# Mechanical Engineering Series

**Series Editor**

Francis A. Kulacki, Department of Mechanical Engineering,
University of Minnesota, Minneapolis, MN, USA

The Mechanical Engineering Series presents advanced level treatment of topics on the cutting edge of mechanical engineering. Designed for use by students, researchers and practicing engineers, the series presents modern developments in mechanical engineering and its innovative applications in applied mechanics, bioengineering, dynamic systems and control, energy, energy conversion and energy systems, fluid mechanics and fluid machinery, heat and mass transfer, manufacturing science and technology, mechanical design, mechanics of materials, micro- and nano-science technology, thermal physics, tribology, and vibration and acoustics. The series features graduate-level texts, professional books, and research monographs in key engineering science concentrations.

More information about this series at http://www.springer.com/series/1161

Zhuomin M. Zhang

# Nano/Microscale Heat Transfer

Second Edition

Zhuomin M. Zhang
Marietta, GA, USA

ISSN 0941-5122 ISSN 2192-063X (electronic)
Mechanical Engineering Series
ISBN 978-3-030-45041-0 ISBN 978-3-030-45039-7 (eBook)
https://doi.org/10.1007/978-3-030-45039-7



This Springer imprint is published by the registered company Springer Nature Switzerland AG
The registered company address is: Gewerbestrasse 11, 6330 Cham, Switzerland

*To my wife Lingyun*

# Preface

Right after the first edition was published in 2007, I was invited to give seminars at a number of universities in several countries. At that time, many people considered nanotechnology research to be new and emerging; however, quite a few also wondered what it was about and questioned its real-world applications. Nowadays, the benefits of nanotechnology have entered our daily lives; it is hard for a teenager or a young adult to imagine living in an environment without modern mobile devices, internet, big data, etc. Many universities have since used my book in teaching graduate and undergraduate courses related to the micro/nanoscale thermal transport. Remarkable progress has been made in the last decade on the simulations, measurements, and applications of nanoscale thermal science and engineering. The second edition is an update of the first edition, covering the recent advances in this field. The structure and chapters are not changed, and the revision follows the same philosophy: to put the readers first and to make it easy to understand. More advanced topics are covered as overviews with pertinent references so that readers can seek further details from the literature and other resources.

Over the past thirty years, there have been tremendous developments in microelectronics, microfabrication technology, MEMS and NEMS, quantum structures (*e.g.*, superlattices, nanowires, nanotubes, graphene and other two-dimensional materials, and nanoparticles), optoelectronics and lasers including ultrafast lasers, and molecular- to atomic-level imaging techniques (such as high-resolution electron microscopy, scanning tunneling microscopy, atomic force microscopy, near-field optical microscopy, and scanning thermal microscopy). The field is fast moving into scaling up and systems engineering to explore the unlimited potential that nanoscience and nanoengineering may offer to restructure the technologies in the new millennia. When the characteristic length becomes comparable to the mechanistic length scale, continuum assumptions that are often made in conventional thermal analysis may break down. Similarly, when the characteristic time becomes comparable to the mechanistic time scale, traditional equilibrium approaches may not be appropriate. Understanding the energy transport mechanisms in small dimensions and short timescale is crucial for the future advancement of nanotechnology. In recent years, a growing number of research publications have been in nano/microscale thermophysical engineering. Timely dissemination of the knowledge gained from contemporary research to educate future scientists and engineers is of emerging significance. For this reason, more and more universities are offering courses in microscale/nanoscale thermal transport. A self-contained textbook suitable for engineering students is much needed. Many practicing engineers who have graduated earlier wish to learn what is going on in this fascinating area, but are often frustrated due to the lack of a solid background to comprehend the contemporary literature. A book that does not require prior knowledge in statistical mechanics, quantum mechanics, solid-state physics, and electrodynamics is extremely helpful. Nevertheless, such a book should cover all these subjects in some depth without significant prerequisites.

This book is written for engineering senior undergraduates and graduate students, practicing engineers, and academic researchers who have not been extensively exposed to nanoscale sciences but wish to gain a solid background in the thermal phenomena occurring at small length scales and short timescales. The basic philosophy behind this book is to logically integrate the traditional knowledge in thermal engineering and physics with newly developed theories in an easy-to-understand approach, with ample examples and homework problems. The materials have been used in the graduate courses and undergraduate electives that I have taught for a number of times at two universities since 1999. While this book can be used as a text for a senior elective or an entry-level graduate course, it is not expected that all the materials will be covered in a one-semester course. The instructors have the freedom to select materials from the book according to students' backgrounds and interests. Some chapters and sections can also be used to integrate with traditional thermal science courses in order to update the current undergraduate and graduate curricula with nanotechnology contents.

The content of this book includes microscopic descriptions and approaches, as well as their applications in thermal science and engineering, with an emphasis on energy transport in gases and solids by conduction (diffusion) and radiation (with or without a medium), as well as convection in micro/nanofluidics. Following the introduction of Chap. 1, an in-depth overview of the foundation of macroscopic thermodynamics, heat transfer, and fluid mechanics is given in Chap. 2. Chapter 3 summarizes the well-established theories in statistical mechanics, including classical and quantum statistics. Thermal properties of ideal gases are described in the content of statistical thermodynamics, followed by a concise presentation of quantum mechanics. Chapter 4 focuses on microfluidics and introduces the Boltzmann transport equations. The heat transfer and microflow regimes from continuous flow to free molecule flow are described. In Chaps. 5–7, heat transfer in solid nanostructures is discussed. Chapter 5 presents the classical and quantum size effects on specific heat and thermal conductivity without involving detailed solid-state physics, which are introduced in Chap. 6. This arrangement allows a more intuitive learning experience. Chapter 7 focuses on transient as well as nonequilibrium energy transport processes in nanostructures. The next three chapters deal with thermal radiation at nanoscales. Chapter 8 provides the fundamental understanding of electromagnetic waves and the dielectric functions of various materials, including metamaterials with exotic properties. Theories of blackbody radiation, radiation thermometry, and radiation entropy are present. Radiative properties of bulk materials and their relationships are discussed. Chapter 9 describes interference effects of thin films and multilayers, the band structure of photonic crystals, diffraction from surface-relief gratings, scattering from rough surfaces, as well as plasmonics and surface polaritons. Chapter 10 explores evanescent waves and the coupling phenomena in the near field for energy transfer. Recent advances in nanophotonics and nanoscale radiative heat transfer are also summarized. In the second edition, significant enhancements have been made to heat conduction in solids and nanostructures as well as to nanoscale thermal radiation and radiative properties. The dual nature of particles and waves is emphasized throughout the book in explaining the energy carriers, such as molecules in ideal gases, electrons in metals, phonons in dielectric crystalline materials, and photons for radiative transfer. Examples in the text and end-of-chapter homework problems should enhance the understanding of how to apply the formulations and methodology to develop problem-solving skills. Selected homework solutions will be posted on the author's website (http://zhang-nano.gatech.edu/), which also contains author's contact information.

I am deeply in debt to Dr. Markus Flik, my doctoral father, who brought me to the micro/nano world through the three intense and fruitful years at MIT. After graduation, late Professor and Chancellor of Berkeley, Chang-Lin Tien, my academic grandfather, offered immense support and encouragement for me to write this book. I am grateful to my master thesis advisors, Profs. Xin-Shi Ge and Yifang Wang of the University of Science and Technology of China (Hefei), for giving me early research training in thermal radiation. I wish to thank my postdoctoral mentors Dr. Raju Datla and Prof. Dennis Drew (University of Maryland) for providing me a valuable opportunity to do research at NIST, where I also benefited from working with many outstanding researchers, including Drs. Leonard Hansson, Jack Hsia, Joe Rice, Ben Tsai, and late Prof. Dave DeWitt (Purdue).

I am grateful to my colleagues and collaborators at both University of Florida (UF) and Georgia Institute of Technology (GT). I have been fortunate to have very supportive supervisors from Prof. William Tiederman (former Department Chair at UF) to Profs. Ward Winer and William Wepfer (former School Chairs at GT), to Prof. Samuel Graham (current School Chair at GT). I greatly enjoyed the collaboration with Prof. David Tanner (Physics, UF), as well as the valuable interactions with Profs. C. K. Hsieh, Yogi Goswami, Sherif Sherif, Jacob Chung, James Klausner, and David Hahn while at UF. I cherish the friendship and collaboration with my colleagues Profs. Yogendra Joshi, G. P. "Bud" Peterson, Peter Hesketh, William King (now at UIUC), Bara Cola, Peter Loutzenhiser, Devesh Ranjan, and Shannon Yee at Georgia Tech.

I have also benefited greatly from the support, encouragement, and friendship of a large number of peers and colleagues in the heat transfer and thermophysics community; too many to list here. I wish to thank members of the ASME Heat

Transfer Division's K-9 Committee on Nanoscale Thermal Transport for inspiring discussions and comments.

This book would not have been possible without my graduate students' hard work and dedication. Many of them have taken my classes and proofread different versions of the manuscripts. Some materials in the last three chapters are generated based on their thesis research. I would like to thank my former graduate students at UF Ravi Kumar, Brian Johnson, Donghai Chen, David Pearson, Yihui Zhou, Ferdinand Rosa, Yu-Jiun Shen, Jorge Garcia, and Linxia Gu for helping me establish my early academic career. My first Ph.D. students graduated at GT, Qunzhi Zhu (currently at Shanghai University of Electric Power) and Ceju Fu (Peking University), made my transition from UF to GT smoother. They were followed by many wonderful Ph.D. graduates: Hyunjin Lee (Kookmin University), Yu-Bin Chen (National Tsing Hua University), Keunhan Park (University of Utah), Bong Jae Lee (KAIST), Soumya Basu (PsiQuantum Ltd.), Xiaojia Wang (University of Minnesota), Liping Wang (Arizona State University), Andrew McNamara (AMD), Trevor Bright (Aerospace Corp.), Richard Z. Zhang (University of North Texas), Jesse Watjen (Knolls Atomic Power Lab), Xianglei Liu (Nanjing University of Aeronautics and Astronautics), Bo Zhao (post-doc Stanford University), Peiyan Yang (Apple Inc.), and Eric Tervo (NREL director's post-doc fellow). I am glad to see that most of them have developed their own independent research and academic careers and become excellent teachers. My current Ph.D. students Dudong Feng, Chuyang Chen, Shin Young Jeong, and Chiyu Yang have also provided great help during the revision. I am also thankful to many visiting scholars, post-doc researchers, visiting students, master's students, and undergraduate students who have worked with me. Many graduate and undergraduate students who have taken my classes also provided constructive suggestions. I enjoyed working with all of them.

I wish to thank the Thermal Transport Program of NSF for the continuous support of my research and educational endeavor since 1998 and the Program Directors Drs. Ashley Emery, Richard Smith, Alfonso Ortega, Patrick Phelan, Theodore Bergman, Sumanta Acharya, José Lage, and Ying Sun. The ongoing grant number is CBET-1603761. I also gratefully appreciate the Physical Behavior of Materials Program of DOE (Basic Energy Science) and the Program Manager Dr. Refik Kortan for the confidence and support in the past decade. The ongoing grant number is DE-SC0018369. I do take full responsibility for any inadvertent errors or mistakes.

I must thank the Chief Editor of Springer Mechanical Engineering Series, Prof. Francis Kulacki of the University of Minnesota for his encouragement, patience, and valuable comments during the past two years. The Editor at Springer, Michael Luby, and the editorial team are acknowledged for their hard work putting this book to print.

Finally, I thank my family for their understanding and support throughout the writing journey. I can't thank enough my parents and parents-in-law for their unselfish love and support to me and my family. My children Emmy, Angie, and Bryan, now grown-ups, have given me great happiness and made my life meaningful. This book is dedicated to my wife Lingyun for the unconditional love and meticulous care she has provided to me and to our children.

Marietta, GA, USA
January 2020

Zhuomin M. Zhang

# Contents

# List of Symbols

| | |
|---|---|
| $A$ | Area, m$^2$; Helmholtz free energy, J |
| $A_c$ | Cross-sectional area, m$^2$ |
| $A'_\lambda$ | Spectral, directional absorptance of a semitransparent material |
| $\mathbf{a}$ | Acceleration, m/s$^2$ |
| $a$ | Lattice constant, m; magnitude of acceleration, m/s$^2$ |
| $a_0$ | Bohr radius, 0.0529 nm |
| $a_\lambda$ | Absorption coefficient, m$^{-1}$ |
| $\mathbf{B}$ | Magnetic induction or magnetic flux density, T (tesla) or Wb/m$^2$ |
| $C$ | Volumetric heat capacity ($\rho c_p$), J/K m$^3$ |
| $c$ | Phase speed of electromagnetic wave, m/s |
| $c_0$ | Speed of light in vacuum, $2.998 \times 10^8$ m/s |
| $c_p$ or $c_v$ | Specific heat for constant pressure or constant volume, J/kg K |
| $\mathbf{D}$ | Dynamic matrix; electric displacement, C/m$^2$ |
| $D$ | Density of states, m$^{-3}$; diameter, m |
| $D_{AB}$ | Binary diffusion coefficient, m$^2$/s |
| $d$ | Diameter or film thickness, m |
| $\mathbf{E}$ | Electric field vector, N/C or V/m |
| E | Energy, J; magnitude of electric field, V/m |
| $E_F$ | Fermi energy, J |
| $E_g$ | Bandgap energy, J |
| $e$ | Electron charge (absolute value), $1.602 \times 10^{-19}$ C |
| $e_b$ | Blackbody emissive power, W/m$^2$ |
| $\mathbf{F}$, $F$ | Force, N |
| $F$ | Normalized distribution function |
| $f$ | Distribution function |
| $\mathbf{G}$ | Reciprocal lattice vector, m$^{-1}$; dyadic Green's function |
| $G$ | Gibbs free energy, J; electron-phonon coupling constant, W/m$^3$ K |
| $g$ | Degeneracy |
| $\bar{g}$ | Molar specific Gibbs free energy, J/kmol |
| $\mathbf{H}$ | Magnetic field vector, A/m |

| | |
|---|---|
| $H$ | Enthalpy, J; magnetic field strength, A/m |
| $h$ | Mass specific enthalpy, J/kg; convective heat transfer coefficient, W/m$^2$ K; Planck's constant, $6.626 \times 10^{-34}$ J s |
| $h_{\mathrm{m}}$ | Mass transfer coefficient, m/s |
| $\hbar$ | Planck's constant divided by $2\pi$, $h/2\pi$ |
| $\bar{h}$ | Molar specific enthalpy, J/kmol |
| $\mathbf{I}$ | Unit matrix; unit dyadic |
| $I$ | Moment of inertia, kg m$^2$; intensity or radiance, W/m$^2$ μm sr; current, A |
| i | $\sqrt{-1}$ |
| $i, j, k$ | Indices used in series |
| $\mathbf{J}$ or $J$ | Flux vector or magnitude (quantity transferred per unit area per unit time) |
| $\mathbf{J}$ or $\mathbf{J}_{\mathrm{e}}$ | current density (or electric charge flux), A/m$^2$ |
| $\mathbf{K}$ | Block wavevector, m$^{-1}$ |
| $K$ | Spring constant, N/m; Thomson's coefficient, V/K |
| $\mathbf{k}$ | Wavevector, m$^{-1}$ |
| $k$ | Magnitude of the wavevector, m$^{-1}$ |
| $k_{\mathrm{B}}$ | Boltzmann's constant, $1.381 \times 10^{-23}$ J/K |
| $L$, $l$ | Length or characteristic length, m |
| $L_0$ | Average distance between molecules or atoms, m |
| $L_\lambda$ | Radiation entropy intensity, W/K m$^2$ μm sr |
| $l$, $m$, $n$ | Index number |
| $M$ | Molecular weight, kg/kmol |
| $m$ | Mass of a system or a single particle, kg |
| $m_{\mathrm{r}}$ | Reduced mass, kg |
| $m^*$ | Effective mass, kg |
| $N$ | Number of particles; number of phonon oscillators |
| $N_{\mathrm{A}}$ | Avogadro's constant, $6.022 \times 10^{26}$ kmol$^{-1}$; acceptor concentration, m$^{-3}$ |
| $N_{\mathrm{D}}$ | Donor concentration, m$^{-3}$ |
| $\dot{N}$ | Particle flow rate, s$^{-1}$ |
| $n$ | Number density, m$^{-3}$; quantum number; index number; refractive index; or real part of the complex refractive index |
| $\bar{n}$ | Amount of substance, kmol |
| $\tilde{n}$ | Complex refractive index |
| $\mathbf{P}$ | Propagation matrix; polarization vector or dipole moment per unit volume, C/m$^2$ |
| $P$ | Pressure, Pa (N/m$^2$ or J/m$^3$) |
| $P_{ij}$ | Momentum flux component, Pa |
| $\mathbf{p}$ | Momentum vector ($m\boldsymbol{v}$ or $\hbar\mathbf{k}$), kg m/s |
| $p$ | Momentum ($mv$ or $\hbar k$), kg m/s; probability; specularity |
| $Q$ | Heat, J; quality factor of a resonance |
| $\dot{Q}$ | Heat transfer rate, W |
| $q$ | Number of coexisting phases; number of atoms per molecule |

| | |
|---|---|
| $\dot{q}$ | Thermal energy generation rate, W/m$^3$ |
| $\mathbf{q}''$ | Heat flux vector, W/m$^2$ |
| $q''$ | Heat flux, W/m$^2$ |
| $R$ | Gas constant, J/kg K; electric resistance, Ω or V/A |
| $R'$ | Directional–hemispherical reflectance |
| $R''_{\mathrm{b}}$ | Thermal boundary resistance, m$^2$ K/W |
| $R''_{\mathrm{t}}$ | Thermal resistance, m$^2$ K/W |
| $\bar{R}$ | Universal gas constant, 8314.4 J/kmol K |
| $r$ | Distance or radius, m; reflection coefficient |
| $r_{\mathrm{e}}$ | Electrical resistivity, Ω m |
| $\tilde{r}$ | Complex Fresnel's reflection coefficient |
| $\mathbf{S}$ | Poynting vector, W/m$^2$ |
| $S$ | Entropy, J/K |
| $S_j$ | Strength of the *j*th phonon oscillator |
| $\dot{S}$ | Entropy transfer rate, W/K |
| $\dot{S}_{\mathrm{gen}}$ | Entropy generation rate, W/K |
| $s$ | Specific entropy J/kg K; entropy density, J/m$^3$ K |
| $\bar{s}$ | Molar specific entropy, J/kmol K |
| $\dot{s}_{\mathrm{gen}}$ | Volumetric entropy generation rate, W/m$^3$ K |
| $s''$ | Entropy flux, J/m$^2$ K |
| $T$ | Temperature, K |
| $T'$ | Directional–hemispherical transmittance |
| $t$ | Time, s; transmission coefficient |
| $\tilde{t}$ | Complex Fresnel's transmission coefficient |
| $U$ | Internal energy, J; potential, J |
| $\mathbf{u}_{\mathrm{d}}$ | Drift velocity, m/s |
| $u$ | Specific internal energy, J/kg; energy density, J/m$^3$ |
| $\bar{u}$ | Molar specific internal energy, J/kmol |
| $V$ | Volume, m$^3$; voltage, V |
| $\mathbf{v}$ | Velocity, m/s |
| $\mathbf{v}_{\mathrm{B}}$ | Bulk or mean velocity, m/s |
| $\mathbf{v}_{\mathrm{R}}$ | Random or thermal velocity, m/s |
| $v$ | Speed, m/s; specific volume, m$^3$/kg |
| $v_{\mathrm{a}}$ | Speed of sound or average speed of phonons, m/s |
| $v_{\mathrm{F}}$ | Fermi velocity, m/s |
| $v_{\mathrm{g}}$ | Magnitude of group velocity $(\mathrm{d}\omega/\mathrm{d}k)$, m/s |
| $v_l, v_t$ | Longitudinal, transverse phonon speed, m/s |
| $v_{\mathrm{p}}$ | Phase speed $(\omega/k)$, m/s |
| $v_x, v_y, v_z$ | Velocity components, m/s |
| $\bar{v}$ | Average speed, m/s; molar specific volume, m$^3$/kmol |
| $W$ | Work, J; width, m |
| $x, y, z$ | Coordinates, m |
| $Z$ | Partition function |

## Dimensionless Parameters

| | |
|---|---|
| $Kn$ | Knudsen number, $\Lambda/L$ |
| $Le$ | Lewis number, $D_{AB}/\alpha = Pr/Sc$ |
| $Lz$ | Lorentz number, $\kappa/\sigma T$ |
| $Ma$ | Mach number, $v/v_a$ |
| $Nu$ | Nusselt number, $hL/\kappa$ |
| $Pe$ | Peclet number, $RePr = v_\infty L/\alpha$ |
| $Pr$ | Prandtl number, $\nu/\alpha$ |
| $Re$ | Reynolds number, $\rho v_\infty L/\mu$ |
| $Sc$ | Schmidt number, $\nu/D_{AB}$ |
| $ZT$ | Dimensionless figure of merit for thermoelectricity |

## Greek Symbols

| | |
|---|---|
| $\alpha$ | Thermal diffusivity, $m^2/s$; some constant; polarizability |
| $\alpha$ and $\beta$ | Lagrangian multipliers; indices |
| $\alpha_T$ | Thermal accommodation coefficient |
| $\alpha_v$ | (Tangential) momentum accommodation coefficient |
| $\alpha_{v'}$ | Normal momentum accommodation coefficient |
| $\alpha'_\lambda$ | Spectral, directional absorptivity or absorptance |
| $\beta$ | Parallel wavevector component, $m^{-1}$; various coefficients |
| $\beta_P$ | Isobaric thermal expansion coefficient, $K^{-1}$ |
| $\beta_T$ | $2\gamma(2-\alpha_T)Kn/[\alpha_T(\gamma+1)Pr]$ |
| $\beta_v$ | $(2-\alpha_v)Kn/\alpha_v$ |
| $\Gamma_{ij}$ | Hemispherical transmissivity for phonons from media $i$ to $j$ |
| $\Gamma_S$ | Seebeck's coefficient, V/K |
| $\gamma$ | Specific heat ratio; scattering rate ($1/\tau$), rad/s |
| $\gamma_s$ | Sommerfeld constant, J/kg $K^2$ |
| $\delta$ | Differential small quantity; delta function; boundary layer thickness, m |
| $\delta_\lambda$ | Radiation penetration depth, m |
| $\varepsilon$ | Energy of a particle or quasiparticle, J; electric permittivity, F/m; relative permittivity; emittance; or emissivity |
| $\tilde{\varepsilon}$ | Complex dielectric function or relative permittivity |
| $\varepsilon'_\lambda$ | spectral, directional emissivity (or emittance) |
| $\zeta$ | Dummy variable; perpendicular wavevector component, $m^{-1}$ |
| $\eta$ | Various efficiencies; imaginary part of perpendicular wavevector, $m^{-1}$ |
| $\eta_H$ | Hall coefficient $E_y/J_xB$, $m^3/C$ |
| $\Theta$ | Characteristic temperature, K; mean energy of Planck's oscillator |
| $\Theta_D$ | Debye temperature, K |
| $\theta$ | Zenith angle, rad |
| $\theta_B$ | Brewster's angle, rad |

| | |
|---|---|
| $\theta_c$ | Critical angle, rad |
| $\kappa$ | Thermal conductivity, W/m K; extinction coefficient or imaginary part of the refractive index |
| $\kappa_T$ | Isothermal compressibility, $Pa^{-1}$ |
| $\Lambda$ | Mean free path, m; period of a grating or photonic crystal, m |
| $\Lambda_a$ | Average collision distance, m |
| $\lambda$ | Wavelength in vacuum, m (often expressed in μm) |
| $\mu$ | (Dynamic) viscosity, N s/$m^2$; chemical potential, J; electron or hole mobility, $m^2/V$ s; magnetic permeability, N/$A^2$; relative magnetic permeability |
| $\mu_F$ | Fermi energy, J |
| $\nu$ | Kinematic viscosity, $m^2/s$; frequency, Hz |
| $\bar{\nu}$ | Wavenumber, $m^{-1}$ (often expressed in $cm^{-1}$) |
| $\xi$ | Energy transmission coefficient or photon tunneling probability; certain coordinate or variable |
| $\Pi$ | Peltier's coefficient, V |
| $\rho$ | Density, kg/$m^3$ |
| $\rho_e$ | Charge density, C/$m^3$ |
| $\rho'$ | Directional–hemispherical reflectance |
| $\rho'_\lambda$ | Spectral, directional reflectivity |
| $\sigma$ | Electrical conductivity, $(\Omega\,m)^{-1}$; standard deviation |
| $\sigma_{rms}$ | Root-mean-square surface roughness, m |
| $\sigma_{SB}$ | Stefan–Boltzmann constant, $5.67 \times 10^{-8}$ W/$m^2$ $K^4$ |
| $\sigma'_{SB}$ | Phonon Stefan–Boltzmann constant, W/$m^2$ $K^4$ |
| $\tau$ | Relaxation time, s; transmission coefficient |
| $\tau'$ | Directional–hemispherical transmittance |
| $\Phi$ | Scattering phase function; viscous dissipation function; potential function |
| $\phi$ | Number of degrees of freedom; azimuthal angle, rad; intermolecular potential |
| $\chi$ | susceptibility |
| $\Psi$ | Schrödinger's wave function; various functions |
| $\psi$ | Molecular quantity; wave function; work function, J; phase shift, rad |
| $\Omega$ | Solid angle, sr; thermodynamic probability |
| $\omega$ | Angular frequency, rad/s |
| $\omega_p$ | Plasma frequency, rad/s |
| $\varpi$ | Velocity space, $d\varpi = dv_x dv_y dv_z$ |

## Subscripts

| | |
|---|---|
| 0 | Vacuum or free space |
| 1,2,3 | Medium 1,2,3 |
| b | Blackbody; boundary |
| $E$ | Energy |
| e | Electron, electric |
| h | Hole |
| i | Incident |
| m | Bulk or mean; maximum; medium |
| mp | Most probable |
| n | normal direction |
| $n$ or $p$ | $n$-type or $p$-type semiconductor |
| $p$ | TM wave, $p$ (parallel) polarization |
| r | reflected; rotational |
| s | Lattice; scattered; solid; surface |
| $s$ | TE wave, $s$ (perpendicular) polarization |
| t | Transmitted; translational |
| th | Thermal |
| v | Vibrational |
| w | Wall |
| $\infty$ | Free steam |
| $\lambda$, $\nu$, or $\omega$ | Spectral quantity in terms of wavelength, frequency, or angular frequency |

# Chapter 1
# Introduction

Improvement of performance and shrinkage of device sizes in microelectronics have been major driving forces for scientific and economic progress over the past 40 years. Developments in semiconductor processing and surface sciences have allowed precise control over critical dimensions with desirable properties for solid-state devices. In the past 30 years, there have been tremendous developments in micro- and nanoelectromechanical systems (MEMS and NEMS), microfluidics and nanofluidics, quantum structures and devices, photonics and optoelectronics, nanomaterials for molecular sensing and biomedical diagnosis, and scanning probe microscopy for measurement and manipulation at the molecular and atomic levels.

Nanotechnology opens new frontiers in science and engineering, and has also become an integral part of almost all natural science and engineering disciplines. Back in 2007, about 10% of the faculty members at Georgia Tech were conducting some research related to nanoscience and nanoengineering. The number of faculty and research projects related to micro/nanoscales has grown significantly. The same can be said for most major research universities in the United States and in many other countries. Furthermore, the study of nanoscience and nanoengineering requires and has resulted in close interactions across the boundaries of many traditional disciplines. Knowledge of physical behavior at the molecular and atomic levels has played and will continue to play an important role in our understanding of the fundamental processes occurring in the macro world. This will enable us to design and develop novel devices and machines, ranging from a few nanometers all the way to the size of automobiles and airplanes [1, 2]. Ten to fifteen years ago, many people either had never heard about the word "nanotechnology" or had doubts about the usefulness in real world. Today, we depend on and enjoy nanotechnology in our daily life, including cell phones, laptops, computers, internet, medicine and medical devices, energy harvesting, transportation, lighting, batteries, smart clothes, and so on, you name it. The advancement of nano/microscale science and engineering will continue to restructure the technologies currently used in manufacturing, energy production and utilization, communication, transportation, space exploration, and medicine.



Z. M. Zhang, *Nano/Microscale Heat Transfer*, Mechanical Engineering Series,
https://doi.org/10.1007/978-3-030-45039-7_1

A key issue associated with miniaturization is the tremendous increase in the heat dissipation per unit volume. Micro/nanostructures may enable engineered materials with unique thermal properties to allow significant enhancement or reduction of the heat flow rate. Therefore, knowledge of thermal transport from the micrometer scale down to the nanometer scale and thermal properties of micro/nanostructures is of critical importance to future technological growth. Solutions to more and more problems in small devices and systems require a solid understanding of the heat (or more generally, energy) transfer mechanisms in reduced dimensions and/or short time scales, because classical equilibrium and continuum assumptions are not valid anymore. Examples are the thermal analysis and design of micro/nanodevices, thermal management in flexible electronics, ultrafast laser interaction with materials, micromachined thermal sensors and actuators, thermoelectricity in nanostructures, photonic crystals, microscale thermophotovoltaic devices, battery thermal management, and so on [3, 4].

This book was motivated by the need to understand the thermal phenomena and heat transfer processes in micro/nanosystems and at very short time scales for solving problems occurring in contemporary and future technologies. Since the first publication in 2007, many universities have offered micro/nanoscale heat transfer courses and used it as either the textbook or major reference. Significant progress has been made in the last decade and this second edition reflects a major update.

## 1.1 Limitations of the Macroscopic Formulation

As an ancient Chinese philosopher put it, suppose you take a foot-long wood stick and cut off half of it each day; you will never reach an end even after thousands of years, as shown in Fig. 1.1. Modern science has taught us that, at some stage, one would reach the molecular level and even the atomic level, below which the physical and chemical properties are completely different from those of the original material. The wooden stick or slice would eventually become something else that is not distinguishable from the other constituents in the atmosphere. Basically, properties of materials at very small scales may be quite different from those of the corresponding bulk materials. Note that 1 nm (nanometer) is one-billionth of a meter. The diameter of a hydrogen atom H is on the order of 0.1 nm, and that of a hydrogen molecule $H_2$ is approximately 0.3 nm. Using the formula $l_n = 0.3048/2^{n-1}$ m, where $n$ is number of days, we find $l_{30} = 5.7 \times 10^{-10}$ m (or 0.57 nm) after just a month, which is already near the diameter of a hydrogen atom (about 0.1 nm).

While atoms can still be divided with large and sophisticated facilities, our ability to observe, manipulate, and utilize them is very limited. On the other hand, most biological processes occur at the molecular level. Many novel physical phenomena happen at the length scale of a few nanometers and can be integrated into large systems. This is why a nanometer is a critical length scale for the realization of practically important new materials, structures, and phenomena. For example, carbon nanotubes with diameters ranging from 0.4 to 50 nm or so have dramatically different

**Fig. 1.1** The length of the wood stick: $l_1 = 1$ ft in day 1, $l_2 = 1/2$ ft in day 2, and $l_n = 1/2^{n-1}$ ft in day $n$

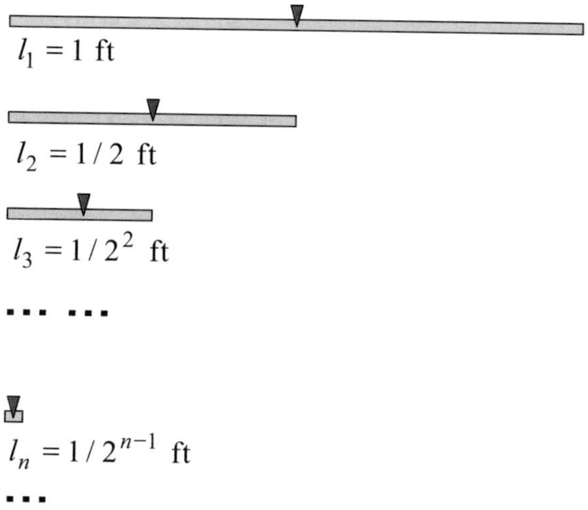

$l_1 = 1$ ft

$l_2 = 1/2$ ft

$l_3 = 1/2^2$ ft

... ...

$l_n = 1/2^{n-1}$ ft

...

properties. Some researchers have shown that these nanotubes hold promise as the building block of nanoelectronics. Others have found that the thermal conductivity of single-walled carbon nanotubes at room temperature could be an order of magnitude higher than that of copper. Therefore, carbon nanotubes have been considered as a candidate material for applications that require a high heat flux.

In conventional fluid mechanics and heat transfer, we treat the medium as a *continuum,* that is, indefinitely divisible without changing its physical nature. All the intensive properties can be defined locally and continuously. For example, the local density is defined as

$$\rho = \lim_{\delta V \to 0} \frac{\delta m}{\delta V} \tag{1.1}$$

where $\delta m$ is the mass enclosed within a volume element $\delta V$. When the characteristic dimension is comparable with or smaller than that of the *mechanistic* length—for example, the molecular *mean free path,* which is the average distance that a molecule travels between two collisions—the continuum assumption will break down. The density defined in Eq. (1.1) will depend on the size of the volume, $\delta V$, and will fluctuate with time even at macroscopic equilibrium. Noting that the mean free path of air at standard atmospheric conditions is about 70 nm, the continuum assumption is well justified for many engineering applications until the submicrometer regime or the nanoscale is reached. Nevertheless, if the pressure is very low, as in an evacuated chamber or at a high elevation, the mean free path can be very large; and thus, the continuum assumption may break down even at relatively large length scales.

Within the macroscopic framework, we calculate the temperature distribution in a fluid or solid by assuming that the medium under consideration is not only a continuum but also at thermodynamic equilibrium everywhere. The latter condition is called the *local-equilibrium* assumption, which is required because temperature can be defined only for stable-equilibrium states. With extremely high temperature gradients at sufficiently small length scales and/or during very short periods of time, the assumption of local equilibrium may be inappropriate. An example is the interaction between short laser pulses and a material. Depending on the type of laser, the pulse duration or width can vary from a few tens of nanoseconds down to several femtoseconds (1 fs = $10^{-15}$ s). In the case of ultrafast laser interaction with metals, free electrons in the metal could gain energy quickly to arrive at an excited state corresponding to an effective temperature of several thousand kelvins, whereas the crystalline lattices remain near room temperature. After an elapse of time represented by the electron relaxation time, the excess energy of electrons will be transferred to phonons, which are energy quanta of lattice vibration, thereby causing a heating effect that raises the temperature or changes the phase of the material under irradiation.

Additional mechanisms may affect the behavior of a system as the physical dimensions shrink or as the excitation and detection times are reduced. A scale-down of the theories developed from macroscopic observations often proves to be unsuitable for applications involving micro/nanoscale phenomena. Examples are reductions in the conductivity of thin films or thin wires due to boundary scattering (size effect), discontinuous velocity and temperature boundary conditions in microfluidics, wave interferences in thin films, and tunneling of electrons and photons through narrow gaps. In the quantum limit, the thermal conductance of a nanowire will reach a limiting value that is independent of the material that the nanowire is made of. At the nanoscale, the radiation heat transfer between two surfaces can exceed that calculated from the Stefan-Boltzmann law by several orders of magnitude. Another effect of miniaturization is that surface forces (such as shear forces) will scale down with $L^2$, where $L$ is the characteristic length, while volume forces (such as buoyancy) will scale down with $L^3$. This will make surface forces predominant over volume forces at the microscale.

## 1.2 The Length Scales

It is instructive to compare the length scales of different phenomena and structures, especially against the wavelength of the electromagnetic spectrum. Figure 1.2 compares the wavelength ranges with some characteristic dimensions. One can see that MEMS generally produce micromachining capabilities from several millimeters down to a few micrometers. Currently, the smallest feature of integrated circuits is well below 100 nm. The layer thickness of thin films ranges from a few nanometers up to several micrometers. The wavelengths of the visible light are in the range from approximately 380 to 760 nm. On the other hand, thermal radiation covers a part of the ultraviolet, the entire visible and infrared, and a portion of the microwave region. The thickness of human hair is between 50 and 100 μm, while the diameter of red blood cells is about 6–8 μm. A typical optical microscope can magnify 100 times with a resolution of 200–300 nm, which is about half the wavelength and is limited due

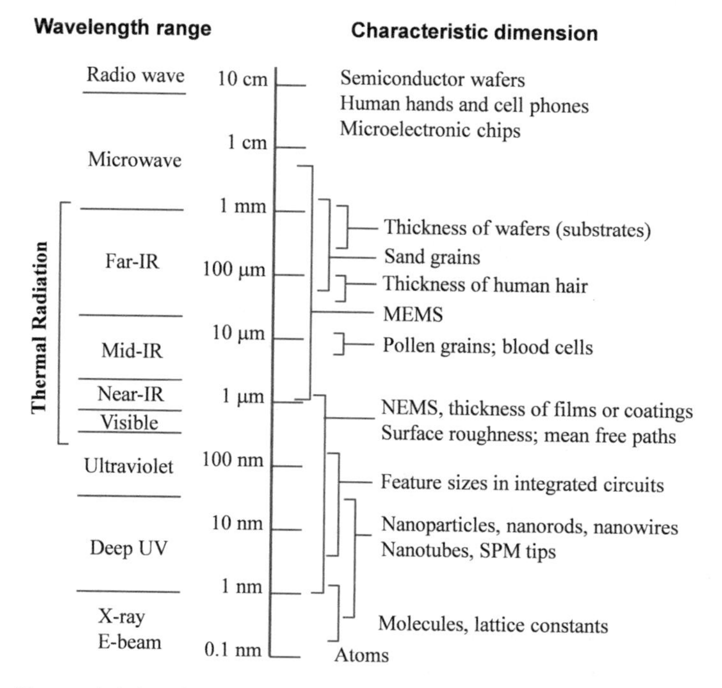


**Fig. 1.2** Characteristic length scales as compared with the wavelength of electromagnetic spectrum

to the diffraction of light. Therefore, optical microscopy is commonly used to study micrometer-sized objects. On the other hand, atoms and molecules are on the order of 1 nm, which falls in the x-ray and electron-beam wavelength region. Therefore, x-ray and electron microscopes are typically used for determining crystal structures and defects, as well as for imaging nanostructures. The development of scanning probe microscopes (SPMs) and near-field scanning optical microscopes (NSOMs) in the 1980s enabled unprecedented capabilities for the visualization and manipulation of nanostructures, such as nanowires, nanotubes, nanocrystals, single molecules, individual atoms, and so forth, as will be discussed in Sect. 1.3.4. Figure 1.2 also shows that the mean free path of heat carriers (e.g., molecules in gases, electrons in metals, and phonons or lattice vibration in dielectric solids) often falls in the micrometer to nanometer scales, depending on the material, temperature, and type of carrier.

A brief historical retrospective is given next on the development of modern science and technologies, with a focus on the recent technological advances leading to nanotechnology. The role of thermal engineering in this technological advancement process is outlined.

## 1.3 From Ancient Philosophy to Contemporary Technologies

Understanding the fundamentals of the composition of all things in the universe, their movement in space and with time, and the interactions between one and another is a human curiosity and the inner drive that makes us different from other living beings on the earth. The ancient Chinese believed that everything was composed of the five elements: metal, wood, water, fire, and earth (or soil) that generate and overcome one another in certain order and time sequence. These simple beliefs were not merely used for fortune-telling but have helped the development of traditional Chinese medicine, music, military strategy, astronomy, and calendar. In ancient Greece, the four elements (fire, earth, air, and water) were considered as the realm wherein all things existed and whereof all things consisted. These classical element theories prevailed in several other countries in somewhat different versions for over 2000 years, until the establishment of modern atomic theory that began with John Dalton's experiment on gases some 200 years ago. In 1811, Italian chemist Amedeo Avogadro introduced the concept of the molecule, which consists of stable systems or bound state of atoms. A molecule is the smallest particle that retains the chemical properties and composition of a pure substance. The first periodic table was developed by Russian chemist Dmitri Mendeleev in 1869. Although the original meaning of atom in Greek is "indivisible," subatomic particles have since been discovered. For example, electrons as a subatomic particle were discovered in 1897 by J. J. Thomson, who won the 1906 Nobel Prize in Physics. An atom is known as the smallest unit of one of the 118 confirmed elements.

The first industrial revolution began in the late eighteenth century and boosted the economy of western countries from manual labor to the machine age by the introduction of machine tools and textile manufacturing. Following the invention of the steam engine in the mid nineteenth century, the second industrial revolution had an even bigger impact on human life through the development of steam-powered ships and trains, along with the internal combustion engines, and the generation of electrical power. Newtonian mechanics and classical thermodynamics have played an indispensable role in the industrial revolutions. The development of machinery and the understanding of the composition of matter have allowed unprecedented precision of experimental investigation of physical phenomena, leading to the establishment of modern physics in the early twentieth century.

The nature of light has long been debated. In the seventeenth century, Isaac Newton formulated the corpuscular theory of light and observed with his prism experiment that sunlight is composed of different colors. In the early nineteenth century, the discovery of infrared and ultraviolet radiation and Young's double-slit experiment confirmed Huygens' wave theory, which was overshadowed by Newton's corpuscular theory for over 100 years. With the establishment of Maxwell's equations that fully describe the electromagnetic waves and Michelson's interferometric experiment, the wave theory of radiation had been largely accepted by the end of the nineteenth century. While the wave theory was able to explain most of the observed phenomena, it could not explain thermal emission over a wide spectrum, nor was it able to explain the photoelectric effect. Max Planck in 1900 used the hypothesis of radiation quanta, or oscillators, to successfully derive the blackbody spectral distribution function. It appears that the energy of light is not indefinitely divisible but must exists in multiples of the smallest massless quanta that we now call photons. In 1905, Albert Einstein explained the photoelectric effect based on the concept of radiation quanta. To knock out an electron from the metal surface, the energy of each incoming phonon ($h\nu$) must be sufficiently large because one electron can absorb only one photon. This explained why photoemission could not occur at frequencies below the threshold value, no matter how intense the incoming light might be. In 1924, Louis de Broglie hypothesized that particles should also exhibit wavelike characteristics. With the electron diffraction experiment, it was found that electrons indeed can behave like waves with a wavelength inversely proportional to the momentum. Electron microscopy was based on the principle of electron diffraction. The wave–particle duality was essential to the establishment of quantum mechanics in the early twentieth century. Quantum mechanics describes the phenomena occurring in minute particles, structures, and their interaction with radiation, for which classical mechanics and electrodynamics are not applicable. The fundamental scientific understanding gained during the first half of the twentieth century has facilitated the development of contemporary technologies that have transformed from the industrial economy to the knowledge-based economy and from the machine age to the information age. The major technological advancements in the last half of the century are highlighted in the following sections.

### *1.3.1 Microelectronics and Information Technology*

In his master's thesis at MIT published in 1940, Claude Shannon (1916–2001) used the Boolean algebra and showed how to use TRUE and FALSE to represent function of switches in electronic circuits. Digital computers were invented during the 1940s in several countries, including the IBM Mark I which is 2.4 m high and 16 m long. In 1948, while working at Bell Labs, Shannon published an article, "A Mathematical Theory of Communication," which marked the beginning of the modern communication and information technology [5]. In that paper, he laid out the basic principles of underlying communication of information with two symbols, 1 and 0, and coined the term "bit" for a binary digit. His theory made it possible for digital storage and transmission of pictures, sounds, and so forth.

In December 1947, scientists at Bell Labs invented the semiconductor point-contact transistor with germanium. The earlier computers and radios were based on bulky vacuum tubes that generated a huge amount of heat. The invention of transistor by William Shockley, John Bardeen, and Walter Brattain was recognized through the Nobel Prize in Physics conferred on them in 1956. There had been intensive research on semiconductor physics using the atomic theory and the mechanism of point contact for the fabrication of transistor to become possible. The invention of transistors ushered the information age with a whole new industry.

In 1954, Gordon Teal at Texas Instruments built the first silicon transistor. The native oxide of silicon appeared to be particularly suitable as the electric insulator. In 1958, Jack Kilby (1923–2005) at Texas Instruments was able to cramp all the discrete components onto a silicon base and later onto one piece of germanium. He filed a patent application the next year on "Miniaturized Electronic Circuits," where he described how to make integrated circuits and connect the passive components via gold wires. Working independently, Robert Noyce at Fairchild Electronics in California found aluminum to adhere well to both silicon and silicon oxide and filed a patent application in 1959 on "Semiconductor Device-and-Lead Structure." Kilby and Noyce are considered the co-inventors of integrated circuits. Noyce was one of the founders of Intel and died in 1990. Kilby was awarded half of the Nobel Prize in Physics in 2000 "for his part in the invention of the integrated circuit." The other half was shared by Zhores Alferov and Herbert Kroemer for developing semiconductor heterogeneous structures used in optoelectronics, to be discussed in the next section.

In 1965, around 60 transistors could be packed on a single silicon chip. Seeing the fast development and future potential of integrated circuits, Gordon Moore, a co-founder of Intel, made a famous prediction that the number and complexity of semiconductor devices would double every year [6]. This is Moore's law, well-known in the microelectronics industry [7]. In the mid-1970s, the number of transistors on a chip increased from 60 to 5000. By 1985, the Intel 386 processor contained a quarter million transistors on a chip. In 2001, the Pentium 4 processor reached 42 million transistors. The number has now exceeded 1 billion per chip in 2006. When the device density is plotted against time in a log scale, the growth almost follows a straight line, suggesting that the packaging density has doubled approximately every 18 months till recent years [7, 8]. Reducing the device size and increasing the packaging density have several advantages. For example, the processor speed increases by reducing the distance between transistors. Furthermore, new performance features can be added into the chip to enhance the performance. The cost for the same performance also reduces. Advanced supercomputer systems have played a critical role in enabling modeling and understanding micro/nanoscale phenomena.

The process is first to grow high-quality silicon crystals and then dice and polish into wafers. Devices are usually made on $SiO_2$ layer that can be grown by heating the wafer to sufficiently high temperatures in a furnace with controlled oxygen partial pressure. The wafers are then patterned using photolithographic techniques combined with etching processes. Donors and acceptors are added to the wafer to form *n*- and *p*-type regions by ion implantation and then annealed in a thermal environment. Metals or heavily doped polycrystalline silicon are used as gates with proper coverage and patterns through lithography. A schematic of metal-oxide-semiconductor field-effect transistor (MOSFET) is shown in Fig. 1.3. Billions of transistors can be packed into the size of a fingernail with several layers through very-large-scale integration (VLSI) with the smallest features on the order of 5 nm. As mentioned earlier, managing heat dissipation is a challenge especially as the device dimension continues to shrink. Local heating or hot spots on the size of 10 nm could cause device failure if not

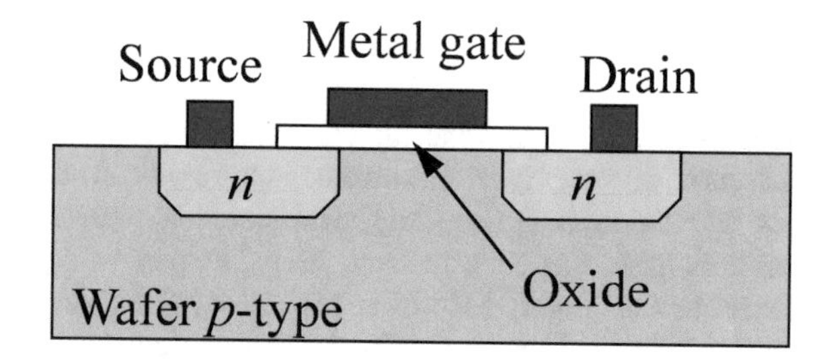


**Fig. 1.3** Schematic of a metal-oxide-semiconductor field-effect transistor (MOSFET)

probably handled. The principles governing the heat transfer at the nanoscale are very different from those at large scales. A fundamental understanding of the phonon transport is required for device-level thermal analysis. Furthermore, understanding heat transfer in microfluidics is necessary to enable reliable device cooling at the micro- and nanoscales.

The progress in microelectronics is not possible without the advances in materials such as crystal growth and thermal processing during semiconductor manufacturing, as well as the deposition and photolithographic technologies. Rapid thermal processing (RTP) is necessary during annealing and oxidation to prevent ions from deep diffusion into the wafer. Thermal modeling of RTP must consider the combined conduction, convection, and radiation modes. Lightpipe thermometer is commonly used to monitor the temperature of the wafer. In an RTP furnace, the thermal radiation emitted by the wafer is collected by the light pipe and then transmitted to the radiometer for inferring the surface temperature [9]. In some cases, the wafer surface is rough with anisotropic features. A better understanding of light scattering by anisotropic rough surfaces is also necessary.

As the process node continues to shrink, high-intensity Ar or Xe arc lamps with millisecond optical pulses are considered as a suitable annealing tool following ion implantation in ultra-shallow junction fabrication. Because the optical energy is absorbed within milliseconds, thermal diffusion cannot distribute heat uniformly across the wafer surface. Therefore, temperature uniformity across the nanometer-patterned wafer is expected to be a critical issue. To reduce the feature size further, deep-UV lithography and x-ray lithography have also been developed. It is inevitable that Moore's law will reach its limit, when the critical dimensions would be less than a few nanometers. Further reduction will be subjected to serious barriers due to problems associated with gate dielectrics and fabrication difficulties. Beyond Moore's law, there are continuous challenges in improving the energy efficiency, overall performance, stability, flexibility, and cost efficiency. Two-dimensional (2D) and 3D very large-scale integration (VLSI) architectures using stacked or sequential integrated systems/circuits may offer future technological solutions. Molecular nanoelectronics using self-assembly has been sought for as an alternative, along with quantum computing. Therefore, nanoelectronics and quantum computing are anticipated to brighten the electronics and computer future.

### 1.3.2 Lasers, Optoelectronics, and Nanophotonics

It is hard to imagine what the current technology would look like without lasers. Lasers of different types have tremendous applications in metrology, microelectronics fabrication, manufacturing, medicine, and communication. Examples are laser printers, laser bar code readers, laser Doppler velocimetry, laser machining, and laser corneal surgery for vision correction. The concept of laser was demonstrated in late 1950s independently in the United States and the Soviet Union during the cold war. The Nobel Prize in Physics of 1964 recognized the fundamental contributions in the field of quantum electronics by Charles Townes, Nicolay Basov, and Aleksandr Prokhorov. The first working laser was Ruby laser built by Theodore Maiman at Hughes Aircraft Company in 1960. The principle of laser dates back to 1917, when Einstein elegantly depicted his conception of stimulated emission of radiation by atoms. Unlike thermal emission and plasma emission, lasers are coherent light sources and, with the assistance of optical cavity, lasers can emit nearly monochromatic light and point to the same direction with little divergence. Lasers enabled a branch of nonlinear optics, which is important to understand the fundamentals of light–matter interactions, communication, as well as optical computing. In 1981, Nicolaas Bloembergen and Arthur Schawlow received the Nobel Prize in Physics for their contributions in laser spectroscopy. There are a variety of nonlinear spectroscopic techniques, including Raman spectroscopy, as reviewed by Fan and Longtin [10]. Two-photon spectroscopy has become an important tool for molecular detection [11]. Furthermore, two-photon 3D lithography has also been developed for microfabrication [12, 13].

Gas lasers such as He–Ne (red) and Ar (green) have been extensively used for precision alignment, dimension measurements, and laser Doppler velocimetry due to their narrow linewidth. On the other hand, powerful Nd:YAG and $CO_2$ lasers are used in thermal manufacturing, where the heat transfer processes include radiation, phase change, and conduction [14, 15]. Excimer lasers create nanosecond pulses in ultraviolet and have been extensively used in materials processing, ablation, eye surgery, dermatology, as well as photolithography in microelectronics and microfabrication. High-energy nanosecond pulses can also be produced by $Q$-switching, typically with a solid-state laser such as Nd:YAG laser at a wavelength near 1 μm. On the other hand, mode-locking technique allows pulse widths from picoseconds down to a few femtoseconds. Pulse durations less than 10 fs have been achieved since 1985. Ultrafast lasers have enabled the study of reaction dynamics and formed a branch in chemistry called *femtochemistry*. Ahmed Zewail of Caltech received the 1999 Nobel Prize in Chemistry for his pioneering research in this field. In 2005, John Hall and Thoedor Hänsch received the Nobel Prize in Physics for developing laser-based precision spectroscopy, in particular, the frequency comb technique. Short-pulse lasers can facilitate fabrication, the study of electron–phonon interaction in the nonequilibrium process, measurement of thermal properties including interface resistance, nondestructive evaluation of materials, and so forth [16–20].

Room-temperature continuous-operation semiconductor lasers were realized in May 1970 by Zhores Alferov and co-workers at the Ioffe Physical Institute in Russia, and independently by Morton Panish and Izuo Hayashi at Bell Labs a month later. Alferov received the Nobel Prize in Physics in 2000, together with Herbert Kroemer who conceived the idea of double-heterojunction laser in 1963 and was also an earlier pioneer of molecular beam epitaxy (MBE). Invented in 1968 by Alfred Cho and John Arthur at Bell Labs and developed in the 1970s, MBE is a high-vacuum deposition technique that enables the growth of highly pure semiconductor thin films with atomic precision. The name heterojunction refers to two layers of semiconductor materials with different bandgaps, such as GaAs/$Al_xGa_{1-x}As$ pair. In a double-heterojunction structure, a lower-bandgap layer is sandwiched between two higher-bandgap layers [21]. When the middle layer is made thin enough, on the order of a few nanometers, the structure is called a quantum well because of the discrete energy levels and enhanced density of states. Quantum well lasers can have better performance with a smaller driving current. Multiple quantum wells (MQWs), also called superlattices, that consist of periodic structures can also be used to further improve the performance. In a laser setting, an optical cavity is needed to confine the laser bandwidth as well as enhance the intensity at a desired wavelength with narrow linewidth. Distributed Bragg reflectors (DBRs) are used on both ends of the quantum well (active region). DBRs are the simplest photonic crystals made of periodic dielectric layers of different refractive indices; each layer thickness is equal to a quarter of the wavelength in that medium ($\lambda/n$). DBRs are dielectric mirrors with nearly 100% reflectance, except at the resonance wavelength $\lambda$, where light will eventually escape from the cavity. Figure 1.4 illustrates a vertical cavity surface emitting laser (VCSEL), where light is emitted through the substrate (bottom of the structure). The energy transfer mechanisms through phonon waves and electron waves have been extensively investigated [22]. Further improvement in the laser efficiency and control of the wavelength has been made using quantum wires and quantum dots (QDs) [21].

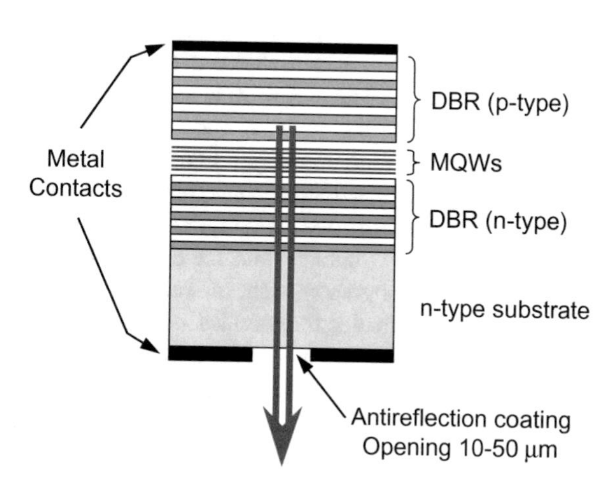


**Fig. 1.4** Schematic of a VCSEL laser made of heterogeneous quantum well structure. The smaller layer thickness can be 3 nm, and there can be as many as several hundred layers

Semiconductor lasers are the most popular lasers (in quantity), and several hundred-million units are sold each year. Their applications include CD/DVD reading/writing, optical communication, laser pointers, laser printers, bar code readers, and so forth. A simpler device is the light-emitting diode (LED), which emits incoherent light with a two-layer *p-n* junction without DBRs. LEDs have been used for lighting, including traffic lights with improved efficiency and decorating lights. The development of wide-bandgap materials, such as GaN and AlN epitaxially grown through metal-organic chemical vapor deposition (MOCVD), allows the LED and semiconductor laser wavelength to be pushed to the blue and ultraviolet. For their invention of efficient blue LEDs, Isamu Akasaki, Hiroshi Amano, and Shuji Nakamura were recognized by the 2014 Nobel Prize in Physics. Organic light-emitting diodes (OLEDs) based on electroluminescence are being developed as a promising candidate for the next-generation computer and TV displays.

Alongside the development of light sources, there have been continuous development and improvement in photodetectors, mainly in focal plane arrays, charge-coupled devices (CCDs), quantum well detectors, readout electronics, data transfer and processing, compact refrigeration and temperature control, and so forth. On the other hand, optical fibers have become an essential and rapidly growing technology in telecommunication and computer networks. The optical fiber technology for communication was developed in the 1970s along with the development of semiconductor lasers. In 1978, Nippon Telegraph and Telephone (NTT) demonstrated the transmission of 32 Mbps (million-bits-per-second) through 53 km of graded-index fiber at 1.3-μm wavelength. By 2001, $3 \times 10^{11}$ m of fiber-optic wires have been installed worldwide; this is a round-trip from the earth to the sun. In March 2006, NEC Corporation announced a 40-Gbps optical-fiber transmission system. The 2009 Nobel Prize in Physics was conferred to Charles K. Kao for his achievements concerning the transmission of light in fibers for optical communications and to Willard S. Boyle and George E. Smith for the invention of CCD sensor. Optical fibers have also been widely applied as sensors for biochemical detection as well as temperature and pressure measurements. Fiber drawing process involves complicated heat transfer and fluid dynamics at different length scales and temperatures [23–25].

Nanophotonics (or nano-optics) is an emerging frontier that integrates photonics with physics, chemistry, biology, materials science, manufacturing, and nanotechnology. The foundation of nanophotonics is to study interactions between light and matter, to explore the unique characteristics of nanostructures for utilizing light energy, and to develop novel nanofabrication and sensing techniques. Recent studies have focused on photonic crystals, nanocrystals, plasmonic waveguides, nanofabrication and nanolithography, light interaction with organic materials, biophotonics, biosensors, quantum electrodynamics, nanocavities, quantum dot and quantum wire lasers, solar cells, and so forth. In the field of thermal radiation, new workshops have been established [26–28] and new experimental discoveries have been made [29–31].

### 1.3.3 Microfabrication and Nanofabrication

Richard Feynman, one of the best theoretical physicists of his time and a Nobel Laureate in Physics, delivered a visionary speech at Caltech in December 1959, entitled "There's plenty of room at the bottom." At that time, lasers had never existed and integrated circuits had just been invented and were not practically useful, and a single computer that is not as fast as a present-day handheld calculator would occupy a whole classroom with enormous heat generation. Feynman envisioned the future of controlling and manipulating things on very small scales, such as writing (with an electron beam) the whole 24 volumes of *Encyclopedia Britannica* on the head of a pin and rearranging atoms one at a time [32]. Many of the things Feynman predicted were once considered scientific fictions or jokes but have been realized in practice by now, especially since 1980s. In 1983, Feynman gave a second talk about the use of swimming machine as a medical device: the surgeon that you could swallow, as well as quantum computing [33]. In the 1990s, micromachining and MEMS emerged as an active research area, with a great success by the commercialization of the micromachined accelerometers in the automobile airbag. Using the etching and lithographic techniques, engineers were able to manufacture microscopic machines with moving parts, as shown in Fig. 1.5, such as gears with a size less than the cross-section of human hair. The technologies used in microfabrication have been extensively discussed in the text of Madou [34]. These MEMS devices were later developed as tools for biological and medical diagnostics, such as the so-called lab-on-a-chip, with pump, valve, and analysis sections on the 10–100 μm scale. In aerospace engineering, an application is to build micro air vehicles or microflyers, with sizes ranging from a human hand down to a bumblebee, that could be used for surveillance and reconnaissance under extreme conditions. Microchannels and microscale heat pipes have also been developed and tested for electronic cooling applications. The study

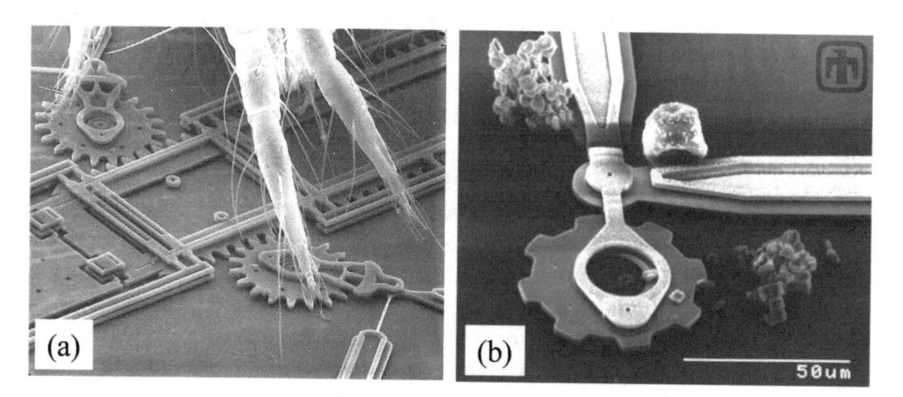


**Fig. 1.5** MEMS structures. **a** A dust mite on a microfabricated mirror assembly, where the gears are smaller than the thickness of human hair. **b** Drive gear chain with linkages, where coagulated red blood cells are on the upper left and the lower right and a grain of pollen is on the upper right. *Courtesy of Sandia National Laboratories,* https://www.sandia.gov/mesa/mems/

of microfluidics has naturally become an active research area in mechanical engineering. The development of SPM and MEMS technologies, together with materials development through self-assembly and other technologies, lead to further development of even smaller structures and the bottom-up approach of nanotechnology. Laser-based manufacturing, focused ion beam (FIB), and electron-beam lithography have also been developed to facilitate nanomanufacturing. In NEMS, quantum behavior becomes important and quantum mechanics is inevitable in understanding the behavior.

Robert Curl, Harold Kroto, and Richard Smalley were winners of the Nobel Prize in Chemistry in 1996 for their discovery of fullerenes in 1985 at Rice University, during a period Kroto visited from University of Sussex. The group used pulsed laser irradiation to vaporize graphite and form carbon plasma in a pressurized helium gas stream. The result as diagnosed by time-of-flight mass spectroscopy suggested that self-assembled $C_{60}$ molecules were formed and would be shaped like a soccer ball with 60 vertices made of the 60 carbon atoms [35]. The results were confirmed later to be $C_{60}$ molecules indeed with a diameter on the order of 1 nm with wave–particle duality. This type of carbon allotrope is called a buckminsterfullerene, or fullerene, or buckyball, after the famous architect Buckminster Fuller (1895–1983) who designed geodesic domes. In 1991, Sumio Iijima of NEC Corporation synthesized carbon nanotubes (CNTs) using arc discharge. Soon his group and an IBM group were able to produce single-walled carbon nanotubes (SWNTs) with a diameter on the order of 1 nm. There have been intensive studies of CNTs for hydrogen storage, nanotransistors, field emission, light emission and absorption, quantum conductance, nanocomposites, and high thermal conductivity. Figure 1.6a shows CNTs growth at a room-temperature environment by chemical vapor deposition on a heated cantilever tip with a size around 5 μm [36]. Figure 1.6b shows the synthesized SWNTs with encapsulated metallofullerenes of Gd:$C_{82}$ (i.e., a gadolinium inside a fullerene molecule). The high-resolution transmission electron microscope (TEM) image suggests that the diameter of the SWNT is from 1.4 to 1.5 nm [37]. It should be noted that electron microscopes, including SEM and TEM, have become a powerful tool for imaging micro/nanoscale objects with a magnification up to 2 million. The first electron microscope was built by Ernst Ruska and Max Knoll in Germany during the early 1930s, and Ruska shared the Nobel Prize in Physics in 1986 for his contributions to electron optics and microscopy.

Various nanostructured materials have been synthesized, such as silicon nanowires, InAs/GaAs QDs, and Ag nanorods. Figure 1.6c shows some images for nanohelices or nanosprings made of ZnO nanobelts or nanoribbons using a solid-vapor process [38, 39]. These self-assembled structures under controlled conditions could be fundamental to the study of electromagnetic coupled nanodevices for use as sensors and actuators, as well as the growth dynamics at the nanoscale.

Since 2004, graphene and other two-dimensional (2D) sheet materials have received great attention due to their amazing and unusual properties. The combination of these materials with micro/nanofabrication holds enormous potentials to revolutionize current microelectronic, optoelectronic, and photonic devices as well as

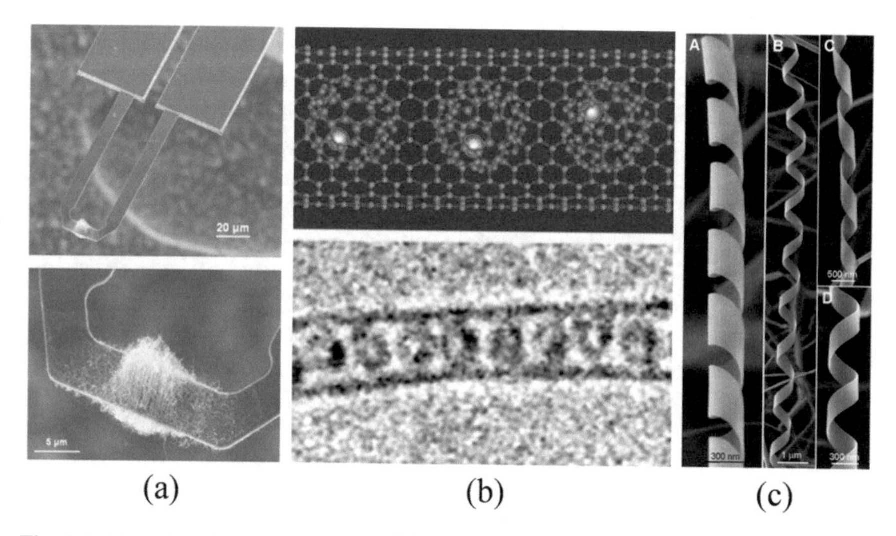


**Fig. 1.6** Examples of nanostructures. **a** SEM image of CNTs grown on heated cantilever tip. *Reprinted with permission from Sunden* et al. [36]; *copyright (2006) American Institute of Physics.* **b** Buckyballs inside a SWNT (the lower is a TEM image in which the nanotube diameter is 1.4–1.5 nm). *Reprinted with permission from Hirahara* et al. [37]; *copyright (2000) American Physical Society.* **c** TEM images of ZnO nanobelts that are coiled into nanohelices or nanosprings. *Reprinted with permission from Gao* et al. [38]; *copyright (2005) AAAS (image courtesy of Prof. Z. L. Wang, Georgia Tech)*

energy harvesting systems [40–48]. For their groundbreaking experiments exfoliating graphene and charactering its properties, Andre Geim and Konstantin Novoselov were awarded the 2010 Nobel Prize in Physics. As a layered 2D material with carbon atoms arranged in a honeycomb lattice, graphene has unique electronic, thermal, mechanical, and optical properties. Unlike conventional metals, free electrons in graphene are massless quasi-particles that exhibit a linear energy-momentum dispersion governed by the Dirac equation for 2D relativistic fermions. As such, graphene offers certain exotic characteristics such as the extremely high mobility, large thermal conductivity, a universal conductance in the optical frequency region, and unique plasmonic characteristics with 2D graphene patches and ribbons [40–42]. Furthermore, the infrared conductance of graphene can be tuned by chemical doping or voltage gating, leading to promising high-speed photodetectors, transistors, solar cells, as well as optical modulators [43, 44]. More recently, a large number of 2D materials have been synthesized chemically or isolated using mechanical or liquid-phase exfoliation from their layered crystalline forms. These 2D materials and their heterostructures have great potentials for photodetectors, nanophotonics, transparent electrodes, and energy conversion and storage [45–48].

One of the successful technologies that operate in the regime of quantum mechanical domain is the giant magnetoresistive (GMR) head and hard drive. The GMR head is based on ferromagnetic layers separated by an extremely thin (about 1 nm) nonferromagnetic spacer, such as Fe/Cr/Fe and Co/Cu/Co. MBE enabled the metallic film growth with required precision and quality. The electrical resistance of GMR materials depends strongly on the applied magnetic field, which affects the spin states of electrons. IBM first introduced this technology in 1996, which was only about 10 years after the publication of the original research results [49, 50]. GMR materials have been extensively used in computer hard drive and read/write head. Albert Fert and Peter Grünberg won the 2007 Nobel Prize in Physics for this discovery. Overheating due to friction with the disk surface can render the data unreadable for a short period until the head temperature stabilizes; such an effect is called *thermal asperity*. Yang et al. [51, 52] performed a detailed thermal characterization of Cu/CoFe superlattices for GMR head applications using MEMS-based thermal metrology tools. Infrared near-field transducer is a key element for heat-assisted magnetic recording (HAMR) to achieve a density of 1 TB/in$^2$. Datta and Xu [53] designed different nanostructures for near-field transducer to boost the coupling efficiency.

### 1.3.4 Probe and Manipulation of Small Structures

Tunneling by elementary particles is a quantum mechanical phenomenon or wavelike behavior. Quantum tunneling refers to the penetration of a particle through a potential barrier whose height (potential energy that the particle would have at the top of the barrier) is greater than the total energy of the particle. When the barrier width is thin enough, quantum tunneling can occur and particles can transmit through the barrier, as if a tunnel is dug through a mountain. An example is the tunneling of electrons through an insulator between two metal strips. Trained in mechanical engineering, Ivar Giaever performed the first tunneling experiment with superconductors in 1960 at the General Electric Research Laboratory and received the 1973 Nobel Prize in Physics, together with Leo Esaki of IBM and Brian Josephson. Esaki made significant contributions in semiconductor tunneling, superlattices, and the development of MBE technology. He invented a tunneling diode, called the Esaki diode, which is capable of very fast operation in the microwave region. Josephson further developed the tunneling theory and a device, called a Josephson junction, which is used in the superconducting quantum interface devices (SQUIDs), for measuring extremely small magnetic fields. SQUIDs are used in magnetic resonance imaging (MRI) for medical diagnostics.

In 1981, Gerd Binnig and Heinrich Rohrer of IBM Zurich Research Laboratory developed the first scanning tunneling microscope (STM) based on electron tunneling through vacuum [54]. This invention has enabled the detection and manipulation of surface phenomena at the atomic level and, thus, has largely shaped the nanoscale science and technology through further development of similar instrumentation [55]. Binnig and Rohrer shared the Nobel Prize in Physics in 1986, along with Ruska who developed the first electron microscope as mentioned earlier. STM uses a sharp-stylus-probe tip and piezoelectricity for motion control. When the tip is near 1 nm from the surface, electron can tunnel through the tip to the conductive substrate. The tunneling current is very sensitive to the gap. Therefore, by maintaining the tip in position and scanning the substrate in the $x$–$y$ direction with a constant current (or distance), the height variation can be obtained with extremely good resolution (0.02 nm). Using STM, Binnig et al. [56] soon obtained the real-space reconstruction of the $7 \times 7$ unit cells of Si (111). In 1993, another group at IBM Almaden Research Center was able to manipulate iron atoms to create a 48-atom quantum corral on a copper substrate [57]. The images have appeared in the front cover of many magazines, including *Science* and *Physics Today*. STM can also be used to assemble organic molecules and to study DNA molecules [2].

In 1986, Gerd Binnig, Calvin Quate, and Christoph Gerber developed another type of SPM, that is, the atomic force microscope (AFM) that can operate without a vacuum environment and for electrical insulators [58]. AFM uses a tapered tip at the end of a cantilever and an optical position sensor, as shown in Fig. 1.7. The position sensor is very sensitive to the bending of the cantilever (with a 0.1-nm vertical resolution). When the tip is brought close to the surface, there exist intermolecular forces (repulsive or attractive) between the tip and the atoms on the underneath surface. In the contact mode, the cantilever is maintained in position using the servo signal from the position-sensing diode to adjust the height of the sample, while it scans in the lateral direction. Surface topographic data can be obtained in an ambient environment for nonconductive materials. Other SPMs have also been developed and the family of SPMs is rather large nowadays. Wickramasinghe and co-workers first investigated thermal probing by attaching a thermocouple to the cantilever tip [59–61]. Later, Arun Majumdar's group developed several types of scanning thermal microscope (SThM) for nanoscale thermal imaging of heated samples, including microelectronic devices and nanotubes [62]. Researchers have also modified SThM for measuring and mapping thermoelectric power at nanoscales [63, 64].

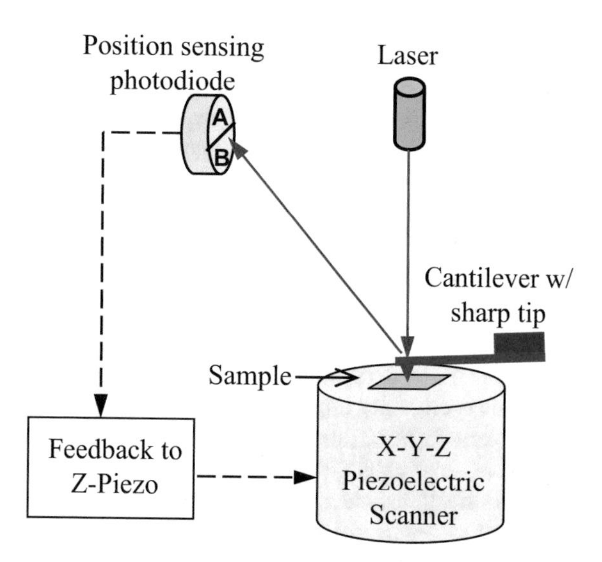


**Fig. 1.7** Schematic of an atomic force microscope (AFM)

Because of its simplicity, AFM has become one of the most versatile tools in nanoscale research, including friction measurements, nanoscale indentation, dip-pen nanolithography, and so forth. Heated cantilever tips were proposed for nanoscale indentation or writing on the polymethyl methacrylate (PMMA) surface, either using a laser or by heating the cantilever legs [65, 66]. The method was further developed to concentrate the heat dissipation to the tip by using heavily doped legs as electrical leads, resulting in writing (with a density near 500 Gb/in$^2$) and erasing (with a density near 400 Gb/in$^2$) capabilities. The temperature signal measured by the tip resistance can also be used to read the stored data due to the difference in heat loss as the tip scans the area [67, 68]. In an effort to improve the data-writing speed, IBM initiated the "millipede" project in 2000 and succeeded in making $32 \times 32$ heated-cantilever array for which each cantilever was separately controlled [69, 70]. Obviously, heat transfer and mechanical characteristics are at the center of these systems. The heated AFM cantilever tips have been used as a local heating source for a number of applications, including the above-mentioned CVD growth of CNTs locally and thermal dip-pen nanolithography [71].

### *1.3.5 Energy Conversion and Storage*

Nanostructures may have unique thermal properties that can be used to facilitate heat transfer for heat removal and thermal management applications. An example was mentioned earlier to utilize nanotubes with high thermal conductivity, although nanotube bundles often suffer from interface resistance and phonon scattering by defects and boundaries. There have been a large number of studies on nanofluids, which are liquids with suspensions of nanostructured solid materials, such as nanoparticles, nanofibers, and nanotubes with diameters on the order of 1–100 nm [72]. Enhanced thermal conductivity and increased heat flux have been demonstrated with a wide range of applications from solar energy harvesting to medical applications, and from electronic cooling to fuel cell thermal management [73–75].

Thermoelectricity utilizes the irreversible thermodynamics principle for thermal-electrical conversion and can be used for cooling in microelectronics as well as miniaturized power generation. A critical issue is to enhance the figure of merit of performance, with a reduced thermal conductivity. Multilayer heterogeneous structures create heat barriers due to size effects and the boundary resistance. These structures have been extensively studied in the literature and demonstrate enhanced performances. Understanding the thermal and electrical properties of heterogeneous structures is critically important for future design and advancement [76–78].

Nanostructures can also help increase the energy conversion efficiency and reduce the cost of solar cells [79]. Furthermore, nanomaterials have been used to develop novel photovoltaic devices. Figure 1.8 shows the device structure of a ZnO-nanowire array for dye-sensitized solar cells [80]. This structure can greatly enhance the absorption or quantum efficiency over nanoparticle-based films. Improvement of photon-to-electron conversion efficiency may be achieved using photonic crystals [81]. In recent

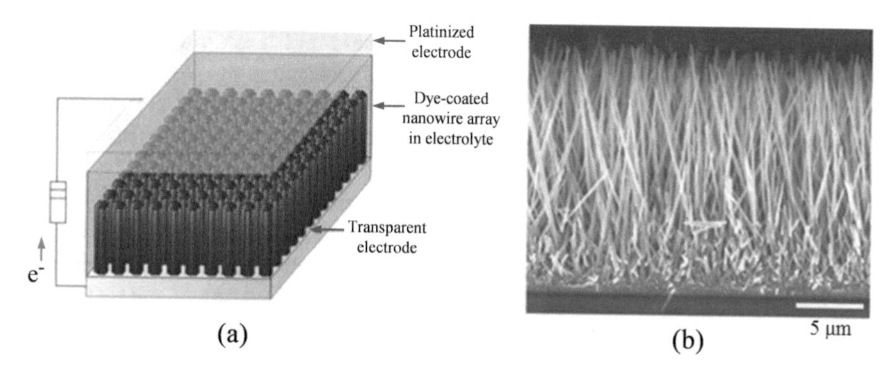


**Fig. 1.8** ZnO nanowires for dye-sensitized solar cells. *Reprinted with permission from Law et al.* [80]; copyright (2005) Macmillan Publishers Ltd. The height of the wires is near 16 μm and their diameters vary between 130 and 200 nm. **a** Schematic of the cell with light incident through the bottom electrode. **b** SEM image of a cleaved nanowire array

years, solar cells based on organic–inorganic halide perovskites have emerged with rapidly increased efficiency and various fabrication processes and structure designs [82].

Fast-depleting reserves of conventional energy sources have resulted in an urgent need for increasing energy conversion efficiencies and recycling of waste heat. One of the potential candidates for fulfilling these requirements is thermophotovoltaic devices, which generate electricity from either the complete combustion of different fuels or the waste heat of other energy sources, thereby saving energy. The thermal radiation from the emitter is incident on a photovoltaic cell, which generates electrical currents. Applications of such devices range from hybrid electric vehicles to power sources for microelectronic systems. At present, thermophotovoltaic systems suffer from low conversion efficiency. Nanostructures have been extensively used to engineer surfaces with designed absorption, reflection, and emission characteristics. Moreover, at the nanoscale, the radiative energy transfer can be greatly enhanced due to tunneling and enhanced local density of states. A viable solution to increase the thermophotovoltaic efficiency is to apply microscale radiation principles in the design of different components to utilize the characteristics of thermal radiation at small distances and in microstructures [83].

Concentrated solar power (CSP) has regained the interest with significant governmental investments in countries like the Spain, United States, Australia, China, United Arab Emirates, India, and so on. Solar energy is reflected by a large array of mirrors to a central receiver (power tower) to create a high-temperature source that can be used to heat a working fluid and then used to generate electricity through a steam turbine or a gas turbine power plant [84]. CSP may be combined with thermal storage system for operation during night or bad weathers; therefore, it can potentially offer a high-efficiency and cost-effective renewable energy solution. Challenges remain in the selected materials for energy storage, receiver efficiency, and system integration [85]. Various spectrally selective absorber/emitter that can operate at high temperatures are being developed using multilayers and nanostructures [86, 87]. Thermochemical cycle with redox reactions may be used as the high-temperature storage to further boost the efficiency of CSP [88, 89]. Further research is needed to understand the materials properties, develop and identify suitable storage substances, as well as develop high-temperature high-pressure power systems [84, 85].

Alternatives to traditional turbomachinery have also been developed to improve the cost effectiveness [85]. One of such methods is called solar thermophotovoltaic (STPV) systems [90, 91]. As illustrated in Fig. 1.9, an intermediate absorber/emitter assembly converts the solar radiation to a relatively lower temperature and then emits with a larger area to a PV cell, which then generates electricity. The prototype by Lenert et al. [90] used an carbon nanotube array to fully absorb the solar irradiation and a 1D $Si/SiO_2$ photonic crystal as the selective emitter to match with the bandgap of 0.55 eV ($\lambda = 2.26\,\mu m$). An efficiency of 3.2% was demonstrated and later an efficiency of 6.8% has been achieved using an improved PC design with a bandpass

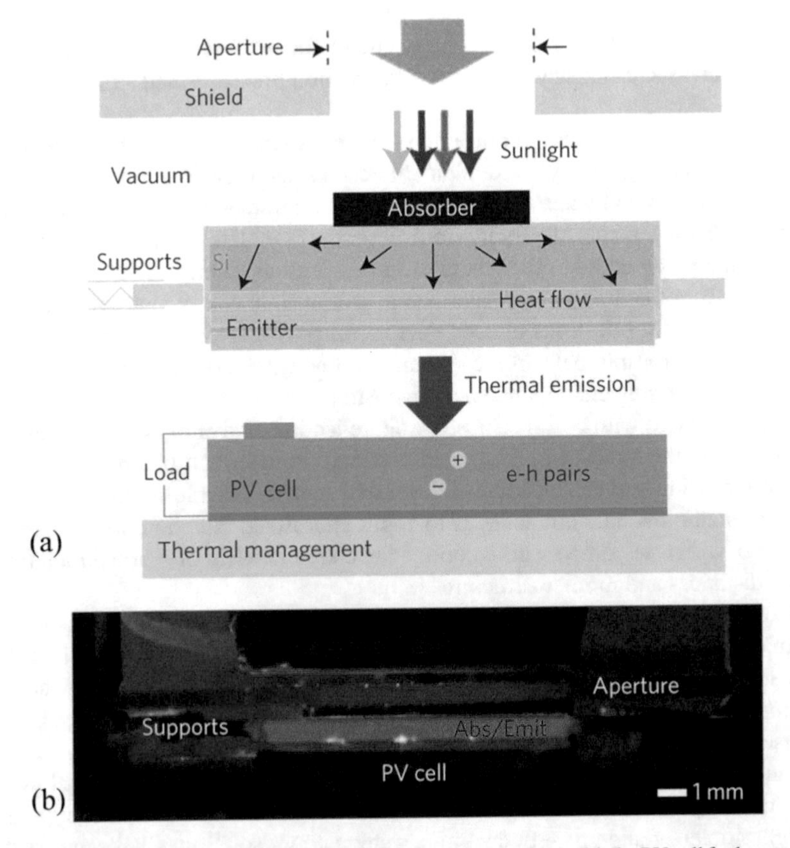


**Fig. 1.9** A nanophotonic solar TPV: **a** Schematic of absorber/emitter with the PV cell for harvesting concentrated solar power; **b** Microscopic image of cross-section. *Reprinted with permission from Lenert et al.* [90]; *copyright (2014) Springer Nature*

filter [91]. These devices are scalable and significant improvement may be made in order to approach the theoretical conversion efficiency limit of about 60%.

Hydrogen technologies are being considered and actively pursued as the energy source of the future [92]. There are two ways in which hydrogen $H_2$ may be used: one is in a combustion heat engine where hydrogen reacts with oxygen intensively while releasing heat; the other is in a fuel cell where electrochemical reaction occurs quietly to generate electricity just like a battery. Because the only reaction product is water, hydrogen-powered automobiles can be made pollution free in principle. Grand challenges exist in generation, storage, and transport of hydrogen. If all hydrogen is obtained from fossil fuels, there will be no reduction in either the fossil fuel consumption or the carbon dioxide emission, except that the emission is centralized in the hydrogen production plant. Alternatively, hydrogen may be produced from water with other energy sources, such as renewable energy sources. Nanomaterials are being developed for several key issues related to hydrogen technologies, such as hydrogen storage using nanoporous materials, effective hydrogen generation by harvesting solar energy with inexpensive photovoltaic materials, and fuel cells based on nanostructure catalysts [93, 94]. Effective thermal management and cooling are also very important to improve the performance and reliability of the fuel cell technology [75, 95].

Lithium-ion batteries are commonly used in cell phones, laptops, and electric cars. Moreover, competing technologies have been developed and commercialized. The 2019 Nobel Prize in Chemistry honors John B. Goodenough, M. Stanley Whittingham, and Akira Yoshino for the pioneering research toward the development of Li-ion batteries during 1970s and 1980s. Overheating or thermal runaway has been known to cause device failure as well as fire disasters. Therefore, understanding the thermal properties and thermal transport at the microscale is critically important to improving the performance and reliability [96, 97]. Nanobatteries and nanogenerators have also been actively explored [98, 99].

### 1.3.6 Biomolecule Imaging and Molecular Electronics

Optical microscopy has played an instrumental role in medical diagnoses because it allows us to see bacteria and blood cells. Optical wavelength is more desirable than x-ray or electron beam because of the less invasiveness and the more convenience. However, the resolution of a traditional microscope is on the order of half the wavelength due to the diffraction limit. While the concept of near-field imaging existed in the literature before 1930, it has been largely forgotten because of the inability in building the structures and controlling their motion. With the microfabrication and precision-positioning capabilities, near-field scanning optical microscopes (NSOMs, also called SNOMs) were realized in the early 1980s by different groups and extensively used for biomolecule imaging with a resolution of 20–50 nm [100]. The principle is to bring the light through an aperture of a tapered fiber of very small diameter at the end or to bring the light through an aperture of very small diameter. The beam out from the fiber tip or aperture will diverge quickly if the sample is placed in the far field, that is, away from the aperture. However, high resolution can be achieved by placing the sample in close proximity to the aperture within a distance much less than the wavelength, that is, in the near field, such that the beam size is almost the same as the aperture. An apertureless metallic tip can be integrated with an SPM to guide the electromagnetic wave via surface plasmon resonance with a spatial resolution as high as 10 nm, for high-resolution imaging and processing. There have since been extensive studies on near-field interactions between electromagnetic waves and nanostructured materials, from semiconductor QDs, metallic nanoaperture and nanohole arrays, to DNA and RNA structures. The 2014 Nobel Prize in Chemistry was awarded to Eric Betzig, Stefan W. Hell, and William E. Moerner for the development of super-resolved fluorescence microscopy for imaging individual molecules. The 2017 Nobel Prize in Chemistry recognized the development of cryo-electron microscopy by Jacques Dubochet, Joachim Frank, and Richard Henderson for probing biomolecules with high resolution.

Nanoparticles are among the earliest known nanostructures that have been used for centuries in making stained glass with gold or other metallic nanoparticles as well as photographic films with silver nanoparticles. A QD has a spherical core encapsulated in a shell made of another semiconductor material, such as a CdSe core in a ZnS shell. The outer shell is only several monolayers thick, and the diameters of QDs range from 2 to 10 nm. The material for the inner core has a smaller bandgap. Quantum confinement in the core results in size-dependent fluorescent properties. Compared with molecular dyes conventionally used for fluorescent labeling in cellular imaging, the emission from QD fluorophores is brighter with a narrower spectral width. QDs also allow excitation at shorter wavelengths, making it easier to separate the fluorescent signal from the scattered one, and are resistive to photobleaching that causes dyes to lose fluorescence. Furthermore, the emission wavelength can be selected by varying the core size of QDs to provide multicolor labeling. It was first demonstrated in 1998 that QDs could be conjugated to biomolecules such as antibodies, peptides, and DNAs, enabling surface passivation and water solubility. In recent years, significant development has been made to employ QDs for in vivo and in vitro imaging, labeling, and sensing [101, 102].

CMOS technology is a top-down semiconductor fabrication process, in which patterns are created by first making a mask and then printing the desired features onto the surface of the wafer via lithography. Integrated circuits have dominated the technological and economic progress in the past 40 years, and complex and high-density devices have been manufactured on silicon wafers. However, this technology is coming to a limit, as the smallest feature size is less than a few nanometers or just about ten unit cells. While opportunities still remain in semiconductor technology as discussed previously [8], molecular electronics is considered as a promising alternative [103]. A 3D assembly with short interconnect distances would greatly increase the information storage density and transfer speed with reduced power consumption and amount of heat being dissipated. Self-assembly means naturally occurring processes, from biological growth to the galaxy formation. In materials synthesis, self-assembly implies that the end products or structures are formed under favorable conditions and environments. An example is the growth of bulk crystals from a seed. Fullerenes and nanotubes are formed by self-assembling, not by slicing a graphite piece and then rolling and bending it to the shape of a tube or a shell. Self-assembly is referred to as a bottom-up process, like constructing an airplane model with LEGO pieces. Biological systems rely on self-assembly and self-replication to develop. Since 2000, CNT-based transistors have been built by several groups and found to be able to outperform Si-based ones. Transistors have also been created using a single molecule of a transition-metal organic complex nanobridge between two electrodes [104]. Because of the small dimensions, quantum mechanics should govern the electrical and mechanical behaviors [105]. Figure 1.10 illustrates an engineered DNA strand between metallic atoms, noting that the width of a DNA strand

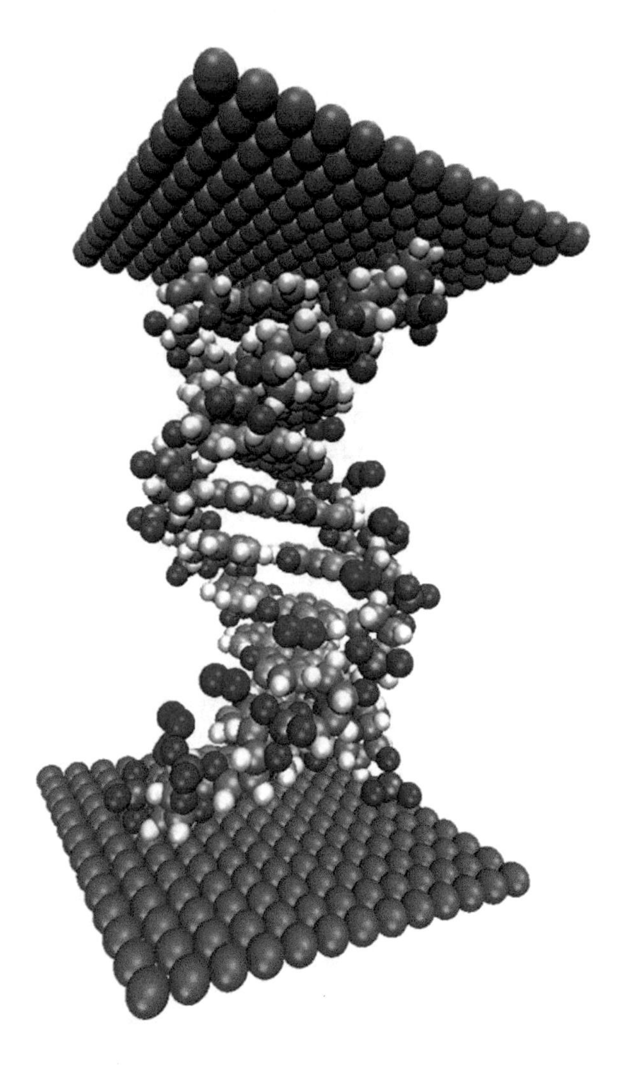

**Fig. 1.10** An engineered DNA strand between metal and atom contacts that could function as a molecular electronics device. *Courtesy of NASA Ames Center of Nanotechnology*

is around 2 nm. Such a structure could function as a sensor and other electronic components. Molecular electronics, while still at its infancy, is expected to revolutionize electronics industry and to enable continuous technological progress through the twenty-first century.

Nano/microscale research and discoveries have been instrumental to the development of technologies used today in microelectronics, photonics, communication, manufacture, and biomedicine. However, systematic and large-scale government investment toward nanoscience and engineering did not start until late 1990s, when the Interagency Working Group on Nanoscience, Engineering, and Technology (IWGN) was formed under the National Science and Technology Council (NSTC). The first report was released in fall 1999, entitled "Nanostructure Science and Technology," followed by the report, "Nanotechnology Research Directions." In July 2000, NSTC published the "National Technology Initiative (NNI)." A large number of nanotechnology centers and nanofabrication facilities have been established since then; see www.nano.gov. In the United States, the government spending on nanotechnology R&D exceeded $1 billion in 2005, as compared to $464 million in 2001 and approximately $116 million in 1997. The total government investment worldwide was over $4 billion in 2005, and Japan and European countries invested similar amount of money as the United States did. Over 60 countries have launched nanotechnology research programs. The NNI funding has totaled near $29 billion from 2001 to 2020. Recognizing the increasing impact on engineering and science, the American Society of Mechanical Engineers established the ASME Nanotechnology Institute in mid-2001 and sponsored a large number of international conferences and workshops. Understanding the thermal transport and properties at the nanoscale is extremely important as mentioned earlier. In 2008, Professor "Bob" D. Y. Tzou initiated the ASME International Conference of Microscale/Nanoscale Heat and Mass Transfer (MNHMT) in Tainan, which was followed by five successive MNHMTs in Shanghai, Atlanta, Hong Kong, Singapore, and Dalian [106–110]. These conferences have provided a highly interactive forum for researchers, educators, and practitioners around the world to exchange and promote the knowledge and new advances on the state-of-the-art research and development in this interdisciplinary field. The ASME Heat Transfer Division established the committee on Nanoscale Thermal Transport in 2012 and organized many focused research sessions at various ASME conferences.

Engineers have the responsibility to transfer the basic science findings into technological advances, to design and develop better materials with desired functions, to build systems that integrate from small to large scales, to perform realistic modeling and simulation that facilitate practical realization of improved performance and continuously reduced cost, and to conduct quantitative measurements and tests that determine the materials properties and system performance. Like any other technology, nanotechnology may also have some adverse effects, such as toxic products and biochemical hazards, which are harmful to human health and the environment. There are also issues and debates concerning security, ethics, and religion. Governmental and industrial standard organizations, as well as universities, have paid great attention to the societal implications and education issues in recent years. Optimists believe that we can continue to harness nanobiotechnology to improve the quality of human life and benefit social progress, while overcoming the adverse effects, like we have done with electricity, chemical plants, and space technology.

## 1.4 Objectives and Organization of This Book

Scientists, engineers, entrepreneurs, and lawmakers must work together for the research outcomes to be transferred into practical products that will advance the technology and benefit the society. Nanotechnology is still in the early stage and holds tremendous potentials; therefore, it is important to educate a large number of engineers with a solid background in nanoscale analysis and design so that they will become tomorrow's leaders and inventors. There is a growing demand of educating mechanical engineering students at both the graduate and undergraduate levels with a background in thermal transport at micro/nanoscales. Micro/nanoscale heat transfer courses have been introduced in a number of universities; however, most of these courses are limited at the graduate level. While an edited book on *Microscale Energy Transport* has been available since 1998 [3], it is difficult to use as a textbook due to the lack of examples, homework problems, and sufficient details on each subject. Some universities have introduced nanotechnology-related courses to the freshmen and sophomores, with no in-depth coverage on the fundamentals of physics. A large number of institutions have introduced joint mechanical-electrical engineering courses on MEMS/NEMS, with a focus on device-level manufacturing and processing technology. To understand the thermal transport phenomena and thermophysical properties at small length scales, learning the concepts and principles of quantum mechanics, solid-state physics, and electrodynamics are inevitable while being difficult for engineering students.

The aim of this book is to introduce the much needed physics knowledge without overwhelming mathematical operators or notions that are unfamiliar to engineering students. Therefore, this book can be used as the textbook not only in a graduate-level course but also in an elective for senior engineering undergraduates. While the book contains numerous equations, the math requirement mostly does not exceed engineering calculus including series, differential and integral equations, and some vector and matrix algebra. The reason to include such a large number of equations is to provide necessary derivation steps, so that readers can follow and understand clearly. This is particularly helpful for practicing engineers who do not have a large number of references at hand. The emphasis of this book is placed on the fundamental understanding of the phenomena and properties: that is, why do we need particular equations and how can we apply them to solve thermal transport problems at the prescribed length and time scales? Selected and refined examples are provided that are both practical and illustrative. At the end of each of the remaining nine chapters, a large number of exercises are given at various levels of complexity and difficulty. Numerical methods are not presented in this book. Most of the problems can be solved with a personal computer using a typical software program or spreadsheet. Some open source codes are accessible and downloadable from the author's website.

For course instructors, the solutions of many homework problems can be obtained from the author.

The field of micro/nanoscale heat transfer was cultivated and fostered by Professor Chang-Lin Tien beginning in late 1980s, along with the rapid development in microelectronics, MEMS, and nanotechnology. His long-lasting and legendary contributions to the thermal science research have been summarized in a volume of *Annual Review of Heat Transfer* [111]. As early as in the 1960s, Professor Tien investigated the fundamentals of the radiative properties of gas molecules, the size effect on the thermal conductivity of thin films and wires, and radiation tunneling between closely spaced surfaces. In 1971, he authored with John H. Lienhard a book, entitled *Statistical Thermodynamics,* which provides inspiring discussions on early quantum mechanics and models of thermal properties of gases, liquids, and crystalline solids. While thermodynamics is a required course for mechanical engineering students, the principles of thermodynamics cannot be understood without a detailed background in statistical thermodynamics. Statistical mechanics and kinetic theory are also critical for understanding thermal properties and transport phenomena.

Chapter 2 provides an overview of equilibrium thermodynamics, heat transfer, and fluid mechanics. Built up from the undergraduate mechanical engineering curricula, the materials are introduced in a quite different sequence to emphasize thermal equilibrium, the second law of thermodynamics, and thermodynamic relations. The concept of entropy is rigorously defined and applied to analyze conduction and convection heat transfer problems in this chapter. It should be noted that, in Chap. 8, an extensive discussion is given on the entropy of radiation.

Chapter 3 introduces statistical mechanics and derives the classical (Maxwell-Boltzmann) statistics and quantum (i.e., Bose–Einstein and Fermi-Dirac) statistics. The first, second, and third laws of thermodynamics are presented with a microscopic interpretation, leading to the discussion of Bose–Einstein condensate and laser cooling of atoms. The classical statistics is extensively used to obtain the ideal gas equation, the velocity distribution, and the specific heat. A concise presentation of elementary quantum mechanics is then provided. This will help students gain a deep understanding of the earlier parts of this chapter. For example, the quantization of energy levels and the energy storage mechanisms by translation, rotation, and vibration for modeling the specific heat of ideal polyatomic gases. The combined knowledge of quantum mechanics and statistical thermodynamics is important for subsequent studies. The concept of photon as an elementary particle and how it interacts with an atom are discussed according to Einstein's 1917 paper on the atomic absorption and emission mechanisms. Finally, the special theory of relativity is briefly introduced to help understand the limitation of mass conservation and the generality of the law of energy conservation.

Chapter 4 begins with a very basic kinetic theory of dilute gases and provides a microscopic understanding of pressure and shear. With the help of mean free path and average collision distance, the transport coefficients such as viscosity, thermal conductivity, and mass diffusion coefficient are described. Following a discussion of intermolecular forces, the detailed Boltzmann transport equation (BTE) is presented to fully describe hydrodynamic equations as well as Fourier's law of heat conduction, under appropriate approximations. In the next section, the regimes of microflow are described based on the Knudsen number, and the current methods to deal with microfluidics are summarized. The heat transfer associated with slip flow and temperature jump is presented in more detail with a simple planar geometry. Then, gas conduction between two surfaces under free molecule flow is derived. These examples, while simple, capture some of the basics of microfluidics. No further discussion is given on properties of liquids or multiphase fluids. It should be noted that several books on microflow already exist in the literature.

The next three chapters provide a comprehensive treatment of nano/microscale heat transfer in solids, with an emphasis on the physical phenomena as well as material properties. The materials covered in Chap. 5 are based on simple free-electron model, kinetic theory, and BTE without a detailed background of solid-state physics, which is discussed afterward in Chap. 6. This not only helps students comprehend the basic, underlying physical mechanisms but also allows the instructor to integrate Chap. 5 into a graduate heat conduction course. For an undergraduate elective, Chap. 6 can be considered as reading materials or references without spending too much time going through the details in class. In Chap. 5, the theory of specific heat is presented with a detailed treatment on the quantum size effect. Similarly, the theory of thermal conductivity of metals and dielectric solids is introduced. Because of the direct relation between electrical and thermal conductivities and the importance of thermoelectric effects, irreversible thermodynamics and thermoelectricity are also introduced. The classical size effect on thermal conductivity due to boundary scattering is elaborated. Finally, the concept of quantum conductance (both electric and thermal) is introduced.

Chapter 6 introduces the electronic band structures and phonon dispersion relations in solids. It helps understand semiconductor physics and some of the difficulties of free-electron model for metals. Photoemission, thermionic emission, and electron tunneling phenomena are introduced. The electrical transport in semiconductors is described with applications in energy conversion and optoelectronic devices. Chapter 7 focuses on nonequilibrium energy transport in nanostructures, including non-Fourier equations for transient heat conduction. The equation of phonon radiative transfer is presented and solved for thin-film and multilayer structures. The phenomenon of thermal boundary resistance is studied microscopically. A regime map is developed in terms of the length scale and the time scale from macroscale to microscale to nanoscale heat conduction. Additional reading materials regarding multiscale modeling, atomistic modeling, and thermal metrology are provided as references.

The last three chapters give comprehensive discussion on nano/microscale radiation with extensive background on the fundamentals of electromagnetic waves, the optical and thermal radiative properties of materials and surfaces, and the recent advancement in nanophotonics and nanoscale radiative transfer. Chapter 8 presents the Maxwell equations of electromagnetic waves and the derivation of Planck's law and radiation entropy. The electric and magnetic properties of the newly developed class of materials, that is, negative-refractive-index materials are also discussed.

More extensive discussion of the radiative properties of thin films, gratings, and rough surfaces is given in Chap. 9. The wave interference, partial coherence, and diffraction phenomena are introduced with detailed formulations. Furthermore, various types of surface polaritons and localized excitations in nanostructures and 2D materials are extensively discussed. The focus of Chap. 10 is on near-field thermal radiation, with formulations of simple semi-infinite parallel plates to complicated systems. In addition, Chap. 10 reviews contemporary numerical simulation methods for computing nanoscale thermal radiation and recent experimental techniques for measuring nanoscale radiative transfer. These advancements will continue and are expected to have a huge impact on the energy conversion devices, sensors, and nanoscale photothermal manufacturing.

It is hoped that the present text can be used either as a whole in a one-semester course, or in part for integration into an existing thermal science course for several weeks on a particular topic. Examples are graduate-level thermodynamics (Chaps. 2 and 3), convection heat transfer (Chap. 4), conduction heat transfer (Chaps. 5–7), and radiation heat transfer (Chaps. 8–10). Selected materials may also be used to introduce nanoscale thermal sciences in undergraduate heat transfer and fluid mechanics courses. Some universities offer a second course on thermodynamics at the undergraduate level for which statistical thermodynamics and quantum theory can also be introduced. This text can also be self-studied by researchers or practicing engineers, graduated from a traditional engineering discipline. A large effort is given to balance the depth with the breadth so that it is easy to understand and contains sufficient coverage of both the fundamentals and advanced developments in the field. Readers will gain the background necessary to understand the contemporary research in nano/microscale thermal engineering and to solve a variety of practical problems using the approaches presented in the text, along the codes accessible from author's website [112].

## References

1. C.P. Poole Jr., F.J. Owens, *Introduction to Nanotechnology* (Wiley, New York, 2003)
2. E.L. Wolf, *Nanophysics and Nanotechnology—An Introduction to Modern Concepts in Nanoscience* (Wiley-VCH, Weinheim, Germany, 2004)
3. C.L. Tien, A. Majumdar, F.M. Gerner (eds.), *Microscale Energy Transport* (Taylor & Francis, Washington, DC, 1998)
4. G. Chen, *Nanoscale Energy Transport and Conversion* (Oxford University Press, New York, 2005)
5. C.E. Shannon, A mathematical theory of communication. Bell Syst. Tech. J. **27**, 379–423, 623–656, July & October 1948, http://cm.bell-labs.com/cm/ms/what/shannonday/paper.html
6. G.E. Moore, Cramming more components onto integrated circuits. Electronics **38**(8), 114–117 (1965)
7. G.E. Moore, Progress in digital integrated electronics. IEEE Tech. Digest (International Electron Devices Meeting), 11–13 (1975), www.intel.com/technology/mooreslaw
8. M. Mitchell Wardrop, More than Moore. Nature **530**, 144–147 (2016)
9. Z.M. Zhang, B.K. Tsai, G. Machin, *Radiometric Temperature Measurements: I. Fundamentals; II. Applications* (Academic Press/Elsevier, Amsterdam, 2009)

10. C.-H. Fan, J.P. Longtin, Radiative energy transport at the spatial and temporal micro/nano scales. In: *Heat Transfer and Fluid Flow in Microscale and Nanoscale Structures,* M. Faghri, B. Sunden (eds.) (WIT Press, Southampton, UK, 2003), pp. 225–275
11. W. Denk, J.H. Stricker, W.W. Webb, Two-photon laser scanning fluorescence microscopy. Science **248**, 73–76 (1990)
12. T. Yu, C.K. Ober, S.M. Kuebler, W. Zhou, S.R. Marder, J.W. Perry, Chemically-amplified positive resist system for two-photon three-dimensional lithography. Adv. Mat. **15**, 517–521 (2003)
13. S.M. Kuebler, K.L. Braun, W. Zhou et al., Design and application of high-sensitivity two-photon initiators for three-dimensional microfabrication. J. Photochem. Photobio. A: Chemistry **158**, 163–170 (2003)
14. M.F. Modest, H. Abakians, Heat-conduction in a moving semi-infinite solid subject to pulsed laser irradiation. J. Heat Transfer **108**, 597–601 (1986)
15. M.F. Modest, H. Abakians, Evaporative cutting of a semi-infinite body with a moving cw laser. J. Heat Transfer **108**, 602–607 (1986)
16. C.L. Tien, T.Q. Qiu, P.M. Norris, Microscale thermal phenomena in contemporary technology. Thermal Sci. Eng. **2**, 1–11 (1994)
17. R.J. Stoner, H.J. Maris, Kapitza conductance and heat flow between solids at temperatures from 50 to 300 K. Phys. Rev. B **48**, 16373–16387 (1993)
18. W.S. Capinski, H.J. Maris, T. Ruf, M. Cardona, K. Ploog, D.S. Katzer, Thermal-conductivity measurements of GaAs/AlAs superlattices using a picosecond optical pump-and-probe technique. Phys. Rev. B **59**, 8105–8113 (1999)
19. P.M. Norris, A.P. Caffrey, R. Stevens, J.M. Klopf, J.T. McLeskey, A.N. Smith, Femtosecond pump-probe nondestructive evaluation of materials. Rev. Sci. Instrum. **74**, 400–406 (2003)
20. R.J. Stevens, A.N. Smith, P.M. Norris, Measurement of thermal boundary conductance of a series of metal-dielectric interfaces by the transient thermoreflectance techniques. J. Heat Transfer **127**, 315–322 (2005)
21. O. Manasreh, *Semiconductor Heterojunctions and Nanostructures* (McGraw-Hill, New York, 2005)
22. G. Chen, Heat transfer in micro- and nanoscale photonic devices. Annu. Rev. Heat Transfer **7**, 1–18 (1996)
23. Y. Jaluria, Thermal processing of materials: From basic research to engineering. J. Heat Transfer **125**, 957–979 (2003)
24. X. Cheng, Y. Jaluria, Optimization of a thermal manufacturing process: drawing of optical fiber. Intl. J. Heat Mass Transfer **48**, 3560–3573 (2005)
25. C. Chen, Y. Jaluria, Modeling of radiation heat transfer in the drawing of an optical fiber with multi-layer structure. J. Heat Transfer **129**, 342–352 (2007)
26. Z.M. Zhang, S. Maruyama, A. Sakurai, M.P. Menguç, Special issue on nano- and micro-scale radiative transfer. J. Quant. Spectrosc. Radiat. Transfer **132**, 1–2 (2014)
27. Z.M. Zhang, L.-H. Liu, Q.Z. Zhu, M.P. Menguç, Special issue on the second international workshop on micro-nano thermal radiation. J. Quant. Spectrosc. Radiat. Transfer **158**, 1–2 (2015)
28. B.J. Lee, Y. Shuai, M. Francoeur, M.P. Mengüç, Special issue on the third international workshop on nano-micro thermal radiation. J. Quant. Spectrosc. Radiat. Transfer **237**, 106592 (2019)
29. K. Kim, B. Song, V. Fernández-Hurtado et al., Radiative heat transfer in the extreme near field. Nature **528**, 387–391 (2015)
30. M. Lim, J. Song, S.S. Lee, B.J. Lee, Tailoring near-field thermal radiation between metallo-dielectric multilayers using coupled surface plasmon polaritons. Nat. Commun. **9**, 4302 (2018)
31. J. DeSutter, L. Tang, M. Francoeur, A near-field radiative heat transfer device. Nat. Nanotech. **14**, 751–755 (2019)
32. R.P. Feynman, There's plenty of room at the bottom. J. Microelectromechanical Syst. **1**, 60–66 (1992)

33. R.P. Feynman, Infinitesimal machinery. J. Microelectromechanical Syst. **2**, 4–14 (1993), www.zyvex.com/nanotech/feynman.html
34. M.J. Madou, *Fundamentals of Microfabrication: The Science of Miniaturization*, 2nd edn. (CRC Press, Boca Raton, FL, 2002)
35. H.W. Kroto, J.R. Heath, S.C. O'Brien, R.F. Curl, R.E. Smalley, C60: Buckminsterfullerene. Nature **318**, 162–163 (1985)
36. E.O. Sunden, T.L. Wright, J. Lee, W.P. King, S. Graham, Room-temperature chemical vapor deposition and mass detection on a heated atomic force microscope cantilever. Appl. Phys. Lett. **88**, 033107 (2006)
37. K. Hirahara, K. Suenaga, S. Bandow, et al., One-dimensional metallofullerene crystal generated inside single-walled carbon nanotubes. Phys. Rev. Lett. **85**, 5384 (2000). Also see Phys. Rev. Focus, 19 December 2000, http://focus.aps.org/story/v6/st27
38. P.X. Gao, Y. Ding, W.J. Mai, W.L. Hughes, C.S. Lao, Z.L. Wang, Conversion of zinc oxide nanobelt into superlattice-structured nanohelices. Science **309**, 1700–1704 (2005)
39. X.Y. Kong, Y. Ding, R. Yang, Z.L. Wang, Single-crystal nanorings formed by epitaxial self-coiling of polar nanobelts. Science **309**, 1348–1351 (2004)
40. A.K. Geim, K.S. Novoselov, The rise of graphene. Nat. Mater. **6**, 183–191 (2007)
41. V. Singh, D. Joung, L. Zhai, S. Das, S.I. Khondaker, S. Seal, Graphene based material: past, present and future. Prog. Mater Sci. **56**, 1178–1271 (2011)
42. E. Pop, V. Varshney, A.K. Roy, Thermal properties of graphene: fundamentals and applications. MRS Bull. **37**, 1273–1281 (2012)
43. F.H. Koppens, D.E. Chang, F.J. Garcia de Abajo, Graphene plasmonics: a platform for strong light-matter interactions. Nano Lett. **11**, 3370–3377 (2011)
44. D.N. Basov, M.M. Fogler, A. Lanzara, F. Wang, Y. Zhang, Colloquium: graphene spectroscopy. Rev. Mod. Phys. **86**, 959–994 (2014)
45. K.S. Novoselov, A. Mishchenko, A. Carvalho, A.H. Castro Neto, 2D materials and van der Waals heterostructures. Science **353**, aac9439 (2016)
46. C. Tan, X. Cao, X.J. Wu et al., Recent advances in ultrathin two-dimensional nanomaterials. Chem. Rev. **117**, 6225–6331 (2017)
47. C. Shao, X. Yu, N. Yang, Y. Yue, H. Bao, A review of thermal transport in low- dimensional materials under external perturbation: effect of strain, substrate, and clustering. Nanoscale Microscale Thermophys. Eng. **21**, 201–236 (2017)
48. X. Li, L. Tao, Z. Chen, H. Fang, X. Li, X. Wang, J.-B. Xu, H. Zhu, Graphene and related two-dimensional materials: structure-property relationships for electronics and optoelectronics. Appl. Phys. Rev. **4**, 021306 (2017)
49. P. Grünberg, R. Schreiber, Y. Pang, M.B. Brodsky, H. Sowers, Layered magnetic structures: evidence for antiferromagnetic coupling of Fe layers across Cr interlayers. Phys. Rev. Lett. **57**, 2442–2445 (1986)
50. M.N. Baibich, J.M. Broto, A. Fert, F. Nguyen Van Dau, F. Petroff, P. Etienne, G. Creuzet, A. Friederich, J. Chazelas, Giant magnetoresistance of (001)Fe/(001)Cr magnetic superlattices. Phys. Rev. Lett. **61**, 2472–2475 (1988)
51. Y. Yang, W. Liu, M. Asheghi, Thermal and electrical characterization of Cu/CoFe superlattices. Appl. Phys. Lett. **84**, 3121–3123 (2004)
52. Y. Yang, R.M. White, M. Asheghi, Thermal characterization of Cu/CoFe multilayer for giant magnetoresistive (GMR) head applications. J. Heat Transfer **128**, 113–120 (2006)
53. A. Datta, X. Xu, Infrared near-field transducer for heat-assisted magnetic recording. IEEE Trans. Magnet. **53**, 3102105 (2017); ibid, Optical and thermal designs of near field transducer for heat assisted magnetic recording. Japan. J. Appl. Phys. **57**, 09TA01 (2018)
54. G. Binnig, H. Rohrer, Scanning tunneling microscopy. Helv. Phys. Acta **55**, 726–735 (1982)
55. G. Binnig, H. Rohrer, Ch. Gerber, E. Weibel, Surface studies by scanning tunneling microscopy. Phys. Rev. Lett. **49**, 57–61 (1982)
56. G. Binnig, H. Rohrer, Ch. Gerber, E. Weibel, $7 \times 7$ reconstruction on Si(111) resolved in real space. Phys. Rev. Lett. **50**, 120–123 (1983)

57. M.F. Crommie, C.P. Lutz, D.M. Eigler, Confinement of electrons to quantum corrals on a metal surface. Science **262**, 218–220 (1993)
58. G. Binnig, C.F. Quate, Ch. Gerber, Atomic force microscope. Phys. Rev. Lett. **56**, 930–933 (1986)
59. C.C. Williams, H.K. Wickramasinghe, Scanning thermal profiler. Appl. Phys. Lett. **49**, 1587–89 (1986)
60. J.M.R. Weaver, L.M. Walpita, H.K. Wickramasinghe, Optical absorption microscopy with nanometer resolution. Nature **342**, 783–85 (1989)
61. M. Nonnenmacher, H.K. Wickramasinghe, Optical absorption spectroscopy by scanning force microscopy. Ultramicroscopy **42–44**, 351–354 (1992)
62. A. Majumdar, Scanning thermal microscopy. Annu. Rev. Mater. Sci. **29**, 505–585 (1999)
63. H.-K. Lyeo, A.A. Khajetoorians, L. Shi et al., Profiling the thermoelectric power of semiconductor junctions with nanometer resolution. Science **303**, 818–820 (2004)
64. Z. Bian, A. Shakouri, L. Shi, H.-K. Lyeo, C.K. Shih, Three-dimensional modeling of nanoscale Seebeck measurement by scanning thermoelectric microscopy. Appl. Phys. Lett. **87**, 053115 (2005)
65. H.J. Mamin, D. Rugar, Thermomechanical writing with an atomic force microscope tip. Appl. Phys. Lett. **61**, 1003–1005 (1992)
66. H.J. Mamin, Thermal writing using a heated atomic force microscope tip. Appl. Phys. Lett. **69**, 433–435 (1996)
67. G. Binnig, M. Despont, U. Drechsler et al., Ultrahigh-density atomic force microscopy data storage with erase capability. Appl. Phys. Lett. **74**, 1329–1331 (1999)
68. W.P. King, T.W. Kenny, K.E. Goodson et al., Atomic force microscope cantilevers for combined thermomechanical data writing and reading. Appl. Phys. Lett. **78**, 1300–1302 (2001)
69. U. Dürig, G. Cross, M. Despont, et al. 'Millipede'—an AFM data storage system at the frontier of nanotechnology. Tribology Lett. **9**, 25–32 (2000)
70. P. Vettiger, G. Cross, M. Despont et al., The 'millipede'—nanotechnology entering data storage. IEEE Trans. Nanotechnol. **1**, 39–55 (2002)
71. P.E. Sheehan, L.J. Whitman, W.P. King, B.A. Nelson, Nanoscale deposition of solid inks via thermal dip pen nanolithography. Appl. Phys. Lett. **85**, 1589–1591 (2004)
72. J.A. Eastman, S.R. Phillpot, S.U.S. Choi, P. Kablinski, Thermal transport in nanofluids. Annu. Rev. Mater. Res. **34**, 219–246 (2004)
73. J. Buongiorno, D.C. Venerus, N. Prabhat et al., A benchmark study on the thermal conductivity of nanofluids. J. Appl. Phys. **106**, 094312 (2009)
74. R. Taylor, S. Coulombe, T. Otanicar, P. Phelan, A. Gunawan, W. Lv, G. Rosengarten, R. Prasher, H. Tyagi, Small particles, big impacts: a review of the diverse applications of nanofluids. J. Appl. Phys. **113**, 011301 (2013)
75. M.H. Esfe, M. Afrand, A review on fuel cell types and the application of nanofluid in their cooling. J. Therm. Anal. Calorim. **140**, 1633–1654 (2020) https://doi.org/10.1007/s10973-019-08837-x
76. G. Chen, A. Shakouri, Heat transfer in nanostructures for solid-state energy conversion. J. Heat Transfer **124**, 242–252 (2002)
77. A.J. Minnich, M.S. Dresselhaus, Z.F. Ren, G. Chen, Bulk nanostructured thermoelectric materials: current research and future prospects. Energy Environ. Sci. **2**, 466–479 (2009)
78. S. LeBlanc, S.K. Yee, M.L. Scullin, C. Dames, K.E. Goodson, Material and manufacturing cost considerations for thermoelectrics. Renew. Sustain. Energy Rev. **32**, 313–327 (2014)
79. H.A. Atwater, A. Polman, Plasmonics for improved photovoltaic devices. Nat. Mater. **9**, 205–213 (2010)
80. M. Law, L.E. Greene, J.C. Johnson, R. Saykally, P. Yang, Nanowire dye-sensitized solar cells. Nature Mater. **4**, 455–459 (2005)
81. S. Guldin S. Hüttner, M. Kolle et al., Dye-sensitized solar cell based on a three-dimensional photonic crystal. Nano Lett. **10**, 2303–2309 (2010)

82. M.A. Green, A. Ho-Baillie, H.J. Snaith, The emergence of perovskite solar cells. Nat. Photon. **8**, 506–515 (2014)
83. S. Basu, Y.-B. Chen, Z.M. Zhang, Microscale radiation in thermophotovoltaic devices—a review. Intl. J. Ener. Res. **31**, 689–716 (2007)
84. O. Behar, A. Khellaf, K. Mohammedia, A review of studies on central receiver solar thermal power plants. Renew. Sustain. Energy Rev. **23**, 12–39 (2013)
85. L.A. Weinstein, J. Loomis, B. Bhatia, D.M. Bierman, E.N. Wang, G. Chen, Concentrated solar power. Chem. Rev. **115**, 12797−12838 (2015)
86. H. Wang, V.P. Sivan, A. Mitchell, G. Rosengarten, P.E. Phelan, L.P. Wang, Highly efficient selective metamaterial absorber for high-temperature solar thermal energy harvesting. Sol. Energy Mater. Sol. Cells **137**, 235–242 (2015)
87. Y. Li, C. Lin, D. Zhou et al., Scalable all-ceramic nanofilms as highly efficient and thermally stable selective solar absorbers. Nano Energy **64**, 103947 (2019)
88. P.G. Loutzenhiser, A. Meier, A. Steinfeld, Review of the two-step $H_2O/CO_2$-splitting solar thermochemical cycle based on Zn/ZnO redox reactions. Materials **3**, 4922–4938 (2010)
89. A.J. Schrader, A.P. Muroyama, P.G. Loutzenhiser, Solar electricity via an air Brayton cycle with an integrated two-step thermochemical cycle for heat storage based on $Co_3O_4$/CoO redox reactions: thermodynamic analysis. Sol. Energy **118**, 485–495 (2015)
90. A. Lenert, D.M. Bierman, Y. Nam, W.R. Chan, I. Celanović, M. Soljačić, E.N. Wang, A nanophotonic solar thermophotovoltaic device. Nat. Nanotech. **9**, 126–130 (2014)
91. D.M. Bierman, A. Lenert, W.R. Chan, B. Bhatia, I. Celanović, M. Soljačić, E.N. Wang, Enhanced photovoltaic energy conversion using thermally based spectral shaping. Nat. Energy **1**, 16068 (2016)
92. G. Crabtree, M. Dresselhaus, M. Buchanan, The hydrogen economy. Phys. Today, 39–44, December 2004
93. B.C.H. Steele, A. Heinzel, Materials for fuel-cell technologies. Nature **414**, 345–352 (2001)
94. Z. Gao, L.V. Mogni, E.C. Miller, J.G. Railsback, S.A. Barnett, A perspective on low-temperature solid oxide fuel cells. Energy Environ. Sci. **9**, 1602–1644 (2016)
95. S.M. Senn, D. Poulikakos, Laminar mixing, heat transfer and pressure drop in tree-like microchannel nets and their application for thermal management in polymer electrolyte fuel cells. J. Power Sources **130**, 178–191 (2004)
96. T.M. Bandhauer, S. Garimella, T.F. Fuller, A critical review of thermal issues in lithium-ion batteries. J. Electrochem. Soc. **158**, R1–R25 (2011)
97. R. Kantharaj, A.M. Marconnet, Heat generation and thermal transport in lithium-ion batteries: a scale-bridging perspective. Nanoscale Microscale Thermophys. Eng. **23**, 128–156 (2019)
98. I. Valov, E. Linn, S. Tappertzhofen, S. Schmelzer, J. van den Hurk, F. Lentz, R. Waser, Nanobatteries in redox-based resistive switches require extension of memristor theory. Nat. Commun. **4**, 1771 (2013)
99. Z.L. Wang, J. Chen, L. Lin, Progress in triboelectric nanogenerators as a new energy technology and self-powered sensors. Energy Environ. Sci. **8**, 2250–2282 (2015)
100. A. Lewis, H. Taha, A. Strinkovski et al., Near-field optics: from subwavelength illumination to nanometric shadowing. Nat. Biotechnol. **21**, 1378–1386 (2003)
101. X. Michalet, F.F. Pinaud, L.A. Bentolila et al., Quantum dots for live cells, in vivo imaging, and diagnostics. Science **307**, 538–544 (2005)
102. I.L. Medintz, H.T. Uyeda, E.R. Goldman, H. Mattoussi, Quantum dot bioconjugates for imaging, labelling and sensing. Nature Mater. **4**, 435–446 (2005)
103. B. Yu, M. Meyyappan, Nanotechnology: role in emerging nanoelectronics. Solid-State Electron. **50**, 536–544 (2006)
104. C. Joachim, J.K. Gimzewski, A. Aviram, Electronics using hybrid-molecular and mono-molecular devices. Nature **408**, 541–548 (2000)
105. A. Vilan, D. Aswal, D. Cahen, Large-area, ensemble molecular electronics: motivation and challenges. Chem. Rev. **17**, 4248–4286 (2017)
106. P. Cheng, S. Choi, Y. Jaluria, D. Q. Li, P. M. Norris, D. Y. Tzou, Special issue on micro/nanoscale heat transfer, Part I. J. Heat Transfer **131**, 030301 (2009); Part II, ibid, **131**, 040301 (2009)

107. P. Cheng, Foreword to special issue on micro/nanoscale heat and mass transfer. J. Heat Transfer **134**, 050301 (2012)
108. Z.M. Zhang, P.M. Norris, G.P. Peterson, Foreword to special issue on micro/nanoscale heat and mass transfer. J. Heat Transfer **135**, 090501 (2013)
109. L.Q. Wang, Y. Jaluria, Foreword to special issue on advances in micro/nanoscale heat and mass transfer. J. Heat Transfer **137**, 090301 (2015)
110. Z.M. Zhang, C. Yang, D.Y. Tzou, Foreword to special issue on micro/nanoscale heat and mass transfer, Part I, J. Heat Transfer **139**, 050301 (2017); Part II, ibid, **140**, 010301 (2018)
111. V. Prasad, Y. Jaluria, G. Chen (eds.), *Annual Review of Heat Transfer*, vol. 14 (Begell House, New York, 2005)
112. Z. M. Zhang, http://zhang-nano.gatech.edu/

# Chapter 2
# Overview of Macroscopic Thermal Sciences

This chapter provides a concise description of the basic concepts and theories underlying classical thermodynamics and heat transfer. Different approaches exist in presenting the subject of thermodynamics. Most engineering textbooks first introduce temperature, then discuss energy, work, and heat, and define entropy afterward. Callen developed an axiomatic structure using a simple set of abstract postulates to combine the physical information that is included in the laws of thermodynamics [1]. Continuing the effort pioneered by Keenan and Hatsopoulos [2], Gyftopoulos and Beretta [3] developed a logical sequence to introduce the basic concepts with a rigorous definition of each thermodynamic term. Their book has been a great inspiration to the present author in comprehending and teaching thermodynamics. Here, an overview of classical thermodynamics is provided that is somewhat beyond typical undergraduate textbooks [4, 5]. Details on the historic development of classical thermodynamics can be found from Bejan [6] and Kestin [7], and references therein. The basic phenomena and governing equations in energy, mass, and momentum transfer will be presented subsequently in a self-consistent manner without invoking microscopic theories.

## 2.1 Fundamentals of Thermodynamics

A *system* is a collection of constituents (whose amounts may be fixed or varied within a specified range) in a defined space (e.g., a container whose volume may be fixed or varied within a specified range), subject to other external forces (such as gravitational and magnetic forces) and constraints. External forces are characterized by *parameters*. An example is the volume of a container, which is a parameter associated with the forces that confine the constituents within a specified space. Everything that is not included in the system is called the *environment* or *surroundings* of the system.


Z. M. Zhang, *Nano/Microscale Heat Transfer*, Mechanical Engineering Series,
https://doi.org/10.1007/978-3-030-45039-7_2

Quantities that characterize the behavior of a system at any instant of time are called *properties* of the system. Properties must be measurable, and their values are independent of the measuring devices. Properties supplement constituents and parameters to fully characterize a system. At any given time, the system is said to be in a *state*, which is fully characterized by the type and amount of constituents, a set of parameters associated with various types of external forces, and a set of properties. Two states are identical if the amount of each type of constituents and values of all the parameters and properties are the same. A system may experience a *spontaneous change of state*, when the change of state does not involve any interaction between the system and its environment. If the system changes its state through interactions with other systems in the environment, it is said to experience an *induced change of state*. If a system can experience only spontaneous changes of state, it is said to be an *isolated system*, that is, the change of state of the system does not affect the environment of the system. The study of the possible and allowed states of a system is called *kinematics*, and the study of the time evolution of the state is called *dynamics*.

The relation that describes the change of state of a system as a function of time is the *equation of motion*. In practice, the complete equations of motion are often not known. Therefore, in thermodynamics, the description of the change of state is usually given in terms of the end states (i.e., the initial and final states) and the *modes of interaction* (for example, work and heat, which are discussed later). The end states and the modes of interaction specify a *process*. A spontaneous change of state is also called a *spontaneous process*. A process is *reversible* if there is at least one way to restore both the system and its environment to their initial states. Otherwise, the process is *irreversible*; i.e., it is not possible to restore both the system and its environment to their initial states. A *steady state* is one that does not change as a function of time despite interactions between the system and other systems in the environment.

### 2.1.1 The First Law of Thermodynamics

*Energy* is a property of every system in any state. The first law of thermodynamics states that *energy can be transferred to or from a system but can be neither created nor destroyed*. The energy balance for a system can be expressed as

$$\Delta E = E_2 - E_1 = E_{\text{net,in}} \tag{2.1a}$$

where $\Delta$ denotes a finite change, subscripts 1 and 2 refer to the initial and final states, respectively, and $E_{\text{net,in}} = E_{\text{in}} - E_{\text{out}}$ is the net amount of energy transferred into the system. For an infinitesimal change, the differential form of the energy balance is

$$\mathrm{d}E = \delta E_{\text{net,in}} \tag{2.1b}$$

Here, $d$ is used to signify a differential change of the property of a system, and $\delta$ is used to specify a differentially small quantity that is not a property of any system. Clearly, the energy of an isolated system is conserved. Energy is an additive property, i.e., the energy of a composite system is the sum of the energies of all individual subsystems. Examples are kinetic energy and potential energy, as defined in classical mechanics, and internal energy, which will be discussed later. A similar expression for mass balance can also be written.

The term *mechanical effect* is used for the kind of processes described in mechanics, such as the change of the height of a weight in a gravitational field, the change of the relative positions of two charged particles, the change of the velocity of a point mass, the change of the length of a spring, or a combination of such changes. All mechanical effects are equivalent in the sense that it is always possible to arrange forces and processes that annul all the mechanical effects except one that we choose. It is common to choose the rise and fall of a weight in a gravity field to represent this kind of process.

A *cyclic process* (also called a *cycle*) is one with identical initial and final states. A *perpetual-motion machine of the first kind* (PMM1) is any device (or system) undergoing a cyclic process that produces no external effects but the rise or fall of a weight in a gravity field. A PMM1 violates the first law of thermodynamics, and hence, it is impossible to build a PMM1. Perpetual motion, however, may exist as long as it produces zero net external effect. Examples of perpetual motion are a lossless oscillating pendulum, an electric current through a superconducting coil, and so forth.

### 2.1.2 Thermodynamic Equilibrium and the Second Law

An *equilibrium* state is a state that cannot change spontaneously with time. There are different types of equilibrium: unstable, stable, and metastable. A *stable-equilibrium state* is a state that cannot be altered to a different state without leaving any net effect on the environment. In the following, a stable-equilibrium state is frequently referred to as a state at *thermodynamic equilibrium*.

The *stable-equilibrium-state principle*, or *state principle*, can be phrased as follows: *Among all states of a system with a given set of values of energy, parameters, and constituents, there exists one and only one stable-equilibrium state*. In other words, in a stable-equilibrium state, all properties are uniquely determined by the amount of energy, the value of each parameter, and the amount of each type of constituents. This principle is an integral part of the second law of thermodynamics [2, 3, 7]. It is important for the thermodynamic definition of temperature and the derivation of thermodynamic relations in stable-equilibrium states. Another aspect of the second law of thermodynamics is the definition of an important property, called *entropy*, as discussed next.

Entropy is an additive property of every system in any state. The second law of thermodynamics asserts that, *in an isolated system, entropy cannot be destroyed* but can either be created (in an irreversible process) or remain the same (in a reversible process). The entropy produced as time evolves during an irreversible process is called the *entropy generation* ($S_{\text{gen}}$) due to *irreversibility*. Like energy, entropy can be transferred from one system to another. One can write the entropy balance as follows (keeping in mind that entropy generation must not be negative):

$$\Delta S = S_2 - S_1 = S_{\text{net,in}} + S_{\text{gen}}\,, \quad \text{with } S_{\text{gen}} \geq 0 \tag{2.2a}$$

or

$$\mathrm{d}S = \delta S_{\text{net,in}} + \delta S_{\text{gen}}\,, \quad \text{with } \delta S_{\text{gen}} \geq 0 \tag{2.2b}$$

Here again, $\delta$ is used to indicate an infinitesimal quantity that is *not* a property of any system. For a system with fixed values of energy ($E$), parameters, and constituents, the entropy of the system is the largest in the stable-equilibrium state. This is *the highest entropy principle*. Applying this principle to an isolated system for which the energy is conserved, the entropy of the system will increase until a thermodynamic equilibrium is reached. Spontaneous changes of state are usually irreversible and accompanied by entropy generation.

The second law of thermodynamics can be summarized with the following three statements: (1) There exists a unique stable-equilibrium state for any system with given values of energy, parameters, and constituents. (2) Entropy is an additive property, and for an isolated system, the entropy change must be nonnegative. (3) Among all states with the same values of energy, parameters, and constituents, the entropy of the stable-equilibrium state is the maximum.

The energy of a system with volume ($V$) as its only parameter (neglecting other external forces) is called the *internal energy* ($U$). The state principle implies that there are $r + 2$ (where $r$ is the number of different constituents) independent variables that fully characterize a stable-equilibrium state of such a system. Therefore, in a stable-equilibrium state, all properties are functions of $r + 2$ independent variables. Since entropy is a property of the system, we have

$$S = S(U, V, N_1, N_2, \ldots, N_r) \tag{2.3}$$

where $N_i$ is the number of particles of the $i$th species (or type of constituents). This function is continuous and differentiable [1, 3], and furthermore, it is a monotonically increasing function of energy for fixed values of $V$ and $N_{j's}$. Equation (2.3) can be uniquely solved for $U$ so that

$$U = U(S, V, N_1, N_2, \ldots, N_r) \tag{2.4}$$

which is also continuous and admits partial derivatives of all orders. Each first-order partial derivative of Eqs. (2.3) or (2.4) represents a property of the stable-equilibrium state. For example, *temperature* and *pressure* are properties of a system at thermodynamic equilibrium. The (absolute) temperature is defined by

$$T = \left(\frac{\partial U}{\partial S}\right)_{V,N_{j's}} \tag{2.5a}$$

and the pressure is defined by

$$P = -\left(\frac{\partial U}{\partial V}\right)_{S,N_{j's}} \tag{2.5b}$$

The partial derivative with respect to the $i$th type of constituents defines its chemical potential of that species,

$$\mu_i = \left(\frac{\partial U}{\partial N_i}\right)_{S,V,N_{j's}(j \neq i)} \tag{2.5c}$$

Equation (2.3) or (2.4) is called the *fundamental relation* for states at thermodynamic equilibrium. The differential form of Eq. (2.4) is the Gibbs relation:

$$\mathrm{d}U = T\mathrm{d}S - P\mathrm{d}V + \sum_{i=1}^{r} \mu_i \mathrm{d}N_i \tag{2.6}$$

where Eqs. (2.5a, 2.5b and 2.5c) have been used. The above equation may be rearranged into the form

$$\mathrm{d}S = \frac{1}{T}\mathrm{d}U + \frac{P}{T}\mathrm{d}V - \sum_{i=1}^{r} \frac{\mu_i}{T}\mathrm{d}N_i \tag{2.7}$$

Therefore,

$$\frac{1}{T} = \left(\frac{\partial S}{\partial U}\right)_{V,N_{j's}}, \frac{P}{T} = \left(\frac{\partial S}{\partial V}\right)_{U,N_{j's}}, \text{ and } \frac{\mu_i}{T} = -\left(\frac{\partial S}{\partial N_i}\right)_{U,V,N_{j's}(j \neq i)} \tag{2.8}$$

An interaction between two systems that results in a transfer of energy without net exchanges of entropy and constituents is called a *work interaction*. The amount of energy transferred in such an interaction is called *work* ($W$). An interaction that has only mechanical effects is a work interaction, but a work interaction may involve nonmechanical effects. A process that involves only work interaction is called an *adiabatic* process. Another kind of a typical interaction is *heat interaction*, in which both energy and entropy are transferred without net exchanges of constituents and parameters between two systems. The amount of energy transferred in a heat interaction is called *heat* ($Q$). Furthermore, the amount of entropy transferred ($\delta S$) is equal to the amount of energy transferred ($\delta Q$) divided by the boundary temperature ($T_\mathrm{b}$) at which the heat interaction happens, i.e., $\delta S = \delta Q / T_\mathrm{b}$. If a system cannot exchange constituents with other systems, it is said to be a *closed* system; otherwise, it is an *open* system.

Reversible processes are considered as the limiting cases of real processes, which are always accompanied by a certain amount of irreversibility. Such an ideal process is called a *quasi-equilibrium* (or *quasi-static*) process, in which each stage can be made as close to thermodynamic equilibrium as possible if the movement is frictionless and very slow. In an ideal process, a finite amount of heat can be transferred reversibly from one system to another at a constant temperature. In practice, heat transfer can only happen when there is a temperature difference, and the process is always irreversible.

A *perpetual-motion machine of the second kind* (PMM2) is a cyclic device that interacts with a system at thermodynamic equilibrium and produces no external effect other than the rise of a weight in a gravity field, without changing the value of any parameter or the amount of any constituent of the system. Historically, there exist different statements of the second law of thermodynamics: The Kelvin–Planck statement of the second law is that *it is impossible to build a* PMM2. The Clausius statement of the second law is that *it is not possible to construct a cyclic machine that will produce no effect other than the transfer of heat from a system at lower temperature to a system at higher temperature*. These statements can be proved using the three statements of the second law of thermodynamics given earlier in this chapter.

**Example 2.1** *Criteria for thermodynamic equilibrium*. Consider a moveable piston (adiabatic and impermeable to matter) that separates a cylinder into two compartments (systems A and B), as shown in Fig. 2.1. We learned from mechanics that a mechanical equilibrium requires a balance of forces on both sides of the piston, that is to say, the pressure of system A must be the same as that of system B (i.e., $P_\mathrm{A} = P_\mathrm{B}$). If the piston wall is made of materials that are diathermal (allowing heat transfer) and permeable to all species, under what conditions will the composite system C consisting of systems A and B be at stable equilibrium?

**Solution** Assume system C is isolated from other systems, and each of the subsystems A and B is at a thermodynamic equilibrium state, whose properties are solely determined by its internal energy, volume, and amount of constituents: $U_\mathrm{A}$, $V_\mathrm{A}$, $N_{j's,\mathrm{A}}$ and $U_\mathrm{B}$, $V_\mathrm{B}$, $N_{j's,\mathrm{B}}$, respectively. There exist neighboring states for both subsystems with small differences in $U$, $V$, and $N_{j's}$, but the values of the composite system must be conserved, i.e., $\mathrm{d}U_\mathrm{A} = -\mathrm{d}U_\mathrm{B}, \mathrm{d}V_\mathrm{A} = -\mathrm{d}V_\mathrm{B}$,

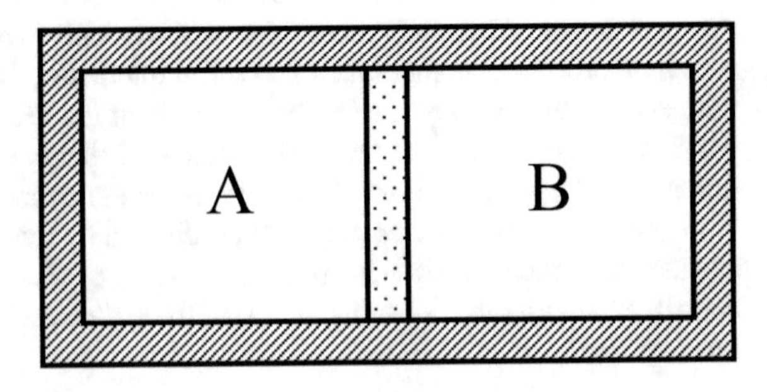


**Fig. 2.1** Illustration of two systems that may exchange work, heat, and species

and $dN_{i,A} = -dN_{i,B}$ ($i = 1, 2, \ldots r$). The differential entropy of system C can be expressed as

$$\begin{aligned} dS_C &= dS_A + dS_B \\ &= \frac{1}{T_A}dU_A - \frac{P_A}{T_A}dV_A + \sum_{i=1}^{r}\frac{\mu_{i,A}}{T_A}dN_{i,A} + \frac{1}{T_B}dU_B - \frac{P_B}{T_B}dV_B \\ &+ \sum_{i=1}^{r}\frac{\mu_{i,B}}{T_B}dN_{i,B} \\ &= \left(\frac{1}{T_A} - \frac{1}{T_B}\right)dU_A - \left(\frac{P_A}{T_A} - \frac{P_B}{T_B}\right)dV_A + \sum_{i=1}^{r}\left(\frac{\mu_{i,A}}{T_A} - \frac{\mu_{i,B}}{T_B}\right)dN_{i,A} \end{aligned} \tag{2.9}$$

If system C is in a stable-equilibrium state, its entropy is maximum and $dS_C = 0$. Since the values of $dU_A$, $dV_A$, and $dN_{i,A}$ are arbitrary, we must have

$$\frac{1}{T_A} = \frac{1}{T_B},\ \frac{P_A}{T_A} = \frac{P_B}{T_B} \text{ and } \frac{\mu_{i,A}}{T_A} = \frac{\mu_{i,B}}{T_B}\ (i = 1,\ 2,\ \ldots\ r)$$

or

$$T_A = T_B,\ P_A = P_B \text{ and } \mu_{i,A} = \mu_{i,B}\ (i = 1,\ 2,\ \ldots\ r) \tag{2.10}$$

These conditions correspond to thermal equilibrium, mechanical equilibrium, and chemical equilibrium, respectively. The combination forms the criteria for thermodynamic equilibrium.

**Discussion**. In the case when the piston is diathermal but rigid and impermeable to matter, the entropy change of system C must be nonnegative, that is,

$$dS_C = dS_A + dS_B = \left(\frac{1}{T_A} - \frac{1}{T_B}\right)dU_A \geq 0 \tag{2.11}$$

The above expression implies that $dU_A \leq 0$ for $T_A > T_B$, and $dU_A \geq 0$ for $T_A < T_B$. Spontaneous heat transfer can occur only from regions of higher temperature to regions of lower temperature. This essentially proves the Clausius statement of the second law of thermodynamics.

The concept of thermal equilibrium provides the physical foundation for *thermometry*, which is the science of temperature measurement. The temperature of a system at a thermodynamic equilibrium state is measured through changes in resistance, length, volume, or other physical parameters of the sensing element used in the thermometer, which is brought to thermal equilibrium with the system. Based on the inclusive statement of the second law of thermodynamics given previously, it can be inferred that two systems are in thermal equilibrium with each other if they are separately in thermal equilibrium with a third system. This is sometimes referred to as the *zeroth law of thermodynamics* [6], especially in the thermometry literature.

**Table 2.1** Two-phase points and the triple point of water

| | Temperature | |
|---|---|---|
| | (K) | (°C) |
| Ice point[a] | 273.15 | 0 |
| Triple point[b] | 273.16 | 0.01 |
| Steam point[c] | 373.124 | 99.974 |

[a]Solid and liquid phases are in equilibrium at a pressure of 1 atm (101.325 kPa)
[b]Solid, liquid, and vapor phases are in equilibrium
[c]Liquid and vapor phases are in equilibrium at 1 atm

The International Temperature Scale of 1990 (ITS-90) was adopted by the International Committee of Weights and Measures in 1989 [8]. The unit of thermodynamic temperature is kelvin (K), which is defined as 1/273.16 of the thermodynamic temperature of the triple point of water. The Celsius temperature is defined as the difference in the thermodynamic temperature and 273.15 K (the ice point). A difference in temperature may be expressed in either kelvins or degrees Celsius (°C). Although earlier attempts were made to define a temperature scale consistent with the original Celsius temperature scale (i.e., 0 °C for the ice point and 100 °C for the steam point), a 0.026 °C departure arose from more accurate measurements of the steam point, as shown in Table 2.1 [9]. The steam point is therefore no longer used as a defining fixed point in the ITS-90. More accurate Steam Tables were developed in the 1990s.

The ITS-90 defines 17 fixed points, which are determined by primary thermometry with standard uncertainties less than 0.002 K below 303 K and up to 0.05 K at the freezing point of copper ($\approx$1358 K). Cryogenic thermometry is essentially based on ideal gas thermometers (up to about 20 K). Platinum resistance thermometers, calibrated at specified sets of fixed points, are used to define the temperature scale from the triple point of hydrogen ($\approx$13.8 K) to the freezing point of silver ($\approx$1235 K). Platinum resistance thermometers have been chosen because of their excellent reproducibility, even though they are not primary thermometers. Radiation thermometers based on Planck's law of thermal radiation are used to define the temperature scale above 1235 K.

It should be noted that the International System of Units (SI) is currently under revision, and the SI units are being redefined based on the fundamental constants without using any materials or prototypes, as documented in the 26th meeting of the General Conference on Weights and Measures (CGPM) [10].

### 2.1.3 The Third Law of Thermodynamics

For each given set of values of constituents and parameters, there exists a unique stable-equilibrium state with *zero absolute temperature* (though not physically attainable). Furthermore, the entropy of any pure substance (in the form of a crystalline solid) vanishes at this state (zero absolute entropy). This is the third law of thermodynamics, also called *the Nernst theorem* after Walther Nernst who received the Nobel Prize in Chemistry in 1920. The energy is the lowest at this state, which is called the *ground-state energy* ($E_g > 0$). The ground-state energy of a system consisting of independent particles may be related to its mass ($m$) using the relativistic theory, i.e., $E_g = mc^2$, where $c$ is the speed of light. Although absolute energy and entropy can be defined according to the third law of thermodynamics, in practice, reference states are often chosen so that the relative values of energy and entropy can be tabulated with respect to those of the reference states.

After reviewing the laws of thermodynamics, it is instructive to give a pictorial presentation to illustrate some of the fundamental concepts in thermodynamics as done by Gyftopoulos and Beretta [3]. For a system that contains a single type of constituents (i.e., pure substance) with fixed values of parameters and amount of constituents, the stable-equilibrium states can be represented as a *convex E–S* curve, whose slope $T = \partial E/\partial S$ defines the temperature of each state on the curve, as shown in Fig. 2.2. The stable-equilibrium-state curve intersects the vertical axis at the ground state, whose energy is the ground-state energy ($E_g$) and whose absolute entropy is zero. Furthermore, the temperature at the ground state is 0 K. This provides a graphical illustration of the third law of thermodynamics. Along the stable-equilibrium-state curve, temperature increases with increasing energy or entropy. The vertical axis above $E_g$ represents *zero-entropy states*, which are not at stable equilibrium (except when $E = E_g$). These are states defined in mechanics, where entropy is not a concern. A spontaneous change of state can be illustrated with this graph as a horizontal line, e.g., from $A_1$ to $A_{10}$, where $A_{10}$ corresponds to the stable-equilibrium state that has the same values of energy, parameters, and constituents as those of $A_1$. No states exist below the stable-equilibrium-state curve because this would violate the highest entropy principle. Each point in the shaded area may correspond to some

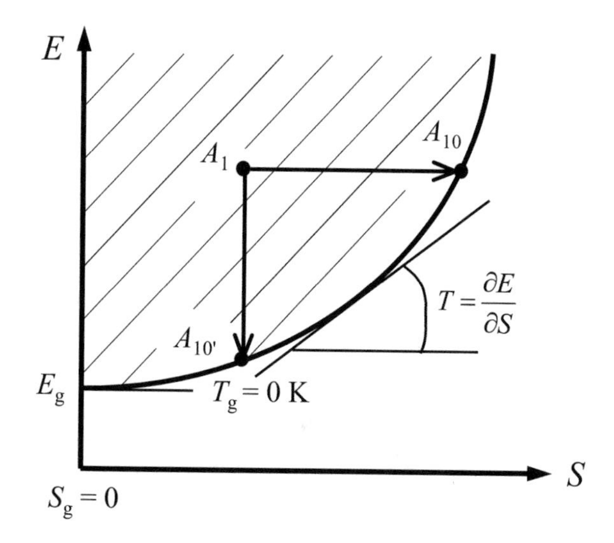


**Fig. 2.2** The *E–S* graph for a pure substance with fixed values of parameters and amount of constituents

states that are not at thermodynamic equilibrium, for which macroscopic properties (such as temperature and pressure) may not be rigorously defined. A nonequilibrium state in general cannot be uniquely determined by the values of its energy (or entropy) and parameters and the amount of constituents. The *lowest energy principle* is expressed as follows: Among all states with the same values of entropy and parameters and the amount of constituents, there exists a stable-equilibrium state whose energy is the lowest. Starting with any state that is not at stable equilibrium, there exists a reversible adiabatic process, in which work can be done by the system until it reaches a stable-equilibrium state. This process is illustrated in the $E - S$ graph by a vertical line from $A_1$ to $A_{10'}$. The corresponding work, which is equal to the energy difference between $A_1$ and $A_{10'}$, is called the *adiabatic availability* [3]. It defines the largest amount of work that can be extracted from a system without any other net effect on the environment of the system.

## 2.2 Thermodynamic Functions and Properties

Several additional properties defined in this section are important in the study of states at thermodynamic equilibrium. The functional relations are derived based on the fundamental relation and are useful under specific circumstances. The phase equilibrium is summarized with an emphasis on pure substances. The concepts of specific heat and latent heat are then introduced. Combining the specific heat and the equation of state, we can evaluate the internal energy and entropy for ideal gases and incompressible solids and liquids.

### *2.2.1 Thermodynamic Relations*

When dealing with substances within the container, the volume is a parameter that characterizes external forces, i.e., the interaction between the system and the wall of the container. If the constituents are confined within a surface, then the surface area will be a parameter instead of the volume. Parameters associated with other external forces (such as gravitational and magnetic forces) can also be included, if necessary. For simplicity, we assume that volume is the only parameter of the systems under investigation, unless otherwise specified.

Enthalpy is defined as $H = U + PV$, and thus we have $\mathrm{d}H = \mathrm{d}U + P\mathrm{d}V + V\mathrm{d}P$. From Eq. (2.6), we obtain

$$\mathrm{d}H = T\mathrm{d}S + V\mathrm{d}P + \sum_{i=1}^{r} \mu_i \mathrm{d}N_i \tag{2.12a}$$

The significance of Eq. (2.12a) is that enthalpy can be expressed as a function of $S$, $P$, and $N_{j's}$,

$$H = H(S, P, N_1, N_2, \ldots, N_r) \tag{2.12b}$$

Furthermore,

$$T = \left(\frac{\partial H}{\partial S}\right)_{P,N_{j's}}, V = \left(\frac{\partial H}{\partial P}\right)_{S,N_{j's}}, \text{ and } \mu_i = \left(\frac{\partial H}{\partial N_i}\right)_{S,P,N_{j's}(j\neq i)} \tag{2.12c}$$

Note that the subscripts in Eq. (2.12c) are different from those in Eqs. (2.5a, 2.5b and 2.5c). Enthalpy $H(S, P, N_{j's})$ is said to be a *characteristic function*, since it allows us to find out all the information about a stable-equilibrium state. A large number of characteristic functions may be defined. Depending on the particular situation and measurements available, it is advantageous to choose the most convenient one. Two other characteristic functions are now introduced. The first one is called *Helmholtz free energy* $A(T, V, N_{j's})$, defined as $A = U - TS$. It follows that

$$\mathrm{d}A = -S\mathrm{d}T - P\mathrm{d}V + \sum_{i=1}^{r} \mu_i \mathrm{d}N_i \tag{2.13a}$$

and

$$S = -\left(\frac{\partial A}{\partial T}\right)_{V,N_{j's}}, P = -\left(\frac{\partial A}{\partial V}\right)_{T,N_{j's}}, \text{ and } \mu_i = \left(\frac{\partial A}{\partial N_i}\right)_{T,V,N_{j's}(j\neq i)} \tag{2.13b}$$

The second is *Gibbs free energy* $G(T, P, N_{j's})$: $G = U + PV - TS = H - TS = A + PV$. It follows that

$$\mathrm{d}G = -S\mathrm{d}T + V\mathrm{d}P + \sum_{i=1}^{r} \mu_i \mathrm{d}N_i \tag{2.14a}$$

and

$$S = -\left(\frac{\partial G}{\partial T}\right)_{P,N_{j's}}, V = \left(\frac{\partial G}{\partial P}\right)_{T,N_{j's}}, \text{ and } \mu_i = \left(\frac{\partial G}{\partial N_i}\right)_{T,P,N_{j's}(j\neq i)} \tag{2.14b}$$

Characteristic functions supplement the fundamental relation and are very useful in the evaluation of the properties of systems under thermodynamic equilibrium.

In a stable-equilibrium state, $T$, $P$, and $\mu_i$ ($i = 1, 2, \ldots r$) must be uniform everywhere in the system. If the system is divided into $k$ equal-volume subsystems, the energy, entropy, and the amount of each type of constituents of the system are the sums of these quantities in all subsystems. If the energy and the amount of each type of constituents in every subsystem are the same, then all subsystems are exactly identical to each other. If this is the case, the system is said to be in a *homogeneous* state; otherwise, it is *heterogeneous*. Examples of homogeneous states are air (which is a mixture of many different kinds of gases) and a well-mixed solution. Examples of heterogeneous states are ice water and water–steam mixture in a boiler.

A system that experiences only homogeneous states is called a *simple system*. In a simple system, $T$, $P$, and $\mu_{j's}$ of each subsystem are the same as those of the system itself and independent of $k$; hence, they are called *intensive properties*. Taking $T$ as an example, we have

$$T\left(\frac{U}{k}, \frac{V}{k}, \frac{N_1}{k}, \frac{N_2}{k}, \ldots \frac{N_r}{k}\right) = T(U, V, N_1, N_2, \ldots, N_r) \tag{2.15}$$

The left-hand side of Eq. (2.15) is the temperature of the subsystem, while the right-hand side is the temperature of the whole system. Unlike temperature and pressure, properties such as $U$, $S$, $V$, and $N$ of each subsystem are inversely proportional to $k$:

$$S\left(\frac{U}{k}, \frac{V}{k}, \frac{N_1}{k}, \frac{N_2}{k}, \ldots, \frac{N_r}{k}\right) = \frac{1}{k}S(U, V, N_1, N_2, \ldots, N_r) \tag{2.16}$$

Properties whose values are proportional to the total amount of constituents are called *extensive properties*. Therefore, $U$, $V$, $S$, and $H$ are extensive properties. Notice that $k$ cannot be arbitrarily large because of the continuum requirement.

The ratio or derivative of two extensive properties is an intensive property, e.g., the density (the ratio of mass to volume) is an intensive property and uniform in a simple system. Note that temperature, pressure, and chemical potentials are derivatives of two extensive properties. The properties $T$, $P$, and $\mu_{j's}$ distinguish themselves from other intensive properties in that they are uniform in both homogeneous and heterogeneous states, whereas others may or may not be uniform in a heterogeneous state. A *specific property* is the ratio of an extensive property to the total amount of constituents (expressed as mass, mole, or number). For example, the mass specific enthalpy is the enthalpy per kilogram of the substance. Specific properties are intensive properties.

For simple systems, the Gibbs relation given in Eq. (2.6) can be integrated to obtain

$$U = TS - PV + \sum_{i=1}^{r} \mu_i N_i \tag{2.17}$$

which is the *Euler relation*. By differentiating Eq. (2.17) and then subtracting Eq. (2.6) from it, we obtain the *Gibbs–Duhem relation*:

$$S\mathrm{d}T - V\mathrm{d}P + \sum_{i=1}^{r} N_i \mathrm{d}\mu_i = 0 \tag{2.18}$$

The Euler relation for a system containing only one type of constituents ($r = 1$) is

$$G = U + PV - TS = \mu N$$

or

$$\mu(T, P) = \frac{G}{N} = g(T, P) \tag{2.19}$$

Hence, the chemical potential of a pure substance is nothing but the specific Gibbs free energy. For a system containing two or more types of constituents, Eq. (2.14b) relates the chemical potential to the partial derivative of the Gibbs free energy with respect to $N_i$ for fixed $T$ and $P$, which is called the *partial* Gibbs free energy of the $i$th type of constituents.

### 2.2.2 The Gibbs Phase Rule

In a heterogeneous state, we consider a subdivision of the system into subsystems, each being a simple system. The collection of all subsystems that have the same values of all intensive properties is called a *phase*. Solid, liquid, and gas (or vapor) are the three distinct phases. The boundary between subsystems of different phases is called an *interface*. Different phases may appear to be clearly separated or well mixed. In space, liquid water droplets could be dispersed throughout water vapor, whereas on the earth, the liquid would occupy the lower part of the container due to gravity.

Assume that there are $q$ coexisting phases, called a $q$-phase heterogeneous state. We can write the Gibbs–Duhem relation for each phase, and thus reduce the independent variables for $T,\ P,\ \mu_i\ (i = 1,\ 2,\ \ldots\ r)$ by $q$. The number of independent variables among $T,\ P,\ \mu_i'$s is determined by the Gibbs phase rule:

$$\phi = r + 2 - q \tag{2.20}$$

For a pure substance, Eq. (2.20) implies that, for a single-phase state, there are only two independent variables among the three intensive properties $T$, $P$, and $\mu$. If $T$ and $P$ are chosen as the independent variables, then all other intensive properties are functions of $T$ and $P$, e.g., specific internal energy $u = u(T, P)$, specific enthalpy $h = h(T, P)$, and specific entropy $s = s(T, P)$. Extensive properties can be determined from the specific properties if the total mass or volume is specified. For a two-phase mixture, such as ice and water or water and steam, only one of $T,\ P,$ and $\mu$ is independent. If $T$ is chosen as the variable, then $P$ and $\mu$ can be expressed as functions of $T$, i.e., $P = P(T)$ and $\mu = \mu(T)$. In order to completely describe the state, however, we will also need to know the amount of constituents in each phase (which may be expressed by the total mass and a mass fraction $x$ of one phase). For example, the specific entropy of a mixture can be expressed as $s = s(T, x)$ or $s = s(P, x)$. In a three-phase mixture, $T$, $P$, and $\mu$ are all fixed. For a pure substance, the solid, liquid, and vapor phases can only coexist at fixed temperature and pressure, which are called *triple-point* properties. Taking water as an example, we have $T_{\text{t.p.}} = 0.01$ °C and $P_{\text{t.p.}} = 0.61$ kPa. One needs to know the amount of constituents in each phase to completely characterize the state. No more than three phases can coexist for any pure substance. It should be noticed that a substance can have different solid phases, e.g., diamond and graphite are allotropes of carbon but with distinct differences in their physical and chemical properties; silicon dioxide can exist in the forms of crystalline quartz or fused silica (glass).

Figure 2.3 shows regions of solid, liquid, and vapor in a *P*–*T* diagram. The S–L, S–V, and L–V lines indicate the coexistence of solid–liquid, solid–vapor, and liquid–vapor phases in thermodynamic equilibrium. The three lines merge to the triple point where all three phases can coexist in thermodynamic equilibrium. There are two S–L lines: the solid line represents a material that expands upon melting, and the dashed line represents a material that contracts upon melting (such as water). There exists a *critical point* or a critical state; the temperature and the pressure at the critical state are called *critical temperature* ($T_c$) and *critical pressure* ($P_c$). The distinction between liquid and vapor phases disappears beyond the critical point. This can be seen clearer in the *T*–*v* diagram shown in Fig. 2.4. The S–L line in Fig. 2.3 becomes an S–L region in Fig. 2.4; the L–V line becomes a dome, called *the saturation dome*. Starting from a solid state, in a constant pressure (isobaric) heating process with $P_{\text{t.p.}} < P < P_c$, the temperature increases until melting starts. As more energy is added to the system, the fraction of solid decreases, whereas the fraction of liquid increases, at a constant temperature. The amount of heat needed to completely melt a unit mass of solid to liquid is called the *specific latent heat of melting*. Once all the substance is in the liquid phase, the temperature rises again with increasing energy until a saturated liquid state is reached. Hereafter, vaporization occurs at constant

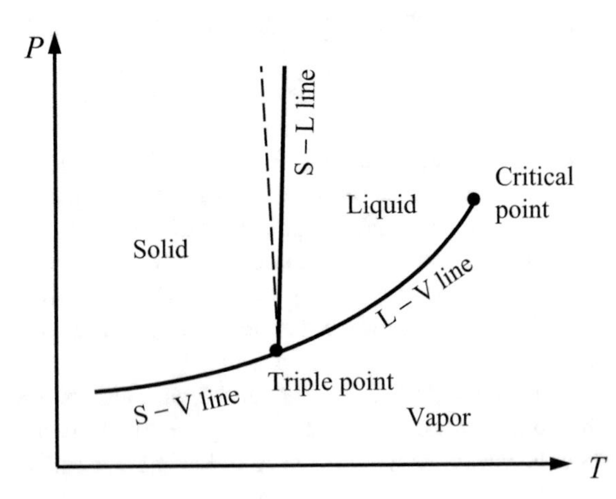


**Fig. 2.3** Schematic of a *P*–*T* diagram for a pure substance

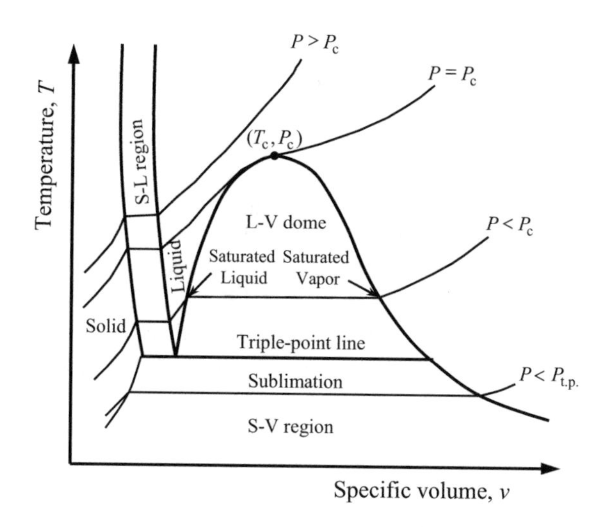


**Fig. 2.4** Schematic of a $T$–$v$ diagram for a material that expands upon melting

temperature (saturation temperature) until it reaches the right side of the saturation dome, which is a saturated vapor state. The amount of energy needed to vaporize a unit mass of a substance is called the *specific latent heat of vaporization*. When the pressure is higher than the critical pressure, however, no vaporization can happen. The liquid and gas forms of aggregation differ in degree rather than in kind. At a pressure lower than the triple-point pressure, the change from solid to vapor can occur without passing through a liquid phase. Such a process is called sublimation. An example is the sublimation of dry ice into cold $CO_2$ gas at room temperature and atmospheric pressure. It can be used to create some theatrical effects such as haze, fog, or smoke.

## 2.2.3 Specific Heats

Specific heats are properties of a system (at stable equilibrium). The *specific heat at constant volume* ($c_v$) and the *specific heat at constant pressure* ($c_p$) are defined as

$$c_v = \left(\frac{\partial u}{\partial T}\right)_V = T\left(\frac{\partial s}{\partial T}\right)_V \tag{2.21a}$$

and

$$c_p = \left(\frac{\partial h}{\partial T}\right)_P = T\left(\frac{\partial s}{\partial T}\right)_P \tag{2.21b}$$

where subscripts $V$ and $P$ signify fixed volume and fixed pressure, respectively. The *heat capacity* is the product of the corresponding specific heat and the mass of the system. Note that only in a reversible process, the amount of heat transferred to a system is $\delta Q = T\mathrm{d}S$. The heat capacity at constant volume of a closed system can be measured in terms of the total amount of energy supplied to it divided by its temperature rise in a constant volume process. On the other hand, the heat capacity at constant pressure of a closed system (such as in a piston–cylinder arrangement) can be measured in terms of the amount of energy per unit mass supplied to the system, excluding the volume work done by the system ($\delta W = p\mathrm{d}V$), divided by the temperature rise in an isobaric process. For example, in a reversible isobaric process, $\mathrm{d}U = \delta Q - p\mathrm{d}V$ and $\mathrm{d}H = \delta Q$. Therefore, $c_p = \frac{1}{m}\frac{\mathrm{d}H}{\mathrm{d}T} = \frac{1}{m}\frac{\delta Q}{\delta T}$.

Specific heats are not defined for all equilibrium states. For example, enthalpy of a two-phase mixture can vary at a constant pressure, such as in a vaporization process, without any change in temperature. This means that the constant pressure specific heat approaches infinity in these states. In fact, the discontinuity in $c_p(T)$ suggests some kind of phase transformation.

A *heat reservoir* is an idealized system that experiences only reversible heat interactions. For any finite amount of energy transfer, its temperature remains unchanged. Therefore, the heat capacity of a reservoir is infinitely large. For a reservoir at temperature $T_\mathrm{R}$, the change of the reservoir energy is proportional to its entropy change:

$$E_{\mathrm{R},2} - E_{\mathrm{R},1} = T_\mathrm{R}(S_{\mathrm{R},2} - S_{\mathrm{R},1}) \tag{2.22a}$$

This suggests that a reservoir can be represented by a straight line in the $E$–$S$ graph. Furthermore, the amount of heat transferred to the reservoir from state 1 to state 2 is given by

$$Q = E_{\mathrm{R},2} - E_{\mathrm{R},1} \tag{2.22b}$$

For a pure substance in a single phase, temperature and pressure are independent, and all other properties can be expressed as functions of $T$ and $P$. The relation among temperature, pressure, and specific volume is called the *equation of state*, which can be expressed as

$$f(T, P, v) = 0 \text{ or } v = v(T, P) \tag{2.23}$$

This equation does not contain information about the internal energy or entropy. However, we can use the function $c_p = c_p(T, P)$, in addition to the equation of state, to fully determine all intensive properties. For example, $\mathrm{d}s = \left(\frac{\partial s}{\partial T}\right)_P \mathrm{d}T + \left(\frac{\partial s}{\partial P}\right)_T \mathrm{d}P$. Using $\left(\frac{\partial s}{\partial T}\right)_P = \frac{c_p(T,P)}{T}$, from the definition of specific heat, and the Maxwell relation $\left(\frac{\partial s}{\partial P}\right)_T = -\left(\frac{\partial v}{\partial T}\right)_P$ (see Problem 2.11), we obtain

$$\mathrm{d}s = \frac{c_p(T, P)}{T}\mathrm{d}T - \left(\frac{\partial v}{\partial T}\right)_P \mathrm{d}P \tag{2.24}$$

Furthermore,

$$dh = c_p(T, P)dT + \left[v(T, P) - T\left(\frac{\partial v}{\partial T}\right)_P\right]dP \tag{2.25}$$

Under certain circumstances, the equation of state is rather simple, and the specific heats can be assumed as functions of the temperature only, i.e., independent of the pressure. These ideal behaviors will be discussed in the next section.

**Example 2.2** *Specific heat and latent heat.* A system consists of 10 kg of $H_2O$ in a closed container that is maintained at a constant pressure of 100 kPa. Initially, the system is at −40 °C (ice), and it is heated to 130 °C (vapor). How much energy must be provided to the system? What is the entropy change of the system? The specific heats of $H_2O$ in the solid, liquid, and vapor states are $c_{p,\mathrm{s}} = 2$ kJ/kg K, $c_{p,\mathrm{f}} = 4.2$ kJ/kg K, and $c_{p,\mathrm{g}} = 2$ kJ/kg K, respectively. The specific latent heats of melting and evaporation are $h_{\mathrm{sf}} = 334$ kJ/kg and $h_{\mathrm{fg}} = 2257$ kJ/kg.

**Solution** From the first law of the closed system in an isobaric process, $\Delta U = Q - W$. Since $\Delta P = 0$, $W = P\Delta V$. Hence, $Q = \Delta H = H_2 - H_1$. Let $T_1 = 233.2$ and $T_2 = 403.2$ K be the initial and final temperatures, respectively, and $T_{\mathrm{sat,m}} = 273.2$ and $T_{\mathrm{sat}} = 373.2$ K be the saturation temperatures. Based on the definition of specific heats, we obtain

$$Q = H_2 - H_1 = m[c_{p,\mathrm{s}}(T_{\mathrm{sat,m}} - T_1) + h_{\mathrm{sf}} + c_{p,\mathrm{f}}(T_{\mathrm{sat}} - T_{\mathrm{sat,m}}) + h_{\mathrm{fg}} + c_{p,\mathrm{g}}(T_2 - T_{\mathrm{sat}})]$$

which gives $Q = 31.51$ MJ. In the single-phase regions, the entropy difference can be evaluated by integrating Eq. (2.21b) or (2.24) since $P$ is fixed. During the phase change, $\Delta S = \Delta H/T$ since the temperature is a constant.

$$S_2 - S_1 = m\left[c_{p,\mathrm{s}}\ln\left(\frac{T_{\mathrm{sat,m}}}{T_1}\right) + \frac{h_{\mathrm{sf}}}{T_{\mathrm{sat,m}}} + c_{p,\mathrm{f}}\ln\left(\frac{T_{\mathrm{sat}}}{T_{\mathrm{sat,m}}}\right) + \frac{h_{\mathrm{fg}}}{T_{\mathrm{sat}}} + c_{p,\mathrm{g}}\ln\left(\frac{T_2}{T_{\mathrm{sat}}}\right)\right]$$

which gives $\Delta S = 90.6$ kJ/K.

**Discussion**. From the Steam Table or software accompanied with common thermodynamics text [4, 5], we can find the specific properties of water as follows: $h_1 = -411.7$ kJ/kg; $s_1 = -1.532$ kJ/kg K; $h_2 = 2737$ kJ/kg; $s_2 = 7.517$ kJ/kg K. Therefore, $Q = \Delta H = m(h_2 - h_1) = 31.49$ MJ; $\Delta S = m(s_2 - s_1) = 90.5$ kJ/K. The negligibly small difference is caused by the assumption of constant specific heat in each phase.

## 2.3 Ideal Gas and Ideal Incompressible Models

The amount of constituents is commonly expressed in terms of the amount of matter in mole. The *mole* is the amount of substance of a system that contains as many elementary entities as there are atoms in 0.012 kg of carbon 12. One mole of substance contains $6.022 \times 10^{23}$ molecules, atoms, or other particles. This value is called the

Avogadro's constant, i.e., $N_A = 6.022 \times 10^{26}$ kmol$^{-1}$. Quantities like molecules and particles do not appear in the units. The mass of the system is $m = \bar{n}M$, where $\bar{n}$ is the amount of constituents (in kmol) and $M$ is called the molecular weight. For example, $M = 18.012$ kg/kmol for water.

### 2.3.1 The Ideal Gas

At relatively high temperature and sufficiently low pressure, most substances behave as a single-phase fluid, in which the interactions between its molecules are generally negligible. The equation of state can be expressed as

$$P\bar{v} = \bar{R}T \text{ or } PV = \bar{n}\bar{R}T \tag{2.26a}$$

where $\bar{v} = V/\bar{n}$ is the molar specific volume in m$^3$/kmol, and $\bar{R}$= 8314 J/kmol K is the *universal gas constant*. Equation (2.26a) is called the ideal gas equation since it can be considered as the definition of an ideal gas. Under *standard conditions* (temperature of 25 °C and pressure of 1 atm), 1 kmol of an ideal gas occupies a volume of 22.5 m$^3$. Dry air can be treated as an ideal gas with an average molecular weight of $M = 29$ kg/kmol. The ideal gas equation of state can be written in terms of the mass quantities for a given substance, i.e.,

$$Pv = RT \text{ or } PV = mRT \tag{2.26b}$$

In the above equation, $v = V/m$ is the specific volume, and $R = \bar{R}/M$ is called the gas constant of the particular substance. The Boltzmann constant is defined as $k_B = \bar{R}/N_A = 1.381 \times 10^{-23}$ J/K. It can be considered as the universal gas constant in terms of particles. Furthermore, if we denote the number density (number of particles per unit volume) as $n$, then the ideal gas equation can be written as $P = nk_BT$ since $n = N_A\bar{n}/V$.

For ideal gases, both $c_p$ and $c_v$ are independent of the pressure, as will be shown from statistical thermodynamics in Chap. 3, but are generally dependent on temperature. The specific internal energy and enthalpy are functions of temperature only, therefore,

$$du = c_v(T)dT \text{ and } dh = c_p(T)dT \tag{2.27}$$

The specific heats $c_p$ and $c_v$ are related by the Mayer relation as

$$\bar{c}_p - \bar{c}_v = \bar{R} \text{ or } c_p - c_v = R \tag{2.28}$$

If $c_v(T) =$ const., which is sometimes referred to as *perfect gas* behavior, then Eq. (2.27) can be integrated to yield

$$u_2 - u_1 = c_v(T_2 - T_1) \tag{2.29a}$$

and

$$h_2 - h_1 = c_p(T_2 - T_1) \tag{2.29b}$$

where subscripts 1 and 2 can be any two (thermodynamic equilibrium) states. The specific entropy depends on both the temperature and the pressure, i.e.,

$$\mathrm{d}s = c_p \frac{\mathrm{d}T}{T} - R\frac{\mathrm{d}P}{P} \tag{2.30a}$$

Integrating the above equation from state 1 to state 2 yields

$$s_2 - s_1 = \int_1^2 \frac{c_p(T)}{T}\mathrm{d}T - R\ln(P_2/P_1) \tag{2.30b}$$

In an isentropic process ($\mathrm{d}s = 0$) of a perfect gas, it can be shown that $Pv^\gamma =$ const., where $\gamma = c_p/c_v$ is the *specific heat ratio*. Note that $Pv =$ const. in an isothermal process.

**Example 2.3** A cylinder contains 0.01 kmol of $N_2$ gas (0.28 kg), which may be modeled as an ideal diatomic gas with $c_v = 2.5R$. A piston maintains the gas at constant pressure, $P_0 = 100$ kPa. The cylinder interacts with a cyclic machine, which in turn interacts with a reservoir at $T_\mathrm{R} = 1000$ K. The cylinder, the reservoir, and the machinery cannot interact with any other systems. The cyclic machine may produce work $W$ (which cannot be negative). A process brings the volume of the cylinder from $V_1 = 0.224$ to $V_2 = 0.448$ m$^3$.

(a) What is the least amount of energy that must be transferred out from the reservoir? In such a case, how much work does the cyclic machine produce? How much entropy is generated in the process?
(b) Find the maximum work that the cyclic machine can produce.

**Analysis**. A schematic drawing is made first as shown in Fig. 2.5. From the ideal gas equation, $T_1 = P_1V_1/\bar{n}\bar{R} = 269.4$ K and $T_2 = 538.8$ K. The initial and final states of the cylinder are fully prescribed. The work done by the cylinder is $W_\mathrm{B} = \int P\mathrm{d}V = P(V_2 - V_1) = 22.4$ kJ, which is also fixed. By applying the first law to the cylinder in an isobaric process, $Q_\mathrm{B} = m(h_2 - h_1) = mc_p(T_2 - T_1) = 0.01 \times 3.5 \times P(V_2 - V_1) = 78.4$ kJ. The work done by the cyclic machine is $W = Q_\mathrm{R} - Q_\mathrm{B}$. Because $Q_\mathrm{B}$ is prescribed and $W \geq 0$, the least amount of energy that must be transferred from the reservoir is when $W = 0$ and $Q_\mathrm{R} = Q_\mathrm{B}$.

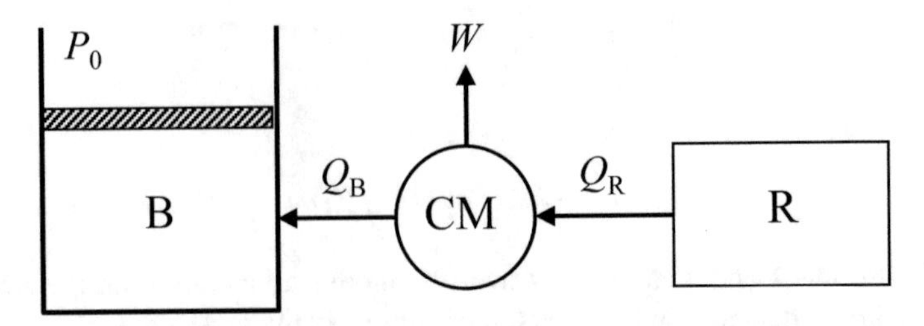


**Fig. 2.5** Schematic drawing for Example 2.3

**Solution**

(a) $Q_R = Q_B = 78.4$ kJ and $W = 0$. We can evaluate the entropy change of the combined system by the following:

$$\begin{aligned}\Delta S &= m(s_2 - s_1) + \Delta S_{CM} + (-Q_R/T_R) \\ &= m[c_p \ln(T_2/T_1) - R\ln(P_2/P_1)] + 0 - 78.4/1000 \\ &= (201.7 - 78.4)\text{ kJ/K} = 123.3\text{ J/K}\end{aligned}$$

Since the system does not have any interactions with any other systems, the entropy change is caused solely by entropy generation.

(b) The maximum work that can be produced is through a reversible process (*not a Carnot cycle since the temperature of the cylinder is not constant*). By setting $\Delta S = m(s_2 - s_1) - Q_R/T_R = 0$, we find $Q_R = T_R m c_p \ln(T_2/T_1) = 201.7$ kJ. The maximum amount of work is therefore $W_{max} = Q_R - Q_B = 123.3$ kJ.

### 2.3.2 Incompressible Solids and Liquids

The assumption for *ideal incompressible* behavior is $v = \text{const.}$, which is the equation of state for incompressible solids and liquids. It can be shown that in this case $c_p = c_v$ and, to a good approximation, the specific heat depends on temperature only. It is common to use $c_p$ for the specific heat of solids and liquids. Using Eqs. (2.24) and (2.25), we obtain the specific internal energy, enthalpy, and entropy for an ideal incompressible solid or liquid as follows:

$$\mathrm{d}u = c_p(T)\mathrm{d}T \tag{2.31}$$

$$\mathrm{d}s = c_p(T)\frac{\mathrm{d}T}{T} \tag{2.32}$$

and

$$dh = c_p(T)dT + vdP \tag{2.33}$$

Notice that while the internal energy and the entropy are functions of temperature only, the enthalpy depends on both temperature and pressure as can be seen from Eq. (2.33). Sometimes only one of the terms on the right-hand side of Eq. (2.33) needs to be considered if the other term is much smaller. For example, if the pressure change is small, the second term can be dropped. Examples when the pressure effect can be neglected are (1) a solid under the normal pressure range and (2) a liquid that flows through a pipeline in a heat exchanger without significant pressure drop. An example when the temperature effect is negligible is pumping water in a reversible adiabatic process, where the enthalpy change between the outlet and inlet of the pump is proportional to the pressure change.

**Example 2.4** In a Rankine cycle, water at 15 °C, 100 kPa is compressed through a pump to 10 MPa before entering the boiler. Model the water as an incompressible liquid with a constant specific heat $c_p = 4.2$ kJ/kg K. What is the least amount of work required to pump 1 kg of water? What is the exit temperature of the water? If the pump efficiency is 80%, what is the actual specific work and exit temperature of the pump?

**Solution** Take $v = 0.001$ m$^3$/kg as an approximation. The least amount of work is needed in a reversible process. It has been shown that the reversible work *done by the system* between bulk flow states is $\delta w = -vdP$. Hence, the work needed in a reversible process is $w_{rev} = h_{2s} - h_1 = 0.001(10,000 - 100) = 9.9$ kJ/kg

Because it is an adiabatic and reversible process, it must be isentropic or $s_{2s} - s_1 = c_p \ln(T_{2s}/T_1) = 0$. Hence, $T_{2s} = T_1 = 15$ °C. Actual work $w = w_{rev}/\eta_p = 12.38$ kJ/kg. Since $w = h_2 - h_1 = c_p(T_2 - T_1) + v(P_2 - P_1)$,

$$T_2 = T_1 + \frac{h_2 - h_1}{c_p} - \frac{v}{c_p}(P_2 - P_1) = T_1 + \frac{w - w_{rev}}{c_p} = 15.59\,°\mathrm{C}$$

which is less than 1 K higher. The entropy generation is $s_{gen} = c_p \ln(T_2/T_1) = 8.6$ J/K kg.

**Discussion**. We can use the Steam Table and notice that all states are compressed liquid. The properties at state 1 can be evaluated at $T_1 = 15$ °C and $P_1 = 100$ kPa, at state 2s (reversible) can be evaluated at $P_{2s} = 10$ MPa and $s_{2s} = s_1$, and at state 2 can be evaluated at $P_2 = 10$ MPa and $h_2 = h_1 + w$. Hence, $w_{rev} = 9.88$ kJ/kg, $T_{2s} = 15.11$ °C, $w = 12.35$ kJ/kg, $T_2 = 15.67$ °C, and $s_{gen} = 8.2$ J/K kg. The differences are negligibly small compared with those obtained from the incompressible assumption. Note that the temperature change in the pump is usually very small. On a $T$–$s$ diagram, it is difficult to distinguish states 1, 2s, and 2. In fact, state 2 crosses the saturated liquid line to overlap with a two-phase-mixture state at $T_2$ and $s_2$. This is because $T$ and $s$ together cannot uniquely determine a stable-equilibrium state.

## 2.4 Heat Transfer Basics

Classical thermodynamics focuses on the changes of mass, energy, and entropy of a system between equilibrium states, and establishes the required balance equations between end states during a given process. For example, we have learned that spontaneous transfer of energy can occur only from a higher temperature to a lower temperature. In thermodynamics, heat interaction is defined as the transfer of energy at the mutual (interface) temperature between two systems. Heat transfer is a subject that extends the thermodynamic principles to detailed energy transport processes that occur as a consequence of temperature differences. Heat transfer phenomena are abundant in our everyday life and play an important role in many industrial, environmental, and biological processes. Examples include energy conversion and storage, electrical power generation, combustion processes, heat exchangers, building-temperature regulation, thermal insulation, refrigeration, microelectronic cooling, materials processing, manufacturing, global thermal budget, agriculture, food industry, and biological systems. Based on the local-equilibrium assumption, heat transfer analysis deals with the rate of heat transfer and/or the temperature distributions (steady state or transient) for given geometries, materials, and initial and boundary conditions. Thermal design, on the other hand, determines the necessary geometric structure and materials for use to achieve optimum performance for a specific task, such as a heat exchanger.

Heat conduction refers to the transfer of heat in a stationary (from the macroscopic point of view) medium, which may be a solid, a liquid, or a gas. Energy can also be transferred between objects by the emission and absorption of electromagnetic waves without any intervening medium; this is called *thermal radiation*, such as the radiation from the sun. When the transfer of heat involves fluid motion, we call it *convection heat transfer*, or simply, *convection*. Examples of convection are cooling with a fan, hot water flowing in a pipe, and cold air blowing outside the wall of a building. The basic macroscopic formulations of conduction, convection, and radiation heat transfer are summarized in this section. The microscopic mechanisms, such as the effects of small dimension and short duration on the thermal transfer processes, will be the subject of the remaining chapters.

### 2.4.1 *Conduction*

In a stationary medium, heat transfer occurs if the medium is not at thermal equilibrium. The assumption of local equilibrium allows us to define the temperature at each location. Fourier's law states that the heat flux (or heat transfer rate per unit area) $\mathbf{q}''$ is proportional to the temperature gradient $\nabla T$, i.e.,

$$\mathbf{q}'' = -\kappa \nabla T \tag{2.34}$$

where $\kappa$ is called *thermal conductivity*, which is a material property that may depend on temperature. Notice that $\mathbf{q}''$ is a vector and its direction is always perpendicular to the isotherms and opposite to the temperature gradient. In an anisotropic medium, such as a thin film or a thin wire, the thermal conductivity depends on the direction along which it is measured.

By doing a control volume analysis using energy balance, a differential equation can be obtained for the transient temperature distribution $T(t, \mathbf{r})$ in a homogeneous isotropic medium; that is [11, 12]

$$\nabla \cdot (\kappa \nabla T) + \dot{q} = \rho c_p \frac{\partial T}{\partial t} \tag{2.35}$$

where $\nabla\cdot$ is the divergence operator, $\dot{q}$ is the volumetric thermal energy generation rate, and $\rho c_p$ can be considered as volumetric heat capacity. Equation (2.35) is called the heat diffusion equation or heat equation. Note that the concept of thermal energy generation is very different from the concept of entropy generation. Thermal energy generation refers to the conversion of other types of energy (such as electrical, chemical, or nuclear energies) to the internal energy of the system, while the total energy is always conserved. Entropy need not be conserved, and entropy generation refers to the creation of entropy by an irreversible process. If there is no thermal energy generation and the thermal conductivity can be assumed to be independent of temperature, Eq. (2.35) reduces to $\nabla^2 T = 0$ at steady state, where $\nabla^2 T = \frac{\partial^2 T}{\partial x^2} + \frac{\partial^2 T}{\partial y^2} + \frac{\partial^2 T}{\partial z^2}$ in the Cartesian coordinates. With the prescribed initial temperature distribution and boundary conditions, the heat equation can be solved analytically for simple cases and numerically for more complex geometries as well as initial and boundary conditions. Typical boundary conditions include (a) constant temperature, (b) constant heat flux, (c) convection, and (d) radiation.

Generally speaking, metals with high electric conductivities and some crystalline solids have very high thermal conductivities ranging from 100 to 1000 W/m K; alloys and metals with low electric conductivities have slightly lower thermal conductivities ranging from 10 to 100 W/m K; water, soil, glass, and rock have thermal conductivities from 0.5 to 5 W/m K; thermal insulation materials usually have a thermal conductivity on the order of 0.1 W/m K; and gases have the lowest thermal conductivity, e.g., the thermal conductivity of air at 300 K is 0.026 W/m K. Notice that thermal conductivity generally depends on temperature. A comprehensive collection of thermal-property data can be found from Touloukian and Ho [13]. At room temperature, Diamond IIa has the highest thermal conductivity, $\kappa = 2300$ W/m K among all natural materials. Researchers have shown that single-walled carbon nanotubes can have even higher thermal conductivity at room temperature. More detailed discussion about the mechanisms of thermal conduction and thermal properties of nanostructures will be provided in subsequent chapters.

**Example 2.5** Consider the steady-state heat conduction through a solid rod, whose sides are insulated, between a constant temperature source at $T_1 = 600$ K and a constant temperature sink at $T_2 = 300$ K. Assume the thermal conductivity of the rod

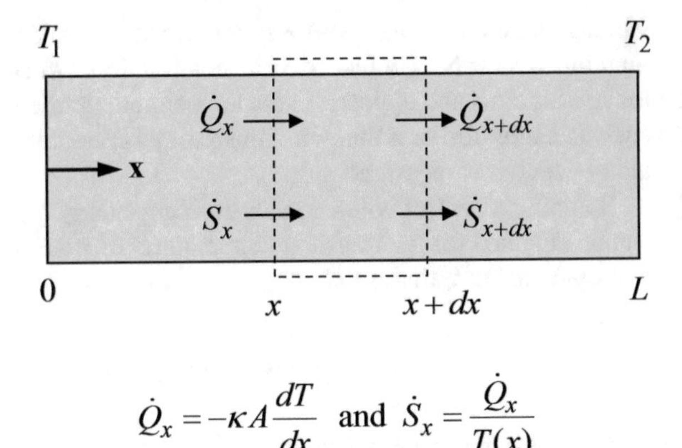


**Fig. 2.6** Illustration of the control volume for energy and entropy balances in a solid rod with heat conduction

is independent of temperature, $\kappa = 150$ W/m K. The rod has a length $L = 0.2$ m and cross-sectional area $A = 0.001$ m$^2$. Show that the temperature distribution along the rod is linear. What is the heat transfer rate? What is the volumetric entropy generation rate? What is the total entropy generation rate?

**Solution** This is a 1D heat conduction problem with no thermal energy generation, as shown in Fig. 2.6. Fourier's law can be written as $\dot{Q}_x = -\kappa A(\mathrm{d}T/\mathrm{d}x)$. At steady state, the heat transfer rate $\dot{Q}_x$ is independent of $x$, since there is no thermal energy generation. Because both $\kappa$ and A are constant, $\mathrm{d}T/\mathrm{d}x$ must not be a function of $x$. Hence, the spatial temperature distribution is a straight line. From the boundary conditions $T(0) = T_1$ and $T(L) = T_2$, we have $T(x) = T_1 + (T_2 - T_1)(x/L)$. Furthermore, $\dot{Q}_x = \kappa A(T_1 - T_2)/L = 225$ W. To evaluate the entropy generation rate, we can apply Eq. (2.2b) to the control volume $A\mathrm{d}x$ to obtain $\dot{s}_{\text{gen}}(x)A\mathrm{d}x$. The net entropy transferred to the control volume is $\dot{S}_x - \dot{S}_{x+\mathrm{d}x} = -\mathrm{d}\left(\frac{\dot{Q}_x}{T}\right)$. The sum of the entropy generation and entropy transferred is equal to the entropy change, which is zero at steady state. Therefore, $\dot{s}_{\text{gen}}(x) = q_x'' \frac{\mathrm{d}(1/T)}{\mathrm{d}x} = \frac{\kappa}{T^2}\left(\frac{\mathrm{d}T}{\mathrm{d}x}\right)^2$, where $q_x'' = \frac{\dot{Q}_x}{A}$ is the heat flux. To calculate the total entropy generation rate, we can integrate $\dot{s}_{\text{gen}}(x)$ over the whole rod. Alternatively, we can perform an entropy balance for the rod as a whole, which gives the rate of entropy generation for a heat transfer rate $\dot{Q}_x$ from $T_1$ to $T_2$ as $\dot{S}_{\text{gen}} = \dot{Q}_x\left(\frac{1}{T_2} - \frac{1}{T_1}\right) = 0.375$ W/K. This example shows that the entropy generation occurs in a finite volume, while the entropy flows through the interface. The amount of entropy flux increases with $x$ as more and more entropy is generated through the irreversible process. More discussion on the entropy generation in heat transfer and fluid flow processes can be found in Bejan [14].

*Contact resistance* is important in microelectronics thermal management and cryogenic heat transfer. A large thermal resistance may exist due to imperfect contact, such as surface roughness. The result is a large temperature difference across the interface. The value of contact resistance depends on the surface conditions, adjacent materials, and contact pressure. As an example, assume a contact resistance between two stainless steel plates to be $R_c'' = 0.001$ m$^2$ K/W and the thermal conductivity of the stainless steel $\kappa = 50$ W/m K. If the thickness of each plate is $L = 5$ mm and the area of the plate is $A = 0.01$ m$^2$, the total thermal resistance is then $R_t = L/(\kappa A) + R_c''/A + L/(\kappa A) = (0.01 + 0.1 + 0.01)$ K/W $= 0.12$ K/W, which is mostly due to the contact resistance. Interfacial fluids and interstitial (filler) materials can be applied to reduce the contact resistance in some cases. Even with a perfect contact, thermal resistance exists between dissimilar materials due to acoustic mismatch, which is especially important at low temperatures [15].

## 2.4.2 Convection

Convection heat transfer refers to the heat transfer from solid to fluid near the boundary when the fluid is in bulk motion relative to the solid. The combination of the bulk motion, known as *advection*, with the random motion of the fluid molecules (i.e., diffusion) is the key for convection heat transfer. Examples are flows over an object or inside a tube, a spray leaving a nozzle that is impinged on a microelectronic component for cooling purposes, and boiling in a pan. The velocity and temperature distributions for a fluid flowing over a heated flat plate are illustrated in Fig. 2.7. A *hydrodynamic boundary layer* or *velocity boundary layer* (VBL) is formed near the surface, and the fluid moves at the free-stream velocity outside the boundary layer. Similarly, a *thermal boundary layer* (TBL) is developed near the surface of the plate where a temperature gradient exists. When the flow speed is not very high and the density of the fluid not too low, the average velocity of the fluid is zero, and the fluid temperature equals the wall temperature in the vicinity of the wall, i.e., $v_x(y = 0) = 0$ and $T(y = 0) = T_w$. For Newtonian fluids, a linear relationship exists between the stress components and the velocity gradients. Many common fluids like air, water, and oil belong to this catalog. The shear stress in the fluid is

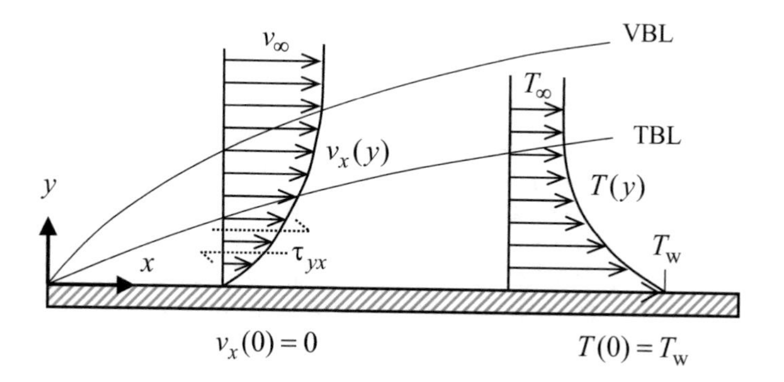


**Fig. 2.7** Illustration of the velocity boundary layer and the thermal boundary layer

$$\tau_{yx} = -\mu \frac{\partial v_x}{\partial y} \tag{2.36}$$

where $\mu$ is the viscosity. Throughout this book, we will use $v_x$, $v_y$, and $v_z$ (or $v_i$ with $i = 1, 2,$ and 3) for the velocity components in the $x$-, $y$-, and $z$-directions, respectively. When Eq. (2.36) is evaluated at the boundary $y = 0$, it gives the force per unit area exerted to the fluids by the wall and is used to calculate the *friction factor* in fluid mechanics [16].

The heat flux between the solid and the fluid can be predicted by applying Fourier's law to the fluid at the boundary; thus,

$$q''_{\mathrm{w}} = -\kappa \left.\frac{\partial T}{\partial y}\right|_{y=0} \tag{2.37}$$

where $\kappa$ is the thermal conductivity of the fluid. Equation (2.37) shows that the basic heat transfer mechanism for convection is the same as that for conduction, i.e., both are caused by heat diffusion and governed by the same equation. Without bulk motion, however, the temperature gradient at the boundary would be smaller. Therefore, advection generally increases the heat transfer rate. Newton's law of cooling is a phenomenological equation for convection. It states that the convective heat flux is proportional to the temperature difference, therefore,

$$q''_{\mathrm{w}} = h(T_{\mathrm{w}} - T_\infty) \tag{2.38}$$

where $h$ is called the *convection heat transfer coefficient*, or *convection coefficient*, $T_{\mathrm{w}}$ is the surface temperature, and $T_\infty$ is the fluid temperature. From Eqs. (2.37) and (2.38), we have

$$h = \frac{-\kappa}{T_{\mathrm{w}} - T_\infty} \left.\frac{\partial T}{\partial y}\right|_{y=0} \tag{2.39}$$

Although $h$ depends on the location, the average convection coefficient is often used in heat transfer calculations. The convection coefficient depends on the fluid thermal conductivity, velocity, and flow conditions (laminar versus turbulent flow, internal versus external flow, and forced versus free convection). Convection can also happen with phase change, such as boiling, which usually causes vigorous fluid motion and enhanced heat transfer. Convection correlations are recommended in most heat transfer textbooks to determine the convection coefficient. For laminar flow over a flat plate of length $L$ with a free-stream velocity $v_\infty$, the following equation correlates the average Nusselt number to the Reynolds number at $x = L$ and the Prandtl number [11]:

$$\overline{Nu_L} = \frac{\bar{h}_L L}{\kappa} = 0.664 Re_L^{1/2} Pr^{1/3}, \text{ for } Pr > 0.6 \text{ and } Re_L < 5 \times 10^5 \tag{2.40}$$

The Reynolds number, defined as $Re_L = \rho v_\infty L/\mu$, is key to the study of hydrodynamics. The Prandtl number $Pr = \nu/\alpha$ is the ratio of *kinematic viscosity* $\nu = \mu/\rho$, which is also known as the *momentum diffusivity*, to the thermal diffusivity $\alpha = \kappa/(\rho c_p)$ of the fluid. A detailed understanding of the fluid flow and convection heat transfer requires the solution of the conservation equations, as summarized in the following.

The differential form of the continuity equation or mass conservation is

$$\frac{\mathrm{D}\rho}{\mathrm{D}t} + \rho\nabla\cdot\mathbf{v} = 0 \tag{2.41}$$

where $\frac{\mathrm{D}}{\mathrm{D}t} = \left(\frac{\partial}{\partial t} + \mathbf{v}\cdot\nabla\right)$ is called the substantial derivative or material derivative. Notice that for an incompressible fluid, the continuity equation reduces to $\nabla\cdot\mathbf{v} = 0$.

Using Stokes' hypothesis that relates the second coefficient of viscosity to the viscosity for Newtonian fluids, the Navier–Stokes equation that describes the momentum conservation can be expressed as follows [16]:

$$\frac{\mathrm{D}\mathbf{v}}{\mathrm{D}t} = -\frac{\nabla P}{\rho} + \mathbf{a} + \nu\nabla^2\mathbf{v} + \frac{\nu}{3}\nabla(\nabla\cdot\mathbf{v}) \tag{2.42}$$

where $\mathbf{a}$ is the body force per unit mass exerted on the fluid, i.e., the acceleration vector.

Energy equation for constant thermal conductivity without thermal energy generation for a moving fluid can be expressed as

$$\rho\frac{\mathrm{D}u}{\mathrm{D}t} = \kappa\nabla^2 T - P\nabla\cdot\mathbf{v} + \mu\Phi \tag{2.43a}$$

where $u$ is the specific internal energy ($\mathrm{d}u = c_v\mathrm{d}T$) and the last term accounts for the viscous dissipation, which is

$$\begin{aligned}\Phi &= 2\left[\left(\frac{\partial v_x}{\partial x}\right)^2 + \left(\frac{\partial v_y}{\partial y}\right)^2 + \left(\frac{\partial v_z}{\partial z}\right)^2\right] + \left(\frac{\partial v_x}{\partial y} + \frac{\partial v_y}{\partial x}\right)^2 + \left(\frac{\partial v_y}{\partial z} + \frac{\partial v_z}{\partial y}\right)^2 + \\ &\quad + \left(\frac{\partial v_z}{\partial x} + \frac{\partial v_x}{\partial z}\right)^2 - \frac{2}{3}(\nabla\cdot\mathbf{v})^2\end{aligned} \tag{2.43b}$$

in the Cartesian coordinates. Equation (2.41) through (2.43a, 2.43b) is usually simplified for specific conditions and solved analytically or numerically using computation fluid dynamics software. In Chap. 4, we will show that the conservation equations can also be derived from the microscopic theories, which are also applicable for rarefied flows and microfluidics.

### 2.4.3 Radiation

Thermal radiation refers to the electromagnetic radiation in a broad wavelength range from approximately 100 nm to 1000 μm. It includes a portion of the ultraviolet region, the entire visible (380–760 nm) region, and the infrared region. Monochromatic radiation refers to radiation at a single wavelength (or a very narrow spectral band), such as lasers and some atomic emission lines. Radiation emitted from a thermal source, such as the sun, an oven, or a blackbody cavity, covers a broad spectral region and can be considered as the spectral integration of monochromatic radiation. In contrast to conduction or convection heat transfer, radiative energy propagates in the form of electromagnetic waves that do not require an intervening medium. Regardless of its wavelength, an electromagnetic wave travels in vacuum at the speed of light, $c_0 = 2.998 \times 10^8$ m/s. Radiation can also be viewed as a collection of particles, called photons, whose energy is proportional to the frequency of radiation. Starting with the definition of intensity and its linkage to the radiative energy flux, radiative transfer between surfaces and in participating media will be briefly described later in this section. More detailed treatment of the mechanism of thermal radiation, radiative properties, and radiative transfer at small length scales will be given in Chaps. 8, 9, and 10.

The *spectral intensity* or *radiance* is defined as the radiative power received within a solid angle, a unit projected area, and a unit wavelength interval; hence [11],

$$I_\lambda(\lambda, \theta, \phi) = \frac{\mathrm{d}\dot{Q}}{\mathrm{d}A \cos\theta \; \mathrm{d}\Omega \; \mathrm{d}\lambda} \tag{2.44}$$

where $(\theta, \phi)$ is the direction of propagation, measured with respect to the surface normal, $\mathrm{d}A \ \cos\theta$ is therefore the projected area, and $\mathrm{d}\Omega$ is an element solid angle. It is convenient to describe the relationship between intensity and radiative power using the spherical coordinates, as shown in Fig. 2.8, where an element area $dA$ whose surface normal is in the $z$-direction is placed at the origin. Note that $r = (x^2 + y^2 + z^2)^{1/2}$, $\theta = \cos^{-1}(z/r)$, $\phi = \tan^{-1}(y/x)$. The solid angle, defined as $\mathrm{d}\Omega = \mathrm{d}A_\mathrm{n}/r^2$, can be expressed as $\mathrm{d}\Omega = (r\mathrm{d}\theta)(r \sin\theta \mathrm{d}\phi)/r^2 = \sin\theta \mathrm{d}\theta \mathrm{d}\phi$.

The spectral heat flux from an element surface d$A$ to the upper hemisphere can be obtained by integrating Eq. (2.44), i.e.,

$$q''_\lambda(\lambda) = \int_0^{2\pi} \int_0^{\pi/2} I_\lambda(\lambda, \theta, \phi) \cos\theta \sin\theta \mathrm{d}\theta \mathrm{d}\phi \tag{2.45}$$

The total heat flux is equal to the heat flux integrated over all wavelengths:

$$q''_\mathrm{rad} = \int_0^\infty q''_\lambda(\lambda) \mathrm{d}\lambda \tag{2.46}$$

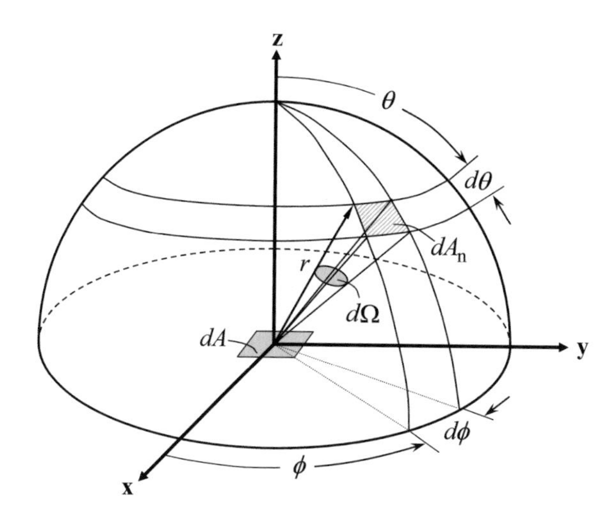


**Fig. 2.8** Illustration of the solid angle in spherical coordinates

We can also define the total intensity as the integral of the spectral intensity over all wavelengths, $I(\theta, \phi) = \int_0^\infty I(\lambda, \theta, \phi)\mathrm{d}\lambda$. An equation similar to Eq. (2.45) holds between the total heat flux and the total intensity. If the radiation is emitted from a surface, the radiative heat flux $q''_{\mathrm{rad}}$ is termed as the (hemispherical) *emissive power*. When the intensity is same in all directions, the surface is said to be diffuse, and Eq. (2.45) can be integrated to obtain the relation, $q''_\lambda = \pi I_\lambda(\lambda, \theta, \phi)$. Similarly, we can obtain $q'' = \pi I$.

The maximum power that can be emitted by a thermal source at a given temperature is from a blackbody. A blackbody is an ideal surface which absorbs all incoming radiation and gives out the maximum emissive power. Radiation inside an isothermal enclosure behaves like a blackbody. In practice, a blackbody cavity is made with a small aperture on an isothermal cavity. The emissive power of a blackbody is given by the Stefan–Boltzmann law, also proportional to the absolute temperature to the fourth power, viz.,

$$e_{\mathrm{b}}(T) = \pi I_{\mathrm{b}}(T) = \sigma_{\mathrm{SB}} T^4 \tag{2.47}$$

where $\sigma_{\mathrm{SB}} = 5.67 \times 10^{-8}$ W/m$^2$ K$^4$ is the Stefan–Boltzmann constant. A blackbody is also a diffuse emitter, i.e., its intensity is independent of the direction. The spectral distribution of blackbody emission is described by Planck's law, which gives the spectral intensity as a function of temperature and wavelength as follows:

$$I_{\mathrm{b},\lambda}(\lambda, T) = \frac{e_{\mathrm{b},\lambda}(\lambda, T)}{\pi} = \frac{2hc^2}{\lambda^5(\mathrm{e}^{hc/k_{\mathrm{B}}\lambda T} - 1)} \tag{2.48}$$

where $h = 6.626 \times 10^{-34}$ J s is the Planck constant, $c$ is the speed of light, and $k_{\mathrm{B}}$ is the Boltzmann constant. The derivation of Planck's law will be given in Chap. 8.

The ratio of the emissive power of a real material to that of the blackbody defines the (total hemispherical) *emissivity* (or *emittance*), $\varepsilon(T) = e(T)/e_{\rm b}(T)$. The spectral directional emissivity is defined as the spectral intensity emitted by the surface to $I_{{\rm b},\lambda}$, i.e.,

$$\varepsilon'_\lambda(\lambda, \theta, \phi, T) = \frac{I_\lambda(\lambda, \theta, \phi, T)}{e_{{\rm b},\lambda}(\lambda, T)/\pi} \tag{2.49}$$

Using $e(T) = \int_0^\infty {\rm d}\lambda \left[\int_0^{2\pi}\int_0^{\pi/2} I_\lambda(\lambda, \theta, \phi, T)\cos\theta\sin\theta{\rm d}\theta{\rm d}\phi\right]$, we have

$$\varepsilon(T) = \frac{\pi}{\sigma T^4}\int_0^\infty e_{{\rm b},\lambda}(\lambda, T){\rm d}\lambda\left[\int_0^{2\pi}\int_0^{\pi/2} \varepsilon'_\lambda(\lambda, \theta, \phi, T)\cos\theta\sin\theta{\rm d}\theta{\rm d}\phi\right] \tag{2.50}$$

This equation suggests that the relationship between the total hemispherical emissivity and the spectral directional emissivity is rather complicated in general. For a gray surface, the spectral emissivity is not a function of the wavelength. For a diffuse surface, the intensity emitted by the surface is independent of the direction. For a diffuse-gray surface, Eq. (2.49) reduces to a simple form $\varepsilon = \varepsilon'_\lambda$, because the emissivity is independent of wavelength and the direction.

Real materials also reflect radiation in contrast to a blackbody. The reflection may be specular for mirrorlike surfaces and more diffuse for rough surfaces. Some window material and thin films are semitransparent. Generally speaking, reflection and transmission are highly dependent on the wavelength, angle of incidence, and polarization status of the incoming electromagnetic wave. The *absorptance*, *reflectance*, and *transmittance* of a material can be defined as the fraction of the absorbed, reflected, and transmitted radiation. The (spectral) directional absorptance, directional–hemispherical reflectance, and directional–hemispherical transmittance are related by

$$A'_\lambda + R'_\lambda + T'_\lambda = 1 \tag{2.51}$$

For an opaque material, the transmittance $T'_\lambda = 0$. It is common to use absorptivity $\alpha'_\lambda$ and reflectivity $\rho'_\lambda$ for opaque materials with smooth surfaces. Note that $\alpha'_\lambda + \rho'_\lambda = 1$. However, the distinction between words ending with "-tivity" and "-tance" is not always clear and both endings are used interchangeably in the literature. The complete nomenclature of radiative quantities and properties can be found from Siegel and Howell [17]. Further discussion about the mechanisms and applications of radiation heat transfer will be provided in Chap. 8.

Kirchhoff's law states that the spectral directional emissivity is always the same as the spectral directional absorptivity, i.e., $\varepsilon'_\lambda = \alpha'_\lambda$. For diffuse-gray surfaces, it can also be shown that $\varepsilon = \alpha$, which may not be generally true for surfaces that are not diffuse-gray, unless they are in thermal equilibrium with the surroundings.

**Example 2.6** Find the net radiative heat flux between two, large parallel surfaces. Surface 1 at $T_1 = 600$ °C has an emissivity $\varepsilon_1 = 0.8$, and surface 2 at $T_2 = 27$ °C has an emissivity $\varepsilon_2 = 0.5$.

**Solution** Assume that the medium in between is transparent, and both surfaces are opaque and diffuse-gray. Note that radiation from one surface to another will be partially absorbed and partially reflected back. Furthermore, the reflected radiation will continue to experience the absorption/reflection processes between the two surfaces. Surface 1 emits $\varepsilon_1 \sigma_{SB} T_1^4$ radiation toward surface 2. The fraction of this emitted radiation that is absorbed by surface 2 can be calculated by tracing the rays between the two surfaces, which is $\varepsilon_2 + (1-\varepsilon_2)(1-\varepsilon_1)\varepsilon_2 + (1-\varepsilon_2)^2(1-\varepsilon_1)^2\varepsilon_2 + \cdots$ since the reflectivity is one *minus* the emissivity. The radiative heat flux from surface 1 to surface 2 is

$$q''_{1\to 2} = \frac{\varepsilon_1 \varepsilon_2 \sigma_{SB} T_1^4}{1-(1-\varepsilon_1)(1-\varepsilon_2)} = \frac{\sigma_{SB} T_1^4}{1/\varepsilon_1 + 1/\varepsilon_2 - 1},$$

and that from surface 2 to surface 1 is

$$q''_{2\to 1} = \frac{\sigma_{SB} T_2^4}{1/\varepsilon_1 + 1/\varepsilon_2 - 1}.$$

Subsequently, the net radiative flux from surface 1 to surface 2 is

$$q''_{12} = q''_{1\to 2} - q''_{2\to 1} = \frac{\sigma_{SB}(T_1^4 - T_2^4)}{1/\varepsilon_1 + 1/\varepsilon_2 - 1} \tag{2.52}$$

Plugging in $T_1 = 873.2$ K, $T_2 = 300.2$ K, and other numerical values, we obtain $q''_{12} = 14.4$ kW/m$^2$.

Gas emission, absorption, and scattering are important for atmospheric radiation and combustion. When radiation travels through a cloud of gas, some of the energy may be absorbed. The absorption of photons raises the energy levels of individual molecules. At sufficiently high temperatures, gas molecules may spontaneously lower their energy levels and emit photons. These changes in energy levels are called *radiative transitions*, which include bound–bound transitions (between nondissociated molecular states), bound–free transitions (between nondissociated and dissociated states), and free–free transitions (between dissociated states). Bound–free and free–free transitions usually occur at very high temperatures (greater than about 5000 K) and emit in the ultraviolet and visible regions. The most important transitions for radiative heat transfer are bound–bound transitions between vibrational energy levels coupled with rotational transitions. The photon energy (or frequency) must be exactly the same as the difference between two energy levels in order for the photon to be absorbed or emitted; therefore, the quantization of the energy levels results in discrete spectral lines for absorption and emission. The rotational lines superimposed on a vibrational line give a band of closely spaced spectral lines, called the vibration–rotation spectrum. Additional discussion will be given in Chap. 3 about quantized transitions in atoms and molecules.

Particles can also scatter electromagnetic waves or photons, causing a change in the direction of propagation. In the early twentieth century, Gustav Mie developed a solution of Maxwell's equations for the scattering of electromagnetic waves by spherical particles, known as the Mie scattering theory. This solution can be used to predict the scattering phase function. In the case when the particle sizes are small compared with the wavelength, the formulation reduces to the simple expression obtained earlier by Lord Rayleigh. The phenomenon is called Rayleigh scattering, in which the scattering efficiency is inversely proportional to the wavelength to the fourth power. The wavelength-dependent characteristic of light scattering by small particles helps explain why the sky is blue and why the sun appears red at sunset. For spheres whose diameters are much greater than the wavelength, geometric optics can be applied by treating the surface as specular or diffuse.

The spectral intensity in a *participating medium*, $I_\lambda = I_\lambda(\xi, \Omega, t)$, depends on the location (the coordinate $\xi$), its direction (the solid angle $\Omega$), and time $t$. In a time interval $dt$, the beam travels from $\xi$ to $\xi + \mathrm{d}\xi$ ($\mathrm{d}\xi = c dt$), and the intensity is attenuated by absorption and out-scattering, but enhanced by emission and in-scattering. The macroscopic description of the radiation intensity is known as the *equation of radiative transfer* (ERT) [17].

$$\frac{1}{c}\frac{\partial I_\lambda}{\partial t} + \frac{\partial I_\lambda}{\partial \xi} = a_\lambda I_{b,\lambda}(T) - (a_\lambda + \sigma_\lambda) I_\lambda + \frac{\sigma_\lambda}{4\pi}\int_{4\pi} I_\lambda(\xi, \Omega', t)\Phi_\lambda(\Omega', \Omega)\,\mathrm{d}\Omega' \tag{2.53}$$

where $a_\lambda$ and $\sigma_\lambda$ are the absorption and scattering coefficients, respectively, and $\Phi_\lambda(\Omega', \Omega)$ is the *scattering phase function* $\Phi_\lambda = 1$, which satisfies the equation: $\frac{1}{4\pi}\int_{4\pi}\Phi_\lambda(\Omega', \Omega)\mathrm{d}\Omega' \equiv 1$. For isotropic scattring, $\Phi_\lambda = 1$. The right-hand side of Eq. (2.53) is composed of three terms: the first accounts for the contribution of emission (which depends on the local gas temperature $T$); the second is the attenuation by absorption and out-scattering; and the third is the contribution of in-scattering from all directions (solid angle $4\pi$) to the direction $\Omega$.

Unless ultrafast laser pulses are involved, the transient term is negligible. The ERT for the steady state can be simplified as

$$\frac{\partial I_\lambda(\zeta_\lambda, \Omega)}{\partial \zeta_\lambda} + I_\lambda(\zeta_\lambda, \Omega) = (1 - \eta_\lambda) I_{b,\lambda} + \frac{\eta_\lambda}{4\pi}\int_{4\pi} I_\lambda(\zeta_\lambda, \Omega')\Phi_\lambda(\Omega', \Omega)\mathrm{d}\Omega' \tag{2.54}$$

where $\zeta_\lambda = \int_0^\xi (a_\lambda + \sigma_\lambda)\mathrm{d}\xi$ is the *optical path length*, and $\eta_\lambda = \sigma_\lambda/(a_\lambda + \sigma_\lambda)$ is called the *scattering albedo*. This is an integrodifferential equation, and its right-hand side is called the source function. The integration of the spectral intensity over all wavelengths and all directions gives the radiative heat flux. Unless the temperature field is prescribed, Eq. (2.54) is coupled with the heat conduction equation in a macroscopically stationary medium and the energy conservation equation in a fluid with convection.

Analytical solutions of the ERT rarely exist for applications with multidimensional and nonhomogeneous media. Approximate models have been developed to deal with special types of problems, including Hottel's *zonal method*, the *differential and moment methods* (often using the spherical harmonics approximation), and the *discrete ordinates method*. The statistical model using the Monte Carlo method is often used for complicated geometries and radiative properties [17]. Analytical solutions can be obtained only for limited simple cases.

**Example 2.7** A gray, isothermal gas at a temperature $T_g = 3000$ K occupies the space between two, large parallel blackbody surfaces. Surface 1 is heated to a temperature $T_1 = 1000$ K, while surface 2 is maintained at a relatively low temperature by water cooling. It is desired to know the amount of heat that must be removed from surface 2. If the scattering is negligible, calculate the heat flux at surface 2 for $a_\lambda L = 0.01, 0.1, 1$, and 10, where $L$ is the distance between the two surfaces.

**Solution** For a gray medium without scattering, Eq. (2.53) becomes $\frac{dI(\theta)}{a_\lambda d\xi} + I(\theta) = I_b(T_g)$, where $\theta$ is the angle between $\xi$ and $x$. With $I_b(T_g) = \sigma_{SB}T_g^4/\pi$ and $I(0) = I_b(T_1) = \sigma_{SB}T_1^4/\pi$, the ERT can be integrated from $x = 0$ to $x = L$. The result is $I(\theta)|_{x=L} = \frac{\sigma_{SB}}{\pi}\left[T_1^4 e^{-a_\lambda L/\cos\theta} + T_g^4(1 - e^{-a_\lambda L/\cos\theta})\right]$. The radiative flux at $x = L$ can be obtained by integrating the intensity over the hemisphere, i.e.,

$$\begin{aligned} q''(a_\lambda L) &= \int_0^{2\pi}\int_0^{\pi/2} \frac{\sigma_{SB}}{\pi}\left[T_g^4 - (T_g^4 - T_1^4)e^{-a_\lambda L/\cos\theta}\right]\cos\theta \sin\theta d\theta d\phi \\ &= \sigma_{SB}T_g^4 - 2\sigma_{SB}(T_g^4 - T_1^4)E_3(a_\lambda L) \end{aligned}$$

where $E_3(\zeta) = \int_0^1 e^{-\zeta/\mu} d\mu$ is called the *exponential integral function of the third kind* and can be numerically evaluated. The final results are tabulated as follows:

| $a_\lambda L$ | 0.01 | 0.1 | 1 | 10 |
|---|---|---|---|---|
| $E_3(a_\lambda L)$ | 0.49 | 0.416 | 0.11 | $3.48 \times 10^{-6}$ |
| $q''$ (W/m$^2$) | $1.474 \times 10^5$ | $8.187 \times 10^5$ | $3.595 \times 10^6$ | $4.593 \times 10^6$ |

**Discussion.** In the optically thick limit ($a_\lambda L \gg 1$), $q'' \approx \sigma_{SB}T_g^4$, and all radiation leaving surface 1 will be absorbed by the gas before reaching surface 2. On the other hand, the heat flux is much greater than $\sigma_{SB}T_1^4 = 56.7$ kW/m$^2$ at $a_\lambda L = 0.01$. The gas absorption can be neglected in the optically thin limit; however, its emission contributes significantly to the radiative flux at surface 2. This is because the gas temperature is much higher than that of surface 1 and $L/\cos\theta$ can be much longer than $L$ for large $\theta$ values.

## 2.5 Summary

This chapter provided an overview of classical thermodynamics, derived following logical steps and on a general basis, as well as the functional relations and thermodynamic properties of simple systems and ideal pure substances. The basic heat transfer modes were elaborated in a coherent way built upon the foundations of thermodynamics. Entropy generation is inevitably associated with any heat transfer process. The connection between heat transfer and entropy generation, which has been omitted by most heat transfer textbooks, was also discussed. The introduction of thermal radiation not only covered most of the undergraduate-level materials but also linked to some basic graduate-level materials. This chapter should serve as a bridge or a reference to the rest of the book, dealing with energy transfer processes in micro/nanosystems and/or from a microscopic viewpoint of macroscopic phenomena.

## Problems

2.1. Give examples of steady state. Give examples of thermodynamic equilibrium state. Give an example of spontaneous process. Is the growth of a plant a spontaneous process? Give an example of adiabatic process.

2.2. What is work? Describe an experiment that can measure the amount of work. What is heat? Describe an apparatus that can be used to measure heat. Are work and heat properties of a system?

2.3. Expand Eqs. (2.1a, 2.1b) and (2.2a) in terms of the rate of energy and entropy change of an open system, which is subjected to work output, heat interactions, and multiple inlets and outlets of steady flow.

2.4. Discuss the remarks of Rudolf Clausius in 1867:

(a) The energy of the universe is constant.
(b) The entropy of the universe strives to attain a maximum value.

2.5. For a cyclic device experiencing heat interactions with reservoirs at $T_1, T_2, \ldots$, the Clausius inequality can be expressed as $\sum_i \frac{\delta Q_i}{T_i} \leq 0$ or $\oint \frac{\delta Q}{T} \leq 0$, regardless of whether the device produces or consumes work. Note that $\delta Q$ is positive when heat is received by the device. Prove the Clausius inequality by applying the second law to a closed system.

2.6. In the stable-equilibrium states, the energy and the entropy of a solid are related by $E = 3 \times 10^5 \exp(\frac{S-S_0}{1000})$, where $E$ is in J, $S$ is in J/K, and $S_0$ is the entropy of the solid at a reference temperature of 300 K. Plot this relation in an $E$–$S$ graph. Find expressions for $E$ and $S$ in terms of its temperature $T$ and $S0$.

2.7. For an isolated system, give the mathematical expressions of the first and second laws of thermodynamics. Give graphic illustrations using $E$–$S$ graph.

2.8. Place two identical metal blocks A and B, initially at different temperatures, in contact with each other but without interactions with any other systems. Assume thermal equilibrium is reached quickly and let system C represents the combined system of both A and B.

   (a) Is the process reversible or not? Which system has experienced a spontaneous change of state? Which systems have experienced an induced change of state?
   (b) Assume that the specific heat of the metal is independent of temperature, $c_p = 240$ J/kg K, the initial temperatures are $T_{A1} = 800$ K and $T_{B1} = 200$ K, and the mass of each block is 5 kg. What is the final temperature? What is the total entropy generation in this process?
   (c) Show the initial and final states of systems A, B, and C in a $u$–$s$ diagram, and indicate which state is not an equilibrium state. Determine the adiabatic availability of system C in the initial state.

2.9. Two blocks made of the same material with the same mass are allowed to interact with each other but isolated from the surroundings. Initially, block A is at 800 K and block B at 200 K. Assuming that the specific heat is independent of temperature, show that the final equilibrium temperature is 500 K. Determine the maximum and minimum entropies that may be transferred from block A to block B.

2.10. A cyclic machine receives 325 kJ heat from a 1000 K reservoir and rejects 125 kJ heat to a 400 K reservoir in a cycle that produces 200 kJ work. Is this cycle reversible, irreversible, or impossible?

2.11. If $z = z(x, y)$, then $\mathrm{d}z = f\mathrm{d}x + g\mathrm{d}y$, where $f(x, y) = \partial z/\partial x$, $g(x, y) = \partial z/\partial y$. Therefore, $\frac{\partial f}{\partial y} = \frac{\partial^2 z}{\partial y \partial x} = \frac{\partial^2 z}{\partial x \partial y} = \frac{\partial g}{\partial x}$. The second-order derivatives of the fundamental equation and each of the characteristic function yield a Maxwell relation. Maxwell's relations are very useful for evaluating the properties of a system in the stable-equilibrium states. For a closed system without chemical reactions, we have $\mathrm{d}N_i \equiv 0$. Show that $\left(\frac{\partial T}{\partial V}\right)_S = -\left(\frac{\partial P}{\partial S}\right)_V$, $\left(\frac{\partial T}{\partial P}\right)_S = \left(\frac{\partial V}{\partial S}\right)_P$, $\left(\frac{\partial S}{\partial V}\right)_T = \left(\frac{\partial P}{\partial T}\right)_V$, and $\left(\frac{\partial S}{\partial P}\right)_T = -\left(\frac{\partial V}{\partial T}\right)_P$.

2.12. The *isobaric volume expansion coefficient* is defined as $\beta_P = \frac{1}{v}\left(\frac{\partial v}{\partial T}\right)_P$, the isothermal compressibility is $\kappa_T = -\frac{1}{v}\left(\frac{\partial v}{\partial P}\right)_T$, and the *speed of sound* is $v_a = \sqrt{\left(\frac{\partial P}{\partial \rho}\right)_s}$. For an ideal gas, show that $\beta_P = 1/T$, $\kappa_T = 1/P$, and $v_a = \sqrt{\gamma RT}$.

2.13. For a system with single type of constituents, the fundamental relation obtained by experiments gives $S = \alpha(NVU)^{1/3}$, where $\alpha$ is a positive constant, and $N$, $V$, $S$, and $U$ are the number of molecules, the volume, the entropy, and the internal energy of the system, respectively. Obtain expressions of the temperature and the pressure in terms of $N$, $V$, $U$, and $\alpha$. Show that $S = 0$ at zero temperature for constant $N$ and $V$.

2.14. For blackbody radiation in an evacuated enclosure of uniform wall temperature $T$, the energy density can be expressed as $u_\nu = \frac{U}{V} = \frac{4}{c}\sigma_{SB}T^4$, where $U$ is the internal energy, $V$ the volume, $c$ the speed of light, and $\sigma_{SB}$ the Stefan–Boltzmann constant. Determine the entropy $S(T, V)$ and the pressure $P(T, V)$, which is called the *radiation pressure*. Show that the radiation pressure is a function of temperature only and negligibly small at moderate temperatures. Hint: $S = \int_0^T \frac{1}{T}\left(\frac{\partial U}{\partial T}\right)_V \mathrm{d}T$ and $P = T\left(\frac{\partial S}{\partial V}\right)_T - \left(\frac{\partial U}{\partial V}\right)_T$.

2.15. A cyclic machine can only interact with two reservoirs at temperatures $T_A = 298$ K and $T_B = 77.3$ K, respectively.

(a) If heat is extracted from reservoir A at a rate of $\dot{Q} = 1000$ W, what is the maximum rate of work that can be generated ($\dot{W}_{max}$)?
(b) If no work is produced, what is the rate of entropy generation ($\dot{S}_{gen}$) of the cyclic machine?
(c) Plot $\dot{S}_{gen}$ versus $\dot{W}$ (the power produced).

2.16. An engineer claimed that it requires much more work to remove 0.1 J of heat from a cryogenic chamber at an absolute temperature of 0.1 K than to remove 270 J of heat from a refrigerator at 270 K. Assuming that the environment is at 300 K, justify this claim by calculating the minimum work required for each refrigeration task.

2.17. A solid block [$m = 10$ kg and $c_p = 0.5$ kJ/kg K], initially at room temperature ($T_{A,1} = 300$ K) is cooled with a large tank of liquid–gas mixture of nitrogen at $T_B = 77.3$ K and atmospheric pressure.

(a) After the block reaches the liquid nitrogen temperature, what is the total entropy generation $S$gen?
(b) Given the specific enthalpy of evaporation of nitrogen, $h_{fg} = 198.8$ kJ/kg, what must be its specific entropy of evaporation $s_{fg}$ in kJ/kg K, in order for the nitrogen tank to be modeled as a reservoir? Does $h_{fg} = T_{sat} \times s_{fg}$ always hold?

2.18. Two same-size solid blocks of the same material are isolated from other systems [specific heat $c_p = 2$ kJ/kg K; mass $m = 5$ kg]. Initially, block A is at a temperature $T_{A1} = 300$ K and block B at $T_{B1} = 1000$ K.

(a) If the two blocks are put together, what will be the equilibrium temperature ($T_2$) and how much entropy will be generated ($S_{gen}$)?
(b) If the two blocks are connected with a cyclic machine, what is the maximum work that can be obtained ($W_{max}$)? What would be the final temperature of the blocks ($T_3$) if the maximum work was obtained?

2.19. A rock [density $\rho = 2800$ kg/m$^3$ and specific heat $c_p = 900$ J/kg K] of 0.8 m$^3$ is heated to 500 K using solar energy. A heat engine (cyclic machine) receives heat from the rock and rejects heat to the ambient at 290 K. The rock therefore cools down.

(a) Find the maximum energy (heat) that the rock can give out.
(b) Find the maximum work that can be done by the heat engine, $W_{max}$.
(c) (c) In an actual process, the final temperature of the rock is 330 K and the work output from the engine is only half of $W_{max}$. Determine the entropy generation of the actual process.

2.20. Consider three identical solid blocks with a mass of 5 kg each, initially at 300, 600, and 900 K, respectively. The specific heat of the material is $c_p =$ 2000 J/kg K. A cyclic machine is available that can interact only with the three blocks.

(a) What is the maximum work that can be produced? What are the final temperatures of each block? Is the final state in equilibrium?
(b) If no work is produced, i.e., simply putting the three blocks together, what will be the maximum entropy generation? What will be the final temperature?
(c) If the three blocks are allowed to interact via cyclic machine but not with any other systems in the environment, what is the highest temperature that can be reached by one of the blocks?
(d) If the three blocks are allowed to interact via cyclic machine but not with any other systems in the environment, what is the lowest temperature that can be reached by one of the blocks?

2.21. Electrical power is used to raise the temperature of a 500 kg rock from 25 to 500 °C. The specific heat of the rock material is $c_p = 0.85$ kJ/kg K.

(a) If the rock is heated directly through resistive (Joule) heating, how much electrical energy is needed? Is this process reversible? If not, how much entropy is generated in this process?
(b) By using cyclic devices that can interact with both the rock and the environment at 25 °C, what is the minimum electrical energy required?

2.22. An insulated cylinder of 2 $m^3$ is divided into two parts of equal volume by an initially locked piston. Side A contains air at 300 K and 200 kPa; side B contains air at 1500 K and 1 MPa. The piston is now unlocked so that it is free to move and it conducts heat. An equilibrium state is reached between the two sides after a while.

(a) Find the masses in both A and B.
(b) Find the final temperatures, pressures, and volumes for both A and B.
(c) Find the entropy generation in this process.

2.23. A piston–cylinder contains 0.56 kg of $N_2$ gas, initially at 600 K. A cyclic machine receives heat from the cylinder and releases heat to the environment at 300 K. Assume that the specific heat of $N_2$ is $c_p = 1.06$ kJ/kg K, and the pressure inside the cylinder is maintained at 100 kPa by the environment.

What is the maximum work that can be produced by the machine? What is the thermal efficiency (defined as the ratio of the work output to the heat received)? The thermodynamic efficiency can be defined as the ratio of the actual work produced to the maximum work. Plot the thermodynamic efficiency as a function of the entropy generation. What is the maximum entropy generation?

2.24. An airstream [$c_p = 1$ kJ/kg K and $M = 29.1$ kg/kmol] flows through a power plant. The stream enters a turbine at $T_1 = 750$ K and $P_1 = 6$ MPa, and exits at $P_2 = 1.2$ MPa into a recovery unit, which can exchange heat with the environment at 25 °C and 100 kPa. The stream then exits the recovery unit to the environment. The turbine is thermally insulated and has an efficiency $\eta_t = 0.85$.

(a) Find the power per unit mass flow rate produced by the turbine.
(b) Calculate the entropy generation rate in the turbine.
(c) Determine the largest power that can be produced by the recovery unit.

2.25. Water flows in a perfectly insulated, steady-state, horizontal duct of variable cross-sectional area. Measurements were taken at two ports, and the data were recorded in a notebook as follows. For port 1, speed $\xi_1 = 3$ m/s, pressure $P_1 = 50$ kPa, and temperature $T_1 = 40$ °C; for port 2, $\xi_2 = 5$ m/s and $P_2 = 45$ kPa. Some information was accidentally left out by the student taking the notes. Can you determine $T2$ and the direction of the flow based on the available information? Hint: Model the water as an ideal incompressible liquid with $c_p = 4.2$ kJ/kg) and specific volume $v = 10^{-3}$ m$^3$/kg.

2.26. An insulated rigid vessel contains 0.4 kmol of oxygen at 200 kPa separated by a membrane from 0.6 kmol of carbon dioxide at 400 kPa; both sides are initially at 300 K. The membrane is suddenly broken and, after a while, the mixture comes to a uniform state (equilibrium).

(a) Find the final temperature and pressure of the mixture.
(b) Determine the entropy generation due to irreversibility.

2.27. Pure $N_2$ and air (21% $O_2$ and 79% $N_2$ by volume), both at 298 K and 120 kPa, enter a chamber at a flow rate of 0.1 and 0.3 kmol/s, respectively. The new mixture leaves the chamber at the same temperature and pressure as the incoming streams.

(a) What are the mole fractions and the mass fractions of $N_2$ and $O_2$ at the exit?
(b) Find the enthalpy change in the mixing process. Find the entropy generation rate of the mixing process.
(c) Consider a process in which the flow directions are reversed. The chamber now contains necessary devices for the separation, and it may transfer heat to the environment at 298 K. What is the minimum amount of work per unit time needed to operate the separation devices?

2.28. A Carnot engine receives energy from a reservoir at $T_H$ and rejects heat to the environment at $T_0$ via a heat exchanger. The engine works reversibly between $T_H$ and $T_L$, where $T_L$ is the temperature of the higher temperature side of the heat exchanger. The *product* of the area and the heat transfer coefficient of the heat exchanger is $\alpha$. Therefore, the heat that must be rejected to the environment through the heat exchanger is $\dot{Q}_L = \alpha(T_L - T_0)$. Given $T_H = 800$ K, $T_0 = 300$ K, and $\alpha = 2300$ W/K. Determine the value of $T_L$ so that the heat engine will produce maximum work, and calculate the power production and the entropy generation in such a case.

2.29. To measure the thermal conductivity, a thin film electric heater is sandwiched between two plates whose sides are well insulated. Each plate has an area of 0.1 m$^2$ and a thickness of 0.05 m. The outside of the plates are exposed to air at $T_\infty = 25$ °C with a convection coefficient of $h = 40$ W/m$^2$ K. The electric power of the heat is 400 W and a thermocouple inserted between the two plates measures a temperature of $T_1 = 175$ °C at steady state. Determine the thermal conductivity of the plate material. Find the total entropy generation rate. Comment on the fraction of entropy generation due to conduction and convection.

2.30. An electric current, $I = 2$ A, passes through a resistive wire of diameter $D =$ 3 mm with a resistivity $r_e = 1.5 \times 10^{-4}\ \Omega$ m. The cable is placed in ambient air at 27 °C with a convection coefficient $h = 20$ W/m$^2$ K. Assume a steady state has been reached and neglect radiation. Determine the radial temperature distribution inside the wire. Determine the volumetric entropy generation rate $\dot{s}_{gen}$ as a function of radius. Determine the total entropy generation rate per unit length of the cable. Hint: For steady-state conduction, $\dot{s}_{gen} = \frac{1}{T}\nabla \cdot \mathbf{q}'' - \frac{1}{T^2}\mathbf{q}'' \cdot \nabla T$. [Hint: Consider $\kappa = 10$ W/m K and $\kappa = 1$ W/m K.]

2.31. Find the thermal conductivity of intrinsic (undoped) silicon, heavily doped silicon, quartz, glass, diamond, graphite, and carbon from 100 to 1000 K from Touloukian and Ho [13]. Discuss the variations between different materials, crystalline structures, and doping concentrations.

2.32. Find the thermal conductivity of copper from 1 to 1000 K from Touloukian and Ho [13]. Discuss the general trend in terms of temperature dependence, and comment on the effect of impurities.

2.33. For laminar flow over a flat plate, the velocity and thermal boundary layer thicknesses can be calculated by $\delta(x) = 5x/\sqrt{Re_x}$ and $\delta_t(x) = 5x Re_x^{-1/2} Pr^{-1/3}$, respectively. Use room temperature data to calculate and plot the boundary layer thicknesses for air, water, engine oil, and mercury for different values of $U_\infty$. Discuss the main features. Hint: Property data can be found from Incropera and DeWitt [11].

2.34. Air at 14 °C and atmospheric pressure is in parallel flow over a flat plate of $2 \times 2$ m$^2$. The air velocity is 3 m/s, and the surface is maintained at 140 °C. Determine the average convection coefficient and the rate of heat transfer from the plate to air. (For air at 350 K, which is the average temperature between the surface and fluid, $\kappa = 0.03$ W/m K, $\nu = 20.9 \times 10^{-6}$ m$^2$/s, and $Pr = 0.7$.)

2.35. Plot the blackbody intensity (Planck's law) as a function of wavelength for several temperatures. Discuss the main features of this function. Show that in the long-wavelength limit, the blackbody function can be approximated by $e_{b,\lambda}(\lambda, T) \approx 2\pi c k_B T/\lambda^4$, which is the Rayleigh–Jeans formula.

2.36. Calculate the net radiative heat flux from the human body at a surface temperature of $T_s = 308$ K, with an emissivity $\varepsilon = 0.9$, to the room walls at 298 K. Assume air is at 298 K and has a natural convection coefficient of 5 W/m$^2$ K. Neglect evaporation, calculate the natural convection heat flux from the person to air. Comment on the significance of thermal radiation.

2.37. Combustion occurs in a spherical enclosure of diameter $D = 50$ cm with a constant wall temperature of 600 K. The temperature of the combustion gases may be approximated as uniform at 2300 K. The absorption coefficient of the gas mixture is $a_\lambda = 0.01$ cm$^{-1}$, which is independent of wavelength. Assuming that the wall is black and neglecting the scattering effect, determine the net heat transfer rate between the gas and the inner wall of the sphere.

## References


1. B. Callen, *Thermodynamics and an Introduction to Thermostatistics*, 2nd edn. (Wiley, New York, 1985)
2. G.N. Hatsopoulos, J.H. Keenan, *Principles of General Thermodynamics* (Wiley, New York, 1965); J.H. Keenan, *Thermodynamics* (Wiley, New York, 1941)
3. E.P. Gyftopoulos, G.P. Beretta, *Thermodynamics: foundations and Applications* (Macmillan, New York, 1991); Also see the augmented edition (Dover Publications, New York, 2005)
4. R.E. Sonntag, C. Borgnakke, G.J. van Wylen, *Fundamentals of Thermodynamics*, 5th edn. (Wiley, New York, 1998)
5. M.J. Moran, H.N. Shapiro, *Fundamentals of Engineering Thermodynamics*, 4th edn. (Wiley, New York, 2000)
6. A. Bejan, *Advanced Engineering Thermodynamics*, 2nd edn. (Wiley, New York, 1997)
7. J. Kestin (ed.), *The Second Law of Thermodynamics* (Dowden, Hutchinson & Ross Inc, Stroudsburg, PA, 1976)
8. H. Preston-Thomas, The international temperature scale of 1990 (ITS-90). Metrologia **27**, 3–10 (1990)
9. Z.M. Zhang, Surface temperature measurement using optical techniques. Annu. Rev. Heat Transfer **11**, 351–411 (2000)
10. BIPM. https://www.bipm.org/en/cgpm-2018/, Accessed 29 Jan 2019
11. T.L. Bergman, A.S. Lavine, F.P. Incropera, D.P. DeWitt, *Fundamentals of Heat and Mass Transfer*, 7th edn. (Wiley, New York, 2011)
12. M.N. Özişik, *Heat Conduction*, 2nd edn. (Wiley, New York, 1993)
13. Y.S. Touloukian, C.Y. Ho (eds.), *Thermophysical Properties of Matter—The TPRC Data Series* (13 volumes compilation of data on thermal conductivity, specific heat, linear expansion coefficient, thermal diffusivity, and radiative properties) (Plenum Press, New York, 1970–1977)
14. A. Bejan, *Entropy Generation Minimization* (CRC Press, Boca Raton, FL, 1996)
15. R.F. Barron, *Cryogenic Heat Transfer* (Taylor & Francis, Philadelphia, PA, 1999)
16. M.C. Potter, D.C. Wiggert, *Mechanics of Fluids* (Prentice Hall, New Jersey, 1991)
17. J.R. Howell, M.P. Mengüç, R. Siegel, *Thermal Radiation Heat Transfer*, 6th edn. (CRC Press, New York, 2016)

# Chapter 3
# Elements of Statistical Thermodynamics and Quantum Theory

Classical statistical mechanics is based on the assumption that all matters are composed of a myriad of small discrete particles, such as molecules and atoms, in any given macroscopic volume [1–5]. There are about $N = 2.5 \times 10^{16}$ molecules per cubic millimeter of air at standard conditions (25 °C and 1 atm). These particles are in continuous random motion, which generally obeys the laws of classical mechanics. A complete microscopic description of a system requires the identification of the position $\mathbf{r}_i(t)$ and velocity $\mathbf{v}_i(t)$ of each particle (here, subscript $i$ indicates the $i$th particle) at any time. For a simple system of $N$ molecules in a box of volume $V$, one can write Newton's law of motion for each molecule as

$$\sum_j \mathbf{F}_{ij}(\mathbf{r}_i, \mathbf{r}_j, t) = m_i \frac{\mathrm{d}\mathbf{v}_i}{\mathrm{d}t}, \quad i = 1, 2, \ldots N \tag{3.1}$$

where $\mathbf{F}_{ij}$ is the intermolecular force that the $j$th molecule exerts on the $i$th molecule, and $m_i$ is the mass of the $i$th molecule. The initial position and velocity, as well as the nature of collisions among particles and that between particles and the walls of the box, must be specified in order to solve the $N$ equations. Although this approach is straightforward, there are two major barriers. First, the intermolecular forces or potentials are often complicated and difficult to determine. Second, the solution of Eq. (3.1) requires significant computer resources even for rather simple problems. Statistical methods are often used instead to obtain microscopic descriptions that are related to macroscopic behaviors. *Statistical mechanics* aims at finding the equilibrium distribution of certain type of particles in the velocity space. It provides a linkage between macroscopic thermodynamic properties and the microscopic behavior and a means to evaluate some thermodynamic properties. *Kinetic theory*, on the other hand, deals with nonequilibrium processes. It gives a microscopic description of transport phenomena and helps predict some important transport properties, as will be seen in Chap. 4.



Z. M. Zhang, *Nano/Microscale Heat Transfer*, Mechanical Engineering Series,
https://doi.org/10.1007/978-3-030-45039-7_3

Along with the rapid development in computing speed and memory, *molecular dynamics* (MD) simulation has become a powerful tool for the investigation of phenomena occurring in nanostructures and/or at very short time scales. In the MD method, the location and the velocity of every particle are calculated at each time step by applying Eq. (3.1) with a suitable potential function [6, 7]. Thermodynamic properties are then evaluated using statistical mechanics formulation. Further discussion about the application of the MD simulation to predict the thermal properties of nanostructures will be given in Chap. 7.

This chapter starts with a statistical model of independent particles and a brief introduction to the basic principles of quantum mechanics. The necessary mathematical background is summarized in Appendix B. It is highly recommended that one review the materials covered in the appendix before studying this chapter. The three important distributions are derived based on the statistics for different types of particles. The microscopic descriptions and results are then linked to macroscopic quantities and the laws of thermodynamics. The application to ideal gases is presented in this chapter, while the applications to blackbody radiation, lattice vibration, free electrons in metals, and electrons and holes in semiconductors will be deferred to later chapters.

## 3.1 Statistical Mechanics of Independent Particles

We say particles are independent when their energies are independent of each other and the total energy is the sum of the energies of individual particles. Consider a system that has $N$ independent particles of the same type confined in a volume $V$. The total internal energy of the system is $U$, which is the sum of the energies of all particles. Particles may have different energies and can be grouped according to their energies. It is of interest to know how many particles are there within certain energy intervals. We can subdivide energy into a large number of discretized energy levels. As illustrated in Fig. 3.1, there are $N_i$ particles on the $i$th energy level, each with energy exactly equal to $\varepsilon_i$.

From classical mechanics point of view, it appears that the increment between adjacent energy levels can be indefinitely small. The particles are distinguishable, and there is no limit on the number of particles on each energy level. Quantum mechanics predicts that the energy levels are indeed discretized with finite increments between adjacent energy levels, and the particles are unidentifiable (*indistinguishable*) according to quantum statistics. Readers are referred to Appendix B.3 for further discussion about the statistical distinguishability and the resulting different permutation and combination theories. The conservation equations for the system shown in Fig. 3.1 are

$$\sum_{i=0}^{\infty} N_i = N \tag{3.2}$$

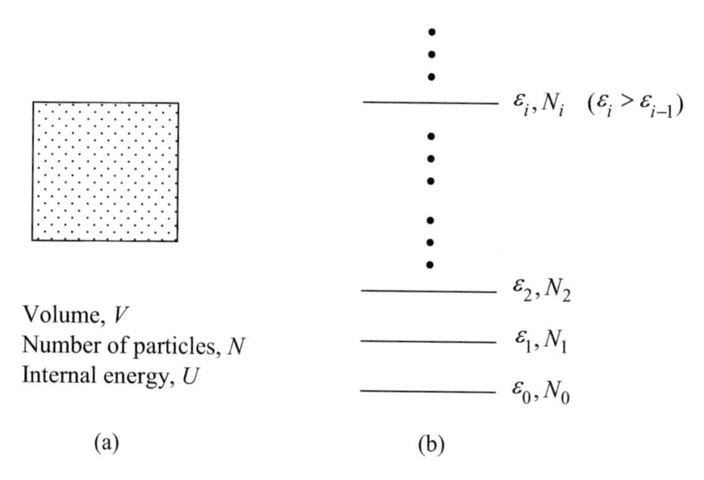


**Fig. 3.1** Illustration of **a** a simple system of independent particles and **b** energy levels

and

$$\sum_{i=0}^{\infty} \varepsilon_i N_i = U \tag{3.3}$$

## 3.1.1 Macrostates Versus Microstates

The thermodynamic state may be viewed in terms of the gross behavior that ignores any differences at the molecular or atomic level, or in terms of the individual particles. A *macrostate* is determined by the values of $N_0, N_1, N_2, \ldots$ for a given volume (which somehow confines the quantized energy levels) though two different macrostates can have the same energy. Each macrostate may be made up of a number of microscopic arrangements; each microscopic arrangement is called a *microstate.* In statistical mechanics, all microstates are assumed *equally probable*. There may be a large number of microstates that correspond to the same macrostate. The number of microstates for each macrostate is termed the *thermodynamic probability* $\Omega$ of that macrostate. Unlike the stochastic probability that lies between 0 and 1, the thermodynamic probability $\Omega$ is usually a very large number. One of the principles underlying statistical mechanics is that the stable-equilibrium state corresponds to the *most probable macrostate*. Therefore, for given values of $U$, $N$, and $V$, the thermodynamic probability is the largest in the stable-equilibrium state. We will use the following example to illustrate the concepts of microstate and macrostate.

**Example 3.1** There are four distinguishable particles in a confined space, and there are two energy levels. How many macrostates are there? How many microstates are there for the macrostate with two particles on each energy level?

**Solution** There are five macrostates in total with $(N_1, N_2) = (0, 4), (1, 3), (2, 2), (3, 1)$, and $(4, 0)$, respectively. Because the particles are distinguishable, the microstates will be different only if the particles from different energy levels are interchanged. Using the combination theory, we can figure out that $\Omega(N_1, N_2) = N!/(N_1!N_2!) = 4!/(2!2!) = 6$. Hence, there are six microstates for the macrostate with two particles on each energy level. It can be shown that this is also the most probable macrostate.

### 3.1.2 Phase Space

The phase space is used to describe all possible values of position and momentum variables that can be used to fully characterize the state of a mechanical system at any given time. It is an important concept in classical and quantum statistics. The *phase space* is a six-dimensional "space" formed by three coordinates for the position $\mathbf{r}$ and three coordinates for the momentum $\mathbf{p} = m\mathbf{v}$ or velocity $\mathbf{v}$. Each point in the phase space defines the exact location and momentum of an individual particle. If both the space and the momentum are described with the Cartesian system, then a volume element in the phase space is $\mathrm{d}x\mathrm{d}y\mathrm{d}z\mathrm{d}p_x\mathrm{d}p_y\mathrm{d}p_z$. Figure 3.2 shows a phase space projected to the $x - p_x$ plane. The three coordinates $(p_x, p_y, p_z)$ form a *momentum space*. One may choose to use $(v_x, v_y, v_z)$ to form a *velocity space*. If the momentum space is described in spherical coordinates, the volume element is $\mathrm{d}p_x\mathrm{d}p_y\mathrm{d}p_z = p^2 \sin\theta \mathrm{d}p\mathrm{d}\theta\mathrm{d}\phi$. The volume contained in a spherical shell from $p$ to $p + \mathrm{d}p$ is $4\pi p^2\mathrm{d}p$. Figure 3.3 illustrates the momentum space projected to the $p_x - p_y$ plane, with a spherical shell.

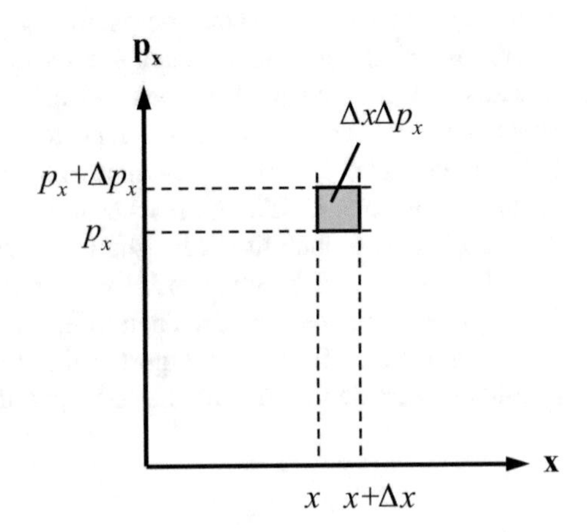


Fig. 3.2 Phase space projected to the $x - p_x$ plane, where $\Delta x \Delta p_x$ is an area element

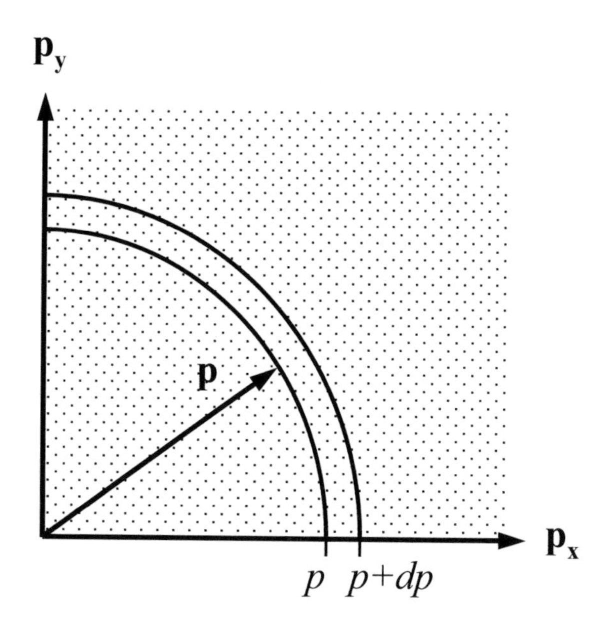


**Fig. 3.3** The $p_x - p_y$ plane of the momentum space, showing a spherical shell

### *3.1.3 Quantum Mechanics Considerations*

The principles of quantum mechanics are important for the advancement of statistical thermodynamics, especially when dealing with particles that cannot be treated with classical statistics. An introduction to the basic principles of quantum mechanics is given in this section and a more detailed introduction of the quantum theory is given in Sect. 3.5. The origin of quantum theory can be traced back to about 100 years ago when Planck first used a discrete set of energies to describe the electromagnetic radiation, and thus obtained Planck's distribution (details to be presented in Sect. 8.2). For any given frequency of radiation $\nu$, the smallest energy increment is given by $h\nu$, where $h = 6.626 \times 10^{-34}$ J · s is called Planck's constant. Radiation can be alternatively viewed as electromagnetic waves or traveling energy quanta. The corpuscular theory treats radiation as a collection of energy quanta, called *photons*. The energy of a photon is given by

$$\varepsilon = h\nu \tag{3.4}$$

From the wave theory, the speed of light $c$ is related to the wavelength $\lambda$ and the frequency by

$$c = \lambda\nu \tag{3.5}$$

In a medium with a refractive index of $n$, $c = c_0/n$ and $\lambda = \lambda_0/n$, where subscript 0 is used to indicate quantities in vacuum with $n = 1$. The speed of light in vacuum is

$c_0 = 299,792,458\,\mathrm{m/s}$, which is a defined quantity as given in Appendix A. Note that the frequency of an electromagnetic wave does not change from one medium to another.

Based on the relativistic theory, the rest energy $E_0$ of a particle with mass $m$ is

$$E_0 = mc^2 \tag{3.6}$$

The momentum of the particle traveling with speed $v$ is $p = mv$. Since the energy of a photon is $h\nu$ and its speed is $c$, the momentum of a (massless) photon is (see Sect. 3.7)

$$p = \frac{h\nu}{c} = \frac{h}{\lambda} \tag{3.7}$$

Another hypothesis of quantum theory is that the motion of matter may be wave-like, with characteristic wavelength and frequency. Therefore, for a particle moving with velocity $v \ll c$.

$$\lambda_{\mathrm{DB}} = \frac{h}{p} = \frac{h}{mv} \quad \text{and} \quad \nu_{\mathrm{DB}} = \frac{mc^2}{h} \tag{3.8}$$

which are called *de Broglie wavelength* and *de Broglie frequency*, respectively. In 1923, Louis de Broglie postulated that matter may also possess wave characteristics and thereafter resolved the controversy as per the nature of radiation. Note that the phase speed of the wave defined by Eq. (3.8) is $c^2/v$, which is greater than the speed of light. The discovery of electron diffraction confirmed de Broglie's hypothesis. For this prediction, de Broglie received the Nobel Prize in Physics in 1929. Seven years later, the 1937 Nobel Prize in Physics was shared by Clinton J. Davisson and George P. Thomson for their independent experiments that demonstrated diffraction of electrons by crystals.

**Example 3.2** Calculate the frequency in Hz and photon energy in eV of an ultraviolet (UV) laser beam at a wavelength of $\lambda = 248\,\mathrm{nm}$ and a microwave at $\lambda = 10\,\mathrm{cm}$. Calculate the de Broglie wavelength of a He atom at 200 °C, using the average speed of 1717 m/s, and an electron traveling with a speed of $10^6$ m/s.

**Solution** The equations are $\nu = c/\lambda$ and $\varepsilon = hc/\lambda$. Assume the refractive index is 1. For the UV beam at $\lambda = 248\,\mathrm{nm}$, $\nu = 1.2\times10^{15}\,\mathrm{Hz}$ and $\varepsilon = 8.01\times10^{-19}\,\mathrm{J} = 5\,\mathrm{eV}$. For $\lambda = 10\,\mathrm{cm}$, $\nu = 3\times10^9\,\mathrm{Hz} = 3\,\mathrm{GHz}$ and $\varepsilon = 2\times10^{-24}\,\mathrm{J} = 1.24\times10^{-5}\,\mathrm{eV} = 124\,\mathrm{meV}$. The mass of a He atom is $m = M/N_{\mathrm{A}} = 6.64\times10^{-27}\,\mathrm{kg}$. Hence, $\lambda_{\mathrm{DB}} = h/mv = 5.8\times10^{-11}\,\mathrm{m} = 58\,\mathrm{pm}$. From Appendix A, $m_{\mathrm{e}} = 9.11\times10^{-31}\,\mathrm{kg}$, therefore, $\lambda_{\mathrm{DB}} = 7.3\times10^{-10}\,\mathrm{m} = 0.73\,\mathrm{nm}$, which is in the x-ray region.

The foundation of quantum mechanics is the Schrödinger equation, which is a partial-differential equation of the time-space dependent complex *probability density function*. More details can be found from the texts of Tien and Lienhard [1], Carey [5],

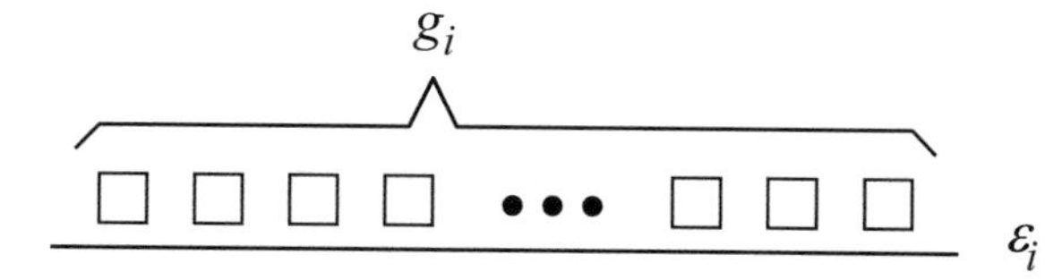


**Fig. 3.4** The degeneracy of the $i$th energy level

and Griffiths [8]. The solutions of the Schrödinger equation support the dual nature of wave and matter, and result in discrete quantized energy levels. Furthermore, there are usually more than one distinguishable *quantum states* at each energy level, i.e., the energy levels may be degenerate. The number of quantum states for a given energy level is called the *degeneracy*, denoted by $g_i$ for the $i$th energy level, as shown in Fig. 3.4.

The *uncertainty principle* states that the position and momentum of a given particle cannot be measured simultaneously with arbitrary precision. The limit is given by

$$\Delta x \Delta p_x \geq h/4\pi \tag{3.9}$$

This result implies that we cannot locate the exact position of a particle in the phase space; all we can say is that the particle is somewhere in a domain whose volume is around $h^3$. The uncertainty principle is one of the cornerstones of quantum mechanics and was formulated in 1927 by Werner Heisenberg, a Nobel laureate in Physics.

For certain particles, such as electrons, each quantum state cannot be occupied by more than one particle. This is *the Pauli exclusion principle,* discovered by Nobel laureate Wolfgang Pauli in 1925. The result, as we will see, is the Fermi-Dirac statistics that can be used to describe the behavior of free electrons. The collection of free electrons in metals is sometimes called the free electron gas, which exhibits very different characteristics from ideal molecular gases.

### *3.1.4 Equilibrium Distributions for Different Statistics*

The characteristics of various types of particles can be described by different statistics. In this section, we will first introduce three statistics and then apply them to obtain the distribution functions, i.e., the number of particles on each energy level. Applications of the distribution functions to various particle systems will also be explained.

- *The Maxwell-Boltzmann (MB) statistics*: Particles are distinguishable and there is no limit for the number of particles on each energy level. From Eq. (B.22) in Appendix B, the thermodynamic probability for the distribution shown in Fig. 3.1b is

$$\Omega = \frac{N!}{N_0!N_1!N_2!\cdots} = \frac{N!}{\prod_{i=0}^{\infty} N_i!}$$

If degeneracy is included as shown in Fig. 3.4, then

$$\Omega_{\text{MB}} = N! \prod_{i=0}^{\infty} \frac{g_i^{N_i}}{N_i!} \tag{3.10}$$

- *The Bose-Einstein (BE) statistics*: Particles are indistinguishable and there is no limit for the number of particles in each quantum state; there are $g_i$ quantum states on the $i$th energy level. From Eq. (B.23), the number of ways of placing $N_i$ indistinguishable objects to $g_i$ distinguishable boxes is $\frac{(g_i+N_i-1)!}{(g_i-1)!N_i!}$. Therefore, the thermodynamic probability for BE statistics is

$$\Omega_{\text{BE}} = \prod_{i=0}^{\infty} \frac{(g_i + N_i - 1)!}{(g_i - 1)!N_i!} \tag{3.11}$$

- *The Fermi-Dirac (FD) statistics*: Particles are indistinguishable and the energy levels are degenerate. There are $g_i$ quantum states on the $i$th energy level, and each quantum state can be occupied by no more than one particle. Using Eq. (B.21), we obtain the thermodynamic probability for FD statistics as

$$\Omega_{\text{FD}} = \prod_{i=0}^{\infty} \frac{g_i!}{(g_i - N_i)!N_i!} \tag{3.12}$$

The three statistics are very important for understanding the molecular, electronic, crystalline, and radiative behaviors that are essential for energy transport processes in both small and large scales. MB statistics can be considered as the limiting case of BE or FD statistics. The thermodynamic relations and the velocity distribution of ideal molecular gases can be understood from MB statistics. BE statistics is important for the study of photons, phonons in solids, and atoms at low temperatures. It is the basis of Planck's law of blackbody radiation, the Debye theory for the specific heat of solids, and the Bose-Einstein condensation, which is important for superconductivity, superfluidity, and laser cooling of atoms. FD statistics can be used to model the electron gas and the electron contribution to the specific heat of solids. It is important for understanding the electronic and thermal properties of metals and semiconductors.

**Example 3.3** Four indistinguishable particles are to be placed in two energy levels, each with a degeneracy of 3. Evaluate the thermodynamic probability of all arrangements, considering BE and FD statistics separately. What are the most probable arrangements?

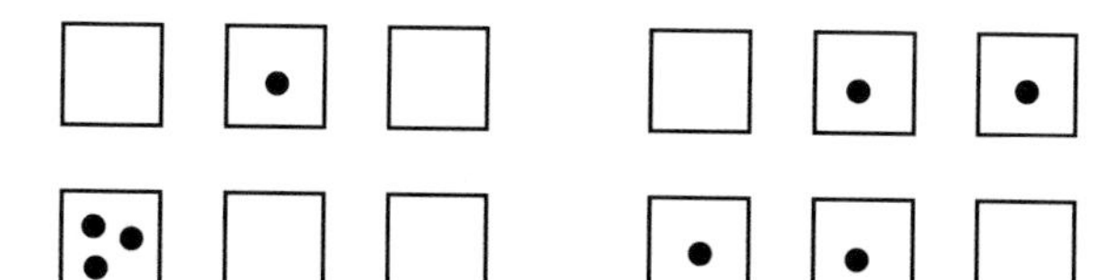

**Fig. 3.5** Illustration of the arrangement for four particles on two energy levels, each with a degeneracy of 3. **a** Bose-Einstein statistics. **b** Fermi-Dirac statistics

**Solution** There are two energy levels, $g_0 = g_1 = 3$ and the total number of particles $N = 4$. The thermodynamic probability is $\Omega = \Omega_0 \times \Omega_1$, which depends on $N_0$ and $N_1 (N_0 + N_1 = 4)$. Figure 3.5 shows specific cases of the BE and FD distributions.

For BE statistics, we have

$$\Omega_{\mathrm{BE}} = \frac{(N_0 + g_0 - 1)!}{(g_0 - 1)!N_0!} \times \frac{(N_1 + g_1 - 1)!}{(g_1 - 1)!N_1!} = \frac{(N_0 + 2)(N_0 + 1)}{2} \times \frac{(6 - N_0)(5 - N_0)}{2}$$

For FD statistics, we must have $N_i \le g_i$; therefore, $1 \le N_0 \le 3$, and

$$\Omega_{\mathrm{FD}} = \frac{g_0!}{(g_0 - N_0)!N_0!} \times \frac{g_1!}{(g_1 - N_1)!N_1!} = \frac{6}{(3 - N_0)!N_0!} \times \frac{6}{(N_0 - 1)!(4 - N_0)!}$$

The results are summarized in the following table. Clearly, the most probable arrangement for both statistics in this case is $N_0 = N_1 = 2$.

| $N_0$ | 0 | 1 | 2 | 3 | 4 |
|---|---|---|---|---|---|
| $N_1$ | 4 | 3 | 2 | 1 | 0 |
| $\Omega_{\mathrm{BE}}$ | 15 | 30 | 36 | 30 | 15 |
| $\Omega_{\mathrm{FD}}$ | – | 3 | 9 | 3 | – |

For a given simple thermodynamics system of volume $V$, internal energy $U$, and total number of particles $N$, we wish to find the state (identified by the distribution $N_0, N_1, N_2, \ldots$) that maximizes $\Omega$ or $\ln \Omega$, under constrains given by Eqs. (3.2) and (3.3), based on the method of Lagrange multipliers (Appendix B). For MB statistics with degeneracy, from Eq. (3.10),

$$\ln \Omega = \ln N! + \sum_{i=0}^{\infty} N_i \ln g_i - \sum_{i=0}^{\infty} \ln N_i!$$

For a large number of particles, the Stirling formula gives $\ln N! \approx N \ln N - N$ from Eq. (B.11). The above equation can be approximated as

$$\ln \Omega = N \ln N - N + \sum_{i=0}^{\infty} N_i \ln g_i - \sum_{i=0}^{\infty} (N_i \ln N_i - N_i) = N \ln N - N + \sum_{i=0}^{\infty} N_i (\ln \frac{g_i}{N_i} + 1)$$

Notice that $N$ and $g_{i's}$ are fixed and only $N_{i's}$ are variables, therefore,

$$\mathrm{d}(\ln \Omega) = \sum_{i=0}^{\infty} \frac{\partial (\ln \Omega)}{\partial N_i} \mathrm{d}N_i = \sum_{i=0}^{\infty} \left( \ln \frac{g_i}{N_i} + 1 - N_i \frac{1}{N_i} \right) \mathrm{d}N_i = \sum_{i=0}^{\infty} \ln \frac{g_i}{N_i} \mathrm{d}N_i = 0 \tag{3.13}$$

From the constraint equations, Eqs. (3.2) and (3.3), we have

$$-\alpha \sum_{i=0}^{\infty} \mathrm{d}N_i = 0 \tag{3.14a}$$

and

$$-\beta \sum_{i=0}^{\infty} \varepsilon_i \mathrm{d}N_i = 0 \tag{3.14b}$$

where $\alpha$ and $\beta$ are Lagrangian multipliers and $\varepsilon_i$'s are treated as constants. Conventionally, negative signs are chosen because $\alpha$ and $\beta$ are generally nonnegative for molecular gases. By adding Eqs. (3.14a) and (3.14b) to Eq. (3.13), we obtain

$$\sum_{i=0}^{\infty} \left( \ln \frac{g_i}{N_i} - \alpha - \beta \varepsilon_i \right) \mathrm{d}N_i = 0$$

Because $\mathrm{d}N_i$ can be arbitrary, the above equation requires that $\ln(g_i/N_i) - \alpha - \beta\varepsilon_i = 0$. Hence,

$$N_i = \frac{g_i}{\mathrm{e}^{\alpha} \mathrm{e}^{\beta \varepsilon_i}} \tag{3.15a}$$

or

$$\frac{N_i}{N} = \frac{g_i \mathrm{e}^{-\alpha} \mathrm{e}^{-\beta \varepsilon_i}}{\sum_{i=0}^{N} g_i \mathrm{e}^{-\alpha} \mathrm{e}^{-\beta \varepsilon_i}} \tag{3.15b}$$

This is the MB distribution. The physical meanings of $\alpha$ and $\beta$ will be discussed later. Using the same procedure described above, we can obtain the following for BE statistics,

$$N_i = \frac{g_i}{\mathrm{e}^{\alpha} \mathrm{e}^{\beta \varepsilon_i} - 1} \tag{3.16}$$

which is the BE distribution. For FD statistics, we can obtain the FD distribution as follows

$$N_i = \frac{g_i}{\mathrm{e}^{\alpha}\mathrm{e}^{\beta\varepsilon_i} + 1} \tag{3.17}$$

The results for all the three statistics are summarized in Table 3.1. Note that $N_i/g_i$ signifies how many particles occupy a quantum state or the probability for a quantum state to be occupied, which is called the *mean occupation number*.

**Example 3.4** Derive the BE distribution step by step. Under which condition can it be approximated by the MB distribution?

**Solution** Using the thermodynamic probability of BE statistics in Eq. (3.11), we have

$$\begin{aligned}
\ln\Omega &= \sum_{i=0}^{\infty}[\ln(g_i+N_i-1)! - \ln(g_i-1)! - \ln N_i!] \\
&\approx \sum_{i=0}^{\infty}[(g_i+N_i-1)\ln(g_i+N_i-1) - (g_i+N_i-1) - (g_i-1)\ln(g_i-1) + (g_i-1) - N_i\ln N_i + N_i] \\
&= \sum_{i=0}^{\infty}[(g_i+N_i-1)\ln(g_i+N_i-1) - (g_i-1)\ln(g_i-1) - N_i\ln N_i]
\end{aligned}$$

Hence,

$$\begin{aligned}
\frac{\partial\ln\Omega}{\partial N_i} &= \ln(g_i+N_i-1) + (g_i+N_i-1)\frac{1}{g_i+N_i-1} - \ln N_i - N_i\frac{1}{N_i} \\
&= \ln\left(\frac{g_i+N_i-1}{N_i}\right) \approx \ln\left(\frac{g_i}{N_i}+1\right), \quad \text{since } N_i \gg 1
\end{aligned}$$

To maximize $\Omega$, we set $\mathrm{d}(\ln\Omega) = \sum_{i=0}^{\infty}\frac{\partial(\ln\Omega)}{\partial N_i}\mathrm{d}N_i \approx \sum_{i=0}^{\infty}\ln\left(\frac{g_i}{N_i}+1\right)\mathrm{d}N_i = 0$.

By adding Lagrangian multipliers, Eq. (3.14a), (3.14b), we get $\sum_{i=0}^{\infty}[\ln(g_i/N_i+1) - \alpha - \beta\varepsilon_i]\mathrm{d}N_i = 0$. Hence, $N_i = g_i/(\mathrm{e}^{\alpha}\mathrm{e}^{\beta\varepsilon_i} - 1)$, which is the BE distribution given in Eq. (3.16) and Table 3.1.

**Discussion**. If $\exp(\alpha+\beta\varepsilon_i) \gg 1$, both Eqs. (3.16) and (3.17) reduce to the MB distribution, Eq. (3.15a), (3.15b). Under the limiting case of $g_i \gg N_i \gg 1$, we have

$$\frac{(g_i+N_i-1)!}{(g_i-1)!\,N_i!} = \frac{\overbrace{(g_i+N_i-1)\cdots(g_i+1)g_i}^{N_i\text{ terms}}}{N_i!} \xrightarrow{g_i \gg N_i \gg 1} \frac{g_i^{N_i}}{N_i!}$$

and

$$\frac{g_i!}{(g_i-N_i)!\,N_i!} = \frac{\overbrace{g_i(g_i-1)\cdots(g_i-N_i+1)}^{N_i\text{ terms}}}{N_i!} \xrightarrow{g_i \gg N_i \gg 1} \frac{g_i^{N_i}}{N_i!}$$

**Table 3.1** Summary of the three statistics

| Statistics | Maxwell-Boltzmann (MB) | Bose-Einstein (BE) | Fermi-Dirac (FD) |
|---|---|---|---|
| Name of particles | Boltzons | Bosons | Fermions |
| Examples | Ideal gas molecules and in the limit of bosons and fermions | Photons and phonons | Electrons and protons |
| Distinguishability | Distinguishable | Indistinguishable | Indistinguishable |
| Degeneracy | Degenerated | Degenerated | Degenerated |
| Particles per quantum state | Unlimited | Unlimited | One |
| Thermodynamic probability ($\Omega$) | $N!\prod_{i=0}^{\infty}\frac{g_i^{N_i}}{N_i!}$ | $\prod_{i=0}^{\infty}\frac{(g_i+N_i-1)!}{(g_i-1)!N_i!}$ | $\prod_{i=0}^{\infty}\frac{g_i!}{(g_i-N_i)!N_i!}$ |
| In the limit of $g_i \gg N_i$ | $\Omega_{\text{MB}}$ (given above) | $\Omega_{\text{MB}}/N!$ | $\Omega_{\text{MB}}/N!$ |
| $\ln\Omega$ | $N\ln N - N + \sum_{i=0}^{\infty} N_i[\ln(g_i/N_i)+1]$ | $\sum_{i=0}^{\infty}[(g_i+N_i-1)\ln(g_i+N_i-1) - N_i\ln N_i - (g_i-1)\ln(g_i-1)]$ | $\sum_{i=0}^{\infty}[g_i\ln g_i - N_i\ln N_i - (g_i-N_i)\ln(g_i-N_i)]$ |
| $\mathrm{d}(\ln\Omega)$ | $\sum_{i=0}^{\infty}\ln(\frac{g_i}{N_i})\,\mathrm{d}N_i$ | $\sum_{i=0}^{\infty}\ln(\frac{g_i}{N_i}+1)\,\mathrm{d}N_i$ | $\sum_{i=0}^{\infty}\ln(\frac{g_i}{N_i}-1)\,\mathrm{d}N_i$ |
| $-\alpha\sum_{i=0}^{\infty}\mathrm{d}N_i - \beta\sum_{i=0}^{\infty}\varepsilon_i\mathrm{d}N_i$ | $\ln(\frac{g_i}{N_i}) - \alpha - \beta\varepsilon_i = 0$ | $\ln(\frac{g_i}{N_i}+1) - \alpha - \beta\varepsilon_i = 0$ | $\ln(\frac{g_i}{N_i}-1) - \alpha - \beta\varepsilon_i = 0$ |
| Distribution function ($Ni$) | $\frac{g_i}{e^{\alpha}e^{\beta\varepsilon_i}}$ | $\frac{g_i}{e^{\alpha}e^{\beta\varepsilon_i}-1}$ | $\frac{g_i}{e^{\alpha}e^{\beta\varepsilon_i}+1}$ |
| Applications | Ideal gases; Maxwell's velocity distribution; limiting cases of BE and FD statistics | Planck's law; Bose-Einstein condensation; specific heat of solids | Electron gas; Fermi level; electron specific heat in metals |

We see that the thermodynamic probability for both the BE and FD statistics reduces to the MB statistics divided by $N!$, which is caused by the assumption of indistinguishable particles. Therefore,

$$\Omega_{\text{MB,corrected}} = \prod_{i=0}^{\infty} \frac{g_i^{N_i}}{N_i!}$$

is called the "corrected" MB statistics. For ideal molecular gases at reasonably high temperatures, $g_i \gg N_i$. For this reason, the MB distribution may be considered as the limiting case of the BE or FD distribution; see Table 3.1.

## 3.2 Thermodynamic Relations

The thermodynamic properties and relations can be understood from the microscopic point of view. This includes the concept of heat and work, entropy, and the third law of thermodynamics. The partition function is key to the evaluation of thermodynamic properties.

### 3.2.1 *Heat and Work*

From Eq. (3.3), we have

$$dU = \sum_{i=0}^{\infty} \varepsilon_i \, dN_i + \sum_{i=0}^{\infty} N_i \, d\varepsilon_i \tag{3.18}$$

The first term on the right is due to a redistribution of particles among the energy levels (which is related to a change in entropy), while the second is due to a shift in the energy levels associated with, e.g., a volume change. Consider a reversible quasi-equilibrium process for a closed system (such as a piston/cylinder arrangement); the work is associated to the volume change that does not change the entropy of the system, while heat transfer changes entropy of the system without affecting the energy levels. Therefore,

$$\delta Q = \sum_{i=0}^{\infty} \varepsilon_i \, dN_i \quad \text{and} \quad \delta W = -\sum_{i=0}^{\infty} N_i \, d\varepsilon_i \tag{3.19}$$

In writing the above equation, $\delta Q$ is positive for heat transferred to the system, and $\delta W$ is positive for work done by the system. They are related to macroscopic quantities for simple system by $\delta Q = T dS$ and $\delta W = P dV$. Hence, we obtain the expression of the first law for a closed system, $\mathrm{d}U = \delta Q - \delta W$. If the system is an open system, then $\sum_{i=0}^{\infty} \varepsilon_i \, \mathrm{d}N_i = \mathrm{d}U + \delta W \neq \delta Q$.

### 3.2.2 *Entropy*

The macroscopic property entropy is related to the thermodynamic probability by

$$S = k_\mathrm{B} \ln \Omega \tag{3.20}$$

where $k_\mathrm{B}$ is the Boltzmann constant. Consider two separate systems A and B, and their combination as a system C. At a certain time, both A and B are individually in thermodynamic equilibrium. Denote the states as $\mathrm{A}_1$ and $\mathrm{B}_1$, and the combined system as state $\mathrm{C}_1$. The thermodynamic probability of system C at state $\mathrm{C}_1$ is related to those of $\mathrm{A}_1$ and $\mathrm{B}_1$ by

$$\Omega_1^\mathrm{C} = \Omega_1^\mathrm{A} \times \Omega_1^\mathrm{B}$$

The entropy of $\mathrm{C}_1$ is then

$$S_1^\mathrm{C} = k_\mathrm{B} \ln \Omega_1^\mathrm{C} = k_\mathrm{B} \ln(\Omega_1^\mathrm{A} \times \Omega_1^\mathrm{B}) = k_\mathrm{B} \ln \Omega_1^\mathrm{A} + k_\mathrm{B} \ln \Omega_1^\mathrm{B} = S_1^\mathrm{A} + S_1^\mathrm{B}$$

Therefore, this definition of entropy meets the additive requirement.

The largest entropy principle states that the entropy of an isolated system will increase until it reaches a stable-equilibrium state (thermodynamic equilibrium), i.e., $\Delta S_\mathrm{isolated} \geq 0$. The microscopic understanding is that entropy is related to the probability of occurrence of a certain macrostate. For a system with specified $U$, $N$, and $V$, the macrostate that corresponds to the thermodynamic equilibrium is the most probable state and, hence, its entropy is the largest. Any states, including those that deviate very slightly from the stable-equilibrium state, will have a much smaller thermodynamic probability. After the equilibrium state is reached, it is not possible for any macrostate, whose thermodynamic probability is much less than that of the equilibrium state, to occur within an observable amount of time.

### 3.2.3 *The Lagrangian Multipliers*

For all the three types of statistics, $\mathrm{d}(\ln \Omega) = \alpha \sum_{i=0}^{\infty} \mathrm{d}N_i + \beta \sum_{i=0}^{\infty} \varepsilon_i \, \mathrm{d}N_i$, where the first term is the change in the total number of particles and the second can be related to the net heat transfer for a closed system; therefore, $\mathrm{d}(\ln \Omega) = \alpha \mathrm{d}N + \beta \delta Q$. In a reversible process in which the total number of particles do not change (closed system), $\mathrm{d}N = 0$, $\mathrm{d}(\ln \Omega) = \mathrm{d}S/k_\mathrm{B}$, and $\delta Q = T\mathrm{d}S$. Hence, we have for all the three statistics

$$\beta \equiv 1/k_{\mathrm{B}}T \tag{3.21}$$

To evaluate $\alpha$, we must allow the system to change its composition. In this case,

$$\mathrm{d}(\ln \Omega) = \alpha \sum_{i=0}^{\infty} \mathrm{d}N_i + \beta \sum_{i=0}^{\infty} \varepsilon_i \mathrm{d}N_i = \alpha \mathrm{d}N + \beta(\mathrm{d}U + P\mathrm{d}V)$$

or

$$T\mathrm{d}S = k_{\mathrm{B}}T\alpha\mathrm{d}N + \mathrm{d}U + P\mathrm{d}V$$

Substituting the above equation into the definition of the Helmholtz function, $\mathrm{d}A = \mathrm{d}(U - TS) = \mathrm{d}U - T\mathrm{d}S - S\mathrm{d}T$, we have

$$\mathrm{d}A = -S\mathrm{d}T - P\mathrm{d}V - k_{\mathrm{B}}T\alpha\mathrm{d}N$$

Noting that the chemical potential $\mu = \left(\frac{\partial A}{\partial N}\right)_{T,V} = -k_{\mathrm{B}}T\alpha$, we obtain

$$\alpha = -\frac{\mu}{k_{\mathrm{B}}T} \tag{3.22}$$

Here, $\mu$ is expressed in molecular quantity, and if $\mu$ is expressed in molar quantity we have $\alpha = -\mu/\overline{R}T$.

### 3.2.4 Entropy at Absolute Zero Temperature

The third law of thermodynamics states that the entropy of any pure substance vanishes at the ground state (with absolute zero temperature); see Sect. 2.1.3. For BE statistics, we have

$$N = N_0 + N_1 + N_2 + \cdots = \frac{g_0}{\mathrm{e}^{\alpha+\beta\varepsilon_0} - 1} + \frac{g_1}{\mathrm{e}^{\alpha+\beta\varepsilon_1} - 1} + \frac{g_2}{\mathrm{e}^{\alpha+\beta\varepsilon_2} - 1} + \cdots$$

At very low temperatures ($T \to 0$), $\beta = 1/k_{\mathrm{B}}T \to \infty$. Since $\varepsilon_0 < \varepsilon_1 < \varepsilon_2 < \cdots$,

$$\frac{N_i}{N_0} \approx \frac{g_i}{g_0}\mathrm{e}^{-\beta(\varepsilon_i - \varepsilon_0)} \to 0 \text{ as } T \to 0 \text{ for } i \geq 1 \tag{3.23}$$

Hence, $N \approx N_0$; that is, all particles will be at the lowest energy level (ground state). If $g_0 = 1$, as it is the case for a pure substance, then $\Omega = 1$ and $S = k_{\mathrm{B}} \ln \Omega = 0$ as $T \to 0$; this is consistent with the third law of thermodynamics. The occurrence for particles that obey BE statistics (bosons) to collapse to the ground state at sufficiently low temperatures is called the *Bose-Einstein condensation*. Such a state of matter is called the *Bose-Einstein condensate*, in which quantum effects dominate the macroscopic behavior.

Some important applications of the Bose-Einstein condensation are superfluidity and superconductivity. Liquid helium ($^4$He) becomes a superfluid with no viscosity at temperatures below the λ-transition ($T \approx 2.17$ K). The specific heat of helium at this temperature becomes infinitely large, suggesting that a phase transition occurs. Bose-Einstein condensate of atoms has been observed with laser cooling and trapping techniques [9]. Photons from the laser collide with the atoms. The absorption can be tuned using the Doppler shift so that only atoms traveling toward the laser can absorb the photons, resulting in reduced momentums in these atoms. Furthermore, the excited atoms will emit photons spontaneously in all directions. The net effect is a decrease in the velocity of the atoms, resulting in a kinetic temperature down to the nanokelvin range. From 1996 to 2003, the Nobel Prize in Physics was awarded for works related to Bose-Einstein condensation for four times: 1996, 1997, 2001, and 2003.

Although electrons are fermions (particles that obey FD statistics) that generally do not condense at zero temperature, they can form pairs at sufficiently low temperatures that behave like bosons. Below the critical temperature, pairs of electrons, called the Cooper pairs can travel freely without any resistance. This is the phenomenon called superconductivity, which was discovered at the beginning of the twentieth century. A large number of elements and compounds can be made superconducting at very low temperatures. Furthermore, some oxides become superconducting at temperatures above 90 K [10]. Superconductors have important applications in magnetic resonance imaging, high-speed and low-noise electronic devices, infrared sensors, and so forth. A similar phenomenon is the superfluidity in helium isotope $^3$He, which undergoes a phase transition at very low temperatures. The fermionic $^3$He atoms pair up to form bosonic entities that experience Bose-Einstein condensation at 3 mK.

For FD statistics, from Eqs. (3.17), (3.21), and (3.22), we have

$$\frac{N_i}{g_i} = \frac{1}{e^{(\varepsilon_i - \mu)/k_B T} + 1} \tag{3.24}$$

As $T \to 0$, it is found that the occupation number $N_i/g_i = 1$ for all energy levels with $\varepsilon_i < \mu$ and $N_i/g_i = 0$ for energy levels with $\varepsilon_i > \mu$. That is, all quantum states are filled or occupied for $i = 0, 1, 2, \ldots, j$ (with $\varepsilon_j < \mu$ ), and all quantum states are empty for $i = j + 1, j + 2, \ldots$ (with $\varepsilon_{j+1} > \mu$ ), as schematically shown in Fig. 3.6. More discussions will be given in Chap. 5 on the behavior of free electrons. For now, it is sufficient to say that the thermodynamic probability $\Omega = 1$ for FD statistics at absolute zero temperature. Therefore, the entropy $S = 0$ at $T \to 0$ K for both the BE and FD statistics. However, MB statistics does not satisfy the third law and is not applicable to very low temperatures.

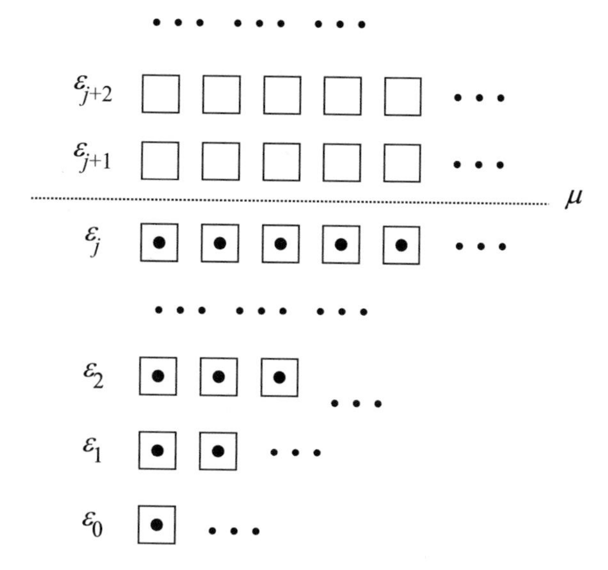


**Fig. 3.6** Schematic of the Fermi-Dirac distribution at 0 K

### 3.2.5 *Macroscopic Properties in Terms of the Partition Function*

The *partition function* is an important quantity in statistical thermodynamics. Unlike the characteristics functions (such as the Helmholtz free energy and the Gibbs free energy defined in Chap. 2) used in macroscopic thermodynamics, the physical meaning of the partition function is not immediately clear. However, the introduction of the partition function allows the calculation of macroscopic thermodynamic properties from the microscopic representation. There are different types of partition functions. For MB statistics, the partition function is defined as

$$Z = N\mathrm{e}^{\alpha} = \sum_{i=0}^{\infty} g_i \mathrm{e}^{-\varepsilon_i/k_\mathrm{B}T} \tag{3.25}$$

Therefore,

$$N_i = \frac{N}{Z} g_i \mathrm{e}^{-\varepsilon_i/k_\mathrm{B}T} \tag{3.26}$$

Since

$$\left[\frac{\partial(\ln Z)}{\partial T}\right]_{V,N} = \frac{1}{Z}\left(\frac{\partial Z}{\partial T}\right)_{V,N} = \frac{\sum_{i=0}^{\infty} g_i \mathrm{e}^{-\varepsilon_i/k_B T}\left(\frac{\varepsilon_i}{k_B T^2}\right)}{\sum_{i=0}^{\infty} g_i \mathrm{e}^{-\varepsilon_i/k_B T}} = \frac{\frac{U\mathrm{e}^{\alpha}}{k_B T^2}}{N\mathrm{e}^{\alpha}} = \frac{U}{N k_B T^2}$$

we have

$$U = N k_B T^2 \left[\frac{\partial(\ln Z)}{\partial T}\right]_{V,N} \tag{3.27}$$

Using the corrected MB statistics by dividing Eq. (3.10) by $N!$, we can express the entropy as

$$S = k_B \ln(\Omega_{MB}/N!) = k_B \sum_{i=0}^{\infty} N_i \left(1 + \ln \frac{g_i}{N_i}\right)$$
$$= k_B \sum_{i=0}^{\infty} N_i \left(1 + \ln \frac{Z}{N} + \beta\varepsilon_i\right) = N k_B + N k_B \ln \frac{Z}{N} + k_B \beta U \tag{3.28a}$$

Had we not divided $\Omega_{MB}$ by $N!$, we would get $S = N k_B \ln Z + k_B \beta U$, which differs from Eq. (3.28a), (3.28b) by a constant. After substituting $\beta$ and $U$ into Eq. (3.28a), (3.28b), we obtain

$$S = N k_B \left\{1 + \ln \frac{Z}{N} + T\left[\frac{\partial(\ln Z)}{\partial T}\right]_{V,N}\right\} \tag{3.28b}$$

The Helmholtz free energy is

$$A = U - TS = -N k_B T \left(1 + \ln \frac{Z}{N}\right) \tag{3.29}$$

The pressure is

$$P = -\left(\frac{\partial A}{\partial V}\right)_{T,N} = N k_B T \left[\frac{\partial(\ln Z)}{\partial V}\right]_{T,N} \tag{3.30}$$

The enthalpy $H$ and the Gibbs free energy $G$ can also be obtained. The partition function is now related to the macroscopic thermodynamic properties of interest for simple substances.

## 3.3 Ideal Molecular Gases

An important application of statistical mechanics is to model and predict the thermal properties of materials. In this section, the application of MB statistics to obtain the equation of state and the velocity distributions for ideal molecular gases is presented. The microscopic theories of the specific heat for ideal monatomic and polyatomic gases are given subsequently.

### 3.3.1 *Monatomic Ideal Gases*

For a monatomic ideal gas at moderate temperatures, MB statistics can be applied, and the translational energies are

$$\varepsilon = \frac{1}{2}m(v_x^2 + v_y^2 + v_z^2) = \frac{1}{2}m\mathbf{v}^2 \tag{3.31}$$

Consider a volume element in the phase space, $\mathrm{d}x\mathrm{d}y\mathrm{d}z\mathrm{d}p_x\mathrm{d}p_y\mathrm{d}p_z$, where $\mathbf{p} = m\mathbf{v}$ is the momentum of a molecule. The accuracy of specifying the momentum and the displacement is limited by $\Delta x \Delta p_x \sim h$, given by the uncertainty principle. The degeneracy, which is the *number of quantum states* (boxes of size $h^3$) in a volume element of the phase space, is given by

$$\mathrm{d}g = \frac{\mathrm{d}x\mathrm{d}y\mathrm{d}z\mathrm{d}p_x\mathrm{d}p_y\mathrm{d}p_z}{h^3} = \frac{m^3}{h^3}\mathrm{d}x\mathrm{d}y\mathrm{d}z\mathrm{d}v_x\mathrm{d}v_y\mathrm{d}v_z \tag{3.32}$$

Many useful results were obtained before quantum mechanics by assuming that $h^3$ is some constant. A more rigorous proof of Eq. (3.32) will be given in Sect. 3.5. When the space between energy levels are sufficiently close, the partition function can be expressed in terms of an integral as $Z_\mathrm{t} = \int \mathrm{e}^{-\varepsilon/k_\mathrm{B}T}\mathrm{d}g$ or

$$Z_\mathrm{t} = \iiint \mathrm{d}x\mathrm{d}y\mathrm{d}z \iiint \frac{m^3}{h^3}\exp\left[-\frac{m}{2k_\mathrm{B}T}\left(v_x^2 + v_y^2 + v_z^2\right)\right]\mathrm{d}v_x\mathrm{d}v_y\mathrm{d}v_z \tag{3.33}$$

The space integration yields the volume $V$, and the velocity integration can be individually performed, viz.

$$\int_{-\infty}^{\infty} \exp\left(-\frac{mv_x^2}{2k_\mathrm{B}T}\right)\mathrm{d}v_x = \sqrt{\frac{2\pi k_\mathrm{B}T}{m}} \tag{3.34}$$

Hence,

$$Z_t = V\left(\frac{2\pi mk_B T}{h^2}\right)^{3/2} \tag{3.35}$$

Therefore,

$$e^{\alpha} = \frac{V}{N}\left(\frac{2\pi mk_B T}{h^2}\right)^{3/2} \tag{3.36}$$

which is indeed much greater than unity at normal temperatures for most substances, suggesting that the MB statistics is applicable for ideal molecular gases. At extremely low temperatures, intermolecular forces cannot be neglected and the molecules are not independent anymore.

From Eq. (3.30), we have $P = Nk_B T\left[\frac{\partial(\ln Z)}{\partial V}\right]_{T,N} = Nk_B T/V$; thus,

$$PV = Nk_B T \quad \text{or} \quad P = nk_B T \tag{3.37}$$

where $n = N/V$ is the number density. The Boltzmann constant is the ideal (universal) gas constant on the molecular basis, $k_B = \overline{R}/N_A$. The internal energy, the specific heats, and the absolute entropy can also be evaluated.

$$U = Nk_B T^2\left[\frac{\partial(\ln Z)}{\partial T}\right]_{V,N} = \frac{3}{2}Nk_B T \tag{3.38}$$

which is not a function of pressure. The molar specific internal energy is $\bar{u} = \frac{3}{2}\overline{R}T$, and the molar specific heats are

$$\bar{c}_v = \left(\frac{\partial \bar{u}}{\partial T}\right)_V = \frac{3}{2}\overline{R} \tag{3.39}$$

and

$$\bar{c}_p = \left(\frac{\partial \bar{h}}{\partial T}\right)_P = \frac{5}{2}\overline{R} \tag{3.40}$$

The above equations show that the specific heats of monatomic gases are independent of temperature, except at very high temperatures when electronic contributions become important. The molar specific heats do not depend on the type of molecules, but the same is not true for mass specific heats. Using Eq. (3.28a), the absolute entropy can be expressed as

$$S = Nk_B\left\{\frac{5}{2} + \ln\left[\frac{V}{N}\left(\frac{2\pi mk_B T}{h^2}\right)^{3/2}\right]\right\}$$

Therefore, the molar specific entropy is a function of $T$ and $P$,

$$\bar{s}(T, P) = \overline{R}\left\{\frac{5}{2} + \ln\left[\frac{k_{\mathrm{B}}T}{P}\left(\frac{2\pi m k_{\mathrm{B}}T}{h^2}\right)^{3/2}\right]\right\} \tag{3.41}$$

This is the *Sackur-Tetrode equation.*

### 3.3.2 Maxwell's Velocity Distribution

Rewrite $N_i = g_i \mathrm{e}^{-\alpha}\mathrm{e}^{-\varepsilon_i/k_{\mathrm{B}}T}$ as $\mathrm{d}N = \mathrm{d}g\mathrm{e}^{-\alpha}\mathrm{e}^{-\varepsilon/k_{\mathrm{B}}T}$. In a volume $V$ and from $\mathbf{v}$ to $\mathbf{v} + d\mathbf{v}$ (i.e., $v_x$ to $v_x + \mathrm{d}v_x$, $v_y$ to $v_y + \mathrm{d}v_y$, and $v_z$ to $v_z + \mathrm{d}v_z$ ), the number of molecules $\mathrm{d}N$ per unit volume may be expressed as

$$\frac{\mathrm{d}N}{V} = \frac{m^3}{h^3}\mathrm{d}v_x\mathrm{d}v_y\mathrm{d}v_z\frac{N}{V}\left(\frac{h^2}{2\pi m k_{\mathrm{B}}T}\right)^{3/2}\exp\left(-\frac{m}{2k_{\mathrm{B}}T}\mathbf{v}^2\right) \tag{3.42}$$

or

$$f(\mathbf{v})\mathrm{d}\mathbf{v} = \frac{\mathrm{d}N}{V} = n\left(\frac{m}{2\pi k_{\mathrm{B}}T}\right)^{3/2}\exp\left(-\frac{m\mathbf{v}^2}{2k_{\mathrm{B}}T}\right)\mathrm{d}\mathbf{v} \tag{3.43}$$

where $f(\mathbf{v})$ is the Maxwell velocity distribution in a unit volume. Notice that

$$F(\mathbf{v}) = \frac{f(\mathbf{v})}{n} = \left(\frac{m}{2\pi k_{\mathrm{B}}T}\right)^{3/2}\exp\left(-\frac{m\mathbf{v}^2}{2k_{\mathrm{B}}T}\right) \tag{3.44}$$

which is a Gaussian distribution. Notice that $\mathbf{v}^2 = \mathbf{v}\cdot\mathbf{v} = v^2 = v_x^2 + v_y^2 + v_z^2$. The distribution of velocity component is also Gaussian, such that

$$F(\mathbf{v}) = F(v_x)F(v_y)F(v_z) \tag{3.45}$$

Taking the $x$-component as an example, we can write

$$F(v_x) = \left(\frac{m}{2\pi k_{\mathrm{B}}T}\right)^{1/2}\exp\left(-\frac{mv_x^2}{2k_{\mathrm{B}}T}\right) \tag{3.46}$$

The speed distribution may be obtained from the following by integrating the velocity distribution in a spherical shell (over the solid angle of $4\pi$).

$$F(v)\mathrm{d}v = \iint\limits_{4\pi} F(\mathbf{v})\mathrm{d}\mathbf{v} = \iint\limits_{4\pi}\left(\frac{m}{2\pi k_{\mathrm{B}}T}\right)^{3/2}\exp\left(-\frac{mv^2}{2k_{\mathrm{B}}T}\right)v^2\sin\theta\mathrm{d}v\mathrm{d}\theta\mathrm{d}\phi$$

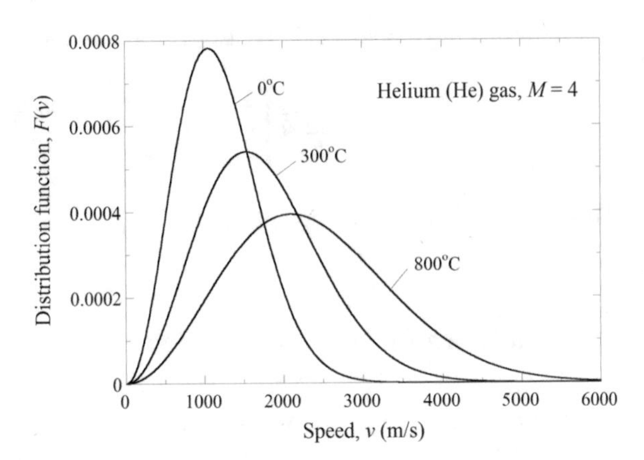


**Fig. 3.7** Speed distribution for helium gas at different temperatures

Therefore,

$$F(v) = 4\pi\left(\frac{m}{2\pi k_B T}\right)^{3/2} v^2 \exp\left(-\frac{mv^2}{2k_B T}\right) \tag{3.47}$$

Figure 3.7 plots the speed distribution of He gas at 0, 300, and 800 °C. When evaluating $k_B T$, we must convert $T$ to absolute temperature. It can be seen that more molecules will be at higher speeds as the temperature increases. It should be noted that $F(v = 0) = 0$ but $F(\mathbf{v})$ is maximum at $v = 0$. In the speed coordinate, an interval between $v$ and $v + dv$ corresponds to a spherical shell in the velocity space. Even though $F(\mathbf{v})$ is maximum at $v = 0$, the probability of finding a molecule per unit speed interval decreases to 0 as $v \to 0$, which is caused by the associated decrease in the volume of the spherical shell.

**Example 3.5** Find the average speed and the root-mean-square speed for a He gas at 200 °C at 100 kPa. What if the pressure is changed to 200 kPa? What are the most probable velocity and the most probable speed?

**Solution** The average speed may be obtained from either the velocity distribution or the speed distribution. That is

$$\bar{v} = \iiint vF(\mathbf{v})\,d\mathbf{v} = \int_0^\infty vF(v)dv = \sqrt{\frac{8k_B T}{\pi m}} \tag{3.48}$$

The average of $v^2$ is (see Appendix B.5)

$$\overline{v^2} = \iiint v^2 F(\mathbf{v})\,d\mathbf{v} = \int_0^\infty v^2 F(v)dv = \frac{3k_B T}{m} \tag{3.49a}$$

Therefore the root-mean-square speed is

$$v_{\text{rms}} = \sqrt{\overline{v^2}} = \sqrt{\frac{3k_\text{B}T}{m}} \tag{3.49b}$$

Plugging in the numerical values, we have $\bar{v} = 1582\,\text{m/s}$ and $v_{\text{rms}} = 1717\,\text{m/s}$ for He gas at 200 °C. We also notice that the pressure has no effect on the speed distribution, unless it is so high that intermolecular forces cannot be neglected.

The most probable velocity $\mathbf{v}_{\text{mp}} = 0$ because of the symmetry in the Gaussian distribution. We can obtain the most probable speed by setting $F'(v) = 0$, viz.

$$2v\exp\left(-\frac{mv^2}{2k_BT}\right) - v^2\left(\frac{mv}{k_BT}\right)\exp\left(-\frac{mv^2}{2k_BT}\right) = 0$$

The solution gives the most probable speed as $v_{\text{mp}} = \sqrt{2k_\text{B}T/m}$. For He gas at 200 °C, it gives $v_{\text{mp}} = 1402\,\text{m/s}$. Note that $v_{\text{mp}} : \bar{v} : v_{\text{rms}} = \sqrt{2} : \sqrt{8/\pi} : \sqrt{3} \approx 1.4 : 1.6 : 1.7$.

**Comment**. An important consequence for Eqs. (3.49a), (3.49b) is that temperature is related to the mean kinetic energy of the molecule such that

$$\frac{1}{2}m\overline{v_x^2} = \frac{1}{2}m\overline{v_y^2} = \frac{1}{2}m\overline{v_z^2} = \frac{1}{2}k_\text{B}T \tag{3.50}$$

Thus, the internal energy of a monatomic gas given in Eq. (3.38) is the sum of the kinetic energy of all molecules.

### 3.3.3 *Diatomic and Polyatomic Ideal Gases*

Additional degrees of freedom or energy storage modes must be considered for diatomic and polyatomic molecules, besides translation. The molecule may rotate about its center of gravity, and atoms may vibrate with respect to each other. For a molecule consisting of $q$ atoms, each atom may move in all three directions, and there will be a total of $3q$ modes. Consider the translation of the molecule as a whole; there are three translational degrees of freedom or modes: $\phi_\text{t} = 3$. For diatomic molecules or polyatomic molecules whose atoms are arranged in a line (such as $CO_2$), as shown in Fig. 3.8, there are two rotational degrees of freedom or modes: $\phi_\text{r} = 2$. Therefore, there are $\phi_\text{v} = 3q - 5$ vibrational degrees of freedom or modes. For polyatomic molecules whose atoms are not aligned (such as $H_2O$ and $CH_4$, see Fig. 3.9), there are three rotational degrees of freedom or $\phi_\text{r} = 3$. The number of vibrational degrees of freedom or modes are thus $\phi_\text{v} = 3q - 6$.

The total energy of a molecule may be expressed as the sum of translational, rotational, and vibrational energies: $\varepsilon = \varepsilon_\text{t} + \varepsilon_\text{r} + \varepsilon_\text{v}$. For simplicity, we have neglected contributions from the electronic ground state and chemical dissociation, which can be included as additional terms in evaluating the internal energy and the entropy [1].

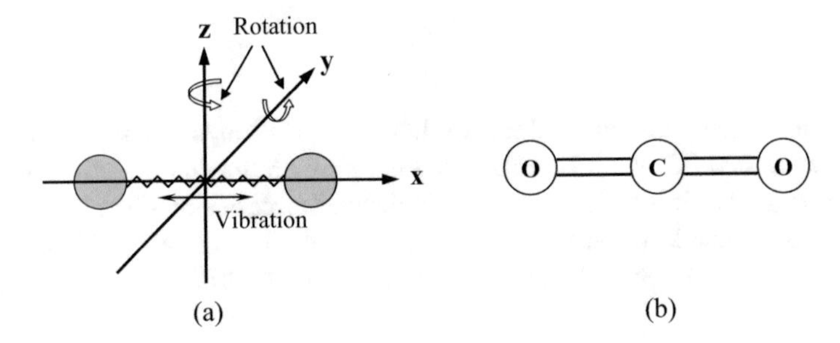


**Fig. 3.8** **a** A diatomic molecule, showing two rotational and one vibrational degrees of freedom. **b** $CO_2$ molecule, where the atoms are aligned

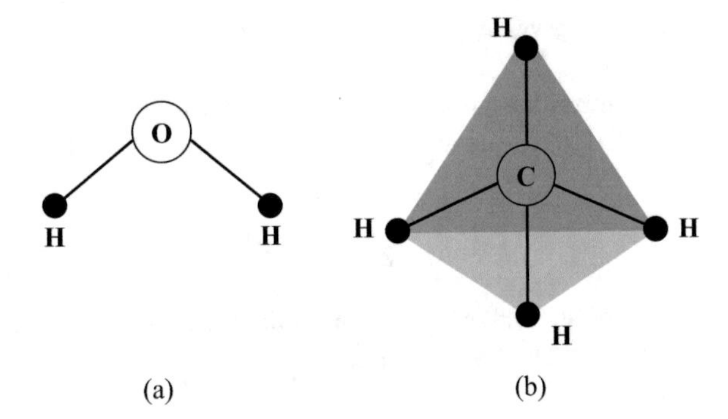


**Fig. 3.9** **a** $H_2O$ molecule, for which the atoms are not aligned. **b** The tetrahedral methane ($CH_4$) molecule

At high temperatures, the vibration mode can be coupled with the rotation mode. Here, however, it is assumed that these modes are independent. The partition function can be written as

$$Z = Z_t Z_r Z_v = \left(\sum g_t e^{-\varepsilon_t/k_B T}\right)\left(\sum g_r e^{-\varepsilon_r/k_B T}\right)\left(\sum g_v e^{-\varepsilon_v/k_B T}\right) \tag{3.51}$$

For polyatomic atoms, Eqs. (3.31) through (3.36) hold for the translational modes. $Z_r$ and $Z_v$ are internal contributions that do not depend on volume; therefore, Eq. (3.37) also holds. Since the degrees of freedom are independent of each other, Maxwell's velocity and speed distributions discussed in Sect. 3.3.2 still hold for polyatomic gases. The problem now is to determine the rotational and vibrational energy levels and degeneracies. Generally speaking, there exists certain characteristic temperature associated with each degree of freedom. The characteristic temperature for translation is very low for molecular gases. On the other hand, the characteristic temperature for rotation is slightly higher, and that for vibration is usually very high, as can be seen from Table 3.2 for selected diatomic molecules. If the temperature is much less than the characteristic temperature of a certain mode, then the contribution of that mode to the energy storage is negligible. For the temperatures

**Table 3.2** Characteristic temperatures of rotation and vibration for some diatomic molecules

| Substance | Symbol | $\Theta_r$ (K) | $\Theta_v$ (K) |
|---|---|---|---|
| Hydrogen | $H_2$ | 87.5 | 6320 |
| Deuterium | $D_2$ | 43.8 | 4490 |
| Hydrogen chloride | HCl | 15.2 | 4330 |
| Nitrogen | $N_2$ | 2.86 | 3390 |
| Carbon monoxide | CO | 2.78 | 3120 |
| Nitric oxide | NO | 2.45 | 2745 |
| Oxygen | $O_2$ | 2.08 | 2278 |
| Chloride | $Cl_2$ | 0.35 | 814 |
| Sodium vapor | $Na_2$ | 0.08 | 140 |

much greater than the characteristic temperature, however, there often exist some asymptotic approximations.

**Rotation**. A quantum mechanical analysis of a rigid rod, to be derived in Sect. 3.5.3, shows that the rotational energy levels are given by

$$\frac{\varepsilon_l}{k_B T} = l(l+1)\frac{\Theta_r}{T} \tag{3.52}$$

Here, $\Theta_r$ is the characteristic temperature for rotation given by $\Theta_r = h^2/(8\pi^2 k_B I)$, where $I$ is the moment of inertia of the molecule about the center of mass. The larger the value of $I$, the smaller the characteristic temperature will be. This is clearly shown in Table 3.2. The degeneracy of rotational energy levels is

$$g_l = \frac{2l+1}{\sigma} \tag{3.53}$$

where $\sigma$ is a symmetry number that arises from molecular symmetry: $\sigma = 1$ if the atoms are of different types (such as in a NO or CO molecule), and $\sigma = 2$ if the atoms are the same (such as in a $O_2$ or $N_2$ molecule).

$$Z_r = \sum_{l=0}^{\infty} \frac{2l+1}{\sigma} \exp\left[-l(l+1)\frac{\Theta_r}{T}\right] \tag{3.54}$$

This series converges very fast for $\Theta_r/T < 0.5$, since

$$Z_r = \frac{1}{\sigma}\left[1 + 3\exp\left(-\frac{2\Theta_r}{T}\right) + 5\exp\left(-\frac{6\Theta_r}{T}\right) + 7\exp\left(-\frac{12\Theta_r}{T}\right) + \cdots\right]$$

For $T/\Theta_r > 1$, Eq. (3.54) may be expanded to give (see Problem 3.26)

$$Z_r = \frac{T}{\Theta_r \sigma}\left[1 + \frac{1}{3}\left(\frac{\Theta_r}{T}\right) + \frac{1}{15}\left(\frac{\Theta_r}{T}\right)^2 + \frac{4}{315}\left(\frac{\Theta_r}{T}\right)^3 + \cdots\right] \tag{3.55}$$

At temperatures much higher than the characteristic temperature of rotation, $T/\Theta_r \gg 1$, the above equation reduces to

$$Z_r = \frac{T}{\sigma \Theta_r} \tag{3.56}$$

Under this limit, the contribution of the rotational energy to the internal energy becomes

$$U_r \approx Nk_B T \tag{3.57}$$

The contribution to the molar specific heat by the two rotational degrees of freedom is

$$\bar{c}_{v,r} = \overline{R} \tag{3.58}$$

**Vibration**. The vibration in a molecule can be treated as a harmonic oscillator. For each vibration mode, the quantized energy levels are given in Sect. 3.5.5 as

$$\varepsilon_{v,i} = (i + \tfrac{1}{2})h\nu,\ i = 0, 1, 2, \ldots \tag{3.59}$$

where $\nu$ is the natural frequency of vibration, and the ground-state energy is $\frac{1}{2}h\nu$. The vibrational energy levels are not degenerated; thus, $g_{v,i} = 1$. Subsequently, we can write

$$Z_v = \sum_{i=0}^{\infty} e^{-(i+1/2)h\nu/k_B T} = e^{-\Theta_v/2T} \sum_{i=0}^{\infty} e^{-i\Theta_v/T}$$

where $\Theta_v = h\nu/k_B$ is a characteristic temperature for vibration and is listed in Table 3.2 for several diatomic molecules. The vibrational partition function becomes

$$Z_v = \frac{e^{-\Theta_v/2T}}{1 - e^{-\Theta_v/T}} = \frac{e^{\Theta_v/2T}}{e^{\Theta_v/T} - 1} \tag{3.60}$$

Its contribution to the internal energy and the specific heat can be written as

$$U_v = Nk_B\Theta_v\left(\frac{1}{2} + \frac{1}{e^{\Theta_v/T} - 1}\right) \tag{3.61}$$

and

$$\bar{c}_{v,\mathrm{v}} = \overline{R}\frac{\Theta_\mathrm{v}^2}{T^2}\frac{\mathrm{e}^{\Theta_\mathrm{v}/T}}{(\mathrm{e}^{\Theta_\mathrm{v}/T}-1)^2} \tag{3.62}$$

At $T \ll \Theta_\mathrm{v}$, the vibrational mode contributes to the internal energy but not to the specific heat. At $T > 1.5\Theta_\mathrm{v}$, $U_\mathrm{v}$ almost linearly depends on $T$ and $\bar{c}_{v,\mathrm{v}} \approx \overline{R}$. In classical statistical mechanics, it is believed that each degree of freedom contributes to the stored thermal energy with an amount of $\frac{1}{2}k_\mathrm{B}T$ and results in a specific heat of $\frac{1}{2}k_\mathrm{B}$ on the particle base. This is called the *equipartition principle*. The contribution of each vibrational mode is $\overline{R}$ not $\overline{R}/2$, due to the fact that each vibrational mode includes a kinetic component and a potential component for energy storage. For this reason, each vibrational mode is equivalent to two degrees of freedom in terms of the energy storage when it is fully excited. It should be noted that the equipartition principle is only applicable at sufficiently high temperatures. Because energy is additive, we can write

$$\bar{c}_v = \bar{c}_{v,\mathrm{t}} + \bar{c}_{v,\mathrm{r}} + \bar{c}_{v,\mathrm{v}} \tag{3.63}$$

The result is schematically shown in Fig. 3.10. One can see that for a diatomic ideal gas,

$$\bar{c}_v = 2.5\overline{R} \quad \text{if} \quad \Theta_\mathrm{r} \ll T \ll \Theta_\mathrm{v} \tag{3.64}$$

which happens to be near room temperature for many gases such as nitrogen and carbon monoxide; see Table 3.2. Figure 3.11 plots the specific heat for several real gases at sufficiently low pressure so that the ideal gas model is applicable. It should be noted that, for hydrogen, nuclear spin is important and Eq. (3.54) needs to be modified to account for the spin degeneracy [1, 2]. However, Eqs. (3.57) and (3.58) predict the right trend and are applicable at temperatures much higher than $\Theta_\mathrm{r}$. At extremely high temperatures (say 3000 K), electronic contributions and the coupling between rotation and vibration become important. Although Eq. (3.63) is the correct expression for the specific heat at moderate temperatures, two additional partition

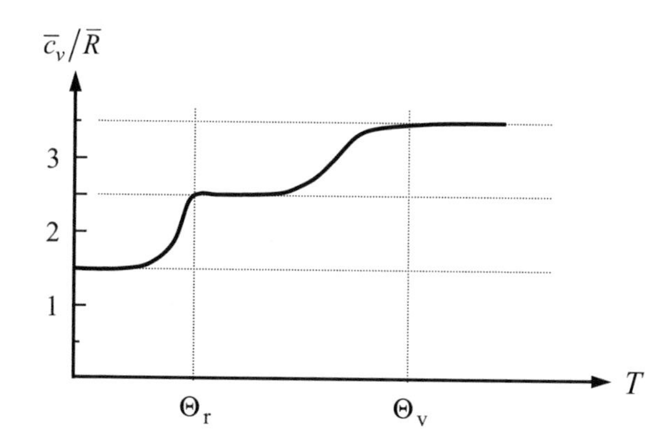


**Fig. 3.10** Typical specific heat curve of a diatomic ideal gas

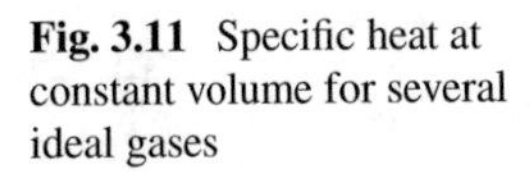
**Fig. 3.11** Specific heat at constant volume for several ideal gases

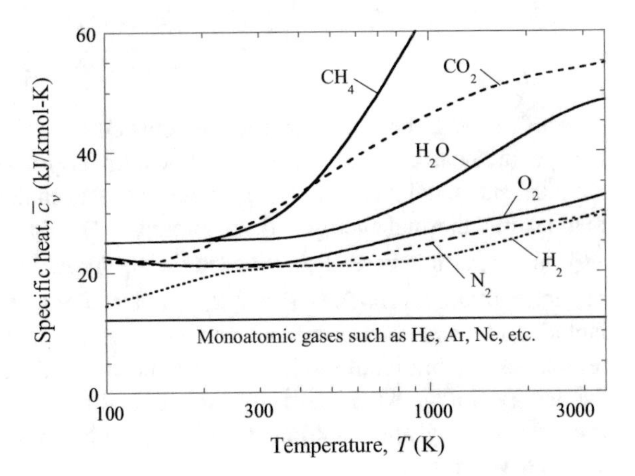


functions must be included to correctly evaluate the internal energy and the entropy (see Problem 3.22). We limit the derivations to the specific heat, which is closely relevant to heat transfer calculations.

The characteristic temperature for rotation is usually very small for polyatomic molecules because of their large moments of inertia. Therefore, the rotational degrees of freedom can be assumed as fully excited in almost any practical situations. Each rotational degree of freedom will contribute $\overline{R}/2$ to the molar specific heat. For molecules whose atoms are aligned (such as $CO_2$), the rotational contribution to the specific heat is $\overline{R}$, and

$$\bar{c}_v = \frac{5}{2}\overline{R} + \overline{R}\sum_{i=1}^{3q-5} \frac{\zeta_i^2 e^{\zeta_i}}{(e^{\zeta_i} - 1)^2}, \quad \zeta_i = \Theta_{v,i}/T \tag{3.65}$$

If $T \gg \Theta_{v,i}$, then $\bar{c}_v \to \overline{R}(3q-2.5)$. For molecules whose atoms are not aligned (such as $H_2O$ and $CH_4$),

$$\bar{c}_v = 3\overline{R} + \overline{R}\sum_{i=1}^{3q-6} \frac{\zeta_i^2 e^{\zeta_i}}{(e^{\zeta_i} - 1)^2} \tag{3.66}$$

In this case, $\bar{c}_v \to \overline{R}(3q - 3)$ at $T \gg \Theta_{v,i}$. Again, electronic contribution may be significant at very high temperatures. Table 3.3 lists the vibrational frequencies for several commonly encountered gases. The unit of frequency is given in inverse centimeter ($cm^{-1}$), which is often used in spectroscopic analyses. Note that $\Theta_v = h\nu/k_B = hc\bar{\nu}/k_B$, where $\bar{\nu}$ is the wavenumber in $cm^{-1}$ if we take $c = 2.998 \times 10^{10}$ cm/s, giving $\Theta_v$ (K) $= 1.44\,\bar{\nu}$ ($cm^{-1}$). One can use this table to estimate the specific heat of these gases based on Eq. (3.65) or (3.66).

**Table 3.3** Vibrational modes of several gases, where the integer in the parentheses indicates the number of degenerate modes

| Type | $cm^{-1}$ | $cm^{-1}$ | $cm^{-1}$ | $cm^{-1}$ | Total $f_v$ |
|---|---|---|---|---|---|
| $CO_2$ | 667 (2) | 1343 | 2349 | – | 4 |
| $H_2O$ | 1595 | 3657 | 3756 | – | 3 |
| $CH_4$ | 1306 (3) | 1534 (2) | 2916 | 3019 (3) | 9 |

In reality, vibration-rotation interactions result in multiple absorption lines around each vibration mode, which can be observed through infrared absorption spectroscopy. Figure 3.12 shows the molecular absorption spectra of $CO_2$ and $H_2O$ measured with a Fourier-transform infrared spectrometer. The absorption spectra were obtained by comparing the spectrum when the measurement chamber is open with that when the chamber is purged with a nitrogen gas, which does not absorb in the mid-infrared region. The concentrations of $H_2O$ and $CO_2$ in the experiments were not controlled since the purpose is to demonstrate the infrared absorption frequencies only. While the resolution of 1 $cm^{-1}$ is not high enough to resolve very fine features, the absorption bands near 670 $cm^{-1}$ due to degenerate bending modes and near 2350 $cm^{-1}$ due to asymmetric stretching mode in $CO_2$ can be clearly seen. Note that the symmetric vibration mode of $CO_2$ at 1343 $cm^{-1}$ is infrared inactive. Hence, it does not show up in the absorption spectrum but can be observed with Raman spectroscopy. Furthermore, the vibration-rotation interactions cause multiple lines in the water vapor absorption bands from 1300 to 2000 $cm^{-1}$ and from 3500 to 4000 $cm^{-1}$.

**Example 3.6** How many rotational degrees of freedom are there in a silane ($SiH_4$) molecule? If a low-pressure silane gas is raised to a temperature high enough to completely excite its rotational and vibrational modes, find its specific heats.

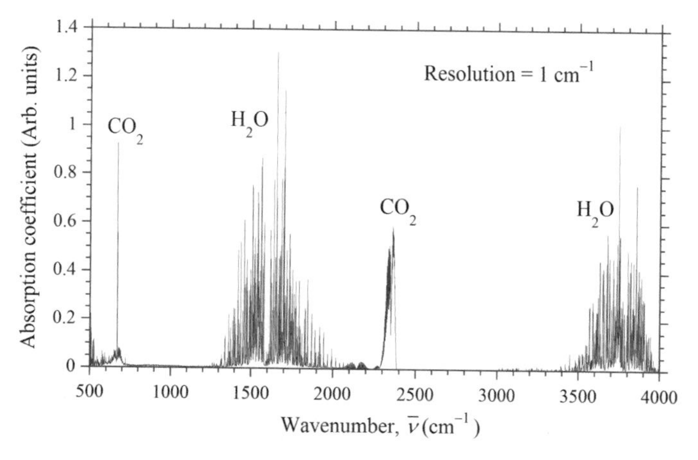


**Fig. 3.12** Infrared absorption spectrum of ambient air obtained with a Fourier-transform infrared spectrometer

**Solution** For $SiH_4$, there will be three translational degrees of freedom $\phi_t = 3$, three rotational degrees of freedom $\phi_r = 3$, and $\phi_v = 3q - 6 = 9$ vibrational degrees of freedom. If all the modes are excited, the specific heat for constant volume will be $c_v = 1.5R + 1.5R + 9R = 12R$. Given that $M = 32$, we find $c_v = 3.12\,\text{kJ/kg K}$, $c_p = 3.38\,\text{kJ/kg K}$, and $\gamma = 13/12 = 1.083$. The actual specific heats would be much smaller at moderate temperatures.

## 3.4 Statistical Ensembles and Fluctuations

We have finished the discussion about statistical thermodynamics of independent particles without mentioning ensembles. In a system of independent particles, there is no energy associated with particle-particle interactions or the configuration of the particles. For dependent particles or dense fluids, the previous analysis can be extended by using *statistical ensembles*, which was pioneered by J. Willard Gibbs (1839–1903) in the late nineteenth century in his 1902 book, *Elementary Principles of Statistical Mechanics.* Statistical ensembles are a large set of macroscopically similar systems. When the properties are averaged over a properly chosen ensemble, the macroscopic properties can be considered as the same as the time-averaged quantity of the same system. There are three basic types of ensembles: microcanonical ensemble, canonical ensemble, and grand canonical ensemble [1, 5].

A *microcanonical ensemble* is composed of a large set of identical systems. Each system in the ensemble is isolated from others by rigid, adiabatic, and impermeable walls. The energy, volume, and number of particles in each system are constant. The results obtained using the microcanonical ensemble for independent particles are essentially the same as what we have obtained in previous sections. It is natural to ask the question as to what extent the statistical mechanics theory presented in previous sections will be valid for nanosystems. If the equilibrium properties are defined based on a large set of microcananical ensembles and considered as the time-averaging properties of the system, there will be sufficiently large number of particles in the whole ensemble to guarantee the basic types of statistics, and the thermodynamics relations derived in Sects. 3.1 and 3.2 are still applicable. On the other hand, the difference between the energy levels due to quantization may be large enough to invalidate the substitution of summation with integration. We will discuss the energy level quantization further in Sect. 3.5. In deriving the properties of ideal gases in Sect. 3.3, the consideration of the translational, rotational, and vibrational degrees of freedom is on the basis of individual molecules. Therefore, the conclusions should be applicable to systems under thermodynamic equilibrium.

In a *canonical ensemble,* each system is separated from others by rigid and impermeable walls, which are diathermal. All systems have the same volume and number of particles. However, the systems can exchange energy. At the equilibrium, the temperature $T$ will be the same for all systems. An important result of applying the canonical ensemble is that the energy fluctuation (i.e., the standard deviation of energy of the system) is proportional to $1/\sqrt{N}$, where $N$ is the total number of independent particles.

In a *grand canonical ensemble,* each system is separated from others by rigid, diathermal, and permeable walls. While the volume is fixed and is the same for each system, the number of particles as well as the energy of each system can vary. The temperature and the chemical potential must be the same for all systems at equilibrium. This allows the study of density fluctuations for each system. The result for monatomic molecules yields that the density fluctuation is also proportional to $1/\sqrt{N}$.

The canonical and grand canonical ensembles are essential for the study of complex thermodynamic systems, such as mixture, chemical equilibrium, dense gases, and liquids, which will not be further discussed in this text. Interested readers can find more details from Refs. [1, 5]. A simple theory based on independent particles of phonons and electrons will be discussed in Chap. 5. While the partition function can also be used to study the thermodynamic relations of solids, the approach used in solid state physics will be adopted in a detailed study of the properties of solids presented in Chap. 6.

## 3.5 Basic Quantum Mechanics

So far we have largely avoided the derivations and equations involving quantum mechanics, by using the conclusions from quantum theory on a need basis without proof. In this section, we shall present the basics of quantum mechanics to enhance the understanding of the materials already presented and to provide some background for future chapters.

In classical mechanics, the state of a system is completely described by giving the position and the momentum of each particle in the system at any given time. The equation of motion is given in Eq. (3.1), which is also the basis for molecular dynamics. The position and the momentum of each particle are determined using the initial values and precisely the forces exerted on it afterward. According to the wave-particle duality, particles also have wave characteristics. The results are described in quantum mechanics by the Schrödinger wave equation. The solution of the Schrödinger equation is given in the form of a *wavefunction*, which describes the probabilities of the possible outcome rather than the exact position and momentum of the particle. Another important aspect in quantum mechanics is the use of operators in mathematical manipulations.

### *3.5.1 The Schrödinger Equation*

Consider the following equation that describes a wave in the $x$ direction (see Appendices B.6 and B.7):

$$\Psi(x,t) = \widetilde{A}\mathrm{e}^{\mathrm{i}(2\pi x/\lambda - 2\pi \nu t)} \tag{3.67}$$

where $\widetilde{A} = A' + \mathrm{i}A''$ is a complex constant, $\lambda$ is the wavelength, and $\nu$ is the frequency. One can take the real part of $\Psi$:

$$\mathrm{Re}(\Psi) = A' \cos(2\pi x/\lambda - 2\pi \nu t) - A'' \sin(2\pi x/\lambda - 2\pi \nu t)$$

which is a cosine function of $x$ for any given $t$. The complex notation is convenient for obtaining derivatives. If Eq. (3.67) is used to describe a moving particle, with a mass $m$ and a momentum $p$, it can be shown that

$$-\mathrm{i}\hbar \frac{\partial}{\partial x}\Psi = \frac{h}{\lambda}\Psi = p\Psi \tag{3.68a}$$

$$-\frac{\hbar^2}{2m}\frac{\partial^2}{\partial x^2}\Psi = \frac{p^2}{2m}\Psi = E_{\mathrm{K}}\Psi \tag{3.68b}$$

and

$$\mathrm{i}\hbar \frac{\partial}{\partial t}\Psi = h\nu\Psi = \varepsilon\Psi \tag{3.68c}$$

where $\hbar$ is the Planck constant divided by $2\pi$, $E_{\mathrm{K}}$ is the kinetic energy of the particle, and $\varepsilon$ is the total energy of the particle. In writing Eqs. (3.68a), (3.68b), (3.68c), we have applied the concept of wave-particle duality to relate $p = h/\lambda$ and $\varepsilon = h\nu$. If the particle possesses only the kinetic and potential energies, we have

$$\varepsilon = E_{\mathrm{K}} + E_{\mathrm{P}} = \frac{p^2}{2m} + \Phi(\mathbf{r}) \tag{3.69a}$$

where $\Phi(\mathbf{r}) = \Phi(x, y, z)$ is the potential function that depends on the position of the particle. Define the Hamiltonian operator in the three-dimensional (3D) case as

$$\widehat{H} = -\frac{\hbar^2}{2m}\nabla^2 + \Phi(\mathbf{r}) \tag{3.69b}$$

It can be seen that $\widehat{H}\Psi = \varepsilon\Psi$. Hence,

$$-\frac{\hbar^2}{2m}\nabla^2\Psi + \Phi(\mathbf{r})\Psi = \mathrm{i}\hbar\frac{\partial \Psi}{\partial t} \tag{3.70}$$

which is the time-dependent Schrödinger equation [8]. From $\varepsilon\Psi = \mathrm{i}\hbar\frac{\partial\Psi}{\partial t}$, one can obtain

$$\Psi(\mathbf{r}, t) = \Psi_0(\mathbf{r})\mathrm{e}^{-\mathrm{i}\varepsilon t/\hbar} \tag{3.71a}$$

The general time dependence for different energy eigenvalues can be written as a summation:

$$\Psi(\mathbf{r}, t) = A_1\Psi_{01}(\mathbf{r})e^{-i\varepsilon_1 t/\hbar} + A_2\Psi_{02}(\mathbf{r})e^{-i\varepsilon_2 t/\hbar} + \cdots \tag{3.71b}$$

Therefore, the key to solve the Schrödinger equation becomes how to obtain the initial wavefunctions. For this reason, Eq. (3.70) can be rewritten as follows:

$$-\frac{\hbar^2}{2m}\nabla^2\Psi + \Phi(\mathbf{r})\Psi = \varepsilon\Psi \tag{3.72}$$

which is called the time-independent Schrödinger equation. The solution gives the wavefunction $\Psi(\mathbf{r})$, which is often expressed in terms of a set of eigenfunctions, $\Psi_1, \Psi_2, \Psi_3, \ldots$, each with an eigenvalue energy, $\varepsilon_1, \varepsilon_2, \varepsilon_3, \ldots$, respectively. The solution, or the wavefunction, must satisfy

$$\int_V \Psi\Psi^* \, dV = 1 \tag{3.73}$$

where the subscript * denotes the complex conjugate since the wavefunction is in general complex, and the integration is over the whole volume. The physical significance is that the probability of finding the particle in the volume must be 1. The wavefunction is also called a state function because it describes the quantum state of the particle, and $\Psi\Psi^*$ is called the probability density function. The average or expectation value of any physical quantity $\eta$ is calculated by

$$\langle\eta\rangle = \int_V \Psi^*\hat{\eta}\Psi \, dV \tag{3.74}$$

where $\hat{\eta}$ signifies an operator of $\eta$. For example, the average energy of the particle is

$$\langle\varepsilon\rangle = \int_V \Psi^*\widehat{H}\Psi \, dV \tag{3.75}$$

Several examples are discussed in the following sections to show how to obtain the wavefunctions and the physical significance of the solutions.

### *3.5.2 A Particle in a Potential Well or a Box*

The one-dimensional (1D) potential well is illustrated in Fig. 3.13a, where a particle is confined within a physical space between $0 < x < L$ and the particle can move

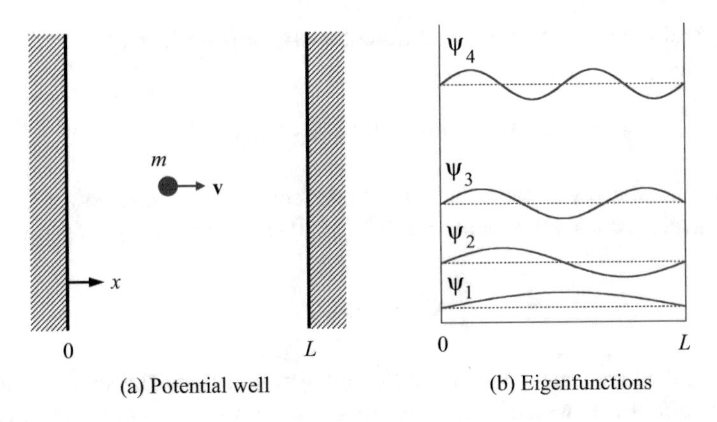


**Fig. 3.13** Illustration of **a** a 1D potential well and **b** the eigenfunctions

parallel to the $x$ axis only. This is equivalent of saying that the potential energy is zero inside and infinite outside the potential well. Hence,

$$\Phi(x) = \begin{cases} 0, & \text{for } 0 < x < L \\ \infty, & \text{at } x = 0 \text{ or } x = L \end{cases} \tag{3.76}$$

The Schrödinger equation becomes

$$-\frac{\hbar^2}{2m}\nabla^2\Psi = \varepsilon\Psi \tag{3.77}$$

whose solutions are $\Psi(x) = A\cos(kx) + B\sin(kx)$, where $k = \sqrt{2m\varepsilon/\hbar^2}$. Because the particle is confined inside the well, the wavefunction must be zero outside the potential well. Another requirement for the wavefunction is that it must be continuous. Thus, we must have $\Psi(0) = \Psi(L) = 0$. This requires that $A = 0$ and, by taking only the positive $k$ values, we have

$$kL = n\pi, \ n = 1, 2, 3\ldots \tag{3.78}$$

Therefore, the eigenfunctions are $\Psi_n(x) = B_n \sin\left(\frac{n\pi x}{L}\right)$, which can be normalized by letting $\int_0^L \Psi_n(x)\Psi_n^*(x)\,\mathrm{d}x = 1$ to get

$$\Psi_n(x) = \sqrt{\frac{2}{L}}\sin\left(\frac{n\pi x}{L}\right) \tag{3.79}$$

The solution requires the particle to possess discretized energy values, i.e., its energy cannot be increased continuously but with finite differences between neighboring states. It can easily be seen that

$$\varepsilon_n = \frac{h^2 n^2}{8mL^2} \tag{3.80}$$

The quantized energy eigenvalues are called energy levels for each quantum state, and the index $n$ is called a quantum number. The eigenfunctions are standing waves as shown in Fig. 3.13b for the first four quantum states. For molecules, the difference between energy levels is very small and the energy distribution can often be approximated as a continuous distribution. For electrons at very small distances, $L \to 10$ nm for example, quantization may be important. The effects of quantum confinement take place when the quantum well thickness becomes comparable to the de Broglie wavelength of the particle, such as electrons or holes in a semiconductor. Quantum wells can be formed by a sandwiched structure of heterogeneous layers, such as AlGaAs/GaAs/AlGaAs. The bandgap of the two outer layers is larger than that of the inner layer to form an effective potential well. These structures are used for optoelectronic applications such as lasers and radiation detectors. The thickness of the active region can be a few nanometers. In some cases, multiple quantum wells are formed with periodic layered structures, called *superlattices*, which have unique optical, electrical, and thermal properties.

**Example 3.7** Derive the uncertainty principle. Suppose the wavefunction is given by Eq. (3.79) for a particle with energy $\varepsilon_n$ given in Eq. (3.80).

**Solution** To find the average position of the particle, we use $\langle x \rangle = \int_0^L \Psi^* x \Psi \mathrm{d}x = \frac{2}{L}\int_0^L x \sin^2\left(\frac{n\pi x}{L}\right)\mathrm{d}x = \frac{L}{2}$. The variance of $x$, $\sigma_x^2 = \langle x - \langle x \rangle \rangle^2 = \langle x^2 \rangle - 2\langle x \rangle^2 + \langle x \rangle^2 = \langle x^2 \rangle - \langle x \rangle^2$.

With $\langle x^2 \rangle = \frac{2}{L}\int_0^L x^2 \sin^2\left(\frac{n\pi x}{L}\right)\mathrm{d}x = \frac{L^2}{3} - \frac{L^2}{2n^2\pi^2}$, we obtain the standard deviation of $x$ as $\sigma_x = L\left(\frac{1}{12} - \frac{1}{2n^2\pi^2}\right)^{1/2}$. For the momentum, we use the operator $p \to -\mathrm{i}\hbar\frac{\partial}{\partial x}$. Hence, $\langle p \rangle = \int_0^L \Psi^*\left(-\mathrm{i}\hbar\frac{\mathrm{d}\Psi}{\mathrm{d}x}\right)\mathrm{d}x = -\mathrm{i}\hbar\frac{2n\pi}{L^2}\int_0^L \sin\left(\frac{n\pi x}{L}\right)\cos\left(\frac{n\pi x}{L}\right)\mathrm{d}x = 0$ and $\langle p^2 \rangle = \int_0^L \Psi^*(-\hbar^2)\frac{\mathrm{d}^2\Psi}{\mathrm{d}x^2}\mathrm{d}x = \left(\frac{n\pi\hbar}{L}\right)^2$. We have $\sigma_p = \frac{n\pi\hbar}{L}$ and obtain the following expression:

$$\sigma_x \sigma_p = \frac{\hbar}{2}\left(\frac{\pi^2 n^2}{3} - 2\right)^{1/2} \tag{3.81}$$

Taking the smallest quantum number, $n = 1$, we get $\sigma_x \sigma_p \approx 0.5678\hbar > \hbar/2$, which is a proof of the uncertainty principle given in Eq. (3.9).

Next, consider a free particle in a 3D box, $0 < x < a,\ 0 < y < b,\ 0 < z < c$. It can be shown that the (normalized) eigenfunctions are

$$\Psi_{x,y,z} = \sqrt{\frac{8}{abc}}\sin\left(\frac{n_x \pi x}{a}\right)\sin\left(\frac{n_y \pi y}{b}\right)\sin\left(\frac{n_z \pi z}{c}\right) \tag{3.82}$$

with the energy eigenvalues:

$$\varepsilon_{x,y,z} = \frac{h^2}{8m}\left(\frac{n_x^2}{a^2} + \frac{n_y^2}{b^2} + \frac{n_z^2}{c^2}\right) \tag{3.83}$$

where $n_x, n_y, n_z = 1, 2, 3, \ldots$ When $a = b = c = V^{1/3}$, Eq. (3.83) can be simplified as

$$\varepsilon_{x,y,z} = \frac{h^2}{8mV^{2/3}}(n_x^2 + n_y^2 + n_z^2) \tag{3.84}$$

Let $\eta = (n_x^2 + n_y^2 + n_z^2)^{1/2}$, then we can evaluate the number of quantum states between $\eta$ and $\eta + \mathrm{d}\eta$, which is nothing but the degeneracy. For sufficiently large $V$, the quantum states are so close to each other that the volume within the spherical shell between $\eta$ and $\eta + \mathrm{d}\eta$ is equal to the number of quantum states. Only one-octant of the sphere is considered in Eq. (3.84) because $n_x > 0,\ n_y > 0,\ n_z > 0$. The total volume is therefore one-eighth of the spherical shell; hence,

$$\mathrm{d}g = \frac{1}{8}4\pi\eta^2\mathrm{d}\eta = \frac{2\pi V(2m)^{3/2}}{h^3}\varepsilon^{1/2}\mathrm{d}\varepsilon \tag{3.85}$$

With $\varepsilon = \frac{1}{2}mv^2$ and $\mathrm{d}\varepsilon = mv\mathrm{d}v$, we obtain

$$\mathrm{d}g = \frac{m^3V}{h^3}4\pi v^2\mathrm{d}v \tag{3.86}$$

This equation is essentially the same as Eq. (3.32), with $\mathrm{d}x\mathrm{d}y\mathrm{d}z = V$ and $\mathrm{d}v_x\mathrm{d}v_y\mathrm{d}v_z = 4\pi\mathrm{d}v$. Equation (3.86) provides a rigid proof of Eq. (3.32), which is the translational degeneracy. It should be noted that the classical statistical mechanics results in the same expression for $U$ and $p$, as well as the Maxwell velocity distribution for ideal gases. However, the constant $h$ must be included to correctly express $S$ as in Eq. (3.41). Equation (3.86) will also be used in Chap. 5 to study the free electron gas in metals. When using the momentum $p = mv$ as the variable, we have

$$\mathrm{d}g = \frac{V}{h^3}4\pi p^2\mathrm{d}p \tag{3.87}$$

Because Eq. (3.87) does not involve mass, it is also applicable to phonons and photons as will be discussed in Chaps. 5 and 8.

### 3.5.3 A Rigid Rotor

The rigid rotor model can be used to study the rotational movement of diatomic molecules as well as the electron around the orbit in a hydrogen atom. Consider two particles separated by a fixed distance $r_0 = r_1 + r_2$ as shown in Fig. 3.14. The masses

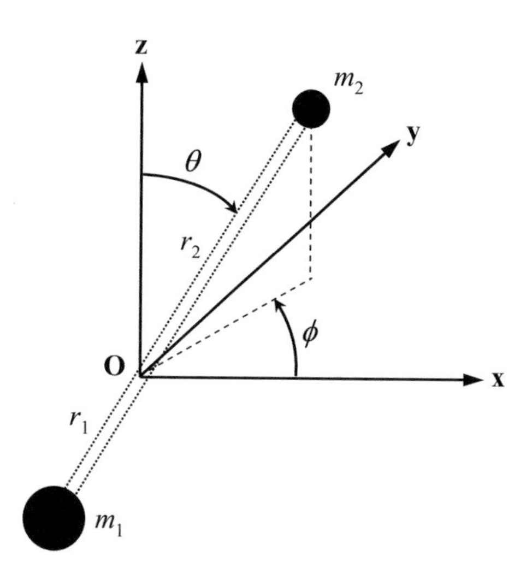


**Fig. 3.14** Schematic of a rotor consisting of two particles

of the particles are $m_1$ and $m_2$, respectively. Since the center of mass is at the origin, we have $m_1 r_1 = m_2 r_2$. The moment of inertia is

$$I = m_1 r_1^2 + m_2 r_2^2 = m_r r_0^2 \tag{3.88}$$

where $m_r = m_1 m_2/(m_1 + m_2)$ is the reduced mass. We can study the rotational movement of the particles by considering a particle with a mass of $m_r$ that rotates around at a fixed distance $r = r_0$ from the origin in the $\theta$ and $\phi$ directions. In the spherical coordinates,

$$\nabla^2 = \frac{1}{r^2}\frac{\partial}{\partial r}\left(r^2\frac{\partial}{\partial r}\right) + \frac{1}{r^2 \sin\theta}\frac{\partial}{\partial\theta}\left(\sin\theta\frac{\partial}{\partial\theta}\right) + \frac{1}{r^2\sin^2\theta}\frac{\partial^2}{\partial\phi^2} \tag{3.89}$$

Because $r \equiv r_0$, the derivative with respect to $r$ vanishes. The potential energy is zero for free rotation. By setting the mass to be $m_r$ and $\Phi = 0$ in Eq. (3.72) and noticing that $m_r r_0^2 = I$, we obtain

$$\frac{1}{\sin\theta}\frac{\partial}{\partial\theta}\left(\sin\theta\frac{\partial\Psi}{\partial\theta}\right) + \frac{1}{\sin^2\theta}\frac{\partial^2\Psi}{\partial\phi^2} = -\frac{2I\varepsilon}{\hbar^2}\Psi \tag{3.90}$$

This partial differential equation can be solved by separation of variables. We get two ordinary differential equations by letting $\Psi(\theta,\phi) = P(\theta)\psi(\phi)$,

$$\frac{\mathrm{d}^2\psi}{\mathrm{d}\phi^2} = -m^2\psi \tag{3.91}$$

and

$$\frac{1}{\sin\theta}\frac{\mathrm{d}}{\mathrm{d}\theta}\left(\sin\theta\frac{\mathrm{d}P}{\mathrm{d}\theta}\right)+\left(\frac{2I\varepsilon}{\hbar^2}-\frac{m^2}{\sin^2\theta}\right)P=0 \tag{3.92}$$

Here, $m$ is a new eigenvalue, and the periodic boundary conditions shall be applied to $P$ and $\psi$, respectively. The solution of Eq. (3.91) is readily obtained as

$$\psi(\phi)=A\mathrm{e}^{\mathrm{i}m\phi} \tag{3.93}$$

with $m = 0, \pm 1, \pm 2, \ldots$, to satisfy the periodic boundary conditions: $\psi(\phi) = \psi(2\pi+\phi)$. A transformation, $\cos\theta = \xi$, can be used so that Eq. (3.92) becomes

$$(1-\xi^2)\frac{\mathrm{d}^2P}{\mathrm{d}\xi^2}-2\xi\frac{\mathrm{d}P}{\mathrm{d}\xi}+\left(\frac{2I\varepsilon}{\hbar^2}-\frac{m^2}{1-\xi^2}\right)P=0 \tag{3.94}$$

Because θ is defined from 0 and $\pi$, we have $-1 \le x \le 1$. In order for Eq. (3.94) to have solutions that are bounded at $x = \pm 1$, $\frac{2I\varepsilon}{\hbar^2} = l(l+1)$, where $l$ is an integer that is greater than or at least equal to the absolute value of $m$. Therefore, the energy eigenvalues are

$$\varepsilon_l=\frac{\hbar^2}{2I}l(l+1),\ l=|m|,\ |m|+1,\ |m|+2,\ \text{etc.} \tag{3.95}$$

Equation (3.94) is called the associated Legendre differential equation. The solutions are the associated Legendre polynomials given as

$$P_l^m(\xi)=\frac{(1-\xi^2)^{m/2}}{l!2^l}\frac{\mathrm{d}^{m+l}}{\mathrm{d}\xi^{m+l}}(\xi^2-1)^l \tag{3.96}$$

Finally, after normalization, the standing wavefunctions can be expressed as

$$\Psi_l^m(\theta,\phi)=\frac{1}{\sqrt{2\pi}}\left[\frac{(2l+1)(l-m)!}{2(m+1)!}\right]^{1/2}P_l^m(\cos\theta)\mathrm{e}^{\mathrm{i}m\phi} \tag{3.97}$$

It can be seen that Eq. (3.95) is identical to Eq. (3.52). The energy level is determined by the principal quantum number $l$. On the other hand, for each $l$, there are $2l+1$ quantum states corresponding to each individual $m$, because $m$ can take $0, \pm 1, \pm 2$ up to $\pm l$. This means that the degeneracy $g_l = 2l+1$. When the two atoms are identical, such as in a nitrogen molecule, the atoms are indistinguishable when they switch positions. The degeneracy is reduced by a symmetry number, as given in the expression of Eq. (3.53). It should be noted that the nuclear spin degeneracy is important for hydrogen (also see Problem 3.27).

### 3.5.4 Atomic Emission and the Bohr Radius

A hydrogen atom is composed of a proton and an electron. Since the mass of the proton is much greater than that of the electron, it can be modeled as the electron moving around the nucleus. The mass of the electron is $m_e = 9.11 \times 10^{-31}$ kg, and the position of the electron can be described in the spherical coordinates as $\mathbf{r} = (r, \theta, \phi)$. The force exerted on the electron is Coulomb's force, which gives a potential field

$$\Phi(r) = -\frac{C_1}{r} \tag{3.98}$$

where $C_1 = e^2/4\pi\varepsilon_0 = 2.307 \times 10^{-28}\,\mathrm{N\,m^2}$, with the electron charge $e = 1.602 \times 10^{-19}\,\mathrm{C}$ and the dielectric constant $\varepsilon_0 = 8.854 \times 10^{-12}\,\mathrm{F/m}$. Let $\Psi(r, \theta, \phi) = R(r)P(\theta)\psi(\phi)$. In doing the separation of variables, we notice that the potential $\Phi$ is independent of $\theta$ and $\phi$, and the total energy is equal to the sum of the rotational energy and the energy associated with $r$. The dependence of rotational energy is given in Eq. (3.45). Using Eqs. (3.72) and (3.89), we can write the equation for $R(r)$ as follows:

$$\frac{\hbar}{2m_e r^2}\frac{\mathrm{d}}{\mathrm{d}r}\left(r^2\frac{\mathrm{d}R}{\mathrm{d}r}\right) + \left(\frac{C_1}{r} + \varepsilon - \frac{l(l+1)\hbar^2}{2I}\right)R = 0 \tag{3.99}$$

which is the associated Laguerre equation, and its solutions are the associated Laguerre polynomials. The solutions give the energy eigenvalues as [5, 8]

$$\varepsilon_n = -\frac{m_e C_1^2}{2\hbar^2 n^2} \tag{3.100}$$

where the negative values are used for convenience to show that the energy increases with the principal quantum number $n$. For $n = 1$, $-m_e C_1^2/2\hbar^2 = -13.6$ eV, as shown in Fig. 3.15. Note that 1 eV $= 1.602 \times 10^{-19}$ J. When the electron is in a higher energy state, it has a tendency of relaxing to a lower energy state by spontaneously emitting a photon, with precisely the same energy as given by the energy difference between the two energy levels:

$$h\nu = \varepsilon_i - \varepsilon_j = \frac{m_e C_1^2}{2\hbar^2}\left(\frac{1}{n_j^2} - \frac{1}{n_i^2}\right) \tag{3.101}$$

The emission or absorption of photons by electrons is called *electronic transitions.* When $i = 3$ and $j = 1$, we have $h\nu = 12.1$ eV, corresponding to the wavelength of 102.6 nm (ultraviolet), which is the second line in the Lyman series. When $i = 3$ and $j = 2$, we have $h\nu = 1.89$ eV, corresponding to the wavelength of 656.4 nm (red),

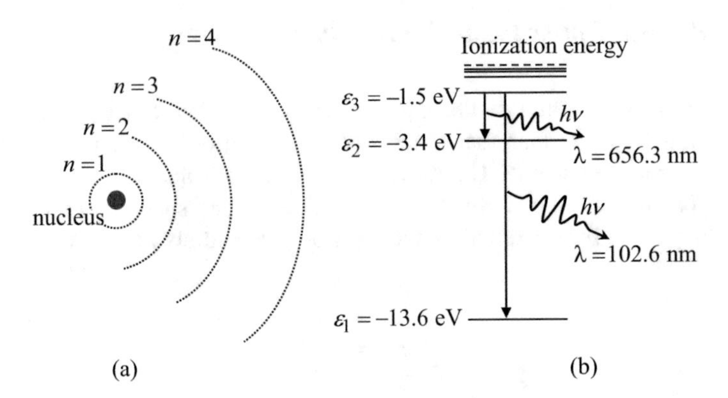


**Fig. 3.15** **a** Electron orbits and **b** energy levels in a hydrogen atom. The ionization energy is the energy required for an electron to escape the orbit

which is the first line in the Balmer series. A more detailed description of the atomic emission lines can be found from Sonntag and van Wylen [2].

The next question is: What is the radius of a particular electron orbit? This is an important question because it gives us a sense of how small an atom is. When a particle is in an orbit, the classical force balance gives that

$$\frac{C_1}{r^2} = m_e\left(\frac{v^2}{r}\right) \tag{3.102}$$

which is to say that $E_K = m_e v^2/2 = C_1/2r$, and the sum of the kinetic and potential energies is

$$\varepsilon = E_K + E_P = \frac{C_1}{2r} - \frac{C_1}{r} = -\frac{C_1}{2r} \tag{3.103}$$

Equations (3.100) and (3.103) can be combined to give discrete values of the radius of each orbit in the following:

$$r_n = \frac{\hbar^2}{m_e C_1} n^2 = a_0 n^2 \tag{3.104}$$

where the electron is in the innermost orbit, with the radius given by $a_0 = \frac{\varepsilon_0 h^2}{\pi m_e e^2} =$ 0.0529 nm, which is called the *Bohr radius*. Niels Bohr (1885–1962) was a Danish physicist who received the Nobel Prize in Physics in 1922 for his contributions to the understanding of the structure of atoms and quantum physics. Therefore, the hydrogen atom in its ground state can be considered as having a diameter of approximately 1 Å (Angstrom), or 0.1 nm. One should accept the quantum interpretation of the electron radius as a characteristic length, not the exact distance that the electron would rotate around the nucleus in the same manner a planet rotates around a star.

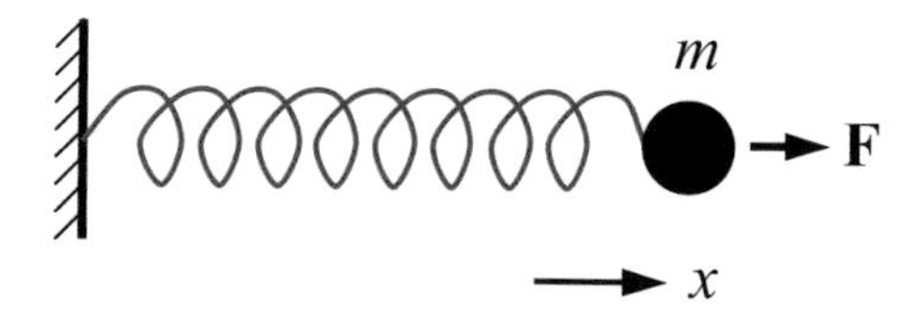


**Fig. 3.16** A linear spring

### 3.5.5 A Harmonic Oscillator

The last example of quantum mechanics is the linear spring as shown in Fig. 3.16. Consider a 1D oscillator with a mass $m$ and the spring force $F(x) = -Kx$. The origin can be selected such that $F(0) = 0$. It can be shown that the potential is

$$\Phi(x) = -\int_0^x F(x)\mathrm{d}x = \frac{1}{2}Kx^2 \quad (3.105)$$

From classical mechanics, we can solve Newton's equation $m\ddot{x} + Kx = 0$ to obtain the solution

$$x = A\sin(\omega t + \phi_0) \quad (3.106)$$

where constant $A$ is the amplitude, constant $\phi_0$ is the initial phase, and parameter $\omega = \sqrt{K/m}$ is the angular resonance frequency. It can be shown that the total energy $\varepsilon = E_\mathrm{K} + E_\mathrm{P} = KA^2/2$ is a constant and the maximum displacement is $A$. The velocity is the largest at $x = 0$ and zero at $x = \pm A$.

The Schrödinger wave equation can be written as

$$\frac{\hbar^2}{2m}\frac{\mathrm{d}^2\Psi}{\mathrm{d}x^2} + \left(\varepsilon - \frac{Kx^2}{2}\right)\Psi = 0 \quad (3.107)$$

with the boundary condition being $\Psi(x) = 0$ at $x \to \pm\infty$. The constants can be grouped by using $\alpha = 2m\varepsilon/\hbar^2$ and $\beta = \sqrt{Km}/\hbar$. Then Eq. (3.107) can be transformed by using $\xi = \sqrt{\beta}x$ and $\Psi(x) = Q(\xi)\exp(-\xi^2/2)$ to

$$\frac{\mathrm{d}^2Q}{\mathrm{d}\xi^2} - 2\xi\frac{\mathrm{d}Q}{\mathrm{d}\xi} + \left(\frac{\alpha}{\beta} - 1\right)Q = 0 \quad (3.108)$$

This is the Hermite equation, and the solutions are Hermite polynomials given by

$$H_n(\xi) = (-1)^n\mathrm{e}^{\xi^2}\frac{\mathrm{d}^n}{\mathrm{d}\xi^n}\left(\mathrm{e}^{-\xi^2}\right) \quad (3.109)$$

when $\alpha$ and $\beta$ must satisfy the eigenvalue equation:

$$\frac{\alpha}{\beta} - 1 = 2n, \quad n = 0, 1, 2, \ldots \tag{3.110}$$

The normalized wavefunctions can be written as

$$\Psi_n(x) = \left(\frac{\sqrt{\beta/\pi}}{n!2^n}\right)^{1/2} H_n(\beta^{1/2}x)\exp\left(-\frac{\beta x^2}{2}\right) \tag{3.111}$$

The energy eigenvalues can be obtained from Eq. (3.110) as

$$\varepsilon_n = (n + \tfrac{1}{2})\hbar\sqrt{K/m} = (n + \tfrac{1}{2})\hbar\omega \tag{3.112}$$

The above equation was used to study the vibrational contributions in diatomic molecules; see Eq. (3.59). The 1/2 term was not included in Planck's original derivation of the blackbody radiation function. The significance lies in that if the ground-state energy is zero, both its kinetic energy and potential energy must be zero, suggesting that both the position and the momentum must be zero. This would violate the uncertainty principle. As mentioned earlier, in classical mechanics, the particle is limited to the region $-A < x < A$, where $A$ is the amplitude given in Eq. (3.106). This is not the case in the quantum theory, as shown in Fig. 3.17, for the first few energy levels and the associated wavefunctions. Notice that probability density function $\Psi^2$ is nonzero even though the absolute value of $x$ exceeds $\sqrt{2\varepsilon/K}$.

The application of quantum theory allows us to predict the specific heat of ideal gases. In deriving the equations shown in Sect. 3.3.3, we have largely neglected nonlinear and anharmonic vibration, electronic contribution, and dissociation. These factors may become important at very high temperatures. The degeneracy due to the coupling of rotation and vibration can cause multiple absorption/emission lines in the infrared in polyatomic molecular gases, as shown in Fig. 3.12.

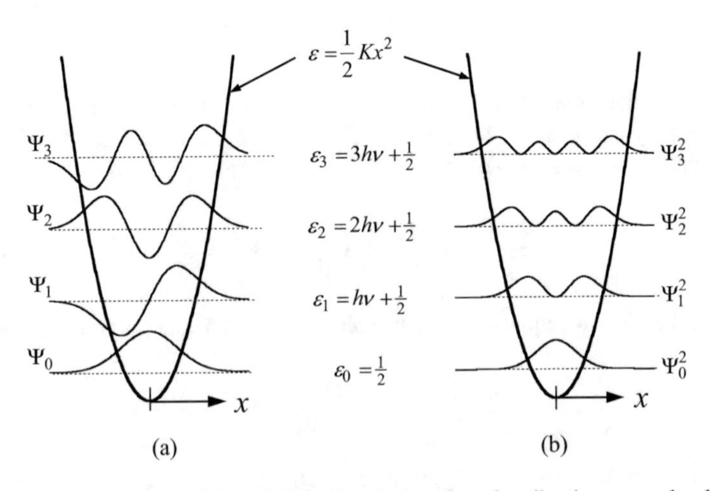


**Fig. 3.17** **a** Wavefunctions and **b** probability density functions for vibration energy levels

## 3.6 Emission and Absorption of Photons by Molecules or Atoms

We have learned that the emission of photons is associated with transitions from a higher energy level to a lower energy level that reduces the total energy of the molecular system. The reverse process is the absorption of photons that increases the energy of the system through transitions from lower energy levels to higher energy levels. As discussed earlier, an electronic transition requires a large amount of energy, and the emitted or absorbed photons are at frequencies from deep ultraviolet ($\lambda \approx 100$ nm) to slightly beyond the red end of the visible region ($\lambda \approx 1$ μm). On the other hand, vibration or rotation-vibration modes lie in the mid-infrared (2.5 μm $< \lambda <$ 25 μm), while their overtones or higher-order harmonics lie in the near-infrared region (0.8 μm $< \lambda <$ 2.5 μm). Rotational modes alone may be active in the far-infrared and microwave regions ($\lambda >$ 25 μm). Transitions between different energy levels of the molecules or atoms are called *bound-bound transitions*, because these energy states are called *bound states*. Bound-bound transitions happen at discrete frequencies due to quantization of energy levels. Dissociation or ionization can also occur at high temperatures. The difference between adjacent energy levels is very small because the electrons can move freely (i.e., not bound to the atom or the molecule). Therefore, *free-free* or *bound-free transitions* happen in a broadband of frequencies. In gases, these broader transitions occur only at extremely high temperatures.

If a molecule at elevated energy states were placed in a surrounding at zero absolute temperature (i.e., empty space), it would lower its energy states by emitting photons in all directions until reaching its ground state. However, the emission processes should occur spontaneously regardless of the surroundings. Suppose the molecule is placed inside an isothermal enclosure, after a long time, the energy absorbed must be equal to that emitted to establish a thermal equilibrium with its surroundings. The thermal fluctuation of oscillators is responsible for the equilibrium distribution of photons governed by Planck's law developed in 1900. Albert Einstein examined how matter and radiation can achieve thermal equilibrium in a fundamental way and published a remarkable paper, "On the quantum theory of radiation" in 1917 [11]. The interaction of radiation with matter is essentially through emission or absorption at the atomistic dimension, although solids or liquids can reflect radiation and small particles can scatter radiation. Einstein noticed that spontaneous emission and pure absorption (i.e., transition from a lower level to a higher level by absorbing the energy from the incoming radiation) alone would not allow an equilibrium state of an atom to be established with the radiation field. He then hypothesized the concept of *stimulated* or *induced emission,* which became the underlying principle of lasers. In a stimulated emission process, an incoming photon interacts with the atom: the interaction results in a transition from a higher energy state to a lower energy state by the emission of another photon of the same energy toward the same direction as the incoming photon. In other words, the stimulated photon is a "clone" of the stimulating photon with the same energy and momentum. Depending on the probability of each event, the incoming photons could either be absorbed or stimulate another photon or pass by without any effect on the atom. Understanding the emission and absorption processes is important not only for coherent emission but also for thermal radiation [12]. While more detailed treatments will be given in later chapters, it is important to gain a basic understanding of the quantum theory of radiative transitions and microscopic description of the radiative properties.

Consider a canonical ensemble of single molecules or atoms, with two nondegenerate energy levels, $\varepsilon_1$ and $\varepsilon_2$ ($\varepsilon_1 < \varepsilon_2$ ), in thermal equilibrium with an enclosure or cavity at temperature $T$. Suppose the total number of particles is $N$, and let $N_1$ and $N_2$ be the number of particles at the energy level corresponding to $\varepsilon_1$ and $\varepsilon_2$, respectively. These particles do not interact with each other at all. The concept of canonical ensemble can be understood as if each cavity has only one atom but there are $N$ single-atom cavities with one atom in each cavity. As shown in Fig. 3.18, there are three possible interaction mechanisms: spontaneous emission, stimulated emission, and *stimulated or induced absorption.* Here, stimulated absorption refers to the process that the energy of the photon is absorbed, and consequently, the transition occurs from the lower energy level to the higher energy level. In a stimulated absorption process, the number of photons before the process is 1 and after the process is $1 - 1 = 0$. In a stimulated emission process, the number of photons beforehand is 1 and afterward is $1 + 1 = 2$. Therefore, stimulated emission is regarded also as *negative absorption*. Each of the photons involved in this process will have an energy equal to $h\nu = \varepsilon_2 - \varepsilon_1$ and a momentum $h\nu/c$.

Transition from the higher energy level to the lower energy level cannot take place if the population of atoms on the higher energy level, $N_2 = 0$, and vice versa. Einstein further assumed that the probability of transition is proportional to the population at the initial energy level, and spontaneous transition should be independent of the radiation field. Hence, the rate of transition from $\varepsilon_2$ to $\varepsilon_1$ due to spontaneous emission can be written as

$$\left(\frac{\mathrm{d}N_1}{\mathrm{d}t}\right)_\mathrm{A} = -\left(\frac{\mathrm{d}N_2}{\mathrm{d}t}\right)_\mathrm{A} = AN_2 \tag{3.113}$$

where $A$ is *Einstein's coefficient* of spontaneous emission. On the other hand, the transition rate due to stimulated emission should also be proportional to the energy

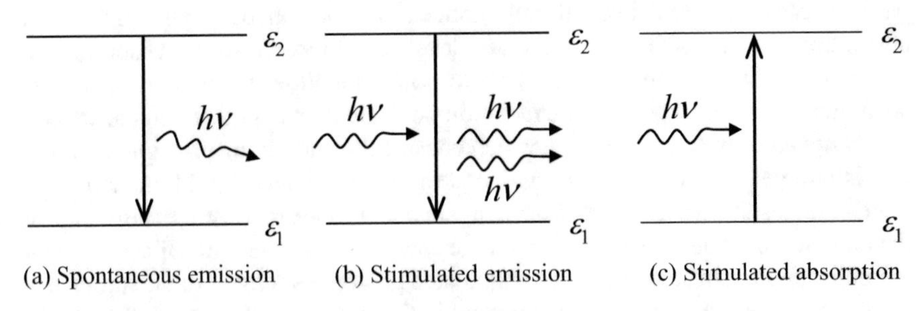


**Fig. 3.18** Illustration of the emission and absorption processes. **a** Spontaneous emission. **b** Stimulated emission. **c** Stimulated absorption

density of the radiation field $u(\nu, T)$. Thus,

$$\left(\frac{\mathrm{d}N_1}{\mathrm{d}t}\right)_\mathrm{B} = BN_2u(\nu, T) \tag{3.114}$$

Stimulated absorption will cause a transition rate that is proportional to $N_1$ and $u(\nu, T)$:

$$\left(\frac{\mathrm{d}N_1}{\mathrm{d}t}\right)_\mathrm{C} = -CN_1u(\nu, T) \tag{3.115}$$

In Eqs. (3.114) and (3.115), constants $B$ and $C$ are Einstein's coefficients of stimulated emission and absorption, respectively. The combination of these processes must maintain a zero net change of the populations at equilibrium. Thus,

$$AN_2 + BN_2u(\nu, T) - CN_1u(\nu, T) = 0 \tag{3.116}$$

Atoms or molecules in a thermal equilibrium are described by the Maxwell-Boltzmann statistics of molecular gases given by Eq. (3.26): $N_1/N_2 = \mathrm{e}^{(\varepsilon_2-\varepsilon_1)/k_\mathrm{B}T} = \mathrm{e}^{h\nu/k_\mathrm{B}T}$. Therefore, Eq. (3.116) can be rewritten as

$$u(\nu, T) = \frac{A/B}{(C/B)e^{h\nu/k_\mathrm{B}T} - 1} \tag{3.117}$$

Comparing this equation with Planck's distribution, Eq. (8.41) in Chap. 8, we see that $B = C$ and $A/B = 8\pi h\nu^3/c^3$. The two-level system can easily be generalized to arbitrary energy levels to describe the fundamental emission and absorption processes. The emission and absorption processes not only exchange energy between the field and the atom but also transfer momentum. How will an atom move inside a cavity? The phenomenon of a molecule or atom in a radiation field is like the Brownian motion, in which the radiation quanta exert forces on the molecule or the atom as a result of momentum transfer during each emission or absorption process. Consequently, the molecule or the atom will move randomly following Maxwell's velocity distribution at the same temperature as the radiation field. The equilibrium radiation field, which obeys the quantum statistics (i.e., BE statistics) that was not realized until 1924, and the motion of a molecular gas, which obeys classical statistics, can be coupled to each other at mutual equilibrium. Einstein also asserted that each spontaneously emitted photon must be directional, while the probability of spontaneous emission should be the same in all directions. In fact, Einstein's 1917 paper complemented Planck's 1900 paper on radiation energy quanta and his own 1905 paper on photoelectric emission and, thus, provided a complete description of the quantum nature of photons, although the name "photon" was not coined until 1928 [12].

At moderate temperatures, the population at higher energy states is too small for stimulated emission to be of significance for optical and thermal radiation. Thus, the absorption comes solely from induced absorption. When stimulated emission is important, the contributions of stimulated emission and stimulated absorption cannot be separated by experiments. The effect is combined to give an effective absorption coefficient by taking stimulated emission as negative absorption, whereas the emission of radiation includes solely the spontaneous emission [12]. The effective absorption coefficient is proportional to the population difference, $N_1 - N_2$. On the other hand, if a *population inversion* can be created and maintained such that $N_2 > N_1$, the material is called a *gain medium* or *active medium*. In an active medium, stimulated emission dominates stimulated absorption so that more and more photons will be cloned and the radiation field be amplified coherently. The principle of stimulated emission was applied in 1950s and early 1960s for the development of maser, which stands for *microwave amplification by stimulated emission of radiation*, and laser, which stands for *light amplification by stimulated emission of radiation* [13]. Lasers have become indispensable to modern technologies and daily life.

## 3.7 Energy, Mass, and Momentum in Terms of Relativity

Special theory of relativity or *special relativity* predicts that energy and mass can be converted to each other. If we retain the definition of mass as in the classical theory, only energy conservation is the fundamental law of physics. The mass does not have to be conserved. On the other hand, for processes that do not involve changes below the atomic level or inside the nuclei, the mass can indeed be considered as conserved. According to the special relativity, the rest energy of a free particle is related to its mass and the speed of light by

$$E_0 = mc^2 \tag{3.118}$$

The rest energy is simply the energy when the particle is not moving relative to the reference frame. Suppose the free particle is moving at a velocity $v$ in a given reference frame, then its momentum is given by [14]

$$p = \frac{mv}{\sqrt{1 - v^2/c^2}} \tag{3.119}$$

When $v \ll c$, Eq. (3.119) reduces to the classical limit $p = mv$. It can be seen that for a particle with nonzero mass, its momentum would increase as $v \to c$ without any bound. There is no way we could accelerate a particle to the speed of light. If there is anything that travels with the speed of light, it has to be massless, i.e., $m = 0$. An example of massless particles is the light quanta or photons. The kinetic energy can be evaluated by integrating the work needed to accelerate a particle, $E_\mathrm{K} = \int_0^x F\mathrm{d}x = \int_0^x \frac{\mathrm{d}p}{\mathrm{d}t}\mathrm{d}x = \int_0^x \frac{\mathrm{d}p}{\mathrm{d}v}\frac{\mathrm{d}v}{\mathrm{d}t}\mathrm{d}x = \int_0^v \frac{\mathrm{d}p}{\mathrm{d}v}v\mathrm{d}v$. Using Eq. (3.119), we find that

$$E_K = \frac{mc^2}{\sqrt{1 - v^2/c^2}} - mc^2 \tag{3.120}$$

When $v \ll c$, we have $1/\sqrt{1 - v^2/c^2} \approx 1 + v^2/2c^2$ so that $E_K = mv^2/2 = p^2/2m$ in the low-speed limit. In the relativistic limit, however, $E_K$ will be on the order of $mc^2$. Because energy is additive, the total energy of a moving free particle is

$$E = E_K + E_0 = \frac{mc^2}{\sqrt{1 - v^2/c^2}} \tag{3.121}$$

Obviously, the energy of a particle would become infinite if its speed approaches the speed of light, unless its mass goes to zero. It can be shown that $E^2 - E_0^2 = \frac{m^2c^4}{1-v^2/c^2} - m^2c^4 = p^2c^2$, where $p$ is given in Eq. (3.119). This gives another expression of energy in terms of the rest energy, the momentum, and the speed of light as follows:

$$E^2 = m^2c^4 + p^2c^2 \tag{3.122}$$

It should be noted that, in general, $pc$ is not equal to the kinetic energy. For $v \ll c$, the total energy is approximately the same as the rest energy. Comparing Eqs. (3.119) and (3.121), we notice that $E = pc(c/v)$. Therefore, when $v \to c$, we see that $E \to pc$ (which is unbounded unless $m = 0$). For a photon that travels at the speed of light, in order for the above equations to be meaningful, we must set its mass to zero. From Eq. (3.122), we have for photons that

$$p = \frac{E}{c} = \frac{h\nu}{c} \tag{3.123}$$

which is the same as Eq. (3.7) in Sect. 3.1.3. By noting that $\lambda\nu = c$, we obtain

$$\lambda = \frac{h}{p} \tag{3.124}$$

The kinetic energy of a photon is $pc$ or $h\nu$ since its rest energy is zero. One should not attempt to calculate the kinetic energy of a photon by $\frac{1}{2}mc^2$, because photons are not only massless but also *relativistic particles,* for which the energy and momentum must be evaluated according to Eqs. (3.122) and (3.123), respectively. While photons do not have mass, it has been observed that photons can be used to create particles with nonzero mass or vice versa, as in *creation* or *annihilation reactions*. High energy physics has proved that mass is not always conserved. It is commonly said that mass and energy can be interconverted. For example, in a nuclear reaction, a small amount of mass can be converted into a large amount of energy. In these statements, the term energy is used in the classical sense (such as kinetic energy and the emission of high-energy photons). When the rest energy $E_0$ is included, the total energy is always conserved.

## 3.8 Summary

This chapter started with very basic independent particle systems to derive the three major statistics, namely, the Maxwell-Boltzmann, Bose-Einstein, and Fermi-Dirac statistics. The classical and quantum statistics were then applied to thermodynamic systems, providing microscopic interpretations of the first, second, and third laws of thermodynamics, as well as Bose-Einstein condensate. The velocity distribution and specific heat of ideal gases were explained based on the semi-classical statistics, followed by a brief description of quantum mechanics to understand the quantization of translational, rotational, and vibrational modes. The fundamental emission and absorption processes of molecules or atoms were discussed along with the concept of stimulated emission. Finally, matter-energy conversion was described within the framework of special relativity. While most of the explanations in this chapter are semi-classical and somehow oversimplified, it should provide a solid background to those who do not have a comprehensive knowledge and background in statistical mechanics and quantum physics. The materials will be frequently referenced in the rest of the book.

## Problems

3.1 For a rectangular prism whose three sides are $x$, $y$, and $z$. If $x + y + z = 9$, find the values of $x$, $y$, and $z$ so that the volume of the prism is maximum.

3.2 Make a simple computer program to evaluate the relative error of Stirling's formula: $\ln x! \approx x \ln x - x$ for $x = 10$, 100, and 1000.

3.3 For each of the following cases, determine the number of ways to place 25 books on 5 shelves (distinguishable by their levels). The order of books within an individual shelf is not considered.

(a) The books are distinguishable, and there is no limit on how many books can be put on each shelf.
(b) Same as (a), except that all the books are the same (indistinguishable).
(c) The books are distinguishable, and there are 5 books on each shelf.
(d) The books are distinguishable, and there are 3 books on the 1st shelf, 4 on the 2nd, 5 on the 3rd, 6 on the 4th, and 7 on the 5th.

3.4 For each of the following cases, determine the number of ways to put 4 books on 10 shelves (distinguishable by their levels). Disregard their order on each shelf.

(a) The books are distinguishable, and there is no limit on how many books you can place on each shelf.
(b) Same as (a), but there is a maximum of 1 book on any shelf.
(c) Same as (a), except that the books are identical (indistinguishable).
(d) Same as (b), except that the books are identical.

3.5 A box contains 5 red balls and 3 black balls. Two balls are picked up randomly. Determine the following:

(a) What's the probability that the second ball is red?
(b) What's the probability that both are red?
(c) If the first one is black, what is the probability that the second is red?

3.6 Suppose you toss two dice, what's the probability of getting a total number (a) equal to 5 and (b) greater than 5?

3.7 Draw 5 cards from a deck of 52 cards.

(a) What is the probability of getting a royal flush?
(b) What is the probability of getting a full house? [A royal flush is a hand with A, K, Q, J, and 10 of the same suit. A full house is a hand with three of one kind and two of another (a pair).]

3.8 For a Gaussian distribution function, $f(x) = a\mathrm{e}^{-(x-\mu)^2}$, where $a$ and $\mu$ are positive constants.

(a) Find the normalized distribution function $F(x)$.
(b) Show that the mean value $\bar{x} = \mu$.
(c) Determine the variance $u_{\mathrm{var}}$ and the standard deviation $\sigma$.

3.9 The speed distribution function for $N$ particles in a fixed volume is given by $f(V) = \frac{AV(B-V)}{B^3}$, where $V$ (>0) is the particle speed, and $A$ and $B$ are positive constants. Determine:

(a) The probability density function $F(V)$.
(b) The number of particles $N$ in the volume.
(c) The minimum speed $V\mathrm{min}$ and maximum speed $V\mathrm{max}$.
(d) The most probable speed where the probability density function is the largest.
(e) The average speed $\overline{V}$ and the root-mean-square average speed $V_{\mathrm{rms}} = \sqrt{\overline{V^2}}$.

3.10 Six bosons are to be placed in two energy levels, each with a degeneracy of two. Evaluate the thermodynamic probability of all arrangements. What is the most probable arrangement?

3.11 Four fermions are to be placed in two energy levels, each with a degeneracy of four. Evaluate the thermodynamic probability of each arrangement. What is the most probable arrangement?

3.12 Derive the Fermi-Dirac distribution step by step. Clearly state all assumptions. Under which condition, can it be approximated by the Maxwell-Boltzmann distribution?

3.13 What is the Boltzmann constant and how is it related to the universal gas constant? Show that the ideal gas equation can be written as $P = nk_{\mathrm{B}}T$. What is the number density of air at standard conditions (1 atm and 25 °C)?

3.14 How many molecules are there per unit volume (number density) for the nitrogen gas at 200 K and 20 kPa? How would you estimate the molecular spacing (average distance between two adjacent molecules)?

3.15 Use Eq. (3.28a), (3.28b) and $\frac{1}{T} = \left(\frac{\partial S}{\partial U}\right)_{V,N}$ to show that $\beta = \frac{1}{k_B T}$.

3.16 Show that $\beta = 1/k_B T$ and $\alpha = -\mu/k_B T$ for all the three statistics. [Hint: Follow the lecture note with a few more steps.]

3.17 Consider 10 indistinguishable particles in a fixed volume that obey the Bose-Einstein statistics. There are three energy levels with $\varepsilon_0 = 0.5$ eu, $\varepsilon_1 = 1.5$ eu, and $\varepsilon_2 = 2.5$ eu, where "eu" refers to certain energy unit. The degeneracies are $g_0 = 1$, $g_1 = 3$, and $g_2 = 5$, respectively.

(a) If the degeneracy were not considered, in how many possible ways could you arrange the particles on the three energy levels?
(b) You may notice that different arrangements may result in the same energy. For example, both the arrangement with $N_1 = 9$, $N_2 = 0$, $N_3 = 1$ and the arrangement with $N_1 = 8$, $N_2 = 2$, $N_3 = 0$ yield an internal energy $U = 7$ eu. How many arrangements are there with $U = 9$ eu? Calculate the thermodynamic probability for all macrostates with $U = 9$ eu.
(c) The ground state refers to the state corresponding to the lowest possible energy of the system. Determine the ground-state energy and entropy. What is the temperature of this system at the ground state?
(d) How many microstates are there for the macrostate with $U = 25$ eu?

3.18 Consider a system of a single type of constituents, with $N$ particles (distinguishable from the statistical point of view) and only two energy levels $\varepsilon_0 = 0$ and $\varepsilon_1 = \varepsilon$ (nondegenerate).

(a) What is the total number of microstates in terms of $N$. How many microstates are there for the macrostate that has energy $U = (N - 1)\varepsilon$? Show that the energy of the most probable macrostate is $N\varepsilon/2$.
(b) What are the entropies of the states with $U = 0$ and $U = (N - 1)\varepsilon$. Sketch $S$ as a function of $U$. Comment on the negative temperature, $1/T = (\partial S/\partial U)_{V,N} < 0$. Is it possible to have a system with a negative absolute temperature?

3.19 A system consists of six indistinguishable particles that obey Bose-Einstein statistics with two energy levels. The associated energies are $\varepsilon_0 = 0$ and $\varepsilon_1 = \varepsilon$, and the associated degeneracies are $g_0 = 1$ and $g_1 = 3$. Answer the following questions:

(a) How many possible macrostates are there? How many microstates corresponding to the macrostate with three particles on each energy level?
(b) What is the most probable macrostate, and what are its corresponding energy $U$ and thermodynamic probability $\Omega$?
(c) Show that at 0 K, both the energy and the entropy of this system are zero. Also, show that for this system the entropy increases as the energy increases.

3.20 From the Sackur-Tetrode equation, show that $s_2 - s_1 = c_p \ln(T_2/T_1) - R\ln(P_2/P_1)$.

3.21 Write $U$, $p$, $A$, and $S$ in terms of the partition function $Z$. Express $H$ and $G$ in terms of the partition function $Z$. For an ideal monatomic gas, express $H$ and $G$ in terms of $T$ and $P$.

3.22 For an ideal diatomic gas, the partition function can be written as $Z = Z_t Z_r Z_v Z_e Z_D$, where $Z_e = g_{e0}$ is the degeneracy of the ground electronic level, and $Z_D = \exp(-D_0/k_B T)$ is the chemical partition function that is associated with the reaction of formation. Here, $g_{e0}$ and $D_0$ can be regarded as constants for a given material. Contributions to the partition function beside the translation are due to internal energy storage and thus are called the *internal contribution*, $Z_{int} = Z_r Z_v Z_e Z_D$. Find the expressions of $U$, $P$, $A$, $S$, $H$, and $G$ in terms of $N$, $T$, and $P$ (or $V$) with appropriate constants, assuming that $\Theta_r \ll T \sim \Theta_v$.

3.23 For an ideal molecular gas, derive the distribution function $f(\varepsilon)$ in terms of the kinetic energy $\varepsilon = mv^2/2$.

3.24 Prove Eqs. (3.48), (3.49a), (3.49b) and (3.50).

3.25 Evaluate and plot the Maxwell speed distribution for Ar gas at 100, 300, and 900 K. Tabulate the average speed, the most probable speed, and the rms speed at these temperatures.

3.26 A special form of the Euler-Maclaurin summation formula is

$$\sum_{j=a}^{\infty} f(j) = \int_a^{\infty} f(x)\mathrm{d}x + \frac{1}{2}f(a) - \frac{1}{12}f'(a) + \frac{1}{720}f^{(3)}(a) - \frac{1}{30,240}f^{(5)}(a) + \cdots$$

Consider the rotational partition function, $Z_r = \sum_{l=0}^{\infty}(2j+1)\exp[-j(j+1)\Theta_r/T]$, and show that $Z_r \approx \frac{T}{\Theta_r}\left[1 + \frac{1}{3}\frac{\Theta_r}{T} + \frac{1}{15}\left(\frac{\Theta_r}{T}\right)^2 + \cdots\right]$, which is Eq. (3.55) for $\sigma = 1$.

3.27 Because of the nuclear spin degeneracy, hydrogen $H_2$ gas is consistent of two different types: *ortho-hydrogen* and *para-hydrogen*. The rotational partition functions can be written, respectively, as

$$Z_{r,ortho} = 3\sum_{l=1,3,5...}(2l+1)\exp\left[-l(l+1)\frac{\Theta_r}{T}\right]$$

and

$$Z_{r,para} = \sum_{l=0,2,4...}(2l+1)\exp\left[-l(l+1)\frac{\Theta_r}{T}\right]$$

so that $Z_{r,H_2} = 3\sum_{l=1,3,5...}(2l+1)\exp\left[-l(l+1)\frac{\Theta_r}{T}\right] + \sum_{l=0,2,4...}(2l+1)\exp\left[-l(l+1)\frac{\Theta_r}{T}\right]$. Evaluate the temperature-dependent specific heat of each of the two types of hydrogen, which can be separated and stay separated for a long time before the equilibrium distribution is restored. Calculate the specific heat of hydrogen in the equilibrium distribution as a function of temperature. The ratio $Z_{r,ortho}/Z_{r,para}$ is the same as the equilibrium ratio of the two types and varies from 0 at very low temperatures to 3 near room temperature.

3.28 Calculate the specific heat and the specific heat ratio $\gamma = c_p/c_v$ for nitrogen $N_2$ at 30, 70, 300, and 1500 K. Assume the pressure is sufficiently low for it to be an ideal gas.

3.29 Calculate the specific heat and the specific heat ratio $\gamma = c_p/c_v$ for oxygen $O_2$ at 50, 100, 300, and 2000 K. Assume the pressure is sufficiently low for it to be an ideal gas.

3.30 Estimate the mole and mass specific heats of CO gas at 100, 300, and 3000 K. Show in a specific heat versus temperature graph the contributions from different modes.

3.31 (a) How many rotational degrees of freedom are there in a $CO_2$ molecule and in a $H_2O$ molecule?
(b) If the temperature of a low-pressure $CO_2$ gas is raised high enough to completely excite its rotational and vibrational modes, what will be its specific heats $c_v$ and $c_p$? Express answers in both kJ/kg K and kJ/kmol K.

3.32 Compute and plot the temperature-dependent specific heat for the following ideal gases and compare your results with tabulated data or graphs: (a) $CO_2$, (b) $H_2O$, and (c) $CH_4$.

3.33 Write down a few sentences to discuss each of the following topics: (a) the significance of partition functions, (b) the different types of statistical ensembles, and (c) statistical fluctuations.

3.34 We have discussed the translational degeneracy d$g$ in a 3D space with a volume $V$, as given in Eq. (3.85). Consider the situation when the particle is confined in a 2D square potential well. Find the proper wavefunctions and the energy eigenvalues. Assuming the area $A$ is very large, find the translational degeneracy d$g$ in terms of $A, m, \varepsilon,$ and d$\varepsilon$.

3.35 Estimate the speed an electron needs in order to escape from the ground state of a hydrogen atom. What is the de Broglie wavelength of the electron at the initial speed? If a photon is used to knock out the electron in the ground state, what would be the wavelength of the photon? Why is it inappropriate to consider the electron movement in an atom analog with the movement of the Mars in the solar system?

3.36 For the harmonic oscillator problem discussed in Sect. 3.5.5. Show that Eq. (3.111) is a solution for Eq. (3.107) for $n = 0, 1,$ and 2. Plot $\Psi_0^2, \Psi_1^2,$ and $\Psi_2^2$ and discuss the differences between classical mechanics and quantum mechanics.

## References

1. C.L. Tien, J.H. Lienhard, *Statistical Thermodynamics* (Hemisphere, New York, 1985)
2. R.E. Sonntag, G.J. van Wylen, *Fundamentals of Statistical Thermodynamics* (Wiley, New York, 1966)
3. J.E. Lay, *Statistical Mechanics and Thermodynamics of Matter* (Harper Collins Publishers, New York, 1990)
4. C.E. Hecht, *Statistical Thermodynamics and Kinetic Theory* (W.H. Freeman and Company, New York, 1990)
5. V.P. Carey, *Statistical Thermodynamics and Microscale Thermophysics* (Cambridge University Press, Cambridge, UK, 1999)
6. F.C. Chou, J.R. Lukes, X.G. Liang, K. Takahashi, C.L. Tien, Molecular dynamics in microscale thermophysical engineering. Annu. Rev. Heat Transfer **10**, 144–176 (1999)
7. S. Maruyama, Molecular dynamics method for microscale heat transfer, in *Advances in Numerical Heat Transfer*, vol. 2, ed. by W.J. Minkowycz, E.M. Sparrow (Taylor and Francis, New York, 2000), pp. 189–226
8. D.J. Griffiths, *Introduction to Quantum Mechanics*, 2nd edn. (Prentice Hall, New York, 2005)
9. H.J. Metcalf, P. van der Straten, *Laser Cooling and Trapping* (Springer, New York, 1999)
10. G. Burns, *High-Temperature Superconductivity: An Introduction* (Academic Press, Boston, MA, 1992)
11. A. Einstein, Zur quantentheorie der strahlung. Phys. Z. **18**, 121–128 (1917); English translation in *Sources of Quantum Mechanics*, B.L. Van der Waerden (ed.), (North-Holland Publishing Company, Amsterdam, the Netherlands, 1967)
12. H.P. Baltes, On the validity of Kirchhoff's law of heat radiation for a body in a nonequilibrium environment. Progress Opt. **13**, 1–25 (1976)
13. J.P. Gordon, H.J. Zeiger, C.H. Townes, The maser—New type of microwave amplifier, frequency standard, and spectrometer. Phys. Rev. **95**, 1264–1274 (1955); A.L. Schawlow, C.H. Townes, Infrared and optical masers. Phys. Rev. **112**, 1940–1949 (1958)
14. R. Wolfson, J.M. Pasachoff, *Physics with Modern Physics for Scientists and Engineers*, 3rd edn. (Addison-Wesley, Reading, MA, 1999)

# Chapter 4
# Kinetic Theory and Micro/Nanofluidics

Statistical mechanics involves determination of the most probable state and equilibrium distributions, as well as evaluation of the thermodynamic properties in the equilibrium states. Kinetic theory deals with the local average of particle properties and can be applied to nonequilibrium conditions to derive transport equations [1–8]. Kinetic theory, statistical mechanics, and molecular dynamics are based on the same hypotheses; they are closely related and overlap each other in some aspects. Knowledge of kinetic theory is important to understanding gas dynamics, as well as electronic and thermal transport phenomena in solid materials.

In this chapter, we first introduce the simple kinetic theory of ideal gases based on the mean-free-path approximation. While it can help us to obtain the microscopic formulation of some familiar transport equations and properties, the simple kinetic theory is limited to local equilibrium and works well only for time durations much longer than the mechanistic timescale, called the relaxation time. The advanced kinetic theory is based on the Boltzmann transport equation (BTE), which will also be presented in this chapter [7, 8]. The BTE is an integro-differential equation of the distribution function in terms of space, velocity, and time. It takes into account changes in the distribution function caused by external forces and collisions between particles. Many macroscopic phenomenological equations, such as Fourier's law of heat conduction, the Navier–Stokes equation for viscous flow, and the equation of radiative transfer for photons and phonons, can be derived from the BTE, under the assumption of local equilibrium. Finally, in the last section of this chapter, we present the application of kinetic theory to the flow of dilute gases in micro/nanostructures and the associated heat transfer. The application of kinetic theory to heat conduction in metals and dielectrics will be discussed in forthcoming chapters.



Z. M. Zhang, *Nano/Microscale Heat Transfer*, Mechanical Engineering Series,
https://doi.org/10.1007/978-3-030-45039-7_4

## 4.1 Kinetic Description of Dilute Gases

In this section, we will introduce the simple kinetic theory of ideal molecular gases. The purpose is to provide a step-by-step learning experience leading to more advanced topics. There are several hypotheses and assumptions in kinetic theory of molecules.

- *Molecular hypothesis*: Matter is composed of small discrete particles (molecules or atoms); any macroscopic volume contains a large number of particles. At 25 °C and 1 atm, 1-$\mu$m$^3$ space of an ideal gas contains 27-million molecules.
- *Statistic hypothesis*: Time average is often used since any macroscopic observation takes much longer than the characteristic timescale of molecular motion (such as the average time lapse between two subsequent collisions of a given molecule).
- *Kinetic hypothesis*: Particles obey the laws of classical mechanics.
- *Molecular chaos*: The velocity and position of a particle are uncorrelated. The velocities of any two particles are not correlated.
- *Ideal gas assumptions*: Molecules are rigid spheres resembling billiard balls. Each molecule has a diameter $d$ and a mass $m$. All collisions are elastic and conserve both energy and momentum. Molecules are widely separated in space (i.e., a dilute gas). Intermolecular forces are negligible except during molecular collisions. The duration of collision is negligible compared with the time between collisions. No collision can occur with more than two particles.

The general molecular *distribution function* is $f(\mathbf{r}, \mathbf{v}, t)$, which is a function of space, velocity, and time. The distribution function gives the particle (number) density in the phase space at any time. Therefore, the number of particles in a volume element of the phase space is

$$\mathrm{d}N = f(\mathbf{r}, \mathbf{v}, t)\mathrm{d}x\mathrm{d}y\mathrm{d}z\mathrm{d}v_x\mathrm{d}v_y\mathrm{d}v_z = f(\mathbf{r}, \mathbf{v}, t)\mathrm{d}V\mathrm{d}\varpi \tag{4.1}$$

where we have used $\varpi$ for the velocity space ($\mathrm{d}\varpi = \mathrm{d}v_x\mathrm{d}v_y\mathrm{d}v_z$). Integrating Eq. (4.1) over the velocity space gives the number of particles per unit volume, or the number density, as

$$n(\mathbf{r}, t) = \frac{\mathrm{d}N}{\mathrm{d}V} = \int_{\varpi} f(\mathbf{r}, \mathbf{v}, t)\mathrm{d}\varpi \tag{4.2}$$

Note that the density is $\rho(\mathbf{r}, t) = m \cdot n(\mathbf{r}, t)$, where $m$ is the mass of a particle. The total number of particles inside the volume $V$ as a function of time is then

$$N(t) = \iint_{V,\varpi} f(\mathbf{r}, \mathbf{v}, t)\mathrm{d}V\mathrm{d}\varpi \tag{4.3}$$

In a thermodynamic equilibrium state,

$$f(\mathbf{r}, \mathbf{v}, t) = f(\mathbf{v}) \tag{4.4}$$

which is independent of space and time. Any intensive property will be the same everywhere.

### 4.1.1 Local Average and Flux

Let $\psi = \psi(\mathbf{r}, \mathbf{v}, t)$ be any additive property of a single molecule, such as kinetic energy and momentum. Note that $\psi$ may be a scalar or a vector. The *local average* or simply the *average* of the property $\psi$ is defined as

$$\bar{\psi} = \frac{\int_{\varpi} f\psi \mathrm{d}\varpi}{\int_{\varpi} f \mathrm{d}\varpi} = \frac{1}{n}\int\limits_{\varpi} f\psi \mathrm{d}\varpi \tag{4.5}$$

which is a function of $\mathbf{r}$ and $t$. The *ensemble average* is the average over the phase space:

$$\langle\psi\rangle = \frac{1}{N}\iint\limits_{V,\varpi} f\psi \mathrm{d}V \mathrm{d}\varpi \tag{4.6}$$

For a uniform gas, the local average and the ensemble average are the same.

The transfer of $\psi$ across an area element d$A$ per unit time per unit area is called the *flux* of $\psi$. As shown in Fig. 4.1, particles having velocities between $\mathbf{v}$ and $\mathbf{v} + \mathrm{d}\mathbf{v}$ that will pass through the area d$A$ in the time interval $dt$ must be contained in the inclined cylinder, whose volume is $\mathrm{d}V = v\mathrm{d}t \cos\theta \, \mathrm{d}A = \mathbf{v} \cdot \mathbf{n} \, \mathrm{d}A\mathrm{d}t$. It is assumed that d$t$ is sufficiently small such that particle–particle collisions can be neglected. The number of particles with velocities between $\mathbf{v}$ and $\mathbf{v} + \mathrm{d}\mathbf{v}$ within the inclined

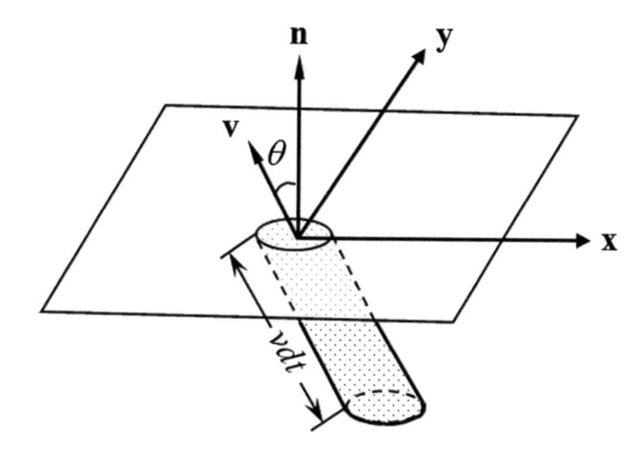


**Fig. 4.1** Illustration of the flux of particles and quantities through a surface

cylinder can be calculated by

$$f(\mathbf{r},\mathbf{v},t)\mathrm{d}V\mathrm{d}\varpi = f(\mathbf{r},\mathbf{v},t)\,\mathbf{v}\cdot\mathbf{n}\,\mathrm{d}A\mathrm{d}t\mathrm{d}\varpi \tag{4.7}$$

The flux of the property $\psi$ is then

$$\text{flux of } \psi \text{ within } \mathrm{d}\varpi = \frac{\psi f(\mathbf{r},\mathbf{v},t)\,\mathbf{v}\cdot\mathbf{n}\,\mathrm{d}\varpi\,\mathrm{d}A\,\mathrm{d}t}{\mathrm{d}A\,\mathrm{d}t}$$

Integrating over all velocities yields the total flux of $\psi$:

$$J_\psi = \int\limits_{\varpi} \psi f \mathbf{v}\cdot\mathbf{n}\,\mathrm{d}\varpi \tag{4.8}$$

Equation (4.8) gives the net flux since it is evaluated for all $\varpi$, or over a solid angle of $4\pi$ in the spherical coordinates. Very often the integration is performed over the hemisphere with $\mathbf{v}\cdot\mathbf{n} = \cos\theta > 0$ for positive flux or $\mathbf{v}\cdot\mathbf{n} = \cos\theta < 0$ for negative flux. When $\psi = 1$, Eq. (4.8) gives the particle flux:

$$J_N = \int\limits_{\varpi} f \mathbf{v}\cdot\mathbf{n}\,\mathrm{d}\varpi \tag{4.9}$$

In an equilibrium state, this integration can be evaluated using the spherical coordinates. Noting that $\mathbf{v}\cdot\mathbf{n} = v\cos\theta$ and $f = f(\mathbf{v})$, which is independent of the direction (isotropic), we can obtain the particle flux in the positive $z$-direction by integrating over the hemisphere in the velocity space,

$$J_N = \int\limits_{v=0}^{\infty}\int\limits_{\phi=0}^{2\pi}\int\limits_{\theta=0}^{\pi/2} f(\mathbf{v})v^3\cos\theta\sin\theta\,\mathrm{d}\theta\mathrm{d}\phi\mathrm{d}v = \pi\int\limits_{0}^{\infty} f(\mathbf{v})v^3\mathrm{d}v \tag{4.10}$$

In writing Eq. (4.10), we have kept the vector variable in $f(\mathbf{v})$ to signify that it is a velocity distribution. One should bear in mind that the last expression is based on the fact that $f(\mathbf{v})$ is not a function of $\theta$ and $\phi$. For an ideal molecular gas, $f(\mathbf{v})$ is given by the Maxwell velocity distribution, i.e., Eq. (3.43) in Chap. 3. If the integration in Eq. (4.10) is performed over the whole sphere with $\theta$ from 0 to $\pi$, we would obtain the net flux of particles, which is zero in the equilibrium case. The average speed can be evaluated using Eq. (4.5); hence,

$$\bar{v} = \frac{1}{n}\int\limits_{\varpi} f(\mathbf{v})v\,\mathrm{d}\varpi = \frac{1}{n}\iiint\limits_{v,\phi,\theta} f(\mathbf{v})v^3\sin\theta\,\mathrm{d}\theta\mathrm{d}\phi\mathrm{d}v = \frac{4\pi}{n}\int\limits_{0}^{\infty} f(\mathbf{v})v^3\mathrm{d}v \tag{4.11}$$

Here, we have assumed an isotropic distribution function to obtain the last expression. The above equation is evaluated over the solid angle of $4\pi$ to obtain the average of all velocities. Comparing Eqs. (4.10) and (4.11), we can see that

$$J_N = \frac{n\bar{v}}{4} \tag{4.12a}$$

For an ideal gas, since $f(\mathbf{v})$ is given by the Maxwell velocity distribution, Eq. (3.44), we obtain

$$J_N = \frac{n\bar{v}}{4} = n\sqrt{\frac{k_B T}{2\pi m}} \tag{4.12b}$$

Because each particle has the same mass, the mass flux is given by

$$J_m = m \int_{\varpi} f\mathbf{v} \cdot \mathbf{n}\, d\varpi = \frac{\rho\bar{v}}{4} \tag{4.13}$$

Substituting $\psi = mv^2/2$ into Eq. (4.8), one obtains the kinetic energy flux $J_{KE}$. In an equilibrium state with an isotropic distribution, the kinetic energy flux in the positive $z$-direction is $J_{KE} = \frac{\pi m}{2} \int_0^\infty f(\mathbf{v}) v^5 dv$, whereas the net kinetic energy flux is zero. Note that Eq. (4.8) is a general equation that is also applicable to nonequilibrium and anisotropic distributions.

When $\psi = m\mathbf{v}$, the momentum flux is a vector, which is often handled by considering individual components. Note that the rate of transfer of momentum across a unit area is equal to the force that the area must exert upon the gas to sustain the equilibrium. Furthermore, the surface may be projected to three orientations, yielding a nine-component tensor in the momentum flux:

$$P_{ij} = \int_{\varpi} (mv_j) f v_i\, d\varpi, \text{ where } i,\ j = 1, 2, 3 \tag{4.14a}$$

Here, $(v_1, v_2, v_3)$ and $(v_x, v_y, v_z)$ are used interchangeably. Let $P = \rho\overline{v_i^2}$, which is always positive, and $\tau_{ij} = \rho\overline{v_j v_i}$ for $i \neq j$. We can rewrite the above equation as

$$P_{ij} = n\overline{mv_j v_i} = \rho\overline{v_j v_i} = P\delta_{ij} + \tau_{ij} \tag{4.14b}$$

where $\delta_{ij}$ is the Kronecker delta, which is equal to 1 when $i = j$ and 0 when $i \neq j$. It can be seen that $P$ is the normal stress or static pressure and $\tau_{ij}(i \neq j)$ is the shear stress, which is zero in a uniform stationary gas (without bulk motion). Notice that the velocity distribution in the vicinity of the wall is the same as that away from the wall because of the reflection by the wall. The pressure is now related to the momentum flux, i.e., $3P = \rho(\overline{v_x^2} + \overline{v_y^2} + \overline{v_z^2}) = \rho\overline{v^2}$, or

$$\frac{P}{\rho} = \frac{1}{3}\overline{v^2} \tag{4.15}$$

which is Boyle's law. Compared with the ideal gas equation, the right-hand side must be related to temperature. In kinetic theory, temperature is associated to the mean translational kinetic energy of the molecule according to

$$\frac{3}{2}k_B T = \frac{1}{2}m\overline{v^2} = \frac{1}{2}m\overline{v_x^2} + \frac{1}{2}m\overline{v_y^2} + \frac{1}{2}m\overline{v_z^2} \tag{4.16}$$

We have derived the above equations from statistical mechanics in Chap. 3. The temperature defined based on the kinetic energy of the particles is sometimes referred to as the *kinetic temperature.* Combining Eqs. (4.15) and (4.16), we get the ideal gas equation, $P = nk_B T$, as expected. From the above discussion, one can see clearly how the macroscopic properties such as pressure and temperature are related to the particle distribution function. For ideal gases at equilibrium, we have derived the Maxwell velocity and speed distributions in Chap. 3.

**Example 4.1** Show that $P = \rho\overline{v_n^2}$, where $v_n$ is the velocity component normal to the wall, and $P = \rho\overline{v^2}/3$ for equilibrium distribution.

**Solution** Consider the horizontal plane shown in Fig. 4.1 as the wall, where at the bottom is a gas in equilibrium. Multiplying Eq. (4.7) by $m\mathbf{v}$ gives the momentum of the particles with velocities between $\mathbf{v}$ and $\mathbf{v} + \mathrm{d}\mathbf{v}$, impinging on the wall: $m\mathbf{v}\, f(\mathbf{v})\mathbf{v} \cdot \mathbf{n}\, \mathrm{d}A\mathrm{d}t\mathrm{d}\varpi$, which of course is equal to the impulse on the wall: $\mathrm{d}\mathbf{F}\mathrm{d}t$. The normal component $v_n = \mathbf{v} \cdot \mathbf{n} = v\cos\theta$ contributes to an impulse on the wall: $mv_n^2 f(\mathbf{v})\mathrm{d}A\mathrm{d}t\mathrm{d}\varpi$, that is always positive regardless of the sign of $v_n$. However, the contributions of all parallel components cancel out due to isotropy. The pressure can be evaluated by integrating over all velocities, $P = \int_\varpi mv_n^2 f(\mathbf{v})\mathrm{d}\varpi = mn\overline{v_n^2} = \rho\overline{v_n^2}$. We have used the definition of local average given by Eq. (4.5). If the distribution is isotropic, then $P = m\int_0^\infty \int_0^{2\pi} \int_0^\pi f(\mathbf{v})v^4 \cos^2\theta \sin\theta\, \mathrm{d}\theta\mathrm{d}\phi\mathrm{d}v = \frac{4\pi m}{3}\int_0^\infty f(\mathbf{v})v^4\, \mathrm{d}v$. Compared with $\overline{v^2} = \frac{1}{n}\int_0^\infty \int_0^{2\pi} \int_0^\pi f(\mathbf{v})v^4 \sin\theta\, \mathrm{d}\theta\mathrm{d}\phi\mathrm{d}v = \frac{4\pi}{n}\int_0^\infty f(\mathbf{v})v^4\mathrm{d}v$, we obtain $P = \frac{1}{3}mn\overline{v^2} = \frac{1}{3}\rho\overline{v^2}$. The distribution function is uniform inside the container; hence, the wall may be a physical wall or merely an imaginary one since pressure exists everywhere in the fluid.

### 4.1.2 The Mean Free Path

The *mean free path,* defined as the average distance the particle travels between two subsequent collisions, is a very important concept. It is often used to determine whether a given phenomenon belongs to the macroscale (continuum) regime or otherwise falls in the microscale regime when the governing equations derived under the

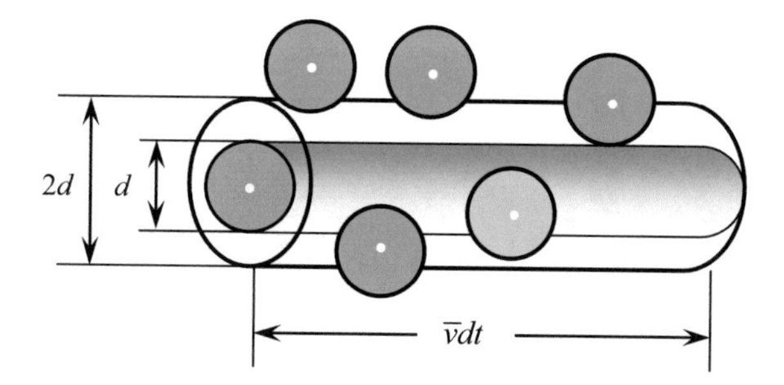


**Fig. 4.2** Schematic used for a simple derivation of the mean free path

assumption of local equilibrium break down. One of the applications is in microfluidics, to be discussed later in this chapter, and another is in the electrical and heat conduction in solids, which will be studied in Chap. 6.

Consider the case shown in Fig. 4.2: a particle of diameter $d$ moving at an average velocity $\bar{v}$ (assuming all other particles are at rest). During a time interval $\mathrm{d}t$, the volume swept by the particle within $d$ from the centerline is $\mathrm{d}V = (\pi d^2)\bar{v}\mathrm{d}t$. The $n\mathrm{d}V$ particles, whose centers are inside this volume element, will collide with the moving particle. Therefore, the frequency of collisions or number of collisions per unit time is $n(\pi d^2)\bar{v}$. The time between two subsequent collisions, $\tau$, is the inverse of the frequency of collision. The mean free path $\Lambda$ is the average distance that a particle travels between two subsequent collisions and is equal to the ratio of the average velocity to the frequency of collision. Therefore,

$$\Lambda = \bar{v}\tau \approx (n\pi d^2)^{-1} \tag{4.17}$$

and depends only on the particle size and the number density. The average time between two subsequent collisions $\tau$ is termed the *relaxation time,*, and the average frequency of collision $\tau^{-1}$ is the *scattering rate* or *collision rate.* The scattering rate is the average number of collisions an individual particle experiences per unit time.

For electrons whose diameters are negligible compared with that of the other particles that scatter them, the mean free path is

$$\Lambda_{\text{electron}}(\text{or } \Lambda_{\text{photon}}) = \frac{1}{nA_{\text{c}}} \tag{4.18}$$

where $A_{\text{c}}$ is the scattering cross-sectional area and $n$ is the number density of the scatter, such as phonons or defects. Equation (4.18) also applies to the case of photons that can be scattered by particles, such as molecules in the atmosphere. The photon mean free path is also called the *radiation penetration depth,* as will be discussed in Chap. 8.

When the relative movement of particles is considered based on the Maxwell velocity distribution, Eq. (4.17) is modified slightly for an ideal gas as follows:

**Table 4.1** Molecular diameter for selected molecules [1]

| Gas type | Molecular weight, $M$ (kg/kmol) | Diameter, $d$ ($10^{-10}$ m, or Å) |
|---|---|---|
| $H_2$ | 2 | 2.74 |
| He | 4 | 2.19 |
| $O_2$ | 32 | 3.64 |
| $N_2$ | 28 | 3.78 |
| Air | 29 | 3.72 |
| $CH_4$ | 16 | 4.14 |
| $NH_3$ | 17 | 4.43 |
| $H_2O$ | 18 | 4.58 |
| $CO_2$ | 44 | 4.64 |

$$\Lambda \approx \frac{1}{\sqrt{2}\pi n d^2} = \frac{k_B T}{\sqrt{2}\pi P d^2} \tag{4.19}$$

The scattering rate, or the collision frequency, is

$$\tau^{-1} = \bar{v}/\Lambda \tag{4.20}$$

Notice that the relaxation time $\tau$ is an important characteristic time. It tells how quickly the system will restore to equilibrium (at least locally), if disturbed. Table 4.1 lists the diameters for some typical molecules.

**Example 4.2** Calculate the mean free path for air at 25 °C and 1 atm. How does it compare with the average spacing between molecules? Find the relaxation time and the number of collisions a molecule experiences per second. What is the speed of sound in air? Explain why we can smell odor far away from its source quickly.

**Solution** $n = P/(k_B T) = 1.0133 \times 10^5/1.381 \times 10^{-23}/298.15 = 2.46 \times 10^{25}\ \text{m}^{-3}$. The average spacing between molecules can be calculated from $L_0 = n^{-1/3} = 3.4$ nm. The mean free path calculated from Eq. (4.19) is $\Lambda = (\sqrt{2}\pi n d^2)^{-1} = 66$ nm (d = 0.37 from Table 4.1), which is about 20 times longer than the molecular spacing. The speed of sound can be calculated from $v_a = \sqrt{\gamma RT} = 345$ m/s using $\gamma = c_p/c_v = 1.4$ and $R = k_B/m$. The average speed is $\bar{v} = \sqrt{8k_B T/\pi m} = 466$ m/s. Therefore, the relaxation time is $\tau = \Lambda/\bar{v} = 0.14$ ns. On the average, each molecule experiences $\tau^{-1}$ or more than 7 billion collisions per second. Although the mean free path is very small, molecules may travel for a long (absolute) distance because of the high average speed. It does not take many molecules for the nose to detect an odor. The odor source usually contains numerous individual molecules.

Let $p(\xi)$ be the probability that a molecule travels at least $\xi$ between collisions. The probability for the particle to collide within an element distance $d\xi$ is $d\xi/\Lambda$. Thus, the probability for a free path greater than $\xi + d\xi$ is less than $p(\xi)$ by the probability of collision between $\xi$ and $\xi + d\xi$. We can write the probability for a molecule to travel at least $\xi + d\xi$ as

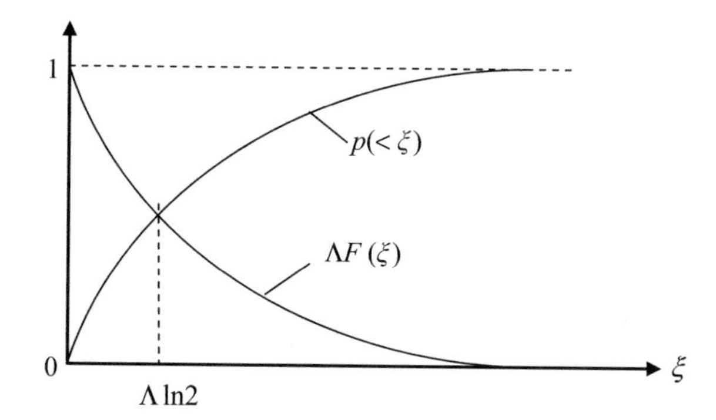


**Fig. 4.3** Free-path distribution functions

$$p(\xi + \mathrm{d}\xi) = p(\xi)\left(1 - \frac{\mathrm{d}\xi}{\Lambda}\right) \tag{4.21}$$

Therefore, $\frac{\mathrm{d}p(\xi)}{p(\xi)} = -\frac{\mathrm{d}\xi}{\Lambda}$. Since $p(0) = 1$, integrating from 0 to $\xi$ yields

$$p(\xi) = \mathrm{e}^{-\xi/\Lambda} \tag{4.22}$$

The probability density function (PDF) for the free path is given by

$$F(\xi) = -\frac{\mathrm{d}p(\xi)}{\mathrm{d}\xi} = \frac{1}{\Lambda}\mathrm{e}^{-\xi/\Lambda} \tag{4.23}$$

One can verify that $\int_0^\infty F(\xi)\mathrm{d}\xi = 1$ and $\bar{\xi} = \int_0^\infty F(\xi)\xi\mathrm{d}\xi = \Lambda$. Therefore, Eq. (4.23) is indeed the free-path PDF. The probability for molecules to have a free path less than $\xi$ is given as

$$p(<\xi) = 1 - p(\xi) = \int_0^\xi F(\xi)\mathrm{d}\xi = 1 - \mathrm{e}^{-\xi/\Lambda} \tag{4.24}$$

Figure 4.3 shows the free-path distribution functions. Equation (4.24) is an exponentially decaying function. In dealing with radiation or photons, the mean free path is called the radiation penetration depth. Radiation will decay exponentially with as the distance increases in an absorbing medium. The fraction of photons that will transmit through a distance equal to the penetration depth is $\mathrm{e}^{-1} \approx 37\%$.

## 4.2 Transport Equations and Properties of Ideal Gases

Consider a molecular gas at steady state but not at equilibrium, with a 1D gradient of some macroscopic properties. Under the assumption of local equilibrium,

$$f(\mathbf{r}, \mathbf{v}, t) = f(\xi, \mathbf{v}) \tag{4.25}$$

where $\xi$ is the coordinate along which the gradient occurs. The *average collision distance* $\Lambda_a$ is defined as the separation of the planes at which particles, on the average, across a plane located at $\xi_0$ will experience the next collision, as shown in Fig. 4.4a. It may be assumed that particles that will cross the plane before the next collision are located in a hemisphere of radius equal to the mean free path $\Lambda$. The problem is how to obtain the average projected length ($\Lambda \cos\theta$) in the $\xi$-coordinate, as shown in Fig. 4.4b. A simple calculation yields

$$\Lambda_a = \frac{\int_{\phi=0}^{2\pi} \int_{\theta=0}^{\pi/2} \Lambda \cos\theta \mathrm{d}A \cos\theta \sin\theta \mathrm{d}\theta \mathrm{d}\phi}{\int_{\phi=0}^{2\pi} \int_{\theta=0}^{\pi/2} \mathrm{d}A \cos\theta \sin\theta \mathrm{d}\theta \mathrm{d}\phi} = \frac{2\pi(\Lambda/3)\mathrm{d}A}{\pi \mathrm{d}A} = \frac{2}{3}\Lambda \tag{4.26}$$

Note that the projected area $\mathrm{d}A \cos\theta$ is used to account for the particle flux. One can consider the free-path distribution and integrate over all free paths. The resulting $\Lambda_a/\Lambda$ is the same [4].

### 4.2.1 *Shear Force and Viscosity*

Consider a gas flowing in the $x$-direction with a velocity gradient in the $y$-direction, as shown in Fig. 4.5. Here, $v_B$ is the *average* or *bulk velocity,* which has a nonzero component only in the $x$-direction. The velocity due to random motion is sometimes

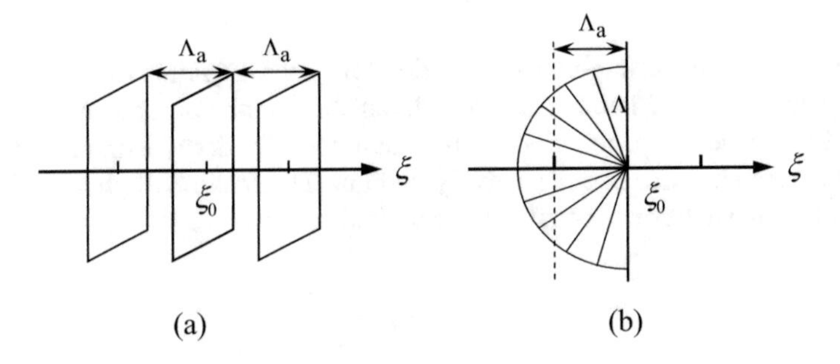


**Fig. 4.4** Illustration of the concepts of **a** average planes of collision and **b** average collision distance $\Lambda_a$, with respect to the mean free path

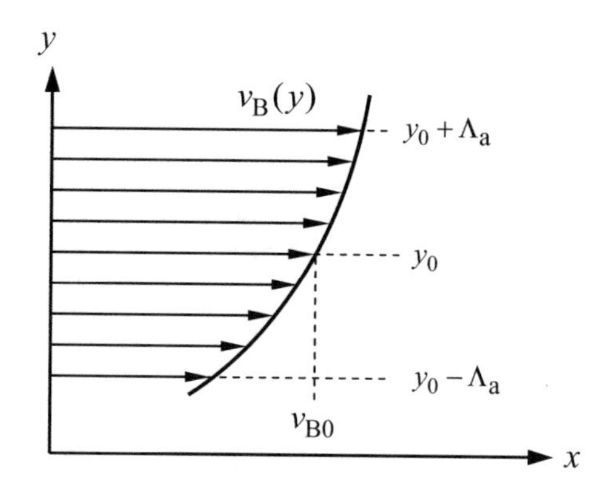


**Fig. 4.5** Schematic of a fluid moving with a bulk velocity $v_B(y)$ that varies in the $y$-direction

called *thermal velocity,* which follows certain equilibrium distribution with an average equal to zero. The fact that the equilibrium distribution is followed everywhere is based on the assumption of local equilibrium. Molecular random motion will cause an exchange of momentum between the upper layer and the lower layer. The net effect is a tendency to accelerate the flow in the upper layer and decelerate the flow in the lower layer. In other words, the flow below the $y = y_0$ plane will exert a shear force to the flow above the $y = y_0$ plane and vice versa. The average momentum of the particles is a function of $y$ only, i.e., $p_x(y) = mv_B(y)$. The momentum flux across the $y = y_0$ plane can be evaluated using the concept of mean planes above and below $y = y_0$. It may be assumed that all the molecules going upward across the $y = y_0$ plane are from the $y = y_0 - \Lambda_a$ plane. Therefore, the momentum flux in the positive $y$-direction is

$$J_p^+ = \frac{n\bar{v}}{4} m \left( v_{B0} - \Lambda_a \left. \frac{\mathrm{d}v_B}{\mathrm{d}y} \right|_{y_0} \right) \tag{4.27a}$$

where $n\bar{v}/4$ is the molecular flux, with $\bar{v}$ as the average speed without considering bulk motion, and $v_{B0} = v_B(y_0)$. Similarly, the momentum flux downward is

$$J_p^- = \frac{n\bar{v}}{4} m \left( v_{B0} + \Lambda_a \left. \frac{\mathrm{d}v_B}{\mathrm{d}y} \right|_{y_0} \right) \tag{4.27b}$$

The net momentum flux, which is equal to the shear force $P_{yx} = \tau_{yx} = J_p^+ - J_p^-$, is therefore

$$\tau_{yx} = -\frac{1}{3} \rho \bar{v} \Lambda \left. \frac{\mathrm{d}v_B}{\mathrm{d}y} \right|_{y_0} \tag{4.28}$$

Comparing Eq. (4.28) with Newton's law of shear stress in Eq. (2.36), $\tau_{yx} = -\mu \left. \frac{\mathrm{d}v_B}{\mathrm{d}y} \right|_{y_0}$, dynamic viscosity is obtained from the simple kinetic theory as

$$\mu = \frac{1}{3}\rho\bar{v}\Lambda \tag{4.29}$$

The above equation provides an order-of-magnitude estimate. While the density is proportional to pressure, the mean free path is inversely proportional to the number density, or density. The average velocity is a function of temperature only. Therefore, the viscosity depends only on temperature and the type of molecules, but not on pressure. The result from more detailed calculations and experiments suggests that Eq. (4.29) be multiplied by 3/2 to give

$$\mu = \frac{1}{2}\rho\bar{v}\Lambda = \frac{m\bar{v}}{2\sqrt{2}\pi d^2} = \frac{1}{\pi d^2}\sqrt{\frac{mk_B T}{\pi}} \tag{4.30}$$

Equation (4.30) is recommended for use in the exercises to estimate the viscosity. It should be noted that the above discussion is based on the simple ideal gas model that each molecule is a rigid (or hard) sphere and all collisions are elastic. Additional modifications have been made to correctly account for the temperature dependence. These models will not be discussed here, and interested readers can find them in the literature [1–8].

## 4.2.2 *Heat Diffusion*

Heat conduction is due to the temperature gradient inside the medium. In an ideal molecular gas, the random motion of molecules transports thermal energy from place to place. Sometimes, we call the particles that are responsible for thermal energy transport *heat carriers*. Similar to the argument for momentum transfer, it is straightforward to illustrate heat diffusion in a 1D temperature gradient system at steady state and under local equilibrium, using Fig. 4.6. The net energy flux across the $x = x_0$ plane is given by

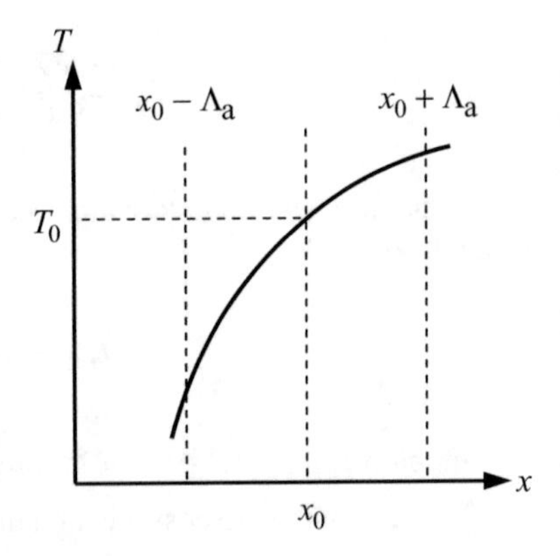


**Fig. 4.6** One-dimensional heat diffusion

$$J_E = J_E^+ - J_E^- = \frac{n\bar{v}}{4}[\bar{\varepsilon}(x_0 - \Lambda_a) - \bar{\varepsilon}(x_0 + \Lambda_a)] = -\frac{1}{3}n\bar{v}\Lambda \left.\frac{d\bar{\varepsilon}}{dx}\right|_{x_0} \tag{4.31}$$

where $\bar{\varepsilon}$ is the average thermal energy per molecule and, hence, is a function of temperature. Based on the definition of specific heat, $n \left.\frac{d\bar{\varepsilon}}{dx}\right|_{x_0} = n \left.\frac{d\bar{\varepsilon}}{dT}\frac{dT}{dx}\right|_{x_0} = nmc_v \left.\frac{dT}{dx}\right|_{x_0}$. The heat flux is related to the temperature gradient as

$$q_x'' = J_E = -\frac{1}{3}\rho c_v \bar{v}\Lambda \left.\frac{dT}{dx}\right|_{x_0} \tag{4.32}$$

which is (1D) Fourier's law with the thermal conductivity given as

$$\kappa = \frac{1}{3}\rho c_v \Lambda \bar{v} \tag{4.33}$$

Because $c_v$ and $\bar{v}$ are functions of temperature only and $\Lambda$ is inversely proportional to $\rho$, the thermal conductivity of a given ideal gas is a function of temperature and independent of pressure.

Comparing Eq. (4.30) with Eq. (4.33), we have $\kappa = 0.667\mu c_v$. The calculated results are consistently lower than the tabulated values for real gases. The reason is the assumption that the average collision distance is the same for both momentum transport and energy transfer. Generally speaking, molecules with a larger speed travel farther than those with a smaller speed. Once the molecules pass the mean plane, they will persist a little while before collision. The persistence effect is larger for energy transfer because the translational kinetic energy of a molecule is proportional to the square of the speed, while that of momentum is proportional to the velocity components. In gases, the average collision distance is greater for energy transfer and depends on the type of gas. Extensive studies of the similarity between $\mu$ and $\kappa$ have resulted in a more accurate expression for calculating the thermal conductivity of ideal gases than the one given in Eq. (4.33). Eucken's formula relates the Prandtl number $Pr \equiv \frac{\nu}{\alpha} = \frac{c_p\mu}{\kappa}$ to the specific heat ratio as follows [7]:

$$Pr = \frac{4\gamma}{9\gamma - 5} \tag{4.34}$$

Based on Eucken's formula, the following equation is recommended to replace Eq. (4.33) in predicting the thermal conductivity of ideal gases:

$$\kappa = \frac{9\gamma - 5}{4}c_v\mu \tag{4.35}$$

where $\mu$ can be calculated from Eq. (4.30). For a monatomic gas, $\gamma = 5/3$ and

$$\kappa = 2.5\mu c_v = 1.25\rho c_v \Lambda \bar{v} \tag{4.36a}$$

For a diatomic gas at intermediate temperatures when the translational and rotational modes are fully excited but no vibrational modes have been excited, we have $\gamma = 1.4$ and

$$\kappa = 1.9\mu c_v = 0.95\rho c_v \Lambda \bar{v} \tag{4.36b}$$

The results calculated from Eq. (4.35) agree reasonably well with the tabulated thermal conductivity values of typical gases. Additional corrections are required when the temperature deviates significantly from the room temperature. More complicated formulations are needed to better account for the temperature dependence [5–8].

**Example 4.3** Calculate the viscosity and the thermal conductivity of air at 300 K and 100 kPa. How will your answers change if the temperature is increased to 306 K and the pressure is decreased to 50 kPa?

**Solution** From Eq. (4.30), we have $\mu = \frac{1}{\pi d^2}\sqrt{\frac{mk_BT}{\pi}} = \frac{1}{\pi(3.72\times10^{-10})^2}\left(\frac{29}{6.022\times10^{26}}\cdot\frac{1.381\times10^{-23}\times300}{\pi}\right)^{1/2} = 1.83\times10^{-5}\ \text{N}\cdot\text{s/m}^2$. It is within 1% of the measured value. From Eq. (4.36b) and $c_v = R/(\gamma - 1) = 716.6$ J/kg K, $\kappa = 1.9\mu c_v = 0.025$ W/m K. This is within 5% of the measured value. Notice that $\mu$ and $\kappa$ depend on temperature only. If the change in specific heat is neglected, then $\kappa \propto \mu \propto \sqrt{T}$. When the temperature is increased by 2%, both $\kappa$ and $\mu$ will increase by 1%.

### 4.2.3 *Mass Diffusion*

Consider a small duct linking two gas tanks containing different types of ideal gases at the same temperature and pressure, as shown in Fig. 4.7a. The total number density of the mixture in the system is conserved such that $n = n_A + n_B$, as illustrated in Fig. 4.7b. Therefore,

$$\frac{dn_A}{dx} = -\frac{dn_B}{dx} \tag{4.37}$$

Fick's law states that

$$J_{N,A} = -D_{AB}\frac{dn_A}{dx} \tag{4.38a}$$

or

$$J_{m,A} = -D_{AB}\frac{d\rho_A}{dx} \tag{4.38b}$$

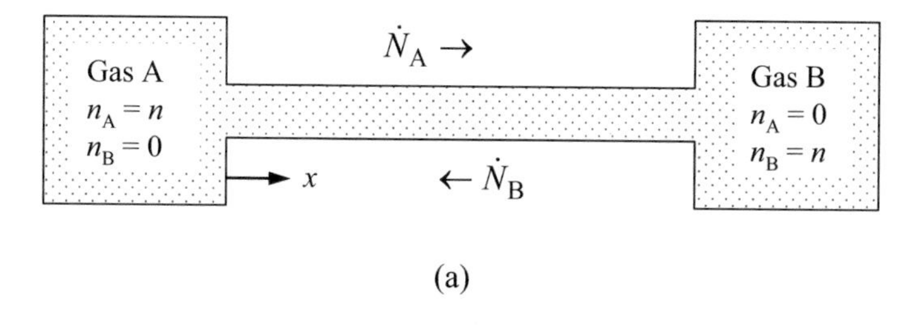


(a)

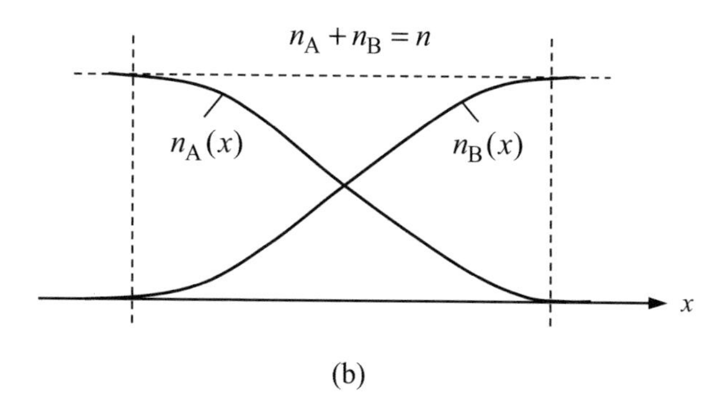


(b)

**Fig. 4.7** Schematic of binary diffusion between ideal gases: **a** Two reservoirs of different types of gas molecules connected through a duct. **b** Concentration distributions in terms of the number densities

where $D_{AB}$ in m$^2$/s is called the *binary diffusion coefficient* or *diffusion coefficient* between A and B. Notice that the molecular transfer rate $\dot{N}_A = J_{N,A}A_c$, where $A_c$ is the cross-sectional area, and the mass transfer rate $\dot{m}_A = J_{m,A}A_c$. Similarly, we can write Fick's law for type B molecules as

$$J_{N,B} = -D_{BA}\frac{dn_B}{dx} \tag{4.39a}$$

or

$$J_{m,B} = -D_{BA}\frac{d\rho_B}{dx} \tag{4.39b}$$

Because the flux of type B molecules must balance that of type A molecules to maintain a uniform pressure, we have

$$J_{N,A} = -J_{N,B} \tag{4.40}$$

Equations (4.37) through (4.40) imply that $D_{BA} = D_{AB}$.

Using the microscopic descriptions of mass diffusion, one can write the positive and negative flux at a certain location $x_0$ using the average distance concept discussed earlier. Hence,

$$J_{N,\mathrm{A}} = \frac{\bar{v}}{4}\left[n_\mathrm{A}(x_0) - \Lambda_\mathrm{a}\left.\frac{\mathrm{d}n_\mathrm{A}}{\mathrm{d}x}\right|_{x_0}\right] - \frac{\bar{v}}{4}\left[n_\mathrm{A}(x_0) + \Lambda_\mathrm{a}\left.\frac{\mathrm{d}n_\mathrm{A}}{\mathrm{d}x}\right|_{x_0}\right]$$

The result is

$$J_{N,\mathrm{A}} = \frac{1}{3}\Lambda\bar{v}\left.\frac{\mathrm{d}n_\mathrm{A}}{\mathrm{d}x}\right|_{x_0} \tag{4.41}$$

Comparing Eq. (4.41) with Eq. (4.38a), we have

$$D_\mathrm{AB} = \frac{1}{3}\Lambda\bar{v} \tag{4.42}$$

In the case of similar molecules (such as isotopes), $D_{\mathrm{AA'}} = \frac{2}{3}\frac{1}{n\pi d^2}\sqrt{\frac{k_\mathrm{B}T}{\pi m}}$, which is often called the self-diffusion coefficient. The calculation for the mean free path and the average velocity for a mixture of dissimilar molecules is certainly more involved. However, a simple expression can be obtained using the central distance $\bar{d} = (d_\mathrm{A} + d_\mathrm{B})/2$ and the reduced mass $\bar{m} = m_\mathrm{A}m_\mathrm{B}/(m_\mathrm{A} + m_\mathrm{B})$; that is

$$D_\mathrm{AB} = \frac{3}{8}\frac{1}{n\bar{d}^2}\sqrt{\frac{k_\mathrm{B}T}{2\pi\bar{m}}} \tag{4.43}$$

Equation (4.43) is recommended for calculation of the binary diffusion coefficient. Recall that the Schmidt number is the ratio of the momentum diffusivity to the mass diffusivity, i.e.,

$$Sc \equiv \frac{\nu}{D_\mathrm{AB}} \tag{4.44}$$

The Lewis number is defined as the ratio of the mass diffusivity to the thermal diffusivity as follows:

$$Le \equiv \frac{D_\mathrm{AB}}{\alpha} = \frac{Pr}{Sc} \tag{4.45}$$

Heat and mass transfer analogy provides a convenient way to calculate convective mass transfer in a boundary layer. The mass transfer rate is related to the convective mass transfer coefficient $h_m$ by

$$\dot{m}_\mathrm{B} = h_m A_\mathrm{s}(\rho_{\mathrm{B,s}} - \rho_{\mathrm{B},\infty}) \tag{4.46}$$

where $A_\mathrm{s}$ is the surface area, $\rho_\mathrm{B}$ is the density of species B, and subscripts s and $\infty$ signify that the quantity is at the surface and in the free stream, respectively. Heat and mass transfer analogy gives

$$h_m = \frac{h}{\rho c_p} Le^{-2/3} \tag{4.47}$$

Equations (4.46) and (4.47) are very useful for calculating the heat transfer during evaporation demonstrated in the following example.

**Example 4.4** Dry air at 30 °C flows at a speed of 2 m/s over a flat plate, with an area of $3 \times 3$ m$^2$, which is maintained at 24 °C. A thin layer of water is formed on the top surface where convection occurs. Determine the heat transfer rate from the plate to the air. For water at 24 °C, the saturation pressure $P_{\text{sat}} = 3$ kPa and the latent heat of evaporation $h_{\text{fg}} = 2445$ kJ/kg.

**Solution** Neglect the temperature gradient inside the water layer and radiative heat transfer. We first evaluate air properties at 300 K and 100 kPa, as in Examples 4.2 and 4.3. The results are $\rho = P/RT = 1.163$ kg/m$^3$, $\mu = 1.83 \times 10^{-5}$ N s/m$^2$, $c_p = 1003$ J/kg K, $Pr = 0.737$, and $\kappa = 0.025$ W/m K. Hence, $Re_L = 3.8 \times 10^5$. From Eq. (2.40), $\bar{h} = 0.664 \frac{\kappa}{L} Re_L^{1/2} Pr^{1/3} = 3.08$ W/m$^2 \cdot$ K and $\dot{Q}_{\text{conv}} = hA_s(T_s - T_\infty) = -166.2$ W. The negative sign indicates that the convection heat transfer is from the air to the surface.

To calculate the mass transfer rate, we assume that $\rho_{\text{B},\infty} = 0$ (dry air) and $\rho_{\text{B,s}} = P_{\text{sat}} M_{\text{H}_2\text{O}}/(\bar{R}T_s) = 3000 \times 18/(8314 \times 297) = 0.022$ kg/m$^3$ (saturated water vapor). Using Eq. (4.43) with $T = 300$ K, we can estimate the binary diffusion coefficient between air and water to be $D_{\text{AB}} = 1.7 \times 10^{-5}$ m$^2$/s, which is about two-thirds of the measured value: $D_{\text{AB}} = 2.56 \times 10^{-5}$ m$^2$/s. Considering the simplifications made in deriving the diffusion coefficient, the agreement is reasonable. Using the measured $D_{\text{AB}}$, we find $Le = D_{\text{AB}}/\alpha = 1.2$ and $h_m = 0.00234$ m/s from Eq. (4.47). The mass transfer rate $\dot{m}_{\text{B}} = h_m A_s \rho_{\text{B,s}} = 0.46$ g/s, and the heat transfer rate by evaporation is $\dot{Q}_{\text{evap}} = \dot{m}_{\text{B}} h_{\text{fg}} = 1124.7$ W. The total heat transfer rate is the sum of evaporation and convection, i.e., $\dot{Q}_{\text{evap}} + \dot{Q}_{\text{conv}} = 1124.7 - 166.2 = 958.5$ W. This example suggests that evaporative cooling is an important mechanism of heat transfer at wetted surfaces.

## 4.3 Intermolecular Forces

Although the mean-free-path method is simple and can predict the temperature and pressure dependence of the transport coefficient correctly, the rigid-elastic-sphere model does not represent the actual collision process. Collision between molecules does not necessarily occur by contact, as in the case with billiard balls. It is a force field described by the intermolecular potential that governs the collision process between molecules, since the force is the gradient of the potential function. It should be noted that electromagnetic forces are responsible for all intermolecular interactions.

*Intermolecular forces* are also responsible for the deviation from ideal gas model as well as the existence of different phases (solid, liquid, or gas) of a substance.

Molecules are held together by chemical bonds. The forces between atoms within a molecule are called *intramolecular forces*. Typical examples are ionic bonds and covalence bonds. In an ionic compound (such as CsI or MgO), valence electrons are transferred from the metal to the nonmetal. The positively charged ion (cation) and negative charged ion (anion) attract each other by electrostatic forces. Covalence bonds are usually formed between nonmetals where the valence electrons are shared between the atoms, such as $N_2$, $Cl_2$, SiC, etc. When the electronegativity of the two atoms is different, a covalent bond is said to be a polar covalence bond since there is a dipole moment pointing from the partial positive side to the partial negative side. In polar molecules, such as SiC or $SiO_2$, the electrons are neither completely transferred from one atom to another nor evenly shared as in homonuclear diatomic molecules, such as $N_2$ or $Cl_2$. Note that $CH_4$, $CO_2$, and $C_2H_4$ are nonpolar molecules due to structural symmetry. It should be mentioned that while all ionic compounds are inherently polar, they are not usually referred to as polar molecules.

Repulsive forces also exist between atoms and molecules when the separation distance becomes very small, on the order of 0.1 nm, due to the overlap of electronic orbits (or electron clouds) when the atoms get very close to each other. While the internuclear forces can be predicted by the quantum mechanical theory, it is difficult to find an explicit expression [9]. Generally speaking, the repulsive forces are short-ranged and increases sharply as the two atoms or molecules approach each other to a distance below 0.5 nm. The repulsive potentials are typically modeled empirically. A common form is the power-law potential, given as

$$\phi(r) = (\sigma/r)^n \tag{4.48}$$

where $\sigma$ is a constant, $r$ is the center-to-center distance between the atoms or molecules, and $n$ is an integer that is typically from 9 and 16. Note that Eq. (4.48) becomes the hard sphere potential when $n \to \infty$, since the potential is infinite for $r < \sigma$ and zero for $r > \sigma$. This describes the rigid sphere assumption in which the force is zero until the two atoms or molecules having a diameter $\sigma$ touch each other. Exponential potentials with the form $\phi(r) = ae^{-r/\sigma_0}$, where $a$ and $\sigma_0$ are constants, have also been used to model the repulsive force [9]. Note that Eq. (4.48) can be used to model both intramolecular and intermolecular repulsive forces.

### 4.3.1 Intermolecular Attractive Forces

Attractive forces between molecules usually are weaker and have a longer range than those between atoms inside the molecule. They can be grouped into ion–dipole forces, hydrogen bonds, and van der Waals forces that may be subdivided into three categories. The details are given in the following.

The ion–dipole force is an electrostatic attraction between a charged ion and a dipole. A cation attracts the partially negative end of a polar molecule, while an anion attracts the partially positive end of a polar molecule. Ion–dipole intermolecular forces are important in many solids, liquids, and solutions, especially for ionic compounds in polar liquids. According to Coulomb's law, the force and potential between two charges scale with $1/r^2$ and $1/r$, respectively. For a charge and a fixed dipole, the attractive force scales with $1/r^3$, while the potential scales with $1/r^2$. Furthermore, for a charge and a freely rotating dipole, the force scales with $1/r^5$, while the potential scales with $1/r^4$ [9].

The hydrogen bond may be considered as a special type of dipole–dipole attractive force that is weaker than the ion–dipole force but stronger than ordinary dipole–dipole forces. Hydrogen bonding occurs when two conditions are met: (1) the hydrogen atom is bonded to a highly electronegative atom (such as O, N, and F); (2); there exists another electronegative atom with a lone pair of electrons in the vicinity of the hydrogen atom. In essence, the hydrogen atom in a hydrogen bond is shared by two electronegative atoms. Hydrogen bonds may occur within a molecule (in some organic compounds) or between molecules. In a water molecule, however, the intramolecular bonds between hydrogen and oxygen atoms are covalence bonding with an interatomic distance of about 0.1 nm. In the liquid water, the distance between O and H atoms in adjacent molecules is approximately 0.176 nm, which is a hydrogen bond [9]. Hydrogen bonding is also responsible for the hydrophobic effect. There is no simple equation for the interaction potential for hydrogen bonding, typically, the potential scales with $1/r^2$ as for charge-fixed dipole interaction. The strength of most hydrogen bonds, measured by the bonding energy, is more than an order of magnitude than a typical van der Waals bond.

In 1873, Dutch theoretical physicist Johannes D. van der Waals studied real gas behavior and attributed the non-ideality to intermolecular interactions. Furthermore, he developed an equation of state named after him, i.e., the van der Waals equation, which applies to both gases and liquid. This work won him the Nobel Prize in Physics in 1910. Generally speaking, van der Waals forces are weak attractive forces between molecules, and the origin can be categorized into three groups: dipole–dipole interactions, dipole–induced dipole interactions, and dispersion forces.

The attractive force between permanent dipoles is due to the Keesom interaction, named after Willem H. Keesom, who developed the first mathematical description of dipole–dipole interactions in 1921. He received a doctoral degree from the University of Amsterdam under van der Waals and worked with Kamerlingh Onnes at the University of Leiden, where he made important contributions to cryogenics and helium liquefaction in the early twentieth century. Keesom's results of dipole–dipole interactions show that the intermolecular potential varies with the inverse sixth power of the distance between the molecules. Examples of Keesom interactions are two HCl molecules, two CO molecules, or an HCl and a CO molecule.

When an ionic molecule or polar molecule is placed near a nonpolar molecule, the molecule with a permanent dipole can induce a dipole in the neighboring molecule to cause mutual attraction. The polarizability of atoms and molecules that arises from such electronic displacements is known as the electronic polarizability. This type of force is called the Debye force, named after the Dutch-American physicist and physical chemist Peter J. W. Debye (a Nobel laureate in Chemistry in 1936). The Debye interaction or the induction interaction also gives an attractive potential that scales with the inverse sixth power of the distance [9].

There exists another type of attractive force between all atoms and molecules, even totally neutral ones such as nitrogen ($N_2$), methane ($CH_4$), and helium (He). These forces are called the London dispersion forces or dispersion forces, which are long-range forces and can have an effect up to 10 nm. They are always present regardless of the type of molecules and play an important role in many phenomena not only between atoms and molecules but also between surfaces and small particles. The physical origin lies in the random fluctuation of electron density or electron clouds that give rise to temporal dipoles or instantaneous dipoles. These fluctuating dipoles can induce other dipoles in the neighborhood. The London dispersion force is named after the German-American physicist Fritz London, who used a quantum mechanical theory, called the second-order perturbation theory, to provide the first explanation of the forces between noble gas atoms in 1930. London derived the interaction potential between two dissimilar atoms as

$$\phi_{\text{disp}}(r) \propto \frac{I_1 I_2}{I_1 + I_2} \frac{\alpha_1 \alpha_2}{r^6} \tag{4.49}$$

where $I_1$ and $I_2$ are the first ionization potentials of the two atoms, respectively, and $\alpha_1$ and $\alpha_2$ are the electronic polarizabilities of the respective atoms. Depending on the polarizability, dispersion forces can be as large as or even greater than the dipole–dipole force. It is worth emphasizing that the potential of all three types of interactions that contribute to the total van der Waals interaction vary with $1/r^6$.

### 4.3.2 *Total Intermolecular Pair Potentials*

As discussed previously, the van der Waals forces due to dipole–dipole, dipole–induced dipole, and fluctuating dipole–induced dipole interactions between two molecules generally vary with $1/r^7$ for sufficiently large $r$, where $r$ is the center-to-center distance between the two molecules. The combination of the attractive and repulsive potentials described previously leads to different semi-empirical functions for the total intermolecular pair potential. There exist a large number of intermolecular potentials with varying complexity for atomistic simulations [10]. One of the most common pair potentials is the Lennard-Jones <6, 12> potential, which was first proposed in 1924 by British mathematician and physical chemist John E. Lennard-Jones. The Lennard-Jones potential can be expressed in the following:

$$\phi_{ij}(r_{ij}) = -4\varepsilon_0 \left[ \left( \frac{r_0}{r_{ij}} \right)^6 - \left( \frac{r_0}{r_{ij}} \right)^{12} \right] \tag{4.50}$$

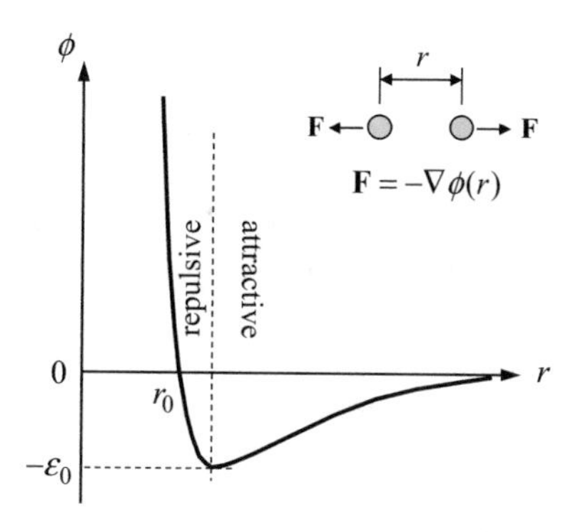


**Fig. 4.8** Illustration of the intermolecular potential $\phi(r)$ as a function of the distance $r$ between two molecules. The subscripts $i$ and $j$ used in Eqs. (4.50) and (4.51) are dropped for simplicity

where $\phi_{ij}$ is the intermolecular potential, $r_{ij}$ is the center-to-center distance between the $i$th and $j$th particles, $\varepsilon_0$ is a constant, and $r_0$ is a characteristic length. Notice that the potential has a minimum $\phi_{ij} = -\varepsilon_0$ at $r_{ij} \approx 1.12r_0$, where the attractive and repulsive forces balance each other. For typical gas molecules, $r_0$ ranges from 0.25 to 0.4 nm. The potential function is illustrated in Fig. 4.8. The force between the molecules is the negative gradient of the potential, i.e.,

$$\mathbf{F}_{ij} = -\nabla\phi_{ij} = \frac{24\varepsilon_0}{r_0}\left[2\left(\frac{r_0}{r_{ij}}\right)^{13} - \left(\frac{r_0}{r_{ij}}\right)^{7}\right]\frac{\mathbf{r}_{ij}}{r_{ij}} \tag{4.51}$$

The combination of Eq. (4.51) with Eq. (3.1) allows a computer simulation of the trajectory of each molecule when the initial position and velocity are prescribed. Although molecular dynamics is a powerful tool for dense phases and for the study of phase change problems, it is not very effective in dealing with dilute gases. The direct simulation Monte Carlo (DSMC) method is an alternative to the deterministic method and has been used extensively in gas dynamics. Additional discussions about these numerical techniques will be given in Sect. 4.4 on microfluidics. In the next section, a more sophisticated kinetic theory based on the Boltzmann transport equation will be presented.

## 4.4 The Boltzmann Transport Equation

In addition to the rigid sphere assumption, the simple kinetic theory is based on local equilibrium and cannot be used to study nonequilibrium processes that happen at a timescale much less than the relaxation time or at a length scale less than the mean free path. The Boltzmann transport equation (BTE) is the basis of classical transport theories of molecular and atomic systems. It is not limited to local equilibrium and can be applied to small length scales and small timescales. The equation formulated by Ludwig Boltzmann in his original investigation of the dynamics of gases over 130 years ago has been extended to the study of electron and phonon transport in solids, as well as radiative transfer in gases. Macroscopic conservation and rate equations can be derived from the BTE under appropriate assumptions. A brief introduction of the BTE is given in this section. More detailed coverage of the history, formulation, and solution techniques of the BTE can be found from Chapman and Cowling [6], Tien and Lienhard [7], and Cercignani [8].

Suppose at time $t$, a particle at the spatial location $\mathbf{r}$ moves with a velocity $\mathbf{v}$. At $t + \mathrm{d}t$, without collision, the particle will move to $\mathbf{r} + d\mathbf{r} = \mathbf{r} + \mathbf{v}dt$ and its velocity becomes $\mathbf{v} + d\mathbf{v} = \mathbf{v} + \mathbf{a}dt$. Here, $\mathbf{a} = \mathbf{F}/m$ is the acceleration in a body force field. Therefore, in the absence of collision, the probability of finding a particle in the phase space does not change with time. Therefore,

$$\frac{f(\mathbf{r} + \mathbf{v}\mathrm{d}t, \mathbf{v} + \mathbf{a}\mathrm{d}t, t + \mathrm{d}t) - f(\mathbf{r}, \mathbf{v}, t)}{\mathrm{d}t} = \frac{\partial f}{\partial t} + \mathbf{v} \cdot \frac{\partial f}{\partial \mathbf{r}} + \mathbf{a} \cdot \frac{\partial f}{\partial \mathbf{v}} = 0 \quad (4.52)$$

where $\frac{\partial f}{\partial \mathbf{r}} = \nabla f = \left( \frac{\partial f}{\partial x} \; \frac{\partial f}{\partial y} \; \frac{\partial f}{\partial z} \right) \begin{pmatrix} \hat{\mathbf{x}} \\ \hat{\mathbf{y}} \\ \hat{\mathbf{z}} \end{pmatrix}$ is the gradient, and $\frac{\partial f}{\partial \mathbf{v}} = \nabla_{\mathbf{v}} f = \left( \frac{\partial f}{\partial v_x} \; \frac{\partial f}{\partial v_y} \; \frac{\partial f}{\partial v_z} \right) \begin{pmatrix} \hat{\mathbf{x}} \\ \hat{\mathbf{y}} \\ \hat{\mathbf{z}} \end{pmatrix}$ can be considered as the gradient defined in the velocity space.

Equation (4.52) is the Liouville equation in classical mechanics. In the absence of both body force and collision, the substantial derivative of the distribution function is

$$\frac{\mathrm{D}f}{\mathrm{D}t} \equiv \frac{\partial f}{\partial t} + \mathbf{v} \cdot \frac{\partial f}{\partial \mathbf{r}} = 0 \quad (4.53)$$

Generally speaking, particles in random motion collide with each other at very high frequencies unless the density is extremely low. A major advance in the kinetic theory of gases is the introduction of the collision term proposed by Boltzmann in the 1870s. The BTE can be written as

$$\frac{\partial f}{\partial t} + \mathbf{v} \cdot \frac{\partial f}{\partial \mathbf{r}} + \mathbf{a} \cdot \frac{\partial f}{\partial \mathbf{v}} = \left[ \frac{\partial f}{\partial t} \right]_{\mathrm{coll}} \quad (4.54)$$

where the collision term can be separated into a source term and a sink term such that

$$\left[ \frac{\partial f}{\partial t} \right]_{\mathrm{coll}} = \Gamma_+ - \Gamma_- = \sum_{\mathbf{v}'} \left[ W(\mathbf{v}, \mathbf{v}') f(r, \mathbf{v}', t) - W(\mathbf{v}', \mathbf{v}) f(r, \mathbf{v}, t) \right] \quad (4.55)$$

Here, $W(\mathbf{v}, \mathbf{v}')$ is called the scattering probability, which can be understood as the fraction of particles with a velocity $\mathbf{v}'$ that will change their velocity to $\mathbf{v}$ per unit time due to collision. The function $W$ depends on the nature of the scatters and is usually a complicated nonlinear function of the velocities.

The BTE is a nonlinear integro-differential equation that cannot be solved exactly. Approximations are usually used to facilitate the solution for given applications. The *relaxation time approximation* provides an easier way to solve the BTE under conditions not too far away from the equilibrium. It gives a linear collision term:

$$\left[\frac{\partial f}{\partial t}\right]_{\text{coll}} = \frac{f_0 - f}{\tau(\mathbf{v})} \tag{4.56}$$

where $f_0$ is the equilibrium distribution and the relaxation time $\tau$ is often treated as independent of the velocity. The solution of Eq. (4.56) gives $f(t) - f_0 = [f(t_1) - f_0]\exp[-(t - t_1)/\tau]$ for $t \geq t_1$, where $t_1$ is the initial time when the system deviates somewhat from the equilibrium. This suggests that an equilibrium will be reached at a timescale $\Delta t = t - t_1$ on the order of $\tau$. Furthermore, it is collision that restores a system from a nonequilibrium state to an equilibrium state. David Enskog proposed a successive approximation method to include higher-order scattering terms by introducing a small perturbation to the equilibrium distribution. This is the well-known Chapman–Enskog method [6–8].

### 4.4.1 Hydrodynamic Equations

The continuity, momentum, and energy equations can be derived from the BTE. Multiplying the BTE by a molecular quantity $\psi$ and integrating it over all velocities, we have

$$\int_{\varpi} \psi \frac{\partial f}{\partial t} \mathrm{d}\varpi + \int_{\varpi} \psi \mathbf{v} \cdot \frac{\partial f}{\partial \mathbf{r}} \mathrm{d}\varpi + \int_{\varpi} \psi \mathbf{a} \cdot \frac{\partial f}{\partial \mathbf{v}} \mathrm{d}\varpi = \int_{\varpi} \psi (\Gamma_+ - \Gamma_-) \mathrm{d}\varpi \tag{4.57}$$

Using the definition of local average $\bar{\psi} = \frac{1}{n} \int_{\varpi} f\,\psi\,\mathrm{d}\varpi$ from Eq. (4.5), the first term in the above equation becomes

$$\int_{\varpi} \psi \frac{\partial f}{\partial t} \mathrm{d}\varpi = \frac{\partial}{\partial t} \int_{\varpi} \psi f \mathrm{d}\varpi - \int_{\varpi} f \frac{\partial \psi}{\partial t} \mathrm{d}\varpi = \frac{\partial (n\bar{\psi})}{\partial t} - n\overline{\frac{\partial \psi}{\partial t}} \tag{4.58a}$$

Note that $\nabla \cdot (\psi \mathbf{v}) = \mathbf{v} \cdot \nabla \psi + \psi \nabla \cdot \mathbf{v} = \mathbf{v} \cdot \nabla \psi$ since the velocity components are independent variables in the phase space. For the second term, we have

$$\int_{\varpi} \psi \mathbf{v} \cdot \nabla f \mathrm{d}\varpi = \nabla \cdot \int_{\varpi} \psi \mathbf{v} f \mathrm{d}\varpi - \int_{\varpi} f \nabla \cdot (\psi \mathbf{v}) \mathrm{d}\varpi = \nabla \cdot (n\overline{\psi \mathbf{v}}) - n\overline{\mathbf{v} \cdot \nabla \psi} \tag{4.58b}$$

The third term is

$$\int_{\varpi} \psi \mathbf{a} \cdot \frac{\partial f}{\partial \mathbf{v}} \mathrm{d}\varpi = \mathbf{a} \cdot \left[ (\psi f)\Big|_{v_x,v_y,v_z=-\infty}^{v_x,v_y,v_z=\infty} - \int_{\varpi} f \frac{\partial \psi}{\partial \mathbf{v}} \mathrm{d}\varpi \right] = -n\mathbf{a} \cdot \overline{\frac{\partial \psi}{\partial \mathbf{v}}} \tag{4.58c}$$

Substituting Eq. (4.58) into Eq. (4.57), we obtain

$$\frac{\partial}{\partial t}(n\bar{\psi}) + \nabla \cdot (n\overline{\psi \mathbf{v}}) - n\left( \overline{\frac{\partial \psi}{\partial t}} + \overline{\mathbf{v} \cdot \nabla \psi} + \mathbf{a} \cdot \overline{\frac{\partial \psi}{\partial \mathbf{v}}} \right) = \Phi_+ - \Phi_- \tag{4.59}$$

where the right-hand side contains a source term and a sink term. When $\psi$ is proportional to the velocity to the $j$th power ($j = 0, 1, 2$), or the $j$th moment, the source and sink terms in Eq. (4.59) cancel out when reaction is not considered, and the gas particles can be treated as rigid spheres.

We can substitute $\psi = m$, the zeroth moment, into Eq. (4.59) to get the mass balance as

$$\frac{\partial \rho}{\partial t} + \nabla \cdot (\rho \mathbf{v}_\mathrm{B}) = 0 \quad \text{or} \quad \frac{\mathrm{D}\rho}{\mathrm{D}t} + \rho \nabla \cdot \mathbf{v}_\mathrm{B} = 0 \tag{4.60}$$

where $\mathbf{v}_\mathrm{B} = \bar{\mathbf{v}}$ is the *bulk velocity.* This is exactly the same as Eq. (2.41). One can extend the above derivation to a system of multiple gas species involving chemical reaction. For the $i$th species, it can be shown that

$$\frac{\mathrm{D}\rho_i}{\mathrm{D}t} + \rho_i \nabla \cdot \mathbf{v}_{i,\mathrm{B}} = \Phi_{i,\mathrm{net}} \tag{4.61}$$

where $\Phi_{i,\mathrm{net}}$ represents the net rate of creation due to reaction.

To derive the momentum equation, substitute the first moment $\psi = m\mathbf{v}$ into Eq. (4.59). The first term becomes $\partial(\rho \mathbf{v}_\mathrm{B})/\partial t$. The second term is more complicated. We can separate the velocity as $\mathbf{v} = \mathbf{v}_\mathrm{B} + \mathbf{v}_\mathrm{R}$, where $\mathbf{v}_\mathrm{R}$ is due to the random motion and is called *thermal velocity,* whose average is zero. Therefore, $\overline{\mathbf{v}\mathbf{v}} = \mathbf{v}_\mathrm{B}\mathbf{v}_\mathrm{B} + \overline{\mathbf{v}_\mathrm{R}\mathbf{v}_\mathrm{R}}$, where $\overline{\mathbf{v}_\mathrm{R}\mathbf{v}_\mathrm{R}}$ is a dyadic whose array is a second-order tensor. In fact, $\rho\overline{\mathbf{v}_\mathrm{R}\mathbf{v}_\mathrm{R}}$ is nothing but the stress tensor given in Eq. (4.14). Because $\psi = m\mathbf{v}$ and the velocity is an independent variable, both $\partial\psi/\partial t$ and $\nabla\psi$ vanish. The last term is simply $\rho\mathbf{a}$. The combination of all the terms gives

$$\rho \frac{\partial \mathbf{v}_\mathrm{B}}{\partial t} + \mathbf{v}_\mathrm{B} \frac{\partial \rho}{\partial t} + \mathbf{v}_\mathrm{B} \nabla \cdot (\rho \mathbf{v}_\mathrm{B}) + \rho(\mathbf{v}_\mathrm{B} \cdot \nabla)\mathbf{v}_\mathrm{B} + \nabla \cdot \{P_{ij}\} - \rho\mathbf{a} = 0$$

Applying the mass balance equation, we can simplify the momentum equation as

$$\frac{\mathrm{D}\mathbf{v}_\mathrm{B}}{\mathrm{D}t} = -\frac{1}{\rho} \nabla \cdot \{P_{ij}\} + \mathbf{a} \tag{4.62}$$

The stress tensor can be obtained from Eq. (4.14). When Stokes' hypothesis is used to simplify the constitutive relations between the stresses and the velocity gradients of a viscous fluid, we have

$$P_{ij} = \begin{cases} P - 2\mu\frac{\partial v_i}{\partial x_i} + \frac{2}{3}\mu\nabla \cdot \mathbf{v}_\mathrm{B},\ i = j \\ -\mu\left(\frac{\partial v_i}{\partial x_j} + \frac{\partial v_j}{\partial x_i}\right),\ i \neq j \end{cases} \tag{4.63}$$

where $x_i = x, y, z$ (for $i = 1, 2, 3$), and $v_i (i = 1, 2, 3)$ is the velocity component of the bulk velocity $\mathbf{v}_\mathrm{B}$. Substituting Eq. (4.63) into Eq. (4.62), one obtains exactly the same result as Eq. (2.42). The derivation is left as an exercise (see Problem 4.12).

Next, we derive the energy equation for viscous flow of a monatomic gas, using the second moment. $\psi = \varepsilon = \frac{1}{2}m\mathbf{v}_\mathrm{R}^2$ because only the random motion contributes to the internal energy. The first term in Eq. (4.59) becomes $\partial(\rho u)/\partial t$, where $u$ is the mass specific internal energy. The second term $\nabla \cdot (n\overline{\psi\mathbf{v}}) = \frac{1}{2}\nabla \cdot (\rho\mathbf{v}_\mathrm{B}\overline{\mathbf{v}_\mathrm{R}^2}) + \frac{1}{2}\nabla \cdot (\rho\overline{\mathbf{v}_\mathrm{R}\mathbf{v}_\mathrm{R}^2})$ $=\nabla \cdot (\rho u\mathbf{v}_\mathrm{B}) + \nabla \cdot \mathbf{J}_E$, where $\mathbf{J}_E = n\int_\varpi f\mathbf{v}_\mathrm{R}\varepsilon \mathrm{d}\varpi$ is the energy flux vector to be discussed further in Sect. 4.3.2. Notice that $\overline{\partial(\frac{1}{2}m\mathbf{v}_\mathrm{R}^2)/\partial t} = 0$ and $n\overline{\mathbf{v} \cdot \nabla(\frac{1}{2}m\mathbf{v}_\mathrm{R}^2)} = \rho\overline{\mathbf{v} \cdot [\mathbf{v}_\mathrm{R} \cdot \nabla(\mathbf{v} - \mathbf{v}_\mathrm{B})]} = -\rho\overline{\mathbf{v} \cdot (\mathbf{v}_\mathrm{R} \cdot \nabla\mathbf{v}_\mathrm{B})} = \{P_{ij}\} : \nabla\mathbf{v}_\mathrm{B}$, which can be considered as the product of the momentum flux and the bulk velocity gradient. This tensor product can be calculated according to $\{P_{ij}\} : \nabla\mathbf{v}_\mathrm{B} = \sum_i\sum_j P_{ij}\frac{\partial v_i}{\partial x_j}$. For the force term, we have $n\overline{\mathbf{a}\partial(\frac{1}{2}m\mathbf{v}_\mathrm{R}^2)/\partial\mathbf{v}} = \rho\mathbf{a} \cdot \overline{\mathbf{v}_\mathrm{R}} = 0$. The energy conservation equation can be expressed as

$$\frac{\partial}{\partial t}(\rho u) + \nabla \cdot (\rho u\mathbf{v}_\mathrm{B}) + \nabla \cdot \mathbf{J}_E + \{P_{ij}\} : \nabla\mathbf{v}_\mathrm{B} = 0$$

After it is simplified using the continuity equation, we have

$$\rho\frac{\mathrm{D}u}{\mathrm{D}t} = -\nabla \cdot \mathbf{J}_E - \{P_{ij}\} : \nabla\mathbf{v}_\mathrm{B} \tag{4.64}$$

The left-hand side consists of the transient term and the advection term. Among the two terms on the right-hand side, the first one corresponds to the energy transfer by heat diffusion, and the second one includes the pressure effect as well as the viscous dissipation. It can be shown that Eq. (4.64) is the same as Eq. (2.43) (see Problem 4.13). In a stationary medium with $\mathbf{v}_\mathrm{B} = 0$, Eq. (4.64) reduces to the heat diffusion equation, $\kappa\nabla^2 T = \rho c_v\frac{\partial T}{\partial t}$ (see Problem 4.14). In the earlier derivations, the velocity $\mathbf{v}$ is taken as an independent variable in the distribution function. Another way of deriving the macroscopic conservation equations is to take the random velocity $\mathbf{v}_\mathrm{R}$ as the independent variable and modify the distribution function to a new one, $f(\mathbf{r}, \mathbf{v}_\mathrm{R}, t)$; see Ref. [6] for example.

In deriving the macroscopic conservation equations, it is assumed that $f(\mathbf{r}, \mathbf{v}, t)$ obeys certain equilibrium distribution at any given location. This is the local-equilibrium assumption, which is only valid when the mean free path is much smaller than the characteristic length. For systems with dimensions comparable to or smaller than the mean free path, the local-equilibrium assumption breaks down, as will be discussed in Sect. 4.4 and forthcoming chapters.

### 4.4.2 Fourier's Law and Thermal Conductivity

The transport equations and coefficients can be obtained based on the BTE. Here, as an example, the 1D Fourier's law will be derived. When the characteristic time $t_c$ is much greater than the relaxation time and the length scale is much greater than the mean free path, we may write the BTE under the relaxation time approximation using Eqs. (4.54) and (4.56) as

$$\frac{\partial f}{\partial t} + \mathbf{v} \cdot \frac{\partial f}{\partial \mathbf{r}} = \frac{f_0 - f}{\tau(\mathbf{v})} \tag{4.65}$$

Assume that the temperature gradient is in the $x$-direction and the medium is stationary. If the medium moves with a bulk velocity, we can set the coordinate to move at the bulk velocity so that the local average velocity is zero. The distribution function will vary with $x$ only, and at steady state, we have $v_x \frac{\partial f}{\partial x} = \frac{f_0 - f}{\tau}$. We further assume that $f$ is not very far away from equilibrium so that $\frac{\partial f}{\partial x} \approx \frac{\partial f_0}{\partial x}$, which is the condition of *local equilibrium*. Therefore,

$$f \approx f_0 - \tau v_x \frac{\partial f_0}{\partial T} \frac{\mathrm{d}T}{\mathrm{d}x} \tag{4.66}$$

The heat flux in the $x$-direction is

$$J_{E,x} = q_x'' = \int_{\varpi} f \varepsilon v_x \mathrm{d}\varpi = \int_{\varpi} \left( f_0 - \tau v_x \frac{\partial f_0}{\partial T} \frac{\mathrm{d}T}{\mathrm{d}x} \right) \varepsilon v_x \mathrm{d}\varpi \tag{4.67a}$$

Let us use Maxwell's velocity distribution Eq. (3.43) as an example to explain the distribution function and equilibrium distribution. The distribution function $f$ can be viewed as a function of $v_x$, $v_y$, and $v_z$ for a given $T$ and other parameters when integrating over the velocity space. On the other hand, it can be viewed as a function of $T$ by fixing $v_x$, $v_y$, $v_z$, and all other parameters. This allows us to obtain $\partial f_0/\partial T$, which, in turn, can be viewed as a function of $v_x$, $v_y$, and $v_z$ in order to carry out the integration. Note that $\int_{\varpi} f_0 \varepsilon v_x \mathrm{d}\varpi = 0$ because $f_0$ is the equilibrium distribution. More discussion will be given in Chap. 7. It should also be noted that the integration over $v_x^2$ is the same as the integration over $v_y^2$ or $v_z^2$. Hence, the integration over $v_x^2$ equals one-third of the integration over $v^2$. After some manipulations, we can write

$$q_x'' = -\kappa \frac{\mathrm{d}T}{\mathrm{d}x} \tag{4.67b}$$

which is Fourier's law with the thermal conductivity expressed as

$$\kappa = \frac{1}{3}\int_{\varpi} \frac{\partial f_0}{\partial T}\tau v^2 \varepsilon \mathrm{d}\varpi \tag{4.68a}$$

The above integral is often converted to integration over the energy, which gives

$$\kappa = \frac{1}{3}\int_0^{\infty} \frac{\partial f_0}{\partial T}\tau v^2 \varepsilon D(\varepsilon)\mathrm{d}\varepsilon \tag{4.68b}$$

where $D(\varepsilon)$ is the density of states, which can be considered as the volume in the velocity space per unit energy interval. If we take both the relaxation time $\tau$ and the velocity $v$ as their average values that can be moved out of the integral, we have $\kappa \approx \frac{1}{3}\tau\bar{v}^2\rho c_v \approx \frac{1}{3}\Lambda\bar{v}\rho c_v$, which is identical to Eq. (4.33). If we assume only $\tau$ is independent of frequency, we can use Maxwell's velocity distribution Eq. (3.43) to evaluate $\kappa = \frac{\tau}{3}\int_{\varpi} \frac{\partial f_0}{\partial T} v^2 \varepsilon \mathrm{d}\varpi$ for a monatomic gas $\left(\varepsilon = \frac{1}{2}m\mathbf{v}^2\right)$ (see Problem 4.15). The result $\kappa = 1.31\Lambda\bar{v}\rho c_v$ is in good agreement with Eq. (4.36a), considering the assumption of a constant relaxation time.

Under local-equilibrium assumption and by applying the relaxation time approximation, we can write the 3D Fourier's law as

$$\mathbf{q}'' = \mathbf{J}_E = \int_{\varpi} f\mathbf{v}\varepsilon \mathrm{d}\varpi = -\kappa \nabla T \tag{4.69}$$

where $\kappa$ is already given in Eq. (4.68a) and $\mathbf{v}$ is the thermal velocity. Eq. (4.69) proves that the first term on the right-hand side of Eq. (4.64) is indeed associated with heat diffusion.

## 4.5 Micro/Nanofluidics and Heat Transfer

A large number of microdevices involving fluid flow in microstructures have been designed and built since the late 1980s. Examples are microsensors, actuators, valves, heat pipes, and microducts used in heat engines and heat exchangers [11–13]. Micro/nanofluidics research is an active area with applications in biomedical diagnosis (lab-on-a-chip) and drug delivery, MEMS/NEMS sensors and actuators, micropumps for ink-jet printing, and microchannel heat sinks for electronic cooling. Many researchers are also studying fluid flow inside nanostructures, such as nanotubes, and developing unique devices, such as nanojets.

Under the continuum assumption, matter is continuous and indefinitely divisible. Properties are defined as the average over elements much larger than the microscopic structure of the fluid but much smaller than the macroscopic device scale. For flow inside micro/nanostructures, the mean free path of the fluid molecules may be comparable to or smaller than the characteristic dimensions. The continuum assumption is often not valid since the interaction between the molecules and the solid surfaces becomes important. In his seminal paper in 1946, H.-S. Tsien drew the attention of aerodynamicists to the study of noncontinuous fluid mechanics, for applications in high-altitude flights and vacuum systems, which subsequently formed the field of *rarefied gas dynamics* [14]. In the same paper, he delineated the realms from conventional gas dynamics (i.e., continuum regime): *slip flow,* "blank" (which was later called *transition flow*), and *free molecule flow* based on the ratio of the mean free path to the characteristic length, i.e., the Knudsen number as will be discussed in the next section.

Some of the earlier studies are still valid and can help understand fluid flow in microstructures [15, 16]. On the other hand, there are several aspects that are unique to microfluidics, making it distinctly different from the rarefied fluid dynamics. In microstructures, surface-to-volume ratio is much greater than that in macrostructures, and hence, surface forces become dominant over body forces. One of the direct impacts is a significant pressure drop and a greater mass flow rate than that predicted with the continuum theory [12, 13]. Because of the large pressure drop, the velocity is usually not very high. The Reynolds number is significantly smaller due to the small dimensions and relatively low velocity. The axial heat conduction, which is negligible for macroflow, may become important for micro/nanoflow. Due to the large pressure drop, compressibility is another issue that needs to be considered even though the speed is much less than the speed of sound. A change in the density further complicates the pressure distribution, making it nonlinear along the streamline. Liquid is also used in many applications such as microchannel cooling. Furthermore, the phase change by evaporation and condensation is another important aspect in a number of microdevices, such as micro-heat pipes.

Although measurements in micro/nanoflow are challenging, a large number of miniaturized flow and temperature sensors have been developed and integrated into the microdevices to perform measurements with a high spatial resolution. Submicron polysilicon hot-wire anemometers, hot-film shear-stress sensors, piezoresistive and diaphragm-type pressure sensors, and submicron thermocouples are some examples [12]. For flow visualization, both X-ray and caged-dye techniques have been used to image the flow field. Micro-particle image velocimetry (PIV) is a powerful technique for flow visualization and sometimes for thermal measurements. In micro-PIV, small particles imbedded into the fluid scatter pulsed laser light. A microscopic system allows the illumination and collection of the scattered light into a CCD camera. The flow is illuminated at two times, and the velocity vectors are determined based on the displacement of particles. The temperature field can be determined based on the Brownian motion, i.e., random fluctuation of the particles [16].

The next section focuses on gas flow, which can be categorized into different regimes based on the range of the Knudsen number. Examples of slip flow and free molecule conduction are provided to illustrate the effect of rarefaction. More detailed research on microfluidics and microflow devices can be found from the monographs [16, 17]. Reviews of recent studies on the heat transfer in microstructures involving liquids, evaporation, and condensation can be found from Peterson et al. [18], Garimella and Sobhan [19], and Poulikakos et al. [20].

### 4.5.1 The Knudsen Number and Flow Regimes

The continuum model is no longer valid when one of the geometric dimensions, called the characteristic dimension $L$, is comparable to the mechanistic length, such as the mean free path $\Lambda$. This can happen when the gas is at very low pressure (rarefied) or when the characteristic dimension is extremely small: from a few micrometers down to several nanometers in micro- and nano-channels. As a result, boundary scattering becomes significant, and the gas molecules have a large chance to collide with the wall as compared to the collision between molecules.

The ratio of the mean free path to the characteristic length defines an important dimensionless parameter, called the Knudsen number:

$$Kn \equiv \frac{\Lambda}{L} \tag{4.70}$$

The Reynolds number for flow over an object with a characteristic length of $L$ is $Re_L = \rho v_\infty L/\mu$. The Mach number is the ratio of the free steam velocity $v_\infty$ to the speed of sound such that $Ma = v_\infty/v_\mathrm{a}$, where $v_\mathrm{a} = \sqrt{\gamma RT}$. From Eq. (4.30), $\mu = \rho\Lambda\sqrt{2RT/\pi}$. Based on the characteristic length of $L$, Eq. (4.70) can be written as

$$Kn = \sqrt{\frac{\pi\gamma}{2}}\frac{Ma}{Re_L} \tag{4.71a}$$

Within the boundary layer, however, the characteristic length should be the boundary layer thickness $\delta_x$ at location $x$ along the flow direction. Consider a parallel flow over a flat plate in the case of laminar flow, $\delta_x \sim x/Re_x^{1/2}$. Hence, it can be shown that in the boundary layer,

$$Kn_x = \frac{\Lambda}{\delta_x} \sim \frac{Ma}{\sqrt{Re_x}} \tag{4.71b}$$

When internal flow is considered, $v_\infty$ should be replaced by the bulk velocity $v_\mathrm{m}$. The characteristic length can be taken as the diameter of the pipe or the width of a microchannel.

The physics of fluid flow depends much on the magnitude of $Kn$. The local $Kn$ determines the degree of rarefaction and the degree of deviation from the continuum assumption. The regimes are divided based on $Kn$ in Table 4.2. The regime

**Table 4.2** Flow regimes based on the Knudsen number [13]

| Regime | Method of calculation | $Kn$ range |
|---|---|---|
| Continuum | Navier–Stokes and energy equations with no-slip/no-jump boundary conditions | $Kn \leq 0.001$ |
| Slip flow | Navier–Stokes and energy equations with slip/jump boundary conditions, DSMC | $0.001 < Kn \leq 0.1$ |
| Transition | BTE, DSMC | $0.1 < Kn \leq 10$ |
| Free molecule | BTE, DSMC | $Kn > 10$ |

boundaries are instructive rather than exact because they depend on more parameters of the fluid conditions. A small $Kn$ generally corresponds to a continuum flow ($Kn < 0.001$). In this regime, the Navier–Stokes equations are applicable, the velocity of the fluid at the boundary is the same as that of the wall, and the temperature of the fluid adjacent to the wall is the same as the surface temperature. Care must be taken in regard to the compressibility. Conventionally, the flow can be assumed incompressible if $Ma < 0.3$. However, in some microdevices where pressure changes drastically, density change can be significant and thus compressibility must be taken into consideration.

When $Kn$ is increased from about 0.001 to 0.1, noncontinuum (slip) boundary conditions must be applied. Slip flow refers to the situation when the velocity of the fluid at the wall is not the same as the wall velocity, as shown in Fig. 4.9 for fully developed internal flow. In the heat transfer problem, the temperature of the fluid adjacent to the wall is different from that of the wall, as shown in the right of Fig. 4.9. This is called *temperature jump*. In the slip/jump regime, the Navier–Stokes equations can still be used for the flow with modified boundary conditions, as will be discussed in the next section.

If $Kn > 10$, the flow is called a free molecule flow that is dominated by ballistic scattering between the molecules and the surfaces. The continuum assumption breaks down completely. No local velocity or temperature of the gas can be defined for the

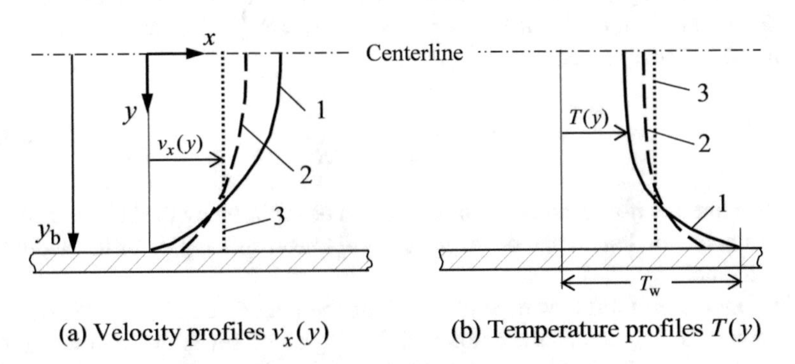


**Fig. 4.9** Illustration of **a** the velocity profile and **b** the temperature profile for internal flow, in the three regimes: 1: continuum, 2: velocity slip and temperature jump, and 3: free molecule

fluid. The "slip" velocity is the same as the velocity of the mainstream, i.e., the fluid velocity will be the same regardless of the distance from the wall, as shown clearly in Fig. 4.9. The same is true for the fluid temperature: no gradient exists near the wall even though there is heat transfer between the wall and the gas. Molecular-based models, such as the BTE or the DSMC, are the best to solve problems in this regime, as well as in the transition regime between the slip flow and the free molecule flow [21].

In the continuum regime, numerical solution techniques include finite element method, finite difference method, boundary element method, and so forth. In recent years, flexible mesh schemes, such as the unstructured grids or mesh-free technique, have become popular. Commercial computational fluid dynamics (CFD) software is often available and can be applied to complex geometries. For numerical solutions of the Boltzmann equations and modeling fluid flow at the molecular level, both deterministic and stochastic methods have been developed. The challenge lies in how to handle the collision terms. Relaxation time approximation and higher order approximations with nonlinear terms have been applied. Lattice Boltzmann (LB) method based on mesoscopic kinetic equations has emerged as a promising numerical technique for simulating single-phase and multiphase flows involving complex interfacial dynamics and geometries [22]. In the LB method, each grid is a volume element that consists of a collection of particles described by the Boltzmann distribution function. The fluid particles collide with each other as they move under the applied force at each discrete time step. By developing simplified version of the kinetic equation, the LB method avoids solving the full BTE and thus reduces computational time and memory. Direct simulation of the molecular movements can be carried out in two ways, as discussed in the following.

Molecular dynamics (MD) considers the position and the velocity of each particle at any time by using a deterministic approach. The molecules are assumed to obey Newton's laws of motion in Eq. (3.1), and their interactions are governed by the intermolecular potentials. An example is the Lennard-Jones $< 6, 12 >$ potential given in Eq. (4.51) that is commonly used directly or with some modifications. In the MD simulation, the first step is called *initialization,* which randomly assigns $N$ molecules in a region of space and sets their velocities according to some equilibrium distribution. After the initial statistical assignment, all the rest steps are deterministic. The time evolution of the position and the velocity of each particle can be found by numerically integrating Newton's equations of motion using small time steps. Periodic boundary conditions are often used to simulate the inlet and the outlet of the flow. Statistical averaging, called ensemble averaging, is used to calculate the internal energy, effective temperature, pressure, and other properties at a given time. The internal energy is the sum of the total kinetic and potential energies. The temperature is based on the average kinetic energy (for monatomic gases). The pressure is calculated using the *virial theorem* [20]. Usually, the simulation time step is on the order of femtoseconds, and it requires thousands of time steps to simulate a process for a few picoseconds in real time. The required computational time is proportional to the square of the number of particles $N$ in the simulation. Therefore, the MD method provides complete information about the trajectories of all particles at a great computational expense. This method is best suited for dense gases and liquids where molecular interactions are less frequent. The MD method is particularly useful at the nanoscale as the number of particles becomes reasonably small and the total time steps are manageable. It can also be used to simulate boiling and vaporization, as well as the ablation process. Note that the MD method is often the only method available for the study of some nanoscale phenomena because no experiments could be conducted at that time.

Considering the inefficiency in modeling dilute gases using the MD method, G. A. Bird [21] in the 1960s established a statistical technique to model rarefied gas flow and transport processes. This method is called the direct simulation Monte Carlo (DSMC) method that has matured as a powerful simulation tool, especially for transition flow and free molecule flow. Some have combined it with continuum models to form a hybrid method for multiscale simulation [17]. The principle of the DSMC method is the same as that of the MD method; however, intermolecular interactions are dealt with entirely on a probabilistic basis rather than the deterministic basis. In the DSMC method, the space is divided into cells, each with a large number of molecules that mimic but do not follow exactly the motion of real molecules. The motion of particles and collisions between them are simulated via a probabilistic process using a time step smaller than the relaxation time. The interaction between the molecules and the boundary is also simulated according to certain statistical models. Since only a small portion of particles are actually simulated at each time step to represent the actual molecules, the computational time is proportional to $N$ rather than $N^2$ as in the MD method. This greatly reduces the required computational resources. Note that the DSMC method is not so efficient for a low $Kn$ flow, where the continuum theory or a direct solution of the BTE is more effective.

### 4.5.2 *Velocity Slip and Temperature Jump*

The interaction between the gas molecules and the wall plays a critical role when the gas becomes rarefied. However, a fundamental understanding of such interaction is often not available. When a molecule impinges on the wall, it will be reflected (or reemitted) after collision with the molecules near the surface of the wall (if adsorption is neglected). If the reflection is specular, the tangential momentum (or velocity) will remain the same, whereas the normal momentum will be reversed. If all the molecules are specularly reflected, there will not be any shear force or friction between the gas and the wall. However, this is not the case for most engineering applications. Another extreme is the diffuse reflection case, in which the molecule will acquire mutual equilibrium with the wall and be reemitted randomly into the hemisphere. For a stream of molecules, the effect is such that the reflected molecules will follow the Maxwell velocity distribution at the wall temperature. The *momentum accommodation coefficients* can be defined as [2, 3]

$$\alpha_v = \left.\frac{p_\mathrm{i} - p_\mathrm{r}}{p_\mathrm{i} - p_\mathrm{w}}\right)_{\parallel}, \quad \text{for tangential components} \tag{4.72a}$$

and

$$\alpha_{v'} = \left.\frac{p_\mathrm{i} - p_\mathrm{r}}{p_\mathrm{i} - p_\mathrm{w}}\right)_\perp, \text{ for normal components} \tag{4.72b}$$

where $p = mv$ is the momentum, the subscripts $i$ and $r$ represent the incident and the reflected, and the subscript $w$ refers to the Maxwell velocity distribution corresponding to the surface temperature $T_\mathrm{w}$. Clearly, $\alpha_v = \alpha_{v'} = 0$ for specular reflection, and $\alpha_v = \alpha_{v'} = 1$ for diffuse reflection. Similarly, the *thermal accommodation coefficient* can be defined based on the ratio of energy differences as

$$\alpha_T = \frac{\varepsilon_\mathrm{i} - \varepsilon_\mathrm{r}}{\varepsilon_\mathrm{i} - \varepsilon_\mathrm{w}} \tag{4.73a}$$

where $\varepsilon$ is the average energy of a molecule and $\varepsilon_\mathrm{w}$ is the energy when the molecules are in thermal equilibrium with the wall. For diffuse reflection, the molecule is completely accommodated by the wall such that $\varepsilon_\mathrm{r} = \varepsilon_\mathrm{w}$ and $\alpha_T = 1$. On the other hand, if the reflection is specular, the molecule is not accommodated at all and the reflected energy will be the same as the incident energy so that $\alpha_T = 0$. For monatomic molecules, thermal accommodation coefficient involves translational kinetic energy only, and the kinetic energy is proportional to the absolute temperature. Hence, we can write the thermal accommodation coefficient in terms of temperatures as

$$\alpha_T = \frac{T_\mathrm{i} - T_\mathrm{r}}{T_\mathrm{i} - T_\mathrm{w}} \tag{4.73b}$$

For polyatomic molecules, it is reasonable to think that the accommodation coefficients for translational, rotational, and vibrational degrees of freedom may be different. However, due to the lack of information on the nature of interaction between the gas molecules and the wall, usually no distinction is made between the accommodation coefficients for different degrees of freedom. In addition, Eq. (4.73b) is often extended to polyatomic molecules with the assumption that the temperature difference is sufficiently small for the specific heat to be independent of temperature. The thermal accommodation coefficient depends on the nature of the molecules, the molecular structure of the solid wall, the surface roughness and cleanness, the temperature, and the degree of rarefaction. Saxena and Joshi [23] provide a comprehensive review and data compilation of earlier works. The values of $\alpha_T$ for air–aluminum and air–steel systems range from 0.87 to 0.97. However, $\alpha_T$ can be less than 0.02 between pure He gas and clean metallic surfaces. Earlier measurements showed that for most engineering surfaces, $\alpha_v$ ranges from 0.87 to 1 for air. Arkilic et al. [24] measured tangential momentum accommodation coefficients for $N_2$, Ar, and $CO_2$ in silicon microchannels and found that $\alpha_v$ is between 0.75 and 0.85. This is possibly due to the relatively smooth crystalline silicon surfaces. Generally speaking, $\alpha_{v'}$ is not very important and can be assumed the same as $\alpha_v$.

Slip flow is an important regime for microchannel flows and MEMS devices. The velocity slip and temperature-jump boundary conditions are presented in this section, together with some analytical solutions for simple cases. If the wall is not moving, the slip boundary condition based on the geometry shown in Fig. 4.9 reads

$$v_x(y_\mathrm{b}) = -\frac{2-\alpha_v}{\alpha_v}\Lambda\left(\frac{\partial v_x}{\partial y}\right)_{y_\mathrm{b}} + 3\sqrt{\frac{R}{8\pi T}}\Lambda\left(\frac{\partial T}{\partial x}\right)_{y_\mathrm{b}} \tag{4.74}$$

All the derivatives and the fluid properties are evaluated at $y = y_\mathrm{b}$. The first term on the right is proportional to the velocity gradient perpendicular to the flow direction, and the second term is known as *thermal creep* due to the temperature gradient along the flow direction. It should be noted that the net mass transfer (creep) is from cold region to hot region. It can be shown that the first term goes with *Kn* and the second term goes with the square of *Kn*. Higher order terms can be included by expressing them as *Kn* raised to higher powers and higher order derivatives [16]. The temperature-jump boundary condition reads

$$T(y_\mathrm{b}) - T_\mathrm{w} = -\frac{2-\alpha_T}{\alpha_T}\frac{2\gamma}{\gamma+1}\frac{\Lambda}{Pr}\left(\frac{\partial T}{\partial y}\right)_{y_\mathrm{b}} + \frac{v_x^2(y_\mathrm{b})}{4R} \tag{4.75}$$

Equation (4.75) suggests that the temperature of the fluid at the wall will not be the same as the wall temperature, as shown in Fig. 4.9. The second term on the right is due to viscous dissipation caused by the slip velocity and is usually negligibly small.

Let us consider the fluid flow through a channel between two fixed parallel plates, i.e., the Poiseuille flow, as shown in Fig. 4.10. It is assumed that $W \geq 2H$ and the edge effect can be neglected. When $Kn = \Lambda/2H$ is less than 0.1 or so, slip flow with temperature-jump boundaries can be applied together with the Navier–Stokes equation and the energy equation to obtain the velocity and temperature distributions. For simplicity, assume the fluid is incompressible and fully developed with constant properties. The momentum equation can be written as

$$\frac{\mathrm{d}^2 v_x}{\mathrm{d}\eta^2} = \frac{H^2}{\mu}\frac{\mathrm{d}P}{\mathrm{d}x} \tag{4.76}$$

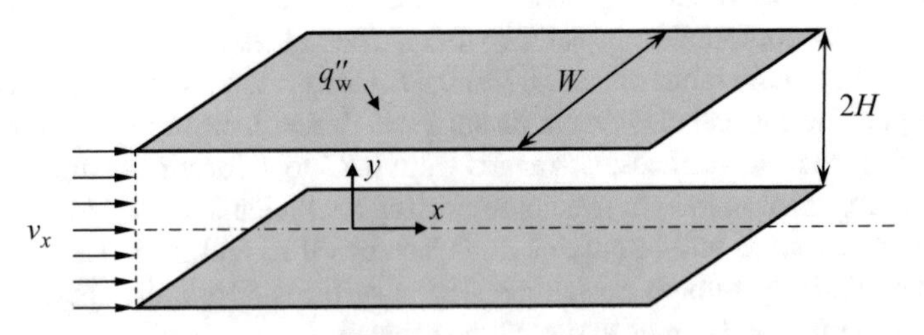


**Fig. 4.10** Micro/nanoscale Poiseuille flow with heat transfer

where $\eta = y/H$. The symmetry requires $\left.\frac{\mathrm{d}v_x}{\mathrm{d}\eta}\right)_{\eta=0} = 0$. The slip condition given by Eq. (4.74) can be simplified by neglecting thermal creep and higher order terms, giving

$$v_x(\eta = 1) = -2\beta_v \left.\frac{\mathrm{d}v_x}{\mathrm{d}\eta}\right)_{\eta=1} \tag{4.77}$$

where

$$\beta_v = \frac{2 - \alpha_v}{\alpha_v} Kn \tag{4.78}$$

The solution gives the fully developed velocity distribution as

$$\frac{v_x(\eta)}{v_\mathrm{m}} = \frac{3}{2}\frac{1 + 4\beta_v - \eta^2}{1 + 6\beta_v} \tag{4.79}$$

where $v_\mathrm{m}$ is the bulk or mean velocity, which can be expressed as

$$v_\mathrm{m} = \int_0^1 v_x(\eta)\mathrm{d}\eta = (1 + 6\beta_v)\frac{H^2}{3\mu}\left(-\frac{\mathrm{d}P}{\mathrm{d}x}\right) \tag{4.80}$$

Define the *velocity slip ratio* $\zeta = \frac{v_x(\eta=1)}{v_\mathrm{m}} = \frac{6\beta_v}{1+6\beta_v}$, which is the ratio of the velocity of the fluid at the wall to the bulk velocity. The velocity distribution can be rewritten as

$$\frac{v_x(\eta)}{v_\mathrm{m}} = \frac{3 - \zeta}{2} - \frac{3(1 - \zeta)}{2}\eta^2 \tag{4.81}$$

The energy equation is simplified based on Eq. (2.43) without dissipation as follows:

$$\rho c_p v_x \frac{\partial T}{\partial x} = \kappa\left(\frac{\partial^2 T}{\partial x^2} + \frac{\partial^2 T}{\partial y^2}\right) \tag{4.82}$$

Let us consider the case with a uniform wall heat flux $q''_\mathrm{w}$ at both plates. For thermally full development, $\partial T/\partial x$ must not depend on $x$ and $y$; hence, the term $\partial^2 T/\partial x^2$ can be dropped out. Applying the energy balance for an elementary controlled volume inside the fluids, we can rewrite Eq. (4.82) after some tedious derivations as follows:

$$\frac{\partial^2 \Theta}{\partial \eta^2} = \frac{v_x}{v_\mathrm{m}} \tag{4.83}$$

where $\Theta(\eta) = \frac{\kappa}{H} \frac{T - T_w}{q''_w}$ is a dimensionless temperature. Integrating Eq. (4.83) yields

$$\Theta(\eta) = \frac{3-\zeta}{4}\eta^2 - \frac{1-\zeta}{8}\eta^4 + C_1\eta + C_2 \tag{4.84}$$

The symmetry at $\eta = 0$ requires that $\Theta'(\eta = 0) = 0$. When the second term on the right of Eq. (4.75) due to viscous dissipation is neglected, the nondimensionalized boundary condition becomes

$$\Theta(\eta = 1) = -2\beta_T \left.\frac{\partial \Theta}{\partial \eta}\right)_{\eta=1} \tag{4.85}$$

where

$$\beta_T = \frac{2-\alpha_T}{\alpha_T} \frac{2\gamma}{\gamma+1} \frac{Kn}{Pr} \tag{4.86}$$

Applying the boundary conditions, we obtain $C_1 = 0$ and $C_2 = (\zeta - 5)/8 - 2\beta_T$. From Eq. (4.85) and Fig. 4.10, the heat flux from the surface to the fluid can be expressed as

$$q''_w = \kappa \left.\frac{\partial T}{\partial y}\right)_{y=H} = \kappa \frac{T_w - T(y=H)}{2\beta_T H} \tag{4.87}$$

Here, $2\beta_T H$ is called the *temperature-jump distance,* which can be thought as an effective length for heat conduction between the wall and the fluid. With the assumption of constant properties, the dimensionless bulk temperature can be calculated by

$$\Theta_m = \int_0^1 \frac{v_x(\eta)}{v_m} \Theta(\eta) d\eta \tag{4.88}$$

The Nusselt number is defined based on the hydraulic diameter $D_h = 4H$ for parallel plates; therefore,

$$Nu = \frac{hD_h}{\kappa} = \frac{q''_w}{T_w - T_m} \frac{4H}{\kappa} = -\frac{4}{\Theta_m} \tag{4.89}$$

Substituting the integration of Eq. (4.88), one obtains after some manipulations [25]

$$Nu = \frac{140}{17 - 6\zeta + (2/3)\zeta^2 + 70\beta_T} \tag{4.90}$$

The above equation approaches to $Nu = 140/17$ when both $\zeta$ and $\beta_T$ become negligibly small, i.e., in the continuum limit. Furthermore, the Nusselt number decreases monotonically as $\beta_T$ increases. Note that $\beta_T$ will be increased if the mean free path increases or if $\alpha_T$ decreases. On the other hand, the Nusselt number increases slightly as $\zeta$ increases, e.g., with a smaller $\alpha_v$. In any case, both $\zeta$ and $\beta_T$ should be much smaller than unity for the slip–jump conditions to hold. If one of the plates is insulated, while the other plate is maintained at a uniform heat flux $q''_{\mathrm{w}}$, the velocity distribution is the same. The Nusselt number can be calculated from Inman [25] as

$$Nu = \frac{140}{26 - 3\zeta + (1/3)\zeta^2 + 70\beta_T}, \text{ with one insulated wall} \tag{4.91}$$

Because there is no heat transfer, temperature jump does not occur at the insulated surface. For a circular tube of inner diameter $D$, Sparrow and Lin [26] derived the Nusselt number for constant heat flux, which can be expressed as

$$Nu_D = \frac{q''_{\mathrm{w}}}{T_{\mathrm{w}} - T_{\mathrm{m}}}\frac{D}{\kappa} = \frac{48}{11 - 6\zeta + \zeta^2 + 48\beta_T} \tag{4.92}$$

where $\zeta = 8\beta_v/(1 + 8\beta_v)$. The expressions for $\beta_v$ and $\beta_T$ are the same as in the case with parallel plates, i.e., Equations (4.78) and (4.86), except that $Kn = \Lambda/D$ for a circular tube.

Figure 4.11 illustrates the variation of Nusselt number as the Knudsen number changes, for air at near room temperature with a uniform heat flux, assuming different accommodation coefficients. Note that $Kn = \Lambda/2H$ for Poiseuille

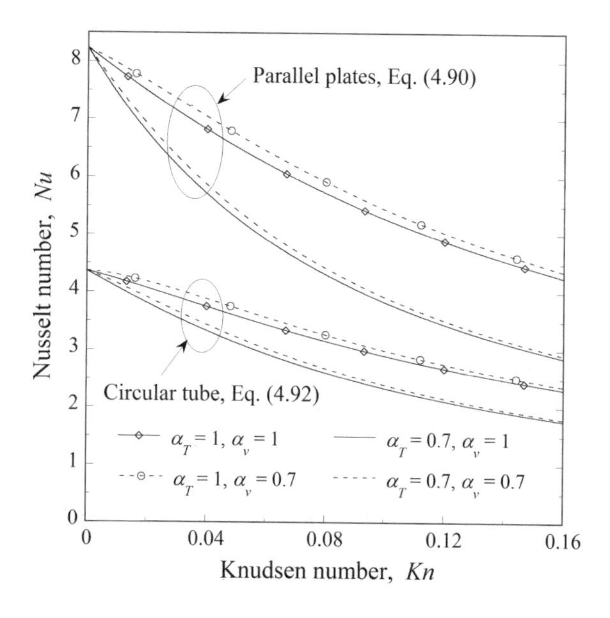


**Fig. 4.11** Calculated Nusselt number as a function of the Knudsen number for air ($\gamma = 1.4$ and $Pr = 0.7$) with different accommodation coefficients

flow, and $Kn = \Lambda/D$ for a circular tube. The change in the Knudsen number can be considered as the combined effect of the pressure and the channel dimension. It should be noted that the slip–jump conditions impose an upper limit on the velocity or the temperature gradient near the boundary. In the continuum limit, the shear stress and the Nusselt number are infinite at the entrance and decrease with $x$ until the flow is fully developed. Assume that the velocity and the temperature are uniform at the entrance. From Eqs. (4.77) and (4.85), we obtain correspondingly $\tau_s = -\mu \left.\frac{\partial v_x}{\partial y}\right)_{y=H} \leq \frac{\mu v_m}{2H\beta_v} = \tau_{s,\max}$ and $Nu \leq 2/\beta_T = Nu_{\max}$, which are the values at the entrance. For a circular tube, it can be shown that $Nu_{\max} = \beta_T^{-1}$ at the entrance (see Problem 4.22).

Yu and Ameel [27] presented analytical solutions for a rectangular channel with constant wall heat flux on all surfaces using an integral transform method. Hadjiconstantinou and Simek [28] provided an extensive review of the literature dealing with slip channel flow with constant wall temperature. Most of the works did not consider the effect of axial conduction. This assumption is good only for large values of the *Peclet number*, defined as the product of the Reynolds number and the Prandtl number ($Pe = RePr$). As the channel dimensions become very small, $Re$ will decrease but $Kn$ will increase. It is possible for $Kn$ to be large enough for slip and jump to occur at a relatively small $Re$. Axial conduction enhances the heat transfer between the fluid and the wall, and thus increases $Nu$ especially when $Kn$ is small. In the no-slip case when $Kn = 0$, it is well known that $Nu = 7.54$ for parallel plates and $Nu = 3.66$ for circular tube without axial conduction ($Pe \to \infty$). In the extreme when $Pe \to 0$, $Nu$ becomes 8.12 (7.7% increase) and 4.18 (14.2% increase), respectively. These values are much closer to the case of constant heat flux, i.e., $Nu = 8.23$ for parallel plates and 4.36 for circular tubes as shown in Fig. 4.11. When both $\alpha_v$ and $\alpha_T$ are unity, it can be shown that the Nusselt number is reduced to about 50% of the value when $Kn$ is varied from 0 to about 0.16, similar to the constant heat flux case. The Nusselt number goes down significantly with decreasing $\alpha_T$ and goes up somewhat with decreasing $\alpha_v$. The lack of sufficient knowledge of the actual behavior of fluids near the wall makes it difficult to precisely determine the accommodation coefficients. Many of the surfaces used in earlier systems are quite different from those used in MEMS and NEMS, where highly pure crystalline dielectric surfaces are commonly used.

### 4.5.3 *Gas Conduction—From the Continuum to the Free Molecule Regime*

Free molecule flow is important for flight at high altitudes and often associated with chemical reactions and shock waves. The heat transfer aspects of high-speed flow can be found from Rohsenow and Choi [4] or Eckert and Drake [15]. In this section, we use a simple case to illustrate the heat transfer regimes for gas conduction. Consider the conduction by gas between two large plates at temperatures $T_1$ and $T_2$,

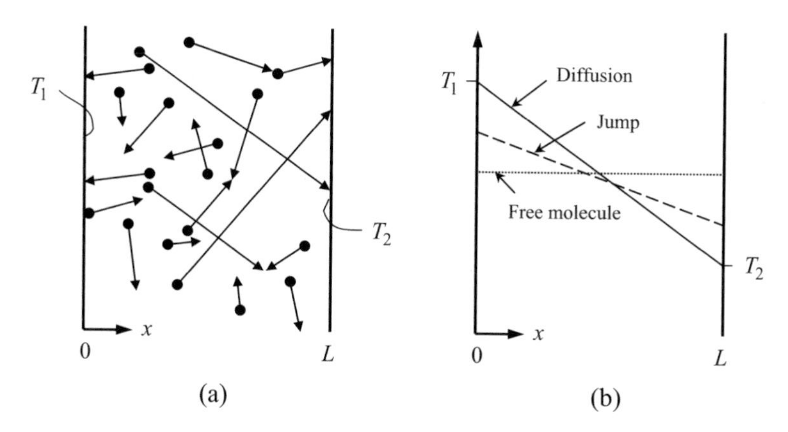


**Fig. 4.12** Heat conduction between two large parallel surfaces filled with an ideal gas. **a** Schematic of the gas molecules. **b** Illustration of the temperature distributions

respectively. The plates are separated at a distance $L$, and the space in between is filled with an ideal gas, as shown in Fig. 4.12. Neglect radiative and convective heat transfer (bulk motion). The heat conduction of a gas is dominated by diffusion, when $Kn = \Lambda/L \gg 1$, where $\Lambda$ is obtained at some effective mean temperature between $T_1$ and $T_2$. In this case, the heat flux can be calculated by applying Fourier's law,

$$q''_{\mathrm{DF}} = \kappa \frac{T_1 - T_2}{L} \tag{4.93}$$

where $\kappa$ can be evaluated using Eq. (4.35) at an effective mean temperature defined as

$$T_{\mathrm{m,DF}} = \left( \frac{2}{3} \frac{T_1^{3/2} - T_2^{3/2}}{T_1 - T_2} \right)^2 \tag{4.94}$$

The above equation takes into consideration the fact that $\kappa \propto \sqrt{T}$, with the assumption that the specific heat is a constant at temperatures between $T_1$ and $T_2$. As long as the density is sufficiently low for the ideal gas model to be valid, the thermal conductivity does not depend on the pressure. The temperature distribution can be obtained by integrating $q'' = -\kappa(T)\frac{\mathrm{d}T}{\mathrm{d}x}$ to obtain

$$T(x) = \left[ T_1^{3/2} - (T_1^{3/2} - T_2^{3/2}) \frac{x}{L} \right]^{2/3} \tag{4.95}$$

which deviates somewhat from a linear relationship. When $Kn = \Lambda/L \gg 1$, however, the chance for molecules to collide with the wall is much larger than that for them to collide with each other. The actual distance a molecule can travel will be less than the mean free path due to collision with the boundary. In the extreme case, one can completely neglect the collisions between molecules and analyze the heat transfer by the molecules, bouncing back and forth between the two plates. The molecules can be sorted into a forward flux and a backward flux, each at a certain equilibrium temperature, determined by the thermal accommodation coefficients. Assume that the thermal accommodation coefficients $a_T$ are the same at both walls. The flux temperatures are

$$T_{1'} = \frac{T_1 + (1-\alpha_T)T_2}{2-\alpha_T} \text{ and } T_{2'} = \frac{T_2 + (1-\alpha_T)T_1}{2-\alpha_T} \tag{4.96}$$

The effective mean temperature of the gas in the free molecule regime is defined as

$$T_{\text{m,FM}} = \frac{4T_{1'}T_{2'}}{\left(\sqrt{T_{1'}} + \sqrt{T_{2'}}\right)^2} \tag{4.97}$$

The net heat flux between the two plates can be expressed as [2]

$$q''_{\text{FM}} = \frac{T_1 - T_2}{\frac{(2-\alpha_T)\sqrt{8\pi RT_{\text{m,FM}}}}{\alpha_T(\gamma+1)c_v P}} \tag{4.98}$$

In the free molecule regime, the heat flux is proportional to the pressure $P$ but independent of $L$ for the given boundary temperatures. This is because the heat transfer rate is proportional to the number density of particles. For intermediate values of $Kn$, the two equations derived under the extreme cases can be combined by adding the thermal resistances such that

$$q'' = \frac{\kappa(T_1 - T_2)}{L\left(1 + Kn\frac{2-\alpha_T}{\alpha_T}\frac{9\gamma-5}{\gamma+1}\sqrt{\frac{T_{\text{m,FM}}}{T_{\text{m,DF}}}}\right)} \tag{4.99}$$

In writing the above equation, we have applied Eq. (4.35) with $\mu = 2\Lambda P/\sqrt{2\pi RT_{\text{m,DF}}}$ from Eq. (4.30). The mean free path $\Lambda$ used in $\kappa$ and $Kn$ should be evaluated at $T_{\text{m,DF}}$. When the temperature difference between the surfaces is smaller than the absolute temperature of the cooler surface, $T_{\text{m,DF}} \approx T_{\text{m,FM}} \approx (T_1 + T_2)/2$. The physical interpretation of Eq. (4.99) is a temperature jump near the surfaces, due to ballistic interaction of the particles with each surface, and a diffusive middle layer, due to particle–particle collisions. For this reason, Eq. (4.99) is called the temperature-jump approximation, which approaches the diffusion limit when $Kn << 1$ and the free molecule limit when $Kn >> 1$. In the transition region, when $\Lambda$ is on the same order as $L$, Eq. (4.99) may be explained by the reduction in the mean free path due to boundary scattering that yields a decrease in the thermal conductivity $\kappa$ from the bulk or diffusion value. This approach will be further explored in the study of the size effect on the thermal conductivity of thin solid films in the next chapter.

**Example 4.5** Calculate the heat flux per 1 K temperature difference near room temperature between two large parallel plates filled with air, assuming $\alpha_T = 0.9$. Plot the results as a function of distance $L$ and pressure $P$. How will you determine the effective thermal conductivity?

**Solution** Let $T_1 = 300.5$ K and $T_2 = 299.5$ K. The effective temperatures calculated from Eqs. (4.94) and (4.97) are very close to the arithmetic mean temperature of 300 K. Note that $\gamma = 1.4$ for air at room temperature. Because $\kappa$ is independent of pressure, we have $\kappa = 0.025$ W/m K from Example 4.3. The mean free path obtained from Eq. (4.19) is $\Lambda/\Lambda_0 = P_0/P$, where $\Lambda_0 = 66.5$ nm at 300 K and the atmospheric pressure $P_0 = 1$ atm. The effective thermal conductivity can be defined as $\kappa_{\text{eff}} = \frac{q''L}{T_1 - T_2}$; hence, Eq. (4.99) reduces to $\kappa_{\text{eff}} = \kappa\left(1 + Kn\frac{2-\alpha_T}{\alpha_T}\frac{9\gamma-5}{\gamma+1}\right)^{-1} = \kappa\left(1 + 3.87\frac{\Lambda_0 P_0}{LP}\right)^{-1}$. It can be seen that $\kappa_{\text{eff}}$ depends on the product $LP$ (which is proportional to $Kn^{-1}$). However, the same cannot be said for the heat flux. The calculated results are shown in Fig. 4.13 for the effective thermal conductivity and heat flux as a function of the separation distance. In the diffusion limit, $\kappa_{\text{eff}}$ is independent of the distance and the pressure, whereas $q''$ increases as $L$ is reduced (proportional to $1/L$). At 1 atm, microscale heat transfer becomes important when $L < 1.5\,\mu\text{m}$ (or $Kn > 0.03$), as $\kappa_{\text{eff}}$ starts to drop, and the dependence of $q''$ on $1/L$ becomes nonlinear. In the free molecule limit, $\kappa_{\text{eff}}$ decreases linearly with both $L$ and $P$ (i.e., the $Kn$), whereas $q''$ is independent of $L$ but depends linearly on $P$. Note that there exists an upper limit of $q''$ for any given pressure. These trends are clearly demonstrated in Fig. 4.13.

The heat transfer calculation mentioned above is important to cryogenic and low-pressure applications. In recent years, atomic force microscopy has become a versatile tool for probing and manipulating, lithography, thermal manufacturing, and measurements at the nanoscales. The heat transfer between the tip and the surface at several

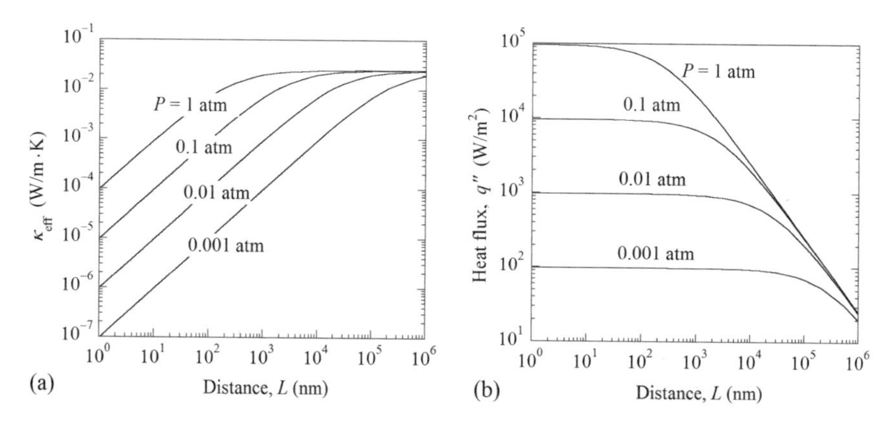


**Fig. 4.13** Distance dependence of **a** the effective thermal conductivity and **b** the heat flux, at different pressures

nanometers may be governed by free molecule flow even at ambient conditions (see Problem 4.27). Radiation heat transfer may increase tremendously when the spacing is less than the characteristic wavelength, which is about 10 μm at 300 K. Hence, radiative heat transfer may be a dominating effect. More details on the nanoscale radiative heat transfer will be given in Chap. 10.

## 4.6 Summary

The simple kinetic theory was introduced based on the ideal gas model, providing a microscopic description of the transport coefficients, such as viscosity, thermal conductivity, and mass diffusion coefficient. This allows one to gain an intuitive understanding of the macroscopic phenomenological or semi-empirical equations, which are important for heat conduction and convection. The complete Boltzmann transport equation (BTE) was then presented from the microscopic point of view. It was shown that the classical transport equations, such as Fourier's law and Navier–Stokes equations can be derived from the BTE under appropriate assumptions. Similar derivations also apply to electron and phonon systems, which will be studied in Chap. 5. The effect of the Knudsen number on the microchannel flow with ideal gases was discussed. The equations for slip flow were solved for simple geometries to provide modified convection heat transfer correlations. Finally, ballistic heat conduction in free molecule flow was described, and a simplified equation was presented that links the continuum region to free molecule flow in the case of conduction between solid walls filled with an ideal gas. The pressure and distance effects on the thermal conductivity and heat flux were clearly demonstrated. The principles discussed in this chapter not only have applications to microfluidics and convection heat transfer but also are important to the subsequent chapters on heat transfer in solid micro/nanostructures.

## Problems

4.1. (a) Determine the mean free path $\Lambda$, average molecular spacing $L_0$, and the frequency of collision $\tau^{-1}$ for air at sea level (15 °C and 1 atm).
(b) Determine the root-mean-square free path.
(c) What is the probability of finding a free path greater than $4\Lambda$?
(d) Calculate $\Lambda$, $L_0$, and $\tau^{-1}$ for air at 200 miles above sea level with $M = 17.3$, $P/P_0 = 5.9 \times 10^{-11}$, and $\rho/\rho_0 = 10^{-11}$, where the subscript 0 signifies properties at sea level. Note: The average size of air molecules may be taken as $3.0 \times 10^{-10}$ m at this altitude.
(e) What is the kinetic temperature at this altitude? Explain the reason why $M$ changes with the altitude.

4.2. Use the mean-free-path distribution to answer the following questions:

(a) What is the root-mean-square free path in terms of the mean free path $\Lambda$.
(b) What is the most probable free path?
(c) What is the probability of finding a free path greater than $\Lambda$?

4.3. Air is pumped to a pressure $P = 10$ Pa at 25 °C. Calculate the following quantities: the average distance between adjacent molecules $L_0$, the molecular mean free path $\Lambda$, the number of collisions that a molecule experiences every second, the molecule flux $J_N$ on any surface, the most probable speed of the molecules, the most probable velocity of the molecules, and the average kinetic energy of each molecule.

4.4. Hydrogen gas is cooled to 100 K, while the pressure is reduced to 0.1 Pa. Determine the mean free path $\Lambda$ and the average frequency of collision. What are the rms speed and the average kinetic energy of a molecule? What is the momentum flux of the gas on the container? What are the most probable free path, the most probable speed, and the most probable velocity?

4.5. What is the dependence of $\mu$ and $\kappa$ on pressure and temperature? How does $D_{\rm AB}$ depend on pressure? For water vapor and air, $D_{\rm AB} = 2.56 \times 10^{-5}$ m$^2$/s at 298 K and 100 kPa. Plot $D_{\rm AB}$ as a function of temperature at $P = 10$, 20, 50, and 100 kPa.

4.6. Calculate $\mu$, $c_v$, $\kappa$, and $Pr$ for oxygen and nitrogen at 100, 300, and 1000 K, and 1 atm. Compare your calculated results with the values tabulated in most heat transfer textbooks to estimate the relative differences.

4.7. A chamber containing $O_2$ at 100 K and $10^{-3}$ atm is placed in the outer space. The oxygen leaks to the outer space through a small hole, 1 μm diameter, in the chamber wall. What is the mass flow rate?

(a) Estimate the number of molecules that escape from the container per unit time.
(b) What is the mass flux?
(c) Evaluate the flux of kinetic energy, $J_{\rm KE}$ using Eq. (4.8). How is your answer compared with $J_N \times m\overline{v^2}/2$? Why are the results different?
(d) If the diameter of the hole is increased to 1 cm, is the basis of your calculation still valid?

4.8. A tube connects a $CH_4$ line to the air. Assuming both ends of the tube are at 1 atm and 25 °C, calculate the binary diffusion coefficient between $CH_4$ and air. Find the mass flow rate of $CH_4$ to the air and that of air to the $CH_4$ line, given the tube has an inner diameter of 5 mm and a length of 7 m. Sketch the concentration distributions in the tube line.

4.9. A tube connects an $O_2$ container to a $N_2$ container. Assume that the temperature is 200 °C and the pressure is 2 atm inside the containers and the tube. Calculate the mass exchange rates of $O_2$ and $N_2$ from one container to the other, assuming that the tube has an inner diameter of 5 mm and a length of 3 m.

4.10. Dry air at 34 °C flows over a flat plate of length $L = 0.1$ m with a velocity of 15 m/s. The width of the plate is 1 m. The surface of the plate is covered with a thin soaked fabric, and electric power is applied to the plate to maintain its surface temperature at 20 °C.

(a) Assuming that the bottom of the plate is insulated, determine the required electric power.
(b) After a long period of operation, the fabric is completely dry. Neglect the changes in the convection coefficient and the electric power. What will be the steady-state surface temperature?
(c) Is it a good assumption to neglect the radiative heat transfer?

4.11. Use $r_0 = 2.869 \times 10^{-10}$ m and $\varepsilon_0/k_B = 10.22$ K for He to calculate and plot the Lennard-Jones potential. Set one molecule at a fixed (pinned) position on the $x$-axis, say at $x = 5$ nm. The other molecule starts at the origin with an initial velocity $\mathbf{v}_0 = v_0(\hat{\mathbf{x}} \cos\beta + \hat{\mathbf{y}} \sin\beta)$, where $\beta$ is a small angle between $\mathbf{v}_0$ and the $x$-axis. Develop a computer program to calculate the trajectory of the moving particle in the $x$-$y$ plane, for various $v_0$ and $\beta$, based on (a) the rigid-elastic-sphere assumption and (b) the intermolecular force field. Comment on the differences between the results obtained from the two models.

4.12. Using Eq. (4.63), show that Eq. (4.62) is identical to Eq. (2.42). Hint: $\nabla \cdot \{P_{ij}\} = \left(\frac{\partial P_{xx}}{\partial x} + \frac{\partial P_{yx}}{\partial y} + \frac{\partial P_{zx}}{\partial z}\right)\hat{\mathbf{x}} + \left(\frac{\partial P_{xy}}{\partial x} + \frac{\partial P_{yy}}{\partial y} + \frac{\partial P_{zy}}{\partial z}\right)\hat{\mathbf{y}} + \left(\frac{\partial P_{xz}}{\partial x} + \frac{\partial P_{yz}}{\partial y} + \frac{\partial P_{zz}}{\partial z}\right)\hat{\mathbf{z}}$.

4.13. Derive the viscous dissipation term in Eq. (2.43) based on Eq. (4.64).

4.14. From Eq. (4.64), derive the heat diffusion equation: $\kappa \nabla^2 T = \rho c_v \frac{\partial T}{\partial t}$.

4.15. Assuming $\tau$ is independent of the frequency, use the Maxwell velocity distribution, Eq. (3.43), to evaluate $\kappa = \frac{\tau}{3} \int_{\varpi} v^2 \varepsilon \frac{\partial f_0}{\partial T} d\varpi$ for a monatomic gas, where $\varepsilon = \frac{1}{2} m \mathbf{v}^2$.

4.16. Consider an isothermal gas flow in the $x$-direction with a bulk velocity distribution $v_B(y) = \overline{v_x}(y)$ as shown in Fig. 4.5. The velocity distribution is not very far from the equilibrium so that $f = f_0 - \tau v_y \frac{\partial f_0}{\partial v_B} \frac{dv_B}{dy}$. Find an expression of the dynamic viscosity $\mu$. Hint: $\tau_{yx} = \int_{\varpi} (m v_x) v_y f d\varpi$ according to Eq. (4.13); the answer is $\frac{1}{4d^2}\left(\frac{m k_B T}{\pi}\right)^{1/2}$.

4.17. What is the continuum assumption, and when does the continuum assumption break down? Define the Knudsen number, and what is its physical significance? What are the unique issues related to microfluidics? What are the applications of microfluidics?

4.18. What happens at the boundary layer for a fluid moving over a large plate during slip flow? Describe both the velocity distribution and the temperature distribution near the wall. Write the slip-flow boundary conditions, and discuss the significance of each term.

4.19. Integrate Eq. (4.88) to find the dimensionless bulk temperature $\Theta_m$; and then use the definition of Nusselt number to prove Eq. (4.90).

4.20. Find the temperature distribution for slip flow between two parallel plates when the bottom plate is insulated and the top plate is heated at a uniform heat flux. Continue on to verify Eq. (4.91).

4.21. Find the velocity and temperature distributions for slip flow through a circular tube with a uniform wall heat flux. Continue on to verify Eq. (4.92).

4.22. For slip flow with temperature jump in a circular tube, show that there exists a maximum Nusselt number at the entrance, given by $Nu_{\max} = 1/\beta_T$.

4.23. For Poiseuille flow with velocity slip, calculate the friction coefficient $C_\mathrm{f} = \tau_\mathrm{s}/(\rho v_\mathrm{m}^2/2)$ at the entrance and for fully developed gas flow.

4.24. For fully developed gas flow in a circular tube, develop an expression for the ratio of the required pump powers with slip and without slip.

4.25. A heat sink contains 100 microchannels, each 1 mm long with a 1 μm × 30 μm cross section. Cold air at 22 °C flows in at 2 atm with a velocity of 4 m/s. The sides of the channel are well insulated, and a constant wall flux $q''_\mathrm{w} = 40\ \mathrm{W/m^2}$ is removed by the flow. Neglecting the entry region, what will be the exit temperature of the air? What will be the wall temperature at the exit? (Assume that $\alpha_v = \alpha_T = 0.8$.)

4.26. For the same fluid, entrance conditions, and wall heat flux as in Problem 4.25, estimate the convection coefficient for fully developed flow in a circular tube as a function of the tube diameter. Take $D = 300$ nm, 3 μm, and 300 μm.

4.27. Model the cantilever tip of an atomic force microscope (AFM) as a flat disk, with a diameter of 100 nm, that is above a flat surface at 300 K. If the tip is heated to 400 K, calculate the heat flux from the tip to the surface when the distance varies from 10 to 1 nm, assuming that the tip and the sample surface are surrounded by dry air at ambient pressure. How will your calculation change if the pressure is reduced to 1 torr? [1 torr = 1 mmHg = 133.3 Pa.]

4.28. Team Project 1: Derive the Nusselt number for constant wall temperature for a laminar slip flow either in a circular tube or between two parallel plates.

4.29. Team Project 2: Develop a computer program to evaluate the Nusselt number in the entry region for uniform wall heat flux in a circular flow.

4.30. Team Project 3: Perform a simulation using the DSMC method for gas conduction between two plates, with different Knudsen numbers. Compare your results with Eq. (4.99).

## References

1. J.H. Jeans, *The Dynamical Theory of Gases*, 4th edn. (Cambridge University Press, Cambridge, UK, 1925)
2. E.H. Kennard, *Kinetic Theory of Gases* (McGraw-Hill, New York, 1938)
3. S.A. Schaaf, P.L. Chambre, *Flow of Rarefied Gases* (Princeton University Press, Princeton, NJ, 1961)
4. E.H. Rohsenow, H.Y. Choi, *Heat, Mass, and Momentum Transfer* (Prentice-Hall, NJ, 1961)
5. J.O. Hirschfelder, C.F. Curtiss, R.B. Bird, *Molecular Theory of Gases and Liquids*, 2nd edn. (Wiley, New York, 1964)

6. S. Chapman, T.G. Cowling, *The Mathematical Theory of Non-uniform Gases*, 3rd edn. (Cambridge University Press, London, UK, 1971)
7. C.L. Tien, J.H. Lienhard, *Statistical Thermodynamics* (Hemisphere, New York, 1985)
8. C. Cercignani, *The Boltzmann Equations and Its Applications* (Springer, New York, 1988)
9. J.N. Israelachvili, *Intermolecular and Surface Forces*, 3rd edn. (Elsevier, Amsterdam, 2011)
10. J.A. Harrison, J.D. Schall, S. Maskey, P.T. Mikulski, M.T. Knippenberg, B.H. Morrow, Review of force fields and intermolecular potentials used in atomistic computational materials research. Appl. Phys. Rev. **5**, 031104 (2018)
11. M.J. Madou, *Fundamentals of Microfabrication: The Science of Miniaturization*, 2nd edn. (CRC Press, Boca Raton, FL, 2002)
12. C.-M. Ho, Y.-C. Tai, Micro-electro-mechanical-systems (MEMS) and fluid flows. Annu. Rev. Fluid Mech. **30**, 579–612 (1998)
13. M. Gad-el-Hak, The fluid mechanics of microdevices—the freeman scholar lecture. J. Fluids Eng. **121**, 5–33 (1999)
14. H.-S. Tsien, Superaerodynamics, mechanics of rarefied gases. J. Aeronautical Sci. **13**, 653–664 (1946)
15. E.R.G. Eckert, R.M. Drake, *Analysis of the Heat and Mass Transfer* (McGraw-Hill, New York, 1972)
16. N.-T. Nguyen, S.T. Wereley, *Fundamentals and Applications of Microfluidics* (Artech House, Boston, MA, 2002)
17. G.E. Karniadakis, A. Beskok, *Micro Flows: Fundamentals and Simulation* (Springer-Verlag, New York, 2002)
18. G.P. Peterson, L.W. Swansoon, F.M. Gerner, Micro heat pipes, in *Microscale Energy Transport,* C.L. Tien, A. Majumdar, F.M. Gerner (eds.), Taylor & Francis, Washington DC, pp. 295–338 (1998)
19. S.V. Garimella, C.B. Sobhan, Transport in microchannels—a critical review. Annu. Rev. Heat Transfer **13**, 1–50 (2003)
20. D. Poulikakos, S. Arcidiacono, S. Maruyama, Molecular dynamics simulation in nanoscale heat transfer: a review. Microscale Thermophys. Eng. **7**, 181–206 (2003)
21. G.A. Bird, *Molecular Gas Dynamics and the Direct Simulation of Gas Flows* (Oxford University Press, Oxford, UK, 1994)
22. S. Chen, G.D. Doolen, Lattice Boltzmann method for fluid flows. Annu. Rev. Fluid Mech. **30**, 329–364 (1998)
23. S.C. Saxena, R.K. Joshi, *Thermal Accommodation and Adsorption Coefficients of Gases* (Hemisphere, New York, 1989)
24. E.B. Arkilic, K.S. Breuer, M.A. Schmidt, Mass flow and tangential momentum accommodation in silicon micromachined channels. J. Fluid Mech. **437**, 29–43 (2001)
25. R. Inman, *Laminar Slip Flow Heat Transfer in a Parallel Plate Channel or a Round Tube with Uniform Wall Heating,* NASA TN D-2393, 1964
26. E.M. Sparrow, S.H. Lin, Laminar heat transfer in tubes under slip-flow conditions. J. Heat Transfer **84**, 363–369 (1962)
27. S. Yu, T.A. Ameel, Slip flow convection in isoflux rectangular microchannels. J. Heat Transfer **124**, 346–355 (2002)
28. N.G. Hadjiconstantinou, O. Simek, Constant-wall-temperature Nusselt number in micro and nano-channels. J. Heat Transfer **124**, 356–364 (2002)

# Chapter 5
# Thermal Properties of Solids and the Size Effect

One of the thrust areas of research in micro/nanoscale heat transfer is related to properties and transport processes in solid materials and devices. In the early 1990s, much research was conducted to identify the regimes when the microscale effect must be considered in dealing with problems occurring at small length and/or timescales [1, 2]. Significant progress has been made in the past decades on understanding the fundamental thermal transport properties of solids and nanostructures. Cahill et al. [3, 4] provided comprehensive surveys on the thermal phenomena and measurement techniques associated with solid-state devices across the nano-, micro-, and macro-length scales and in a large temperature range. The critical dimensions of integrated circuits have continued to shrink during the past few decades, and feature sizes smaller than 10 nm have been reached in recent years. Overheating caused by thermal energy generation is a major source of device failure, and it often occurs in very small regions, known as hot spots. A remarkable number of micro/nanostructured materials and systems have temperature-dependent figures of merit. Therefore, understanding the thermophysical properties, thermal transport physics, and thermal metrology from the micrometer down to the nanometer length scales is critically important for the future development of microelectronic devices and nanobiotechnology.

This chapter focuses on simple phonon theory and electronic theory of the specific heat, thermal conductivity, and thermoelectricity of metals and insulators. The Boltzmann transport equation (BTE) has been used to facilitate the understanding of microscopic behavior, together with the quantum statistics of phonons and electrons. The quantum size effect on phonon specific heat is extensively covered. Examples are given to analyze direct thermoelectric conversion for temperature measurement, power generation, and refrigeration. Furthermore, a detailed treatment of classical size effect on thermal conductivity is presented. Finally, the concepts of quantum electrical conductance and thermal conductance are introduced.



Z. M. Zhang, *Nano/Microscale Heat Transfer*, Mechanical Engineering Series,
https://doi.org/10.1007/978-3-030-45039-7_5

## 5.1 Specific Heat of Solids

In this section, simple models of the specific heat of bulk solids are described considering the contribution of lattice vibrations as well as free electrons in metals. The purpose is to understand macroscopic behavior from a microscopic point of view and to prepare students for further study on the quantum size effect to be discussed in subsequent sections.

### *5.1.1 Lattice Vibration in Solids: The Phonon Gas*

The atoms in solids are close to each other, and interatomic forces keep them in position. Atoms cannot move around except for vibrations near their equilibrium positions. In crystalline solids, atoms are organized into periodic arrays, and each identical structural unit is called a lattice. Lattice vibrations contribute to thermal energy storage and heat conduction. In metals, electrons are responsible for electrical transport and heat conduction but are less important for storing thermal energy except at very low temperatures.

The simple oscillator model treats each atom as a harmonic oscillator, which vibrates along all three axes as shown in Fig. 5.1. If the vibrational degrees of freedom were completely excited, we would expect the high-temperature limit of the specific heat of elementary (monatomic) solids to be

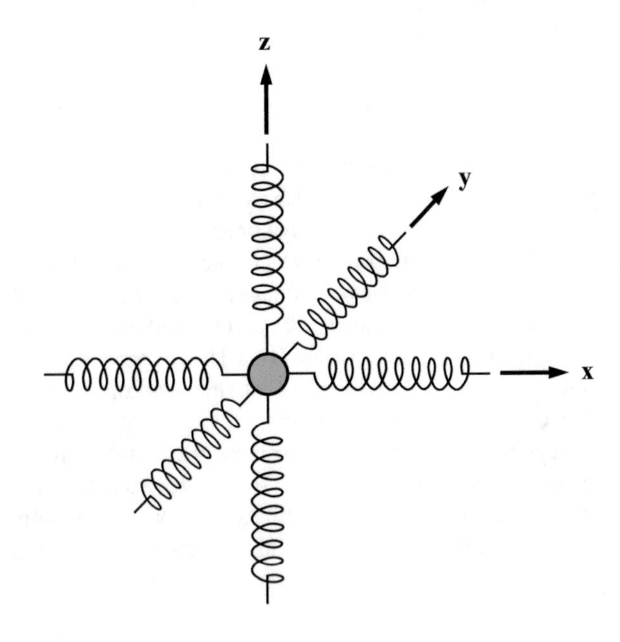


**Fig. 5.1** The harmonic oscillator model of an atom in a solid

$$\bar{c}_v = 3\bar{R} \tag{5.1}$$

This is called the Dulong–Petit law, named after Pierre-Louis Dulong and Alexis-Thérèse Petit in 1819. The Dulong–Petit law can be understood in terms of the equipartition principle in classical statistics. However, it cannot predict low-temperature behavior; even above room temperature, the model significantly overpredicts the specific heats for diamond, graphite, and boron.

Einstein in 1907 proposed a simple harmonic oscillator model and its quantized energy levels $(i + \frac{1}{2})h\nu$, $i = 1, 2,\ldots$, to obtain the specific heat as a function of temperature. Here, the frequency $\nu$ is a characteristic vibration frequency of the solid material. The procedure is similar to the analysis of vibration energies for diatomic gas molecules, e.g., Eqs. (3.59)–(3.62). The resulting specific heat for a monatomic solid is

$$\bar{c}_v(T) = 3\bar{R}\frac{\Theta_{\mathrm{E}}^2}{T^2}\frac{\mathrm{e}^{\Theta_{\mathrm{E}}/T}}{\left(\mathrm{e}^{\Theta_{\mathrm{E}}/T} - 1\right)^2} \tag{5.2}$$

where the factor 3 accounts for oscillation in all three directions and $\Theta_{\mathrm{E}} = h\nu/k_{\mathrm{B}}$ is called the Einstein temperature [5, 6]. It can be shown that $\bar{c}_v \to 0$ as $T \to 0$ and $\bar{c}_v \to 3\bar{R}$ at $T \gg \Theta_{\mathrm{E}}$. In the intermediate temperature range, however, the Einstein specific heat is significantly lower than the experimental data. This can be seen from Fig. 5.2, where the experimental results of the constant-pressure specific heat are taken from Ashcroft and Mermin [6]. It should be noted that $c_p = c_v$ for a solid under the incompressible assumption. The reduced temperature is the ratio of the temperature to the characteristic temperature (either the Einstein temperature or Debye temperature depending on the model). The experimental data were plotted using the Debye temperature given in Table 5.1. The reason that the specific heat of diamond is far from $3\bar{R}$ near room temperature is because of its

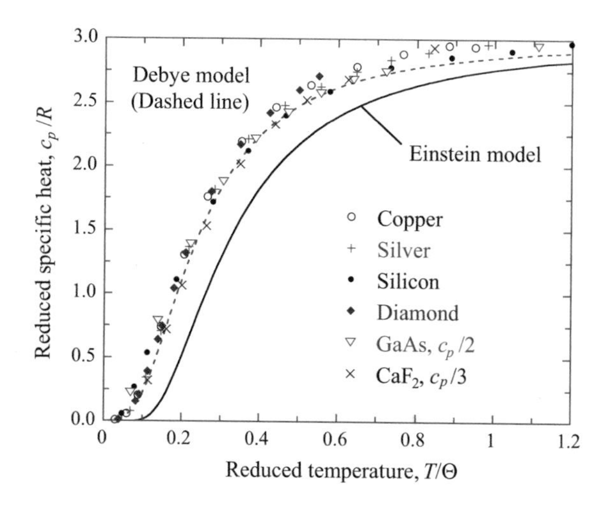


**Fig. 5.2** Comparison of model predictions with experimental data of the specific heat for several crystalline solids

**Table 5.1** The Debye temperature, melting temperature, and other properties for selected solids. The data are mainly taken from Kittel [5] and Ashcroft and Mermin [6]. The reported densities are for 22 °C except for Ar

| Element or compound | Symbol or formula | M (kg/kmol) | $\Theta_D$ (K) | $T_{melt}$ (K) | $n_a$ ($10^{28}$ m$^{-3}$) | $\rho$ ($10^3$ kg/m$^3$) |
|---|---|---|---|---|---|---|
| Argon | Ar | 40 | 92 | 84 | 2.66 (4 K) | 1.77 (4 K) |
| Mercury | Hg | 200.6 | 72 | 234 | 4.26 | 14.26 |
| Sodium | Na | 23 | 158 | 371 | 2.65 | 1.013 |
| Lithium | Li | 6.9 | 344 | 454 | 4.7 | 0.542 |
| Lead | Pb | 207 | 105 | 601 | 3.3 | 11.34 |
| Zinc | Zn | 65.4 | 327 | 692 | 6.55 | 7.13 |
| Magnesium | Mg | 24.3 | 400 | 922 | 4.30 | 1.74 |
| Aluminum | Al | 27 | 428 | 934 | 6.03 | 2.7 |
| Calcium | Ca | 40 | 230 | 1113 | 2.30 | 1.53 |
| Silver | Ag | 108 | 225 | 1235 | 5.85 | 10.5 |
| Copper | Cu | 63.5 | 340 | 1358 | 8.45 | 8.93 |
| Gold | Au | 197 | 165 | 1338 | 5.90 | 19.3 |
| Iron | Fe | 56 | 470 | 1811 | 8.50 | 7.87 |
| Silicon | Si | 28 | 645 | 1687 | 5.0 | 2.33 |
| Diamond | C | 12 | 2000 | 3620 | 17.6 | 3.52 |
| Potassium bromide | KBr | 119 | 177 | 1007 | | 2.75 |
| Sodium chloride | NaCl | 58.5 | 281 | 1074 | | 2.17 |
| Gallium arsenide | GaAs | 144.6 | 360 | 1511 | | 5.32 |
| Calcium fluoride | $CaF_2$ | 78 | 474 | 1696 | | 3.18 |

very high characteristic temperature (or frequency of vibration) compared to other materials as shown in Table 5.1.

In the Einstein model, each atom is treated as an independent oscillator and all atoms are assumed to vibrate at the same frequency. In 1912, Max Born and Theodore von Kármán first realized that the bonding in a solid prevents independent vibrations. Therefore, a collection of vibrations must be considered under the force–spring interactions of the nearby atoms. To avoid the complicated calculations, Peter Debye in 1912 simplified the model by assuming that the velocity of sound is the same in all crystalline directions and for all frequencies. In addition, there is a high-frequency cutoff, and no vibration can occur beyond this frequency. As to be seen from subsequent sections, the Debye model is a great success and has prevailed even though more advanced and realistic theories have been developed.

### 5.1.2 The Debye Specific Heat Model

The Debye model for the specific heat of solids includes a large number of closely spaced modes (or vibration frequencies) up to a certain upper bound $\nu_m$, which is determined by the total number of vibration modes 3*N*, where *N* is the number of atoms. The high-frequency limit is indeed plausible because the shortest wavelength of the lattice wave should be on the order of the interatomic distances, or the lattice constants. Rather than treating each atom as an individual oscillator, the Debye model assumes that vibrations are inside the whole crystal just like standing waves. For elastic vibrations, there are longitudinal waves (e.g., sound waves) and transverse waves (with two polarizations) in a crystal. In analogy to electromagnetic waves and photons, the quanta of lattice waves are called *phonons*. The energy of a phonon is $\varepsilon = h\nu$, where $\nu$ is the vibration frequency. The momentum of a phonon is $p = h\nu/v_p = h/\lambda$, where $\nu$ is the frequency, $\lambda$ is the wavelength, and $v_p = \lambda\nu$ is the speed of propagation (or phase speed) for the given phonon mode. It should be noticed that the propagation speeds of longitudinal and transverse acoustic waves are different. So far, we have related lattice vibrations to lattice waves and to the translational movement of the phonon gas, which follows the Bose–Einstein statistics. However, the total number of phonons is not conserved since it depends on temperature. Thus, we do not need to apply the constraint given in Eq. (3.2) and can simply set $\alpha = 0$ in Eq. (3.16). The result is

$$\frac{N_i}{g_i} = \frac{1}{e^{\varepsilon_i/k_B T} - 1} \tag{5.3}$$

Suppose the energy levels are closely spaced; we can write Eq. (5.3) in terms of a continuous function called the Bose–Einstein distribution function at a given temperature $T$ as

$$f_{BE}(\nu) = \frac{dN}{dg} = \frac{1}{e^{h\nu/k_B T} - 1} \tag{5.4}$$

The *degeneracy* for phonons is the number of quantum states per unit volume in the phase space. For a given volume $V$ and within a spherical shell in the momentum space (from $p$ to $p + dp$), we have from Eq. (3.87) that $dg = 4\pi V p^2 dp/h^3 = 4\pi V \nu^2 d\nu/v_p^3$. Hence,

$$\frac{dg}{V} = \frac{g(\nu)d\nu}{V} = D(\nu)d\nu = \frac{4\pi\nu^2}{v_p^3}d\nu \tag{5.5}$$

Here, we have introduced the *density of states* (DOS) of phonons, $D(\nu)$, which is the number of quantum states per unit volume per unit frequency or energy ($h\nu$)

interval. Equation (5.4) gives the *mean occupation number*, i.e., the average number of bosons per quantum state at frequency $\nu$. The phonon number density in terms of the DOS can be expressed as

$$n = \int_0^\infty f_{\mathrm{BE}}(\nu) D(\nu) \mathrm{d}\nu \tag{5.6}$$

Because there exist one longitudinal and two transverse waves, the phonon DOS in a large spherical shell of the momentum space can be written as

$$D(\nu) = 4\pi \nu^2 \left( \frac{1}{v_l^3} + \frac{2}{v_t^3} \right) = \frac{12\pi \nu^2}{v_{\mathrm{a}}^3} \tag{5.7}$$

where $v_l$ is the speed of the longitudinal wave, $v_t$ is the speed of the transverse wave, and $v_{\mathrm{a}}$ is a weighted average defined in the above equation. The total number of quantum states must be equal to 3*N*, since each quantum state corresponds to a harmonic oscillator. Using integration in place of summation, we have

$$\frac{3N}{V} = \int_0^\infty D(\nu) \mathrm{d}\nu = \int_0^{\nu_{\mathrm{m}}} \frac{12\pi \nu^2}{v_{\mathrm{a}}^3} \mathrm{d}\nu \tag{5.8}$$

where $\nu_{\mathrm{m}}$ is an upper limit of the frequency that can be obtained from Eq. (5.8) as

$$\nu_{\mathrm{m}} = \left( \frac{3n_{\mathrm{a}}}{4\pi} \right)^{1/3} v_{\mathrm{a}} \tag{5.9}$$

Here, $n_{\mathrm{a}} = N/V$ is the number density of atoms.

The Debye temperature is defined as

$$\Theta_{\mathrm{D}} = \frac{h\nu_{\mathrm{m}}}{k_{\mathrm{B}}} = \frac{h}{k_{\mathrm{B}}} \left( \frac{3n_{\mathrm{a}}}{4\pi} \right)^{1/3} v_{\mathrm{a}} \tag{5.10}$$

The Debye temperature and the number density for various solids are listed in Table 5.1 together with some other properties. The listed values of the Debye temperature were based on the experimentally measured specific heat at very low temperatures, rather than that calculated from the speed of sound. The result of the Debye specific heat theory agrees fairly well with the experimental data for several crystalline solids in a large temperature range, as can be seen from Fig. 5.2. The high-temperature limit of the specific heat is $6\bar{R}$ for GaAs and $9\bar{R}$ for $CaF_2$, because the number of atoms in a unit cell of the lattice is 2 and 3, respectively.

**Example 5.1** The average speed of the longitudinal waves is $v_l = 8970\,\mathrm{m/s}$ and that of the transverse waves is $v_t = 5400\,\mathrm{m/s}$ in silicon. Find the average propagation speed, the maximum frequency, the Debye temperature, and the minimum wavelength $\lambda_{\min}$. How does $\lambda_{\min}$ compare with the average distance between atoms?

**Solution** Since $v_\mathrm{a}^3 = 3/(v_l^{-3} + 2v_t^{-3})$, we have $v_\mathrm{a} = 5972\,\mathrm{m/s}$. Given $n_\mathrm{a} = 5.0 \times 10^{28}\,\mathrm{m}^{-3}$, we obtain $\nu_\mathrm{m} = 1.36 \times 10^{13}\,\mathrm{Hz} = 13.6\,\mathrm{THz}$ from Eq. (5.9) and $\Theta_\mathrm{D} = 655\,\mathrm{K}$ from Eq. (5.10), which is a little bit higher than the experimental value of 645 K listed in Table 5.1. The experimental value was obtained by fitting the low-temperature specific heat with the Debye model. The minimum wavelength is estimated by $\lambda_{\min} = v_\mathrm{a}/\nu_\mathrm{m} = 0.44\,\mathrm{nm} = 4.4\,\text{Å}$. The average spacing between atoms can be estimated by $L_0 = n_\mathrm{a}^{-1/3} = 0.27\,\mathrm{nm}$ or $2.7\,\text{Å}$, suggesting that $\lambda_{\min} \approx 2L_0$. The maximum wavelength of the lattice wave will be twice the extension of the solid. For a cubic solid with each side $L$, we have $\lambda_{\max} \approx 2L$. The lattice waves are illustrated in Fig. 5.3 in a 1D case.

The distribution function for phonons can now be written as

$$f(\nu) = \frac{1}{V}\frac{\mathrm{d}N}{\mathrm{d}\nu} = D(\nu) f_\mathrm{BE}(\nu) = \frac{12\pi\nu^2}{v_\mathrm{a}^3(e^{h\nu/k_\mathrm{B}T} - 1)} = \frac{9n_\mathrm{a}\nu^2}{\nu_\mathrm{m}^3(e^{h\nu/k_\mathrm{B}T} - 1)},\ \nu \le \nu_\mathrm{m} \tag{5.11}$$

The vibration contribution to the internal energy can be written as

$$U - U_0 = \int_0^\infty f(\nu) h\nu \mathrm{d}\nu \tag{5.12a}$$

where $U_0$ is the internal energy at 0 K when no vibration modes are excited. The result after some manipulation becomes

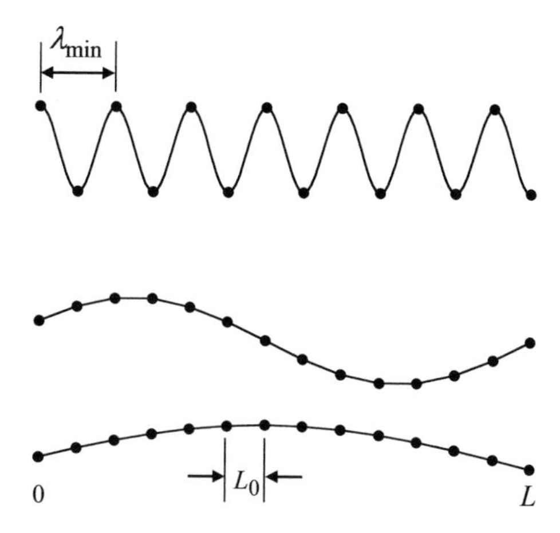


**Fig. 5.3** Illustration of the minimum wavelength $\lambda_{\min} = 2L_0$ and the maximum wavelength $\lambda_{\max} = 2L$ associated with lattice vibrations in a solid with a dimension $L$ and with a periodic array of atoms (dots)

$$U - U_0 = 9Nk_B T \left(\frac{T}{\Theta_D}\right)^3 \int_0^{x_D} \frac{x^3}{e^x - 1} dx \tag{5.12b}$$

where $x_D = \Theta_D/T$. The molar specific heat is then

$$\bar{c}_v(T) = \left(\frac{\partial \bar{u}}{\partial T}\right)_V = 9\bar{R}\left(\frac{T}{\Theta_D}\right)^3 \int_0^{x_D} \frac{x^4 e^x}{(e^x - 1)^2} dx \tag{5.13}$$

The specific heat predicted by the Debye theory agrees very well with experimental data of many solids, as shown in Fig. 5.2. Notice that $\int_0^{x_D} x^4 e^x (e^x - 1)^{-2} dx = 4\int_0^{x_D} x^3 (e^x - 1)^{-1} dx - x_D^4/(e^{x_D} - 1)$. When $T \gg \Theta_D$, $x_D \to 0$ and $e^x - 1 \approx x$. Thus, $\int_0^{x_D} x^3 (e^x - 1)^{-1} dx \to x_D^3/3$, and the Debye specific heat approaches $3\bar{R}$ in the high-temperature limit. The relative difference is about 5% at $T = \Theta_D$. Using Eq. (B.9), it can be shown that at $T \ll \Theta_D$, Eq. (5.13) can be approximated by

$$\bar{c}_v(T) = \frac{12\pi^4}{5}\bar{R}\left(\frac{T}{\Theta_D}\right)^3 \propto T^3 \tag{5.14}$$

which is known as the $T^3$ law, and it agrees with experiments for many solid materials within a few percent for $T/\Theta_D < 0.1$ [7].

In essence, the Einstein specific heat theory assumed that all oscillations are at the same frequency, and it implied that the DOS has a sharp peak at that frequency and is zero at all other frequencies. On the other hand, the Debye theory is based on a parabolic function, $D(\nu) \propto \nu^2$. More detailed studies have revealed that the actual phonon density of states is a complicated function of the frequency [6, 8], as illustrated in Fig. 5.4 for aluminum and copper according to neutron scattering measurements. There are different phonon branches in a real crystal that affect the DOS in different frequency regions. A detailed discussion will be deferred to Chap. 6 when we take

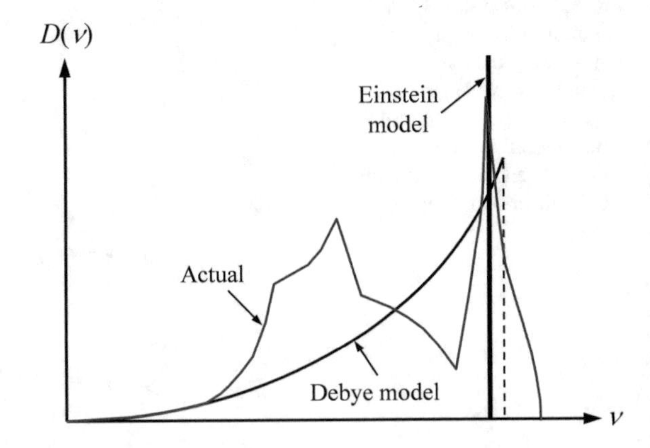


**Fig. 5.4** Illustration of the phonon density of states in the Einstein model and the Debye model as compared with the actual behavior of metals

a deeper look into the crystalline structures and phonon dispersion relations. In general, the Debye theory predicts correctly the low-temperature behavior when only the low-frequency phonon modes are excited; this is probably the most significant contribution of the Debye model. At higher temperatures, the Debye model can be considered as a first-order approximation, as shown in Fig. 5.2.

### 5.1.3 *Free-Electron Gas in Metals*

The translational motion of free electrons within metals is largely responsible for their electrical and thermal conductivities. Sometimes, the free electrons are called electron gas, drawing an analogy between electrons and monatomic molecules. However, there are distinct differences between electrons in a solid and molecules in an ideal gas. The number of free electrons is on the order of the number of atoms. For Au, Cu, and Ag, we shall assume there is one (1) free electron per atom, but there are three (3) electrons per atom for Al and four (4) electrons per atom for Pb (see Table 5.2). Electrons obey the Fermi–Dirac distribution given in Eq. (3.24). A continuous function called the *Fermi function* can be defined as

$$f_{\mathrm{FD}}(\varepsilon) = \frac{\mathrm{d}N}{\mathrm{d}g} = \frac{1}{\mathrm{e}^{(\varepsilon-\mu)/k_{\mathrm{B}}T} + 1} \tag{5.15}$$

The Fermi function is plotted in Fig. 5.5a, where $\mu_{\mathrm{F}} = \mu$ at $T = 0$ K is called the *Fermi energy* (or Fermi level). It will be shown later that $\mu$ changes little when the temperature is not very high. At the absolute temperature of 0 K, $f_{\mathrm{FD}} = 1$ when $\varepsilon < \mu_{\mathrm{F}}$, and $f_{\mathrm{FD}} = 0$ when $\varepsilon > \mu_{\mathrm{F}}$. This suggests that each quantum state whose energy is below the Fermi energy is occupied by one electron. All quantum states whose energies exceed the Fermi energy are not occupied. As the temperature increases, the function falls less sharply. Hence, the quantum states slightly below the Fermi level are still filled, and those slightly above the Fermi level remain empty. However, the quantum states are only partially filled around the Fermi level.

The degeneracy for electrons is further increased by 2, due to the existence of positive and negative spins. In a volume $V$ of a spherical shell in the momentum space, we have $\mathrm{d}g = 8\pi V(m_{\mathrm{e}}/h)^3 v^2 \mathrm{d}v$ from Eq. (3.86) by considering the spin degeneracy. Hence, the distribution function in terms of the electron speed is

$$f(v) = \frac{1}{V}\frac{\mathrm{d}N}{\mathrm{d}v} = 8\pi\left(\frac{m_{\mathrm{e}}}{h}\right)^3 \frac{v^2}{\mathrm{e}^{(\varepsilon-\mu)/k_{\mathrm{B}}T} + 1} \tag{5.16}$$

Using $f(v)\mathrm{d}v = f(\varepsilon)\mathrm{d}\varepsilon$ and $\varepsilon = m_{\mathrm{e}}v^2/2$, we obtain the distribution function in terms of the kinetic energy of electrons as

$$f(\varepsilon) = \frac{1}{V}\frac{\mathrm{d}N}{\mathrm{d}\varepsilon} = 4\pi\left(\frac{2m_{\mathrm{e}}}{h^2}\right)^{3/2} \frac{\sqrt{\varepsilon}}{\mathrm{e}^{(\varepsilon-\mu)/k_{\mathrm{B}}T} + 1} \tag{5.17}$$

**Table 5.2** Electronic properties of selected metals; data mainly from Kittel [5]

| | Li | Na | K | Cu | Ag | Au | Mg | Ca | Zn | Al | Pb |
|---|---|---|---|---|---|---|---|---|---|---|---|
| $\mu_F$ (eV) | 4.72 | 3.23 | 2.12 | 7.0 | 5.51 | 5.5 | 7.13 | 4.68 | 9.39 | 11.6 | 9.37 |
| $n_e(10^{28}\ m^{-3})$ Valence electrons | 4.7 (1) | 2.65 (1) | 1.4 (1) | 8.45 (1) | 5.85 (1) | 5.90 (1) | 8.60 (2) | 4.60 (2) | 13.1 (2) | 18.1 (3) | 13.2 (4) |
| $r_e(\mu\Omega\,cm)$ at 22 °C | 9.32 | 4.75 | 7.19 | 1.70 | 1.61 | 2.20 | 4.30 | 3.60 | 5.92 | 2.74 | 21.0 |

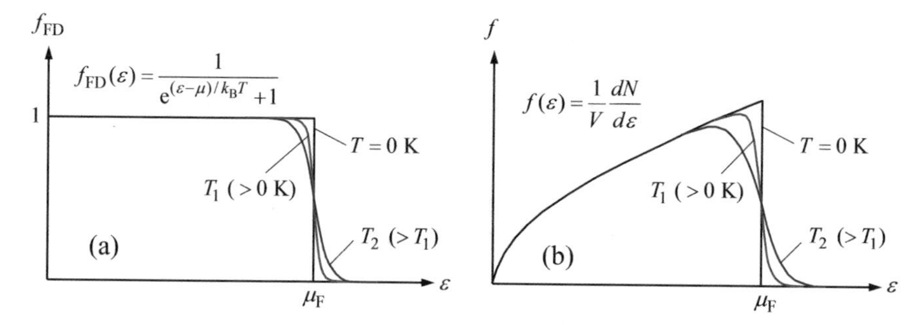


**Fig. 5.5** **a** The Fermi function and **b** the distribution function of free electrons in a metal

This equation is plotted in Fig. 5.5b. Note that $f(\varepsilon) = f_{FD}(\varepsilon)D(\varepsilon)$, where $D(\varepsilon)$ is the density of states for free electrons and is expressed as

$$D(\varepsilon) = 4\pi\left(\frac{2m_e}{h^2}\right)^{3/2}\sqrt{\varepsilon} \tag{5.18}$$

Now, we are ready to evaluate the Fermi energy $\mu_F$. At $T \to 0$, the number density of electrons becomes

$$n_e = \frac{N_e}{V} = \lim_{T\to 0}\int_0^{\mu} 4\pi\left(\frac{2m_e}{h^2}\right)^{3/2}\frac{\sqrt{\varepsilon}}{e^{(\varepsilon-\mu)/k_BT}+1}d\varepsilon \tag{5.19}$$

which gives

$$\mu_F = \frac{h^2}{8m_e}\left(\frac{3n_e}{\pi}\right)^{2/3} \tag{5.20}$$

Typical values of $\mu_F$ range from 2 to 12 eV. Table 5.2 lists the Fermi energy, the electron number density, the number of electrons per atom, and the electrical resistivity of various metals. The temperature dependence of $\mu$ for electrons is given by the Sommerfeld expansion [6]:

$$\mu(T) = \mu_F\left[1 - \frac{1}{3}\left(\frac{\pi k_BT}{2\mu_F}\right)^2 + \cdots\right] \tag{5.21a}$$

It can be seen that $\mu(T) \approx \mu_F$ at moderate temperatures. Arnold Sommerfeld (1868–1951) was a German physicist and one of the founders of quantum mechanics. As a professor at the University of Munich, he advised a large number of doctorate students who became famous in their own right, including Peter Debye, Wolfgang Pauli, and Werner Heisenberg, among others. Sommerfeld applied the FD statistics to study free electrons in metals and resolved the difficulty in the classical theory for electron specific heat. The Sommerfeld expansion for the integration involving the FD function is derived in Appendix B.8. As discussed in Chap. 3, electrons tend to fill all the quantum states up to a certain energy level. In many texts, $\mu(T)$ is called the Fermi level or the Fermi energy, which is temperature dependent. As the temperature increases, only those electrons near the Fermi level will be redistributed. By noticing that the difference between $\mu(T)$ and $\mu_F$ is small, we can use Eqs. (B.74) and (B.78) to express the electron number density as follows:

$$n_e = \int_0^\infty D(\varepsilon) f_{FD}(\varepsilon, T) d\varepsilon \approx \int_0^{\mu_F} D(\varepsilon) d\varepsilon + (\mu - \mu_F) D(\mu_F) + \frac{\pi^2 (k_B T)^2}{6} D'(\mu_F)$$

where the first term is the same as the right-hand side of Eq. (5.19). Since the number density is independent of temperature, we must have

$$(\mu - \mu_F) D(\mu_F) + \frac{\pi^2 (k_B T)^2}{6} D'(\mu_F) = 0 \tag{5.21b}$$

which proves Eq. (5.21a) since $D(\varepsilon)/D'(\varepsilon) = 2\varepsilon$.

**Example 5.2** Calculate $\mu$ at 300 and 10,000 K for copper using $\mu_F = 7\,\text{eV}$. Find the maximum speed (Fermi velocity) and the average speed of electrons for copper at 0 K. How will the Fermi velocity change if the temperature is changed to $T =$ 300 K?

**Solution** Note that $k_B = 1.381 \times 10^{-23}/1.602 \times 10^{-19} = 8.62 \times 10^{-5}$ eV/K. Let us calculate the relative changes of $\mu$ at a given temperature $T$. From Eq. (5.21a), we have

$$\frac{\mu(T) - \mu_F}{\mu_F} \approx -\frac{1}{3}\left(\frac{\pi k_B T}{2\mu_F}\right)^2 = -1.24 \times 10^{-10} T^2$$

which is about 0.0011% at 300 K and 1.2% at 10,000 K. The change in $\mu$ is indeed very small. At $T = 0$, $\mu_F = \frac{1}{2} m_e v_{max}^2 = \frac{1}{2} m_e v_F^2$. Hence,

$$v_{max} = v_F = \sqrt{2\mu_F / m_e} \tag{5.22a}$$

$$\bar{\varepsilon} = \frac{1}{2} m_e \overline{v^2} = \frac{U}{N} = \int_0^{\mu_F} f(\varepsilon)\varepsilon d\varepsilon \bigg/ \int_0^{\mu_F} f(\varepsilon) d\varepsilon = \frac{3}{5}\mu_F \tag{5.22b}$$

$$v_{rms} = \sqrt{\frac{2\bar{\varepsilon}}{m_e}} = \sqrt{\frac{6\mu_F}{5m_e}} \tag{5.22c}$$

Electrons are constantly moving even at absolute zero temperature. For copper, we get $v_F = 1.57 \times 10^6$ m/s and $v_{rms} = 1.22 \times 10^6$ m/s, which is about three-quarters of $v_F$. The classical model based on the equipartition principle or the Maxwell–Boltzmann distribution would give $\frac{3}{2}k_B T = \frac{1}{2}m_e\overline{v^2}$ or $v_{rms} = \sqrt{3k_B T/m_e} = 0$ at absolute zero temperature. Because $\mu$ changes little from 0 to 300 K, the Fermi velocity at 300 K is essentially the same as that obtained at 0 K.

**Discussion:** If we use the rms velocity to calculate the de Broglie wavelength as in Example 3.2, we obtain $\lambda_{DB} = 0.6$ nm. If an electron is accelerated in vacuum to 50 keV, the velocity will be greater than one-third of that of light, and the de Broglie wavelength will be extremely small ($\lambda_{DB} \approx 0.0066$ nm). The resolutions in conventional optical microscopy and photolithography are usually limited by $\lambda/2$ (the diffraction limit), which is on the order of 200 nm for visible light. Electron microscopy can have a much higher resolution (~0.1 nm), and e-beam nanolithography allows the manufacturing of features just a few nanometers.

In order to find out the specific heat of electrons, we first calculate the internal energy:

$$U = V\int_0^\infty \varepsilon f_{FD}(\varepsilon)D(\varepsilon)d\varepsilon \tag{5.23a}$$

Because the distribution function does not vary significantly except near $\varepsilon = \mu$, the Sommerfeld expansion can be used to express the integration [see Eq. (B.78) in Appendix B]. Hence,

$$\frac{U}{V} \approx \int_0^{\mu_F} \varepsilon D(\varepsilon)d\varepsilon + \mu_F(\mu - \mu_F)D(\mu_F) + \frac{(\pi k_B T)^2}{6}[\mu_F D'(\mu_F) + D(\mu_F)]$$

One can see from Eq. (5.21b) that the two middle terms on the right side cancel out. It should also be noted that $D(\mu_F) = 3n_e/2\mu_F$. Therefore,

$$U \approx \frac{3}{5}N\mu_F\left[1 + \frac{5\pi^2}{12}\left(\frac{k_B T}{\mu_F}\right)^2 + \cdots\right] \tag{5.23b}$$

The specific heat of free electrons can then be obtained as

$$\bar{c}_{v,e} = \left(\frac{\partial \bar{u}}{\partial T}\right)_V = \frac{\pi^2 k_B T}{2\mu_F}\bar{R} \tag{5.24}$$

which is much smaller than $\frac{3}{2}\bar{R}$ as we would obtain if electrons were behaving as an ideal monatomic molecular gas. Another way of obtaining Eq. (5.24) is to use integration, which is left as an exercise (see Problem 5.6). Electronic contribution to the specific heat of solids is negligible except at very low temperatures (a few kelvins or less). The specific heat of metals at very low temperatures can thus be expressed as

$$c_v(T) = \gamma T + BT^3 \tag{5.25}$$

where the linear term is the electronic contribution and the cubic term is the lattice contribution for which $B$ can be obtained from Eq. (5.14). The coefficient $\gamma$ is known as the *Sommerfeld constant.* The experimental values of $\gamma$ generally agree with those predicted by the free-electron model given in Eq. (5.25) for most alkali metals (e.g., Na, K) and noble metals (e.g., Cu, Ag, Au). For transition metals with magnetic properties, such as Fe and Mn, the measured $\gamma$ value can be an order of magnitude greater than the predicted. On the other hand, for semimetals like Bi, the measured $\gamma$ value can be an order of magnitude smaller than the predicted. Further discussions can be found from the text of Ashcroft and Mermin [6].

**Example 5.3** Calculate and plot the specific heat of copper, and compare with the data in Touloukian and Buyco [7]. Discuss the contribution of electrons and lattice vibrations.

**Solution** From Table 5.1, the Debye temperature for Cu is $\Theta_D = 340\,\text{K}$. At $T < 30$ K, we can apply the $T^3$ law given in Eq. (5.14) to find the coefficient $B$ in Eq. (5.25) to be $5.95 \times 10^{-6}\,\bar{R}\,[\text{K}^{-3}]$. Using $\mu_F = 7\,\text{eV}$ from Table 5.2, the Sommerfeld coefficient can be calculated from Eq. (5.24) as $\gamma = 6.08 \times 10^{-5}\,\bar{R}\,[\text{K}^{-1}]$. Therefore, the two contributions will be equal at $T = 3.2$ K. The results are plotted in Fig. 5.6a at temperatures below 10 K. At higher temperatures, as shown in Fig. 5.6b, the electronic contribution is much smaller compared with the lattice specific heat: about 0.3% at 100 K, 0.6% at 300 K, and 2% at 1000 K. The data show much higher specific heat values than those predicted by the Debye model. The addition of the electronic contributions cannot fully account for the difference. Noting that $R = \bar{R}/M = 130.9$ J/kg K at 1000 K, the specific heat calculated from the Debye model of $c_v = 390.6$ J/kg K is 99.5% of $3R$ given by the Dulong–Petit law. There are several reasons that may be responsible for the deviation between the Debye model

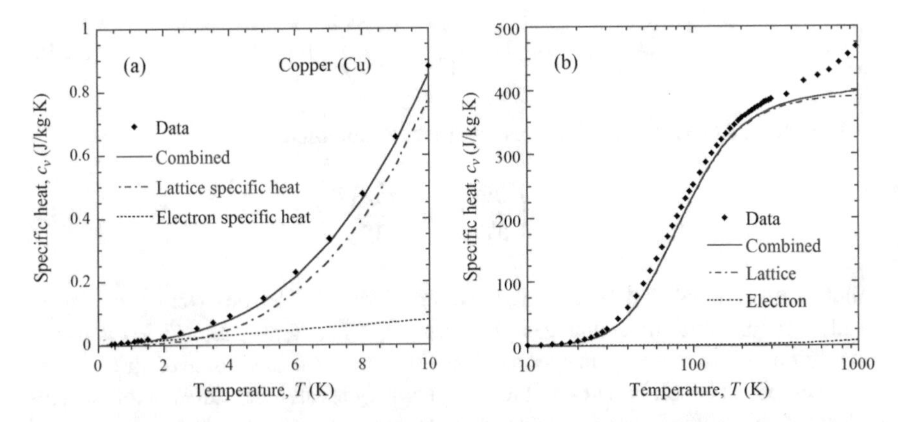


**Fig. 5.6** Electron and lattice contributions to the specific heat of Cu **a** at low temperatures and **b** from 10 to 1000 K

and measurements at high temperatures. The first is the anharmonic vibration that was not considered in the simple models with harmonic vibrations. The contribution of anharmonic vibrations becomes more important at higher temperatures since the amplitude of vibration increases with temperature. Secondly, thermal expansion cannot be ignored at high temperatures. The variation of the distance between atoms may change the potential function and thus increase the specific heat. Additionally, when thermal expansion is not negligibly small, the specific heat at constant pressure (that is measured) may be greater than that at constant volume (that is predicted). Interested readers are referred to the literature for further discussions [9, 10].

## 5.2 Quantum Size Effect on Specific Heat

The above discussion assumes that the physical dimensions are much larger than the lattice constant. In nanoscale devices and structures, such as 2D thin films or superlattices, 1D nanowires or nanotubes, or 0D quantum dots or nanocrystals, substitution of summation by integration is no longer appropriate. Note that a 2D thin film is confined in one dimension, a 1D wire is confined in two dimensions, and a 0D quantum dot is confined in all three dimensions. In nanostructures, it is necessary to consider quantization of the energy levels. The specific heat becomes a function of the actual dimensions. Experimental demonstrations of quantum size effect on specific heat have been made on Pb particles [11], carbon nanotubes [12], and titanium dioxide nanotubes [13], to name a few. To analyze the quantum size effect on the lattice specific heat, we begin with a wavelike treatment of the vibrational modes in this section.

### 5.2.1 *Periodic Boundary Conditions*

Consider a 1D chain of $N + 1$ atoms as sketched in Fig. 5.3, where the end nodes are fixed in position. The solution should be a standing wave with the following eigenfunctions:

$$\sin\left(\frac{\pi x}{L}\right),\quad \sin\left(\frac{2\pi x}{L}\right),\quad \sin\left(\frac{3\pi x}{L}\right),\ldots,\quad \sin\left(\frac{\pi x}{L_0}\right) \tag{5.26}$$

where $L/L_0 = N$, which is the total number of vibration modes within a length of $L$. Another approach is based on the Born–von Kármán periodic boundary conditions [6]. Instead of treating the solid as a bounded specimen whose atoms are fixed at each boundary, the Born–von Kármán lattice model takes the medium as an infinite extension with periodic boundary conditions. For a solid whose dimensions are $L_x$, $L_y$, $L_z$ in the Cartesian coordinates, the standing wave solutions are

$$\exp(ik_x x), \quad \exp(ik_y y), \quad \exp(ik_z z) \tag{5.27}$$

where $\mathbf{k} = (k_x, k_y, k_z)$ is called the *lattice wavevector* with $k^2 = k_x^2 + k_y^2 + k_z^2$. The allowed discretized values are

$$k_x = 0, \quad \pm\frac{2\pi}{L_x}, \quad \pm\frac{4\pi}{L_x}, \quad \pm\frac{6\pi}{L_x}, \ldots, \quad \pm\frac{(N_x - 1)\pi}{L_x}, \quad +\frac{N_x\pi}{L_x} \tag{5.28a}$$

$$k_y = 0, \quad \pm\frac{2\pi}{L_y}, \quad \pm\frac{4\pi}{L_y}, \quad \pm\frac{6\pi}{L_y}, \ldots, \quad \pm\frac{(N_y - 1)\pi}{L_y}, \quad +\frac{N_y\pi}{L_y} \tag{5.28b}$$

$$k_z = 0, \quad \pm\frac{2\pi}{L_z}, \quad \pm\frac{4\pi}{L_z}, \quad \pm\frac{6\pi}{L_z}, \ldots, \quad \pm\frac{(N_z - 1)\pi}{L_z}, \quad +\frac{N_z\pi}{L_z} \tag{5.28c}$$

where the last term only has "+" term and should only be included if the number of atoms along each direction $N_x$, $N_y$, or $N_z$ is an even number. The central distance between adjacent atoms is $L_x/N_x$, $L_y/N_y$, or $L_z/N_z$ in the given direction. The individual components of the lattice wavevector may be negative or zero in this case. In the 1D case, it can be seen that the total number of modes is the same as the total number of atoms along the 1D chain. However, the infinite medium representation with periodic boundary conditions is advantageous not only in mathematical derivations but also for the physical interpretation of lattice dynamics.

### 5.2.2 General Expressions of Lattice Specific Heat

The general expression of the lattice vibrational energy in a solid is given as

$$u(T) = u_0 + \sum_P \sum_K \hbar\omega \left( \frac{1}{e^{\hbar\omega/k_B T} - 1} + \frac{1}{2} \right) \tag{5.29}$$

where $u_0$ accounts for the static energy at absolute zero temperature, the first term in the parenthesis is the Bose–Einstein distribution $f_{BE}(\omega, T)$ given in Eq. (5.4), and the second term in the parenthesis corresponds to the *zero-point energy* that is associated with the $\frac{1}{2}h\nu$, due to *quantum fluctuation* or *vacuum fluctuation,* in the vibrational energy levels. We use $h\nu$ and $\hbar\omega$ interchangeably whichever is more convenient. The summation is over all phonon branches in terms of the wavevector index $K$ and the polarization index $P$. A phonon branch (sometimes also called a phonon mode) describes the behavior of a type of phonons with a continuous frequency rather than a discrete frequency. The concept of phonon branches will be presented in detail in the subsequent chapter. The lattice specific heat can be expressed as [5]

$$c_v(T) = \sum_P \sum_K \hbar\omega \frac{\partial f_{BE}(\omega, T)}{\partial T} \tag{5.30}$$

Upon introducing the DOS, we can replace the summation over $k$-space with an integration as follows:

$$c_v(T) = \sum_P \int_0^\infty \hbar\omega \frac{\partial f_{\mathrm{BE}}}{\partial T} D(\omega)\mathrm{d}\omega$$
$$= k_{\mathrm{B}} \sum_P \int_0^\infty \left(\frac{\hbar\omega}{k_{\mathrm{B}}T}\right)^2 \frac{\mathrm{e}^{\hbar\omega/k_{\mathrm{B}}T}}{(\mathrm{e}^{\hbar\omega/k_{\mathrm{B}}T}-1)^2} D(\omega)\mathrm{d}\omega \tag{5.31}$$

Since the DOS is expressed as the number of modes per unit volume, Eq. (5.31) gives the specific heat per unit volume. Neutral scattering and Raman scattering are common ways of determining the DOS from the relationship between $\omega$ and the lattice wavevector $\mathbf{k}$ along selected crystal directions. The function $\omega = \omega(\mathbf{k})$ is called a *dispersion relation.* If discretized values are expressed using the Delta functions in the expression of $D(\omega)$, Eq. (5.31) is equivalent to Eq. (5.30), and both the equations can be considered as the general expressions of the specific heat due to lattice vibrations. For a nanostructure with very few atoms in a particular direction, Eq. (5.30) may be more convenient to use. On the other hand, in directions with a large number of atoms, Eq. (5.31) would be the preferable choice.

### 5.2.3 Dimensionality

The method of periodic boundary conditions allows one to determine the density of states for simple dispersion relations easily. Figure 5.7 shows the $k$-space, or the *reciprocal lattice space,* in the 2D case. Each individual block of area $4\pi^2/(L_xL_y)$

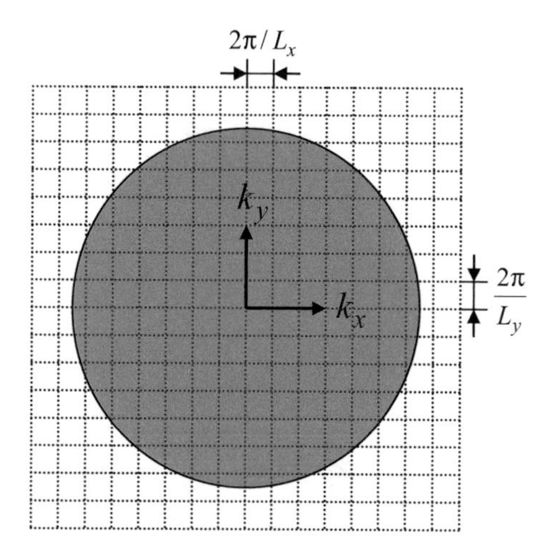


**Fig. 5.7** Schematic of the reciprocal lattice space, or $k$-space

represents a mode, and the number of modes up to a certain value of $k$ is equal to the total number of blocks inside the circle. One can also use this graph to visualize the 3D case. Each box of volume $8\pi^3/(L_x L_y L_z)$ represents a mode. The number of modes for a given upper limit $k$ is equal to the total number of boxes within a sphere of radius $k$, hence,

$$N = \frac{4\pi k^3}{3} \frac{L_x L_y L_z}{8\pi^3} = \frac{V k^3}{6\pi^2} \tag{5.32}$$

When the dimensions are large enough, the DOS can be expressed as

$$D(\omega) = \frac{1}{V} \frac{\mathrm{d}N}{\mathrm{d}\omega} = \frac{k^2}{2\pi^2} \frac{\mathrm{d}k}{\mathrm{d}\omega} \tag{5.33}$$

Assume the dispersion relation is linear, then

$$\omega = v_\mathrm{a} k \tag{5.34}$$

where $v_\mathrm{a}$ is the average speed of the longitudinal and transverse waves as in Eq. (5.7). We can rewrite Eq. (5.33) as

$$D(\omega) = \frac{\omega^2}{2\pi^2 v_\mathrm{a}^3} \tag{5.35}$$

This expression is equivalent to Eq. (5.7) for a single polarization. Equations (5.32) and (5.34) can be combined to obtain the high-frequency limit by setting $N$ equal to the number of atoms. The result is the same as Eq. (5.9). When Eq. (5.35) is substituted into Eq. (5.31), the Debye expression of the specific heat given in Eq. (5.13) is readily obtained.

If the number of atoms is very small in a particular direction, there will only be a few values for the particular wavevector component. The dimensionality will be reduced, and the wavevector component can be assumed as zero in that direction. For a 2D solid (such as a thin film or a quantum well), the DOS is defined as the number of quantum states per unit area. By assuming a linear dispersion relation, we obtain

$$N = \frac{\pi k^2}{4\pi^2/(L_x L_y)} = \frac{A k^2}{4\pi} \tag{5.36}$$

and

$$D(\omega) = \frac{1}{A} \frac{\mathrm{d}N}{\mathrm{d}\omega} = \frac{k}{2\pi} \frac{\mathrm{d}k}{\mathrm{d}\omega} = \frac{\omega}{2\pi v_\mathrm{a}^2} \tag{5.37}$$

For a 1D solid (such as a nanowire or a nanotube), by noting that $N = 2k\big/(2\pi/L_x) = Lk/\pi$, we find the DOS to be

$$D(\omega) = 1/(\pi v_a) \tag{5.38}$$

which is independent of the frequency. It can be shown that, in the low-temperature limit, the specific heat for a 2D solid is proportional to $T^2$ and that for a 1D solid is proportional to $T$ [14, 15]. Experimental evidence of the dimensionality change has been known for a long time in graphite, which has a layered lattice structure with a strong bonding between atoms within each layer and a weak interactive force between layers. The specific heat of graphite is approximately proportional to $T^2$ at low temperatures [16]. On the other hand, the linear temperature dependence of specific heat has been observed in carbon nanotubes [12].

It can be seen from Eq. (5.31) that when $\hbar\omega \gg k_B T$, the integrand approaches zero. Therefore, the contribution to the specific heat is negligibly small when the phonon energy is much higher than $k_B T$. The speed of lattice waves ranges from 1000 to 10,000 m/s, and the phonon wavelength corresponding to $k_B T$ is called *thermal phonon wavelength,* which can be calculated from $\lambda_{th} = v_a h / k_B T$. At room temperature, $\lambda_{th}$ is approximately 0.3 nm for $v_a = 2000\,\text{m/s}$ and 1 nm for $v_a = 6000\,\text{m/s}$. At 10 K, $\lambda_{th} \approx 10\,\text{nm}$ for $v_a = 2000\,\text{m/s}$, and $\lambda_{th} \approx 30\,\text{nm}$ for $v_a = 6000\,\text{m/s}$. It is expected that the quantum size effect will become more significant at low temperatures, because the thermal phonon wavelength may be greater than the smallest physical length, such as the thickness of the film and the diameter of the wire.

### 5.2.4 Thin Films and Nanowires

Thin films, or quantum wells, are important components for microelectronic and photonic devices. We will use the following example to elucidate the effect of film thickness and temperature on the specific heat of thin films.

**Example 5.4** Evaluate the low-temperature behavior of the specific heat of a thin film made of a monatomic solid. Assume that the film thickness is $L$, which has $q$ monatomic layers, i.e., $L = qL_0$. The average acoustic speed $v_a$ may be assumed to be independent of temperature. Values of silicon given in Example 5.2 may be used in the numerical evaluation.

**Solution** The molar specific heat can be expressed as

$$\bar{c}_v(T) = \frac{3V\bar{R}}{Nk_B} \sum_{k_x, k_y, k_z} \hbar\omega \frac{\partial f_{BE}}{\partial T} \tag{5.39}$$

where the number 3 accounts for the three phonon polarizations. Assume the dimension perpendicular to the film is the $z$-direction. The allowable modes in the $z$-direction are given by $k_z = 0, \pm 2\pi/L, \pm 4\pi/L, \ldots$ In order for the total number of modes in the $z$-direction to be equal to $q$ for all $q$ values, we shall use the following limits:

$$k_z = \begin{cases} 0, \pm\frac{2\pi}{L}, \pm\frac{4\pi}{L}, \ldots \pm\frac{(q-1)\pi}{L}, \text{ for } q = 1, 3, 5\ldots \\ 0, \pm\frac{2\pi}{L}, \ldots \pm\frac{(q-2)\pi}{L}, +\frac{q\pi}{L}, \text{ for } q = 2, 4, 6\ldots \end{cases} \tag{5.40}$$

Assume that the lattice is infinitely extended in the directions parallel to the film. We can substitute the summation with integration in the parallel directions using cylindrical coordinates. Therefore,

$$\bar{c}_v(T) = \frac{3V\bar{R}}{(2\pi)^3 N k_B} \sum_{k_z} \left( \int_0^{\beta_D} \hbar\omega \frac{\partial f_{BE}}{\partial T} 2\pi\beta \mathrm{d}\beta \right) \Delta k_z \tag{5.41}$$

where $\beta^2 = k_x^2 + k_y^2$, $\Delta k_z = 2\pi/L$, and $\beta_D = \sqrt{k_D^2 - k_z^2}$. The cutoff value $k_D$ is determined by setting the total number of modes equal to the number of atoms per unit area. Equation (5.36) can be used to evaluate the number of modes for each $k_z$ and then summed up over all $k_z$ values. Hence,

$$\frac{N}{A} = \sum_{k_z} \frac{\beta_D^2}{4\pi} = \sum_{k_z} \frac{k_D^2 - k_z^2}{4\pi} \tag{5.42a}$$

Note that $N = AL/L_0^3 = Aq/L_0^2$ and there are $q$ terms in the summation according to the $k_z$ values given in Eq. (5.40). We can solve Eq. (5.42a) to obtain

$$k_D = \sqrt{\frac{4\pi}{L_0^2} + \sum_{k_z} \frac{k_z^2}{q}} \tag{5.42b}$$

In the limit of a single atomic layer, $k_D = 2\sqrt{\pi}/L_0 \approx 3.54 L_0^{-1}$; when $q \to \infty$, $k_D \approx 3.98 L_0^{-1}$, which is very close to the 3D value of $k_D = (6\pi^2)^{1/3}/L_0 \approx 3.90 L_0^{-1}$. Note that the value of $k_D$ normalizes the specific heat so that $c_v$ approaches to the high-temperature limit of $3R$. At low temperatures, when the quantum size effect is significant, a slight difference in $k_D$ does not alter the results much.

Using the linear dispersion relation, $\omega = v_a k = v_a\sqrt{\beta^2 + k_z^2}$, we see that $2\beta \mathrm{d}\beta = k\mathrm{d}k$ for fixed $k_z$. Therefore, Eq. (5.41) can be recast to the following:

$$\bar{c}_v(T) = \frac{3\bar{R}}{2\pi N/A} \left( \frac{k_B T}{\hbar v_a} \right)^2 \sum_{k_z} \int_{x_z}^{x_D} \frac{x^3 e^x}{(e^x - 1)^2} \mathrm{d}x \tag{5.43}$$

where $x = \frac{\hbar v_a k}{k_B T}$, $x_D = \frac{\hbar v_a k_D}{k_B T}$, and $x_z = \frac{\hbar v_a |k_z|}{k_B T}$. The $T^2$ dependence at low temperatures is evident when $q = 1$ or $k_z = 0$ only. The modes associated with $k_z = 0$ are

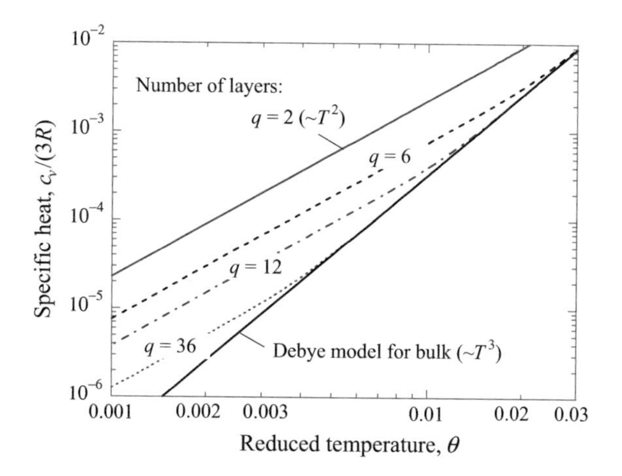


**Fig. 5.8** Quantum size effect on the specific heat of thin films, where the reduced temperature is defined as $\theta = Tk_B L_0/hv_a$

parallel to the interface and are called *planar modes*. We have carried out a numerical evaluation of Eqs. (5.42) and (5.43) for different values of $N$ to see when the departure from bulk behavior will occur. The results are plotted in Fig. 5.8 as a function of the reduced temperature, defined as $\theta = Tk_B L_0/hv_a = L_0/\lambda_{th}$. It can be shown that $hv_a/k_B L_0 \approx 1.61\Theta_D$, where $\Theta_D$ is the Debye temperature. When $q$ and $T$ are sufficiently large, the result from Eq. (5.43) is the same as that predicted by the Debye model for bulk materials. The departure occurs at low temperatures and especially for small $q$ values.

As mentioned earlier, due to the layerlike structure of graphite, its specific heat exhibits 2D solid behavior at low temperatures [16]. The procedure used for this example is essentially the same as that used by Prasher and Phelan [17], except that we have considered the planar modes ($k_z = 0$) in evaluating Eq. (5.43). The result is an increase in the specific heat in the microscopic regime, as discussed in detail in Ref. [15]. Hence, planar modes are critically important when the thickness is small, especially at low temperatures.

A similar formulation can be derived for a nanowire with a square or rectangular cross section [15, 16]. When axial modes are allowed for a wire parallel to the $z$-axis, $k_x = k_y = 0$, the specific heat of this single mode dominates all other modes at sufficiently low temperatures and varies linearly with the temperature $T$ [15]. Another way to show the linear temperature dependence is to combine the DOS given in Eq. (5.38) with Eq. (5.31) as mentioned previously [14]. Note that the discussion here assumes that the linear dispersion relation, Eq. (5.34), holds for the average phonon speed with only three phonon modes. The actual phonon modes and dispersion relations can be rather complicated, and rather sophisticated tools are required to model the thermal properties of nanostructured materials. Several studies have been conducted for silicon nanowires [18–20].

### 5.2.5 *Nanoparticles or Nanocrystals*

For 3D confinement or 0D structures, consider cuboidal nanoparticles of dimensions $(L_x, L_y, L_z)$ with the number of atoms in each direction as $(N_x, N_y, N_z)$. The molar specific heat can be written as a summation using Eq. (5.30) as follows:

$$\bar{c}_v(\theta) = \frac{3\bar{R}}{N_x N_y N_z} \sum_{j,m,n} \frac{\eta^2 e^\eta}{(e^\eta - 1)^2} \tag{5.44}$$

where

$$\eta = \frac{1}{\theta}\left(\frac{j^2}{N_x^2} + \frac{m^2}{N_y^2} + \frac{n^2}{N_z^2}\right)^{1/2} \quad \text{with } \theta = \frac{T k_B L_0}{h\nu_a} \tag{5.45}$$

Note that the summation indices are based on the wavevector values given in Eq. (5.28). For nanoparticles, the aspect ratio and shape can affect the specific heat characteristics. Therefore, $T^n$ (where $1 < n < 3$) behavior may occur for cuboids below the Debye temperature but not at very low temperatures [15].

When the temperature is very low, only the mode(s) with the lowest frequency can be excited and a *second quantum size effect* will occur. Consider a cubic nanocrystal with $q$ atoms in each dimension. Among the $q^3$ total modes, we are left with only six axial modes, which are $\mathbf{k} = (\pm 2\pi/L, 0, 0)$, $(0, \pm 2\pi/L, 0)$, and $(0, 0, \pm 2\pi/L)$. These modes have the longest phonon wavelength. From Eq. (5.44), the specific heat can be expressed as

$$c_v(T \to 0) = \frac{a}{T^2} \exp\left(-\frac{b}{T}\right) \tag{5.46}$$

where $a$ and $b$ are positive constants. Because Eq. (5.46) converges to zero faster than $T^3$, the second quantum size effect will reduce the specific heat at extremely low temperatures [21, 22]. Experiments were made in the early 1970s on lead particles as small as 2.2-nm diameter [11]. At temperatures below 15 K, the specific heat of these particles is much greater than that for the bulk material. However, as the temperature is reduced to about 2 K, the difference diminishes. Below 2 K, the specific heat of the nanoparticles decreases much rapidly than that of the bulk. Note that Eq. (5.46) only applies to the 3D confined case because for 1D or 2D confined cases, the wavevector in the unconfined direction is not restricted.

For bulk solids, as discussed previously, the Born–von Kármán periodic boundary condition is equivalent to the Dirichlet or Neumann boundary conditions [6]. However, for nanocrystals, the applied boundary conditions can affect the model predictions significantly [15, 22]. Let us consider cuboidal nanoparticles. The Dirichlet boundary condition fixes the value at the boundary and is called a clamped boundary condition. The eigenfunctions are given in Eq. (5.26), allowing only positive wavevectors. The specific heat may still be computed using Eq. (5.44), while Eq. (5.45) is modified as

$$\eta = \frac{1}{2\theta}\left(\frac{j^2}{N_x^2} + \frac{m^2}{N_y^2} + \frac{n^2}{N_z^2}\right)^{1/2}, \; j = 1, 2, \ldots, N_x; \; m = 1, 2, \ldots, N_y; \; n = 1, 2, \ldots, N_z \tag{5.47a}$$

For a cubic nanoparticle with $N_x = N_y = N_z = q$, the lowest phonon mode corresponds to $(j, m, n) = (1, 1, 1)$ or $k_x = k_y = k_z = \pi/L$. The result is a minimum wavevector $k_{\min} = \sqrt{3}\pi/L$ and $\eta_{\min} = \sqrt{3}/(2\theta q)$. For the periodic boundary conditions as discussed before, $k_{\min} = 2\pi/L$ and $\eta_{\min} = (\theta q)^{-1}$. While the minimum wavevector is slightly smaller, the resulting $\eta_{\min}$ for the lowest phonon mode is greater with Dirichlet boundary conditions. Hence, the predicted specific heat using the Dirichlet boundary conditions is always lower than that of the corresponding bulk material which is independent of the boundary conditions. This is also true for spherical particles using spherical Bessel functions [22].

On the other hand, when Neumann free-surface boundary conditions are applied, the eigenfunctions are cosine functions and the indices in Eq. (5.47a) should be modified to allow zero indices as long as at least one of them is nonzero, that is

$$j = 0, 1, \ldots, (q_x - 1); \; m = 0, 1, \ldots, (q_y - 1); \; n = 0, 1, \ldots, (q_z - 1) \tag{5.47b}$$

where $j$, $m$, and $n$ cannot be simultaneous zero. In this case, we see from Eq. (5.47a) that $k_{\min} = \pi/L$ and $\eta_{\min} = (2\theta q)^{-1}$. Therefore, the phonon frequency of the lowest mode is half of that in the case of periodic boundary conditions. The reduction of phonon frequency toward low temperatures is called *phonon softening*. Phonon softening results in an enhancement of the specific heat of nanoparticles at low temperatures until the temperature becomes sufficiently low when the second quantum size effect described by Eq. (5.46) will dominate the specific heat behavior. The result for cubic nanoparticles is in general consistent with that of spherical nanoparticles based on the Neumann boundary conditions [15, 22].

### 5.2.6 Graphite, Graphene, and Carbon Nanotubes

Unlike diamond, which contains 3D tetrahedral structures, graphite crystallizes in the hexagonal system with sheetlike structures. While diamond and graphite are each a polymorph of the element carbon, they exhibit dramatically different properties due to their different crystalline structures. Diamond is hard, transparent, and an electrical insulator. On the contrary, graphite is quite soft, opaque, and a good electrical conductor. Graphene is a single atomic layer of carbon atoms packed into a periodic benzene-ring structure. Carbon nanotubes may be considered as rolled from a graphene sheet into a hollow cylinder, with one or both of its ends capped with half a fullerene molecule. The discovery of $C_{60}$ and other fullerenes by Robert Curl, Harold Kroto, and Richard Smalley was recognized through the 1996 Nobel Prize in Chemistry conferred on them. The diameter of single-walled carbon nanotubes (SWNTs) can be as small as 0.4 nm with a typical diameter 1–2 nm and as long as 100 μm or so. Multi-walled carbon nanotubes (MWNTs) and nanotube ropes can have a diameter from 10 to 200 nm.

As mentioned earlier, graphite has a 2D structure and exhibits $T^2$ dependence at low temperatures [16]. For an isolated graphene sheet, the in-plane or parallel transverse acoustic phonon mode or branch has a velocity of $v_{\text{TA-p}} = 15,000\,\text{m/s}$ and the longitudinal acoustic phonon mode has a velocity of $v_{\text{LA}} = 24,000\,\text{m/s}$. On the other hand, the out-of-plane or perpendicular transverse phonon branch is described by a quadratic dispersion relation, $\omega \propto k^2$, which is the dominant mode for the specific heat at low temperatures. Considering the dimensionality and the dispersion relation, the specific heat of a graphene sheet depends almost linearly on $T$ at lower temperatures (see Problem 5.11) and on $T^2$ as the temperature is raised above 100 K or so.

The four acoustic phonon modes or branches are expected to be the dominant contributions to the specific heat of isolated SWNTs at low temperatures. These include two (degenerate) transverse modes, one longitudinal mode, and a twisting mode or torsional mode associated with the rigid rotation around the nanotube axis. The dispersion relation is linear for all four modes at low frequencies [23]. Therefore, because of the 1D structure, the specific heat is expected to be linearly dependent on temperature. As the temperature is raised, however, higher frequency modes are excited and the 2D characteristics of carbon nanotubes come into play. Watt de Heer has written an elegant article on this topic [24]. There are significant differences between SWNTs, MWNTs, and nanotube ropes or bundles; the actual temperature dependence can be more complicated and dependent on the diameter [12, 23–25].

In nanostructures, the electron DOS is also subject to quantization. The theory for the electronic contribution to the specific heat is more complicated. The electron–electron and electron–phonon interactions as well as the distribution of energy levels and the Fermi energy need to be considered in a detailed model [26, 27]. The electronic specific heat of small particles is still a linear function of temperature. Generally speaking, the electronic contribution to the specific heat is negligibly small unless the temperature is below about 1 K. Therefore, we will not discuss the electronic size effect on the specific heat any further.

## 5.3 Electrical and Thermal Conductivities of Solids

In this section, we use kinetic theory to study the electron and phonon transport properties of metals and insulators in the bulk form. The coupling between electrical current and heat flux due to electric field and temperature gradient will be studied in the next section, followed by a discussion of the size effect on the electrical and thermal conductivities.

### 5.3.1 *Electrical Conductivity*

We start with the simple kinetic theory approach based on the Drude free-electron model, also known as the Drude–Lorentz theory. As shown in Fig. 5.9, the electrical resistance of a resistor is $R = r_e L/A_c = L/(\sigma A_c)$, where $r_e$ is the resistivity; its inverse $\sigma$ is the conductivity, $L$ is the length, and $A_c$ is the cross-sectional area. Ohm's law relates the voltage drop $\Delta V$ and the current $I$ by $\Delta V = IR$, which can be rearranged as

$$\frac{I}{A_c} = \sigma \frac{\Delta V}{L} \tag{5.48}$$

Notice that $J = I/A_c$ is the *current density* (charge per unit cross-sectional area per unit time), and $E = \Delta V/L$ is the electric field (note that the electric field is in the direction of decreasing voltage). Rewriting it in the vector form, we have

$$\mathbf{J} = \sigma \mathbf{E} \tag{5.49}$$

The above equation may be considered as the microscopic Ohm's law. An electron of charge $-e$ is accelerated in an electric field according to Newton's law as

$$\mathbf{F} = -e\mathbf{E} = m_e \frac{d\mathbf{v}}{dt} \tag{5.50}$$

Due to collisions, electrons cannot move completely freely. The velocity change of an electron during a relaxation time $\tau$ (the average traveling time between collisions) due to an external field is called the *drift velocity* $\mathbf{u}_d$. The probability that a traveling particle will collide with another particle or a defect during an infinitesimal time $dt$ is given by $dt/\tau$. The acceleration term in Eq. (5.50) can then be approximated by

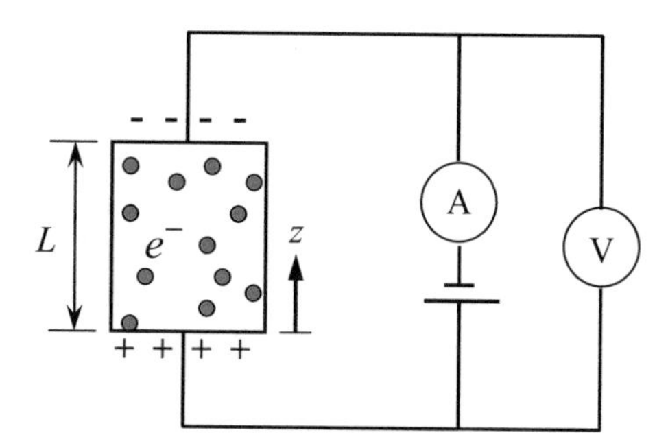


**Fig. 5.9** Illustration of electrical conduction

$\mathbf{u}_d/\tau$. Alternatively, one can also consider a damping force that is proportional to the drift velocity given as $m_e\gamma\mathbf{u}_d$, where the damping coefficient $\gamma$ happens to be the electron scattering rate $1/\tau$. At steady state, the damping force must balance the external electrical force, i.e., $-e\mathbf{E} = m_e\mathbf{u}_d/\tau$ [28]. The current density is related to the drift velocity by $\mathbf{J} = -en_e\mathbf{u}_d$, hence,

$$\mathbf{J} = \frac{n_e e^2 \tau}{m_e}\mathbf{E} \tag{5.51}$$

Comparing the above equation with Eq. (5.49), we obtain the Drude–Lorentz expression:

$$\sigma = \frac{n_e e^2}{m_e}\tau \tag{5.52}$$

The preceding equation is often used to obtain the relaxation time $\tau$ from the measured electrical conductivity $\sigma$. At moderate temperatures, it can be assumed that the characteristic velocity of electrons is the Fermi velocity $v_F$, and the mean free path of electrons can be written as

$$\Lambda_e = v_F\tau \tag{5.53}$$

The electron scattering mechanisms are illustrated in Fig. 5.10. Electron–electron scattering is inelastic and usually negligible compared with electron–phonon scattering, which is also inelastic. Because lattice vibrations are enhanced as temperature increases, electron–phonon scattering is expected to be dominant at high temperatures. Defect or impurity scattering, on the other hand, is important at low temperatures. For bulk materials that are large enough, boundary scattering is negligible. According to Matthiessen's rule, the scattering rate of independent scattering events can be added to yield the total scattering rate. For a bulk material, we have

$$\frac{1}{\tau} = \frac{1}{\tau_{e-e}} + \frac{1}{\tau_{e-ph}} + \frac{1}{\tau_{e-d}} \approx \frac{1}{\tau_{e-ph}} + \frac{1}{\tau_{e-d}} \tag{5.54}$$

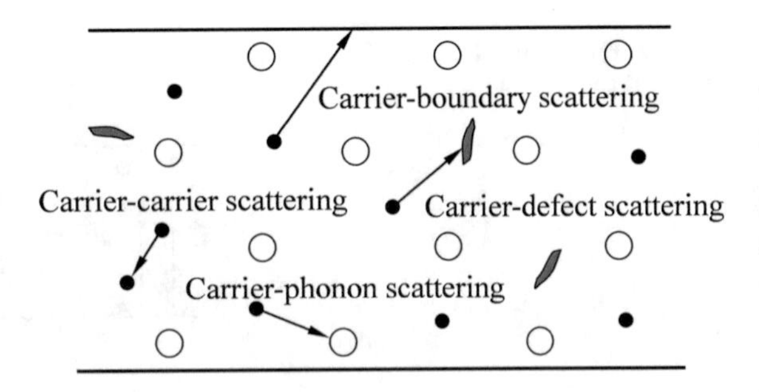


**Fig. 5.10** Schematic of various carrier scattering mechanisms

where the subscripts e–e, e–ph, and e–d are for electron–electron, electron–phonon, and electron–defect scattering. It should be noted that the free-electron description is often applied to model the electrical conductivity of (doped) semiconductors, in which both electrons and holes (positive charges) can carry currents. For some metals, as to be explained in Chap. 6, the charge carriers are actually holes rather than electrons. Using Eq. (5.53), we can write Eq. (5.54) in terms of the mean free path as follows:

$$\frac{1}{\Lambda_{\mathrm{e}}} = \frac{1}{\Lambda_{\mathrm{e-ph}}} + \frac{1}{\Lambda_{\mathrm{e-d}}} \tag{5.55}$$

Different scattering mechanisms can be considered separately. Boundary scattering becomes important when the characteristic dimension $L_{\mathrm{c}}$ is comparable to the mean free path of the bulk material $\Lambda_{\mathrm{e}}$. Here, $L_{\mathrm{c}}$ can be the thickness of a thin film or the diameter of a thin wire. An effective mean free path can be defined for the evaluation of the scattering rate and the conductivity:

$$\frac{1}{\Lambda_{\mathrm{e,eff}}} = \frac{1}{\Lambda_{\mathrm{e}}} + \frac{1}{\Lambda_{\mathrm{e-b}}} \tag{5.56}$$

where the subscript e–b is for electron–boundary scattering. It can be seen that when boundary scattering is important, the effective mean free path will be suppressed, or the scattering rate will increase. The electrical conductivity will be reduced, and the reduction is size dependent. This is similar to the molecular heat transfer discussed in Chap. 4 when the *Kn* number, i.e., the ratio of the mean free path to the characteristic length ($\Lambda/L_{\mathrm{c}}$), is comparable or greater than 1. Further discussion of the size effect on the conductivities of solids will be given in Sect. 5.5.

The Bloch formula for electrical resistivity due to electron–phonon scattering gives

$$r_{\mathrm{e-ph}} = 4r_0\left(\frac{T}{\Theta}\right)^5 \int_0^{\Theta/T} \frac{x^5 \mathrm{e}^x}{(\mathrm{e}^x - 1)^2}\mathrm{d}x \tag{5.57}$$

where $r_0$ is a constant, and $\Theta$ is a characteristic temperature that is very close to the Debye temperature [29]. The derivation of the above equation requires a careful treatment of the electron–phonon interaction within the framework of the electron band theory considering both the *N* process and the *U* process, which will be discussed in Chap. 6. The Bloch formula predicts that the electrical resistivity approaches zero as the temperature approaches absolute zero for a pure metal. When $T \ll \Theta$, the low-temperature approximation of the lattice resistivity can be written as

$$r_{\mathrm{e-ph}} \approx 498 r_0 T^5/\Theta^5 \tag{5.58}$$

Because of impurities, electron–defect scattering gives a residual resistivity $r_{e-d}$ that is important at low temperatures, and its value is independent of temperature. Adding the scattering rates using Matthiessen's rule, the electrical resistivity is obtained as [29]

$$r_e = r_{e-ph} + r_{e-d} \tag{5.59}$$

Figure 5.11 compares the model with the electrical resistivity data recommended for high-purity bulk metals after annealing [30]. Taking the electrical resistivity of gold as an example, it can be seen that phonon scattering dominates the electrical resistivity at high temperatures and results in $r_{e-ph} \approx r_0 T/\Theta$, which is proportional to $T$. It should be noted that $\Theta_D$ listed in Table 5.1 can be used to approximate $\Theta$ in most cases. The constant $r_0$ can be determined using the resistivity values at 22 °C, or 295 K, given in Table 5.2. At very low temperatures, $r_e \approx r_{e-d}$, which is independent of temperature but depends strongly on the impurity concentration.

**Example 5.5** Consider a large copper specimen of high purity with a very small defect scattering rate of $\tau_{e-d}^{-1} = 5 \times 10^8$ rad/s at the liquid helium temperature of 4.2 K. Find the electrical resistivity, the electron relaxation time,, and the mean free path of this specimen at 1, 295, and 590 K.

**Solution** We first use Eq. (5.52) to evaluate the residual resistivity at 1 K by assuming that the scattering rate is the same at 4.2 and 1 K. This yields an electrical resistivity $r_e \approx r_{e-d} = \frac{m_e}{n_e e^2}\frac{1}{\tau_{e-d}} = 2.1 \times 10^{-5}\ \mu\Omega\,\text{cm}$ or conductivity $\sigma = 4.76 \times 10^{12}\ (\Omega\,\text{m})^{-1}$. The electrical resistivity at 295 K is given in Table 5.2 to be $r_{e-ph} \approx r_e = 1.7\ \mu\Omega\,\text{cm}$. Because the Debye temperature for Cu is 340 K, we can approximate the resistivity

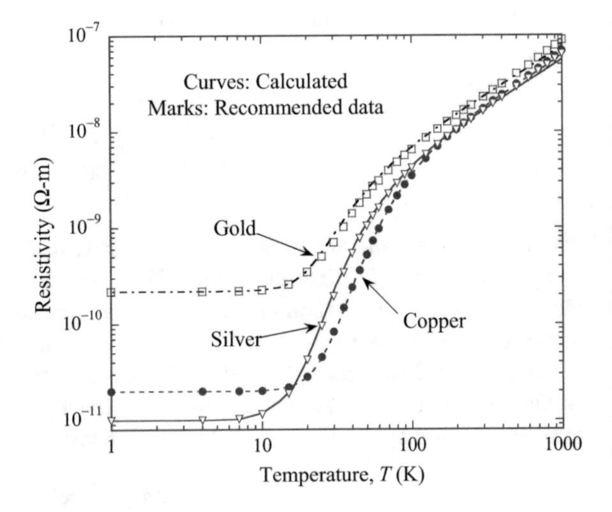


**Fig. 5.11** Comparison of the measured electrical resistivity data [30] of 99.999% pure copper, gold, and silver with the model considering electron–phonon scattering and electron–defect scattering using Eq. (5.59)

at 590 K to be twice that of the resistivity at 295 K, or 3.4 μΩ cm. The relaxation time is approximately $2 \times 10^{-9}$ s at 1 K, $2.47 \times 10^{-14}$ s at 295 K, and $1.24 \times 10^{-14}$ s at 590 K since the number density is assumed to be temperature independent. Using Eq. (5.53) and the Fermi velocity of $v_F = 1.57 \times 10^6$ m/s from Example 5.2, we have the mean free path $\Lambda_e = 3.14$, 38.8, and 19.4 nm at 1, 295, and 590 K, respectively. The conductivity of a copper film with a thickness of less than 100 nm may be affected by boundary scattering. At low temperatures, however, boundary scattering may be dominant for low-dimensional structures even at the micrometer length scale.

For metals, electrons are also responsible for thermal transport. Knowledge of the electrical transport is critical to the understanding of thermal properties. The effect of boundary scattering on transport properties is called the classical size effect [1, 2]. Quantum size effect can modify the DOS of electrons and hence the electrical and thermal properties, as will be discussed in Sect. 5.6.

### 5.3.2 *Thermal Conductivity of Metals*

In metals, free electrons are the main thermal energy carriers. As discussed in Chap. 4, kinetic theory predicts that the thermal conductivity is

$$\kappa = \frac{1}{3}\rho c_{v,e} v_F \Lambda_e \tag{5.60}$$

where $\rho = n_e m_e$ is the mass of electrons per unit volume and $c_{v,e}$ is the mass specific heat of the electrons. Note that $\rho c_{v,e}$ is the volumetric specific heat of electrons and can be expressed as $\rho c_{v,e} = \frac{n_e \pi^2 k_B^2 T}{2\mu_F}$ using the electron specific heat formula given in Eq. (5.24). Substituting the expression for $\rho c_{v,e}$ and $v_F \Lambda_e = v_F^2 \tau \approx 2\mu_F \tau / m_e$ into Eq. (5.60), we obtain the thermal conductivity of a given metal as follows:

$$\kappa = \frac{n_e \pi^2 k_B^2 T}{3 m_e} \tau \tag{5.61}$$

which is proportional to $\tau T$. The Wiedemann–Franz law can be obtained by comparing this equation with the expression for the electrical conductivity given in Eq. (5.52), viz.,

$$Lz \equiv \frac{\kappa}{\sigma T} = \frac{1}{3}\left(\frac{\pi k_B}{e}\right)^2 = 2.44 \times 10^{-8}\ \mathrm{W\,\Omega/K^2} \tag{5.62}$$

where $Lz$ is called *the Lorentz number*. The measured $Lz$ value for most conductors is between 2.2 and $2.7 \times 10^{-8}$ W Ω/K$^2$ at room temperature. The derivations given above were based on the simple kinetic theory, which is consistent with the solution of the BTE under the assumptions of local equilibrium and the relaxation time approximation. The actual scattering process may result in some differences in the effectiveness of transferring momentum and energy during electron–phonon scattering. More detailed theories and experiments have shown that the thermal conductivity of metals is independent of temperature at moderate and high temperatures [29]. The Wiedemann–Franz law is therefore valid near and above room temperature for most metals. As the temperature is lowered, electron–phonon scattering yields a thermal resistance (or $1/\kappa$ ) that is proportional to $T^2$, not $T^4$ as one would obtain by combining Eqs. (5.57) and (5.62). Recall that in the intermediate region, approximately between 10 and 100 K, the Wiedemann–Franz law is not valid. At very low temperatures, defect scattering dominates and, because defect scattering is elastic, the Wiedemann–Franz law is valid again so that $\kappa \propto T$. Therefore, the thermal conductivity at cryogenic temperatures can be expressed as

$$\frac{1}{\kappa(T)} = \frac{A}{T} + BT^2 \tag{5.63}$$

where $A$ and $B$ are positive constants. The first term on the right-hand side dominates at very low temperatures, when the thermal conductivity is proportional to $T$. As the temperature increases, the thermal conductivity reaches a peak and then falls down proportional to $T^{-2}$. As the temperature approaches the room temperature, the thermal conductivity changes little with temperature until the melting point is reached. Figure 5.12 plots the measured thermal conductivity of copper with different impurity concentrations [31]. The highest purity annealed copper has a residual resistivity of $5.79 \times 10^{-12}\ \Omega\,\text{m}$. Oxygen-free high conductivity (OFHC) copper is commonly used in absolute cryogenic radiometers to build the cavity receiver. Even 0.5% impurity concentration will make the conductivity to dramatically decrease at lower temperatures. On the other hand, the thermal conductivity is less sensitive to impurity at temperatures above 100 K and changes little until the melting temperature

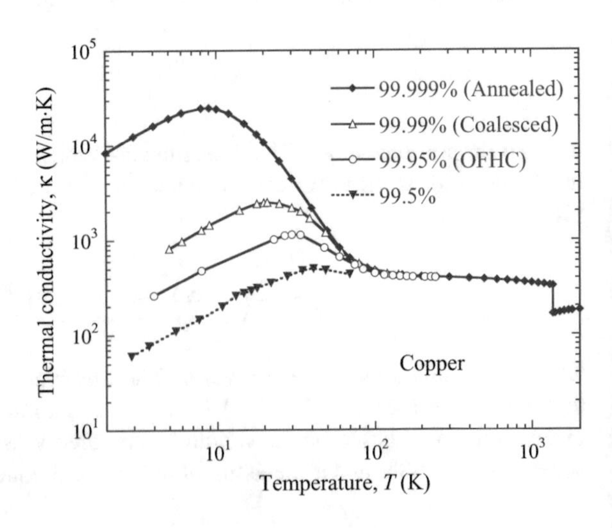


**Fig. 5.12** Thermal conductivity of copper with different purity levels [31]

of 1358 K. Beyond the melting temperature, the thermal conductivity values are for liquid copper.

### 5.3.3 Derivation of Conductivities from the BTE

So far, we have used simple kinetic theory to discuss the electrical and thermal conductivities of metals. It is hoped that these discussions have provided some insights into basic phenomena. To understand the detailed mechanisms, we now present the approaches based on the BTE under two assumptions: local equilibrium and relaxation time approximation. Recall from Chap. 4 that the distribution function can be expressed in terms of $f(\mathbf{r}, \mathbf{v}, t)$ or $f(\mathbf{r}, \mathbf{p}, t)$, where $\mathbf{p} = m_e\mathbf{v}$ for electrons. In describing the phonon specific heat, we have extensively used the phonon wavevector $\mathbf{k}$ as well as the $k$-space. The advanced theory based on the electronic band structure, to be discussed in Chap. 6, is also based on the $k$-space. Using the magnitude relations, $p = h/\lambda$ and $k = 2\pi/\lambda$, we have $\mathbf{k} = \mathbf{p}/\hbar$. Therefore, the distribution function can be written in terms of $\mathbf{k}$ or $f(\mathbf{r}, \mathbf{k}, t)$. The energy of an electron is related to its wavevector by $\varepsilon = \frac{\hbar\mathbf{k}\cdot\hbar\mathbf{k}}{2m_e} = \frac{\hbar^2k^2}{2m_e}$. Under the local-equilibrium condition, the distribution function can be written in terms of temperature $T(\mathbf{r}, t)$ and energy $\varepsilon$ such that

$$f(\mathbf{r}, \mathbf{k}, t)\mathrm{d}\mathbf{k} = f_1(\varepsilon, T)\frac{\mathrm{d}\mathbf{k}}{\mathrm{d}\varepsilon}\mathrm{d}\varepsilon = f_1(\varepsilon, T)D(\varepsilon)\mathrm{d}\varepsilon \tag{5.64}$$

where $D(\varepsilon) = \mathrm{d}\mathbf{k}/\mathrm{d}\varepsilon$ is the DOS, and $f_1(\varepsilon, T)$ is such that $n(\mathbf{r}, t) = \int_0^\infty f_1(\varepsilon, T)D(\varepsilon)\mathrm{d}\varepsilon$ and $\bar{\varepsilon}(\mathbf{r}, t) = \int_0^\infty \varepsilon f_1(\varepsilon, T)D(\varepsilon)\mathrm{d}\varepsilon$. For the equilibrium distribution of free electrons, $f_1(\varepsilon, T)$ is nothing but the Fermi–Dirac function given in Eq. (5.15). When the distribution function is isotropic in the $k$-space, the DOS is given in Eq. (5.18) since $\mathrm{d}\mathbf{k} = \mathrm{d}k_x\mathrm{d}k_y\mathrm{d}k_z = 4\pi k^2\mathrm{d}k$ and $\mathrm{d}\varepsilon = \hbar^2k\mathrm{d}k/m_e$. As discussed earlier, free electrons will occupy all the quantum states below the Fermi level. The Fermi level corresponds to a maximum $k$ in all directions in the $k$-space, which is a spherical surface. All the electron quantum states are included in this *Fermi sphere*. The argument is similar to the Debye model of phonons, where there is an upper bound of the wavevector and the distribution is assumed to be isotropic. We will see in Chap. 6 that the Fermi surface even for monatomic solids with the simplest crystalline structures is not exactly spherical. This is because the electrons in solids are not really independent particles. For simplicity, a spherical Fermi surface is assumed in this section.

Suppose there is a constant electric field $E$ along with a temperature gradient in the $z$-direction. The function $f_1(\varepsilon, T)$ is a nonequilibrium distribution that depends on $z$. At steady state under the relaxation time approximation, we can rewrite Eq. (4.54) as follows:

$$f_1(\varepsilon, T) = f_0(\varepsilon, T) + \tau(\varepsilon)\left(\frac{eE}{m_e}\frac{\partial f_1}{\partial \varepsilon}\frac{\partial \varepsilon}{\partial v_z} - v_z\frac{\partial f_1}{\partial T}\frac{dT}{dz}\right) \quad (5.65)$$

where $f_0(\varepsilon, T)$ corresponds to the equilibrium distribution, which for electrons is the Fermi–Dirac function $f_{FD}$. The relaxation time is not taken as a constant; rather, it is assumed to be dependent on the wavevector or the energy. Note that $\frac{\partial \varepsilon}{\partial v_z} = \frac{\partial \varepsilon}{\partial v}\frac{\partial v}{\partial v_z} = m_e v \frac{v_z}{v} = m_e v_z$. As discussed in Chap. 4, under local equilibrium, we also assume that

$$\frac{\partial f_1}{\partial \varepsilon} \approx \frac{\partial f_0}{\partial \varepsilon} \quad \text{and} \quad \frac{\partial f_1}{\partial T} \approx \frac{\partial f_0}{\partial T} \quad (5.66)$$

Note that this should be viewed as a simplified notation that is valid only when the partial derivatives are substituted into the integration over the $k$-space. We will consider the effect of applied field and temperature gradient separately. When there is no temperature gradient, the current density can be written as

$$J_e = -eJ_N = -e\int_0^\infty v_z\left(f_{FD} + \tau v_z eE\frac{\partial f_{FD}}{\partial \varepsilon}\right)D(\varepsilon)d\varepsilon \quad (5.67a)$$

The first term $-\int_0^\infty ev_z f_{FD}(\varepsilon, T)D(\varepsilon)d\varepsilon$ is zero; and therefore,

$$J_e = -e^2E\int_0^\infty \tau(\varepsilon)v_z^2\frac{\partial f_{FD}}{\partial \varepsilon}D(\varepsilon)d\varepsilon \quad (5.67b)$$

Because the integration is over the equilibrium distribution, it is one-third of the integration if $v_z^2$ is replaced by $v^2 = 2\varepsilon/m_e$. The electrical conductivity can be expressed as

$$\sigma = -\frac{2e^2}{3m_e}\int_0^\infty \frac{\partial f_{FD}}{\partial \varepsilon}\tau(\varepsilon)\varepsilon D(\varepsilon)d\varepsilon \quad (5.68)$$

Note that $\partial f_{FD}/\partial \varepsilon \approx -\delta(\varepsilon - \mu)$, where $\delta(\varepsilon - \mu)$ is the Dirac delta function with a sharp peak at $\varepsilon = \mu$ and essentially zero when $\varepsilon \neq \mu$. Furthermore, $\int_\infty^\infty f(x)\delta(x-a)dx = f(a)$. Consequently, the only active electrons are those around the Fermi level. This small fraction of electrons, however, is responsible for the conduction of electricity and heat in metals. We have by assuming $\mu(T) \approx \mu_F$ that

$$\sigma = \frac{2e^2}{3m_e}\tau_F\mu_F D(\mu_F) \quad (5.69)$$

which is the same as Eq. (5.52) since $D(\mu_F) = 3n_e/(2\mu_F)$ according to Eqs. (5.18) and (5.20). The relaxation time is not the average of all electrons but the average of only those electrons near the Fermi surface.

To evaluate the thermal conductivity, we set the applied field to be zero. Note that for an open system of fixed volume, $dU = \delta Q - \mu dN$, i.e., the heat flux is equal to the energy flux *minus* the product of the chemical potential and the particle flux. Hence,

$$q_z'' = J_E - \mu J_N = \int_0^\infty v_z(\varepsilon - \mu)\left(f_{FD}(\varepsilon, T) - \tau(\varepsilon)v_z \frac{\partial f_{FD}}{\partial T}\frac{dT}{dz}\right)D(\varepsilon)d\varepsilon \quad (5.70)$$

Note again that the integration of the equilibrium distribution function in Eq. (5.70) is zero. Furthermore, the integration for $v_z^2$ can be converted into the integration for $v^2 = 2\varepsilon/(3m_e)$. After some manipulations, it can be shown that the thermal conductivity is

$$\kappa = \frac{2}{3m_e}\int_0^\infty \tau(\varepsilon)(\varepsilon - \mu)\varepsilon \frac{\partial f_{FD}}{\partial T} D(\varepsilon)d\varepsilon \quad (5.71a)$$

Using Eq. (B.82) from Appendix B.8, i.e., $\frac{\partial f_{FD}}{\partial T} = -\frac{\partial f_{FD}}{\partial \varepsilon}\left(\frac{\varepsilon - \mu}{T}\right)$, we obtain after applying Eq. (B.80) that

$$\begin{aligned} \kappa &= -\frac{2}{3m_e T}\int_0^\infty \tau(\varepsilon)(\varepsilon - \mu)^2 \varepsilon \frac{\partial f_{FD}}{\partial \varepsilon} D(\varepsilon)d\varepsilon \\ &= \frac{2}{3m_e T}\tau(\mu_F)\mu_F D(\mu_F)\frac{\pi^2 (k_B T)^2}{3} \end{aligned} \quad (5.71b)$$

This is essentially the same expression as in Eq. (5.61) for the electron thermal conductivity obtained from simple kinetic theory. The discussion above based on the Fermi–Dirac distribution not only confirms the simple kinetic theory but also explains why $v_F$ should be used in Eqs. (5.53) and (5.60) rather than the rms velocity of electrons. A familiarity with the BTE will help the study of the classical size effect due to boundary scattering and thermoelectricity phenomena to be discussed in subsequent sections.

The derivation above has confirmed the electrical conductivity and thermal conductivity expressions. This also explains that the scattering rate corresponds to electrons with energy equal to the Fermi energy. Therefore, the Wiedemann–Franz law is also confirmed since the scattering rates for the electron (momentum) transport and that for energy transport cancel each other. Electron–phonon scattering must satisfy the energy and momentum conservations. When the amount of energy change of electrons before and after collision is comparable with $k_B T$, the scattering is inelastic, and thus the two scattering processes can differ significantly. This happens at intermediately lower temperatures since $k_B T$ is small. At very low temperatures, since electron–defect scattering is elastic, the transport of electron momentum is as effective as the transport of energy. As discussed earlier, the result in the intermediate low-temperature region for electron–phonon scattering is such that the electrical resistivity follows $T^5$, while $1/\kappa$ follows $T^2$. In order for Eqs. (5.60) and (5.61) to be valid, it is often thought as if the relaxation time for thermal conductivity is somewhat different from that for electrical conductivity. Actually, it is not because the relaxation times are different; it is because the relaxation time approximation is not valid. By using two relaxation times, one can simplify the scattering process. The relaxation time for momentum transfer retains its meaning of the relaxation time, as in Eq. (5.52) for the electrical conductivity. On the other hand, the relaxation time for thermal transport given in Eq. (5.61) is sometimes called the *energy relaxation time*, which is taken as a weighted average to approximate the difference in the scattering effectiveness for energy exchange [6, 29].

## 5.3.4 Thermal Conductivity of Insulators

Heat conduction in electrical insulators is dominated by lattice waves or phonons. This class of materials includes diamond, quartz, sapphire, and silicon carbide, as well as semiconductor materials like silicon, germanium, and gallium arsenide. Kinetic theory predicts the thermal conductivity of dielectric materials or electrical insulators as follows:

$$\kappa = \frac{1}{3}\rho c_v v_a \Lambda_{ph} \tag{5.72}$$

where $\rho c_v$ is the lattice volumetric specific heat, $v_a$ is the average speed of corresponding acoustic waves or phonons, and $\Lambda_{ph}$ is the phonon mean free path and is related to the scattering rate by $\Lambda_{ph} = v_a \tau$. When $v_a$ is used, it is often assumed that the dispersion relation is linear, i.e., $v_g = v_p$. For crystalline solids, the acoustic speed is on the order of 5000 m/s and depends little on temperature; however, it may depend on the polarization. The density decreases slightly as temperature increases due to thermal expansion, but the change is negligibly small. The specific heat $cv$ is a function of temperature as predicted by the Debye theory, and it is nearly constant at temperatures close to or higher than the Debye temperature. The mean free path can be evaluated based on phonon–phonon scattering and phonon–defect scattering.

The BTE for phonons was first derived by Rudolf E. Peierls in 1929. In some publications, it is referred to as the Boltzmann–Peierls or Peierls–Boltzmann equation. Here, we use a simplified model to derive Eq. (5.72) from the relaxation time approximation of the BTE, based on the Debye theory. The assumption is that the phonon velocity can be taken as a constant that is averaged over all three modes according to Eq. (5.7), as described by the DOS. For phonons, the distribution function can be conveniently converted into the frequency $\nu$ domain. Suppose there is a temperature gradient in the $z$-direction; using the similar procedure as done in the previous section, the thermal conductivity can be expressed as

$$\kappa = \iiint\limits_{\nu,\phi,\theta} v_z h\nu\tau v_z \frac{\partial f_{\mathrm{BE}}}{\partial T} \frac{D(\nu)}{4\pi} \sin\theta \mathrm{d}\theta \mathrm{d}\phi \mathrm{d}\nu \tag{5.73}$$

where $D(\nu)/4\pi$ can be viewed as the density of states per unit solid angle. Noting that $v_z = v_{\mathrm{a}} \cos\theta$ and the distribution function is independent of the direction, we can integrate Eq. (5.73) over all angles first to get $\int_0^{2\pi} \int_0^{\pi} \cos^2\theta \sin\theta \mathrm{d}\theta \mathrm{d}\phi = 4\pi/3$. With the upper limit of frequency $\nu_{\mathrm{m}}$ determined by Eq. (5.9), we can rewrite Eq. (5.73) in the following:

$$\kappa = \frac{1}{3} \int_0^{\nu_{\mathrm{m}}} \tau v_{\mathrm{a}}^2 h\nu \frac{\partial f_{\mathrm{BE}}}{\partial T} D(\nu) \mathrm{d}\nu \tag{5.74}$$

The integration over the spherical coordinates offers a different way for deriving the 1/3 term in the kinetic expression of thermal conductivity obtained earlier for a molecular gas and an electron gas. In addition to the assumption that the acoustic velocity is independent of the frequency, we further assume that the scattering rate is independent of the frequency. Hence, both $\tau$ and $v_{\mathrm{a}}$ can be taken out of the integrand. The remaining part is the specific heat per unit volume, defined in Eq. (5.31). It is clear that Eq. (5.72) can be obtained based on the assumption that phonon speed, relaxation time, and mean free path are independent of frequency.

Using Matthiessen's rule, the phonon mean free path can be expressed as

$$\frac{1}{\Lambda_{\mathrm{ph}}} = \frac{1}{\Lambda_{\mathrm{ph-ph}}} + \frac{1}{\Lambda_{\mathrm{ph-d}}} \tag{5.75}$$

where ph–ph and ph–d stand for phonon–phonon scattering and phonon–defect scattering, respectively. The inverse of the mean free path can be added because they are proportional to the number of collisions per unit time (or scattering rate). The scattering rate due to phonon–phonon scattering is inversely proportional to temperature at relatively high temperatures, i.e., $\Lambda_{\mathrm{ph-ph}}$ decreases as temperature increases. This causes a reduction in thermal conductivity as temperature goes up. Thus, in the high-temperature limit, the thermal conductivity can be modeled as inversely proportional to temperature in a first-order approximation.

At low temperatures, defect scattering dominates and the scattering rate is more or less constant. The thermal conductivity depends on the specific heat and should also vary with $T^3$. The size of the sample affects the mean free path and hence the thermal conductivity. Also, as the temperature is reduced, phonons with lower frequencies play an important role in the thermal transport and storage. Thus, boundary scattering is expected to be more important at low temperatures. Similar to that for electron scattering, the effective mean free path including boundary scattering can be defined as

$$\frac{1}{\Lambda_{\mathrm{ph,eff}}} = \frac{1}{\Lambda_{\mathrm{ph}}} + \frac{1}{\Lambda_{\mathrm{ph-b}}} \tag{5.76}$$

Figure 5.13 shows the thermal conductivity of silicon with different impurity concentrations. For highly pure single-crystal silicon, the thermal conductivity is comparable with a good electrical conductor such as aluminum. As the impurity concentration increases, the scattering rate increases and the mean free path decreases, resulting in a reduction in the thermal conductivity. The contribution of free electrons or holes to the thermal conductivity of semiconductors is insignificant as compared to that of lattice vibration. Therefore, the temperature dependence of thermal conductivity for other crystalline insulators is similar to that of Si. At very low temperatures, $\kappa \propto T^3$ due to the temperature dependence of the specific heat; at high temperatures, $\kappa \propto T^{-1}$ due to the increased phonon–phonon scattering rate. Diamond has the highest thermal conductivity (as high as 2200 W/m K at room temperature) among all bulk materials, due to its large sound velocity and mean free path.

**Example 5.6** Estimate the mean free path and the phonon scattering rate of pure silicon at 5, 10, 20, 100, 300, and 1000 K. Also, calculate the corresponding thermal diffusivity $\alpha = \kappa/\rho c_p$.

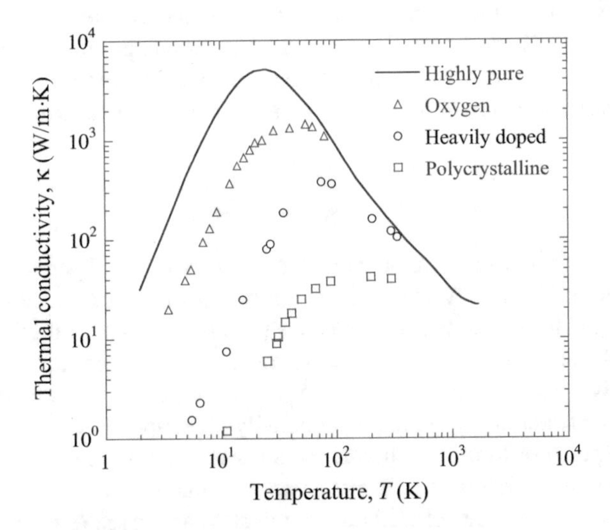


**Fig. 5.13** Data of thermal conductivity of silicon taken from Touloukian et al. [31]. The fitted curve is for a highly pure silicon with a dopant concentration less than $10^{16}$ cm$^{-3}$; triangles are for a *p*-type single-crystal silicon with an oxygen concentration of $2 \times 10^{17}$ cm$^{-3}$; circles are for a heavily doped *n*-type silicon with a phosphorus concentration of $2 \times 10^{19}$ cm$^{-3}$; and squares are for a *p*-type polycrystalline silicon with a boron concentration of $3 \times 10^{20}$ cm$^{-3}$

**Solution** The purpose of this example is to give some quantitative information about the mean free path and its temperature dependence. The calculation is straightforward using Eq. (5.74) by assuming that the density and the phonon velocity are independent of temperature. From Example 5.1, we have $v_a \approx 6000$ m/s, and the density is 2330 kg/m$^3$ (from Table 5.1). The specific heat can be calculated from the Debye model, and the thermal conductivity of intrinsic Si can be found from Fig. 5.13. The computed results are tabulated in the following table.

| Temperature (K) | 5 | 10 | 20 | 100 | 300 | 1000 |
|---|---|---|---|---|---|---|
| Thermal conductivity $\kappa$ (W/m K) | 424 | 2110 | 4940 | 884 | 148 | 31.2 |
| Specific heat $c_p$ (J/kg K) | 0.034 | 0.28 | 3.43 | 260 | 712 | 921 |
| Mean free path $\Lambda$ (m) | $2.7 \times 10^{-3}$ | $1.6 \times 10^{-3}$ | $3.1 \times 10^{-4}$ | $7.3 \times 10^{-7}$ | $4.5 \times 10^{-8}$ | $7.3 \times 10^{-9}$ |
| Scattering rate $1/\tau$ (rad/s) | $2.2 \times 10^{6}$ | $3.7 \times 10^{6}$ | $1.9 \times 10^{7}$ | $8.2 \times 10^{9}$ | $1.3 \times 10^{11}$ | $8.3 \times 10^{11}$ |
| Thermal diffusivity $\alpha$(m$^2$/s) | 5.4 | 3.3 | 0.62 | $1.5 \times 10^{-3}$ | $8.9 \times 10^{-5}$ | $1.5 \times 10^{-6}$ |

The mean free path and thermal diffusivity increase dramatically as the temperature is lowered. Because the crystal is highly pure, there is very little scattering at low temperatures. The decrease in conductivity is caused by the reduction in the specific heat. At high temperatures, the specific heat of Si does not change significantly. Hence, the decrease in thermal conductivity is due to the increase of the phonon–phonon scattering rate. It should be mentioned that at very high temperatures, thermally activated free electrons and holes will also increase the impurity scattering.

When the phonon mean free path is comparable with the smallest dimension so that $Kn \equiv \Lambda/L_c > 1$, boundary scattering or the classical size effect should be considered, as will be discussed in Sect. 5.5. When $Kn \gg 1$, ballistic or phonon–boundary scattering becomes dominant compared with phonon–phonon and phonon–defect scattering. As in the case of free molecule flow, Fourier's law is applicable only in the diffusion limit. When ballistic scattering is significant, the temperature at the boundary is discontinuous. The heat transfer process by phonons is more radiative than conductive, as in the case of thermal radiation through a transparent medium. Even at steady state, the 1D temperature distribution without heat generation is nonlinear. We will study the equation of phonon radiative transfer (EPRT) in Chap. 7 along with other equations that should be used for small timescales or length scales, where Fourier's law of heat conduction breaks down. This is especially important at low temperatures, for small structures, and/or in rapid processes such as during a short laser pulse.

So far, we have studied the basics of phonon contributions to the thermal conductivity under the relaxation time approximation for a gray medium, i.e., by assuming that $\tau$ is independent of the vibration frequency. Furthermore, we have taken the average acoustic velocity and assumed that it is also independent of the vibration frequency. A further assumption is that the phonon dispersion relations are isotropic and linear up to a maximum frequency. Real crystals behave very differently from the simple pictures just presented. To understand this, we must study the phonon dispersion relations for all phonon branches, with different polarizations and along different crystal directions. While the study of crystalline structures and phonon dispersion relations will be deferred to Chap. 6, we can write the general expression for thermal conductivity under the local-equilibrium condition in two forms. The summation form reads as

$$\kappa(\hat{\mathbf{n}}) = \sum_{P}\sum_{K} \hbar\omega(\mathbf{k})\frac{\partial f_{\mathrm{BE}}}{\partial T}\tau(\mathbf{k})v_{\mathrm{g,n}}^{2}(\mathbf{k}) \tag{5.77}$$

where the summation is over the wavevector index $K$ and the polarization index $P$. Note that $v_{\mathrm{g,n}}(\mathbf{k})$ is the phonon group velocity for the given polarization in the direction $\hat{\mathbf{n}}$ along which the thermal conductivity is to be evaluated. The integration form reads

$$\kappa(\hat{\mathbf{n}}) = k_{\mathrm{B}}\sum_{P}\int_{0}^{\infty} \tau(\omega)v_{\mathrm{g,n}}^{2}(\omega)\left(\frac{\hbar\omega}{k_{\mathrm{B}}T}\right)^{2}\frac{\mathrm{e}^{\hbar\omega/k_{\mathrm{B}}T}}{(\mathrm{e}^{\hbar\omega/k_{\mathrm{B}}T}-1)^{2}}D(\omega)\mathrm{d}\omega \tag{5.78}$$

where $D(\omega)$ is the DOS for an individual polarization. If the DOS is properly handled so that it contains information about a particular microstructure, Eq. (5.78) would be identical to Eq. (5.77). Otherwise, Eq. (5.78) is the approximation of Eq. (5.77) for large systems. For a large system with isotropic dispersion in the $k$-space, we have

$$D(\omega) = \frac{1}{(2\pi)^{3}}\frac{\mathrm{d}\mathbf{k}}{\mathrm{d}\omega} = \frac{1}{2\pi^{2}}\frac{k^{2}}{\mathrm{d}\omega/\mathrm{d}k} = \frac{\omega^{2}}{2\pi^{2}v_{\mathrm{p}}^{2}v_{\mathrm{g}}} \tag{5.79}$$

where $v_{\mathrm{p}} = \omega/k$ and $v_{\mathrm{g}} = \mathrm{d}\omega/\mathrm{d}k$ are the phase and group speeds for the corresponding polarization and can be calculated if the dispersion relation $\omega = \omega(\mathbf{k})$ is known.

Therefore,

$$\kappa = \frac{k_{\mathrm{B}}}{6\pi^{2}}\left(\frac{k_{\mathrm{B}}T}{\hbar}\right)^{3}\sum_{P}\int_{0}^{x_{\mathrm{m}}} \tau(x)\frac{v_{\mathrm{g}}(x)}{v_{\mathrm{p}}^{2}(x)}\frac{x^{4}\mathrm{e}^{x}}{(\mathrm{e}^{x}-1)^{2}}\mathrm{d}x \tag{5.80}$$

where the upper limit corresponds to the maximum frequency of each phonon polarization or branch. Equation (5.79) helps us understand low-temperature behavior of thermal conductivity of insulators.

For the same frequency, while the energy of a phonon is the same as that of a photon $h\nu$, the acoustic wave has a much shorter wavelength than the electromagnetic wave because of the small propagation speed $v_a$ compared to the speed of light. Thus, the momentum of a phonon is much greater than that of a photon of the same frequency. As an example, our ears sense sound waves in the frequency range from 20 to 20,000 Hz. Assume $v_a = 1000$ m/s; then, the wavelength range is 50 m to 5 cm. In solids, however, the most important frequencies for thermal energy transfer are much higher and temperature dependent. The smallest vibration wavelength is roughly $\lambda_{min} = 2L_0 \approx 0.5$ nm. With a typical velocity of $v_a = 5000$ m/s in crystalline solids, the highest frequency $\nu_m$ is on the order of 10 THz or $10^{13}$ Hz. Compared with the electromagnetic wave spectrum, this frequency falls in the mid-infrared spectral region. Therefore, electromagnetic radiation can interact with such phonons, and the resulted absorption is called lattice absorption or phonon absorption. High-frequency phonons are called *optical phonons.* On the other hand, the frequency of acoustic phonons ranges from 0 to 10 THz. By setting $k_B T = h\nu$, we find that the frequency corresponding to the thermal energy of translational motion of a particle is on the order of $\nu = k_B T/h = 6$ THz at 300 K (where $k_B T = 26$ meV). The thermal phonon wavelength $\lambda_{th}$ is therefore on the order of 1 nm with $v_a \approx 5000$ m/s. On the other hand, low-frequency phonons are responsible for energy storage and transfer in crystalline solids at cryogenic temperatures. The shift in the dominant frequency for phonon transport resembles Wien's displacement law for blackbody radiation because phonons and photons are governed by the same statistics. The phonon wave effect and quantum size effect are expected to become important when the characteristic dimension is on the order of the thermal wavelength, as illustrated earlier in the study of specific heat of solids.

For amorphous and disordered solids that are poor electric conductors, periodic lattice structure does not exist and phonons if they exist cannot propagate very far. Cahill et al. [32] extended the work of Albert Einstein in 1911 by assuming that the mean free path for the $i$th phonon mode $\Lambda_i = \tau_i v_i$ is limited to half of the phonon wavelength. That is to say that the relaxation time is half of the period, $\tau_i = \pi/\omega$. Some earlier works used the lattice constant or the phonon wavelength as the minimum mean free path [32–34]. By substituting $\tau = \pi/\omega$ and $v_p = v_g = v_i$ into Eq. (5.80), the minimum thermal conductivity can be expressed as

$$\kappa_{min} = \frac{k_B}{6\pi}\left(\frac{k_B T}{\hbar}\right)^2 \sum_i \int_0^{x_i} \frac{1}{v_i} \frac{x^3 e^x}{(e^x - 1)^2} dx \tag{5.81a}$$

where $x_i = \Theta_i/T$ and $\Theta_i$ can be calculated from (5.10) by substituting $v_i$ for $v_a$. Using Eq. (5.10), Eq. (5.81) may be expressed as follows according to Ref. [32]:

$$\kappa_{\min} = k_{\mathrm{B}}\left(\frac{\pi}{6}n_{\mathrm{a}}^{2}\right)^{1/3}\sum_{i} v_i \left(\frac{T}{\Theta_i}\right)^2 \int_0^{x_i} \frac{x^3 \mathrm{e}^x}{(\mathrm{e}^x - 1)^2}\mathrm{d}x \tag{5.81b}$$

At room temperature, the thermal conductivity of most amorphous solids falls in the range 0.2–5 W/m K [34]. Note that there are no fitting parameters in Eq. (5.81b) as long as the number density of the atoms and the acoustic velocities for the transverse and longitudinal phonons are given. Overall, Eq. (5.81b) agrees well with the measured thermal conductivity for a large number of disordered solids, though some materials exhibit even lower thermal conductivities than predicted $\kappa_{\min}$. While Eq. (5.81a) or (5.81b) removes the relaxation time and mean free path, in disordered materials when some of the vibration eigenstates are localized, definition of phonon velocities and wavevectors is questionable.

Another approach was developed by Allen and Feldman [35] by extending the Kubo–Greenwood formulation, which is a quantum mechanical theory for electron transport based on the linear response theory, to the thermal conductivity of disordered solids. The key is to relate the conductivity to the heat current operator matrix, which under the harmonic assumption can be related to the mode diffusivity without defining the group velocity or scattering rate. The obtained conductivity formula is expected to be applicable to disordered media where the wavevectors of the carriers can hardly be defined [35–38]. The temperature-dependent thermal conductivity is thus expressed in terms of a summation [35]:

$$\kappa = \frac{1}{V}\sum_{i} C(\omega_i) D_{\mathrm{dif}}(\omega_i) \tag{5.82a}$$

where $C(\omega)$ is the specific heat of the harmonic oscillator,

$$C(\omega) = \hbar\omega \frac{\partial f_{\mathrm{BE}}}{\partial T} = \frac{k_{\mathrm{B}} x^2 \mathrm{e}^x}{(\mathrm{e}^x - 1)^2}, \text{ with } x = \frac{\hbar\omega}{k_{\mathrm{B}} T} \tag{5.82b}$$

and the mode diffusivity is expressed as

$$D_{\mathrm{dif}}(\omega_i) = \frac{\pi V^2}{3\hbar^2 \omega_i^2} \sum_{j(\neq i)} \left|S_{ij}\right|^2 \delta(\omega_i - \omega_j) \tag{5.82c}$$

The heat current operator is a measure of the coupling strength between vibration mode $i$ and $j$ and can be calculated from harmonic lattice dynamic theory [35–37]. Some discussions on how to obtain semi-classical expressions of the diffusivity will be given later.

In a follow-up study of amorphous silicon, Allen et al. [36] divided the heat carriers in crystals (*vibrons*) into *propagons* that have a larger mean free path than the lattice constant and are propagating modes, *diffusons* that are most popular and largely responsible for heat transfer but are not propagating, and *locons* that are localized modes that do not contribute to heat transport. It should be noted that when anharmonicity is considered, the locon's contribution to heat transfer cannot be neglected [39, 40]. Based on atomistic simulation of amorphous silicon, Allen et al. [36] assigned 4% of the modes to propagons with frequencies less than 3 THz. The region between propagons and diffusons is called the Ioffe–Regel crossover where the mean free path is about the same as the atomic distance. When the frequency is further increased, the mean free path and wavevector cannot be rigorously defined. Locons are at frequencies higher than 17 THz and comprise about 3% of the modes; this was determined by finding the decay lengths, inverse participation ratios, and coordination numbers of the participating atoms [36]. It was shown that propagons dominate the thermal transport at low temperatures, while diffusons contribute to about 2/3 of the thermal conductivity at ambient temperature. A recent theoretical study [41] of amorphous silicon based on lattice and molecular dynamics showed that the propagon–diffuson transition frequencies could be as high as 5–10 THz and propagons might consist of 24% of all modes, suggesting that most heat is carried by elastic waves in amorphous silicon at temperatures from 100 K to 500 K.

We may rewrite Eq. (5.82a) in an integral form using the density of states [38]:

$$\kappa = \int_0^\infty C(\omega) D_{\mathrm{dif}}(\omega) D(\omega) \mathrm{d}\omega \tag{5.83}$$

For propagons, by comparing Eq. (5.74) with (5.83), we see that $D_{\mathrm{dif}} = \tau v_{\mathrm{a}}^2/3 = \Lambda v_{\mathrm{a}}/3$ and the upper limit in Eq. (5.83) can be set as the high-frequency limit of propagons. For diffusion, if the frequency-dependent mode diffusivity is obtained, Eq. (5.83) can be applied to calculate the thermal conductivity. Allen et al. [36] found a temperature independent $D_{\mathrm{dif}}(\omega) \sim \omega^{-2}$ for amorphous silicon in the intermediate temperature range and predicted a low-temperature plateau of thermal conductivity between 10 and 30 K by combining the contributions of propagons and diffusons. Assuming diffusons travel stepwise following a random walk with two steps per period of oscillation, Agne et al. [38] obtained an expression of the mode diffusivity as follows:

$$D_{\mathrm{dif}} = \frac{\omega}{3\pi} n_{\mathrm{a}}^{-2/3} p \tag{5.84}$$

where $p$ is the probability of a successful jump that may be taken as 1 for diffusons. Plugging Eq. (5.84) into Eq. (5.83) and setting the integration maximum $\omega_{\mathrm{m}}$ according to Eq. (5.9), they obtained a minimum thermal conductivity expression, which may be applicable near room temperatures. At lower temperatures, propagons are responsible for the heat transfer, and the minimum thermal conductivity may be predicted with Eq. (5.81b). It should be noted that the minimum thermal conductivity using the combination of Eqs. (5.83) and (5.84) depends on the upper integration limit that is a function of the sound velocity.

Most polymers are disordered due to their complex morphologies and long chains structures. Typically, they are also electrical insulators and have relatively low thermal conductivities [42]. Due to the advancement of flexible electronics, energy harvesting, and biophysics, thermal transport in polymeric materials has received growing attention lately [43–47]. At room temperature, the thermal conductivity of commonly used amorphous polymers is mostly around 0.1–0.5 W/m K, though it can be as low as 0.06 W/m K or as high as 0.67 W/m K [42–44]. Like inorganic amorphous materials, as the temperature goes down from room temperature, the thermal conductivity of amorphous polymers monotonically decreases with a plateau-like behavior. Choy [42] used the BTE and the Debye DOS, Eq. (5.74), to describe the thermal transport by assuming that the mean free path $\Lambda$ (or the scattering rate) has an inverse frequency dependence ($\tau \sim \omega^{-1}$) at low frequencies, and $\Lambda = \Lambda_{\min} = L_0$ when the frequency exceeds a certain threshold value to describe the localized modes. Kommandur and Yee [43] used the Allen–Feldman model, given in Eq. (5.83) with the Debye DOS, and considered the mode diffusivity as frequency dependent according to

$$D_{\text{dif}}(\omega) = \frac{1}{3} a v_{\text{a}}^2 \omega^{-n} \tag{5.85}$$

where $n$ is taken as an adjustable parameter approximately between 1 and 2. The parameters $a$ and $n$ for propagons and diffusons, as well as the crossover frequency, were taken as adjustable parameters to fit the thermal conductivity data, since the contribution of locons is negligibly small. Both the BTE approach and the Allen–Feldman model can fit the temperature dependency reasonably well [42, 43]. While the starting points of the BTE and Kubo's linear response theory are conceptually different, it appears that the final model relations are well correlated though with different interpretations of the physical significance of the parameters.

In general, the properties of polymeric materials depend on the morphology, crystallinity, and chain orientation and alignment [45–47]. Polymers can have skeletal structures, planar molecular structures, or 1D linear macromolecules; subsequently, the specific heat may follow the general cubic, quadratic, and linear temperature dependence [48]. Furthermore, there exist glass transitions and other phase transitions that can give spikes in the specific heat of polymeric materials at the transition temperatures [48]. Filling the polymer with highly conductive nanostructured materials can increase the thermal conductivity of the polymer composite significantly (by more than an order or magnitude) to 10–20 W/m K at room temperature [45–47]. Furthermore, aligned polymer chains and nanofibers are expected to have very high thermal conductivities. Molecular dynamics modeling of single polyethylene chains has shown a converging thermal conductivity up to 350 W/m K and a value over 100 W/m K for polyethylene nanofibers [45, 49].

## 5.4 Thermoelectricity

Solid-state energy conversion devices are very important, and it is hoped that nanotechnology may offer solutions for improving the efficiency of these devices, such as thermoelectric refrigerators and power generators. An understanding of thermoelectricity is useful for further development of these solid-state energy conversion devices. To illustrate the *thermoelectric effect,* assume an electric field **E** and a temperature gradient $\nabla T$ exist along the $z$-direction of a conductor. The right-hand side of Eqs. (5.69a) or (5.70) needs to be modified to consider the existence of both an electric field and a temperature gradient. This can be done by applying Eqs. (5.65) and (5.66). By dropping the integration for the equilibrium distribution and using Appendix B.8, we can write the 3D vector forms of the current density and the heat flux as

$$\mathbf{J}_{\mathrm{e}} = L_{11}\left(\mathbf{E} + \frac{\nabla\mu}{e}\right) - L_{12}\nabla T \tag{5.86}$$

and

$$\mathbf{q}'' = L_{21}\left(\mathbf{E} + \frac{\nabla\mu}{e}\right) - L_{22}\nabla T \tag{5.87}$$

where

$$L_{11} = -e^2\Psi_0 \quad L_{12} = \frac{e}{T}\Psi_1 \; L_{21} = TL_{12} = e\Psi_1, \text{ and } L_{22} = -\frac{1}{T}\Psi_2 \tag{5.88}$$

Here, the function $\Psi_n$ is defined as

$$\Psi_n = \frac{1}{3}\int_0^\infty (\varepsilon - \mu)^n \tau v^2 \frac{\partial f_{\mathrm{FD}}}{\partial \varepsilon} D(\varepsilon)\mathrm{d}\varepsilon \tag{5.89}$$

In writing this equation, we have used Eq. (B.81) and converted $(\mathrm{d}\mu/\mathrm{d}T)\nabla T = \nabla\mu$ in order to consider the spatial dependence of $\mu$. The detailed derivation of the preceding equations is left as an exercise (Problem 5.21). Let

$$\mathbf{E} + \frac{\nabla\mu}{e} = -\nabla\Phi \tag{5.90}$$

where $\Phi$ is called the *electrochemical potential* because it is the combination of the electrostatic potential and the chemical potential. For metals at low or intermediate temperatures, the variation in $\mu$ is relatively small, and the terms involving $\nabla\mu$ in Eqs. (5.86) and (5.87) can be dropped out. For semiconductors, changing the dopant or impurity concentration as well as the temperature may cause a large gradient of $\mu$, and thus $\nabla\mu$ cannot be neglected. When there is no temperature gradient, we can easily find the electrical conductivity of metals to be

$$\sigma = L_{11} \tag{5.91}$$

The thermal conductivity is defined according to $\mathbf{q}'' = -\kappa \nabla T$ when no electric current flows. By setting $J_e = 0$ and combining Eqs. (5.86) and (5.87), we find that the thermal conductivity is related to the coefficients by

$$\kappa = L_{22} - L_{12}L_{21}/L_{11} \tag{5.92}$$

For metals, the second term on the right-hand side is much smaller than the first one, so that we can approximate $\kappa \approx L_{22}$, as already discussed in Eq. (5.71a).

## 5.4.1 The Seebeck Effect and Thermoelectric Power

If a temperature gradient exists, according to Eq. (5.86), there will be a current flow even in the absence of an external field. In the case of open circuit when the current flow is zero, there will be a voltage across the rod whose ends are held at different temperatures. The *Seebeck effect,* as it was first noticed by T. J. Seebeck in 1821, can be used to directly produce electric power from a temperature difference. The *Seebeck coefficient,* also called *thermopower* or *thermoelectric power,* is defined as the induced thermoelectric voltage across a material of unit length per unit temperature difference. Therefore,

$$\Gamma_S = \frac{-\nabla \Phi}{\nabla T} = \frac{L_{12}}{L_{11}} \tag{5.93}$$

which has units V/K. To calculate $L_{12}$ for a metal, we can use Eq. (B.79) to evaluate $\Psi_1$ in Eq. (5.89). The simplest approach is to assume that $\tau$ does not change much near the Fermi surface. The result gives (see Problem 5.22)

$$\Gamma_S \approx -\frac{\pi^2 k_B}{2e} \frac{k_B T}{\mu_F} \tag{5.94}$$

For metals, the Seebeck coefficient is negative, and its magnitude will increase as temperature goes up. From Table 5.2, $\mu_F = 7$ eV for copper. We have from Eq. (5.94) that $\Gamma_S = -1.6\,\mu$V/K at 300 K and $-3.2\,\mu$V/K at 600 K. However, the experimental values are positive with 1.83 $\mu$V/K at 300 K and 3.33 $\mu$V/K at 600 K [50, 51]. This sign error is due to the simplification used to evaluate $\Psi_1$, and it is an indication that the nearly free-electron model may not capture all the fundamental physics of metals. A proper quantum mechanical evaluation based on the actual band structure is rather complicated but has been carried out in some studies [6, 52]. Higher values of the Seebeck coefficient can exist in some alloys and semiconductors.

Generally speaking, the Seebeck coefficient is positive for $p$-type semiconductors whose majority carriers are holes and negative for $n$-type semiconductors whose majority carriers are electrons.

For a wire whose ends are at different temperatures $T_1$ and $T_2$, as in the open circuit shown in Fig. 5.14a, there will be a voltage difference between 1 and 2 according to the relation $V_2 - V_1 = -\int_{T_1}^{T_2} \Gamma_S(T)\mathrm{d}T$. For $n$-type semiconductors, $\Gamma_S(T)$ is negative and electrons at the higher temperature end tend to diffuse toward the lower temperature end. An electrostatic potential will be built up to balance the diffusion process. Hence, the voltage is higher at the higher temperature end. Thermoelectric voltage cannot be measured with the same type of wires because the electrostatic potentials would cancel out each other. To measure the thermoelectric power, a junction is formed with two types of wires having different Seebeck coefficients, type I (+) and type II (−), as shown in Fig. 5.14b. The leads can be a third type of wire or the same as one of the thermocouple wires. This is of course the familiar thermocouple arrangement for temperature measurement. A reference temperature ($T_1$) is needed because a thermocouple can only measure the temperature difference. The voltage output can be expressed as

$$\Delta V = \int_{T_1}^{T_2} \left[\Gamma_{S,I}(T) - \Gamma_{S,II}(T)\right]\mathrm{d}T = \Gamma_{I,II}\Delta T \tag{5.95}$$

In thermocouple practice, the difference $\Gamma_{I,II}$ is called the Seebeck coefficient or thermopower, and the potential difference $\Delta V$ is called the electromotive force (emf). Because the Seebeck coefficient is zero when a material becomes superconducting ($\sigma \to \infty$), superconductors have been used to establish an absolute scale of thermoelectric power [51]. In thermometry, a wire with a positive Seebeck coefficient and another with a negative Seebeck coefficient are combined to form a thermocouple junction. For example, a type-E thermocouple is made of a nickel–chromium alloy (chromel) and a copper–nickel alloy (constantan); on the other hand, a type-J thermocouple is made of copper and constantan. Historically, galvanometers were used

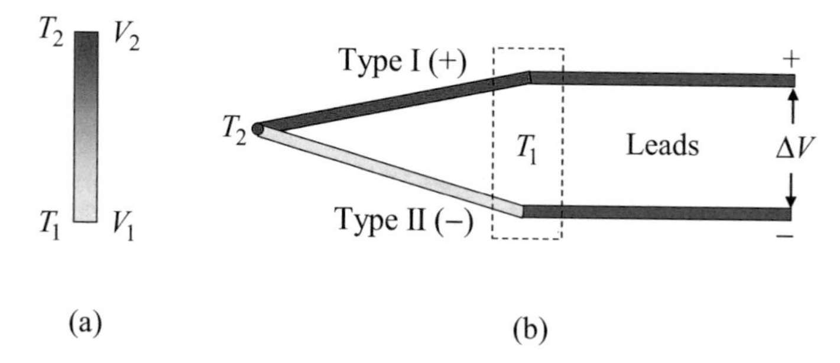


**Fig. 5.14** Illustration of the Seebeck effect. **a** Single wire with a temperature difference between the two ends. **b** A thermocouple made of two different materials

to accurately measure the electric current in a potentiometer. The DC voltage can now be measured quickly and very accurately with a digital voltmeter/multimeter (DVM). Detailed discussions about the fundamentals and practice of thermoelectric thermometry based on metallic and alloy wires can be found from Bentley [50].

### 5.4.2 The Peltier Effect and the Thomson Effect

Equations (5.86) and (5.87) can be combined to eliminate the potential term so that

$$\mathbf{q}'' = \frac{L_{21}}{\sigma}\mathbf{J}_\mathrm{e} - \kappa\nabla T \tag{5.96}$$

This equation suggests that there will be a heat flux in a material due to an external electric current, even without any temperature difference. This phenomenon, first discovered by Jean Peltier in 1834, is called the *Peltier effect,* which can be used for refrigeration (known as *thermoelectric cooling*) by passing through an electric current through a material. The coefficient $L_{21}/\sigma$ is called the *Peltier coefficient*. It can be seen from Eqs. (5.88), (5.91), and (5.93) that

$$\Pi = L_{21}/\sigma = T\Gamma_\mathrm{S} \tag{5.97}$$

This quantitative relationship between the Seebeck coefficient and the Peltier coefficient was revealed by William Thomson (Lord Kelvin) in the 1850s. Thomson's thermodynamic derivation led him to discover a third thermoelectric effect, known as the *Thomson effect,* which states that heat can be *released* or *absorbed* when current flows in a material with a temperature gradient. The energy received by a volume element for prescribed $\mathbf{J}_\mathrm{e}$ and $\nabla T$ can be expressed as follows:

$$\mathbf{J}_\mathrm{e}\cdot(-\nabla\Phi) - \nabla\cdot\mathbf{q}'' = \frac{J_\mathrm{e}^2}{\sigma} + \nabla\cdot(\kappa\nabla T) - \left(T\frac{\mathrm{d}\Gamma_\mathrm{S}}{\mathrm{d}T}\right)\mathbf{J}_\mathrm{e}\cdot\nabla T \tag{5.98}$$

Notice that the common term $\Gamma_\mathrm{S}\mathbf{J}_\mathrm{e}\cdot\nabla T$ in both $\mathbf{J}_\mathrm{e}\cdot(-\nabla\Phi)$ and $\nabla\cdot\mathbf{q}''$ cancels out. In Eq. (5.98), the first term is the heat generated by the Joule heating, the second term is the heat transferred into the control volume due to the temperature gradient, and the third term is caused by the Thomson effect. The last term on the right-hand side is nonzero when there is a current flow with a temperature gradient, unless the Seebeck coefficient is independent of temperature. It should be noted that, like the Seebeck effect and the Peltier effect, the Thomson effect is also a reversible process per se. The *Thomson coefficient* $K$ is defined as the rate of the absorbed heat divided by the product of the current density and the temperature gradient. Thus,

$$K = T\frac{\mathrm{d}\Gamma_\mathrm{S}}{\mathrm{d}T} \tag{5.99}$$

Equation (5.98) has provided a way to determine $d\Gamma_S/dT$, after $\sigma$ and $\kappa$ are measured at different temperatures. This allows for the absolute thermopower to be determined for certain materials at higher temperatures, since superconductivity can only occur at very low temperatures. A systematic study has resulted in the determination of absolute thermoelectric power for lead and platinum, which can then be used as reference materials to determine the absolute thermoelectric power for other materials [51]. It should be noted that, before the discovery of high-temperature superconductors, the highest temperature that a material could be made superconducting was 23 K in an alloy. Superconductivity at temperatures above 35 K was discovered in a ceramic material in 1986 and, shortly afterward, superconductivity above the boiling temperature of liquid nitrogen (78 K) was made possible.

**Example 5.7** Consider a $p$-type semiconductor rod of diameter $d = 1$ mm and length $L = 2$ mm. One end of the rod is in contact with a heat sink at $T_L = 300$ K, and the other end is in contact with a heat source at $T_H = 350$ K. What is the open-circuit voltage? If a current $I = 0.8$ A is allowed to flow from the cold end to the hot end, what is the heat transfer rate to the heat sink? Neglect the temperature dependence of the thermal conductivity, the electrical resistivity, and the Seebeck coefficient by using $\kappa = 1.1$ W/m K, $r_e = 19\,\mu\Omega$ m, and $\Gamma_S = 220\,\mu$V/K, respectively.

**Solution** Assume there is no heat transfer via the side of the rod. For an open circuit, the electric potential is higher at the cold end, and the voltage across the rod is $V_{open} = \Gamma_S(T_H - T_L) = 11$ mV. The rate of heat transfer to the heat sink by conduction from the heat source is $q_C = (\pi d^2/4)\kappa(T_2 - T_1)/L = 21.6$ mW.

When an electric current is running from the cold end to the hot end, the Joule heating is generated uniformly inside the rod. The dissipated heat must reach both ends equally by conduction. The additional heat transfer to the heat sink is $q_J = I^2R/2 = 15.5$ mW, where $R = 48.4$ m$\Omega$ is the resistance of the rod. On the other hand, the Peltier effect results in cooling, or heat removal from the heat sink. From Eq. (5.84), we have $q_P = -T_1\Gamma_S I = -52.8$ mW. The combination of the three terms gives the heat transfer rate as $q = q_C + q_J + q_P = -15.7$ mW. The negative sign indicates that heat is removed from the heat sink.

This example demonstrates the Peltier effect for thermoelectric refrigeration. It can be seen that a smaller thermal conductivity will decrease the heat transfer between the two ends: a smaller electrical resistivity will reduce the Joule heating,, and a larger Seebeck or Peltier coefficient will enhance the heat removal. For most metals, the thermal conductivity is too high, and the Seebeck coefficient is too small for refrigeration application. Some insulators can have a large Seebeck coefficient but their electrical resistivity is too high for them to be used in thermoelectric devices.

### 5.4.3 *Thermoelectric Generation and Refrigeration*

The study of thermoelectric generation and refrigeration has become an active research area since the 1950s, along with the development of semiconductor materials

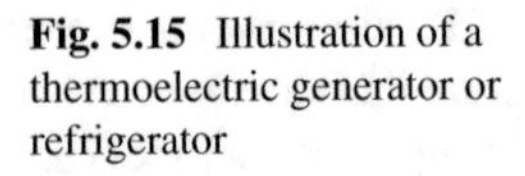
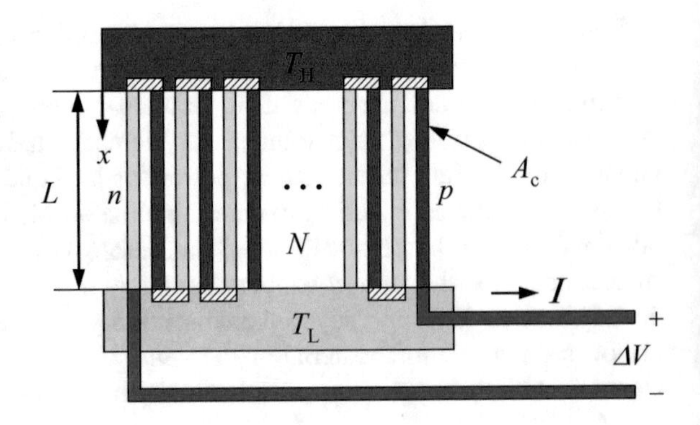


**Fig. 5.15** Illustration of a thermoelectric generator or refrigerator

or *p*-*n* junctions. Heavily doped semiconductors exhibit large Seebeck coefficients. Alternative *n*-type or *p*-type semiconductors (or semimetals) are used as thermoelectric materials or *thermoelectric elements.* These include antimony tellurium (SbTe), bismuth tellurium (BiTe), and silicon germanium (SiGe) compounds. More recently, nanostructured materials are investigated as candidates to increase the performance of thermoelectric devices [53].

With the understanding of the Seebeck effect, the Peltier effect, and the Thomson effect, we are ready to perform a thermodynamic analysis of thermoelectric generators or refrigerators as illustrated in Fig. 5.15. There are $N$ pairs of junctions that are connected electrically in series by metallic interconnects and thermally in parallel between the two heat sinks. To simplify the analysis, contact resistances are neglected, and it is assumed that all the thermoelectric elements have the same length $L$ and the same cross-sectional area $A_c$. Furthermore, heat transfer by other modes is neglected except conduction by thermoelectric elements. Because contact electrical resistance is neglected, heat generation by the Joule heating happens due to resistance of the thermoelectric elements only. A load resistance $R_L$ is used to evaluate the electric power output of the generator. A further assumption is that the thermal and electrical conductivities, as well as the Seebeck coefficient, are independent of temperature. This assumption is reasonable when the temperature difference between the two heat reservoirs is very small.

Consider a thermoelectric generator. In this case, heat is taken from the high-temperature reservoir $T_H$ at the rate $q_H$, and some heat is released to the low-temperature reservoir $T_L$ at the rate $q_L$. The generated thermoelectric power is

$$P = I\Delta V = q_H - q_L \tag{5.100}$$

The temperature distribution along the thermoelectric element is not linear, i.e., the temperature gradient is not constant. The steady-state temperature distribution along a single thermoelectric element can be solved by setting Eq. (5.98) to zero. Because of the assumption of constant values of $I$, $\kappa$, $\sigma$, and $\Gamma_S$, the Thomson coefficient also becomes zero. Therefore, we obtain

$$T(x) = \frac{J_e^2}{2\sigma\kappa}(L - x)x - \frac{x}{L}(T_H - T_L) + T_H \tag{5.101}$$

The resulting heat transfer rates due to temperature gradient are

$$-\kappa A_c \frac{dT}{dx}\bigg|_{x=0} = \kappa A_c \frac{T_H - T_L}{L} - \frac{I^2 L}{2\sigma A_c} \tag{5.102a}$$

and

$$-\kappa A_c \frac{dT}{dx}\bigg|_{x=L} = \kappa A_c \frac{T_H - T_L}{L} + \frac{I^2 L}{2\sigma A_c} \tag{5.102b}$$

Clearly, half of the Joule heating goes to the heat source, and the other half goes to the heat sink, as noticed in Example 5.7. Substituting Eq. (5.102) into Eq. (5.96) and using the subscripts $n$ and $p$ for different thermoelectric elements, we have

$$q_H = NI\Gamma_{np}T_H + NA_c\kappa_{np}\frac{\Delta T}{L} - N\frac{I^2 L}{2A_c\sigma_{np}} \tag{5.103a}$$

$$q_L = NI\Gamma_{np}T_L + NA_c\kappa_{np}\frac{\Delta T}{L} + N\frac{I^2 L}{2A_c\sigma_{np}} \tag{5.103b}$$

where $\Gamma_{np} = \Gamma_{S,p} - \Gamma_{S,n}$, $\kappa_{np} = \kappa_n + \kappa_p$, $\Delta T = T_H - T_L$, and $\sigma_{np} = \left(1/\sigma_n + 1/\sigma_p\right)^{-1}$. The output power is therefore

$$P = I\Delta V = q_H - q_L = NI\Gamma_{np}\Delta T - I^2 R_0 \tag{5.104}$$

where $R_0 = NL/(A_c\sigma_{np})$ is the resistance of all thermoelectric elements. The voltage is solely caused by the Seebeck effect, i.e., $\Delta V = N\Gamma_{np}\Delta T$. Assuming the load resistance is $R_L$, we have

$$I = \frac{\Delta V}{R_0 + R_L} = \frac{N\Gamma_{np}\Delta T}{R_0 + R_L} \tag{5.105}$$

Substituting Eq. (5.105) into Eq. (5.104), we see that the electric power is indeed $P = I^2 R_L$. The thermal efficiency can be calculated as follows:

$$\eta = \frac{P}{q_H} = \frac{\frac{R_L}{R_0}\frac{\Delta T}{T_H}}{\frac{1}{Z^* T_H}\left(1 + \frac{R_L}{R_0}\right)^2 + \left(1 + \frac{R_L}{R_0}\right) - \frac{\Delta T}{2T_H}} \tag{5.106}$$

where

$$Z^* = \frac{NL}{A_c\kappa_{np}}\frac{\Gamma_{np}^2}{R_0} = \frac{\sigma_{np}\Gamma_{np}^2}{\kappa_{np}} \tag{5.107}$$

is independent of the geometry [54]. When $1/Z^*T_H \ll 1$ and $R_0/R_L \ll 1$, we have $\eta \to 1 - T_L/T_H$, which is exactly the Carnot efficiency. Increasing $Z^*$ will improve the efficiency. Hence, minimizing the thermal conduction, reducing the electrical resistance, and increasing the Seebeck coefficient of the thermoelectric elements are essential to improve the performance. A similar analysis can be done for thermoelectric cooling, which is left as an exercise (see Problem 5.25). In general, the *figure of merit* of thermoelectricity is defined as

$$Z = \frac{\sigma\Gamma_S^2}{\kappa} \tag{5.108}$$

which has units of $K^{-1}$ and can be nondimensionalized by multiplying the temperature $T$. The resulting dimensionless parameter $ZT$ (zee-tee) is often quoted as the figure of merit for thermoelectric materials or devices. This applies to both thermoelectric generation and refrigeration (see Problems 5.23 and 5.25).

Because of the compromise between a large electrical conductivity and a small thermal conductivity, along with the requirement of a large Seebeck coefficient, it has turned out that semiconductors are the best choice for thermoelectric applications. After an extensive pursuit in the 1950s, materials with $ZT$ values between 0.5 and 1 near room temperature have been developed using $Bi_xSb_{2\text{-}x}Te_3$ and $Bi_2Se_yTe_{3\text{-}y}$. These materials are essentially doped V-VI semiconductors $Sb_2Te_3$ or $Bi_2Te_3$. In the past 25 years, intensive theoretical and experimental research has been conducted to increase the thermoelectric device performance by using nanostructured materials. Mildred Dresselhaus and coworkers predicted that multiple quantum wells or superlattices may enhance $ZT$ values due to quantum confinement as well as a reduction in the phonon thermal conductivity; the idea has also been extended to PbTe/PbSe superlattice nanowires [55]. Superlattices made of SiGe/Si and GaAs/AlAs have also been considered. Since 2001, several groups have demonstrated $ZT$ values exceeding 2 [53, 56]. Gang Chen's group has performed extensive investigations on the phonon and electron transport in nanostructured materials related to low-dimensional thermoelectricity, as discussed in a recent review [57]. The reduction in thermal conductivity may come from a combination of a number of factors including the mean-free-path reduction by boundary scattering, thermal resistance associated with acoustic mismatch or phonon scattering at the interface of dissimilar materials, and/or quantum confinement of the phonon DOS.

Before moving to the discussion of size effects on thermal conductivity, let us give an overview of irreversible thermodynamics and a brief introduction to nonequilibrium thermodynamics.

### 5.4.4 Onsager's Theorem and Irreversible Thermodynamics

The set of coupled equations given in Eqs. (5.86) and (5.87) is an example of *irreversible thermodynamics,* pioneered by Lars Onsager in the 1930s, alternatively known as the thermodynamics of irreversible processes or Onsager's theorem. Onsager described the phenomenological relations of interrelated or coupled transport processes using the following equation [58]:

$$\mathbf{J}_i = \sum_j \alpha_{ij}\mathbf{F}_j \tag{5.109}$$

where $\mathbf{J}_i$ is the flux of a physical quantity $X_i$ with $J_i = \mathrm{d}X_i/\mathrm{d}t$, $\alpha_{ij}$ is called the *Onsager kinetic coefficient,* and $\mathbf{F}_i$ is the $i$th generalized driving force or *affinity.* In an equilibrium state, all $\mathbf{F}_i$'s are zero. Furthermore, the entropy of a system can be expressed as [59]

$$\mathrm{d}s = \sum_i f_i \mathrm{d}X_i \tag{5.110}$$

where $f_i$ is a property that is related to $\mathbf{F}_i$ such that $\mathbf{F}_i$ is proportional to the gradient of $f_i$. The entropy flux is thus

$$\mathbf{s}'' = \sum_i f_i \mathbf{J}_i \tag{5.111}$$

If an infinitesimal control volume is chosen, the continuity equation can be written as

$$\frac{\partial X_i}{\partial t} + \nabla \cdot \mathbf{J}_i = 0 \tag{5.112}$$

The entropy balance becomes

$$\frac{\partial s}{\partial t} = \dot{s}_{\text{gen}} - \nabla \cdot \mathbf{s}'' \tag{5.113}$$

where $\frac{\partial s}{\partial t} = \sum_i f_i \frac{\partial X_i}{\partial t}$ and $\nabla \cdot \mathbf{s}'' = \sum_i \nabla f_i \cdot \mathbf{J}_i + \sum_i f_i \nabla \cdot \mathbf{J}_i$. Using the continuity equation, we obtain the *volumetric entropy generation rate*:

$$\dot{s}_{\text{gen}} = \sum_i \nabla f_i \cdot \mathbf{J}_i \tag{5.114}$$

Furthermore, the *Onsager reciprocity* is expressed as follows [58, 59]:

$$\alpha_{ij} = \alpha_{ji} \tag{5.115}$$

Lars Onsager (1903–1976) received the Nobel Prize in Chemistry in 1968 "for the discovery of the reciprocal relations bearing his name, which are fundamental for the thermodynamics of irreversible processes." The Onsager reciprocity was even considered by some researchers as the *fourth law of thermodynamics.*

**Example 5.8** Determine the Onsager kinetic coefficients and the volumetric entropy generation rate for a conductor with constant current and temperature gradient.

**Solution** It should be noted that in thermoelectricity, $\mathbf{J}_1 = \mathbf{J}_e$, $\mathbf{J}_2 = \mathbf{q}''$, $\mathbf{F}_1 = -(1/T)\nabla\Phi$, and $\mathbf{F}_2 = \nabla(1/T) = -(1/T^2)\nabla T$. Thus, the Onsager relations are expressed as

$$\mathbf{J}_e = \alpha_{11}\frac{-\nabla\Phi}{T} - \alpha_{12}\frac{\nabla T}{T^2} \tag{5.116}$$

$$\mathbf{q}'' = \alpha_{21}\frac{-\nabla\Phi}{T} - \alpha_{22}\frac{\nabla T}{T^2} \tag{5.117}$$

Comparing the above expressions with Eqs. (5.86) and (5.87), we find that

$$\alpha_{11} = TL_{11}\ \alpha_{12} = \alpha_{21} = T^2L_{12}\ \text{and}\ \alpha_{22} = T^2L_{22} \tag{5.118}$$

The entropy generation rate can be calculated by using Eq. (5.99). Note that

$$\mathrm{d}s = \frac{\delta Q - \mu \mathrm{d}N}{TV} = \mathbf{q}'' \cdot \nabla\left(\frac{1}{T}\right) + \mathbf{J}_e \cdot \left(-\frac{\nabla\Phi}{T}\right) \tag{5.119}$$

In the steady state, the energy equation, Eq. (5.98), becomes

$$\mathbf{J}_e \cdot (-\nabla\Phi) - \nabla \cdot \mathbf{q}'' = 0 \tag{5.120}$$

Therefore, the volumetric entropy generation rates for 3D and 1D cases, respectively, are

$$\dot{s}_{\text{gen}} = \mathbf{q}'' \cdot \nabla\left(\frac{1}{T}\right) + \frac{1}{T}\nabla \cdot \mathbf{q}''\ \text{and}\ \dot{s}_{\text{gen}} = \frac{q''}{T^2}\frac{\mathrm{d}T}{\mathrm{d}x} + \frac{1}{T}\frac{\mathrm{d}q''}{\mathrm{d}x} \tag{5.121}$$

These results are consistent with the analysis in Chap. 2 (see Example 2.5 and Problem 2.29). Furthermore, Eq. (5.121) suggests that the Thomson effect is a reversible process that does not cause any entropy generation. The same can be said for both the Seebeck effect and the Peltier effect, which are reversible thermoelectric effects. In addition to thermoelectricity, irreversible thermodynamics has found applications in multicomponent diffusion, nonisothermal diffusion (when both a temperature gradient and a concentration gradient exist), and some magnetic processes [59]. A further advancement in nonequilibrium thermodynamics was made by Ilya Prigogine (1917–2003) who was awarded the Nobel Prize in Chemistry in 1977.

Prigogine's study extended irreversible thermodynamics to systems that are far from equilibrium and are allowed to exchange energy, mass, and entropy with their surroundings. Prigogine and colleagues demonstrated that ordered dissipative systems can be formed from disordered systems, when the systems are far from equilibrium, and dubbed this theory *dissipative structure,* which led to pioneering research in self-organization or self-assembly. The formation of ordered structures from disordered structures has diverse applications in chemical, biological, and social systems [60, 61]. It is beyond the scope of this book to go into the details of this theory.

## 5.5 Classical Size Effect on Conductivities

When the characteristic length, such as the thickness of a film, the diameter of a wire, or the size of a grain (for polycrystalline solids), is comparable to the mechanistic length, i.e., the mean free path, boundary or interface scattering becomes important. Subsequently, the thermal conductivity (as well as other transport coefficients) becomes size dependent and can also be anisotropic [62, 63]. Because the mean free paths of electrons and phonons tend to increase as temperature goes down, size effects are usually more important at low temperatures. The criteria are also different for different materials due to the different carrier types and scattering mechanisms. In the following section, we will study the effect of boundary scattering on electrical and thermal conductivities based on simple geometric considerations as well as derivations using the BTE.

### 5.5.1 *Simple Geometric Considerations*

The simple expression of thermal conductivity based on the kinetic theory is $\kappa = \frac{1}{3}(\rho c_v)v\Lambda_b$ for either electrons or phonons. Here, $\Lambda_b$ is called the *bulk mean free path,* which is the mean free path when the material is infinitely extended. While the specific heat and the velocity are also size dependent, especially for phonons, let us focus on the size dependence of the mean free path. The main objective of this section is to illustrate how boundary scattering affects the thermal conductivity by reducing the mean free path. The argument is also applicable to the electrical conductivity, since it is also proportional to the mean free path. Shown in Fig. 5.16 are two geometric configurations to be considered: (*a*) and (*b*) for a thin film and (*c*) for a thin wire or rod.

In the ballistic transport limit when $d \ll \Lambda_b$, we assume that the mean free path in the film is the same as the thickness $d$, $\Lambda_f = d$. Thus, the conductivity ratio can be obtained as

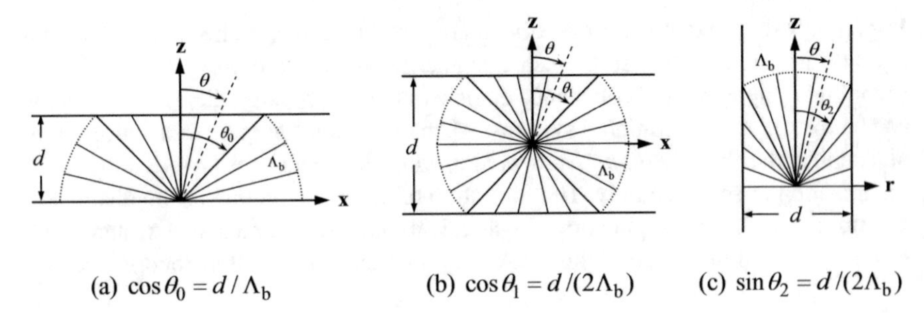


**Fig. 5.16** Illustration of free-path reduction due to boundary scattering. **a** A thin film for paths originated from the surface. **b** A thin film for paths originated from the center. **c** A thin wire for paths originated from the center

$$\frac{\kappa_f}{\kappa_b} = \frac{\Lambda_f}{\Lambda_b} = \frac{1}{Kn} \tag{5.122}$$

where $Kn = \Lambda_b/d$ is the Knudsen number for electrons or phonons, borrowed from the definition used in rarefied gas dynamics. In the intermediate region, we can apply Matthiessen's rule as suggested in Eqs. (5.56) and (5.76) such that

$$\frac{1}{\Lambda_{eff}} = \frac{1}{\Lambda_b} + \frac{1}{\Lambda_f} \tag{5.123}$$

Accordingly,

$$\frac{\kappa_{eff}}{\kappa_b} = \frac{\Lambda_{eff}}{\Lambda_b} = \frac{1}{1 + Kn} \tag{5.124}$$

The result calculated from Eq. (5.124) is plotted in Fig. 5.17 to illustrate the size dependence of the effective thermal conductivity. It appears that this simple formula overpredicts the reduction in thermal conductivity, as compared with the more realistic models to be discussed next.

As early as 1901, J. J. Thomson first considered the size effect on the electrical conductivity of thin films. His argument was extended by K. Fuchs in 1938 based on the BTE. The geometric argument assumes that boundary scattering is diffuse and inelastic, i.e., the electrons are fully accommodated after scattering by a boundary. The concept of accommodation is the same as that used for ideal gas particles in the free molecule flow regime discussed in Sect. 4.4. However, for simplicity, the distribution of free paths is not taken into consideration. In other words, all paths are assumed to be the same as the bulk mean free path. When $d \ll \Lambda_b$, we may assume that all energy carriers originate from the boundary. From Fig. 5.16a, we see that

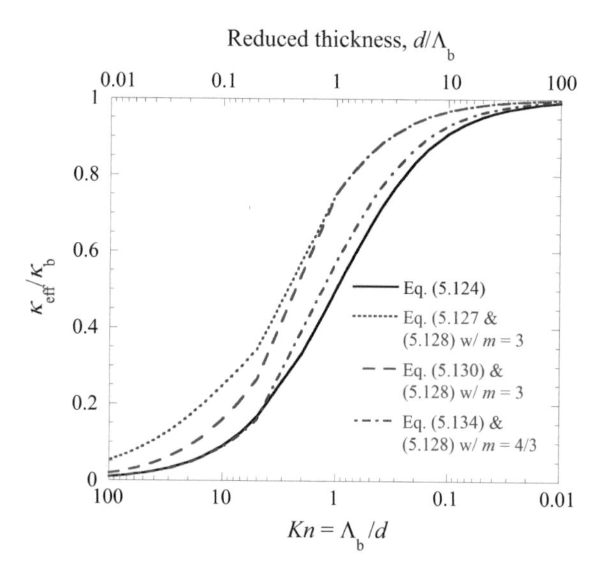


**Fig. 5.17** Reduction in thermal conductivity due to boundary scattering. Note that Eq. (5.128) was used with different $m$ values for $Kn < 1$, and interpolation was used for $1 < Kn < 5$

$$\Lambda(\theta) = \begin{cases} d/\cos\theta, & 0 < \theta < \theta_0 \\ \Lambda_b, & \theta_0 < \theta < \pi/2 \end{cases} \quad (5.125)$$

The free paths should be averaged over the hemisphere, and the weighted average can be written and evaluated as follows:

$$\frac{\Lambda_f}{\Lambda_b} = \frac{\int_0^{2\pi}\int_0^{\pi/2}\Lambda(\theta)\sin\theta d\theta d\phi}{\int_0^{2\pi}\int_0^{\pi/2}\Lambda_b\sin\theta d\theta d\phi} = \frac{\ln(Kn)+1}{Kn} \quad (5.126)$$

Applying Matthiessen's rule again, we have

$$\frac{\kappa_{eff}}{\kappa_b} = \frac{\Lambda_{eff}}{\Lambda_b} = \left(1 + \frac{Kn}{\ln(Kn)+1}\right)^{-1} \quad (5.127)$$

This equation, however, cannot be applied for small values of $Kn$ since $\ln(Kn)$ becomes negative. Let us assume Eq. (5.127) is applicable for $Kn > 5$. When $Kn < 1$, we may use

$$\frac{\kappa_{eff}}{\kappa_b} = \frac{\Lambda_{eff}}{\Lambda_b} = \left(1 + \frac{Kn}{m}\right)^{-1} \quad (5.128)$$

where $m \approx 3$ for thin films [62, 63]. Equation (5.124) can be considered as a special case of Eq. (5.128) with $m = 1$. The results based on Eqs. (5.127) and (5.128) are plotted in Fig. 5.17 for comparison. The thermal conductivity in the intermediate region for $1 < Kn < 5$ is linearly interpolated based on the values at $Kn = 1$ and 5.

Equations (5.126) and (5.127) do not consider the direction of transport and cannot capture the anisotropic feature due to the size effect. Flik and Tien [63] employed a weighted average of the free-path components in the parallel and normal directions of thin films. Their work was extended to different geometries by Richardson and Nori [64]. For the $z$-direction, the projected mean free path is $\Lambda_z = \Lambda(\theta)\cos\theta$; hence, the weighted average becomes

$$\frac{\Lambda_z}{\Lambda_{\mathrm{b},z}} = \frac{\int_0^{2\pi}\int_0^{\pi/2} \Lambda(\theta)\cos\theta\sin\theta \mathrm{d}\theta\mathrm{d}\phi}{\int_0^{2\pi}\int_0^{\pi/2} \Lambda_\mathrm{b}\cos\theta\sin\theta \mathrm{d}\theta\mathrm{d}\phi} = \frac{2}{Kn} - \frac{1}{Kn^2} \tag{5.129}$$

The use of Matthiessen's rule allows us to obtain

$$\frac{\kappa_{\mathrm{eff},z}}{\kappa_\mathrm{b}} = \left(1 + \frac{Kn}{2 - Kn^{-1}}\right)^{-1} \text{ for } Kn > 5 \tag{5.130}$$

For $Kn < 1$, Eq. (5.128) should be used with $m = 3$, which can be obtained by integrating over the film when $Kn \ll 1$ [63]. The result from Eq. (5.130) is also shown in Fig. 5.17. For transport along the $x$-direction, one may assume that all the electrons originate from the center of the film for simplicity. The component of the free path is $\Lambda_x = \Lambda(\theta)\sin\theta\cos\phi$, where $\phi$ is the azimuthal angle. Due to symmetry, the integration can be carried out in a single octant only. It can be seen from Fig. 5.16b that $\Lambda(\theta) = d/(2\cos\theta)$ for $0 \le \theta < \theta_1$, and $\Lambda(\theta) = \Lambda_\mathrm{b}$ for $\theta_1 \le \theta < \pi/2$, where $\theta_1 = \cos^{-1}(d/2\Lambda_\mathrm{b})$. Subsequently,

$$\frac{\Lambda_x}{\Lambda_{\mathrm{b},x}} = \frac{\int_0^{\pi/2}\int_0^{\pi/2} \Lambda(\theta)\sin^2\theta\cos\phi \mathrm{d}\theta\mathrm{d}\phi}{\int_0^{\pi/2}\int_0^{\pi/2} \Lambda_\mathrm{b}\sin^2\theta\cos\phi \mathrm{d}\theta\mathrm{d}\phi} \tag{5.131}$$

After evaluation of the above integral, we obtain

$$\frac{\kappa_{\mathrm{eff},x}}{\kappa_\mathrm{b}} = \frac{2}{\pi Kn}\ln[2Kn(1+\sin\theta_1)] + 1 - \frac{2\theta_1}{\pi} - \frac{\sin\theta_1}{\pi Kn} \tag{5.132}$$

In the ballistic limit, i.e., $Kn \gg 1$, Eq. (5.132) reduces to $\kappa_{\mathrm{eff},x}/\kappa_\mathrm{b} \approx (2/\pi)\ln(4Kn)/Kn$. If the free paths were to originate from the boundary, the result could be obtained by replacing $Kn$ with $Kn/2$ in Eq. (5.132). While it is perfectly logical to assume that all the carriers originate from the surface for the $z$-component in the ballistic limit. For thermal transport along a film with a temperature gradient in the $x$-direction, carriers must originate from a cross section or $y$-$z$ plane inside the film. The transport process along the film is essentially diffusion-like with significant boundary scattering contributions. Anisotropy may arise between $\kappa_{\mathrm{eff},x}$ and $\kappa_{\mathrm{eff},z}$ due to boundary scattering. A simple argument is that paths with large polar angles are more important for parallel conduction, whereas paths with smaller polar angles are more important for normal conduction. Based on the geometry, it can be seen that paths with smaller polar angles are more likely to be scattered by the boundary. Another reason that causes $\kappa_{\text{eff},x}$ to be greater than $\kappa_{\text{eff},z}$ is that scattering tends to be more specular for larger incidence angles. Specular reflection or elastic scattering does not reduce the conductivity because the incident particles only change the direction without any exchange of energy with the surface. Crystal anisotropy is another major reason for anisotropic conduction, sometimes the dominant reason, as in high-temperature superconducting $YBa_2Cu_3O_7$ films [63]. Grain boundaries can strongly influence the thermal conductivity in polycrystalline films [62]. For chemical–vapor-deposited polycrystalline diamond films, depending on the crystal orientation, $\kappa_x$ may be greater or smaller than $\kappa_z$ [65].

For circular wires, considering the conduction along a thin wire as shown in Fig. 5.16c, we have $\Lambda_z(\theta) = \Lambda_\text{b} \cos\theta$ for $0 < \theta < \theta_2$, and $\Lambda_z(\theta) = d \cot\theta/2$ for $\theta_2 < \theta < \pi/2$, where $\theta_2 = \sin^{-1}(d/2\Lambda_\text{b})$. Thus,

$$\frac{\Lambda_{\text{w},z}}{\Lambda_{\text{b},z}} = \frac{\int_0^{2\pi}\int_0^{\pi/2} \Lambda_z(\theta)\sin\theta \text{d}\theta \text{d}\phi}{\int_0^{2\pi}\int_0^{\pi/2} \Lambda_b \cos\theta \sin\theta \text{d}\theta \text{d}\phi} = \frac{1}{Kn} - \frac{1}{4Kn^2} \tag{5.133}$$

Applying Matthiessen's rule yields

$$\frac{\kappa_{\text{eff,w}}}{\kappa_\text{b}} = \frac{4Kn - 1}{4Kn^2 + 4Kn - 1} \tag{5.134}$$

which can be applied for $Kn > 5$ and approaches to Eq. (5.124) at large $Kn$. For $Kn < 1$, studies have shown that Eq. (5.128) is a good approximation with $m = 4/3$ [66, 67]. The reduction in thermal conductivity for thin wires is also indicated in Fig. 5.17, where values for $1 < Kn < 5$ are again based on a simple interpolation between the two expressions. Due to geometric confinement, the reduction in the mean free path is more severe for thin wires than for thin films. The geometric argument is easy to understand and may help gain a physical intuition of the size effect due to boundary scattering. In consideration of the classical size effect, it is assumed that Fourier's law is still applicable with a modified thermal conductivity. The size effect on the electron or phonon transport properties can also be formulated using the BTE for thin films and wires, as presented in the following.

### 5.5.2 *Conductivity Along a Thin Film Based on the BTE*

In Sect. 5.3.3, we derived electrical and thermal conductivities based on the BTE for bulk materials. The relaxation time approximation was adopted, and the distribution function was assumed to be not too far away from equilibrium, i.e., under the local-equilibrium conditions. To determine the size effect on the conductivities along thin films, the same assumptions will be applied. Consider the geometry shown in Fig. 5.16a, with a temperature gradient and an electric field in the $x$-direction only. Because of the finite thickness in the $z$-direction, the distribution function should also be an explicit function of $z$, viz.,

$$f_1(\varepsilon, T, z) \approx f_0(\varepsilon, T) + \tau(\varepsilon)\left(\frac{eE}{m_e}\frac{\partial f_0}{\partial \varepsilon}\frac{\partial \varepsilon}{\partial v_x} - v_x\frac{\partial f_0}{\partial T}\frac{\mathrm{d}T}{\mathrm{d}x} - v_z\frac{\partial f_1}{\partial z}\right) \quad (5.135\mathrm{a})$$

Compared with Eq. (5.65), the last term was added because $f_1$ depends also on $z$. Here, the electric field and the temperature gradient are along the $x$-direction. In Eq. (5.135a), we have already replaced $\partial f_1/\partial \varepsilon$ with $\partial f_0/\partial \varepsilon$ and $\partial f_1/\partial T$ with $\partial f_0/\partial T$. We can rearrange Eq. (5.135a) as follows:

$$-\frac{eE}{m_e}\frac{\partial f_0}{\partial \varepsilon}\frac{\partial \varepsilon}{\partial v_x} + v_x\frac{\partial f_0}{\partial T}\frac{\mathrm{d}T}{\mathrm{d}x} + v_z\frac{\partial f_1}{\partial z} = -\frac{f_1 - f_0}{\tau(\varepsilon)} \quad (5.135\mathrm{b})$$

which is nothing but the steady-state BTE under the relaxation time approximation. The general solution can be expressed as

$$f_1 = f_0 + \tau v_x\left(eE\frac{\partial f_0}{\partial \varepsilon} - \frac{\partial f_0}{\partial T}\frac{\mathrm{d}T}{\mathrm{d}x}\right)\left[1 - \psi(\mathbf{v})\exp\left(-\frac{z}{\tau v_z}\right)\right], \; v_z > 0 \quad (5.136\mathrm{a})$$

and

$$f_1 = f_0 + \tau v_x\left(eE\frac{\partial f_0}{\partial \varepsilon} - \frac{\partial f_0}{\partial T}\frac{\mathrm{d}T}{\mathrm{d}x}\right)\left[1 - \psi(\mathbf{v})\exp\left(-\frac{d-z}{\tau v_z}\right)\right], \; v_z < 0 \quad (5.136\mathrm{b})$$

where $\psi(\mathbf{v})$ is an arbitrary function that accounts for the accommodation and scattering characteristics. If perfect accommodation is assumed with inelastic and diffuse scattering, then $\psi(\mathbf{v}) = 1$. Let us consider electrical conduction without any temperature gradient. For diffuse scattering only with $\psi(\mathbf{v}) = 1$, it can be shown that

$$f_1 = f_0 + \tau v_x eE\frac{\partial f_0}{\partial \varepsilon}\left[1 - \exp\left(-\frac{z}{\tau v_z}\right)\right], \quad v_z > 0 \quad (5.137\mathrm{a})$$

and

$$f_1 = f_0 + \tau v_x eE\frac{\partial f_0}{\partial \varepsilon}\left[1 - \exp\left(-\frac{d-z}{\tau v_z}\right)\right], \quad v_z < 0 \quad (5.137\mathrm{b})$$

We must substitute the distribution function into Eq. (5.67a) and integrate over $(v_x, v_y, v_z)$, or over $(v, \theta, \phi)$ or $(\varepsilon, \theta, \phi)$, in spherical coordinates, to obtain $J_e(z) = -eJ_N(z)$ along the film. Therefore,

$$J_e(z) = -e^2E\int_0^\infty \tau\frac{\partial f_{\mathrm{FD}}}{\partial \varepsilon}\mathrm{d}\varepsilon\int_0^{2\pi}\mathrm{d}\phi\left\{\int_0^{\pi/2} v_x^2\left[1 - \exp\left(-\frac{z}{\tau v\cos\theta}\right)\right]v^2\sin\theta\,\mathrm{d}\theta\right.$$

$$+\int_{\pi/2}^{\pi} v_x^2\left[1-\exp\left(-\frac{d-z}{\tau v\cos\theta}\right)\right]v^2\sin\theta d\theta\Bigg\} \quad (5.138)$$

Putting $v_x = v\sin\theta\cos\phi$, the integration over $\phi$ can be carried out independently. The average current flux $J_{e,avg} = (1/d)\int_0^d J_e(z)dz$ can also be obtained. The properties of the Fermi integral allow the integration over $\varepsilon$ to be carried out and expressed in terms of the properties at the Fermi surface, i.e., $\tau(\mu_F)$ and $v_F$. Notice that $\Lambda_b = \tau(\mu_F)v_F$, and let $J_{e,avg} = \sigma_f E$, where $\sigma_f$ is the effective electrical conductivity of the film. After normalization of the electrical current density based on Eqs. (5.67a) and (5.68), we obtain the following relation:

$$\frac{\sigma_f}{\sigma_b} = F(Kn) \quad (5.139a)$$

where

$$\begin{aligned} F(Kn) &= \frac{3}{4d}\int_0^{\pi/2}\sin^3\theta\int_0^d\left[1-\exp\left(-\frac{z}{\Lambda_b\cos\theta}\right)\right]dz d\theta \\ &+ \frac{3}{4d}\int_{\pi/2}^{\pi}\sin^3\theta\int_0^d\left[1-\exp\left(-\frac{d-z}{\Lambda_b\cos\theta}\right)\right]dz d\theta \\ &= \frac{3}{2d}\int_0^{\pi/2}\sin^3\theta\left\{d-\Lambda_b\cos\theta\left[1-\exp\left(-\frac{d}{\Lambda_b\cos\theta}\right)\right]\right\}d\theta \\ &= 1-\frac{3Kn}{8}+\frac{3Kn}{2}\int_1^{\infty}\left(\frac{1}{t^3}-\frac{1}{t^5}\right)\exp\left(-\frac{t}{Kn}\right)dt \end{aligned} \quad (5.139b)$$

Note that the $m$th-order *exponential integral* is defined as $E_m(x) = \int_1^{\infty} e^{-xt}t^{-m}dt$ or $E_m(x) = \int_0^1 \eta^{m-2}e^{-x/\eta}d\eta$, which has the relation $E_{m+1}(x) = m^{-1}\left[e^{-x} - xE_m(x)\right]$. Equation (5.139b) can also be expressed as

$$F(Kn) = 1-\frac{3Kn}{8}+\frac{3Kn}{2}\left[E_3\left(\frac{1}{Kn}\right)-E_5\left(\frac{1}{Kn}\right)\right] \quad (5.139c)$$

The asymptotic relations are

$$\frac{\sigma_f}{\sigma_b} \approx 1-\frac{3Kn}{8} \text{ for } Kn \ll 1 \quad (5.140a)$$

and

$$\frac{\sigma_{\mathrm{f}}}{\sigma_{\mathrm{b}}} \approx \frac{3\ln(Kn)}{4Kn} \text{ for } Kn \gg 1 \tag{5.140b}$$

which is close to Eq. (5.132) for $Kn \gg 1$. The derivation using the BTE presented here inherently assumes that the electrons are originated from the film rather than from the boundaries.

For thermal conductivity, we can substitute Eq. (5.136a, 5.136b) with $\psi(\mathbf{v}) = 1$ into Eq. (5.70) and follow a similar procedure to obtain $\kappa_{\mathrm{f}}/\kappa_{\mathrm{b}} = F(Kn)$, where $F(Kn)$ is given in Eq. (5.139b) or (5.139c). At very low temperatures or near room temperature, the Wiedemann–Franz law is applicable, and the reduction in electrical and thermal conductivities is essentially the same. In the intermediate region, one could use different scattering rates or mean free paths for the bulk thermal and electrical conductivities to determine the size effect individually based on Eq. (5.139a). Another way to obtain $\kappa_{\mathrm{f}}/\kappa_{\mathrm{b}}$ is to calculate the heat flux using Eq. (5.136a, 5.136b) with a finite temperature gradient as done by Kumar and Vradis [68]. They obtained complicated expressions and showed that the results are similar to $\sigma_{\mathrm{f}}/\sigma_{\mathrm{b}}$ in a large range.

According to the discussion of thermoelectricity in Sect. 5.4, we could in principle quantify the size effect on other coefficients. If the same assumptions are used, to the first-order approximation, $L_{12}$ and $L_{21}$ are subject to boundary scattering and will also be reduced according to Eq. (5.139a). Because the thermoelectric power is the ratio of the two coefficients, the Seebeck coefficient along the film should be expected to remain the same regardless of boundary scattering. One should be cautious about this conclusion because the assumption of a spherical Fermi surface and the free-electron model are questionable when modeling thermoelectricity, as mentioned previously.

The above discussion can be extended to scattering with a specular component. Let parameter $p$, which is called *specularity,* represent the probability of scattering being elastic and specular. For specular and elastic scattering, carriers will continue to exchange energy and momentum inside the film after reflection by the boundary. Therefore, these scattering events do not cause any reduction in the effective mean free paths or conductivities along the film. If $p$ is assumed to be independent of the incident direction, the function $\psi(\mathbf{v})$ in Eq. (5.136a, 5.136b) becomes

$$\psi(\mathbf{v}) = \frac{1-p}{1-p\exp(-d/\tau v_z)} \tag{5.141}$$

The function given in Eq. (5.139b) may be modified after some tedious derivations as follows:

$$F(Kn,p) = 1 - \frac{3(1-p)Kn}{2}\int_1^{\infty}\left(\frac{1}{t^3}-\frac{1}{t^5}\right)\frac{1-\exp(-t/Kn)}{1-p\exp(-t/Kn)}\mathrm{d}t \tag{5.142}$$

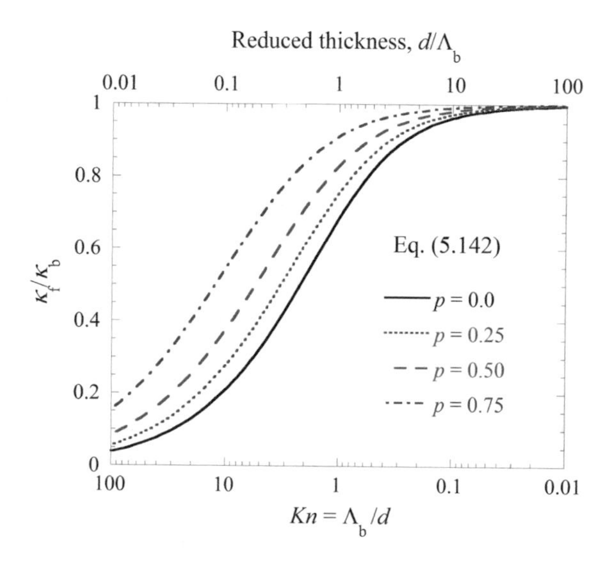


**Fig. 5.18** Size effect on thermal conductivity along the film of thickness $d$, as predicted by the BTE with different specularities.

The effects of $p$ and $Kn$ on the effective conductivity are shown in Fig. 5.18. The trends with respect to $Kn$ are very similar to those in Fig. 5.17 obtained from simple geometric considerations. For electron transport, since the de Broglie wavelength of electrons is less than 1 nm, boundary scattering can usually be considered diffuse, i.e., $p = 0$. For phonons, the wavelength may vary from the atomistic scale up to the size of the crystal. Therefore, the size effect needs to be considered for different phonon frequencies. The parameter $p$ can be estimated based on the rms surface roughness $\sigma_{\text{rms}}$ and the wavelength $\lambda$ of the carrier by

$$p = \exp\left(-\frac{16\pi^2\sigma_{\text{rms}}^2\cos^2\theta_{\text{i}}}{\lambda^2}\right) \tag{5.143}$$

where $\theta_{\text{i}}$ is the angle of incidence. This equation can be derived from the wave scattering theory [69]. Generally speaking, $p \ll 1$ when $\lambda \leq \sigma_{\text{rms}}$. When $\lambda > 10\sigma_{\text{rms}}$, the specular reflection cannot be neglected. Furthermore, the specularity $p$ increases with the incidence angle. The actual scattering distribution often consists of a broad specular lobe, and the nonspecular component is not perfectly diffuse. This is similar to light scattering by rough surfaces for which an in-depth discussion will be given in Chap. 9. Feng et al. [70] studied the effect of specularity and grain boundary scattering on the thermal conductivity of thin metal films. Their model for the reduction of thermal conductivity of copper and gold films agrees well with experimental values over a large temperature range.

As can be seen from Figs. 5.17 and 5.18, when $Kn = \Lambda_{\text{b}}/d > 0.1$, i.e., when $d < 10\Lambda_{\text{b}}$, the size effect may be significant, and boundary scattering dominates when $d < 0.1\Lambda_{\text{b}}$. Note that Examples 5.5 and 5.6 provide typical numerical values of the bulk mean free paths of electrons in a noble metal and of phonons in silicon. At room temperature, the electron mean free path of a metal is on the order of tens of nanometers, and thus one would expect some size effect when $d$ is less than 300 nm. For a highly pure metal at very low temperatures, however, the electron mean free path could be on the order of millimeters. In this case, even when $d$ of the film is on the order of micrometers, boundary scattering would dominate the scattering process. Note that the method presented in this section is only for heat conduction along a film. For phonon conduction across a film, the BTE may be simplified using the equation of phonon radiative transfer to be discussed in Chap. 7.

### *5.5.3 Conductivity Along a Thin Wire Based on the BTE*

The above discussion can be extended to conduction along a thin wire. For wires with circular cross sections, the effective conductivity can be expressed as [66, 67]

$$\frac{\kappa_{\mathrm{w}}}{\kappa_{\mathrm{b}}} \text{ or } \frac{\sigma_{\mathrm{w}}}{\sigma_{\mathrm{b}}} = 1 - \frac{12}{\pi} \int_0^1 \sqrt{1-\xi^2} \int_1^\infty \exp\left(-\frac{\xi t}{Kn}\right) \frac{\sqrt{t^2-1}}{t^4} \mathrm{d}t \, \mathrm{d}\xi \tag{5.144}$$

In particular, the asymptotic approximations with $\approx$1% accuracy are

$$\frac{\kappa_{\mathrm{w}}}{\kappa_{\mathrm{b}}} \text{ or } \frac{\sigma_{\mathrm{w}}}{\sigma_{\mathrm{b}}} \approx 1 - \frac{3}{4}Kn + \frac{3}{8}Kn^3 \text{ for } Kn < 0.6 \tag{5.145a}$$

and

$$\frac{\kappa_{\mathrm{w}}}{\kappa_{\mathrm{b}}} \text{ or } \frac{\sigma_{\mathrm{w}}}{\sigma_{\mathrm{b}}} \approx \frac{1}{Kn} - \frac{3(\ln Kn + 1)}{8Kn^2} - \frac{2}{15Kn^3} \text{ for } Kn > 1 \tag{5.145b}$$

If the scattering is not completely diffuse, a specularity parameter $p$ similar to that for thin films can be introduced, and the expression becomes

$$\frac{\kappa_{\mathrm{w}}}{\kappa_{\mathrm{b}}} \text{ or } \frac{\sigma_{\mathrm{w}}}{\sigma_{\mathrm{b}}} = 1 - \frac{12(1-p)^2}{\pi} \sum_{m=1}^{\infty} m p^{m-1} G(Kn, m) \tag{5.146a}$$

where

$$G(Kn, m) = \int_0^1 \sqrt{1-\xi^2} \int_1^\infty \exp\left(-\frac{m\xi t}{Kn}\right) \frac{\sqrt{t^2-1}}{t^4} \mathrm{d}t \, \mathrm{d}\xi \tag{5.146b}$$

Again, different mean free paths and $Kn$ numbers should be used for thermal and electrical conductivities in the region where the Wiedemann–Franz law is not applicable.

### 5.5.4 *Size Effects on Crystalline Insulators*

For crystalline solids, phonons are the principal heat carriers. Simple geometric arguments can also be applied to give qualitative results of the size effect using the phonon mean free path. The BTE or Boltzmann–Peierls equation can be used to more rigorously predict the thermal conductivity of bulk solids as well as the size effect on the thermal conductivity. The distribution function of phonons depends on the frequency or the wavevector, which are related by the dispersion relation. The group velocity can be calculated from the dispersion curve for a given phonon mode or branch. In general, the scattering rate is frequency dependent. The Boltzmann–Peierls equation at a given frequency under the relaxation time approximation for steady-state 1D conduction can be expressed as follows:

$$v_x \frac{\partial f_{\mathrm{BE}}}{\partial T}\frac{\mathrm{d}T}{\mathrm{d}x} + v_z \frac{\partial f_1}{\partial z} = -\frac{f_1 - f_{\mathrm{BE}}}{\tau(\omega)} \tag{5.147}$$

where $v_x$ and $v_z$ are the components of the group velocity that depend on the frequency. The solution is similar to Eq. (5.136a, 5.136b), especially for the $z$-dependence. Following the discussions in Sect. 5.3.4 on phonon thermal conductivity, in conjunction with the average heat flux along the film, we can rewrite Eq. (5.80) as follows:

$$\kappa_{\mathrm{f}} = \frac{k_{\mathrm{B}}^4 T^3}{6\pi^2 \hbar^3} \sum_P \int_0^{x_{\mathrm{m}}} \frac{\tau(x) v_{\mathrm{g}}(x)}{v_{\mathrm{p}}^2(x)} \frac{x^4 \mathrm{e}^x}{(\mathrm{e}^x - 1)^2} F(Kn_x, p)\mathrm{d}x \tag{5.148}$$

where $x = \hbar\omega/k_{\mathrm{B}}T$ is a reduced frequency and the Knudsen number, $Kn_x = \tau(x)v_{\mathrm{g}}(x)/d = \Lambda(x)/d$, is thus a function of the frequency $\omega$. In this equation, the summation index $P$ accounts for all phonon polarizations, the upper bound of the integration is the cutoff frequency for each polarization, and the function $F(\xi, p)$ can be calculated from Eq. (5.142). If an average $Kn$ that is independent of the frequency can be used, combining Eqs. (5.148) with (5.80) gives $\kappa_{\mathrm{f}}/\kappa_{\mathrm{b}} = F(Kn, p)$ as expected. A similar equation can be developed for thin wires [67, 71].

For semiconductors, such as silicon, the phonon mean free path is on the order of tens of nanometers at room temperature. Therefore, the size effect can be neglected for a 1-μm-thick silicon film above room temperature. However, as temperature is lowered, the size effect becomes more and more significant. Numerical calculations dealing with the conductivity reduction are left as exercises. Kenneth Goodson's group has experimentally demonstrated the size effect on the thermal conductivity of both intrinsic and doped silicon films with thicknesses from a few micrometers down to 20 nm [72]. The thermal transport properties of silicon nanowires and other semiconductor nanowires have been extensively studied both experimentally and theoretically for thermoelectric applications [4, 53, 73–79].

### 5.5.5 Mean-Free-Path Distribution

As previously discussed, the scattering rate and the mean free path both depend on the frequency. If all modes are combined, the thermal conductivity can be expressed as an integration with respect to frequency $\kappa = \int_0^\infty \kappa_\omega(\omega)\mathrm{d}\omega$. The integrand is the distribution function of thermal conductivity in terms of frequency (i.e., a frequency spectrum). To facilitate understanding and analysis of the experimental data, it has been proposed to use the mean free path (MFP) as the independent variable such that the bulk thermal conductivity can be expressed as follows [80, 81]:

$$\kappa_{\text{bulk}} = \int_0^\infty \kappa_\Lambda(\Lambda)\mathrm{d}\Lambda \tag{5.149a}$$

where

$$\kappa_\Lambda(\Lambda) = \kappa_\omega(\omega)\left(\frac{\mathrm{d}\Lambda}{\mathrm{d}\omega}\right)^{-1} = -\sum_P \int_0^\infty \frac{1}{3}C(\omega)v\Lambda\left(\frac{\mathrm{d}\Lambda}{\mathrm{d}\omega}\right)^{-1} \tag{5.149b}$$

Here, $C(\omega)$ is the volumetric specific heat as given in Eq. (5.82b). The group velocity $v$ and mean free path $\Lambda$ for each phonon branch are also functions of frequency. We can interpret Eq. (5.149b) as the thermal conductivity per unit MFP, i.e., the thermal conductivity distribution function in terms of the mean free path. It is simply referred to as the *MFP distribution* or *MFP spectrum* [81], unless noted otherwise, $\Lambda$ refers to the bulk mean free path.

With nanostructures, due to boundary scattering, the $\kappa_\Lambda(\Lambda)$ at a given bulk MFP is reduced by a factor that depends on the Knudsen number. Therefore, we can write

$$\kappa_{\Lambda,\text{nano}}(\Lambda) = \kappa_\Lambda(\Lambda)F_{\text{nano}}(Kn, p) \tag{5.150}$$

where $F_{\text{nano}}$ is a structure-dependent function of $Kn$ and the specularity $p$. It may be thought as the ratio of the MFP in the nanostructure to that of the bulk, i.e., $F_{\text{nano}} = \Lambda_{\text{nano}}/\Lambda$. The Knudsen number depends on the characteristic length and varies with the bulk MFP. Thus, the thermal conductivity of the nanostructure can be written as

$$\kappa_{\text{nano}} = \int_0^\infty \kappa_{\Lambda,\text{nano}}(\Lambda)\mathrm{d}\Lambda = \int_0^\infty \kappa_\Lambda(\Lambda)F_{\text{nano}}(Kn, p)\mathrm{d}\Lambda \tag{5.151}$$

The cumulative distribution function (CDF) can also be defined for thermal conductivity in terms of the mean free path as follows:

$$K_\Lambda(\Lambda) = \frac{1}{\kappa_{\text{nano}}} \int_0^\Lambda \kappa_\Lambda(\xi) F_{\text{nano}}(Kn, p) \mathrm{d}\xi \tag{5.152}$$

For bulk materials, simply set $\kappa_{\text{nano}} = \kappa$ and $F_{\text{nano}} = 1$ in Eq. (5.152). This equation is called the (normalized) thermal conductivity accumulation function.

The mean-free-path spectrum and the thermal conductivity accumulation function have been experimentally and theoretically investigated for both bulk and nanostructured materials [78–82]. These studies have significantly improved our understanding of heat conduction in bulk and nanostructured solid materials and devices. Further discussion is postponed to Chap. 7.

## 5.6 Quantum Conductance and the Landauer Formalism

In the above discussion, the Fourier law was assumed to hold under the local-equilibrium approximation, with reduced thermal conductivities to include the effect of boundary scattering. Many works have employed ab initio techniques, lattice dynamics, and equilibrium or nonequilibrium molecular dynamics to study thermal transport at the nanoscale [4, 82–84]. In heterogeneous structures, such as superlattices, when thermal transport across the multiple layers is considered, the local-equilibrium assumption breaks down in the ballistic regime. Further discussion of non-Fourier conduction, especially for transient processes, will be deferred to Chap. 7. For superlattice nanowires, both lateral and longitudinal confinements exist, so each element is like a quantum dot confined in all three dimensions. When the quantum confinement becomes significant, the relaxation time approximation used to solve the BTE is not applicable. Landauer's formalism is presented here for modeling certain nonequilibrium and ballistic transport phenomena. This section also introduces the quantum size effect on electrical and thermal transport processes, with an emphasis on the concept of quantum conductance and its implications.

Quantum size effect on the lattice specific heat was discussed in Sect. 5.2. Attention is now paid to the electrical conductance of metallic materials and thermal conductance of dielectric materials. For bulk solids, the DOS for electrons $D(\varepsilon)$ is proportional to $\sqrt{\varepsilon}$, as given in Eq. (5.18) and illustrated in Fig. 5.5b. Note that for phonons or photons, the energy $\varepsilon = \hbar\omega$ is proportional to the frequency and $D(\omega)$ is proportional to $\omega^2$ when the dispersion is linear; see Eq. (5.35) and Fig. 5.4. For electrons or holes, $\varepsilon = \frac{p^2}{2m^*} = \frac{\hbar^2 k^2}{2m^*}$, where $k$ is the wavevector and $m^*$ is the effective mass. For the electron gas in a 2D solid, the density of states becomes $D(\varepsilon) = 2 \times \frac{k}{2\pi}\frac{\mathrm{d}k}{\mathrm{d}\varepsilon} = \frac{m^*}{\pi\hbar^2}$, which can be derived using Eq. (5.37) considering the spin degeneracy. In a quantum well of thickness $L$, the energy levels are quantized in the normal or $z$-direction according to Eq. (3.80), i.e., $\frac{n^2 h^2}{8m^* L^2}$, where $n$ is a positive integer. The combined energies can be expressed as

$$\varepsilon_n(k) = \frac{n^2\hbar^2}{2m^*L} + \frac{\hbar^2(k_x^2 + k_y^2)}{2m^*} \tag{5.153}$$

and the resulting DOS is given by

$$D(\varepsilon) = \frac{nm^*}{\pi\hbar^2}, \quad \text{for } \varepsilon_n \leq \varepsilon < \varepsilon_{n+1} \text{ and } n = 1, 2, \ldots \tag{5.154}$$

which is a staircase function, as depicted in Fig. 5.19a, along with the bulk DOS. The reason that the DOS for the $n$th subband is multiplied by $n$ is because $k_{z,n} = n\pi/L$, where $k_{\min} = \pi/L$. Before applying Eqs. (5.36) and (5.37), we must multiply the total number of modes $N$ by $k_{z,n}L/\pi$. For 1D quantum wires confined in both $y$- and $z$-directions (assuming a rectangular shape of $L_y \times L_z$), the energy levels are given by

$$\varepsilon_{l,n} = \frac{l^2\hbar^2}{2m^*L_y} + \frac{n^2\hbar^2}{2m^*L_z} \tag{5.155}$$

For each subband $(l, n)$, the DOS becomes

$$D(\varepsilon) = \frac{nl}{\pi\hbar}\sqrt{\frac{2m^*}{\varepsilon - \varepsilon_{l,n}}} \tag{5.156}$$

which has an inverse square-root dependence of energy and a singularity at $\varepsilon_{l,n}$, as shown in Fig. 5.19b. For 3D confined quantum dots, the energy levels are completely discrete; subsequently, the DOS becomes isolated delta functions (not shown in Fig. 5.19).

The quantization of electron energy levels or phonon frequencies in small structures suggests that the resulting transport properties may also be quantized. For example, the electrical conductance may depend on the applied current or force for the nanocontact in a stepwise manner. The thermal conductance of insulators can also be quantized due to limited available phonon modes in small structures and

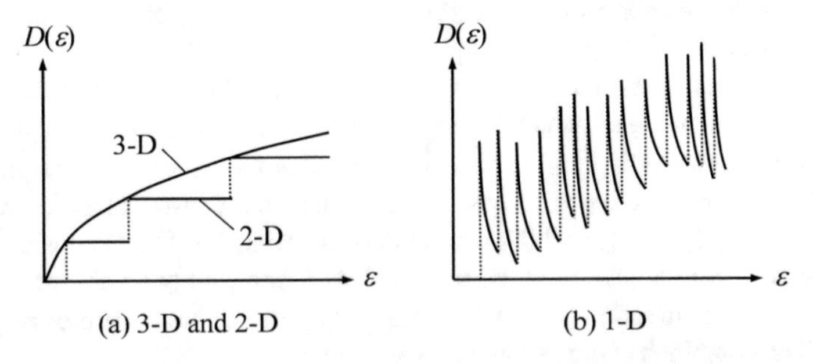


**Fig. 5.19** Electron density of states due to quantum confinement. **a** 2D quantum wells versus 3D bulk solids. **b** 1D quantum wires.

at low temperatures. In this section, we use conductance rather than conductivity for reasons to be explained soon. Long before the quantization of conductance was experimentally observed, physicists had formulated different theories to understand the transport phenomena in the quantum or ballistic regimes. Landauer and collaborators [85] have developed a formalism to treat electrical current flow as a transmission probability when carriers are scattered coherently and the resulting ballistic transport behaves quantum mechanically. Landauer's formalism can easily be applied to the 1D case for conductance through a narrow channel, as illustrated in Fig. 5.20a. Suppose ballistic transmission exists in the channel connecting two reservoirs of different electrochemical potentials; there will be a current flow from 1 to 2 and reversely from 2 to 1. In the absence of losses by scattering and reflection, the net current flow can be expressed as

$$J_{\mathrm{e}} = J_{1\to 2} - J_{2\to 1} = -ev_{\mathrm{F}}(\mu_1 - \mu_2)D(\varepsilon) \tag{5.157}$$

where $\mu$ is the chemical potential. The derivation can be easily generalized to include the electrostatic potential. Note that the DOS in the 1D case is $D(\varepsilon) = (\pi\hbar v_{\mathrm{F}})^{-1}$ considering the electronic spin degeneracy. Because the voltage drop is $V_1 - V_2 = -(\mu_1 - \mu_2)/e$, the electrical conductance for complete transmission becomes

$$g_{\mathrm{e}0} = \frac{J_{\mathrm{e}}}{V_1 - V_2} = \frac{e^2}{\pi\hbar} \text{ or } \frac{2e^2}{h} \tag{5.158}$$

which gives a universal constant with a value of $7.75 \times 10^{-5}\ \Omega^{-1}$ or a resistance value of 12.91 kΩ. This is the quantum conductance for an ideal 1D conductor, in which there is no resistance or voltage drop associated with the channel itself. Instead, the voltage drop is associated with the perturbation at each end of the channel as it interacts with the reservoir [85]. In the above derivation, we assumed that the Fermi distribution function can be approximated as a step function (i.e., at absolute zero temperature). By introducing a transmission coefficient $\xi_{12}$ and using the actual

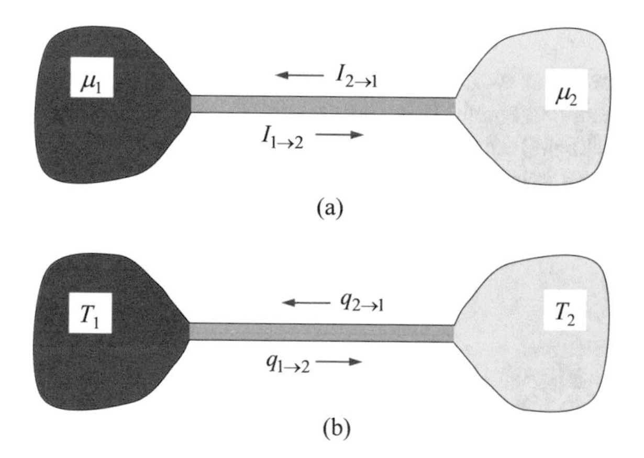


**Fig. 5.20** Illustration of quantum conductance. **a** Electrical current flow through a narrow metallic channel due to different electrochemical potentials. **b** Heat transfer between two heat reservoirs through a narrow dielectric channel

distribution function, one can modify Eq. (5.157) to the following [85, 86]:

$$J_e = \int_0^\infty (-ev_F)\xi_{12}(\varepsilon)[f_{FD}(\varepsilon, \mu_1) - f_{FD}(\varepsilon, \mu_2)]D(\varepsilon)d\varepsilon \tag{5.159}$$

For small potential differences, using the following approximation,

$$\frac{f_{FD}(\varepsilon, \mu_1) - f_{FD}(\varepsilon, \mu_2)}{\mu_2 - \mu_1} = -\frac{\partial f_{FD}(\varepsilon, \mu)}{\partial \mu} = \frac{\partial f_{FD}(\varepsilon, \mu)}{\partial \varepsilon}$$

we obtain the expression of the electrical conductance:

$$g_e = -\frac{2e^2}{h}\int_0^\infty \xi_{12}(\varepsilon)\frac{\partial f_{FD}}{\partial \varepsilon}d\varepsilon \tag{5.160}$$

which reduces to Eq. (5.158) at absolute zero temperature when $\xi_{12}(0)$ is taken to be 1. The transmission coefficient or probability is given by a scattering matrix (the $S$-matrix) based on a solution of Schrödinger's equation. The solution is in the form of eigenvalues called eigenchannels, each with a transmission coefficient $\tau_i$ between 0 and 1. Thus, the expression of conductance is reduced to

$$g_e = \frac{2e^2}{h}\sum_i \tau_i \tag{5.161}$$

Depending on how many propagation modes at the Fermi level are excited, the conductance varies in a discontinuous manner. Conductance quantization has been realized in metallic nanocontacts, nanowires, and carbon nanotubes [86–89], even at room temperature, and has also been predicted by molecular dynamics simulations [90, 91]. These discoveries are very important for the development of single-electron transistors, nanoelectromechanical systems, nanotribology, and quantum computing.

The ballistic thermal transport process resembles electromagnetic radiation between two blackbodies separated by a vacuum. For a 1D photon gas, the Stefan–Boltzmann law reads $q'' \propto T^2$ rather than $q'' \propto T^4$. In a solid nanostructure (channel) that links two heat reservoirs, as illustrated in Fig. 5.20b, the ballistic heat conduction can be treated in a similar way so that

$$q_{1\to 2} = \frac{1}{2\pi}\sum_P \int_{\omega_P}^{\omega_D} \xi_P(\omega)\hbar\omega f_{BE}(\omega, T_1)d\omega \tag{5.162a}$$

and

$$q_{2\to 1} = \frac{1}{2\pi}\sum_P \int_{\omega_P}^{\omega_D} \xi_P(\omega)\hbar\omega f_{BE}(\omega, T_2)d\omega \tag{5.162b}$$

where $\xi_P(\omega)$ is the transmission coefficient (or probability) of the polarization branch $P$, which accounts for both scattering in the channel and reflection from the junctions. Here, the upper bound $\omega_D$ approaches infinity at very low temperatures, and the lower bound is the cutoff frequency for the phonon mode $P$. This cutoff frequency is determined by the width of the channel and the order of propagating phonon modes, like in a waveguide. More specifically, if a rectangular cross section is considered whose dimensions are $L_x$ and $L_y$, the cutoff frequency for the ($m$,$n$) mode is given by

$$k_{mn} = \frac{\omega_{mn}}{v_s} = \sqrt{\left(\frac{m\pi}{L_x}\right)^2 + \left(\frac{n\pi}{L_y}\right)^2} \tag{5.163}$$

Apparently, a narrow channel enables a large cutoff wavenumber. Note that the zeroth-order mode always exists because it has a zero cutoff frequency. If the integration in Eq. (5.162) is expressed in terms of the wavevector, there will be a group velocity $v_g$ term. In writing Eq. (5.162), we have assumed $v_g = v_p$ for a linear dispersion relation. The net heat transfer is calculated by $q_{12} = q_{1\to 2} - q_{2\to 1}$, which is commonly done in radiation heat transfer. Assuming that the temperature difference is small, we obtain the thermal conductance as

$$g_T = \frac{q_{12}}{T_1 - T_2} = \frac{1}{2\pi}\sum_P \int_{\omega_P}^{\omega_D} \xi_P(\omega)\hbar\omega\frac{\partial f_{BE}(\omega, T)}{\partial T}d\omega \tag{5.164a}$$

or

$$g_T = \frac{k_B^2 \bar{T}}{h}\sum_P \int_{x_P}^{x_D} \xi_P(x)\frac{x^2 e^x}{(e^x - 1)^2}dx \tag{5.164b}$$

Note that $\bar{T}$ represents the average temperature. At sufficiently low temperatures, only the lowest phonon branches, whose cutoff frequency equals zero, may contribute to the conductance. If the transmission coefficient is assumed to be unity, each of the lowest phonon modes will contribute to the thermal conductance by

$$g_{T0} = \frac{\pi k_B^2 T}{6\hbar} \quad \text{or} \quad \frac{\pi^2 k_B^2 T}{3h} \tag{5.165}$$

which has a value $g_{T0}/T = 0.947\ \text{pW/K}^2$ and is another universal constant that can be viewed as the Stefan–Boltzmann constant in 1D space for each mode. If the above derivation is repeated to obtain electron thermal conductance, we will end up with $2g_{T0}$ due to the electronic spin degeneracy. Therefore, the Lorentz number $Lz = \frac{\kappa}{\sigma T} = \frac{g_T}{g_e T}$ in the ballistic regime remains the same as given in Eq. (5.62) for the diffusive regime [92]. Roukes and collaborators [93] have experimentally demonstrated quantum thermal conductance using a 60-nm-thick silicon nitride membrane. They reported a $16g_{T0}$ behavior at temperatures below 0.6 K since the structure was suspended by four narrow bridges (channels). Each bridge or channel acts like a wire with four phonon modes (two transversal, one longitudinal, and one torsional). Murphy and Moore [74] used Landauer's formalism to study phonon transport in silicon nanowires considering temperature dependence and the effect of diffusive and localized modes on the frequency-dependent transmission coefficient.

Carbon nanotubes (CNTs) have been known for a while, especially for its associated large thermal conductivities [94–98]. Single-walled carbon nanotubes can be made essentially free from defect scattering and boundary scattering due to atomistic smoothness. Their diameters can be made as small as a few nanometers, while their lengths can be several micrometers. Thermal conductivities of single-walled and multi-walled nanotubes have been measured with suspended MEMS bridges and are found to exceed that of diamond at room temperature [97]. The thermal conductivity was calculated from the measured thermal conductance based on an effective cross-sectional area. Above room temperature, phonon–phonon anharmonic interactions may provide a means for diffusive conduction behavior. Nanotube bundles, on the other hand, are subject to various scattering mechanisms and possess a lower thermal conductivity; yet they may still behave like good thermal conductors ($\kappa$ values from 50–300 W/m K). Furthermore, the contact may be attributed to the reduction in conductance. Contact resistance due to interface scattering needs to be further addressed in order to realize the potential of nanotubes for use in heat transfer enhancement [99]. Mingo and Broido calculated the thermal conductance of carbon nanotubes in the ballistic limit [100]; for semiconductor nanotubes at sufficiently low temperatures, the thermal conductance becomes $4g_{T0}T$ due to the four lowest phonon modes regardless of the length and the cross-sectional area. In this regime, the thermal conductivity of CNTs increases with length. As the temperature increases from cryogenic temperatures, the thermal conductivity of CNTs first increases due to the increased specific heat and reaches a peak around 300–400 K, and then decreases due to phonon–phonon scattering. In the diffusion limit, the conductivity is independent of the length and diminishes as temperature further increases. For nanotubes whose band structures are metal-like, such as with (6,0) and (18,0) chiral numbers, electron ballistic transport may be important; however, electron–phonon scattering will dominate at sufficiently high temperatures.

## 5.7 Summary

This chapter began with lattice vibrations (i.e., phonons) in solids and discussed the dimensionality and the quantum size effect on the lattice specific heat. Free-electron theory was applied, assuming a spherical Fermi surface, to predict the electronic specific heat, as well as electrical and thermal conductivities of solids. The Boltzmann transport equation under the relaxation time approximation and the local-equilibrium assumption was used to derive the electrical and thermal conductivities as well as the thermoelectric coefficients within the framework of irreversible thermodynamics. A brief discussion of the efficiency of thermoelectric power and refrigeration systems was then provided. The classical size effect on electrical and thermal conductivities was presented using both geometric arguments and the BTE, followed by a discussion on the mean-free-path distribution and the thermal conductivity accumulation function. Finally, the concept of conductance quantization for both electrical current and heat flow was introduced using Landauer's formalism. The properties were discussed with examples of representative materials, such as noble metals, semiconductors, quantum wells, superlattices, nanowires, and carbon nanotubes. In the next chapter, the band theory for electrons and phonons will be introduced as an advanced topic of the transport theory of solids.

## Problems

5.1 Calculate the specific heat of lead, using both the Einstein model and the Debye model, for temperatures equal to 2, 10, 20, 50, 100, 200, 300, 600, and 800 K. Use $\Theta_D = 88$ K and $\Theta_E = 65$ K since the specific heats calculated with these values agree with the data well for the whole temperature range. Compare your answer with the values from Touloukian and Buyco [7]. Explain the low-temperature and high-temperature behavior.

5.2 In the first stage of designing a refrigeration system that will cool 1 kg of Pb from 300 to 2 K. Assume the Debye model can be used to calculate the temperature-dependent specific heat of lead (with $\Theta_D = 88$ K). Answer the following questions:

(a) How much energy must be removed from Pb?
(b) How much entropy must be transferred out from Pb?
(c) Assuming that the environment is at 300 K, what is the least amount of work necessary to perform this refrigeration task?
(d) Consider the refrigeration in three temperature ranges: (1) from 300 to 100 K; (2) from 100 to 20 K; and (3) from 20 to 2. What is the least amount of work needed in each temperature range?

5.3 Plot the Fermi function $f_{\mathrm{FD}}$ versus $\varepsilon$ for $T = 0$, 500, and 5000 K. Plot the distribution function of free electrons in metal $f(\varepsilon)$ as a function of $\varepsilon$. Discuss the main features of these plots. [Use eV as the unit for energy.]

5.4 The Fermi energy (at 0 K) of copper is $\mu_{\mathrm{F}} = 7.07$ eV. What is $\mu(T)$ of Cu at 1000 and 10,000 K? Determine the maximum and root-mean-square free-electron speeds in copper at 0 K. Plot the electron distribution functions in terms of the speed and the kinetic energy for $T = 0$, 300, and 4000 K.

5.5 The Fermi energy of silver is $\mu_{\mathrm{F}} = 5.51$ eV. Calculate $\mu(T)$ of Ag at $T = 400$ and 4000 K. What is the rms speed of electrons at 0 K? What is the Fermi velocity? Plot the Fermi function at 0 and 4000 K in one graph and discuss the differences.

5.6 For $k_{\mathrm{B}}T \ll \mu_{\mathrm{F}}$, the specific heat of free-electron gas in metal may be expressed as $\bar{c}_{v,\mathrm{e}} = \frac{\bar{R}}{n_{\mathrm{e}}k_{\mathrm{B}}} \int_0^\infty \frac{\partial f_{\mathrm{FD}}}{\partial T} D(\varepsilon)\varepsilon d\varepsilon$. Evaluate this integration to obtain Eq. (5.24) by referring Appendix B.8.

5.7 Calculate the Fermi energy of silver using the molecular weight and density. Estimate the spacing between the adjacent atoms of Ag. Calculate and plot the electron specific heat and the lattice specific heat of Ag at temperatures from 0 to 1000 K. Show in a separate graph the low-temperature behavior. How do your calculated values agree with experimental data found in a heat transfer text?

5.8 Calculate the Fermi energy $\mu_{\mathrm{F}}$ for copper based on the molecular weight and density. What is the rms speed of free electrons in Cu at 0 and 300 K? Find the electronic specific heat and the lattice specific heat in J/kg K of Cu at 0.1, 1, 10, 30, and 500 K. When can you apply the $T^3$ law, and when can you use the Delong–Petit law?

5.9 Calculate the electronic specific heat and the lattice specific heat of gold at 1, 10, 100, 300, and 1000 K. Sketch their temperature dependence. At what temperature is the electronic and lattice contributions the same? How does your calculated result compare with the value given in a heat transfer text?

5.10 The Mayer relation for the specific heat can be written as $c_p - c_v = \frac{T\beta_P^2}{\rho\kappa_T}$, where $\beta_P = \frac{1}{v}\left(\frac{\partial v}{\partial T}\right)_P$ is the isobaric volume expansion coefficient, $\kappa_T = -\frac{1}{v}\left(\frac{\partial v}{\partial P}\right)_T$ is the isothermal compressibility, and $\rho$ is the density. Noting that the sound speed $v_{\mathrm{a}}$ is defined according to $v_{\mathrm{a}}^2 = \left(\frac{\partial P}{\partial \rho}\right)_s = \frac{c_p}{c_v \rho \kappa_T}$, we can write $\frac{c_p - c_v}{c_v} = \frac{T\beta_P^2 v_{\mathrm{a}}^2}{c_p}$. A simple estimate of the relative difference between the specific heats is readily obtained by assuming that $v_{\mathrm{a}}$ is independent of temperature, $c_p$ on the right-hand side is approximately $3R$, and $\beta_P = 3\alpha$, where $\alpha$ is the linear thermal expansion coefficient. For silicon, $\alpha \approx 4.6 \times 10^{-6}\ \mathrm{K}^{-1}$ at 1000 K and $v_{\mathrm{a}} \approx 5000$ m/s. For copper, $\alpha \approx 2.2 \times 10^{-5}\ \mathrm{K}^{-1}$ and $v_{\mathrm{a}} \approx 2500$ m/s. Estimate the relative difference between $c_p$ and $c_v$ at 1000 K for silicon and copper.

5.11 Graphene is a single sheet of carbon atoms arranged in hexagonal pattern. The phonon mode with the lowest speed is the out-of-plane transverse acoustic mode, when the atoms vibrate perpendicular to the plane. It has a dispersion relation $\omega(k) = ak^2$, with $a = 6 \times 10^{-7}\ \mathrm{m^2 s}$. It is expected that this mode is the dominate mode for the lattice specific heat at low temperatures (below 100 K). Using the 2D solid model with the quadratic dispersion to show that $c_v(T) \propto T$ at low temperatures, i.e., $T \ll \Theta_D$.

5.12 Evaluate the specific heat of a thin GaAs film of two different thicknesses: $L = 2$ and 10 nm. Plot the calculated specific heat with and without planar modes. Compare your results with that predicted by the Debye model for the bulk GaAs at $T \ll \Theta_D$.

5.13 Develop a computer program to calculate the lattice specific heat of CdS or $ZnO_2$ cubic nanocrystals with different sizes: $L = 2$, 10, and 20 nm. Discuss the low-temperature behavior in terms of Eqs. (5.43) and (5.44).

5.14 For a nanowire of diameter $d = 5$ nm, show that $c_v(T) \propto T$ at low temperatures for a linear dispersion. If the length of the nanowire is $L = 10d$, what is the lowest temperature asymptote of the specific heat due to the second quantum size effect?

5.15 Calculate the electron scattering rate and the mean free path of copper at 295 K. Use the linear relations for the electrical resistivity and the Wiedemann–Franz law to calculate the thermal conductivity at 200, 400, 600, and 800 K. Compare the calculated results with data from a heat transfer textbook.

5.16 Calculate the electron scattering rate $1/\tau$, the mean free path $\Lambda$, the electrical conductivity $\sigma$, and the thermal conductivity $\kappa$ of aluminum near room temperature. If the temperature is increased by 5%, how will $1/\tau$, $\Lambda$, $\sigma$, and $\kappa$ change? Express the scattering rate in both rad/s and Hz. Discuss why one should divide it by $2\pi$ to express $1/\tau$ in Hz.

5.17 Sketch the thermal conductivity versus temperature from 0 to 1000 K for silver. What is the dependence of $\kappa$ on $T$, as the temperature approaches absolute zero? How does the thermal conductivity change above 300 K?

5.18 Find the data for the electrical and thermal conductivities of a good conductor in a large temperature range, and evaluate when the Wiedemann–Franz law is valid. Show the low-temperature and high-temperature asymptotes for both $\sigma$ and $\kappa$.

5.19 In the text, we stated that $\partial f_{FD}/\partial \varepsilon$ is a Dirac delta function and used it to obtain the electrical conductivity in Eq. (5.63). Prove that when $k_B T \ll \mu_F$, the integral $\int_0^\infty G(\varepsilon)\frac{\partial f_{FD}}{\partial \varepsilon} d\varepsilon \approx -G(\mu_F)$, where $G(x)$ is an analytical function of $x$. Then, derive Eq. (5.49) from Eq. (5.63).

5.20 Sketch the thermal conductivity of germanium (relatively pure) as a function of temperature [31]. Explain the trend of thermal conductivity at very low temperatures and at above room temperature. Can you assume that the thermal conductivity is independent of temperature near room temperature?

5.21 Derive Eqs. (5.74) through (5.80). Show that in Eq. (5.80), the second term is much smaller than the first term for metals.

5.22 Prove Eq. (5.82a, 5.82b, 5.82c), and calculate the Seebeck coefficient for Ag at 300 and 600 K. The measured Seebeck coefficient of Ag is 1.51 μV/K at 300 K and 3.72 μV/K at 600 K. On the other hand, the Seebeck coefficient for Pt is −5.28 μV/K at 300 K and −11.66 μV/K at 600 K. If an Ag-Pt thermocouple is formed with a junction temperature $T_2 = 600$ K and a reference temperature $T_1 = 300$ K, find the output voltage (see Fig. 5.14b).

5.23 For given values of $T_L$, $T_H$, and $Z^*$, there exists an optimal ratio $R_L/R_0$ for achieving the maximum efficiency of the thermoelectric generator given in Eq. (5.94). Show that

$$\eta_{max} = \frac{\Delta T}{T_H} \frac{\sqrt{1+Z^*T_M} - 1}{\sqrt{1+Z^*T_M} + T_L/T_H},$$

where $T_M = (T_H + T_L)/2$. Calculate the maximum efficiency, normalized to the Carnot efficiency, for $T_L = 300$ K and $T_H = 800$ K as a function of the dimensionless parameter $Z^*T_M$. Plot it for $Z^*T_M$ from 0.3 to 3. Discuss the significance of $ZT$ in thermoelectric devices.

5.24 Consider a thermoelectric generator made of two semiconductors working between $T_L = 300$ K and $T_H = 600$ K. The $p$-type material is made of $Bi_{0.5}Sb_{1.5}Te_3$, and the $n$-type material is made of $Bi_2Se_{0.75}Te_{2.25}$, with the following average properties: $\kappa_p = 1.2$ W/m K, $\kappa_n = 1.3$ W/m K, $r_{e,p} = 15\ \mu\Omega$ m, $r_{e,n} = 13\ \mu\Omega$ m, $\Gamma_p = 210\ \mu$V/K, and $\Gamma_n = -190\ \mu$V/K. Assume that the length $L = 0.8$ cm and the cross section $A_c = 0.3\ cm^2$ for both materials. A generator with a diameter of 10 cm contains 100 pairs ($N = 100$). Find the power output at the maximum efficiency (see Problem 5.23).

5.25 Perform a thermodynamic analysis of the thermoelectric cooling using the same configuration as in Fig. 5.15. By noting that no load resistance is needed and the voltage supplied $\Delta V = N\Gamma_{np}\Delta T + IR_0$, show that the coefficient of performance of a thermoelectric refrigeration is

$$COP = \frac{|q_L|}{P} = \frac{I\Gamma_{np}A_c\sigma_{np}T_L - I^2L/2 - A_c^2\sigma_{np}\kappa_{np}\Delta T/L}{I\Gamma_{np}A_c\sigma_{np}\Delta T + I^2L}.$$

The maximum $COP$ can be obtained by setting the derivative with respect to $I$ equal to zero. Show that

$$COP_{max} = \frac{T_L}{\Delta T} \frac{\sqrt{1+Z^*T_M} - T_H/T_L}{\sqrt{1+Z^*T_M} + 1},$$

where $T_M = (T_H + T_L)/2$.

5.26 Estimate the thermal conductivity along a copper film with various thicknesses: $d = 400$, 100, and 50 nm at 300 K. What if the temperature is reduced to 1 K?

5.27 Estimate the thermal conductivity along a copper wire with various diameters: $d = 400$, 100, and 50 nm at 1 and 300 K, respectively. Compare simple geometric averaging of free paths with the BTE. What are the electron de Broglie wavelengths at these temperatures? If the surface roughness parameter $\sigma_{rms} = 2$ nm, will the scattering be mostly diffuse or specular at each temperature?

5.28 At 5 K, calculate the thermal conductivity, perpendicular ($\kappa_{\text{eff},z}$) and parallel ($\kappa_{\text{eff},x}$) to the plane, for a 200-nm-thick gold film. Calculate the effective thermal conductivity $\kappa_{\text{eff,w}}$ of a gold wire of 5-μm thickness. Hint: use the bulk resistivity value from Fig. 5.11.

5.29 In Example 5.6, we have calculated the properties of a single-crystal silicon at various temperatures. Use simple relations with $p = 0$ to estimate the thermal conductivities of silicon from 5 to 1000 K along a 50-nm-thick thin film and a 100-nm-thick thin wire. Assume the surface roughness $\sigma_{\text{rms}} = 2$ nm. Will the diffuse model be a good assumption? For the thin film, redo the calculation using the specularity $p$ estimated based on the thermal phonon wavelength $\lambda_{\text{th}}$.

5.30 The diameter of a carbon nanotube is determined by its chiral numbers $(m, n)$ according to $d = 0.07834\sqrt{m^2 + mn + n^2}$. What is the diameter of (10,10) single-walled nanotubes? Assume that the wall thickness (unit atomic layer) is 0.34 nm. What is the cross-sectional area? Calculate the phonon thermal conductivity $\kappa$ in the ballistic limit considering the four phonon modes at 100 K for (10,10) nanotubes with length $L = 100$ nm, 1 μm, and 10 μm. Will the ballistic limit of thermal conduction hold at room temperature and above?

## References


1. M.I. Flik, B.I. Choi, K.E. Goodson, Heat transfer regimes in microstructures. J. Heat Transf. **114**, 666–674 (1992)
2. C.L. Tien, G. Chen, Challenges in microscale conductive and radiative heat transfer. J. Heat Transf. **116**, 799–807 (1994)
3. D.G. Cahill, K. Goodson, A. Majumdar, Thermometry and thermal transport in micro/nanoscale solid-state devices and structures. J. Heat Transf. **124**, 223–241 (2002)
4. D.G. Cahill, W.K. Ford, K.E. Goodson, G.D. Mahan, A. Majumdar, H.J. Maris, R. Merlin, S.R. Phillpot, Nanoscale thermal transport, J. Appl. Phys. **93**, 793–818 (2003); D.G. Cahill, P.V. Braun, G. Chen, D.R. Clarke, S. Fan, K. E. Goodson, P. Keblinski, W.P. King, G.D. Mahan, A. Majumdar, H.J. Maris, S.R. Phillpot, E. Pop, L. Shi, Nanoscale thermal transport. II. 2003–2012. Appl. Phys. Rev. **1**, 011305 (2014)
5. C. Kittel, *Introduction to Solid State Physics*, 7th edn. (Wiley, New York, 1996)
6. N.W. Ashcroft, N.D. Mermin, *Solid State Physics* (Harcourt College Publishers, Fort Worth, TX, 1976)
7. Y.S. Touloukian, E.H. Buyco (eds.), *Thermophysical Properties of Matter*, Vol. 4: Specific Heat – Metallic Elements and Alloys; Vol. 5: Specific Heat – Nonmetallic Solids (IFI/Plenum, New York, 1970)
8. G. Nilsson, S. Rolandson, Lattice dynamics of copper at 80 K. Phys. Rev. B **7**, 2393–2400 (1973)
9. A.J.E. Foreman, Anharmonic specific heat of solids. Proc. Phys. Soc. (London) **79**, 1124–1141 (1962)
10. R.A. MacDonald, W.M. MacDonald, Thermodynamic properties of fcc metals at high temperatures. Phys. Rev. B **24**, 1715–1724 (1981)
11. V. Novotny, P.P.M. Meincke, J.H.P. Watson, Effect of size and surface on the specific heat of small lead particles. Phys. Rev. Lett. **28**, 901–903 (1972); V. Novotny, P.P.M. Meincke, Thermodynamic lattice and electric properties of small particles. Phys. Rev. B **8**, 4186–4199 (1973)

12. W. Yi, L. Lu, D.-L. Zhang, Z.W. Pan, S.S. Xie, Linear specific heat of carbon nanotubes. Phys. Rev. B **59**, R9015–R9018 (1999)
13. C. Dames, B. Poudel, W.Z. Wang, J.Y. Huang, Z.F. Ren, Y. Sun, J.I. Oh, C. Opeil, M.J. Naughton, G. Chen, Low-dimensional phonon specific heat of titanium dioxide nanotubes. Appl. Phys. Lett. **87**, 031901 (2005)
14. A.A. Valladares, The Debye specific heat in *n* dimensions. Am. J. Phys. **43**, 308–311 (1975)
15. A.J. McNamara, B.J. Lee, Z.M. Zhang, Quantum size effect on the lattice specific heat of nanostructures. Nanoscale Microscale Thermophys. Eng. **14**, 1–20 (2010)
16. W. DeSorbo, W.W. Tyler, The specific heat of graphite from 13 to 300 K. J. Chem. Phys. **21**, 1660–1663 (1953)
17. R.S. Prasher, P.E. Phelan, Size effect on the thermodynamic properties of thin solid films. J. Heat Transf. **120**, 1078–1081 (1998); R.S. Prasher, P.E. Phelan, Non-dimensional size effects on the thermodynamic properties of solids. Int. J. Heat Mass Transf. **42**, 1991–2001 (1999)
18. Y. Zhang, J.X. Cao, Y. Xiao, X.H. Yan, Phonon spectrum and specific heat of silicon nanowires. J. Appl. Phys. **102**, 104303 (2007)
19. Y. Zhou, X. Zhang, M. Hu, Nonmonotonic diameter dependence of thermal conductivity of extremely thin Si nanowires: competition between hydrodynamic phonon flow and boundary scattering. Nano Lett. **17**, 1269–1276 (2017)
20. Z. Rashid, L. Zhu, W. Li, Effect of confinement on anharmonic phonon scattering and thermal conductivity in pristine silicon nanowires. Phys. Rev. B **97**, 075441 (2018)
21. H.P. Baltes, E.R. Hilf, Specific heat of lead grains. Solid State Commun. **12**, 369–373 (1973)
22. R. Lautenschlager, Improved theory of the vibrational specific heat of lead grains. Solid State Commun. **16**, 1331–1334 (1975)
23. M.S. Dresselhaus, P.C. Eklund, Phonons in carbon nanotubes. Adv. Phys. **49**, 705–814 (2000)
24. J. Hone, B. Batlogg, Z. Benes, A.T. Johnson, J.E. Fischer, Quantized phonon spectrum of single-wall carbon nanotubes. Science **289**, 1730–1733 (2000); W.A. de Heer, A question of dimensions. Science **289**, 1702–1703 (2000)
25. J. Zimmermann, P. Pavone, G. Cuniberti, Vibrational modes and low-temperature thermal properties of graphene and carbon nanotubes: minimal force-constant model. Phys. Rev. B **78**, 045410 (2008)
26. R. Denton, B. Muhlschlegel, D.J. Scalapino, Thermodynamic properties of electrons in small metal particles. Phys. Rev. B **7**, 3589–3607 (1973)
27. W.P. Halperin, Quantum size effects in metal particles. Rev. Mod. Phys. **58**, 533–606 (1986)
28. Z.M. Zhang, Clarification of the relation between drift velocity and relaxation time. J. Thermophys. Heat Transf. **26**, 189–191 (2012)
29. J.M. Ziman, *Electrons and Phonons* (Oxford University Press, Oxford, UK, 1960); reprinted in the Oxford Classics Series, 2001
30. R.A. Matula, Electrical resistivity of copper, gold, palladium, and silver. J. Phys. Chem. Ref. Data **8**, 1147–1298 (1979)
31. Y.S. Touloukian, R.W. Powell, C.Y. Ho, P.G. Klemens (eds.), *Thermophysical Properties of Matter,* Vol. 1: Thermal Conductivity – Metallic Elements and Alloys; Vol. 2: Thermal Conductivity – Nonmetallic Solids (IFI/Plenum, New York, 1970)
32. D.G. Cahill, S.K. Watson, R.O. Pohl, Lower limit to the thermal conductivity of disorderd crystals. Phys. Rev. B **46**, 6131–6140 (1992)
33. G.A. Slack, The thermal conductivity of nonmetallic crystals. Solid State Phys. **34**, 1–71 (1979)
34. M.C. Wingert, J. Zheng, S. Kwon, R. Chen, Thermal transport in amorphous materials: a review. Semicond. Sci. Technol. **31**, 113003 (2016)
35. P.B. Allen, J.L. Feldman, Thermal conductivity of disordered harmonic solids. Phys. Rev. B **48**, 12581–12588 (1993)
36. P.B. Allen, J.L. Feldman, J. Fabian, F. Wooten, Diffusons, locons and propagons: Character of atomic vibrations in amorphous Si. Philos. Mag. B **79**, 1715–1731 (1999)
37. J.M. Larkin, A.J.H. McGaughey, Thermal conductivity accumulation in amorphous silica and amorphous silicon. Phys. Rev. B **89**, 144303 (2014)

38. M.T. Agne, R. Hanus, G.J. Snyder, Minimum thermal conductivity in the context of diffuson-mediated thermal transport. Energy Environ. Sci. **11**, 609–616 (2018)
39. D.M. Leitner, Vibrational energy transfer and heat conduction in a one-dimensional glass. Phys. Rev. B **64**, 094201 (2001)
40. W. Lv, H. Asegun, Non-negligible contributions to thermal conductivity from localized modes in amorphous silicon dioxide. Sci. Rep. **6**, 35720 (2016)
41. J. Moon, B. Latour, A.J. Minnich, Propagating elastic vibrations dominate thermal conduction in amorphous silicon. Phys. Rev. B **97**, 024201 (2018)
42. C.L. Choy, Thermal conductivity of polymers. Polymer **18**, 984–1004 (1977)
43. S. Kommandur, S.K. Yee, An empirical model to predict temperature-dependent thermal conductivity of amorphous polymers. J. Polymer Sci. B: Polymer Phys. **55**, 1160–1170 (2017)
44. X. Xie, K. Yang, D. Li, T.-H. Tsai, J. Shin, P.V. Braun, D.G. Cahill, High and low thermal conductivity of amorphous macromolecules. Phys. Rev. B **95**, 035406 (2017)
45. A. Henry, Thermal transport in polymers. Ann. Rev. Heat Transf. **17**, 485–520 (2014)
46. H. Chen, V.V. Ginzburg, J. Yang, Y. Yang, W. Liu, Y. Huang, L. Du, B. Chen, Thermal conductivity of polymer-based composites: fundamentals and applications. Prog. Polymer Sci. **59**, 41–85 (2016)
47. X. Xu, J. Chen, J. Zhou, B. Li, Thermal conductivity of polymers and their nanocomposites. Adv. Mater. **30**, 1705544 (2018)
48. M. Pyda, A. Boller, J. Grebowicz, H. Chuah, B.V. Lebedev, B. Wunderlich, Heat Capacity of Poly(trimethylene terephthalate). J. Polymer Sci. B: Polymer Phys. **36**, 2499–2511 (1998)
49. S. Shen, A. Henry, J. Tong, R. Zheng, G. Chen, Polyethylene nanofibres with very high thermal conductivities. Nat. Nanotech. **5**, 251–255 (2010)
50. R.E. Bentley, *Theory and Practice of Thermoelectric Thermometry* (Springer, Singapore, 1998)
51. R.B. Roberts, The absolute scale of thermoelectricity II. Phil. Mag. B **43**, 1125–1135 (1981)
52. O. Dreirach, The electrical resistivity and thermopower of solid noble metals. J. Phys. F: Met. Phys. **3**, 577–584 (1973)
53. C.J. Vineis, A. Shakouri, A. Majumdar, M.G. Kanatzidis, Nanostructured thermoelectrics: big efficiency gains from small features. Adv. Mater. **22**, 3970–3980 (2010)
54. S.L. Soo, *Direct Energy Conversion* (Prentice-Hall, Englewood Cliffs, NJ, 1968)
55. L.D. Hicks, M.S. Dresselhaus, Effect of quantum-well structures on the thermoelectric figure of merit. Phys. Rev. B **47**, 12727–12731 (1993); M.S. Dresselhaus, Y.-M. Lin, O. Rabin, G. Dresselhaus, Bismuth nanowires for thermoelectric applications. Microscale Thermophys. Eng. **7**, 207–219 (2003); Y.-M. Lin, M. S. Dresselhaus, Thermoelectric properties of superlattice nanowires. Phys. Rev. B **68**, 075304 (2003)
56. J. He, T.M. Tritt, Advances in thermoelectric materials research: looking back and moving forward. Science **357**, eaak9997 (2017)
57. Z. Tian, S. Lee, G. Chen, Comprehensive review of heat transfer in thermoelectric materials and devices. Ann. Rev. Heat Transf. **17**, 425–483 (2014)
58. L. Onsager, Reciprocal relations in irreversible processes. I & II. Phys. Rev. **37**, 405–426; **38**, 2265–2279 (1931)
59. H.B. Callen, *Thermodynamics and an Introduction to Thermostatistics*, 2nd edn. (Wiley, New York, 1985)
60. D. Kondepudi, I. Prigogine, *Modern Thermodynamics: From Heat Engines to Dissipative Structures* (Wiley, New York, 1998)
61. D. Jou, G. Lebon, J. Casas-Vázquez, *Extended Irreversible Thermodynamics*, 4th edn. (Springer, Berlin, 2010)
62. C.R. Tellier, A.J. Tosser, *Size Effects in Thin Films* (Elsevier, Amsterdam, 1982)
63. M.I. Flik, C.L. Tien, Size effect on the thermal conductivity of high-$T_c$ thin-film superconductors. J. Heat Transf. **112**, 872–881 (1990)
64. R.A. Richardson, F. Nori, Transport and boundary scattering in confined geometrics: analytical results. Phys. Rev. B **48**, 15209–15217 (1993)

65. J.E. Graebner, S. Jin, G.W. Kammlott, J.A. Herb, C.F. Gardinier, Large anisotropic thermal conductivity in synthetic diamond films. Nature **359**, 401–403 (1992)
66. D. Stewart, P.M. Norris, Size effect on the thermal conductivity of thin metallic wires: Microscale implications. Microscale Thermophys. Eng. **4**, 89–101 (2000)
67. S.G. Walkauskas, D.A. Broido, K. Kempa, T.L. Reinecke, Lattice thermal conductivity of wires. J. Appl. Phys. **85**, 2579–2582 (1999)
68. S. Kumar, G.C. Vradis, Thermal conductivity of thin metallic films. J. Heat Transfer **116**, 28–34 (1994)
69. P. Beckman, A. Spizzichino, *The Scattering of Electromagnetic Waves from Rough Surfaces* (Artech House Inc, Norwood, MA, 1987)
70. B. Feng, Z. Li, X. Zhang, Effect of grain-boundary scattering on the thermal conductivity of nanocrystalline metallic films. J. Phys. D Appl. Phys. **42**, 055311 (2009)
71. J. Zou, A. Balandin, Phonon heat conduction in a semiconductor nanowire. J. Appl. Phys. **89**, 2932–2938 (2001)
72. M. Asheghi, M.N. Touzelbaev, K.E. Goodson, Y.K. Leung, S.S. Wong, Temperature-dependent thermal conductivity of single-crystal silicon layers in SOI substrates. J. Heat Transf. **120**, 30–36 (1998); M. Asheghi, K. Kurabayashi, R. Kasnavi, K.E. Goodson, Thermal conduction in doped single-crystal silicon films. J. Appl. Phys. **91**, 5079–5088 (2002); W. Liu, M. Asheghi, Thermal conductivity measurements of ultra-thin single crystal silicon layers. J. Heat Transf. **128**, 75–83 (2006)
73. D. Li, Y. Wu, P. Kim, L. Shi, P. Yang, A. Majumdar, Thermal conductivity of individual silicon nanowires. Appl. Phys. Lett. **83**, 2934–2936 (2003)
74. P.G. Murphy, J.E. Moore, Coherent phonon scattering effects on thermal transport in thin semiconductor nanowires. Phys. Rev. B **76**, 155313 (2007)
75. A.I. Hochbaum, R. Chen, R.D. Delgado, W. Liang, E.C. Garnett, M. Najarian, A. Majumdar, P. Yang, Enhanced thermoelectric performance of rough silicon nanowires. Nature **451**, 163–167 (2008)
76. P. Martin, Z. Aksamija, E. Pop, U. Ravaioli, Impact of phonon-surface roughness scattering on thermal conductivity of thin Si nanowires. Phys. Rev. Lett. **102**, 125503 (2009)
77. H. Kim, I. Kim, H.-J. Choi, W. Kim, Thermal conductivities of $Si_{1-x}Ge_x$ nanowires with different germanium concentrations and diameters. Appl. Phys. Lett. **96**, 233106 (2010)
78. G. Xie, Y. Guo, X. Wei, K. Zhang, L. Sun, J. Zhong, G. Zhang, Y.-W. Zhang, Phonon mean free path spectrum and thermal conductivity for $Si_{1-x}Ge_x$ nanowires. Appl. Phys. Lett. **104**, 233901 (2014)
79. A. Malhotra1, M. Maldovan, Impact of phonon surface scattering on thermal energy distribution of Si and SiGe nanowires. Sci. Rep. **6**, 25818 (2016)
80. A.J. Minnich, J.A. Johnson, A.J. Schmidt, K. Esfarjani, M.S. Dresselhaus, K.A. Nelson, G. Chen, Thermal conductivity spectroscopy technique to measure phonon mean free paths. Phys. Rev. Lett. **107**, 095901 (2011)
81. F. Yang, C. Dames, Mean free path spectra as a tool to understand thermal conductivity in bulk and nanostructures. Phys. Rev. B **87**, 035437 (2013)
82. T. Shiga, D. Aketo, L. Feng, J. Shiomi, Harmonic phonon theory for calculating thermal conductivity spectrum from first-principles dispersion relations. Appl. Phys. Lett. **108**, 201903 (2016)
83. P.K. Schelling, S.R. Phillpot, P. Keblinski, Comparison of atomic-level simulation methods for computing thermal conductivity. Phys. Rev. B **65**, 144306 (1999)
84. A.J. Kulkarni, M. Zhou, Size-dependent thermal conductivity of zinc oxide nanobelts. Appl. Phys. Lett. **88**, 141921 (2006)
85. R. Landauer, Spatial variation of currents and fields due to localized scatters in metallic conduction. IBM J. Res. Develop. **1**, 223–231 (1957); R. Landauer, Conductance determined by transmission: probes and quantized constriction resistance. J. Phys.: Condens. Matter **1**, 8099–8110 (1989); Y. Imry, R. Landauer, Conductance viewed as transmission. Rev. Mod. Phys. **71**, S306–S312 (1999)

86. G. Rubio, N. Agraït, S. Vieira, Atomic-sized metallic contacts: mechanical properties and electronic transport. Phys. Rev. Lett. **76**, 2302–2305 (1996)
87. N. Agraït, A.L. Yeyati, J.M. van Ruitenbeek, Quantum properties of atomic-sized conductors. Phys. Rep. **377**, 81–279 (2003)
88. L. Chico, L.X. Benedict, S.G. Louie, M.L. Cohen, Quantum conductance of carbon nanotubes with defects. Phys. Rev. B **54**, 2600–2606 (1996)
89. S. Frank, P. Poncharal, Z.L. Wang, W.A. de Heer, Carbon nanotube quantum resistors. Science **280**, 1744–1746 (1998)
90. U. Landman, W.D. Luedtke, N.A. Burnham, R.J. Colton, Atomistic mechanisms and dynamics of adhesion, nanoindentation, and fraction. Science **248**, 454–461 (1990)
91. U. Landman, W.D. Leudtke, B.E. Salisbury, R.L. Whetten, Reversible manipulations of room temperature mechanical and quantum transport properties in nanowire junctions. Phys. Rev. Lett. **77**, 1362–1365 (1996)
92. A. Greiner, L. Reggiani, T. Kuhn, L. Varani, Thermal conductivity and Lorenz number for one-dimensional ballistic transport. Phys. Rev. Lett. **78**, 1114–1117 (1997)
93. K. Schwab, E.A. Henriksen, J.M. Worlock, M.L. Roukes, Measurement of the quantum of thermal conductance. Nature **404**, 974–977 (2000); K. Schwab, J.L. Arlett, J.M. Worlock, M.L. Roukes, Thermal conductance through discrete quantum channels. Physica E **9**, 60–68 (2001)
94. J. Hone, M. Whitney, C. Piskoti, A. Zettl, Thermal conductivity of single-walled carbon nanotubes. Phys. Rev. B **59**, 2514–2516 (1999)
95. S. Berber, Y.-K. Kwon, D. Tománek, Unusually high thermal conductivity of carbon nanotubes. Phys. Rev. Lett. **84**, 4613–4616 (2000)
96. S. Maruyama, A molecular dynamics simulation of heat conduction of a finite length single-walled carbon nanotube. Microscale Thermophys. Eng. **7**, 41–50 (2003)
97. P. Kim, L. Shi, A. Majumdar, P.L. McEuen, Thermal transport measurements of individual multiwalled nanotubes. Phys. Rev. Lett. **87**, 215502 (2001)
98. C. Yu, L. Shi, Z. Yao, D. Li, A. Majumdar, Thermal conductance and thermopower of an individual single-wall carbon nanotube. Nano Lett. **5**, 1842–1846 (2005)
99. A.J. McNamara, Y. Joshi, Z.M. Zhang, Characterization of nanostructured thermal interface materials – a review. Int. J. Thermal Sci. **62**, 2–11 (2012)
100. N. Mingo, D.A. Broido, Carbon nanotube ballistic thermal conductance and its limits. Phys. Rev. Lett. **95**, 096105 (2005); N. Mingo, D.A. Broido, Length dependence of carbon nanotube thermal conductivity and the problem of long waves. Nano Lett. **5**, 1221–1225 (2005)

# Chapter 6
# Electron and Phonon Transport

In the preceding chapter on solid properties, we relied on the Drude–Sommerfeld model, which assumes that electrons are completely free and the Fermi surface is spherical and isotropic in all directions of the wavevector. While the concepts of electronic band structures and phonon dispersion in real solids were often mentioned, we have deliberately avoided any details. It is hoped that the free-electron model will help readers gain an intuitive picture of electrons without a deep knowledge of solid-state physics. Note that the free-electron model described in Sect. 5.1.3 is applicable only for metals, usually good conductors, and cannot be applied to semiconductors. The Sommerfeld theory, albeit successful in quantitatively describing electronic transport for certain metals, does not touch on the fundamental mechanisms of electron scattering and the shape of the Fermi surface. The free-electron model also fails to explain certain phenomena including thermoelectricity. The Hall effect and magnetoresistance, to be discussed in the following section, provide further evidence of the inadequacy of the free-electron model.

This chapter introduces electronic band theory after a brief discussion of electronic structures in atoms, binding in crystals, and crystal lattices. The phonon dispersion relations are presented subsequently and explained in terms of different branches of acoustic and optical phonons. Subsequently, the electron and phonon scattering mechanisms are outlined. The next section addresses electronic emission and tunneling phenomena, including photoelectric effect, thermionic emission, field emission, as well as electron tunneling through a potential. A significant portion of this chapter is then devoted to semiconductor materials and devices, with an emphasis on optoelectronic applications such as solar cells, thermophotovoltaic systems, light-emitting diodes (LEDs), and semiconductor lasers including quantum well lasers.



Z. M. Zhang, *Nano/Microscale Heat Transfer*, Mechanical Engineering Series,
https://doi.org/10.1007/978-3-030-45039-7_6

## 6.1 The Hall Effect

When a conductor carrying electric current is placed in a magnetic field perpendicular to the current flow, there is a Lorentz force acting on the conductor according to $\mathbf{F} = \sum q\mathbf{u}_\mathrm{d} \times \mathbf{B} = l\mathbf{I} \times \mathbf{B}$, where $q$ is the charge of each carrier, $\mathbf{u}_\mathrm{d}$ is the drift velocity of the carrier, $\mathbf{B}$ is the magnetic induction, $\mathbf{I}$ is the current in the conductor, and $l$ is the length of the conductor. This principle was used in the electromagnetic motor invented by Michael Faraday in 1821. Because electric current is always defined in the direction of the applied electric field $\mathbf{E} = -\nabla V$, the force acting on the conductor is independent of the nature of the carriers (electrons or holes). Microscopically, however, there is a subtle difference that can be distinguished by the experiment first performed by Edwin Hall in 1878 when he was a graduate student at Johns Hopkins University. As shown in Fig. 6.1, an electric current passes through a metal foil in the $x$-direction, while the electrons are drifted opposite to the $x$-direction. When a uniform magnetic field $\mathbf{B}$ is applied in the $z$-direction, the electrons are subjected to a force toward the negative $y$-direction. Gradually, an electric field is built up across the foil as manifested by a nonzero voltage $V_\mathrm{H}$, which is called the Hall voltage. The electric potential in the $y$-direction eventually balances the magnetic force such that the electrons drift in the $x$-direction only. This effect is called the *Hall effect*. By setting the $y$-component of the Lorentz force $\mathbf{F} = q(\mathbf{E} + \mathbf{u}_\mathrm{d} \times \mathbf{B})$ to zero, one obtains

$$V_\mathrm{H} = \frac{IB}{nqd} \tag{6.1}$$

where $n$ is the number density of the carrier and $d$ is the thickness of the conductor [1, 2]. The *Hall coefficient* is defined as follows:

$$\eta_\mathrm{H} = \frac{V_\mathrm{H} d}{IB} = \frac{1}{nq} \tag{6.2}$$

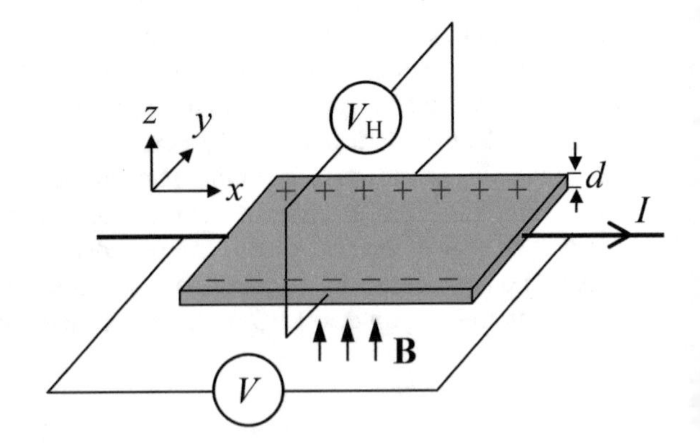


**Fig. 6.1** Illustration of the Hall effect experiment

The *Hall resistance* can be defined as $R_H = V_H/I = B\eta_H/d$, and its inverse is called the *Hall conductance*. Similarly, the *Hall resistivity* is given by $r_H = B\eta_H = E_y/J_x$, where $E_y$ is the electric field in the $y$-direction and $J_x$ is the current density. For metals, $q = -e$ and $n = n_e$, the number density of free electrons, and one would expect a negative Hall resistance.

**Example 6.1** Find the Hall coefficient and the Hall voltage for a copper foil of 2 × 2 cm$^2$ area, with a thickness of 10 μm. Given the electrical current $I = 0.5$ A and the magnetic induction $B = 1.0$ T (tesla) = 1.0 Wb/m$^2$, what is the voltage drop along the current flow direction?

**Solution** Based on the previous chapter, the number density of electrons in copper is $n_e = 8.45 \times 10^{28}$ m$^{-3}$. From Eq. (6.2), we obtain $\eta_H = -7.4 \times 10^{-5}$ cm$^3$/C, and from Eq. (6.1) we find $V_H = -3.7$ μV, which is a very small voltage but can be measured accurately. Using the resistivity of copper $r_e = 1.7 \times 10^{-8}$ Ω m, we see that $V = 850$ μV, which is much larger than the Hall voltage. The Hall coefficient is much larger for semiconductors because of their usually much lower carrier densities.

Before the discovery of the Hall effect, many people, including James Clerk Maxwell, believed that the force acted only on the conductor but not on the current carriers [3]. Measurement of the Hall coefficient allows the determination of the sign of the charge carriers as well as the carrier concentration. This is important especially for semiconductor materials. The Hall coefficient is positive for $p$-type semiconductors, but negative for $n$-type semiconductors. In reality, the Hall coefficient depends also on the applied magnetic field although such a dependence cannot be predicted by the Drude free-electron model. For some common metals like Al, Be, Cd, In, W, and Zn, the Hall coefficient can even become positive. Therefore, the Hall effect cannot be fully accounted by the free-electron model. It is necessary to understand the electronic band structures.

Magnetoresistance is the change in resistance of a material under an applied magnetic field. The magnetoresistance may be transverse, when the applied magnetic field is perpendicular to the current flow, and longitudinal, when the applied magnetic field is parallel to the current flow. In the free-electron theory, resistance is expected to be independent of the strength of the applied transverse magnetic field. In reality, most materials exhibit transverse magnetoresistance that depends on the magnetic field strength. In the late 1980s, researchers observed a giant magnetoresistive (GMR) effect, also called giant magnetoresistance, with extremely thin films of ferromagnetic and metallic layers. The GMR effect has been applied to read heads for magnetic hard disk drives [4] and the discovery is recognized by the 2007 Nobel Prize in Physics.

Klaus von Klitzing and coworkers in 1980 measured the Hall voltage of a 2D electron gas using a metal-oxide-semiconductor field-effect transistor (MOSFET), at very low temperatures ($T \approx 1.5$ K) with a high magnetic field ($B > 15$ T), at the Grenoble High Magnetic Field Laboratory in France [5]. They found that the Hall conductance is quantized and increases with the applied magnetic field by steps in a staircase sequence. The Hall conductance is a multiple of a fundamental constant, $1/R_K$, where

$$R_K = h/e^2 = 25812.807449 \pm 0.000086\,\Omega \tag{6.3}$$

is called the von Klitzing constant. Note that $e^2/h$ is proportional to the fine-structure constant, which is related to the strength of light–matter interaction in quantum electrodynamics. For this work, von Klitzing was awarded the Nobel Prize in Physics in 1985. The remarkable precision and gauge invariance of quantized conductance allowed the definition of a resistance standard used worldwide since 1990 [6]. As discussed in Chap. 5, quantized conductance has also been observed between nanocontacts and nanostructures with an increment of $2/R_K$. The discovery of the fractional quantum Hall effect in 1982, on the other hand, rendered three physicists (Robert Laughlin, Horst Störmer, and Daniel Tsui) the 1998 Nobel Prize. This has led to a breakthrough in our fundamental understanding of the physical world. For example, in a 2D system, electrons may switch between Fermi–Dirac statistics and Bose–Einstein statistics, continuously [7]. Strohm et al. [8] reported phonon Hall effect by applying a magnetic field perpendicular to the heat flow in a paramagnetic dielectric material at low temperatures. A transverse temperature difference was measured, which reverses sign when the magnetic field is inversed. Researchers have also observed magnon Hall effect and photonic spin Hall effect.

## 6.2 General Classifications of Solids

There are several ways to classify solids. Based on their electrical conductivities, solids may be classified as insulators, semiconductors, or conductors. They may exist in different forms, such as amorphous or crystalline phases, depending on how the atoms in the solids are arranged. A general introduction is given in this section considering chemical bonds and electrical properties of solids. Let us first take a look at the electron configuration in atoms because it is directly related to physical and chemical properties.

### *6.2.1 Electrons in Atoms*

The periodic table of elements is arranged sequentially according to atomic number, which is determined by the number of protons inside the nucleus and equal to the number of electrons orbiting the nucleus, since an atom itself is charge neutral. The electrons occupy different quantum states, which are fully described by the Schrödinger wave equation as discussed in Chap. 3. By solving the wave equation in spherical coordinates [9, 10], the number of quantum states can be determined and identified using indices $n$, $l$, and $m$. The first or principal quantum number $n =$

1, 2, 3, 4, ... corresponds to different shells, denoted as $K, L, M, N, O, \ldots$ In each shell, there are $n$ subshells defined by the orbital number $l = 0, 1, 2, \ldots, (n-1)$. The corresponding symbols are $s, p, d, f, g, h$, and so forth. For each $l$, the magnetic quantum number $m = 0, \pm 1, \pm 2, \ldots, \pm l$, which gives a total of $2l + 1$ orbits (or *orbitals* since electrons do not follow an exact path that can be described by classical mechanics) for each subshell. Hence, there are a total of $n^2$ orbitals in the $n$th shell. When spin degeneracy is considered, the total allowable quantum states are $2n^2$ in the $n$th shell. In other words, there are 2, 8, 18, and 32 quantum states in the first ($K$), second ($L$), third ($M$), and fourth ($N$) shells, respectively. On the other hand, there are $2(2l + 1)$ quantum states in the $l$th ($l < n$) subshell. For example, the $s, p, d$, or $f$ subshell contains correspondingly 2, 6, 10, or 14 quantum states. According to Pauli's exclusion principle, each quantum state can have no more than one electron, i.e., at most only two electrons (one with $+\frac{1}{2}$ and the other with $-\frac{1}{2}$ spin) can share the same orbital.

According to the Aufbau principle, electrons will fill the lowest energy states first. The electron configuration of an atom is expressed by the numbers in each subshell. For example, we can write for aluminum and calcium, respectively,

$$^{13}\text{Al: } 1s^2 2s^2 2p^6 3s^2 3p^1 \text{ and } ^{20}\text{Ca: } 1s^2 2s^2 2p^6 3s^2 3p^6 4s^2$$

Note that the $4s$ orbitals are filled before the $3d$ orbitals because the associated energy level of a $3d$ orbital is higher than that of a $4s$ orbital. However, the electron configuration for $^{29}Cu$ is

$$1s^2 2s^2 2p^6 3s^2 3p^6 4s^1 3d^{10} \text{ rather than } 1s^2 2s^2 2p^6 3s^2 3p^6 4s^2 3d^9$$

This is due to the fact that a half-filled or filled $d$-subshell is more stable than the $s$ shell of the next level [10]. Similarly, the outermost shells for chromium ($^{24}Cr$) are $4s^1 3d^5$ not $4s^2 3d^4$, and those for gold ($^{79}Au$) are $6s^1 4f^{14} 5d^{10}$ not $6s^2 4f^{14} 5d^9$. The properties of an element depend largely on the filled state of the outermost orbitals. Alkali metals, such as $^{3}Li$, $^{11}Na$, and $^{19}K$, have one electron in the outermost orbital and can easily lose it, especially when interacting with halogens whose outermost orbitals can be filled by adding only one electron each. The result is the formation of chemically stable compounds such as NaCl and CsF. The outermost electrons are called *valence electrons*. The $4s^1$ electron in copper is largely responsible for its high electrical conductivity because it can leave the atom relatively easily. When the outermost orbitals are completely filled, as in noble gases like He and Ne, the atoms are very stable and reluctant to react with others. Noble gases are also called inert gases since they are monatomic gases at ambient conditions. At the atmospheric pressure, helium must be cooled to 4.2 K for it to condense into liquid. The general sequence of electron configuration in order of increasing energy is schematically given in the following:

$$1s^2 \,|\, 2s^2\, 2p^6 \,|\, 3s^2\, 3p^6 \,|\, 4s^2\, 3d^{10}\, 4p^6 \,|\, 5s^2\, 4d^{10}\, 5p^6 \,|\, 6s^2\, 4f^{14} 5d^{10} 6p^6 \,|\, 7s^2\, 5f^{14} 6d^{10}$$

For convenience, each dashed line indicates the electron configuration of an inert gas listed underneath that line. Each noble gas contains a completely filled $p$ subshell (with the exception of He which has a filled $K$ shell) before the next $s$ subshell. In atomic physics, *ionization energy* is the energy required to separate an electron from the atomic nucleus. The ionization energy varies periodically according to the atomic number: Alkali metals have the lowest ionization energy because of the single electron in the outermost $s$-orbitals. On the other hand, inert gases have the highest ionization energy. Helium is the most stable element with an ionization energy of 24.6 eV. The ionization energy of lithium is only 5.4 eV. For a hydrogen atom, the ionization energy is 13.6 eV as discussed in Chap. 3.

### 6.2.2 Insulators, Conductors, and Semiconductors

The picture of free-electron gas depicted in Chap. 5 is an oversimplified version in which the electron energies are limited to a nearly continuous band from the zero energy level up to the Fermi energy or Fermi level. Only those near the Fermi surface contribute to electronic transport properties. Electrons in a single atom are in various discrete energy levels, which are well predicted by quantum mechanics. In real solids, atoms are arranged in close proximity; hence, electrons interact strongly with one another as well as with the crystal lattices, resulting in complex wavefunctions as manifested by their band structures. There exist a large number of *allowable bands* that may be occupied by electrons. Between two consecutive allowable bands, there exists a *forbidden band* that cannot be occupied by any electron. Electrons occupy broad bands with allowable energy states up to the Fermi level. The distinction between insulators and metals can be understood by looking at the electronic states near the Fermi surface as illustrated in Fig. 6.2. A brief qualitative description is given here, whereas more detailed theories are deferred to subsequent sections.

For insulators, the highest occupied band is completely filled as shown in Fig. 6.2a. This is called a *valence band* due to the contribution of valence electrons. The next higher band is a *conduction band* which is completely empty. There exists a large energy gap between the valance band and the conduction band, usually between 5 and 15 eV. Examples are $E_g \approx 8$ eV for fused silica ($SiO_2$) and $E_g \approx 14$ eV for LiF. The Fermi level lies in the middle of the forbidden band. Because the valence band is completely filled, electrons are not free to move around (i.e., change from one quantum state to another) under the influence of an electric field. An electrical insulator is also called a dielectric. Pure crystalline dielectrics are transparent to visible light because their valence electrons cannot be excited unless the incoming radiation frequency is high enough that the photon energy exceeds the bandgap energy. Note that a photon energy of $h\nu = 2$ eV corresponds to a visible wavelength

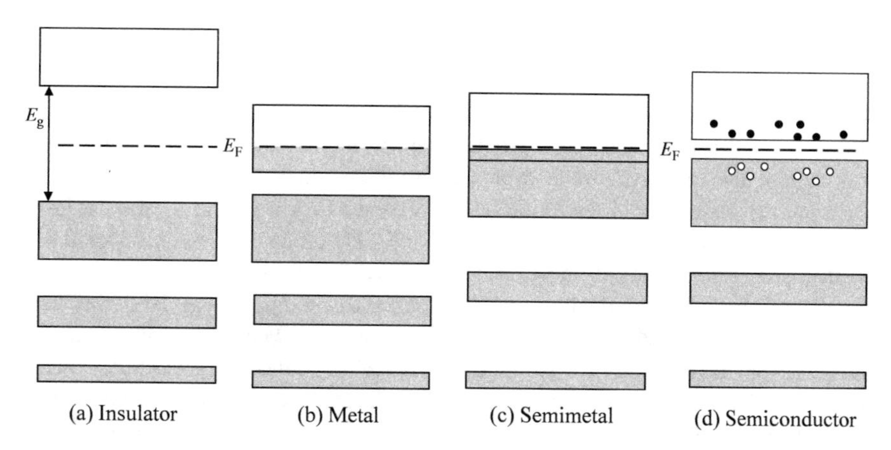


**Fig. 6.2** Schematic of the energy band for different materials, where $E_g$ is the bandgap energy and $E_F$ is the Fermi energy. **a** An insulator has a completely filled valance band and a completely empty conduction band, with a wide bandgap between the two. **b** A metal has a partially filled conduction band and the Fermi level lies in this band. **c** A semimetal, also called a metal, has a conduction band that overlaps the filled valence band. **d** A semiconductor is like an insulator but with a much smaller bandgap and may conduct electricity at elevated temperatures due to thermally excited electrons and holes. Doping or impurities in a semiconductor can result in a large electrical conductance

$\lambda = 620$ nm, and that of 10 eV corresponds to $\lambda = 124$ nm, which lies in the deep ultraviolet. On the other hand, lattice vibrations or phonons in dielectric materials often yield absorption of radiation in the mid infrared.

A metal has a partially filled conduction band, which is the highest occupied band, as shown in Fig. 6.2b. The Fermi level lies inside this allowable band. For some metals like Bi and Sn, the conduction band overlaps the valence band as illustrated in Fig. 6.2c. These metals are sometimes called semimetals since their electrical conductivities are not as high as the alkali or noble metals. Because the energy states within the conduction band are continuous, the uppermost electrons in the partially filled conduction band or the top of the valence band can be excited to a higher unoccupied energy level by an arbitrarily applied field. Over 80% of the elements in the periodic table are metals (or semimetals). All group Ia (alkali, excluding hydrogen), group IIa (alkaline earth), group IIIa (except boron), and transition (all b groups from columns 3 to 12 of the periodic table) elements are metals. The interaction between electromagnetic radiation and a material is much like applying an electric field to the material, except that the frequency of the applied field is very high. Note that the frequency of red light at $\lambda = 632$ nm is $\nu = c/\lambda = 475$ THz. Because of their relatively free electrons, metals interact with electromagnetic radiation strongly. This is manifested by the strong absorption by thin metallic films and the high reflection from polished bulk metals. The strong interaction of metals with microwaves can easily be demonstrated by placing a piece of aluminum foil in a microwave oven and then observing the noises and sparkles as the oven is turned on. At shorter wavelengths in the visible spectrum and in the ultraviolet, additional absorption mechanisms emerge that may be better explained by the particle nature of light.

Semiconductors have band structures similar to those of insulators, except that the energy bandgap $E_g$ is much narrower, i.e., on the order of 1 eV. For example, diamond has a bandgap of 5.5 eV and is usually classified as an insulator, whereas silicon has a bandgap of 1.1 eV at room temperature and is a semiconductor. Some semiconductors can have a relatively large bandgap and hence are called *wideband semiconductors*. Examples are the III–V semiconductor GaN (3.4 eV) and the II–VI semiconductors CdS (2.4 eV) and ZnS (3.7 eV). Diamond may be considered as a wideband semiconductor because of its crystal structure similar to those of Si and Ge. Pure or *intrinsic* semiconductors are insulators at low temperatures. At higher temperatures, as illustrated in Fig. 6.2d, some electrons (dots) can be *thermally excited* from the valence band to the conduction band, leaving holes (circles) in the valence band. Subsequently, electrical current may flow through, although with a large resistance as compared to metals. Bandgap absorption is essential for the interaction of semiconductors with optical radiation. When the photon energy exceeds the energy gap, strong absorption occurs. This is why a silicon wafer looks dark and is opaque to visible light.

By doping the semiconductor with impurities, the charge distribution can be significantly changed, while, at the same time, the bandgap and the Fermi level are slightly modified. The semiconductor becomes *extrinsic*, meaning that the number of electrons is no longer the same as that of holes. A group V element, such as phosphorous with five valance electrons, may substitute a small fraction of silicon atoms. The extra valence electrons can be thermally excited to the conduction band via ionization of the impurities. The phosphorus atom is said to be a *donor*, and the doped semiconductor becomes *n*-type since majority of its carriers are electrons. The electron concentration can be significantly increased to enhance the electrical conductivity. From the band structure point of view, the donated electrons form a filled impurity band right below the conduction band. The difference in energy between the conduction band and the impurity band is called *ionization energy*, which is on the order of 0.05 eV. The ionization energy of a semiconductor has a different meaning from the ionization energy required to separate an electron from the atomic nucleus discussed earlier. Likewise, when impurities from a group III element such as boron with three valance electrons are introduced, additional holes are created such that the silicon semiconductor becomes a *p*-type semiconductor because of the additional positive charge carriers. The boron atoms are called *acceptors*, which form an empty impurity band right above the valence band [11]. The energy difference between these two bands is also called the ionization energy. Doping can strongly affect the infrared properties of semiconductors because of free-carrier absorption. Furthermore, impurities and defects tend to increase phonon scattering and reduce thermal conductivity since thermal transport in semiconductors is mainly by lattice vibration.

### 6.2.3 Atomic Binding in Solids

Two or more atoms can combine to form a molecule, mainly through the electrons in the outermost orbitals (i.e., valence electrons), since the electrons in the inner shells remain tightly bonded to their nuclei. The wavefunctions of the valance electrons are significantly modified as compared with those of the individual atoms. There are five major kinds of chemical bonds: the ionic, covalent, molecular, and hydrogen bonds for insulators and the metallic bond for conductors. Solids with identical chemical composition can have different stable forms or phases, which exhibit distinct differences in their appearances as well as electrical, mechanical, and thermal properties. A notable example is carbon, which may exist in the form of diamond, graphite, carbon black (amorphous carbon), or the fullerene family. A crystal contains periodic and densely packed atoms or lattices, whereas an amorphous solid does not have well-organized lattice structures. The atoms in an amorphous solid are disordered and irregular, like those in a liquid, except that they are firmly bonded together. Therefore, a crystal is usually denser and harder than the amorphous phase of the same composition. A crystal usually exhibits distinct facets along the crystalline planes and has a sharp transition between solid and liquid at a fixed melting point. An amorphous solid does not have clear facets when broken. When heated up, an amorphous solid is first softened and then gradually it melts over a wide temperature range. An example is quartz versus fused silica (glass), both made of $SiO_2$. For a given composition, the thermal conductivity is usually much higher in the crystal form because of lattice vibrations.

Alkali metals and alkaline earth metals have one and two valance electrons, respectively, that are loosely bonded. A metal atom can lose its outermost electrons to become a positive ion. On the other hand, the elements in groups VIIa and VIa tend to gain additional electrons to fill the outermost orbitals and become negative ions. The positive and negative ions attract each other by electrostatic force and form an *ionic bond*, which is quite strong. *Ionic crystals*, such as NaCl, CsCl, KBr, $CaF_2$, and MgO, are hard and usually have high melting points (above 1000 K). They are insulators because the ions cannot move around freely and are transparent in the visible spectrum because of the large bandgap. Nevertheless, some of these crystals are soluble and can be dissolved in water. The solution becomes conductive because of the ions. The positive and negative ions form an electrical dipole and can absorb infrared radiation through lattice vibrations. These solids belong to the group of *polar materials*, in terms of polarizability. Note that the elements in groups Ib (noble metals) and IIb (Zn, Cd, and Hg) resemble those in groups Ia and IIa because of the outermost $s$-orbital electrons. The difference is that groups Ib and IIb also have filled $d$-subshells. Therefore, II–VI semiconductors such as ZnSe and CdTe are largely ionic bonded.

The main contribution to the binding energy is the electrostatic or Madelung energy [2]. The long-range electrostatic force between two ions with charges $q_1$ and $q_2$ is proportional to $q_1 q_2 / r^2$, where $r$ is the separation distance measured from the center of the ion cores. Depending on the sign of the charges, either attractive or repulsive force may occur. The ions arrange themselves in a way that gives the strongest attractive interaction, which is balanced by the short-range repulsive force between atoms. The contribution of the Coulomb attraction to the total energy of the system is roughly proportional to $-1/r$. As atoms are brought very close to each other, the charge distributions or the electron orbitals begin to overlap with each other. Pauli's exclusion principle requires some of the electrons move to higher quantum states, resulting in an increased total energy of the system. Associated with the increased energy is a repulsive force between the atoms. The magnitude of this repulsive force varies with $1/r^{m+1}$ (where $m$ is between 6 and 10 for alkali halides with NaCl structure), and thus is negligible at large distances but increases rapidly when the distance is less than 0.5 nm [1]. The repulsive force contributes to the energy of the system by $1/r^m$. There exists a minimum energy or equilibrium position of the system when all the repulsive and attractive forces balance each other. Readers are reminded about the similar discussion in Sect. 4.2.4 on the intermolecular force and potential; see Eq. (4.51) and Fig. 4.8. Understanding the binding energy or the interatomic potential is very important for atomic scale simulations, e.g., those using molecular dynamics.

*Covalent bonds* are formed between nonmetallic elements when the electrons in the outermost orbitals are shared by more than one atom. Covalent binding is important for gaseous molecules like $Cl_2$, $N_2$, and $CO_2$. When the atoms are brought close enough, the electron orbitals overlap, allowing them to share one or more electrons. Covalent interactions result in attractive forces, and the binding of atoms is associated with a reduced total energy. *Covalent crystals* consist of an infinite network of atoms joined together by covalent bonds. Examples are diamond, silicon, SiC, and quartz ($SiO_2$). The whole crystal is better viewed as a large molecule or supermolecule. In diamond structure, each atom is bonded to four neighboring atoms, which form a tetrahedron. In a SiC crystal, each silicon atom is bonded to four carbon atoms and vice versa. In a $SiO_2$ crystal, while each silicon atom is bonded to four oxygen atoms at tetrahedral angles, each oxygen atom is bonded only to two silicon atoms. Covalent solids are usually very hard with a high melting point and thermal conductivity. The melting points of quartz and silicon are 1920 K and 1690 K, respectively. Diamond has the highest melting point (3820 K) among all known materials. At room temperature, the thermal conductivity of diamond is 2300 W/m K, which is the highest of all known bulk materials. Pure diamond and intrinsic silicon do not absorb radiation at frequencies lower than that of the corresponding bandgap energy. Because of its wide bandgap, diamond is clear in the visible region and transparent throughout the whole infrared and microwave regions.

Some solids have both ionic and covalent characteristics. Examples are the III–V semiconductors such as GaN, GaAs, and InSb. II–VI materials such as ZnO and CdS have a large proportion (30%) of covalent bond characteristics. Even SiC has some ionic bond characteristics because of the dipoles formed due to different attractive forces by different atoms. Therefore, SiC is also a polar material that can absorb and emit infrared radiation through lattice vibrations.

Inert gases can be solidified at very low temperatures via *molecular bonds*. At atmospheric pressure, argon becomes liquid at temperatures between 84 and 87 K.

At temperatures below 84 K, it crystallizes into a dense solid, called a *molecular crystal*. Van der Waals' force caused by induced dipole moments between atoms is responsible for the attraction and binding of atoms. The van der Waals weak interaction gives a long-range potential that is proportional to $-1/r^6$, as discussed in Sect. 4.3.1. The repulsive potential for inert gas is proportional to $1/r^{12}$. Molecular bonds are also important for many organic molecules.

Hydrogen has only one electron per atom and can form a covalent bond with another to form $H_2$ molecule. When interacting with other atoms, a hydrogen atom may be attracted to form a *hydrogen bond*. The hydrogen bond and the resulting electrostatic attraction are important for $H_2O$ molecules, with many striking physical properties in its vapor, water, and ice phases. Hydrogen bonds and molecular bonds are essential to organic molecules and polymers.

*Metallic bonds* are responsible for the binding energy in metals. Pure metals can form densely packed periodic lattices or crystals. Metals often exist in *polycrystalline* form in which small grains of crystals are joined together randomly, or in *alloy* form in which the atoms are arranged irregularly like an amorphous insulator. Unlike in a covalent crystal where atoms share a few electrons, in a metallic crystal, some valence electrons leave the ion cores completely and are shared by all the ions in the crystal. This is consistent with the picture of free-electron gas and describes well the behavior of alkali metals. Transition metals, including the noble metals, contain electrons in the $d$ subshell. The metallic bonds are supplemented by covalent and molecular bonds. Due to the relatively free-electron gas, metals have high thermal and electrical conductivities. Metallic crystals are also more flexible than nonmetallic crystals, which are usually brittle. The melting points of metals vary significantly. Examples are Hg (234 K), Ga (303 K), Au (1338 K), and W (3695 K). As mentioned in previous chapters, the physical properties would change significantly as the structure is reduced down to hundreds, tens, or even a few atomic layers in one, two, or three dimensions. Examples are carbon nanotubes, silicon nanowires, ZnO nanobelts, and CdSe-ZnS quantum dots. In order to further understand the properties of solids, let us examine the crystal structures more closely in the following section.

## 6.3 Crystal Structures

A crystal is constructed by the continuous repetition in space of an identical structural unit. Geometrically speaking, a crystal is a 3D periodic array, or network, of lattices. All lattice points are identical to one another. For a crystal made of only one type of element, each lattice point may be treated as a single atom or ion. However, this is not necessary as will be illustrated later. In general, each lattice point represents a set of atoms, ions, or molecules, located in its neighborhood. This set of atoms, ions, or molecules is called a *basis*. A *unit cell* of a crystal structure contains both the lattice and the basis, and can be repeated by translations to cover the whole crystal.

It has long been hypothesized that crystalline materials must have some periodicity in their microstructures. In 1913, W. L. Bragg and his father W. H. Bragg used x-rays to provide microscopic evidence of the existence of periodic lattice structures. This was a giant step because the distances between atoms are on the order of 0.1 nm. X-ray crystallography provided a powerful tool for the determination of the microscopic structure of solids. The Braggs received the Nobel Prize in Physics in 1915, when Lawrence Bragg was only 25 years old. It was not until 1983 that atomic images were obtained in real space using a scanning tunneling microscope (STM) as discussed in Chap. 1. The physical properties of crystalline solids are largely determined by the arrangement of atoms in a unit cell, in addition to the chemical bonds between atoms. It is of great importance to know the structure of a crystal first in order to understand its electrical, thermal, mechanical, and optical properties.

### *6.3.1 The Bravais Lattices*

In three dimensions, crystal lattices can be grouped into 14 different types as required by translational symmetry. These are called Bravais lattices, named after French physicist Auguste Bravais (1811–1863), who showed that there are only 14 unique Bravais lattices from the point of view of symmetry. Bravais lattices are then categorized into seven crystal systems, resulting in seven types of *conventional unit cells*, namely, cubic, tetragonal, orthorhombic, hexagonal, rhombohedral, monoclinic, and triclinic, as illustrated in Fig. 6.3.

There are three cubic lattices: the simple cubic with lattice points only on its apexes, the body-centered cubic (bcc) with one additional lattice at the center, and the face-centered cubic (fcc) with one additional lattice at each face, as shown in Figs. 6.3a1, a2, and a3. To illustrate the difference between bcc and fcc lattices clearly, Fig. 6.4 displays the top views of these two structures with the same $a$, which is called the *lattice constant*. Some practical examples will be given soon. If one looks at Fig. 6.4b along the diagonals, the face-centered structure becomes body-centered. However, the lattice constant would become $a/\sqrt{2}$ along the lateral directions but remains $a$ in the vertical direction. Such a structure is a special case of the tetragonal, because one side is not the same as the other two. There are two tetragonal Bravais lattices, the simple and the body-centered, because a face-centered tetragonal lattice can simply be rotated by 45° to become a body-centered one. A tetragonal lattice can be thought of as a cubic lattice stretched in one direction.

In the orthorhombic lattices shown in Fig. 6.3c, the three lattice constants, $a$, $b$, and $c$, are not equal to each other. Besides the simple, body-centered, and face-centered orthorhombic lattices, there exists a base-centered lattice structure, in which two additional lattices are placed at the center of the top and bottom faces. An orthorhombic lattice can be thought of as a corresponding tetragonal lattice stretched along one side of its square. To produce the additional two, one can simply rotate the tetragonal by 45° and then stretch it.

A hexagonal lattice contains equal triangular or honeycomb-layered structures (see Fig. 6.3d). The next three types of Bravais lattices have inclined faces (see Fig. 6.3e–g). The rhombohedral (or trigonal) has equal sides, whereas the triclinic

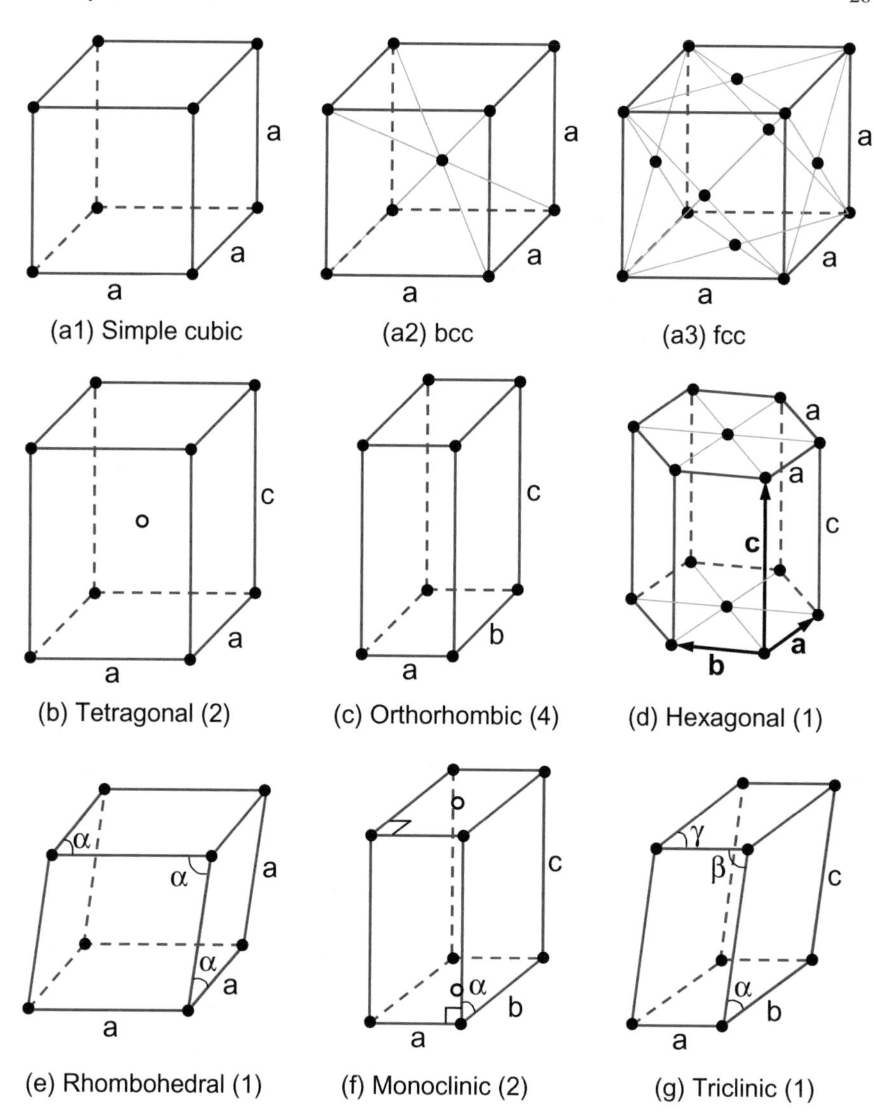


**Fig. 6.3** The seven crystal systems with a total of 14 Bravais lattices, where each point is called a lattice point. The number in parentheses refers to the number of Bravais lattices in the crystal system. **a** Three types of the cubic: simple cubic, body-centered cubic (bcc), and face-centered cubic (fcc). **b** Tetragonal: either simple or body-centered as represented by the empty circle at the center. **c** Orthorhombic: simple, body-centered, face-centered, or base-centered. d Hexagonal. **e** Rhombohedral (also trigonal). **f** Monoclinic: simple or base-centered as represented by the empty circles on the opposite faces. **g** Triclinic

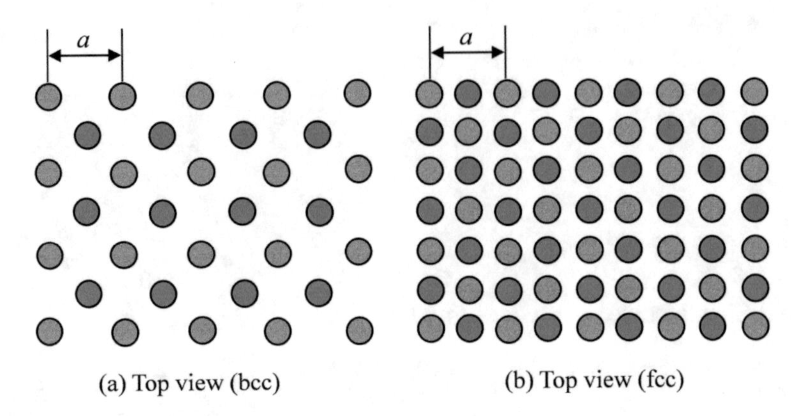


**Fig. 6.4** Top views of **a** body-centered cubic and **b** face-centered cubic Bravais lattices. The two different filling patterns (hatched and shaded) represent lattice points on alternative layers as in Fig. 6.3a2, a3

has three different sides and angles. Both contain six parallelogram faces. The monoclinic, on the other hand, has four rectangular faces and two parallelogram faces.

**Example 6.2** Copper is an fcc lattice. Estimate the lattice constant and the distance between nearest copper atoms (or ion cores to be exact) from the density and the molecular weight of copper.

**Solution** From Table 5.1, we have for Cu that $\rho = 8930\ \text{kg/m}^3$ and $M = 63.5$ kg/kmol. The number density of Cu atoms is $n = \rho N_A/M = 8.47 \times 10^{28}\ \text{m}^{-3}$. If the structure were simple cubic, we would easily find that $a = n^{-1/3} = 0.228$ nm, which would also be the closest distance between atoms. For an fcc lattice, there are eight corner points and six face points. If each lattice point is made of one atom, each corner point contains one-octant of an atom and each face point contains half of an atom inside the cube. Therefore, there are four atoms inside each fcc unit cell. The number of unit cells per unit volume becomes $n/4$ and the calculated lattice constant is $a = 0.361$ nm for Cu. The closest distance between atoms is $a/\sqrt{2} = 0.256$ nm. If we assume that all the atoms are rigid spheres that are closely packed (touching one another), then we can calculate the packing density or the fraction of occupied space. Assume that the diameter of an atom is $d$. For a simple cubic lattice, $a = d$ and there is only one atom per lattice. Hence, $f = (1/6)\pi d^3/a^3 = 0.52$. For an fcc lattice, $a = d\sqrt{2}$ and $f = 4(1/6)\pi d^3/a^3 = 0.74$. What is the packing density for a bcc lattice then?

Some solids with bcc or fcc lattices are listed in Table 6.1, along with others that form a hexagonal close-packed (hcp) lattice. An hcp lattice can be considered as two Bravais hexagonal lattices that are interlocked at $c/2$. Each lattice point is surrounded by, at equal distances, 12 neighboring points: 3 above, 3 below, and 6 at the same height. Imagine that atoms are rigid spheres with a diameter $d$; it can

**Table 6.1** Crystal structures and lattice constants of common elements [1, 2]. Room temperature values unless otherwise indicated. Note that 1 Å = 0.1 nm

| fcc | | bcc | | hcp | | |
|---|---|---|---|---|---|---|
| Element | $a$ (Å) | Element | $a$ (Å) | Element | $a$ (Å) | $c$ (Å) |
| Ar (4.2 K) | 5.26 | Ba | 5.02 | H (4 K) | 3.75 | 6.12 |
| Ag | 4.09 | Cr | 2.88 | Be | 2.27 | 3.59 |
| Al | 4.05 | Cs (78 K) | 6.05 | Cd | 2.98 | 5.62 |
| Au | 4.08 | Fe | 2.87 | Er | 3.56 | 5.59 |
| Ca | 5.58 | K (5 K) | 5.23 | Gd | 3.64 | 5.78 |
| Ce | 5.16 | Li (78 K) | 3.49 | Mg | 3.21 | 5.21 |
| Cu | 3.61 | Mo | 3.15 | Ti | 2.95 | 4.69 |
| Pb | 4.95 | Na (5 K) | 4.23 | Tl | 3.46 | 5.53 |
| Pd | 3.89 | Nb | 3.30 | Y | 3.65 | 5.73 |
| Pt | 3.92 | V | 3.03 | Zn | 2.66 | 4.95 |
| Yb | 5.49 | W | 3.16 | Zr | 3.23 | 5.15 |

be shown that $a = d$ and $c = d\sqrt{8/3}$ for an hcp lattice. Each sphere is in contact with 12 others. It can be seen from Table 6.1 that these hcp crystals follow the ratio $c/a = \sqrt{8/3} \approx 1.633$ within $\pm 16\%$.

### 6.3.2 *Primitive Vectors and the Primitive Unit Cell*

A set of *primitive vectors* can be defined for Bravais lattices **a**, **b**, and **c** so that the vector between any two lattice points can be expressed by the *lattice translation vector* (or operator)

$$\mathbf{R} = m\mathbf{a} + n\mathbf{b} + l\mathbf{c} \tag{6.4}$$

where $m$, $n$, and $l$ are integers. For a simple cubic lattice, we can simply assign $\mathbf{a} = a\hat{\mathbf{x}}$, $\mathbf{b} = a\hat{\mathbf{y}}$, $\mathbf{c} = a\hat{\mathbf{z}}$, as can be seen from Fig. 6.3$a$1. However, the assignment of primitive vectors is not unique. The parallelepiped formed by the three vectors is called a *primitive unit cell*, whose volume $V_{\text{uc}} = \mathbf{a} \times \mathbf{b} \cdot \mathbf{c}$ remains the same no matter how the primitive vectors are chosen. Taking the bcc lattice as an example, we may choose the primitive vectors as either

$$\mathbf{a} = a\hat{\mathbf{x}}, \mathbf{b} = a\hat{\mathbf{y}}, \mathbf{c} = 0.5a(\hat{\mathbf{x}} + \hat{y} + \hat{z}) \tag{6.5a}$$

or

$$\mathbf{a} = 0.5a(-\hat{x} + \hat{\mathbf{y}} + \hat{z}), \mathbf{b} = 0.5a(\hat{\mathbf{x}} - \hat{y} + \hat{z}), \mathbf{c} = 0.5a(\hat{\mathbf{x}} + \hat{y} - \hat{z}) \tag{6.5b}$$

From Eq. (6.5b), we see that $\mathbf{a}+\mathbf{b}+\mathbf{c}$ points to the center point and $\mathbf{a}+\mathbf{b}=a\hat{\mathbf{z}}$. Either way, we end up with $V_{\text{uc}} = 0.5a^3$, suggesting that the Bravais bcc lattice is not a primitive cell. In fact, only the simple Bravais lattices are primitive unit cells. Of course, there are other ways of choosing the primitive vectors. For a Bravais fcc lattice, we can write

$$\mathbf{a} = 0.5a(\hat{\mathbf{y}} + \hat{z}), \mathbf{b} = 0.5a(\hat{\mathbf{x}} + \hat{z}), \mathbf{c} = 0.5a(\hat{\mathbf{x}} + \hat{y}) \tag{6.6}$$

Each vector conveniently ends at the three face-centered points. The total volume of the primitive cell becomes $V_{\text{uc}} = 0.25a^3$, as expected. Each lattice point is associated with a basis of atoms whose locations relative to the lattice point can be specified by $\mathbf{r}_j = x_j\mathbf{a} + y_j\mathbf{b} + z_j\mathbf{c}$ with $0 \le x_j, y_j, z_j \le 1$ for the $j$th atom.

Another way of choosing a unit cell is to follow the two steps: (1) Draw lines to connect a given lattice point to all nearby lattice points. (2) At the midpoint and normal to these lines draw new lines or planes. The smallest volume enclosed in this way is called the *Wigner–Seitz primitive cell*, as illustrated in Fig. 6.5. The Wigner–Seitz cell for a 2D lattice becomes a hexagon whose opposite sides are parallel, and that for an fcc lattice is a rhombic dodecahedron. The longer diagonal of each rhombic face is $\sqrt{2}$ times that of the shorter diagonal. There are six apexes where four surfaces meet and eight apexes where three surfaces meet. The distance between opposite axes joined by four faces is exactly the Bravais lattice constant $a$. The axes, $x$, $y$, and $z$, pass through these six apexes as well as the center. Each Wigner–Seitz cell contains only one lattice point, and it has been proven to be a primitive cell.

It is convenient to describe the orientation of the crystal plane by the Miller indices, which are three integers $h$, $k$, and $l$, without common factors, and denoted by ($hkl$). These numbers give a vector $h\mathbf{a} + k\mathbf{b} + l\mathbf{c}$ that is perpendicular to the plane. For example, if **a**, **b**, and **c** are along the $x$-, $y$-, and $z$-axes, respectively, the six surfaces of the cubic unit cell are represented by (001), $(00\bar{1})$, (010), $(0\bar{1}0)$, (100), and $(\bar{1}00)$,

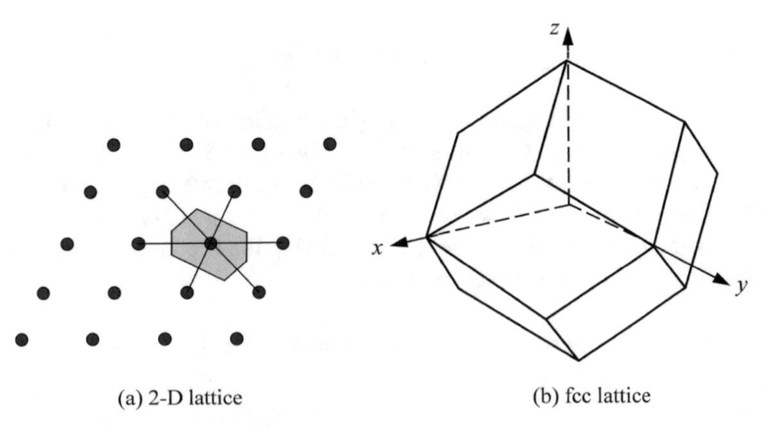


**Fig. 6.5** The Wigner–Seitz cells: **a** For a 2D lattice as shown by the shaded region and **b** for an fcc lattice as shown by the rhombic dodecahedron

where negative sign is denoted by a bar on top of the number. The whole set of surfaces can be denoted by $\{100\}$ due to symmetry. Most commercial semiconductor wafers are (100) oriented and some (111). The way to find the smallest $h, k, l$ of any specified crystal facet is first to extend the plane so that it intersects the axes formed by the lattice vectors. Find the intercepts on each axis in terms of multiples of the unit cell vector, e.g., (2, 4, −6); the numbers must be integers for any specified crystal plane. Take the reciprocals of these numbers, which are $(\frac{1}{2}, \frac{1}{4}, -\frac{1}{6})$. Multiply them by the least common multiple, which is 12 in this example. Then put into the Miller indices $(6, 3, \bar{2})$. All parallel planes are characterized by the same set of Miller indices.

**Example 6.3** Find all angles between the (100), (111), and (311) surfaces in a cubic lattice.

**Solution** For two vectors **a** and **b**, $\mathbf{a} \cdot \mathbf{b} = ab \cos\alpha = x_a x_b + y_a y_b + z_a z_b$. Thus, the angle between (100) and (111) planes is $\alpha = \cos^{-1}\left(1/\sqrt{3}\right) = 54.7°$; that between (100) and (311) planes is $\alpha = \cos^{-1}\left(3/\sqrt{11}\right) = 25.2°$; and that between (111) and (311) planes is $\alpha = \cos^{-1}\left(\frac{3+1+1}{\sqrt{11\times 3}}\right) = 29.5°$.

### 6.3.3 *Basis Made of Two or More Atoms*

With respect to the primitive vector and basis, a bcc lattice can be thought of as a simple cubic with a basis made of two atoms, one at (0, 0, 0) and the other at $a(\frac{1}{2}, \frac{1}{2}, \frac{1}{2})$. Each of the eight lattice points contains the same basis by translation, according to Eq. (6.4), and the unit vectors along the three orthogonal sides of the cubic. The simple cubic lattice having a basis of two atoms, however, breaks some of the symmetry of the Bravais cubic lattice and is called a non-Bravais lattice. Lattices with a basis consisting of more than one atom have important practical applications as discussed in the following. The cesium chloride structure is made of two types of elements, each forming a simple Bravais lattice, as shown in Fig. 6.6a. The two Bravais lattices can be thought of as being placed in identical positions first, and then one is moved by $a\left(\frac{1}{2}, \frac{1}{2}, \frac{1}{2}\right)$ so that the point at the origin is translated to the center of the other. It is not a body-centered cubic lattice. Rather, the crystal structure can be viewed as a simple cubic with a base of two ions, Cs at (0, 0, 0) and Cl at $a\left(\frac{1}{2}, \frac{1}{2}, \frac{1}{2}\right)$. The sodium chloride structure is more common. In this case, it can be considered as two fcc lattices made of different ions. The two fcc lattices are then translated exactly the same way as in the CsCl structure. The resulting structure is shown in Fig. 6.6b, where each ion is surrounded by six ions of the other type. The lattice constants of some common crystals are listed in Table 6.2. It can be seen that most ionic crystals form NaCl or CsCl structures.

The crystal structures of diamond and zincblende semiconductors are also derivatives of the cubic structure. The zincblende structure is formed from two fcc lattices

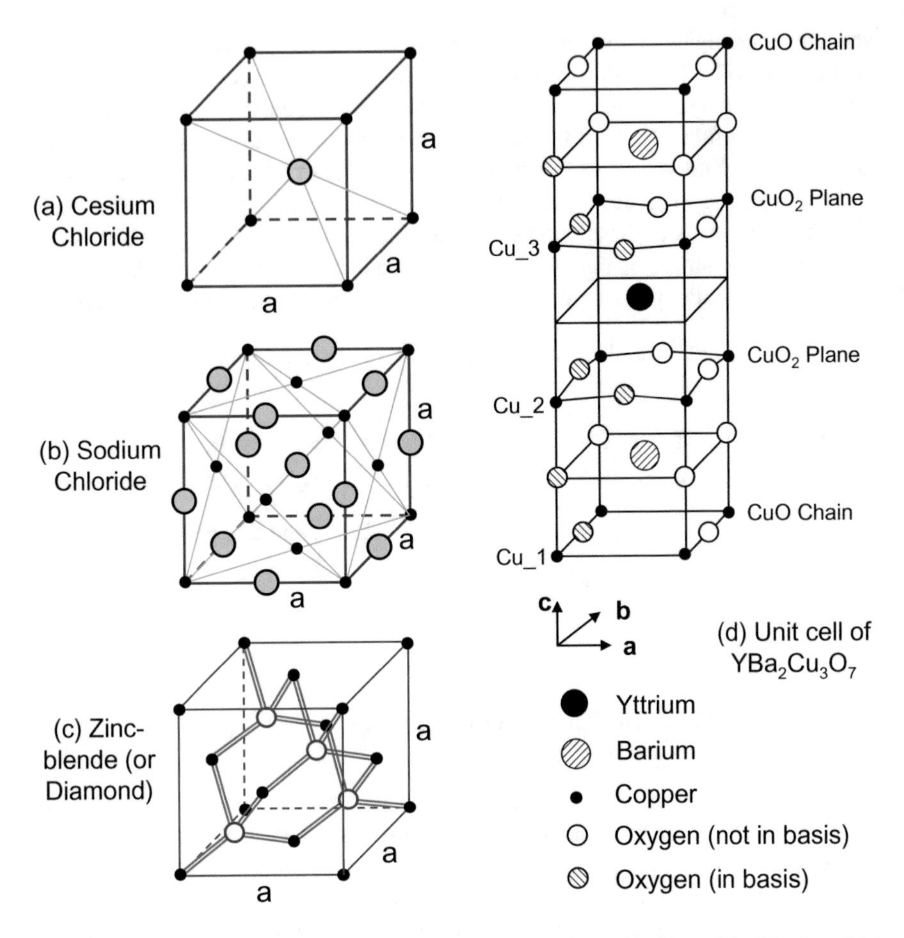


**Fig. 6.6** Crystalline structures. **a** Cesium chloride; **b** Sodium chloride. **c** Zincblende, which becomes a diamond structure when the atoms in the empty circles are the same as the filled ones. **d** $YBa_2Cu_3O_7$ superconductor whose lattice constants are approximately $a = 0.38$, $b = 0.39$, and $c = 1.17$ nm

with different types of atoms, displaced along the body diagonal by one-quarter the length of the diagonal. Specifically, the basis is made of one atom at (0, 0, 0) and the other atom at $a\left(\frac{1}{4}, \frac{1}{4}, \frac{1}{4}\right)$, as shown in Fig. 6.6c. A total of four atoms are moved completely inside the cube, and each atom has a covalent bond with each of the four adjacent atoms, which together form a tetrahedron. Examples of zincblende structure are GaAs, SiC, and so forth. A diamond structure can be viewed as a special case of a zincblende structure for which there is only one type of element, such as C, Si, or Ge. The outermost subshell of Si is $3s^2 3p^2$, and the $s$ subshell is filled. By promoting an $s$-electron to a $p$-orbital to form $sp^3$ hybrid orbitals, four covalent bonds can be formed. This is also true for C and Ge. In essence, the diamond lattice can be thought of as an fcc lattice with a basis containing two identical atoms: one is on

**Table 6.2** Crystal properties of some compounds and semiconductors at room temperature [1, 2]. For semiconductors, the bandgap energy is indicated, and "i" in parentheses denotes an indirect bandgap

| Compound | | Semiconductors | | |
|---|---|---|---|---|
| Composition | $a$ (Å) | Composition | $a$ (Å) | $E_g$ (eV) |
| Sodium chloride structure | | Diamond structure | | |
| LiF | 4.02 | C | 3.57 | 5.47 (i) |
| LiCl | 5.13 | Si | 5.43 | 1.11 (i) |
| NaBr | 5.97 | Ge | 5.66 | 0.66 (i) |
| NaCl | 5.64 | | | |
| KBr | 6.60 | Zincblende structure | | |
| KCl | 6.29 | BN | 3.62 | 7.5 (i) |
| CsF | 6.01 | CdS | 5.82 | 2.42 |
| AgCl | 5.55 | CdSe | 6.05 | 1.70 |
| AgBr | 5.77 | CdTe | 6.48 | 1.56 |
| MgO | 4.21 | GaAs | 5.65 | 1.42 |
| MgS | 5.20 | GaN (w) | 5.45 | 3.36 |
| CaO | 4.81 | GaP | 5.45 | 2.26 (i) |
| CaS | 5.69 | GaSb | 6.43 | 0.72 |
| CaSe | 5.91 | HgTe | 6.04 | <0 |
| BaTe | 6.99 | InAs | 5.87 | 0.36 |
| | | InP | 6.48 | 1.35 |
| Cesium chloride structure | | InSb | 4.35 | 0.17 |
| CsCl | 4.12 | SiC | 4.63 | 2.36 |
| CsBr | 4.29 | ZnO | 5.41 | 3.35 |
| CsI | 4.57 | ZnS | 5.67 | 3.68 |
| TlBr | 3.97 | ZnSe | 6.09 | 2.58 |

the corner and the other on the body diagonal at a distance of one-quarter diagonal. Table 6.2 also presents commonly used diamond and zincblende semiconductors with associated lattice constants and bandgap energies. Notice that GaN crystal is wurtzite in its stable form with a hexagonal symmetry. This is also the case for AlN and InN, which are not shown in the table. The III-nitride materials have a wide-band, and thus are important for UV-blue-green LEDs and lasers. On the other hand, ZnS, ZnO, CdS, and CdSe can also be wurtzite. HgTe is a semimetal with a negative bandgap and can be mixed with the wideband semiconductor CdTe to form the ternary compound of $Hg_{1-x}Cd_xTe$, which can be used as infrared detectors, namely, mercury–cadmium–telluride (MCT) detectors.

Yttrium–barium–copper oxide ($YBa_2Cu_3O_7$) is a high-temperature superconductor, which becomes superconducting at temperatures below 91 K [12]. It belongs to the cuprate-perovskite family and is a ceramic material when one oxygen atom is removed from the unit cell to form $YBa_2Cu_3O_6$ [13]. The crystal structure of $YBa_2Cu_3O_7$ is a simple orthorhombic lattice, whose basis contains 13 atoms, as shown in Fig. 6.6d. The structure is very close to a tetragonal one since $a \approx b$. The properties of $YBa_2Cu_3O_7$ are highly anisotropic in the $c$-axis direction. Superconductivity is found in the $a$–$b$ plane, which is presumed due to the $CuO_2$ planes above and below the yttrium atom. Other examples of Bravais lattices include As, Sb, and Bi with rhombohedral lattices; In and Sn with tetragonal lattices; and Ga, Cl, Br, and S (rhombic) with orthorhombic lattices [1].

Graphite is a form of carbon made of layered structures as shown in Fig. 6.7. When separated from others, each individual layer or sheet is called a graphene. In the graphite structure, each carbon atom is covalently bonded to three others in the plane and loosely bonded between planes. There are relatively free electrons, and hence graphite is a conductor along the plane. The layer of graphite has honeycomb shape, and at first sight, it may be difficult to link it with the arrays of triangles in the hexagonal lattice. It becomes more obvious, however, if a basis is chosen to contain two atoms so that a hexagon with all diagonals can be seen by the dashed lines in Fig. 6.7a. In this way, graphite can be considered as a hexagonal Bravais lattice. The primitive unit cell of graphite is a rhombic prism (with six surfaces) formed using three layers, as illustrated by the dashed lines in Fig. 6.7b. Each unit cell contains a total of four carbon atoms.

The structure of carbon nanotubes (CNTs) can be understood based on the graphene structure and the chiral vector,

$$\mathbf{C}_\mathrm{h} = m\mathbf{a}_1 + n\mathbf{a}_2 \tag{6.7}$$

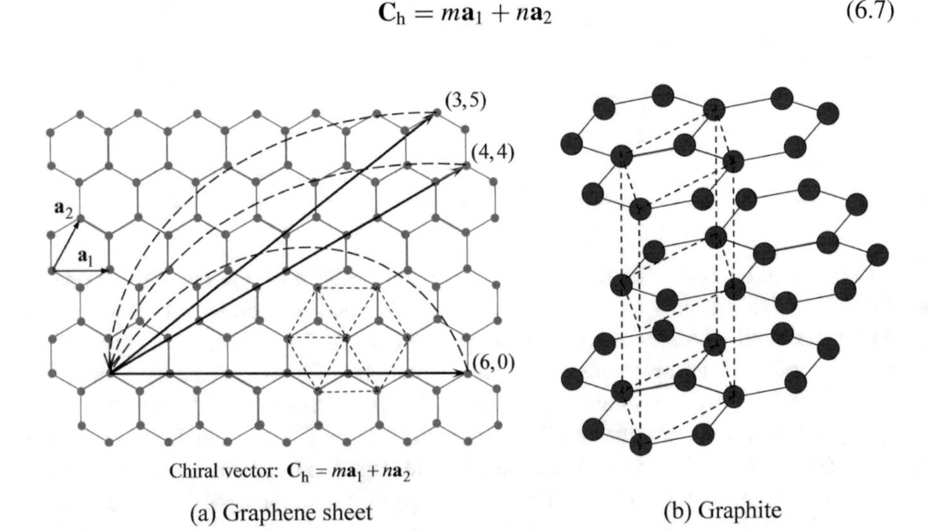


**Fig. 6.7** Crystal structures of **a** graphene layer and **b** graphite. Carbon nanotubes can be viewed as rolling a graphene sheet in a direction perpendicular to the chiral vector

Different CNTs are based on rolling in the chiral vector so that the axis is perpendicular to the chiral vector and the magnitude of the chiral vector becomes the perimeter of the tube. The diameter of the tube becomes

$$d_t = \frac{C_h}{\pi} = \frac{a_{C-C}}{\pi}\sqrt{3(m^2 + mn + n^2)} \tag{6.8}$$

where $a_{C-C} = 0.1421$ nm is the nearest distance between the carbon atoms in graphene [14]. Notice that the chiral vector has a magnitude $a = a_{C-C}\sqrt{3} = 0.246$ nm. In calculating the cross-sectional surface area of a single-walled nanotube (SWNT), one could use $a$ as the wall thickness and obtain

$$A_c = \pi d_t a = 3(a_{C-C})^2\sqrt{(m^2 + mn + n^2)} \tag{6.9}$$

Take the (20, 20) SWNT as an example, we have $d_t = 2.7$ nm and $A_c = 2.1\ \text{nm}^2$. Some researchers suggested using a layer thickness equal to the separation of graphite as 0.335 nm, which gives $A_c = \pi d_t * 0.335 = 2.9\ \text{nm}^2$. Note that $A_c = \pi D^2/4 = 5.8\ \text{nm}^2$.

## 6.4 Electronic Band Structures

The behavior of electrons in solid is complicated because the solution of wave functions involves a rather complicated many-body problem. Electrons in solids can be thought of as in a periodic potential due to the periodic arrays of atoms. Electronic band structures are functions that describe the electron states in the energy versus wavevector space. Let us first look at the reciprocal lattice in three dimensions.

### *6.4.1 Reciprocal Lattices and the First Brillouin Zone*

The reciprocal lattice of a crystal structure is defined in the **k**-space (wavevector space). Since a crystal is a periodic array of lattices in real space, the reciprocal lattice can be obtained by performing a spatial Fourier transform of the crystal. For a simple orthorhombic lattice with the primitive vectors $\mathbf{a} = a\hat{\mathbf{x}}$, $\mathbf{b} = b\hat{\mathbf{y}}$, and $\mathbf{c} = c\hat{\mathbf{z}}$, the reciprocal lattice can be defined by the three vectors $\mathrm{A} = \frac{2\pi}{a}\hat{\mathbf{x}}$, $\mathbf{B} = \frac{2\pi}{b}\hat{\mathbf{y}}$, and $\mathbf{C} = \frac{2\pi}{c}\hat{\mathbf{z}}$, which define another orthorhombic. The product of the volumes of the unit lattice and the reciprocal lattice is $8\pi^3$. Some of this aspect was discussed in Chap. 5. In general, the reciprocal primitive vectors can be generated by

$$\mathrm{A} = 2\pi\frac{\mathbf{b}\times\mathbf{c}}{\mathbf{a}\cdot(\mathbf{b}\times\mathbf{c})};\ \mathbf{B} = 2\pi\frac{\mathbf{c}\times\mathbf{a}}{\mathbf{a}\cdot(\mathbf{b}\times\mathbf{c})};\ \mathbf{C} = 2\pi\frac{\mathbf{a}\times\mathbf{b}}{\mathbf{a}\cdot(\mathbf{b}\times\mathbf{c})} \tag{6.10}$$

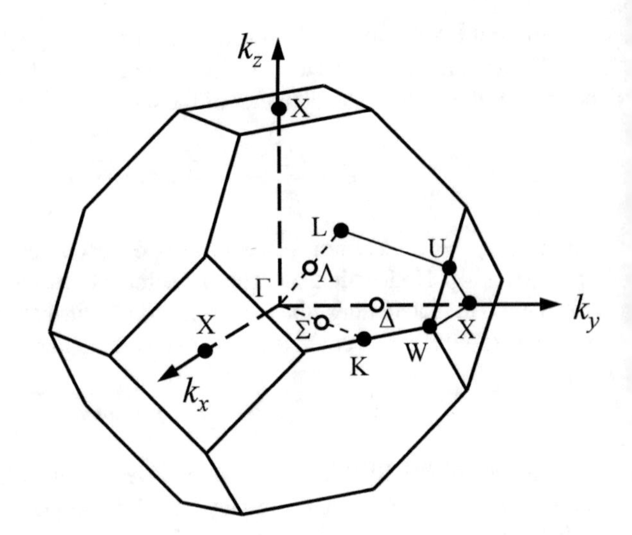


**Fig. 6.8** The first Brillouin zone of a face-centered cubic structure. The shape is a truncated octahedron with eight hexagons and six squares. This is also the Wigner–Seitz cell for a bcc lattice, whose first Brillouin zone has the same shape as the Wigner–Seitz cell for an fcc lattice shown in Fig. 6.5b

In solid-state physics, a Brillouin zone is defined as a Wigner–Seitz cell in the reciprocal lattice and the smallest of which is called the first Brillouin zone. The definition of the Brillouin zone gives a vivid geometric interpretation of the Bragg diffraction condition and thus is of importance in the study of electron and phonon states in crystals, as well as their interactions with electromagnetic waves. Figure 6.8 shows the first Brillouin zone of a face-centered cubic lattice. The directions $k_x$, $k_y$, and $k_z$ are called the [100], [010], and [001] directions, respectively. The center of the Brillouin zone is called the Γ-point, and the intersection of the three axes with the zone edge is called the X-point. The *body diagonal*, or the [111] direction, meets the zone edge at the L-point. Other points of interest such as K, W, and U at the zone edges and Δ, Λ, and Σ, located halfway between the zone center and an edge, can also be defined.

## 6.4.2 Bloch's Theorem

The total potential in a crystal includes the core–core, electron–electron, and electron–core Coulomb interactions. For solving electron wave functions subjected to such a potential, one would have to deal with a many-body problem, which turned out to be very difficult in mathematics. However, this problem can be simplified using the so-called *nearly free-electron model*, in which each electron moves in the average field created by the other electrons and ions. This is also called the one-electron model. The Hamiltonian operator $H$ for the one-electron model is given as

$$H = \frac{p_{\mathrm{e}}^2}{2m_{\mathrm{e}}} + U(\mathbf{r}) \tag{6.11}$$

where $p_e$ and $m_e$ are the momentum and the mass of the electron, respectively, and $U(\mathbf{r})$ is a periodic potential function resulted from both the electron–electron and electron–core interactions. The one-electron Schrödinger equation is (see Sect. 3.5.1) given as follows:

$$\left[-\frac{\hbar^2}{2m_e}\nabla^2 + U(\mathbf{r})\right]\psi(\mathbf{r}) = E\psi(\mathbf{r}) \tag{6.12}$$

where $E$ is the electron energy and $\psi(\mathbf{r})$ is the electron wave function. The periodicity of the lattice structure yields the boundary condition,

$$U(\mathbf{r}) = U(\mathbf{r} + \mathbf{R}) \tag{6.13}$$

where $\mathbf{R}$ is the vector between two lattice points, called the periodic potential, which ensures that $U(\mathbf{r})$ can be expanded as a Fourier series in terms of the reciprocal lattice vector $\mathbf{G}$ as follows:

$$U(\mathbf{r}) = \sum_{\mathbf{G}} U_{\mathbf{G}} e^{i\mathbf{G}\cdot\mathbf{r}} \tag{6.14}$$

The reciprocal lattice vector can be expressed as $\mathbf{G} = l_1\mathbf{A} + l_2\mathbf{B} + l_3\mathbf{C}$, where $\mathbf{A}$, $\mathbf{B}$, and $\mathbf{C}$ are primitive vectors of the reciprocal lattic e as given in Eq. (6.10), and the integers $l_1, l_2$, and $l_3$ are indices. In Eq. (6.14), $U'_{\mathbf{G}}s$ are complex Fourier expansion coefficients for a given set of $l_1, l_2$, and $l_3$.

According to the Bloch theorem, the wave function of an electron in a periodic potential must have the form:

$$\psi(\mathbf{r}) = e^{i\mathbf{k}\cdot\mathbf{r}} u_{\mathbf{k}}(\mathbf{r}) \tag{6.15}$$

where $u_{\mathbf{k}}(\mathbf{r})$ is a periodic function with the periodicity of the lattice, similar to Eq. (6.13), and thus $\psi(\mathbf{r} + \mathbf{R}) = e^{i\mathbf{k}\cdot\mathbf{R}}\psi(\mathbf{r})$. The wave function $\psi(\mathbf{r})$ can also be expressed as a Fourier series summed over all values of the permitted wavevector such that

$$\psi(\mathbf{r}) = \sum_{\mathbf{k}} C_{\mathbf{k}} e^{i\mathbf{k}\cdot\mathbf{r}} \tag{6.16}$$

The summation is over all wavevectors $\mathbf{k}$'s. From Eq. (6.16), we have

$$\nabla^2\psi(\mathbf{r}) = \sum_{\mathbf{k}} C_k (ik)^2 e^{i\mathbf{k}\cdot\mathbf{r}} = -\sum_{\mathbf{k}} k^2 C_k e^{i\mathbf{k}\cdot\mathbf{r}} \tag{6.17}$$

The combination of Eqs. (6.14) and (6.16) gives

$$U(\mathbf{r})\psi(\mathbf{r}) = \sum_{\mathbf{k}}\sum_{\mathbf{G}} U_{\mathbf{G}} C_{\mathbf{k}} e^{i(\mathbf{k}+\mathbf{G})\cdot\mathbf{r}} \tag{6.18}$$

Using Eqs. (6.16) through (6.18), we can rewrite the Schrödinger equation as follows:

$$\sum_{\mathbf{k}} \frac{\hbar^2 k^2}{2m_e} C_{\mathbf{k}} e^{i\mathbf{k}\cdot\mathbf{r}} + \sum_{\mathbf{k}}\sum_{\mathbf{G}} U_{\mathbf{G}} C_{\mathbf{k}} e^{i(\mathbf{k}+\mathbf{G})\cdot\mathbf{r}} = \sum_{\mathbf{k}} E C_{\mathbf{k}} e^{i\mathbf{k}\cdot\mathbf{r}} \tag{6.19}$$

The coefficients of each Fourier component must be equal on both sides of the equation. Thus,

$$\left(\frac{\hbar^2 k^2}{2m_e} - E\right) C_{\mathbf{k}} + \sum_{\mathbf{G}} U_{\mathbf{G}} C_{\mathbf{k}-\mathbf{G}} = 0 \tag{6.20}$$

where $C_{\mathbf{k}-\mathbf{G}}$ is the coefficient for the term with $\mathbf{k} - \mathbf{G}$ in the exponent, i.e., $\exp[i(\mathbf{k} - \mathbf{G}) \cdot \mathbf{r}]$ in Eq. (6.16). Equation (6.20) is paramount in the electronic band theory of crystals, and it is, therefore, called *the central equation* [2]. When $U(\mathbf{r}) \equiv 0$, Eq. (6.20) reduces to $E_k^0 = \hbar^2 k^2/(2m_e)$ by noting that $\hbar k = p_e$ for free electrons, as used in the Sommerfeld theory. Under the influence of a periodic potential, the relationship becomes more complex because it is a set of linear equations for infinite numbers of coefficients. Because the equation is homogeneous, the determinant of the characteristic matrix must be zero. In some cases, the terms can be significantly reduced to yield simple solutions with insightful physics.

Consider the 1D case when the Fourier components are relatively small compared with the kinetic energy of electrons at the zone boundary. This is the weak-potential assumption. At the first Brillouin zone boundaries, we have

$$k = G/2 = \pi/a \tag{6.21}$$

Because there are only two values of $k$ and $G$, Eq. (6.20) reduces to the following two equations due to symmetry:

$$(E_\mu^0 - E)C_\mu + UC_{-\mu} = 0 \tag{6.22a}$$

and

$$(E_{-\mu}^0 - E)C_{-\mu} + UC_\mu = 0 \tag{6.22b}$$

where $\mu = \frac{1}{2}G$ is introduced merely for the convenience of notation. These equations have solutions only when the determinant becomes zero, i.e.,

$$\begin{vmatrix} E_\mu^0 - E & U \\ U & E_{-\mu}^0 - E \end{vmatrix} = 0 \tag{6.23}$$

Because $E_\mu^0 = E_{-\mu}^0 = \hbar^2\mu^2/(2m_e)$, the two roots are then obtained as

$$E = E_\mu^0 \pm U = \frac{(\hbar\pi/a)^2}{2m_e} \pm U \tag{6.24}$$

The two solutions at the zone edge, i.e., $k = \pi/a$, are actually on two $E(k)$ curves. When $k$ is near the zone edge, we can express the central equation, Eq. (6.20), as the following two equations [1, 2]:

$$(E_k^0 - E)C_k + UC_{k-G} = 0 \tag{6.25a}$$

and

$$(E_{k-G}^0 - E)C_{k-G} + UC_k = 0 \tag{6.25b}$$

By setting its determinant to zero, we obtain

$$E(k) = \tfrac{1}{2}(E_k^0 + E_{k-G}^0) \pm \left[\tfrac{1}{4}(E_k^0 - E_{k-G}^0)^2 + U^2\right]^{1/2} \tag{6.26}$$

which gives two branches near the zone edge, as shown in Fig. 6.9. A bandgap of $2U$ is formed at the first Brillouin zone edge. The corresponding wave functions at the zone edge are

$$\psi_{1,2}(x) = \frac{1}{\sqrt{2L}}\left(e^{i\pi x/a} \pm e^{-i\pi x/a}\right) \tag{6.27a}$$

where $L$ is the length of the crystal. This forms two standing waves:

$$\psi_1(x) = \sqrt{2/L}\cos(\pi x/a) \quad \text{and} \quad \psi_2(x) = i\sqrt{2/L}\sin(\pi x/a) \tag{6.27b}$$

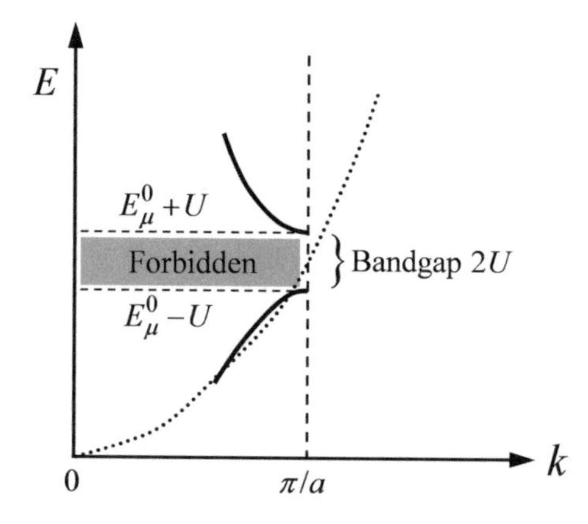


**Fig. 6.9** Illustration of the energy bands, where the solid curves are calculated from Eq. (6.26). The lower and upper bands correspond to the choice of the minus and plus signs, respectively. When $k = G/2 = \pi/a$, the two bands are separated by a gap of magnitude $2U$. The dashed line, on the other hand, represents the free-electron behavior according to $E \propto k^2$

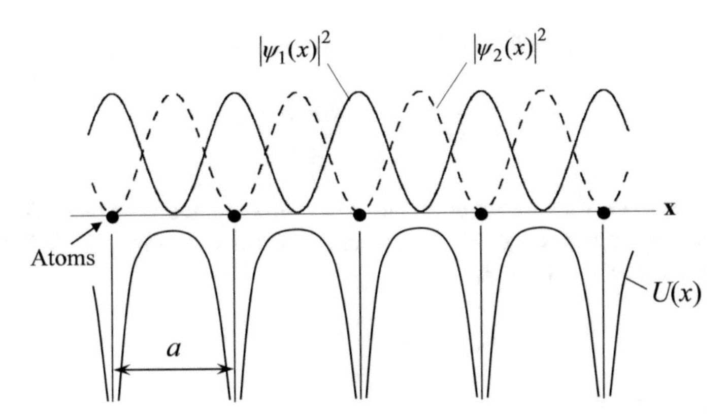


**Fig. 6.10** The upper part of the figure plots the probability density $|\psi|^2$ in a 1D weak potential at the edge of the first Brillouin zone; the lower part of the figure illustrates the actual potential $U(x)$ of electrons

As shown in Fig. 6.10, the lower energy state $E_\mu^0 - U$ corresponds to $\psi_1$ with a probability density $|\psi_1|^2$ peaked at core sites. The probability density function describes electrons that are piled up close to the core site. The upper energy state $E_\mu^0 + U$ corresponds to $\psi_2$ with a probability density $|\psi_2|^2$ that distributes electrons between the cores. The energy difference between these two states is the origin of formation of the gap at the Brillouin zone edge. On the other hand, away from the zone edge, the electron wave functions can be expressed as

$$\psi(x) \approx L^{-1/2} \mathrm{e}^{\pm \mathrm{i}kx} \tag{6.28}$$

which are propagating waves that characterize the wavelike behavior of free electrons [2, 15].

When all the Brillouin zones and their associated Fourier components are included, the result is a set of curves, as those shown in Fig. 6.11a. An easier way to show this is to use the Kronig–Penney model, first formulated in 1931, in which the potential is assumed to be a square-well array [2, 9]. The details are left as an exercise (see Problem 6.12). The allowable bands are illustrated by the solid curves in Fig. 6.11. If the electrons were completely free, then $E(k) = E_k^0 = \hbar^2 k^2/(2m_\mathrm{e})$ would be a parabola, as illustrated by the dashed curve in Fig. 6.11a, without any bandgap. It is useful to plot all the energy levels in the first Brillouin zone. This can be done by folding the branches in Fig. 6.11a, which is known as the *extended-zone scheme*, using the reciprocal lattice vector. The result is shown in Fig. 6.11b, which is called the *reduced-zone scheme* for the representation of the electronic bands.

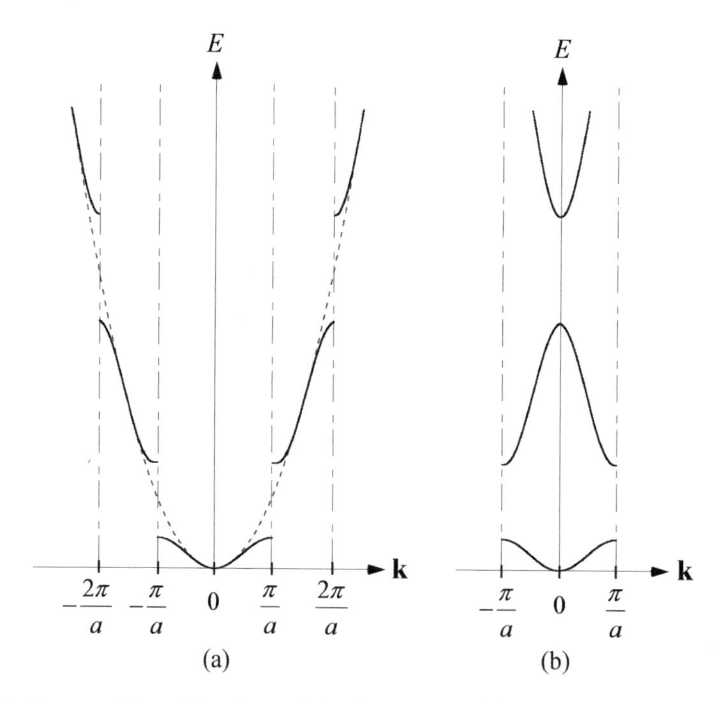


**Fig. 6.11** Representation of the electronic band structure. **a** The extended-zone scheme. **b** The reduced-zone scheme

### 6.4.3 Band Structures of Metals and Semiconductors

The nearly free-electron model described in the previous section assumes a weak potential and cannot predict the behavior of electrons in the inner orbitals or near the nuclei. A simple way to calculate the electronic structure of inner electrons, such as those in the *d* subshells, is the *tight-binding method*, which assumes that the potential is so large that electrons can hardly move out of the ion core. Due to the complicated 3D structure and the multiple numbers of outermost electrons in each atom, the actual electronic band structures are rather complicated. More advanced methods include the augmented plane wave (APW) method, the Korringa–Kohn–Rostoker (KKR) Green function method, and the pseudopotential method. More details can be found from Ashcroft and Mermin [1], Kittel [2], and Omar [15], and references therein.

It can be shown that the number of orbitals in a band in the first Brillouin zone is the same as the number of unit cells in the crystal, *N*. According to the Pauli exclusion principle, the number of electrons that can occupy a band is 2*N*. For copper, the outermost electron configuration is $4s^1 3d^{10}$. The *s*- and *d*-subshell electrons result in six bands (with some overlap), as can be seen from Fig. 6.12, along the direction according to the first Brillouin zone depicted in Fig. 6.8 [16–18]. The *d* bands are from 2 to 5.5 eV below the Fermi level and are completely filled. The *s* band, illustrated by the thicker line segments, is interrupted by the *d* bands. The *s* band is only half filled

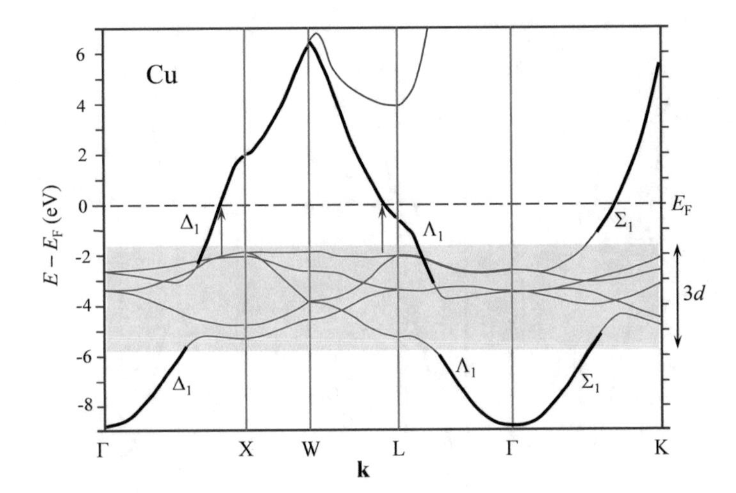


**Fig. 6.12** Calculated energy band structure of copper, *adapted from Refs.* [16–18] *with permission*

and half empty. For alkali metals, there is only one valence electron and the *s* band is continuous. Electrons in the *s* band can be easily excited from below the Fermi level to above the Fermi level within the same band. This explains why copper is a conductor. When radiation is incident on a copper surface, because of the relatively high frequency, free electrons have an inductive characteristic and tend to reflect the radiation. The absorption of photons will cause the electrons in the *s* band to reach a higher level within the same band. If the phonon energy exceeds 2 eV, transition from the top *d* band to the *s* band right above the Fermi level is possible, as indicated by the two arrows in Fig. 6.12. The *interband transitions* result in strong absorption as well as a reduction in reflection of copper at wavelengths shorter than about 0.6 μm. Pure copper has a red-brown color because it does not reflect blue and violet colors. Gold has a similar interband transition that absorbs short-wavelength visible light. On the other hand, for silver, the interband transition occurs at a much shorter wavelength. Thus, silver can reflect light in the whole visible spectrum.

The Fermi surface is anisotropic and not spherical for real crystals. For alkali metals with bcc lattices, such as Na and K, the Fermi surface is nearly spherical lying inside the first Brillouin zone [1]. The Fermi surface of Al is close to the free-electron surface for an fcc lattice with three conduction electrons per atom. For noble metals, due to the effect of *d* bands, the Fermi surface is characterized by a sphere that bulges out in the eight <111> directions.

The electronic band structures of Si and GaAs in the first Brillouin zone are shown in Fig. 6.13, along reciprocal lattice directions [19–21]. Si and GaAs are chosen here because these two types of semiconductors have distinct energy gap features that can represent a wide range of semiconductor materials. Degeneracy causes additional subbands within the conduction and valence bands. *Intraband transitions* refer to the excitation or relaxation of electrons between subbands. For intrinsic semiconductors, the Fermi level lies right in the middle between the bottom of the conduction band

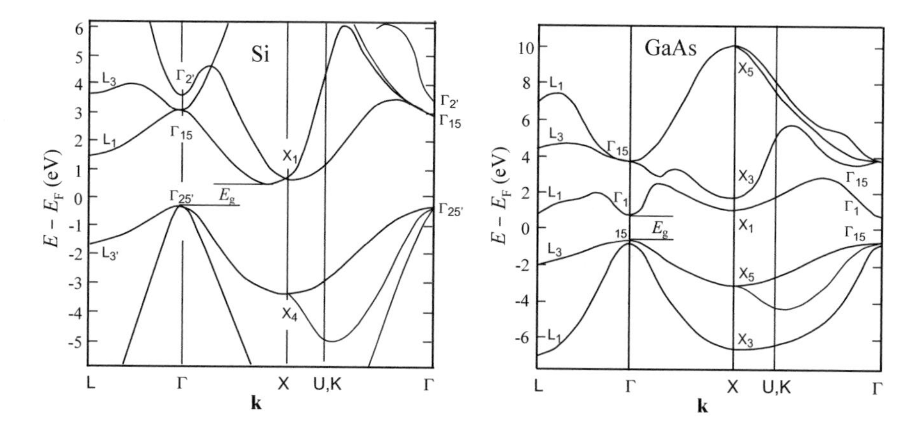


**Fig. 6.13** Calculated energy band structure of silicon (left) and gallium arsenide (right), *adapted from Refs.* [19–21] *with permission*

and the top of the valence band. The valence bands are formed by the bonded valence electrons, and they are completely filled at low temperatures. The electrons in the conduction band are dissociated from the atom and hence become free charges. The bandgap energy, or energy gap, $E_g$ is the difference between the energies at the top of the valence band ($E_V$) and the bottom of the conduction band ($E_C$). The values of $E_g$ for some semiconductors are included in Table 6.2. For Si, as shown in Fig. 6.13a, the bottom of the conduction band and the top of the valence band do not occur at the same $k$. This type of semiconductor is called an *indirect gap* semiconductor. For a *direct gap* semiconductor, such as GaAs, the bottom of the conduction band and the top of the valence band occur at the same value of $k$ at the $\Gamma$-point, as shown in Fig. 6.13b. The mechanism for electron transition between the valence band and the conduction band in a direct gap semiconductor is different from that in an indirect gap semiconductor. Additional discussion about radiation absorption processes will be given in Chap. 8.

At absolute zero temperature, there are no electrons in the conduction band and the valence band is completely filled. When the temperature increases or there exist optical excitations, electrons in the valence band can transit to the conduction band, leaving behind some vacancies in the valence band. The vacancies left in the valence band are called *holes*, which carry opposite charge as electrons. Usually the electrons are found almost exclusively in levels near the conduction band minima, while the holes are found in the neighborhood of the valence band maxima. Therefore, the energy versus wavevector relations for the carriers can generally be approximated by quadratic forms in the neighborhood of such extrema, i.e.,

$$E_e(k) = E_C + \frac{\hbar^2 k^2}{2m_e^*} \quad \text{and} \quad E_h(k) = E_V - \frac{\hbar^2 k^2}{2m_h^*} \tag{6.29}$$

where subscript e and h are for electrons and holes, respectively, $E_C$ is the energy at the bottom of the conduction band and $E_V$ is the energy at the top of the valence band. In the 1D case, the effective mass $m^*$ for electrons and holes is defined as

$$\frac{1}{m_e^*} = \frac{1}{\hbar^2}\frac{d^2E_e}{dk^2} \quad \text{and} \quad \frac{1}{m_h^*} = -\frac{1}{\hbar^2}\frac{d^2E_h}{dk^2} \tag{6.30}$$

where the negative sign is assigned to make the effective mass of the hole positive at the top of the valance band. Effective mass is defined based on the quantum mechanical description of the group velocity and the acceleration of charge carriers, respectively, as

$$v_g = \frac{1}{\hbar}\frac{\partial E}{\partial k} \quad \text{and} \quad a = \frac{dv_g}{dt} = \frac{1}{\hbar}\frac{\partial^2 E}{\partial k^2}\frac{dk}{dt} = \frac{1}{\hbar^2}\frac{\partial^2 E}{\partial k^2}F \tag{6.31}$$

where $F = \frac{dp_e}{dt} = \hbar\frac{dk}{dt}$ is the force exerted on the charge carrier due to an electric field. In 3D case, the effective mass depends also on the direction and is a $3 \times 3$ tensor [15]. Note that the above definition of effective mass is for parabolic bands only according to Eq. (6.29) and hence does not apply to 2D solids such as graphene, to be discussed next.

### 6.4.4 Electronic Properties of Graphene

As a layered 2D material with carbon atoms arranged in a honeycomb lattice, as shown in Fig. 6.7a, graphene has unique electronic, mechanical, thermal, and optical properties. Due to its large carrier mobility and electrical conductivity, along with the feasibility of controlling the carrier density by a gate voltage, graphene is a promising material for the next generation of transistors and 2D flexible nanoelectronics [22, 23]. Graphene can be synthesized chemically (e.g., by chemical vapor deposition on a metal surface) or isolated using mechanical or liquid-phase exfoliation from graphite [24]. As discussed in Chap. 5, the thermal conductivity of graphene can be as high as or even higher than that of diamond [25]. Graphene and related materials also hold promise for energy conversion and storage [26]. In addition, graphene exhibits unique optical and infrared properties [27] for optoelectronics and photonics applications (to be discussed in Chap. 9). Knowledge of the electronic structure of carbon and its related materials is critical for understanding the unique electronic and other properties of graphene.

A carbon atom has six electrons configured as $^6$C: $1s^22s^22p^2$. There are four electrons in the second shell, two in the $s$-orbital and two in $p$-orbitals. However, this is merely the ground-state configuration without excitation. Note that the electron cloud for the $s$-orbital is isotropic or spherical shaped. Each of the three $p$-orbitals shapes like a dumbbell (or the number 8), identified as $p_x$, $p_y$, and $p_z$ with the direction along the orbital axis. In methane ($CH_4$), a carbon atom is bonded to four hydrogen atoms, forming a tetrahedral molecular geometry (referring to Fig. 3.9b). The hydrogen atoms are at the vertices of the regular tetrahedron while the carbon atom is at the centroid. All four bonds between the C and H are equally spaced with equal strength measured by the bond energy. The underlying mechanism can be explained by orbital hybridization, which may be explained in two steps. Firstly, one electron is promoted from the $2s$ orbital to the $2p$ orbital due to excitation, such that each of the orbital in the outmost shell of carbon $(s,\ p_x,\ p_y, \text{and } p_z)$ is occupied by one electron; secondly, the four electron orbitals combine and rearrange themselves so that each contains 25% of $s$, $p_x$, $p_y$, and $p_z$ components. The hybrid orbitals look like asymmetric dumbbell, and each points out in one of the tetrahedral directions to form a chemical bond with an H atom. This is called the $sp^3$ hybridization and each hybrid orbital possesses 25% $s$-orbital and 75% $p$-orbital characteristics. The orbital hybridization theory was originally developed in 1931 by Linus Carl Pauling, who received the Nobel Prize in Chemistry in 1954. The $sp^3$ hybridization is also responsible to the diamond (or crystalline silicon) structure where each carbon (or silicon) atom is bonded to four other carbon (silicon) atoms.

In an ethylene (or ethene) molecule ($C_2H_4$), as shown in Fig. 6.14a, each carbon atom is bonded to two hydrogen atoms and the carbon atoms form a double bond with each other. In this case, the $s$-orbital electron is hybridized with the electrons in the $p_x$ and $p_y$ orbital to form a $sp^2$ hybridization (with 1/3 $s$-orbital component). The hybrid electrons form a sigma ($\sigma$) bond with each hydrogen ($1s$ electron) as well as between the carbon atoms. All six atoms lie in a plane and the angle between H–C–H bonds is close to 120°. Nevertheless, for each carbon atom, there is one lonely electron in the $p_z$ orbital whose axis is perpendicular to the plane. The two lonely electrons form a pi ($\pi$) bond to share the orbital. In essence, the double bond between the carbon atoms contains a $\sigma$ bond and a $\pi$ bond with very different characteristics. Note that $\sigma$ bonds are the strongest type of covalence bond. In a $\sigma$ bond, the atomic orbitals overlap with each other in a head-on position, so that their orbitals are symmetrical with respect to rotation about the bond axis. On the other hand, in a $\pi$ bond, the orbital axes are perpendicular to the bond axis between the two atoms. In acetylene (or ethyne) molecule ($C_2H_2$), all four atoms are aligned in the $x$-direction, as shown in Fig. 6.14b. The electrons in the $s$ and $p_x$ orbitals form $sp$ hybridization, which is responsible for the strong $\sigma$ bond between C and H as well as between C and C atoms. Furthermore, the electrons in the $p_y$ (or $p_z$) orbital of each carbon atom form a $\pi$ bond whose orbital axis is in the $y$ (or $z$) direction. The triple bond between the carbon atoms consists of one $\sigma$ bond and a pair of $\pi$ bonds; the latter is much weaker than the former.

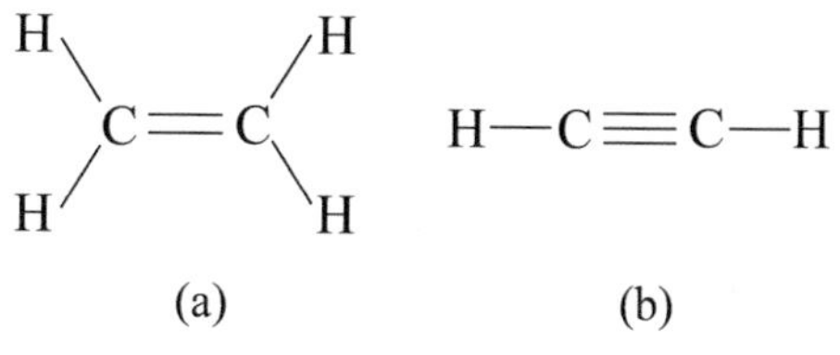

**Fig. 6.14** The chemical structure of **a** ethylene and **b** acetylene

H—C≡C—H

(a) (b)

The $sp^2$ hybridization is also responsible to the carbon bonds in graphene, where each carbon is bonded to three neighborhood carbon atoms via $\sigma$ bonds. These electrons are also called $\sigma$ electrons. The unhybridized electron in the $p_z$ orbital forms a $\pi$ bond with another carbon atom's $p_z$ orbital. In a way, the carbon–carbon bonding in graphene contains alternating single and double bonds, forming a conjugated system like benzene. In contrary to benzene, the $\pi$ electrons in graphene are shared by the atoms and are highly mobile along the graphene sheet, like free electrons in metal. What is more, there are some unique properties of graphene that can only be explained quantum mechanically, with the help of the electron band structures.

Figure 6.15 shows the band structure of graphene calculated in the first Brillouin zone based on the density-functional theory (DFT), when the Fermi energy is set to zero [28, 29]. The bands below (or above) the Fermi level are completely filled (or empty). The filled bands, called $\pi$ or $\sigma$ bands, are associated with electrons in the $\pi$ and $\sigma$ bonds. The unfilled bands are associated with $\pi^*$ or $\sigma^*$ antibonding orbitals. It can be seen that the $\sigma$ and $\sigma^*$ bands are spaced far away from the Fermi level. The transition from $\pi \rightarrow \pi^*$ is responsible for nearly all electronic and optical properties of graphene, except with high-energy excitations (> 5 eV) such as irradiation by photons in the deep ultraviolet and x-ray regions [27]. The most striking feature of 2D graphene is the gapless feature of the bands at the K point (one of the six Dirac points), as indicated by the dashed box. Furthermore, the $\pi$ and $\pi^*$ bands are conical as shown in Fig. 6.16. For 3D semiconductor materials, there is a bandgap between the conduction band and the valence band and the band structures are parabolic in a 2D diagram or parabola in a 3D diagram.

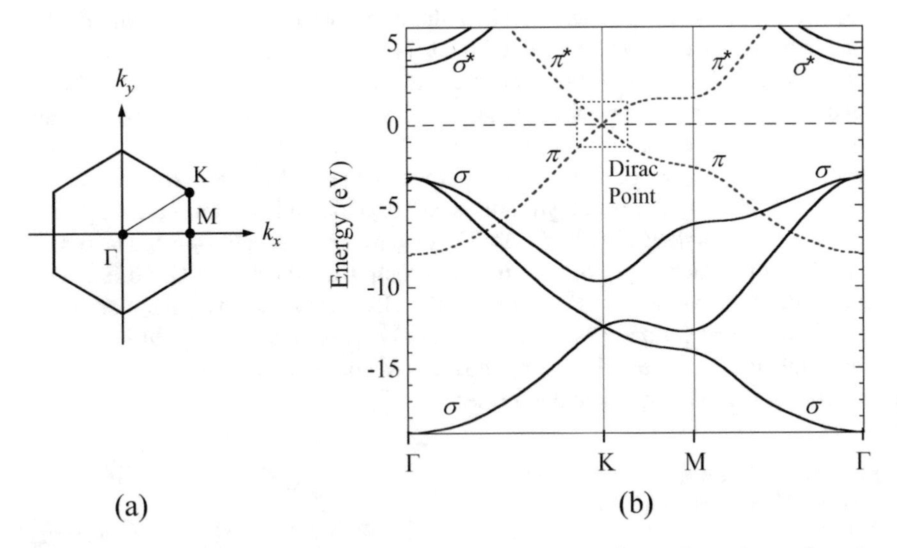


**Fig. 6.15** **a** First Brillouin zone of graphene; **b** band structure of a graphene sheet, where the dashed curves are for $\pi$ or $\pi^*$ bands, and solid curves are for $\sigma$ or $\sigma^*$ bands. *Adapted from* [29] *with permission of American Physical Society*

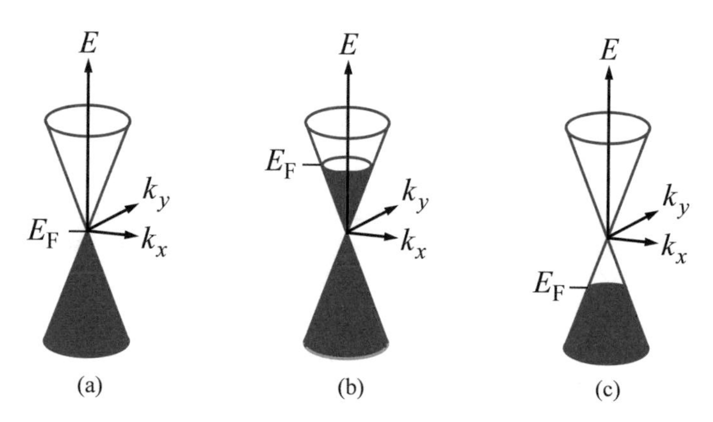


**Fig. 6.16** Band shape near the Dirac point with different Fermi levels: **a** $E_F = 0$; **b** $E_F > 0$; **c** $E_F < 0$

According to the tight-binding model with some approximation [22], the bands can be described by

$$E = \pm\gamma_0\sqrt{3 + 2\cos(K_y a) + 4\cos(\sqrt{3}K_x a/2)\cos(K_y a/2)} \tag{6.32}$$

where $E = 0$ corresponds to the Dirac point, $\gamma_0$ is a constant about 2.8 eV, and $a \approx 0.246$ nm is the magnitude of the chiral vector defined in Fig. 6.7a.

Equation (6.32) predicts conical band structures at the six Dirac points when the energy is within about 1 eV from the apex. For example, at the K point, $K_x = 2\pi/(\sqrt{3}a)$ and $K_y = 2\pi/(3a)$. This can be schematically shown in Fig. 6.16 with different Fermi levels. It can be seen that

$$E = \pm v_F\hbar\sqrt{k_x^2 + k_y^2} = \pm v_F\hbar k \tag{6.33}$$

Here, $v_F \approx 1 \times 10^6$ m/s is the Fermi velocity, and the wavevector **k** at the $K$ point is set to zero. In 1928, Paul Dirac derived a relativistic wave equation that modifies the Schrödinger equation and can be applied to massless particles. According to Eq. (3.122) as discussed in Sect. 3.7, $E^2 = m^2c^4 + p^2c^2$ for a particle with mass $m$ and momentum $p$ that travels at the speed $c$. Therefore, for a massless particle, $E$ is the product of the momentum and speed. Note that in Eq. (6.33), $\hbar k$ is the momentum of the electron and $v_F$ is essentially a constant. Therefore, it can be said that the electrons in pure graphene at low temperatures ($E_F = 0$) are massless Dirac fermions. Gating with a voltage across the graphene sheet allows the Fermi level to be changed, as shown in Fig. 6.16b, c, similar to chemically doping with a small amount of impurities. The cyclotron effective mass is given as [22, 30].

$$m_c = \hbar k_F/v_F = |E_F|/v_F^2 \tag{6.34}$$

which has been experimentally determined by measuring the cyclotron frequency $\omega_c = eB/m_c$ by applying a magnetic field $B$. Note that Eq. (6.34) resembles Einstein's equation, $E = mc^2$. Measurements under applied magnetic field have revealed another exotic behavior in graphene: the anomalous (half-integer) quantum Hall effect at room temperature [22, 30].

From the semi-classical Boltzmann theory, the electrical conductivity of 2D graphene (unit: S or $\Omega^{-1}$) can be expressed in terms of the relaxation time (at the Fermi level) [31]:

$$\sigma_{2D} = \frac{e^2 v_F^2}{2} D(E_F)\tau \tag{6.35a}$$

where $\tau$ is the relaxation time and $D$ is the density of states (DOS) per unit area. For a relatively pure graphene sheet, the DOS can be approximated as

$$D(E) = \frac{2|E|}{\pi \hbar^2 v_F^2} \tag{6.35b}$$

Substituting Eq. (6.35b) into Eq. (6.35a), we obtain

$$\sigma_{2D} = \frac{e^2 |E_F|}{\pi \hbar^2} \tau \tag{6.36}$$

Furthermore, the carrier concentration $n$ (number per unit surface area) and the wavevector at the Fermi level are related by $k_F = \sqrt{\pi |n|}$ or $|E_F| = \hbar v_F \sqrt{\pi |n|}$, which can be used to estimate the carrier concentration. Equation (6.36) breaks down $E_F \to 0$ or $n \to 0$. Due to the gapless feature, there exists a universal minimum conductivity theoretically given as [22]

$$\sigma_{\min} = \frac{4e^2}{\pi h} \approx 4.932 \times 10^{-5}\ \mathrm{S} \tag{6.37}$$

which has been experimentally observed with somewhat higher values due to impurity, size, and other experimental factors [32].

Using the relation $\sigma_{2D} = e|n|\mu$, the electron (or hole) mobility in graphene can be calculated from

$$\mu = \frac{e v_F}{\hbar \sqrt{|n|\pi}} \tau \tag{6.38}$$

where the relaxation time $\tau$ is on the order of $10^{-14} - 10^{-12}$ s for impurity concentrations in the range $10^{11} - 10^{12}\,\mathrm{cm}^{-2}$ [31]. Note that the carrier concentration is different from the impurity concentration and can be tuned either way by a gating voltage. The carrier mobility $\mu$ in graphene typically ranges from 2,000 to 20, 000 $\mathrm{cm}^2/\mathrm{V\,s}$ [32], but could be as high as 200, 000 $\mathrm{cm}^2/\mathrm{V\,s}$ in suspended graphene, even at room temperature [33].

The success in obtaining single or few layers of graphene and measuring their exotic properties have spurred growing interest in a variety of other 2D or quasi-2D materials, whose properties are dramatically distinctive from their 3D counterparts [34–38]. Examples are hexagonal boron nitride (hBN), transition metal dichalcogenides (TMD) such as $MoS_2$ and $WSe_2$, black phosphorus (or phosphorene), layered $Bi_2Te_3$ and GeSe, etc. The intensive theoretical and experimental investigations of these materials systems expand the current understanding of the electrical, mechanical, optical, and thermal properties of existing materials and may provide pathways for new technologies in novel electronic devices, energy systems, nanophotonics, and biomedical applications [26, 34–39].

## 6.5 Phonon Dispersion and Scattering

In the above discussion of electronic band structures, it is assumed that the cores of atoms are fixed. In a real crystal, however, the cores of atoms are vibrating about their equilibrium positions and the vibration of atoms has an important influence on energy storage and transport in crystals. Lattice vibration causes elastic waves to propagate in crystalline solids. Phonons are the energy quanta of lattice waves. For a given vibration frequency $\omega$, the energy of a phonon $\hbar\omega$ is the smallest discrete value of energy. Thermal vibrations in crystals are thermally excited phonons, like the thermally excited photons in a blackbody cavity.

### *6.5.1 The 1D Diatomic Chain*

Phonon dispersion describes the relationship between the vibration frequency and the phonon wavevector. A simple example is given first for a diatomic chain of linear spring–mass arrays, as shown in Fig. 6.17. It is assumed that the spring constant $K$ is the same between the nearest-neighbor atoms. The spring is a conceptual representation of the combined attractive and repulsive forces, which can be assumed linear if the displacement is sufficiently small. Anharmonic vibrations may become

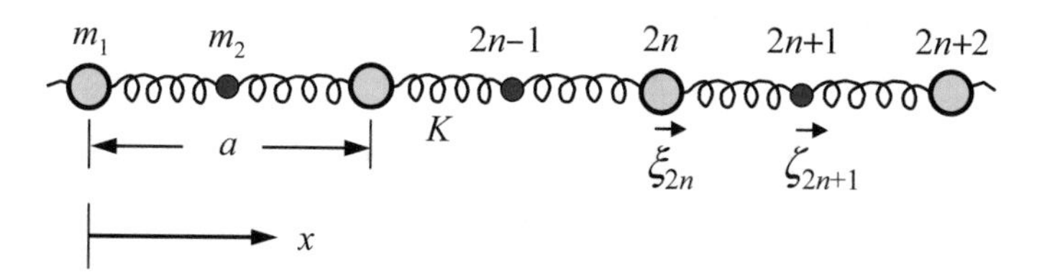


**Fig. 6.17** A chain of two atoms with different masses $m_1$ and $m_2$ linked by springs of the same spring constant $K$, where $\xi$ and $\zeta$ denote the displacements of individual atoms from their equilibrium positions

significant at high temperatures. Another assumption of the nearest-neighbor model is that the forces on an atom come from the nearest neighbors only [40]. The equation of motion of the atoms can be written as follows:

$$m_1 \frac{d^2 \xi_{2n}}{dt^2} = K(\zeta_{2n+1} + \zeta_{2n-1} - 2\xi_{2n}) \tag{6.39a}$$

and

$$m_2 \frac{d^2 \zeta_{2n+1}}{dt^2} = K(\xi_{2n+2} + \xi_{2n} - 2\zeta_{2n+1}) \tag{6.39b}$$

where $\xi_{2n}$ is the displacement of the atom with mass $m_1$ indexed by an even number and $\zeta_{2n+1}$ is the displacement of the atom with mass $m_2$ indexed by an odd number [41, 42]. To solve these equations, we substitute the general solutions $\xi_{2n} = A_1 \exp[i(nka - \omega t)]$ and $\zeta_{2n+1} = A_2 \exp[i(n + 1/2)ka - i\omega t]$ into Eq. (6.39a). After some manipulations, we can obtain

$$(2K - m_1\omega^2)A_1 - 2K\cos(ka/2)A_2 = 0 \tag{6.40a}$$

$$(2K - m_2\omega^2)A_2 - 2K\cos(ka/2)A_1 = 0 \tag{6.40b}$$

The determinant of Eq. (6.40a) must be zero, that is,

$$m_1 m_2 \omega^4 - 2K(m_1 + m_2)\omega^2 + 4K^2[1 - \cos^2(ka/2)] = 0 \tag{6.41}$$

The two roots for $\omega^2$ can be expressed as

$$\omega_{1,2}^2 = K\left(\frac{1}{m_1} + \frac{1}{m_2}\right) \pm K\left[\left(\frac{1}{m_1} + \frac{1}{m_2}\right)^2 - \frac{4\sin^2(ka/2)}{m_1 m_2}\right]^{1/2} \tag{6.42}$$

The resulting $\omega - k$ curves are the dispersion relations, as shown in Fig. 6.18. Two branches are formed when $m_1 \neq m_2$. The upper branch that corresponds to the plus sign is called the *optical phonon branch*, or simply *optical branch*, because it is important for infrared activities in ionic solids. The lower branch that corresponds to the minus sign is called the *acoustic branch*. At very low frequencies, the atoms in the unit cell move in phase with each other. Such a behavior is characteristic for a sound wave.

It can be seen that the dispersion curves vary periodically with $k$. The results outside the first Brillouin zone merely reproduce lattice dynamics that can be fully described by the dispersion curves in the first Brillouin zone. Due to the periodicity of the solution in terms of $k$, we may treat a value of $k$ outside the first Brillouin zone by subtracting an appropriate integer times the reciprocal lattice constant $2\pi/a$ to give a value of $k$ within the limits of the first Brillouin zone. Given that $|k| \leq \pi/a$,

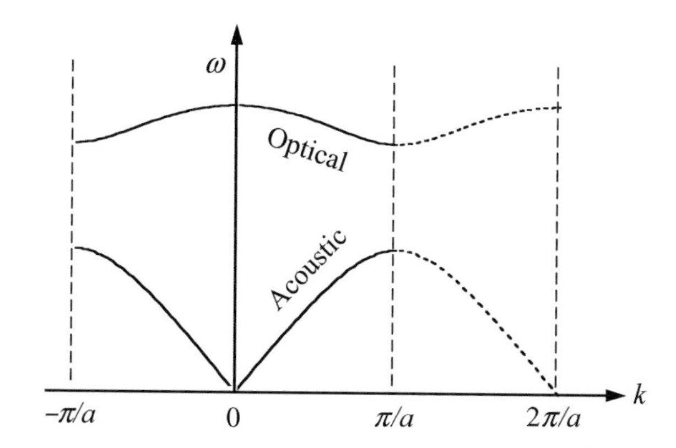


**Fig. 6.18** Phonon dispersion of the linear diatomic chain, calculated by the nearest-neighbor model. The first Brillouin zone is between $-\pi/a$ and $\pi/a$

the phonon wavelength is specified by

$$\lambda = \frac{2\pi}{k},\ 2a \leq \lambda < \infty \tag{6.43}$$

This makes perfect sense as the wavelength should not be smaller than the lattice constants, as explained in previous chapter; see Fig. 5.3. For solids with small dimensions, there is also a limit of the maximum wavelength $2L$. For $k << \pi/a$, the acoustic branch gives $\omega \propto k$, which is a linear dispersion relation. At $k = \pi/a$, $\omega = \sqrt{2K/m_1}$ and $\sqrt{2K/m_2}$, and the two branches are separated when $m_1 \neq m_2$. In this case, it should be noticed that the group velocity $v_g \equiv d\omega/dk = 0$. Only standing waves exist. If $m_1 = m_2$, then the upper and lower branches will be continuous at $k = \pi/a$ and the slope is not zero. However, the lattice constant needs to be modified to $a/2$ in Fig. 6.17, and thus the range of the first Brillouin zone is between $-2\pi/a$ and $2\pi/a$. If the upper branch is not folded at $k = \pi/a$, it will connect smoothly with the lower branch and extend to $2\pi/a$. Detailed discussion can be found from Ref. [41].

### 6.5.2 *Dispersion Relations for Real Crystals*

The above discussion can be extended to 3D systems, in which lattice vibrations allow both transverse and longitudinal modes. For the case of two atoms per primitive cell, there are one longitudinal and two transverse branches for both acoustic and optical vibration modes. The phonon dispersion relations for silicon and silicon carbide are shown in Fig. 6.19 [43–46]. Experimental determination of the phonon dispersion curves was made with neutron scattering [43, 44] for Si and Raman scattering for SiC [46]. Because $m_1 = m_2$ for Si, the longitudinal optical (LO) and longitudinal acoustic (LA) branches meet at the zone edge and thus the group velocity is not equal to zero there. For SiC, on the other hand, the two roots in Eq. (6.42) are different

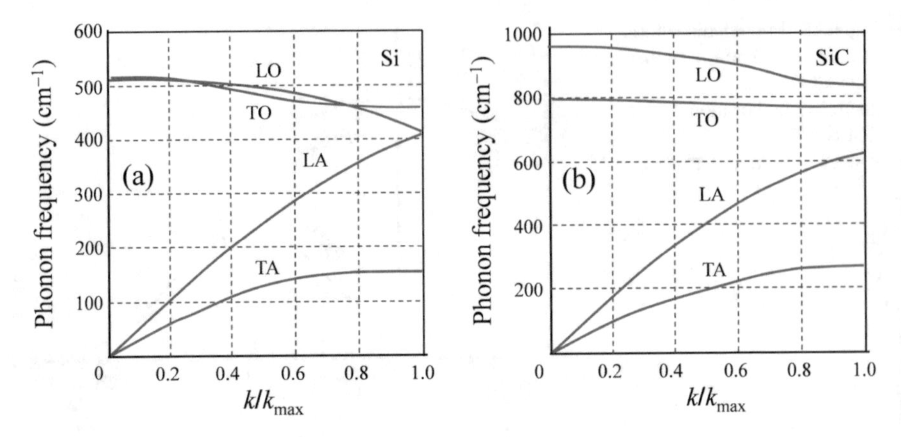


**Fig. 6.19** Optical and acoustical branches of phonon dispersion. **a** Si [101] direction. *Adapted with permission from Refs.* [43–45]. **b** SiC. *Adapted with permission from Ref.* [46]

because $m_1 \neq m_2$. There exists a frequency gap between the LO and LA branches at the zone edge. The frequency gap is forbidden for propagating waves, i.e., no phonons can propagate at frequencies within the gap, similar to the bandgap for electrons. The group velocities of LO and LA phonon modes are zero at the zone edge; this can be seen by the flat dispersion curves. One should not worry about the negative or positive sign of the group velocity as it is merely a result of folding the dispersion curves. The group velocity is always in the direction of energy transfer. It should be mentioned that the speed of sound and the phonon propagation speed refer to the group velocity, not to the phase velocity.

According to the wave–particle duality, a phonon with energy $\hbar\omega$ should also have an associated momentum, given by

$$\mathbf{p} = \hbar\mathbf{k} \tag{6.44}$$

where $\mathbf{k}$ is the wavevector of the phonon. There is a distinction between phonons and photons. Phonons do not carry any physical momentum because the physical momentum associated with lattice vibration is zero, except when all lattices are in phase. On the other hand, when interacting with other elementary particles, such as electrons or photons, the wavevector must follow the selection rule such that it looks as if a phonon has a real momentum given by Eq. (6.44). This momentum is often called the *crystal momentum* [1, 2].

The group velocity of phonons in the optical branches is usually small, and subsequently optical phonons contribute little to the thermal conduction in solids. On the other hand, optical phonons can interact or scatter with acoustic phonons, especially at elevated temperatures, to reduce the thermal conductivity [41]. Although LA phonons have higher group velocities than TA phonons, one must also consider the frequency distribution of phonons since phonons obey Bose–Einstein statistics; see Eq. (5.77) and discussions in Chap. 5. At low temperatures, the TA phonon mode is the dominant contributor to both the thermal conductivity and the specific heat of insulators and semiconductors. As the temperature goes up, LA phonons also play a significant role. While optical phonons contribute little to the heat conduction, they contribute about half of the heat capacity above room temperature. This is because group velocity does not enter the equation for specific heat; refer to Eq. (5.30). In general, if there are $q$ atoms in the primitive cell or basis, there will be one longitudinal and two transverse acoustic branches, and $q - 1$ longitudinal and $2(q - 1)$ transverse optical branches. However, degeneracy of the transverse branches may occur due to symmetry [40–42]. An example of complex materials is the family of zeolites, which are hydrated aluminosilicate minerals that exhibit nanoporous crystalline structures. Zeolites have important applications as filters, catalysts, solar collector, and adsorption refrigeration. Greenstein et al. [47] studied the thermal properties of MFI zeolite films considering phonon dispersion. MFI is a special type of zeolite that has ordered channel directions and an average pore size of 0.6 nm. The calculation of specific heat and thermal conductivity involved summation over 864 polarizations (phonon branches) over all wavevectors in the first Brillouin zone. The modeling results were in reasonable agreement with experiments [47].

In recent years, lots of studies have been done on the phonon transport in graphene and other 2D materials, as well as the interface between 2D materials and substrates [25, 36, 48]. For example, recent measurement and simulation have shown that a thermal conductivity in monolayer of hBN can be as high as 750 W/m K [37].

Recently, researchers have demonstrated experimentally very high thermal conductivity (900–1300 W/m K) in boron arsenide (BAs) crystals at room temperature; the result verified the previous predictions by first-principles simulations in 2013 and 2017; see Ref. [49, 50] and works cited therein. Cubic BAs has since replaced cubic BN (around 740 W/m K) to become the bulk material with the second highest (next to diamond) thermal conductivity at room temperature [49].

Another important aspect of phonon transport is scattering. The mean free path of phonons is often small compared with the size of crystals. For nanostructures, on the contrary, the mean free path can be larger than the characteristic length, resulting in boundary scattering. Some qualitative discussions have been given in the previous chapter. A summary of the characteristics of phonon and photon is given in Table 6.3. In most situations, phonons are treated as particles, especially in dealing with interactions among phonons themselves as well as with electrons, photons, and defects. For long-wavelength phonons, lattice vibration can also be described by a sound wave or an acoustic wave of three polarizations. To analyze the acoustic wave behavior, the crystal is viewed as a continuous medium because the individual vibration of atoms is not of interest. A brief discussion of the microscopic conservation (or selection rules) during scattering events involving phonons and/or electrons is presented next.

**Table 6.3** Comparison of the characteristics of phonon and photon

| Phonon | Photon |
|---|---|
| Bose–Einstein statistics | Bose–Einstein statistics |
| Massless | Massless |
| Energy: $\varepsilon = h\nu$ | Energy: $\varepsilon = h\nu$ |
| Phase speed: $v_\mathrm{p} = \omega/k$ | Phase speed: $v_\mathrm{p} = \lambda\nu$ |
| Mechanical vibration (existence in solids and some liquids, such as liquid helium) | Electromagnetic waves (existence in any medium as well as in vacuum) |
| Both transverse and longitudinal | Transverse only |
| Crystal momentum: $\mathbf{p} = \hbar\mathbf{k}$ | Physical momentum: $\mathbf{p} = \hbar\mathbf{k}$ |
| Frequency: less than $\approx$ 50 THz | Frequency: no limit |
| Group velocity: $<\approx 2 \times 10^4$ m/s | Group velocity: order of $10^8$ m/s |
| Mean free path: $\approx$ 10 to 100 nm (except at very low temperatures and in nanotubes) | Mean free path: no limit (largely dependent on the medium) |

### 6.5.3 *Scattering Mechanisms*

Phonon scattering governs the thermal transport properties of dielectric and semiconductor materials. Proper modeling of phonon scattering is important for the application of the Boltzmann transport equation (BTE) or Peierls–Boltzmann equation, considering the frequency-dependent scattering rate. The anharmonic nature of the interatomic potential offers a coupling mechanism for phonon–phonon interactions, which was not included in the linear oscillator model. The phonon–phonon scattering is inelastic because the phonon frequency before the scattering event is different from that after the event. The energy conservation requires the scattering to involve at least three phonons. A three-phonon process is mostly common since the probability is usually much larger than the values for processes involving four or more phonons. In a three-phonon process, either two phonons interact to form a third one or one phonon breaks into two others. The phonon energy and crystal momentum are conserved as given by [1, 2]

$$\hbar\omega_1 + \hbar\omega_2 = \hbar\omega_3 \quad \text{or} \quad \hbar\omega_1 = \hbar\omega_2 + \hbar\omega_3 \tag{6.45}$$

$$\hbar\mathbf{k}_1 + \hbar\mathbf{k}_2 = \hbar\mathbf{k}_3 \quad \text{or} \quad \hbar\mathbf{k}_1 = \hbar\mathbf{k}_2 + \hbar\mathbf{k}_3 \tag{6.46}$$

In Eqs. (6.45) and (6.46), the left-hand side terms are for phonon(s) before scattering and the right-hand side terms are for phonon(s) after scattering. The processes just described are called *normal (or N) processes*, in which the wavevectors of phonons are inside the first Brillouin zone. Since both the energy and the momentum are conserved, *N*-processes do not alter the direction of energy flow. Hence, *N*-processes make no contribution to the thermal resistance and do not affect the thermal conductivity.

Scattering is also permitted when two phonons interact to form a third one, whose wavevector is outside the Brillouin zone. This can be understood by the equivalence of phonons with the same energy but with different wavevectors $\mathbf{k}'$ and $\mathbf{k}$ that follow the relationship:

$$\mathbf{k}' = \mathbf{k} + \mathbf{G} \tag{6.47}$$

where $\mathbf{G}$ is a reciprocal lattice vector. The reverse process is also possible with the assistance of $\mathbf{G}$ so that one phonon is annihilated to create two others. The momentum relations given in Eq. (6.46) need to be modified as follows after dropping $\hbar$ in all terms:

$$\mathbf{k}_1 + \mathbf{k}_2 = \mathbf{k}_3 + \mathbf{G} \quad \text{or} \quad \mathbf{k}_1 + \mathbf{G} = \mathbf{k}_2 + \mathbf{k}_3 \tag{6.48}$$

These equations, combined with the energy conservation described by Eq. (6.45), describe *the umklapp (or U) processes*. The net momentum is not conserved in the *U*-processes, which introduce thermal resistance and thus reduce the thermal conductivity. Figure 6.20 schematically shows the relationship between the wavevectors for an *N*-process and a *U*-process. An *N*-process can be viewed as the general case of a *U*-process when $\mathbf{G} = 0$.

Above room temperature, *U*-processes dominate and the thermal conductivity decreases linearly as temperature increases. This is because the scattering rate $\gamma = 1/\tau$ between acoustic phonons due to *U*-processes can be described by [41]

$$\gamma_U = \left(A\omega + B\omega^2\right)T \tag{6.49}$$

where $A$ and $B$ are positive constants. When the temperature is reduced, the *U*-process becomes weaker because of the shift in phonon distribution function toward longer wavelengths. Scattering of phonons by defects becomes important. As shown in Fig. 5.13, as the temperature is decreased below room temperature, the thermal conductivity increases to a maximum and then decreases due to the reduction in

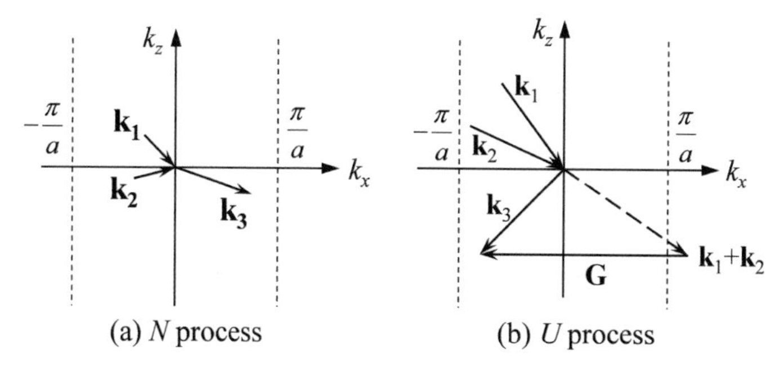


**Fig. 6.20** Schematic illustrations of phonon–phonon scattering processes: **a** the *N*-process and **b** the *U*-process

the specific heat. Four-phonon processes are also possible. Four-phonon scattering includes the annihilation of two phonons to create two others, the annihilation of one phonon to create three others, and the annihilation of three phonons to create another. The calculation of the probability of scattering is more involved [41]. Ecsedy and Klemens [51] estimated the scattering rate due to four-phonon processes to be

$$\gamma_{\text{Four}} \propto \omega^2 T^2 \tag{6.50}$$

Their simplified model also suggests that the probability of four-phonon processes in the temperature range from 300 to 1000 K is negligibly small compared with the three-phonon $U$-processes [41, 51]. Recent first-principles calculations by Feng et al. [50] have shown that the four-phonon process cannot be neglected even for common materials like silicon and diamond above room temperature. At 1000 K, the inclusion of four-phonon scattering could reduce the predicted $\kappa$ by 30%, resulting in excellent agreement with the experimental value. Furthermore, the predicted $\kappa$ value for single-crystal BAs with and without considering four-phonon processes is 2200 W/m K and 1400 W/m K, respectively [50]. Later, the value around 1000 W/m K was experimentally demonstrated by several groups in 2018 [49]. According to the study by Feng et al. [50], the four-phonon scattering rate should scale with

$$\gamma_{\text{Four}} \propto \omega^4 T^2 \tag{6.51}$$

In addition to phonon–phonon interactions, phonons may also interact with defects (such as impurities, vacancies, or dislocations) and boundaries. These scattering processes can also influence the mean free path of phonons. Scattering of phonons by defects is elastic since the phonon energy remains the same. At temperatures near and above the Debye temperature, phonon–phonon interactions are dominant. As the temperature drops, the dominant wavelengths of phonons become comparable to the size of defects; therefore, scattering of phonons by defects becomes important. The scattering rate for phonon-defect scattering is independent of temperature but dependent on the phonon wavelength. This can be modeled using the Rayleigh scattering theory for small particles such that the scattering rate due to defects is inversely proportional to the fourth power of the phonon wavelength $\lambda$, viz.,

$$\gamma_{\text{ph-d}} \propto \lambda^{-4} \quad \text{or} \quad \omega^4 \tag{6.52}$$

When the bulk mean free path is comparable or greater than the characteristic dimension, such as the thickness of the film or the diameter of the wire, scattering of phonons by boundaries becomes important. Boundary scattering is important for nanostructure materials and at low temperatures when the phonon mean free path is large, as extensively discussed in the previous chapter.

In metals and semiconductors, electronic transport becomes important. The scattering of charge carriers controls the electric conduction in solids and dominates the thermal conduction in metals. Carrier–carrier inelastic scattering is negligible except for highly conductive materials, such as a high-temperature superconductor. Since lattice vibrations are enhanced with increasing temperature, electron–phonon scattering usually dominates the scattering process at high temperatures, while at low temperatures, lattice vibrations are weak and defect scattering becomes important. The vibration of lattice ions causes deviations from the perfect periodic lattice and distorts the carrier wave function. This is more easily visualized as the scattering of electrons by phonons. Both the acoustic branch and the optical branch can scatter electrons. Usually, the energy of acoustic phonons can be neglected compared with the electron energy. Therefore, scattering by acoustic phonons is essentially elastic. Scattering by optical phonons is inelastic because the exchange of energy between the carriers and the phonons can be significant. This process facilitates the energy transfer between electrons and phonons, which is associated with Joule heating. For materials with two different atoms per primitive cell, the asymmetric charge distribution in the chemical bond forms a dipole. Scattering by optical phonons in these materials is called *polar scattering*, which can effectively scatter electrons or holes. The energy and momentum conservations for carrier–phonon scattering can be written as

$$E_{\rm f} = E_{\rm i} \pm \hbar\omega_{\rm phonon} \tag{6.53a}$$

and

$$\mathbf{k}_{\rm f} + \mathbf{G} = \mathbf{k}_{\rm i} \pm \mathbf{k}_{\rm phonon} \tag{6.53b}$$

where subscripts i and f indicate the initial and final states of the carrier, the minus sign corresponds to phonon emission, and the plus sign corresponds to phonon absorption. The momentum of an electron is similar to that of a phonon and is also referred to as the *crystal momentum*. If $\mathbf{G}$ is set to zero, the process is an *N*-process; otherwise, it is a *U*-process as in phonon–phonon scattering. In semiconductors at low temperatures, only *N*-processes are energized. In metals and semiconductors, the electron–phonon scattering rate typically ranges from $10^{12}$ to $10^{14}$ Hz at room temperature. Near or above the Debye temperature, the specific heat is almost a constant and the number of phonons increases linearly with temperature. Hence, the electron–phonon scattering rate is proportional to temperature in metals, resulting in nearly temperature-independent thermal conductivity, while the electrical resistance is proportional to temperature.

An electron or hole in a periodic lattice does not really collide with ions. The transport of free carriers can be viewed as the propagation of a wave in a periodic potential created by the ions. In addition to lattice vibrations, defects or impurities may break the periodicity of the potential or alter its amplitude. Kinetic theory gives the defect scattering rate as

$$\gamma_{\rm e-d} = n_{\rm d}\sigma_{\rm d}v_{\rm e} \tag{6.54}$$

where $n_{\rm d}$ and $\sigma_d$ are the defect number density and scattering cross section, respectively, and $v_{\rm d}$ is the average carrier velocity. The scattering cross-section is an effective area related to the scattering probability and not the actual geometric cross-sectional area. For metals, the electron velocity is the Fermi velocity $v_F$, which is on the order of $10^6$ m/s. For semiconductors, the random velocity of electrons or holes can be calculated by

$$v_{th} = (3k_B T/m^*)^{1/2} \tag{6.55}$$

which is called the *thermal velocity* and is on the order of $10^5$ m/s at room temperature.

In semiconductors, the interband transition requires the conservation of both energy and momentum. This can occur by electronic transitions when interacting with the incident radiation. For indirect gap semiconductors, however, the photon itself cannot provide a large enough change in momentum. Therefore, a phonon is either emitted or absorbed for momentum conservation. The energy and momentum conservation equations are, respectively,

$$E_f - E_i = \hbar\omega_{photon} \pm \hbar\omega_{phonon} \tag{6.56a}$$

and

$$\mathbf{k}_f - \mathbf{k}_i = \mathbf{k}_{photon} \pm \mathbf{k}_{phonon} \tag{6.56b}$$

where the plus and minus signs correspond to phonon absorption and emission, respectively. This kind of transition is called the indirect interband transition. For a direct interband transition, there is no need to emit or absorb a phonon and, thus, the last term in both Eqs. (6.56a) and (6.56b) should be dropped out. The interaction of photons with solids will be left to Chap. 8 (Sect. 8.4) for a more detailed discussion about the absorption and emission processes.

In addition to the absorption and the emission, photons may be scattered by phonons, causing a nonlinear effect. There exists inelastic scattering when photons are scattered by phonons, resulting in x-ray scattering, neutron scattering, Raman scattering, and Brillouin scattering. In Raman scattering, the creation (emission) and annihilation (absorption) of a phonon cause a shift in the frequency of the radiation, namely, the Stokes and anti-Stokes shifts, as shown in Fig. 6.21. The energy conservation equations are

$$\hbar\omega_s = \hbar\omega_i - \hbar\omega_{ph}, \text{ for a Stokes shift} \tag{6.57a}$$

and

$$\hbar\omega_s = \hbar\omega_i + \hbar\omega_{ph}, \text{ for an anti-Stokes shift} \tag{6.57b}$$

where subscripts i, s, and ph are for incident photon, scattered photon, and phonon, respectively. Because the interaction involved two photons and one phonon, the

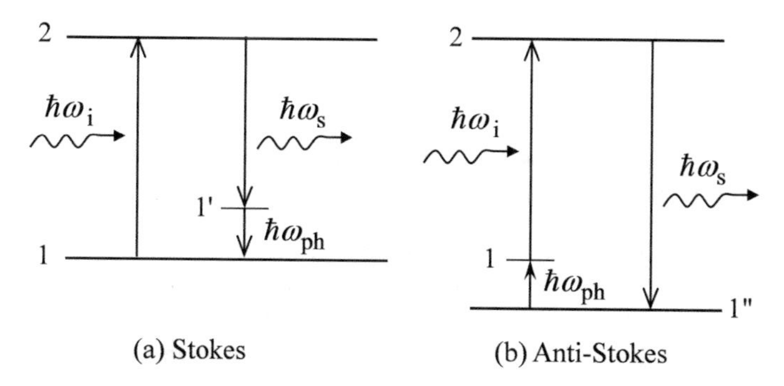


**Fig. 6.21** Illustration of Raman scattering and **a** the Stokes and **b** anti-Stokes processes

momentum of the phonon is restricted to small values. The Raman effect, or the Raman scattering, was named after Indian physicist C. V. Raman (1888–1970), who won the Nobel Prize in Physics in 1930 for the discovery. The intensity of the anti-Stokes shift is usually much weaker than that of the Stokes shift. In certain cases, however, the phonons generated by the Stokes process can subsequently participate in the anti-Stokes process, causing a strong excitation to the anti-Stokes component. It is interesting to note that the anti-Stokes component actually pumps energy out from the material, resulting in a radiative cooling effect.

Note that the resulting photon can interact with the phonon again, creating a cascade process that emits $m$ phonons. The photon energy is reduced by $m$ times the energy per phonon. The probability decreases as the order increases. Raman spectroscopy has become a major analytical instrument for the study of solids. High-intensity lasers, high-resolution spectrometers, and sensitive detectors such as photomultiplier tubes (PMTs) are often employed to measure narrow Raman lines. The Raman intensity and intensity ratio depend upon temperature, as illustrated in Fig. 6.22. The ratio of the Raman intensities can be expressed by

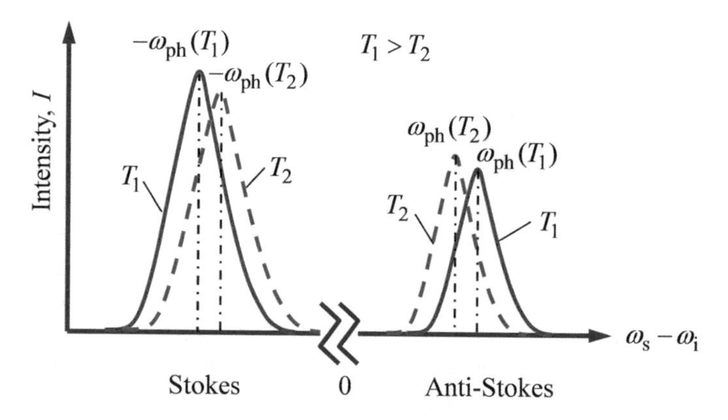


**Fig. 6.22** Raman intensity for the Stokes and anti-Stokes scattering at two different temperatures

$$\frac{I_{\text{anti-Stokes}}}{I_{\text{Stokes}}} = \left(\frac{\omega_{\text{i}} - \omega_{\text{ph}}}{\omega_{\text{i}} + \omega_{\text{ph}}}\right)^2 \exp\left(-\frac{\hbar\omega_{\text{ph}}}{k_{\text{B}}T}\right) \tag{6.58}$$

which can be used for surface temperature measurements in microelectronics and microcantilever heaters [52, 53].

**Example 6.4** Neutron scattering by phonons is important for measuring the dispersion relations. Express the energy conservation and the momentum conservation during the neutron–phonon scattering in terms of the wavevector and the mass of the neutron, and the wavevector and the frequency of the phonon. Assume the process involves one phonon only.

**Solution** A neutron has a mass $m_{\text{ne}} = 1.673 \times 10^{-27}$ kg that is 1834 times that of an electron. Based on the wave–particle duality, the kinetic energy of a neutron can be expressed as $E_{\text{ne}} = \frac{p^2}{2m_{\text{ne}}} = \frac{\hbar^2 k^2}{2m_{\text{ne}}}$; thus, the energy conservation becomes

$$\frac{\hbar^2 k_{\text{s}}^2}{2m_{\text{ne}}} = \frac{\hbar^2 k_{\text{i}}^2}{2m_{\text{ne}}} \pm \hbar\omega_{\text{ph}} \tag{6.59}$$

where $k_{\text{i}}$ and $k_{\text{s}}$ are the magnitude of wavevector of the incident and scattered neutrons. The wavevector selection rule gives

$$\mathbf{k}_{\text{s}} + \mathbf{G} = \mathbf{k}_{\text{i}} \pm \mathbf{k}_{\text{ph}} \tag{6.60}$$

These relations characterize the inelastic scattering of neutrons by phonons. The plus and minus signs refer to the process that absorbs or releases phonons, respectively.

### 6.5.4 Phononics and Coherent Phonons

Phonons are quantized lattice vibration waves. The wavelength of phonons that are important for thermal transport falls in the region from 1 to 10 nm at room temperature and can be longer at cryogenic temperatures. While phonons are typically treated as particles, their wave nature may become important in 1D superlattices and 2D or 3D periodic structures due to interference effects, especially at low temperatures. These effects may enable tailoring the thermal transport properties of semiconductors and insulators, resulting in phonon engineering that belongs to a research branch called *phononics*. Phonon engineering may allow us to control heat transfer in unprecedented manner, such as thermal diode, thermal transistors, thermal memory, topological phonon hall effect, heat cloaking, thermal lens for heat focusing, to name a few [54–56]. The development of graphene and 2D van der Waals materials increases the possibilities and broadens the applications of phonon engineering [25, 57]. Furthermore, with femtosecond to picosecond excitations, phonons may exhibit coherent nature. Resonance lattice vibrations or coherent phonons have been experimentally observed not only in bulk solids but also in superlattices [58–62].

In a solid crystal, valence electrons are confined and their properties are manifested by the band structure. The electron wave functions can be analyzed based on the Bloch theory. Given the periodicity of typical solids from 0.1 to 1 nm, the crystalline structure also affects the phonon dynamics, resulting in phonon dispersion. Since phonons can extend to longer wavelengths than electrons do, periodic structures with a period of 1–200 nm can affect the phonon band structure (dispersion relations), forming *phononic crystals* that can modify the DOS and group velocity of phonons, thereby reducing the thermal conductivity due to the coherence effect [54, 63]. One of the promising applications is in thermoelectricity where the confinement reduced the thermal conductivity dramatically with little effect on the Seebeck coefficient or electrical conductivity. This is desired to improve the figure of merit of thermoelectric devices as discussed in Chap. 5. It is worth mentioning here that *photonic crystals* with a period from 100 nm to 10 μm can exhibit unique optical properties in the visible and infrared regions, as to be discussed in Chap. 9.

Luckyanova et al. [64] demonstrated coherent phonon transport on thermal conductivity using GaAs/AlAs superlattices fabricated by metal-organic chemical vapor deposition. Each period contains 12 nm GaAs and 12 nm AlAs films, as confirmed by transmission electron microscopic images. The cross-plane thermal conductivities of five samples whose periods are 1, 3, 5, 7, and 9 were measured using a time-domain thermoreflectance (TDTR) technique at temperatures from 30 to 300 K. For periodic layers in the incoherent regime, interface scattering often dominates the thermal resistance, which can be modeled as a combination of the internal and interface resistances. The resulting cross-plane thermal resistance increases with the number of layers, and subsequently the effective thermal conductivity of superlattices with the same period is independent of the number of layers. When interface resistance is small and the phonon mean free path in the bulk material is much longer than the total thickness of the superlattice, the thermal resistance is dominated by boundary resistance at the front and back of the superlattice. This situation resembles ballistic phonon transport in a homogeneous film, when the thermal conductivity decreases linearly as the thickness is reduced as already discussed in Chap. 5. Luckyanova et al. [64] observed a linear dependence of the thermal conductivity at temperatures below 150 K, and thereby demonstrated coherent phonon contributions to thermal conductivity. As the temperature increases, due to phonon scattering at interfaces, the linearity breaks down. The phonon mean free path ($\Lambda$) distribution for an infinitely extended superlattice was evaluated using first-principles calculation. It was found that phonons whose mean free path exceed 216 nm (the total thickness of the superlattices used in the experiments) contribute to the thermal conductivity about 87% at 100 K and 71% even at 300 K. Due to the lack of internal scattering and interface scattering, these low-frequency phonons can be described by the dispersion relations based on the superlattice band structures. The results unambiguously demonstrate coherent phonon transport in superlattices even at room temperature [64].

In the incoherence limit, the thermal conductivity of a superlattice decreases as the interface density increases or period decreases in superlattices made of the same materials with similar interfaces. If the interfaces are made very smooth to avoid scattering, when the period is smaller than the phonon coherence length (such that coherent phonons can play a significant role), the thermal conductivity increases as the period is reduced due to the modification of the phonon dispersion and DOS. The combined effects give rise to a minimum when the thermal conductivity is plotted against the period of the superlattice or the interface density. Such a minimum thermal conductivity separates the particle regime and wave regime of heat conduction by phonons. The predicted minimum thermal conductivity was experimentally demonstrated by Ravichandran et al. [65] using perovskite superlattices made of either $SrTiO_3/CaTiO_3$ or $SrTiO_3/BaTiO_3$ at room temperature. They used molecular epitaxy deposition to grow (001) oxides with atomically sharp interfaces. The layer thickness varies from a single unit cell to 40 nm for $SrTiO_3/CaTiO_3$. For this kind of superlattices with a 50:50 volume ratio between the two materials and a total thickness of 200 nm, a minimum cross-plane thermal conductivity was observed when the period is between 2 and 3 nm. The valley gets deeper when the temperature is lowered from 307 to 142 K and then to 84 K, with a gradual shift toward large period as the temperature is lowered [65]. The experimental results are consistent with theoretical predictions of the crossover between the wave–particle behaviors and the value of the minimum thermal conductivity. The fact that the thermal conductivity can be manipulated by changing the superlattice period may have extensive applications in thermoelectric devices.

Phononic nanomesh structure refers to periodically perforated thin membranes or films, which form a 2D phononic crystal [54–57]. Significant reduction of the in-plane thermal conductivity at room temperature has been observed in Si nanomeshes. Yu et al. [66] fabricated nanomeshes on 22-nm-thick doped epitaxial Si film with a period of a few tens of nanometers and positioned the nanomesh structure between two suspended membranes (one for heating and one for sensing) to measure the thermal conductance at steady state. They reported a nanomesh thermal conductivity near 1.9 W/m K at temperatures from 150 to 280 K, which then decreases to approximately 1.4 W/m K at 100 K. Hopkins et al. [67] fabricated relatively large nanomesh structures with 500-nm-thick single crystalline Si. The period was varied from 500 to 800 nm while the diameter of the holes is 300 or 400 nm. The thermal conductivity of suspended nanomesh structure was measured with a TDTR setup to be as low as 6.8 W/m K, more than an order of magnitude lower than the value of 148 W/m K obtained for bulk Si. While low-frequency coherent or partially coherent phonons with a longer mean free path may play a role in the thermal transport of nanomeshes [68], later theoretical studies [69, 70] and recent experiments [71] suggest that coherent phonons may not play a significant role due to the relatively large surface roughness of microfabricated structures (1–3 nm) as compared with the phonon wavelength at room temperature. Classical phonon scattering including backscattering and native oxide layers may be largely responsible for the thermal conductivity reduction at near room temperature [57, 70, 71].

Nevertheless, in the sub-kelvin temperature range where the dominant thermal phonons are in the gigahertz frequency region, coherent phonon effects on thermal conduction have been observed in suspended silicon-nitride nanomeshes with a period of 970 nm and 2425 nm, respectively [72]. At such low temperatures, the dominant phonon wavelength is in the micrometer range, much larger than the surface roughness. In this regime, a strong modification of the phonon band structure can affect the thermal transport as demonstrated experimentally [72]. Maire et al. [73] fabricated ordered and disordered 1D and 2D hole arrays with a nominal period of 300 nm on a 145 nm single crystalline Si layer. The Si nanobeam or membrane with a nominal hole diameter of 150 nm was patterned on a silicon-on-insulator wafer and etched into suspended structures. The roughness of the surface was measured with an atomic force microscope (AFM) to be 0.5 nm, while the roughness of the inner surface of the hole was estimated to be 2.5 nm from scanning electron microscopic images. Measurements with a micro-TDTR setup suggested that the thermal decay rates for ordered and disordered samples remain the same at temperatures higher than about 10 K. Below 10 K, especially at 4 K, disorder can significantly increase the decay rate [73]. Coherent phonons with frequencies less than 200 GHz are expected to play a significant role at 4 K, resulting in a decreased decay rate (or thermal conductivity) in the ordered structure. Their results are consistent with coherent acoustic phonon spectroscopy performed with femtosecond pump-probe thermoreflectance experiments [74].

## 6.6 Atomistic Simulation of Lattice Thermal Properties

In Chap. 5, we extensively discussed the relaxation time approximation in modeling thermal conductivity based on BTE, where the Debye model is often assumed in predicting the phonon DOS. While the results can predict the temperature dependence of thermal conductivity as well as classical size effects, the accuracy is rather limited. Furthermore, the parameters that enter the model are typically obtained by fitting the measurement data. In Sect. 6.5.3, we gave a brief introduction of phonon dispersion and scattering mechanisms. In the last 20 years, especially since the publication of the first edition of this book in 2007, more and more researchers have applied atomistic simulations rooted in quantum mechanics to predict lattice thermal properties for semiconductors, superlattices, and low-dimensional materials (including polymer chains) [75–81]. Classical molecular dynamics (MD) simulations have also been extensively used to predict the thermal conductivity of solid materials based on empirical potentials (refer to Sect. 4.3) using either the linear response theory (i.e., the Green–Kubo relation) or the direct method (i.e., nonequilibrium approach) [82]. Phonon properties can be extracted by analyzing the wave-packet dynamics [82, 83]. Furthermore, the phonon scattering rate or lifetime can also be determined from MD simulations by analyzing the normal modes [83–85]. The interatomic potentials can also be obtained from the first-principles calculations and used in MD simulations [78, 80]. This section overviews the ab initio simulation methodology and how it can be applied to obtain phonon properties and calculate lattice thermal conductivity. The equilibrium and nonequilibrium MD methods will also be introduced.

### 6.6.1 Interatomic Force Constants (IFCs)

Nowadays, most first-principles calculations for materials properties are based on the density-functional theory (DFT) [86, 87], which has been referred to or cited by over one hundred-thousand papers per year in the last decade according to Google Scholar. While the term "first principles" can have different meanings in different fields, it generally refers to a basic set of postulates or physical laws that do not rely on any other assumptions or fitting parameters that are frequently used in empirical modeling. Therefore, solving the Schrödinger equation to obtain the electronic structure of a crystal under suitable approximations without using parameters obtained from fitting the experimental data is a first-principles approach. Some of the methods were mentioned in Sect. 6.4.3, but they are limited to a few atoms in the unit cell. Rather than solving the many-body Schrödinger equation to obtain the wave functions, DFT treats it as a vibrational problem to obtain the electron densities by minimizing the energy functional using the method of Lagrange multipliers as described in Appendix B.2. The details are rather complicated [86, 87]; thus, only a conceptual description is given next.

The many-electron wave function can be expressed as a product of the single-electron wavefunctions as

$$\Psi(\mathbf{r}_1, \mathbf{r}_2, \ldots \mathbf{r}_N) = \psi_1(\mathbf{r}_1)\psi_2(\mathbf{r}_2)\ldots\psi_N(\mathbf{r}_N) \tag{6.61}$$

where $N$ is the number of electrons in the system. Each of the wave function satisfies the Schrödinger equation:

$$\left[-\frac{\hbar^2}{2m_e}\nabla^2 + V_{\text{ext}}(\mathbf{r}) + \Phi_{\text{e}}(\mathbf{r})\right]\psi_i(\mathbf{r}) = \varepsilon_i\psi_i(\mathbf{r}) \tag{6.62}$$

where $\psi_i$ and $\varepsilon_i$ are, respectively, the wave function and orbital energy of the $i$th electron, $V_{\text{ext}}$ is the potential due to all nuclei that is determined by the structure and elemental composition of the system, and $\Phi_e$ is the potential due to the existence of other electrons (the Coulomb potential). Most solution methods start with a suitable potential $V_{\text{ext}}$ to obtain the wave function $\Psi(\mathbf{r}_1, \mathbf{r}_2, \ldots \mathbf{r}_N)$. From there, all observables can be obtained, and among them, the electron density can be expressed as

$$n(\mathbf{r}) = N\int d\mathbf{r}_2 \int d\mathbf{r}_3 \ldots \int d\mathbf{r}_N \Psi^*\Psi \tag{6.63}$$

Hohenberg and Kohn [86] proved that all ground-state properties can be expressed as functionals of the electron density $n(\mathbf{r})$ and the total energy is such a functional that attains its minimum for the correct ground-state density. In essence, DFT is a variational approach in which the electron density is obtained first and then used to obtain the many-electron wave function, along with the potential. All other observables can be obtained consequently. Compared with traditional first-principles methods, the computational cost of DFT is relatively low. This advantage becomes more obvious especially with more and more complex systems.

The local-density approximation (LDA) could be used for a slowly varying density so that the ground-state energy can be obtained in terms of single-particle equations for the interacting system [86]. The Kohn–Sham equations provide a means to obtain the electron density by solving $N$ noninteracting one-electron Schrödinger's equations [86, 87]. Before 1990, DFT was not considered as an accurate method in band structure computations [21]. Since then, some approximations have been greatly refined to better model the exchange and correlation potentials [88]. The 1998 Nobel Prize in Chemistry was bestowed on Walter Kohn for his establishment of DFT and John A. Pople for his development of computational methods in quantum chemistry including the implementation of DFT [89]. More recently, DFT has been widely used to predict the chemical, electronic, structural, lattice dynamics, and even magnetic properties of materials from the atomic scale. A number of software packages are available for solid-state simulations based on DFT [88, 90]. It should be noted that the term ab initio (from the beginning) is often used with the same meaning as from first principles, especially in computational chemistry based on quantum mechanics.

DFT can be used to predict phonon properties by using the theory of lattice dynamics [88]. The Born–Oppenheimer approximation assumes that the motion of atomic nuclei can be treated separately from that of electrons. This is the so-called adiabatic approximation, which allows us to decouple the vibrational degrees of freedom from the electronic degrees of freedom [91]. A similar Hamiltonian for a crystal can be written in terms of the interatomic force constants (IFCs) and then used in lattice dynamics to obtain phonon properties.

The density-functional perturbation theory (DFPT) is a linear response theory that aims at obtaining IFCs through a small perturbation of the system from its equilibrium [91]. Suppose $\hat{H}$ is a Hamiltonian and write the Schrödinger equation of the $i$th particle as follows:

$$\hat{H}\psi_i = \varepsilon_i \psi_i \tag{6.64}$$

Then by applying a small perturbation (parameter $\xi$), we can write the perturbed Hamiltonian, wave function, and energy of the particle as [92]

$$\hat{H} = \hat{H}^{(0)} + \xi \hat{H}^{(1)} + \xi^2 \hat{H}^{(2)} + \cdots \tag{6.65a}$$

$$\psi_i = \psi_i^{(0)} + \xi \psi_i^{(1)} + \xi^2 \psi_i^{(2)} + \cdots \tag{6.65b}$$

and

$$\varepsilon_i = \varepsilon_i^{(0)} + \xi \varepsilon_i^{(1)} + \xi^2 \varepsilon_i^{(2)} + \cdots \tag{6.65c}$$

By substituting Eq. (6.65a) into Eq. (6.64) and equalizing terms with the same order on both sides, we obtain a series of equations based on the order of perturbation.

The first equation is

$$\hat{H}^{(0)}\psi_i^{(0)} = \varepsilon_i^{(0)}\psi_i^{(0)} \tag{6.66}$$

Apparently, this is for the unperturbed system or the system at equilibrium. The first-order perturbation gives

$$\hat{H}^{(0)}\psi_i^{(1)} + \hat{H}^{(1)}\psi_i^{(0)} = \varepsilon_i^{(0)}\psi_i^{(1)} + \varepsilon_i^{(1)}\psi_i^{(0)} \tag{6.67}$$

and the second-order perturbation gives

$$\hat{H}^{(2)}\psi_i^{(0)} + \hat{H}^{(1)}\psi_i^{(1)} + \hat{H}^{(0)}\psi_i^{(2)} = \varepsilon_i^{(2)}\psi_i^{(0)} + \varepsilon_i^{(1)}\psi_i^{(1)} + \varepsilon_i^{(0)}\psi_i^{(2)} \tag{6.68}$$

The third-order and higher order perturbation relations can also be applied when necessary. It should be noted that

$$\varepsilon_i^{(n)} = \frac{1}{n!}\frac{\mathrm{d}^n\varepsilon_i}{\mathrm{d}\xi^n}\bigg|_{\xi=0} \quad \text{and} \quad \psi_i^{(n)} = \frac{1}{n!}\frac{\mathrm{d}^n\psi_i}{\mathrm{d}\xi^n}\bigg|_{\xi=0} \tag{6.69}$$

The Hellmann and Feynman theorem gives $\varepsilon_i^{(1)}$ in terms of $\psi_i^{(0)}$ and $\hat{H}^{(1)}$, which are supposed to be known. The second-order energy derivative and higher order derivatives can also be obtained [92]. The $(2n+1)$ theorem states that if one knows the wave functions up to order of $n$, i.e., $\psi_i^{(n)}$, one can deduce the energy derivative up to the order $(2n+1)$, i.e., $\varepsilon_i^{(2n+1)}$. The actual multivariable perturbation problem is very complicated and readers are referred to [91, 92] and references therein.

Let us assume that, based on the elementary composition of the system, the unit cell parameters and equilibrium atomic positions can be obtained from DFT. If the atoms are allowed to have small displacement around their equilibrium position, then the potential energy of the system $U$ can be expressed as a Taylor expansion [80, 81]:

$$U = U_0 + \sum_i\sum_\alpha \Pi_i^\alpha \zeta_i^\alpha + \frac{1}{2!}\sum_{i,j}\sum_{\alpha,\beta}\Phi_{ij}^{\alpha\beta}\zeta_i^\alpha\zeta_j^\beta + \frac{1}{3!}\sum_{i,j,k}\sum_{\alpha,\beta,\gamma}\Psi_{ijk}^{\alpha\beta\gamma}\zeta_i^\alpha\zeta_j^\beta\zeta_k^\gamma + O(\zeta^4) \tag{6.70}$$

Here, $i, j, k \ldots = 1, 2, \ldots N$ are the atom indices, and $\alpha, \beta, \gamma \ldots = 1, 2,$ or 3 are the coordinate indices ($x$, $y$, or $z$ in the Cartesian coordinates system). The IFCs are given as follows:

$$\Pi_i^\alpha = \frac{\partial U}{\partial \zeta_i^\alpha} = -F_i^\alpha \tag{6.71}$$

where $F_i^\alpha$ is the force component on the $i$th atom and it is zero at equilibrium. Thus, the first summation on the right-hand side of Eq. (6.70) disappears. The second-order derivatives are called the harmonic force constants,

$$\Phi_{ij}^{\alpha\beta} = \frac{\partial^2 U}{\partial \zeta_i^\alpha \partial \zeta_j^\beta} \tag{6.72}$$

This will become clear in the next section. The third-order derivatives are known as the cubic force constants, which are related to the anharmonic processes due to three-phonon interactions:

$$\Psi_{ijk}^{\alpha\beta\gamma} = \frac{\partial^3 U}{\partial \zeta_i^\alpha \partial \zeta_j^\beta \partial \zeta_k^\gamma} \tag{6.73}$$

The next anharmonic constants are related to four-phonon processes [50]. Once the IFCs are obtained using DFPT [78, 91, 92], lattice dynamics can be used to obtain phonon dispersion relations, as well as the wavevectors and wave functions of the normal modes (phonon modes) from harmonic lattice dynamics. The higher order terms have little impact on the phonon dispersion, which determines the group velocity and DOS of phonons. However, the anharmonic terms are related and can be used to obtain phonon–phonon scattering rates, which are temperature dependent. Besides DFPT, other methods, such as the real-space small displacement method and frozen-phonon method, have also been used to extract the IFCs for various materials [78, 93–96]. When the phonon properties are fully determined, the data can be combined with the Peierls–Boltzmann equation. The solution allows the determination of the thermal conductivity from first principles as discussed in the following two sections [75, 76, 79].

### 6.6.2 Lattice Dynamics and Fermi's Golden Rule

The phonon properties can be obtained once the IFCs are calculated. The harmonic force constants allow the determination of phonon dispersion and normal modes according to lattice dynamics theory. The anharmonic force constants can be used to obtain the scattering rate for three-phonon or four-phonon interactions using Fermi's golden rule. This section provides a brief coverage of lattice dynamics and the golden rule.

In a solid crystal, due to periodicity, the atom index is expressed in terms of double indices, that is, $i \to lb$, $j \to l'b'$, $k \to l''b''$, etc. Here, $l$ is the index of the unit cell and $b$ is the index of the atom inside the unit cell $l$. Note that the lattice translation vector of the $l$th unit cell can be represented by $\mathbf{R}_l$ according to Eq. (6.4), though $l$ means all three indices here. Another vector $\mathbf{r}_b$ can be used to indicate the position of the atom with respect to the lattice point. With these notations, the Hamiltonian for the crystal with potential energy $U$ becomes

$$\hat{H} \approx \sum_{lb} \frac{\hat{p}_{lb}^2}{2m_b} + \frac{1}{2!} \sum_{lb,l'b'} \sum_{\alpha,\beta} \Phi_{lb,l'b'}^{\alpha\beta} \zeta_{lb}^{\alpha} \zeta_{l'b'}^{\beta} + \frac{1}{3!} \sum_{lb,l'b',l''b''} \sum_{\alpha,\beta,\gamma} \Psi_{lb,l'b',l''b''}^{\alpha\beta\gamma} \zeta_{lb}^{\alpha} \zeta_{l'b'}^{\beta} \zeta_{l''b''}^{\gamma} \tag{6.74}$$

Note that $\hat{p}_{lb}$ is the momentum operator for the $b$th atom in the $l$th cell and $m_b$ is the mass of the atom. The other parameters are defined in Eqs. (6.69–6.72). The fourth-order term is not given for simplicity. The equation of motion of atom $b$ in unit cell $l$ can be expressed as follows:

$$m_b \frac{\partial^2 \zeta_{lb}^{\alpha}}{\partial t^2} = -\sum_{l'b'} \sum_{\beta} \Phi_{lb,l'b'}^{\alpha\beta} \zeta_{l'b'}^{\beta} \tag{6.75}$$

The dynamic equation may be solved using a Fourier transform and expressed in terms of a series of Fourier components, each one is a plane wave with a wavevector $\mathbf{k}$ and angular frequency $\omega$,

$$\zeta_{lb}^{\alpha} = \frac{1}{\sqrt{m_b}} \sum_{\mathbf{k}} \eta_b^{\alpha}(\mathbf{k}) \mathrm{e}^{\mathrm{i}(\mathbf{k}\cdot\mathbf{R}_l - \omega t)} \tag{6.76}$$

Note that $\mathbf{R}_l$ is the equilibrium position vector of the $l$th unit cell and the coefficient $\eta_b^{\alpha}(\mathbf{k})$ is independent of $l$. Substituting Eq. (6.76) into Eq. (6.75) gives

$$\omega^2 \eta_b^{\alpha}(\mathbf{k}) = \sum_{b',\beta} D_{bb'}^{\alpha\beta}(\mathbf{k}) \eta_{b'}^{\beta}(\mathbf{k}) \tag{6.77}$$

where the dynamic matrix is expressed as

$$D_{bb'}^{\alpha\beta}(\mathbf{k}) = \frac{1}{\sqrt{m_b m_{b'}}} \sum_{l'} \Phi_{0b,l'b'}^{\alpha\beta} \mathrm{e}^{\mathrm{i}\mathbf{k}\cdot(\mathbf{R}_0 - \mathbf{R}_{l'})} \tag{6.78}$$

In writing Eq. (6.78), the translational invariance has been used [40, 41]. For periodic boundary conditions, the dynamic matrix is essentially a Fourier transform of the harmonic force constant matrix. The determinant of the following matrix must be zero, viz.,

$$\left| D_{bb'}^{\alpha\beta}(\mathbf{k}) - \omega^2 \delta_{\alpha\beta} \delta_{bb'} \right| = 0 \tag{6.79a}$$

which can be written in a matrix form,

$$\det\left[\mathbf{D}(\mathbf{k}) - \omega^2 \mathbf{I}\right] = 0 \tag{6.79b}$$

This allows the determination of phonon dispersion curves for all polarizations in the first Brillouin zone. The analysis of a 1D chain with two atoms was illustrated in Sect. 6.5.1. If a unit cell has one atom, there will be three polarizations for each wavevector; if a unit cell has two atoms, there will be six polarizations (or branches), and so forth. For 2D and 3D crystals, the dispersion curves are rather complicated since they are different along different directions between the lattice points [45], similar to the electron band structures. Note that Fig. 6.19a only plots Si dispersion curves along the direction from the zone center Γ to point X in the first Brillouin zone, while Fig. 6.19b is for several SiC polytypes along the direction with a maximum magnitude of wavevector at the zone edge.

Equation (6.77) may be written in terms of the eigenvectors $\mathbf{e}_{b,\lambda}$ of the normal modes, where $\lambda$ specifies a (phonon) mode, in the following [41]:

$$\omega^2 e^{\alpha}_{b,\lambda}(\mathbf{k}) = \sum_{b',\beta} D^{\alpha\beta}_{bb'}(\mathbf{k}) e^{\beta}_{b',\lambda}(\mathbf{k}) \tag{6.80}$$

For a system with $N$ atoms, there exists $3N$ discrete eigenvectors that are orthogonal and normalized. These normal modes or Bloch modes represent $3N$ harmonic vibrational modes or phonons with specific wavevectors. The corresponding wavevectors and frequencies (eigenvalues) can be obtained from the dynamic matrix. All vibrational motion can be expressed as a superposition of the normal modes. As discussed previously, the phonon DOS, group velocity, and specific heat can be calculated using the dispersion relations. While the harmonic vibrations are for low temperatures, anharmonic vibrations do not affect the dispersion significantly unless the temperature is very high (e.g., close to the melting temperature). Anharmonic lattice dynamics can be used to predict the frequency shift, linewidth, and lifetime by including anharmonic perturbation or using the self-consistent phonon formulation [78, 97].

*Fermi's golden rule*, or simply the golden rule, was derived from the time-dependent perturbation theory in quantum physics to express the scattering rate or decay rate. The general form of the Fermi golden rule can be applied to nuclear decay, atomic electron transitions, interband and intraband transitions, electron scattering, phonon scattering, etc. For a system that undergoes a transition from an initial state described by the wavevector $\mathbf{k}$ to a final state denoted by $\mathbf{k}'$, the golden rule can be expressed as [42, 90]

$$\gamma_{\mathbf{k}\to\mathbf{k}'} = \sum_{\mathbf{k}'} \frac{2\pi}{\hbar} M^2_{\mathbf{k}\mathbf{k}'} \delta(E_{\mathbf{k}'} - E_{\mathbf{k}} \mp \hbar\omega) \tag{6.81}$$

where $\gamma$ is the transition rate (probability of transition per unit time) and its inverse is called the lifetime or relaxation time, $M_{\mathbf{k}\mathbf{k}'}$ is the interaction matrix element for transition from state $\mathbf{k}$ to $\mathbf{k}'$, and the Dirac delta function signifies energy conservation. Note that momentum conservation is also necessary for the transition to occur as discussed in Sect. 6.5.3. As an example, for electron–phonon interaction, $\omega$ is the frequency of the emitted or absorbed phonon, referring to Eq. (6.45). The term $\hbar\omega$ can be dropped when electrons are scattered elastically such as in the case of electron-defect scattering. Note that $M_{\mathbf{kk}'}$ can be expressed in terms of a suitable Hamiltonian that represents the scattering potential (or strength) so that

$$M_{\mathbf{kk}'} = \frac{1}{V} \int_V \Psi_{\mathbf{k}'}^* \hat{H}_{\text{scat}} \Psi_{\mathbf{k}} d\mathbf{r} \tag{6.82}$$

The Fermi golden rule can be expressed in terms of phonon–phonon scattering involving three or four phonons [50, 90]. This is very important for modeling anharmonic scattering considering three-phonon and four-phonon processes. In the evaluation of the interaction matrix elements, the IFCs of the third order allow the determination of the three-phonon processes (both $N$-process and $U$-process) and fourth-order IFC allows the determination of four-phonon processes [50, 81]. Formulations for defect scattering and interfacial scattering can also be obtained based on Fermi's golden rule [78]. For heavily doped semiconductors, phonon–electron scattering may also play a role in reducing the lattice thermal conductivity [81, 98]. The detailed formulations and computational methodology can be found from the cited literature.

### 6.6.3 Evaluation of Thermal Conductivity

This section describes several methods used to calculate thermal conductivity of insulators and semiconductors. These methods can be used for both bulk and nanostructured materials if boundary scattering is included. They can also be used to study thermal transport in inhomogeneous media and thermal transport across interfaces, though the focus of this chapter is mostly on the intrinsic properties of solids.

If Fourier's law of heat conduction is extended to include anisotropy, we can write

$$q''_\beta = -\sum_\alpha \kappa_{\alpha\beta} \frac{\partial T}{\partial \alpha} \tag{6.83}$$

Here again, $\alpha$ or $\beta$ specifies a coordinate axis. The Peierls–Boltzmann equation or phonon BTE can be solved using relaxation time approximation for each mode. The thermal conductivity can be obtained by comparing the BTE solution with Eq. (6.83) [75, 78]:

$$\kappa_{\alpha\beta} = \kappa_{\alpha\beta} = \sum_P \sum_K c(\mathbf{k}) v_\alpha(\mathbf{k}) v_\beta(\mathbf{k}) \tau(\mathbf{k}) \tag{6.84}$$

Here, $c(\mathbf{k}) = h\omega(\mathbf{k})\partial f_{\text{BE}}/\partial T$, and $v_\alpha$ and $v_\beta$ are the $\alpha$ and $\beta$ components of the group velocity, respectively. These quantities can be obtained from the dispersion relations. The scattering rate $\tau(\mathbf{k})$ for each mode $\lambda$ (in terms of the wavevector index $K$ and polarization index $P$) is obtained from the golden rule as discussed previously. Equation (6.84) is written in a way that is consistent with Eq. (5.71) but including anisotropy. The thermal conductivity tensor becomes a diagonal tensor $\bar{\bar{\kappa}} = \text{diag}(\kappa_{xx}, \kappa_{yy}, \kappa_{zz})$ in the principal coordinates and becomes $\bar{\bar{\kappa}} = \kappa\mathbf{I}$ for an isotropic medium. Since 2007, this method has been used to obtain the thermal conductivity of numerous semiconductor materials and oxide [78–81].

In the past 20 years, molecular dynamics simulations have been extensively used in modeling thermal transport and thermal properties of bulk and nanostructured materials. A brief introduction was given in Sect. 4.5.1. Molecular dynamics is advantageous for modeling crystals whose basis contains a large number of atoms, unstructured and disordered materials (such as amorphous solids and liquid phase), as well as soft matters (such as polymer and organic compounds) [76, 82–85, 99]. It is particularly suitable for modeling nanostructured materials such as nanotubes, nanowires, and fullerenes and related materials. Classical MD simulations are based on empirical or semi-empirical intermolecular or interatomic potentials. The number of atoms that can be simulated is rather limited by the earlier computational capabilities. The computational speed and capabilities have been significantly improved in the last two decades. Nowadays, advanced large-scale molecular dynamics simulation packages, e.g., LAMMPS [100], are accessible. Various functional forms obtained by comparison with experiments and from ab initio methods are available to model the interatomic potentials of different types of materials. The combination of these functions and parameters is often called a *force field* in molecular modeling. Another advantage of MD simulation of thermal transport is that it inherently includes anharmonicity and is particularly suitable at high temperatures. The wave characteristics of lattice vibration (phonon) cannot be easily observed since MD is a time-domain simulation. This problem has been addressed using various post-processing techniques such as Fourier analysis of the time variation of the locations and velocities of particles, wave-packet analysis, and mode decomposition [80, 82–84].

There are two MD simulation methods in studying thermal transport. One is called the direct method or nonequilibrium molecular dynamics (NEMD) and the other is often called the Green–Kubo method or equilibrium molecular dynamics (EMD). NEMD is more intuitive since it is based on 1D Fourier's law to determine the thermal conductivity:

$$k_x = -\frac{q_x''}{\partial T/\partial x} \tag{6.85}$$

In the direct simulation, either the temperatures at both ends of the structure (heat baths) are preset or the heat flux is preset. The simulation is then run for a sufficient duration of time using millions of time steps, each at subpicosecond timescale, to determine the unknown heat flux or the temperature gradient, respectively [85]. Periodic boundary conditions may be used to model bulk solids. Equation (6.85) is then used to calculate the thermal conductivity. In the MD simulation, the temperature is calculated according to [82]

$$T = \frac{1}{3Nk_{\mathrm{B}}}\left\langle \sum_{i}^{N} m_i v_i^2 \right\rangle \tag{6.86}$$

In the study where stress is considered, the pressure can be calculated by the virial equation [83],

$$P = \frac{Nk_{\mathrm{B}}T}{V} + \frac{1}{3V}\left\langle \sum_{i}\sum_{j>i} \mathbf{r}_{ij} \cdot \mathbf{F}_{ij} \right\rangle \tag{6.87}$$

In general, the temperature distribution is not linear due to temperature jumps near the two heat baths. However, the temperature profile can be approximated as linear in the middle section. The size effect and length effect on thermal conductivity of nanostructures can be easily modeled [76]. NEMD can also be used to study interface thermal resistance as well as non-Fourier heat conduction using the transient method [85]. These topics will be further discussed in Chap. 7.

In EMD, the Green–Kubo linear response theory is used. Statistical fluctuations always exist at temperatures exceeding absolute zero, causing random motion of atoms. If the sampling time is sufficiently small, the summation of the magnitude (or the square of the fluctuation term) depends on the temperature and the properties of the system. EMD is based on the Green–Kubo fluctuation-dissipation theorem and the thermal conductivity is computed from the *fluctuating heat current* as follows [82–84]:

$$k_{\alpha\beta} = \frac{1}{Vk_{\mathrm{B}}T^2}\int_0^\infty \langle J_\alpha(t) J_\beta(0)\rangle \mathrm{d}t \tag{6.88}$$

where $V$ is the simulation volume, and $J_\alpha$ or $J_\beta$ is the *heat current* component in the $\alpha$ or $\beta$ direction, respectively, since the heat current is a vector. Note that $\alpha = \beta$ along the principle thermal conductivity coordinates. The operator <> signifies ensemble average. The fluctuating heat current vector can be expressed as follows:

$$\mathbf{J} = \frac{\mathrm{d}}{\mathrm{d}t}\sum_i \varepsilon_i \mathbf{r}_i \tag{6.89a}$$

where $\varepsilon_i$ is the total energy of the $i$th particle that includes both kinetic and potential energies. Note that the heat current defined here has a unit of heat flux multiplied by volume or [W m]. If only pairwise potentials are considered, we have

$$\mathbf{J} = \sum_i E_i \mathbf{v}_i + \frac{1}{2}\sum_{i,j} (\mathbf{F}_{ij} \cdot \mathbf{v}_i)\mathbf{r}_{ij} \tag{6.89b}$$

Three-body interaction terms can also be introduced and may be necessary to describe the force field, such as in the Stillinger–Weber potential commonly used in modeling diamond structure semiconductors [82].

By analyzing the EMD simulation in the reciprocal lattice space or *k*-space (sometimes called phonon space) and using the theory of lattice dynamics, phonon relaxation time and mean free paths can be extracted [80, 83, 84]. The phonon DOS can be obtained by a Fourier transform of the velocity autocorrelation function [80]. The spectral partition ratio and spatial energy distribution can be obtained using the DOS spectrum. These quantities are important to assess phonon confinement and localization especially in determination of the minimum thermal conductivity. The ensemble average of the heat current is called the heat current autocorrelation function, which is key to perform phonon space analysis in the first Brillouin zone. The analysis can provide information on the phonon spectral energy distribution and relaxation time. Anharmonic effects can also be examined [84]. The uncertainties associated with EMD calculations have also been systematically investigated [101].

First-principles calculations can be used to obtain harmonic and anharmonic force field and then applied to calculate the thermal conductivity [102, 103]. Recently, ab initio MD simulations have also been applied to model thermal conductivity [104–106].

## 6.7 Electron Emission and Tunneling

In all the discussions given so far, electrons are confined to the solid. Emission or discharge of electrons from a solid surface to vacuum or through a barrier (such as in a metal–insulator–metal multilayer structure) is possible, under the influence of an incident electromagnetic wave, an electric field, or a heating effect. Because of the importance of electron emission and tunneling to fundamental physics and device applications, the basic concepts are described in this section.

### *6.7.1 Photoelectric Effect*

In 1887, Heinrich Hertz observed the *photoelectric effect* or *photoemission*. Shortly afterward, the phenomenon was experimentally studied by several others, including J. J. Thomson, who discovered electron as a subatomic particle. When radiation is incident on a metal plate, the electrons in the metal can be excited by absorbing the energy of the electromagnetic wave to escape the surface, as illustrated in Fig. 6.23a. The actual apparatus used for measuring the ejected photoelectrons was to use another electrode and measure the current flow via a closed circuit. This is similar to the arrangement shown in Fig. 6.23b for thermionic emission, but with photons incident on the left plate without heating up any of the plates. If the frequency of the incident radiation is not high enough, no electrical current can be measured no matter how

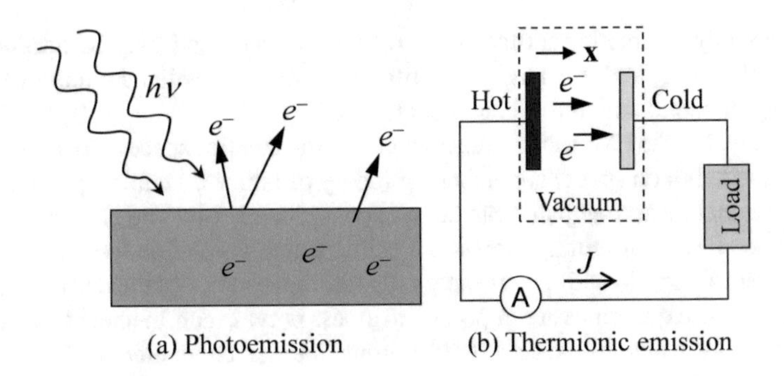


**Fig. 6.23** Illustration of **a** the photoelectric effect and **b** the thermionic emission

intense the incident radiation is. Saying in other words, there appears to be a threshold frequency for photoemission to occur in a given material. The photoelectric effect was explained in 1905 by Albert Einstein with the concept of light quanta, postulated by Max Planck a few years earlier. Although Einstein also made seminal contributions to the theory of relativity and Brownian motion, he was awarded the Nobel Prize in Physics in 1921 mainly for his discovery of the law that governs the photoelectric effect.

From the Fermi–Dirac distribution function of free-electron gas, we can see that at low temperatures, electrons fill all energy levels up to the Fermi energy $E_F$. Note that we use $E$ and $E_F$ as the relative electron energy and, thus, they can be either positive or negative. Because of the binding of the electron with the rest of the solid, an additional energy, called the *work function* $\psi$, must be provided to the electron for it to escape from the solid. For Ag, Al, Au, Cu, Fe, Pb, and W, the work function ranges from 4 to 5 eV, which corresponds to a wavelength in the ultraviolet region from 250 to 300 nm. For Na, K, Cs, and Ca, the work function ranges from 2 to 3 eV, which falls in the visible spectrum. Because a photon can interact with only one electron at a time, the photon energy $h\nu$ must exceed the work function in order for the incident radiation to eject electrons from the surface. If $h\nu > \psi$, the photon energy may be absorbed by an electron right at the Fermi level. Subsequently, the electron will have a kinetic energy of

$$\tfrac{1}{2}m_e v_{e,\max}^2 = h\nu - \psi \tag{6.90}$$

after leaving the surface. If an electron is below the Fermi level, the kinetic energy of the ejected electron will be smaller than that given by Eq. (6.90). Therefore, Eq. (6.90) predicts the *maximum kinetic energy* of an electron for the prescribed photon frequency. A direct method for the determination of the work function is to measure the kinetic energy distribution of the photoelectrons, for a given frequency of the incident radiation.

One of the applications of photoemission is to measure the electron binding energy using the x-ray photoelectron spectroscopy (XPS), which is also called the electron spectroscopy for chemical analysis (ESCA). The basic principle for XPS is

$$E_{\mathrm{bd}} = h\nu - \tfrac{1}{2}m_{\mathrm{e}}v_{\mathrm{e}}^2 - \psi \tag{6.91}$$

where $E_{\mathrm{bd}}$ stands for the binding energy with respect to the Fermi energy. The high-energy photons from an x-ray source (200–2000 eV) can interact with the inner electrons and eject them out of the surface. The photoelectron intensity can be plotted as a function of the electron kinetic energy using an electron energy analyzer. The intensity peaks are associated with the binding energies of the particular atomic structures. Comparing with the recorded photoelectron spectra, XPS allows the determination of the chemical composition of the substance near the surface. Swedish physicist Kai Siegbahn shared the Nobel Prize in Physics in 1981 for his contribution leading to the practical application of XPS. Furthermore, ultraviolet photoemission spectroscopy (UPS) with photon energies ranging from 5 to 100 eV, often from a synchrotron radiation source, has been used to study the band structures of crystalline solids [107].

### 6.7.2 *Thermionic Emission*

The charge emission from hot bodies was independently discovered by British scientist Frederich Guthrie in 1873, with a heated iron ball, and Thomas Edison in 1880, while working on his incandescent bulbs. *Thermionic emission* was extensively studied in the early 1900s by Robert Millikan, Nobel Laureate in Physics in 1923; Owen Richardson, Nobel Laureate in Physics in 1928; and Irving Langmuir, Nobel Laureate in Chemistry in 1932, among others.

With the understanding of the work function as the threshold energy that an electron must gain to escape the solid, it becomes straightforward to explain the emission of electrons from a heated metal. We use metal here to illustrate thermionic emission because good conductors can be better approximated by the Sommerfeld theory. The distribution function of a free-electron gas has been extensively discussed in Chap. 5 (Sect. 5.1.3). At absolute zero temperature, all states below the Fermi level are filled by electrons and all states above the Fermi level are empty. Note that this picture is consistent with the electronic band theory. At elevated temperatures, the distribution function is modified as illustrated in Fig. 5.5. Some electrons will have energies above $E_{\mathrm{F}}$ (or $\mu_{\mathrm{F}}$ as was used in Chap. 5). Because the distribution function becomes zero only when $E \to \infty$, a small fraction of electrons must occupy energy levels exceeding $E_{\mathrm{F}} + \psi$. We wish to quantitatively evaluate the current density or the charge flux from the hot plate to the cold plate, as illustrated in Fig. 6.23b. Let the electron flow be along the $x$-direction.

From Eq. (5.16), the number of electrons per unit volume between $\mathbf{v}$ and $\mathbf{v} + \mathrm{d}\mathbf{v}$ is

$$n(\mathbf{v})\mathrm{d}v_x\mathrm{d}v_y\mathrm{d}v_z = 2\left(\frac{m_{\mathrm{e}}}{h}\right)^3 \frac{\mathrm{d}v_x\mathrm{d}v_y\mathrm{d}v_z}{\mathrm{e}^{(E-E_{\mathrm{F}})/k_{\mathrm{B}}T} + 1} \tag{6.92}$$

where $E = \frac{1}{2}m_e(v_x^2 + v_y^2 + v_z^2)$ is the kinetic energy of an electron. The current density in the $x$-direction is given by

$$J_x = (1 - r') \iiint (-e)v_x n(\mathbf{v})\mathrm{d}v_x \mathrm{d}v_y \mathrm{d}v_z \tag{6.94}$$

where $r'$ is the electron reflection coefficient or the fraction of electron reflected by the receiver.

The integration is from $-\infty$ to $\infty$ in both the $y$- and $z$-directions. In order for an electron to escape in the $x$-direction, the following criterion must be satisfied:

$$v_x > v_{x,0} = \sqrt{2(E_F + \psi)/m_e} \tag{6.95}$$

This equation suggests that the integration is carried out only in the tail of the distribution function, where the $x$ velocity is positive and the kinetic energy is sufficiently large, i.e., $E - E_F > \psi$, which is on the order of several electron volts. Note that $k_B T = 0.086$ eV at 1000 K and 0.026 eV at 300 K. When $\psi/k_B T = 4$, dropping the unity term in the denominator of Eq. (6.92) causes less than 2% error. The error becomes even smaller at a larger $v_x$ so that its impact on the integration is negligibly small. For this reason, it appears safe to substitute the Fermi–Dirac distribution by the Maxwell–Boltzmann distribution, viz.,

$$J_x = -2e(1 - r')\left(\frac{m_e}{h}\right)^3 \mathrm{e}^{E_F/k_B T} \int_{v_{x,0}}^{\infty} v_x \exp\left(-\frac{m_e v_x^2}{2k_B T}\right)\mathrm{d}v_x$$
$$\times \int_{-\infty}^{\infty}\int_{-\infty}^{\infty} \exp\left(-\frac{m_e v_y^2}{2k_B T} - \frac{m_e v_z^2}{2k_B T}\right)\mathrm{d}v_y \mathrm{d}v_z \tag{6.95}$$

The result is the famous *Richardson–Dushman equation* for the current density [1, 2]:

$$J = A_{RD}(1 - r')T^2 \mathrm{e}^{-\psi/k_B T} \tag{6.96}$$

where $A_{RD} = 4\pi m_e e k_B^2/h^3 = 1.202 \times 10^6$ A/m$^2$ K$^2$ is called the *Richardson constant*, and the direction of $J$ is as shown in Fig. 6.23b. The heat transfer associated with the electron flow can be evaluated by considering the kinetic energy associated with each electron, $\frac{1}{2}m_e v^2 \approx \frac{1}{2}m_e v_x^2$, so that

$$q_x'' = (1 - r') \iiint v_x\left(\tfrac{1}{2}m_e v_x^2\right) f(\mathbf{v})\mathrm{d}v_x \mathrm{d}v_y \mathrm{d}v_z = (\psi + E_F + k_B T)\frac{J_x}{e} \tag{6.97}$$

This equation suggests that the average energy of the "hot electron" is $\psi + E_F + k_B T$, as expected. Vacuum tubes operate based on the principle of thermionic emission. Vacuum tubes had wide applications in the mid-twentieth century in radio, TV, and computer systems, but have largely been replaced by transistors nowadays. Thermionic generators produce electricity without any moving parts and belong to the category of direct energy converters. Extensive discussion of the thermodynamics and efficiency of thermionic converters can be found from Hatsopoulos and Gyftopoulos [108].

In some applications, a voltage can be applied between the electrodes. Furthermore, a semiconductor can be used to form a Schottky barrier between a metal and a semiconductor [11]. The applied voltage changes the potential distribution so that it gradually decreases inside the barrier. Furthermore, the work function can be significantly reduced. Assuming the transmission coefficient is unity, Eq. (6.96) can be modified to the following for the net charge transfer:

$$J_{\text{net}} = A^* T^2 e^{-\psi^*/k_B T} (e^{e\Delta V/k_B T} - 1) \tag{6.98}$$

where $A^*$ should be calculated according to the effective mass, $\psi^*$ is the effective work function, and $\Delta V$ is the applied voltage [11]. In deriving Eq. (6.98), we assumed that hot electrons from the cathode will go through the barrier through ballistic processes. This means that the electron mean free path must be larger than the thickness of the semiconductor film. Otherwise, the electron transport is governed by diffusion because of collisions with phonons or impurities. When diffusion occurs, the electron transport under the influence of a temperature difference is described by the thermoelectric effect, based on irreversible thermodynamics, as discussed in Chap. 5. When the barrier thickness is extremely small, another phenomenon called quantum tunneling may occur such that an electron whose energy is lower than the potential barrier has a chance to transmit through the barrier. Tunneling effect will be discussed in the next subsection. Mahan and coworkers pointed out that, for the thermionic phenomenon to be the dominant transport mechanism, the electron mean free path in the barrier must be greater than the thickness of the barrier [109]. Furthermore, the latter must exceed the characteristic length, below which tunneling becomes significant. Thermionic emission in semiconductor heterogeneous structures has been extensively studied in the last decades for both refrigeration and power generation [109–111]. The refrigeration process is a reversed thermionic power generation process. In *thermionic refrigeration*, the cold cathode emits electrons to the room-temperature anode as a result of the applied voltage. In order to achieve any cooling effect, energy that is carried through by the electrical current must be greater than that by heat conduction via lattice vibration from the hot electrode to the cold electrode. The nonequilibrium electron and phonon transport phenomena have also been investigated. In some cases, both thermionic and thermoelectric effects may show up [111, 112]. In other cases, thermionic and tunneling effects can work together or against each other [113, 114].

### 6.7.3 Field Emission and Electron Tunneling

From the above discussion, we have noticed that thermionic emission may be enhanced or even reversed (from a colder cathode to a hotter anode) by an applied electric field. Some thermal excitation is necessary for part of the electrons to occupy energy levels above the Fermi level by a finite amount, prescribed by the work function. This is commonly referred to as a *potential barrier* or a *potential hill*. An electron must acquire sufficient energy for it to surmount the barrier. When the field strength is very high, however, electrons at energy levels lower than the height of the barrier can tunnel through the potential hill. The word *tunneling* gives a vivid (but inaccurate) picture of the tunneling phenomenon as if a hole were drilled for the electrons without sufficient energy to pass through a potential hill, without climbing to its top first. This phenomenon of electron emission at high applied field is called *field emission*, which can occur at very low temperatures. The applied electric field can exceed several billion volts per meter. Because of the high field, field emission can occur only in ultrahigh vacuum (UHV); otherwise, ionization of the gas molecules would occur that can cause *discharge glow*. In essence, field emission is a form of *quantum tunneling*, which cannot be understood within the framework of classical mechanics. The electron motion is governed by Schrödinger's wave equation, and the transmission can be predicted by the probability of finding an electron on the other side of the potential hill, as illustrated in Fig. 6.24.

In 1928, Fowler and Nordheim [115] provided the first quantum mechanical derivation of the field emission current density $J$ as follows:

$$J = C\left(\frac{\Delta V}{L}\right)^2 \exp\left(-\frac{\alpha\psi^{3/2}}{\Delta V/L}\right) \tag{6.99}$$

This is called the Fowler–Nordheim equation, in which $\Delta V/L$ is the electric field, $C$ and $\alpha$ are two positive constants, and $\psi$ is the work function defined previously.

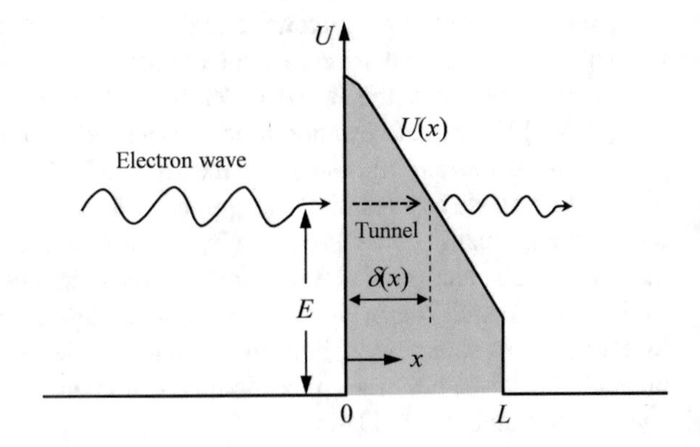


**Fig. 6.24** Illustration of quantum tunneling through a potential barrier by an electron wave

The WKB approximation is commonly used to find the transmission probability $\tau'$ of tunneling. WKB (also KWB or BWK) stands for Wentzel, Kramers, and Brillouin, although a fourth person Jeffreys was also included in some literature—so, the abbreviation appeared as JWKB. The main assumption in the WKB approximation is that the potential $U(x)$ is a slow function of $x$ [116]. In the region where the electron energy $E$ is greater than $U(x)$, the wave function is of the form

$$\Psi(x,t) = A\exp(-\,\mathrm{i}\omega t)\exp\left[\pm\frac{\mathrm{i}}{\hbar}\sqrt{2m_\mathrm{e}(E-U)}\right] \tag{6.100}$$

where $A$ is the amplitude of the electron wave. In the region where $E < U(x)$, the wave function is of the form

$$\Psi(x,t) = A\exp(-\,\mathrm{i}\omega t)\exp\left[\pm\frac{1}{\hbar}\sqrt{2m_\mathrm{e}(U-E)}\right] \tag{6.101}$$

The transmission probability or transmission coefficient can be approximated as

$$\tau'(E) = \exp\left[-\frac{2}{\hbar}\int_0^\delta \mathrm{d}x\sqrt{2m_\mathrm{e}(U-E)}\right] \tag{6.102}$$

where $\delta$ is the width of the potential at $E$ [116].

**Example 6.5** Consider a potential barrier in the region $0 \le x \le L$ whose potential is the highest but linear decreases with $x$ according to $U(x) = \psi - \frac{x}{L}e\Delta V$ and $\delta(E) = \frac{\psi - E}{e\Delta V}L$. Find the transmission coefficient.

**Solution** For the triangular barrier shown in Fig. 6.24, we note that

$$\int_0^\delta \mathrm{d}x\sqrt{U-E} = \int_0^\delta \mathrm{d}x\sqrt{\psi - E - e\Delta V x/L} = \frac{(\psi-E)^{3/2}}{e\Delta V/L}\int_1^0\sqrt{1-u}\mathrm{d}u = \frac{2(\psi-E)^{3/2}}{3e\Delta V/L}$$

Substituting this equation into Eq. (6.102), we obtain

$$\tau'(E) = \exp\left[-\frac{4(\psi-E)^{3/2}}{3\hbar\Delta V/L}\sqrt{2m_\mathrm{e}}\right] \tag{6.103}$$

When $E << \psi$, we see that $\tau' \approx \exp\left(-\frac{\alpha\psi^{3/2}}{\Delta V/L}\right)$, where $\alpha = \frac{4\sqrt{2m_\mathrm{e}}}{3\hbar}$. At elevated temperatures, however, we need to consider the energy distribution of electrons. Esaki and coworkers demonstrated that the resonant tunneling of electron waves may allow the transmission coefficient to approach unity in superlattice and double-barrier structures [117]. Electron tunneling is similar to photon tunneling of electromagnetic waves, to be discussed in Chap. 10.

The tunneling current density can be calculated by

$$J_t = \int_{E_{min}}^{E_{max}} e\tau'(E)n(E)\mathrm{d}E \tag{6.104}$$

where $E$ is the kinetic energy in the $x$-direction, $E_{max}$ corresponds to the energy at the top of the potential barrier, $E_{min}$ is a reference energy, and $n(E)\mathrm{d}E$ is the number of available electrons, with energy between $E$ and $E + \mathrm{d}E$, per unit area per unit time, given as

$$n(E) = \frac{m_e k_B T}{2\pi^2\hbar^3} \ln\left[1 + \exp\left(-\frac{E - E_F}{k_B T}\right)\right] \tag{6.105}$$

Some analytical expressions similar to Eq. (6.100) have been presented [118, 119] to approximate the integration of Eq. (6.104).

The energy transfer during field emission or electron tunneling can also be evaluated [113, 120, 121]. A salient difference between thermionic emission and field emission is that thermionic emission always gives out energy as the electrons are emitted and transfer the energy to the other side of the barrier. This is because the emitted electrons are in the high-energy tail of the distribution function, called hot electrons, with a much higher effective temperature than the equilibrium cathode temperature. On the other hand, field emission allows electrons with energies much lower than that corresponding to the equilibrium temperature to escape the surface. Since the replacement electrons have a higher average energy than the emitted electrons, a heating effect occurs that increases the cathode temperature. Depending on the geometry, temperature, transmission coefficient, and energy distribution, both heating and cooling of the cathode are made possible by field emission. This is known as the Nottingham effect originally published in 1941.

Some applications of quantum tunneling in semiconductors and superconductors were discussed in Chap. 1. One of the applications of electron tunneling was the invention of scanning tunneling microscope (STM). Xu et al. [120] developed a model for the energy exchange by the tunneling electrons and made a comparison with STM measurements. They considered the Nottingham effect on both electrodes, as well as resistive heating. At short distances, thermionic emission, field emission, and photon tunneling could occur simultaneously. Photon tunneling will be studied in Chap. 10. Fisher and Walker [121] analyzed the energy transport in nanoscale field emission processes by considering the geometry of the emission tip. Quantum size effect may play a role in modifying some of the critical parameters. Field emission by nanotubes has been proposed for nanoscale manufacturing and thermal writing [122]. Wong et al. [123] performed a detailed thermal analysis during electron beam heating and laser processing. Carbon nanotube field emission display (CNT-FED) has been demonstrated. While CNT-FEDs resemble the cathode-ray tubes (CRTs) in many ways, it can be made thin and flat with a much lower applied voltage [124].

## 6.8 Electrical Transport in Semiconductor Devices

Semiconductors are the most important materials for microelectronics, MEMS, and optoelectronics. Much of the discussions in Chap. 5 and the previous sections of this chapter are applicable to semiconductors, especially for the energy storage and transport by phonons. This section focuses on the basics of electrical transport and properties for some common semiconductor devices used in optoelectronics.

### *6.8.1 Number Density, Mobility, and the Hall Effect*

The calculation of the number density of electrons and holes at any given temperature $T$ is very important for the determination of the electrical, optical, and thermal properties of semiconductor materials and devices. The free-electron gas model can be modified to describe the electron and hole distributions and the transport in semiconductors. The Fermi–Dirac distribution function is applicable to electrons and holes according to

$$f_{\mathrm{e}}(E) = \frac{1}{\mathrm{e}^{(E-E_{\mathrm{F}})/k_{\mathrm{B}}T}+1} \quad \text{and} \quad f_{\mathrm{h}}(E) = \frac{1}{\mathrm{e}^{(E_{\mathrm{F}}-E)/k_{\mathrm{B}}T}+1} \tag{6.106}$$

Note that $f_{\mathrm{e}}(E) + f_{\mathrm{h}}(E) \equiv 1$. The number density of electrons or holes is given by

$$n_{\mathrm{e}} = \int_{E_{\mathrm{C}}}^{\infty} \frac{D_{\mathrm{e}}(E)\mathrm{d}E}{\mathrm{e}^{(E-E_{\mathrm{F}})/k_{\mathrm{B}}T}+1} \quad \text{and} \quad n_{\mathrm{h}} = \int_{-\infty}^{E_{\mathrm{V}}} \frac{D_{\mathrm{h}}(E)\mathrm{d}E}{\mathrm{e}^{(E_{\mathrm{F}}-E)/k_{\mathrm{B}}T}+1} \tag{6.107}$$

where $D_{\mathrm{e}}(E)$ and $D_{\mathrm{h}}(E)$ are the densities of states in the conduction and valence bands, respectively. With the approximated quadratic forms of the conduction and valence bands, Eq. (6.29), the densities of states can be written as

$$D_{\mathrm{e}}(E) = M_{\mathrm{C}} \frac{\mathrm{d}\mathbf{k}}{\mathrm{d}E}\bigg|_{\mathrm{C}} = \frac{M_{\mathrm{C}}}{2\pi^2}\left(\frac{2m_{\mathrm{e}}^*}{\hbar^2}\right)^{3/2}(E-E_{\mathrm{C}})^{1/2} \tag{6.108}$$

and

$$D_{\mathrm{h}}(E) = \frac{\mathrm{d}\mathbf{k}}{\mathrm{d}E}\bigg|_{\mathrm{V}} = \frac{1}{2\pi^2}\left(\frac{2m_{\mathrm{h}}^*}{\hbar^2}\right)^{3/2}(E_{\mathrm{V}}-E)^{1/2} \tag{6.109}$$

where $M_{\mathrm{C}}$ is the number of equivalent minima in the conduction band. Equations (6.108) and (6.109) are derived based on the parabolic shape near the bottom of the conduction band for electrons or the top of the valance band for holes.

The effective mass of electrons is a geometric average over the three major axes because the effective mass of silicon depends on the crystal direction. The effective mass of holes is an average of heavy holes and light holes because there exist different subbands [11]. At moderate temperatures, $E_C - E_F >> k_B T$ and $E_F - E_V >> k_B T$ are satisfied; subsequently, $f_e$ and $f_h$ can be approximated with the classical Maxwell–Boltzmann distribution:

$$f_e(E) \approx e^{(E_F - E)/k_B T} \quad \text{and} \quad f_h(E) \approx e^{(E - E_F)/k_B T} \tag{6.110}$$

We can carry out the integrations in Eq. (6.107) and thus obtain

$$n_e = N_C e^{-(E_C - E_F)/k_B T} \tag{6.111}$$

and

$$n_h = N_V e^{-(E_F - E_V)/k_B T} \tag{6.112}$$

where

$$N_C = 2M_C \left( \frac{m_e^* k_B T}{2\pi \hbar^2} \right)^{3/2} \quad \text{and} \quad N_V = 2 \left( \frac{m_h^* k_B T}{2\pi \hbar^2} \right)^{3/2} \tag{6.113}$$

are called the *effective density of states* in the conduction band and in the valance band, respectively. The combination of Eqs. (6.111) and (6.112) gives, in terms of $E_g = E_C - E_V$,

$$n_e n_h = N_{th}^2 = N_C N_V e^{-E_g/k_B T} \propto T^3 e^{-E_g/k_B T} \tag{6.114}$$

This expression does not involve the Fermi energy. Therefore, it holds for both intrinsic and doped semiconductors. The number density $N$th can be viewed as thermally excited electron–hole pairs per unit volume. It is also referred to as the number density of intrinsic carriers because $n_e = n_h = N_{th}$, in an intrinsic semiconductor. It can be seen that the number densities increase with temperature so that the electrical conductivity of an intrinsic semiconductor increases with temperature. The Fermi energy for an intrinsic semiconductor can be obtained by setting $n_e = n_h$ in Eqs. (6.111) and (6.112), yielding

$$E_F = \frac{E_C + E_V}{2} + \frac{k_B T}{2} \ln\left( \frac{N_V}{N_C} \right) \approx \frac{E_C + E_V}{2} \tag{6.115}$$

The Fermi energy for an intrinsic semiconductor is expected to lie in the middle of the forbidden band or the bandgap. The requirement for the approximate distributions given in Eq. (6.110) to hold with less than 2% error is $E_g/k_B T > 8$, such that $\exp[-E_g/(2k_B T)] < 0.02$. For $E_g > 0.8$ eV, we have $T < 1150$ K. One should keep in mind that $E_g$ reduces as temperature increases. For silicon, $E_g \approx 1.11$ eV at 300 K and $\approx 0.91$ eV at 900 K.

When impurities of either donors or acceptors or both are involved, the calculation of Fermi energy and number densities becomes more involved [11, 15]. Let $N_D$ and $N_A$ stand, respectively, for the number densities (i.e., doping concentrations) of donors (e.g., P and As) and acceptors (e.g., B and Ga). In brief, the energy level of donors $E_D$ is usually lower but very close to $E_C$. As a result, the Fermi energy $E_F$ goes up but is always below $E_D$. The difference $E_C - E_D$ is called the ionization energy of donors, which is required for the donors to become ionized. The ionization of donors increases the number of free electrons, and the semiconductor is said to be of $n$-type. For the semiconductor Si, the ionization energy for P is 45 meV and that of As is 54 meV. Likewise, the energy level of acceptors $E_A$ is slightly above $E_V$, and $E_A - E_V$ is called the ionization energy of acceptors. The ionization of acceptors increases the number of holes, and the semiconductor is said to be of $p$-type. For the semiconductor Si, the ionization energy for B is 45 meV and that of Ga is 72 meV. Note that there are $5.0 \times 10^{22}$ cm$^{-3}$ (atoms per cubic centimeters) for silicon. For $n$-type silicon with an arsenic doping concentration of $N_D = 5.0 \times 10^{16}$ cm$^{-3}$, the impurities occupy one atomic site per million. Because of the change in Fermi energy, most of the impurities are ionized at room temperature when the doping concentration is less than $5.0 \times 10^{17}$ cm$^{-3}$. For fully ionized impurities, the charge neutrality requires that

$$n_e + N_A = n_h + N_D \tag{6.116}$$

If the impurities are partially ionized, $N_D$ and $N_A$ in Eq. (6.116) should be replaced by the ionized donor and acceptor concentrations, respectively.

**Example 6.6** For boron-doped Si with $N_A = 2.0 \times 10^{16}$ cm$^{-3}$, find $N_{th}$, $n_e$, and $n_h$ at temperatures from 300 to 1000 K; compare your answers with the values for intrinsic silicon. Assume $m_e^* = 0.3m_e$ and $m_h^* = 0.6m_e$. Use $M_C = 6$ and $E_g(T) = 1.155 - 0.000473T^2/(T + 636)$ eV where $T$ is in kelvin.

**Solution** This is a $p$-type semiconductor with $N_D = 0$, and from Eq. (6.116), we have $n_h = n_e + N_A$. Substituting it into Eq. (6.114), we have $n_e^2 + N_A n_e = N_{th}^2 = N_C N_V e^{-E_g/k_B T}$. The solution gives

$$n_e = \frac{1}{2}\left(\sqrt{N_A^2 + 4N_{th}^2} - N_A\right) \text{ and } n_h = \frac{1}{2}\left(\sqrt{N_A^2 + 4N_{th}^2} + N_A\right) \tag{6.117}$$

The calculated values of $N_C = 2.47 \times 10^{19}$ cm$^{-3}$ and $N_V = 1.17 \times 10^{19}$ cm$^{-3}$ at 300 K are somewhat lower than the recommended values of $N_C = 2.86 \times 10^{19}$ cm$^{-3}$ and $N_V = 2.66 \times 10^{19}$ cm$^{-3}$ [11]. The results are plotted in Fig. 6.25 for comparison. In the extrinsic region when $T < 700$ K, the majority carriers are holes, and $n_h \approx N_A$ depends little on temperature. In the intrinsic region when $T > 800$ K, $n_e \approx n_h \approx N_{th}$, due to thermal excitation. It should be mentioned that at very low temperatures,

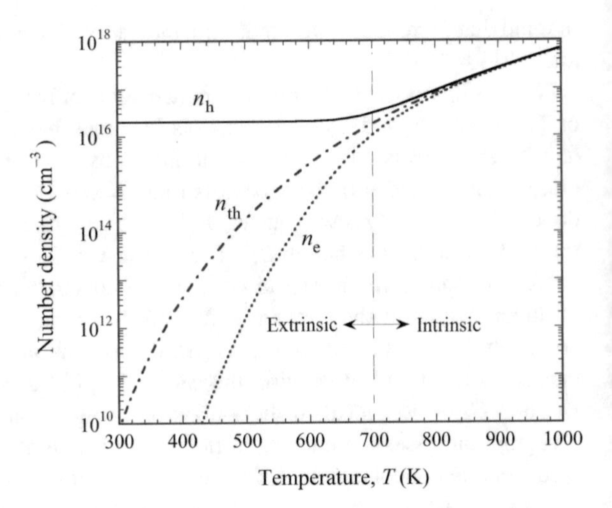


**Fig. 6.25** Calculated number densities for Example 6.6

i.e., $T < 100$ K, ionization is not very effective and $n_h << N_A$. Therefore, the low-temperature region is called freeze-out zone, which is not shown in the plot.

The Drude free-electron model predicts $\sigma = \tau n_e e^2/m_e$, as given in Eq. (5.52), which can be applied for both electrons and holes, using proper effective masses and relaxation times. In semiconductor physics, the electron or hole mobility is defined based on the effective mass:

$$\mu_e = \frac{e\tau_e}{m_e^*} \quad \text{and} \quad \mu_h = \frac{e\tau_h}{m_h^*} \tag{6.118}$$

The physical significance is that mobility is the drift velocity per unit applied field, i.e.,

$$\mathbf{u}_{d,e} = -\mu_e \mathbf{E} \quad \text{and} \quad \mathbf{u}_{d,h} = \mu_h \mathbf{E} \tag{6.119}$$

The electrical conductivity of a semiconductor is thus

$$\sigma = en_e\mu_e + en_h\mu_h \tag{6.120}$$

Depending on the impurity and temperature range, one term may be dominant, or both the terms may be comparable. It is crucial to understand the scattering mechanism in semiconductors. In metals, all the conducting electrons are near the Fermi surface, and their average energy cannot be described by the classical statistics because $\bar{\varepsilon} = \frac{1}{2}m_e\overline{v_e^2} \approx \frac{3}{5}\mu_F \neq \frac{3}{2}k_BT$ (see Example 5.2). For semiconductors, on the other hand, the Boltzmann distribution given in Eq. (6.110) suggests that $\bar{\varepsilon} = \frac{1}{2}m_e^*\overline{v_e^2} \approx \frac{3}{2}k_BT = \frac{1}{2}m_e^*v_{th}^2$, and the classical statistics is applicable to a large temperature range. Thermal velocity is the velocity of electrons or holes at the equilibrium temperature and was given in Eq. (6.55). At sufficiently high temperatures when phonon scattering dominates, the electron mean free path $\Lambda_e \propto T^{-1}$. Based on the relation $\tau_e = \Lambda_e/\bar{v}_e$, we have

$$\mu_{\text{ph}} \propto T^{-3/2} \tag{6.121}$$

where $\mu_{\text{ph}}$ is the contribution of carrier–phonon scattering. Equation (6.121) describes intrinsic semiconductor without defects. The scattering by impurities results in a mobility given by

$$\mu_{\text{d}} \propto T^{3/2}/N_{\text{d}} \tag{6.122}$$

where $N_d$ stands for the concentration of the ionized impurities. The combination gives the mobility for either electron or hole as follows:

$$\frac{1}{\mu} = \frac{1}{\mu_{\text{ph}}} + \frac{1}{\mu_{\text{d}}} \tag{6.123}$$

For intrinsic semiconductor, the electrical conductivity is very small and proportional to $\exp\left[-E_{\text{g}}/(2k_{\text{B}}T)\right]$ so that the electrical conductivity increases with temperature. For intermediately doped semiconductors, there exists a maximum value of the mobility below room temperature due to the opposite temperature dependence of $\mu_{\text{ph}}$ and $\mu_{\text{d}}$. At that temperature, the electrical conductivity is maximum. As the temperature goes up beyond room temperature, the conductivity decreases due to the increased phonon scattering. When the semiconductor reaches the intrinsic region, the number density suddenly increases and the conductivity increases again with temperature.

The Hall effect is very useful in measuring the mobility of semiconductors. In the extrinsic region, the Hall effect allows measurement of the type and concentration of the carriers. The measurements are usually carried out with the van der Pauw method, which is a four-probe technique for determining the electrical resistance and the Hall coefficient. The data of electrical resistivity and number density allow the extraction of the mobility, based on the effective mass determined using cyclotron resonance technique.

When both the carriers are significant to the transport properties, the situation is rather interesting. Referring to Fig. 6.1, when current flows to the positive $x$-direction, we have $v_{\text{e},x} < 0$ and $v_{\text{h},x} > 0$. The magnetic force drives both the electrons and the holes toward the negative $y$-direction, such that $v_{\text{e},y} < 0$ and $v_{\text{h},y} < 0$ if $E_y = 0$. At steady state, a finite $E_y$, known as the Hall field, may exist. Since there is no net current flow in the $y$-direction, we must have

$$J_x = -ev_{\text{e},x}n_{\text{e}} + ev_{\text{h},x}n_{\text{h}} \tag{6.124}$$

$$J_y = -ev_{\text{e},y}n_{\text{e}} + ev_{\text{h},y}n_{\text{h}} = 0 \tag{6.125}$$

In general, both $v_{e,y}$ and $v_{h,y}$ are not zero. The Lorentz force $\mathbf{F} = q(\mathbf{E}+\mathbf{u}_d \times \mathbf{B})$ in the $y$-direction is related to the drift velocities for electrons or holes by

$$-eE_y + ev_{e,x}B = ev_{e,y}/\mu_e \tag{6.126}$$

and

$$eE_y - ev_{h,x}B = ev_{h,y}/\mu_h \tag{6.127}$$

Rewrite Eqs. (6.126) and (6.127) as $n_e\mu_e(E_y - v_{e,x}B) = -n_e v_{e,y}$ and $n_h\mu_h(E_y - v_{h,x}B) = n_h v_{h,y}$, respectively. Compared with Eq. (6.122), we notice that $n_e\mu_e(E_y - v_{e,x}B) + n_h\mu_h(E_y - v_{h,x}B) = 0$, or

$$\frac{E_y}{B} = \frac{n_e\mu_e v_{e,x} + n_h\mu_h v_{h,x}}{n_e\mu_e + n_h\mu_h}$$

Combining it with Eq. (6.121), we obtain the Hall coefficient as follows:

$$\eta_H = \frac{E_y}{J_x B} = \frac{n_e\mu_e v_{e,x} + n_h\mu_h v_{h,x}}{e(n_e\mu_e + n_h\mu_h)(-n_e v_{e,x} + n_h v_{h,x})} \tag{6.128}$$

Substituting $v_{e,x} = -\mu_e E_x$ and $v_{h,x} = \mu_h E_x$ into the previous equation, we obtain

$$\eta_H = \frac{n_h\mu_h^2 - n_e\mu_e^2}{e(n_h\mu_h + n_e\mu_e)^2} \tag{6.129}$$

after canceling $E_x$. The Hall coefficient for semiconductors may be positive or negative, and becomes zero when $n_h\mu_h^2 = n_e\mu_e^2$. The drift velocities in the $y$-direction, however, cannot be zero unless $B = 0$ or $J_x = 0$.

### 6.8.2 Generation and Recombination

The generation, recombination, and diffusion processes are directly related to the charge transport in semiconductors and optoelectronic devices. This section takes photoconductivity as an example to illustrate the generation and recombination processes, followed by a brief discussion of luminescence.

Much has been said previously about absorption of light that causes a transition in the electronic states in solids. The bandgap absorption of Si, Ge, and GaAs corresponds to the wavelengths in the visible and near-infrared spectral regions. The excitation of electrons from the valence band to the conduction band by the absorption of radiation increases the conductivity of the semiconductor dramatically. This is known as *photoconductivity* and can be used for sensitive radiation detectors. For some semiconductors, the bandgap is very narrow so that transitions can happen at longer wavelengths. For example, the bandgap energy of $Hg_{0.8}Cd_{0.2}Te$ is 0.1 eV at 77 K (liquid nitrogen temperature), and the material can be used as infrared detectors, which are commonly referred to as MCT detectors. At very low temperatures, impurities cannot be ionized thermally even though the ionization energy is very small. For boron-doped germanium, the acceptor ionization energy $E_A - E_V \approx 10\,\text{meV}$ corresponds to a wavelength of about 120 μm [125]. Therefore, Ge:B can be used as far-infrared radiation detectors. There are two groups of radiation detectors. The first group is called thermal or bolometric detectors, which rely on the temperature change of the detector as a result of the absorbed radiation. The temperature change can be monitored by a temperature-dependent property, such as the electrical resistance. An example is the superconductive bolometer, which relies on the drastic change in resistance with temperature, near the superconducting-to-normal-state transition or critical temperature $T_c$. The second group is called nonthermal, nonbolometric, or nonequilibrium detectors. An example is the photoconductive detector in which the conductivity changes as a result of the direct interaction of electrons with photons.

Before the radiation is incident on the photoconductive detector, the conductivity can be expressed as $\sigma_0 = en_{e,0}\mu_e + en_{h,0}\mu_h$ at thermal equilibrium. Under the influence of an incident radiation with photon energies greater than the bandgap, additional electron–hole pairs are created so that the concentration is increased by $\Delta n$ for both types of carriers. The relative change in the electrical conductance $\Delta\sigma/\sigma_0$ can be expressed as

$$\frac{\Delta\sigma}{\sigma_0} = \frac{\Delta n(\mu_e + \mu_h)}{n_{e,0}\mu_e + n_{h,0}\mu_h} \tag{6.130}$$

Here, $\Delta n$ is the net increase in carrier concentration as a result of both generation and recombination. The *generation* is associated with the absorbed radiation and depends on the intensity of the incident light and the *quantum efficiency*, which is wavelength dependent. The quantum efficiency is the percentage of the incoming photons that generate an electron–hole pair. The *recombination* is a relaxation process because the excess charges are not at thermal equilibrium. If the incident radiation is blocked off, the semiconductor will quickly reach an equilibrium with the conductivity $\sigma_0$. The characteristic time of the recombination process is called the *recombination lifetime* or *recombination time* $\tau_{rc}$. While it is also related to electron scattering, lattice scattering, and/or defect scattering, the recombination time is usually much longer than the relaxation time used in charge transport processes. The net rate of change can be expressed as the rate of generation (creation) minus the rate of recombination (annihilation), viz.,

$$\frac{dn}{dt} = \dot{n}_g - \frac{n - n_0}{\tau_{rc}} \tag{6.131}$$

Under a steady-state incident radiation, we can set $dn/dt = 0$ so that $\Delta n = n - n_0 = \tau_{rc}\dot{n}_g$. Suppose that the incoming photon is of frequency $\nu$ in Hz with a spectral irradiance $I_\nu$ in W/m$^2$ Hz, and the detector has an effective area $A$, thickness $d$, and absorptance $\alpha_\nu$. We obtain the rate of generation:

$$\dot{n}_{\mathrm{g}} = \frac{\alpha_\nu I_\nu A}{h\nu A d} = \frac{\alpha_\nu I_\nu}{h\nu d} \tag{6.132}$$

Substituting into Eq. (6.130), we obtain the *sensitivity* of a photoconductive detector as follows:

$$\frac{1}{I_\nu}\frac{\Delta\sigma}{\sigma_0} = \frac{\alpha_\nu \tau_{\mathrm{rc}}(\mu_{\mathrm{e}} + \mu_{\mathrm{h}})}{h\nu d(n_{\mathrm{e},0}\mu_{\mathrm{e}} + n_{\mathrm{h},0}\mu_{\mathrm{h}})} \tag{6.133}$$

Increasing the recombination time $\tau_{\mathrm{rc}}$ improves the sensitivity but decreases the speed or response time of the detector. Photoconductivity requires that $h\nu > E_{\mathrm{g}}$ for bandgap absorption to occur. However, the sensitivity decreases toward higher frequencies, or shorter wavelengths, because there are fewer photons per unit radiant power. Consequently, the sensitivity of a photoconductive detector increases with wavelength first and then suddenly drops to zero close to the band edge. For thin films, the absorptance depends on the film thickness when the photon penetration depth is comparable to the thickness.

In photoconductivity, the recombination is not associated with the emission of radiation, and therefore it is said to be nonradiative. The Auger effect and multiphonon emission are two common processes of nonradiative recombination. In the Auger effect, the energy released by a recombining electron–hole pair is absorbed by another electron in the conduction band, which subsequently relaxes to the equilibrium condition by the emission of phonons. In a multiphonon emission process, the recombination of an electron–hole pair is associated with the release of a cascade of phonons, each having a much lower energy. More details on the recombination process and how to calculate the associated lifetime can be found from the texts of Sze [11].

Radiative recombination can also occur and is very important for light-emitting applications, such as *luminescence*, which is essentially the inverse process of absorption. The excitation of electrons may be accomplished by passing through an electrical current. An example is the semiconductor light-emitting diode, in which the electronic transition from the conduction band to the valence band can result in optical radiation. Photoluminescence is often referred to as *fluorescence*, when the emission occurs at the same time as the absorption, or *phosphorescence*, when the emission continues for a while after the excitation.

### *6.8.3 The* p-n *Junction*

The *p-n* junction is familiar to every reader although many of us are unfamiliar with the underlying physics. Let us first take a look at the charge transport by diffusion, which is a very important process in semiconductor applications. Diffusion takes place when there is a spatial nonuniformity in the carrier concentration. The principle is the same as the diffusion of ideal gas molecules described in Sect. 4.2.3. Using Fick's law, we can write the current densities resulting from the diffusion of electrons and holes as follows:

$$J_\mathrm{e} = eD_\mathrm{e}\frac{\mathrm{d}n_\mathrm{e}}{\mathrm{d}x} \quad \text{and} \quad J_\mathrm{h} = -eD_\mathrm{h}\frac{\mathrm{d}n_\mathrm{h}}{\mathrm{d}x} \tag{6.134}$$

where the diffusion coefficient for electrons and holes can be related to the mean free path and the average velocity by $D_\mathrm{e} = \frac{1}{3}\Lambda_\mathrm{e}\bar{v}_\mathrm{e}$ and $D_\mathrm{h} = \frac{1}{3}\Lambda_\mathrm{h}\bar{v}_\mathrm{h}$, according to Eq. (4.42). Assuming $\bar{v}_\mathrm{e} \approx \bar{v}_\mathrm{h} \approx v_\mathrm{th}$, we have $D_\mathrm{e} = \frac{1}{3}\Lambda_\mathrm{e}v_\mathrm{th} = \frac{1}{3}\tau_\mathrm{e}v_\mathrm{th}^2$. Combined with $\frac{1}{2}m_\mathrm{e}^*v_\mathrm{th}^2 = \frac{3}{2}k_\mathrm{B}T$, we obtain

$$D_\mathrm{e} = \frac{\tau_\mathrm{e}k_\mathrm{B}T}{m_\mathrm{e}^*} = \frac{\mu_\mathrm{e}k_\mathrm{B}T}{e} \tag{6.135}$$

which is known as the *Einstein relation*. A similar equation holds also for the holes. In transient heat conduction, the *thermal diffusion length* is usually calculated by $l_\mathrm{th} = \sqrt{\alpha t}$, where $\alpha = \kappa/\rho c_p$ is the thermal diffusivity and $t$ is a characteristic time. The *diffusion length* for electrons is defined as $l_\mathrm{dif} = \sqrt{D_\mathrm{e}\tau_\mathrm{e}}$, which is proportional to $\tau_\mathrm{e}$ and $v_\mathrm{th}$. The diffusion velocity is sometimes defined as $v_\mathrm{dif} = l_\mathrm{dif}/\tau_\mathrm{e} = v_\mathrm{th}/\sqrt{3}$. The factor of $\sqrt{3}$ reduction arises because the diffusion velocity is the average thermal velocity along one direction only. In semiconductors, charge transfer is a combined effect of the carrier drift and diffusion. Electron diffusion is not important for metals because of the large drift velocity given by the Fermi velocity, which changes little with temperature at moderate temperatures.

Through oxidation, lithography, diffusion and ion implantation, and metallization, semiconductor *p*-*n* junctions can be fabricated with microelectronics manufacturing technology [11]. A *p*-*n* junction consists of a *p*-type semiconductor, with a high hole concentration, joined with an *n*-type semiconductor, with a high electron concentration, as shown in Fig. 6.26.

If one compares Fig. 6.26a with Fig. 4.7b, the process looks similar to a binary diffusion. Because of the concentration gradient, holes will diffuse right and electrons will diffuse left. Diffusion causes the region near the interface to be depleted, so that there are fewer free holes on the left side and fewer free electrons on the right side of the depletion region. Keep in mind that electrons and holes are charged particles. As they leave the host material, ions of opposite charges are left behind. This results in a charge accumulation, as shown in Fig. 6.26a, that leads to a built-in potential in the depletion region that will inhibit further diffusion. As a consequence of this built-in potential, the energy in the *p*-doped region is raised relative to that in the *n*-doped region, as shown in Fig. 6.26b. The Fermi level is the same everywhere, and it is closer to the conduction band for the *n*-type and the valence band for the *p*-type.

**Example 6.7** Prove that the Fermi energy in a *p*-*n* junction is independent of $x$ at thermal equilibrium, as shown in Fig. 6.26b.

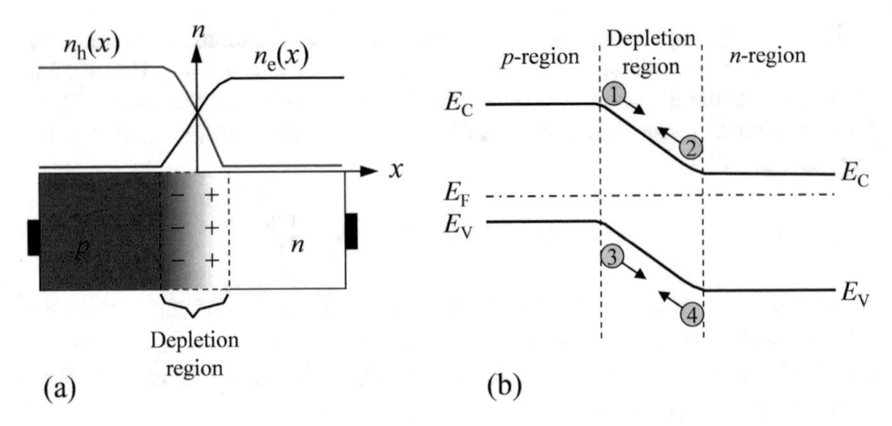


**Fig. 6.26** Schematic of a *p-n* junction at thermal equilibrium. **a** The device and carrier concentrations, including the charge distribution in the depletion region. The width of the depletion region is exaggerated for clarity. **b** The energy band diagram for the $p - n$ junction near the depletion region. The dash-dotted line is the Fermi level. The four processes are (1) electron drift, (2) electron diffusion, (3) hole diffusion, and (4) hole drift

**Solution** Without any externally applied voltage, the current densities become

$$J_e = J_{e,drif} + J_{e,diff} = -en_e\mu_e\frac{dV_0}{dx} + eD_e\frac{dn_e}{dx} \tag{6.136}$$

$$J_h = J_{h,drif} + J_{h,diff} = -en_h\mu_h\frac{dV_0}{dx} - eD_h\frac{dn_h}{dx} \tag{6.137}$$

where $V_0$ is the built-in potential and the electric field is $-dV_0/dx$. Because a high potential $V_0$ means a smaller electron kinetic energy, we have

$$\frac{dV_0}{dx} = -\frac{1}{e}\frac{dE_C}{dx} = -\frac{1}{e}\frac{dE_V}{dx} \tag{6.138}$$

From Fig. 6.26, we see that the built-in electric field in the depletion region points toward the negative $x$-direction and thus $dV_0/dx > 0$; consequently, $dE_C/dx < 0$ and $dE_V/dx < 0$. Employing Eq. (6.110), we notice that

$$\frac{1}{n_e}\frac{dn_e}{dx} = \frac{1}{k_BT}\left(\frac{dE_F}{dx} - \frac{dE_C}{dx}\right) \text{ and } \frac{1}{n_h}\frac{dn_h}{dx} = \frac{1}{k_BT}\left(\frac{dE_V}{dx} - \frac{dE_F}{dx}\right) \tag{6.139}$$

Substituting Eqs. (6.135), (6.138), and (6.139) into Eq. (6.136) and setting $J_e = 0$, we end up with

$$\frac{dE_F}{dx} = 0 \tag{6.140}$$

This equation can also be derived using Eq. (6.137) in the $p$-region; hence, the Fermi energy $E_F$ is independent of $x$ at thermal equilibrium.

A popular application of $p$-$n$ junction is as a diode rectifier, which allows current to flow easily with a forward bias but becomes highly resistive when the bias is reversed. For the configuration shown in Fig. 6.26, a forward bias means that the electrical field is in the positive $x$-direction, opposite to the built-in field. Qualitatively, this can be understood as a forward bias removes the barrier for holes to diffuse right and for electrons to diffuse left. On the other hand, a reverse bias creates an even stronger barrier for these diffusion processes. Quantitatively, it can be shown that for an externally applied voltage $V$ (positive for forward bias and negative for reverse bias), the current density can be expressed as

$$J = J_0\left[\exp\left(\frac{eV}{k_B T}\right) - 1\right] \tag{6.141}$$

where $J_0$ is the saturation current density, which depends on the diffusion coefficient, scattering time, number density, and other factors. Noting that

$$\frac{dJ}{dV} = \frac{eJ_0}{k_B T}\exp\left(\frac{eV}{k_B T}\right) \tag{6.142}$$

The electrical conductance increases with $V$ for forward bias, and decreases to zero as $V \to -\infty$. It should be noted that in practice, the width of the depletion region is often less than 0.5 μm and the built-in potential may be around 1 V through the depletion region. There is actually a very large built-in field.

Heterojunction is a junction of dissimilar semiconductors with different bandgap energies. The energy band diagram can be very different from that shown in Fig. 6.26b. The Fermi energies can be different on each side. Bipolar transistors were invented in 1947 at Bell Labs. It is based on two $p$-$n$ junctions arranged in a $p$-$n$-$p$ or $n$-$p$-$n$ configuration. Field-effect transistors (FETs) work on a different principle. Referring to Fig. 1.3 in Chap. 1, free electrons cannot move from the source to the drain because of the lack of free carriers in the $p$-type wafer. If a negative voltage is applied to the gate, electrons below the gate will be pushed even further, and there is still little chance for the electron to flow from the source to the drain. However, as soon as a positive voltage is applied to the gate, electrons will be attracted to the region below it and form a path for electricity to flow from the source to the drain. Furthermore, a transistor can amplify the signal since only a weak signal is necessary to the gate. Metal-oxide-semiconductor field-effect transistors (MOSFETs) have become the most important device in contemporary integrated circuits. Thermal management is important for such devices because of the local heating or hot spots where Fourier's law often fails to predict the temperature history. More discussion on nonequilibrium heat conduction will be given in Chap. 7. A brief discussion on photovoltaic devices will be given next.

### 6.8.4 Optoelectronic Applications

The photovoltaic effect is a direct energy conversion process in which electromagnetic radiation, incident upon a *p-n* junction, generates electron–hole pairs. The built-in electric field in the *p-n* junction tends to push the generated holes to the *p*-region and the generated electrons to the *n*-region, resulting in a reverse photocurrent. Solar cells and photovoltaic detectors have been developed and applied for over half a century. Thermophotovoltaic (TPV) devices have also been considered as energy conversion systems that allow recycling of the waste heat [126]. Figure 6.27 shows a typical TPV cell and the associated electrical circuit. When the incident radiation with a photon energy greater than the bandgap energy $E_g$ of the cell material strikes the *p-n* junction, an electron–hole pair is generated at the location as each photon is absorbed. Carriers generated in the depletion region are swept by the built-in electric field and then collected by the electrodes at the ends of the cell, resulting in a drift current. For radiation absorbed near the depletion region, the minority carriers (electrons in the *p*-region and holes in the *n*-region) tend to diffuse toward the depletion region, yielding a diffusion current. If the load resistance $R_L$ is zero, i.e., in the case of a short circuit, there is a photocurrent $I_{sc}$ flowing in the circuit due to the combination of the diffusion and drift of charge carriers. The direction of this current is indicated

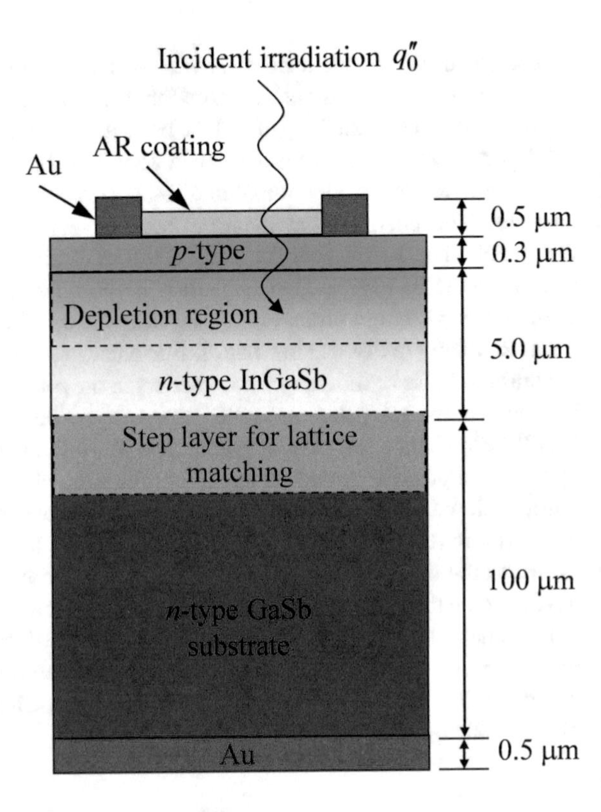


**Fig. 6.27** Schematic of a typical TPV cell [126]

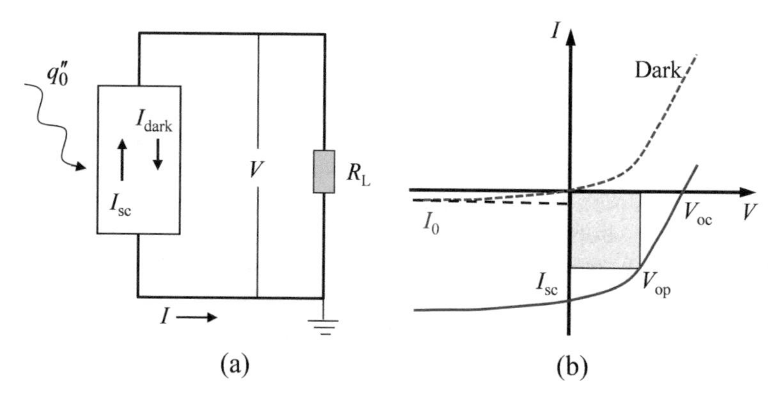


**Fig. 6.28** Schematic of **a** the circuit diagram and **b** the $I$–$V$ characteristics of a PV cell

on Fig. 6.28a. If the circuit is open, or the load resistance $R_L$ approaches infinite, a positive open-circuit voltage is built up due to irradiation. This gives the maximum voltage $V = V_{max}$, when no current flows through the load, i.e., $I = 0$. When the load has a finite resistance $R_L$, a voltage $V$ is developed, not only across the load but also across the photovoltaic cell. This voltage reduces the built-in potential of the cell as if a forward bias is applied to the *p-n* junction. Subsequently, the diffusion of minority carriers produces a forward current, which is called the *dark current* in photovoltaic devices. The current $I$ flowing through the load resistor becomes

$$I = -I_{sc} + I_0\left[\exp\left(\frac{eV}{k_B T}\right) - 1\right] \tag{6.143}$$

The first term on the right is the photocurrent, or the short-circuit current, which depends on the incident photon flux, quantum efficiency, as well as the transport properties. The second term on the right is the dark current $I_{dark}$ with $I_0$ being the saturation current, as previously given in Eq. (6.141) based on the current density. The dark current is zero, when $V = 0$, as shown in Fig. 6.28b. Consider the photovoltaic cell, shown in Figs. 6.27 and 6.28a. If the total (net) irradiance $q_0'' = 0$, then without bias voltage, $I = I_{sc} = 0$, which is the condition of thermal equilibrium. When $q_0'' \neq 0$, which can be the situation where the photovoltaic cell is exposed to a high-temperature emitter, the $I$–$V$ curve is shifted down by $I_{sc}$ as shown in Fig. 6.28b. The output power is determined by the product of $|IV|$, which can be optimized to yield the maximum output at the optimal point. Basu et al. [126] provided an extensive review of the operation principle and the state of the art in TPV technology, as well as the potential application of microscale radiative heat transfer for performance improvement. Near-field TPV will be further discussed in Chap. 10.

Light-emitting diodes (LEDs) are based on *p-n* junctions as well but with direct gap semiconductors. At low forward bias voltages, the recombination processes are essentially nonradiative. At high forward bias voltages, however, radiative recombination results in the emission of photons; this phenomenon is called electroluminescence. The emission is a spontaneous process and is therefore incoherent. Depending on the materials used and their bandgaps, LEDs can emit in the ultraviolet, visible, and infrared regions. It should be noted that electroluminescence can also be employed to create a refrigeration effect [127].

Semiconductor lasers are based on the stimulated emission process, as discussed in Chap. 3, and have numerous important applications due to their small size, portability, and ease of operation. Semiconductor lasers have been used in laser printers, optical fiber communication, CD reading/writing, and so forth. The key is to create population inversion so that lasing can occur. Quantum well lasers, based on quantum confinement, offer significant advantages over conventional semiconductor lasers, such as low threshold current, high output power, high speed, and so forth. Further explanation of the optical and electronic characteristics of semiconductor lasers can be found from the books of Sze [11] and Zory [128], for example.

## 6.9 Summary

This chapter began with an introduction to the atomic structures, chemical bonds, and crystal lattices. Emphasis was placed on electronic band structures and phonon dispersion relations, allowing one to gain a deeper knowledge of solid-state physics, beyond the previous chapter. Electronic band structures of metal, semiconductor, and 2D materials (especially graphene) are introduced. Phonon dispersion and scattering mechanisms are presented, along with some coverage of coherent phonons. The basic concepts of atomistic simulations based on first principles and molecular dynamics methods are delineated with extensive references of recent literature. Photoelectric effect, thermionic emission, and field emission were described in subsequent sections to stress the interrelation between these phenomena. The basic electrical transport processes in semiconductors, such as number density, mobility, electrical conductivity, charge diffusion, and photoconductivity, were explained. The *p-n* junction was discussed along with applications, such as photovoltaic cells, thermophotovoltaics, LEDs, luminescent refrigeration, and semiconductor lasers.

## Problems

6.1. Consider a phosphorus-doped 250-μm-thick silicon wafer, with a doping concentration of $10^{17}$ cm$^{-3}$. The applied current is 10 mA and the magnetic induction is 0.5 T.

(a) Determine the Hall coefficient and the Hall voltage, assuming there is only one type of carriers.

(b) For a chip of area $1 \times 1$ cm$^2$ and resistivity 0.075 Ω cm, what is the voltage drop along the direction of current flow?

6.2. Consider the Hall experiment arranged in Fig. 6.1, under steady-state operation and with uniform magnetic field. Assume a current is flowing in the $y$-direction.

(a) Show that $v_x = -(e\tau/m_e)E_x - \omega_c \tau v_y$ and $v_y = -(e\tau/m_e)E_y + \omega_c \tau v_x$, when the current is carried by electrons. Here, $\tau$ is the relaxation time, $v_x$ and $v_y$ are the electron drift velocities in the $x$- and $y$-directions, respectively, and $\omega_c = eB/m_e$ is called the *cyclotron frequency*.
(b) Prove Eq. (6.1) by setting $v_y = 0$.

6.3. Express the electron configurations for Ag and Au. Based on the orbital occupation of outer electrons, discuss the similarities in their chemical and electrical properties.

6.4. Express the electron configurations for Ca and Zn. Based on the orbital occupation of outer electrons, discuss the similarities in their chemical and electrical properties.

6.5. Give a general discussion of insulators, semiconductors, and metals. Explain why glass ($SiO_2$) is transparent, silicon wafers appear dark, and aluminum foils look bright. What are the types of chemical bonds in $SiO_2$, Si, and Al?

6.6. How many billiard balls can you pack in a basket with a volume of 0.25 m$^3$? Assume that the balls are rigid spheres with a diameter $d$ = 43 mm and mass $m$ = 46 g. Arrange the spheres in a crystal lattice according to the diamond, simple cubic, bcc, fcc, and hcp structures. What is the total weight for each arrangement? [Hint: Show that for close-packed spheres, the fraction of volume occupied by the spheres is $\sqrt{3}\pi/16 \approx 0.340$ (diamond), $\pi/6 \approx 0.524$ (simple cubic), $\sqrt{3}\pi/8 \approx 0.680$ (bcc), and $\sqrt{2}\pi/6 \approx 0.740$ (fcc or hexagonal close - packed).]

6.7. (a) Count the number of atoms inside a unit cell of $YBa_2Cu_3O_7$ as shown in Fig. 6.6d, and confirm that it is the same as that in the basis.
(b) Find the density of $YBa_2Cu_3O_7$ crystal based on the dimensions of the unit cell, noting that the molecular weight $M$ = 88.9 (Y), 137.3 (Ba), 63.6 (Cu), and 16.0 (O) kg/kmol.

6.8 (a) Calculate the diameter and cross-sectional area for CNTs with chiral indices $(m, n)$ = (5, 5), (8, 8), (10, 10), (10, 20), and (20, 40).
(b) Take (40, 40) SWNTs of 10-μm length, with a thermal conductivity $\kappa$ = 3200 W/m K at room temperature. Align sufficient nanotubes to make a bundle with a diameter of 1 μm; how many CNTs are needed?
(c) Neglect the effect of interface and defects on the thermal conductivity. What is the heat transfer rate if the temperatures at both ends are 320 and 300 K?
(d) Compare the heat transfer rate if the CNT is replaced by a Si nanowire of 1-μm diameter and 10-μm length.

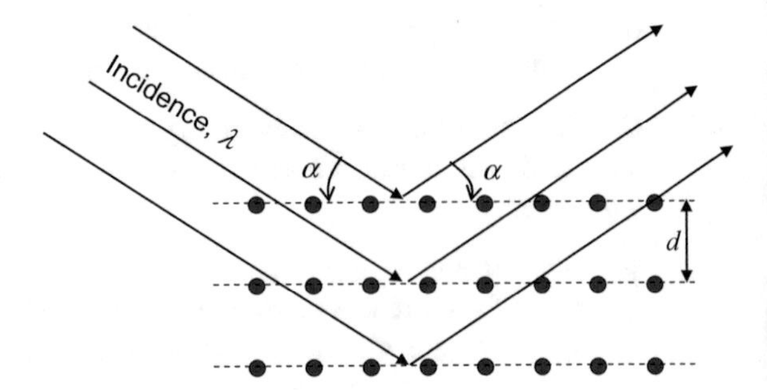


**Fig. 6.29** Schematic of Bragg's x-ray diffraction experiment

6.9. The interatomic potential for a KBr crystal can be expressed as $\phi(r) = -\frac{\alpha_M e^2}{4\pi\varepsilon_0 r} + C\left(\frac{a}{r}\right)^m$, where $\alpha_M$ is the Madelung constant, which is 1.748 for crystals with NaCl structure, $\varepsilon_0 = 8.854 \times 10^{-12}$ $C^2$/J m is the electric permittivity of vacuum, $a = 6.60 \times 10^{-10}$ m is the lattice constant, $m = 8.85$, and $C = 2.65 \times 10^{-21}$ J for KBr. Note that $r$ is in meter.

(a) Plot the attractive potential, the repulsive potential, and the combined potential in eV as a function of $r$ in Å.
(b) Find the equilibrium distance, which should be the nearest distance between $K^+$ and $Br^-$ ions.
(c) At the equilibrium distance, what are the attractive and repulsive forces between each ion pair?

6.10. Bragg's x-ray diffraction formula relates the angle $\alpha$ of diffraction maximum and the x-ray wavelength $\lambda$ as follows: $2d \sin\alpha = n\lambda$, where $n$ is the refractive index that can be taken as unity in the x-ray region, $d$ is the spacing between adjacent layers of atoms, and $\alpha$ is measured between the incidence and the crystal plane, as shown in Fig. 6.29. This formula can be understood by the constructive interference between the two layers.

(a) To measure a spacing $d = 3.12$ Å, what is the maximum wavelength $\lambda$ that can still be used to perform the experiment successfully?
(b) In an x-ray experiment, $\lambda = 1.5$ Å. Assume that the errors in $\lambda$ and $n$ are negligible. How accurately must one determine $\alpha$ in order to measure the spacing with an uncertainty of 0.01 Å?

6.11. Using Eq. (6.10) to show that the reciprocal lattice of a hexagon is also a hexagon, as shown in Fig. 6.3d. Calculate the volumes of the direct and reciprocal lattices in terms of $a$ and $c$.

6.12. Use the Kronig–Penney model to solve the Schrödinger equation for an electron in a square-well array. Referring to Fig. 6.10, assume that the potential function is $U(x) = 0$ at $0 \le x \le (a-b)/2$ and $(a+b)/2 \le x \le a$, and $U(x) = U_0 > 0$ at $(a-b)/2 \le x \le (a+b)/2$. Note that $x = 0$ at the core of atom location and the potential is periodic. Find the conditions for the solutions to exist. For simplicity, you may now assume $b \to 0$ and $U_0 \to \infty$

to obtain the relation for $E(k)$. Plot this function to illustrate the electronic band structure.

6.13. Discuss the difference between interband transitions and transitions that occur within a band for copper. Explain why copper appears reddish brown.

6.14. What is the difference between a direct-bandgap semiconductor and an indirect bandgap semiconductor? Why are Si and GaAs wafers opaque to the visible light?

6.15. Plot the band structure of graphene according to Eq. (6.32) and discuss the main features.

6.16. Why we say electrons in pure graphene or graphene whose Fermi level is zero are massless? How does the free-electron mass in graphene changes with bias voltage?

6.17. Derive Eqs. (6.41) and (6.42) first. Then, plot the phonon dispersion curves for a diatomic chain with mass ratio $m_1/m_2$ equal to 1, 2, 3, and 4. What happens when $m_1/m_2 = 1$?

6.18. Approximate $k_{\max} = \pi/a$, and find the group velocities of LA and TA phonons for Si and SiC at $k = 0.3k_{\max}$, using Fig. 6.19. What is the phase speed at $k = 0.3k_{\max}$ for LO phonon in SiC? [Hint: Convert the unit of $\omega$ from $\text{cm}^{-1}$ to rad/s first.]

6.19. Perform a literature search to discuss phonon–phonon scattering mechanism. When will four-phonon scattering be important? When can four-phonon scattering be neglected?

6.20. Prove Eqs. (6.96) and (6.97). Assume that $\psi = 0.4$ eV and $E_F = 3$ eV, estimate the error in Eq. (6.96) caused by approximating the Fermi–Dirac distribution with the Maxwell–Boltzmann distribution in the numerical evaluation.

6.21. Clearly explain the differences between thermionic emission and field emission.

6.22. For a gallium-doped silicon with $N_A = 5 \times 10^{16}\ \text{cm}^{-3}$, use the information from Example 6.6 to calculate the number density of electrons and holes from 300 to 1000 K. Assume the effect of impurity on the mobility can be neglected, so that $\mu_e = 1450\ \text{cm}^2/\text{V s}$ for electrons, and $\mu_h = 500\ \text{cm}^2/\text{V s}$ for holes at 300 K. Determine the electrical resistivity of the doped silicon from 300 to 1000 K.

6.23. For a single-type doped silicon with $\mu_e = 1350\ \text{cm}^2/\text{V s}$ and $\mu_h = 450\ \text{cm}^2/\text{V s}$ at 400 K, the Hall coefficient is zero. Is this semiconductor $n$-type or $p$-type? What is the impurity concentration? [Hint: Use the parameters given in Example 6.6.]

6.24. For a single-type doped silicon with $\mu_e = 1350\ \text{cm}^2/\text{V s}$, $\mu_h = 450\ \text{cm}^2/\text{V s}$, and $N_{th} = 2\times 10^{10}\ \text{cm}^{-3}$, calculate and plot the Hall coefficient for $p$-type doping, with $N_A$ ranging from 0 to $2 \times 10^{12}\ \text{cm}^{-3}$. Discuss, without calculation, the trend with $n$-type doping.

6.25. For a phosphorus-doped silicon, $N_D = 2 \times 10^{15}\ \text{cm}^{-3}$, $\mu_e = 1350\ \text{cm}^2/\text{V s}$, and $\mu_h = 450\ \text{cm}^2/\text{V s}$ at 300 K. Use the parameters from Example 6.6 as needed.

(a) Calculate the thermal velocity and the diffusion length for the electrons and holes at room temperature.
(b) Find the electrical conductivity at room temperature.
(c) Plot the thermal velocity and wavelength as a function of temperature.

6.26. Plot the $J$–$V$ curve of a $p$-$n$ junction based on Eq. (6.141), using dimensionless groups $J/J_0$ and $eV/k_BT$ as the axes. Discuss the meaning of saturation current (density).

6.27. Based on the $I$–$V$ curve for a photovoltaic cell shown in Fig. 6.28b, explain how to determine the open voltage. How to determine the optimal operating point? The $I$–$V$ curve with irradiation is shifted downward, what if the cell is facing a heat sink at lower temperatures?

## References


1. N.W. Ashcroft, N.D. Mermin, *Solid State Physics* (Harcourt College Publishers, Fort Worth, TX, 1976)
2. C. Kittel, *Introduction to Solid State Physics*, 7th edn. (Wiley, New York, 1996)
3. J. E. Avron, D. Osadchy, R. Seiler, A topological look at the quantum Hall effect. Phys. Today, 38–42 (2003)
4. G.A. Prinz, Magnetoelectronics. Science **282**, 1660–1663 (1998)
5. K. von Klitzing, G. Dorda, M. Pepper, New method for high-accuracy determination of the fine-structure constant based on quantized hall resistance. Phys. Rev. Lett. **45**, 494–497 (1980); K. von Klitzing, The quantized Hall effect. Rev. Mod. Phys. **58**, 519–530 (1986)
6. A. Hartland, The quantum Hall effect and resistance standards. Metrologia **29**, 175–190 (1992)
7. J.P. Eisenstein, H.L. Strörmer, The fractional quantum Hall effect. Science **248**, 1510–1516 (1990)
8. C. Strohm, G.L.J.A. Rikken, P. Wyder, Phenomenological evidence for the phonon Hall effect. Phys. Rev. Lett. **95**, 155901 (2005)
9. J.E. Lay, *Statistical Mechanics and Thermodynamics of Matter* (Harper Collins Publishers, New York, 1990)
10. F. Yang, J.H. Hamilton, *Modern Atomic and Nuclear Physics* (McGraw-Hill, New York, 1996)
11. S.M. Sze, *Physics of Semiconductor Devices*, 2nd edn. (Wiley, New York, 1981); S.M. Sze, *Semiconductor Devices: Physics and Technology*, 2nd edn. (Wiley, New York, 2002)
12. R.J. Cava, Structure chemistry and the local charge picture of copper oxide superconductors. Science **247**, 656–662 (1990)
13. B.I. Choi, Z.M. Zhang, M.I. Flik, T. Siegrist, Radiative properties of Y-Ba-Cu–O films with variable oxygen content. J. Heat Transfer **114**, 958–964 (1992)
14. M.S. Dresselhaus, P.C. Eklund, Phonons in carbon nanotubes. Adv. Phys. **49**, 705–814 (2000)
15. M.A. Omar, *Elementary Solid State Physics: Principles and Applications* (Addison-Wesley, New York, 1975)
16. B. Segall, Fermi surface and energy band of copper. Phys. Rev. **125**, 109–122 (1962)
17. G.A. Burdick, Energy band structure of copper. Phys. Rev. **129**, 138–150 (1963)
18. R.E. Hummel, *Electronic Properties of Materials* (Springer, Berlin, 1993)
19. M.L. Cohen, T.K. Bergstresser, Band structure and pseudopotential from factors for fourteen semiconductors of the diamond and zinc-blende structures. Phys. Rev. **141**, 789–796 (1966)
20. F. Herman, W.E. Spicer, Spectral analysis of photoemission yields in GaAs and related crystals. Phys. Rev. **174**, 906–908 (1968)

21. M.L. Cohen, J.R. Chelikowsky, *Electronic Structure and Optical Properties of Semiconductors* (Springer, Berlin, 1988)
22. A.H. Castro Neto, F. Guinea, N.M.R. Peres, K.S.Novoselov, A.K. Geim, The electronic properties of graphene. Rev. Mod. Phys. **81**, 109–162 (2009)
23. D. Akinwande, N. Petrone, J. Hone, Two-dimensional flexible nanoelectronics. Nat. Commun. **5**, 5678 (2014)
24. P. Avouris, C. Dimitrakopoulos, Graphene: synthesis and applications. Mater. Today **15**, 86–97 (2012)
25. A.A. Balandin, Thermal properties of graphene and nanostructured carbon materials. Nat. Mater. **10**, 569–581 (2011)
26. F. Bonaccorso, L. Colombo, G. Yu, M. Stoller, V. Tozzini, A.C. Ferrari, R.S. Ruoff, V. Pellegrini, Graphene, related two-dimensional crystals, and hybrid systems for energy conversion and storage. Science **347**, 1246501 (2015)
27. X. Luo, T. Qiu, W. Lu, Z. Ni, Plasmons in graphene: recent progress and applications. Mater. Sci. Eng. R **74**, 351–376 (2013)
28. S.B. Trickey, F. Müller-Plathe, G.H. Diercksen, J.C. Boettger, Interplanar binding and lattice relaxation in a graphite delayer. Phys. Rev. B **45**, 4460–4468 (1992)
29. D.W. Boukhvalov, M.I. Katsnelson, A.I. Lichtenstein, Hydrogen on graphene: electronic structure, total energy, structural distortions and magnetism from first-principles calculations. Phys. Rev. B **77**, 035427 (2008)
30. K.S. Novoselov, A.K. Geim, S.V. Morozov, D. Jiang, M.I. Katsnelson, I.V. Grigorieva, S.V. Dubonos, A.A. Firsov, Two-dimensional gas of massless Dirac fermions in graphene. Nature **438**, 197–200 (2005)
31. T. Stauber, N.M.R. Peres, F. Guinea, Electronic transport in graphene: a semiclassical approach including midgap states. Phys. Rev. B **76**, 205423 (2007)
32. Y.-W. Tan, Y. Zhang, K. Bolotin, Y. Zhao, S. Adam, E.H. Hwang, S. Das Sarma, H.L. Stormer, P. Kim, Measurement of scattering rate and minimum conductivity in graphene. Phys. Rev. Lett. **99**, 246803 (2007)
33. S.V. Morozov, K.S. Novoselov, M.I. Katsnelson, F. Schedin, D.C. Elias, J.A. Jaszczak, A.K. Geim, Giant intrinsic carrier mobilities in graphene and its bilayer. Phys. Rev. Lett. **100**, 016602 (2008)
34. M. Xu, T. Liang, M. Shi, H. Chen, Graphene-like two-dimensional materials. Chem. Rev. **113**, 3766–3798 (2013)
35. M. Ashton, J. Paul, S.B. Sinnott, R.G. Hennig, Topology-scaling identification of layered solids and stable exfoliated 2D materials. Phys. Rev. Lett. **118**, 106101 (2017)
36. X. Gu, Y. Wei, X. Yin, B. Li, R. Yang, Colloquium: phononic thermal properties of two-dimensional materials. Rev. Mod. Phys. **90**, 041002 (2018)
37. Q. Cai, D. Scullion, W. Gan et al., High thermal conductivity of high-quality monolayer boron nitride and its thermal expansion. Sci. Adv. **5**, eaav0129 (2019)
38. F. Xia, H. Wang, D. Xiao, M. Dubey, A. Ramasubramaniam, Two-dimensional material nanophotonics. Nat. Photon. **8**, 899–907 (2014)
39. D. Chimene, D.L. Alge, A.K. Gaharwar, Two-dimensional nanomaterials for biomedical applications: emerging trends and future prospects. Adv. Mater. **27**, 7261–7284 (2015)
40. M. Born, K. Huang, *Dynamic Theory of Crystal Lattices* (Oxford University Press, London, 1954)
41. G.P. Srivastava, *The Physics of Phonons* (Adam Hilger, Bristol, 1990)
42. J.M. Ziman, *Electrons and Phonons* (Oxford University Press, Oxford, 1960)
43. B.N. Brockhouse, Lattice vibration in silicon and germanium. Phys. Rev. Lett. **2**, 256–258 (1959)
44. G. Dolling, in *Second Symposium on Inelastic Scattering of Neutrons in Solids and Liquids*, vol. II, Chalk River, Canada (IAEA, Vienna, 1963), p. 37
45. R. Tubino, L. Piseri, G. Zerbi, Lattice dynamics and spectroscopic properties by a valence force potential of diamondlike crystals: C, Si, Ge, and Sn. J. Chem. Phys. **56**, 1022–1039 (1972)

46. D.W. Feldman, J.H. Parker Jr., W.J. Choyke, L. Patrick, Phonon dispersion curves by Raman scattering in SiC, polytypes 3C, 4H, 6H, 15R, and 21R. Phys. Rev. **173**, 787–793 (1968)
47. A.M. Greenstein, S. Graham, Y.C. Hudiono, S. Nair, Thermal properties and lattice dynamics of polycrystalline MFI zeolite films. Nanoscale Microscale Thermophys. Eng. **10**, 321–331 (2006)
48. S.L. Shindé, G.P. Srivastava (eds.), *Length-Scale Dependent Phonon Interactions* (Springer, New York, 2014)
49. C. Dames, Ultrahigh thermal conductivity confirmed in boron arsenide. Science **361**, 549–550 (2018)
50. T.L. Feng, L. Lindsay, X.L. Ruan, Four-phonon scattering significantly reduces intrinsic thermal conductivity of solids. Phys. Rev. B **96**, 161201(R) (2017)
51. D.J. Ecsedy, P.G. Klemens, Thermal conductivity of dielectric crystals due to four-phonon processes and optical modes. Phys. Rev. B **15**, 5957–5962 (1977)
52. Z.M. Zhang, Surface temperature measurement using optical techniques. Annu. Rev. Heat Transfer **11**, 351–411 (2000)
53. J. Lee, T. Beechem, T.L. Wright, B.A. Nelson, S. Graham, W.P. King, Electrical, thermal, and mechanical characterization of silicon microcantilever heaters. J. Microelectromech. Sys. **15**, 1644–1655 (2006)
54. M. Maldovan, Sound and heat revolutions in phononics. Nature **503**, 209–217 (2013); M. Maldovan, Narrow low-frequency spectrum and heat management by thermocrystals. Phys. Rev. Lett. **110**, 025902 (2014); M. Maldovan, Phonon wave interference and thermal bandgap materials. Nat. Mater. **14**, 667–674 (2015)
55. R. Anufriev, A. Ramiere, J. Maire, M. Nomura, Heat guiding and focusing using ballistic phonon transport in phononic nanostructures. Nat. Commun. **8**, 15505 (2017)
56. G. Xie, D. Ding, G. Zhang, Phonon coherence and its effect on thermal conductivity of nanostructures. Adv. Phys: X **3**, 1480417 (2018)
57. Y. Xiao, Q. Chen, D. Ma, N. Yang, Q. Hao, Phonon transport within periodic porous structures—from classical phonon size effects to wave effects. ES Mater. Manuf. **5**, 2–18 (2019)
58. R. Merlin, Generating coherent THz phonons with light pulses. Solid State Commun. **102**, 207–220 (1997)
59. C. Colvard, R. Merlin, M.V. Klein, A.C. Gossard, Observation of folded acoustic phonons in a semiconductor superlattice. Phys. Rev. Lett. **45**, 298–301 (1980)
60. Y. Ezzahri, S. Grauby, J.M. Rampnoux, H. Michel, G. Pernot, W. Claeys, S. Dilhaire, C. Rossignol, G. Zeng, A. Shakouri, Coherent phonons in Si/SiGe superlattices. Phys. Rev. B **75**, 195309 (2007)
61. A.A. Maznev, F. Hofmann, A. Jandl, K. Esfarjani, M.T. Bulsara, E.A. Fitzgerald, G. Chen, K.A. Nelson, Lifetime of sub-THz coherent acoustic phonons in a GaAs-AlAs superlattice. Appl. Phys. Lett. **102**, 041901 (2013)
62. F. He, W. Wu, Y. Guo, Direct measurement of coherent thermal phonons in $Bi_2Te_3/Sb_2Te_3$ superlattice. Appl. Phys. A **122**, 777 (2016)
63. R. Anufriev, M. Nomura, Reduction of thermal conductance by coherent phonon scattering in two-dimensional phononic crystals of different lattice types. Phys. Rev. B **93**, 045410 (2016); *ibid.* Heat conduction engineering in pillar-based phononic crystals. Phys. Rev. B **95**, 155432 (2017)
64. M.N. Luckyanova, J. Garg, K. Esfarjani, A. Jandl, M.T. Bulsara, A.J. Schmidt, A.J. Minnich, S. Chen, M.S. Dresselhaus, Z. Ren, E.A. Fitzgerald, G. Chen, Coherent phonon heat conduction in superlattices. Science **338**, 936–939 (2012)
65. J. Ravichandran, A.K. Yadav, R. Cheaito, P.B. Rossen, A. Soukiassian, S.J. Suresha, J.C. Duda, B.M. Foley, C.-H. Lee, Y. Zhu, A.W. Lichtenberger, J.E. Moore, D.A. Muller, D.G. Schlom, P.E. Hopkins, A. Majumdar, R. Ramesh, M.A. Zurbuchen, Crossover from incoherent to coherent phonon scattering in epitaxial oxide superlattices. Nat. Mater. **13**, 168–172 (2014)
66. J.K. Yu, S. Mitrovic, D. Tham, J. Varghese, J.R. Heath, Reduction of thermal conductivity in phononic nanomesh structures. Nat. Nanotech. **5**, 718–721 (2010)

67. P.E. Hopkins, C.M. Reinke, M.F. Su, R.H. Olsson III, E.A. Shaner, Z.C. Leseman, J.R. Serrano, L.M. Phinney, I. El-Kady, Reduction in the thermal conductivity of single crystalline silicon by phononic crystal patterning. Nano Lett. **11**, 107–112 (2011)
68. E. Dechaumphai, R. Chen, Thermal transport in phononic crystals: the role of zone folding effect. J. Appl. Phys. **111**, 073508 (2012)
69. A. Jain, Y.J. Yu, A.J.H. McGaughey, Phonon transport in periodic silicon nanoporous films with feature sizes greater than 100 nm. Phys. Rev. B **87**, 195301 (2013)
70. N.K. Ravichandran, A.J. Minnich, Coherent and incoherent thermal transport in nanomeshes. Phys. Rev. B. **89**, 205432 (2014)
71. J. Lee, W. Lee, G. Wehmeyer, S. Dhuey, D.L. Olynick, S. Cabrini, C. Dames, J.J. Urban, P.D. Yang, Investigation of phonon coherence and backscattering using silicon nanomeshes. Nat. Commun. **8**, 14054 (2017)
72. N. Zen, T.A. Puurtinen, T.J. Isotalo, S. Chaudhuri, I.J. Maasilta, Engineering thermal conductance using a two-dimensional phononic crystal. Nat. Commun. **5**, 3435 (2014)
73. J. Maire, R. Anufriev, R. Yanagisawa, A. Ramiere, S. Volz, M. Nomura, Heat conduction tuning by wave nature of phonons. Sci. Adv. **3**, e1700027 (2017)
74. M.R. Wagner, B. Graczykowski, J.S. Reparaz, A. El Sachat, M. Sledzinska, F. Alzina, C.M.S. Torres, Two-dimensional phononic crystals: disorder matters. Nano Lett. **16**, 5661–5668 (2016)
75. D. Broido, M. Malorny, G. Birner, N. Mingo, D. Stewart, Intrinsic lattice thermal conductivity of semiconductors from first principles. Appl. Phys. Lett. **91**, 231922 (2007)
76. T. Luo, G. Chen, Nanoscale heat transfer—from computation to experiment. Phys. Chem. Chem. Phys. **15**, 3389–3412 (2013)
77. G. Chen, Multiscale simulation of phonon and electron thermal transport. Annu. Rev. Heat Transfer **17**, 1–8 (2014)
78. K. Esfarjani, J. Garg, G. Chen, Modeling heat conduction from first principles. Annu. Rev. Heat Transfer **17**, 9–47 (2014)
79. L. Lindsay, First principles Peierls-Boltzmann phonon thermal transport: a topical review. Nanosc. Microsc. Thermophys. Eng. **20**, 67–84 (2016)
80. H. Bao, J. Chen, X. Gu, B. Cao, A review of simulation methods in micro/nanoscale heat conduction. ES Energy Environ. **1**, 16–55 (2018)
81. A.J.H. McGaughey, A. Jain, H.Y. Kim, B. Fu, Phonon properties and thermal conductivity from first principles, lattice dynamics, and the Boltzmann transport equation. J. Appl. Phys. **125**, 011101 (2019)
82. P.K. Schelling, S.R. Phillpot, P. Keblinski, Comparison of atomic-level simulation methods for computing thermal conductivity. Phys. Rev. B **65**, 144306 (2002); ibid, Phonon wave-packet dynamics at semiconductor interfaces by molecular dynamics simulation. Appl. Phys. Lett. **80**, 2484–2487 (2002)
83. A.J.H. McGaughey, M. Kariany, Phonon transport in molecular dynamics simulations: formulation and thermal conductivity prediction. Adv. Heat Transfer **39**, 169–255 (2006)
84. A.J.H. McGaughey, J.M. Larkin, Prediction of phonon properties from equilibrium molecular dynamics simulations. Annu. Rev. Heat Transfer **17**, 49–87 (2014)
85. J. Shiomi, Nonequilibrium molecular dynamics method for lattice heat conduction calculations. Annu. Rev. Heat Transfer **17**, 177–203 (2014)
86. P. Hohenberg, W. Kohn, Inhomogeneous electron gas. Phys. Rev. **136**, B864–B871 (1964); W. Kohn, L.J. Sham, Self-consistent equation including exchange and correlation effects. Phys. Rev. **140**, A1133–A1138 (1965)
87. R.O. Jones, Density functional theory: its origins, rise to prominence, and future. Rev. Mod. Phys. **87**, 897–923 (2015)
88. P.J. Hasnip, K. Refson, M.I.J. Probert, J.R. Yates, S.J. Clark, C.J. Pickard, Density functional theory in the solid state. Phil. Trans. R. Soc. A **372**, 20130270 (2014)
89. B.G. Johnson, P.M.W. Gill, J.A. Pople, The performance of a family of density functional methods. J. Chem. Phys. **98**, 5612–5626 (1993)
90. M. Kaviany, *Heat Transfer Physics*, 2nd edn. (Cambridge University Press, Cambridge, 2014)

91. S. Baroni, S. de Gironcoli, A. Dal Corso, Phonons and related crystal properties from density-functional perturbation theory. Rev. Mod. Phys. **73**, 515–562 (2001)
92. X. Gonze, J.-P. Vigneron, Density-functional approach to nonlinear-response coefficients of solids. Phys. Rev. B **39**, 13120–13128 (1989)
93. D. Vanderbilt, S.H. Taole, S. Narasimhan, Anharmonic elastic and phonon properties of Si. Phys. Rev. B **40**, 5657–5668 (1989)
94. K. Parlinski, Z.Q. Li, Y. Kawazoe, First-principles determination of the soft mode in cubic $ZrO_2$. Phys. Rev. Lett. **78**, 4063–4066 (1997)
95. K. Esfarjani, H.T. Stokes, Method to extract anharmonic force constants from first principles calculations. Phys. Rev. B **77**, 144112 (2008)
96. A. Togo, I. Tanaka, First principles phonon calculations in materials science. Scr. Mater. **108**, 1–5 (2015); A. Togo, L. Chaput, I. Tanaka, Distributions of phonon lifetimes in Brillouin zones. Phys. Rev. B **91**, 094306 (2015)
97. T. Tadano, S. Tsuneyuki, First-principles lattice dynamics method for strongly anharmonic crystals. J. Phys. Soc. Jpn. **87**, 041015 (2018); *ibid*, Self-consistent phonon calculations of lattice dynamical properties in cubic $SrTiO_3$ with first-principles anharmonic force constants. Phys. Rev. B **92**, 054301 (2015)
98. B. Liao, B. Qiu, J. Zhou, S. Huberman, K. Esfarjani, G. Chen, Significant reduction of lattice thermal conductivity by the electron-phonon interaction in silicon with high carrier concentrations: a first-principles study. Phys. Rev. Lett. **114**, 115901 (2015)
99. Y.Z. Liu, Y.H. Feng, Z. Huang, X.X. Zhang, Thermal conductivity of 3D boron-based covalent organic frameworks from molecular dynamics simulations. J. Phys. Chem. C **120**, 17060–17068 (2016)
100. LAMMPS Molecular Dynamics Simulator, Sandia National Laboratory. https://lammps.sandia.gov/, Accessed 13 Dec 2019
101. Z. Wang, S. Safarkhani, G. Lin, X. Ruan, Uncertainty quantification of thermal conductivities from equilibrium molecular dynamics simulations. Int. J. Heat Mass Transfer **112**, 267–278 (2017)
102. B.-L. Huang, M. Kaviany, Ab initio and molecular dynamics predictions for electron and phonon transport in bismuth telluride. Phys. Rev. B **77**, 125209 (2008)
103. K. Esfarjani, G. Chen, H.T. Stokes, Heat transport in silicon from first-principles calculations. Phys. Rev. B **84**, 085204 (2011)
104. A. Marcolongo, P. Umari, S. Baroni, Microscopic theory and quantum simulation of atomic heat transport. Nat. Phys. **12**, 80–84 (2015)
105. C. Carbogno, R. Ramprasad, M. Scheffler, Ab initio Green-Kubo approach for the thermal conductivity of solids. Phys. Rev. Lett. **118**, 175901 (2017)
106. J.S. Tse, N.J. English, K. Yin, T. Iitaka, Thermal conductivity of solids from first-principles molecular dynamics calculations. J. Phys. Chem. **122**, 10682–10690 (2018)
107. S. Hüfner, *Photoelectron Spectroscopy: principles and Applications*, 3rd edn. (Springer, Berlin, 2003)
108. G. N. Hatsopoulos, E.P. Gyftopoulos, *Thermionic Energy Conversion*, vol. 1 (1973); Vol. 2. (MIT Press, Cambridge, MA, 1979)
109. G.D. Mahan, Thermionic refrigeration. J. Appl. Phys. **76**, 4362–4366 (1994); G.D. Mahan, L.M. Woods, Multilayer thermionic refrigeration. Phys. Rev. Lett. **80**, 4016–4019 (1998); G.D. Mahan, J.O. Sofo, M. Barkowiak, Multilayer thermionic refrigerator and generator. J. Appl. Phys. **83**, 4683–4689 (1998)
110. A. Shakouri, J.E. Bowers, Heterostructure integrated thermionic coolers. Appl. Phys. Lett. **71**, 1234–1236 (1997); A. Shakouri, C. LaBounty, J. Piprek, P. Abraham, J.E. Bowers, Thermionic emission cooling in single barrier heterostructures. Appl. Phys. Lett. **74**, 88–89 (1999)
111. D. Vashaee, A. Shakouri, Nonequilibrium electrons and phonons in thin film thermionic coolers. Microscale Thermophys. Eng. 8, 91–100 (2004); D. Vashaee, A. Shakouri, Electronic and thermoelectric transport in semiconductor and metallic superlattices. J. Appl. Phys. 95, 1233–1245 (2004)

112. T. Zeng, G. Chen, Interplay between thermoelectric and thermionic effects in heterostructures. J. Appl. Phys.**92**, 3152–3161 (2002); T. Zeng, G. Chen, Nonequilibrium electron and phonon transport and energy conversion in heterostructures. Microelectron. J. **34**, 201–206 (2003)
113. Y. Hishinuma, T.H. Geballe, B.Y. Moyzhes, T.W. Kenny, Refrigeration by combined tunneling and thermionic emission in vacuum: use of nanometer scale design. Appl. Phys. Lett. **78**, 2572–2574 (2001); Y. Hishinuma, T.H. Geballe, B.Y. Moyzhes, T.W. Kenny, Measurements of cooling by room-temperature thermionic emission across a nanometer gap. J. Appl. Phys. **94**, 4690–4696 (2003)
114. T. Zeng, Thermionic-tunneling multilayer nanostructures for power generation. Appl. Phys. Lett. **88**, 153104 (2006)
115. R.H. Fowler, L. Nordheim, Electron emission in intense electric field. Proc. Royal Soc. Lond. A **119**, 173–181 (1928)
116. D.J. Griffiths, *Introduction to Quantum Mechanics*, 2nd edn, Chap. 8 (Pearson Prentice Hall, Upper Saddle River, NJ, 2005)
117. R. Tsu, L. Esaki, Tunneling in a finite superlattice. Appl. Phys. Lett. **22**, 562–564 (1973); L.L. Chang, L. Esaki, R. Tsu, Resonant tunneling in semiconductor double barriers. Appl. Phys. Lett. **24**, 593–595 (1974)
118. J.W. Gadzuk, E.W. Plummer, Field emission energy distribution (FEED). Rev. Mod. Phys. **45**, 487–548 (1973)
119. L. Nilsson, O. Groening, P. Groening, O. Kuettel, L. Schlapbach, Characterization of thin film electron emitters by scanning anode field emission spectroscopy. J. Appl. Phys. **90**, 768–780 (2001)
120. J.B. Xu, K. Läuger, R. Möller, K. Dransfeld, I.H. Wilson, Energy-exchange processes by tunneling electrons. Appl. Phys. A **59**, 155–161 (1994)
121. T.S. Fisher, Influence of nanoscale geometry on the thermodynamics of electron field emission. Appl. Phys. Lett. **79**, 3699–3701; T.S. Fisher, D.G. Walker, Thermal and electrical energy transport and conversion in nanoscale electron field emission process. J. Heat Transfer **124**, 954–962 (2002)
122. C. Trinkle, P. Kichambare, R.R. Vallance, B. Wong, M.P. Mengüç, B. Sadanadan, et al., Thermal transport during nanoscale machining by field emission of electrons from carbon nanotubes. J. Heat Transfer **125**, 546 (2003); R.R. Vallance, A.M. Rao, M.P. Mengüç, Processes for nanomachining using carbon nanotubes. US Patent No. 6,660,959, 9 Dec 2003
123. B.T. Wong, M.P. Mengüç, R.R. Vallance, Nano-scale machining via electron beam and laser processing. J. Heat Transfer **126**, 566–576 (2004)
124. W.I. Milne, K.B.K. Teo, G.A.J. Amaratunga, P. Legagneux, L. Gangloff, J.-P. Schnell, V. Semet, V. Thien Binh, O. Groening, Carbon nanotubes as field emission sources. J. Mater. Chem. **14**, 933–943 (2004)
125. R.J. Keyes (ed.), *Optical and Infrared Detectors* (Springer, Berlin, 1980)
126. S. Basu, Y.-B. Chen, Z.M. Zhang, Microscale radiation in thermophotovoltaic devices—A review. Intl. J. Ener. Res. **31** (2007)
127. T.P. Xiao, K. Chen, P. Santhanam, S. Fan, E. Yablonovitch, Electroluminescent refrigeration by ultra-efficient GaAs light-emitting diodes. J. Appl. Phys. **123**, 173104 (2018)
128. P.S. Zory Jr. (ed.), *Quantum Well Lasers* (Academic Press, San Diego, 1993)

# Chapter 7
# Nonequilibrium Energy Transfer in Nanostructures

Fourier's law and the associated heat diffusion equation comprise one of the most celebrated models in mathematical physics. Joseph Fourier in 1824 wrote: *Heat, like gravity, penetrates every substance of the universe; its rays occupy all parts of space. …The theory of heat will hereafter form one of the most important branches of general physics.* Soon afterward, heat transfer also became an important engineering field, essential to the second industrial revolution and the development of modern technologies.

Recall the discussion of heat interaction and heat transfer in Chap. 2. We have treated heat conduction as a diffusion process based on the concept of local thermal equilibrium. This allows us to define and determine the equilibrium temperature at each location in a body instantaneously, under the continuum assumption described in Chap. 1. The local-equilibrium condition breaks down at the microscale when the characteristic length $L$ is smaller than a mechanistic length scale, such as the mean free path $\Lambda$. For conduction by molecules, consider a rarefied gas between two parallel plates at different temperatures. If the mean free path is much greater than the separation distance, i.e., the Knudsen number $Kn = \Lambda/L >> 1$, the gas is in the free molecule regime and its velocity distribution cannot be described by Maxwell's distribution function. Furthermore, the transport becomes ballistic rather than diffusive. Nonequilibrium energy transfer refers to the situation when the assumption of local equilibrium does not hold. This can occur in solid nanostructures even at room temperature and in a steady state, or in bulk solids under the influence of short pulse heating.

For heat conduction *across* a dielectric thin film, when the thickness is much smaller than the phonon mean free path, which increases as the temperature goes down, the condition of local equilibrium is not satisfied. Hence, the phonon statistics at a given location cannot be described by the equilibrium distribution function at any given temperature. Strictly speaking, temperature cannot be defined inside the medium. However, an *effective* temperature is typically adopted, based on the statistical average of the particle energies. In the case of heat transfer across a thin



Z. M. Zhang, *Nano/Microscale Heat Transfer*, Mechanical Engineering Series,
https://doi.org/10.1007/978-3-030-45039-7_7

dielectric film or between two plates separated by a rarefied molecular gas, the effective temperature distribution cannot be described by the heat diffusion theory derived from Fourier's law, using the concept of equilibrium temperature without considering the temperature jumps at the boundaries. The concept of temperature jump was introduced in Chap. 4 (e.g., Fig. 4.12). Consider a metal or a superconductor that is subjected to ultrafast pulsed laser heating, in which the pulse duration may range from several femtoseconds to a few nanoseconds. The electrons gain energy quickly to reach a state that is far from equilibrium with the crystal lattice or the phonon system. The transport processes during and immediately after the laser pulse become nonequilibrium both temporally and spatially. Conventional Fourier's law cannot be directly applied.

In Chap. 5, we have considered the size effect on thermal transport in solids. Two approaches have been used under different situations. In the first situation, we apply Matthiessen's rule to account for the reduction in mean free path by assuming that Fourier's law is still applicable but with a size-dependent thermal conductivity. In the second situation, where the transport is completely ballistic, we use the concept of quantum conductance based on the Landauer formulation to solve the problem in a straightforward manner. The definition of an effective thermal conductivity is particularly useful for the study of transport processes *along* a thin film or a thin wire, when the length in the direction of transport is much greater than the mean free path. In this case, a local equilibrium can be established, and thus, the energy transfer is well described by Fourier's law, even though the thickness is less than the mean free path. Here, the only microscale effect is the classical size effect, which arises from boundary scattering of electrons in a metal or phonons in an insulator or a semiconductor. For energy transport across a thin film or in a multilayer structure, on the other hand, the local-equilibrium condition breaks down when the film thickness is much smaller than the mean free path. Furthermore, thermal boundary resistance (TBR) may become significant at the interfaces. Because of the wave-particle duality, the electron wave or phonon wave effect may need to be considered in some cases. For nonmetallic crystalline materials, the most commonly used method to study thermal transport is based on the Boltzmann transport equation (BTE) of phonons. Various assumptions and techniques have been developed to solve the phonon BTE. In very small structures, such as nanotubes or nanowires, molecular dynamics (MD) and other atomistic simulation methods may be more suitable.

This chapter begins with a description of the phenomenological theories in which the energy transport processes are represented by a single differential equation or a set of differential equations that can be solved with appropriate initial and boundary conditions. These equations are often called non-Fourier heat equations, which can be considered as extensions of the conventional heat diffusion equation based on Fourier's law. The limitations of the phenomenological theories are discussed. While the BTE, Monte Carlo method, and MD simulations have been presented in previous chapters, this chapter stresses the application in solid nanostructures, including thermal boundary resistance (TBR) and multilayer structures. The equation of phonon radiative transfer (EPRT) is introduced and used to delineate the diffusive and ballistic heat conduction regimes in thin films. A heat conduction regime with respect to length and time scale is presented, followed by a summary of the contemporary methods for measuring thermal transport properties of solids, thin films, and nanostructures.

## 7.1 Phenomenological Theories

A fundamental difficulty of Fourier's heat conduction theory was thought to be that a thermal disturbance in one location of the medium would cause a response at any other location instantaneously, as required by the mathematical solution of the diffusion equation. In theory, the speed of heat propagation appears to be unlimited; this has been viewed by some as a direct violation of the principle of causality. Let us begin with an example of 1D transient heating of a semi-infinite medium. Assume that the medium is homogeneous, with constant thermal properties, and is initially at a uniform temperature $T(x, 0) = T_i$. The thermal diffusivity of the medium is $\alpha = \kappa/(\rho c_p)$, where $\kappa$, $\rho$, and $c_p$ are the thermal conductivity, density, and specific heat of the material, respectively. The wall at $x = 0$ is heated with a constant heat flux $q_0''$ at $0 < t \le t_p$, where $t_p$ is the width of the step heating, and insulated at $t > t_p$. The solution of the temperature distribution $T(x, t)$ can be found from Refs. [1, 2] as follows:

$$T(x,t) - T_i = 2q_0'' \frac{\sqrt{\alpha t}}{\kappa} F(\xi) \text{ at } 0 < t \le t_p \tag{7.1a}$$

$$T(x,t) - T_i = 2q_0'' \frac{\sqrt{\alpha t}}{\kappa} \left[ F(\xi) - \eta F\left(\frac{\xi}{\eta}\right) \right] \text{ at } t > t_p \tag{7.1b}$$

where $\xi = x/\sqrt{4\alpha t}$, $\eta = \sqrt{1 - t_p/t}$, and $F(\xi) = \pi^{-1/2} \exp(-\xi^2) - \xi \mathrm{erfc}(\xi)$ with erfc being the complementary error function as given in Appendix B.1.2. While $F(10) = 2.1 \times 10^{-44}$ and the right-hand sides of both Eqs. (7.1a) and (7.1b) are essentially negligible when $x > 3\sqrt{\alpha t}$, the paradox is that a nonzero response must not occur faster than the speed of the thermal energy carriers, such as the Fermi velocity in metals or the speed of sound in dielectrics. In reality, this rarely causes any problem because a signal that is below the noise level cannot be detected by any physical instrument, as will be discussed in the example next.

**Example 7.1** A thick plate of fused silica $SiO_2$, initially at room temperature, is heated at one surface by a heat flux of $10^5$ W/m$^2$ for 5 s and then insulated. Treat the heated surface to be at $x = 0$, and assume the other surface is at $x \to \infty$. Plot the temperature distributions at various times. Imagine a temperature sensor is placed at certain locations with instantaneous response and zero additional heat capacity. Estimate the time for the thermometer to sense the temperature rise as a function of the location $x$. Assume that the thermophysical properties of the glass are constant, $\kappa = 1.4$ W/m K, and $\alpha = 8.5 \times 10^{-7}$ m$^2$/s.

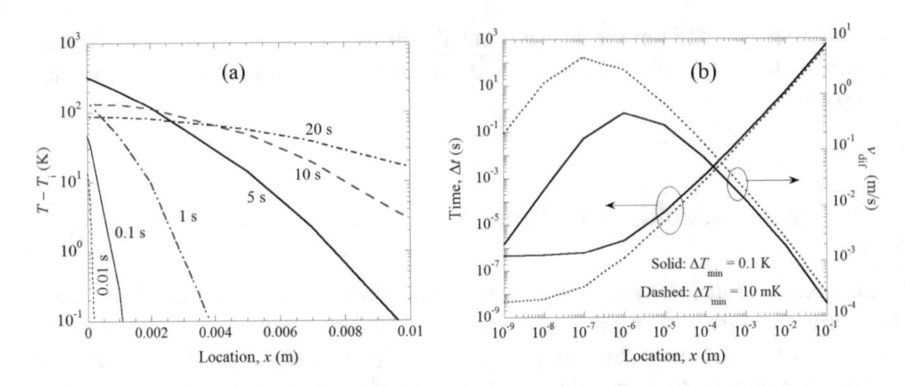


**Fig. 7.1** **a** The temperature distributions at various times. **b** The time required for a given location to acquire a minimum temperature rise and the estimated thermal diffusion speed

**Solution** The temperature distribution is shown in Fig. 7.1a at $t = 0.01, 0.1, 1, 5,$ 10, and 20 s. During the heating, the temperature monotonically increases with time and the heat flux is always positive. After the heat input is stopped when $t = 5$ s, the temperature near the surface decreases but is still the highest and the temperature decreases toward increasing $x$. While the predicted temperature rises everywhere instantaneously, the magnitude may be too small to be observed practically. We can calculate the time $\Delta t$ required for a minimum temperature rise $\Delta T_{\min}$, specified by the thermometer sensitivity. Let us choose $\Delta T_{\min} = 10$ mK and 0.1 K for illustration. The *average thermal diffusion speed* can be estimated by $v_{\text{dif}}(x) = x/\Delta t$, for any given location $x$. The results are shown in Fig. 7.1b. In reality, diffusion is often a slow process near room temperature. For the example given here, $v_{\text{dif}}$ for $\Delta T_{\min} = 10$ mK is between 1 and 5 m/s, for 5 nm $< x <$ 5 μm, and goes down rapidly at $x >$ 5 μm. At $x = 10$ mm, $v_{\text{dif}}$ is only 2–3 mm/s. On the other hand, the speed of sound in glass is on the order of 5 km/s, which is several orders of magnitude greater than the average thermal diffusion speed.

Recall that the uncertainty principle in quantum mechanics states that $\Delta E \Delta t > \hbar$, suggesting that we cannot measure time and energy simultaneously with unlimited precision. From statistical mechanics, the distribution function allows a small fraction of particles to have a very high speed or to travel a very large distance without collision, although the probability may be extremely low. Based on the uncertainty principle and statistical mechanics, it seems convincing that Fourier's law, in its applicable regime, does not violate the principle of causality. What is physically problematic and practically impossible is to provide a temperature impulse to the surface or at any given location instantaneously. We further conclude that the heat diffusion equation does not produce an infinite speed of thermal energy propagation; rather, it is often a very slow process. Microscopically, Fourier's law fails when a local equilibrium is not established, as explained earlier. At the same time, the concept of an equilibrium temperature cannot be applied. It is critically important for the technological advancement to establish and apply thermal transport theories, both microscopically and macroscopically, under nonequilibrium conditions.

Several phenomenological theories have been developed to describe transient heat transfer processes in solids and micro/nanostructures. Applications of transient and ultrafast heating include laser processing, nanothermal fabrication, and the measurement of thermophysical properties. In the literature, there appears to be controversial experimental evidence on the existence of certain phenomena predicted by the hyperbolic heat conduction [3]. Furthermore, a large division exists as regards the formulation and the interpretation of the theories of non-Fourier conduction. While the intention is to provide a clear and objective presentation, the discussion will inevitably reflect the author's personal views and limitations at the time when the manuscript was prepared. This section should help readers gain a general understanding of the basic concepts and phenomena related to non-Fourier heat conduction. Although relatively few papers out of a large number of publications are cited in the text and the reference section, interested readers can easily trace the relevant literature from the cited sources, especially Refs. [3–6].

### *7.1.1 Hyperbolic Heat Equation*

Several earlier studies have pointed out that the instantaneous response may be an indication of a nonphysical feature of the Fourier heat theory. Carlo Cattaneo in 1948 used kinetic theory of gas to derive a rate equation given by

$$\mathbf{q}''(\mathbf{r}, t) + \tau_{\mathrm{q}} \frac{\partial \mathbf{q}''(\mathbf{r}, t)}{\partial t} = -\kappa \nabla T(\mathbf{r}, t) \tag{7.2}$$

which is a *modified Fourier equation* called *Cattaneo's equation*. The historical contributions by James Clerk Maxwell in 1867 and Pierre Vernotte in 1958 have been extensively reviewed by Joseph and Preziosi [4] and will not be repeated here. In Eq. (7.2), $\tau_{\mathrm{q}}$ is a kind of relaxation time, originally thought to be the same as $\tau$, i.e., the average time between collisions. The energy equation for heat conduction involving an internal source or volumetric heat generation rate $\dot{q}(\mathbf{r}, t)$ is

$$\dot{q}(\mathbf{r}, t) - \nabla \cdot \mathbf{q}''(\mathbf{r}, t) = \rho c_p \frac{\partial T(\mathbf{r}, t)}{\partial t} \tag{7.3}$$

The divergence of Eq. (7.2) and the time derivative of Eq. (7.3) give two new equations, which can be combined with Eq. (7.3) to eliminate the heat flux terms. The resulting differential equation for constant properties can be written as

$$\frac{\dot{q}}{\kappa} + \frac{\tau_{\mathrm{q}}}{\kappa} \frac{\partial \dot{q}}{\partial t} + \nabla^2 T = \frac{1}{\alpha} \frac{\partial T}{\partial t} + \frac{\tau_{\mathrm{q}}}{\alpha} \frac{\partial^2 T}{\partial t^2} \tag{7.4}$$

This is the *hyperbolic heat equation*, in contrast to the heat diffusion equation or parabolic heat equation. Without heat generation, we can rewrite Eq. (7.4) as

$$\nabla^2 T = \frac{1}{\alpha}\frac{\partial T}{\partial t} + \frac{1}{v_{\text{tw}}^2}\frac{\partial^2 T}{\partial t^2} \tag{7.5}$$

which is a telegraph equation or a damped wave equation. The solution of the hyperbolic heat equation results in a propagating wave, the amplitude of which decays exponentially as it travels. The speed of this *temperature wave* in the high-frequency limit, or the short-time limit, is given by

$$v_{\text{tw}} = \sqrt{\alpha/\tau_{\text{q}}} \tag{7.6}$$

The amplitude of the temperature wave decays according to $\exp(-t/\tau_{\text{q}})$ due to the damping caused by the first-order time-derivative term $(1/\alpha)(\partial T/\partial t)$, which is also called the diffusion term. For an insulator, from the simple kinetic theory we have $\kappa = \frac{1}{3}(\rho c_v)v_{\text{g}}^2\tau$. Noting that $c_v = c_p$ for an incompressible solid and assuming $\tau_{\text{q}} = \tau$, we get

$$v_{\text{tw}} = v_{\text{g}} \Big/ \sqrt{3} \tag{7.7}$$

Equation (7.7) relates the speed of the temperature wave to the speed of sound in an insulator. The square root of three can be understood as due to the randomness of thermal fluctuations in a 3D medium, just like the relation between the velocity and its components, $\overline{v^2} = \overline{v_x^2} + \overline{v_y^2} + \overline{v_z^2}$, in kinetic theory. Earlier experiments at cryogenic temperatures have demonstrated a second sound propagating at the velocity $v_{\text{2nd}} = v_{\text{g}} \Big/ \sqrt{3}$ in liquid helium and some solids [4].

Equation (7.5) sets a limit on the heat propagation speed, which is manifested by a sharp wavefront that travels at $v_{tw}$ inside the medium for a sudden temperature change at the boundary. As a wave equation, the solution of the temperature has an amplitude and a phase. Theoretically, the temperature wave can be reflected by another boundary and can interfere, constructively or destructively, with a forward propagating wave. The interaction between the temperature waves may also result in a resonance effect, a typical wave phenomenon. Numerous analytical and numerical predictions have been made [6–10]. It should be noted that the terms *heat wave* [4] and *thermal wave* [7] have also been frequently used in the literature to describe the temperature wave behavior. The term "temperature wave" is used in this chapter for the wavelike behavior associated with the hyperbolic-type heat equations, because "heat wave" might be confused with the calamitous weather phenomenon and "thermal wave" might be confused with the diffusion wave used in photoacoustic techniques. Bennett and Patty [11] clarified: *The term thermal wave interference is used to mean the superposition of simple harmonic solutions of the thermal diffusion equation. Although wavelike in nature there are important differences between thermal waves arising from a differential equation that is of the first order in time*

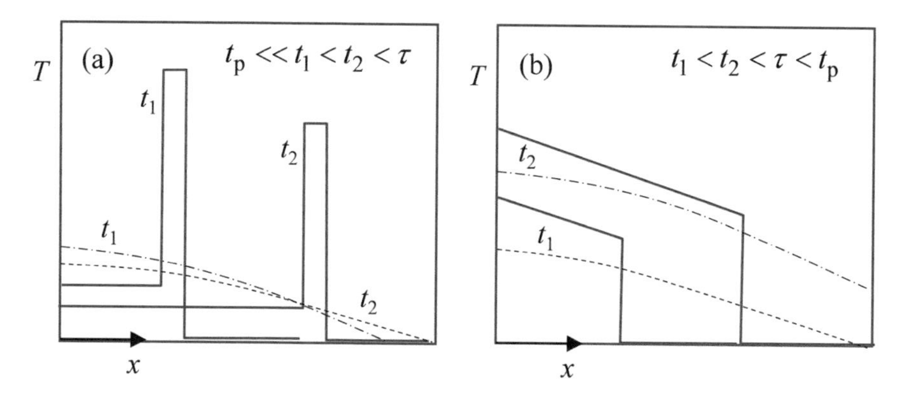


**Fig. 7.2** (Not to scale) Illustration of the solution of the hyperbolic heat equation at short timescales. **a** A short pulse, $t_p << \tau$. **b** A long pulse, $t_p > \tau$. The solid curves are the solutions of the hyperbolic heat equation (7.5), and the dash-dotted and dashed curves are the solutions of the heat diffusion equation (7.1a, 7.1b)

*and waves that are solution to a wave equation that is of the second order in time.* In the heat transfer literature, thermal wave often refers to periodic-heating techniques used widely for thermophysical property measurements [12].

Let us consider an example of a semi-infinite solid under a constant heat flux at the surface. Figure 7.2 illustrates the solutions for a small $t_p$ and a large $t_p$, compared with $\tau$. Here again, we have assumed $\tau_q = \tau$. The propagation speed is equal to $v_{tw}$, and the pulse wavefront is given by $x_1 = v_{tw}t_1$ and $x_2 = v_{tw}t_2$. Hence, $x_1 < x_2 < \Lambda$, where $\Lambda = v_g\tau$ is the mean free path. In the case of a short pulse, the temperature pulse propagates and its height decays by dissipating its energy to the medium as it travels. The parabolic heat equation, on the other hand, predicts a continuous temperature distribution without any wavefront (see Fig. 7.2).

As time passes on, the first-order time derivative, or the diffusion term, in Eq. (7.5) dominates. If the relative change of $\partial T/\partial t$ or $\mathbf{q}''$ during one $\tau_q$ is large, then the wave feature is important. This should happen immediately after a sudden thermal disturbance that results in a temporal nonequilibrium, as well as a spatial nonequilibrium near the heat pulse or the wavefront. After a sufficiently long time, usually 5–10 times $\tau_q$, a local equilibrium will be reestablished, and the thermal field can be described by the parabolic heat equation. At steady state, the hyperbolic and parabolic equations predict the same results. While Eq. (7.4) is mathematically more general than the heat diffusion equation, it should not be taken as a correction, or a more realistic theory than the Fourier conduction model, because Cattaneo's equation has not been justified on a fundamental basis, nor has it been validated by any plausible experiments.

Many researchers have investigated the hyperbolic heat equation based on the second law of thermodynamics [13–15]. It has been found that the hyperbolic heat equation sometimes predicts a negative entropy generation and even allows energy to be transferred from a lower temperature region to a higher temperature region. The entropy generation rate for heat conduction without an internal source can be calculated by [15]

$$\dot{s}_{\text{gen}} = -\frac{1}{T^2}\mathbf{q}'' \cdot \nabla T = \frac{1}{\kappa T^2}\mathbf{q}'' \cdot \left(\mathbf{q}'' + \tau_{\text{q}}\frac{\partial \mathbf{q}''}{\partial t}\right) \tag{7.8a}$$

The above equation was obtained by setting the energy and entropy balances as follows:

$$\rho\frac{\partial u}{\partial t} = -\nabla \cdot \mathbf{q}'' \quad \text{and} \quad \rho\frac{\partial s}{\partial t} = -\nabla \cdot \left(\frac{\mathbf{q}''}{T}\right) + \dot{s}_{\text{gen}} \tag{7.8b}$$

Note that $du = T ds$. A negative entropy generation can easily be numerically demonstrated from Eq. (7.5) during the temperature wave propagation. Here, a negative entropy generation does not constitute a violation of the second law of thermodynamics because the concept of "temperature" in the hyperbolic heat equation cannot be interpreted in the conventional sense, due to the lack of local thermal equilibrium. Extended irreversible thermodynamics has been proposed by Jou et al. [16] by modifying the definition of entropy such that it is not a property of the system anymore but also depends on the heat flux vector. The theory of extended irreversible thermodynamics is self-consistent but has not been fully validated by experiments; hence, it cannot be taken as a generalized thermodynamic theory. Similarly, the hyperbolic heat equation should not be treated as a more general theory over Fourier's heat conduction theory [17].

**Example 7.2** Derive the modified Fourier equation, or Cattaneo's equation, based on the BTE under the relaxation time approximation.

**Solution** Tavernier [18] first showed that Cattaneo's equation could be derived for phonons and electrons using the relaxation time approximation of the BTE. As done in Sect. 4.3.2, where we have derived Fourier's law based on the BTE, let us start by assuming that the temperature gradient is in the $x$-direction only. The transient 1D BTE under the relaxation time approximation can be written as follows:

$$\frac{\partial f}{\partial t} + v_x\frac{\partial f}{\partial x} = \frac{f_0 - f}{\tau} \tag{7.9}$$

A further assumption is made such that $\frac{\partial f}{\partial x} \approx \frac{\partial f_0}{\partial x} = \frac{\partial f_0}{\partial T}\frac{\partial T}{\partial x}$, which is the condition of local equilibrium. Substitute the local-equilibrium condition into Eq. (7.9) and multiply each term by $\tau\varepsilon v_x$. We can then perform integration of each term over the momentum space to obtain

$$\int\limits_{\varpi} \tau\varepsilon v_x\frac{\partial f}{\partial t}d\varpi + \int\limits_{\varpi} \tau\varepsilon v_x^2\frac{\partial f}{\partial x}d\varpi = \int\limits_{\varpi} \varepsilon v_x f_0 d\varpi - \int\limits_{\varpi} \varepsilon v_x f d\varpi \tag{7.10a}$$

By treating the relaxation time as a constant, applying the local-equilibrium condition to the second term, and noting that the first term on the right-hand side is zero, we have

$$\tau\frac{\partial q_x''}{\partial t}+\kappa\frac{\partial T}{\partial x}=-q_x'' \text{ or } q_x''+\tau\frac{\partial q_x''}{\partial t}=-\kappa\frac{\partial T}{\partial x} \tag{7.10b}$$

This equation can be generalized to the 3D case as given in Eq. (7.2), after replacing $\tau$ with $\tau_q$.

The derivation given in this example, however, does *not* provide a microscopic justification of the hyperbolic heat equation, because it is strictly valid only under the local-equilibrium assumption with an averaged relaxation time. The local-equilibrium assumption prohibits application of the derived equation to length scales comparable to or smaller than the mean free path [19, 20]. Suppose a thermal disturbance occurs at a certain time and location; after a duration that is much longer than the relaxation time, Fourier's law and the parabolic heat equation are well justified because both the spatial and temporal local-equilibrium conditions are met. On the other hand, if we wish to use the modified Fourier equation to study the transient behavior at a timescale less than $\tau$, then the disturbance will propagate by a distance shorter than the mean free path, as shown in Fig. 7.2. Therefore, the derivation based on the BTE, under local-equilibrium and relaxation time approximations, is not a microscopic proof of the hyperbolic heat equation, which is meaningful only in a nonequilibrium situation. To this end, it appears that Maxwell in 1867 made the right choice in dropping terms involving the relaxation time in the paper, by assessing that the rate of conduction will rapidly establish itself [3, 4].

Rigorously speaking, the local-equilibrium condition can be expressed in terms of integration, i.e.,

$$\left|\int_{\varpi}\tau v_x^2\frac{\partial}{\partial x}(f-f_0)\varepsilon \mathrm{d}\varpi\right| << \left|\kappa\frac{\partial T}{\partial x}\right| \tag{7.11}$$

In deriving Eq. (7.10b), we have loosely assumed $\int_{\varpi}\tau\varepsilon v_x\frac{\partial f}{\partial t}\mathrm{d}\varpi = \frac{\partial}{\partial t}\int_{\varpi}\tau\varepsilon v_x f\mathrm{d}\varpi=\tau\frac{\partial q_x''}{\partial t}$. After a careful examination of the derivations, Zhang et al. [17] noted that Eq. (7.10a) can be rearranged to obtain the following expression:

$$\int_{\varpi}\tau\varepsilon v_x\left(\frac{\partial}{\partial t}+v_x\frac{\partial}{\partial x}\right)(f-f_0)\mathrm{d}\varpi+\kappa\frac{\partial T}{\partial x}=-q_x'' \tag{7.12}$$

One may define a new local-equilibrium condition as follows:

$$\left|\int_{\varpi}\tau\varepsilon v_x\frac{\mathrm{D}}{\mathrm{D}t}(f-f_0)\mathrm{d}\varpi\right| << \left|\kappa\frac{\partial T}{\partial x}\right| \tag{7.13}$$

where the operator $\frac{\mathrm{D}}{\mathrm{D}t} = \frac{\partial}{\partial t} + v_x \frac{\partial}{\partial x}$ in the 1D case and can be generalized to 3D cases. With the new local-equilibrium condition, Eq. (7.12) becomes Fourier's law. Therefore, we can derive Fourier's law directly from the BTE even in the transient situation [17]. Based on the above discussion, Fourier's law is not an approximation of Cattaneo's equation. Hence, one should not treat Cattaneo's equation as a generalized Fourier's law. It may be more appropriate to name Eq. (7.2) and the like as modified Fourier's equations.

Without knowing the heat carrier types and statistics, it is impossible to compare Eqs. (7.11) and (7.13). Both assumptions will break down when the smallest geometric dimension is on the same order or smaller than the mean free path. The basic assumption in the relaxation time approximation is that the distribution function is not too far from equilibrium. For a heat pulse with a duration less than $\tau_\mathrm{p}$, the relaxation time approximation should generally be applied when the time duration $t > 3\tau_\mathrm{p}$, regardless of whether we are dealing with a thin film or a semi-infinite medium. Atomistic simulations, based on molecular dynamics and the lattice Boltzmann method, have provided further evidence that the hyperbolic heat equation is not applicable at very short timescales or in the nonequilibrium regime, where the applicability of the relaxation time approximation is also questionable [21, 22]. Nevertheless, after some modifications, there exist a number of special cases when the modified heat equation becomes physically plausible and practically applicable. The modified equation does not produce sharp wavefronts like those illustrated in Fig. 7.2.

### 7.1.2 *Dual-Phase-Lag Model*

Chester [23] first related Cattaneo's equation with a lagging behavior, specifically, there exists a finite buildup time after a temperature gradient is imposed on the specimen for the onset of a heat flow, which does not start instantaneously but rather grows gradually during the initial period on the order of the relaxation time $\tau$. Conversely, if the thermal gradient is suddenly removed, there will be a *lag* in the disappearance of the heat current. Gurtin and Pipkin [24] introduced the memory effect to account for the delay of the heat flux with respect to the temperature gradient. They expressed the heat flux as an integration of the temperature gradient over time, in analogy with the stress–strain relationship of viscoelastic materials with instantaneous elasticity. The linearized constitutional equation reads

$$\mathbf{q}''(\mathbf{r}, t) = -\int_{-\infty}^{t} K(t - t')\nabla T(\mathbf{r}, t')\mathrm{d}t' \tag{7.14}$$

where $K(\xi)$ is a kernel function. Equation (7.14) reduces to Fourier's law when $K(\xi) = \kappa\delta(\xi)$ and to Cattaneo's equation when $K(\xi) = (\kappa/\tau_\mathrm{q})\mathrm{e}^{-\xi/\tau}$. By assuming

$$K(\xi) = \kappa_0 \delta(\xi) + \frac{\kappa_1}{\tau_q} e^{-\xi/\tau} \tag{7.15}$$

Joseph and Preziosi [4] showed that the heat flux can be separated into two parts:

$$\mathbf{q}''(\mathbf{r}, t) = -\kappa_0 \nabla T - \frac{\kappa_1}{\tau_q} \int_{-\infty}^{t} \exp\left(-\frac{t - t'}{\tau_q}\right) \nabla T(\mathbf{r}, t') \mathrm{d}t' \tag{7.16a}$$

Hence,

$$\mathbf{q}'' + \tau_q \frac{\partial \mathbf{q}''}{\partial t} = -\kappa \nabla T - \tau_q \kappa_0 \frac{\partial}{\partial t} \nabla T \tag{7.16b}$$

where $\kappa = \kappa_0 + \kappa_1$ is the steady-state thermal conductivity, as can be seen from Eq. (7.16a). Combined with Eq. (7.3), the heat equation becomes a partial differential equation,

$$\nabla^2 T + \tau_T \frac{\partial}{\partial t} \nabla^2 T = \frac{1}{\alpha} \frac{\partial T}{\partial t} + \frac{\tau_q}{\alpha} \frac{\partial^2 T}{\partial t^2} \tag{7.17}$$

where $\tau_T = \tau_q \kappa_0 / \kappa$ is known as the *retardation time*. Unless $\tau_T = 0$ or $\kappa_0 = 0$, Eq. (7.17) maintains the diffusive feature and produces an instantaneous response, albeit small, throughout the medium for an arbitrary thermal disturbance.

In a series of papers published in the 1990s, Tzou extended the lagging concept to a dual-phase-lag model, as described in his monograph first published in 1997 and the second edition in 2015 [6]. The starting point of the dual-phase-lag model is the constitutive relationship,

$$\mathbf{q}''(\mathbf{r}, t + \tau_q) = -\kappa \nabla T(\mathbf{r}, t + \tau_T) \tag{7.18}$$

The introduction of a delay time $\tau_T$ in Eq. (7.18) implies the existence of a lag in the temperature gradient, with respect to the heat flux driven by an internal or external heat source. The rationale of the phenomenological equation given in Eq. (7.18) was that, in some cases, the heat flux might be viewed as the result of a preceding temperature gradient; in other cases, the temperature gradient might be viewed as the result of a preceding heat flux. The heat flux and the temperature gradient can switch roles in the relationship between "cause" and "effect." Moreover, both lags might occur simultaneously in certain materials under dramatic thermal disturbances, such as during short-pulse laser heating [6, 7]. These primitive arguments should not be scrutinized rigorously; rather, they are merely thinking instruments to help us gain an intuitive understanding of the heat flux and temperature gradient relationship. After applying the Taylor expansion to both sides of Eq. (7.18) and using the first-order approximation, one immediately obtains

$$\mathbf{q}'' + \tau_q \frac{\partial \mathbf{q}''}{\partial t} = -\kappa \nabla T - \tau_T \kappa \frac{\partial}{\partial t} \nabla T \tag{7.19}$$

which is mathematically identical to Eq. (7.16b), with the substitution of $\tau_q \kappa_0 = \tau_T \kappa$. Applying the first-order approximation of Eq. (7.18), one may end up with $\mathbf{q}'' + (\tau_q - \tau_T)\frac{\partial \mathbf{q}''}{\partial t} = -\kappa \nabla T$, or $\mathbf{q}'' = -\kappa \nabla T - (\tau_T - \tau_q)\frac{\partial}{\partial t}\nabla T$, or even $\mathbf{q}'' + \left(\tau_q - \frac{\tau_T}{3}\right)\frac{\partial \mathbf{q}''}{\partial t} = -\kappa \nabla T - \frac{2\tau_T}{3}\kappa \frac{\partial}{\partial t}\nabla T$. These equations are merely special cases of Eq. (7.19), after regrouping $\tau_q$ and $\tau_T$. The only requirement for Eq. (7.19) to make logical sense is that both $\tau_q$ and $\tau_T$ are nonnegative. The reason that a lag in time has been called a phase lag is perhaps because the temperature field can be viewed as a Fourier transform: $T(\mathbf{r}, t) = \int_{-\infty}^{\infty} \tilde{T}(\mathbf{r}, \omega) \mathrm{e}^{-\mathrm{i}\omega t} \mathrm{d}\omega$, where $\tilde{T}(\mathbf{r}, \omega)$ is the Fourier component at frequency $\omega$. The actual phase lag $\omega \tau_T$ (or $\omega \tau_q$ for heat flux) depends on the frequency. Equation (7.19) is mathematically more general and has some advantages over Cattaneo's equation. From now on, Eq. (7.17) will be called the *lagging heat equation*. It is straightforward to include the source terms in the lagging heat equation, as well as to treat thermophysical properties as temperature dependent. The solution, however, becomes more and more difficult as the complexity increases. Numerous studies have appeared in the literature on analytical solutions and numerical methods [4, 25–28].

It should be noted that in Eq. (7.15), $\kappa_0$ and $\kappa_1$ denote the effective and elastic conductivities, respectively, and are supposed to be nonnegative [4]. Therefore, $\tau_T$ must not be greater than $\tau_q$. In fact, the ratio $\eta = \kappa_0/(\kappa_0 + \kappa_1)$ is a direct indication of whether thermal behavior can be described by heat diffusion (when $\eta = 1$ and $\kappa_1 = 0$) or the hyperbolic heat equation (when $\eta = 0$ and $\kappa_0 = 0$). In general, $0 \le \eta \le 1$ and the thermal process lies somewhere between the two extremes prescribed by Fourier's law and Cattaneo's equation. In other words, there will be wavelike features in the solution, which is superimposed by an instantaneous diffusive response throughout the medium. The diffusive response here, as well as in Fourier's law, does not correspond to an infinite speed of propagation. Rather, it is well justified by quantum statistics as explained previously.

The dual-phase-lag model relaxes the requirement of $\tau_T \le \tau_q$; but in the meantime, it produces a negative thermal conductivity component, i.e., $\kappa_1 < 0$ according to Eq. (7.15). This drawback has long been overcome by Tzou [6], who proposed a new memory function in accordance with Eq. (7.19) as follows:

$$\mathbf{q}''(\mathbf{r}, t) = -\frac{\kappa}{\tau_q} \int_{-\infty}^{t} \exp\left(-\frac{t - t'}{\tau_q}\right) \left[\nabla T(\mathbf{r}, t') + \tau_T \frac{\partial}{\partial t'} \nabla T(\mathbf{r}, t')\right] \mathrm{d}t' \tag{7.20}$$

Equation (7.20) suggests that the heat flux depends not only on the history of the temperature gradient but also on the history of the time derivative of $\nabla T$. When $\tau_T = 0$, Eq. (7.20) becomes Cattaneo's equation. When $\tau_T = \tau_q$, Eq. (7.20) reduces to Fourier's law. However, $\tau_T > \tau_q$ is theoretically permitted because Eq. (7.20) does not presume that the thermal conductivity is composed of an effective conductivity and an elastic conductivity. The inclusion of $\tau_T > \tau_q$ makes Eq. (7.19) more general than

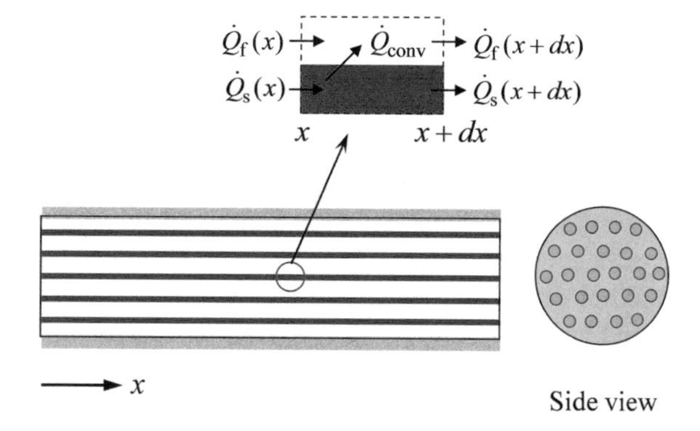


**Fig. 7.3** Illustration of heat transfer in a solid–fluid heat exchanger, where long solid rods are immersed in a fluid inside a sealed pipe, which is insulated from the outside

the Eq. (7.16a) since this allows the lagging heat equation to describe the behavior of parallel heat conduction that can occur in a number of engineering situations.

Sometimes a microscale phenomenon can be understood easily if a macroscale analogy can be drawn. For this reason, let us consider the solid–fluid heat exchanger shown in Fig. 7.3. Assume that a fluid is stationary inside a sealed pipe, filled with long solid rods. The pipe is insulated from the outside. If the rods are sufficiently thin, we may use the average temperature in a cross section and assume that heat transfer takes place along the $x$-direction only. Let us denote the temperatures of the solid rods and the fluid by $T_s(x, t)$ and $T_f(x, t)$, respectively, and take their properties $\kappa_s$, $C_s = (\rho c_p)_s$, $\kappa_f$, and $C_f = (\rho c_p)_f$ to be constant. Note that $C_s$ and $C_f$ are the *volumetric heat capacities*. Given the rod diameter $d$, the number of rods $N$, and the inner diameter $D$ of the pipe, the total surface area per unit length is $P = N\pi D$, and the total cross-sectional areas of the rods and the fluid are $A_c = N\pi d^2/4$ and $A_f = (\pi/4)(D^2 - Nd^2)$, respectively. Assume the average convection coefficient is $h$. The energy balance equations can be obtained using the control volume analysis as follows:

$$C_s \frac{\partial T_s}{\partial t} = \kappa_s \frac{\partial^2 T_s}{\partial x^2} - G(T_s - T_f) \tag{7.21a}$$

and

$$C_f' \frac{\partial T_f}{\partial t} = G(T_s - T_f) \tag{7.21b}$$

where $G = hP/A_c$ and $C_f' = C_f A_f/A_c$. In writing Eq. (7.21b), we have assumed that $\kappa_f << \kappa_s$ and dropped the term $\kappa_f \frac{\partial^2 T_f}{\partial x^2}$. Equations (7.21a) and (7.21b) are coupled equations that can be solved for the prescribed initial and boundary conditions. These are completely macroscopic equations governed by Fourier's law of heat conduction. Nevertheless, we can combine Eqs. (7.21a) and (7.21b) to eliminate $T_f$ and, consequently, obtain the following differential equation for $T_s$:

$$\frac{\partial^2 T_s}{\partial x^2} + \tau_T \frac{\partial}{\partial t}\left(\frac{\partial^2 T_s}{\partial x^2}\right) = \frac{1}{\alpha}\frac{\partial T_s}{\partial t} + \frac{\tau_q}{\alpha}\frac{\partial^2 T_s}{\partial t^2} \tag{7.22}$$

where $\alpha = \frac{\kappa_s}{C_s+C_f'}$, $\tau_T = \frac{C_f'}{G}$, and $\tau_q = \frac{C_s\tau_T}{C_s+C_f'} < \tau_T$. The same equation can also be obtained for the fluid temperature $T_f$. Here, $\tau_q$ does not have the meaning of relaxation time. Equation (7.22) is completely physical but should not be viewed as a wave equation; rather, it describes a parallel or coupled heat diffusion process. The concept of dual phase lag can still be applied. It should be noted that, due to the initial temperature difference between the rod and the fluid, a local equilibrium is not established at any $x$ inside the pipe, until after a sufficiently long time.

Although no fundamental physics can be gained from this example, it can help us appreciate that the lagging heat equation may be useful for describing the behavior in inhomogeneous media. Minkowycz et al. [29] studied the heat transfer in porous media by considering the departure from local thermal equilibrium and obtained higher order differential equations similar to Eq. (7.22). On the other hand, Kaminski [30] made an experimental attempt to determine $\tau_q$ in the hyperbolic heat equation, by measuring the time interval between when the heat source was turned on and when a temperature signal was detected. The heat source and the thermometer used were long needles, placed in parallel and separated by a gap of 5–20 mm. What the experiment actually measured was the average thermal diffusion speed $v_{dif}$ if the cylindrical geometry and the initial conditions were properly taken into consideration in the analysis. The main problem with this frequently cited paper and similar studies in the 1990s was that most researchers did not realize that the hyperbolic heat equation is physically unjustified to be superior to the parabolic heat equation; instead, some researchers took the parabolic equation as a special case of the more general hyperbolic equation [3]. While many researchers have expressed doubt about the applicability of the hyperbolic heat equation, few have realized that an instantaneous response is merely a mathematical solution that does not affect the application of the diffusion equation in macroscopic problems. Electron gas and phonon gas in solids are quantum mechanical particles, which do not have memory of any kind. Ideal molecular gases obey classical statistics and do not have memory either, unless the deposited energy is too intense to cause ionization or reaction.

Does the temperature wave exist? What is a temperature wave anyway? In the early 1940s, Russian theoretical physicist Lev Landau (1908–1968) used a two-fluid model to study the behavior of quasiparticles in superfluid helium II and predicted the existence of a second sound, propagating at a speed between $v_g/\sqrt{3}$ and $v_g$, depending on the temperature. Note that the group velocity is the same as the phase velocity for a linear dispersion. Above the $\lambda$-point, where superfluidity is lost, the second sound should also disappear. Landau was awarded the Nobel Prize in Physics in 1962 for his pioneering theories of condensed matter at low temperatures. He authored with his students a famous book series in mechanics and physics. Landau's prediction was validated experimentally by Russian physicists in the 1940s. The existence of a second sound in crystals was also postulated when scattering by defects becomes minimized. However, it was not until the mid-1960s that the second sound associated with heat pulse propagation was observed in solid helium (below 1 K) and other crystals at low temperatures (below 20 K). The second sound can occur only at very low temperatures when the mean free path of phonons in the *U*-processes, in which the total momentum is not conserved, is longer than the specimen size; while at the same time, the scattering rate of the *N*-processes, in which the total momentum is conserved, is high enough to dominate other scattering processes. It should be noted that while the *N*-processes have a much shorter mean free path than the size of the specimen, scattering by *N*-processes does not dissipate heat (see Sect. 6.5.3). Callaway [31] simplified the BTE for phonon systems by a two-relaxation-time approximation, which should be applicable when $t > \tau_N$:

$$\frac{\partial f}{\partial t} + \mathbf{v} \cdot \frac{\partial f}{\partial \mathbf{r}} = \frac{f_0 - f}{\tau} + \frac{f_1 - f}{\tau_N} \tag{7.23}$$

where $\tau$ stands for the relaxation time for the *U*-processes, $\tau_N$ is the relaxation time for the *N*-processes, and $f_0$ and $f_1$ are the associated equilibrium distribution functions. Guyer and Krumhansl [32] solved the linearized BTE and derived the following equation for the phonon effective temperature:

$$\nabla^2 T + \frac{9\tau_N}{5} \frac{\partial}{\partial t} \nabla^2 T = \frac{3}{\tau v_a^2} \frac{\partial T}{\partial t} + \frac{3}{v_a^2} \frac{\partial^2 T}{\partial t^2} \tag{7.24}$$

where $v_a$ is the average phonon speed. Assuming a linear dispersion, it can be evaluated using Eq. (5.10). Substituting $\alpha = \tau v_a^2/3$, $\tau_q = \tau$, and $\tau_T = 9\tau_N/5$, we see that Eq. (7.24) is identical to Eq. (7.17). The condition $t > \tau_N$ can be satisfied even at $t < \tau$ since $\tau_N << \tau$. The significance of Eq. (7.24) lies in that the temperature wave or the second sound is not universal, but rather, requires strict conditions to be met [32]. When the condition $\tau_N << \tau$ is satisfied, we have $\tau_T << \tau_q$ and the energy transfer is dominated by wave propagation. At higher temperatures, the scattering rate for the *U*-processes is usually very high, and the *N*-processes contribute little to the heat conduction or thermal resistance, as discussed in Chap. 6. Therefore, the reason why temperature waves have not been observed in insulators at room temperature is not because of the small $\tau$, in the range from $10^{-10}$ to $10^{-13}$ s, but because of the lack of mechanisms required for a second sound to occur. No experiments have ever shown a second sound in metals, as suggested by the hyperbolic heat equation.

Shiomi and Maruyama [33] performed molecular dynamics simulations of the heat conduction through (5,5) single-walled carbon nanotubes, 25 nm in length, for several femtoseconds. They found that the wavelike behavior could be fitted by the lagging heat equation, but could not be described by the hyperbolic heat equation due to local diffusion. The ballistic nature of heat propagation in nanotubes has already been explained in Chap. 5. They suspected that optical phonons might play a major role in the non-Fourier conduction process [33]. Tsai and MacDonald [34] studied the strong anharmonic effects at high temperature and pressure using molecular dynamics. Their work predicted a second sound response. The coupling of elastic and thermal effects was thought to be important. Studies on thermomechanical effects such as thermal expansion, thermoelasticity, and shock waves can be found from Tzou [6] and Wang and Xu [35, 36], and will not be discussed further.

Tang and Araki [26] clearly delineated four regimes in the lagging heat equation, according to the ratio $\eta = \tau_T/\tau_q$. (1) When $\eta = 0$, it is a damped wave, i.e., hyperbolic heat conduction. (2) When $0 < \eta < 1$, it is wavelike diffusion, for which wave features can be clearly seen if $\eta << 1$. (3) When $\eta = 1$, it is pure diffusion or diffusion, i.e., Fourier's conduction. (4) When $\eta > 1$, it is called over-diffusion, which makes the dimensionless temperature decay faster than pure diffusion would. In the next section, we will discuss a microscopic theory on short-pulse laser heating of metals, which falls in the regime of over-diffusion, or parallel conduction.

### *7.1.3 Two-Temperature Model*

With a short laser pulse, 5 fs–500 ps, free electrons absorb radiation energy and the absorbed energy excites the electrons to higher energy levels. The "hot electrons" move around randomly and dissipate heat mainly through electron–phonon interactions. In the 1970s, Anisimov et al. [37] proposed a *two-temperature model*, which is a pair of coupled nonlinear equations governing the effective temperatures of electrons and phonons. This model was experimentally confirmed in the 1980s by researchers at the Massachusetts Institute of Technology [38, 39]. The two-temperature model was introduced to the heat transfer community by Qiu and Tien [40, 41] in early 1990s. In a series of papers [40–42], Qiu and Tien analyzed the size effect due to boundary scattering and performed experiments with thin metallic films. In the two-temperature model, it is assumed that the electron and phonon systems are each at their own local equilibrium, but not in mutual equilibrium. The electron temperature could be much higher than the lattice (or phonon) temperature due to absorption of pulse heating. Therefore,

$$C_e \frac{\partial T_e}{\partial t} = \nabla \cdot (\kappa \nabla T_e) - G(T_e - T_s) + \dot{q}_a \tag{7.25a}$$

$$C_s \frac{\partial T_s}{\partial t} = G(T_e - T_s) \tag{7.25b}$$

Here, the subscript e and s are for the electron and phonon systems, respectively, $C$ is the volumetric heat capacity, $G$ is the electron–phonon coupling constant, and $\dot{q}_a$ is the source term that represents the absorbed energy rate per unit volume during the laser pulse and drops to zero after the pulse. Heat conduction by phonons is neglected, and thus, the subscript e is dropped in the thermal conductivity $\kappa$. Note that $\mathbf{q}'' = -\kappa \nabla T_e$, according to Fourier's law. We have already given a macroscopic example of parallel heat transfer, as shown in Fig. 7.3, which should ease the understanding of the phenomenological relations given in Eqs. (7.25a), (7.25b). Equations (7.25a), (7.25b) originate from microscopic interactions between photons, electrons, and phonons. In order to examine the parameters in Eqs. (7.25a), (7.25b) and their dependence on $T_e$ and $T_s$, let us assume that the lattice temperature is near or above the Debye temperature for simplicity. In such a case, electron–electron scattering and electron–defects scattering are insignificant compared with electron–phonon scattering. It is expected that the electron relaxation time is inversely proportional to the lattice temperature, i.e., $\tau \approx \tau_{\text{e-ph}} \propto T_s^{-1}$. The meaning of the relaxation time is that the electron system can be assumed to be at internal local equilibrium when $t > \tau$, which is the condition for Eqs. (7.25a), (7.25b) to be applicable. Boundary scattering may play a role for very thin films or in polycrystalline materials. An effective mean free path can be introduced to modify the scattering rate [40, 43, 44]. The volumetric heat capacity for the lattice or phonons is $C_s = \rho c_p$ is a weak function of the lattice temperature; the volumetric heat capacity of electrons, from Eq. (5.25), becomes

$$C_e = \frac{n_e \pi^2 k_B^2}{2\mu_F} T_e = \gamma T_e \tag{7.26}$$

Recall that $C_e$ is relatively small compared with $C_s$, even at several thousand kelvins. From simple kinetic theory, the thermal conductivity is

$$\kappa = \frac{\pi^2 n_e k_B^2}{3m_e} \tau T_e \approx \frac{\kappa_{eq}}{T_s} T_e \tag{7.27}$$

where $\kappa_{eq}$ is the thermal conductivity when $T_e = T_s$, which can be set at room temperature value. The term $T_e$ in Eq. (7.27) comes from the heat capacity. The size effect can be included using an effective relaxation time. Theoretically, the coupling constant can be estimated by

$$G = \frac{\pi^2 m_e n_e v_a^2}{6\tau T_s} \text{ or } G = \frac{\pi^4 (n_e v_a k_B)^2}{18\kappa_{eq}} \tag{7.28}$$

which is independent of temperature, when boundary scattering is not important, but proportional to the square of the speed of sound in the metal. With the speed of sound in the low-frequency limit, the dispersion is linear; thus, we do not have to worry about the difference between the phase velocity and the group velocity. From Eq. (5.10), we have

$$v_a = \frac{k_B \Theta_D}{h} \left( \frac{4\pi}{3n_a} \right)^{1/3} \tag{7.29}$$

When boundary scattering is included, $G$ is expected to increase from the bulk value and depend on the lattice temperature. Using the Debye temperature and for $n_a = n_e$, we have

$$G = \frac{\pi^2}{12 \times \sqrt[3]{4}} \frac{n_e k_B^2 \Theta_D^2}{\tau T_s \mu_F} \approx 0.518 \frac{n_e k_B^2 \Theta_D^2}{\tau T_s \mu_F} \tag{7.30}$$

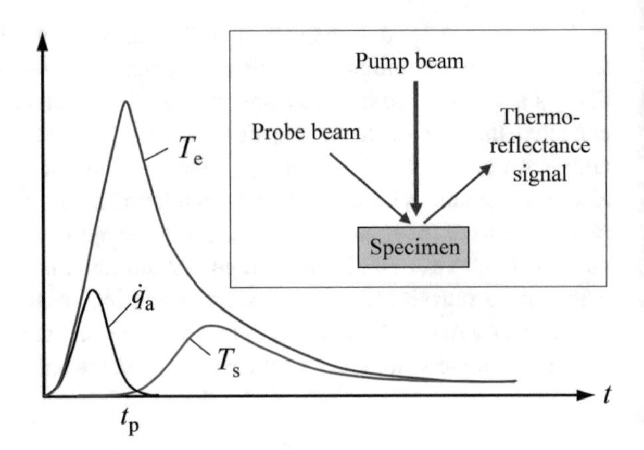


**Fig. 7.4** Illustration (not to scale) of ultrafast thermoreflectance experiments and the associated electron and phonon temperatures near the surface, during a short pulse

Typical values of $G$ are on the order of $10^{16}$ W/K m$^3$, e.g., $G \approx 2.9 \times 10^{16}$ W/K m$^3$ for gold. The behavior of the electron and phonon temperatures near the surface is shown in Fig. 7.4, for a short pulse. The electron temperature rises quickly during the pulse and begins to decrease afterward; in the meantime, the lattice temperature gradually begins to increase until the electron and lattice systems reach a thermal equilibrium. Both the temperatures will go down as heat is carried away from the surface. Note that the electron temperature can rise very high due to its small heat capacity, but the lattice or solid may be just slightly above room temperature. If the temperatures of the electron and lattice were the same, Eqs. (7.25a), (7.25b) would reduce to the simple Fourier heat conduction equation. This would lead to a prediction of a much lower temperature rise, because the heat capacity of the lattice is much greater than that of the electrons.

Given such a short timescale and the nonequilibrium nature between electrons and phonons locally, no contact thermometer could possibly measure the effective electron temperature. Experiments are usually performed by the femtosecond or picosecond thermoreflectance technique, also known as the pump-and-probe method, shown in the inset of Fig. 7.4. The reflectance of the surface depends on the electron temperature $T_e$. The experimental setup is rather involved and the details will be given in Sect. 7.4.3. The procedure is to send a pump pulse train that is synchronized with a probe pulse train at a fixed delay time. The electron temperature change near the surface is related to the reflectance as a function of the delay time. Electron–phonon coupling, boundary scattering, and thermal boundary resistance can all affect the thermoreflectance signal. Comparing with the model described in Eqs. (7.25a), (7.25b), along with the dependence of the reflectance on the electron temperature, the microscopic characteristics can be analyzed. Ultrafast thermoreflectance techniques have become an important thermal metrology tool for the study of electron–phonon interactions, TBR, and thermophysical properties [40–49]. Thermionic emission can also occur from the surface, especially when the electrons are excited to higher energy states [50].

Similar to what has been done for Eqs. (7.21a) and (7.21b), Eqs. (7.25a) and (7.25b) can be combined to formulate partial differential equations for either the electron or phonon temperature. Neglecting the temperature dependence of the parameters, one obtains the following differential equations for the electron temperature and the phonon temperature, respectively,

$$\nabla^2 T_\mathrm{e} + \tau_\mathrm{T}\frac{\partial}{\partial t}\nabla^2 T_\mathrm{e} + \frac{\dot{q}_a}{\kappa} + \frac{\tau_T}{\kappa}\frac{\partial \dot{q}_\mathrm{a}}{\partial t} = \frac{1}{\alpha}\frac{\partial T_\mathrm{e}}{\partial t} + \frac{\tau_\mathrm{q}}{\alpha}\frac{\partial^2 T_\mathrm{e}}{\partial t^2} \tag{7.31a}$$

$$\nabla^2 T_\mathrm{s} + \tau_\mathrm{T}\frac{\partial}{\partial t}\nabla^2 T_\mathrm{s} + \frac{\dot{q}_\mathrm{a}}{\kappa} = \frac{1}{\alpha}\frac{\partial T_\mathrm{s}}{\partial t} + \frac{\tau_\mathrm{q}}{\alpha}\frac{\partial^2 T_\mathrm{s}}{\partial t^2} \tag{7.31b}$$

where $\alpha = \frac{\kappa}{C_\mathrm{e}+C_\mathrm{s}}$, $\tau_\mathrm{T} = \frac{C_\mathrm{s}}{G}$, and $\tau_\mathrm{q} = \frac{\tau_\mathrm{T} C_\mathrm{e}}{C_\mathrm{e}+C_\mathrm{s}} \approx \frac{C_\mathrm{e}}{G} << \tau_\mathrm{T}$. These equations are identical to the lagging heat equations and can be solved with appropriate boundary conditions. The results again belong to the regime of over-diffusion, or parallel conduction, without any wavelike features. Cooling caused by thermionic emission is usually neglected, and the surface under illumination can be assumed adiabatic. A 1D approximation further simplifies the problem. The solution follows the general trends depicted in Fig. 7.4. The situation will be completely changed if a phase change occurs or if the system is driven to exceed the linear harmonic behavior [6, 35].

The term $\tau_\mathrm{q}$ is clearly not the same as the relaxation time $\tau$ due to collision. The resulting solution is more diffusive than wavelike. In the literature, $\tau_\mathrm{q}$ is commonly referred to as the *thermalization time*. The physical meaning of $\tau_\mathrm{q}$ is a *thermal time constant* for the electron system to reach an equilibrium with the phonon system. For noble metals at room temperature, the relaxation time $\tau$ is on the order of 30–40 fs, the thermalization time $\tau_\mathrm{q}$ is 0.5–0.8 ps, and the *retardation time* $\tau_\mathrm{T}$ is 60–90 ps. In practice, we need to consider the temperature dependence of the parameters in Eqs. (7.25a, 7.25b), as mentioned earlier. Some numerical solutions, considering temperature dependence, and comparisons with experiments can be found from Smith et al. [51] and Tzou and Chiu [27]. Given that the two-temperature model cannot be applied to $t < \tau$, due to the limitation of Fourier's law, one may prefer to use a pulse width $t_\mathrm{p}$ between 100 and 200 fs and measure the response during several picoseconds until the thermalization process is complete, i.e., the electron and phonon temperatures become the same. This first-stage measurement allows the determination of the coupling constant $G$. In the case of a thin film, the TBR sets a barrier for heat conduction between the film and the substrate. The time constant of the film can range from several tens to hundreds of picoseconds. Therefore, the TBR between the film and the substrate can be determined by continuing the observation of thermoreflectance signals for 1–2 ns after each pulse. Fitting the curves in the second-stage measurement allows an estimate of the TBR. Of course, one could use a longer pulse width $t_\mathrm{p}$ to determine the TBR. Most advanced femtosecond research laboratories are equipped with Ti:sapphire lasers whose pulse widths range from 50 to 500 fs. Femtosecond lasers with a pulse width of 25 fs have also been used in some studies; see for example Li et al. [52]. For $t_\mathrm{p}$ below 50 fs, Eq. (7.25a) is not applicable during the heating, at least for noble metals. The relaxation time for Cr is about 3 fs, and Eqs. (7.25a, 7.25b) can be safely applied even with $t_p = 10$ fs. However, the processes below 20 fs may largely involve electron–electron inelastic scattering, thermionic emission, ionization, phase transformation, chemical reaction, and so forth. Other difficult issues associated with the reduced pulse width include widened frequency spectrum, increased pulse intensity, decreased pulse energy, and so forth. A simple hyperbolic formulation cannot properly address these issues. One must investigate the physical and chemical processes occurring at this timescale in order to develop a physically plausible model, with or without the concept of effective temperatures. Femtosecond laser interactions with dielectric materials have also been extensively studied (see Jiang and Tsai [53, 54] and references therein). Recently, Ma [55] proposed a two-parameter heat conduction model for analyzing transient heat conduction data for dielectric materials and for thermal interfaces based on both frequency-domain and time-domain measurements. In the two-parameter model, a nonequilibrium (effective) temperature is defined and used to obtain a nondiffusive phenomenological equation. A thermal conductivity (that describes diffusive thermal transport) and a ballistic heat transfer length (that is the product of a ballistic relaxation time and the speed of carriers) are taken as the fitting parameters [55]. It should be noted that the formulation of the two-parameter model differs significantly from the ballistic-diffusion heat conduction equations proposed by Chen [56] in 2001.

Let us reiterate some major points presented in this section: (a) Fourier's law, which is limited to local-equilibrium conditions, does not predict an infinite speed of heat diffusion, nor does it violate the principle of causality [3]. An instantaneous response at a finite distance is permitted by quantum statistics although the probability of such a response sharply approaches zero as the distance increases. An instantaneous temperature change or heat flux at a precise location is not physically possible. Only under the continuum assumption, we can use the concept of sudden change of temperature at the boundary. (b) Heat diffusion is usually a very slow process, compared with the speed of sound. The temperature wave, or the second sound, has been observed only in helium and some very pure dielectric crystals, at low temperatures, where the $U$-processes are ballistic and the $N$-processes have a very high scattering rate. (c) Both Fourier's law and Cattaneo's equation can be derived from the BTE under slightly different approximations [17]. Fourier's law is not an approximation of Cattaneo's equation and, hence, Cattaneo's equation is not more general than Fourier's law. Nevertheless, the introduction of an additional parameter (the relaxation time) in Cattaneo's equation may allow the hyperbolic heat equation to better fit some experiments in inhomogeneous medium with coupled phenomena [4]. (c) All kinds of non-Fourier equations are based on some sort of effective temperature, which are not measurable using a contact thermometer. The principle of contact thermometry is based on the assumption of thermodynamic equilibrium according to the zeroth law of thermodynamics. The concept of coldness or hotness should be abandoned in reference to nonequilibrium energy transport processes. Noncontact thermometry, on the other hand, relies on certain physical responses to deduce the equilibrium temperature or the effective temperature of the system being measured. (d) The memory hypothesis and the lagging argument are phenomenological models that may be useful in the study of certain nonequilibrium or parallel conduction processes, but are not universally applicable. These and similar equations must be derived and applied on a case-by-case basis. It is important to understand the microscopic processes occurring at the appropriate length scales and timescales in order to develop physically reliable models.

## 7.2 Heat Conduction Across Layered Structures

In Sect. 5.5.2, we have given a detailed discussion on the heat conduction along a thin film using the BTE, under the local-equilibrium assumption. An effective thermal conductivity can be used after taking proper account of boundary scattering. The heat conduction problem can thus be well described by Fourier's law using the effective thermal conductivity. As mentioned earlier, for heat transfer across a film or a superlattice, the condition of local equilibrium breaks down in the acoustically thin limit. The local distribution function cannot be approximated by an equilibrium distribution function at any temperature. Conventional Fourier's law breaks down because it relies on the definition of an equilibrium temperature and the existence of local equilibrium. It is natural to ask the following two questions. (1) Is it possible for us to define an effective temperature? (2) Can Fourier's law still be useful in the nonequilibrium regime, according to the effective temperature? This section presents the equation of phonon radiative transfer (EPRT) and the solution of EPRT for thin films under the relaxation time approximation. A resistance network representation is used to illustrate how Fourier's law of heat conduction may be applied inside the medium, at least approximately, with temperature-jump boundary conditions. Because of the importance of understanding the boundary conditions, this section also discusses models of thermal boundary resistance (TBR) in layered structures.

### 7.2.1 *Equation of Phonon Radiative Transfer (EPRT)*

The phonon BTE under the relaxation time approximation, in a region with heat generation, may be written as

$$\frac{\partial f}{\partial t} + \mathbf{v} \cdot \frac{\partial f}{\partial \mathbf{r}} = \frac{f - f_0}{\tau(\omega, T)} + S_0 \tag{7.32}$$

where the second term $S_0$ on the right-hand side is a source term to model the generation of phonons due to heat dissipation, such as electron–phonon scattering. Phonon–phonon scattering is already included in the first term on the right-hand side. The scattering rate may also include phonon-defect scattering. Many studies have treated phonon transport in analogy to thermal radiative transfer [19, 20, 57–67]. In the following, a simplified case is used to illustrate how to model heat transfer across a thin film as well as multilayer structures. Let us consider a film of thickness

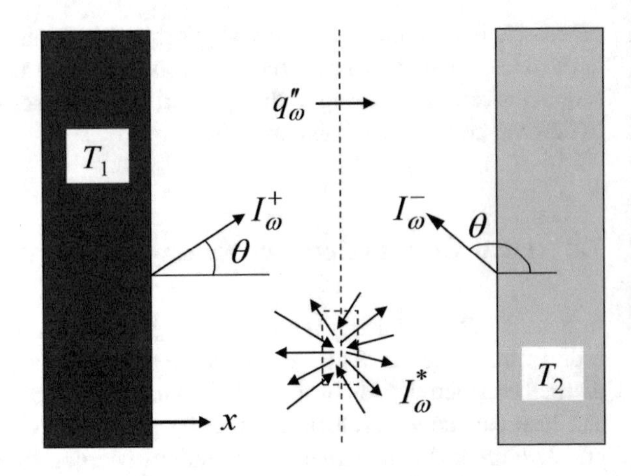


**Fig. 7.5** Schematic of phonon radiative transfer inside a dielectric medium between two walls maintained at temperatures $T_1$ and $T_2$. These walls are like heat reservoirs, but their surfaces are not necessarily blackbodies

$L$ between two boundaries without any internal source. The phonon BTE becomes

$$\frac{\partial f}{\partial t} + v_x \frac{\partial f}{\partial x} = \frac{f_0 - f}{\tau} \tag{7.33}$$

Realizing the nonequilibrium distribution function may be anisotropic, let us define

$$I_\omega(x, \Omega, t) = \frac{1}{4\pi} \sum_P v_g \hbar\omega f D(\omega) \tag{7.34}$$

where $P$ is the index for phonon mode or polarization and $D(\omega)$ is the DOS. Equation (7.34) gives the *phonon intensity*, which is the energy transfer rate in the direction $\Omega$ from a unit area, per unit frequency and per unit solid angle. The geometry of the problem and illustration of the intensity is given in Fig. 7.5. As done before, let $v_g$ and $v_p$ be the group velocity and phase velocity, respectively. Note that $v_x = v_g \cos\theta$, where $\theta$ is the polar angle. Substituting Eq. (7.34) into Eq. (7.33), we obtain

$$\frac{1}{v_g}\frac{\partial I_\omega}{\partial t} + \mu \frac{\partial I_\omega}{\partial x} = \frac{I_\omega^* - I_\omega}{v_g \tau} \tag{7.35}$$

where $\mu = \cos\theta$ and $I_\omega^*(\omega, T)$ is the intensity for equilibrium distribution that is independent of the direction. Equation (7.35) is called the equation of phonon radiative transfer (EPRT) [19, 61]. Comparing the EPRT with the ERT given in Eq. (2.53), we see that the scattering terms are neglected in the EPRT, and the emission and absorption are replaced by the phonon collision terms. The phonon mean free path $\Lambda = v_g \tau$ is also called the phonon penetration depth (see Example 4.2). The inverse of the penetration depth ($1/\Lambda$) corresponds to the absorption coefficient in the ERT. Conversion to the EPRT allows well-established theories and numerical techniques, developed in radiative transfer, to be applied to solve Eq. (7.35) and to interpret the physical significance of the solutions [68, 69]. If $\tau$ does not depend on frequency, we are dealing with a gray medium.

If the phonon Knudsen number $Kn = \Lambda/L << 1$, then most phonons will collide with phonons or defects inside the medium. This regime is called the *acoustically thick limit*, in analogy to the *optically thick limit* for photons. This is also known as the macroscale regime or the local-equilibrium situation. Unless at a very short timescale, when a sudden local disturbance occurs, we expect that Fourier's law is applicable and the heat conduction is by diffusion. On the other hand, if $Kn = \Lambda/L >> 1$, phonons originated from one boundary will most likely reach the other boundary without colliding with other phonons or defects inside the medium. This is the ballistic regime, corresponding to free molecule flow for molecular gases. This regime is called the *acoustically thin limit*, where the phonon distribution inside the medium cannot be characterized by an equilibrium distribution function if the walls are at different temperatures, even in the steady state. Because the BTE is more fundamental than Fourier's law, it is applicable to both limiting cases as well as those between the two limits. It would be very useful if a macroscopic model can also be developed to bridge these two limits. Some basic formulations are given in the following.

Note that $I_\omega^*$ is the equilibrium distribution function, which is independent of the direction. Using Bose–Einstein statistics, we have

$$I_\omega^*(\omega, T) = \sum_P \frac{v_g \hbar\omega}{e^{\hbar\omega/k_B T} - 1} \frac{k^2}{(2\pi)^3} \frac{dk}{d\omega} = \sum_P \frac{\hbar\omega^3}{8\pi^3 v_p^2 (e^{\hbar\omega/k_B T} - 1)} \tag{7.36}$$

This equilibrium distribution is also the distribution function for blackbody radiation with $v_p$ replaced by the speed of light. Integrating Eq. (7.36) over all frequencies gives the total intensity for all three phonon modes:

$$I^*(T) = \int_0^\infty I_\omega^*(\omega, T) d\omega = \frac{3k_B^4 T^4}{8\pi^3 \hbar^3 v_a^2} \int_0^\infty \frac{x^3 dx}{e^x - 1} = \frac{\sigma'_{SB} T^4}{\pi} \tag{7.37}$$

where $\sigma'_{SB} = \pi^2 k_B^4/(40\hbar^3 v_a^2)$ is the phonon Stefan–Boltzmann constant, and $v_a$ is the average phase velocity of the two translational and one longitudinal phonon modes, defined according to Eq. (5.7). Let us consider a solid at temperatures higher than the Debye temperature. The integration can be carried out to an upper limit $\omega_m$ with $x_m = \hbar\omega_m/(k_B T) << 1$. From the discussion following Eq. (5.13), one can easily show that

$$I^*(T) = \int_0^{\omega_m} I_\omega^*(\omega, T) d\omega = \frac{\omega_m^3 k_B}{8\pi^3 v_p^2} T \tag{7.38}$$

This integration is a good approximation, even at temperatures slightly lower than the Debye temperature. When phonons are at equilibrium, the energy flux is

$\pi I^*$, which is obtained by integrating $I^*\cos\theta\,\mathrm{d}\Omega$ over the hemisphere. According to Eq. (4.12), the energy density can be expressed as

$$u(T) = \frac{4\pi}{v_\mathrm{g}} I^*(T) \tag{7.39}$$

Note that the volumetric heat capacity $C = \mathrm{d}u/\mathrm{d}T$ when $u$ is expressed in terms of energy density. We therefore obtain the low-temperature relation of the specific heat, i.e., the $T^3$ law, and the high-temperature relation of the specific heat, i.e., the Dulong–Petit law, as already derived in Sect. 5.1.2. It is important to pay attention to the meaning of $C$ in the kinetic expression of thermal conductivity:

$$\kappa = \frac{1}{3} C v_\mathrm{g}^2 \tau \tag{7.40}$$

At very low temperatures, when $T << \Theta_\mathrm{D}$, $C$ is the volumetric heat capacity of all phonon modes combined because only low-frequency modes or acoustic branches contribute to the specific heat. However, at temperatures close to the Debye temperature, phonons in the optical branches contribute little to the thermal conductivity, as already discussed in Chap. 6. The relative contributions of LA and TA branches are also temperature dependent. The Debye temperature for most materials, except diamond, is not much higher than room temperature (see Table 5.1). Therefore, one may treat the volumetric heat capacity $C$ as a fraction of the volumetric specific heat in dealing with Si, GaAs, Ge, ZnS, or GaN, near room temperature. Also, we must use the appropriate upper limit in the integral in calculating the total energy transfer when applying EPRT. The heat flux per unit frequency interval can thus be expressed as

$$q''_\omega = \int_{4\pi} I_\omega \cos\theta \,\mathrm{d}\Omega = 2\pi \int_{-1}^{1} I_\omega \mu \,\mathrm{d}\mu \tag{7.41}$$

Energy balance at any given location requires that the incoming flux from all directions be the same as the outgoing flux toward all directions, for both steady and transient states, as illustrated in Fig. 7.5. This is the criterion for *radiative equilibrium* [68, 69], which can be expressed as follows [19]:

$$4\pi \int_0^{\omega_\mathrm{m}} \frac{1}{\Lambda_\omega} I_\omega^* \mathrm{d}\omega = 2\pi \int_0^{\omega_\mathrm{m}} \int_{-1}^{1} \frac{1}{\Lambda_\omega} I_\omega \,\mathrm{d}\mu \,\mathrm{d}\omega \tag{7.42}$$

where $\Lambda_\omega$ is the mean free path at $\omega$, $4\pi$ on the left-hand side came from the integration over all solid angles in a sphere, and $2\pi$ on the right-hand side came from integration over the azimuth angles. For a gray medium, $\Lambda_\omega = v_\mathrm{a}\tau$ is independent of the frequency. Equation (7.42) gives a definition of an *effective phonon temperature*

$T^*$ based on the equilibrium distribution: $I^*_\omega(T^*, \omega)$. An equivalent expression can be obtained based on the energy density, viz.

$$u(T^*) - u_0 = \sum_P \sum_K \hbar\omega f_0(T^*, \omega, \Omega) = \sum_P \sum_K \hbar\omega f(\omega, \Omega) \tag{7.43}$$

where $u_0$ is a reference value. Note that the spectral component (integrand) on both sides of Eq. (7.42) may not be equal at all frequencies. Even for a gray medium, in general, one cannot deduce the following from Eq. (7.42):

$$I^*_\omega(T^*, \omega) = \frac{1}{2} \int_{-1}^{1} I_\omega \mathrm{d}\mu \tag{7.44}$$

The physical significance of Eq. (7.44) is that the angular average of the intensity, at a given location and time, can be described by an equilibrium intensity that satisfies the equilibrium distribution function at a certain temperature. As a matter of fact, Eq. (7.44) is equivalent to the *local-equilibrium approximation* [70]. It can be shown that the local-equilibrium approximation is valid only in the acoustically thick limit or the diffusive heat conduction regime.

**Example 7.3** For a dielectric medium of thickness $L = 0.01\Lambda$, the mean free path $\Lambda$ is independent of wavelength. The boundary or wall temperatures are $T_1 = 50$ K at $x = 0$ and $T_2 = 100$ K at $x = L$. Both the temperatures are much lower than the Debye temperature. Assume that reflection at the boundaries is negligible, i.e., the walls can be modeled as blackbodies. At steady state, express the heat flux through the medium and find the effective photon temperature distribution $T^*(x)$.

**Solution** Because $Kn = \Lambda/L >> 1$, the medium is said to be in the acoustically thin limit, in which phonons travel from one wall to another ballistically with little chance of being scattered by other phonons or defects inside the medium. The forward intensity can be expressed as $I^+_\omega = I^*_\omega(T_1, \omega)$ for $\mu > 0$, and the backward intensity $I^-_\omega = I^*_\omega(T_2, \omega)$ for $\mu < 0$. From Eq. (7.41), we have

$$q''_x = \int_0^\infty q''_\omega \mathrm{d}\omega = 2\pi \int_0^\infty \int_0^1 (I^+_\omega - I^-_\omega)\, \mu \mathrm{d}\mu \mathrm{d}\omega = \sigma'_{\mathrm{SB}}(T_1^4 - T_2^4) \tag{7.45}$$

For heat conduction, the above equation is called the Casimir limit [71]. To numerically evaluate this equation, we need data for $v_\mathrm{a}$. From Eq. (7.42), we have

$$\sigma'_{\mathrm{SB}}(T^*)^4 = \frac{\pi}{2} \int_0^{\omega_\mathrm{m}} (I^+_\omega + I^-_\omega) \mathrm{d}\omega = \frac{1}{2}\left(\sigma'_{\mathrm{SB}} T_1^4 + \sigma'_{\mathrm{SB}} T_2^4\right) \tag{7.46}$$

We obtain $T^* = 85.37$ K, which is the effective temperature inside the medium $0 < x < L$ and is independent of $x$. Since $T(0) = T_1$ and $T(L) = T_2$ are the boundary conditions, there is a temperature jump at each boundary similar to Fig. 4.12b in the free molecule regime for gas conduction. If the walls are not black but diffuse-gray with emissivities $\varepsilon_1$ and $\varepsilon_2$, similar to Eq. (2.52), the heat flux becomes

$$q''_x = \frac{\sigma'_{SB} T_1^4 - \sigma'_{SB} T_2^4}{1/\varepsilon_1 + 1/\varepsilon_2 - 1} \tag{7.47}$$

**Comments**: (1) Taking diamond with $v_a = 12,288$ m/s as an example, we have $\sigma'_{SB} = 50.63$ W/m$^2$ K$^4$. The magnitude of the heat flux in the ballistic limit for $T_1 = 50$ K and $T_1 = 100$ K is 4.75 GW/m$^2$, which is quite high. Note that the mean free path of diamond in this temperature region is around 1.3 μm [19, 70]. Thus when $Kn = 100$, the thickness is only 13 nm. If an effective diffusive thermal conductivity is used, $\kappa_{eff} = 1.23$ W/m K, which is much smaller than the bulk thermal conductivity of diamond! In general, ballistic transport in nanostructures results in a restriction to the heat flow as compared with diffuse transport with the same thermal conductivity. (2) Figure 7.6 shows the phonon intensity spectra at $T_1$, $T_2$, and $T^*$ for diamond, calculated from Eq. (7.36) for the sum of the three phonon modes taking the average velocity $v_a$ in place of $v_p$. We notice immediately that Eq. (7.44) cannot be satisfied in the acoustically thin limit. Let us designate

$$I_{avg}(\omega) = \frac{1}{2}\int_{-1}^{1} I_\omega d\mu = \frac{1}{2}(I_\omega^+ + I_\omega^-) \tag{7.48}$$

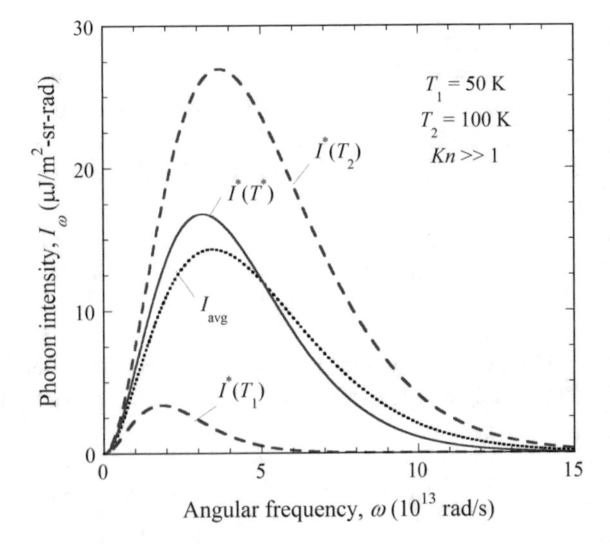


**Fig. 7.6** Phonon intensity spectra for equilibrium distribution at the wall temperatures $T_1$ and $T_2$, and the effective temperature $T^*$. The intensity calculated based on Eq. (7.48) is also plotted for comparison. Note that $I_{avg}(\omega)$ may be considered as a nonequilibrium distribution in terms of the phonon intensity

which is also plotted in Fig. 7.6. It can be seen that $I_\omega^*(\omega) \neq I_{avg}(\omega)$ in general. It is well known that a monochromatic temperature can be defined and is useful in radiation thermometry (refer to Sect. 8.2 for further discussion). Bright and Zhang [70] used the concept of monochromatic phonon temperature to study entropy generation in a thin film from the diffusive regime to the ballistic regime.

## 7.2.2 *Solution of the EPRT*

The two-flux method is very helpful in developing a solution of the EPRT in planar structures, as shown in Fig. 7.5. The equations for the forward and backward intensities, denoted respectively by superscripts (+) and (−), can be separated. Assuming the medium is gray, at steady state, we can rewrite the EPRT given in Eq. (7.35) as follows [68, 69]:

$$\mu \frac{\partial I_\omega^+}{\partial x} = \frac{I_\omega^* - I_\omega^+}{\Lambda}, \text{ when } 0 < \mu < 1 \tag{7.49a}$$

$$\mu \frac{\partial I_\omega^-}{\partial x} = \frac{I_\omega^* - I_\omega^-}{\Lambda}, \text{ when } -1 < \mu < 0 \tag{7.49b}$$

If we further assume that the walls are diffuse and gray, then the boundary conditions become

$$T(0) = T_1 \text{ and } T(L) = T_2 \tag{7.50}$$

Thus,

$$I_\omega^+(0, \mu) = \varepsilon_1 I_\omega^*(T_1) + (1 - \varepsilon_1) I_\omega^-(0, \mu) \tag{7.51a}$$

$$I_\omega^-(L, \mu) = \varepsilon_2 I_\omega^*(T_2) + (1 - \varepsilon_2) I_\omega^+(L, \mu) \tag{7.51b}$$

The solutions of Eqs. (7.49a) and (7.49b) can be expressed as follows:

$$I_\omega^+(x, \mu) = I_\omega^+(0, \mu) \exp\left(-\frac{x}{\Lambda\mu}\right) + \int_0^x I_\omega^*(\xi) \exp\left(-\frac{x - \xi}{\Lambda\mu}\right) \frac{d\xi}{\Lambda\mu} \text{ for } \mu > 0 \tag{7.52a}$$

and

$$I_\omega^-(x, \mu) = I_\omega^-(L, \mu) \exp\left(\frac{L - x}{\Lambda\mu}\right) - \int_x^L I_\omega^*(\xi) \exp\left(-\frac{x - \xi}{\Lambda\mu}\right) \frac{d\xi}{\Lambda\mu} \text{ for } \mu < 0 \tag{7.52b}$$

In Eq. (7.53), the first term represents intensity originated from the left surface, after being attenuated, and the second term is the contribution of generation that is subject to attenuation as well. Equation (7.54) is viewed reversely for intensity from the right to the left. The spectral heat flux, defined in Eq. (7.41), can be obtained

$$q''_\omega = 2\pi \int_0^1 \left[ I_\omega^+(0,\mu)\exp\left(-\frac{x}{\Lambda\mu}\right) - I_\omega^-(L,-\mu)\exp\left(-\frac{L-x}{\Lambda\mu}\right)\right]\mu \mathrm{d}\mu$$
$$+ 2\pi \int_0^x I_\omega^*(\xi) E_2\left(\frac{x-\xi}{\Lambda}\right)\frac{\mathrm{d}\xi}{\Lambda} - 2\pi \int_x^L I_\omega^*(\xi) E_2\left(\frac{\xi-x}{\Lambda}\right)\frac{\mathrm{d}\xi}{\Lambda} \quad (7.53)$$

where $E_m(x) = \int_0^1 \eta^{m-2} e^{-x/\eta}\,\mathrm{d}\eta$ is again the $m$th-order *exponential integral.* If the surface is diffuse, then we have

$$q''_\omega = 2\pi I_\omega^+(0) E_3\left(\frac{x}{\Lambda}\right) - 2\pi I_\omega^-(L) E_3\left(\frac{L-x}{\Lambda}\right)$$
$$+ 2\pi \int_0^x I_\omega^*(\xi) E_2\left(\frac{x-\xi}{\Lambda}\right)\frac{\mathrm{d}\xi}{\Lambda} - 2\pi \int_x^L I_\omega^*(\xi) E_2\left(\frac{\xi-x}{\Lambda}\right)\frac{\mathrm{d}\xi}{\Lambda} \quad (7.54)$$

Energy balance requires that the derivative of the radiative heat flux be zero, viz.

$$\frac{dq''_x}{\mathrm{d}x} = \int_0^{\omega_\mathrm{m}} \frac{\partial}{\partial x} q''_\omega(x,\omega)\mathrm{d}\omega = 0 \quad (7.55)$$

This equation is another form of radiative equilibrium since radiative equilibrium means that the divergence of the radiative heat flux to be zero or $\nabla \cdot \mathbf{q}'' = 0$. Differentiating Eq. (7.54) yields

$$\frac{\partial q''_\omega}{\partial x} = -\frac{2\pi}{\Lambda} I_\omega^+(0) E_2\left(\frac{x}{\Lambda}\right) - \frac{2\pi}{\Lambda} I_\omega^-(L) E_2\left(\frac{L-x}{\Lambda}\right)$$
$$- \frac{2\pi}{\Lambda} \int_0^L I_\omega^*(\xi) E_1\left(\frac{|x-\xi|}{\Lambda}\right)\frac{\mathrm{d}\xi}{\Lambda} + \frac{4\pi}{\Lambda} I_\omega^*(x) \quad (7.56)$$

In radiative transfer, we call $J_1 = \int \pi I_\omega^+(0)\mathrm{d}\omega$ and $J_2 = \int \pi I_\omega^-(L)\mathrm{d}\omega$ the total radiosities at surfaces 1 and 2, respectively, and $e_\mathrm{b}(T) = \int \pi I_\omega^* \mathrm{d}\omega$ the total blackbody emissive power. Substituting Eq. (7.56) into Eq. (7.55), after performing the integration, we obtain

$$2e_{\mathrm{b}}(T(x)) = J_1 E_2\left(\frac{x}{\Lambda}\right) + J_2 E_2\left(\frac{L-x}{\Lambda}\right) + \int_0^L e_{\mathrm{b}}(T(\xi)) E_1\left(\frac{|x-\xi|}{\Lambda}\right)\frac{\mathrm{d}\xi}{\Lambda} \tag{7.57}$$

This is the radiative equilibrium condition and it is always valid if there is no internal generation. Note that Eq. (7.56) becomes zero for all frequencies only in the diffusive limit.

**Example 7.4** Find the temperature distribution, heat flux, and thermal conductivity for a gray medium with diffuse-gray surfaces in the acoustically thick limit, i.e., $Kn << 1$; under two extreme conditions: (i) $T_1, T_2 << \Theta_{\mathrm{D}}$ and (ii) $T_1, T_2 > \Theta_{\mathrm{D}}$.

**Solution** In the thick limit, the first two terms in Eq. (7.53) can be dropped as long as $x$ is not too close to either surface. Applying the first-order Taylor expansion $I_\omega^*(x) = I_\omega^*(\xi) + \frac{\mathrm{d}I_\omega^*}{\mathrm{d}x}(x-\xi) + \ldots$ and letting $z = \frac{x-\xi}{\Lambda}$ in the third and fourth terms, we obtain

$$q_\omega'' = -4\pi\Lambda\frac{\partial I_\omega^*}{\partial x}\int_0^\infty zE_2(z)\mathrm{d}z = -\frac{4\pi}{3}\Lambda\frac{\partial I_\omega^*}{\partial x} \tag{7.58}$$

Since $\int_0^\infty zE_2(z)\mathrm{d}z = 1/3$. In fact, this equation applies to everywhere inside the medium because the spectral heat flux is continuous in the acoustically thick limit. Integrating Eq. (7.58) over the frequencies of interest, we see that under condition (i):

$$q_x'' = -\frac{16\sigma_{\mathrm{SB}}' T^3}{3}\Lambda\frac{\mathrm{d}T}{\mathrm{d}x}, \text{ when } T << \Theta_{\mathrm{D}} \tag{7.59}$$

This is nothing but a heat diffusion equation if we define the thermal conductivity as

$$\kappa(T) = \frac{16}{3}\sigma_{\mathrm{SB}}' T^3 \Lambda \tag{7.60}$$

Comparing Eq. (7.60) with Eq. (7.40), $\kappa(T) = \frac{1}{3}Cv_{\mathrm{g}}\Lambda$, we see that $Cv_{\mathrm{g}} = 16\sigma_{\mathrm{SB}}' T^3$ in this case and it is consistent with the $T^3$ law for the specific heat at low temperatures. In the thick limit, the temperature distribution is continuous at the wall, i.e., $T(0^+) = T(0) = T_1$ and $T(L^-) = T(L) = T_2$. Furthermore, the radiosity at the wall becomes the blackbody emissive power, even though the surface is not black. Hence, we can integrate Eq. (7.59):

$$\int_0^L q_x''\mathrm{d}x = \frac{4\Lambda}{3}\sigma_{\mathrm{SB}}'\int_{T_1}^{T_2} 4T^3\mathrm{d}T \tag{7.61a}$$

which gives

$$q''_x = \frac{4}{3} Kn(\sigma'_{SB} T_1^4 - \sigma'_{SB} T_2^4) \tag{7.61b}$$

as well as the temperature distribution:

$$T(x) = \left[T_1^4 - \frac{x}{L}(T_1^4 - T_2^4)\right]^{1/4} \tag{7.62}$$

This distribution is linear in terms of the fourth power of temperature [69, 70]. From the definition of thermal resistance $q''_x = (T_1 - T_2)/R'_t$, we have

$$R''_t = \frac{3(T_1 + T_2)(T_1^2 + T_2^3)}{4\sigma'_{SB} Kn} \tag{7.63}$$

Under condition (ii), when the temperature is greater than the Debye temperature, we have

$$q''_x = -\frac{\omega_m^3 k_B}{6\pi^2 v_p^2} \Lambda \frac{dT}{dx} \text{ when } T > \Theta_D \tag{7.64}$$

Compared with Eq. (7.40), we obtain

$$C v_g = \frac{\omega_m^3 k_B}{2\pi^2 v_p^2} \tag{7.65}$$

This suggests that the specific heat is independent of temperature in the high-temperature limit as expected. A proper $\omega_m$ should be chosen so that only propagating phonons or acoustic phonons are considered [61]. Assuming that the temperature difference is small so that we can approximate the thermal conductivity as a constant, we have

$$q''_x = \frac{1}{3} C v_g Kn(T_1 - T_2) \tag{7.66}$$

The thermal resistance becomes $R''_t = 3/(C v_g Kn)$, which increases as $L$ increases. The temperature distribution is linear. One should realize that the scattering rate increases with temperature and depends on the frequency, due to phonon–phonon scattering. If we look at the radiative equilibrium condition again, by assuming $T_1 > T_2$, we see that $I_\omega^+ > I_\omega^* > I_\omega^-$. Therefore, local equilibrium is not a stable-equilibrium state. In the thick limit, the difference between $I_\omega^+$ and $I_\omega^-$ is caused by the spatial variation of $I_\omega^*$ as can be clearly seen from Eqs. (7.52a) and (7.52b). Hence, the local-equilibrium approximation given in Eq. (7.44) is valid.

**Comment**. In the acoustically thin limit under the condition (ii) that $T > \Theta_{\mathrm{D}}$, by using the linear temperature relationship given in Eq. (7.38), we can modify Eq. (7.45) to the following,

$$q''_x = \frac{1}{4} C v_{\mathrm{g}} (T_1 - T_2) \tag{7.67}$$

Here, we have used the definition of Eq. (7.65). The effective thermal conductivity in the ballistic limit: $\kappa_{\mathrm{eff}} = 3\kappa_{\mathrm{b}}/(4Kn)$, where $\kappa_{\mathrm{b}}$ is the bulk or diffusive thermal conductivity. It can be seen that in the ballistic regime, the thermal conductivity is inversely proportional to $Kn$.

Although no closed form exists for the solution of the ERT between the thick and thin limits, a number of approximation techniques and numerical methods can be used to provide satisfactory solutions, such as the discrete ordinates method ($S_N$ approximation) and the spherical harmonics method ($P_N$ approximation) [69]. It is important to see that, except in the thick limit, energy transfer occurs inside the medium in two ways: one is through exchange with the walls, and the other is through diffusion. For this reason, a ballistic-diffusion approximation has been developed to solve the EPRT [56]. In general, the temperature distribution looks like that in Fig. 4.12b if $T_2$ is comparable to the Debye temperature. If $T_1 << \Theta_{\mathrm{D}}$, then the temperature distribution can be plotted in terms of $T^4$ so that the distribution looks more or less linear. There exists a temperature jump such that $T(0^+) \neq T(0)$ and $T(L^-) \neq T(L)$, except in the thick limit. Understanding that the temperature is only an effective temperature and given such a temperature distribution, one may assume that there is a thermal resistance at each boundary and an internal thermal resistance, which may be described by Fourier's heat conduction [64]. For thermal radiative transfer in the absence of heat conduction, there exists a radiation slip or radiation jump at the boundary, unless the medium is optically thick. Without a participating medium, photons do not scatter on itself to dissipate heat or transfer heat by diffusion. This is a distinction between photons and phonons. Radiation slip is manifested by a discontinuous change of the intensity at the boundary. The temperature in the medium adjacent to the wall differs from the surface temperature. Such a temperature jump does not exist in classical Fourier's heat conduction theory; however, both velocity slip and temperature jump have already been incorporated in microfluidics research, as discussed in Chap. 4; see Eq. (4.99). The temperature-jump concept was first applied in the study of heat conduction in rarefied gases over 100 years ago. A straightforward approach for phonon transport is to sum up the thermal resistances in the acoustically thin and thick limits. The heat flux at very low temperatures can be expressed as

$$q''_x = \frac{4\Lambda}{3L} \frac{\sigma'_{\mathrm{SB}}(T_1^4 - T_2^4)}{1 + \left(\frac{1}{\varepsilon_1} - \frac{1}{2} + \frac{1}{\varepsilon_2} - \frac{1}{2}\right)\frac{4Kn}{3}} \tag{7.68}$$

Here, we separately write $\left(\frac{1}{\varepsilon_1} - \frac{1}{2}\right)$ and $\left(\frac{1}{\varepsilon_2} - \frac{1}{2}\right)$ to emphasize the thermal resistance due to radiation slip at each boundary. In the thick limit, the temperature jump approaches zero as $Kn \to 0$. Basically, Eq. (7.68) reduces to Eqs. (7.47) and (7.61b), in the extreme cases. If the walls can be treated as blackbodies with $\varepsilon_1 = \varepsilon_2 = 1$, and the temperature difference between $T_1$ and $T_2$ is small, we can approximate the heat flux as follows:

$$q_x'' = \frac{\kappa_b}{L} \frac{\Delta T}{1 + 4Kn/3} = \kappa_{\text{eff}} \frac{\Delta T}{L} \tag{7.69}$$

where $\Delta T = T_1 - T_2 << T_2 < T_1$, the bulk thermal conductivity $\kappa_b(T) = \frac{16}{3}\sigma'_{\text{SB}} T^3 \Lambda$, and the effective conductivity of the film is

$$\kappa_{\text{eff}} = \frac{\kappa_b}{1 + 4Kn/3} \tag{7.70}$$

At relatively high temperatures close to the Debye temperature, from Eqs. (7.66) and (7.67), we can write

$$q_x'' = \frac{\kappa_b}{L} \frac{T_1 - T_2}{1 + \left(\frac{1}{\varepsilon_1} + \frac{1}{\varepsilon_2} - 1\right)\frac{4Kn}{3}} = \kappa_{\text{eff}} \frac{T_1 - T_2}{L} \tag{7.71}$$

where $\kappa_b(T) = \frac{1}{3}Cv_g\Lambda$. Equation (7.71) gives the same conductivity ratio $\kappa_{\text{eff}}/\kappa_b$ as in Eq. (7.70) for blackbody walls. These effective thermal conductivities are on the same order of magnitude as we have derived in Sect. 5.5.5, based on simple geometric arguments and Matthiessen's rule for the mean free path given in Eq. (5.128). In previous chapters, however, we did not elaborate in detail on the nature of nonequilibrium and the necessity of defining an effective temperature. It is interesting that different schools of thought can result in rather consistent results. The heat diffusion equation per se cannot tell us the cause of a temperature jump or how to evaluate it. The phonon BTE enables us to explore the microscopic phenomena and helps to evaluate the parameters and the properties. The microscopic understanding and the macroscopic phenomenological equations can work together to provide an effective thermal analysis tool.

The results presented previously are consistent with the detailed derivation of the temperature jump or the radiation slip, originally formulated by Deissler [72], for thermal radiation in gases not too far from the optically thick limit. Nevertheless, the expressions given here can be approximately applied between the diffusion and ballistic extremes [70]. It should be noted that when the temperature jump is treated as a thermal resistance at the boundary, Fourier's law can be used for the heat conduction inside the medium with bulk thermal conductivity. This is very different from heat conduction along the film.

While the meaning of emissivity for optical radiation is very clear, a question still remains as how to interpret the boundary conditions in the case of phonon conduction,

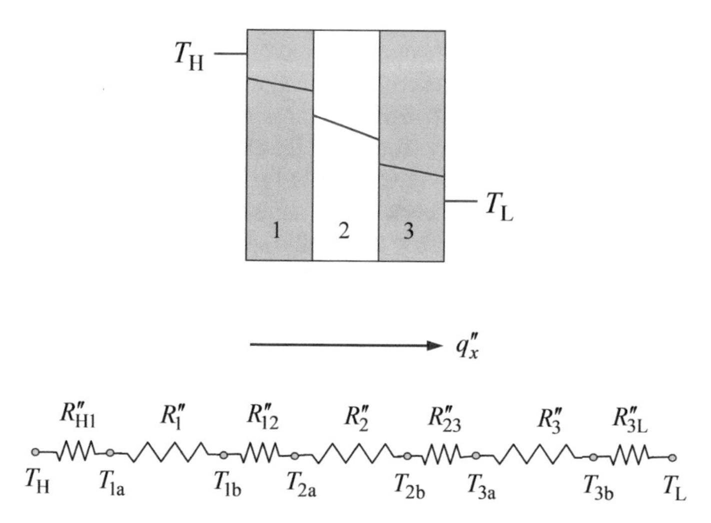


**Fig. 7.7** Temperature distribution in a multilayer structure, with thermal boundary resistance, and the thermal resistance network representation. Here, $R''_i$ is the internal resistance in the *i*th layer due to heat conduction, and $R''_{ij}$ is the thermal boundary resistance between the *i*th and *j*th media. Two temperatures are needed to specify the effective temperature of different media at the interface

since it is not easy to perceive the concepts of phonon emission and emissivity. If a multilayer structure is considered, we need to better understand the reflection and the transmission of phonons at the interfaces between dissimilar materials. A three-layer structure is shown in Fig. 7.7 to illustrate the temperature distribution in a multilayer structure. Depending on the temperature range, we may express the internal thermal resistance using Fourier's law, i.e., $R''_i = L_i/\kappa_i$, where $\kappa_i$ is the effective thermal conductivity of the *i*th layer. For the thermal resistance at the interface inside the layered structures, we could replace the emissivity with a transmissivity $\Gamma_{ij}$ such that [64]

$$R''_{ij} = \frac{4\Lambda_i}{3\kappa_i}\left(\frac{1}{\Gamma_{ij}} - \frac{1}{2}\right) + \frac{4\Lambda_j}{3\kappa_j}\left(\frac{1}{\Gamma_{ji}} - \frac{1}{2}\right) \tag{7.72}$$

At the boundaries, we can still use $R''_{H1} = \frac{4\Lambda_1}{3\kappa_1}\left(\frac{1}{\varepsilon_1} - \frac{1}{2}\right)$ and $R''_{3L} = \frac{4\Lambda_3}{3\kappa_3}\left(\frac{1}{\varepsilon_3} - \frac{1}{2}\right)$.

The heat flux can be estimated by $q''_x = (T_H - T_L)/R''_{tot}$, where $R''_{tot}$ is the sum of all thermal resistances. The effective thermal conductivity of the whole layered structure becomes $\kappa_{eff} = L_{tot}/R''_{tot}$. The details were presented by Chen and Zeng [64], who further considered nondiffuse surfaces and defined equivalent equilibrium temperatures. The assumption is that the deviation from the thick limit is not significant. If we are dealing with the ballistic regime, we might need to consider phonon wave effects as well as the quantum size effect. Recently, Maldovan's group has performed comprehensive studies of phonon transport across superlattices considering surface roughness and various length scales [73, 74]. The thermal resistance network method, however, cannot be easily extended to multidimensional problems or to transient heating by a localized heat source. Statistical models (such as the Monte Carlo method) or atomistic simulations (such as the atomistic Green's function method or molecular dynamics) are necessary. Therefore, the extension of Fourier's law for 1D nonequilibrium heat transfer should be considered only as a special case. It is intriguing to apply the same approach to electron systems for the study of both electrical conductivity and thermal conductivity of metallic solids, as well as metal-dielectric multilayer structures. Further discussion on the classical and advanced models of thermal boundary resistance is given in the next section.

### *7.2.3 Thermal Boundary Resistance (TBR)*

Thermal resistance at the interface between dissimilar materials is very important for heat transfer in heterostructures. Let us first clarify the difference between thermal contact resistance and thermal boundary resistance (TBR). The former refers to the thermal resistance between two bodies, usually with very rough surfaces whose root-mean-square roughness $\sigma_{\mathrm{rms}}$ is greater than 0.5 $\mu$m, brought or joined together mechanically. For thermal contact resistance, readers are referred to a recent comprehensive review by Yovanovich [75]. Originally, TBR refers to the resistance at the interface between two solids or between a liquid and a dielectric at low temperatures. Even when the materials are in perfect contact with each other, reflections occur when phonons travel toward the boundary, because of the difference in acoustic properties of adjacent materials. In practice, the interface can be atomically smooth, or with a roughness ranging from several tenths of a nanometer to several nanometers. The thermal resistance between a solid material and liquid helium is called the Kapitza resistance, first observed by the Russian physicist and 1978 Nobel Laureate Pyotr Kapitza, in the 1940s. The existence of a thermal resistance gives rise to a temperature discontinuity at the boundary and has been modeled, based on the *acoustic mismatch model* (AMM). TBR exists between two dielectrics as well as between a metal and a dielectric. In a thin-film structure, an interface is often accompanied by the formation of an intermediate layer of mixed atoms. An extensive review of earlier studies can be found in the work of Swartz and Pohl [59]. Stoner and Maris [76] used a picosecond thermoreflectance technique to measure the TBR for several metal-dielectric interfaces from 50 to 300 K and observed anomalously large conductance that can be understood as due to the anharmonicity of the metal, resulting in an inelastic channel that facilitated the thermal transport. Phelan and coworkers [77–79] performed extensive research and provided literature survey of TBR of high-temperature superconductors in both the normal and superconducting states, for applications in superconducting electronics and radiation detectors.

Little [60] showed that the heat flux across the boundary of a perfectly joined interface between two solids is proportional to the difference in the fourth power of temperature on each side of the interface. This can be understood based on previous discussions of phonon radiative transfer and blackbody radiation. Consider longitudinal

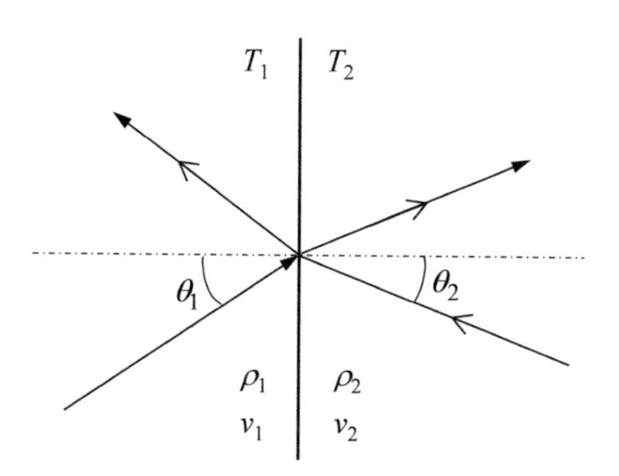


**Fig. 7.8** Schematic of phonon transport across an interface between two semi-infinite media, each at a thermal equilibrium. Note that arrows at the end denote incidence from the left side, while arrows in the middle denote incidence from the right side

phonon modes that follow the linear dispersion in a Debye crystal, and assume that the interface is perfectly smooth. At any given frequency, the transmission coefficients can be written as follows [60, 77]:

$$\tau_{12} = \tau_{21} = \frac{4\rho_1\rho_2 v_{l1} v_{l2}\cos\theta_1\cos\theta_2}{\left(\rho_1 v_{l1}\cos\theta_2 + \rho_2 v_{l2}\cos\theta_1\right)^2} \tag{7.73}$$

where subscripts 1 and 2 denote the media 1 and 2, respectively, $\rho$ is the density, $v_l$ is the propagation speed of longitudinal phonons, and $\theta$ is the polar angle, as illustrated in Fig. 7.8. The scattering is assumed to be purely elastic since the phonon frequency is conserved. An analog of Snell's law can be written as follows:

$$\frac{1}{v_{l1}}\sin\theta_1 = \frac{1}{v_{l2}}\sin\theta_2 \tag{7.74}$$

Assume $v_{l1} > v_{l2}$, for incidence from medium 2 to 1, there exists a critical angle $\theta_c = \sin^{-1}(v_{l2}/v_{l1})$, beyond which all phonons will be reflected. Due to the boundary resistance, there will be a temperature difference across the interface. By assuming that the phonons are at equilibrium on either side, the heat flux from medium 1 to 2 can be expressed as follows:

$$q''_{1\to 2} = \frac{1}{4\pi}\int_0^{\omega_m}\int_0^{2\pi}\int_0^{\pi/2} \hbar\omega v_{l1} f_1(\omega, T_1)\tau_{12} D(\omega)\cos\theta_1\sin\theta_1 \mathrm{d}\theta_1 \mathrm{d}\phi_1 \mathrm{d}\omega \tag{7.75}$$

If the distribution function is isotropic over the hemisphere, we have

$$q''_{1\to 2} = \frac{1}{4}\frac{\Gamma_{12}}{v_{l1}^2}\int_0^{\omega_m} \hbar\omega v_{l1}^3 f_1(\omega, T_1) D(\omega)\mathrm{d}\omega \tag{7.76}$$

where $\Gamma_{12}$ can be viewed as the hemispherical transmissivity that is expressed as

$$\Gamma_{12} = \frac{1}{\pi}\int_0^{2\pi}\int_0^{\pi/2} \tau_{12}\cos\theta_1 \sin\theta_1 \mathrm{d}\theta_1 \mathrm{d}\phi = 2\int_0^{\pi/2} \tau_{12}\cos\theta_1 \sin\theta_1 \mathrm{d}\theta_1 \tag{7.77}$$

It should be noted that

$$\Gamma_{21} = 2\int_0^{\theta_c} \tau_{21}\cos\theta_2 \sin\theta_2 \mathrm{d}\theta_2 = \frac{v_{l2}^2}{v_{l1}^2}\Gamma_{12} \tag{7.78}$$

One can prove Eq. (7.78) by noting that $\tau_{21} = \tau_{12}$ and using Eq. (7.74) and its derivative, i.e., $v_{l1}^{-1}\cos\theta_1 \mathrm{d}\theta_1 = v_{l2}^{-1}\cos\theta_2 \mathrm{d}\theta_2$. The difference between $\Gamma_{21}$ and $\Gamma_{12}$ can be explained as due to total internal reflection since for incidence from medium 2 to 1, portion of the photons will be totally reflected if the incidence angle exceeds the critical angle. For the Debye density of states, we have

$$\frac{1}{4\pi} v_l \hbar\omega f(\omega, T) D(\omega) \mathrm{d}\omega = \frac{\hbar\omega^3}{8\pi^3 v_l^2 (\mathrm{e}^{\hbar\omega/k_B T} - 1)} \tag{7.79}$$

Therefore, the net heat flux across the interface becomes

$$q_x'' = q_{1\to 2}'' - q_{2\to 1}'' = \frac{1}{4}\frac{\Gamma_{12}}{v_{l1}^2}\int_0^{\omega_m} \hbar\omega\left[v_{l1}^3 f_1(\omega, T_1) - v_{l2}^3 f_2(\omega, T_1)\right] D(\omega) \mathrm{d}\omega \tag{7.80a}$$

or

$$q_x'' = \frac{\Gamma_{12}}{v_{l1}^2}\frac{k_B^4}{8\pi^2\hbar^3}\left(T_1^4 \int_0^{x_{m,1}} \frac{x^3 \mathrm{d}x}{\mathrm{e}^x - 1} - T_2^4 \int_0^{x_{m,2}} \frac{x^3 \mathrm{d}x}{\mathrm{e}^x - 1}\right) \tag{7.80b}$$

In the low-temperature limit, we obtain

$$q_x'' = \frac{\Gamma_{12}}{v_{l1}^2}\frac{\pi^2 k_B^4}{120\hbar^3}(T_1^4 - T_2^4) \tag{7.81}$$

After replacing $v_{l1}^{-2}$ with $\sum_j v_{j1}^{-2} = v_{l1}^{-2} + 2v_{t1}^{-2}$, i.e., one longitudinal and two transverse phonon modes, we obtain

$$q_x'' = \frac{\pi^2 k_B^4}{120\hbar^3}(T_1^4 - T_2^4)\Gamma_{12}\sum_j v_{j1}^{-2} \tag{7.82}$$

The TBR can now be obtained as $R_b'' = (T_1 - T_2)/q_x''$. Furthermore, by assuming that the temperature difference is small, we can approximate $R_b''$ by

$$R_b'' = \frac{30\hbar^3 T^{-3}}{\pi^2 k_B^4 \Gamma_{12} \sum_j v_{j1}^{-2}} \tag{7.83}$$

which is inversely proportional to $T^3$. Equations (7.82) and (7.83) are the results of the AMM.

The characteristic wavelength is the most probable wavelength in the phonon distribution function. It can be approximated by

$$\lambda_{mp} \approx a\frac{\Theta_D}{T} \tag{7.84}$$

where $a$ is the lattice constant, on the order of 0.3–0.6 nm [77]. Only when $\lambda_{mp} >> \sigma_{rms}$, we can assume that the scattering is completely specular. Even for atomically smooth interfaces, the characteristic wavelength for phonons will be on the same order of magnitude as the rms surface roughness, when the temperature approaches the Debye temperature. The specularity parameter was introduced in Chap. 5, Eq. (5.143) and repeated here for normal incidence:

$$p = \exp\left(-\frac{16\pi^2\sigma_{rms}^2}{\lambda^2}\right) \tag{7.85}$$

This equation has been wrongly expressed in some literature with $\pi^2$ being mistaken as $\pi^3$ due to a typo in an earlier work. In the high-temperature limit, TBR is expected to be small, especially when compared with conduction in the solids. Other considerations are (a) the interface may not be perfectly smooth, (b) there exists an upper limit of the frequency or a lower limit of wavelength, and (c) phonons on either side of the boundary may not be in a local-equilibrium state. These difficulties post some real challenges in modeling TBR. Nevertheless, we shall present the *diffuse mismatch model* (DMM) that was introduced by Swartz and Pohl [59]. In the DMM, it is assumed that phonons will be scattered according to a probability, determined by the properties of the two media but independent of where the phonons originate from. For phonons coming from medium 1, the transmission and reflection probabilities are related by $\Gamma_{12} + R_{12} = 1$. For phonons originating from medium 2, on the other hand, $\Gamma_{21} = R_{12}$ and $R_{21} = \Gamma_{12}$. Hence, the reciprocity requires that

$$\Gamma_{12} + \Gamma_{21} = 1 \tag{7.86}$$

We can rewrite Eq. (7.78), considering all three polarizations, as follows:

$$\Gamma_{12} \sum_j v_{j1}^{-2} = \Gamma_{21} \sum_j v_{j2}^{-2} \tag{7.87}$$

The combination of Eqs. (7.86) and (7.87) gives

$$\Gamma_{12} = \frac{\sum_j v_{j2}^{-2}}{\sum_j v_{j1}^{-2} + \sum_j v_{j2}^{-2}} \tag{7.88}$$

This is the DMM prediction of the "hemispherical" transmission coefficient. The heat flux can be calculated according to

$$q_x'' = \frac{k_{\mathrm{B}}^4}{8\pi^2\hbar^3}\left(T_1^4\int_0^{x_{\mathrm{m},1}}\frac{x^3\mathrm{d}x}{\mathrm{e}^x-1} - T_2^4\int_0^{x_{\mathrm{m},2}}\frac{x^3\mathrm{d}x}{\mathrm{e}^x-1}\right)\Gamma_{12}\sum_j v_{j1}^{-2} \tag{7.89}$$

Equations (7.88) and (7.89) are the only equations needed to calculate TBR in the DMM. In addition to the Debye temperatures and the speeds of longitudinal and transverse waves, one would need to determine the upper limits of the integrals in Eq. (7.89). Alternatively, Eq. (7.89) can be recast using the volumetric heat capacity and the group velocity to obtain

$$q_x'' = \frac{1}{4}\left(C_{v1}v_{\mathrm{g}1}T_1 - C_{v1}v_{\mathrm{g}1}T_2\right)\Gamma_{12} \tag{7.90}$$

One must be careful in applying the heat capacity in Eq. (7.80) since the heat capacity in the expression of thermal conductivity is different from $\rho c_p$, unless at very low temperatures. Both AMM and DMM assume that phonons on each side of the interface are individually at equilibrium, and do not take into account the nonequilibrium distribution of phonons near the interface. In multilayer thin films, especially in quantum wells and superlattices, when the film thickness is comparable with or smaller than the phonon mean free path, thermal transport inside the film cannot be modeled as pure diffusion anymore. A detailed treatment of temperature-jump conditions and boundary resistance in superlattices can be found from Refs. [61–65]. Majumdar [80] proposed a modified AMM, by modeling interface roughness, using a fractal structure and assuming that the reflection can be approximated by geometric optics which is applicable when the phonon wavelength is smaller than the autocorrelation length of the rough surface. TBR between highly dissimilar materials, metal–metal interface, and metal–dielectric interface has also been extensively studied [47, 48, 81, 82].

### 7.2.4 *Atomistic Green's Function (AGF)*

As mentioned previously, the Monte Carlo method has been used extensively for solving the phonon transport equations [57, 58, 83–85]. The lattice Boltzmann method has also been employed in a number of publications [22, 67, 86]. Equilibrium and nonequilibrium molecular dynamics approaches have also been extensively employed to study thermal transport in nanostructures and TBR [82]. The basics of molecular dynamics simulation of solids have been discussed in Chap. 6 and can be found from the literature [87–95]. Another method called the nonequilibrium Green's function (NEGF) method has been extensively used to model the electron transport in semiconductor nanodevices [96] and has been introduced to study phonon transport across various interfaces, which is called the atomistic Green's function (AGF) method [97–99]. The AGF method is briefly discussed in the following.

The NEGF is an atomic-level quantum mechanical model based on the density matrix that can be obtained from the Hamiltonian matrix. As discussed in Sect. 5.6, the electrical current can be expressed in terms of Landauer's formalism, where the transmission probability can be obtained from the Green's function formulation [96]. Ozpineci and Ciraci [100] developed the Green's function method for thermal conductance in a phononic system that consists of chain of atoms between two reservoirs. Mingo and Yang [97] further developed the AGF approach and used it to study phonon transmission through coated nanowires by neglecting inelastic scattering. This method is further extended to study Si/Ge interfaces using an empirical interatomic potential that includes the strain effect [98]. A plane-wave formulation based on the wavevector space is developed to evaluate the harmonic matrix for a unit cell in the $x$-$y$ plane and multilayers in the $z$-direction across the interface (from left to the right). Green's function is used to represent the response of the dynamic system to an infinitesimal perturbation and can be used to obtain the transmission coefficient, which is a function of the frequency and parallel wavevector.

$$\Xi(\omega, \mathbf{k}_{\parallel}) = \mathrm{Trace}[\Gamma_{\mathrm{L}}\mathbf{G}\Gamma_{\mathrm{R}}\mathbf{G}^{\dagger}] \tag{7.91}$$

Here, $\mathbf{k}_{\parallel}$ is the wavevector parallel to the interface, the matrices $\Gamma_{\mathrm{L}}$ and $\Gamma_{\mathrm{R}}$ represent the phonon escape rate at the left and right contacts, $\mathbf{G}$ is a suitable Green's function matrix, and superscript "†" denotes conjugate transpose. The symbol $\Xi$ is the transmission coefficients for all phonon modes or polarizations. The determination of these matrices requires knowledge of the harmonic matrix and interatomic potentials, and is rather complicated [98, 99]. The heat flux can be obtained based on Landauer's formalism by integration over the frequency and wavevector space, which can be performed through numerical discretization [98]. The thermal conductance can be obtained as the ratio of the heat flux to the temperature difference. Some studies have separately obtained the polarization-dependent transmission coefficients [101, 102].

While lattice dynamics has been applied to calculate the phonon transport across interfaces [103, 104], it is difficult to implement for various geometric and boundary conditions. Both the lattice dynamics and AGF methods treat phonons as waves and study the coherent propagation, reflection, and transmission of lattice waves through thin films, nanostructures, and interfaces. Similar to the lattice dynamics method, the AGF method is based on the harmonic matrix of the system that can be related to the derivatives of the total interatomic potential. In the AGF method, only the equilibrium positions of the atoms and interatomic force constants (IFCs) are needed. As discussed in Sect. 6.6, the IFCs can be obtained from first-principles calculations. Most AGF simulations have only dealt with harmonic vibrations, and thus are applicable at temperatures much lower than the Debye temperatures of the materials involved. Molecular dynamic simulations are inherently time domain. The wave-packet method can be applied to molecular dynamics simulations to extract the lattice vibration parameters, such as the mode-specific transmissivity. Due to the fact that molecular dynamics can inherently include anharmonic scattering, it is mostly suitable at high temperatures. On the other hand, standard molecular dynamics employs the Boltzmann distribution for phonons and therefore is not valid at temperatures much lower than the Debye temperature. Therefore, the AGF method and the MD method each has its own advantages depending on the temperature range of interest. Recently, Sadasivam et al. [105] modeled thermal transport across metal silicide and silicon interface using first-principles AGF, and included the anharmonic phonon scattering by modifying the conventional recursive Green's function approach. The AGF method has been used to model coherent phonon across Si/Ge superlattices [106] as well as TBR across stacked graphene/hexagonal boron nitride (hBN) heterostructures [107].

## 7.3 Heat Conduction Regimes

There has been a continuous effort to delineate the regimes of microscale heat conduction since 1992 as discussed in the previous chapters. Nonequilibrium phonon transport in dimensions less than 100 nm has become an important issue in silicon-on-insulator transistors. Multiscale and multiphysics simulations have been developed and applied to nanoelectronic devices [48, 66, 67, 82, 95, 108–110]. This section presents a regime map for heat conduction in solids by electrons and phonons, as schematically depicted in Fig. 7.9. Here, the timescale $\tau_c$ is known as *effective collision interaction time,* since collision does not occur instantaneously but is through intermolecular potential and force interactions. These forces become important only when the particles become very close to each other. Of course, this is the classical picture of atomic or molecular interactions. Electrons and phonons are quantum mechanical particles; thus, the interaction is via the wavefunctions predicted by Schrödinger's equations. For ultrafast pulse heating, the collision time can be the time required for a photon and an electron to interact. Generally speaking, the relaxation time is much shorter than the relaxation time and neglected in the BTE. The characteristic phonon or electron wavelength $\lambda$ is assumed to be less than the mean free path $\Lambda$.

Region 1 is the macroscale regime where Fourier's law and the heat diffusion equation can be applied, when the timescale is greater than $\tau$ and the length scale is greater than about $10\Lambda$. Region 2 is called the mesoscale or quasi-equilibrium regime, which is characterized by the classical size effect. This region is also known as the first microscale. For heat transfer along with a film or a wire, local-equilibrium assumption is appropriate and boundary scattering reduces the effective mean free path and thermal conductivity. For heat transfer across a film or a multilayer, it is possible

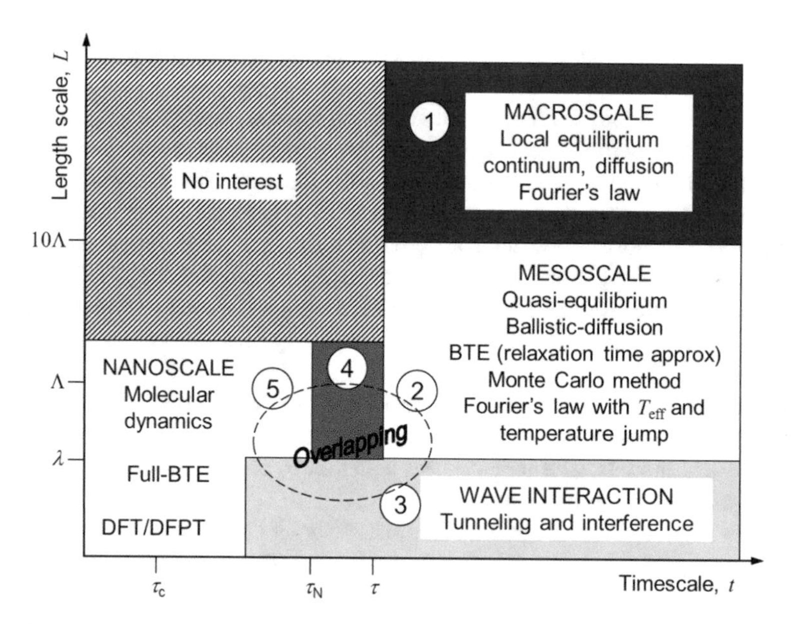


**Fig. 7.9** Heat conduction regimes

to use Fourier's law inside the medium by considering an effective temperature and the temperature-jump boundary condition. It is difficult, if not impossible, to apply Fourier's law to complex geometries or local heating. The two-temperature model for fast laser heating can be in either region 1 or 2, depending on how the length scale is compared with the mean free path. Most of the research on microscale heat transfer between 1990 and 2005 dealt with the microscale phenomena in region 2.

Region 3 is the regime of wave behavior, which is described by Schrödinger's wave equations and where quantum tunneling can occur. Quantum size effect becomes significant on thermal conductivity and specific heat. Quantum conductance is a special case of quantum tunneling, for which the ballistic processes are confined in one dimension through a channel. For very thin layers, wave interference and coherent phonon effects may become important. However, due to the interface roughness, the coherence may be destroyed so that the energy ray method or the particle approach can still be applied at very small length scales. We will give a comprehensive treatment of electromagnetic wave interference and scattering phenomena in subsequent chapters. The region on the upper left is said to be of no interest at short timescales because a thermal disturbance cannot travel that far and affect the temperature field.

Region 4 is designed to represent the wavelike behavior, described by the Jeffreys-type equation, Eq. (7.17). When we say Jeffreys-type equation, we mean that both $\kappa_0$ and $\kappa_1$ in Eq. (7.16a) are positive. As discussed earlier, $\tau_N$ is the second relaxation time for phonon scattering that does not transfer or dissipate thermal energy, as in the $N$ processes. In this regime, the BTE based on the two-relaxation-time approximation may be applied [31, 32]. This regime includes the heat pulse propagation and the second sound in dielectric crystals, at low temperatures. It suffices to say that this region, while of great academic interest, has very limited applications. The pure hyperbolic heat equation, however, predicts a nonphysical wavefront and cannot be applied without the additional diffusion term. Nevertheless, theoretical studies of the hyperbolic heat equation have helped in better understanding heat transfer behavior on short timescales and, subsequently, facilitated the development of more realistic models. While the lagging heat equation can mathematically describe both wavelike behavior and parallel heat conduction, it does not provide much new physics. On the other hand, the memory concept may be related to the anharmonic and nonlinear effects that are inherent to the solid and crystal structures. Study of the thermomechanical and thermoelastic effects, and thermal transport in polymers and inhomogeneous materials, such as biological materials, may require empirical and semiempirical models. The lagging heat equation or similar differential equations may be quite helpful in these applications.

Region 5 belongs to the nanoscale regime, where it is necessary to employ quantum or sometimes classical molecular dynamics to study the underlying phenomena. At the very fundamental level, DFT and DFPT are needed that can be coupled with molecular dynamics or the first-principles-based BTE as discussed in the previous chapter. The dashed ellipse indicates the overlapping between different regions, where molecular dynamics simulation may provide rich information as well as a bridge between different timescales and length scales.

## 7.4 Thermal Metrology

Thermal metrology plays an important role not only in determining the unique properties but also in testing the theoretical predictions and helping to understand the fundamental mechanisms. Thermal metrology includes measurements of temperature (thermometry), specific heat (calorimetry), and heat flux. Thermophysical properties, such as thermal conductivity, thermal diffusivity, and specific heat, can be measured with steady-state, periodically modulated, pulsed, and combined techniques [111–115]. MEMS and NEMS have enabled the fabrication of miniaturized heaters and sensors. Furthermore, optical techniques such as thermoreflectance, Raman spectroscopy, photothermal radiometry, fluorescence, and laser flash techniques have been widely used in the measurement of temperature [116] and thermal properties of nano/microstructured materials [117]. Scanning thermal microscopy and near-field optical microscopy have further improved the spatial resolution [47, 118]. A large number of publications can be found from the bibliography of the previous and present chapters and references therein. A brief overview of selected measurement techniques is given in the following.

### 7.4.1 Microbridge and Suspended Microdevices

The four-point probing microbridge shown in Fig. 7.10 is commonly used for measuring thermal properties. The metal bridge can serve as either a heater or thermometer or both. Platinum (Pt) is mostly used due to its relatively high resistivity, large temperature coefficient of resistance (TCR), and chemical stability. Either steady-state, transient, or periodic-heating methods can be used in the measurements; in some cases, a combined heating method can be used alternatively or simultaneously. The microbridge can be fabricated on a dielectric substrate, a thin insulating film on a substrate, or a suspended membrane, allowing both in-plane and cross-plane thermal transport properties to be measured. The thickness of the metal film is typically several tens of nanometers and the width of the bridge can vary from tens of nanometers to several micrometers. Depending on the applications, the bridge length can vary from tens of micrometers to several millimeters. Extensive discussions on the use of electrothermal techniques for measuring the thermal conductivity and thermal diffusivity can be found from Refs. [47, 112, 119].

As an example, Fig. 7.10b and c display the SEM images of a microfabricated bridge used as a thermometer [120]. The Pt film with a thickness of 35 nm was etched

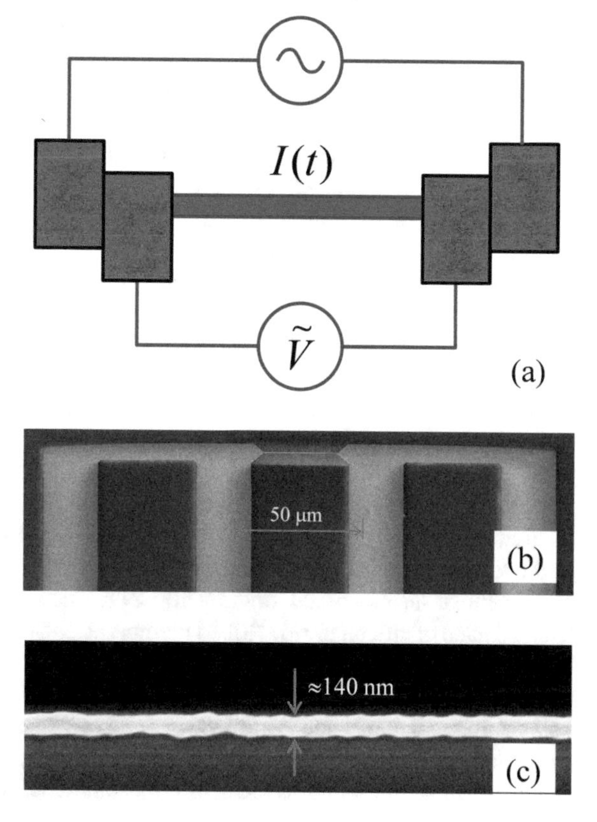


**Fig. 7.10** A patterned heater/thermometer microbridge in a four-terminal sensing scheme. **a** Schematic of the microbridge circuit; **b** a Pt microbridge with a length of 29 μm connected to four electrodes and **c** portion of the microbridge whose width is approximately 140 nm [120]

using focused ion beam (FIB) to a width of 140 nm over 29 μm length. The bridge was fabricated over a $SiO_2$ film on Si substrate for characterizing the heating effect from an AFM cantilever as it approaches and scans the surface. The TCR was calibrated to be near 20% that of bulk Pt, which is approximately 0.0039 $K^{-1}$. The resistivity was about five times that of pure Pt, suggesting grain boundary and geometric boundary scattering effects may play a role in the deposited Pt film and etched microbridge [120]. It is necessary to calibrate the microfabricated thermometers and to determine the TCR curve before performing actual measurements.

Since it often takes a long time to achieve steady states, traditionally, the hot wire and hot strip methods have been developed to measure thermal properties using a step function or a short impulse of electrical power. In the late 1980s, Cahill et al. [121–123] developed the 3-omega or 3-$\omega$ method for measuring thermal conductivity of amorphous solids and thin films using a lock-in amplifier to generate a harmonic oscillating current signal $I \sim \cos(\omega t) = \cos(2\pi f t)$ and measure the voltage signal oscillating at a frequency of $3\omega$. This method greatly reduces the effect of background effects such as thermal radiation and can be used for both cross-plane and in-plane thermal conductivity. The basic principle is that when an alternating current passes through the bridge as illustrated in Fig. 7.10a at a frequency $\omega$, the voltage $\tilde{V}_\omega = \tilde{I}_\omega R$ also oscillates at a frequency of $\omega$. Consequently, the electrical power $\tilde{P}_{2\omega} = \tilde{I}_\omega \tilde{V}_\omega$ is modulated at $2\omega$, which is dissipated as Joule heating to the bridge. The resulting temperature oscillates around the mean temperature at a frequency of $2\omega$ with a phase delay $\phi$ that depends on the properties and geometry of the system. The mean temperature (operating temperature) of the bridge depends on the average heating power. The resistance of the bridge is therefore modulated about its operating point at a frequency of $2\omega$. The lock-in amplifier collects the voltage signal and performs a frequency analysis to extract the $3\omega$ voltage signal $\tilde{V}_{3\omega} = \tilde{I}_\omega \tilde{R}_{2\omega}$. Through careful models of the heat transfer processes and known parameters such as the film thickness and specific heat capacity, the 3-$\omega$ method has become a powerful technique in measuring thermal conductivity, especially for semiconductors and insulators [47]. Dames [124] gave an extensive review with background information of the $3\omega$ methods and its variations. Kommandur and Yee [125] fabricated a microbridge on a suspended semiconducting polymer film and used the $3\omega$ method to measure the in-plane thermal conductivity and to characterize the anisotropy in thermal transport properties.

Shi et al. [126] microfabricated suspended devices for measuring thermal and electrical properties of nanostructures. Kim et al. [127] reported the first thermal conductivity measurements of individual carbon nanotubes (CNTs) using a suspended microdevice. Yu et al. [128] measured the thermal conductance and the Seebeck coefficient of an individual single-wall CNT. The device includes two suspended islands made of silicon nitride ($SiN_x$) membrane and each island is supported by five $SiN_x$ beams as shown in Fig. 7.11. A Pt thin film is coated on the membrane and patterned in serpentine winding on each island. The four beams or leads form four contact points that provide heating power and measure the temperature of the island simultaneously. One of the islands is used as the heater (with its own thermometer) and the other island serves as the heat sink. Nanotubes or nanowires with a length of

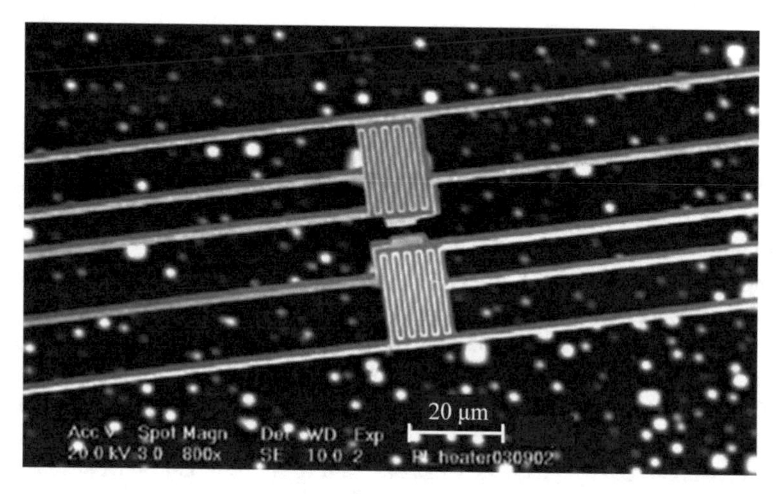


**Fig. 7.11** Schematic of the microfabricated suspended device that has two isolated membranes with patterned Pt resistors supported by silicon nitride beams. *Reprinted with permission from Yu et al.* [128]; *copyright (2005) American Chemical Society*

about 5–20 μm can be laid between the islands. Additional beams may be used to measure the resistance of the suspended nanotubes, nanowires, and nanofilms [126–130]. Both steady-state and transient measurements have been performed. Detailed analysis of the thermal resistance and the effect of contact resistance need to be taken into consideration; see a recent review by Weathers and Shi [130].

Fujii et al. [131] fabricated a suspended T-shape nanosensor to measure the thermal conductivity of individual CNTs of a few micrometers in length. The Pt strip of a length of 5–6 μm, width on the order of 0.5 μm, and thickness a few tens of nanometers, is suspended. The CNT is suspended from the middle of the Pt strip to a heat sink. Under steady-state operation with DC current, the temperature difference between the ends of the CNT and heat flow rate through the CNT can be determined by analyzing the measurement results to determine its thermal conductivity.

Recently, Kim et al. [132] proposed to use four suspended parallel bridges made of Pt strip on $SiN_x$ beams to measure the thermal and thermoelectric properties of nanostructures. Though the analysis involves detailed heat transfer and thermal resistances through the beams, the fabrication is much easier than the suspended islands structures. Furthermore, individual beams can serve as a four-point probe and heater. Contact resistance can also be compensated for through a careful analysis of the thermal resistance network. The setup has been used to measure Si nanowires from 100 to 500 K and BAs microrods from 250 to 350 K [132, 133]. Transient and $3\omega$ sensing schemes may also be employed to measure the thermal and thermoelectric properties of nanowire structures.

### 7.4.2 Scanning Probe Microscopic Techniques

As mentioned in Chap. 1, the family of scanning probe microscopy (SPM) has been established as a powerful toolbox in nanotechnology from manipulating and imaging single atoms to probing the topological, chemical, and thermal profiles near the interfaces. Majumdar [134] reviewed the development and applications of scanning thermal microscopy (SThM) for local temperature mapping with a few tens of nanometer resolution by fabricating a thermocouple or resistance thermometer. The method was developed by Majumdar et al. [135, 136] in the early 1990s to allow surface temperature measurements based on the previous work at IBM [137]. Another method, also pioneered by Majumdar [138] used the thermal expansion principle called the scanning Joule expansion microscopy, which has been further developed to measure the temperature profile with 10 nm resolution for studying the size effect of thermal conductivity [139] as well as imaging the thermal and thermoelectric characteristics at graphene-metal contact [140]. The most frequently used SThM is based on fabricating a thermocouple at the tip. The method has been further developed through the years not only for local temperature measurements but also for thermal conductivity measurement and thermoelectric property characterization as reviewed in Refs. [141, 142].

A representative high-quality SThM with a thermocouple at the tip is shown in Fig. 7.12, which can be used in air [143]. The probe was made of silica with a very low thermal conductivity, and the tip was made to be 12 μm long to minimize the air gap effect. In addition to measuring the thermal profile for a heated sample, the thermal conductivity profile can be obtained by heating the tip with a high-frequency (>100 kHz) AC current such that a steady-state temperature is sensed by the thermocouple whose time constant is greater than 1 ms [143].

The cantilever tip or cantilever can be optically or electrically heated with controllable temperature for thermal processing, nanofabrication, data writing and reading, and for the study of thermal transport at nanoscales [144, 145]. Lee et al. [146, 147] performed a steady-state and frequency-dependent characterization of heated AFM cantilevers over a range of pressures for thermal metrology applications. The temperature distribution in heated Si cantilevers was obtained with micro-Raman spectroscopy with a spatial resolution of 1 μm. Park et al. [148] analyzed the frequency response of heated AFM cantilevers in the frequency range from 10 Hz to 1 MHz, and observed high-order harmonic responses, such as $3\omega$, $5\omega$, and $7\omega$, at frequencies below 100 kHz and impedance effects at higher frequencies. Park et al. [149] also investigated thermal behavior of heated cantilevers at cryogenic temperatures, down to 78 K. By measuring the thermal response at various frequencies, this study extracted the specific heat near the cantilever tip and the thermal conductivity along the heavily doped silicon legs, at temperatures ranging from 80 to 200 K. There appears to be a significant reduction in the thermal conductivity for the free-standing silicon cantilever, with a thickness of 0.59 μm, at low temperatures. The heat transfer between heated AFM microcantilever and substrate has also been investigated [120]. As reviewed by King et al. [150], heated AFM cantilevers have become a useful thermal analysis tool at the micro- and nanoscales.

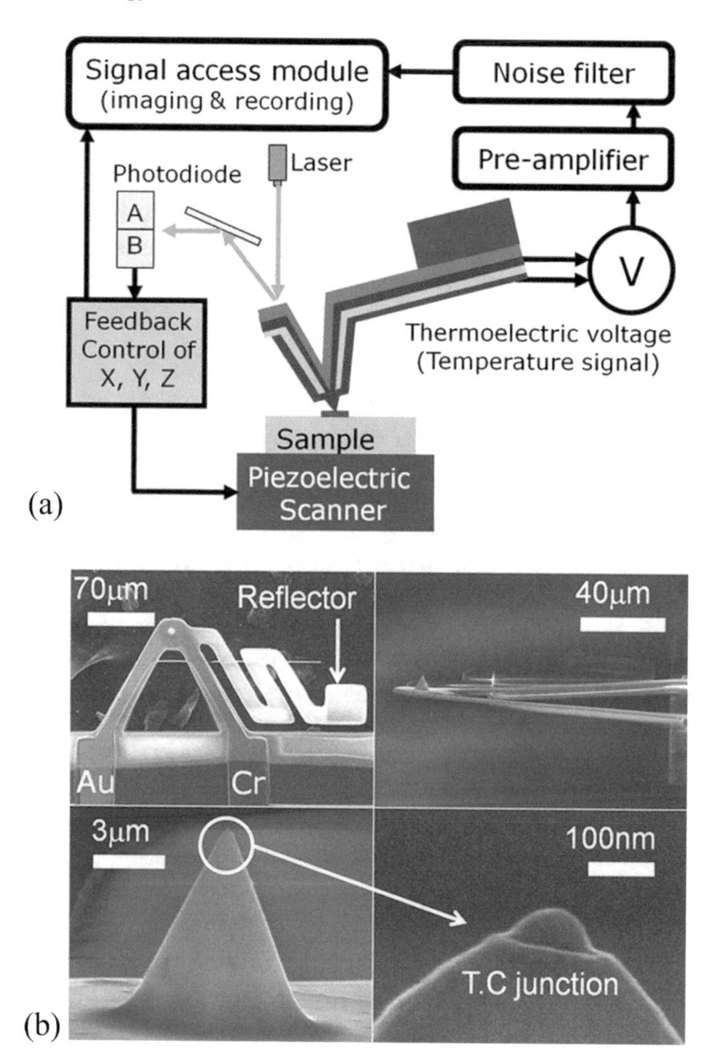


**Fig. 7.12** Scanning thermal microscopy with a special resolution of about 50 nm for temperature and thermal conductivity characteristics.: **a** Schematic of the experimental and the tip layout; **b** SEM images of the probe tip and cantilever. *Reprinted with permission from Kim et al.* [143]; *copyright (2011) American Chemical Society*

### *7.4.3 Noncontact Optical Techniques*

Optical methods are noncontact and can have a large range of temporal resolutions for measuring temperature and thermal properties such as thermal conductivity, diffusivity, specific heat, thermal boundary resistance, and the electron–phonon coupling constant [116, 117]. Femtosecond lasers have become much more affordable and accessible in recent years [49, 151–155]. For measurement of bulk and film properties, a temporal resolution of 10 ns–10 ms is usually sufficient. In order to probe thermal boundary resistance between films or thermal properties of very thin films, a resolution of 100 ps–10 ns is frequently used [156–158]. To measure the electron–phonon coupling, ultrafast lasers are needed since picosecond resolution is required [38–46, 159]. Another advantage of optical methods is that the beam spot size can be made relatively small, down to a few micrometers using an objective lens. Submicron resolution can be achieved with micro-Raman thermometry. Measurements with 50–500 nm spatial resolution can be made possible using near-field optics or fabricated nanostructures [118, 160–162].

Pump-and-probe methods are often employed in which the sample surface is heated by a laser beam (or another optical source) and the thermal responses are measured using one of the variety of probing techniques. Examples are the thermoreflectance method based on the temperature dependence of the reflectance of the surface or film, micro-Raman thermometry based on the Raman shift due to phonon scattering being temperature dependent, the radiometric method based on the thermal emission signal according to the theory of blackbody radiation, and photoacoustic and photodeflection techniques [112, 113, 117]. Measurements are often accomplished either in the time domain, when the transient thermal response after pulsed or step heating is observed, or in the frequency domain, when periodic heating is used and the periodic response with a time delay is measured [47, 112, 117]. The latter is also called the thermal wave method [11, 12].

Figure 7.13 illustrates a time-domain thermoreflectance (TDTR) setup [154] for measuring the thermal conductivity of film or bulk materials as well as TBR. The pump-probe scheme is shown in Fig. 7.13a. The transducer is usually a metal film. The thermoreflectance coefficient is defined as follows

$$C_{\mathrm{TR}} = \frac{1}{R}\frac{\partial R}{\partial T} \quad \text{or} \quad \frac{\Delta R}{R} = C_{\mathrm{TR}}\Delta T \tag{7.92}$$

The temperature and wavelength dependence of $C_{\mathrm{TR}}$ of metal films have been extensively characterized [163]. For Au, due to the interband transition near the wavelength of $\lambda = 500$ nm, the absorptance and thermoreflectance coefficient is relatively large. For Al, the absorptance is high near $\lambda = 800$ nm. The wavelength of pulsed Ti-sapphire lasers ranges from 720 to 880 nm; thus, Al coating is typically used [49, 159]. The pulse duration is typically 90–150 fs, though shorter pulses can also be generated using a mode-locking technique. As shown in Fig. 7.13b, the laser beam after the optical isolator is split into a pump beam (high power) and a probe beam (low power) using a polarizing beamsplitter (PBS). The output of the laser is a pulse train at a typical frequency of 80 MHz. An electro-optic modulator (EOM) is used to reduce the modulation frequency to 1–10 MHz range for measuring thermal properties or TBR. The probe beam goes through a delay stage in order to probe the sample temperature after the pump pulse heating. Both beams are sent to the sample through the objective lens. After averaging over many pulses, the delay stage is moved to vary the delay time. It should be noted this method is different from the traditional

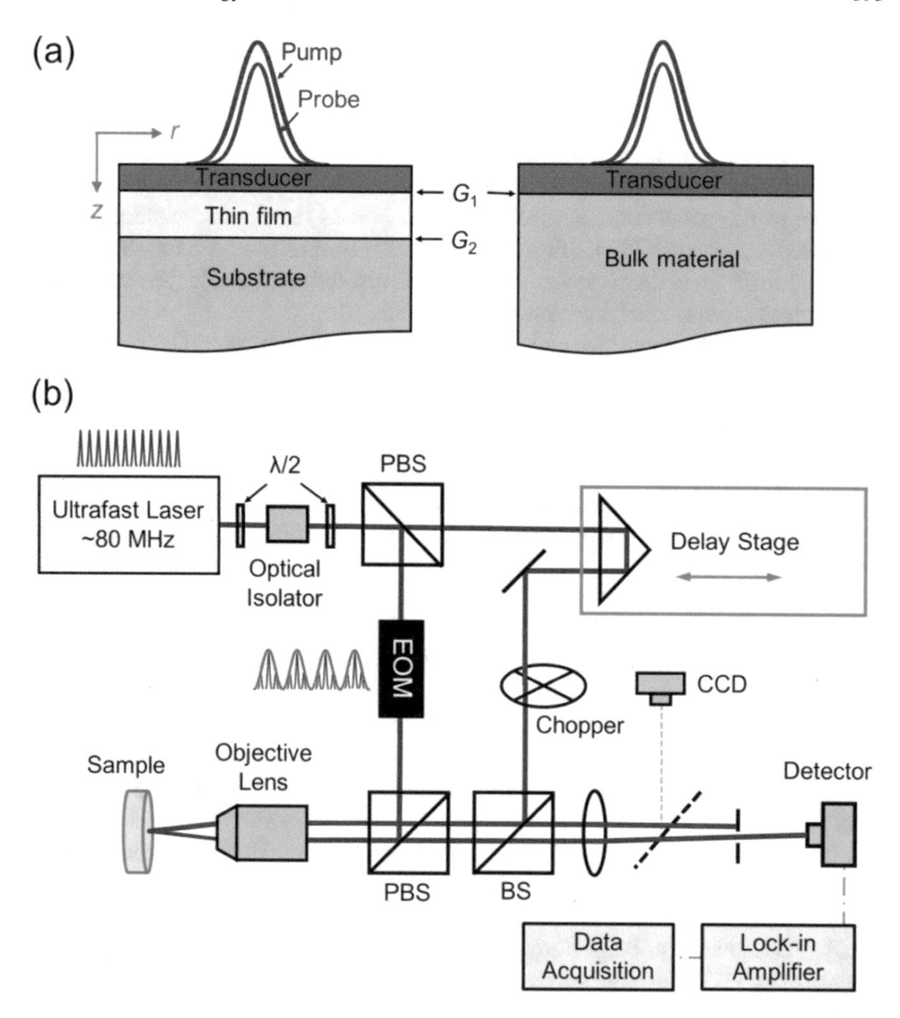


**Fig. 7.13** Schematic of a TDTR: **a** Illustration of the pump and probe beams on a sample; **b** the optical layout. *Reprinted with permission from Jiang et al.* [154]; *copyright (2018) AIP Publishing*

transient response method. For example, in the conventional transient laser heating and relaxation method, the temperature rise and fall after the laser pulse is monitored continuously with a temporal resolution typically from 1 μm to 1 ms [112]. In a TDTR measurement, the response at each delay step is recorded and then plotted as a function of time. The delay step determines the temporal resolution which can be varied from tens of femtoseconds to tens of picoseconds. Since the movement of the delay stage can be precisely controlled with a micrometer resolution, the smallest time delay that can be achieved is less than 10 fs. For example, if the total travel length of the delay stage is 30 cm, the maximum delay time is 2 ns. In some experimental setups, a forward advance is used for the pump beam rather than delaying the probe beam. The chopper may be used to vary the frequency of the probe beam independently for dual-frequency measurement [154, 164]. A second harmonic generator can be used to double the frequency of either the pump beam or probe beam so that its wavelength is changed to the visible range [49]. TDTR methods have been used to measure the thermal conductivity accumulation function in terms of the mean free path [151, 152] as well as the thermal conductivity of perforated membranes [165, 166]. Wagner et al. [166] also used a two-laser setup with a micro-Raman thermometer at submicron resolution to obtain the steady-state temperature profile during continuous laser heating.

By changing the beam size and modulation frequency, it is possible to determine both in-plane and cross-plane thermal conductivity by fitting the model prediction to the experimental data using the least-squares method. Another way to probe the in-plane thermal transport is to use a lateral offset between the pump beam and the probe beam. Wang et al. [167] used both methods to study the thermal conductivity of layered borides. To measure the properties of 2D materials it is critically important to reduce the metal layer thickness. A magneto-optical thin film (on the order of 20 nm) has been used as the transducer. Under a magnetic field due to the Kerr effect, the polarization of the reflected beam is a function of temperature. The method based on time-resolved magneto-optical Kerr effect (TR-MOKE) has been developed and used to measure the anisotropic thermal conductivity of molybdenum disulfide [168] and black phosphorus [153].

The femtosecond laser setup can be used for frequency-domain thermoreflectance (FDTR) with few hardware modifications [169]. The signal reaching the detector, or the reflection of the probing beam, is at the same frequency as the pump beam with a phase delay [151]. The modulation frequency is determined by the EOM and can be varied from 25 kHz to 20 MHz. By fixing the time delay and changing the modulation frequency, one can obtain the frequency response. Theoretical models are necessary to relate the frequency response to the properties being determined [169]. FDTR can also be performed with two continuous-wave lasers [169–171]. Regner et al. [170] developed a two-laser FDTR setup to measure the phonon mean free spectra for crystalline Si, doped Si, amorphous Si, and amorphous $SiO_2$.

A modified setup is used to measure anisotropic thermal conductivity of thin films, as shown in Fig. 7.14 [171]. Two continuous-wave green lasers at slightly different wavelengths are used. The wavelengths of 488 nm and 532 nm match well with the peaks of absorption and thermoreflectance coefficients of the gold film (transducer), respectively [163]. An optical isolator (ISO) is used after each laser to prevent the reflected beam from reentering the laser cavity. After a half-wave plate (HWP), the pump beam is modulated by EOM at frequency *f*, which can be varied in a large range from about 9 kHz to 200 GHz, though only the middle range is useful for the data analysis and parameter reduction. A picomotor mirror, which uses a piezoelectric actuator to fine tune the angular rotation, is used to offset the probe beam position. After the beamsplitter (BS), both beams go through the PBS, a quarter-wave plate (QWP), and the objective lens (OBJ) to focus on the sample surface. Only reflected light at the probe beam frequency is allowed to enter the photodetector (PD) thanks to a bandpass filter (BP). The detector receives a signal

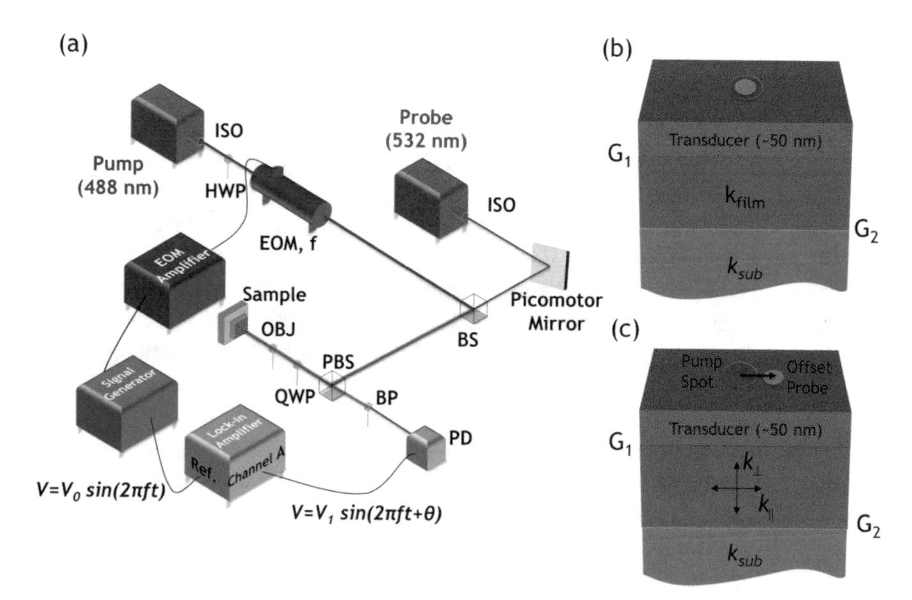


**Fig. 7.14** Illustration of a frequency-domain thermoreflectance measurement system and scenarios. **a** The optical layout of the two-laser FDTR setup; **b** scheme for measuring thermal conductivities of the film, substrate, and TBR, $G_1$ and $G_2$, with concentrated beams; **c** scheme for measuring anisotropic thermal conductivities, both in-plane $k_{\|}$ and across-plane $k_{\perp}$, of the film with an offset beam spot. *Reprinted with permission from Rodin and Yee* [171]; *copyright (2017) AIP Publishing*

due to the temperature change of the sample at the same frequency $f$ but with a phase lag $\phi$, which depends on the materials properties, lateral offset, and the modulation frequency $f$. The phase lag can be measured with a lock-in amplifier using either a heterodyne modulator [170] or a signal generator [171]. Through careful thermal modeling and a least-squares fitting, the desired properties such as the TBR, in-plane and cross-plane thermal conductivities of the film can be determined. The results for crystalline and amorphous $Al_2O_3$ and $SiO_2$, respectively, along with highly oriented pyrolytic graphite have been reported [171].

The femtosecond transient thermal grating (TTG) method has also been used for measuring the thermal conductivity accumulation functions of thin membranes [172–174]. A pulsed laser (wavelength 515 nm and pulse width 60 ps) is split into beams using diffraction optics and then focused to the sample, causing sinusoidal interference patterns on the sample, which can be a thin film or membrane. A continuous laser is either reflected or transmitted through the sample, producing a diffraction signal whose decay is related to the thermal diffusivity. Silicon membranes of thicknesses from 1500 nm down to 17.5 nm have been measured to demonstrate ballistic thermal transport as well as to study the thermal conductivity accumulation function [172, 173]. Transport along perforated silicon membrane has also been examined [174].

## 7.5 Summary

The present chapter, together with Chaps. 5 and 6, provides a comprehensive treatment of thermal properties of and transport processes in micro/nanostructured solid materials. This chapter focused on the transient and nonequilibrium heat conduction, when the local-equilibrium condition is not satisfied to justify the conventional heat diffusion theory, based on Fourier's law. Several modified phenomenological theories were critically reviewed with an emphasis on their application regimes. The phonon BTE was presented using the EPRT, and the solutions were discussed for the nonequilibrium heat transfer across a thin film or a multilayer structure. The basic models of TBR were outlined. A summary on the advanced atomistic scale modeling is provided focusing on the atomistic Green's function method. A heat transfer regime was developed to assist readers in choosing an appropriate methodology for a given situation. Finally, some important thermal measurement techniques are discussed with extensive references.

## Problems

7.1. What are the characteristic lengths for heat conduction along a thin film? Why is local equilibrium a good assumption in this case, even though the film thickness is less than the mean free path of heat carriers? Why does the thermal conductivity depend on the thickness of the film?

7.2. Why do we say that Fourier's law is a fundamental physical law, like Newton's laws in mechanics, but Cattaneo's equation is not? Comment on the paradox of infinite speed of heat diffusion by considering the feasibility of exciting the surface temperature or depositing a heat flux to the surface instantaneously.

7.3. Consider a 1D semi-infinite medium, initially at uniform temperature $T_i$, where the surface temperature is suddenly changed to a constant temperature, $T(0,t) = T_s$. The analytical solution of the heat diffusion equation gives $\theta(x,t) = \frac{T(x,t)-T_i}{T_s-T_i} = \mathrm{erfc}\left(\frac{x}{2\sqrt{\alpha t}}\right)$. For silicon at various temperatures, use the properties given in Example 5.6 to estimate how long it will take for a given location to gain a temperature rise that is $10^{-12}$, or one part per trillion of the maximum temperature difference. Estimate the average thermal diffusion speed in terms of $x$ and $T_i$. Hint: $\mathrm{erfc}(5.042) = 1.00 \times 10^{-12}$.

7.4. Repeat Problem 7.3, using copper instead of silicon as the material, based on the properties given in Example 5.5. Discuss why the average thermal diffusion speed is different under different boundary conditions, i.e., constant heat flux and constant temperature. From an engineering point of view, do you think heat diffusion is a fast or slow process? Why?

7.5. (a) Derive Eq. (7.4), the hyperbolic heat equation from Cattaneo's equation
(b) Derive Eq. (7.14), the lagging heat equation, based on the dual-phase-lag model.

7.6. Take GaAs as an example. How would you compare the speed of sound with the average thermal diffusion speed, at different temperatures and length scales? This problem requires some literature search on the properties.

7.7. Assume the hyperbolic heat equation would work for transient heat transfer in glass (Pyrex), at room temperature. Given $\kappa = 1.4$ W/m K, $\rho = 2500$ kg/m$^3$, $c_p = 835$ J/kg K, and $v_a = 5640$ m/s.

(a) At what speed would the temperature wave propagate?
(b) For an excimer laser with a pulse width $t_p = 10$ ns, 0.1 ns after the pulse starts, could the hyperbolic equation be approximated by the parabolic equation?
(c) Suppose we have an instrument available to probe the timescale below $\tau_q$, will the hyperbolic heat equation be able to describe the observation?

7.8. Derive Eq. (7.13b) from Eq. (7.13a). Discuss the conditions for these equations to be reduced to Fourier's law or Cattaneo's equation.

7.9. Show that Eq. (7.17) satisfies Eq. (7.16). Discuss the conditions for Eq. (7.17) to represent Fourier's law or Cattaneo's equation.

7.10. Derive Eqs. (7.18a), (7.18b), and (7.18c).

7.11. Derive Eqs. (7.27a) and (7.27b). Calculate $\tau$, $\tau_q$, and $\tau_T$ of copper, for $T_e =$ 300, 1000, and 5000 K, assuming the lattice temperature $T_s = 300$ K.

7.12. Calculate the electron–phonon coupling constant $G$ for aluminum, copper, gold, and silver, near room temperature. Discuss the dependence of $\kappa$ and $G$ upon the electron and lattice temperatures $T_e$ and $T_s$.

7.13. At $T_e = 1000$, 3000, and 6000 K, estimate the energy transfer by thermionic emission from the copper surface, assuming that the electrons obey the equilibrium distribution function at $T_e$.

7.14. Based on Example 7.3, evaluate the heat flux in a thin silicon film. How thin must it be in order for it to be considered as in the radiative thin limit? Calculate the medium temperature $T$. Plot the left-hand side and the right-hand side of Eq. (7.43). Furthermore, assuming Eq. (7.43) to be true for each frequency, find a frequency-dependent temperature $T(\omega)$ of the medium. At what frequency does $T(\omega) = T$? Is there any physical significance of $T(\omega)$?

7.15. Derive Eq. (7.53), using Eqs. (7.38), (7.49a), (7.49b), and (7.50).

7.16. In principle, one should be able to study nonequilibrium electrical and thermal conduction in the direction perpendicular to the plane and use the BTE to determine the effective conductivities. This could be a team project, for a few students, to formulate the necessary equations. As an individual assignment, describe how to set up the boundary conditions, as well as the steps you plan to follow, without actually deriving the equations.

7.17. For a diamond type IIa film, $v_l = 17,500$ m/s, $v_t = 12,800$ m/s, and $\kappa = 3,300$ W/m K, near 300 K. Assume that the boundaries can be modeled as blackbodies for phonons. For boundary temperatures $T_1 = 350$ K and $T_2 = 250$ K, calculate and plot the heat flux $q''_x$ and the effective thermal conductivity $\kappa_{\text{eff}}$ across the film of thickness $L$, varying from 0.05 to 50 μm.

7.18. Calculate the TBR between high-temperature superconductor $YBa_2Cu_3O_{7-\delta}$ and MgO substrate, at an average temperature between 10 and 90 K, using both the AMM and the DMM without considering the electronic effect. The following parameters are given for $YBa_2Cu_3O_{7-\delta}$: $v_l = 4780$ m/s, $v_t = 3010$ m/s, $\rho = 6338$ kg/m$^{-3}$, and $\Theta_D = 450$ K; and for MgO: $v_l = 9710$ m/s, $v_t = 6050$ m/s, $\rho = 3576$ kg/m$^{-3}$, and $\Theta_D = 950$ K.

7.19. Evaluate the effective thermal conductivity near room temperature of a GaAs/AlAs superlattice, with a total thickness of 800 nm, using the DMM to compute the transmission coefficient. Assume the end surfaces are blackbodies to phonons; consider that (a) each layer is 4 nm thick and (b) each layer is 40 nm thick. The following parameters are given, considering phonon dispersion on thermal conductivity, for GaAs: $C = 880$ kJ/m$^3$ K, $v_g = 1024$ m/s, and $\Lambda = 145$ nm; and for AlAs: $C = 880$ kJ/m$^3$ K, $v_g = 1246$ m/s, and $\Lambda = 236$ nm. How is the result compared with a single layer of either GaAs or AlAs?

7.20. Evaluate the effective thermal conductivity near room temperature of a Si/Ge superlattice, with a total thickness of 1000 nm, using the DMM to compute the transmission coefficient. Assume the end surfaces are blackbodies to phonons; consider that (a) each layer is 5 nm thick and (b) each layer is 50 nm thick. The following parameters are given, considering phonon dispersion on thermal conductivity, for Si: $C = 930$ kJ/m$^3$ K, $v_g = 1804$ m/s, and $\Lambda = 260$ nm; and for Ge: $C = 870$ kJ/m$^3$ K, $v_g = 1042$ m/s, and $\Lambda = 199$ nm. How is the result compared with a single layer of either Si or Ge?

7.21. Make a comparison of the different methods for measuring the thermal conductivity of a thin film.

7.22. Suppose one wishes to measure the thermal conductivity of a graphene sheet of 10 μm × 10 μm, what method(s) would you recommend?

7.23. Suppose one wishes to measure the thermal conductivity of a superlattice Si/Ge nanowire of length 50 μm and diameter 3 nm, what method would you suggest?

7.24. What is the mechanism of transient thermal grating? What properties can be measured by the TTG method?

## References


1. H.S. Carslaw, J.C. Jaeger, *Conduction of Heat in Solids*, 2nd edn. (Clarendon Press, Oxford, 1959)
2. M.N. Özişik, *Heat Conduction,* 2nd ed., Wiley, New York, 1993; *also* D.W. Hahn and M.N. Özişik, *Heat Conduction,* 3rd ed., Wiley, New York, 2012
3. T.J. Bright, Z.M. Zhang, Common misperceptions of the hyperbolic heat equation. J. Thermophys. Heat Transfer **23**, 601–607 (2009)
4. D. D. Joseph, L. Preziosi, Heat waves. Rev. Mod. Phys., **61**, 41–73 (1989)
5. D.D. Joseph, L. Preziosi, Addendum to the paper 'heat waves'. Rev. Mod. Phys. **62**, 375–391 (1990)

6. D.Y. Tzou, *Macro- to Microscale Heat Transfer: The Lagging Behavior*, 2nd edn. (Wiley, New York, 2015)
7. M.N. Özişik, D.Y. Tzou, On the wave theory in heat conduction. J. Heat Transfer **116**, 526–535 (1994)
8. W.K. Yeung, T.T. Lam, A numerical scheme for non-Fourier heat conduction, Part I: one-dimensional problem formulation and applications. Numer. Heat Transfer B **33**, 215–233 (1998)
9. A. Haji-Sheikh, W.J. Minkowycz, E.M. Sparrow, Certain anomalies in the analysis of hyperbolic heat conduction. J. Heat Transfer **124**, 307–319 (2002)
10. J. Gembarovic, J. Gembarovic Jr., Non-Fourier heat conduction modeling in a finite medium. Int. J. Thermophys. **25**, 1261–1268 (2004)
11. C.A. Bennett, R.R. Patty, Thermal wave interferometry: a potential application of the photoacoustic effect. Appl. Opt. **21**, 49–54 (1982)
12. A. Mandelis (ed.), *Photoacoustic and Thermal Wave Phenomena in Semiconductors* (Elsevier, Amsterdam, 1987)
13. M.B. Rubin, Hyperbolic heat conduction and the second law. Int. J. Eng. Sci. **30**, 1665–1676 (1992)
14. C. Bai, A.S. Lavine, On hyperbolic heat conduction and the second law of thermodynamics. J. Heat Transfer **117**, 256–263 (1995)
15. A. Barletta, E. Zanchini, Hyperbolic heat conduction and local equilibrium: a second law analysis. Int. J. Heat Mass Transfer **40**, 1007–1016 (1997)
16. D. Jou, G. Lebon, J. Casas-Vázquez, *Extended Irreversible Thermodynamics*, 4th edn. (Springer, Berlin, 2010)
17. Z.M. Zhang, T.J. Bright, G.P. Peterson, Reexamination of the statistical derivations of Fourier's law and Cattaneo's equation. Nanoscale Microscale Thermophys. Eng. **15**, 220–228 (2011)
18. J. Tavernier, Sur l'équation de conduction de la chaleur. Comptes Rendus Acad. Sci. **254**, 69–71 (1962)
19. A. Majumdar, Microscale heat conduction in dielectric thin films. J. Heat Transfer **115**, 7–16 (1993)
20. A.A. Joshi, A. Majumdar, Transient ballistic and diffusive phonon heat transport in thin films. J. Appl. Phys. **74**, 31–39 (1993)
21. S. Volz, J.-B. Saulnier, M. Lallemand, B. Perrin, P. Depondt, M. Mareschal, Transient Fourier-law deviation by molecular dynamics in solid argon. Phys. Rev. B **54**, 340–347 (1996)
22. J. Xu, X.W. Wang, Simulation of ballistic and non-Fourier thermal transport in ultra-fast laser heating. Phys. B **351**, 213–226 (2004)
23. M. Chester, Second sound in solids. Phys. Rev. **131**, 2013–2015 (1963)
24. M.E. Gurtin, A.C. Pipkin, A general theory of heat conduction with finite wave speeds. Arch. Ration. Mech. Anal. **31**, 113–126 (1968)
25. P.J. Antaki, Solution for non-Fourier dual phase lag heat conduction in a semi-infinite slab with surface heat flux. Int. J. Heat Mass Transfer **41**, 2253–2258 (1998)
26. D.W. Tang, N. Araki, Wavy, wavelike, diffusive thermal responses of finite rigid slabs to high-speed heating of laser-pulses. Int. J. Heat Mass Transfer **42**, 855–860 (1999)
27. D.Y. Tzou, K.S. Chiu, Temperature-dependent thermal lagging in ultrafast laser heating. Int. J. Heat Mass Transfer **44**, 1725–1734 (2001)
28. L.Q. Wang, X.S. Zhou, X.H. Wei, *Heat Conduction: Mathematical Models and Analytical Solutions* (Springer-Verlag, Berlin, 2008)
29. W.J. Minkowycz, A. Haji-Sheikh, K. Vafai, On departure from local thermal equilibrium in porous media due to a rapid changing heat source: the Sparrow number. Int. J. Heat Mass Transfer **42**, 3373–3385 (1999)
30. W. Kaminski, Hyperbolic heat conduction equation for materials with a nonhomogeneous inner structure. J. Heat Transfer **112**, 555–560 (1990)
31. J. Callaway, Model for lattice thermal conductivity at low temperatures. Phys. Rev. **113**, 1046–1951 (1959)

32. R. A. Guyer, J. A. Krumhansl, Solution of the linearized phonon Boltzmann equation. Phys. Rev. **148**, 766–778 (1966); Thermal conductivity, second sound, and phonon hydrodynamic phenomena in nonmetallic crystals. Phys. Rev. **148**, 778–788 (1966)
33. J. Shiomi, S. Maruyama, Non-Fourier heat conduction in a single-walled carbon nanotube: Classical molecular dynamics simulations. Phys. Rev. B **73**, 205420 (2006)
34. D.H. Tsai, R.A. MacDonald, Molecular-dynamics study of second sound in a solid excited by a strong heat pulse. Phys. Rev. B **14**, 4714–4723 (1976)
35. X.W. Wang, X. Xu, Thermoelastic wave induced by pulsed laser heating. Appl. Phys. A **73**, 107–114 (2001)
36. X.W. Wang, Thermal and thermomechanical phenomena in picosecond laser copper interaction. J. Heat Transfer **126**, 355–364 (2004)
37. S.I. Anisimov, B.L. Kapeliovich, T.L. Perel'man, Electron emission from metal surfaces exposed to ultrashort laser pulses. Sov. Phys. JETP **39**, 375–377 (1974)
38. J.G. Fujimoto, J.M. Liu, E.P. Ippen, N. Bloembergen, Femtosecond laser interaction with metallic tungsten and nonequilibrium electron and lattice temperatures. Phys. Rev. Lett. **53**, 1837–1840 (1984)
39. S.D. Brorson, J.G. Fujimoto, E.P. Ippen, Femtosecond electronic heat-transport dynamics in thin gold films. Phys. Rev. Lett. **59**, 1962–1965 (1987)
40. T.Q. Qiu, C.L. Tien, Short-pulse laser heating on metals. Int. J. Heat Mass Transfer **35**, 719–726 (1992)
41. T.Q. Qiu, C.L. Tien, Size effect on nonequilibrium laser heating of metal films. J. Heat Transfer **115**, 842–847 (1993)
42. T.Q. Qiu, T. Juhasz, C. Suarez, W.E. Bron, C.L. Tien, Femtosecond laser heating of multi-layer metals—II. Experiments. Int. J. Heat Mass Transfer **37**, 2799–2808 (1994)
43. J.L. Hostetler, A.N. Smith, D.M. Czajkowsky, P.M. Norris, Measurement of the electron-phonon coupling factor dependence on film thickness and grain size in Au, Cr, and Al. Appl. Opt. **38**, 3614–3620 (1999)
44. S. Link, C. Burda, Z.L. Wang, M.A. El-Sayed, Electron dynamics in gold and gold-silver alloy nanoparticles: The influence of a nonequilibrium electron distribution and the size dependence of the electron-phonon relaxation. J. Chem. Phys. **111**, 1255–1264 (1999)
45. A.N. Smith, P.M. Norris, Influence of intraband transition on the electron thermoreflectance response of metals. Appl. Phys. Lett. **78**, 1240–1242 (2001)
46. R.J. Stevens, A.N. Smith, P.M. Norris, Measurement of thermal boundary conductance of a series of metal-dielectric interfaces by the transient thermoreflectance techniques. J. Heat Transfer **127**, 315–322 (2005)
47. D.G. Cahill, K.E. Goodson, A. Majumdar, Thermometry and thermal transport in micro/nanoscale solid-state devices and structures. J. Heat Transfer **124**, 223–241 (2002)
48. D.G. Cahill, W.K. Ford, K.E. Goodson et al., Nanoscale thermal transport. J. Appl. Phys. **93**, 793–818 (2003)
49. J. Zhu, D.W. Tang, W. Wang, J. Liu, K.W. Holub, R. Yang, Ultrafast thermoreflectance techniques for measuring thermal conductivity and interface thermal conductance of thin films. J. Appl. Phys. **108**, 094315 (2010)
50. D.M. Riffe, X.Y. Wang, M.C. Downer et al., Femtosecond thermionic emission from metals in the space-charge-limited regime. J. Opt. Soc. Am. B **10**, 1424–1435 (1993)
51. A.N. Smith, J.L. Hostetler, P.M. Norris, Nonequilibrium heating in metal films: An analytical and numerical analysis. Numer. Heat Transfer A **35**, 859–874 (1999)
52. M. Li, S. Menon, J.P. Nibarger, G.N. Gibson, Ultrafast electron dynamics in femtosecond optical breakdown of dielectrics. Phys. Rev. Lett. **82**, 2394–2397 (1999)
53. L. Jiang, H.-L. Tsai, Energy transport and nanostructuring of dielectrics by femtosecond laser pulse trains. J. Heat Transfer **128**, 926–933 (2006)
54. L. Jiang, H.-L. Tsai, Plasma modeling for ultrashort pulse laser ablation of dielectrics. J. Appl. Phys. **100**, 023116 (2006)
55. Y. Ma, A two-parameter nondiffusive heat conduction model for data analysis in pump-probe experiments. J. Appl. Phys. **116**, 243505 (2014); *ibid*, Hotspot size-dependent thermal boundary conductance in nondiffusive heat conduction. J. Heat Transfer **137**, 082401 (2015)

56. G. Chen, Ballistic-diffusion heat-conduction equations. Phys. Rev. Lett. **86**, 2297–2300 (2001); *ibid*, Ballistic-diffusive equations for transient heat conduction from nano to macroscales. J. Heat Transfer **124**, 320–328 (2002)
57. T. Klitsner, J.E. VanCleve, H.E. Fischer, R.O. Pohl, Phonon radiative heat transfer and surface scattering. Phys. Rev. B **38**, 7576–7594 (1988)
58. R.B. Peterson, Direct simulation of phonon-mediated heat transfer in a Debye crystal. J. Heat Transfer **116**, 815–822 (1994)
59. E.T. Swartz, P.O. Pohl, Thermal boundary resistance. Rev. Mod. Phys. **61**, 605–668 (1989)
60. W.A. Little, The transport of heat between dissimilar solids at low temperatures. Can. J. Phys. **37**, 334–349 (1959)
61. G. Chen and C.L. Tien, "Thermal conductivity of quantum well structures," *J. Thermophys. Heat Transfer,* **7**, 311–318, 1993
62. G. Chen, Size and interface effects on thermal conductivity of superlattices and periodic thin-film structures. J. Heat Transfer **119**, 220–229 (1997)
63. G. Chen, Thermal conductivity and ballistic-phonon transport in the cross-plane direction of superlattices. Phys. Rev. B **57**, 14958–14973 (1998)
64. G. Chen, T. Zeng, Nonequilibrium phonon and electron transport in heterostructures and superlattices. Microscale Thermophys. Eng. **5**, 71–88 (2001)
65. T. Zeng, G. Chen, Phonon heat conduction in thin films: impacts of thermal boundary resistance and internal heat generation. J. Heat Transfer **123**, 340–347 (2001)
66. S. Sinha, K.E. Goodson, Review: multiscale thermal modeling in nanoelectronics. Int. J. Multiscale Comp. Eng. **3**, 107–133 (2005)
67. R.A. Escobar, S.S. Ghai, M.S. Jhon, C.H. Amon, Multi-length and time scale thermal transport using the lattice Boltzmann method with application to electronics cooling. Int. J. Heat Mass Transfer **49**, 97–107 (2006)
68. E.M. Sparrow, R.D. Cess, *Radiation Heat Transfer*, Augmented edn. (McGraw-Hill, New York, 1978)
69. M.F. Modest, *Radiative Heat Transfer*, 3rd edn. (Academic Press, New York, 2013)
70. T.J. Bright, Z.M. Zhang, Entropy generation in thin films evaluated from phonon radiative transport. J. Heat Transfer **132**, 101301 (2010)
71. H.B.G. Casimir, Note on the conduction of heat in crystal. Physica **5**, 495–500 (1938)
72. R.G. Deissler, Diffusion approximation for thermal radiation in gasses with jump boundary condition. J. Heat Transfer **86**, 240–245 (1964)
73. A. Malhotra, K. Kothari, M. Maldovan, Cross-plane thermal conduction in superlattices: Impact of multiple length scales on phonon transport. J. Appl. Phys. **125**, 044304 (2019)
74. K. Kothari, A. Malhotra, M. Maldovan, Cross-plane heat conduction in III–V semiconductor superlattices. J. Phys. Condens. Matter **31**, 345301 (2019)
75. M.M. Yovanovich, Four decades of research on thermal contact, gap, and joint resistance in microelectronics. IEEE Trans. Compon. Packag. Technol. **28**, 182–206 (2005)
76. R.J. Stoner, H.J. Maris, Kapitza conductance and heat flow between solids at temperatures from 50 to 300 K. Phys. Rev. B **48**, 16373–16387 (1993)
77. R.S. Prasher and P.E. Phelan, "Review of thermal boundary resistance of high-temperature superconductors," *J. Supercond.*, **10**, 473–484, 1997
78. P.E. Phelan, Application of diffuse mismatch theory to the prediction of thermal boundary resistance in thin-film high-$T_c$ superconductors. J. Heat Transfer **120**, 37–43 (1998)
79. L. De Bellis, P.E. Phelan, R.S. Prasher, Variations of acoustic and diffuse mismatch models in predicting thermal-boundary resistance. J. Thermophys. Heat Transfer **14**, 144–150 (2000)
80. A. Majumdar, Effect of interfacial roughness on phonon radiative heat conduction. J. Heat Transfer **113**, 797–805 (1991)
81. A. Majumdar, P. Reddy, Role of electron–phonon coupling in thermal conductance of metal–nonmetal interfaces. Appl. Phys. Lett. **84**, 4768–4770 (2004)
82. A. Giri, P.E. Hopkins, A review of experimental and computational advances in thermal boundary conductance and nanoscale thermal transport across solid interfaces. Adv. Func. Mater. **2019**, 1903857 (2019)

83. S. Mazumdar, A. Majumdar, Monte Carlo study of phonon transport in solid thin films including dispersion and polarization. J. Heat Transfer **123**, 749–759 (2001)
84. Q. Hao, G. Chen, M.-S. Jeng, Frequency-dependent Monte Carlo simulations of phonon transport in two-dimensional porous silicon with aligned pores. J. Appl. Phys. **106**, 114321 (2009)
85. J.-P.M. Péraud, C.D. Landon, N.G. Hadjiconstantinou, Monte Carlo methods for solving the Boltzmann transport equation. Annu. Rev. Heat Transfer **17**, 205–265 (2014)
86. A. Nabovati, D.P. Sellan, C.H. Amon, On the lattice Boltzmann method for phonon transport. J. Comput. Phys. **230**, 5864–5876 (2011)
87. S.R. Phillpot, P.K. Schelling, P. Keblinski, Phonon wave-packet dynamics at semiconductor interfaces by molecular-dynamics simulation. Appl. Phys. Lett. **80**, 2484–2486 (2002); *ibid*, Interfacial thermal conductivity: Insights from atomic level simulation. J. Mater. Sci. **40**, 3143–3148 (2005)
88. C.-J. Twu, J.-R. Ho, Molecular-dynamics study of energy flow and the Kapitza conductance across an interface with imperfection formed by two dielectric thin films. Phys. Rev. B **67**, 205422 (2003)
89. H. Zhong, J.R. Lukes, Interfacial thermal resistance between carbon nanotubes: Molecular dynamics simulations and analytical thermal modeling. Phys. Rev. B **74**, 125403 (2006)
90. R.J. Stevens, L.V. Zhigilei, P.M. Norris, Effects of temperature and disorder on thermal boundary conductance at solid-solid interfaces: non-equilibrium molecular dynamics simulations. Int. J. Heat Mass Transfer **50**, 3977–3989 (2007)
91. E.S. Landry, A.J.H. McGaughey, Thermal boundary resistance predictions from molecular dynamics simulations and theoretical calculations. Phys. Rev. B **80**, 165304 (2009)
92. Y. Chalopin, K. Esfarjani, A. Henry, S. Volz, G. Chen, Thermal interface conductance in Si/Ge superlattices by equilibrium molecular dynamics. Phys. Rev. B **85**, 195302 (2012)
93. S. Merabia, K. Termentzidis, Thermal conductance at the interface between crystals using equilibrium and nonequilibrium molecular dynamics. Phys. Rev. B **86**, 094303 (2012)
94. Z. Liang, M. Hu, Tutorial: Determination of thermal boundary resistance by molecular dynamics simulations. J. Appl. Phys. **123**, 191101 (2018)
95. F. VanGessel, J. Peng, P.W. Chung, A review of computational phononics: the bulk, interfaces, and surfaces. J. Mater. Sci. **53**, 5641–5683 (2018)
96. S. Datta, Nanoscale device modeling: the Green's function method. Superlattices Microstruct. **28**, 253–278 (2000)
97. N. Mingo, L. Yang, Phonon transport in nanowires coated with an amorphous material: an atomistic Green's function approach. Phys. Rev. B **68**, 245406 (2003)
98. W. Zhang, T.S. Fisher, N. Mingo, Simulation of interfacial phonon transport in Si–Ge heterostructures using an atomistic Green's function method. J. Heat Transfer **129**, 483–491 (2007); ibid, The atomistic Green's function method: an efficient simulation approach for nanoscale phonon transport. Numerical Heat Transfer B **51**, 333–349 (2007)
99. S. Sadasivam, Y. Che, Z. Huang, L. Chen, S. Kumar, T.S. Fisher, The atomistic Green's function method for interfacial phonon transport. Annu. Rev. Heat Transfer **17**, 89–145 (2014)
100. A. Ozpineci, S. Ciraci, Quantum effects of thermal conductance through atomic chains. Phys. Rev. B **63**, 125415 (2001)
101. Z.-Y. Ong, G. Zhang, Efficient approach for modeling phonon transmission probability in nanoscale interfacial thermal transport. Phys. Rev. B **91**, 174302 (2015)
102. L. Yang, B. Latour, A.J. Minnich, Phonon transmission at crystalline-amorphous interfaces studied using mode-resolved atomistic Green's functions. Phys. Rev. B **97**, 205306 (2018)
103. D.A. Young, H.J. Maris, Lattice-dynamical calculation of the Kapitza resistance between fcc lattices. Phys. Rev. B **40**, 3685–3693 (1989)
104. H. Zhao, J.B. Freund, Lattice-dynamical calculation of phonon scattering at ideal Si–Ge interfaces. J. Appl. Phys. **97**, 024903 (2005)
105. S. Sadasivam, N. Ye, J.P. Feser, J. Charles, K. Miao, T. Kubis, T.S. Fisher, Thermal transport across metal silicide-silicon interfaces: First-principles calculations and Green's function transport simulations. Phys. Rev. B **95**, 085310 (2017)

106. Z. Tian, K. Esfarjani, G. Chen, Green's function studies of phonon transport across Si/Ge superlattices. Phys. Rev. B **89**, 235307 (2014)
107. Z. Yan, L. Chen, M. Yoon, S. Kumar, Phonon transport at the interfaces of vertically stacked graphene and hexagonal boron nitride heterostructures. Nanoscale **8**, 4037 (2016)
108. J. Lai, A. Majumdar, Concurrent thermal and electrical modeling of sub-micrometer silicon devices. J. Appl. Phys. **79**, 7353–7361 (1996)
109. P.G. Sverdrup, Y.S. Ju, K.E. Goodson, Sub-continuum simulation of heat conduction in silicon-on-insulator transistors. J. Heat Transfer **123**, 130–137 (2001)
110. S. Sinha, E. Pop, R.W. Dutton, K.E. Goodson, Non-equilibrium phonon distribution in sub-100 nm silicon transistors. J. Heat Transfer **128**, 638–647 (2006)
111. C.D.S. Brites, P.P. Lima, N.J.O. Silva, A. Millán, V.S. Amaral, F. Palacio, L.D. Carlos, Thermometry at the nanoscale. Nanoscale **4**, 4799–4829 (2012)
112. X.W. Wang, *Experimental Micro/Nanoscale Thermal Transport* (Wiley, New York, 2012)
113. A.J. McNamara, Y. Joshi, Z.M. Zhang, Characterization of nanostructured thermal interface materials—a review. Int. J. Thermal Sci. **62**, 2–11 (2012)
114. G. Chen, Probing nanoscale heat transfer phenomena. Annu. Rev. Heat Transfer **16**, 1–8 (2013)
115. D. Zhao, X. Qian, X. Gu, S.A. Jajja, R. Yang, Measurement techniques for thermal conductivity and interfacial thermal conductance of bulk and thin film materials. J. Electron. Package **138**, 040802 (2016)
116. Z.M. Zhang, Surface temperature measurement using optical techniques. Annu. Rev. Heat Transfer **11**, 351–411 (2000)
117. B. Abad, D.-A. Borca-Tasciuc, M.S. Martin-Gonzalez, Non-contact methods for thermal properties measurement. Renew. Sustain. Energy Rev. **76**, 1348–1370 (2017)
118. A.C. Jones, B.T. O'Callahan, H.U. Yang, M.B. Raschke, The thermal near-field: coherence, spectroscopy, heat-transfer, and optical forces. Prog. Sur. Sci. **88**, 349–392 (2013)
119. K.E. Goodson, Y.S. Ju, Heat conduction in novel electronic films. Annu. Rev. Mater. Sci. **29**, 261–293 (1999)
120. K. Park, G.L.W. Cross, Z.M. Zhang, W.P. King, Experimental investigation on the heat transfer between a heated microcantilever and a substrate. J. Heat Transfer **130**, 102401 (2008)
121. D. G. Cahill and R. O. Pohl, "Thermal conductivity of amorphous solids above the plateau," *Phys. Rev. B*, **35**, 4067–4073, 1987
122. D.G. Cahill, H.E. Fischer, T. Klitsner, E.T. Swartz, R.O. Pohl, Thermal conductivity of thin films: measurements and understanding. J. Vac. Sci. Technol. A **7**, 1259–1266 (1989)
123. D.G. Cahill, Thermal conductivity measurement from 30 K to 750 K: the 3-omega method. Rev. Sci. Instrum. **61**, 802–808 (1990)
124. C. Dames, Measuring the thermal conductivity of thin films: 3 omega and related electrothermal methods. Annu. Rev. Heat Transfer **16**, 7–49 (2013)
125. S. Kommandur, S.K. Yee, A suspended 3-omega technique to measure the anisotropic thermal conductivity of semiconducting polymers. Rev. Sci. Instrum. **89**, 114905 (2018)
126. L. Shi, D. Li, C. Yu, W. Jang, D. Kim, Z. Yao, P. Kim, A. Majumdar, Measuring thermal and thermoelectric properties of one-dimensional nanostructures using a microfabricated device. J. Heat Transfer **125**, 881–888 (2003)
127. P. Kim, L. Shi, A. Majumdar, P.L. McEuen, Thermal transport measurements of individual multiwalled nanotubes. Phys. Rev. Lett. **87**, 215502 (2001)
128. C. Yu, L. Shi, Z. Yao, D. Li, A. Majumdar, Thermal conductance and thermopower of an individual single-wall carbon nanotube. Nano Lett. **5**, 1842–1846 (2005)
129. A. Mavrokefalos, M.T. Pettes, F. Zhou, L. Shi, Four-probe measurements of the in-plane thermoelectric properties of nanofilms. Rev. Sci. Instrum. **78**, 034901 (2007)
130. A. Weathers, L. Shi, Thermal transport measurement techniques for nanowires and nanotubes. Annu. Rev. Heat Transfer **16**, 101–134 (2013)
131. M. Fujii, X. Zhang, H. Xie, H. Ago, K. Takahashi, T. Ikuta, H. Abe, T. Shimizu, Measuring the thermal conductivity of a single carbon nanotube. Phys. Rev. Lett. **95**, 065502 (2005)

132. J. Kim, E. Ou, D.P. Sellan, L. Shi, A four-probe thermal transport measurement method for nanostructures. Rev. Sci. Instrum. **86**, 044901 (2015)
133. J. Kim, D.A. Evans, D.P. Sellan, O.M. Williams, E. Ou, A.H. Cowley, L. Shi, Thermal and thermoelectric transport measurements of an individual boron arsenide microstructure. Appl. Phys. Lett. **108**, 201905 (2016)
134. A. Majumdar, Scanning thermal microscopy. Annu. Rev. Mater. Sci. **29**, 505–585 (1999)
135. A. Majumdar, J. P. Carrejo, J. Lai, Thermal imaging using the atomic force microscope. Appl. Phys. Lett. **62**, 2501–2503 (1993)
136. A. Majumdar, J. Lai, M. Chandrachood, O. Nakabeppu, Y. Wu, J. Shi, Thermal imaging by atomic force microscopy using thermocouple cantilever probes. Rev. Sci. Instrum. **66**, 3584–3592 (1995)
137. C.C. Williams, H.K. Wickramasinghe, Scanning thermal profiler. Appl. Phys. Lett. **49**, 1587–1589 (1986)
138. A. Majumdar, J. Varesi, Nanoscale temperature distribution measured by scanning Joule expansion microscopy. J. Heat Transfer **120**, 297–305 (1998)
139. S.P. Gurrum, W.P. King, Y.K. Joshi, K. Ramakrishna, Size effect on the thermal conductivity of thin metallic films investigated by scanning Joule expansion microscopy. J. Heat Transfer **130**, 082403 (2008)
140. K.L. Grosse, M.-H. Bae, F. Lian, E. Pop, W.P. King, Nanoscale Joule heating, Peltier cooling and current crowding at graphene-metal contacts. Nat. Nanotech. **6**, 287–290 (2011)
141. T. Borca-Tasciuc, Scanning probe methods for thermal and thermoelectric property measurements. Annu. Rev. Heat Transfer **16**, 211–258 (2013)
142. S. Gomès, A. Assy, P.-O. Chapuis, Scanning thermal microscopy: a review. Phys. Status Solidi A **212**, 477–494 (2015)
143. K. Kim, J. Chung, G. Hwang, O. Kwon, J.S. Lee, Quantitative measurement with scanning thermal microscope by preventing the distortion due to the heat transfer through the air. ACS Nano **11**, 8700–8709 (2011)
144. H.F. Hamann, Y.C. Martin, H.K. Wickramasinghe, Thermally assisted recording beyond traditional limits. Appl. Phys. Lett. **84**, 810–812 (2004)
145. W.P. King, T.W. Kenny, K.E. Goodson et al., Atomic force microscope cantilevers for combined thermomechanical data writing and reading. Appl. Phys. Lett. **78**, 1300–1302 (2001)
146. J. Lee, T. Beechem, T. L. Wright, B. A. Nelson, S. Graham, W. P. King, Electrical, thermal, and, mechanical characterization of silicon microcantilever heaters. J. Microelectromech. Syst. **15**, 1644 (2007)
147. J. Lee, T.L. Wright, M.R. Abel et al., Thermal conduction from microcantilever heaters in partial vacuum. J. Appl. Phys. **101**, 014906 (2007)
148. K. Park, J. Lee, Z.M. Zhang, W.P. King, Frequency-dependent electrical and thermal response of heated atomic force microscope cantilevers. J. Microelectromech. Syst. **16**, 213–222 (2007)
149. K. Park, A. Marchenkov, Z.M. Zhang, W.P. King, Low temperature characterization of heated microcantilevers. J. Appl. Phys. **101**, 094504 (2007)
150. W.P. King, B. Bhatia, J.R. Felts, H.J. Kim, B. Kwon, B. Lee, S. Somnath, M. Rosenberger, Heated atomic force microscope cantilevers and their applications. Annu. Rev. Heat Transfer **16**, 287–326 (2013)
151. A.J. Schmidt, X. Chen, G. Chen, Pulse accumulation, radial heat conduction, and anisotropic thermal conductivity in pump-probe transient thermoreflectance. Rev. Sci. Instrum. **79**, 114802 (2008)
152. A.J. Minnich, Measuring phonon mean free paths using thermal conductivity spectroscopy. Annu. Rev. Heat Transfer **16**, 183–210 (2013)
153. J. Zhu, H. Park, J.-Y. Chen et al., Revealing the origins of 3D anisotropic thermal conductivities of black phosphorus. Adv. Electron. Mater. **2**, 1600040 (2016)
154. P. Jiang, X. Qian, R. Yang, Tutorial: time-domain thermoreflectance (TDTR) for thermal property characterization of bulk and thin film materials. J. Appl. Phys. **124**, 161103 (2018)

155. Z. Cheng, T. Bougher, T. Bai et al., Probing growth-induced anisotropic thermal transport in high-quality CVD diamond membranes by multifrequency and multiple-spot-size time-domain thermoreflectance. ACS Appl. Mater. Interfaces. **10**, 4808–4815 (2018)
156. S. Huxtable, D.G. Cahill, V. Fauconnier, J.O. White, J.-C. Zhao, Thermal conductivity imaging at micrometrescale resolution for combinatorial studies of materials. Nat. Mater. **3**, 298–301 (2004)
157. D.G. Cahill, Analysis of heat flow in layered structures for time-domain thermoreflectance. Rev. Sci. Instrum. **75**, 5119–5122 (2004)
158. J. Jeong, X. Meng, A. K. Rockwell et al., Picosecond transient thermoreflectance for thermal conductivity characterization. Nanoscale Microscale Thermophys. Eng. **23**, 211−221 (2019)
159. P.M. Norris, A.P. Caffrey, R.J. Stevens, J.M. Klopf, J.T. McLeskey, A.N. Smith, Femtosecond pump–probe nondestructive examination of materials. Rev. Sci. Instrum. **74**, 400–406 (2003)
160. K.E. Goodson, M. Asheghi, Near-field optical thermometry. Microscale Thermophys. Eng. **1**, 225–235 (1997)
161. D. Seto, R. Nikka, S. Nishio, Y. Taguchi, T. Saiki, Y. Nagasaka, Nanoscale optical thermometry using a time-correlated single-photon counting in an illumination-collection mode. Appl. Phys. Lett. **110**, 033109 (2017)
162. M.E. Siemens, Q. Li, R. Yang, K.A. Nelson, E.H. Anderson, M.M. Murnane, H.C. Kapteyn, Quasi-ballistic thermal transport from nanoscale interfaces observed using ultrafast coherent soft X-ray beams. Nat. Mater. **9**, 26–30 (2010)
163. T. Favaloro, J.-H. Bahk, A. Shakouri, Characterization of the temperature dependence of the thermoreflectance coefficient for conductive thin films. Rev. Sci. Instrum. **86**, 024903 (2015)
164. C. Wei, X. Zheng, D.G. Cahill, J.-C. Zhao, Invited article: Micron resolution spatially resolved measurement of heat capacity using dual-frequency time-domain thermoreflectance. Rev. Sci. Instrum. **84**, 071301 (2013)
165. P.E. Hopkins, C.M. Reinke, M.F. Su, R.H. Olsson III, E.A. Shaner, Z.C. Leseman, J.R. Serrano, L.M. Phinney, I. El-Kady, Reduction in the thermal conductivity of single crystalline silicon by phononic crystal patterning. Nano Lett. **11**, 107–112 (2011)
166. M.R. Wagner, B. Graczykowski, J.S. Reparaz et al., Two-dimensional photonic crystals: Disorder matters. Nano Lett. **16**, 5661–5668 (2016)
167. X. Wang, T. Mori, I. Kuzmych-Ianchuk, Y. Michiue, K. Yubuta, T. Shishido, Y. Grin, S. Okada, D.G. Cahill, Thermal conductivity of layered borides: The effect of building defects on the thermal conductivity of $TmAlB_4$ and the anisotropic thermal conductivity of $AlB_2$. APL Mater. **2**, 046113 (2014)
168. J. Liu, G.-M. Choi, D.G. Cahill, Measurement of the anisotropic thermal conductivity of molybdenum disulfide by the time-resolved magneto-optic Kerr effect. J. Appl. Phys. **116**, 233107 (2014)
169. A. J. Schmidt, R. Cheaito, M. Chiesa, A frequency-domain thermoreflectance method for the characterization of thermal properties. Rev. Sci. Instrum. **80**, 094901 (2009); ibid, Characterization of thin metal films via frequency-domain thermoreflectance. J. Appl. Phys. **107**, 024908 (2010)
170. K.T. Regner, D.P. Sellan, Z. Su, C.H. Amon, A.J.H. McGaughey, J.A. Malen, Broadband phonon mean free path contributions to thermal conductivity measured using frequency domain thermoreflectance. Nat. Commun. **4**, 1640 (2013)
171. D. Rodin, S.K. Yee, Simultaneous measurement of in-plane and through-plane thermal conductivity using beam-offset frequency domain thermoreflectance. Rev. Sci. Instrum. **88**, 014902 (2017)
172. J. Johnson, A. A. Maznev, J. Cuffe, J. K. Eliason, A. J. Minnich, T. Kehoe, C. M. Sotomayor Torres, G. Chen, K. A. Nelson, Direct measurement of room-temperature nondiffusive thermal transport over micron distances in a silicon membrane. Phys. Rev. Lett. **110**, 025901 (2013)
173. J. Cuffe, J.K. Eliason, A.A. Maznev et al., Reconstructing phonon mean-free-path contributions to thermal conductivity using nanoscale membranes. Phys. Rev. B **91**, 245423 (2015)
174. A. Vega-Flick, R.A. Duncan, J.K. Eliason et al., Thermal transport in suspended silicon membranes measured by laser-induced transient gratings. AIP Adv. **6**, 120903 (2016)

# Chapter 8
# Fundamentals of Thermal Radiation

Radiation is one of the fundamental modes of heat transfer. However, the concepts of thermal radiation are much more complicated and, hence, very difficult to perceive. The main features of radiation that are distinct from conduction and convection are as follows: (a) radiation can transfer energy with and without an intervening medium; (b) the radiant heat flux is not proportional to the temperature gradient; (c) radiation emission is wavelength dependent, and the radiative properties of materials depend on the wavelength and the temperature; (d) the radiant energy exchange and the radiative properties depend on the direction and orientation [1, 2].

The dual theory explains the nature of radiation as either electromagnetic waves or a collection of particles, called photons. Although radiation can travel in vacuum, it originates from matter. All forms of matter emit radiation through complicated mechanisms (e.g., molecular vibration in gases, and electron and lattice vibrations in solids). In most solids and some liquids, radiation emitted from the interior is strongly absorbed by adjoining molecules. Therefore, radiation from or to these materials is often treated as *surface phenomena,* while radiation in gases and some semitransparent solids or liquids has to be treated as *volumetric phenomena.* Nevertheless, one must treat solids or liquids volumetrically as a medium to understand the mechanisms of reflection and emission, to predict the radiative properties of thin films and small particles, and to calculate radiation heat transfer between objects placed in close vicinity. *Thermal radiation* refers to a type of radiation where the emission is directly related to the temperature of the body (or surface).

There are numerous engineering applications where radiation heat transfer is important, such as solar energy, combustion, furnaces, high-temperature materials processing and manufacturing, and insulation in space and cryogenic systems. Even at room temperature, radiative heat transfer may be of the same order of magnitude as convective heat transfer. The study of thermal radiation went along with the study of light phenomena and led to some major breakthroughs in modern physics. It is instructive to give a brief survey of major historical developments related to thermal radiation.



Z. M. Zhang, *Nano/Microscale Heat Transfer*, Mechanical Engineering Series,
https://doi.org/10.1007/978-3-030-45039-7_8

Quantitative understanding of light phenomena began in the seventeenth century with the discoveries of Snell's law of refraction, Fermat's least-time principle of light path, Huygens' principle of constructing the wavefront from secondary waves, and Newton's prism that helped him prove white light consists of many different types of rays. In the dawn of the nineteenth century, Sir Frederick Herschel (1738–1822), a German-born English astronomer, discovered infrared radiation [3]. His original objective was to find a suitable color for a glass filter, which could transmit the most of light but the least amount of heat, for use in solar observations. By moving a thermometer along the spectrum of solar radiation that passed through a prism, Herschel accidently found that the temperature of the thermometer would rise even though it was placed beyond the red end of the visible light. He published several papers in *Philosophical Transactions of the Royal Society of London* in 1800 and called the unknown radiation *invisible light* or *heat-making rays.* Young's double-slit experiment in 1801 demonstrated the interference phenomenon and the wave nature of light. It was followed by extensive studies on polarization and reflection led by French physicist Augustin-Jean Fresnel (1788–1827) who contributed significantly to the establishment of the wave theory of light. In 1803, radiation beyond the violet end of the visible spectrum via chemical effects was also discovered. The ultraviolet, visible, and infrared spectra were thus associated with chemical, luminous, and heating effects, respectively. Yet, the common nature of the different types of radiation was not known until the late nineteenth century.

One of the obstacles of accurately measuring infrared radiation (or heat radiation, as it was called in those days) was the lack of sensitive detectors. In the earlier years, measurements were performed using mercury-in-glass thermometers with blackened bulbs. In 1829, Italian physicists Leopoldo Nobili (1784–1835) and Macedonio Melloni (1798–1854) invented the thermopile made by connecting a number of thermocouples in series that is much more sensitive and faster than the thermometer. Melloni used the device to study the infrared radiation from hot objects and the sun. Gustav Kirchhoff (1824–1887), a German physicist, contributed greatly to the fundamental understanding of spectroscopy and thermal emission by heated objects. In 1862, he coined the term "black body" radiation and established Kirchhoff's law, which states that the emissivity of a surface equals its absorptivity at thermal equilibrium.

Many famous physicists and mathematicians have contributed to electromagnetism. The complete equations of electromagnetic waves were established in 1873 by Scottish physicist James Clerk Maxwell (1831–1879), and later confirmed experimentally by German physicist Heinrich Hertz (1857–1894) through the discovery of radio waves due to electrical vibrations. Before the existence of electrons was proved, Dutch physicist Hendrik Lorentz (1853–1928) proposed that light waves were due to oscillations of electric charges in the atom. His electron theory could explain the phenomenon discovered by his mentee Pieter Zeeman (1865–1943) that the lines in the spectrum can split into several lines under a strong magnetic field (known as the Zeeman effect). They shared the Nobel Prize in Physics in 1902 for their research into the influence of magnetism upon radiation phenomena. The electromagnetic wave theory has played a central role in radio, radar, television, microwave technology, telecommunication, thermal radiation, and physical optics. Albert Einstein arrived at the famous formula $E = mc^2$ in 1905, after connecting the relativity principle with the Maxwell equations.

In 1881, Samuel Langley (1834–1906), the American astronomer, physicist, and aeronautics pioneer, invented a highly sensitive device called a *bolometer* for detecting thermal radiation. The bolometer used two platinum strips, connected in a Wheatstone bridge circuit with a sensitive galvanometer, to read the imbalance of the bridge caused by the exposure of one of the strips to radiation. Langley was the first to make an accurate map of the solar spectrum up to a wavelength of 2.8 $\mu$m. The Stefan–Boltzmann law of blackbody radiation is buit upon the empirical relation obtained by Slovenian physicist Joseph Stefan (1835–1893) in 1879, through careful experimental observation. The theoretical proof was provided by Austrian physicist Ludwig Boltzmann (1844–1906) in 1884, based on the thermodynamic relations of a Carnot cycle with radiation as a working fluid using the concept of radiation pressure. In the late nineteenth century, German physicist Wilhelm Wien (1864–1928) derived the displacement law in 1893 by considering a piston moving within a mirrored empty cylinder filled with thermal radiation. Wien also derived a spectral distribution of blackbody radiation, called Wien's formula, which is applicable to the short-wavelength region of the blackbody spectrum but deviates from experiments toward long wavelengths. Wien received the Nobel Prize in 1911 "for his discoveries regarding the laws governing the radiation of heat." In 1900, Lord Rayleigh (1842–1919), British physicist and Nobel Laureate in Physics in 1904, used the equipartition theorem to show that the blackbody emission should be directly proportional to temperature but inversely proportional to the fourth power of wavelength. Sir James Jeans (1877–1946), a British physicist, astronomer, and mathematician, derived a more complete expression in 1905. The Rayleigh–Jeans formula agreed with experiments at sufficiently high temperatures and long wavelengths, where Wien's formula failed, but disagreed with experiments at short wavelengths. It is noteworthy that Rayleigh has made great contributions to light scattering and wave phenomena, such as the discovery of Rayleigh scattering by small objects that explains why the sky is blue and the sunset glows red and orange. Rayleigh also predicted the existence of *surface waves,* sometimes called *Rayleigh waves,* which propagate along the interface between two different media. The amplitude of the wave, however, diminishes in each media as the distance from the interface increases.

In an effort to obtain a better agreement with measurements at long wavelengths, German physicist Max Planck (1858–1947) in 1900 used the maximum entropy principle, based on Boltzmann's entropy expression, to derive an equation, known as Planck's law, which agrees with experiments in the whole spectral region. Planck obtained his expression independently of Rayleigh's work, while the complete derivation of Rayleigh–Jeans formula was obtained several years later. In his book *The Theory of Heat Radiation,* Planck [4] showed that his formula would reduce to Wien's formula at small $\lambda T$ and Rayleigh–Jeans formula at large $\lambda T$. In his derivation, Planck used a bold assumption that is controversial to classical electrodynamics. His hypothesis was that the energy of linear oscillators is not infinitely divisible but must assume discrete values that are multiples of $h\nu$, where $h$ is a universal constant and $\nu$ is the frequency of the oscillator. This concept would have been easily accepted for a system consisting of particles, like atoms or gas molecules, but not for oscillators that radiate electromagnetic energy. Planck's work opened the door to quantum mechanics. The idea of quantization of radiation was further developed by Einstein, who applied it to explain the photoelectric effect in 1905. Planck was awarded the Nobel Prize in Physics in 1918 for the discovery of energy quanta. In 1924, Indian mathematical physicist Satyendra Nath Bose (1894–1974) modified the Boltzmann statistics of ideal molecular gases, by introducing the concept of different quantum states at each energy level (degeneracy) using the phase space while treating the light quanta as indistinguishable. Subsequently, Bose was able to statistically derive Planck's distribution function without using the semi-classical oscillator concept. With the help of Einstein, Bose's work was published in *Zeitschrift für Physik* in 1924. Einstein further extended Bose's theory to atoms and predicted the existence of a phenomenon, known as Bose–Einstein condensate, as discussed in Chap. 3. It is clear that the journey of questing for the truth in understanding thermal radiation has led to important discoveries in modern physics.

This chapter contains an introduction to the electromagnetic wave theory, blackbody radiation, plane wave reflection, and refraction at the boundary between two semi-infinite media, evanescent waves and total internal reflection, and various models used to study the optical properties of different materials. A brief description of the typical experimental methods used to measure the spectral radiative properties is also presented. The materials covered in the following sections are intended to provide a sound background for more in-depth studies on the applications of thermal radiation to micro/nanosystems in subsequent chapters.

## 8.1 Electromagnetic Waves

In this section, we will study macroscopic Maxwell's equations and electromagnetic (EM) waves in isotropic media from dielectric to dissipative (lossy) to magnetic media. The concepts of polarization, absorption, and evanescent waves are introduced using complex wavevectors. Poynting's theorem describes the energy balance for EM waves including transfer, storage, and dissipation. The complex dielectric function is defined from which the complex refractive index for nonmagnetic materials can be calculated.

### *8.1.1 Maxwell's Equations*

The propagation of electromagnetic waves in any media is governed by a set of equations, first stated together by Maxwell. The macroscopic Maxwell equations can be written in the differential forms as follows [5–9]:

$$\nabla \times \mathbf{E} = -\frac{\partial \mathbf{B}}{\partial t} \tag{8.1}$$

$$\nabla \times \mathbf{H} = \mathbf{J} + \frac{\partial \mathbf{D}}{\partial t} \tag{8.2}$$

$$\nabla \cdot \mathbf{D} = \rho_e \tag{8.3}$$

$$\nabla \cdot \mathbf{B} = 0 \tag{8.4}$$

Here, **E** is the electric field, **H** is the magnetic field, **J** is the electric current density (or charge flux according to the definition in Chap. 4), **D** is the electric displacement, **B** is the magnetic flux density (also called magnetic induction), and $\rho_e$ is the charge density. In the SI units, **E** is in V/m, **H** in C/m s, **J** in A/m$^2$, **D** in C/m$^2$, **B** in Wb/m$^2$, and $\rho_e$ in C/m$^3$. Note that 1 T (T) = 1 Wb/m$^2$ and 1 Wb (Wb) = 1 V s. The charge conservation or continuity equation, $\nabla \cdot \mathbf{J} + \partial \rho_e / \partial t = 0$, is implicitly included in the Maxwell equations, because it can be obtained by taking the divergence of Eq. (8.2) and then applying Eq. (8.3). The constitutive relations for a linear isotropic medium are

$$\mathbf{D} = \varepsilon_m \mathbf{E} \tag{8.5}$$

$$\mathbf{B} = \mu_m \mathbf{H} \tag{8.6}$$

where $\varepsilon_m$ is the electric permittivity in F/m and $\mu_m$ the magnetic permeability of the medium in N/A$^2$. Note that the farad (F) is the SI unit of capacitance: 1 F = 1 C/V. The permittivity and permeability values of free space (vacuum) are $\varepsilon_0 = 8.854 \times 10^{-12}$ F/m and $\mu_0 = 4\pi \times 10^{-7}$ N/A$^2$, respectively. For anisotropic media, $\mu_m$ and $\varepsilon_m$ are dyadic tensors. The microscopic form of Ohm's law gives

$$\mathbf{J} = \sigma \mathbf{E} \tag{8.7}$$

where $\sigma$ is the electric conductivity in A/V m.

A brief discussion on the physical interpretation of Maxwell's equations is given next. Equation (8.1) is an expression of Faraday's law of induction, which states that a time-varying magnetic field produces an electric field in a coil. In other words, through any closed electric field line, there is a time-varying magnetic field. Combining Eq. (8.1) with Green's theorem, Eq. (B.71), we see that the integral of the electric field around a closed loop is equal to the negative of the integral of the time derivative of the magnetic induction, over the area enclosed by the loop. Equation (8.2) is the general Ampere's law, which includes Maxwell's displacement current ($\partial \mathbf{D}/\partial t$). It states that through any closed magnetic field line, there is an electric current density **J** or a displacement current or both. Conversely, circulating magnetic fields are produced by passing an electrical current through a conductor or changing electric fields or both. Equation (8.3) is Gauss's law, which implies that the electric field diverges from electric charges. Using Gauss's theorem, Eq. (B.70), it can be seen from Eq. (8.3) that the integral of the electric field over a closed surface is proportional to the electric charges enclosed by that surface. If there are no electric charges inside a closed surface, there is no net electric field penetrating the surface. Equation (8.4) is an analogy to Gauss's law for the magnetic field. However, isolated magnetic poles (i.e., magnetic monopoles) have not been observed, so the integration of a magnetic field over any closed surface is zero.

The interpretations given in the preceding paragraph are straightforward since all variables and coefficients are considered as real quantities. However, Maxwell's equations are most useful when all quantities are expressed in complex variables. The material properties, such as $\varepsilon_{\rm m}$ and $\mu_{\rm m}$, are generally complex and frequency dependent. To facilitate understanding, we will start with simple cases first and then generalize the theory for more realistic problems.

### 8.1.2 The Wave Equation

Sometimes called free charge density, $\rho_{\rm e}$ in Eq. (8.3) should be treated as excess charges or net charges per unit volume. Because the number of electrons equals the number of protons in the nuclei, in most media, we can assume $\rho_{\rm e} = 0$. For a nonconductive material, $\sigma = 0$. We further assume that $\varepsilon_{\rm m}$ and $\mu_{\rm m}$ are both real and independent of position, time, and the field strength. This is true for a nondissipative (lossless), homogeneous, and linear material. If $\mu_{\rm m} = \mu_0$, the material is said to be nonmagnetic. Therefore, a nonconductive and nonmagnetic material is a dielectric for which only $\varepsilon_{\rm m}$ is needed to characterize its electromagnetic behavior. Materials with both $\varepsilon_{\rm m}$ and $\mu_{\rm m}$ being real but $\mu_{\rm m} \neq \mu_0$ are sometimes called general dielectrics or dielectric-magnetic media. Substituting the constitutive relations into Maxwell's equations and then combining Eqs. (8.1) and (8.2), we obtain

$$\nabla^2 \mathbf{E} = \mu_{\rm m}\varepsilon_{\rm m}\frac{\partial^2 \mathbf{E}}{\partial t^2} \tag{8.8}$$

where the vector identity given in Eq. (B.64), $\nabla \times (\nabla \times \mathbf{E}) = \nabla(\nabla \cdot \mathbf{E}) - \nabla^2\mathbf{E} = -\nabla^2\mathbf{E}$, has been employed. Equation (8.8) is the *wave equation,* which can also be written in terms of the magnetic field. The wave equation has infinite number of solutions (see Problem 8.1). The solution of Eq. (8.8) for a monochromatic plane wave can be written as

$$\mathbf{E} = \mathbf{E}_0 {\rm e}^{-{\rm i}(\omega t - \mathbf{k}\cdot\mathbf{r})} \tag{8.9}$$

where $\mathbf{E}_0$ is the amplitude vector, $\omega$ is the angular frequency, $\mathbf{r} = x\hat{\mathbf{x}} + y\hat{\mathbf{y}} + z\hat{\mathbf{z}}$ is the position vector, and $\mathbf{k} = k_x\hat{\mathbf{x}} + k_y\hat{\mathbf{y}} + k_z\hat{\mathbf{z}}$ is the *wavevector,* which points toward the direction of propagation. In order for Eq. (8.9) to be a solution of Eq. (8.8), the magnitude of $\mathbf{k}$ must be $k = \omega\sqrt{\mu_{\rm m}\varepsilon_{\rm m}}$. The complex form of the electric field is used in Eq. (8.9) to facilitate mathematical manipulation. The actual electric field may be expressed as the real part of Eq. (8.9), viz.,

$$\mathrm{Re}(\mathbf{E}) = \mathrm{Re}(\mathbf{E}_0)\cos\phi + \mathrm{Im}(\mathbf{E}_0)\sin\phi \tag{8.10}$$

where Re or Im stands for taking the real part or the imaginary part, and $\phi = \omega t - \mathbf{k}\cdot\mathbf{r}$ is the phase. Equation (8.9) is a time-harmonic solution at a fixed frequency. Because any time-space-dependent function can be expressed as a Fourier series of many frequency components, we can integrate Eq. (8.9) over all frequencies to obtain the total electric field at any time and position. Therefore, understanding the nature of Eq. (8.9) is very important to the study of electromagnetic wave phenomena.

When Eq. (8.9) is substituted into Maxwell's equations, a time derivative $\partial/\partial t$ can be replaced by a multiplication of $-\mathrm{i}\omega$ and the operator $\nabla$ can be replaced by $\mathrm{i}\mathbf{k}$. Hence, the first two Maxwell equations can be written as

$$\mathbf{k}\times\mathbf{E} = \omega\mu_{\mathrm{m}}\mathbf{H} \tag{8.11a}$$

and

$$\mathbf{k}\times\mathbf{H} = -\omega\varepsilon_{\mathrm{m}}\mathbf{E} \tag{8.11b}$$

The two equations suggest that **E**, **H**, and **k** are orthogonal and form a right-handed triplet, when both $\varepsilon_{\mathrm{m}}$ and $\mu_{\mathrm{m}}$ are positive. On the surface normal to the wavevector **k**, the electric or magnetic field is a function of time only, because $\mathbf{k}\cdot\mathbf{r} = \text{const}$. This surface is called a *wavefront*. In the **k**-direction, the wavefront travels at the speed given by

$$c = \frac{\omega}{k} = \frac{1}{\sqrt{\mu_{\mathrm{m}}\varepsilon_{\mathrm{m}}}} \tag{8.12}$$

which is called *phase speed* and it is the smallest speed at which the wavefront propagates. The phase velocity is the phase speed times the unit wavevector [8].

Figure 8.1 illustrates a plane wave, propagating in the positive $x$-direction, whose electric field is parallel to the $y$-direction and magnetic field parallel to the $z$-direction.

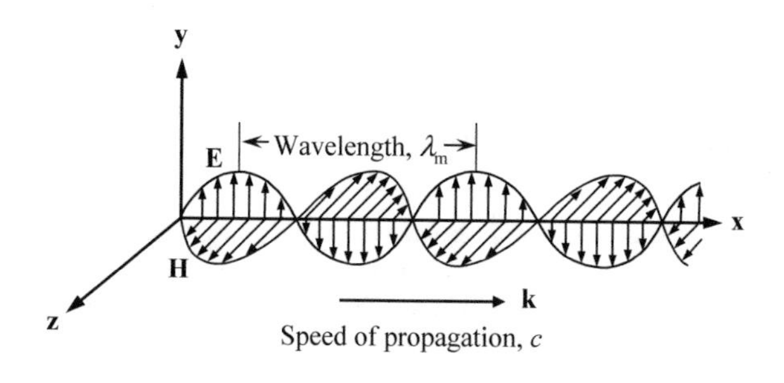


**Fig. 8.1** Illustration of a linearly polarized electromagnetic wave

In such cases, $k = k_x$ and $\mathbf{k} \cdot \mathbf{r} = kx$. The wavefront is perpendicular to the $x$-direction. It can be seen clearly that the wavevector is related to the wavelength $\lambda_\mathrm{m}$ in the medium by $k = 2\pi/\lambda_\mathrm{m}$.

In free space, the speed of an electromagnetic wave is given by $c_0 = 1/\sqrt{\mu_0\varepsilon_0}$. The speed of light in a vacuum was instated as an exact number, $c_0 = 299,792,458$ m/s, by the General Conference on Weights and Measures (abbreviated as CGPM for Conférence Générale des Poids et Mesures) in 1983. The SI base unit meter has since been defined as the distance that light travels in a vacuum during a time interval of $1/299,792,458$ s. The reference on constant, units, and uncertainty can be found from the web page of the National Institute of Standards and Technology (NIST) for detailed discussions about the fundamental physical constants and the base SI units [10]. For most calculations, it suffices to use $c_0 = 2.998 \times 10^8$ m/s. The refractive index of a medium is given as $n = \sqrt{\frac{\mu_\mathrm{m}\varepsilon_\mathrm{m}}{\mu_0\varepsilon_0}} = \frac{c_0}{c}$. Therefore, $c = c_0/n$ and $\lambda_\mathrm{m} = \lambda/n$, where $\lambda$ is the wavelength in vacuum. For nonmagnetic materials $\mu_\mathrm{m}/\mu_0 = 1$; thus, $n = \sqrt{\varepsilon_\mathrm{m}/\varepsilon_0}$.

Notice that $n$ of a medium is a function of frequency (or wavelength) and is, in general, temperature dependent. For polychromatic light, the phase speed usually depends on wavelength because $n = n(\lambda)$ in a dispersive medium. In a vacuum, the energy propagation velocity is the same as the phase velocity. For polychromatic waves in a dispersive medium, the group velocity $\mathbf{v}_\mathrm{g}$ determines the direction and speed of energy flow and is defined as

$$\mathbf{v}_\mathrm{g} = \nabla_k \omega = \frac{\mathrm{d}\omega}{\mathbf{dk}} = \frac{\partial\omega}{\partial k_x}\hat{\mathbf{x}} + \frac{\partial\omega}{\partial k_y}\hat{\mathbf{y}} + \frac{\partial\omega}{\partial k_z}\hat{\mathbf{z}} \tag{8.13}$$

which is the gradient of $\omega$ in the $k$-space. In a homogeneous and isotropic medium, $v_\mathrm{g} = c_0\left(n + \omega\frac{dn}{d\omega}\right)^{-1}$ and the direction of the group velocity will be the same as that of the wavevector $\mathbf{k}$. In a nondispersive medium, where $n$ is not a function of frequency, it is clear that $v_\mathrm{g} = c = c_0/n$. A group front can also be defined based on the constant-amplitude surface of the wave group. In general, it is not parallel to the wavefront, when light is refracted from a nondispersive medium to a dispersive medium; furthermore, the energy flow direction is not necessarily perpendicular to the group front [11]. Notice that the wave equation is also applicable to other types of waves such as acoustic waves, which are matter waves with a longitudinal and two transverse modes, as mentioned in Chap. 5.

### *8.1.3 Polarization*

A simple transverse wave will oscillate perpendicular to the wavevector. Because electromagnetic waves have two field vectors that can change their directions during propagation, the polarization behavior may be complicated. It is important to understand the nature of polarization in order to fully characterize an electromagnetic wave. There are two equivalent ways to interpret a complex vector $\mathbf{A}$. The first method considers it as a vector whose components are complex, i.e.,

$$\mathbf{A} = A_x\hat{\mathbf{x}} + A_y\hat{\mathbf{y}} + A_z\hat{\mathbf{z}} \tag{8.14a}$$

where $A_x$, $A_y$, and $A_z$ are complex numbers:

$$A_x = A'_x + \mathrm{i}A''_x,\ A_y = A'_y + \mathrm{i}A''_y,\ \text{and } A_z = A'_z + \mathrm{i}A''_z \tag{8.14b}$$

The second method decomposes it into two real vectors such that

$$\mathbf{A} = \mathbf{A}' + \mathrm{i}\mathbf{A}'' \tag{8.15a}$$

where $\mathbf{A}'$ and $\mathbf{A}''$ are the real and imaginary parts of the complex vector, given by

$$\mathbf{A}' = A'_x\hat{\mathbf{x}} + A'_y\hat{\mathbf{y}} + A'_z\hat{\mathbf{z}} \text{ and } \mathbf{A}'' = A''_x\hat{\mathbf{x}} + A''_y\hat{\mathbf{y}} + A''_z\hat{\mathbf{z}} \tag{8.15b}$$

In either case, a complex vector has six real scalar terms.

For the time being, let us assume all the material properties to have real values and $\mathbf{k}$ to be a real vector. Both $\mathbf{E}$ and $\mathbf{H}$ are complex, according to Eq. (8.9). To ensure that $\mathbf{k} \cdot \mathbf{E} = 0$ at any time and location, both $\mathrm{Re}(\mathbf{E}_0)$ and $\mathrm{Im}(\mathbf{E}_0)$ must be perpendicular to $\mathbf{k}$. The same is true for the magnetic vector. Because $\mathbf{H}$ can be obtained from Eq. (8.11a), the state of polarization can be based on how the electric field varies in time and along the $\mathbf{k}$-direction in space. In order to study the time dependence of the electric field, rewrite Eq. (8.10) as

$$\mathrm{Re}(\mathbf{E}) = \mathbf{a}\cos(\omega t) + \mathbf{b}\sin(\omega t) \tag{8.16}$$

where $\mathbf{a} = \mathrm{Re}(\mathbf{E}_0 \mathrm{e}^{\mathrm{i}\mathbf{k}\cdot\mathbf{r}})$ and $\mathbf{b} = \mathrm{Im}(\mathbf{E}_0 \mathrm{e}^{\mathrm{i}\mathbf{k}\cdot\mathbf{r}})$ are both real vectors and perpendicular to $\mathbf{k}$. In general, the electric field will vary with time in an ellipse, called the *vibration ellipse*, as shown in Fig. 8.2. If $\mathbf{a}$ and $\mathbf{b}$ are parallel or, equivalently, $\mathrm{Re}(\mathbf{E}_0)$ and

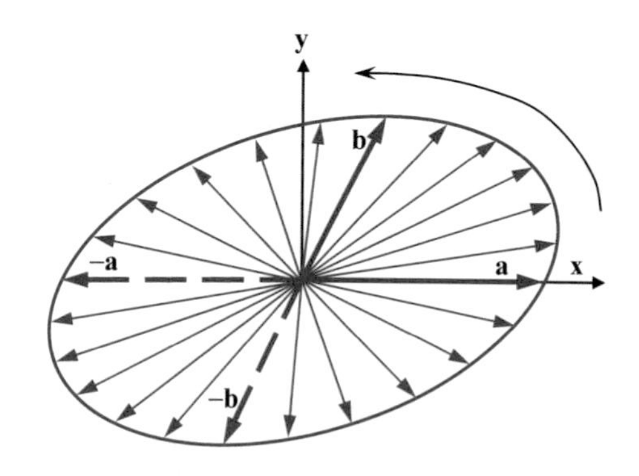


**Fig. 8.2** Illustration of polarization by the vibration ellipse, for a plane wave propagating in the positive $z$-direction (out of the paper). The electric field vector is plotted at an increment of $\omega\Delta t = \pi/12$

$\mathrm{Im}(\mathbf{E}_0)$ are parallel to each other, then the electric field will not change its directions. The wave is said to be *linearly polarized*, and either $\mathbf{a}$ or $\mathbf{b}$ specifies the direction of polarization. An example of a linearly polarized wave is the wave shown in Fig. 8.1. When $\mathbf{a} \perp \mathbf{b}$ and $|\mathbf{a}| = |\mathbf{b}|$, the vibration ellipse is a circle and the wave is said to be *circularly polarized*. In general, a monochromatic wave described by Eq. (8.10) is *elliptically polarized.* For circularly or elliptically polarized light, if $\mathbf{a} \times \mathbf{b}$ is in the same direction as $\mathbf{k}$, the vibration ellipse will rotate counterclockwise (left handed) when viewed toward the light source, and if $\mathbf{a} \times \mathbf{b}$ is opposite to the direction of propagation, the vibration ellipse will rotate clockwise (right handed). Similarly, one can consider the polarization of the electric field at a fixed time and observe the vibration ellipse along the direction of propagation as an exercise (see Problem 8.2).

Because of the random nature of thermal radiation, the Fourier component does not vary with time exactly following $\mathrm{e}^{-\mathrm{i}\omega t}$ but with some fluctuations in the amplitude. The polarization may become completely random, which is said to be *unpolarized*, *randomly polarized*, or *completely uncorrelated.* In any case, the electric field can be decomposed into the two orthogonal directions on the vibration ellipse. This is particularly useful for calculating energy transfer. The polarization status can be fully described by the four Stokes parameters and the response of an optical element can be modeled using the Mueller matrix formulation [6, 7]. For coherent monochromatic light, however, the Jones vector and Jones matrix formulation can provide additional phase information since it is based on the transformation of the electric field rather than the amplitude [12].

### 8.1.4 Energy Flux and Density

The energy conservation for an electromagnetic field can be obtained from Maxwell's equations, according to English physicist John Poynting (1852–1914). To derive Poynting's theorem, one can dot multiply Eqs. (8.1) and (8.2) by –$\mathbf{H}$ and $\mathbf{E}$, respectively, and then add up each side. Using the vector identity in Eq. (B.63), we get $\nabla \cdot (\mathbf{E} \times \mathbf{H}) = (\nabla \times \mathbf{E}) \cdot \mathbf{H} - (\nabla \times \mathbf{H}) \cdot \mathbf{E}$. After some simplifications, we obtain

$$-\nabla \cdot (\mathbf{E} \times \mathbf{H}) = \frac{\partial}{\partial t}\left(\frac{1}{2}\varepsilon_{\mathrm{m}}\mathbf{E} \cdot \mathbf{E} + \frac{1}{2}\mu_{\mathrm{m}}\mathbf{H} \cdot \mathbf{H}\right) + \mathbf{E} \cdot \mathbf{J} \tag{8.17}$$

The left-hand term represents the energy flow into a differential control volume, the first term on the right is the rate of change of the stored energy (associated with the electric and magnetic fields), and the last term is the dissipated electromagnetic work or Joule heating. The *Poynting vector* is defined as

$$\mathbf{S} = \mathbf{E} \times \mathbf{H} \tag{8.18a}$$

The Poynting vector is essentially the energy flux, which gives both the direction and the rate of energy flow per unit projected surface area. Equations (8.17) and (8.18a) can be easily extended to the complex field notation. Although it is easy to write the Poynting vector (which is always real) as $\mathbf{S} = \text{Re}(\mathbf{E}) \times \text{Re}(\mathbf{H})$, it is not very helpful because one would have to evaluate the real parts of $\mathbf{E}$ and $\mathbf{H}$ individually. Besides, the frequency of oscillation is usually too high to be measured. For harmonic fields, the time-averaged Poynting vector can be expressed as

$$\langle \mathbf{S} \rangle = \frac{1}{2}\text{Re}(\mathbf{E} \times \mathbf{H}^*) \tag{8.18b}$$

where * signifies the complex conjugate. Similarly, the time-averaged energy density for time-harmonic fields becomes [5]

$$\langle u \rangle = \frac{1}{4}\varepsilon_\text{m}\mathbf{E} \cdot \mathbf{E}^* + \frac{1}{4}\mu_\text{m}\mathbf{H} \cdot \mathbf{H}^* \tag{8.19}$$

For an absorbing or dissipative medium, different approaches exist regarding the definition and determination of the electromagnetic energy density especially when magnetic materials are involved [13, 14].

**Example 8.1** Prove that Eq. (8.18b) is the time-averaged Poynting vector for time-harmonic fields.

**Solution** Let $\mathbf{E} = \mathbf{E}(\mathbf{r})\text{e}^{-\text{i}\omega t}$ and $\mathbf{H} = \mathbf{H}(\mathbf{r})\text{e}^{-\text{i}\omega t}$, where $\mathbf{E}(\mathbf{r})$ and $\mathbf{H}(\mathbf{r})$ are complex vectors. Integrating the Poynting vector over a period $T$, we have

$$\begin{aligned}
\langle \mathbf{S} \rangle &= \frac{1}{T}\int_T \text{Re}(\mathbf{E}) \times \text{Re}(\mathbf{H})\text{d}t \\
&= \frac{1}{4T}\int_T \left[\mathbf{E}(\mathbf{r})\text{e}^{-\text{i}\omega t} + \mathbf{E}^*(\mathbf{r})\text{e}^{\text{i}\omega t}\right] \times \left[\mathbf{H}(\mathbf{r})\text{e}^{-\text{i}\omega t} + \mathbf{H}^*(\mathbf{r})\text{e}^{\text{i}\omega t}\right]\text{d}t \\
&= \frac{1}{4}\left(\mathbf{E} \times \mathbf{H}^* + \mathbf{E}^* \times \mathbf{H}\right) = \frac{1}{2}\text{Re}\left(\mathbf{E} \times \mathbf{H}^*\right)
\end{aligned}$$

### 8.1.5 Dielectric Function

The conductivity is large at low frequencies for metals, due to free electrons. Even for good conductors, however, the electrons are not completely free but will be scattered by defects and phonons. At high frequencies, the current density $\mathbf{J}$ and the electric field $\mathbf{E}$ are not in phase anymore, suggesting that the conductivity should be a complex number. For insulators such as crystalline or amorphous dielectrics, electromagnetic waves can interact with bound electrons or lattice vibrations to transfer energy to the medium. At optical frequencies, the distinction between a conductor and an insulator becomes ambiguous unless the optical response over a wide frequency region is considered. It is well known that a good conductor is highly reflective in a broad spectral region from the near infrared all the way to radio frequencies. Nevertheless, a dielectric material can also be highly reflective in certain frequency bands, especially in the mid-infrared region. At certain frequencies or in a narrow frequency band, the dielectric function $\varepsilon' + i\varepsilon''$ may appear to be very similar for a metal and a dielectric material.

Let us consider a nonmagnetic material whose conductivity is $\sigma$. The wave equation for $\sigma \neq 0$ and $\mu_m = \mu_0$ has the following form:

$$\nabla^2 \mathbf{E} = \mu_0 \sigma \frac{\partial \mathbf{E}}{\partial t} + \mu_0 \varepsilon_m \frac{\partial^2 \mathbf{E}}{\partial t^2} \tag{8.20}$$

Suppose Eq. (8.9) is a solution of this equation. We can substitute $\partial \mathbf{E}/\partial t = -i\omega \mathbf{E}$, $\partial^2 \mathbf{E}/\partial t^2 = -\omega^2 \mathbf{E}$, and $\nabla^2 \mathbf{E} = -k^2 \mathbf{E}$ into Eq. (8.20) to obtain

$$k^2 = i\omega\mu_0\sigma + \omega^2 \mu_0 \varepsilon_m \tag{8.21}$$

Therefore, the wavevector becomes complex: $\mathbf{k} = \mathbf{k}' + i\mathbf{k}''$, where $\mathbf{k}' = k'_x\hat{\mathbf{x}} + k'_y\hat{\mathbf{y}} + k'_z\hat{\mathbf{z}}$ and $\mathbf{k}'' = k''_x\hat{\mathbf{x}} + k''_y\hat{\mathbf{y}} + k''_z\hat{\mathbf{z}}$ are real vectors. Note that Eq. (8.21) tells us the value of $k^2 = \mathbf{k} \cdot \mathbf{k} = k_x^2 + k_y^2 + k_z^2$, where each wavevector component may be complex, but does not specify the individual components. The *complex dielectric function* is defined as

$$\varepsilon = \varepsilon' + i\varepsilon'' = \frac{\varepsilon_m}{\varepsilon_0} + i\frac{\sigma}{\omega\varepsilon_0} \tag{8.22}$$

For a nonmagnetic material, the *complex refractive index* $\tilde{n} = n + i\kappa$ is related to the complex dielectric function by $\varepsilon = (n + i\kappa)^2$. The imaginary part $\kappa$ of the complex refractive index is called the *extinction coefficient*. By definition, we have

$$\varepsilon' = n^2 - \kappa^2 \quad \text{and} \quad \varepsilon'' = 2n\kappa \tag{8.23}$$

The refractive index $n$ and the extinction coefficient $\kappa$ are also called *optical constants*, although none of them are constant over a large wavelength region for real materials [15]. The dielectric function is also called relative permittivity, with respect to the permittivity of vacuum $\varepsilon_0$. One can consider the $\sigma/\omega$ term in Eq. (8.22) as the imaginary part of the permittivity. Some texts used $\varepsilon = \varepsilon' - i\varepsilon''$ for the dielectric function and $\tilde{n} = n - i\kappa$ for the complex refractive index. In doing so, Eq. (8.9) must be revised to $\mathbf{E} = \mathbf{E}_0 e^{i(\omega t - \mathbf{k}\cdot\mathbf{r})}$. In either convention, $\varepsilon''$ and $\sigma$ must be nonnegative for a passive medium. Equation (8.21) can be rewritten as

$$k = \tilde{n}\omega/c_0 \tag{8.24}$$

For simplicity, we will remove the tilde and simply use $n$ for the complex refractive index, where it can be clearly understood from the context.

By substituting $\mathbf{ik}$ for $\nabla$ and $-\mathrm{i}\omega$ for $\partial/\partial t$, we can rewrite Maxwell's curl equations as

$$\mathbf{k} \times \mathbf{E} = \omega\mu_0\mathbf{H} \tag{8.25}$$

and

$$\mathbf{k} \times \mathbf{H} = -\omega\varepsilon_0\varepsilon\mathbf{E} \tag{8.26}$$

Similar to the definition of the complex dielectric function, one may choose to define a complex conductivity that satisfies Ohm's law at high frequencies, $\mathbf{J} = \tilde{\sigma}\mathbf{E}$, where

$$\tilde{\sigma} = \sigma' + \mathrm{i}\sigma'' = \sigma - \mathrm{i}\omega\varepsilon_\mathrm{m} \tag{8.27}$$

Note that we have assumed that $\sigma$ is the real part of $\tilde{\sigma}$. Therefore,

$$\sigma'' = -\omega\varepsilon_0\varepsilon' \text{ and } \varepsilon'' = \sigma'/\omega\varepsilon_0 \tag{8.28}$$

Equation (8.26) can be recast in terms of the complex conductivity as

$$\mathbf{k} \times \mathbf{H} = -\mathrm{i}\tilde{\sigma}\mathbf{E} \tag{8.29}$$

In the subsequent discussion, we will omit the tilde above $\sigma$, when the context is sufficiently clear. The complex conductivity and the complex dielectric function are related to each other. For a linear, isotropic, and homogeneous nonmagnetic material, only two frequency-dependent functions are needed to fully characterize the electromagnetic response. The function pairs often found in the literature are $(n, \kappa)$, $(\sigma', \varepsilon')$, $(\varepsilon', \varepsilon'')$, and $(\sigma', \sigma'')$. The principle of causality, which states that the effect cannot precede the cause, or no output before an input, imposes additional restrictions on the frequency dependence of the optical properties so that the real and imaginary parts are not completely independent, but related, to each other. In general, the relative permeability, which is complex and frequency dependent, can be expressed as

$$\mu = \mu' + \mathrm{i}\mu'' = \mu_\mathrm{m}/\mu_0 \tag{8.30}$$

The complex refractive index for magnetic materials should be defined as follows:

$$n = \sqrt{\varepsilon\mu} \tag{8.31}$$

The amplitude of the complex wavevector is $k = n\omega/c_0$, same as Eq. (8.24). One can verify that Eq. (8.9) is a solution of the wave equation. The relative permittivity $\varepsilon$ and permeability $\mu$ will be used to formulate the general equations, later in this chapter. In most sections of this chapter, we deal with nonmagnetic materials, such as metals, dielectrics, and semiconductors. However, we will devote the discussion of the optical properties of magnetic materials to Sect. 8.4.6, because of the emerging interest in *metamaterials*, which are synthesized materials with magnetic responses at microwave and higher frequencies (see Problem 8.6, for example).

### 8.1.6 Propagating and Evanescent Waves

In an absorbing nonmagnetic medium, the electric and magnetic fields will attenuate exponentially. As an example, consider a wave that propagates in the positive $x$-direction, with its electric field polarized in the $y$-direction. Then,

$$\mathbf{E} = \hat{\mathbf{y}} E_0 \mathrm{e}^{-\mathrm{i}(\omega t - k'x)} \mathrm{e}^{-k''x} \tag{8.32}$$

where $k' = \omega n / c_0$ and $k'' = \omega \kappa / c_0$ are the real and imaginary parts of the wavevector, respectively, that is, $\mathbf{k} = (k' + \mathrm{i}k'')\hat{\mathbf{x}}$. Equation (8.32) suggests that the amplitude of the electric field will decay exponentially according to $\mathrm{e}^{-(2\pi\kappa/\lambda)x}$. The magnetic field can be obtained from Eq. (8.25) as

$$\mathbf{H} = \hat{\mathbf{z}} \frac{n + \mathrm{i}\kappa}{\mu_0 c_0} E_0 \mathrm{e}^{-\mathrm{i}(\omega t - k'x)} \mathrm{e}^{-k''x} \tag{8.33}$$

By substituting Eqs. (8.32) and (8.33) into Eq. (8.18a), we obtain the time-averaged energy flux in the $x$-direction as

$$\langle S \rangle = \frac{n}{2\mu_0 c_0} E_0^2 \mathrm{e}^{-2k''x} = \frac{n}{2\mu_0 c_0} E_0^2 \mathrm{e}^{-a_\lambda x} \tag{8.34}$$

where $a_\lambda = 4\pi\kappa/\lambda$ is called the *absorption coefficient*. The inverse of $a$ is called the *radiation penetration depth* (or photon mean free path) given by

$$\delta_\lambda = \frac{1}{a_\lambda} = \frac{\lambda}{4\pi\kappa} \tag{8.35}$$

It is the distance through which the radiation power is attenuated by a factor of $\mathrm{e}^{-1}$ ($\approx$37%). (See Problem 8.6 for some typical values of the penetration depth in various materials at different wavelengths and temperatures.)

When $\mathbf{k}$ is complex, the plane normal to $\mathbf{k}'$ is the constant-phase plane and the plane normal to $\mathbf{k}''$ is the constant-amplitude plane because

$$\mathbf{E} = \mathbf{E}_0 \mathrm{e}^{-\mathrm{i}(\omega t - \mathbf{k}' \cdot \mathbf{r})} \mathrm{e}^{-\mathbf{k}'' \cdot \mathbf{r}} \tag{8.36}$$

When $\mathbf{k}' \times \mathbf{k}'' = 0$, the wave is said to be *homogeneous*; otherwise, the constant-phase planes will not be parallel to the constant-amplitude planes, and the wave is said to be *inhomogeneous*. A typical homogeneous wave is given in Eq. (8.32). Now let us

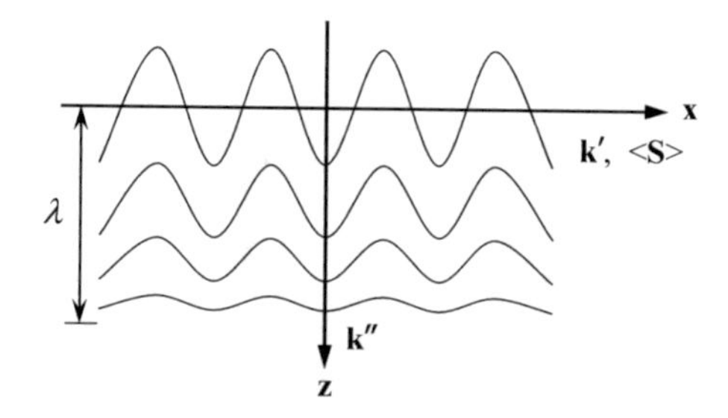


**Fig. 8.3** Schematic of an evanescent wave near the $z = 0$ surface

use an example to illustrate an inhomogeneous wave. Consider a wave, defined in the $z \geq 0$ half plane filled with vacuum, with a wavevector $\mathbf{k} = 2\omega/c_0\hat{\mathbf{x}} + \mathrm{i}\sqrt{3}\omega/c_0\hat{\mathbf{z}}$. The electric field is linearly polarized in the $y$-direction; thus, $\mathbf{E} = \hat{\mathbf{y}}E_0\mathrm{e}^{-\mathrm{i}(\omega t - \mathbf{k}\cdot\mathbf{r})}$. It can be shown that $\mathbf{k}\cdot\mathbf{k} = k^2 = \omega^2/c_0^2$; hence, $\mathbf{k}$ is indeed a valid wavevector in vacuum. The electric field can be written as

$$\mathbf{E} = \hat{\mathbf{y}}E_0\mathrm{e}^{-\mathrm{i}(\omega t - k_x x)}\mathrm{e}^{-\eta z} \tag{8.37}$$

Here, $k_x = 2k = 4\pi/\lambda$, and $\eta = \mathrm{Im}(k_z) = 2\sqrt{3}\pi/\lambda$. Clearly, the wave has a constant phase for any constant-$x$ plane and a constant amplitude for any constant-$z$ plane. Furthermore, the amplitude decays exponentially toward the positive $z$-direction and becomes negligible, when $z > \lambda$, as shown schematically in Fig. 8.3. Such a wave is called an *evanescent wave*, which exists in waveguides and is important for near-field optics and nanoscale radiation heat transfer. It can be shown that the time-averaged Poynting vector is parallel to the $x$-direction so that no energy is transported toward the $z$-direction (see Problem 8.7).

## 8.2 Blackbody Radiation: The Photon Gas

This section deals with Planck's law of blackbody radiation, which is the foundation of far-field radiation heat transfer analysis. After the discussion of radiation thermometry and radiance temperature, we will study radiation (or photon) entropy and pressure. Photon entropy and exergy may be important for analyzing and designing advanced energy harvesting systems from solar radiation to near-field thermal radiative devices. The limitation of Planck's law is also addressed.

### *8.2.1 Planck's Law*

Consider an enclosure of volume $V$, whose walls are at a uniform temperature $T$, as shown in Fig. 8.4a. The enclosure may contain a medium (such as a molecular gas) or

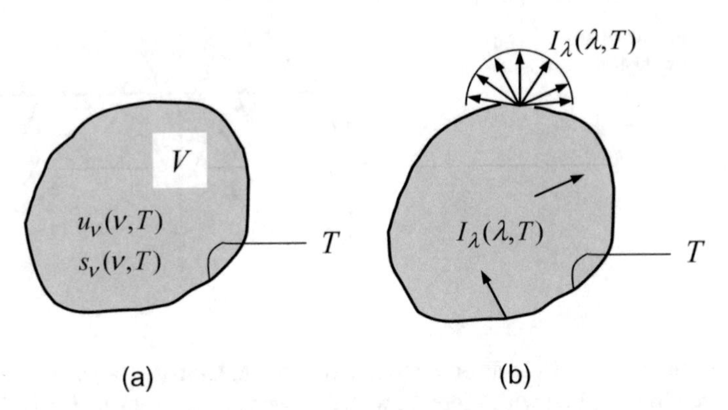


**Fig. 8.4** An isothermal enclosure (blackbody cavity): **a** without an opening and **b** with a small opening on the wall that has little effect on the equilibrium distribution

may be evacuated (vacuum). Inside the enclosure, there exist electromagnetic fields, which may be viewed either as many transverse waves at different frequencies or as a large number of quanta with different energies. The particle theory treats radiation as a collection of photons. The energy and the momentum of each photon are related to the frequency and the speed of light, by $\varepsilon = h\nu$ and $p = h\nu/c$, respectively. We are interested in finding the equilibrium distribution of photons with respect to photon energy or frequency or momentum. Photons obey Bose–Einstein statistics, without requiring the total number be conserved. The number of photons $\mathrm{d}N$ in a frequency interval from $\nu$ to $\nu + \mathrm{d}\nu$ per unit volume is equal to the *mean occupation number* multiplied by the number of quantum states (degeneracy):

$$\mathrm{d}N = f_{\mathrm{BE}}(\nu)\mathrm{d}g = \frac{\mathrm{d}g}{\mathrm{e}^{h\nu/k_{\mathrm{B}}T} - 1} \tag{8.38}$$

The quantum states in the phase space, consisting of a volume $V$ and a spherical shell in the momentum space (from $p$ to $p + \mathrm{d}p$), are given by $\mathrm{d}g = 2V(4\pi p^2 \mathrm{d}p)/h^3$, where the factor 2 accounts for the two polarization states of electromagnetic waves. Thus, we can write the density of states (DOS), which is the number of quantum states per unit volume per unit frequency interval, as

$$D(\nu) = \frac{1}{V}\frac{\mathrm{d}g}{\mathrm{d}\nu} = \frac{8\pi\nu^2}{c^3} \tag{8.39}$$

Notice that $c$ is the speed of light in the medium and it is assumed that the refractive index of the medium is real and independent of the frequency. Because the Bose–Einstein distribution function gives the mean occupation number of each quantum state, the number of photons per unit volume per unit frequency interval is

$$f(\nu) = \frac{1}{V}\frac{\mathrm{d}N}{\mathrm{d}\nu} = f_{\mathrm{BE}}(\nu)D(\nu) = \frac{8\pi\nu^2}{c^3(\mathrm{e}^{h\nu/k_{\mathrm{B}}T} - 1)} \tag{8.40}$$

Integrating the above equation over all frequencies yields the total number of photons at a given temperature per unit volume. Clearly, the number of photons is not conserved in a blackbody cavity with a fixed volume $V$.

Since the energy of a photon is $h\nu$, the spectral energy density (energy per unit volume per unit frequency interval) at a fixed temperature $T$ can be written as

$$u_\nu(\nu) = h\nu f(\nu) = \frac{8\pi h\nu^3}{c^3(\mathrm{e}^{h\nu/k_\mathrm{B}T} - 1)} \tag{8.41}$$

For an area element inside the enclosure, the radiant energy flux is related to the energy density and the speed of light by

$$q''_{rad,\nu} = \frac{u_\nu c}{4} \tag{8.42}$$

If a blackbody is placed inside the enclosure, it will absorb all incoming radiant energy that reaches its surface; at thermal equilibrium, it must emit the same amount of energy. After substituting Eq. (8.41) into Eq. (8.42), we obtain the *spectral emissive power* of a blackbody as a function of frequency and temperature as

$$e_{\mathrm{b},\nu}(\nu, T) = \frac{2\pi h\nu^3}{c^2(\mathrm{e}^{h\nu/k_\mathrm{B}T} - 1)} \tag{8.43}$$

Note that the spectral emissive power is the power emitted per unit area per frequency or wavelength interval. To express the spectral emissive power in terms of the wavelength (in vacuum), we can substitute $c = c_0/n$, $\nu = c_0/\lambda$, $\mathrm{d}\nu = -c_0\mathrm{d}\lambda/\lambda^2$, and $e_{\mathrm{b},\nu}\mathrm{d}\nu = -e_{\mathrm{b},\lambda}\mathrm{d}\lambda$ into Eq. (8.43). Therefore,

$$e_{\mathrm{b},\lambda}(\lambda, T) = \frac{2\pi h c_0^2 n^2}{\lambda^5(\mathrm{e}^{hc/k_\mathrm{B}\lambda T} - 1)} = \frac{C_1 n^2}{\lambda^5(\mathrm{e}^{C_2/\lambda T} - 1)} \tag{8.44}$$

where $C_1 = 3.742 \times 10^8$ W m$^{-2}$ μm$^4$ and $C_2 = 1.439 \times 10^4$ μm K [10] are called the first and second radiation constants. Equations (8.43) and (8.44) are called *Planck's law* or *Planck's distribution* (of blackbody radiation) in terms of the frequency and wavelength, respectively. It should be noted that the blackbody intensity is $I_{\mathrm{b},\lambda}(\lambda, T) = e_{\mathrm{b},\lambda}(\lambda, T)/\pi$, as in Eq. (2.48), and isotropic inside the whole cavity regardless of the radiative properties of the wall. Furthermore, when there is a small opening, the emitted radiation is diffuse and obeys the blackbody distribution, as shown in Fig. 8.4b. The requirement is that the opening should be sufficiently small compared with the size of the enclosure, but large enough compared to the wavelengths of interest. The concept of blackbody cavity was made clear by Wien in his 1911 Nobel lecture, as seen from the excerpt below:

> … there must exist, in a cavity surrounded by bodies of equal temperature, a radiation energy that is independent of the nature of the bodies. If in the walls surrounding this cavity a small aperture is made through which radiation issues, we obtain a radiation which is independent

> of the nature of the emitting body, and is wholly determined by the temperature. The same radiation would also be emitted by a body which does not reflect any rays and which is therefore designated as completely black, and this radiation is called the radiation of a black body or blackbody radiation.

It should be noted that if the refractive index is a weak function of wavelength and absorption by the medium is negligible, $c^3$ in Eqs. (8.40) and (8.41) should be replaced by $c^2 v_g$ where $v_g$ is the group velocity defined in Eq. (8.13). The group velocity $v_g$ should also be used in Eq. (8.42) to replace $c$. Nevertheless, the expressions of the emissive power given in Eqs. (8.43) and (8.44) remain the same. In the following, $n = 1$ or vacuum is assumed for Planck's distribution unless otherwise indicated.

Equation (8.43) or (8.44) can be integrated over the whole spectrum to obtain the Stefan–Boltzmann law: $e_b = \sigma_{SB} T^4$. In Fig. 8.5, $e_{b,\lambda}/\sigma_{SB} T^5$ is plotted as a function of $\lambda T$ so that the area under Planck's distribution (solid curve) is $\int_0^\infty \frac{e_{b,\lambda}(\lambda,T)}{\sigma_{SB} T^5} d(\lambda T) = \frac{1}{\sigma_{SB} T^4} \int_0^\infty e_{b,\lambda}(\lambda, T) d\lambda = 1$. The Planck's distribution has a peak and approaches zero at extremely short and long wavelengths. If $C_2/\lambda T \gg 1$, the right-hand side of Eq. (8.44) can be approximated by $C_1 \lambda^{-5} e^{-C_2/\lambda T}$. This is called Wien's formula, which gives good approximation, even beyond the maximum emissive power, as can be seen from Fig. 8.5. At very long wavelengths, Wien's formula underpredicts the emissive power and asymptotically approaches to $C_1 \lambda^{-5}$, suggesting that the emissive power is independent of temperature. Note that the right-hand side of Eq. (8.44) approaches $C_1 T/(C_2 \lambda^4)$ if $C_2/\lambda T \ll 1$, since $e^x - 1 \approx x$ for $x \ll 1$. This is called the Rayleigh–Jeans formula, which is applicable at very long wavelengths, as shown in Fig. 8.5. The significance of the Rayleigh–Jeans formula is that it correctly predicts the temperature dependence of the blackbody spectrum, at very long wavelengths, where Wien's formula fails. The failure of the Rayleigh–Jeans formula at short wavelengths is called the *ultraviolet catastrophe*. The significance of Planck's formula is more than a unified mathematical formulation. It was derived based on the hypothesis of energy quanta that do not exist in classical Newtonian mechanics or Maxwell's electrodynamics. It should be

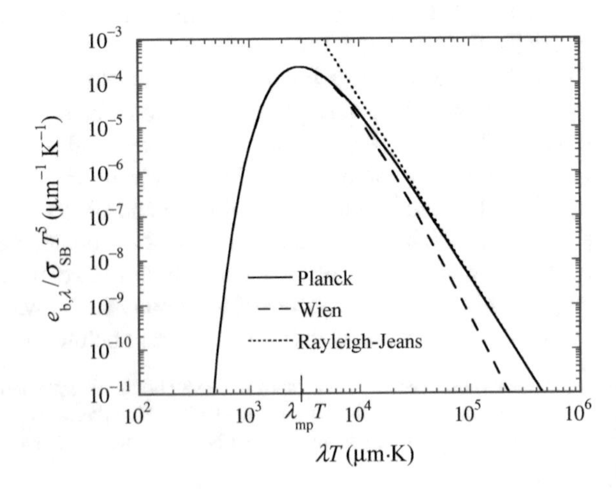


**Fig. 8.5** Planck's law for blackbody emissive power

noted that the preceding derivation is based on statistical thermodynamics, presented in Chap. 3, rather than on Planck's original semi-classical oscillator model [4].

**Example 8.2** Find the wavelength $\lambda_{\text{mp}}$ at which Planck's distribution reaches a maximum. What is the ratio of the energy emitted at $\lambda < \lambda_{\text{mp}}$ to that at $\lambda > \lambda_{\text{mp}}$?

**Solution** By setting the derivative of Eq. (8.44) equal to zero, i.e., $\mathrm{d}e_{\text{b},\lambda}/\mathrm{d}\lambda = 0$, we have

$$\frac{hc}{k_{\text{B}}\lambda T} + 5\exp\left(-\frac{hc}{k_{\text{B}}\lambda T}\right) - 5 = 0$$

This equation can be solved by iteration or numerically to yield

$$\lambda_{\text{mp}}\ [\mu\text{m}] = \frac{2898\ \mu\text{m K}}{T\ [\text{K}]} \tag{8.45}$$

This is *Wein's displacement law*. The location of $\lambda_{\text{mp}}$ is also marked on Fig. 8.5. To find out the ratio of the energy emitted at $\lambda < \lambda_{\text{mp}}$ to that at $\lambda > \lambda_{\text{mp}}$, we can numerically evaluate $\int_0^{\lambda_{\text{mp}}} e_{\text{b},\lambda}(\lambda, T)\mathrm{d}\lambda / \int_{\lambda_{\text{mp}}}^{\infty} e_{\text{b},\lambda}(\lambda, T)\mathrm{d}\lambda$. The numerical result is approximately 1:3 and independent of temperature.

**Example 8.3** Assuming the sun to be a blackbody at 5800 K, calculate the emissive power at the following wavelength intervals: $\lambda < 0.3\ \mu\text{m}$, $0.3\ \mu\text{m} < \lambda < 0.4\ \mu\text{m}$, $0.4\ \mu\text{m} < \lambda < 0.7\ \mu\text{m}$, $0.7\ \mu\text{m} < \lambda < 3\ \mu\text{m}$, and $\lambda > 3\ \mu\text{m}$. Neglect the absorption by the atmosphere. What is the radiant power arriving at the earth's surface from the sun?

**Solution** The total emissive power is $\sigma_{\text{SB}} T_{\text{sun}}^4 = 5.67 \times 10^{-8} \times 5800^4 \approx 64\ \text{MW/m}^2$. We can obtain the emissive power in each spectral region by integrating Eq. (8.44), as listed in the following table:

| $\lambda$ ($\mu$m) | <0.3 | 0.3–0.4 | 0.4–0.7 | 0.7–3 | >3 | Total |
|---|---|---|---|---|---|---|
| $\lambda_2 T$ ($\mu$mK) | 1740 | 2320 | 4060 | 17400 | $\infty$ | – |
| $F_{0\to\lambda_2}$ | 0.03 | 0.12 | 0.49 | 0.98 | 1 | – |
| $F_{\lambda_1\to\lambda_2}$ | 0.03 | 0.09 | 0.37 | 0.49 | 0.02 | 1 |
| $\Delta E_b$ (MW/m$^2$) | 1.9 | 5.8 | 23.7 | 31.4 | 1.3 | 64.1 |

Note that $F_{\lambda_1\to\lambda_2}$ represents the fraction of radiation falling between $\lambda_1$ and $\lambda_2$. The total power emitted by the sun equals the emissive power multiplied by the surface area of the sun. The fraction of the power that reaches the earth equals the solid angle of the earth divided by $4\pi$. Note that the radius of the sun $r_{\text{sun}} = 6.955 \times 10^8$ m, the radius of the earth $r_{\text{earth}} = 6.378 \times 10^6$ m, and the earth–sun distance $R_{\text{earth-sun}} = 1.496 \times 10^{11}$ m. Therefore, the total power that will reach the earth's surface, if the

**Table 8.1** Spectral regions expressed in different units

| | Wavelength $\lambda$ (μm) | Wavenumber $\bar{\nu}$ (cm$^{-1}$) | Frequency $\nu$ (THz) | Angular frequency $\omega$ ($10^{14}$ rad/s) | Photon energy $E$ (eV) |
|---|---|---|---|---|---|
| UV | 0.01–0.38 | $(10-0.26)\times10^5$ | 30,000–790 | 1900–50 | 120–3.3 |
| VIS | 0.38–0.76 | $(2.6-1.3)\times10^4$ | 790–400 | 50–25 | 3.3–1.6 |
| NIR | 0.76–2.5 | $(1.3-0.4)\times10^4$ | 400–120 | 25–7.5 | 1.6–0.5 |
| MIR | 2.5–25 | 4000–400 | 120–12 | 7.5–0.75 | 0.5–0.05 |
| FIR | 25–1000 | 400–10 | 12–0.3 | 0.75–0.019 | 0.05–0.0012 |
| MW | $10^3$–$10^5$ | 10–0.1 | 0.3–0.003 | $(19–0.19)\times10^{-3}$ | $(12–0.12)\times10^{-4}$ |

absorption by the atmosphere is neglected, is

$$\dot{Q} = 4\pi r_{\text{sun}}^2 \cdot \sigma_{\text{SB}} T_{\text{sun}}^4 \cdot \frac{\pi r_{\text{earth}}^2}{4\pi R_{\text{earth-sun}}^2} \approx 1.8\times10^{17}\ \text{W}$$

The average irradiation on the earth is: $G = \dot{Q}/\pi r_{\text{earth}}^2 \approx 1377$ W/m$^2$. This value is very close to the total solar irradiance (TSI), measured outside the earth's atmosphere.

Because of the broad spectral region of electromagnetic waves, alternative units are often used, such as wavelength $\lambda$ (in vacuum), wavenumber $\bar{\nu} = 1/\lambda$, frequency $\nu = c_0/\lambda$, angular frequency $\omega = 2\pi\nu$, and photon energy $E = h\nu$. Generally speaking, optical radiation covers the spectral region including ultraviolet (UV), visible (VIS), near infrared (NIR), mid infrared (MIR), and far infrared (FIR). Table 8.1 outlines the subdivisions of the spectral region in different units from ultraviolet (UV) to microwave (MW). Note that $\bar{\nu}$ [cm$^{-1}$] $= 10{,}000/\lambda$ [μm] and $E$ [eV] $= 1.24/\lambda$ [μm]. Thermal radiation covers part of the UV from $\lambda = 0.1$ μm through some of the MW region.

### *8.2.2 Radiation Thermometry*

The developments of the absolute temperature scale and radiation thermometry are among the most important applications of blackbody radiation [16]. The Stefan–Boltzmann law $e_b = \sigma_{\text{SB}} T^4$ defines an absolute thermodynamic temperature, which is consistent with the one defined by the ideal gas law and the Carnot cycle. While radiation thermometry can serve as a primary standard, most practical radiation thermometers are not absolute instruments because of other considerations such as fast response, easy operation, and low cost. High-temperature furnaces are commonly used as calibration standards. The cavity is a hollow cylinder, made of graphite for example, with a conical ending and a small aperture. The most accurate calibration source is the fixed-point heat pipe blackbody, for which a pure metal is melted outside the graphite cylinder to maintain a constant temperature in a two-phase state. The freezing temperatures are then used to define the temperature scales (1234.93 K for Ag, 1337.33 for Au, and 1357.77 K for Cu).

To measure the absolute temperature of a thermally radiative body, two blackbody cavities at different temperatures would be needed: one serves as the emitter (blackbody source) and the other as the receiver (radiometer). Quinn and Martin [17] used a blackbody source and a cryogenic radiometer to directly determine the thermodynamic temperatures and measure the Stefan–Boltzmann constant. The experimentally obtained Stefan–Boltzmann constant was $(5.66967 \pm 0.00076) \times 10^{-8}\ \mathrm{Wm^{-2}K^{-4}}$. The difference is 0.13% of the theoretical value $(5.67040 \pm 0.00004) \times 10^{-8}\ \mathrm{Wm^{-2}K^{-4}}$, based on Planck's constant, Boltzmann's constant, and the speed of light. Since the early 1990s, NIST has developed a high-accuracy cryogenic radiometer (HACR) facility to serve as the primary standard for optical radiation measurements. A schematic of the original HACR receiver is shown in Fig. 8.6. The receiver is mounted at the bottom of a liquid helium cryostat in an evacuated chamber, and the optical access is through a Brewster window below the cavity. The HACR facility has gone through some major upgrades in recent years. The receiver cavity is made of copper with a high thermal conductivity and low specific heat at cryogenic temperatures. The inner wall of the cavity is coated with a specular black paint to absorb the incident radiation with an effective absorptance greater than 99.998%. The electrical-substitution technique links the radiant power to the electric power to achieve an overall uncertainty within 0.02% for optical power measurements. Detailed descriptions can be found from Pearson and Zhang [18] and references therein. The cosmic radiation background (in the far-infrared and microwave region), measured with cryogenic bolometers, can be fitted to the blackbody distribution at 2.7 K; this is the temperature of the universe at the present time. The discovery of cosmic radiation background in 1964 and the subsequent measurements and theoretical studies have been recognized by the Nobel Prizes in Physics to Arno Penzias and Robert Wilson in 1978 and to John Mather and George Smoot in 2006.

Most radiation thermometers are based on spectral measurements rather than on the measurement of the total irradiance from the target. When a radiation thermometer is used to measure the temperature of a real surface, the unknown emittance of the surface and the influence of the surrounding radiation are the major issues that affect the measurement. Various methods have been developed to deal with these problems, including the creation of a blackbody cavity on the surface, the two-color method, and the use of a controlled reference source. The development of optical fibers has allowed radiometric temperature measurements for surface locations that are otherwise inaccessible by imaging radiometers. The detailed theory and practice of radiation thermometry can be found from the two book volumes compiled by Zhang et al. [19]. A brief discussion of the basic operational principles of spectral radiation thermometry is given in the following.

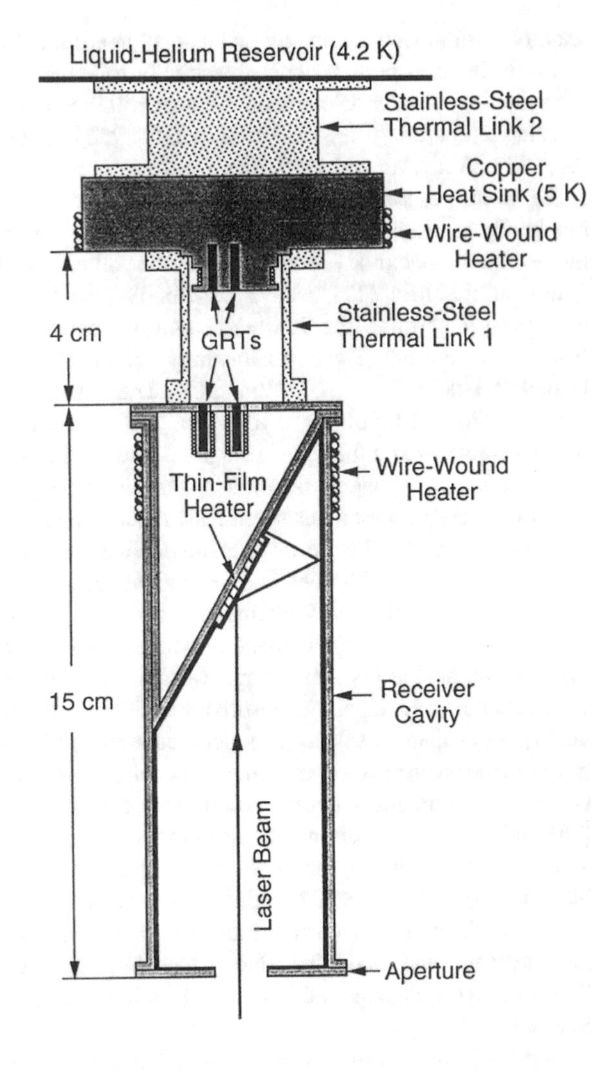


**Fig. 8.6** Schematic of the receiver cavity of an absolute cryogenic radiometer, where GRT stands for germanium resistance thermometer, from Pearson and Zhang [18]

The *measurement equation* of a spectral radiation thermometer can be approximated as follows:

$$V_d = C_I I_{ex,\lambda}(\lambda) \tag{8.46}$$

where $V_d$ is the detector output signal and $C_I$ is an instrument constant that is independent of the target material and temperature. The term $I_{ex,\lambda}(\lambda)$ is called the *exitent spectral radiance*, which includes the radiation emitted by the target and the surroundings, as well as that reflected by the target. The *radiance temperature* $T_\lambda$ (also called the *brightness temperature*) is defined according to

$$I_{b,\lambda}(\lambda, T_\lambda) = I_{ex,\lambda}(\lambda) \tag{8.47}$$

where $I_{b,\lambda}(\lambda, T_\lambda)$ is the blackbody intensity at the wavelength $\lambda$ and temperature $T_\lambda$. If the surrounding emission and absorption can be neglected, the exitent spectral radiance is due only to the emission; therefore,

$$I_{\mathrm{ex},\lambda}(\lambda) = I_{\mathrm{e},\lambda}(\lambda, T) = \varepsilon'_\lambda I_{\mathrm{b},\lambda}(\lambda, T) \tag{8.48}$$

where $\varepsilon'_\lambda$ is the directional-spectral emittance, and $I_{\mathrm{e},\lambda}(\lambda, T)$ is the intensity emitted by the target. By combining Eqs. (8.47) and (8.48) and applying Wien's formula, the surface temperature is related to the radiance temperature by

$$\frac{1}{T} = \frac{1}{T_\lambda} + \frac{\lambda}{C_2} \ln \varepsilon'_\lambda \tag{8.49}$$

The uncertainty in the measured temperature due to an uncertainty in the emittance is

$$\frac{\delta T}{T} = -\frac{\lambda T}{C_2} \frac{\delta \varepsilon'_\lambda}{\varepsilon'_\lambda} \tag{8.50}$$

The impact of emittance on the temperature measurement decreases as $\lambda$ decreases. Equation (8.50) suggests that it may be advantageous to choose a wavelength that is somewhat shorter than the wavelength at which $I_{\mathrm{b},\lambda}(\lambda, T)$ is a maximum as given by Wien's displacement law. If the surrounding radiation is not negligible, $I_{\mathrm{ex},\lambda}(\lambda)$ is the sum of the emitted and reflected spectral radiances. In practice, when choosing the operating wavelength, one should also consider the material's properties and the effect of surrounding radiation, as well as the detector availability and sensitivity. Hence, the choice of a radiation thermometer requires a detailed analysis of different effects in the actual measurements.

**Example 8.4** Rapid thermal processing is a semiconductor single-wafer manufacturing technique. A lightpipe radiation thermometer, operated at $\lambda = 0.95$ μm, is used to measure the temperature of the wafer. The emittance or emissivity of a plain silicon wafer is approximately 0.7 at this wavelength. Neglect the reflected radiation from the wafer. If the wafer is at a temperature of 1200 K, what is the radiance temperature? If the temperature needs to be determined within an uncertainty of 1 K, how much tolerance in the emittance error is acceptable?

**Solution** From Eq. (8.49), $T_\lambda \approx 1167$ K, which differs from the actual temperature by approximately 33 K. One can also solve Eqs. (8.47) and (8.48), using Planck's law, and the result is essentially the same. Based on Eq. (8.50), to obtain a temperature within 1 K, the emittance must be determined within an uncertainty of $\delta\varepsilon'_\lambda = 0.0074$. Zhou et al. [20] developed a model to predict the effective emittance of silicon wafers in rapid thermal processing furnaces and showed that, by using a reflective cavity, the temperature measurement uncertainty can be significantly reduced.

### 8.2.3 Radiation Pressure and Photon Entropy

Like other particles, photon gas also has the property of entropy and can be related to other properties in equilibrium states. Express the energy density in an enclosure of volume $V$, at thermodynamic equilibrium, with a temperature $T$ as $u = \frac{U}{V} = \frac{4}{c}\sigma_{\mathrm{SB}}T^4$. It can be seen that the specific heat at constant volume is $c_v = \left(\frac{\partial u}{\partial T}\right)_V = \frac{16}{c}\sigma_{\mathrm{SB}}T^3$. The radiation entropy or photon entropy can therefore be obtained as

$$S = \int_0^T Vc_v \frac{\mathrm{d}T}{T} = \frac{16}{3c}V\sigma_{\mathrm{SB}}T^3 \tag{8.51a}$$

or

$$s = \frac{16}{3c}\sigma_{\mathrm{SB}}T^3 \tag{8.51b}$$

Note that $T = \left(\frac{\partial U}{\partial S}\right)_V$ is satisfied. The Helmholtz free energy $A = U - TS = -\frac{4}{3c}V\sigma_{\mathrm{SB}}T^4$. Thus, the radiation pressure is

$$P = -\left(\frac{\partial A}{\partial V}\right)_T = \frac{4}{3c}\sigma_{\mathrm{SB}}T^4 \tag{8.52}$$

The force from radiation pressure, albeit small, has some important applications in trapping and manipulating atomic to molecular particles. This technique is called optical traps or optical tweezers; see Lang and Block [21] for a bibliographical review. Arthur Ashkin shared the Nobel Prize in Physics in 2018 "for the optical tweezers and their application to biological systems." Another way to view radiation pressure is that photons or electromagnetic waves carry both energy and momentum. The interaction of electromagnetic waves or photons with matter therefore involves a change of momentum, resulting in a pressure on the object. Radiation pressure was first predicted by J. C. Maxwell in 1873 and experimentally demonstrated by Russian physicist P. Lebedev in 1900 and American physicists E. F. Nichols and G. F. Hull in 1901. Solar radiation pressure plays a role in the formation of a comet's dust tail. Radiation pressure is also important in cosmology concerning the formation and evaluation of the stars and galaxies.

If each photon mode (frequency) is individually considered, the *spectral entropy density* for unpolarized radiation can be expressed as follows [4]:

$$s_\nu(\nu, T) = \frac{8\pi k_{\mathrm{B}}\nu^2}{c^3}\left[\frac{x}{\mathrm{e}^x - 1} + \ln\left(\frac{\mathrm{e}^x}{\mathrm{e}^x - 1}\right)\right] \tag{8.53}$$

where $x = \frac{h\nu}{k_{\mathrm{B}}T}$. Note that $\frac{1}{T} = \left(\frac{\partial s_\nu}{\partial u_\nu}\right)_\nu = \frac{k_{\mathrm{B}}}{h\nu}\ln\left(1 + \frac{8\pi h\nu^3}{c^3 u_\nu}\right)$, which is consistent with Eq. (8.41). Similar to the energy flux (emissive power) and intensity, the *radiation entropy flux* can be obtained by multiplying a factor $c/4$ to Eqs. (8.51b) and (8.53), and the *radiation entropy intensity* can be obtained by dividing the flux by $\pi$, because of the isotropic nature of blackbody radiation. Clearly, electromagnetic radiation carries both energy and entropy.

**Example 8.5** Consider the radiation heat transfer between two parallel plates at $T_1$ and $T_2$, respectively. Assume each plate has an area of $A$ and both plates are blackbodies. The separation distance is much smaller than $\sqrt{A}$ but much greater than the wavelength of thermal radiation.

(a) How much entropy is generated at each plate? Evaluate the ratio of entropy generation assuming that $T_1 = 2T_2$.
(b) If a thermophotovoltaic receiver is mounted on the lower temperature side to convert thermal radiative energy to electricity (work), what is its maximum achievable efficiency?

**Solution**

(a) The net energy flow from plate 1 to 2 is $\dot{Q}_{12} = A\sigma_{SB}(T_1^4 - T_2^4)$. The entropy of plate 1 will decrease at the rate of $\mathrm{d}S_1/\mathrm{d}t = -\dot{Q}_{12}/T_1$, and the entropy of plate 2 will increase at the rate of $\mathrm{d}S_2/\mathrm{d}t = \dot{Q}_{12}/T_2$. On the other hand, the net entropy flow from plate 1 to 2 can be calculated as $\dot{S}_{12} = \frac{4}{3}A\sigma_{SB}\left(T_1^3 - T_2^3\right)$. Therefore, $\dot{S}_{\mathrm{gen},1} = -\frac{\dot{Q}_{12}}{T_1} + \dot{S}_{12} = A\sigma_{SB}\left(\frac{1}{3}T_1^3 - \frac{4}{3}T_2^3 + \frac{T_2^4}{T_1}\right)$, $\dot{S}_{\mathrm{gen},2} = A\sigma_{SB}\left(\frac{1}{3}T_2^3 - \frac{4}{3}T_1^3 + \frac{T_1^4}{T_2}\right)$. The combined total entropy generation is equal to $\dot{Q}_{12}\left(\frac{1}{T_2} - \frac{1}{T_1}\right)$, as expected. It can be shown that the entropy generation at each plate is always greater than zero if $T_1 \neq T_2$, or equal to zero if $T_1 = T_2$. When $T_1 = 2T_2$, the entropy generation by plate 1 is about one-quarter and that by plate 2 is about three-quarters of the total entropy generated.
(b) The available energy or exergy of thermal radiation is defined as the maximum work that can be produced by a system with respect to a large reservoir. In the present example, we may assume that the reservoir is at the same temperature as $T_2$. Suppose an amount of heat is taken from the high-temperature plate. We would like to find out the maximum work that can possibly be produced. Let us consider a reversible heat engine at $T_2$. The radiative energy leaving surface 1 can still be described by $\dot{Q}_1 = A\sigma_{SB}(T_1^4 - T_2^4)$, and the entropy leaving surface 1 is $\dot{S}_1 = \frac{4}{3}A\sigma_{SB}\left(T_1^3 - T_2^3\right)$. Therefore, the entropy generation in plate 1 cannot be eliminated. In other words, it is impossible to achieve the Carnot efficiency of $\eta_{\mathrm{Carnot}} = 1 - T_2/T_1$. The maximum work can be obtained when the irreversibility at the lower temperature plate is negligible and the heat engine is also reversible. It can easily be shown that the maximum work $\dot{W}_{\max} = \dot{Q}_1 - T_2\dot{S}_1$, and the optimal efficiency is given by

$$\eta_{\mathrm{opt}} = \frac{\dot{W}_{\max}}{\dot{Q}_1} = 1 - \frac{4(1 + y + y^2)}{3(1 + y)(1 + y^2)} \tag{8.54a}$$

where $y = T_1/T_2 \geq 0$. When $y = 2$, we obtain an optimal efficiency $\eta_{\mathrm{opt}} = 37.8\%$, which is less than the Carnot efficiency of 50%, because of the unrecoverable irreversibility at plate 1.

Consider a black receiver on the earth's surface that converts solar radiation to electricity. Since the incoming radiation is from a narrow solid angle, the emitted radiation can be assumed to be in equilibrium with the surroundings at $T_0$. Therefore, the received radiant power and entropy flux is $\dot{Q}_1 = \phi A \sigma_{SB} T_s^4$ and $\dot{S}_1 = \frac{4}{3}\phi A \sigma_{SB} T_s^3$, where $T_s$ is the temperature of the sun and $\phi$ is a fraction accounting for the view angle and atmospheric transmittance (neglecting the scattering effect). Assuming a power $\dot{W}$ is developed, the heat transferred to the surroundings is $\dot{Q}_0 = \dot{Q}_1 - \dot{W}$ and the entropy transferred to the surroundings is $\dot{Q}_0/T_0$. In a reversible energy conversion device, the entropy generation must be zero and the maximum efficiency is obtained as

$$\eta_{opt} = \frac{\dot{W}_{max}}{\dot{Q}_1} = 1 - \frac{4}{3}\frac{T_0}{T_s} \tag{8.54b}$$

Different formulas on the optimal efficiency exist for solar energy conversion devices due to the different model assumptions used. For example, we can set $T_1 = T_s$ and $T_2 = T_0$ in Eq. (8.54a) and change the denominator $\dot{Q}_1$ to the absorbed solar radiation $\dot{Q}_{1,s} = A\sigma_{SB}T_1^4$. Then one would obtain Petela's formula [22]:

$$\eta_{opt} = \frac{\dot{W}_{max}}{\dot{Q}_{1,s}} = 1 - \frac{4}{3}\frac{T_0}{T_s} + \frac{1}{3}\left(\frac{T_0}{T_s}\right)^4 \tag{8.54c}$$

Since $T_0/T_s$ is about one-twentieth, the difference between Eqs. (8.54b) and (8.54c) is practically negligible. A comprehensive discussion on energy conversion efficiency can be found from the review of Landsberg and Tonge [23]. This topic is of contemporary interest especially when dealing with near-field radiative energy conversion devices [24].

The next question one may ask is whether temperature can be defined for laser radiation. The answer is *yes,* and the temperature for high-intensity lasers can be very high. An intuitive guess is to define the temperature, based on the intensity $I_\nu$ of the laser or the monochromatic radiation, by setting $I_\nu = I_{b,\nu}(\nu, T_\nu)$. The definitions of entropy and thermodynamic temperature for optical radiation are very important for analyzing optical energy conversion systems, such as solar cells, thermophotovoltaic generators, luminescence devices, and laser cooling apparatus [25, 26]. Assume that the monochromatic radiation is from a thermodynamic equilibrium state, such as a resonance cavity that allows only a single mode to exist. The spectral entropy intensity of unpolarized radiation can be written as follows [4, 25]:

$$L_\nu = \frac{2k_B\nu^2}{c^2}\left[\left(1 + \frac{c^2 I_\nu}{2h\nu^3}\right)\ln\left(1 + \frac{c^2 I_\nu}{2h\nu^3}\right) - \frac{c^2 I_\nu}{2h\nu^3}\ln\left(\frac{c^2 I_\nu}{2h\nu^3}\right)\right] \tag{8.55}$$

Thermodynamically, the *monochromatic radiation temperature* can be defined as $\frac{1}{T_\nu(\nu)} = \left(\frac{\partial L_\nu}{\partial I_\nu}\right)_\nu$ and given as

$$\frac{1}{T_\nu(\nu)} = \left(\frac{\partial L_\nu}{\partial I_\nu}\right)_\nu = \frac{k_B}{h\nu}\ln\left(1 + \frac{2h\nu^3}{c^2 I_\nu}\right) \tag{8.56}$$

This is indeed Planck's distribution of intensity at the same temperature. The expressions can be modified for polarized radiation. When the energy intensity is very high, Eq. (8.56) approaches $T_\nu(\nu) = \frac{c^2 I_\nu}{2k_B\nu^2}$, which is in the Rayleigh–Jeans limit. The radiation temperature will be proportional to the intensity of the monochromatic radiation and can exceed $10^{10}$ K, with a 1–mW He-Ne laser at 632.8 nm wavelength [26]. Therefore, for lasers with a moderate intensity, $T_\nu$ tends to be so high that the entropy is nearly zero; hence, the laser power can be considered as "work." If a collimated beam is randomly scattered by a rough surface, the scattered radiation will have a much lower intensity because of the increase in the solid angle. The process is accompanied with an entropy increase and is thus irreversible. It is not possible to increase the intensity of the scattered light, back to its original intensity, without leaving any net effect on the environment. On the other hand, if a nearly collimated light is split into two beams with a beamsplitter, the transmitted and reflected beams can interfere with each other to reconstruct the original beam. This process is reversible because the two beams are *correlated*. The correlated beams have lower entropy than those with the same intensity at thermodynamic equilibrium. The concept of temperature is applicable only if the maximum entropy state has been reached [25]. While the definition of the monochromatic radiation temperature is similar to that of the radiance temperature, the physical significance is somewhat different. In the definition of radiance temperature, the concepts of entropy and thermodynamic equilibrium do not enter into consideration.

Consider a gray-diffuse body, for which the emissive power is proportional to the blackbody emissive power, at any frequency and angle of emission. The monochromatic temperature calculated from Eq. (8.56), however, is frequency dependent. This is because the emitted radiation, as a whole, cannot be considered as a blackbody at any temperature. Thermal radiation of this type has been called *dilute blackbody radiation* [23]. This simple example shows that photons at any given frequency can be considered as in a thermodynamic equilibrium but not necessarily in equilibrium with photons at other frequencies. When radiation has two linear polarizations with different intensities, the monochromatic temperatures will be different, even for the two polarizations. In general, it is a function of frequency, direction, and polarization. The requirement is that each subsystem be in a thermodynamic equilibrium, even though it is not in equilibrium with other subsystems at the same spatial location. Photons at different frequencies, with different polarization states, or propagating in different directions, can coexist in their own equilibrium states without any interaction with each other. The concept may be called partial equilibrium, as in the case when the two parts of a cylinder were separated by a moveable adiabatic wall. The mechanical equilibrium would be established to maintain the same pressure on each side, but the temperatures may be different from each other because thermal equilibrium is reached only inside each portion but not between them. Another example is in ultrafast laser heating of metals, as discussed in Chap. 7, where the electron and phonon systems can be treated as being in separate equilibrium states but not in equilibrium with each other.

The concept of entropy intensity has recently been applied by Caldas and Semiao [27] to study the entropy generation in an absorbing, emitting, and scattering medium, based on the equation of radiative transfer (ERT) introduced in Sect. 2.4.3. The key is that the change in entropy in an elemental path length equals the change in intensity divided by the radiance temperature. The entropy change at steady state can be obtained from Eq. (2.53) in Chap. 2 as follows:

$$\frac{\mathrm{d}L_\lambda}{\mathrm{d}\xi} = \frac{a_\lambda I_{\mathrm{b},\lambda}}{T_\lambda(I_\lambda)} - \frac{(a_\lambda + \sigma_\lambda)I_\lambda}{T_\lambda(I_\lambda)} + \frac{\sigma_\lambda}{4\pi}\int_{4\pi}\frac{I_\lambda(\Omega')}{T_\lambda(I_\lambda)}\Phi(\Omega',\Omega)\mathrm{d}\Omega' \tag{8.57}$$

Like $I_\lambda$, the entropy intensity $L_\lambda$ is a function of wavelength, location, and direction. Note that $I_{\mathrm{b},\lambda} = I_{\mathrm{b},\lambda}(\lambda, T_\mathrm{g})$, where $T_\mathrm{g}$ is the local temperature. Usually, $T_\lambda(I_\lambda)$ depends not only on the wavelength but also on the direction for a given location. The term $I_\lambda / T_\lambda(I_\lambda)$, however, is not the same as $L_\lambda$. Integrating Eq. (8.57) over the solid angle of $4\pi$ and over all wavelengths yields the entropy change in the volume element due to the intensity field variation. Furthermore, the entropy change in the control volume is equal to the total energy absorbed divided by $T_g$. The energy rate received per unit volume can be expressed as

$$\dot{q} = \int_0^\infty \int_{4\pi} a_\lambda (I_\lambda - I_{\mathrm{b},\lambda})\mathrm{d}\Omega \mathrm{d}\lambda \tag{8.58}$$

The above equation works even with scattering since the integration of in-scattering and out-scattering cancels out. The rate of entropy change of the medium due to the net absorption by the matter is simply $\dot{q}/T_g$, which may be either positive or negative. The sum of the entropy change due to the field and that due to the matter is the total entropy change that is attributed to entropy generation by irreversibility. Therefore, we can express the volumetric entropy generation rate in terms of the absorption, emission, and scattering as follows [27]:

$$\dot{s}_{\mathrm{gen}} = \dot{s}_{\mathrm{abs\,+\,emi}} + \dot{s}_{\mathrm{sca}} \tag{8.59}$$

The entropy generation due to combined absorption and emission is

$$\dot{s}_{\mathrm{abs\,-\,emi}} = \int_0^\infty \int_{4\pi} a_\lambda (I_\lambda - I_{b,\lambda})\left[\frac{1}{T_\mathrm{g}} - \frac{1}{T_\lambda(I_\lambda)}\right]\mathrm{d}\Omega \mathrm{d}\lambda \tag{8.60a}$$

The entropy generation due to scattering is

$$\dot{s}_{\text{sca}} = \int_0^\infty \int_{4\pi} \left[ \int_{4\pi} \frac{I_\lambda(\Omega')}{4\pi T_\lambda(I_\lambda)} \Phi(\Omega', \Omega) \mathrm{d}\Omega' - I_\lambda(\Omega) \right] \sigma_\lambda \mathrm{d}\Omega \mathrm{d}\lambda \tag{8.60b}$$

Note that Eq. (8.60a) should be treated as the combined absorption and emission effect since entropy generation due to absorption and emission processes cannot be separated. Furthermore, the entropy generation due to either absorption–emission or scattering is always greater than or equal to zero. When a surface is involved in radiative heat transfer, the entropy generation rate per unit area can be expressed as

$$s''_{\text{gen}} = \int_0^\infty \int_0^{2\pi} \int_0^{\pi/2} \left[ \frac{I_{\text{in},\lambda} - I_{\text{out},\lambda}}{T_{\text{w}}} - (L_{\text{in},\lambda} - L_{\text{out},\lambda}) \right] \cos\theta \sin\theta \mathrm{d}\theta \mathrm{d}\phi \mathrm{d}\lambda \tag{8.61}$$

where $T_{\text{w}}$ is the wall temperature, and subscripts "in" and "out" signify the energy or entropy intensity to and from the surface, respectively. If the surface is not a blackbody, the outgoing intensity includes both the emitted and reflected intensities. An alternative approach is to integrate the intensity over the whole sphere with a solid angle of $4\pi$. In Eq. (8.61), the entropy intensity is related to the energy intensity by Eq. (8.55), which is recast in terms of wavelength as follows:

$$L_\lambda(\lambda, I_\lambda) = \frac{2k_{\text{B}}c}{\lambda^4} \left[ \left(1 + \frac{\lambda^5 I_\lambda}{2hc^2}\right) \ln\left(1 + \frac{\lambda^5 I_\lambda}{2hc^2}\right) - \frac{\lambda^5 I_\lambda}{2hc^2} \ln\left(\frac{\lambda^5 I_\lambda}{2hc^2}\right) \right] \tag{8.62}$$

The use of Eq. (8.62) may be disputed when multiple reflections occur. The intensity of the emitted radiation is less than that of the blackbody and is reduced by each reflection. The question still remains as to whether the blackbody intensity should be used to calculate the entropy or the actual intensity after each reflection or the combined intensity at any given location. An example is a system of two large parallel plates, separated by vacuum. One of the plates is at a temperature $T_1$ and is diffuse-gray with an emittance of 0.5. The other plate is insulated and is a perfect reflector (i.e., no emission). It is clear that a thermal equilibrium will be established in the cavity after a long time. Again, the separation distance is much larger than the thermal radiation wavelengths. The radiation leaving surface 1 includes the emitted rays, as well as the first-order and higher order reflected rays. An attempt to define the entropy of the emitted ray and each reflected ray will result in a total entropy intensity greater than the entropy intensity calculated based on the blackbody intensity $I_{\text{b},\lambda}(\lambda, T_1)$. Therefore, to apply the previous analysis in a consistent way and to obtain meaningful results, we must make the following hypotheses:

- The intensity at any given location is additive regardless of where it originates from, as long as it falls within the same solid angle and wavelength intervals. While this sounds obvious, it is untrue when interference effects become important. The resulting intensity is called the combined intensity.

- The monochromatic radiation temperature $T_\lambda$, defined in Eq. (8.56), is a function of the combined intensity and is, in general, dependent on the direction and wavelength. The effect of polarization is neglected to simplify the problem. Equation (8.56) must not be applied to each of the reflected or scattered rays. The physical significance is that all the photons, with the same wavevector and frequency, can be considered as a subsystem that is at thermodynamic equilibrium with the temperature $T_\lambda[I_\lambda(\lambda, \theta, \phi)]$.
- The entropy intensity is defined based on the combined intensity, according to Eq. (8.62). While entropy must be additive, the entropy of all individual rays must be calculated based on the monochromatic temperature of the combined intensity. Because the number of photons, intensity, and entropy are additive, the fraction of the entropy of each ray is the same as the ratio of the intensity of that ray to the combined intensity.

With the theories presented in this section, one should be able to perform a second law thermodynamic analysis for a given system, involving radiative transfer of energy. Zhang and Basu [28] investigated entropy flow and generation considering incoherent multiple reflections. Different approximations exist in analyzing the entropy of radiation. For example, the method of dilute blackbody radiation uses a dilution factor and defines an effective temperature for each wavelength [23]. When the process is very complicated, such an effective temperature cannot be easily defined and this definition cannot be applied to multiple reflections. Entropy generation is usually accompanied by the generation of heat, such as heating by friction, electrical resistor, chemical reaction, or the absorption of solar radiation. On the other hand, it appears that entropy generation can occur in radiation without the generation of heat, such as by scattering. The definition of inelastic scattering is based on the conservation of energy (wavelength) and momentum, which does not impose any constraints on the reversibility. Further research is much needed in order to better understand the nature of entropy of radiation and determine the ultimate efficiency of photovoltaic cells and other radiative processes, including laser cooling and trapping. Another area of possible application of radiation entropy is in nanoscale heat conduction using EPRT, as discussed in Chap. 7. The entropy concept may be extended to the phonon system by defining radiation entropy and entropy intensity of phonons. Bright and Zhang [29] extended the concept of radiation entropy in a participating medium to phonon radiation, providing a method to evaluate local entropy generation. The conventional formula for entropy generation in heat diffusion can be derived under the local-equilibrium assumption. Furthermore, the entropy generation mechanism during phonon transport is elucidated as due to the "absorption" of high-frequency phonons and "emission" of lower frequency phonons, arising from the actual phonon scattering processes [29]. There is a need to further develop photon entropy analysis for near-field thermal radiation considering both interferences and photon tunneling [24].

### 8.2.4 Limitations of Planck's Law

The concept that *a blackbody absorbs all radiant energy that is incident upon it* is purely from the geometric-optics point of view, in which light travels in a straight line and cannot interact with an object that does not intercept the light ray. Another example of the geometric-optics viewpoint is that the transmittance of an iris (open aperture) should be 1, meaning that *all radiation incident on the opening will go through but no radiation outside the opening can go through.* However, for an aperture whose diameter is comparable to the wavelength of the incident radiation, diffraction may become important. As a result, the transmittance could exceed 1 in some cases. Due to the diffraction effect, a particle that is sufficiently small compared to the wavelength will interact with the radiation field, according to the scattering and absorption cross sections, which can be greater than the projected surface area. In some cases, it is possible for the object to absorb more energy than the product of the radiant flux and the projection area. The absorptance can be greater than 1 and thus exceeds the limit set by a blackbody. When such an object is placed in an isothermal enclosure, the emitted energy will be greater than that from a blackbody having the same dimensions. This anomaly has been discussed in detail by Bohren and Huffman [9].

The energy density near the surface within a distance less than the wavelength can be much greater than that given by Eq. (8.41) and increases as the distance is further reduced. When two objects are placed at a distance much smaller than the characteristic wavelength of thermal radiation, i.e., in the near field, photon tunneling can occur and cause significant enhancement of the energy transfer. In recent years, there have been numerous studies of light transmission through small apertures, radiation heat transfer at nanometer distances, and light emission from nanostructures [30–33]. Recent studies have also demonstrated that radiation heat transfer can be greatly enhanced for micro/nanostructures even when they are separated by distances longer than the characteristic wavelength [34]. This is still an open field with many new developments and applications. We will study these phenomena and the underlying physics in the following two chapters.

## 8.3 Radiative Properties of Semi-infinite Media

The reflection and refraction (transmission) of a semi-infinite isotropic medium are studied based on Maxwell's equations using suitable boundary conditions at the interface between the incident and transmitting media. Only plane waves with different polarizations are considered. Total internal reflection and the associated Goos–Hänchen phase shift are also introduced. For real materials or interfaces, the bidirectional reflectance distribution function (BRDF) is often needed to fully describe the radiative properties. The emittance can be calculated based on the reflectance by introducing Kirchhoff's law.

### 8.3.1 *Reflection and Refraction of a Plane Wave*

Consider radiation incident from one medium to another at the interface or the boundary. The boundary that separates the media is assumed to be a smooth plane and extends to infinity. Each medium is homogeneous and isotropic such that there is no scattering within the medium. Therefore, the electric response can be characterized by the relative permittivity or dielectric function $\varepsilon$, and the magnetic response can be characterized by the relative permeability $\mu$. For nonmagnetic materials, the refractive index is related to the dielectric function by $n = \sqrt{\varepsilon}$. Keep in mind that these quantities are, in general, complex and frequency dependent. The real and imaginary parts of the refractive index are often called the optical constants. In this section, we present the general formulation for both magnetic and nonmagnetic materials. For certain crystalline and amorphous solids, like quartz and glass, the refractive index is real in a wide spectral region and is the only parameter needed to fully characterize the optical response of the material. In such a case, the expression can be largely simplified and the results can be easily comprehended. The reduced results will also be presented because of their importance to numerous engineering problems.

The incident radiation is a monochromatic plane wave with an angular frequency $\omega$. As shown in Fig. 8.7, the wavevector of the incident wave is $\mathbf{k}_1^+ = (k_{1x}, 0, k_{1z})$, and the surface normal defines the *plane of incidence*, which is the *x*-*z*-plane. The wavevectors of the reflected and transmitted waves must lie in the same plane. The angle of incidence $\theta_1$ is the angle between the incident wavevector and the *z*-direction, i.e., $\sin\theta_1 = k_{1x}/k_1$ and $\cos\theta_1 = k_{1z}/k_1$, where $k_1^2 = k_{1x}^2 + k_{1z}^2 = \mu_1\varepsilon_1\omega^2/c_0^2$. It is common to study the reflection and the refraction for linearly polarized waves, with either the electric or magnetic field being parallel to the *y*-axis, because other polarizations can be decomposed into the two polarization components.

When the electric field is in the *y*-direction, as shown in Fig. 8.7a, the wave is called a transverse-electric (TE) wave or is said to be perpendicularly (*s*) polarized. The incident electric field can be expressed as follows by omitting the time-harmonic term of $e^{-i\omega t}$ hereafter:

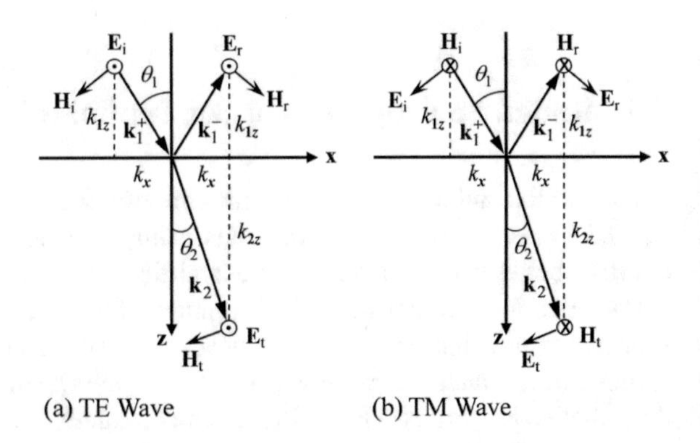


**Fig. 8.7** Illustration of reflection and transmission at an interface: **a** TE wave or *s*-polarization. **b** TM wave or *p*-polarization

$$\mathbf{E}_{\rm i} = \hat{\mathbf{y}} E_{\rm i} {\rm e}^{{\rm i}k_{1z}z + {\rm i}k_{1x}x} \tag{8.63}$$

The boundary conditions state that the tangential components of both **E** and **H** must be continuous at the interface. This implies that the $x$-component of the wavevector must be the same for the incident, reflected, and transmitted waves, i.e., $k_{1x} = k_{2x} = k_x$. Because the angle of reflection must be the same as the angle of incidence (specular reflection), we have $\mathbf{k}_1^- = (k_x, 0, -k_{1z})$. For the transmitted or refracted wave, we have $\mathbf{k}_2 = (k_x, 0, k_{2z})$ and

$$\sin\theta_2 = \frac{k_x}{k_2} = \frac{n_1 \sin\theta_1}{n_2} \tag{8.64}$$

which is called Snell's law. It can be easily visualized by observing the bended image of a chopstick in a bowl of water. Note that $k_{2z}^2 = k_2^2 - k_x^2 = \mu_2\varepsilon_2\omega^2/c_0^2 - k_x^2 = k_2^2\cos^2\theta_2$. Generally speaking, the wavevector components and the refractive indices may be complex. Complex angles can be defined so that Eq. (8.64) is always valid. Near the interface, the nonzero components of the electric and magnetic fields are

$$E_y = \begin{cases} (E_{\rm i}{\rm e}^{{\rm i}k_{1z}z} + E_{\rm r}{\rm e}^{-{\rm i}k_{1z}z}){\rm e}^{{\rm i}k_x x}, & \text{for } z < 0 \\ E_{\rm t}{\rm e}^{{\rm i}k_{2z}z}{\rm e}^{{\rm i}k_x x}, & \text{for } z > 0 \end{cases} \tag{8.65}$$

$$H_x = \begin{cases} -\frac{k_{1z}}{\omega\mu_0\mu_1}(E_{\rm i}{\rm e}^{{\rm i}k_{1z}z} - E_{\rm r}{\rm e}^{-{\rm i}k_{1z}z}){\rm e}^{{\rm i}k_x x}, & \text{for } z < 0 \\ -\frac{k_{2z}}{\omega\mu_0\mu_2}E_{\rm t}{\rm e}^{{\rm i}k_{2z}z}{\rm e}^{{\rm i}k_x x}, & \text{for } z > 0 \end{cases} \tag{8.66}$$

and

$$H_z = \begin{cases} \frac{k_x}{\omega\mu_0\mu_1}(E_{\rm i}{\rm e}^{{\rm i}k_{1z}z} + E_{\rm r}{\rm e}^{-{\rm i}k_{1z}z}){\rm e}^{{\rm i}k_x x}, & \text{for } z < 0 \\ \frac{k_x}{\omega\mu_0\mu_2}E_{\rm t}{\rm e}^{{\rm i}k_{2z}z}{\rm e}^{{\rm i}k_x x}, & \text{for } z > 0 \end{cases} \tag{8.67}$$

where $E_{\rm i}$, $E_{\rm r}$, and $E_{\rm t}$ are, respectively, the amplitudes of the incident, reflected, and transmitted electric fields at the interface. It is further assumed that $k_x$ is real so that the amplitude of the field is independent of $x$. The Fresnel reflection and transmission coefficients for a TE wave are defined as $r_{12,s} = E_{\rm r}/E_{\rm i}$ and $t_{12,s} = E_{\rm t}/E_{\rm i}$, respectively. Boundary conditions require that $E_y$ and $H_x$ be continuous at $z = 0$. From Eqs. (8.65) and (8.66), we obtain $1 + r_{12,s} = t_{12,s}$ and $(k_{1z}/\mu_1)(1 - r_{12,s}) = (k_{2z}/\mu_2)t_{12,s}$; thus,

$$r_{12,s} = \frac{E_{\rm r}}{E_{\rm i}} = \frac{k_{1z}/\mu_1 - k_{2z}/\mu_2}{k_{1z}/\mu_1 + k_{2z}/\mu_2} \tag{8.68a}$$

and

$$t_{12,s} = \frac{E_{\rm t}}{E_{\rm i}} = \frac{2k_{1z}/\mu_1}{k_{1z}/\mu_1 + k_{2z}/\mu_2} \tag{8.68b}$$

which are generally applicable, as long as each medium is homogeneous and isotropic [8]. For nonmagnetic materials, the previous equations can be written as follows:

$$r_{12,s} = \frac{n_1 \cos\theta_1 - n_2 \cos\theta_2}{n_1 \cos\theta_1 + n_2 \cos\theta_2} \tag{8.69a}$$

and

$$t_{12,s} = \frac{2n_1 \cos\theta_1}{n_1 \cos\theta_1 + n_2 \cos\theta_2} \tag{8.69b}$$

The spectral reflectivity, or simply reflectivity, $\rho'_\lambda$ is given by the ratio of the reflected energy flux to the incident energy flux, and the absorptivity $\alpha'_\lambda$ is the ratio of the transmitted energy flux to the incident energy flux, since all the photons transmitted through the interface will be absorbed inside the second medium. Terms ending with "-ivity" are typically used for a perfect interface and those with "-tance" are for general surfaces including smooth and rough surfaces, thin films, as well as layered structures. The energy flux is related to the time-averaged Poynting vector, defined in Eq. (8.18b). From Eqs. (8.65) to (8.67), the $x$- and $z$-components of the Poynting vector at the interface ($z \to 0$) in medium 1 are

$$\langle S_{1x} \rangle = \frac{1}{2}\mathrm{Re}\left[\frac{k_x^*}{\omega\mu_0\mu_1^*}(E_\mathrm{i} + E_\mathrm{r})(E_\mathrm{i}^* + E_\mathrm{r}^*)\right] \tag{8.70a}$$

and

$$\langle S_{1z} \rangle = \frac{1}{2}\mathrm{Re}\left[\frac{k_{1z}^*}{\omega\mu_0\mu_1^*}(E_\mathrm{i} + E_\mathrm{r})(E_\mathrm{i}^* - E_\mathrm{r}^*)\right] \tag{8.70b}$$

It can be seen that, in general, the reflected wave and the incident wave are coupled and the energy flow cannot be separated into a reflected flux and an incident flux. Under the assumption that medium 1 is lossless (nonabsorbing or nondissipative) and $k_x^2 < k_1^2$, we can write

$$\langle S_{1z} \rangle = \langle S_{\mathrm{i}z} \rangle - \langle S_{\mathrm{r}z} \rangle \tag{8.71}$$

where

$$\langle S_{\mathrm{i}z} \rangle = \frac{k_{1z}}{2\omega\mu_0\mu_1}|E_\mathrm{i}|^2 \text{ and } \langle S_{\mathrm{r}z} \rangle = \frac{k_{1z}}{2\omega\mu_0\mu_1}|E_\mathrm{r}|^2 \tag{8.72}$$

If medium 1 is lossy, there will be additional terms associated with $E_\mathrm{i}E_\mathrm{r}^*$ and $E_\mathrm{i}^*E_\mathrm{r}$. In this case, the power flow normal to the interface cannot be separated as forward and backward terms, because of the cross-coupling terms. Therefore, the lossless condition in medium 1 is required in order to properly define the energy or power reflectivity [35]. This is usually not a problem when radiation is incident from air or a dielectric prism onto a medium. The spectral reflectivity can then be obtained based on the $z$-components of the reflected and incident Poynting vectors as

$$\rho'_{\lambda,s}(\theta_1) = |E_\mathrm{r}|^2/|E_\mathrm{i}|^2 = \left|r_{12,s}\right|^2 \tag{8.73}$$

The Poynting vector at the interface in medium 2 can be written as

$$\langle \mathbf{S}_\mathrm{t} \rangle = \frac{1}{2\omega\mu_0} \mathrm{Re}\left( \frac{k_x^* \hat{\mathbf{x}} + k_{2z}^* \hat{\mathbf{z}}}{\mu_2^*} \right) |E_\mathrm{t}|^2 \tag{8.74}$$

which is not parallel to $\mathrm{Re}(\mathbf{k}_2)$ unless $\mathrm{Im}(\mu_2) = 0$. Recall that the plane of constant phase is perpendicular to $\mathrm{Re}(\mathbf{k}_2)$. If medium 2 is dissipative, $\mathrm{Im}(\mathbf{k}_2)$ is parallel to the $z$-axis and the amplitude will vary along the $z$-direction. The wave becomes inhomogeneous in medium 2, except when $k_x = 0$ (normal incidence). The definition of the transmitted energy flux at the interface is based on the projected Poynting vector in the $z$-direction. Hence, the absorptivity is the ratio of the $z$-components of the transmitted and incident Poynting vectors, viz.,

$$\alpha'_{\lambda,s}(\theta_1) = \frac{\mathrm{Re}(k_{2z}/\mu_2)}{\mathrm{Re}(k_{1z}/\mu_1)} \left| t_{12,s} \right|^2 \tag{8.75}$$

Note that $\mathrm{Re}(k_{2z}/\mu_2) = \mathrm{Re}(k_{2z}^*/\mu_2^*)$, and $\mathrm{Re}(k_{1z}/\mu_1) = k_{1z}/\mu_1$ since medium 1 is lossless. It can be shown that $\rho'_{\lambda,s} + \alpha'_{\lambda,s} = 1$, as required by energy conservation: $\langle S_{1z} \rangle = \langle S_{2z} \rangle$ at $z = 0$. For nonmagnetic and nondissipative materials, we have

$$\alpha'_{\lambda,s}(\theta_1) = \frac{n_2 \cos\theta_2}{n_1 \cos\theta_1} \left| t_{12,s} \right|^2 \tag{8.76}$$

The reflection and transmission coefficients for the transverse magnetic (TM) wave or parallel ($p$) polarization are defined as the ratios of the magnetic fields: $r_{12,p} = H_\mathrm{r}/H_\mathrm{i}$ and $t_{12,p} = H_\mathrm{t}/H_\mathrm{i}$, respectively [7]. Hence,

$$r_{12,p} = \frac{H_\mathrm{r}}{H_\mathrm{i}} = \frac{k_{1z}/\varepsilon_1 - k_{2z}/\varepsilon_2}{k_{1z}/\varepsilon_1 + k_{2z}/\varepsilon_2} \tag{8.77a}$$

$$t_{12,p} = \frac{H_\mathrm{t}}{H_\mathrm{i}} = \frac{2k_{1z}/\varepsilon_1}{k_{1z}/\varepsilon_1 + k_{2z}/\varepsilon_2} \tag{8.77b}$$

In the case of nonmagnetic materials, we obtain

$$r_{12,p} = \frac{n_2 \cos\theta_1 - n_1 \cos\theta_2}{n_2 \cos\theta_1 + n_1 \cos\theta_2} \tag{8.78a}$$

and

$$t_{12,p} = \frac{2n_2 \cos\theta_1}{n_2 \cos\theta_1 + n_1 \cos\theta_2} \tag{8.78b}$$

At normal incidence, the reflection coefficients calculated based on Eqs. (8.69a) and (8.78a) are related by

$$r_{12,s} = \frac{n_1 - n_2}{n_1 + n_2} = -r_{12,p} \tag{8.79}$$

When both $n_1$ and $n_2$ are real, for $n_1 < n_2$, the electric field will experience a phase reversal (phase shift of $\pi$) upon reflection but the magnetic field will not. On the other hand, for $n_1 > n_2$, it is the magnetic field that will experience a phase reversal. In fact, based on Maxwell's equations, the electric and magnetic quantities obey a duality when $\rho_e = 0$. They can be interchanged with the following substitutions: $\mathbf{E} \rightarrow \mathbf{H}$ and $\mathbf{H} \rightarrow -\mathbf{E}$. Note that $\varepsilon$ and $\mu$, as well as the polarization states $s$ and $p$, should also be interchanged. The Poynting vector for a TM wave is $\langle \mathbf{S} \rangle = \frac{1}{2\omega\varepsilon_0}\text{Re}\left(\frac{\mathbf{k}}{\varepsilon}\right)\left|H_y\right|^2$, which is not parallel to Re($\mathbf{k}$) when Im($\varepsilon_2$) $\neq 0$. Upon refraction into an absorbing medium, the waves become inhomogeneous and the Poynting vectors for different polarizations may split into different directions [36]. Nevertheless, the constant-amplitude plane is always perpendicular to the $z$-direction because the amplitude cannot change along the $x$-$y$-plane. The reflectivity for $p$-polarization is

$$\rho'_{\lambda,p}(\theta_1) = \left|r_{12,p}\right|^2 \tag{8.80}$$

Hence, the absorptivity becomes

$$\alpha'_{\lambda,p}(\theta_1) = \frac{\text{Re}(k_{2z}/\varepsilon_2)}{\text{Re}(k_{1z}/\varepsilon_1)}\left|t_{12,p}\right|^2 \tag{8.81}$$

For nonmagnetic and nonabsorbing materials, we have

$$\alpha'_{\lambda,p}(\theta_1) = \frac{n_1 \cos\theta_2}{n_2 \cos\theta_1}\left|t_{12,p}\right|^2 \tag{8.82}$$

If the incident wave is unpolarized or circularly polarized, the reflectivity can be obtained by averaging the values for $p$- and $s$-polarized waves, i.e.,

$$\rho'_{\lambda} = \frac{\rho'_{\lambda,p} + \rho'_{\lambda,s}}{2} \tag{8.83}$$

The reflectivity for radiation incident from air ($n_1 \approx 1$) to a dielectric medium ($n_2 = 2$) and that from the dielectric to air are shown in Fig. 8.8 for each polarization as well as for the unpolarized incident radiation. When $n_1 > n_2$, the reflectivity will reach 1 at $\theta_1 = \theta_c = \sin^{-1}(n_2/n_1)$. This angle is called the *critical angle*, and *total internal reflection* occurs at angles of incidence greater than the critical angle. This is the principle commonly used in optical fibers and waveguides, since light is trapped inside the high-index material and propagates along the medium. It can be seen that in total internal reflection, $k_x > k_2$ and $k_{2z}$ becomes purely imaginary. The amplitude of the wave exponentially attenuates in the positive $z$-direction. This is similar to Eq. (8.37) and makes it an evanescent wave, as shown in Fig. 8.3. The time-averaged Poynting vector is zero in the $z$-direction. Hence, no energy is transmitted through the boundary.

For the TE wave, the reflectivity increases monotonically with the angle of incidence and reaches 1 at the grazing angle (90°) or at the critical angle when $n_1 > n_2$.

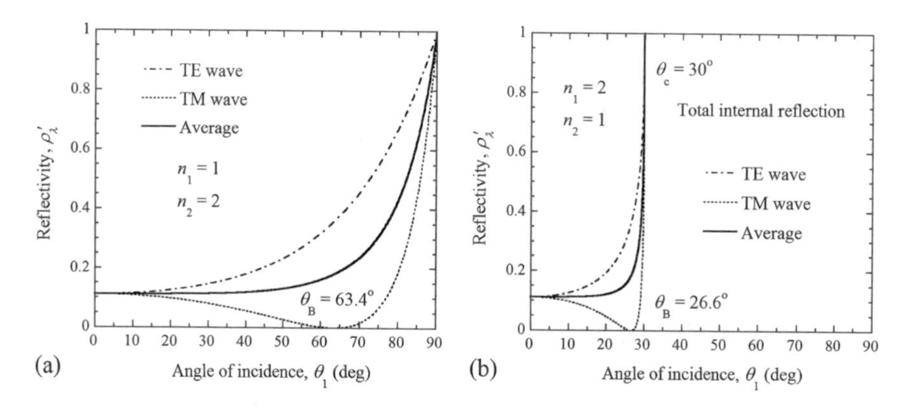


**Fig. 8.8** Reflectivity versus the angle of incidence between air and a dielectric: **a** Incident from air to a medium; **b** Incident from a medium to air

The reflectivity for the TM wave, on the other hand, goes through a minimum that is equal to zero. The angle at which $\rho'_{\lambda,p} = 0$ is called the Brewster angle, given by $\theta_B = \tan^{-1}(n_2/n_1)$ for nonmagnetic materials. For $p$-polarization, all the incident energy will be transmitted into medium 2, without reflection at the Brewster angle. This phenomenon has been used to build polarizers and transmission windows in absolute cryogenic radiometers. The physical mechanism of reflection can also be understood as re-emission by the *induced electric dipoles* in the medium, based on the *Ewald–Oseen extinction theorem*. At the Brewster angle, the electric dipoles induced in the material align in the direction of the reflected wave, and the refracted wave is perpendicular to the reflected wave (i.e., $\theta_1 + \theta_2 = 90°$). The reflective power goes to zero because an electric dipole cannot radiate along its own axis. The situation is changed when magnetic materials are involved, such as a negative index material. The fields radiated by both the induced electric dipoles and *magnetic dipoles* are responsible for the reflection. The Brewster angle can occur for either polarization when the radiated fields cancel each other. A detailed discussion can be found from the publication of Fu et al. [37]. In an absorbing medium, there is a drop in reflectivity for $p$-polarization, but the minimum is not zero. Furthermore, there exists a *principal angle* at which the phase difference between the two reflection coefficients equals 90° and the ratio of the reflectivity for the TM and TE waves is near the reflectivity minimum [8]; see Problem 8.24.

The reflectivity for radiation incident from air ($n_1 \approx 1$) or a vacuum, at normal incidence, becomes

$$\rho'_{\lambda,\mathrm{n}} = \frac{(n_2 - 1)^2 + \kappa_2^2}{(n_2 + 1)^2 + \kappa_2^2} \tag{8.84}$$

for any polarization. It can be seen that the normal reflectivity will be close to 1, when either $n_2 \ll 1$ or $n_2 \gg 1$. The reflectivity is large for most metals in the infrared, because both $n_2$ and $\kappa_2$ are large. The reflectivity of a conventional superconductor approaches 1 when the frequency is lower than that of the superconducting energy gap, since $n_2 \to 0$ in this case. On the other hand, $\rho'_{\lambda,\mathrm{n}} \to 0$ when $n_2 \approx 1$ and $\kappa_2 \ll 1$. This can occur in a dielectric material at certain mid-infrared wavelengths and also for most metals in the x-ray region.

### 8.3.2 Total Internal Reflection and the Goos–Hänchen Shift

Total internal reflection (TIR) occurs when light comes from an optically denser material to another material at incidence angles greater than the critical angle determined by Snell's law. As discussed in the preceding section, the amplitude of the reflection coefficient becomes unity at incidence angles greater than the critical angle. Although no energy is transferred from medium 1 to medium 2, there exists an electromagnetic field in the second medium near the surface. This electromagnetic field can store as well as exchange energy with medium 1 at any instant of time.

While evanescent waves do not carry energy into the second medium, there is a shift in the phase of the reflected wave upon TIR. Consider a plane wave of angular frequency $\omega$ incident from a semi-infinite medium 1 to medium 2, as shown in Fig. 8.9a. The wavevector $\mathbf{k}_1^+ = k_x\hat{\mathbf{x}} + k_{1z}\hat{\mathbf{z}}$, $\mathbf{k}_1^- = k_x\hat{\mathbf{x}} - k_{1z}\hat{\mathbf{z}}$, and $\mathbf{k}_2 = k_x\hat{\mathbf{x}} + k_{2z}\hat{\mathbf{z}}$, since the parallel wavevector component $k_x$ must be the same as required by the phase-matching boundary condition. The magnitudes of the wavevectors are

$$k_1^2 = k_x^2 + k_{1z}^2 = \varepsilon_1\mu_1\omega^2/c^2 \tag{8.85a}$$

and

$$k_2^2 = k_x^2 + k_{2z}^2 = \varepsilon_2\mu_2\omega^2/c^2 \tag{8.85b}$$

where $\varepsilon$ and $\mu$ are the relative (ratio to those of a vacuum) permittivity and permeability, respectively, and $c$ is the speed of light in a vacuum (omitting the subscript 0). Assume that the incident wave is $p$ polarized or a TM wave, so that the only nonzero component of the magnetic field is in the $y$-direction. The magnetic field of the incident wave may be expressed as $\mathbf{H}_\mathrm{i} = (0, H_y, 0)$, where

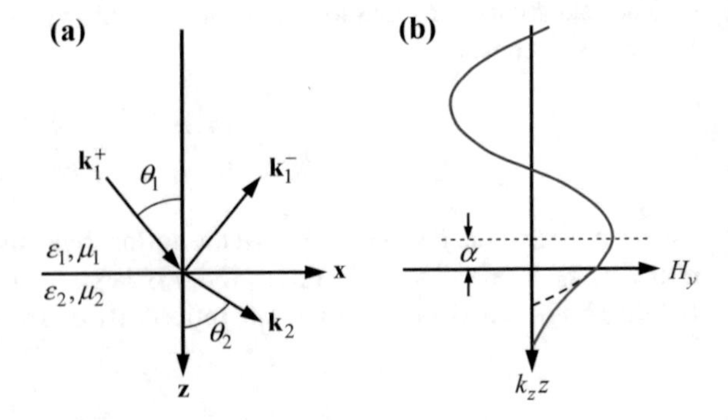


**Fig. 8.9** Illustration of total internal reflection. **a** Schematic of the incident, reflected, and transmitted waves at the interface between two semi-infinite media. **b** The magnetic field distribution for a TM wave when total internal reflection occurs

$H_y(x, y, z, t) = H_i e^{ik_{1z}z + ik_{1x}x - i\omega t}$. For simplicity, let us omit $\exp(-i\omega t)$ from now on. Recall that the Fresnel coefficients for a TM wave are defined as the ratios of the reflected or transmitted magnetic field to the incident magnetic field. For example, the Fresnel reflection coefficient is

$$r_p = \frac{H_r}{H_i} = \frac{k_{1z}/\varepsilon_1 - k_{2z}/\varepsilon_2}{k_{1z}/\varepsilon_1 + k_{2z}/\varepsilon_2} \tag{8.86}$$

The field in medium 1 is composed of the incident and reflected fields, and that in medium 2 is the transmitted field. Therefore,

$$\frac{H_y}{H_i} = \begin{cases} (e^{ik_{1z}z} + r_p e^{-ik_{1z}z})e^{ik_x x}, & \text{for } z \le 0 \\ (1 + r_p)e^{ik_{2z}z}e^{ik_x x}, & \text{for } z > 0 \end{cases} \tag{8.87}$$

The electric fields can be obtained by applying the Maxwell equations. Hence, we can write the electric and magnetic fields in both media as follows:

$$\frac{E_x}{H_i} = \begin{cases} \frac{k_{1z}}{\omega\varepsilon_1\varepsilon_0}(e^{ik_{1z}z} - r_p e^{-ik_{1z}z})e^{ik_x x}, & \text{for } z \le 0 \\ \frac{k_{2z}}{\omega\varepsilon_2\varepsilon_0}(1 + r_p)e^{ik_{2z}z}e^{ik_x x}, & \text{for } z > 0 \end{cases} \tag{8.88}$$

and

$$\frac{E_z}{H_i} = \begin{cases} -\frac{k_x}{\omega\varepsilon_1\varepsilon_0}(e^{ik_{1z}z} + r_p e^{-ik_{1z}z})e^{ik_x x}, & \text{for } z \le 0 \\ -\frac{k_x}{\omega\varepsilon_2\varepsilon_0}(1 + r_p)e^{ik_{2z}z}e^{ik_x x}, & \text{for } z > 0 \end{cases} \tag{8.89}$$

Assume that $\varepsilon$'s and $\mu$'s are real and furthermore, $\varepsilon_1\mu_1 > \varepsilon_2\mu_2 > 0$. From Eq. (8.85b), we have $k_{2z}^2 = \varepsilon_2\mu_2\omega^2/c^2 - k_x^2$. When $\sqrt{\varepsilon_2\mu_2} < k_x c/\omega < \sqrt{\varepsilon_1\mu_1}$, the incidence angle $\theta_1$ is defined but the refraction angle is not, because $k_{2z}$ becomes imaginary. One can write $k_{2z} = i\eta_2$, where $\eta_2 = \sqrt{k_x^2 - \varepsilon_2\mu_2\omega^2/c^2}$ is a real positive number. In this case, $|r_p| = 1$ and

$$r_p = e^{i\delta} = e^{-i2\alpha} \tag{8.90}$$

where $\tan\alpha = (\eta_2/\varepsilon_2)/(k_{1z}/\varepsilon_1)$. Following Haus [38], the magnetic field at $x = 0$ in medium 1 can be written as

$$H_y = 2H_i e^{-i\alpha}\cos(k_{1z}z + \alpha), \ \ z \le 0 \tag{8.91a}$$

Similarly, $H_y$ in medium 2 becomes

$$H_y = 2H_i e^{-i\alpha}\cos(\alpha)e^{-\eta_2 z}, \ \ z > 0 \tag{8.91b}$$

The magnetic field at $x = 0$ is plotted in Fig. 8.9b with respect to $k_z z$, at the instant of time when the phase of $H_i e^{-i\alpha - i\omega t}$ becomes zero. From this figure, one can see that the field decays exponentially in medium 2. As a result, there is a phase shift in medium 1 upon TIR so that the maximum amplitude is shifted from the interface to $k_z z = -\alpha$. The phase angle of the reflection coefficient $\delta = -2\alpha$ is called the *Goos–Hänchen phase shift,* which depends on the incidence angle $\theta_1$ or $k_x$. The difference in $\delta$ for TE and TM waves in a dielectric prism was used to construct a polarizer called *Fresnel's rhomb*, which can change a linearly polarized wave into a circularly polarized wave, or vice versa [7].

**Example 8.6** Calculate the time-averaged Poynting vector near the interface in the case of total internal reflection.

**Solution** Based on Example 8.1, it can be seen that the Poynting vector $\mathbf{S} = \mathrm{Re}(\mathbf{E}) \times \mathrm{Re}(\mathbf{H})$ is, in general, a function of time. The time-dependent terms that oscillate with $2\omega$, however, become zero after integration. The time-averaged Poynting vector is $\langle \mathbf{S} \rangle = \frac{1}{2}\mathrm{Re}(\mathbf{E} \times \mathbf{H}^*)$. For z > 0, $\langle S_z \rangle = \frac{1}{2}\mathrm{Re}(E_x H_y^*) = 0$ because $k_{2z}$ is purely imaginary. It can also be shown that $\langle S_z \rangle = 0$ for $z \leq 0$ (see Problem 8.26). Furthermore,

$$\langle S_x \rangle = -\frac{1}{2}\mathrm{Re}(E_z H_y^*) = \begin{cases} \frac{k_x}{\omega\varepsilon_1\varepsilon_0}|H_i|^2\left[1 + \cos(2k_{1z}z + 2\alpha)\right], & z \leq 0 \\ \frac{k_x}{\omega\varepsilon_2\varepsilon_0}|H_i|^2[1 + \cos(2\alpha)]\mathrm{e}^{-2\eta_2 z}, & z > 0 \end{cases} \tag{8.92}$$

Note that $\langle S_x \rangle$ does not have to be continuous across the interface. Depending on whether $\varepsilon$ is positive or negative, the sign of $\langle S_x \rangle$ may be parallel or antiparallel to $k_x$. It should also be noted that $\langle S_x \rangle$ is a sinusoidal function of z in medium 1 and decays exponentially in medium 2 as z approaches infinity.

Newton conjectured that, when a light beam is reflected at the boundary upon TIR, the light corpuscles would penetrate some distance into the optically rarer medium and then reenter the optically denser medium. In addition, he suspected that the path of the beam would be a parabola with its vertex in the rarer medium and, consequently, the actual reflected beam would be shifted laterally with respect to the geometric-optics prediction. From the Poynting vector formulation given in Eq. (8.92), the energy must penetrate into the second medium to maintain the energy flow parallel to the interface and reenter the first medium so that no net energy is transferred across the interface. The actual beams have a finite extension so that the reflected beam in the far field can be separated from the incident beam since the Poynting vector is parallel to the wavevector. The effect of the parallel energy flow indeed causes the reflected beam to shift forward from that expected by the geometric-optics analysis. F. Goos and H. Hänchen were the first to observe the lateral beam shift through a cleverly devised experiment in 1947. A schematic of this experiment is shown in Fig. 8.10, in which a glass plate was used so that the incident light was multiply reflected by the top and bottom surfaces. In the middle of one or both of the surfaces, a silver strip was deposited. This way, the beam reflected by the silver film (solid line) would essentially follow geometric optics and that by total internal reflection would experience a lateral shift. Although the lateral shift is on the order of the wavelength, a large number of reflections (over 100 times) allowed the shift to be observed by a photographic plate. Lotsch [39] published a series of papers on the comprehensive study of the Goos–Hänchen effect. Puri and Birman [40] provided an elegant review of earlier works, including several methods for analyzing the Goos–Hänchen effect. A quantitative study of the Goos–Hänchen effect is presented next.

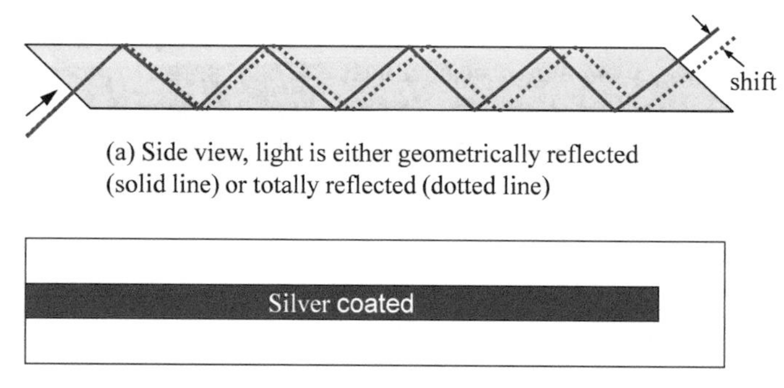


(a) Side view, light is either geometrically reflected (solid line) or totally reflected (dotted line)

(b) Top view, surface coated with a silver strip

**Fig. 8.10** Illustration of the Goos–Hänchen experiment: **a** side view, **b** top view

One way to model the lateral shift is to use a beam of finite width rather than an unbounded plane wave. Another method that is mathematically simpler considers the phase change of an incoming wave packet, which is composed of two plane waves with a slightly different $k_x$. Upon TIR, the phase shift $\delta = -2\alpha$ for a given polarization is a function of $k_x$. The difference in the phase shift will cause the reflected beam to exhibit a lateral shift along the interface (x-direction) given as

$$D = -\frac{d\delta}{dk_x} = \frac{\varepsilon_1}{\varepsilon_2}\frac{2k_x}{\eta_2 k_{1z}}\frac{k_{1z}^2 + \eta_2^2}{k_{1z}^2 + (\eta_2\varepsilon_1/\varepsilon_2)^2}, \text{ for } p \text{ polarization} \tag{8.93}$$

where we have used $\alpha = \tan^{-1}(\eta_2\varepsilon_1/k_{1z}\varepsilon_2)$. In formulating the above equation, $k_x$ is always taken as positive. Equation (8.92) clearly suggests that $\langle S_x \rangle$ and $k_x$ have the same sign when the permittivity is positive and different signs when the permittivity is negative [41]. When $\varepsilon_1$ and $\varepsilon_2$ have different signs, the lateral shift D will be negative, which implies that the lateral shift is opposite to $\langle \mathbf{S} \rangle_x$ of the incident beam. For a TE wave, one can simply replace $\varepsilon$'s by $\mu$'s in Eqs. (8.93). For two dielectrics, we have $\mu_1 = \mu_2 = 1$, $\varepsilon_1 = n_1^2$, and $\varepsilon_2 = n_2^2$, where $n_1$ and $n_2$ are the refractive indices of medium 1 and 2, respectively. Consequently, Eq. (8.93) reduces to the following:

$$D_s = \frac{2\tan\theta_1}{\eta_2} \text{ for a TE wave} \tag{8.94a}$$

and

$$D_p = \frac{2\tan\theta_1}{\eta_2\left(n_1^2\sin^2\theta_1/n_2^2 - \cos^2\theta_1\right)} \text{ for a TM wave} \tag{8.94b}$$

At grazing incidence, $k_{1z} \to 0$, however, the shift in the direction parallel to the beam is $D\cos\theta_1 = (2/\eta_2)(\varepsilon_2/\varepsilon_1)\sin\theta_1$, which approaches a finite value and does not diverge. At the critical angle, $\theta_1 = \theta_c = \sin^{-1}(n_1/n_1)$, $\eta_2 = \delta = 0$, and D approaches infinity. This difficulty can be removed by using the Gaussian beam incidence [42]. Quantum mechanics has also been applied to predict the lateral beam shift [39]. The Goos–Hänchen effect also has its analogy in acoustics and is of contemporary interest in dealing with negative index materials, waveguides, and photon tunneling [41, 43, 44].

### 8.3.3 Bidirectional Reflectance Distribution Function

Real surfaces contain roughness or texture that depends on the processing method. A surface appears to be smooth if the wavelength is much greater than the surface roughness height. A highly polished surface can have a roughness height on the order of nanometers. Some surfaces that appear "rough" to human eyes may appear to be quite "smooth" for far-infrared radiation. The root-mean-square (rms) roughness is a commonly used parameter to describe surface roughness. The power spectral density provides more general information on the vertical and spatial extent of surface irregularities. Zhang et al. [31] gave a detailed discussion on the roughness parameters as well as the instruments used for surface characterization.

The reflection of radiation by rough surfaces is more complicated. For randomly rough surfaces, there often exist a peak around the direction of specular reflection, an off-specular lobe, and a diffuse component. When the surface contains periodic structures, such as patterned or microfabricated surfaces, diffraction effects may become important and several peaks may appear. The bidirectional reflectance distribution function (BRDF), which is a function of the angles of incidence and reflection, fully describes the reflection characteristics from a rough surface at a given wavelength. As illustrated in Fig. 8.11, the BRDF is defined as the reflected radiance (intensity) divided by the incident irradiance (flux) at the surface [45]

$$f_\mathrm{r}(\lambda, \theta_i, \phi_\mathrm{i}, \theta_r, \phi_\mathrm{r}) = \frac{\mathrm{d}I_\mathrm{r}}{I_\mathrm{i} \cos\theta_i \mathrm{d}\Omega_i} \ [\mathrm{sr}^{-1}] \tag{8.95}$$

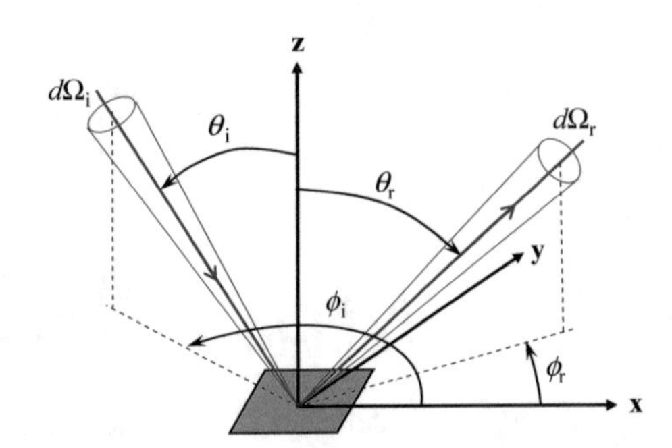


**Fig. 8.11** Geometry of the incident and reflected beams in defining the BRDF

where $(\theta_i, \phi_i)$ and $(\theta_r, \phi_r)$ denote the directions of incident and reflected beams, respectively, $I_i$ is the incident irradiance (radiant power per unit area), and $dI_r$ is the reflected radiance (intensity).

The directional-hemispherical reflectance can be obtained by integrating the BRDF over the hemisphere:

$$\rho'_\lambda = \int_{2\pi} f_r \cos\theta_r \, d\Omega_r \tag{8.96}$$

An important principle of the BRDF is reciprocity, which specifies symmetry of the BRDF, with regard to reflection and incidence angles. In other words, the reflectance for energy incident from $(\theta_i, \phi_i)$ and reflected to $(\theta_r, \phi_r)$ is equal to that for energy incident from $(\theta_r, \phi_r)$ and reflected to $(\theta_i, \phi_i)$. Therefore,

$$f_r(\lambda, \theta_i, \phi_i, \theta_r, \phi_r) = f_r(\lambda, \theta_r, \phi_r, \theta_i, \phi_i) \tag{8.97}$$

The BRDF reciprocity is an extension of the Helmholtz reciprocity principle [46]. While the reciprocity principle holds for most passive medium and surfaces, it does not hold in some nonlinear or magnetic media.

For a *diffuse* or *Lambertian* surface, the BRDF is independent of $(\theta_r, \phi_r)$ and is related to the directional-hemispherical reflectance as $f_{r,dif} = \rho'_\lambda/\pi$. On the other hand, the BRDF for an ideal *specular,* or mirror-like, reflector can be represented as

$$f_{r,spe} = \frac{\rho'_\lambda}{\cos\theta_i}\delta_\theta(\theta_r - \theta_i)\delta_\phi(\phi_r - \phi_i - \pi) \tag{8.98}$$

where the Dirac delta function $\delta(x)$ is zero everywhere, except at $x = 0$. Furthermore, the delta functions are normalized such that

$$\int_{2\pi} \delta_\theta(\theta_r - \theta_i)\delta_\phi(\phi_r - \phi_i - \pi) d\Omega_r = 1 \tag{8.99}$$

In general, the BRDF of a real surface should fall between the two extreme cases. It should be noted that for a perfectly smooth surface, the reflectivity calculated from the Fresnel coefficient, discussed in Sect. 8.3.1, can also be understood as the directional-hemispherical reflectance. Further discussions on BRDF models based on geometric optics and physical optics, as well as rigorous solutions of the Maxwell equations, will be given in Chap. 9, where we will also study the effect of surface microstructures on the BRDF and how to characterize a rough surface.

### 8.3.4 *Emittance (Emissivity) and Kirchhoff's Law*

Real materials have finite thicknesses. The assumption of semi-infinity or opaqueness requires that the thickness be much greater than the radiation penetration depth. This is usually not a problem for a metal in the visible or infrared spectral regions. When this is not the case, we are dealing with a transparent or semitransparent material, like a glass window. The radiative properties of semitransparent layers and thin films will be studied in the next chapter. Laser beams or light from a spectrophotometer do not extend to infinity and are not perfectly collimated. Nevertheless, as long as the diameter of the beam spot is much greater than the wavelength and the beam divergence is not very large, the directional, spectral reflectivity and absorptivity, calculated from the previous section, are applicable to most situations and can be integrated to obtain the properties for finite conic angles or hemispherical properties. For real materials, we use reflectance and absorptance that depend on the nature of surfaces and coatings.

For real surfaces, the ratio of the *emissive power* of the surface to that of a blackbody at the same temperature defines the hemispherical emittance (or emissivity) $\varepsilon_\lambda^{\mathrm{h}}$. The directional emittance (or emissivity) $\varepsilon_\lambda'$ is defined based on the intensity ratio. The total emittance can be evaluated by integrating the spectral emittance over all wavelengths weighted by the blackbody distribution function. A concise discussion of radiative properties can be found from a popular heat transfer textbook [47] and more complete definitions and relations can be found from Howell et al. [1] and Modest [2].

To establish the relationship between the radiative properties, consider an opaque surface at a temperature $T_\mathrm{s}$ inside a vacuum enclosure whose walls are at a temperature $T_\mathrm{w}$, as shown in Fig. 8.12. Whether the surface is inside the enclosure or not, we must have

$$\alpha_\lambda'(\theta, \phi) + \rho_\lambda'(\theta, \phi) = 1 \tag{8.100a}$$

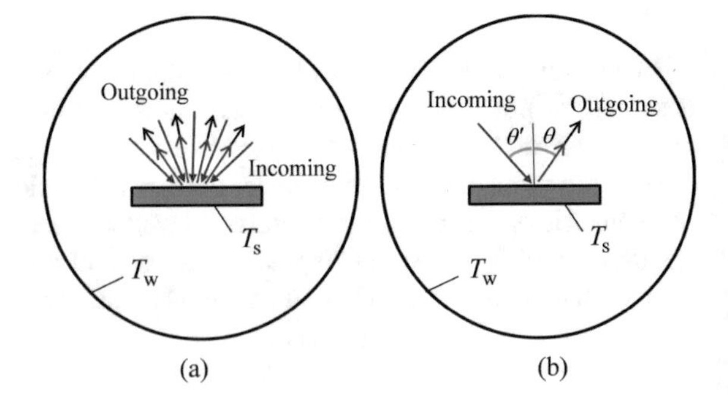


**Fig. 8.12** Schematic of a blackbody enclosure for consideration of **a** hemispherical properties and **b** directional properties. Note that the outgoing radiation has two arrows, one represents emission and the other reflection

Because the incoming radiation will either be absorbed or reflected as long as the surface is opaque, the sum of the directional absorptance and the directional-hemispherical reflectance must be unity. Furthermore, the sum of the hemispherical absorptance and the hemispherical–hemispherical reflectance must also be unity:

$$\alpha_\lambda^{\mathrm{h}} + \rho_\lambda^{\mathrm{h}} = 1 \text{ and } \alpha_{\mathrm{tot}}^{\mathrm{h}} + \rho_{\mathrm{tot}}^{\mathrm{h}} = 1 \tag{8.100b}$$

where the subscript "tot" signifies a total property.

Consider the enclosure at thermal equilibrium where no internal sources or sinks exist in the surface element. The temperatures of the surface element and the wall must be the same. Furthermore, as shown earlier with the blackbody cavity concept, the intensity and emissive power inside the enclosure are independent on the location and direction. Based on Fig. 8.12a for the hemispherical properties, we see that the combination of the emitted and reflected power per unit area must be the same as that of the incident. Energy balance requires that

$$e_{\mathrm{b},\lambda}(\lambda, T_{\mathrm{w}}) = \varepsilon_\lambda^{\mathrm{h}} e_{\mathrm{b},\lambda}(\lambda, T_{\mathrm{s}}) + \rho_\lambda^{\mathrm{h}} e_{\mathrm{b},\lambda}(\lambda, T_{\mathrm{w}}) \tag{8.101a}$$

and

$$\sigma_{\mathrm{SB}} T_{\mathrm{w}}^4 = \varepsilon_{\mathrm{tot}}^{\mathrm{h}} \sigma_{\mathrm{SB}} T_{\mathrm{s}}^4 + \rho_{\mathrm{tot}}^{\mathrm{h}} \sigma_{\mathrm{SB}} T_{\mathrm{w}}^4 \tag{8.101b}$$

By setting $T_{\mathrm{s}} = T_{\mathrm{w}}$ and combining with Eq. (8.100b), we have

$$\varepsilon_\lambda^{\mathrm{h}} = \alpha_\lambda^{\mathrm{h}} \text{ and } \varepsilon_{\mathrm{tot}}^{\mathrm{h}} = \alpha_{\mathrm{tot}}^{\mathrm{h}} \tag{8.102}$$

The equality between emittance and absorptance is called Kirchhoff's law. Note that in the literature, both emissivity and absorptivity are commonly used regardless of the nature of the surfaces. The hemispherical properties depend on the directional and spectral behavior of the surface. Furthermore, the hemispherical absorptance depends on the condition of the incident intensity distribution. Hence, the two equalities given in Eq. (8.102) do not hold in general. Special situations exist under ideal assumptions, for example, if a surface is diffuse, $\varepsilon_\lambda^{\mathrm{h}} = \alpha_\lambda^{\mathrm{h}}$ always holds. Furthermore, if a surface is diffuse-gray, both the equalities in Eq. (8.102) hold without requiring thermal equilibrium [47]. Real surfaces rarely meet these requirements, however.

Equation (8.102) can also be understood by considering the energy balance of the surface, that is, the absorbed radiant power must equal the emitted so that the emittance must be equal to the absorptance. This argument is justifiable for the hemispherical properties. When dealing with directional properties, as can be seen from Fig. 8.12b, the emitted and the reflected components toward the same direction $\theta$ should add up to give the blackbody intensity. Under thermal equilibrium, we can obtain the following expression [46]:

$$\varepsilon_\lambda'(\lambda, \theta, \phi) + \int_{2\pi} f_{\mathrm{r}}(\lambda, \theta', \phi', \theta, \phi) \cos\theta' \mathrm{d}\Omega' = 1 \tag{8.103a}$$

Here, the integration is over the incident hemisphere. The second term in Eq. (8.103a) is the hemispherical-directional reflectance [1, 2]. When BRDF reciprocity holds, it is the same as the directional-hemispherical reflectance $\rho'_\lambda$. For a specular surface, Eq. (8.103a) can be written as

$$\varepsilon'_\lambda(\lambda, \theta, \phi) + \rho'_\lambda(\lambda, \theta', \phi') = 1 \tag{8.103b}$$

where $(\theta', \phi')$ and $(\theta, \phi)$ are a pair of specular incidence and reflection angles, i.e., $\theta' = \theta$ and $\phi' = \phi + 180°$. Zhu and Fan [48] showed that Eq. (8.103b) holds even in nonreciprocal systems such as magneto-optical materials. When BRDF reciprocity holds, the conventional Kirchhoff's law $\varepsilon'_\lambda = \alpha'_\lambda$ for the spectral directional properties can be derived. Therefore, the spectral, directional emittance can be expressed in terms of the spectral, directional-hemispherical reflectance as follows:

$$\varepsilon'_\lambda = 1 - \rho'_\lambda \tag{8.104}$$

When a material is not at thermal equilibrium with its surroundings, its emittance is defined based solely on spontaneous emission and is an intrinsic property of the material that does not depend on the surroundings. On the other hand, the absorptance is defined based on the net absorbed energy by treating stimulated or induced emission as negative absorption. Under appropriate conditions, Kirchhoff's law according to the equality given in Eq. (8.104) is valid for individual polarization with or without thermal equilibrium. The assumptions are: (a) the material under consideration is reciprocal and at local thermal equilibrium, though not necessarily at equilibrium with the surroundings; (b) the external field is not strong enough to alter the material's intrinsic properties or cause a nonlinear effect. We can then compute the directional emittance for an opaque surface or semi-infinite media, from the directional-hemispherical reflectance for incidence from air or a vacuum.

The emittance is typically calculated by averaging over the two polarizations. The hemispherical emittance can then be obtained by integrating the directional emittance so that

$$\varepsilon^{\mathrm{h}}_\lambda = \frac{1}{\pi}\int_0^{2\pi}\int_0^{\pi/2} \varepsilon'_\lambda \cos\theta \sin\theta \mathrm{d}\theta \mathrm{d}\phi \tag{8.105}$$

It can be seen from Fig. 8.8a that, when averaged over the two polarizations, the reflectivity changes little until the Brewster angle and then increases to 1 when the incidence angle approaches 90°. The hemispherical emittance for a nonmetallic surface is about 10% smaller than the normal emittance. On the other hand, the hemispherical emittance for metallic surfaces is about 20% greater than the normal emittance. Diffuse emission is a good first-order approximation, even though the surface is smooth and the reflection is specular. Thus, the hemispherical emittance may be approximated by the normal emittance. In most studies, the emittance is calculated from the indirect method, based on the reflectivity and Kirchhoff's law given in Eq. (8.104). Direct calculations can be accomplished by considering the emission, along with the absorption and transmission, inside the material. According to the fluctuation-dissipation theorem (FDT), thermal emission arises from the induced field that originated from the random charge fluctuation. Wang et al. [49] used the FDT to directly calculate the emittance of a layered structure and demonstrated the equivalence between the direct method and the indirect method based on Eq. (8.104). The fluctuational electrodynamics is essential for the study of near-field radiation and will be carefully discussed in Chap. 10.

The total-hemispherical emittance can be evaluated using Planck's distribution. Therefore,

$$\varepsilon_{\text{tot}}^{\text{h}} = \frac{\int_0^\infty \varepsilon_\lambda^{\text{h}}(\lambda) e_{\text{b},\lambda}(\lambda, T)\text{d}\lambda}{\int_0^\infty e_{\text{b},\lambda}(\lambda, T)\text{d}\lambda} = \frac{\int_0^\infty \varepsilon_\lambda^{\text{h}}(\lambda) e_{\text{b},\lambda}(\lambda, T)\text{d}\lambda}{\sigma_{\text{SB}} T^4} \tag{8.106}$$

The total emittance depends on the surface temperature and the spectral variation of the optical constants. Pure metals usually have a very low emittance, and the emittance can increase due to surface oxidation. Spectrally selective materials that appear to be reflective to visible light may exhibit a large total emittance, greater than 0.9 near ambient temperature; examples are snow and white paint. An earlier compilation of the radiative properties of many engineering materials can be found in Touloukian and DeWitt [50]. The use of surface microstructures to modify emission characteristics will be discussed in the next chapter.

## 8.4 Dielectric Function Models

Unlike in dilute gases where the molecules are far apart, in solids, the closely packed atoms form band structures. Absorption in solids usually happens in a much broader frequency region or band. Free electrons in metals can interact with the incoming electromagnetic waves or photons, and cause broadband absorption from the visible (or even ultraviolet) all the way to the microwave and longer wavelengths. For semiconductors especially with high impurity (doping) concentrations or at elevated temperatures, both the free electrons and holes contribute to the absorption process. The absorption of a photon makes the electron or the hole transit to a higher energy state within the same band. Therefore, free-carrier absorption is caused by *intraband transitions*. In order to conserve momentum, the carriers must also collide with ionized impurities, phonons, other carriers, grain boundaries, interfaces, and so forth. The collisions act as a *damping force* on the motion of carriers. The Drude model describes the oscillatory movement of an electron, driven by a harmonic field, which is subjected to a damping force. The model is simple in form and predicts the dielectric function of some metals fairly well in a broad spectral region, especially in the mid and far infrared.

Absorption by lattice vibrations or bound electrons, which is important for insulators and lightly doped semiconductors, is due to the existence of electric dipoles formed by the lattice. The strongest absorption is achieved when the frequency equals the vibrational mode of the dipole, i.e., the resonance frequency, which is usually in the mid- to far-infrared region of the spectrum. The contribution of bound electrons is often modeled by the Lorentz model.

Interband transition is the *fundamental absorption process* in semiconductors. An electron can be excited from the valence band to the conduction band by absorbing a photon, whose energy is greater than the energy gap $E_g$. Because the absorption by electrons is usually weak in semiconductors, a strong absorption edge is formed near the bandgap. In this transition process, both the energy and the momentum must be conserved.

This section discusses the formulation for different contributions to the dielectric function. It should be noted that the real and imaginary parts of the dielectric function are interrelated according to the causality, which is discussed first. Because all naturally occurring and most of the synthesized materials are nonmagnetic at high frequencies, only nonmagnetic materials are considered so that $\mu = 1$ and $n = \sqrt{\varepsilon}$ in the following, except in Sect. 8.4.6.

### 8.4.1 Kramers–Kronig Dispersion Relations

The real and imaginary parts of an analytic function are related by the Hilbert transform relations. Hendrik Kramers and Ralph Kronig were the first to show that the real and imaginary parts of the dielectric function are interrelated. These relations are called the Kramers–Kronig dispersion relations or K-K relations. The K-K relations can be interpreted as the causality in the frequency domain and are very useful in obtaining optical constants from limited measurements. The principle of causality states that *the effect cannot precede the cause,* or *no output before input.* Some important relations are given here, and a detailed derivation and proofs can be found from Jackson [5], Born and Wolf [8], and Bohren and Huffman [9].

The real part $\varepsilon'$ and the imaginary part $\varepsilon''$ of a dielectric function are related by

$$\varepsilon'(\omega) - 1 = \frac{2}{\pi}\wp\int_0^\infty \frac{\zeta\varepsilon''(\zeta)}{\zeta^2 - \omega^2}\mathrm{d}\zeta \tag{8.107a}$$

and

$$\varepsilon''(\omega) - \frac{\sigma_0}{\varepsilon_0\omega} = -\frac{2\omega}{\pi}\wp\int_0^\infty \frac{\varepsilon'(\zeta) - 1}{\zeta^2 - \omega^2}\mathrm{d}\zeta \tag{8.107b}$$

where $\sigma_0$ is the dc conductivity, $\wp$ denotes the Cauchy principal value of the integral, and $\zeta$ is a dummy frequency variable. These relations can be written in terms of $n$ and $\kappa$ as

$$n(\omega) - 1 = \frac{2}{\pi}\wp \int_0^\infty \frac{\zeta\kappa(\zeta)}{\zeta^2 - \omega^2} d\zeta \tag{8.108a}$$

$$\kappa(\omega) = -\frac{2\omega}{\pi}\wp \int_0^\infty \frac{n(\zeta) - 1}{\zeta^2 - \omega^2} d\zeta \tag{8.108b}$$

Equations (8.107a), (8.107b) and (8.108a), (8.108b) are the K-K relations, which relate the real part of a causal function to an integral of its imaginary part over all frequencies, and vice versa. A number of sum rules can be derived based on the K-K relations and are useful in obtaining or validating the dielectric function of a given material [15]. The K-K relations can be applied to reflectance spectroscopy to facilitate the determination of optical constants from the measured reflectivity of a material. For radiation incident from a vacuum to a medium with a complex refractive index $(n + \mathrm{i}\kappa)$ at normal incidence, the Fresnel reflection coefficient for TE waves is

$$r(\omega) = |r(\omega)|\mathrm{e}^{\mathrm{i}\phi(\omega)} = \frac{1 - n(\omega) - \mathrm{i}\kappa(\omega)}{1 + n(\omega) + \mathrm{i}\kappa(\omega)} \tag{8.109}$$

where $|r|$ is the amplitude and $\phi$ the phase shift upon reflection for the electric field. The reflectivity expressed in terms of $\omega$ is

$$\rho'_\omega(\omega) = rr^* = |r|^2 \tag{8.110}$$

The amplitude and the phase are related, and it can be shown that

$$\phi(\omega) = -\frac{\omega}{\pi}\wp \int_0^\infty \frac{\ln \rho'_\omega(\zeta)}{\zeta^2 - \omega^2} d\zeta \tag{8.111}$$

The refractive index and the extinction coefficient can be calculated, respectively, from

$$n(\omega) = \frac{1 - \rho'_\omega}{1 + \rho'_\omega - 2\cos\phi\sqrt{\rho'_\omega}} \tag{8.112a}$$

and

$$\kappa(\omega) = \frac{2\sin\phi\sqrt{\rho'_\omega}}{1 + \rho'_\omega - 2\cos\phi\sqrt{\rho'_\omega}} \tag{8.112b}$$

### 8.4.2 *The Drude Model for Free Carriers*

The Drude model describes the frequency-dependent conductivity of metals and can be extended to free carriers in semiconductors. In the absence of an electromagnetic field, free electrons move randomly. When an electromagnetic field is applied, free electrons acquire a nonzero average velocity, giving rise to an electric current that oscillates at the same frequency as the electromagnetic field. The collisions with the stationary atoms result in a damping force on the free electrons, which is proportional to their velocity. The equation of motion for a single free electron is then

$$m_e\ddot{\mathbf{x}} = -m_e\gamma\dot{\mathbf{x}} - e\mathbf{E} \tag{8.113}$$

where $e$ is the absolute charge of an electron, $m_e$ is the electron mass, and $\gamma$ denotes the strength of the damping due to collision, i.e., the *scattering rate* or the inverse of the relaxation time $\tau$. Assume the electron motion under a harmonic field $\mathbf{E} = \mathbf{E}_0\mathrm{e}^{-\mathrm{i}\omega t}$ is of the form $\mathbf{x} = \mathbf{x}_0\mathrm{e}^{-\mathrm{i}\omega t}$ so that $\ddot{\mathbf{x}} = -\mathrm{i}\omega\dot{\mathbf{x}}$. We can rewrite Eq. (8.113) as

$$\dot{\mathbf{x}} = \frac{e/m_e}{\mathrm{i}\omega - \gamma}\mathbf{E} \tag{8.114}$$

The electric current density is $\mathbf{J} = -n_e e\dot{\mathbf{x}} = \tilde{\sigma}(\omega)\mathbf{E}$; therefore, the complex conductivity is

$$\tilde{\sigma}(\omega) = \frac{n_e e^2/m_e}{\gamma - \mathrm{i}\omega} = \frac{\sigma_0}{1 - \mathrm{i}\omega/\gamma} \tag{8.115}$$

where $\sigma_0 = n_e e^2\tau/m_e$ is the dc conductivity, as discussed in Chap. 5. Equation (8.115) is called the Drude free-electron model, which describes the frequency-dependent complex conductivity of a free-electron system, in terms of the dc conductivity and the scattering rate, in a rather simple form. The electrical conductivity approaches the dc conductivity at very low frequencies (or very long wavelengths). The dielectric function is related to the conductivity by Eq. (8.28); thus,

$$\varepsilon(\omega) = \varepsilon_\infty - \frac{\sigma_0\gamma}{\varepsilon_0(\omega^2 + \mathrm{i}\gamma\omega)} \tag{8.116}$$

where $\varepsilon_\infty$, which is on the order of 1, is included to account for contributions, other than the contribution of the free electrons, that are significant at high frequencies. There exist several transitions at the ultraviolet and visible regions for metals, such as *interband transitions*. Note that when $\omega \to \infty$, the real part of the dielectric function of all materials should approach unity, as can be seen from Eq. (8.107a). In the low-frequency limit when $\omega \ll \gamma$, $\tilde{\sigma}(\omega \to 0) \approx \sigma_0$ and $\varepsilon'' \gg \varepsilon'$. Therefore,

$$n \approx \kappa \approx \sqrt{\frac{\sigma_0}{2\varepsilon_0\omega}} \tag{8.117}$$

This is the Hagen–Ruben equation and is applicable at very long wavelengths [1]. Both the refractive index and the extinction coefficient will increase with the square root of wavelength in vacuum. It is interesting to note that the radiation penetration depth $\delta_\lambda = \lambda/(4\pi\kappa)$ will also increase with the square root of wavelength. As an example, consider gold at $\lambda = 4$ μm with $\kappa = 25$. The penetration depth is 13 nm at this wavelength. If the wavelength is increased to 4 cm, which is well into the microwave region, the penetration depth would increase to 1.3 μm. Generally speaking, metals are highly reflecting in the infrared wavelength region.

The *plasma frequency* is defined according to $\omega_\mathrm{p}^2 = \frac{\sigma_0 \gamma}{\varepsilon_0} = \frac{n_\mathrm{e} e^2}{m_\mathrm{e}\varepsilon_0}$. Using the plasma frequency, we can write Eq. (8.116) in a more compact form as follows:

$$\varepsilon(\omega) = \varepsilon_\infty - \frac{\omega_\mathrm{p}^2}{\omega(\omega + \mathrm{i}\gamma)} \tag{8.118}$$

If $\omega \gg \gamma$, the dielectric function can be approximated as

$$\varepsilon(\omega) \approx \varepsilon_\infty - \frac{\omega_\mathrm{p}^2}{\omega^2}\left(1 - \mathrm{i}\frac{\gamma}{\omega}\right), \text{ when } \omega \gg \gamma \tag{8.119}$$

The plasma frequency falls in the ultraviolet region for most metals. For example, the wavelength corresponding to the plasma frequency is approximately 80 nm for aluminum and 200 nm for tungsten. When $\omega \gg \omega_\mathrm{p}$, as in the x-ray region, $\varepsilon(\omega) \to 1 + \mathrm{i}\gamma\omega_\mathrm{p}^2/\omega^3$. Thus, metals become highly absorptive and not so reflective. Take tungsten as an example. At $\lambda = 1$ nm, the optical constants are $n \approx 1$ and $\kappa = 4 \times 10^{-4}$. The penetration depth is calculated to be $\delta_\lambda = 200$ nm. Because the refractive index is similar to that of air, the reflection is very weak and most of the incident radiation is absorbed within a depth of 1 μm. Some metal foils become semitransparent, for example, the radiation penetration depth in lithium is close to 100 μm at $\lambda = 1$ nm. The Center for X-Ray Optics at Lawrence Berkeley National Laboratory maintains a website on x-ray properties [51]. If $\omega < \omega_p$, the real part of the dielectric function $\varepsilon'$ becomes negative, and the extinction coefficient is much greater than the refractive index, i.e., $\kappa \gg n$. According to Eq. (8.84), this corresponds to a high reflectivity. A vanishing real part of the refractive index corresponds to a longitudinal collective oscillation of the electron gas, i.e., a *plasma oscillation*. Plasma oscillations originate from a long-range correlation of electrons caused by Coulomb forces.

**Example 8.7** From Table 5.2, calculate the plasma frequency and the electron scattering rate for aluminum. Then calculate its dielectric function and compare the normal reflectivity with data.

**Solution** For aluminum near room temperature, $n_\mathrm{e} = 18.1 \times 10^{28}$ m$^{-3}$ and $\sigma_0 = 1/r_\mathrm{e} = 3.75 \times 10^7$ m/Ω. From Appendix A, $e = 1.602 \times 10^{-19}$ C, $m_\mathrm{e} = 9.109 \times 10^{-31}$ kg, and $\varepsilon_0 = 8.854 \times 10^{-12}$ C$^2$/N m$^2$. Hence, $\gamma = n_\mathrm{e}e^2/m_\mathrm{e}\sigma_0 = 1.4 \times 10^{14}$ rad/s, or the scattering time $\tau = 7.2 \times 10^{-14}$ s, and $\omega_\mathrm{p} = 2.4 \times 10^{16}$ rad/s,

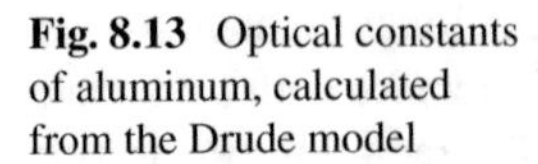

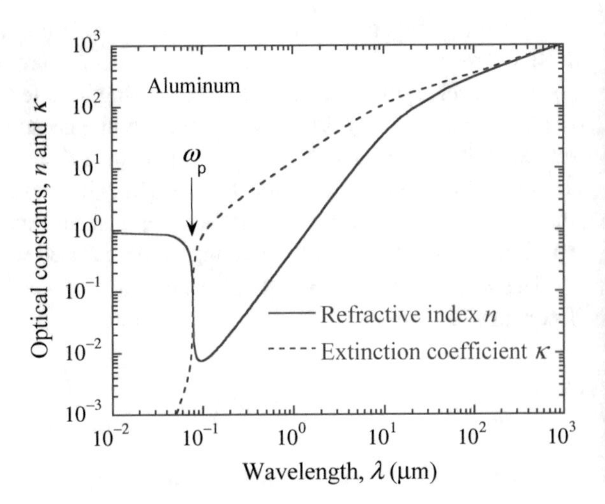


**Fig. 8.13** Optical constants of aluminum, calculated from the Drude model

which corresponds to a wavelength of 79 nm. The exact parameters may differ slightly in different references, and sometimes an effective mass is used which is slightly larger than the electron rest mass. The predicted optical constants are plotted in Fig. 8.13, assuming $\varepsilon_\infty = 1$. It can be seen that as the wavelength exceeds 100 μm, the difference between $n$ and $\kappa$ diminishes. In the region 0.1 μm $< \lambda <$ 200 μm, $n < \kappa$ so that the real part of the dielectric function $\varepsilon' = n^2 - \kappa^2$ becomes negative. A sharp transition occurs at the plasma frequency so that $n \to 1$ and $\kappa$ decreases rapidly toward higher frequencies.

As shown in Fig. 8.14, the reflectivity calculated from Eq. (8.84) is compared with the measured data for an aluminum film, which was prepared by ultrahigh vacuum deposition and measured in high vacuum to avoid oxidation [15]. The results agree very well at wavelengths greater than 2 μm. For $\lambda < 1$ μm, the contribution from the interband transition causes a reduction in the reflectivity. Note that the simple Drude model did not include these effects and is applicable for long wavelengths only. The established optical constants of metals are based on the measured reflectivity in a broad spectral region by using the K-K relations described in Sect. 8.4.1. The results for a large number of samples are tabulated in *Handbook of the Optical Constants of Solids*, with pertinent references [15].

In some studies, the Drude model is modified by considering the temperature and frequency dependence of the scattering rate and the effective mass. While the Drude model predicts well the radiative properties at room temperature or above, caution should be taken at extremely low temperatures. If the electron mean free path becomes comparable to the distance over which the electric field varies, i.e., the field penetration depth, nonlocal effects become important and the Drude theory breaks down. This can occur at cryogenic temperatures, and a more complex theory called the *anomalous skin effect* theory must then be applied [52].

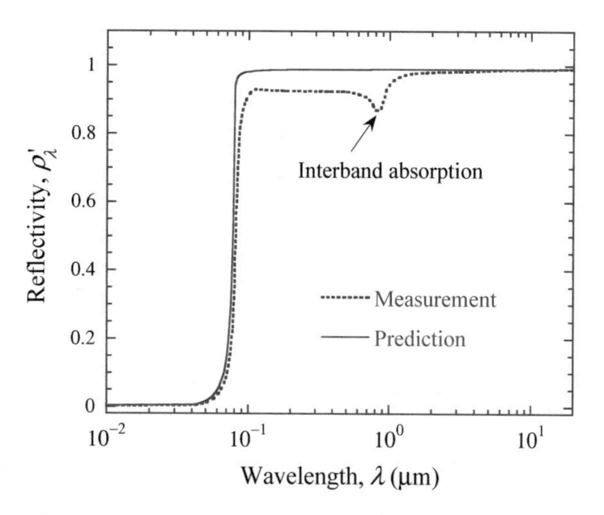


**Fig. 8.14** Normal spectral reflectivity of aluminum

## *8.4.3 The Lorentz Oscillator Model for Phonon Absorption*

Vibrations of lattice ions and bound electrons contribute to the dielectric function in a certain frequency region, often in the mid infrared. The refractive index can be calculated using the Lorentz oscillator model, which assumes that a bound charge $e$ is accelerated by the local electric field **E**, which is assumed to be the same as the applied field here. In contrast to free electrons, a bound charge experiences a restoring force determined by a spring constant $K_j$. The oscillator is further assumed to have a mass $m_j$ and a damping coefficient $\gamma_j$, as shown in Fig. 8.15. The force balance yields the equation of motion for the oscillator:

$$m_j\ddot{\mathbf{x}} + m_j\gamma_j\dot{\mathbf{x}} + K_j\mathbf{x} = e\mathbf{E} \tag{8.120}$$

The solution for a harmonic field $\mathbf{E} = \mathbf{E}_0 \mathrm{e}^{-\mathrm{i}\omega t}$, valid at timescales greater than the relaxation time, is given by

$$\mathbf{x} = \frac{e/m_j}{\omega_j^2 - \mathrm{i}\gamma_j\omega - \omega^2}\mathbf{E} \tag{8.121}$$

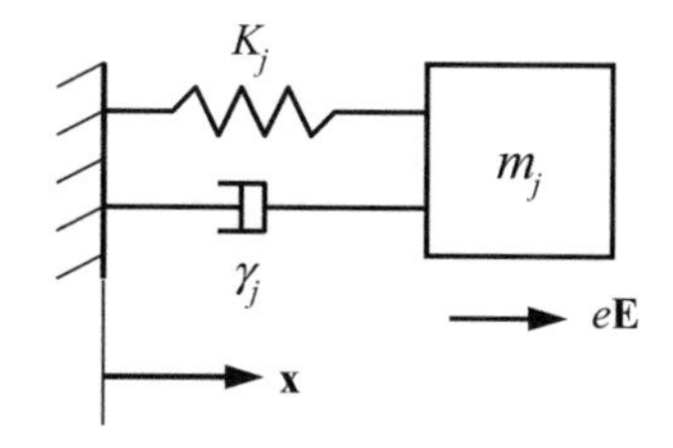


**Fig. 8.15** The classical oscillator model

where $\omega_j = (K_j/m_j)^{1/2}$ is the resonance frequency of the *j*th oscillator. The motion of the single oscillator causes a dipole moment $e\mathbf{x}$. If the number density of the *j*th oscillator is $n_j$, the polarization vector, or the dipole moment per unit volume, is $\mathbf{P} = \sum_{j=1}^{N} n_j e\mathbf{x}$, where $N$ is the total number of infrared active phonon modes (oscillators). The constitutive relation gives the polarization as $\mathbf{P} = (\varepsilon - 1)\varepsilon_0\mathbf{E}$. It can be shown that

$$\varepsilon(\omega) = \varepsilon_\infty + \sum_{j=1}^{N} \frac{S_j\omega_j^2}{\omega_j^2 - \mathrm{i}\gamma_j\omega - \omega^2} \tag{8.122}$$

where $\varepsilon_\infty$ is a high-frequency constant and $S_j = \omega_{pj}^2/\omega_j^2 = n_j e^2/(\varepsilon_0 m_j \omega_j^2)$ is called the *oscillator strength.*

At very low frequencies, $\varepsilon(0) = \varepsilon_\infty + \sum_{j=1}^{N} S_j$, which is called the dielectric constant. The real and imaginary parts of the dielectric function ($\varepsilon'$, $\varepsilon''$) and optical constants ($n$, $\kappa$) for a simple oscillator are illustrated in Fig. 8.16, near the resonance frequency for $\varepsilon_\infty = 1$. It can be seen from Eq. (8.122) and Fig. 8.16 that, for frequencies much lower or much higher than the resonance frequency $\omega_j$, $\varepsilon''$ and $\kappa$ are negligible. Only within an interval of $\gamma_j$ around the resonance frequency is the absorption appreciable. Within the absorption band, the real part of the refractive index decreases with frequency; this phenomenon is called *anomalous dispersion*. It follows that in an interval of width $\gamma_j$ around the resonance frequency, the Lorentz oscillator is highly reflecting and absorbing, while for higher or lower frequencies, it acts as a transparent material. The real part of the dielectric function becomes negative in a frequency region somewhat higher than $\omega_j$. A more complicated treatment based on quantum mechanics yields a four-parameter model [53]. The previous classical oscillator model can be considered as a good approximation when the relaxation times of the longitudinal and transverse optical phonons are close to each other. In some

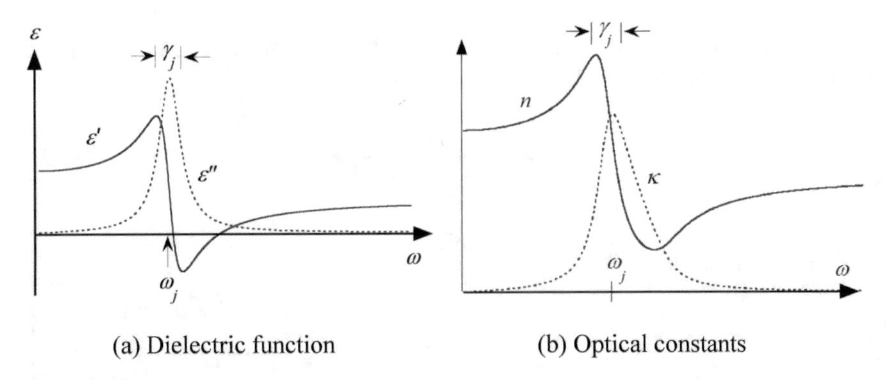


**Fig. 8.16** The dielectric behavior predicted by the Lorentz oscillator model. **a** Real part and imaginary part of the dielectric function. **b** Refractive index and extinction coefficient

studies, frequency- and temperature-dependent scattering rate is also considered to model the infrared spectra.

Due to the large number of parameters involved, it is much more difficult to determine the Lorentz oscillator parameters than to determine the Drude parameters. In practice, the oscillator parameters are often treated as adjustable parameters that are obtained by fitting Eq. (8.122) to the measured reflectivity data. The Lorentz model has been applied to a large number of dielectric materials by fitting the reflectance spectra [15]. The author and collaborators have obtained the Lorentz parameters for several perovskite crystals ($LaAlO_3$, $LaGaO_3$, and $NdGaO_3$), thin polyimide films, $HfO_2$ and $Ta_2O_5$ films, as well as certain ceramic materials [54].

**Example 8.8** The Lorentz model for SiC at room temperature for an ordinary ray is given as follows:

$$\varepsilon(\omega) = \varepsilon_\infty \left[ 1 + \frac{\omega_{\rm LO}^2 - \omega_{\rm TO}^2}{\omega_{\rm TO}^2 - {\rm i}\gamma\omega - \omega^2} \right] \tag{8.123}$$

where $\omega_{\rm LO} = 969\ {\rm cm}^{-1}$ and $\omega_{\rm TO} = 793\ {\rm cm}^{-1}$ are the frequencies corresponding to the longitudinal and transverse optical phonons, respectively, $\gamma = 4.76\ {\rm cm}^{-1}$, and $\varepsilon_\infty = 6.7$ [55]. What are the refractive indices at the high- and low-frequency limits? Calculate the normal reflectivity and compare it with the experimental result.

**Solution** Comparing Eqs. (8.122) and (8.123), we see that the resonance frequency corresponds to the TO phonon frequency, and the oscillation strength is $S_1 = \varepsilon_\infty(\omega_{\rm LO}^2/\omega_{\rm TO}^2 - 1) = 3.3$. The high-frequency limit of the refractive index is $n \approx \sqrt{\varepsilon_\infty} = 2.6$, and the low-frequency limit is $n = \sqrt{\varepsilon_\infty + S_1} = 3.16$. Note that transitions that occur in the visible and ultraviolet regions are not included so that the high-frequency limit is approximately 1 $\mu$m. On the other hand, because there are no other transitions at long wavelengths, the dielectric constant is approximately the same for zero frequency. The normal reflectivity is calculated using Eq. (8.84) and compared with the data, as shown in Fig. 8.17. The agreement is excellent since the Lorentz parameters were fitted to the experimental data [55]. The phonon band causes a large $\kappa$ value and hence a high reflectivity (very low emissivity) between $\omega_{\rm TO} = 793\ {\rm cm}^{-1}$ and $\omega_{\rm LO} = 969\ {\rm cm}^{-1}$. This band is called *reststrahlen band*. The German word "reststrahlen" means "residual rays" and the reststrahlen effect indicates the phenomenon of high reflectance in a dielectric material that is otherwise transparent. At $\omega = 1000\ {\rm cm}^{-1}$, the reflectivity is nearly 0 such that the emissivity is almost 1. This happens at the edge of the reststrahlen band, where the refractive index increases close to 1 and the extinction coefficient decreases to a very small value. This wavelength is called the *Christiansen wavelength*, and the associated phenomenon is called the *Christiansen effect* [9].

The density-functional perturbation theory (DFPT) can be used to perform first-principles calculations of the lattice dynamics. It can provide phonon dispersions as well as the resonance frequencies of different phonon modes [56]. It should be noted that some optical phonons are symmetric and they cannot be detected by

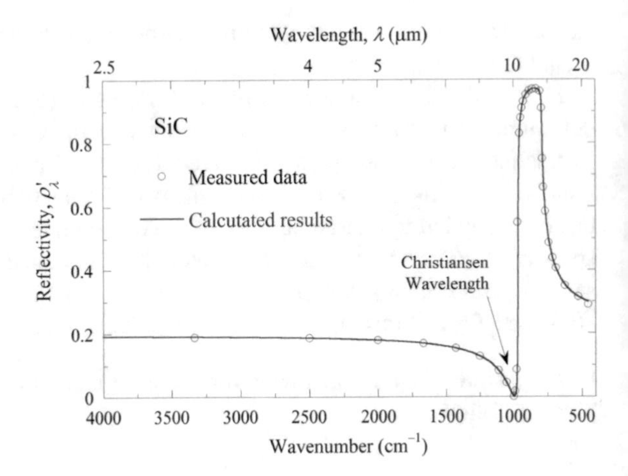


**Fig. 8.17** The calculated and measured normal reflectivity of SiC at room temperature

infrared spectroscopy but can show up in Raman spectroscopy like the phonons in Si and Diamond. Note that both the TO and LO vibrational frequencies can be determined with the DFPT including those that are not infrared active but are Raman active [56, 57]. However, the determination of the scattering rate from the first-principles simulation is more challenging. Bao et al. [58] obtained the phonon lifetime and resonance frequencies using an analysis of the ab initio molecular dynamics (MD) trajectories based on the normal modes and spectral density analysis methods, allowing the calculation of the far-infrared dielectric function of GaAs. This method may also be applied to multiple phonon oscillators.

## *8.4.4 Semiconductors*

The absorption coefficient of lightly doped silicon is shown in Fig. 8.18 to illustrate the contribution of different mechanisms [9, 59]. Let us look at the absorption of silicon in the visible and the infrared first, as shown in Fig. 8.18a. At short wavelengths, photon energies are large enough to excite electrons from the valence band to the conduction band. This interband transition causes the absorption coefficient to rise quickly as the photon energy $h\nu$ is increased above the indirect bandgap, which is approximately $E_g = 1.1$ eV at room temperature and decreases somewhat as temperature increases. As the wavelength further increases beyond the absorption edge, the absorption coefficient is affected by the existence of impurities and defects, absorption by free carriers (i.e., intraband or intersubband transitions by electrons and holes), and absorption by lattice vibrations. While the lattice vibration affects certain regions of the spectrum, the free-carrier contribution increases at longer wavelengths. For intrinsic silicon at low temperatures, the free-carrier concentration is very low, and thus silicon is transparent at wavelengths longer than the bandgap wavelength. Lattice absorption occurs in the mid infrared and introduces some absorption for

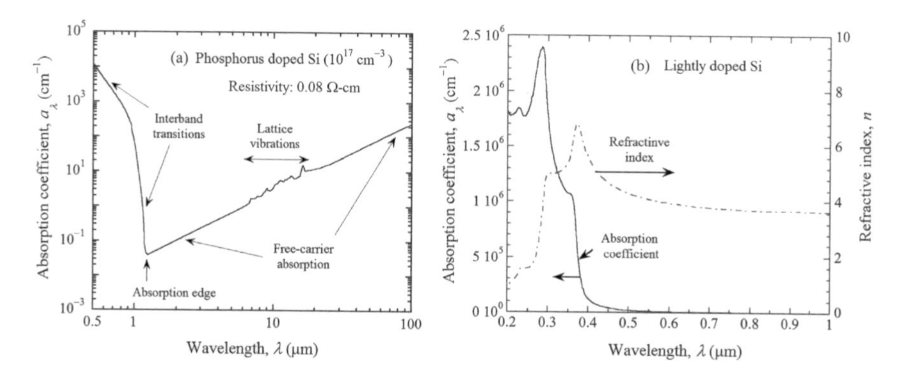


**Fig. 8.18** The absorption coefficient and the refractive index of Si at room temperature. **a** Absorption coefficient in the visible and the infrared. **b** Absorption coefficient and refractive index from the ultraviolet to the near infrared

$6\ \mu m < \lambda < 25\ \mu m$. Free-carrier absorption is important for doped silicon at longer wavelengths. Note that even for intrinsic silicon at high temperatures, thermally excited free carriers dominate the absorption at longer wavelengths; a 0.5-mm-thick silicon wafer is essentially opaque above 1000 K. The free-carrier concentration for intrinsic silicon is about $10^{10}$ cm$^{-3}$ at 300 K and nearly $10^{18}$ cm$^{-3}$ at 1000 K. As shown in Fig. 8.18b, the absorption coefficient continues to increase at shorter wavelengths, due to the interband transition associated with the direct bandgap, which dominates the optical characteristics of silicon in the ultraviolet region. This transition also affects the refractive index of silicon at longer wavelengths. Beyond 500 nm, the refractive index of lightly doped Si decreases somewhat as the wavelength increases.

Modeling the *interband transitions* requires quantum theory. First-principles or ab initio calculations have been performed to study the optical absorption spectrum of semiconductors and insulators, considering electron–hole interactions [60, 61]. In a direct-bandgap semiconductor, shown in Fig. 8.19a, the lowest point of the conduction band occurs at the same wavevector as the highest point of the valence band. An electron can be excited from the top of the valence band to the bottom of the conduction band by absorbing a photon of energy that is at least equal to the bandgap energy. When the valence band and the conduction band are parabola-like, the absorption coefficient due to direct bandgap absorption can be expressed as

$$a_{\text{bg}} = A(\hbar\omega - E_{\text{g}})^{1/2} \tag{8.124}$$

where $A$ is a parameter that depends on the effective masses of the electrons and the holes, and the refractive index of the material.

When a transition requires a change in both energy and momentum, as in the case for an indirect bandgap semiconductor shown in Fig. 8.19b, a phonon is either emitted (process 1) or absorbed (process 2) for momentum conservation because the photon itself cannot provide a change in momentum. This kind of transition is called *indirect interband transition*. With the involvement of phonons, the absorption

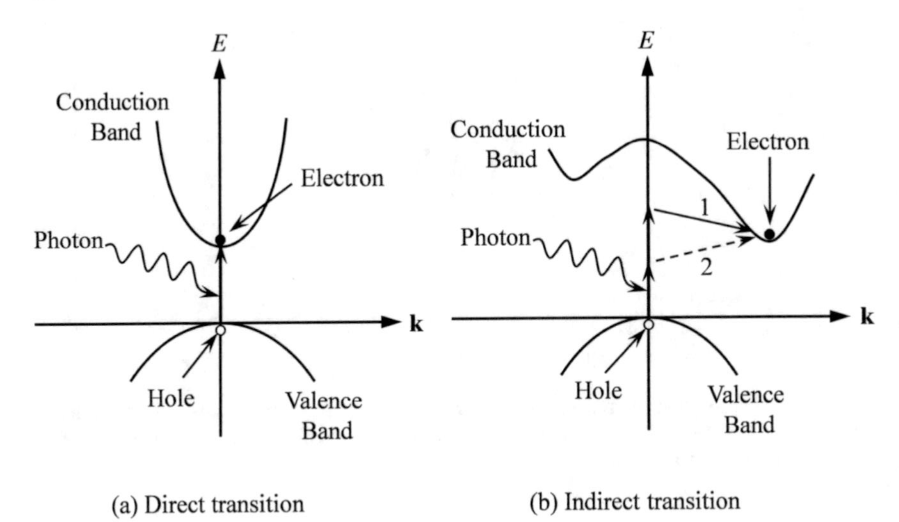


**Fig. 8.19** Interband transitions in semiconductors. **a** Direct transition without involving a phonon. **b** Indirect transition involving the emission or absorption of a phonon

coefficient is given as

$$a_{\rm a}(\omega) = \frac{A(\hbar\omega - E_{\rm g} + \hbar\omega_{\rm ph})^2}{\exp(\hbar\omega_{\rm ph}/k_{\rm B}T) - 1},\quad \hbar\omega > E_{\rm g} - \hbar\omega_{\rm ph} \tag{8.125}$$

and

$$a_{\rm e}(\omega) = \frac{A(\hbar\omega - E_{\rm g} - \hbar\omega_{\rm ph})^2}{1 - \exp(-\hbar\omega_{\rm ph}/k_{\rm B}T)},\quad \hbar\omega > E_{\rm g} + \hbar\omega_{\rm ph} \tag{8.126}$$

where $a_{\rm a}$ and $a_{\rm e}$ correspond to the absorption coefficients for transitions with phonon absorption and emission, respectively, and their values are nonzero only when the photon energy is greater than the bandgap energy subtracted (or added) by the phonon energy. There may be several phonon modes that can cause indirect interband transitions, and their effects on the absorption coefficient can be superimposed [59].

The Drude model can be applied to model the free-carrier contribution for both intrinsic and doped silicon as given in the following [59, 62]:

$$\varepsilon(\omega) = \varepsilon_{\rm bl} - \frac{N_{\rm e}e^2/\varepsilon_0 m_{\rm e}^*}{\omega^2 + i\omega\gamma_{\rm e}} - \frac{N_{\rm h}e^2/\varepsilon_0 m_{\rm h}^*}{\omega^2 + i\omega\gamma_{\rm h}} \tag{8.127}$$

where the first term on the right $\varepsilon_{\rm bl}$ accounts for contributions by transitions across the bandgap and lattice vibrations, the second term is the Drude term for transitions in the conduction band (free electrons), and the last term is the Drude term for transitions in the valence band (free holes). Here, $N_{\rm e}$ and $N_{\rm h}$ are the concentrations, $m_{\rm e}^*$ and $m_{\rm h}^*$ the effective masses, and $\gamma_{\rm e}$ and $\gamma_{\rm h}$ the scattering rates of free electrons and holes, respectively. The effective masses of silicon are taken as $m_e^* = 0.27m_0$ and $m_h^* = 0.37m_0$, where $m_0$ is the electron mass in vacuum.

The value of $\varepsilon_{bl}$ is determined using the refractive index and the extinction coefficient of intrinsic silicon. The refractive index of silicon changes from 3.6 at $\lambda = 1$ μm to 3.42 at $\lambda > 10$ μm at room temperature and increases slightly at higher temperatures. Absorption by lattice vibrations occurs in silicon at wavelengths between 6 and 25 μm. To account for the lattice absorption, the extinction coefficients are taken from the tabulated values in *Handbook of the Optical Constants of Solids* [15]. At elevated temperatures or for heavily doped silicon, the effect of absorption by lattice vibrations is negligible compared to the absorption by free carriers. The carrier concentration and the scattering rate depend on the temperature and dopant concentrations. For bulk silicon, the scattering is caused by the collision of electrons or holes with the lattice (phonons) or ionized dopant sites (impurities or defects). The total scattering rates can be calculated by

$$\gamma_e = \gamma_{e-l} + \gamma_{e-d} \text{ and } \gamma_h = \gamma_{h-l} + \gamma_{h-d} \tag{8.128}$$

where the subscripts l and d stand for lattice and defects, respectively. Generally speaking, increasing the defect concentration or temperature gives rise to a larger scattering rate. For intrinsic silicon, the concentration of the thermally excited free electrons and holes is the same and can be found from the relation:

$$N_{th}^2 = N_C N_V \exp(-E_g/k_B T) \tag{8.129}$$

where $N_C$ and $N_V$ are the effective densities of states in the conduction band and the valence band, respectively, and for silicon, $E_g = 1.17 - 0.000473T^2/(T + 636)$ eV. Note that $N_C = 2.86 \times 10^{19}$ cm$^{-3}$ and $N_V = 2.66 \times 10^{19}$ cm$^{-3}$ at 300 K; however, both increase with temperature proportional to $T^{3/2}$. When the dopant concentrations are not very high, the free-carrier concentrations can be obtained from

$$N_e = \frac{1}{2}\left[N_D - N_A + \sqrt{(N_D - N_A)^2 + 4N_{th}^2}\right] \tag{8.130}$$

and $N_h = N_{th}^2/N_e$ when the majority impurities are $n$-type. When the majority impurities are $p$-type, the equations become $N_h = \frac{1}{2}\left[N_A - N_D + \sqrt{(N_A - N_D)^2 + 4N_{th}^2}\right]$ and $N_e = N_{th}^2/N_h$. Equation (8.130) has been derived based on complete ionization, which does not hold for heavily doped semiconductors or at very low temperatures. Integration is needed to determine the concentration when complete ionization is not expected, as described by Fu and Zhang [62].

The calculated optical constants $n$ and $\kappa$ of silicon, for wavelengths in the range between 1 and 100 μm, are shown in Fig. 8.20 at 300 and 1000 K for $n$-type phosphorus donors. The refractive index changes little for lightly doped silicon, even at high temperatures. The refractive index for heavily doped silicon first decreases and

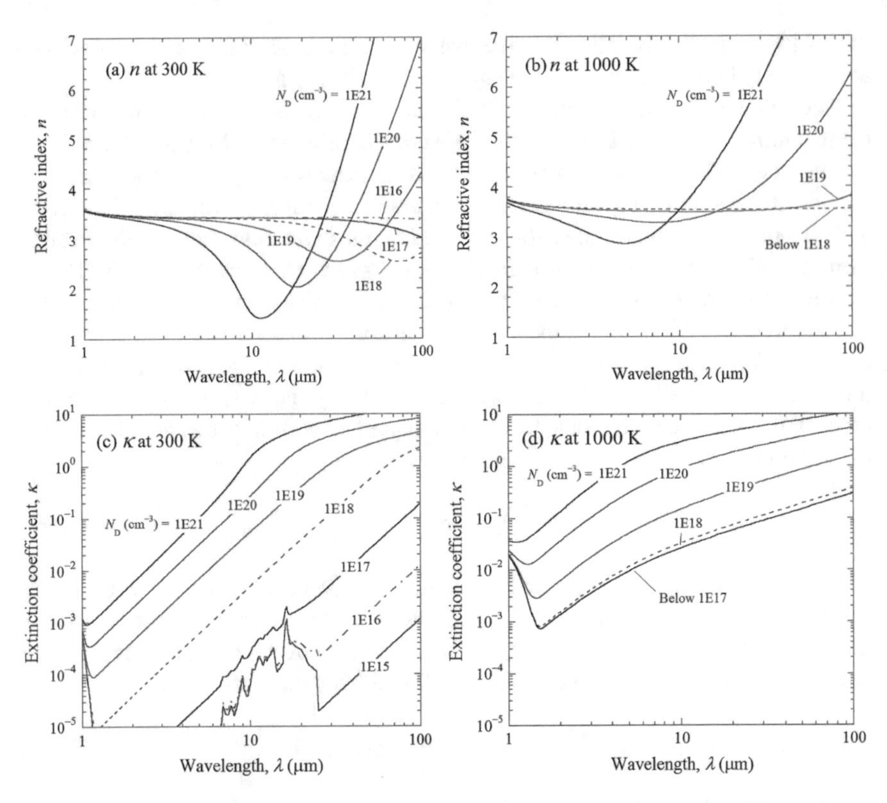


**Fig. 8.20** Optical constants of $n$-type phosphorus-doped silicon, at 300 K and 1000 K, for different dopant concentrations

then increases abruptly toward longer wavelengths. The carrier contribution to the extinction coefficient at 300 K is very small for lightly doped silicon, and the lattice contribution can be clearly seen between 6 and 25 μm. As the doping level exceeds $10^{17}$ cm$^{-3}$, these phonon features are screened out. This is also true for lightly doped silicon at 1000 K as the thermally excited carriers have a concentration of about $10^{18}$ cm$^{-3}$. At 1000 K, $\kappa$ is essentially the same for $N_D \leq 10^{17}$ cm$^{-3}$ and increases with higher dopant concentrations. At 300 K, the calculated $\kappa$ at $\lambda > 1.12$ μm decreases with reducing dopant concentration until $N_D$ is less than $10^{10}$ cm$^{-3}$, when most carriers are from the thermal excitation rather than the doping. The lattice absorption features become prominent when $N_D \leq 10^{16}$ cm$^{-3}$. For doping levels under, $10^{18}$ cm$^{-3}$, $\kappa \ll n$ unless the wavelength is very long, and silicon behaves as a dielectric. The significance is that the radiation penetration depth can be very large in the mid infrared because of the small $\kappa$ values. For heavily doped silicon, on the other hand, the Drude model predicts that $n \approx \kappa$ in the long-wavelength limit, just like in a metal. The accuracy of the simple Drude model is subjected to a number of factors, such as the dependence of the effective mass on temperature, dopant concentration, and even frequency. The scattering rate may be frequency dependent as

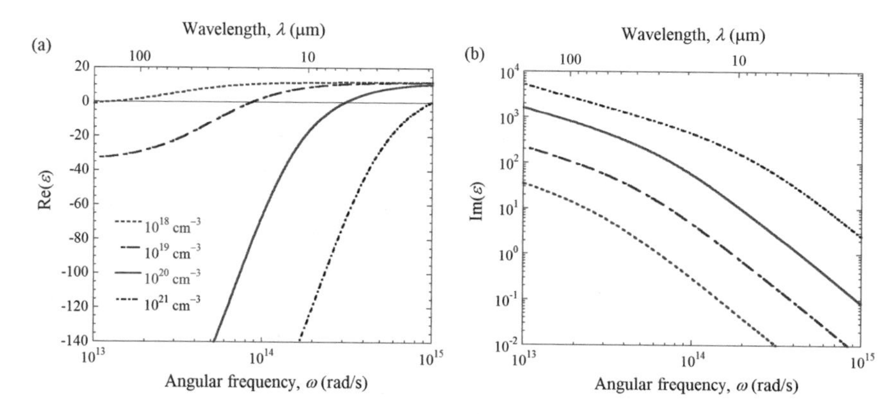


**Fig. 8.21** **a** Real and **b** imaginary parts of the dielectric function for *n*-type silicon at 400 K for different dopant concentrations

well. The band structure may be modified for heavily doped silicon. Nevertheless, this model has captured the essential features of the dielectric function of silicon, for wavelengths greater than 0.5 μm, at temperatures from 300 to 1200 K, and with a doping level up to $10^{19}$ cm$^{-3}$.

Basu et al. [63] performed a comprehensive study on the ionization models and mobility models of doped silicon. It was pointed out that the ionization model adopted in the previous study of Fu and Zhang [62] would underpredict the carrier concentrations for doping concentrations greater than $10^{17}$ cm$^{-3}$. The model recommended by Basu et al. [63] compares well with the mid-infrared transmittance and reflectance spectra of both phosphorus (*p*-type) and boron (*n*-type)-doped silicon films at room temperature. The model can be extended into the temperature range from 250 to 400 K by modeling the temperature-dependent scattering rate using $\gamma(T)/\gamma_0 = (300/T)^{1.5}$, where $\gamma_0$ is the impurity scattering rate at room temperature calculated from the mobility model. Figure 8.21 plots the calculated dielectric function of heavily doped silicon at 400 K for angular frequencies between $10^{13}$ and $10^{15}$ rad/s with different dopant concentrations. An important feature is that the real part of the dielectric function $\varepsilon'$ becomes negative especially for high doping concentrations. Such a metallic behavior due to free carriers can enable surface plasmon resonances and enhance near-field radiative heat transfer as will be discussed in subsequent chapters.

### *8.4.5 Superconductors*

A *superconductor* is a material that exhibits zero resistance and perfect diamagnetism when it is maintained at temperatures below the critical temperature $T_c$, under a bias current less than the critical current and an applied magnetic field less than the critical magnetic field. The discovery of high-temperature superconductors in the late 1980s has generated tremendous excitement in the public because the achievement of superconductivity above the boiling temperature of nitrogen (77 K at atmospheric pressure) offers many technological promises. More and more materials have been found to be superconducting at higher and higher temperatures. Extensive studies have been devoted to the infrared properties of superconducting films for applications such as radiation detectors, optical modulators, and other optoelectronic devices [64]. High-temperature superconducting (HTS) materials are made of ceramic structures, such as $YBa_2Cu_3O_{7\text{-}\delta}$, where $\delta$ is between 0 and 1. The Y-Ba-Cu-O compound behaves as an insulator when $\delta > 0.6$ and as a conductor when $\delta < 0.2$ at room temperature.

In the normal state ($T > T_c$), the dielectric function $\varepsilon(\omega)$ can be modeled as a sum of the free-electron contribution using the Drude model, an intraband absorption that is important for the mid-infrared region by using the Lorentz term, and a high-frequency constant [65]:

$$\varepsilon(\omega) = \varepsilon_\infty + \varepsilon_{\text{Mid-IR}} + \varepsilon_{\text{Drude}} \tag{8.131}$$

The expression of the Drude term is the same as in Eq. (8.116) or (8.118). Although phonon contributions can be neglected compared to the large electronic contributions, a broadband mid-infrared electronic absorption often exists in the HTS materials, which is typically modeled with a Lorentz oscillator that has a large width, or a frequency-dependent scattering rate.

Many properties of superconductors can be explained in terms of a two-fluid model that postulates that *a fluid of normal electrons coexists with a superconducting electron fluid.* These two fluids coexist but do not interact. According to the BCS theory [66], interaction between a pair of free electrons and a phonon (or other thermally generated excitations) leads to the formation of an electron pair, called a *Cooper pair*. The Cooper pairs cannot be scattered by any sources as they move in the lattice structure. In the superconducting state, only a fraction of free electrons $f_s$ is in the condensed phase (or superconducting state) and the remaining electrons are in the normal state. The value of $f_s$ is temperature dependent and goes to zero at $T_c$. The dielectric function in the superconducting state can be modeled by

$$\varepsilon(\omega) = \varepsilon_\infty + \varepsilon_{\text{Mid-IR}} + (1 - f_s)\varepsilon_{\text{Drude}} + f_s\varepsilon_{\text{Sup}} \tag{8.132}$$

The Drude term remains due to the presence of normal electrons with a number density of $(1 - f_s)n_e$. In Eq. (8.132), the dielectric function of the superconducting electrons can be modeled as

$$\varepsilon_{\text{Sup}} = -\frac{\omega_p^2}{\omega^2} + \mathrm{i}\pi\delta(\omega)\frac{\omega_p^2}{\omega} \tag{8.133}$$

where $\delta(\omega)$ is the Dirac delta function. The calculated results are usually fitted with the experimental measurements by adjusting the plasma frequency, the scattering rate, and the fraction of superconducting electrons. Excellent agreement has been observed between the predicted and experimental values of both the transmittance and the reflectance of superconducting films, at temperatures ranging from 300 down to 10 K [65].

### 8.4.6 Metamaterials with a Magnetic Response

The concept of negative refractive index ($n < 0$) was first postulated by Victor Veselago in 1968 for a hypothetical material that has both negative permittivity and permeability in the same frequency region. In this case, the sign of $n$ should be chosen as negative in $n = \pm\sqrt{\varepsilon\mu}$. Many of the unique features associated with negative index materials (NIMs) were summarized in Veselago's original paper, such as negative phase velocity, reversed Doppler effect, and the prediction of a planar lens. As illustrated in Fig. 8.22a, if $n$ is negative, the phase speed will be negative and light incident from a conventional positive index material (PIM) to a NIM will be refracted to the same side as the incidence. This is called *bending light in the wrong way*. Furthermore, if light can be bent differently, then a planar slab of a NIM can focus light as shown in Fig. 8.22b. The lack of simultaneous occurrence of negative $\varepsilon$ and $\mu$ in natural materials hindered further study on NIMs for some 30 years. On the basis of the theoretical work by John Pendry and coworkers in the late 1990s, Shelby et al. [67] first demonstrated that a metamaterial exhibits negative refraction at x-band microwave frequencies. In a NIM medium, the phase velocity of an electromagnetic wave is opposite to its energy flux. The electric field, the magnetic field, and the wavevector form a left-handed triplet. For this reason, NIMs are also called left-handed materials (LHMs). Because both $\varepsilon$ and $\mu$ are simultaneously negative, NIMs are also called double negative (DNG) materials.

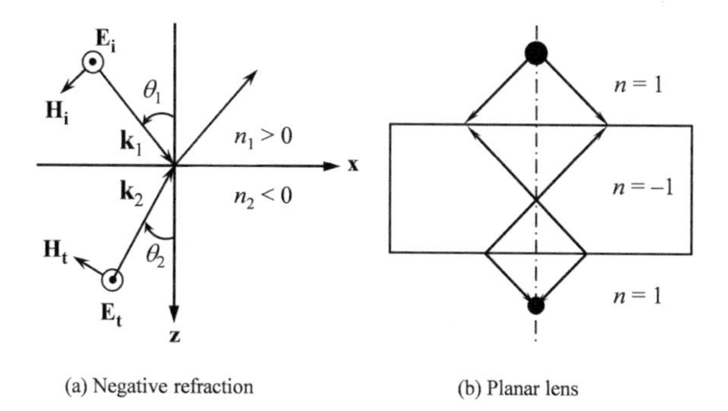


**Fig. 8.22** Unique features of a negative index material (NIM). **a** The refracted ray bends toward the same side as the incidence. **b** A slab of NIM can focus light like a lens does. Arrows indicate the wavevector directions. Note that the energy direction is the opposite of the wavevector direction in a NIM

Pendry [68] conceived that a NIM slab with $\varepsilon = \mu = -1$ would perform the dual function of correcting the phase of the propagating components and amplifying the evanescent components, which only exist in the near field of the object. The combined effects could make a perfect lens that eliminates the limitations on image resolution imposed by diffraction for conventional lenses. Despite the doubt cast by some researchers on the concept of a "perfect lens" and even on negative refraction, both the hypotheses of negative refraction and the ability to focus light by a slab of NIM have been verified by analytical, numerical, and experimental methods. Potential applications of NIMs range from nanolithography to novel Bragg reflectors, phase-compensated cavity resonators, waveguides, and enhanced photon tunneling for microscale energy conversion devices [69–71]. Ramakrishna [70] gave an extensive bibliographic review on the theoretical and experimental investigations into NIMs and relevant materials. There has been growing interest in the study of NIMs because of the promising new applications as well as the intriguing new physics. The search for new ways of constructing NIMs also calls for the development of new materials and processing techniques.

The ideal case, where $\varepsilon = \mu = -1$, cannot exist at more than a single frequency because both $\varepsilon$ and $\mu$ of a NIM must be inherently dependent on the frequency as required by the causality. In addition, real materials possess losses, and hence both $\varepsilon$ and $\mu$ are complex. The negative index can be realized by considering the complex plane, as illustrated in Fig. 8.23. Note that $\varepsilon = r_\varepsilon e^{i\phi_\varepsilon}$ and $\mu = r_\mu e^{i\phi_\mu}$. Then, we have

$$n = r_n e^{i\phi_n} = \sqrt{r_\varepsilon r_\mu} e^{i(\phi_\varepsilon + \phi_\mu)/2} \tag{8.134}$$

Therefore, if both $\varepsilon'$ and $\mu'$ are negative, $n$ will be negative, but $\kappa$ will always be positive. Note that a negative $n$ can be obtained as long as $\phi_n > \pi/2$. Generally speaking, one would like to see all the phase angles be close to $\pi$ so that the loss is minimized. Note that the principal value of the phase is chosen to be from 0 to $2\pi$ in the preceding discussion, rather than from $-\pi$ to $\pi$. If the latter is chosen, one would obtain a negative $\kappa$ and a positive $n$ for a NIM. Many metals and polar dielectrics have a negative $\varepsilon$ in the visible and the infrared. Furthermore, periodic structures of thin metal wires or strips can dilute the average concentration of electrons and shift the plasma frequency to the far-infrared or longer wavelengths. Negative-$\mu$

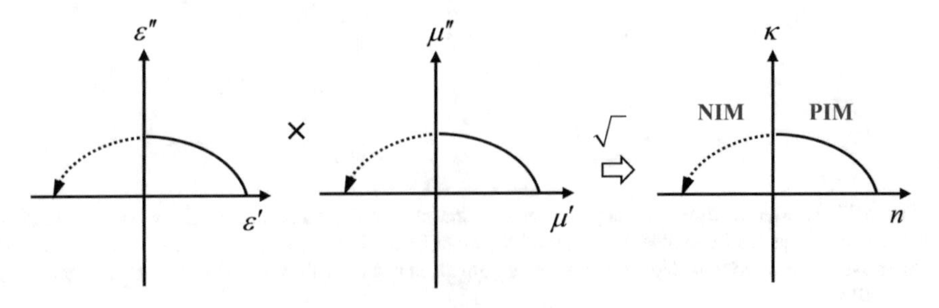


**Fig. 8.23** Illustration of a negative refractive index, using the complex planes

materials rarely exist in nature, at the optical frequencies, but can be obtained using metamaterials consisting of split-ring resonator structures at microwave frequencies. These structures can be scaled down to achieve negative $\mu$ toward higher frequencies. The combination of repeated unit cells of interlocking copper strips and split-ring resonators makes a metamaterial exhibit a negative $\varepsilon$ and $\mu$ simultaneously. Based on an effective-medium approach, the relative permittivity and permeability of a NIM can be expressed as functions of the angular frequency $\omega$ as follows:

$$\varepsilon(\omega) = 1 - \frac{\omega_p^2}{\omega^2 + i\gamma_e\omega} \tag{8.135}$$

and

$$\mu(\omega) = 1 - \frac{F\omega^2}{\omega^2 - \omega_0^2 + i\gamma_m\omega} \tag{8.136}$$

where $\omega_p$ is the effective plasma frequency, $\omega_0$ is the effective resonance frequency, $\gamma_e$ and $\gamma_m$ are the damping terms, and $F$ is the fractional area of the unit cell occupied by the split ring. From Eqs. (8.135) and (8.136), both negative $\varepsilon$ and $\mu$ can be realized in a frequency range between $\omega_0$ and $\omega_p$ for adequately small $\gamma_e$ and $\gamma_m$. Here, the values of $\omega_0$, $\omega_p$, $\gamma_e$, $\gamma_m$, and $F$ depend on the geometry of the unit cell that constructs the metamaterial. These structures can be scaled down to achieve a negative index at higher frequencies.

To illustrate the negative index behavior, Fig. 8.24 shows the calculated refractive index and the extinction coefficient of a hypothetical NIM using the following parameters [72]: $\omega_0 = 0.5\omega_p$, $F = 0.785$, and $\gamma_e = \gamma_m = \gamma = 0.0025\omega_p$. Because of the scaling capability of the metamaterial, the frequency is normalized to $\omega_p$. It can be seen that in the frequency range from $\omega_0$ to $\omega_p$, where the real parts of $\varepsilon$ and $\mu$ are negative, $n$ is negative and $\kappa$ (for small values of $\gamma$) is small at frequencies not

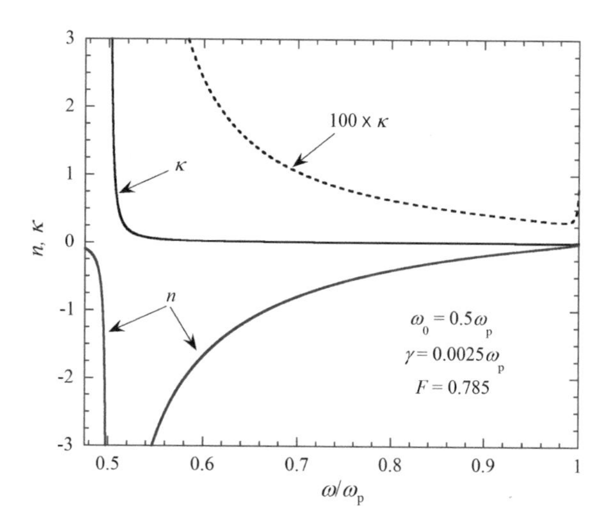


**Fig. 8.24** Calculated refractive index of a hypothetical negative index material (NIM)

too close to $\omega_0$. Further discussions of metamaterials and their radiative properties will be given in Chap. 9.

## 8.5 Experimental Techniques

Measurements of radiative properties (absorptance, emittance, reflectance, and transmittance) of real materials are critical for understanding the physical behavior of materials as well as for thermal analysis and design. The optical constants can be derived from the measured radiative or optical properties. For an opaque object with a smooth surface, measurement of the reflectivity in a broad spectral range can be used through KK relation to determine the complex refractive index if the material is nonmagnetic and isotropic. In the semitransparent region for a smooth slab (thick film), the measured spectral transmittance and reflectance can be used to extract the refractive index and absorption coefficient. For polar materials in the mid infrared (such as $SiO_2$ and $Al_2O_3$), the phonon oscillator parameters can be obtained by fitting the measured reflectance spectrum. Bidirectional reflectance and transmittance measurements are often used to study materials with surface roughness or inhomogeneity (such as porous materials or carbon nanotube arrays) due to surface and volume scattering. On the other hand, integrating spheres allow the diffused light to be collected and can be used to measure directional-hemispherical properties. The methods for measuring radiative properties can be grouped generally into two categories: calorimetric measurements and radiometric measurements, discussed in the following.

In a calorimetric technique, the thermal response of the specimen is used to determine the absorptance or emittance of the sample under investigation. Both the steady-state temperature change and transient temperature history can be used to deduce the radiative properties, though the calorimetric methods often use transient thermal responses during the heating or cooling process. The calorimetric method is well suited for measuring the total-hemispherical emittance of opaque materials [50, 73–75]. The sample is suspended in a large vacuum closure whose walls may be cooled with a cryogen such as liquid nitrogen. Due to the large wall area, the rate of net radiative transfer from the surface can be written as $q_{\mathrm{rad}} = \varepsilon_{\mathrm{tot}}^{\mathrm{h}} A_{\mathrm{s}} \sigma_{\mathrm{SB}} (T^4 - T_{\mathrm{w}}^4)$, where $\varepsilon_{\mathrm{tot}}^{\mathrm{h}}$, $A_{\mathrm{s}}$, and $T$ are the emittance, surface area, and temperature of the sample, and $T_{\mathrm{w}}$ is the wall temperature. The methods have been used to measure $\varepsilon_{\mathrm{tot}}^{\mathrm{h}}$ of certain metals up to about 1100 K [73, 74] as well as some solids and coatings down to 100 K [76]. In laser calorimetry [15, 77–79], the sample is heated up by a laser beam and the sample temperature depends on the laser power and spectral absorptance. Either the heating curve or cooling curve, after a shuttle is opened or closed, can be used to determine the absorptance based on a suitable thermal model. This method is particularly useful for measuring crystals with very low absorption coefficients [78, 79].

Radiometric techniques are based on the measurement of the radiant power reaching the detector (or receiver) from the source (or emitter). A variety of radiometric techniques or optical methods exist for measuring the spectral radiative properties. An optical instrument or measurement system typically includes four parts: the source, detector, optical components, and the sample. Sometimes the sample to be studied can be the source (in emission measurements), detector (in absorption measurements), or part of the components. In general, the source can be a lamp, laser, a blackbody cavity, or a thermal emitter. The detector can be a thermopile or a bolometer that measures the radiation based on a temperature rise or a semiconductor detector that is based on photoconductive or photovoltaic principles as described in Chap. 6. Simple optical components include lenses, mirrors, polarizers, filters and windows, beam-splitters, prisms, gratings, and optical fibers. More complicated components, such as an interferometer and a monochromator, to be discussed in subsequent sessions, may combine several simple components.

Besides calorimetry and radiometry, polarimetry and ellipsometry are commonly used to determine the optical constants, from which the radiative properties can be calculated [15, 80]. These methods are largely based on the phase and amplitude of the electric field component, rather than the radiant power or intensity. Cezairliyan et al. [81] used the division-of-amplitude photopolarimeter (DOAP), which can measure the Stokes parameters, to determine the refractive index and extinction coefficient at $\lambda = 633$ nm for cylindrical specimens heated by a pulsed laser. The normal spectral emissivities of molybdenum and tungsten at temperatures between 2000 and 2800 K measured by the polarimetric technique agree well with those measured by the spectral radiometric technique using high-speed pyrometers [81]. Spectroscopic ellipsometers can nowadays perform measurements not only from the ultraviolet to the near-infrared (wavelengths 150–2500 nm) region, but also from the mid- to far-infrared region up to $\lambda = 120\ \mu\text{m}$ [82].

In the following, we give some general discussions about the sources and detectors, the basics of dispersive instruments and the Fourier-transform spectrometer, along with setups for measuring directional-hemispherical properties with integrating spheres and for measuring spectral, directional emittance of materials at elevated temperatures. Measurements of the bidirectional reflectance and transmittance distribution functions will also be discussed, followed by a section on spectral ellipsometry.

### *8.5.1 Sources*

For thermal radiation, a blackbody is the ideal source since its spectral distribution is well defined, as discussed previously. The radiation from the sun can be approximated as a blackbody at a temperature of about 5800 K. However, sunlight varies with time and atmospheric conditions. Therefore, it cannot be used as a source for quantitative measurements. Because the surface area of the walls must be much greater than the opening, blackbody cavities are bulky and must be carefully designed to maintain a uniform inner wall temperature. This has been done successfully in national metrology laboratories/institutes and used for measuring the Stefan–Boltzmann constant and the spectral distribution that has resulted in the discovery of Planck's law [3, 16–19]. For high-temperature emittance measurements, blackbody cavities can be used as references. For measuring spectral transmittance and reflectance, since reference methods are commonly used, there is no need to precisely know the radiant power or intensity.

Lasers are quite commonly used in optical measurements since they provide well-collimated and nearly monochromatic radiation at discrete wavelengths. Some gas lasers have a very narrow spectral band due to atomic or molecular transitions, such as He-Ne lasers in the visible (633 nm) and near-infrared (1154 nm) or $CO_2$ lasers at wavelengths near 10 μm, especially at 10.6 μm. An optical cavity is made of a Fabry–Pérot resonator with two highly reflecting mirrors: one is opaque and the other partially transparent. The beam is reflected back and forth between the mirrors and makes multiple passes through the gain region (lasing medium) before it exits through the partially transparent mirror. This way, the stimulated emission within the gain region can be amplified. At present, semiconductor-based solid-state lasers are very popular and inexpensive. An example is a laser diode with a *p-n* junction that is similar to a light-emitting diode (LED). However, LEDs are based on spontaneous emission and produce light in a relatively broad spectral band (30–60 nm width). On the other hand, a diode laser is based on the stimulated emission of a *p-i-n* junction in which the active region is the intrinsic region (*i*) sandwiched between the *p*- and *n*-type direct bandgap semiconductors, such as GaAs, GaSb, InP, etc. Detailed descriptions of the mechanisms, types, and performances of various laser systems can be found elsewhere. While the laser is a powerful tool for optical measurements, spectrometer systems can quickly produce continuous spectral measurements in a broad wavelength band and thus are the most common instruments for measuring spectral radiative properties.

Incandescent lamps give out light when the filament is heated to an elevated temperature. Tungsten halogen lamps are perhaps the most popular and inexpensive light source for UV, VIS, and NIR measurements. The tungsten filament is heated to about 2800–3200 K in a mixture of inert gas and a halogen gas (such as bromine). The halogen gas reacts with tungsten that is being evaporated from the filament at high temperatures and redeposits the tungsten atoms back onto the filament; this is called the halogen cycle. The bulb is made of fused silica (or quartz) that is an amorphous $SiO_2$, which has a low coefficient of thermal expansion, high strength, and a high melting temperature. For a regular lamp without a halogen gas, the filament temperature cannot be very high and the tungsten is gradually evaporated and deposited onto the glass wall. At a temperature around 3000 K, the blackbody emission peak is near 1 μm. Since the emissivity of tungsten decreases with increasing wavelength, the emission peak shifts toward a shorter wavelength. Typically, a quartz tungsten-halogen lamp can be used from about 250 to 2500 nm wavelengths. The signal becomes weaker at longer wavelengths due to the reduction of tungsten emissivity. Furthermore, fused silica begins absorbing beyond 3 μm wavelength.

Globar made of SiC with the addition of rare earth oxides has been commonly used for infrared spectroscopy at wavelengths from about 2 to 100 μm. The temperature of the heating element is typically 1300–1650 K. The emissivity of the globar is from 0.82 to 0.94 at wavelengths from 0.65 to 15 μm [83]. The globar source can be exposed to ambient conditions and lasts for a long time without turning it off, especially with nitrogen purging. Nowadays, most commercial mid-infrared spectrometers use globar emitters with different shapes and somewhat different operating temperatures.

A high voltage can ionize the gas molecules placed in between the cathode and anode. The high-temperature plasma generated by the electric discharge gives out arc light. This is the mechanism of lightning and has been used in arc welding. Common gas discharge lamps include the mercury arc lamp, xenon arc lamp, xenon flash lamp, and deuterium lamp. Although a high-voltage pulse (>20 kV) is needed to initiate the discharging process, gas discharge lamps usually operate with a low DC voltage (around 20 V) and a high current. Deuterium lamps emit unidirectional ultraviolet radiation with high stability. The wavelengths can range from as short as 115 nm (10.8 eV) to about 400 nm. They are commonly used in UV spectroscopic applications. The color temperature of a typical xenon lamp is around 6000 K and the emitted radiation is from 185 to 2000 nm. Hence, the xenon lamp has a closer match to the solar spectrum than other artificial sources. For this reason, xenon lamps are often used in solar simulators, UV/VIS spectrophotometers, and microscopes. Mercury arc lamps emit sharp peaks in the UV region with higher intensities than the xenon lamps with the same power consumption. Since the peaks are centered around 254 nm, the effective color temperature may exceed 10000 K. Another application of the mercury arc lamp is in the far infrared from 30 to 1000 μm, where the intensity of globar decreases more rapidly. In the far-infrared applications, the effective radiance temperature of mercury arc lamps is about 5000 K [84].

Synchrotron radiation or a synchrotron light source uses a circular particle accelerator. When electrons are accelerated under the magnetic field in the storage ring to a relativistic speed, electromagnetic radiation is emitted in the broad spectrum from x-ray to microwave with high brightness, collimation, and stability. Both linear and circular polarizations can be produced. Of course, synchrotron radiation is very expensive and available only in limited facilities. Synchrotron radiation is mostly used in x-ray studies since it is the brightest x-ray source and also in THz radiation (0.1–10 THz) studies where there is a lack of intensive sources [85].

### 8.5.2 Detectors

Generally speaking, there are two types of radiation detectors: thermal detectors and photon detectors. In a thermal detector, incident radiation causes a temperature variation that can be measured by a transducer that converts it to an electrical signal. In a photon detector or quantum detector, incident photons interact with the materials such as a semiconductor diode and cause electronic transitions to generate electron–hole pairs. Photodiodes can operate in either the photoconductive (PC) mode or the photovoltaic (PV) mode. The former is based on the change in electric conductivity and the latter is based on the voltage or current output, just like a solar cell, due

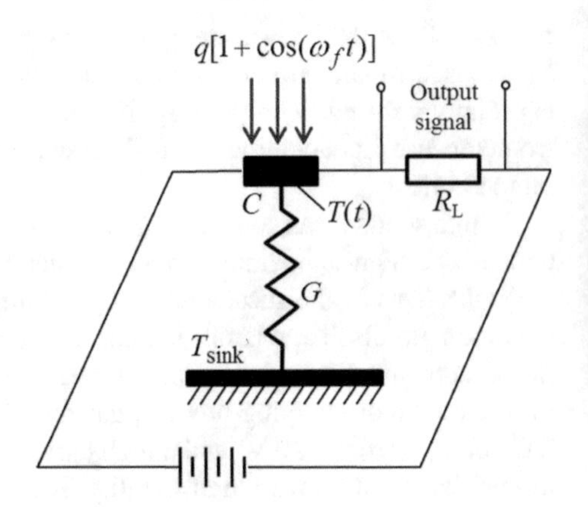


**Fig. 8.25** Illustration of a bolometer coupled with a heat sink and a simple measurement system

to absorbed photons. Photon detectors are generally more sensitive and faster than thermal detectors. On the other hand, thermal detectors typically have a broadband response with good linearity.

A thermopile utilizes the thermoelectric effect and combines many thermocouple junctions in a series: the hot junctions are coated black to receive radiation and cold junctions are maintained at the heat sink temperature. Since their invention in the early 1830s, thermopiles combined with a galvanometer had been successfully used for measuring the solar spectrum and for studying blackbody radiation until the bolometer was invented by S. P. Langley in 1880 [3]. Bolometers are based on the temperature dependence of electrical resistance and can be orders of magnitude more sensitive than thermopiles. In the following, we will use a bolometer to illustrate the figures of merit of a radiation detector, specifically a thermal detector.

As shown in Fig. 8.25, a lumped capacitance model is used in which the detector is assumed to be at a uniform temperature with a heat capacitance $C$ and is linked to a heat sink at temperature $T_{\text{sink}}$ with a thermal conductance $G$. For incidence with a radiant power modulated at an angular frequency $\omega_f$, the transient heat conduction equation may be written as [64, 86]

$$C\frac{\mathrm{d}\theta}{\mathrm{d}t} + G\theta = I^2R + \alpha q[1 + \cos(\omega_f t)] \tag{8.137}$$

Here, $\theta = T - T_{\text{sink}}$ is the reduced temperature, $R$ is the resistance of the detector element which is a function of temperature, $I$ is the bias current which is assumed to be constant, and $\alpha$ is the absorptance of the detector (often coated with a wavelength-independent absorbing layer). The solution of Eq. (8.137) can be expressed as

$$\theta(t) = \frac{\alpha q \cos(\omega_f t - \phi)}{G_{\text{eff}}\left(1 + \omega_f^2\tau^2\right)^{1/2}} + \theta_0 \tag{8.138}$$

where $\phi = \tan^{-1}(\omega_f t)$ is the phase lag, $G_{\text{eff}} = G - I^2(\mathrm{d}R/\mathrm{d}T)$ is the effective thermal conductance, and $\tau = C/G_{\text{eff}}$ is the *time constant*. The last term in Eq. (8.138) can be expressed as $\theta_0 = T_0 - T_{\text{sink}} = (\alpha q + I^2 R_0)/G$, where $T_0$ is the (average) operating temperature and $R_0$ is the resistance at $T_0$. The resistance of the bolometer may be expressed as $R = R_0 + (\mathrm{d}R/\mathrm{d}T)(T - T_0)$. Note that the *temperature coefficient of resistance* (TCR) is defined as $\beta = (\mathrm{d}R/\mathrm{d}T)/R_0$. Usually the bias current is sufficiently small such that $G_{\text{eff}} \approx G$. The time constant determines how high a modulation frequency can be used. Thermal detectors typically have a time constant in the range from milliseconds to seconds.

The responsivity $S$ is the ratio of the output (voltage) signal to the incident power (modulated portion only). For a bolometer based on Eq. (8.138), we see that

$$S = \frac{\alpha\beta I R_0}{G_{\text{eff}}\left(1 + \omega_f^2\tau^2\right)^{1/2}} \tag{8.139}$$

High TCR is critical for a high sensitivity. For this reason, superconductor bolometers have been developed that use the sharp resistance transition just above the critical temperature [64]. Some solids like vanadium dioxide ($VO_2$) experience a phase change above room temperature, with a large negative TCR during the insulator (semiconductor) to metal transition (around 340 K). This phenomenon has been used to build uncooled microbolometer arrays for infrared imaging applications.

Another figure of merit is called the noise equivalent power (*NEP*), which is the noise floor that limits the sensitivity since any signal below *NEP* cannot be distinguished from the noise. The *NEP* of a thermal detector depends on the background fluctuation called background noise, phonon noise due to the random exchange of thermal energy through the conductance *G*, Johnson noises of the detector resistance and load resistance due to random charge fluctuations, and the $1/f$ noise that is inversely proportional to frequency of the electronic signal [64, 86, 87]. By operating at cryogenic temperatures (e.g., using liquid helium), *NEP* can be reduced by orders of magnitude. The detectivity $D^*$ is often used for comparing the sensitivity of different detectors and is defined as

$$D^* = \frac{\sqrt{AB}}{NEP} \tag{8.140}$$

Here, $A$ is the detector area and $B$ is the bandwidth in Hz. The units of $D^*$ are usually expressed in terms of cm $\mathrm{Hz}^{1/2}\ \mathrm{W}^{-1}$.

Another type of thermal detector that is commonly used in infrared spectrometers is the pyroelectric detector, which is based upon the thermally induced polarization change in pyroelectric materials. Commonly used pyroelectric materials are lithium tantalate ($LiTaO_3$), triglycine sulfate (TGS) or deuterated triglycine sulfate (DTGS), and lead zirconate titanate (PZT), which also has a large piezoelectric effect. When oscillating radiation is absorbed by a pyroelectric material, the temperature variation will change the degree of polarization, resulting in an oscillating voltage signal on the load resistor. Room-temperature DTGS detectors have been adopted for many Fourier-transform infrared spectrometers. While it has a relatively low $D^*$ compared to photon detectors, as shown in Fig. 8.26, a DTGS detector has a nearly flat spectral response and can be used in a large spectral range, extending to the far-infrared region (wavelengths up to 1000 μm). In the far infrared, liquid-helium-cooled bolometers are often used since the detectivity can be increased by orders of magnitude. Note that photoconductive mercury–cadmium–telluride (HgCdTe or MCT) detectors with various compositions and bandgaps are often used to achieve higher sensitivity in the mid infrared. The detectivity of a narrower band (higher sensitivity) MCT-1 and a wider band (lower sensitivity) MCT-2 is also shown in Fig. 8.26, along with a more sensitive photovoltaic indium antimonite (InSb) detector that is useful up to 5.5 μm. A photon detector based on a semiconductor diode has a cutoff frequency since the incoming photon energy must exceed the bandgap of the semiconductor material. Furthermore, as the frequency increases, a portion of the photon energy that exceeds the bandgap is lost to heat. It should be noted that MCT PC detectors have been reported with poor linearity and sometimes need to be calibrated and corrected [88].

In the visible and near-infrared region, photodiode detectors such as GaP (150–550 nm), Si (190–1100 nm), Ge (800–1800 nm), and InGaAs (900–2600 nm) can be used in their applicable spectral range with high linearity and detectivity. For measurements with very low light signals, such as measuring scattered light or single photon detection, photomultiplier tubes (PMTs) can be used from UV to NIR due to their extremely high sensitivities. A photomultiplier tube is a quartz vacuum tube where electrons are generated at the photocathode by photoemission as described in Chap. 6. The emitted electrons undergo a set of electrodes (called dynodes) where secondary emission occurs to release additional electrons. More and more electrons

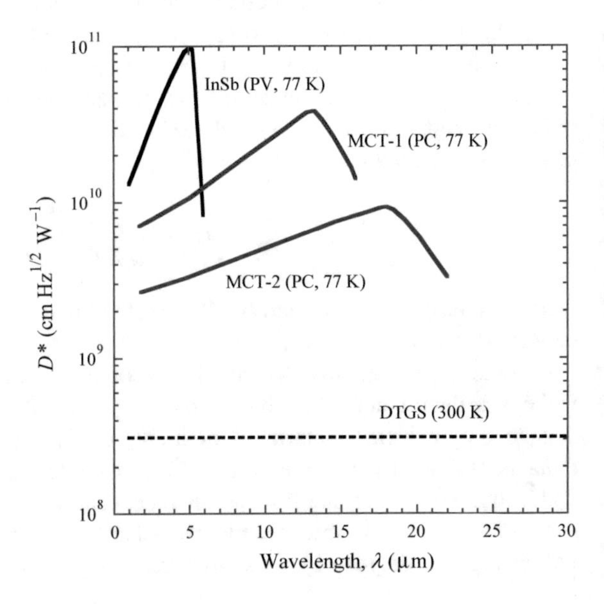


**Fig. 8.26** Detectivity ($D^*$) of several infrared detectors, where the photovoltaic InSb and photoconductive MCT detectors are cooled by liquid nitrogen and the DTGS detector operates at room temperature

are generated and accelerated through multiple dynode stages with higher and higher electrical potential. The electrical current emitted by the photocathode can be amplified by up to eight orders of magnitude by the time the electrons reach the anode where they are collected [86, 87].

Though the original purpose of this invention was to develop better electronic memory for information storage, the charge-coupled device (CCD) has become a major technology in digital images and spectroscopic applications, such as digital cameras and the Hubble space telescope. CCDs are silicon devices that contain an array of metal-oxide-semiconductor capacitors. The charges (electrons or holes) generated by incoming photons through photoelectric effect are first stored in the potential well created by the gate electrodes and then shifted (transferred) to the next capacitor by applying appropriate clock pulses to the gate electrodes. Using a pulse train, the charges stored in each capacitor are eventually transferred to a terminal row to be read out serially and matched up with the location to provide a map of the incoming photon flux. This ingenious conception has led to development of the digital camera technology and certain UV/VIS/NIR spectrometers. In recent years, active pixel sensors based on complementary metal-oxide-semiconductor (CMOS) have been developed as the alternative technology for imaging applications. In a CMOS-based device, each photodetector (pixel) has its own amplifier so that the generated photocurrent is read out simultaneously by an integrated circuit array. At present, most of the commercial infrared focal plane arrays or IR cameras use CMOS technology as the integrated read-out device.

While detectors can be calibrated to measure the actual radiant power as a power meter, radiation detectors are often used for relative measurements according to the ratio of the sample signal to a reference signal. In some applications, such as solar irradiance measurements, radiation thermometry, absolute radiometry, and thermal imaging where accurate radiant power measurements are required, calibrations of the detector responsivity and optical throughput are necessary. Standard instruments such as the electrically self-calibrated radiometers, absolute cryogenic radiometers, and blackbody sources are often employed for these purposes [17, 18].

### *8.5.3 Dispersive Instruments*

Before the laser was invented, most of the light sources were polychromatic. Interference filters, prisms, or gratings are typically used to obtain nearly monochromatic radiation. Multilayered dielectric (and sometimes metallic) films can be coated on a substrate to form interference filters that allow radiation from a narrow spectral band to pass through. An example is the Fabry–Perot interferometry that has sharp transmission peaks as to be further discussed in Chap. 9. A prism can effectively deflect broadband light into different directions, achieving nearly monochromatic radiation in selected directions. However, for spectroscopic applications, most contemporary instruments use surface relief gratings whose surfaces are corrugated periodically.

Polychromatic irradiation on a periodically corrugated surface is diffracted toward different directions (i.e., spatially dispersed according to the wavelength) based on the grating equation [38]:

$$\sin\theta_{\rm d}^{(m)} = \sin\theta_{\rm i} + \frac{m\lambda}{\Lambda} \tag{8.141}$$

where $\theta_{\rm d}^{(m)}$ is the angle of diffraction that depends on the angle of incidence $\theta_{\rm i}$, wavelength $\lambda$, grating period $\Lambda$, and diffraction order $m$ (which is an integer). The zeroth order ($m = 0$) corresponds to specular reflection that is in the same direction for all wavelengths. When $m$ is not equal to zero (positive or negative), the diffraction angle is wavelength dependent and different monochromatic radiation can be spatially dispersed upon reflection. Further discussion of grating theory will be given in Chap. 9.

Figure 8.27 shows the configuration of a Czerny–Turner monochromator with a rotating grating. The grating period must be greater than the maximum measurable wavelength. For a grating spectrometer, depending on the wavelength range, there may be tens to several thousands of grooves per millimeter. Some modern grating monochromators employ a linear array of sensors (CCD) without moving parts. It should be noted that suitable shortwave cutoff filters are needed to prevent unwanted radiation with higher diffraction orders from reaching the detector. This is because the diffraction angle depends on the product, $m\lambda$. Hence, radiation from shorter

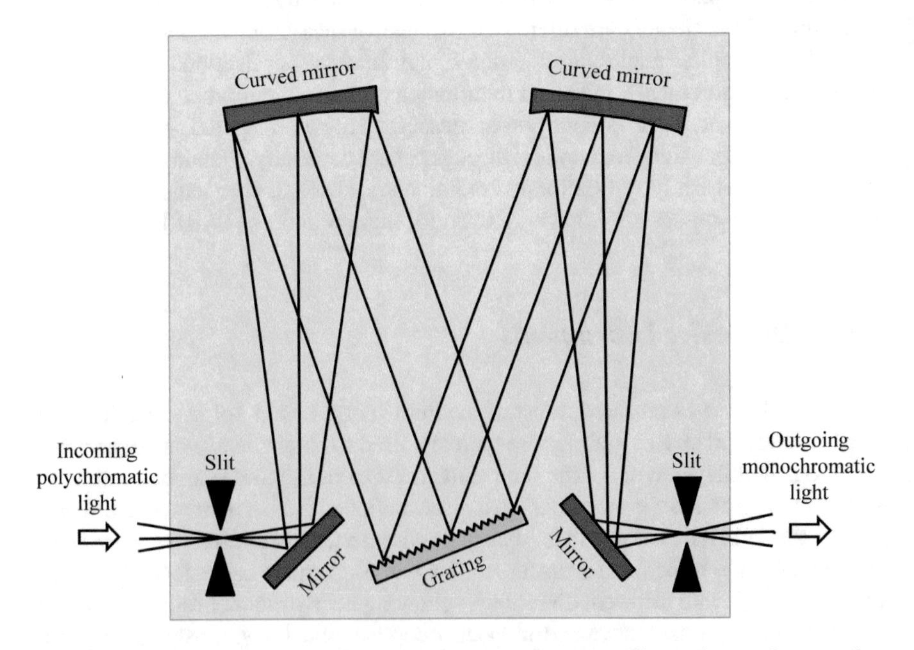


**Fig. 8.27** Illustration of a Czerny–Turner grating monochromator. The grating can be rotated to vary the output wavelength

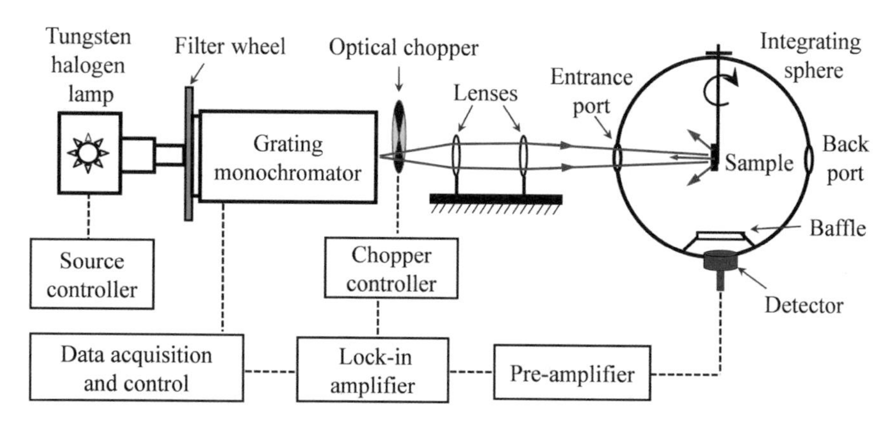


**Fig. 8.28** A setup for measuring directional-hemispherical radiative properties using a monochromator and an integrating sphere [89, 90]

wavelengths and higher orders can have the same $\theta_d$ as for radiation from longer wavelengths and lower orders. The spectral resolution is often determined by the width of the exit slit. However, the ultimate *resolving power* (i.e., the ratio of the wavelength divided by the spectral resolution $\delta\lambda$) is limited by $mN$, where $N$ is the total number of illuminated grooves [38]. A variety of commercial monochromators and spectrophotometers are available to meet the specific requirements for spectral range, sensitivity, and resolution.

Typically, grating spectrophotometers are used from the ultraviolet to the near infrared. At wavelengths beyond 2500 nm, the Fourier-transform infrared spectrometer has become the prevailing choice, as discussed in the subsequent section. A custom-built setup for measuring the directional-hemispherical reflectance and transmittance is shown in Fig. 8.28. The system is composed of a halogen lamp, a grating monochromator with a filter wheel, a chopper, and two lenses, which guide the monochromatic radiation to a 200-mm-diameter integrating sphere [89–91]. The inner wall of the integrating sphere is coated with a polytetrafluoroethylene (PTFE) diffuse reflector. Incident radiation can be focused on a spot size of approximately 6 mm $\times$ 6 mm on the center of the sphere, where the sample is mounted on a rotary holder through the top port. By rotating the sample holder, the beam can be directed either onto the back port of the sphere (covered by a PTFE plate) to obtain the reference signal or onto the sample to obtain the sample signal. A baffle placed above the detector located at the bottom port prevents the direct illumination of the detector by the first reflection of the sample or reference. Two photon detectors can be mounted at the bottom port of the sphere interchangeably: a Si photodiode for wavelengths from 300 to 1050 nm and a Ge photodiode for wavelengths from 1000 to 1800 nm. The detector signal is sent to a transimpedance pre-amplifier that has eight decades of dynamic range with a linear response. Afterward, the voltage signal is collected by a lock-in amplifier at the chopping frequency of 400 Hz. For an opaque sample, the directional-hemispherical reflectance can be obtained from the ratio of the sample signal to the reference signal. Furthermore, the back-mount method can also be used for convenient measurement. In the back-mount configuration, the sample is placed at the back port outside the sphere, interchangeably with the PTFE reference [91]. A number of factors can affect the accuracy of the integrating sphere measurements and, therefore, calibration and corrections are often necessary to reduce the measurement uncertainty [92].

### 8.5.4 Fourier-Transform Infrared Spectrometer

Developed in the late 1960s, Fourier-transform infrared (FTIR) spectrometers have become a versatile tool for infrared spectral characterization of materials, including spectral transmittance, reflectance, absorptance, and emittance [93–97]. As schematically shown in Fig. 8.29, an FTIR system utilizes Michelson interferometer that consists of a beamsplitter, a fixed mirror, and a moving mirror to produce interference effects. The strength of the output optical signal depends on the relative position of the moving mirror. If the path lengths between the beamsplitter and the two mirrors are the same, the situation is identified as zero path difference (ZPD), and the power reaching the detector will be the largest since constructive interferences occur at all wavelengths. For monochromatic incident light, a periodic signal will reach the detector as the moving mirror travels due to the alternating constructive and destructive interferences. For polychromatic incident radiation, the detector receives a time-varying signal called an *interferogram*, which is a Fourier transform of the incident radiation weighted by the spectral efficiency of the optical system and detector responsivity. In general, the interferogram appears somewhat like a sinc function with a peak at the ZPD. Unlike dispersive spectrometers, the FTIR detector receives a time-varying signal that carries information about the radiative power in

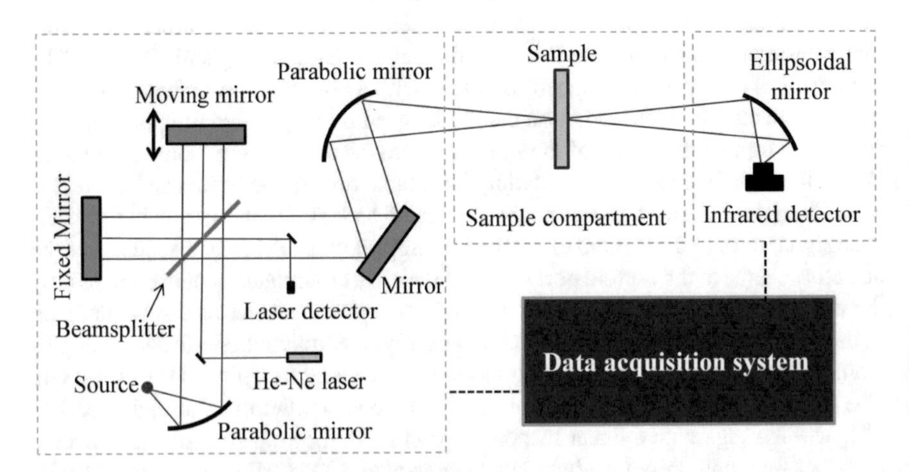


**Fig. 8.29** Illustration of the Fourier-transform infrared spectrometer. Accessories can be introduced in the sample compartment for reflectance measurements

a broadband. Suppose the spectrum of the source is $I(\bar{\nu})$. After passing through the Michelson interferometer, the spectral radiant power arriving at the detector is a periodic function with a dc component:

$$\frac{1}{2}[1 + \cos(2\pi x\bar{\nu})]\xi(\bar{\nu})I(\bar{\nu})$$

where $\bar{\nu}$ is the wavenumber, $x$ is the path length difference, and $\xi(\bar{\nu})$ is the optical efficiency. The one-half term is due to the interferometer since the other half of the energy is reflected back to the source.

Since only the modulated part contributes to the interferogram, the detector output signal for polychromatic incidence is given as follows:

$$V(x) = \frac{1}{2}\int_0^\infty \xi(\bar{\nu})S(\bar{\nu})I(\bar{\nu})\cos(2\pi\bar{\nu}x)\mathrm{d}\bar{\nu} \tag{8.142}$$

where $S(\bar{\nu})$ is the detector responsivity. The data acquisition system performs an inverse fast Fourier transform with a computer to generate a relative spectral response function, typically called the single beam spectrum [93]:

$$I^*(\bar{\nu}) = \int_{-\infty}^{\infty} V(\bar{\nu})\cos(2\pi\bar{\nu}x)\mathrm{d}x \tag{8.143}$$

If the velocity of the moving mirror is $u$, then $x = 2ut$. Hence, the detector receives a time-varying signal. However, the frequency $f = 2u\bar{\nu}$ typically falls in the range from several hundred to several thousand Hz depending on the wavelength of the incident radiation. This frequency can easily be measured by a thermal detector [94].

As shown in Fig. 8.29, a He-Ne laser with a well-characterized wavelength is used to precisely determine the location of the moving mirror with respect to the fixed mirror (i.e., the path length difference). The laser beam goes through the same interferometer to generate a sinusoidal wave that is detected by a photodiode detector. This enables high wavenumber accuracy for the resulting spectrum.

Spectral transmittance can be measured by dividing the spectrum with the sample by the reference spectrum when the sample is moved out of the optical paths as shown in Fig. 8.29. Reflectance accessories can be used both for specular and diffuse reflectance measurements [63, 91, 96]. Various other accessories can be used with FTIR spectrometers including attenuated total reflectance (ATR) that is based on evanescent waves [15, 93].

FTIR spectrometers have several advantages over dispersive spectrophotometers, such as high throughput, high signal-to-noise ratio, high resolution, and short measurement time. They are particularly suitable for measurements at wavelengths beyond 2 μm. While FTIR spectrometers have very high wavelength accuracy, caution must be taken with regard to its radiometric accuracy in order to quantitatively measure the radiative properties [88, 94, 95].

Figure 8.30 shows the optical layout of a custom-built spectral emissometer that allows the heated sample to be rotated to measure the directional emittance for each polarization [97]. A blackbody calibration source is used as the reference and a flip mirror allows the emission signal from either the sample surface or the blackbody to be collected by the FTIR through its side port. The ellipsoidal reflector collects the radiation through a half-cone angle of approximately 3° from the sample surface and focuses it again onto the iris, which is used to limit the collecting area on the sample and to adjust the amount of radiation reaching the spectrometer. The opening of the iris can be matched to the FTIR system to achieve a spectral resolution of 1 $cm^{-1}$. For most measurements, however, a resolution of 4 $cm^{-1}$ is usually sufficient. The parabolic reflector converts the radiation to a nearly collimated beam with a diameter of about 25 mm before sending it to the FTIR. A liquid-nitrogen-cooled InSb detector can also be used, which has a higher detectivity but a narrow spectral range from 2.0 to 5.5 μm. An IR wire-grid polarizer is mounted next to the iris for measuring the emittance with a chosen polarization.

The heater assembly is also shown on the right of Fig. 8.30. The sample was compressed on a copper disk, which was nickel plated to prevent oxidation. The nickel-plated copper disk maintained a uniform temperature underneath the sample. The copper surface was also polished before nickel plating to reduce thermal contact resistance. A coil heater was located at the back of the copper disk with an alumina plate inserted in between for electrical insulation. A K-type thermocouple probe

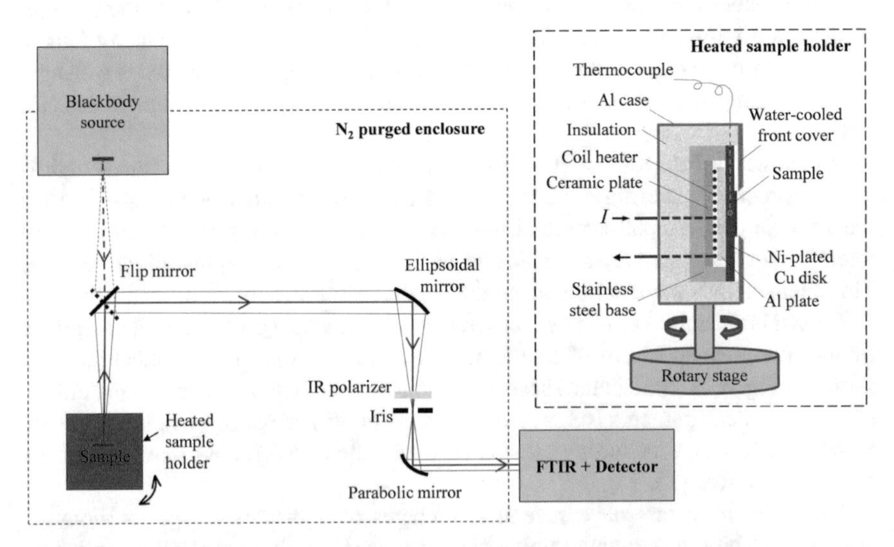


**Fig. 8.30** Schematic of optical layout for the spectral emissometer for measuring angular-dependent emittance for each polarization. The heated sample holder is mounted on a rotary stage as shown on the right [97]

with oxidation-resistive sheathing was embedded inside the copper disk for sample temperature measurement. The thermocouple is also used with a PID temperature controller to set and control the sample temperature. The heater assembly was placed in refractory materials and mounted inside a metal box. The sample temperature can reach 1000 K with a power input around 140 W. The front cover of the heater assembly was water-cooled with an aperture of 25 mm in diameter. The heater assembly was mounted on a rotary stage to change the emission angle. The emissometer has been used to measure a SiC substrate for calibration and the coherent emission from an asymmetric Fabry–Perot planar multilayer structure as well as a metamaterial structure by excitation of magnetic polaritons [97]. More discussions of multilayers and magnetic polaritons will be given in the subsequent chapter.

### 8.5.5 BRDF and BTDF Measurements

To measure BRDF or BTDF, both $(\theta_i, \phi_i)$ and $(\theta_r, \phi_r)$ need to be changed while the distance between the sample and the detector should be fixed [45]. For in-plane measurements, the plane of incidence is the same as the plane of reflection so that the azimuthal angles can be fixed [98]. Figure 8.31 shows a diagram of a laser scatterometer. The laser beam is in a fixed position; rotating the detector allows the change of the polar angle of incidence $\theta_i$, while rotating the detector around the sample using the goniometer allows the change of the reflection angle $\theta_r$. In the actual setup, the detector is allowed to move out of the horizontal plane so that

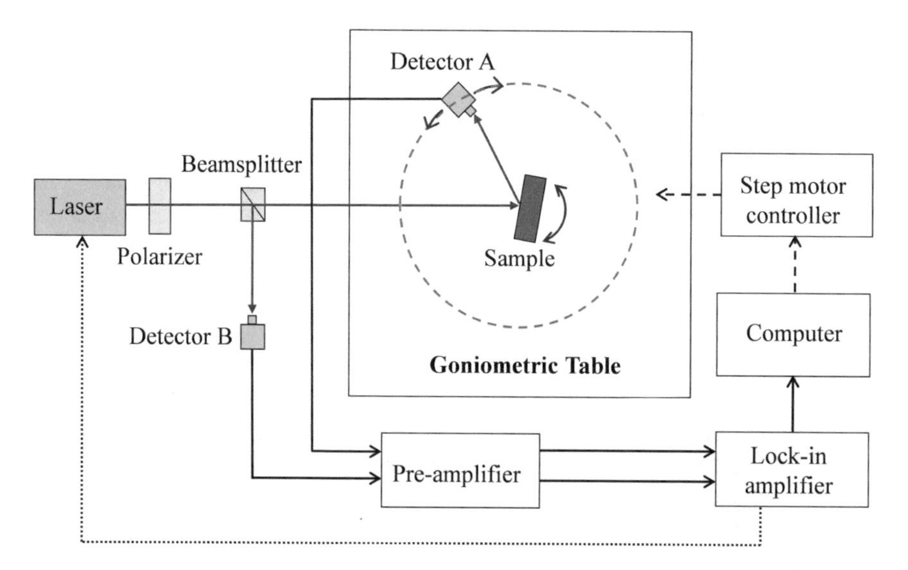


**Fig. 8.31** Schematic of a laser scatterometer for BRDF/BTDF measurements [45]

the azimuthal angle of reflection $\phi_r$ can also be changed. This facility is called the three-axis automated scatterometer (TAAS) and was developed in the author's lab [45]. The three rotary stages are independently controlled by step motors with very high angular precision (better than 0.01°). The incident laser beam is parallel to the optical table and the sample is vertically mounted.

A fiber-coupled diode laser system serves as a collimated light source. A thermoelectric-cooled temperature controller maintains the laser at a constant operation temperature to achieve superior power stability less than 0.2%. The lock-in amplifier provides an alternating current (typically 400 Hz) to the laser controller and measures the detector signal (after the pre-amplifier) at the same modulation frequency to eliminate the effect of background radiation. The wavelength of the laser can be chosen using different laser diodes. A linear polarizer is used to polarize the incident light either parallel or perpendicular to the plane of incidence. Then, the light is split into two paths by a beamsplitter. The majority is transmitted through the beamsplitter to the sample and then reflected/scattered by the sample. The output power is measured with a signal detector A, whose signal is sent to a pre-amplifier. A smaller portion is reflected by the beamsplitter and measured by a reference detector B, whose signal is also sent to the pre-amplifier. Si and Ge photodiode detectors measure the radiant power in the wavelength range from 350 to 1100 nm and from 800 to 1800 nm, respectively. The transimpedance pre-amplifiers convert the current signal from the detectors to a voltage output with resistance values switchable from 10 to $10^9$ $\Omega$ to achieve an eight-order dynamic range. A typical solid angle of the signal detector (with respect to the center of the laser spot on the sample) is $\Delta\Omega_r = 1.84 \times 10^{-4}$ sr, resulting in a half-cone angle of 0.45° [45]. The lock-in amplifier and step motors are connected to a desktop with the LabView environment for data acquisition and automatic rotary-stage control. It should be noted that the BRDF within ± 2.5° of the retroreflection direction cannot be measured since the sample detector would block the incident beam at this position. Laser diodes at 635, 891, 977, and 1550 nm have been employed in several investigations [45, 54, 99].

In the experiment, the detector output signal is proportional to the solid angle $\Delta\Omega_r$. The denominator of Eq. (8.95) gives the incident radiant power reaching the detector. Hence, the BRDF or BTDF can be obtained from the following measurement equation [45, 98]:

$$f_s = \frac{1}{P_i} \frac{P_s}{\cos\theta_r \Delta\Omega_r} \tag{8.144}$$

where $f_s$ refers to either BRDF or BTDF, $P_i$ is the laser power incident on the sample, $P_s$ is the scattered power reaching the signal detector, and $\Delta\Omega_r$ is the solid angle of the detector area viewed from the beam centered on the sample. During the measurements, the beamsplitter ratio is first calibrated. Detector A is rotated to behind the sample while the sample is removed. The ratio of the signal from detector B to that from detector A gives an instrument constant $C_I$. In the measurements, $P_s/P_i = C_I V_A/V_B$, where $V_A$ and $V_B$ are the output voltages from the lock-in amplifier for detectors A and B, respectively. As long as the responses are linear, there is no need to calibrate the detector responses. The use of the reference detector also eliminates the effect of laser power instability during the measurements. The specular reflectance, $R_{sp}$, can also be measured by positioning the signal detector in the specular direction with $\theta_r = \theta_i$ using the equation $R_{sp} = f_r \cos\theta_r \Delta\Omega_r$. The laser beam diameter (FWHM) is 3–5 mm, which is much smaller than the detector aperture whose diameter is 8 mm. This allows the specularly reflected power to be fully captured by the signal detector. In the measurements, $V_A$ and $V_B$ are averaged over many measurements at a given position to reduce the random error and the resulting uncertainty is typically within 5%.

### 8.5.6 Ellipsometry

When linearly polarized light is incident on an isotropic surface with or without a film, if the incident wave is either *p*- or *s*-polarized, the reflected wave is also linearly polarized as shown in Fig. 8.7. The ratio of the Fresnel reflection coefficients can be expressed as

$$\tilde{R} = \frac{r_p}{r_s} = \tan(\Psi)\mathrm{e}^{\mathrm{i}\Delta} \tag{8.145}$$

where $\Psi$ and $\Delta$ are called the ellipsometric angles or parameters. Note that $\tan(\Psi)$ is the amplitude ratio and $\Delta$ is the phase difference between the two Fresnel reflection coefficients. For an opaque and nonmagnetic medium, at given angle of incidence, $\tilde{R}$ is a function of the optical constants $(n, \kappa)$. If both $\tan(\Psi)$ and $\Delta$ can be measured, then the optical constants can be obtained. If a thin dielectric film is coated on an opaque substrate with known optical properties, then the refractive index and the film thickness can be simultaneously determined [80, 96]. It is also possible to use ellipsometry to study anisotropic crystals [82]. This is the principle of ellipsometry and various methods can be used to measure the ellipsometric parameters under oblique incidence [80, 82].

Figure 8.32 shows the rotating analyzer setup where a monochromatic beam (either from a laser or spectrometer) is incident at $\theta_i$, which is usually greater than 60°. The incident wave on the sample is linearly polarized but with both *s*- and

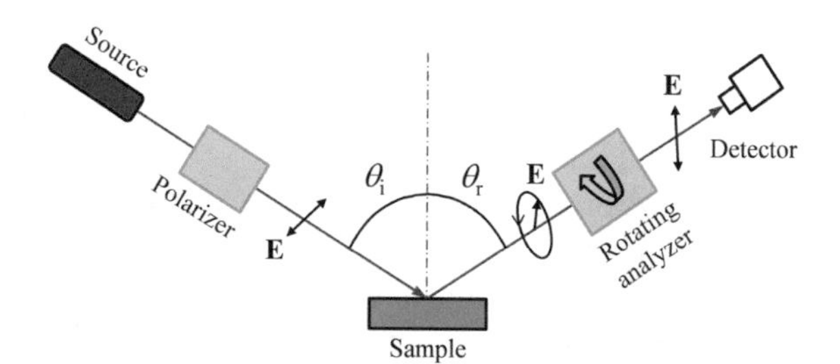


**Fig. 8.32** Schematic of a rotary analyzer ellipsometer

$p$-components. In this case, the reflected wave is, in general, elliptically polarized. The analyzer is another linear polarizer that is rotated to give out a sinusoidal signal to the detector. A quarter-wave compensator can sometimes be used. Furthermore, the incident polarizer can also be rotated to change the incident wave polarization. Through suitable data processing including regression analysis, the two ellipsometric parameters can be determined at each wavelength. With the development of spectral ellipsometers and the extension to the mid- to far-infrared regions, ellipsometry has become a complementary and alternative technique for the study of radiative properties of materials [82, 100].

## 8.6 Summary

In this chapter, we used the macroscopic Maxwell equations to derive the plane wave equation and subsequently defined the optical properties for isotropic materials. Planck's law was derived based on statistical mechanics. After a brief discussion of radiation thermometry, radiation pressure and photon entropy were then introduced. The reflection and refraction of waves at a smooth interface were derived based on the electromagnetic wave theory. This chapter also presented the dielectric functions for metals, dielectrics, semiconductors, superconductors, as well as materials with a magnetic response or metamaterials. The concept of NIM or DNG materials, as well as their unique features, was also explained. The last section surveyed the experimental techniques typically used for measuring radiative properties from ultraviolet to the far infrared. This chapter serves as the foundation of the subsequent chapters, in which we will provide extensive discussions on the radiative properties of semitransparent materials, windows, multilayers, periodic gratings, rough surfaces, as well as evanescent waves, surface polaritons, photon tunneling, and near-field radiative heat transfer.

## Problems

8.1 Write the wave equation in the 1D scalar form as $\frac{\partial^2 \psi}{\partial x^2} = \frac{1}{c^2}\frac{\partial^2 \psi}{\partial t^2}$, where $c$ is a positive constant. Prove that any analytical function $f$ can be its solution as long as $\psi(x, t) = f(x \pm ct)$. Plot $\psi$ as a function of $x$ for two fixed times $t_1$ and $t_2$. Show that the sign determines the direction (either forward or backward) and $c$ is the speed of propagation. Develop an animated computer program to visualize wave propagation.

8.2 Consider an electromagnetic wave propagating in the positive $z$-direction, i.e., $\mathbf{k} = k\hat{\mathbf{z}}$. Plot the vibration ellipse, and compare it with Fig. 8.2 for two cases: (1) $\mathbf{a} = 3\hat{\mathbf{x}}$ and $\mathbf{b} = \hat{\mathbf{x}} + 2\hat{\mathbf{y}}$ and (2) $\mathbf{a} = 3\hat{\mathbf{x}}$ and $\mathbf{b} = -2\hat{\mathbf{x}} + \hat{\mathbf{y}}$. Consider the spatial dependence of the electric field at a given time, say $\omega t = 2\pi m$, where $m$ is an integer. Discuss how $\mathbf{E}$ will change with $kz$ for the following two cases:

(3) $\text{Re}(\mathbf{E}_0) = 3\hat{\mathbf{x}}$ and $\text{Im}(\text{E}_0) = 0$ and (4) $\text{Re}(\mathbf{E}_0) = 3\hat{\mathbf{x}}$ and $\text{Im}(\mathbf{E}_0) = -3\hat{\mathbf{y}}$. The polarization is said to be right handed if the end of the electric field vector forms a right-handed coil or screw in space at any given time. Otherwise, it is said to be left handed. Discuss the handedness for all four cases.

8.3 Integrate Eq. (8.17) over a control volume to show that the energy transferred through the boundary into the control volume is equal to the sum of the storage energy change and energy dissipation. Write an integral equation using Gauss's theorem.

8.4 Derive the wave equation in Eq. (8.20) for a conductive medium; show Eq. (8.9) is a solution if $k$ is complex, as given in Eq. (8.21). Many books use $\mathbf{E} = \mathbf{E}_0 e^{i(\omega t - \mathbf{k}\cdot\mathbf{r})}$ instead of Eq. (8.9) as the solution; how would you modify Eqs. (8.21) and (8.22)? Show that the complex refractive index must be defined as $\tilde{n} = n - i\kappa$, where $\kappa \geq 0$.

8.5 Calculate the refractive index, the absorption coefficient, and the radiation penetration depth for the following materials, based on the dielectric function values at room temperature.

(a) Glass ($SiO_2$): $\varepsilon = 2.1 + i0$ at 1 μm; $\varepsilon = 1.8 + i0.004$ at 5 μm.
(b) Germanium: $\varepsilon = 21 + i0.14$ at 1 μm; $\varepsilon = 16 + i0.0003$ at 20 μm.
(c) Gold: $\varepsilon = -10 + i1.0$ at 0.65 μm; $\varepsilon = -160 + i2.1$ at 2 μm.

8.6 Consider a metamaterial with $\mu = -1 + i0.01$ and $\varepsilon = -2 + i0.01$; determine the refractive index and the extinction coefficient. Calculate the radiation penetration depth. Do a quick Internet search on negative index materials, and briefly describe what you have learned.

8.7 Find the magnetic field $\mathbf{H}$ for the wave given in Eq. (8.37). Show that the time-averaged Poynting vector is parallel to the $x$-axis. That is, the $z$-component of $\langle \mathbf{S} \rangle$ for such a wave vanishes. Briefly describe the features of an evanescent wave.

8.8 Write Planck's distribution in terms of wavenumber $\bar{\nu} = 1/\lambda$, i.e., the emissive power in terms of the wavenumber: $e_{b,\bar{\nu}}(\bar{\nu}, T)$. What is the most probable wavenumber in $cm^{-1}$? Compare your answer with the most probable wavelength obtained from Wien's displacement law in Eq. (8.45). Explain why the constants do not agree with each other. Cosmic background radiation can be treated as blackbody radiation at 2.7 K; what is the wavenumber corresponding to the maximum emissive power?

8.9 Based on the geometric parameters provided in Example 8.3 and neglecting the atmospheric effect, calculate the total intensity of the solar radiation arriving at earth's surface. Calculate the spectral intensity for solar radiation at 628 nm wavelength. A child used a lens to focus solar radiation to a small spot on a piece of paper and set fire this way. Does the beam focusing increase the intensity of the radiation? The lens diameter is 5 cm, and the distance between the lens and the paper is 2.5 cm. What are the focus size and the heat flux at the focus? Neglect the loss through the lens.

8.10 For a surface at $T = 1800$ K, with an emissivity of 0.6, what are the radiance temperatures at $\lambda = 0.65$ μm and 1.5 μm? If a conical hole is formed with a half-cone angle of 15°, what is the effective emittance and the radiance temperature at $\lambda = 0.65$ μm?

8.11 Derive Planck's law for a medium with a refractive index $n \neq 1$ in terms of the medium wavelength $\lambda_m$, $e_{b,\lambda_m}(\lambda_m, T)$ from Eq. (8.43). Assume that $n$ is not a function of frequency (i.e., the medium is nondispersive) in the spectral region of interest. How does it compare with Eq. (8.44)?

8.12 Express Eq. (8.53) in terms of wavelength, $s_\lambda(\lambda, T)$. Find an expression of the entropy intensity for blackbody radiation, $L_\lambda(\lambda, T)$, and show that $L_\lambda(\lambda, T) = \frac{c}{4\pi} s_\lambda(\lambda, T)$.

8.13 Assume that all the blue light at $\lambda$ in the range between 420 and 490 nm of solar radiation is scattered by the atmosphere and uniformly distributed over a solid angle of $4\pi$ sr. What are the monochromatic temperatures of the scattered radiation at $\lambda = 420$ and 490 nm?

8.14 A diode-pumped solid-state laser emits continuous-wave (cw) green light at a wavelength of 532 nm with a beam diameter of 1.1 mm. If the beam divergence is $2 \times 10^{-7}$ sr, what would be the spot size at a distance of 100 m from the laser (without scattering)? If the output optical power is 2 mW and the spectral width is $\delta\lambda = 0.1$ nm (assuming a square function), what is the average intensity of the laser beam? Find the monochromatic radiation temperature of the laser when it is linearly polarized. Suppose the laser hits a rough surface and is scattered into the hemisphere isotropically. Find the radiation temperature of the scattered radiation and the entropy generation caused by scattering.

8.15 In Example 8.5, the two plates are blackbodies. Assume that the plates are diffuse-gray surfaces with emissivities $\varepsilon'_1$ and $\varepsilon'_2$. Calculate the entropy generation rate in each plate per unit area. How will you determine the optimal efficiency for an energy conversion device installed at plate 2? For $T_1 = 1500$ K, $T_2 = 300$ K, and $\varepsilon'_2 = 1$, plot the optimal efficiency versus $\varepsilon'_1$.

8.16 The concept of dilute blackbody radiation can be used as an alternative method to calculate the entropy generation of a two-plate problem as in Problem 8.15. Assume that the multiply reflected rays are at not in equilibrium with each other. Rather, each ray retains its original entropy and can be treated as having an effective temperature of $T_1$ or $T_2$ depending on which plate the ray is emitted from. How would you evaluate the entropy transfer from plate 1 to 2 and the entropy generation by each plate then?

8.17 Calculate the entropy generation rate per unit volume for Example 2.7. Further, calculate the entropy generated at each surface, assuming that surface 2 is at 300 K.

8.18 The conversion efficiency of thermophotovoltaic devices is wavelength dependent, and the optical constants are wavelength dependent as well. Perform a literature search to find some recent publications in this area. Use the entropy concept to determine the ultimate efficiency of a specific design. Based on your analysis, propose a few suggestions for further improvement of the particular design you have chosen.

8.19 Derive the Fresnel reflection coefficient for a TM wave, following the derivation given in the text for a TE wave.

8.20 Show that $\rho'_{\lambda,s} + \alpha'_{\lambda,s} = 1$, where $\rho'_{\lambda,s}$ is given in Eq. (8.73) and $\alpha'_{\lambda,s}$ is given in Eq. (8.75). Discuss why the $z$-component of the time-averaged Poynting vector must be continuous at the boundary but not the $x$-component.

8.21 For nonmagnetic lossy media with $\varepsilon_1 = \varepsilon'_1 + \mathrm{i}\varepsilon''_1$ and $\varepsilon_2 = \varepsilon'_2 + \mathrm{i}\varepsilon''_2$, expand Eq. (8.70b) and compare your results with Eq. (8.71).

8.22 For plane waves incident from air to a nonmagnetic material with $\varepsilon = -2 + \mathrm{i}0$ (negative real), show that the reflectivity is always 1 regardless of the angle of incidence and the polarization. What can you say about $k_{2z}$ and $\langle S_{2z} \rangle$? Is the wave in the medium a homogeneous wave or an evanescent wave?

8.23 The refractive index of glass is approximately 1.5 in the visible region. What is the Brewster angle for glass when light is incident from air? Calculate the reflectance and plot it against the incidence angle for $p$-polarization, $s$-polarization, and random polarization. Redo the calculation for incidence from glass to air, and plot the reflectance against the incidence angle. At what angle does total internal reflection begin and what is this angle called?

8.24 Denote the incidence angle at which the ratio of the reflectance for TM and TE waves is minimized as $\theta_\mathrm{M}$. For radiation incident from air to a medium with $n = 2$ and $\kappa = 1$, determine $\theta_\mathrm{M}$ and compare it with the principle angle $\theta_\mathrm{P}$, at which the phase difference between the two reflection coefficients equals to $\pi/2$. [Hint: Use graphs to prove the existence of $\theta_\mathrm{M}$ and $\theta_\mathrm{P}$.]

8.25 For incidence from glass with $n = 1.5$ to air, calculate the Goos–Hänchen phase shift $\delta$ for both TE and TM waves. Plot $\delta$ as a function of the incidence angle $\theta_1$.

8.26 Show that the normal component of the time-averaged Poynting vector is zero in both the incident and transmitting media when total internal reflection occurs. Furthermore, derive Eq. (8.92).

8.27 Calculate the Goos–Hänchen lateral shift upon total internal reflection from a dielectric with $n = 2$ to air. Plot the lateral shift for both TE and TM waves as a function of $\theta_1$. Discuss the cause and the physical significance of the lateral beam shift.

8.28 A perfect conductor can be understood based on the Drude free-electron model by neglecting the collision term. The dielectric function becomes $\varepsilon(\omega) = 1 - \omega_\mathrm{p}^2/\omega^2$, where $\omega_\mathrm{p}$ is the plasma frequency. For radiation incident from air to a perfect conductor, calculate the phase shift when $\omega = \omega_\mathrm{p}/2$ for TE and TM waves as a function of the incidence angle. Use Eq. (8.93) to calculate the lateral beam shift for a TM wave and modify it for a TE wave. Do you expect a sign difference between the TE and TM waves?

8.29 Calculate and plot the emissivity (averaged over the two polarizations) versus the zenith angle for the materials and wavelengths given in Problem 8.5. Calculate and tabulate the normal and hemispherical emissivities for all cases.

8.30 Calculate the optical constants and the radiation penetration depth for either gold or silver at room temperature, using the Drude model, and plot them as functions of wavelength. In addition, calculate the reflectivity and plot it against wavelength. Compare the results using the Hagen–Ruben equation. How will the scattering rate and the plasma frequency change if the temperature is raised to 600 K?

8.31 Calculate the normal emissivity of MgO from 2000 to 200 cm$^{-1}$ (5 to 50 μm) using the Lorentz model with two oscillators having the following parameters: $\varepsilon_\infty = 3.01$; $\omega_1 = 401$ cm$^{-1}$, $\gamma_1 = 7.62$ cm$^{-1}$, and $S_1 = 6.6$; $\omega_2 = 640$ cm$^{-1}$, $\gamma_2 = 102.4$ cm$^{-1}$, and $S_2 = 0.045$. Can you develop a program to calculate the hemispherical emissivity and plot it against the normal emissivity for a comparison?

8.32 Find the Brewster angles for light incident from air to a NIM with (a) $\varepsilon_2 = -2$ and $\mu_2 = -2$, (b) $\varepsilon_2 = -1$ and $\mu_2 = -4$, and (c) $\varepsilon_2 = -8$ and $\mu_2 = -0.5$.

8.33 Use the online resources posted on the author's webpage [54] to calculate the absorption coefficient and normal reflectivity of intrinsic doped silicon for 0.5 μm $< \lambda <$ 25 μm.

8.34 First reproduce Fig. 8.21 for the dielectric function at 400 K and then calculate the dielectric function at 300 K for the same doping concentrations. Furthermore, calculate the real and imaginary parts of the refractive index of $n$-type doped silicon with a dopant concentration of 10$^{19}$ cm$^{-1}$ and plot them versus angular frequency.

8.35 Suppose a NIM can be described by Eqs. (8.135) and (8.136) with the following parameters: $\omega_p = 4.0 \times 10^{14}$ rad/s (i.e., $\lambda_p = 4.71$ μm), $\omega_0 = 2.0 \times 10^{14}$ rad/s (i.e., $\lambda_0 = 9.42$ μm), $\gamma = 0$, and $F = 0.785$. Assume a wave is propagating in such a medium in the region of $n < 0$ with a wavevector $\mathbf{k} = k_x\hat{\mathbf{x}}$, where $k_x = k = |n|\omega/c_0$. Show that the group velocity is in the negative $x$-direction. Also show that the Poynting vector is in the same direction as the group velocity.

8.36 Suppose a NIM can be described by Eqs. (8.135) and (8.136) with the following parameters: $\omega_p = 4.0 \times 10^{14}$ rad/s (i.e., $\lambda_p = 4.71$ μm), $\omega_0 = 2.0 \times 10^{14}$ rad/s (i.e., $\lambda_0 = 9.42$ μm, and $F = 0.5$. Calculate and plot the refractive index and the extinction coefficient in the spectral region from 2 to 15 μm, for $\gamma = 0$, 10$^{12}$, and 10$^{13}$ rad/s.

8.37 What is a detector? What is a bolometer? What is a radiometer? If you are asked to buy a detector for infrared radiation measurement for the wavelength range between 2 and 16 μm, discuss how you would select a detector and why. [Hint: Do some online search.]

8.38 A bolometer uses a thin YBCO film on a sapphire substrate whose area is 2 mm × 2 mm, operating at 90 K. The thickness of the sapphire plate is 25 μm. The thermal conductance between the detector element and a heat sink is $G = 8.4 \times 10^{-5}$ W/K. The resistance $R_0(90\text{ K}) = 200\ \Omega$ and $\beta = 1.5$ K$^{-1}$. Assume the absorptance $\alpha = 0.7$. Calculate the time constant for different bias currents, $I = 0.1, 0.2$ and $0.3$ mA. Calculate and plot the detector responsivity as a function of modulation frequency $\omega_f$ between 0.1 and 10 Hz for each bias current value given above. Neglect the heat capacity of the YBCO film. The density and specific heat of sapphire at the operating temperature are $\rho = 3970$ kg/m$^3$ and $c_p = 102$ J/kg K.

# References

1. J.R. Howell, M.P. Mengüç, R. Siegel, *Thermal Radiation Heat Transfer*, 6th edn. (CRC Press, New York, 2016)
2. M.F. Modest, *Radiative Heat Transfer*, 3rd edn. (Academic Press, New York, 2013)
3. E.S. Barr, Historical survey of the early development of the infrared spectral region. Am. J. Phys. **28**, 42–54 (1960)
4. M. Planck, *The Theory of Heat Radiation* (Dover Publications, New York, 1959) (originally published in German in 1913 and translated to English by M. Masius in 1914)
5. J.D. Jackson, *Classical Electrodynamics*, 3rd edn. (Wiley, New York, 1998)
6. D.J. Griffiths, Introduction to Electrodynamics, 4th edn. (Cambridge University Press, Cambridge, UK, 2017)
7. A.J. Kong, *Electromagnetic Wave Theory*, 2nd edn. (Wiley, New York, 1990)
8. M. Born, E. Wolf, *Principles of Optics*, 7th edn. (Cambridge University Press, Cambridge, UK, 1999)
9. C.F. Bohren, D.R. Huffman, *Absorption and Scattering of Light by Small Particles* (Wiley, New York, 1983)
10. NIST, http://physics.nist.gov/cuu/index.html. Last Accessed 24 Dec 2018
11. Z.M. Zhang, K. Park, Group front and group velocity in a dispersive medium upon refraction from a nondispersive medium. J. Heat Transfer **126**, 244–249 (2004)
12. P. Yeh, C. Gu, *Optics of Liquid Crystal Displays*, 2nd edn. (Wiley, New York, 2010)
13. T.J. Chui, J.A. Kong, Time-domain electromagnetic energy in a frequency-dispersive left-handed medium. Phys. Rev. B **70**, 205106 (2004)
14. J.M. Zhao, Z.M. Zhang, Electromagnetic energy storage and power dissipation in nanostructures. J. Quant. Spectrosc. Radiat. Transfer **151**, 49–57 (2015)
15. E.D. Palik (ed.), *Handbook of the Optical Constants of Solids*, vol. I, II, and III (Academic Press, San Diego, CA, 1998)
16. Z.M. Zhang, Surface temperature measurement using optical techniques. Annu. Rev. Heat Transfer **11**, 351–411 (2000)
17. T.J. Quinn, J.E. Martin, A radiometric determination of the Stefan-Boltzmann constants and thermodynamic temperatures between –40 °C and +100 °C. Philos. Trans. Royal Soc. London, Ser. A **316**, 85–189 (1985)
18. D.A. Pearson, Z.M. Zhang, Thermal-electrical modeling of absolute cryogenic radiometers. Cryogenics **39**, 299–309 (1999)
19. Z.M. Zhang, B.K. Tsai, G. Machin, *Radiometric Temperature Measurements*, I. Fundamentals & II. Applications (Academic Press, Amsterdam, 2010)
20. Y.H. Zhou, Y.-J. Shen, Z.M. Zhang, B.K. Tsai, D.P. DeWitt, A Monte Carlo model for predicting the effective emissivity of the silicon wafer in rapid thermal processing furnaces. Int. J. Heat Mass Transfer **45**, 1945–1949 (2002)
21. M.J. Lang, S.M. Block, Resource Letter: LBOT-1: Laser-based optical tweezers. Am. J. Phys. **71**, 201–215 (2003)
22. R. Petela, Exergy of heat radiation. J. Heat Transfer **86**, 187–192 (1964)
23. P.T. Landsberg, G. Tonge, Thermodynamic energy conversion efficiencies. J. Appl. Phys. **51**, R1–R20 (1980)
24. E.J. Tervo, E. Bagherisereshki, Z.M. Zhang, Near-field radiative thermoelectric energy converters: a review. Frontiers Energy **12**, 5–21 (2018)
25. C.E. Mungan, Radiation thermodynamics with application to lasing and fluorescent cooling. Am. J. Phys. **73**, 315–322 (2005)
26. C. Essex, D.C. Kennedy, R.S. Berry, How hot is radiation? Am. J. Phys. **71**, 969–978 (2003)
27. M. Caldas, V. Semiao, Entropy generation through radiative transfer in participating media: analysis and numerical computation. J. Quant. Spectrosc. Radiat. Transfer **96**, 423–437 (2005)
28. Z.M. Zhang, S. Basu, Entropy flow and generation in radiative transfer between surfaces. Int. J. Heat Mass Transfer **50**, 702–712 (2007)

29. T.J. Bright, Z.M. Zhang, Entropy generation in thin films evaluated from phonon radiative transport. J. Heat Transfer **132**, 101301 (2010)
30. Z.M. Zhang, C.J. Fu, Q.Z. Zhu, Optical and thermal radiative properties of semiconductors related to micro/nanotechnology. Adv. Heat Transfer **37**, 179–296 (2003)
31. Z.M. Zhang, H. Ye, Measurements of radiative properties of engineered micro-/nanostructures. Annu. Rev. Heat Transfer **16**, 351–411 (2013)
32. X.L. Liu, L.P. Wang, Z.M. Zhang, Near-field thermal radiation: recent progress and outlook. Nanoscale Microscale Thermophys. Eng. **19**, 98–126 (2015)
33. B. Zhao, Z.M. Zhang, Design of optical and radiative properties of solids, in *Handbook of Thermal Science and Engineering: Radiative Heat Transfer,* ed. by F.A. Kulachi (Springer Nature, 2017), pp. 1023–1068
34. D. Thompson, L. Zhu, R. Mittapally, S. Sadat, Z. Xing, P. McArdle, M.M. Qazilbash, P. Reddy, E. Meyhofer, Hundred-fold enhancement in far-field radiative heat transfer over the blackbody limit. Nature **561**, 216–221 (2018)
35. B. Salzberg, A note on the significance of power reflections. Am. J. Phys. **16**, 444–446 (1948); Z.M. Zhang, Reexamination of the transmittance formula of a lamina. J. Heat Transfer **119**, 645–647 (1997)
36. P. Halevi, A. Mendoza-Hernández, Temporal and spatial behavior of the Poynting vector in dissipative media: refraction from vacuum into a medium. J. Opt. Soc. Am. **71**, 1238–1242 (1981)
37. C.J. Fu, Z.M. Zhang, P.N. First, Brewster angle with a negative index material. Appl. Opt. **44**, 3716–3724 (2005)
38. H.A. Haus, *Waves and Fields in Optoelectronics* (Prentice-Hall, Englewood Cliffs, NJ, 1984)
39. H.K.V. Lotsch, Beam displacement at total reflection: The Goos-Hänchen effect, I, II, III, IV. Optik **32**(2), 116–137; (3), 189–204; (4), 299–319; (6), 553–569 (1970–1971)
40. A. Puri, J.L. Birman, Goos-Hänchen beam shift at total internal reflection with application to spatially dispersive media. J. Opt. Soc. Am. A **3**, 543–549 (1986)
41. D.-K. Qing, G. Chen, Goos-Hänchen shifts at the interfaces between left- and right-handed media. Opt. Lett. **29**, 872–874 (2004); X.L. Hu, Y.D. Huang, W. Zhang, D.-K. Qing, J.D. Peng, Opposite Goos-Hänchen shifts for transverse-electric and transverse-magnetic beams at the interface associated with single-negative materials. Opt. Lett., **30**, 899–901 (2005)
42. H.M. Lai, F.C. Cheng, W.K. Tang, Goos-Hänchen effect around and off the critical angle. J. Opt. Soc. Am. A **3**, 550–557 (1986)
43. I.V. Shadrivov, A.A. Zharov, Y.S. Kivshar, Giant Goos-Hänchen effect at the reflection from left-handed metamaterials. Appl. Phys. Lett. **83**, 2713–2715 (2003)
44. X. Chen, C.-F. Li, Lateral shift of the transmitted light beam through a left-handed slab. Phys. Rev. E **69**, 066617 (2004)
45. Y.J. Shen, Q.Z. Zhu, Z.M. Zhang, A scatterometer for measuring the bidirectional reflectance and transmittance of semiconductor wafers with rough surfaces. Rev. Sci. Instrum. **74**, 4885–4892 (2003)
46. W.C. Snyder, Z. Wan, X. Li, Thermodynamic constraints on reflectance reciprocity and Kirchhoff's law. Appl. Opt. **37**, 3464–3470 (1998)
47. T.L. Bergman, A.S. Lavine, F.P. Incropera, D.P. DeWitt, *Fundamentals of Heat and Mass Transfer*, 8th edn. (Wiley, New York, 2017)
48. L. Zhu, S. Fan, Near-complete violation of detailed balance in thermal radiation. Phys. Rev. B **90**, 220301(R) (2014)
49. L.P. Wang, S. Basu, Z.M. Zhang, Direct and indirect methods for calculating thermal emission from layered structures with nonuniform temperatures. J. Heat Transfer **133**, 072701 (2011)
50. Y.S. Touloukian, D.P. DeWitt, Thermal Radiative Properties, in *Thermophysical Properties of Matter,* TPRC Data Series, vol. 7, 8, and 9 ed. by Y.S. Touloukian, C.Y. Ho (IFI Plenum, New York, 1970–1972)
51. LBL-CXRO, http://henke.lbl.gov/optical_constants. Last Accessed 25 Dec 2018
52. W.M. Toscano, E.G. Cravalho, Thermal radiation properties of the noble metals at cryogenic temperatures. J. Heat Transfer **98**, 438–445 (1976)

53. F. Gervais, B. Piriou, Temperature dependence of transverse-and longitudinal-optic modes in $TiO_2$ (rutile). Phys. Rev. B **10**, 1642–1654 (1974)
54. Z.M. Zhang, http://zhang-nano.gatech.edu. Last Accessed 15 Jan 2019
55. W.G. Spitzer, D. Kleinman, D. Walsh, Infrared properties of hexagonal silicon carbide. Phys. Rev. **113**, 127–132 (1959)
56. S. Baroni, S. de Gironcoli, A. Dal Corso, P. Giannozzi, Phonons and related crystal properties from density-functional perturbation theory. Rev. Mod. Phys. **73**, 515–562 (2001)
57. G.-M. Rignanese, X. Gonze, G. Jun, K. Cho, A. Pasquarello, First-principles investigation of high-k dielectrics: Comparison between the silicates and oxides of hafnium and zirconium. Phys. Rev. B **69**, 184301 (2004)
58. H. Bao, B. Qiu, Y. Zhang, X. Ruan, A first-principles molecular dynamics approach for predicting optical phonon lifetimes and far-infrared reflectance of polar materials. J. Quant. Spectrosc. Radiat. Transfer **113**, 1683–1688 (2012)
59. P.J. Timans, The thermal radiative properties of semiconductors, in *Advances in Rapid Thermal and Integrated Processing*, ed. by F. Roozeboom (Kluwer Academic Publishers, The Netherlands, 1996), pp. 35–101
60. L.X. Benedict, E.L. Shirley, R.B. Bohn, Theory of optical absorption in diamond, Si, Ge, and GaAs. Phys. Rev. B **57**, R9385–R9387 (1998)
61. M. Rohlfing, S.G. Louie, Electron-hole excitations and optical spectra from first principles. Phys. Rev. B **62**, 4927–4944 (2000)
62. C.J. Fu, Z.M. Zhang, Nanoscale radiation heat transfer for silicon at different doping levels. Int. J. Heat Mass Transfer **49**, 1703–1718 (2006)
63. S. Basu, B.J. Lee, Z.M. Zhang, Infrared radiative properties of heavily doped silicon at room temperature", J. Heat Transfer, 132, 023301, 2010; "Near-field radiation calculated with an improved dielectric function model for doped silicon. J. Heat Transfer **132**, 023302 (2010)
64. Z.M. Zhang, A. Frenkel, Thermal and nonequilibrium responses of superconductors for radiation detectors. J. Supercond. **7**, 871–884 (1994)
65. A.R. Kumar, Z.M. Zhang, V.A. Boychev, D.B. Tanner, L.R. Vale, D.A. Rudman, Far-infrared transmittance and reflectance of $YBa_2Cu_3O_{7-\delta}$ films on Si substrates. J. Heat Transfer **121**, 844–851 (1999)
66. J. Bardeen, L.N. Cooper, J.R. Schrieffer, Theory of superconductivity. Phys. Rev. **108**, 1175–1204 (1957)
67. R.A. Shelby, D.A. Smith, S. Schultz, Experimental verification of a negative index of refraction. Science **292**, 77–79 (2001)
68. J.B. Pendry, Negative refraction makes a perfect lens. Phys. Rev. Lett. **85**, 3966–3969 (2000)
69. Z.M. Zhang, C.J. Fu, Unusual photon tunneling in the presence of a layer with a negative refractive index. Appl. Phys. Lett. **80**, 1097–1099 (2002)
70. S.A. Ramakrishna, Physics of negative refractive index materials. Rep. Prog. Phys. **68**, 449–521 (2005)
71. X. Zhang, Z. Liu, Superlenses to overcome the diffraction limit. Nature Mat. **7**, 435–441 (2008)
72. C.J. Fu, Z.M. Zhang, D.B. Tanner, Energy transmission by photon tunneling in multilayer structures including negative index materials. J. Heat Transfer **127**, 1046–1052 (2005)
73. K.G. Ramanathan, S.H. Yen, High-temperature emissivities of copper, aluminum, and silver. J. Opt. Soc. Am. **67**, 32–38 (1977)
74. S.X. Cheng, P. Cebe, L.M. Hanssen, D.M. Riffe, A.J. Sievers, Hemispherical emissivity of V, Nb, Ta, Mo, and W from 300 to 1000 K. J. Opt. Soc. Am. B **4**, 351–356 (1987)
75. J. Hameury, B. Hay, J.R. Filtz, Measurement of total hemispherical emissivity using a calorimetric technique. Int. J. Thermophys. **28**, 1607–1620 (2007)
76. D. Giulietti, A. Gozzini, M. Lucchesi, R. Stampacchia, A calorimetric technique for measuring total emissivity of solid materials and coatings at low temperatures. J. Phys. D Appl. Phys. **12**, 2027–2036 (1979)
77. D.A. Pinnow, T.C. Rich, Development of a calorimetric method for making precision optical absorption measurements. Appl. Opt. **12**, 984–992 (1973)

78. H.B. Rosenstock, D.A. Gregory, J.A. Harrington, Infrared bulk and surface absorption by nearly transparent crystals. Appl. Opt. **15**, 2075–2079 (1976)
79. D. Bunimovich, L. Nagli, A. Katzir, Absorption measurements of mixed silver halide crystals and fibers by laser calorimetry. Appl. Opt. **33**, 117–119 (1994)
80. R.M.A. Azzam, N.M. Bashara, *Ellipsometry and Polarized Light* (North Holland, Amsterdam, 1987)
81. A. Cezairliyan, S. Krishanan, J.L. McClure, Simultaneous measurements of normal spectral emissivity by spectral radiometry and laser polarimetry at high temperatures in millisecond-resolution pulse-heating experiments: Application to molybdenum and tungsten. Int. J. Thermophys. **17**, 1455–1473 (1996)
82. S. Schöche, T. Hofmann, R. Korlacki, T.E. Tiwald, M. Schubert, Infrared dielectric anisotropy and phonon modes of rutile $TiO_2$. J. Appl. Phys. **113**, 164102 (2013)
83. C.A. Mitchell, Emissivity of globar. J. Opt. Soc. Am. **52**, 341–342 (1962)
84. M.F. Kimmitt, J.E. Walsh, C.L. Platt, K. Miller, M.R.F. Jensen, Infrared output from a compact high pressure arc source. Infrared Phys. Technol. **37**, 471–477 (1996)
85. R.A. Lewis, A review of terahertz sources. J. Phys. D Appl. Phys. **47**, 374001 (2014)
86. R.H. Kingston, *Detection of Optical and Infrared Radiation* (Springer, Berlin, 1978)
87. R.J. Keyes (ed.), *Optical and Infrared Detectors* (Springer, Berlin, 1980)
88. Z.M. Zhang, C.J. Zhu, L.M. Hanssen, Absolute detector calibration applied to nonlinearity error correction in FT-IR measurements. Appl. Spectrosc. **51**, 576–579 (1997)
89. H. Lee, A. Bryson, Z. Zhang, Measurement and modeling of the emittance of silicon wafers with anisotropic roughness. Int. J. Thermophys. **28**, 918–933 (2007)
90. X.J. Wang, J.D. Flicker, B.J. Lee, W.J. Ready, Z.M. Zhang, Visible and near-infrared radiative properties of vertically aligned multi-walled carbon nanotubes. Nanotechnology **20**, 215704 (2009)
91. P. Yang, C.Y. Chen, Z.M. Zhang, A dual-layer structure with record-high solar reflectance for daytime radiative cooling. Sol. Energy **169**, 316–324 (2018)
92. L.M. Hanssen, K.A. Snail, Integrating spheres for mid- and near-infrared reflection spectroscopy, in *Handbook of Vibrational Spectroscopy*, ed. by J.M. Chalmers, P.R. Griffiths (New York, Wiley, 2002), pp. 1175–1192
93. P.R. Griffiths, J.A. de Haseth, *Fourier Transform Infrared Spectrometer* (Wiley, New York, 1986)
94. M.I. Flik, Z.M. Zhang, Influence of nonequivalent detector responsivity on FT-IR photometric accuracy. J. Quant. Spectrosc. Radiat. Transfer **47**, 293–303 (1992)
95. S.G. Kaplan, L.M. Hanssen, R.U. Datla, Testing the radiometric accuracy of Fourier transform infrared transmittance measurements. Appl. Opt. **36**, 8896–8908 (1997)
96. T.J. Bright, J.I. Watjen, Z.M. Zhang, C. Muratore, A.A. Voevodin, Optical properties of $HfO_2$ thin films deposited by magnetron sputtering: from the visible to the far-infrared. Thin Solid Films **520**, 6793–6802 (2012); T.J. Bright, J.I. Watjen, Z.M. Zhang, C. Muratore, A.A. Voevodin, D.I. Koukis, D.B. Tanner, D.J. Arenas, Infrared optical properties of amorphous and nanocrystalline $Ta_2O_5$ thin films. J. Appl. Phys. **114**, 083515 (2013)
97. L.P. Wang, S. Basu, Z.M. Zhang, Direct measurement of thermal emission from a Fabry-Perot cavity resonator. J. Heat Transfer **134**, 072701 (2012); L.P. Wang, Z.M. Zhang, Measurement of coherent thermal emission due to magnetic polaritons in subwavelength microstructures. J. Heat Transfer **135**, 091014 (2013)
98. P.Y. Barnes, E.A. Early, A.C. Parr, *Spectral Reflectance* (NIST Special Publication, Washington DC, 1998), pp. 250–248
99. Q.Z. Zhu, Z.M. Zhang, Anisotropic slope distribution and bidirectional reflectance of a rough silicon surface. J. Heat Transfer **126**, 985–993 (2004); H.J. Lee, Z.M. Zhang, Measurement and modeling of the bidirectional reflectance of $SiO_2$ coated Si surfaces. Int. J. Thermophys. **27**, 820–839 (2006)
100. E. Franke, C.L. Trimble, M.J. DeVries, J.A. Woollam, M. Schubert, F. Frost, Dielectric function of amorphous tantalum oxide from the far infrared to the deep ultraviolet spectral region measured by spectroscopic ellipsometry. J. Appl. Phys. **88**, 5166–5174 (2000)

# Chapter 9
# Radiative Properties of Nanomaterials

Optical and thermal radiative properties are fundamental physical properties that describe the interaction between electromagnetic waves and matter from deep ultraviolet to far-infrared spectral regions. A large number of studies have been devoted to the measurement, analysis, modeling, and simulation of optical and radiative characteristics of materials in solid, liquid, gas, and plasma phases. The radiative properties of nanostructured materials are critical to the functionality and the performance of many devices, such as semiconductor lasers, radiation detectors, tunable optical filters, waveguides, solar cells, and selective emitters and absorbers. The use of microstructures not only modifies the optical properties for optoelectronic applications and processing control but also facilitates some important energy conversion devices, such as solar cells and thermophotovoltaic applications.

This chapter will start with the radiative properties of a single layer with or without considering the wave interference effect. Partial coherence and the effect of surface scattering will be considered next. The approach will then be generalized to multilayered structures using the 1D matrix formulation. Furthermore, periodic structures such as photonic crystals and gratings will be studied based on the Bloch wave equation. Subsequently, the effective medium formulations will be briefly discussed. Finally, the effect of surface roughness and microstructures on the radiative properties will be presented.

## 9.1 Radiative Properties of a Single Layer

Crystalline films, from a few nanometers to several micrometers thick, have been deposited (by physical vapor deposition, chemical vapor deposition, sputtering, laser ablation, molecular beam epitaxy, rapid thermal processing, and other techniques) onto suitable substrates. These layered structures play important roles in contemporary technologies, such as integrated circuits, semiconductor lasers, quantum well



Z. M. Zhang, *Nano/Microscale Heat Transfer*, Mechanical Engineering Series,
https://doi.org/10.1007/978-3-030-45039-7_9

detectors, superconductor/semiconductor hybrid devices, optical filters, and spectrally selective coatings for solar thermal applications. Radiative energy transport in thin films differs significantly from that at bulk solid surfaces and through thick windows because of multiple reflections and interference effects. The radiative properties of a lamina with smooth and parallel surfaces will be discussed first, with emphasis on different formulations for various applications. At the end of this section, the effect of surface scattering will be considered in the regime where the roughness is much smaller than the wavelength.

### 9.1.1 The Ray-Tracing Method for a Thick Layer

A "thick" layer or slab refers to the case where interference between multiply reflected waves can be neglected. In other words, the waves are *incoherent.* On the contrary, a "thin" film refers to the case where all multiply reflected waves are *coherent* and interfere with each other. The condition for being thick has often been commonly interpreted as that the layer thickness $d$ is much greater than the wavelength. A more rigorous criterion is that the thickness is much greater than the coherence length, which can be much greater than the wavelength. The coherence length depends on the spectral width of the source and the spectral resolution of the spectrophotometer, such as a grating monochromator or a Fourier transfer spectrometer. In addition, beam divergence, surface roughness, and nonparallelism of the surfaces further reduce the degree of coherence. Generally speaking, when the thickness is comparable to the wavelength, wave interference becomes important. However, this does not guarantee complete coherence because of the nature of the source and imperfect surfaces. Let us first consider the radiative properties of a layer or a slab, in the incoherent limit, because of its simplicity.

Either the ray-tracing method or the net radiation method can be applied to find out the transmittance and the reflectance of a thick layer [1]. Consider a slab of thickness $d$, placed in air or vacuum, as shown in Fig. 9.1. The refractive index and the extinction coefficient of the material are $n_2$ and $\kappa_2$, respectively. As mentioned earlier, it is generally required that the thickness be much greater than the wavelength so that the interference effect can be neglected. Because the intensity will attenuate exponentially inside an absorbing medium, the penetration depth $\delta_\lambda = \lambda/4\pi\kappa$ should be greater than the layer thickness in order to have appreciable transmission. For this reason, the extinction coefficient is usually much smaller than the refractive index, i.e., $\kappa \ll n$. Therefore, we can limit our consideration to dielectric materials with a small loss, such as a glass window or a silicon wafer in the semitransparent region. For a given surface reflectivity $\rho'_\lambda$ and an internal transmissivity $\tau'_\lambda$, ray-tracing yields the directional-hemispherical spectral reflectance as

$$R'_\lambda = \rho'_\lambda + \rho'_\lambda(1-\rho'_\lambda)^2\tau'^2_\lambda + \rho'^3_\lambda(1-\rho'_\lambda)^2\tau'^4_\lambda + \rho'^5_\lambda(1-\rho'_\lambda)^2\tau'^6_\lambda + \cdots$$

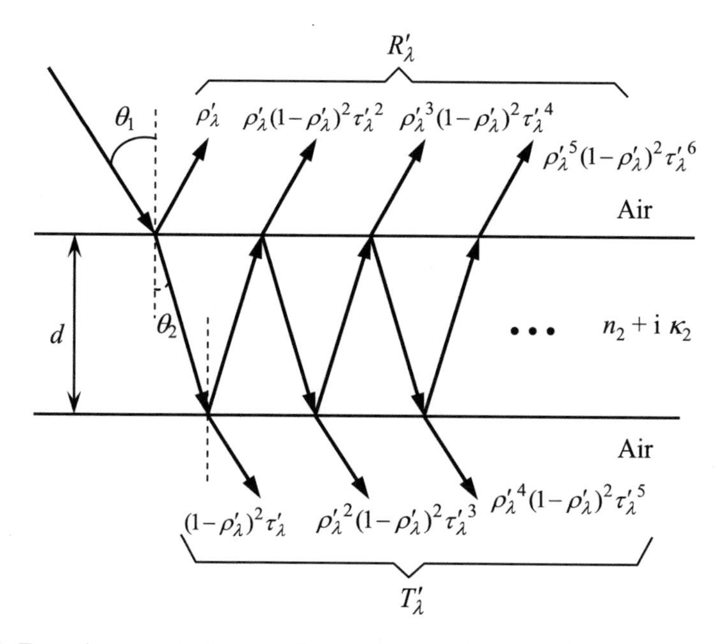


**Fig. 9.1** Transmittance and reflectance of a lamina as a result of multiple reflections

$$= \rho'_\lambda + \frac{\rho'_\lambda(1-\rho'_\lambda)^2\tau'^2_\lambda}{1-\rho'^2_\lambda\tau'^2_\lambda} = \rho'_\lambda\left[1 + \frac{(1-\rho'_\lambda)^2\tau'^2_\lambda}{1-\rho'^2_\lambda\tau'^2_\lambda}\right] \tag{9.1}$$

because the second term and beyond form a geometric series. Similarly, the directional-hemispherical spectral transmittance can be expressed as

$$T'_\lambda = \frac{(1-\rho'_\lambda)^2\tau'_\lambda}{1-\rho'^2_\lambda\tau'^2_\lambda} \tag{9.2}$$

Hence, the directional-spectral absorptance of the lamina at the given direction and wavelength is

$$A'_\lambda = 1 - T'_\lambda - R'_\lambda = \frac{(1-\rho'_\lambda)(1-\tau'_\lambda)}{1-\rho'_\lambda\tau'_\lambda} \tag{9.3}$$

The reflectivity $\rho'_\lambda$ can be calculated from Eqs. (8.73) and (8.80), for each polarization, as a function of the angle of incidence $\theta_1$ and the refractive index. For unpolarized incident radiation, $R'_\lambda$, $T'_\lambda$, and $A'_\lambda$ should be averaged over the two linear polarizations. The influence of $\kappa_2$ on $\rho'_\lambda$ is often negligibly small. On the other hand, $\kappa_2$ affects the absorption through the internal transmissivity $\tau'_\lambda$, defined as

$$\tau'_\lambda = \exp\left(-\frac{4\pi\kappa_2 d}{\lambda\cos\theta_2}\right) \tag{9.4}$$

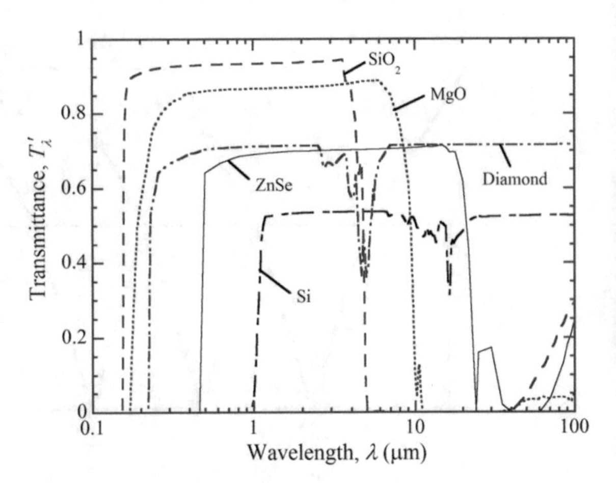


**Fig. 9.2** Normal transmittance of several dielectric materials with 0.5 mm thickness at room temperature

where $\lambda$ is the wavelength in air or vacuum, $\theta_2$ is the refraction angle inside the slab, and $d/\cos\theta_2$ can be considered as the actual path length of the ray inside the layer. From Snell's law, we have $\cos\theta_2 = \sqrt{1 - (1/n_2)^2 \sin^2\theta_1}$. Here again, the effect of $\kappa_2$ is neglected. Figure 9.2 shows the transmittance at normal incidence for several semitransparent materials with a thickness $d = 0.5$ mm, calculated using the tabulated optical constants from Palik [2]. It can be seen that silicon dioxide ($SiO_2$) is transparent in the visible region but opaque to infrared radiation beyond 5 μm wavelength. Here, the $SiO_2$ spectrum is for hydroxyl free fused silica, since common $SiO_2$ glass has a strong absorption band near 3 μm due to hydroxyl (OH) groups. Intrinsic silicon (Si) is opaque for visible light but has a transmittance of about 53% in the far-infrared region.

When there is no absorption, the reflectance and the transmittance are independent of the layer thickness *d*, and for normal incidence, the following simplified equation can be used:

$$T'_\lambda = \frac{2n_2}{n_2^2 + 1} \tag{9.5}$$

For a fused silica ($SiO_2$) window in the visible range, with a refractive index around 1.5, the transmittance is 0.923. For a diamond with a refractive index of about 2.4, the transmittance is 0.71 in the transparent region regardless of the thickness.

As discussed in Chap. 8, lattice vibrations in polar materials tend to give mid-infrared absorption bands. At high frequencies or short wavelengths, the fundamental bandgap absorption becomes important. The region in between is the transparent region, whose cutoff wavelengths depend on the material. Table 9.1 lists the transparent region and the range of the refractive index for a number of crystalline insulators or intrinsic (lightly doped) semiconductors, mostly taken from Ref. [2]. In this spectral region, the refractive index decreases slightly as the wavelength increases (normal dispersion). The simple Cauchy's equation [3] and the Sellmeier equation

**Table 9.1** Refractive index and transparent region for some typical materials near room temperature. The bandgap energy is also listed

| Materials | Symbol | Bandgap (eV) | Refractive index range[a] | Shortwave cutoff[b] (μm) | Longwave cutoff[b] (μm) |
|---|---|---|---|---|---|
| Diamond | C | 5.5 | 2.37–2.74 | 0.25 | – |
| Silicon | Si | 1.11 | 3.42–3.52 | 1.2 | – |
| Germanium | Ge | 0.66 | 4.0–4.1 | 2.0 | 20 |
| Silicon carbide | SiC | 3.05 | 2.45–2.75 | 0.45 | 5.0 |
| Boron nitride | BN | 7.5 | 1.8–2.1 | 0.2 | 5.0 |
| Aluminum nitride | AlN | 6.0 | 1.9–2.1 | 0.3 | 7.0 |
| Gallium arsenide | GaAs | 1.43 | 3.2–3.5 | 0.9 | 17 |
| Gallium antimonite | GaSb | 0.68 | 3.7–3.8 | 2.0 | 28 |
| Zinc selenide | ZnSe | 2.6 | 2.1–2.4 | 0.6 | 19 |
| Cadmium telluride | CdTe | 1.56 | 2.6–2.9 | 0.9 | 27 |
| Calcium fluoride | $CaF_2$ | 12 | 1.3–1.8 | 0.15 | 9.0 |
| Cesium iodide | CsI | 5.4 | 1.6–2.2 | 0.25 | 50 |
| Potassium bromide | KBr | 7.6 | 1.3–1.8 | 0.22 | 28 |
| Fused silica | $SiO_2$ | 9.0 | 1.4–1.8 | 0.15 | 3.6 |
| Magnesium oxide | MgO | 7.8 | 1.5–1.8 | 0.25 | 7.0 |
| Hafnium dioxide | $HfO_2$ | 5.6 | 1.8–2.1 | 0.3 | 7.0 |
| Rutile | $TiO_2$ | 3.05 | 2.5–3.0 | 0.5 | 5.0 |
| Zirconia | $ZrO_2$ | 6.00 | 2.0–2.3 | 0.36 | 5.0 |
| Sapphire | $Al_2O_3$ | 9.5 | 1.6–1.9 | 0.15 | 6.0 |
| Tantalum oxide | $Ta_2O_5$ | 4.0 | 1.8–2.2 | 0.5 | 5.0 |
| Silicon nitride | $Si_3N_4$ | 5.3 | 1.8–2.2 | 0.3 | 5.0 |
| Strontium titanate | $SrTiO_3$ | 3.25 | 1.9–2.2 | 0.4 | 7.0 |
| Water (liquid)[c] | $H_2O$ | | 1.32–1.35 | 0.3 | 1.2 |

[a]Higher refractive index corresponds to the shorter wavelength end and vice versa. Some are anisotropic and some may have different crystalline structures. Numerical values may vary from sample to sample due to different crystalline structures, impurities, and defects

[b]The exact values depend on the thickness and may depend on the preparation methods, defects, impurities, etc

[c]Water is listed for the sake of comparison

[4] are often used to describe the wavelength-dependent refractive index, though other major complicated dispersion relations have also been introduced [5].

For diamond and Si, multiphonon absorption becomes important in the mid-IR region, as shown in Fig. 9.2. Similarly, multiphonon absorption becomes important for Ge from 20 to 40 μm. So the longwave cutoff is listed as 20 μm. In some applications, antireflection coatings are often used to reduce reflectance and enhance transmittance, which will be discussed later for multilayer structures. The list comprises selected elemental semiconductors, group IV compound semiconductors, III–V semiconductors, II–VI semiconductors, a perovskite, as well as some dielectric materials such as oxides, halides, and nitride. Liquid water is also included for reference purpose. Note that the refractive index of ice is close to that of liquid water.

### *9.1.2 Thin Films*

Thin film coatings are of practical importance to the design of spectrally selective surfaces for solar energy utilization and space applications, optical filters, and antireflection coatings. When the wavelength of radiation is comparable to the coherence length, which depends not only on the properties of the film but also on the characteristics of the source and the detector, wave interference becomes important. To consider the interference effect, the amplitude and the phase of the electric field (or the magnetic field) must be traced during multiple reflections. The method is usually referred to as thin-film optics, as illustrated in Fig. 9.3 for a thin film of thickness $d$ between two semi-infinite media [6, 7]. There are several practical configurations

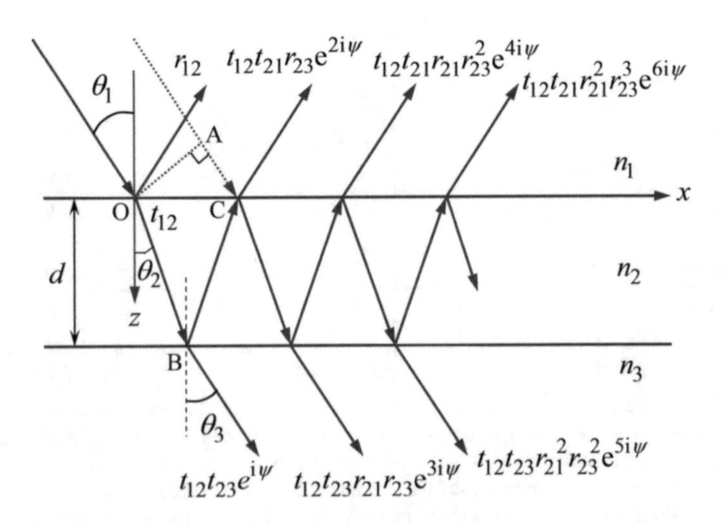


**Fig. 9.3** Illustration of interference between multiple reflections

based on the structure shown in this figure. (a) The first is for a free-standing film in air. (b) The second is for radiation incident from air (medium 1) on a thin film (medium 2) coated onto a semi-infinite substrate (medium 3). (c) In the third configuration, media 1 and 3 are dielectrics but medium 2 is a vacuum. This configuration is important for photon tunneling experiments to be discussed in Chap. 10.

Let us first consider the lossless case, where the refractive indices are all real, and so are the angles of incidence and refraction. It will be seen later that the equations can easily be extended to absorbing media using complex variables. A plane wave with either $p$ or $s$ polarization is incident from medium 1. Note that $t_{jk}$ and $r_{jk}$, where $j, k = 1, 2,$ or 3, are, respectively, the transmission and reflection coefficients between the media $j$ and $k$ for the given polarization. While the multiply reflected waves are illustrated with a spatial displacement, interference occurs at the same time and location between multiply reflected beams. For this reason, upon traversing the film, the wave acquires a phase shift $\psi$ given by

$$\psi = \frac{2\pi d}{\lambda}\sqrt{n_2^2 - n_1^2 \sin^2\theta_1} \tag{9.6}$$

Note again that $\lambda$ is the wavelength in vacuum. This is to say that $\psi = 2\pi(n_2/\lambda)\mathrm{d}\cos\theta_2$. The reason that $\cos\theta_2$ is in the numerator, instead of in the denominator, is because the phase for the same location $x$ is considered when $z$ is changed from 0 to $d$. The phase of the electric field is given by $\mathbf{k}\cdot\mathbf{r}$, and thus, the phase difference is $\psi = k_2\mathrm{d}\cos\theta_2$, where $k_2 = 2\pi n_2/\lambda$. Another way to understand the phase shift is to consider the plane of constant phase, as illustrated in Fig. 9.3 with the line OA. The first reflected wave is the wave from A to C that acquires a phase difference of $(k_1\sin\theta_1)(2\mathrm{d}\tan\theta_2) = (k_2\sin\theta_2)(2\mathrm{d}\tan\theta_2)$ because $k_x = k_j\sin\theta_j$ is the same in all media. The second reflected wave goes through the film twice (from O to B and then from B to C) and gains a phase difference of $2k_2d/\cos\theta_2$. It can easily be shown that the phase shift between the first and the second reflected waves is $2k_2\mathrm{d}\left(\frac{1}{\cos\theta_2} - \frac{\sin^2\theta_2}{\cos\theta_2}\right) = 2\psi$. More detailed discussion can be found from Brewster [8]. After the superposition, the field reflection and transmission coefficients of the film can be expressed as

$$r = r_{12} + \frac{t_{12}t_{21}r_{23}\mathrm{e}^{2\mathrm{i}\psi}}{1 - r_{21}r_{23}\mathrm{e}^{2\mathrm{i}\psi}} \tag{9.7}$$

and

$$t = \frac{t_{12}t_{23}\mathrm{e}^{\mathrm{i}\psi}}{1 - r_{21}r_{23}\mathrm{e}^{2\mathrm{i}\psi}} \tag{9.8}$$

which are known as Airy's formulae [6, 7] . It should be noted that these coefficients are defined based on the electric fields for $s$ polarization and the magnetic fields for $p$-polarization, respectively. The energy reflectance can be calculated by

$$R'_\lambda = rr^* = \left| r_{12} + \frac{t_{12}t_{21}r_{23}\mathrm{e}^{2\mathrm{i}\psi}}{1 - r_{21}r_{23}\mathrm{e}^{2\mathrm{i}\psi}} \right|^2 \tag{9.9}$$

For the incident radiation with random polarization, Eq. (9.9) should be averaged over the two linear polarizations by evaluating Fresnel's coefficients for each polarization separately.

It should be noted that Eqs. (9.6)–(9.9) are not limited to lossless situations as long as the absorption in medium 1 is negligible [9]. When $n_2$ and $n_3$ are complex, the phase shift given in Eq. (9.6) becomes complex. Note that the reflection and transmission coefficients in Eqs. (9.7) and (9.8) are always complex. Waves inside an absorbing medium are typically inhomogeneous because the constant-phase planes are defined by the real part of the wavevector and the constant-amplitude planes are parallel to the interfaces. To determine the direction of energy flow, one needs to carefully evaluate the Poynting vector in medium 3. The expression of the energy transmittance is similar to those for the absorptivity in Eqs. (8.75) and (8.81). If medium 3 is also lossless as for medium 1, we can write the transmittance in terms of the transmission coefficient as in the following:

$$T'_{\lambda,s} = \frac{n_3 \cos\theta_3}{n_1 \cos\theta_1} tt^*, \quad \text{for } s \text{ polarization} \tag{9.10a}$$

and

$$T'_{\lambda,p} = \frac{n_1 \cos\theta_3}{n_3 \cos\theta_1} tt^*, \quad \text{for } p \text{ polarization} \tag{9.10b}$$

which are the exact expressions of transmittance of given polarization for an absorbing film (medium 2). For a free-standing film in air, since $n_1 = n_3 = 1$, the transmittance can be approximated by the following equation when the film is slightly absorbing (i.e., $\kappa_2 \ll n_2$):

$$T'_\lambda = \frac{(1 - \rho'_\lambda)^2 \tau'_\lambda}{1 + \rho'^2_\lambda \tau'^2_\lambda - 2\rho'_\lambda \tau'_\lambda \cos(2\psi)} \tag{9.11}$$

Here, $\psi$ and $\rho'_\lambda$ are calculated by neglecting $\kappa_2$, and $\tau'_\lambda$ is from Eq. (9.4). Note that $\rho'_\lambda$ depends on the polarization so does $T'_\lambda$. The transmittance will oscillate even though the optical constants are unchanged. A change in wavelength, thickness, or refractive index can cause the transmittance to oscillate. The transmittance spectrum has peaks at $\psi = m\pi$ and valleys at $\psi = (m + \frac{1}{2})\pi$, where $m$ is a nonnegative integer. Figure 9.4 shows the calculated normal transmittance for $d = 10$ μm and $n_2 = n + \mathrm{i}\kappa$, with $n = 2$ and $\kappa = 0$, 0.005, and 0.05. The subscript 2 is dropped for convenience. The results are plotted in terms of wavenumber between 750 and 1500 cm$^{-1}$. The *free spectral range* is the frequency interval between two peaks. It is convenient to use the wavenumber instead of frequency. For normal incidence, the free spectral range in terms of wavenumber is given by

**Fig. 9.4** Calculated transmittance of a thin film of 10 μm thickness with $n = 2$ and various $\kappa$ values

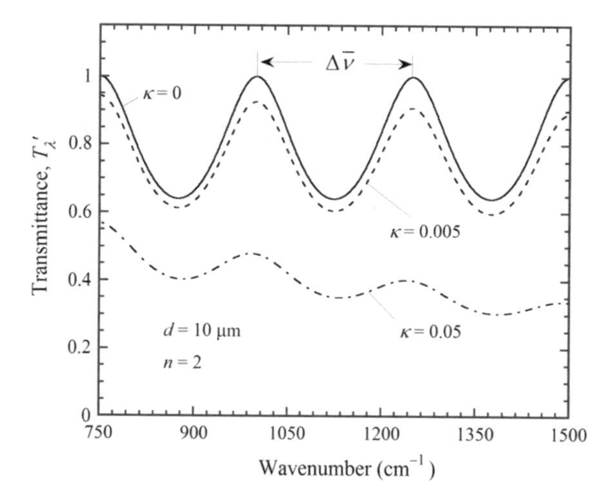


$$\Delta\bar{\nu} = \frac{1}{2nd} \tag{9.12}$$

where $d$ is in cm and $\Delta\bar{\nu}$ is in cm$^{-1}$. When plotted in terms of wavelength, the free spectral range becomes

$$\Delta\lambda = \frac{\Delta\bar{\nu}}{\bar{\nu}^2} = \frac{\lambda^2}{2nd} \tag{9.12a}$$

which increases with wavelength for constant $n$ and $d$. In the absence of absorption, the maximum transmittance is unity. The inclusion of a very small nonzero extinction coefficient $\kappa$ can cause the transmittance to be reduced from the lossless situation, especially at shorter wavelengths. When $\kappa = 0.05$, the internal transmissivity $\tau'_\lambda$ is a strong function of wavelength and the transmittance is significantly reduced. Furthermore, the *fringe contrast* is also reduced due to absorption. The fringe contrast $\Phi$ is defined, based on the maximum transmittance $T_{\max}$ and minimum transmittance $T_{\min}$, as

$$\Phi = \frac{T_{\max} - T_{\min}}{T_{\max} + T_{\min}} \tag{9.13}$$

For broadband or polychromatic radiation, the total transmittance is defined as the fraction of the energy transmitted. Suppose the spectral intensity is $I_\lambda$, then the total transmittance is

$$T'_{\text{tot}} = \int_0^\infty I_\lambda(\lambda) T'_\lambda(\lambda)\mathrm{d}\lambda \bigg/ \int_0^\infty I_\lambda(\lambda)\mathrm{d}\lambda \tag{9.14}$$

In some practice, one needs to integrate the transmittance over a narrow band. An example is the radiation coming through a filter or a spectrometer with a finite resolution. The intensity is nearly constant within the small bandwidth; the transmittance can be averaged over a spectral width $\Delta\lambda$ around $\lambda$ for each wavelength, viz.,

$$\overline{T'_\lambda}(\lambda) = \frac{1}{\Delta\lambda}\int_{\lambda-\Delta\lambda/2}^{\lambda+\Delta\lambda/2} T'_\lambda(\lambda)\mathrm{d}\lambda \tag{9.15}$$

It can be shown that integrating the coherence formula in Eq. (9.11) over a free spectral range $\Delta\lambda = \Delta\bar{\nu}/\bar{\nu}^2$ gives the same result as the incoherence formula in Eq. (9.2). However, the fringe-averaged transmittance is not equal to the arithmetic average of the transmittance maximum and minimum. When $d$ is much greater than the wavelength by a factor of, say, 1000, the free spectral range $\Delta\lambda$ will become so small that most spectrophotometers do not have the sufficient resolution to discern the fringes. Furthermore, a slight variation in the film thickness or the wedge effect will cause the phases of multiple reflections to be canceled out. The measured transmittance will follow Eq. (9.2) without the high-frequency oscillation. That is why Eq. (9.2) has practical importance even though it can be obtained from Eq. (9.11) by spectral averaging. The spectral-averaging method is useful to obtain radiative properties in the partial coherence regime, to be discussed in Sect. 9.1.3.

It should be emphasized that for metallic films, when the extinction coefficient is not much smaller than the refractive index, Eq. (9.11) breaks down and the transmittance of a thin film must be calculated according to Eqs. (9.10a) and (9.10b). Consider a 100 nm gold film with $n_2 = 0.916 + \mathrm{i}1.84$ at the wavelength $\lambda = 0.5\ \mu\mathrm{m}$. The penetration depth is $\delta_\lambda = \lambda/(4\pi\kappa) = 21.6\ \mathrm{nm}$. At normal incidence, $\rho'_\lambda = 0.481$ and $\tau'_\lambda = 0.0098$, and both Eqs. (9.2) and (9.11) reduce to $T'_\lambda \approx (1-\rho'_\lambda)^2\tau'_\lambda = 0.0026$. This result, however, is incorrect because neither equation is applicable for large extinction coefficients. Using Eqs. (9.10a) and (9.10b) and the complex Fresnel coefficients defined in Chap. 8, we have reevaluated the normal transmittance of the gold film to be $T'_\lambda \approx 0.013$ in this case (see Problem 9.5 and Zhang [9] for more discussion).

**Example 9.1** Calculate the reflectance in terms of film thickness $d$ for a dielectric film onto a silicon substrate with a refractive index of $n_3 = 3.44$ near $\lambda = 2.5\ \mu\mathrm{m}$. Assume radiation is incident at normal incidence from air. Consider two different coatings: $n_2 = 1.83$ (SiO) and $n_2 = 4.07$ (Ge).

**Solution** Equation (9.9) can be recast as $R'_\lambda = \left|\frac{r_{12}+r_{23}e^{i2\psi_2}}{1+r_{12}r_{23}e^{i2\psi_2}}\right|^2$, where for normal incidence, $r_{12} = (n_1 - n_2)/(n_1 + n_2)$ and $r_{23} = (n_2 - n_3)/(n_2 + n_3)$. While the Fresnel coefficients are for $s$ polarization, a minus sign for both $r_{12}$ and $r_{23}$ for $p$-polarization will not change the value of $R'_\lambda$. The results are plotted in terms of the dimensionless parameter $\xi = n_2 d/\lambda$ in Fig. 9.5. The reflectance oscillates with a period $\Delta d = \lambda/(2n_2)$. When $n_2 d = \lambda/2, 3\lambda/2, 5\lambda/2$, and so on, the reflectance is reduced to that of silicon without coating: $R'_\lambda = (n_1 - n_3)^2/(n_1 + n_3)^2$. When $n_2 > n_3$ or $n_2 < n_1$, the reflectance is always greater than that without coating and

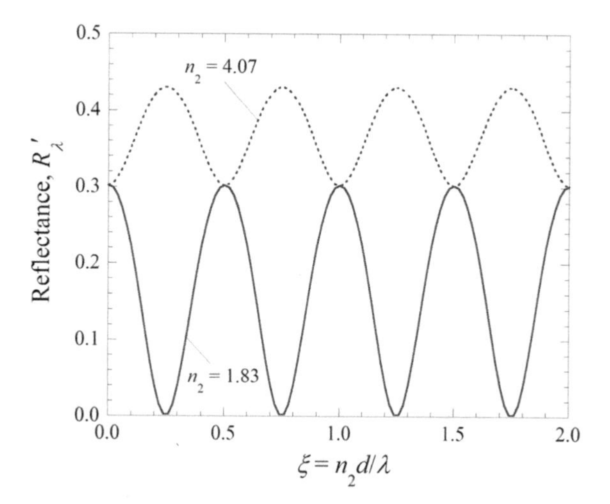


**Fig. 9.5** Reflectance for a silicon monoxide (SiO) or germanium (Ge) film onto a silicon (Si) substrate at $\lambda = 2.5\ \mu\text{m}$

reaches a maximum at $n_2 d = \lambda/4, 3\lambda/4, 5\lambda/4$, and so on. When $n_1 < n_2 < n_3$, the reflectance is always smaller than that without coating and reaches a minimum at $n_2 d = \lambda/4, 3\lambda/4, 5\lambda/4$, and so on. The values are determined by $R'_\lambda = \left[(n_1 n_3 - n_2^2)/(n_1 n_3 + n_2^2)\right]^2$. Note that the reflectance minimum becomes zero when $n_2 = \sqrt{n_1 n_3} = 1.855$. Since the refractive index of SiO is close to this value, a nearly zero reflectance can be obtained. This is called the *antireflection effect* and has numerous applications in many optical systems including eye glasses. In addition, quarter-wave antireflection coatings can be used to improve the energy conversion efficiency for solar energy applications.

### 9.1.3 Partial Coherence

It should be noted that no source is perfectly coherent—even laser or atomic emission has a nonzero line width. Likewise, no source is completely incoherent—even the most chaotic blackbody radiation has a small *coherence length*. The coherence length is related to the distance that light travels within a *coherence time*. The concept of coherence is related to the situation where the wave nature will be preserved. When fluctuations manifest, interference effects disappear, when the time is longer than the coherence time or when waves travel a distance longer than the coherence length [7]. Although complete incoherence and coherence formulae can be applied to a variety of practical problems, there are situations that do not fall in either regime. An example is the measured transmittance spectra of a slab with a spectrometer, such as a grating spectrophotometer or the Fourier-transform infrared (FTIR) spectrometer based on the Michelson interferometer as discussed in the previous chapter. Due to the finite instrument resolution and imperfections of the sample surfaces (not perfectly parallel or smooth), the fringe contrast defined in Eq. (9.13) for transmittance is always less than that predicted by the coherence formula. A similar definition also applies to the reflectance spectrum.

Partial coherence theory was developed before the first laser was invented in 1960s and has gone through significant advancements along with the developments of lasers and quantum optics, including the application to radiometry [10]. A brief introduction is given here with an emphasis on the radiative properties of thin films. The electric field can be expressed in either frequency domain as $E(\nu)$ or time domain as $E(t)$, which are related by Fourier transforms. The *mutual coherence function* of any two waves is defined as

$$\langle E_j(t)E_k^*(t)\rangle = 4\int_0^\infty G_{jk}(\nu)\mathrm{d}\nu \tag{9.16}$$

where the angular bracket <> symbolizes the time-averaging operation according to

$$\langle E_j(t)E_k^*(t)\rangle = \lim_{\tau\to\infty}\frac{1}{2\tau}\int_{-\tau}^{\tau} E_j(t)E_k^*(t)dt \tag{9.17}$$

and $G_{jk}(\nu)$ is the mutual spectral density given by

$$G_{jk} = \lim_{\tau\to\infty}\frac{1}{2\tau}\overline{E_j(\nu)E_k^*(\nu)} \tag{9.18a}$$

where the "long bar" denotes ensemble averaging. The spectral density of a wave is defined by

$$G(\nu) = \lim_{\tau\to\infty}\frac{1}{2\tau}\overline{E(\nu)E^*(\nu)} \tag{9.18b}$$

and the optical intensity, which is proportional to the radiant energy flux in a given medium, is

$$I = \langle E(t)E^*(t)\rangle = 4\int_0^\infty G(\nu)\mathrm{d}\nu \tag{9.19}$$

The complex degree of coherence is defined as

$$\gamma_{jk} = \frac{\langle E_j(t)E_k^*(t)\rangle}{\sqrt{\langle E_j(t)E_j^*(t)\rangle\langle E_k(t)E_k^*(t)\rangle}} \tag{9.20}$$

Note that $|\gamma_{jk}| \le 1$. If there are only two waves, each with an optical intensity of $I_1$ and $I_2$, the combined optical intensity of the two waves is given as follows:

$$I_c = I_1 + I_2 + \sqrt{I_1 I_2}(\gamma_{12} + \gamma_{12}^*) \tag{9.21}$$

Let us use Young's double-slit experiment as an example, where light from a pinhole goes through two slits. Interference patterns will be projected on a screen. When the slits are of very small width and the source is nearly monochromatic, a sine wave pattern will be observed with alternate bright and dark fringes. This is because $\gamma_{12} = \exp(i\delta)$, where $\delta$ is the phase difference between the two beams and varies with the position on the screen. The outcome is completely coherent because $|\gamma_{12}| = 1$. On the other hand, when the source is polychromatic, the pattern will be the brightest at the center because constructive interference occurs for all wavelengths only at the center. The interference fringes will fade away from the center and eventually disappear because of the lack of coherence. In this case, $|\gamma_{12}|$ is position dependent. Partial coherence can also occur as the width of the slit is enlarged. If the slit width is comparable to or larger than the wavelengths, the screen will be evenly illuminated. This corresponds to a complete incoherence with $|\gamma_{12}| = 0$ and $I_c = I_1 + I_2$.

Chen and Tien employed the partial coherence theory to calculate the radiative properties of a layer, by taking the forward propagating field in the film as composed of two components: the first transmitted wave and all the rest that are caused by multiple reflections [11]. Alternatively, the degree of coherence may be defined between any two multiply reflected waves, and the radiative properties in the partial coherence regime can be expressed in an infinite summation. Several factors affect the degree of coherence, such as the beam divergence, the thickness variation, or the finite spectral width of the instrument. The combined effect is that multiple reflections become less and less coherent, because the phase of the wave increases by $2\psi$ each time when it undergoes a round trip inside the film (see Fig. 9.3). Recently, Fu et al. [12] obtained analytic formulae for the reflectance and the transmittance of a thin film using direct spectral integration. The integral averaging of transmittance, calculated from wave optics over a finite frequency interval, yields the same result as the partial coherence formulation does. The spectral averaging of the transmittance can be evaluated by

$$\bar{T}(\nu) = \frac{1}{\delta\nu} \int_{\nu-\delta\nu/2}^{\nu+\delta\nu/2} T(\zeta)\,d\zeta \tag{9.22}$$

where $\zeta$ is a dummy variable and the frequency interval used for the averaging $\delta\nu$ is called the *coherence spectral width* [13]. The directional-hemispherical spectral transmittance is simply expressed as $T(\nu)$ in Eq. (9.22) without any subscript or superscript for clarity. The frequency $\nu$ is most conveniently expressed in $cm^{-1}$ or in terms of wavenumber as done before. It should be emphasized that the spectrally averaged property is still a spectral property rather than a total property. It is inherently assumed that $\delta\nu$ is a small bandwidth within which the source spectral intensity is independent of frequency. Furthermore, $\delta\nu$ is related not only to the effective bandwidth, the resolution, and the sampling interval of the spectrometer but also to the conditions of the specimen. Figure 9.6 illustrates the effect of spectral averaging on the transmittance spectrum for a film with $n = 2$ and $\kappa = 0$, with various $\delta\nu$

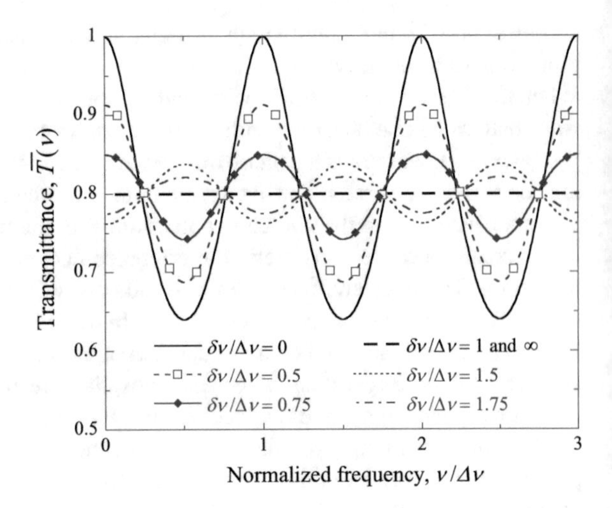


**Fig. 9.6** The effect of coherence spectral width on the spectrally averaged transmittance

values, at normal incidence. Both the frequency $\nu$ and the coherence spectral width $\delta\nu$ are normalized by the free spectral range $\Delta\nu$ so that the curves are independent of the film thickness and the frequency unit used. As $\delta\nu/\Delta\nu$ increases from 0 (the coherent limit), the fringe contrast decreases until $\delta\nu/\Delta\nu = 1$ when all the fringes disappear. When $\delta\nu/\Delta\nu > 1$, however, the fringes reappear but the peaks and the valleys invert from the original. The inversion is largest when $\delta\nu/\Delta\nu = 1.5$. When $\delta\nu/\Delta\nu \gg 1$, the fringe contrast becomes negligible, and the transmittance approximates the incoherent limit when geometric optics is applicable.

Although $\delta\nu = 0$ and $\delta\nu \to \infty$ correspond to the coherent and incoherent limits, respectively, the magnitude of $\delta\nu$ is not directly related to the degree of coherence in the partial coherence regime. For example, $\delta\nu/\Delta\nu = 1.5$ is more coherent than $\delta\nu/\Delta\nu = 1$ (when all fringes disappear). The degrees of coherence are difficult to calculate even for smooth films and not applicable to films with rough surfaces. Lee et al. [13] introduced a coherence function:

$$\phi = \frac{\bar{T}(\nu_{\max}) - \bar{T}(\nu_{\min})}{T_{\text{coh}}(\nu_{\max}) - T_{\text{coh}}(\nu_{\min})} \tag{9.23}$$

where $T_{\text{coh}}$ is the transmittance calculated from the coherence formulation without scattering loss based on thin-film optics, $\bar{T}$ is the spectral averaging of transmittance calculated from Eq. (9.22) to include partial coherence, and $\nu_{\max}$ and $\nu_{\min}$ are the frequencies corresponding to transmittance maximum and minimum, respectively, in the coherent limit [13]. In essence, the denominator equals the difference between transmittance extrema in the coherent limit, and the numerator equals the difference in transmittance extrema, when partial coherence is considered.

The coherence function is plotted in Fig. 9.7 as a function of a dimensionless parameter $\delta\nu/\Delta\nu$ for dielectric thin films. The film thickness is implicitly included in the parameters and does not affect the shape of the curves. The coherence function

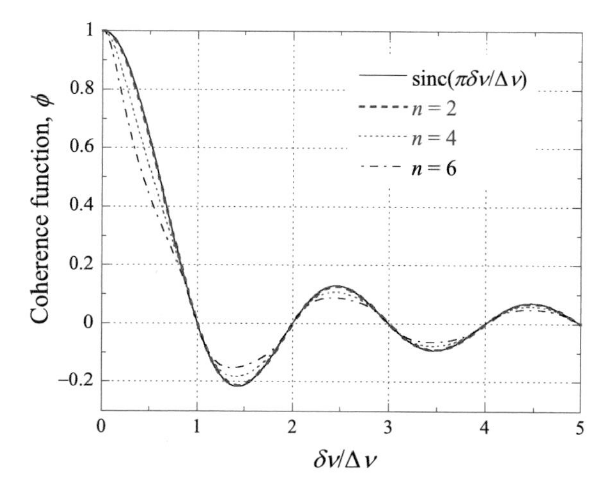


**Fig. 9.7** Coherence function versus the ratio of the coherence spectral width to the free spectral range for different refractive indices

varies within (−1, 1), and its magnitude quantifies the reduction in the fringe contrast from 1 in the coherent limit to 0 in the incoherent limit. The locations where $\phi = 0$ correspond to $\delta\nu = m\Delta\nu$ ($m$ = 1, 2, 3…), when all fringes disappear in the transmittance spectra. When $\phi < 0$, the peaks and the valleys are inverted in the transmittance spectrum, resulting in fringe flipping. When $n \leq 2$, it can be seen from Fig. 9.7 that the coherence function is approximated by the sinc function: $\text{sinc}(x) = \sin(x)/x$. As refractive index increases, however, the coherence function becomes flatter and deviates from the sinc function. The coherence function serves the same role as the degree of coherence that helps determine which approach (i.e., wave optics, partial coherence formulation, or geometric optics) is most suitable for modeling the radiative properties for a particular case. In addition, Eq. (9.23) can also be applied to rough surfaces, as will be discussed in the next section.

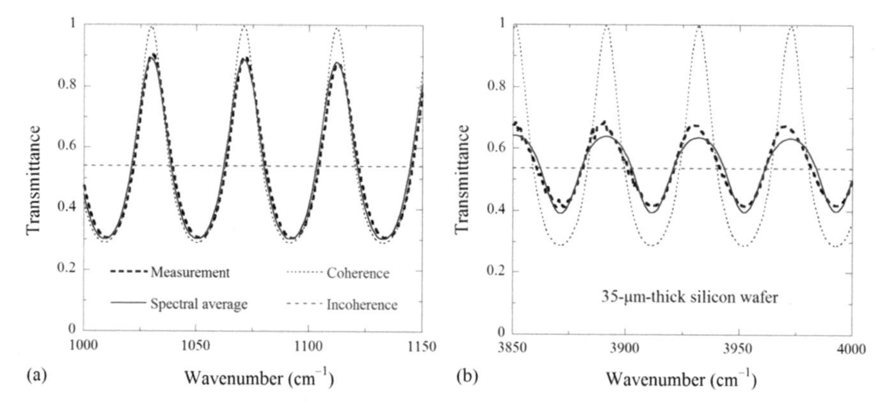


**Fig. 9.8** Normal transmittance of a 35-μm-thick Si wafer in two narrow spectral regions near the wavelengths of **a** 10 μm (1000 cm$^{-1}$) and **b** 2.5 μm (4000 cm$^{-1}$), respectively [13]

Figure 9.8 shows the measured and predicted transmittance for a double-side polished silicon wafer in two narrow spectral regions as functions of the wavenumber. The transmittance spectra in the coherent and incoherent limits are shown for comparison. Because the refractive index of silicon changes less than 1% ($n = 3.432 \pm 0.011$), the free spectral range in wavenumber is $\Delta\bar{\nu} \approx 41.3\,\text{cm}^{-1}$, and the transmittance predicted by the incoherence formula is approximately 0.537. It can be seen that the transmittance is less coherent toward short wavelengths (increasing wavenumber). Therefore, a wavenumber-dependent coherence spectral width was used to fit the data obtained from the FTIR spectrometer [13]. The coherence spectral width $\delta\bar{\nu}$ varies from 10.4 $\text{cm}^{-1}$ at $\bar{\nu} = 1000\,\text{cm}^{-1}$ to 28.7 $\text{cm}^{-1}$ at $\bar{\nu} = 4000\,\text{cm}^{-1}$. The coherence function $\phi$ calculated from Eq. (9.23) changes from 0.84 at $\bar{\nu} = 1000\,\text{cm}^{-1}$ to 0.33 at $\bar{\nu} = 4000\,\text{cm}^{-1}$. The coherence spectral width is much greater than the instrument resolution of 1 $\text{cm}^{-1}$, suggesting that the surfaces of the wafer may be slightly nonparallel. The measured transmittance is also sensitive to the mechanical stress on the wafer.

### 9.1.4 Effect of Surface Scattering

In order to model the losses in the reflectance and transmittance due to scattering at the surfaces, shown in Fig. 9.9, the Fresnel coefficients can be modified by the scattering factors that depend on the rms roughness. Notice that the reflectance and transmittance obtained this way are not directional-hemispherical properties. Because only the reflection and transmission near the specular directions are considered, we will use *specular reflectance* $R'_{\lambda,\text{sp}}$ and *specular transmittance* $T'_{\lambda,\text{sp}}$. The derivation of the scattering factor is based on the assumptions that the surface height follows the Gaussian distribution and the autocovariance function of surface roughness is also Gaussian. When both the rms roughness and the autocorrelation length are much less than the wavelength of the incident radiation, the *scalar scattering theory* may be applied to determine the reflection coefficients, considering

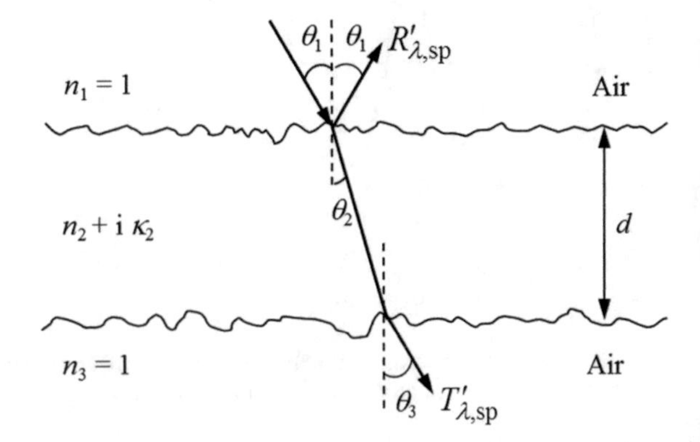


**Fig. 9.9** Geometry of a thin film with rough surfaces, in the model of the specular transmittance and reflectance, when $\kappa_2 \ll n_2$ and $\sigma_{\text{rms}} \ll \lambda$

scattering losses [14]. The modified Fresnel coefficients between the media $j$ and $k$ ($j = 1, 2,$ or 3; $k = j \pm 1$) are given in the following:

$$r'_{jk} = r_{jk} S_{\mathrm{r},jk} \tag{9.24a}$$

and

$$t'_{jk} = t_{jk} S_{\mathrm{t},jk} \tag{9.24b}$$

where the prime refers to the modified Fresnel coefficients for a given polarization, and the scattering factors are defined as follows, based on real refractive indices only:

$$S_{\mathrm{r},jk} = \exp\left[-\frac{1}{2}\left(\frac{4\pi\sigma_{\mathrm{rms}} n_j \cos\theta_j}{\lambda}\right)^2\right] \tag{9.25a}$$

and

$$S_{\mathrm{t},jk} = \exp\left[-\frac{1}{2}\left(\frac{2\pi\sigma_{\mathrm{rms}}(n_j \cos\theta_j - n_k \cos\theta_k)}{\lambda}\right)^2\right] \tag{9.25b}$$

where $\sigma_{\mathrm{rms}}$ is the rms roughness of the interface [14]. It should be noted that some relations of the Fresnel coefficients, such as $r_{jk} = -r_{kj}$ and $1 + r_{jk} = t_{jk}$, do not hold after the modifications, because of scattering losses. The reflectance and the transmittance should be obtained from Eqs. (9.9), (9.10a), and (9.10b). Furthermore, the energy losses due to surface roughness increase toward shorter wavelengths, because of the $\sigma_{\mathrm{rms}}/\lambda$ term in the scattering factors; this yields a reduction in the fringe contrasts and a decrease in the overall transmittance. Even for a nonabsorbing film, the sum of the specular transmittance and reflectance is not equal to 1, because of scattering losses.

**Example 9.2** Calculate the normal transmittance of a 10 μm film with a refractive index $n = 2.4$, when there is no absorption, in the spectral range from 1000 to 3000 cm$^{-1}$. Both surfaces are rough with a roughness $\sigma_{\mathrm{rms}}$ of 0.10 μm. How does the $\sigma_{\mathrm{rms}}$ value affect the transmittance?

**Solution** We can use Eqs. (9.10a) and (9.10b) to calculate the transmittance but with the reflection and transmission coefficients modified by Eqs. (9.24a) and (9.24b). The results are plotted in Fig. 9.10, for $\sigma_{\mathrm{rms}} = 0.05$, 0.10, and 0.20 μm, to examine the effect of roughness on the specular transmittance. It can be seen that surface roughness reduces both the peak transmittance and the fringe contrast. Furthermore, the reduction is more prominent toward shorter wavelengths.

An optically smooth surface has an rms roughness on the order of 10 nm. Some highly polished semiconductor wafers or thin films, grown by molecular beam epitaxy, can have an rms roughness less than 1 nm. On the other hand, chemical vapor

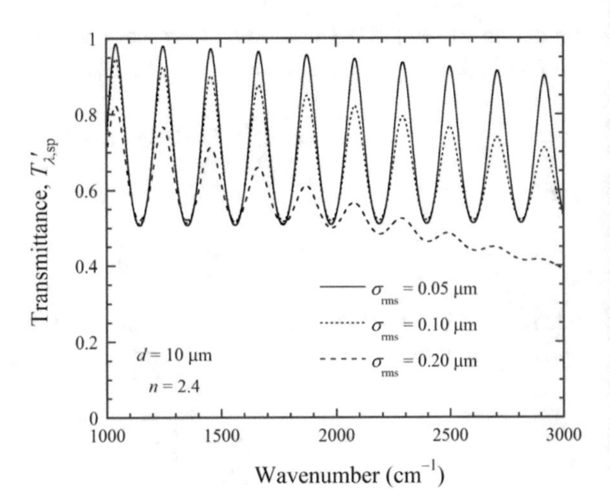


**Fig. 9.10** Transmittance of a dielectric thin film with surface roughness on both sides

deposited (CVD) diamond films and the backside of silicon wafers can have a roughness ranging from 100 nm to 1 μm. The fringe contrast in the measured spectrum is often less than that predicted by wave optics after the modification of the Fresnel coefficient, due to the lack of parallelism between the two surfaces. In other words, when the effect of partial coherence is significant, the scalar scattering theory alone cannot accurately predict the transmittance of thin films. Lee et al. [13] used the fringe-averaging method together with the scalar scattering theory to obtain excellent agreement of the specular transmittance for rough surfaces with FTIR measurements for a CVD diamond film and several silicon wafers. On the other hand, the scalar scattering theory cannot be applied when either the autocorrelation length or the rms roughness is comparable with the wavelength.

## 9.2 Multilayer Structures

For multiple parallel plates that are thick layers, without considering interference, the net radiation method or ray-tracing method can be applied along with recursion technique to obtain the transmittance and reflectance [1]. Since many applications involve a thin film on a substrate or multilayer thin films, expressions of the radiative properties of multilayer structures involving interference are analyzed in this section beginning with few layers of thin films. The transfer matrix method for thin-film multilayer structures will then be described, and its application to films on a thick substrate will also be discussed.

### 9.2.1 *Thin Films with Two or Three Layers*

Examples of two-layer thin films include a metallic coating on a thin dielectric substrate, especially in the long-wavelength region, where interference in the substrate cannot be ignored. The film can also be modeled as a sheet resistance for metallic films in the far-infrared and microwave regions. Nevertheless, thin-film optics is generally applicable to any spectral region and for different materials. The expressions of the reflectance and the transmittance of a thin film-substrate composite in vacuum are

$$R'_{\lambda,\mathrm{F}} = \left| r_\mathrm{a} + \frac{t_\mathrm{a} t_\mathrm{b} r_{\mathrm{S}0} \mathrm{e}^{\mathrm{i}2\psi_\mathrm{s}}}{1 - r_\mathrm{b} r_{\mathrm{S}0} \mathrm{e}^{\mathrm{i}2\psi_\mathrm{s}}} \right|^2 \tag{9.26}$$

$$R'_{\lambda,\mathrm{S}} = \left| r_{0\mathrm{S}} + \frac{t_{0\mathrm{S}} t_{\mathrm{S}0} r_\mathrm{b} \mathrm{e}^{\mathrm{i}2\psi_\mathrm{s}}}{1 - r_\mathrm{b} r_{\mathrm{S}0} \mathrm{e}^{\mathrm{i}2\psi_\mathrm{s}}} \right|^2 \tag{9.27}$$

and

$$T'_\lambda = \left| \frac{t_\mathrm{a} t_{\mathrm{S}0} \mathrm{e}^{\mathrm{i}\psi_\mathrm{s}}}{1 - r_\mathrm{b} r_{\mathrm{S}0} \mathrm{e}^{\mathrm{i}2\psi_\mathrm{s}}} \right|^2 \tag{9.28}$$

where the subscripts F and S indicate whether the incoming radiation is incident on the film or substrate, since the direction of incidence makes a difference for the reflectance, $\psi_\mathrm{s}$ is the complex phase shift inside the substrate, $t_\mathrm{a}$ and $r_\mathrm{a}$ are the transmission and reflection coefficients for incidence from vacuum to the film, when the substrate is assumed semi-infinite, $t_\mathrm{b}$ and $r_\mathrm{b}$ are the transmission and reflection coefficients for incidence from the substrate to the film, and subscripts S0 and 0S refer to the Fresnel coefficients at the substrate-vacuum interface. The reflection and transmission coefficients $r_\mathrm{a}$, $r_\mathrm{b}$, $t_\mathrm{a}$, and $t_\mathrm{b}$ are generally complex and should be calculated from Eqs. (9.7) and (9.8) using the phase shift of the film. The absorptance also depends on which side the radiation is incident from. When there is another coating at the backside of the substrate, one can replace the Fresnel coefficients with the transmission and reflection coefficients of the film.

**Example 9.3** A Fabry–Perot interferometer can be built with two mirrors made by coating highly reflecting materials (e.g., ultrathin metallic films) on both sides of a dielectric thin film, as illustrated on the left of Fig. 9.11. Derive a formula for the transmittance, and show that resonance in transmittance can be obtained within narrow spectral bands.

**Solution** In 1899, Charles Fabry and Alfred Perot constructed a device based on interference effect and published a series of papers on the possible applications in metrology and spectroscopy. This is the Fabry–Perot interferometer, also known as an optical cavity resonator or etalon. Like the Michelson interferometer, the Fabry–Perot interferometer is an important device used in spectroscopy, laser applications, and

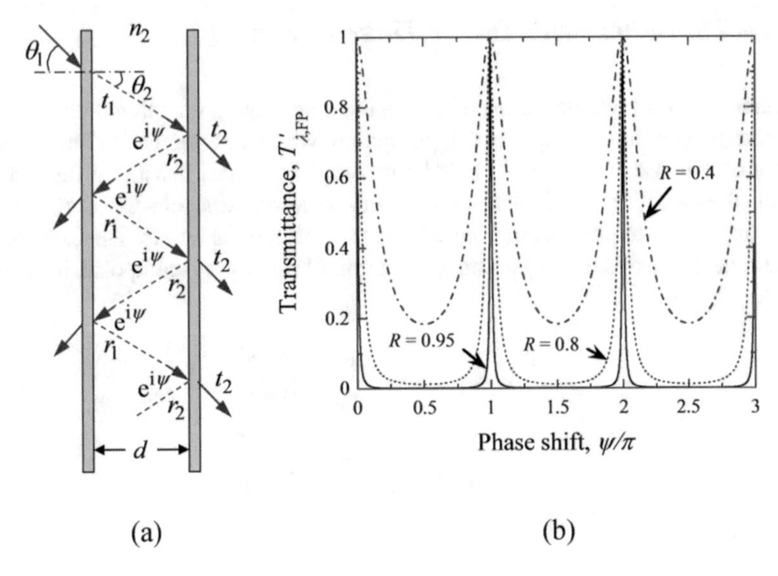


**Fig. 9.11** **a** Schematic of a Fabry–Perot interferometer; **b** the calculated transmittance for different $R$ values

wavelength and frequency standards [14]. By considering the transmission and reflection coefficients $t_1, t_2, r_1,$ and $r_2$, at each boundary of the dielectric film, the overall transmittance coefficient of the Fabry–Perot interferometer, shown in Fig. 9.11, can be expressed as follows:

$$t_{\mathrm{FP}} = \frac{t_1 t_2 \mathrm{e}^{\mathrm{i}\psi}}{1 - r_1 r_2 \mathrm{e}^{\mathrm{i}2\psi}} \tag{9.29}$$

where $\psi = 2\pi n_2 \bar{\nu} d_2 \cos\theta_2$ is the phase shift according to Eq. (9.6). Here, $\bar{\nu} = 1/\lambda$ is the wavenumber in $\mathrm{cm}^{-1}$. The energy transmittance can be written as follows:

$$T'_{\lambda,\mathrm{FP}} = t_{\mathrm{FP}} t^*_{\mathrm{FP}} = \frac{T_1 T_2}{\left(1 - \sqrt{R_1 R_2}\right)^2 + 4\sqrt{R_1 R_2}\sin^2\psi_{\mathrm{t}}} \tag{9.30}$$

where $\psi_{\mathrm{t}} = \psi + \arg(r_1)/2 + \arg(r_2)/2$ is the total phase shift that includes contributions by the interfaces, $T_1 = t_1 t_1^*$ and $T_2 = t_2 t_2^*$ are not exactly the transmittances through the coating, and $R_1 = r_1 r_1^*$ and $R_2 = r_2 r_2^*$ are indeed the reflectances for incidence from the dielectric to the left and right boundaries, respectively. When the loss can be neglected and the structure is symmetric, then $\psi_{\mathrm{t}} = \psi$, $R_1 = R_2 = R$, and $T_1 T_2 = (1 - R)^2$. In such case, Eq. (9.30) can be simplified as

$$T'_{\lambda,\mathrm{FP}} = \frac{(1 - R)^2}{(1 - R)^2 + 4R\sin^2\psi} \tag{9.31}$$

The results for different $R$ values are shown on the right of Fig. 9.11. Clearly, a large $R$ yields sharp transmission peaks at $\psi = m\pi$. Suppose the refractive index of the dielectric is kept constant and the change of the phase shift corresponds to the frequency variation, the free spectral range is the interval between two resonance peaks, given by $\Delta\bar{\nu} = 1/(2n_2 d_2 \cos\theta_2)$, similar to that of Eq. (9.12). The full-width-at-half-maximum (FWHM), $\delta\bar{\nu}$, measures how sharp the peak is. The ratio $Q = \Delta\bar{\nu}/\delta\bar{\nu}$ is called the *finesse* of the interferometer, which determines the resolving power. The finesse is known as the $Q$-factor of the resonator. For a lossless Fabry–Perot cavity, it can be shown that

$$Q = \frac{\Delta\bar{\nu}}{\delta\bar{\nu}} = \frac{\pi\sqrt{R}}{1-R} \tag{9.32}$$

which is 313, when $R = 0.99$. Kumar et al. [15] constructed a Fabry–Perot resonator, based on high-critical-temperature superconducting films on Si substrates, and demonstrated sharp transmission peaks in the far-infrared at cryogenic temperatures, when $YBa_2Cu_3O_{7-\delta}$ becomes superconducting. Wang et al. [16] demonstrated asymmetric Fabry–Perot structures for coherent thermal emission at elevated temperatures.

### 9.2.2 *The Matrix Formulation*

A multilayer structure containing $N$ layers is shown in Fig. 9.12. In this section, the 1D matrix formulation or transfer matrix method is presented in such a way

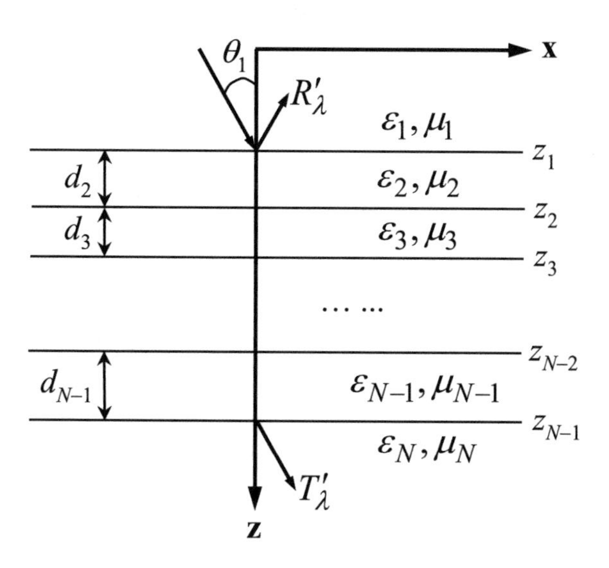


**Fig. 9.12** Schematic illustration of an $N$-layer structure, where the first and last layers are semi-infinite, and each layer is assumed to be homogeneous and isotropic

that magnetic materials can also be included. Each layer is assumed to be isotropic and homogeneous, and it can be fully described by a relative permittivity $\varepsilon_l$ and a relative permeability $\mu_l$ ($l = 1, 2,\ldots, N$). For a monochromatic plane wave originated from layer 1, which is assumed to be lossless, the phase-matching condition requires that $k_{lx} \equiv k_x = \omega n_1 \sin\theta_1/c_0$. Consider a linearly polarized electromagnetic wave, whose plane of incidence is perpendicular to the $y$-axis. For $s$ polarization or TE wave, where the electric field is parallel to the $y$-axis, the electric field in the $l$th layer can be written as $E_l(z)\mathrm{e}^{\mathrm{i}(k_x x-\omega t)}$, where

$$E_1(z) = A_1\mathrm{e}^{\mathrm{i}k_{1z}z} + B_1\mathrm{e}^{-\mathrm{i}k_{1z}z}$$

and

$$E_l(z) = A_l\mathrm{e}^{\mathrm{i}k_{lz}(z-z_{l-1})} + B_l\mathrm{e}^{-\mathrm{i}k_{lz}(z-z_{l-1})},\ l = 2, 3, \ldots, N \tag{9.33}$$

Here, $A_l$ and $B_l$ are the amplitudes of the forward and backward waves at the interface, respectively, $z_l = z_{l-1} + d_l$ ($l = 2, 3, \ldots, N - 1$), and $d_l$ is the layer thickness. The magnetic field can be obtained from the electric field using Maxwell's equations. The expression of the wave component $k_{l_z}$ is calculated from $k_x^2 + k_{lz}^2 = \varepsilon_l\mu_l\omega^2/c_0^2$. The only condition imposed is that the imaginary part of $k_{l_z}$ must not be less than zero. This will ensure that the wave will decay toward positive $z$. After applying boundary conditions at the interface, we can see that the field amplitudes of adjacent layers are related by [17]

$$\begin{pmatrix} A_l \\ B_l \end{pmatrix} = \mathbf{P}_l\mathbf{D}_l^{-1}\mathbf{D}_{l+1}\begin{pmatrix} A_{l+1} \\ B_{l+1} \end{pmatrix},\quad l = 1, 2, \ldots, N - 1 \tag{9.34}$$

In Eq. (9.34), $\mathbf{P}_l$ is the propagation matrix given by

$$\mathbf{P}_l = \mathbf{I} = \begin{pmatrix} 1 & 0 \\ 0 & 1 \end{pmatrix},\quad l = 1$$

and

$$\mathbf{P}_l = \begin{pmatrix} \mathrm{e}^{-\mathrm{i}k_{lz}d_l} & 0 \\ 0 & \mathrm{e}^{\mathrm{i}k_{lz}d_l} \end{pmatrix},\ l = 2, 3, \ldots, N - 1 \tag{9.35}$$

$\mathbf{D}_l$ is called the dynamical matrix, and $\mathbf{D}_l^{-1}$ is its inverse. For $s$ polarization, $\mathbf{D}_l$ is given in terms of $k_{lz}$ and $\mu_l$ as follows:

$$\mathbf{D}_l = \begin{pmatrix} 1 & 1 \\ k_{lz}/\mu_l & -k_{lz}/\mu_l \end{pmatrix},\ l = 1, 2, \ldots, N \tag{9.36}$$

By successively applying Eq. (9.35) to all layers, we have

$$\begin{pmatrix} A_1 \\ B_1 \end{pmatrix} = \mathbf{M} \begin{pmatrix} A_N \\ B_N \end{pmatrix} \tag{9.37}$$

where

$$\mathbf{M} = \begin{pmatrix} M_{11} & M_{12} \\ M_{21} & M_{22} \end{pmatrix} = \prod_{l=1}^{N-1} \mathbf{P}_l \mathbf{D}_l^{-1} \mathbf{D}_{l+1} \tag{9.38}$$

The electric field transmission and reflection coefficients are obtained by setting $B_N = 0$, because the last layer is semi-infinite, and thus there is no backward wave. Simple algebraic manipulations give the expressions of the coefficients as

$$t = A_N / A_1 = 1/M_{11} \tag{9.39}$$

and

$$r = B_1 / A_1 = M_{21}/M_{11} \tag{9.40}$$

Furthermore, the energy reflectance and transmittance are given as follows:

$$R'_\lambda = rr^* = \left| \frac{M_{21}}{M_{11}} \right|^2 \tag{9.41}$$

$$T'_\lambda = \frac{\mathrm{Re}\left(k_{Nz}^*/\mu_N^*\right)}{\mathrm{Re}\left(k_{1z}^*/\mu_1^*\right)} tt^* = \frac{\mathrm{Re}(k_{Nz}/\mu_N)}{\mathrm{Re}(k_{1z}/\mu_1)} \left| \frac{1}{M_{11}} \right|^2 \tag{9.42}$$

For $p$-polarization or TM wave, the magnetic field is parallel to the $y$-axis. Equation (9.33) can be written in terms of the magnetic field. The above procedure can then be applied to derive the transmission and reflection coefficients based on the magnetic fields. Then, the dynamic matrix $\mathbf{D}_l$ given in Eq. (9.36) must be replaced by

$$\mathbf{D}_l = \begin{pmatrix} 1 & 1 \\ k_{lz}/\varepsilon_l & -k_{lz}/\varepsilon_l \end{pmatrix}, \quad l = 1, 2, \ldots, N \tag{9.43}$$

The expression for the reflectance is the same as Eq. (9.41) and that for transmittance for $p$-polarization becomes

$$T'_\lambda = \frac{\mathrm{Re}(k_{Nz}/\varepsilon_N)}{\mathrm{Re}(k_{1z}/\varepsilon_1)} tt^* = \frac{\mathrm{Re}(k_{Nz}/\varepsilon_N)}{\mathrm{Re}(k_{1z}/\varepsilon_1)} \left| \frac{1}{M_{11}} \right|^2 \tag{9.44}$$

The assumption that the first medium is lossless is necessary because the reflectance is ill-defined if the first medium is lossy or dissipative, because of the coupling between the reflected and incident waves [6]. Nevertheless, Eqs. (9.41),

(9.42), and (9.44) are applicable even when the last medium is lossy. Comparing Eq. (9.42) with Eq. (9.44), and Eq. (9.36) with Eq. (9.43), we immediately notice the duality of the electric and magnetic fields, since the only difference is the interchange of $\varepsilon$ and $\mu$ in these equations. Further applications of the matrix formulation will be discussed in subsequent sections as well as in the next chapter.

### 9.2.3 Thin Films on a Thick Substrate

Radiative properties of thin coatings on a substrate are important for a large number of applications, such as a thermal oxide on a Si substrate, antireflection coatings on the lens of glasses, interference filters, metallic coatings, and superconducting films. In these cases, the coating thicknesses are on the order of nanometers and must be considered as thin films. On the other hand, the substrate is usually thick enough to be considered as incoherent, while being semitransparent for energy transfer consideration. Furthermore, the substrate is either lossless or slightly absorbing ($\kappa_s \ll n_s$), as discussed earlier. Figure 9.13 shows the geometry of an incoherent substrate of thickness $d_s$ bounded by multilayer thin films on both sides. The refraction angle $\theta_s$ in the substrate can be calculated from the incidence angle $\theta_1$ by neglecting absorption of the substrate. In Fig. 9.13, $\rho_a$ or $\rho_b$ refer to the reflectance of the first multilayer structure for rays originated from air or the substrate, and $\tau_a$ and $\tau_b$ are the corresponding transmittance. Furthermore, $\rho_s$ and $\tau_s$ represent the reflectance and the transmittance for rays originated from the substrate at the second multilayer structure. Since the coupling between the incident and reflected waves in the substrate is negligible, the transmittance is the same whether the ray is originated from air or the substrate, i.e., $\tau_b = \tau_a$.

In order to calculate these parameters, the transfer matrix method, discussed in the previous section, can be separately applied to the multilayer structures for a given

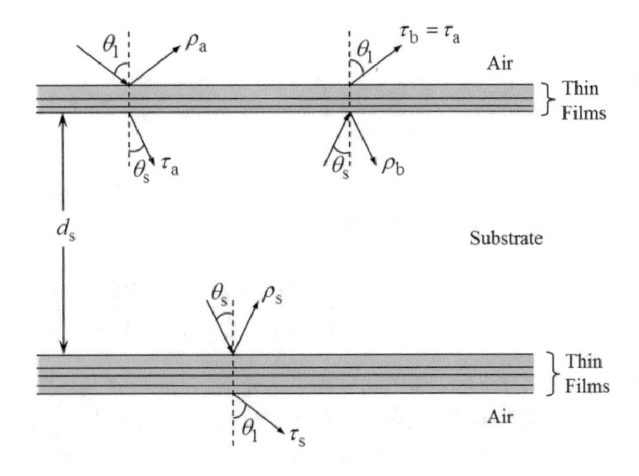


**Fig. 9.13** Radiative properties of multilayer thin films on an incoherent, thick substrate

incident direction. The internal transmittance of the substrate is $\tau = \exp\left(-\frac{4\pi\kappa_s d_s}{\lambda\cos\theta_s}\right)$, where $\lambda$ is the wavelength in vacuum. The reflectance and the transmittance of the multilayer structure can be calculated using the ray-tracing method and expressed as follows:

$$R'_\lambda = \rho_a + \frac{\rho_s\tau_a^2\tau^2}{1-\rho_s\rho_b\tau^2} \tag{9.45}$$

$$T'_\lambda = \frac{\tau_a\tau_s\tau}{1-\rho_s\rho_b\tau^2} \tag{9.46}$$

Radiative properties of arbitrary numbers of thick and thin layers have been derived theoretically [18]. For each thin-film stack, the field reflection and transmission coefficients are obtained first, using the matrix formulation described previously. The power transmittance and reflectance at the interfaces of each thick layer can then be obtained. Using the net radiation method, the energy transmittance and reflectance can be evaluated. Spectral averaging is another and perhaps more powerful technique of obtaining the transmittance and reflectance for systems involving thick and thin layers.

The absorptance of the composite layers can be calculated by subtracting the reflectance and the transmittance from unity. The Poynting vector can be evaluated as a function of $z$ to obtain the radiant energy flux $S(z) = \frac{1}{2}\mathrm{Re}(\mathbf{E}\times\mathbf{H}^*)$. The fraction of energy absorbed between $z_1$ and $z_2$ is given by

$$\alpha_{z_1-z_2} = \frac{S(z_1)-S(z_2)}{S_{iz}} \tag{9.46}$$

where $S_{iz}$ is the incident radiant energy flux in the $z$-direction. From Sect. 8.1.4, one can obtain the local energy density. The energy dissipated per unit volume is given by $-\nabla\cdot\mathbf{S}$.

### 9.2.4 Waveguides and Optical Fibers

Optical fibers and waveguides are essential for optical communication and optoelectronics. There are numerous other applications such as noncontact radiation thermometry, near-field microscopy, and decoration lightings. According to a report in 2000, the total length of optical fiber wires that had been installed worldwide exceeded $3.0 \times 10^{11}$ m, which equals the distance of a round trip from the earth to the sun. Optical fibers usually operate based on the principle of total internal reflection, as shown in Fig. 9.14. The fiber core is usually surrounded by a cladding material with a lower refractive index.

The numerical aperture *NA* is defined according to the half angle $\theta_h$ of the acceptance cone, within which total internal reflection occurs. It can be seen from Fig. 9.14

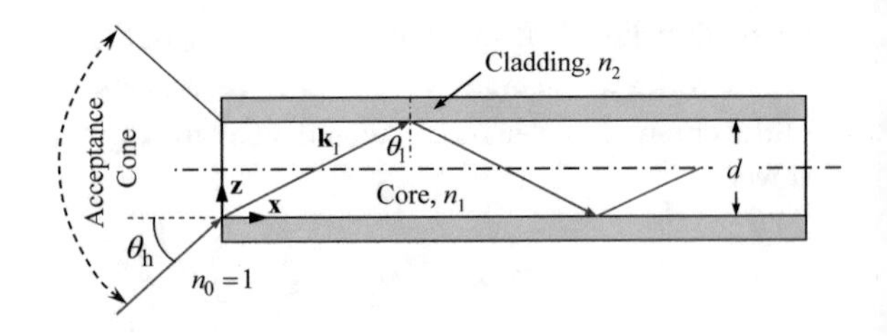


**Fig. 9.14** Schematic of a planar dielectric waveguide

that

$$NA = \sin\theta_h = n_1\cos\theta_c = \sqrt{n_1^2 - n_2^2} \tag{9.47}$$

For example, if $n_1 = 1.53$ and $n_2 = 1.46$, the critical angle $\theta_c = 72.6°$, the maximum cone angle $\theta_h = 27°$, and $NA = 0.46$. There are different types of waveguides, such as graded-index waveguides and metallic waveguides, in addition to the simple dielectric type. The cross-section may be circular, annular, rectangular, or elliptical. In some cases, the diameter of the fiber is much greater than the wavelength and the electromagnetic waves inside the fiber are incoherent. These devices are sometimes called lightpipes, which are used for relatively short distances. Optical fibers in communication technology use very thin wires and transmit light with well-defined *modes*. In the following, the configuration of a 1D dielectric slab between two media is discussed to illustrate the basics of an optical waveguide. More detailed treatments can be found from Haus [19] and Kong [20]. The present author was fortunate to learn optoelectronics and the electromagnetic wave theory through graduate courses taught by these professors.

Consider the planar structure shown in Fig. 9.14 that is infinitely extended in the *y*-direction. When the variation of *d* along the *x*-direction is negligibly small compared to the wavelength, the electromagnetic waves inside the waveguide are coherent. A standing wave pattern must be formed in the *z*-direction. This requires the phase shift in the *z*-direction for the round trip including two reflections at the boundary to be a multiple of $2\pi$, viz,

$$2k_{1z}d + 2\delta = 2m\pi,\ m = 0, 1, 2, \ldots \tag{9.48}$$

where $k_{1_z} = (\omega/c)n_1\cos\theta_1$, and the phase shift upon total internal reflection is

$$\delta = -2\alpha = -2\tan^{-1}\left(g\frac{\sqrt{\sin^2\theta_1 - \sin^2\theta_c}}{\cos\theta_1}\right) \tag{9.49}$$

where $g = 1$ for TE waves and $g = n_1^2/n_2^2$ for TM waves; see Sect. 8.3.2.

The solutions of Eq. (9.48) give discrete values of $\theta_1$ or $k_x = (\omega/c_0)n_1\sin\theta_1$, at which waves can propagate through the fiber for a prescribed frequency. These are called *guided modes* of the optical fiber, and Eq. (9.48) may be regarded as the *mode equation*. The orders of mode are identified as $\text{TE}_0, \text{TE}_1, \ldots, \text{TE}_m$ or $\text{TM}_0, \text{TM}_1, \ldots, \text{TM}_m$ for a 1D waveguide. For a 2D waveguide, the subscripts consist of two indices "*ml*" for each mode. As $\theta_1$ decreases from $\pi/2$ to $\theta_c$, $k_{1z}$ increases and higher order modes can be excited. One might wonder why $\theta_1 = \pi/2$ or $k_{x1} = k_1$ is *not* a guided mode. In this case, the beam would go through the core, the cladding, and air in a straight line. Any bending in the waveguide would result in some loss of energy transfer. On the other hand, the guided modes are much less affected by the bending. This is why an optical fiber can transfer signals to a very long distance while being flexible.

To illustrate the solution in terms of $k_{1z}d$, let us rearrange Eq. (9.49) as follows:

$$\tan\left(\frac{k_{1z}d}{2} - \frac{m\pi}{2}\right) = \tan\alpha = g\frac{\eta_2}{k_{1z}} = g\sqrt{\frac{(k_1 d)^2 - (k_2 d)^2}{(k_{1z}d)^2} - 1} \tag{9.50}$$

The functions on the left and right sides of Eq. (9.50) can be plotted in the same graph against $k_{1z}d$, as shown in Fig. 9.15, for two values of $\omega d$, assuming $\omega_2 d_2 > \omega_1 d_1$. The solid lines are for the left side, which is independent of polarization. The dash-dotted curves are for TE waves, and the dotted curves are for TM waves. The intersections within the circles identify the guided modes. It is noted that fewer modes are permitted with a smaller $\omega d$ or $d/\lambda$. For $\omega_1 d_1$, the possible modes are $\text{TE}_0$, $\text{TE}_1$, $\text{TM}_0$, and $\text{TM}_1$ only. A fiber that supports only a single mode for a given frequency is called a *single-mode fiber*; otherwise, it is called a *multimode fiber*.

**Example 9.4** Determine the range of $d/\lambda$ so that only the $\text{TE}_0$ and $\text{TM}_0$ waves are guided in the planar waveguide with $n_1 = 1.55$ and $n_2 = 1.42$. Moreover, if $d/\lambda = 1000$, how many TE and TM modes may be guided?

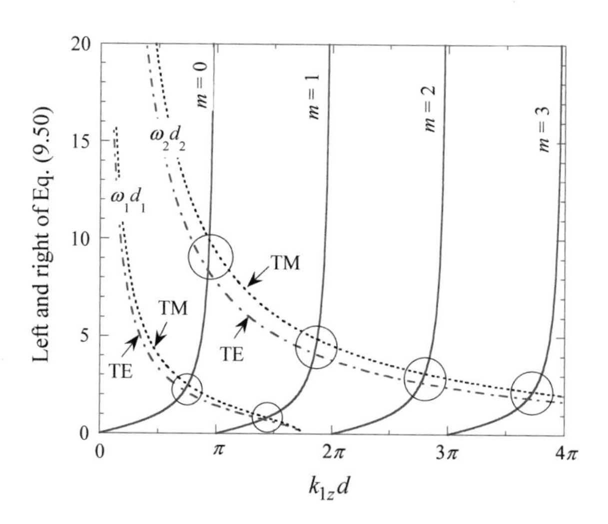


**Fig. 9.15** Solutions of the mode equation, when $\omega_2 d_2 > \omega_1 d_1$. The circles indicate the intersections between the curves described by the left and right sides of Eq. (9.50)

**Solution** Because $d/\lambda$ must be small enough so that the left-hand side of Eq. (9.50) becomes zero at $k_{1z}d = \pi$, we have $(k_1 d)^2 - (k_2 d)^2 = \pi^2$ or $4\pi^2(n_1^2 - n_2^2)(d/\lambda)^2 = \pi^2$. Finally, we find $d/\lambda < 0.5\left(n_1^2 - n_2^2\right)^{-1/2} = 1.3$. Moreover, from Fig. 9.15, we can estimate the highest order mode $M$, using $k_{1z}d = M\pi$ and $\cos\theta_1 = \cos\theta_c$ when $d \gg \lambda$. Hence, $2\pi(d/\lambda)\cos\theta_c = M\pi$ or $M = 2(d/\lambda)\cos\theta_c = 801.8$. There will be 802 TE modes and 802 TM modes including the zeroth-order modes.

Next, we will study the fields in a planar waveguide. Assume that the waveguide contains a core which is medium 1 for $0 \le z \le d$ surrounded by medium 2 that extends to both $+\infty$ and $-\infty$. Let us take a TE wave and write in the more general terms $\varepsilon_1$, $\mu_1$, $\varepsilon_2$, and $\mu_2$. The electric field is nonzero only in the $y$-direction, and the $y$-component of the electric field is given by

$$E_y = \begin{cases} C\mathrm{e}^{\eta_2 z}\mathrm{e}^{\mathrm{i}k_x x}, & z < 0 \\ \left(A\mathrm{e}^{\mathrm{i}k_{1z}z} + B\mathrm{e}^{-\mathrm{i}k_{1z}z}\right)\mathrm{e}^{\mathrm{i}k_x x}, & 0 \le z \le d \\ D\mathrm{e}^{-\eta_2(z-d)}\mathrm{e}^{\mathrm{i}k_x x}, & z > d \end{cases} \tag{9.51}$$

where the time-harmonic term $\exp(-\mathrm{i}\omega t)$ is again omitted for simplicity. Note that $\eta_2 = \mathrm{Im}(k_{z2})$ such that the field decays in medium 2 away from the interfaces with medium 1. The magnetic fields can be obtained as $H_x = -\frac{1}{\mathrm{i}\omega\mu_1\mu_0}\frac{\partial E_y}{\partial z}$ and $H_z = \frac{1}{\mathrm{i}\omega\mu_1\mu_0}\frac{\partial E_y}{\partial x}$. There are four boundary conditions for the tangential components to be continuous at $z = 0$ and $z = d$. We end up with a set of homogeneous linear equations of the coefficients $A$, $B$, $C$, and $D$. The solution exists only when the determinant of the characteristic $4 \times 4$ matrix becomes zero and can be expressed in a combined equation as follows:

$$\tan(k_{1z}d)\left(\frac{k_{1z}^2}{\varepsilon_1^2} - \frac{\eta_2^2}{\varepsilon_2^2}\right) = 2\left(\frac{k_{1z}\eta_2}{\varepsilon_1\varepsilon_2}\right) \tag{9.52}$$

This is an equivalent expression of the mode equation. An easier way to solve Eq. (9.51) is by considering the condition of total internal reflection at the boundaries, i.e.,

$$A = B\mathrm{e}^{\mathrm{i}\delta} \text{ and } B = A\mathrm{e}^{\mathrm{i}(2k_{1z}d+\delta)} \tag{9.53}$$

The combination gives $\mathrm{e}^{\mathrm{i}(2k_{1z}d+2\delta)} = 1$, which is nothing but Eq. (9.48). After substituting $A = B\mathrm{e}^{-\mathrm{i}2\alpha}$ into Eq. (9.51), the boundary conditions require that

$$E_y = \begin{cases} 2\mathrm{e}^{-\mathrm{i}\alpha}B\cos\left(\frac{k_{1z}d}{2} - \frac{m\pi}{2}\right)\mathrm{e}^{\eta_2 z}\mathrm{e}^{\mathrm{i}k_x x}, & z < 0 \\ 2\mathrm{e}^{-\mathrm{i}\alpha}B\cos\left(k_{1z}z - \frac{k_{1z}d}{2} - \frac{m\pi}{2}\right)(k_{1z}z - \alpha)\mathrm{e}^{\mathrm{i}k_x x}, & 0 \le z \le d \\ 2\mathrm{e}^{-\mathrm{i}\alpha}B\cos\left(\frac{k_{1z}d}{2} + \frac{m\pi}{2}\right)\mathrm{e}^{-\eta_2(z-d)}\mathrm{e}^{\mathrm{i}k_x x}, & z > d \end{cases} \tag{9.54}$$

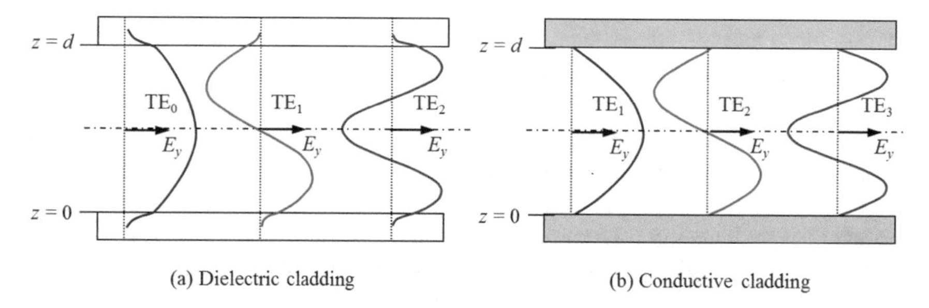


**Fig. 9.16** Electric field distribution $E_y(z)$ in planar waveguides for **a** dielectric cladding and **b** conductive cladding. For the conducting cladding, $\sigma \to \infty$ and the lowest order TE mode is the first order

Figure 9.16a shows the electric field distribution for $\text{TE}_0$, $\text{TE}_1$, and $\text{TE}_2$. The decaying fields inside the cladding are clearly demonstrated. For a cladding with the conductivity $\sigma \to \infty$, the waves will be perfectly reflected at the interface without any phase shift and the electric field must vanish in the cladding. Only the odd $m$'s are guided modes. The first guided mode is $\text{TE}_1$, and the guided mode $\text{TE}_q$ corresponds to $q = (m+1)/2$, with $m = 1, 3, 5, \ldots$ The electric fields for the conducting waveguide modes $\text{TE}_1$, $\text{TE}_2$, and $\text{TE}_3$ are shown in Fig. 9.16b, for comparison with those for the first three modes in the dielectric waveguide. The difference lies in that no fields can penetrate into the conducting waveguide, whereas the fields can penetrate into the dielectric cladding.

**Example 9.5** Determine the energy flux, the phase velocity, and the group velocity of the electromagnetic waves in a planar dielectric waveguide.

**Solution** Obviously, there is no net energy flow in the $z$-direction, and $\langle \mathbf{S} \rangle_x = \frac{1}{2}\text{Re}\big(E_y H_z^* - E_z H_y^*\big)$. The second term on the right becomes zero for a TE wave; thus, $\langle \mathbf{S} \rangle_x = \frac{k_x}{2\omega\mu_0} E_y E_y^*$.Integration of $\langle \mathbf{S} \rangle_x$ from $z = -\infty$ to $+\infty$ gives the power transmitted per unit length in the $y$-direction. Note that a small portion of the energy is transmitted through the cladding. The phase velocity along the $x$-direction is $v_\text{p} = \omega/k_x = c_0/(n_1 \sin\theta_1)$. The group velocity for a given mode is given by $v_\text{g} = (\text{d}k_x/\text{d}\omega)^{-1}$, which requires the solution of Eq. (9.52) accounting for the frequency-dependent refractive index.

Losses cannot be avoided in practical systems. If the conductivity does not approach infinite, there will be losses of absorption in the conductive cladding. The dielectric cladding may have a small extinction coefficient that results in attenuation as waves propagating through the fiber. Furthermore, if there are absorbing particles or another high-index medium outside the cladding dielectric, wave coupling or leakage can occur that results in absorption or attenuation.

In Sect. 9.2.1, we introduced the concept of Fabry–Perot resonant cavities. Two- and three-dimensional optical cavities and microwave cavities support resonance modes, which are standing waves inside the cavity. These devices are important for photonics and optoelectronics. Microcavities have also been used to modify the surface radiative properties. The quality factor, or the $Q$-factor, of a resonator is defined as the ratio of energy storage to the energy dissipation. High $Q$-factors can be achieved with the microfabricated optical cavities for quantum electrodynamics (QED), enhancement and suppression of spontaneous emission, and biological and chemical sensing [21]. A special microcavity is made of spheres or disks, where the resonance is built up around a circumference in the form of a polygon. Total internal reflection traps the light inside the microsphere or the disk. At a particular wavelength, when resonance occurs, light undergoes multiple reflections, and a strong electric field which is confined near the perimeter can be built. This is the so-called *whispering gallery mode* (WGM), named after the whispering gallery at St. Paul's Cathedral in London. A whispering gallery is a circular gallery under a dome where whispers can be heard from the opposite side of the building. Optical fibers or waveguides are commonly used to couple the photon energy to or from the microcavities via evanescent waves. Ultrahigh $Q$-factors can be achieved with WGMs. The energy coupling mechanisms have recently been studied by Guo and Quan using a finite-element method [22].

A recent development in fiber optics is the use of photonic crystals (PCs) to confine the light into a fiber, whose cladding region is made of PCs, rather than a solid low-index material. Note that further discussions on photonic crystals will be given in Sect. 9.3.1. The fiber core may be either solid or hollow, and the PCs in the cladding region may contain air-filled holes in silica. For this reason, these fibers are called *photonic crystal fibers* (PCFs) and some are called *holey fibers* [23] . In the stop band, waves cannot propagate inside the PC and thus effectively confine the propagating wave to the core region, where the modes can be guided, without using total internal reflection. One of the advantages of PCFs over conventional optical fibers is the spectral broadening that enables high-intensity pulses to be transmitted with less distortion or loss of the spectral information; this has important applications such as optical coherence spectroscopy and tomography. Another advantage is that the use of large guiding areas can provide low-loss high-power delivery for imaging, lithography, and astronomy. Other potential applications range from birefringence and nonlinear optics to atomic particle guidance [23].

## 9.3 Photonic Crystals and Periodic Gratings

The unique features of periodic microstructures (i.e., photonic crystals or gratings) can be utilized to engineer the radiative properties for specific applications [23–26]. After introducing photonic crystals, the one-dimensional photonic crystal is used as an example to illustrate the photonic bandgaps and dispersion. The general grating equation is then described with an introduction of the rigorous coupled-wave analysis (RCWA) numerical technique. The effective medium theory for periodic structures will also be introduced.

### 9.3.1 Photonic Crystals

A *photonic crystal* (PC) is a periodic array of unit cells (i.e., photonic lattices in analogy to those in real crystals), which replicate infinitely into one, two, or three dimensions. Figure 9.17a illustrates a 1D PC by placing alternating *Journal of Heat Transfer* and *Journal of Thermophysics and Heat Transfer* issues in author's bookshelf. To have a PC with a period of the order of infrared wavelengths, say 3 μm, the thickness needs to be reduced by a factor of 6000. Figure 9.17b is a photo of a stack of chopsticks in three dimensions. Structures of 3D tungsten PCs have been fabricated with a rod width of 1.2 μm and rod-to-rod spacing of 4.2 μm, for tuning the infrared thermal emission properties [27].

From the analogy of the electron movement in crystals, electromagnetic wave propagation in a PC should also satisfy the Bloch condition, discussed in Chap. 6. Similarly, due to the periodicity, a PC exhibits band structures consisting of pass and stop bands when the frequency is plotted against the wavevector. In the pass band, for instance, waves can propagate inside a PC. Whereas in the stop band, no energy-carrier waves can exist inside a PC, and only oscillating but evanescently decaying

Fig. 9.17 Illustration of **a** 1D and **b** 3D photonic crystal (PC) structures

(a) 1-D structure made by alternating layers

(b) 3-D structure made by stacking rod

fields possibly exist. The existence of stop bands enables a PC to be used in many optoelectronic devices from waveguides to band-pass filters [7, 8, 28]. Most of the 1D PCs are made with alternating layers of two lossless dielectrics; nevertheless, metallodielectric PCs have also been extensively investigated especially in recent years. In some cases, the dimension may be smaller than 100 nm for tuning the visible properties.

While 3D PCs with complicated structures have been fabricated and used in a number of applications, the fundamental physics can be illustrated using 1D PCs and can easily be generalized for 2D or 3D structures. The 1D PC, illustrated in Fig. 9.18, is a periodic multilayer structure, where $\Lambda = d_a + d_b$ is the period of the PC or photonic lattice constant. The unit cell is composed of alternating dielectrics with different refractive indices $n_a$ and $n_b$. It is assumed that all layers are infinitely extended in the $x$-$y$ plane, and the PC is in the positive-$z$ half space starting with $m = 0$ at $z = 0$. From the analogy between wave propagation in a periodic media and the motion of electrons in crystalline materials, the electric field vector in the 1D PC, for a monochromatic electromagnetic wave of angular frequency $\omega$, should satisfy the Bloch condition given by

$$\mathbf{E}(x, y, z, t) = \mathbf{u}(z)e^{iKz}e^{i(k_x x + k_y y - \omega t)} \tag{9.55}$$

where $\mathbf{u}(z + \Lambda) = \mathbf{u}(z)$ is a periodic function of $z$ with a period equal to the lattice constant of the photonic crystal, $k_x$ and $k_y$ are the parallel components of the wavevectors that must be the same in all layers as required by the phase-matching condition, and $K$ represents the Bloch wavevector that is a scalar in the 1D case. Here,

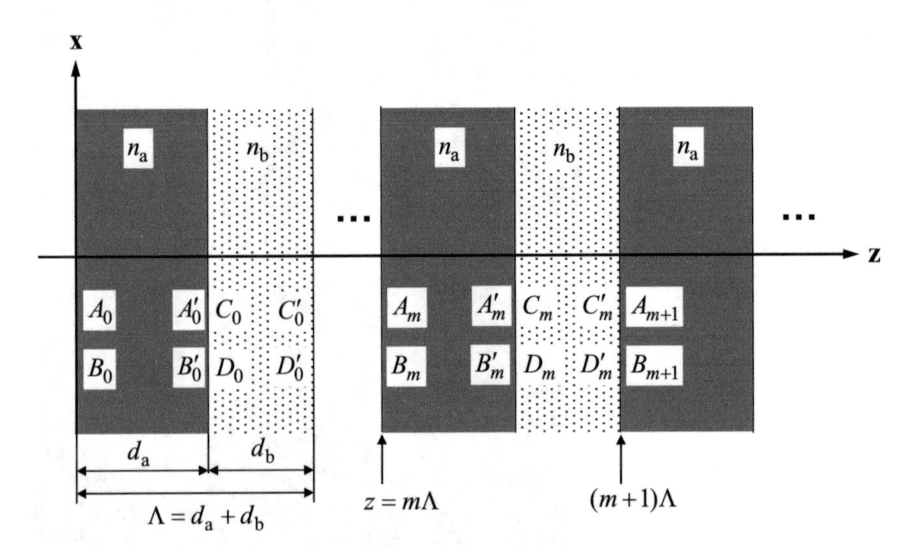


**Fig. 9.18** Amplitudes of the forward and backward waves in a semi-infinite 1D PC, in the right half space. The unit cell of the 1D PC is made of two dielectric layers: type a and type b, and has a period $\Lambda = d_a + d_b$

$K$ is a characteristic parameter of the PC that is the same for all layers. The wavevector components in the $z$-direction are $k_{az}$ and $k_{bz}$ in media a and b, respectively, and are determined by the relations $k_x^2 + k_y^2 + k_{za}^2 = k_a^2 = n_a^2\omega^2/c_0^2$ and $k_x^2 + k_y^2 + k_{zb}^2 = k_b^2 = n_b^2\omega^2/c_0^2$. From the Bloch condition, the electric field in the 1D PC satisfies the following equation:

$$\mathbf{E}(x, y, z + \Lambda, t) = \mathbf{E}(x, y, z, t)e^{iK\Lambda} \quad (9.56)$$

The magnetic field is related to the electric field by Maxwell's equations and must also follow the Bloch condition. Therefore, the fields inside the PC are not periodic functions.

Because of the axial symmetry, the coordinate can always be rotated around the $z$-axis to make $k_y = 0$. For $s$ polarization, the electric field is parallel to the $y$-direction and can be expressed as

$$E_y(x, z) = \left[A_m e^{ik_{za}(z-m\Lambda)} + B_m e^{-ik_{za}(z-m\Lambda)}\right]e^{ik_x x} \quad (9.57a)$$

for $m\Lambda \le z \le (m\Lambda + d_a)$ and

$$E_y(x, z) = \left[C_m e^{ik_{zb}(z-m\Lambda-d_a)} + D_m e^{-ik_{zb}(z-m\Lambda-d_a)}\right]e^{ik_x x} \quad (9.57b)$$

for $(m\Lambda + d_a) \le z \le (m+1)\Lambda$. In Eqs. (9.57a) and (9.57b), the time-dependent term $\exp(-i\omega t)$ is omitted for simplicity, $m$ is an integer, $A_m$ and $C_m$ are the amplitudes of forward waves, and $B_m$ and $D_m$ are the amplitudes of backward waves at the interfaces, as shown in Fig. 9.18 [17]. The amplitudes at the other side of boundary are given by the coefficients: $A'_m = e^{ik_{za}d_a}A_m$, $B'_m = e^{-ik_{za}d_a}B_m$, $C'_m = e^{ik_{zb}d_b}C_m$, and $D'_m = e^{-ik_{zb}d_b}D_m$. Boundary conditions require that the tangential components of the electric field $E_y$ and magnetic field $H_x$ to be continuous at each interface. From the matrix formulation, the coefficients $A_m$ and $B_m$ at $z = m\Lambda$ are related to those at $z = (m + 1)\Lambda$ with the propagation matrix $\mathbf{P}$ and dynamical matrix $\mathbf{D}$ as follows:

$$\begin{pmatrix} A_m \\ B_m \end{pmatrix} = (\mathbf{P}_a\mathbf{D}_a^{-1}\mathbf{D}_b)(\mathbf{P}_b\mathbf{D}_b^{-1}\mathbf{D}_a)\begin{pmatrix} A_{m+1} \\ B_{m+1} \end{pmatrix} \quad (9.58)$$

From Eq. (9.56), the ratio of the electric fields at two points separated by a period $\Lambda$ along the $z$-direction is equal to $\exp(iK\Lambda)$; thus,

$$\begin{pmatrix} A_{m+1} \\ B_{m+1} \end{pmatrix} = e^{iK\Lambda}\begin{pmatrix} A_m \\ B_m \end{pmatrix} \quad (9.59)$$

The Bloch wavevector parameter $K$ can be obtained by solving the eigenvalue equation:

$$\mathbf{M}\begin{pmatrix} A_{m+1} \\ B_{m+1} \end{pmatrix} = \mathrm{e}^{-\mathrm{i}K\Lambda}\begin{pmatrix} A_{m+1} \\ B_{m+1} \end{pmatrix} \tag{9.60}$$

where $\mathrm{M} = (\mathbf{P}_\mathrm{a}\mathbf{D}_\mathrm{a}^{-1}\mathbf{D}_\mathrm{b})(\mathbf{P}_\mathrm{b}\mathbf{D}_\mathrm{b}^{-1}\mathbf{D}_\mathrm{a})$. In general, $K$ depends on the frequency $\omega$ and the parallel wavevector component $k_x$, for a given geometry and refractive indices. Once $K$ is determined, the electric field in the PC can be expressed in the Bloch wave form as

$$E_y(x, z) = u(z)\mathrm{e}^{\mathrm{i}Kz}\mathrm{e}^{\mathrm{i}k_x x} \tag{9.61}$$

where $u(z)$ is a periodic function of $z$. For $m\Lambda \le z \le (m\Lambda + d_\mathrm{a})$,

$$u(z) = \left[A_0\mathrm{e}^{\mathrm{i}k_{z\mathrm{a}}(z-m\Lambda)} + B_0\mathrm{e}^{-\mathrm{i}k_{z\mathrm{a}}(z-m\Lambda)}\right]\mathrm{e}^{-\mathrm{i}K(z-m\Lambda)} \tag{9.62a}$$

and for $(m\Lambda + d_\mathrm{a}) \le z \le (m+1)\Lambda$,

$$u(z) = \left[C_0\mathrm{e}^{\mathrm{i}k_{z\mathrm{b}}(z-m\Lambda-d_\mathrm{a})} + D_0\mathrm{e}^{-\mathrm{i}k_{z\mathrm{b}}(z-m\Lambda-d_\mathrm{a})}\right]\mathrm{e}^{-\mathrm{i}K(z-m\Lambda)} \tag{9.62b}$$

Note that $A_0$ and $B_0$ are amplitudes of the first layer, i.e., at $m = 0$, and

$$\begin{pmatrix} C_0 \\ D_0 \end{pmatrix} = (\mathbf{P}_\mathrm{a}\mathbf{D}_\mathrm{a}^{-1}\mathbf{D}_\mathrm{b})^{-1}\begin{pmatrix} A_0 \\ B_0 \end{pmatrix} \tag{9.63}$$

The expressions for the magnetic field can be obtained from those of the electric field using Maxwell's equations. For $p$-polarization, the magnetic field is parallel to the $y$-axis. The same procedure can be used to determine the magnetic field first and then the electric field. The amplitudes $A_0$ and $B_0$ depend on the boundary condition at $z = 0$, i.e., the interaction of the PC with the medium in the left half space.

For a given PC, the Bloch wavevector can be solved from the eigenvalue problem given in Eq. (9.60), for any real positive values of $\omega$ and $k_x$. In general, $K$ is complex. When $K$ is purely real, i.e., $\mathrm{Im}(K) = 0$, the electric field oscillates in the direction of $z$, and the Bloch wave propagates into the positive $z$-direction, which is called an *extended mode.* When $\mathrm{Im}(K) \ne 0$, on the other hand, the amplitude of the Bloch wave decays exponentially along the positive $z$-direction, and the wave is confined to the first few unit cells of the photonic crystal; this is called a *localized mode* [17, 25]. For the *localized mode,* the field is localized in the vicinity of the defect or the edge. Notice that $K = K(k_x, \omega)$, and the regions with $\mathrm{Im}(K) = 0$ in the $\omega$–$k_x$ plane are called pass bands, and those with $\mathrm{Im}(K) \ne 0$ are called stop bands. Suppose light is incident from air (in the left half space) on the PC at $z = 0$. In the stop band, the PC will act like a perfect mirror, which is also called a Bragg reflector. A diagram in the $\omega$–$k_x$ domain, showing the different regions, allows one to study the band structures of a PC.

**Example 9.6** Consider the 1D PC depicted in Fig. 9.18, with the following parameters: $n_a = 2.4$, $n_b = 1.5$, and $d_a = d_b$. Construct the band structure for both polarizations, and calculate the normal reflectance.

**Solution** The PC is semi-infinite, and the incidence is from air. The unit cell of the 1D PC is defined by the thickness of the unit cell $\Lambda = d_a + d_b$ since $d_a = d_b$. Following the previous discussion, we have calculated the band structure of the 1D PC for either polarization, and the results are shown in Fig. 9.19. Here, the parallel component of the wavevector is $k_x = (\omega/c_0)\sin\theta$, where $\theta$ is the angle of incidence. The band structure is expressed by the reduced frequency and wavevector; hence, it is independent of the period of the PC. For the calculation, it is assumed that the 1D PC is a perfectly periodic structure infinitely extended into $z$-direction (i.e., no defect or edge exists in the PC). The shaded regions represent the stop bands, while unshaded regions are the pass bands. The light line in air, which corresponds to $\theta = 90°$, is plotted as a dash-dot line based on $\omega = k_x c_0$. On the upper left side of this line, propagating waves exist in air and $\theta \leq 90°$. On the lower right side of this line, evanescent waves exist in air since $\theta$ becomes complex. Note that stop bands shrink to zero only for $p$-polarization. The point where the top and bottom band edges merge together corresponds to the Brewster angle between the dielectric of types $a$ and $b$ of the PC. At the Brewster angle, the reflectivity at the interface between two dielectrics is zero; thus, waves or incident energy can propagate into the PC. For the 1D PC considered here, because the Brewster angle is located on the lower right side of the light line, the propagating waves in air will not be affected by the Brewster angle of the constituent dielectrics of the PC.

Figure 9.20 shows the reflectance of the 1D PC structure with different numbers of periods ($N = 30$ and 300), calculated using the 1D matrix formulation. The wavelength is normalized to the period $\Lambda$. The reflectance approaches to unity in the stop band (when $N > 30$). In the pass band, interference effects affect the free spectral range of oscillation and, thus, the oscillation frequency increases with the number of periods of the PC structure. A special type of 1D PCs is the Bragg reflector,

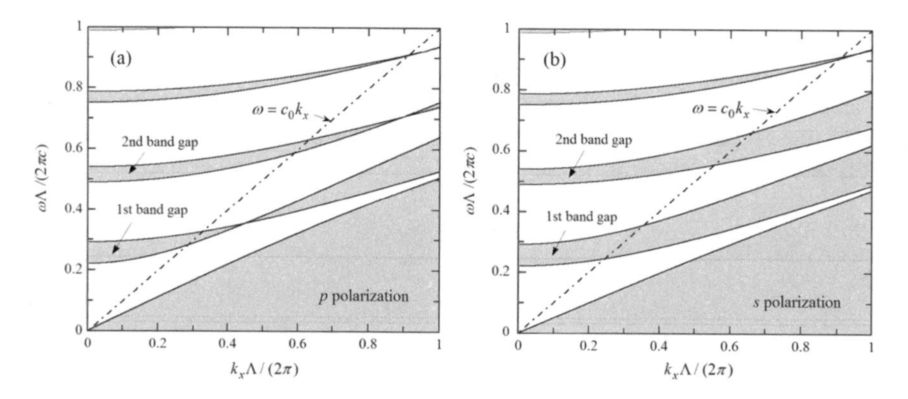


**Fig. 9.19** Band structures of a 1D PC. **a** TM wave ($p$-polarization). **b** TE wave ($s$ polarization)

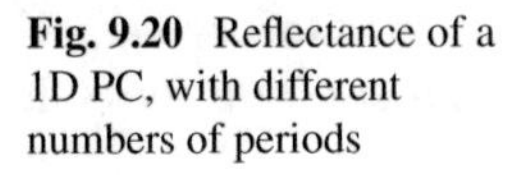

**Fig. 9.20** Reflectance of a 1D PC, with different numbers of periods

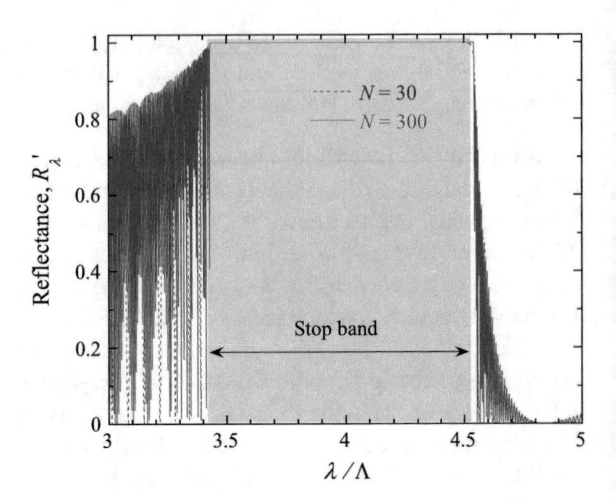


which is composed of alternating high- and low-index films, each at a thickness of one-quarter of the wavelength in the film, i.e., $d_a = \lambda/(4n_a)$ and $d_b = \lambda/(4n_b)$. Further discussion about surface waves and coherent emission characteristics of PC structures will be deferred to the next chapter.

## *9.3.2 Periodic Gratings*

Diffraction grating is considered as one of the simplest and most important devices in optical metrology; subsequently, there have been extensive research works on the effect of gratings on the radiative property modification [29, 30]. Nanoscale diffraction elements fabricated with nanolithography have important applications in biochemical sensing, surface diagnostics, and nanophotonics. Patterned semiconductor microelectronics has periodic structures on the surface with a period below 100 nm [31]. Understanding the radiative properties is essential for thermal processing and modeling in semiconductor manufacturing as the feature size continues to shrink.

In the inhomogeneous region, where the permittivity $\varepsilon$ and the permeability $\mu$ are spatial functions, the monochromatic plane wave equations become more complicated. By assuming the solution is a time-harmonic plane wave, we can rewrite the Maxwell equations as follows [32]:

$$\nabla \times \mathbf{E} = \mathrm{i}\omega\mu\mu_0\mathbf{H} \tag{9.64}$$

$$\nabla \times \mathbf{H} = -\mathrm{i}\omega\varepsilon\varepsilon_0\mathbf{E} \tag{9.65}$$

$$\varepsilon\nabla \cdot \mathbf{E} + \mathbf{E} \cdot \nabla\varepsilon = 0 \tag{9.66}$$

$$\mu\nabla\cdot\mathbf{H}+\mathbf{H}\cdot\nabla\mu=0 \tag{9.67}$$

Since only isotropic media are considered here, both $\varepsilon$ and $\mu$ are scalars. By taking the curl of Eq. (9.64) and applying the vector identities in Appendix B.7 and Eqs. (9.66) and (9.67), we obtain

$$\nabla^2\mathbf{E}+\nabla(\mathbf{E}\cdot\nabla\ln\varepsilon)+\nabla\ln\mu\times(\nabla\times\mathbf{E})+k^2\mu\varepsilon\mathbf{E}=0 \tag{9.68}$$

$$\nabla^2\mathbf{H}+\nabla(\mathbf{H}\cdot\nabla\ln\mu)+\nabla\ln\varepsilon\times(\nabla\times\mathbf{H})+k^2\mu\varepsilon\mathbf{H}=0 \tag{9.69}$$

where $k=\omega/c$ is the wavevector in vacuum.

These equations cannot be solved easily and numerical methods are often required. Among them are rigorous coupled-wave analysis (RCWA), finite-difference time-domain (FDTD), finite-element method (FEM), boundary element method (BEM), as well as the volume integral method. Effective medium formulation is another approach that takes the average field by approximating the inhomogeneous medium with effective homogenous electric and magnetic properties. The concept of RCWA will be presented next because it is an effective tool for calculating the optical properties of the grating geometry with sufficient accuracy.

### *9.3.3 Rigorous Coupled-Wave Analysis (RCWA)*

We begin with a discussion restricted to $s$ polarization and for incidence perpendicular to the gratings, as illustrated in Fig. 9.21. A plane wave is incident on a 1D grating surface from free space, region I with $\varepsilon_{\text{I}}=1$, $n_{\text{I}}=1$, and $\kappa_{\text{I}}=1$. Region II is composed of binary materials $A$ and $B$ so that the dielectric function in region II is a periodic function of $x$ and the period $\Lambda$ is called the grating period. The filling ratio $\phi$ is the volume fraction of material $A$, and the lateral extension of the grating is assumed to be infinite. Region III is the substrate with a dielectric function $\varepsilon_{\text{III}}$ .

The wavevector $\mathbf{k}$ defines the direction of incidence, and the angle between $\mathbf{k}$ and the surface normal $\hat{\mathbf{z}}$ is the angle of incidence $\theta$, also called the polar angle. The grating vector $\mathbf{K}$ is defined in the positive $x$-direction with a magnitude $K=$

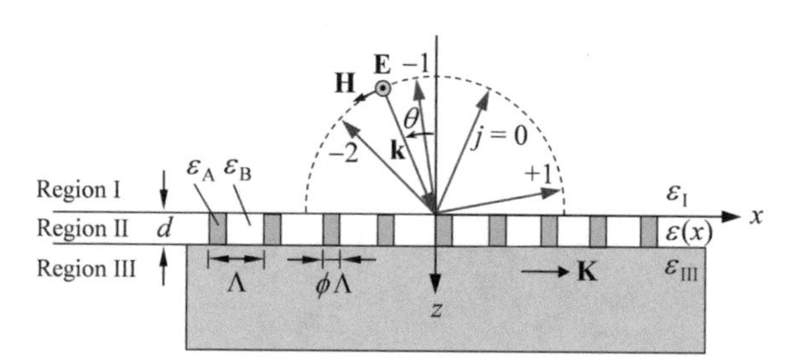


**Fig. 9.21** Schematic drawing for TE wave incident on a grating layer, showing the reflected diffraction orders $j=-2,-1$, 0, and 1

$2\pi/\Lambda$. In the following discussion, it is assumed that the incident wavevector is on the $x$-$z$ plane, i.e., the $y$-component of $\mathbf{k}$ is zero. For $s$ polarization, the electric field $\mathbf{E}$ is parallel to the $y$-direction and perpendicular to the grating vector $\mathbf{K}$. The magnitude of the incident electric field, after normalization, can be expressed as $\exp(\mathrm{i}k_x x + \mathrm{i}k_z z - \mathrm{i}\omega t)$. For simplicity, the time-harmonic term exp(-i$\omega t$) will be omitted hereafter. The magnitude of $\mathbf{k}$ in regions I and III can be expressed as

$$k_{\mathrm{I}} = \frac{2\pi n_{\mathrm{I}}}{\lambda} = \frac{2\pi}{\lambda} = k \quad \text{and} \quad k_{\mathrm{III}} = \frac{2\pi n_{\mathrm{III}}}{\lambda} = n_{\mathrm{III}} k \tag{9.70}$$

where $n_{\mathrm{III}}$ is the refractive index of region III. There exists a phase difference of $2\pi\Lambda\sin\theta/\lambda = k_x\Lambda$ between the incident wave at $(x, z)$ and that at $(x + \Lambda, z)$ due to a path difference of $\Lambda\sin\theta$. This condition must also be satisfied by each diffracted wave, i.e., the magnitude of the $j$th-order reflected wave can be written as $r_j \exp(\mathrm{i}k_{x,j}x - \mathrm{i}k_{\mathrm{I}z}z)$, where $r_j$ is the reflection coefficient, and $k_{x,j}$ is determined from the Bloch–Floquet condition [30, 32]:

$$k_{x,j} = \frac{2\pi}{\lambda}\sin\theta + \frac{2\pi}{\Lambda}j = k_x + K_j \tag{9.71a}$$

This equation can be expressed in terms of the angle of reflection given by

$$\sin\theta_j = \sin\theta + \frac{j\lambda}{\Lambda} \tag{9.71b}$$

where $\theta_j = \sin^{-1}(k_{x,j}/k)$ is the $j$th-order diffraction angle for reflection and Eq. (9.71b) is the well-known *grating equation*. When $k_{x,j} > k_{\mathrm{I}}$, $\sin\theta_j > 1$ and the $j$th-order reflected wave decays exponentially toward the negative $z$-direction. This is an evanescent wave that exists only near the surface, within a distance on the order of the wavelength. Note that the $z$-component of $\mathbf{k}$ for the $j$th-order reflected wave is

$$k_{\mathrm{I}z,j} = \begin{cases} \left(k_{\mathrm{I}}^2 - k_{x,j}^2\right)^{1/2}, & k_{\mathrm{I}} > k_{x,j} \\ \mathrm{i}\left(k_{x,j}^2 - k_{\mathrm{I}}^2\right)^{1/2}, & k_{x,j} > k_{\mathrm{I}} \end{cases} \tag{9.72}$$

Because $k_{x,j}$ must be the same in all media, similar criteria can be applied to the transmitted waves in region III to obtain $k_{\mathrm{III}z,j}$ by replacing I by III in the subscripts in Eq. (9.72).

The electric field in region I is a superposition of the incident and reflected waves; therefore,

$$E_{\mathrm{I}}(x, z) = \exp(\mathrm{i}k_x x + \mathrm{i}k_z z) + \sum_j r_j \exp\left(\mathrm{i}k_{x,j}x - \mathrm{i}k_{\mathrm{I}z,j}z\right) \tag{9.73}$$

The electric field in region III can be obtained by superimposing all transmitted waves as

$$E_{\mathrm{III}}(x,z)=\sum_j t_j \exp\left[\mathrm{i}k_{x,j}x+\mathrm{i}k_{\mathrm{III}z,j}(z-d)\right] \tag{9.74}$$

where $t_j$ is the transmission coefficient for the $j$th-order transmitted wave.

The electric field in region II can be expressed as

$$E_{\mathrm{II}}(x,z)=\sum_j \Psi_j(z)\exp\left(\mathrm{i}k_{x,j}x\right) \tag{9.75}$$

where $\Psi_j(z)$ is the amplitude of the $j$th space-harmonic component. Here, the order $j$ is matched with the diffraction order in regions I and III. Due to the periodic structure, the dielectric function of region II can be expanded in the following Fourier series:

$$\varepsilon(x)=\sum_m \varepsilon_m \exp\left(\mathrm{i}\frac{2m\pi}{\Lambda}x\right),\ m=0,\pm1,\pm2,\ldots \tag{9.76}$$

where $\varepsilon_m$ is the $m$th coefficient that can be calculated from

$$\varepsilon_0=\phi\varepsilon_{\mathrm{A}}+(1-\phi)\varepsilon_{\mathrm{B}} \text{ and } \varepsilon_m=\frac{(\varepsilon_{\mathrm{A}}-\varepsilon_{\mathrm{B}})\sin(m\phi\pi)}{m\pi}\ (m\neq0) \tag{9.77}$$

for rectangular gratings depicted in Fig. 9.21. It should be noted that each $\varepsilon_m$ is not a physical property of the material, and its imaginary part may be negative for a passive medium.

The coupled-wave formulation comes from the wave equation of the total electric field in region II. Due to the factors that $\varepsilon$ is independent of $y$ and $\mathbf{E}$ is parallel to the $y$-axis, we have from Eq. (9.68) that

$$\nabla^2 E_{\mathrm{II}}(x,z)+k^2\varepsilon(x)E_{\mathrm{II}}(x,z)=0 \tag{9.78}$$

A differential equation can be obtained by substituting Eqs. (9.75) and (9.76) into Eq. (9.78) as

$$\sum_j \frac{d^2\Psi_j}{dz^2}\exp\left(\mathrm{i}k_{x,j}x\right)-\sum_j k_{x,j}^2\Psi_j\exp\left(\mathrm{i}k_{x,j}x\right) \\ +k^2\left[\sum_m \varepsilon_m\exp\left(\mathrm{i}\frac{2m\pi}{\Lambda}x\right)\right]\left[\sum_n \Psi_n\exp\left(\mathrm{i}k_{x,n}x\right)\right]=0 \tag{9.79}$$

Equation (9.79) can be rearranged in terms of $\exp(\mathrm{i}k_{x,j}x)$ for the $j$th-order as follows:

$$\sum_j \left( \frac{d^2\Psi_j}{dz^2} - k_{x,j}^2 \Psi_j + k^2 \sum_n \varepsilon_{j-n} \Psi_n \right) \exp(\mathrm{i}k_{x,j}x) = 0 \tag{9.80}$$

In order to satisfy this equation for any value of $x$, the coefficient of $\exp(\mathrm{i}k_{xj}x)$ must be zero for all $j$'s. Hence, Eq. (9.80) is an infinite set of second-order coupled equations. Note that each space-harmonic term is coupled to other components through the harmonics of the grating. The numerical solution is obtained with sufficiently large number of diffraction orders. Suppose $j = 0, \pm 1, \pm 2, \ldots, \pm q$, then there are $N = 2q + 1$ diffraction orders so that $n = 0, \pm 1, \pm 2, \ldots, \pm q$ will also have $N$ terms. Equation (9.80) can be represented by an $N \times N$ matrix. The Fourier expansion of the dielectric function will have $m = 0, \pm 1, \pm 2, \ldots, \pm 2q$, or $4q + 1$, terms. The magnetic field can be obtained from Eq. (9.67) and expressed in terms of $\Psi_j$. The $N$ unknown functions $\Psi_j (j = 0, \pm 1, \pm 2, \ldots, \pm q)$ can be expressed as summations of the eigenfunctions, which have $2N$ unknown coefficients. Together with $r_j$ and $t_j (j = 0, \pm 1, \pm 2, \ldots, \pm q)$, there are $4N$ unknowns. By matching the boundary conditions for the electric field and the tangential component of the magnetic field at the interface between regions I and II and that between regions II and III, the corresponding $4N$ linear equations can be solved using the matrix method. An enhanced, numerically stable transmittance matrix approach was developed and applied to the implementation of RCWA for surface-relief and multilevel gratings, with detailed equations and solution procedures [32]. The derivation of TM wave is more complicated because of the extra term in Eq. (9.69). Nevertheless, a corrected procedure has been proposed by Li [33]. Many researchers have considered the effect of azimuthal angle of incidence on the radiative properties of gratings, i.e., when the incident wavevector **k** is not perpendicular to the grating grooves [34]. RCWA has also been developed and applied for 2D gratings as well as gratings of complicated geometries and anisotropy [35]. Some RCWA codes and other codes for radiative properties and thermal radiation can be found from author's website [36].

Once the reflection and transmission coefficients are obtained, it is possible to compute the fields inside and outside the grating structures, as well as to obtain the grating efficiency for each diffracted wave by calculating the time-averaged Poynting vector. The directional-hemispherical reflectance $\rho'_\lambda$ is the summation of the reflectance of all orders. Furthermore, the directional absorptance can be calculated by $\alpha'_\lambda = 1 - \rho'_\lambda$, assuming region III is semi-infinite.

As an example, the reflectance at normal incidence of a silicon grating for both $p$ and $s$ polarizations is shown in Fig. 9.22. The grating region simulates polycrystalline silicon gates in the 65 nm devices, used in high-performance complementary metal-oxide-semiconductor (CMOS) technology [31]. The grating period is $\Lambda = 240$ nm. The thickness of the grating or the height of the gates is $d = 50$ nm. The width of the gates is 30 nm, yielding a filling ratio of $\phi = 1/6$. The properties of the gates and the substrate are taken from data combined in Palik [2] for single-crystal silicon at room temperature. Comparison has been made to the reflectance of plain silicon, which is independent of the polarization, and that predicted by effective medium formulations

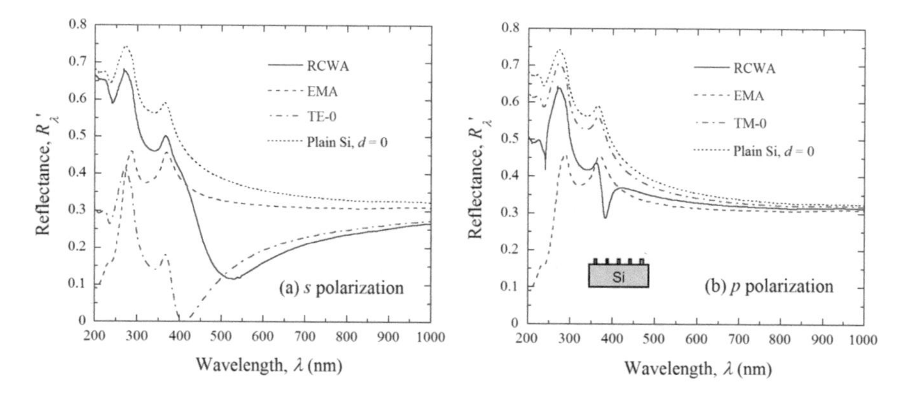


**Fig. 9.22** Calculated reflectance of silicon gratings. **a** TE wave ($s$ polarization). **b** TM wave ($p$-polarization) [31]

(to be discussed later). When compared with plain silicon, the reflectance is significantly reduced by the thin grating layer, and the reduction depends strongly on the wavelength and the polarization. For TE wave, the reduction is largest at $\lambda = 520$ nm; whereas for TM wave, the reduction is more significant at shorter wavelengths and grating anomalies occur at the wavelengths of 240 and 380 nm. Figure 9.22 also shows the calculation based on effective medium formulations as discussed next.

### 9.3.4 *Effective Medium Formulations*

When the grating period is much smaller than the wavelength, i.e., $\Lambda/\lambda < (n_{\text{III}} + \sin\theta)^{-1}$, all the diffracted waves are evanescent waves, except the zeroth-order (specular direction) one. The reflection is similar to a smooth film with an effective uniform dielectric function. This approach is called the *method of homogenization,* and the underlying physics is based on the effective medium theory (EMT). Effective medium formulations have been used widely to describe the optical properties of inhomogeneous media. The EMT was first postulated by Maxwell-Garnett [37] to obtain the effective dielectric function of metallic particles embedded in a dielectric medium. The general assumption is that the spacing separating the particles to be sufficiently large or the filling ratio of the particles is small. Bruggeman [38] developed a different formulation by assuming that two materials are embedded in the effective medium and obtained an expression, which has been successfully applied to study the effect of porosity on refractive index and absorption coefficient of different materials. The dielectric function of the effective medium $\varepsilon_{\text{EMA}}$ is related to those of the two components by

$$\phi\frac{\varepsilon_{\text{EMA}} - \varepsilon_{\text{A}}}{\varepsilon_{\text{A}} + 2\varepsilon_{\text{EMA}}} + (1-\phi)\frac{\varepsilon_{\text{EMA}} - \varepsilon_{\text{B}}}{\varepsilon_{\text{B}} + 2\varepsilon_{\text{EMA}}} = 0 \tag{9.81}$$

where $\phi$ is the volume fraction (filling ratio) of material A. Bruggeman's expression given in Eq. (9.81) is often called the effective medium approximation (EMA). Rytov [39] in 1956 first applied the EMT for a periodic structure by treating a stratified medium as a homogeneous uniaxial crystal and obtained the effective permittivity and permeability tensors. The zeroth order is considered to be applicable when $\Lambda \ll \lambda$ and has been used for designing surfaces with antireflection and selective radiative properties. The expression has been extended to include higher order terms for both 1D and 2D gratings [40]. The effective medium formulation for gratings depends on the polarization. The zeroth-order expressions of the dielectric function for different polarizations are given below:

$$\varepsilon_{\mathrm{TE},0} = \phi\varepsilon_{\mathrm{A}} + (1-\phi)\varepsilon_{\mathrm{B}}, \text{ for TE wave} \tag{9.82}$$

$$\frac{1}{\varepsilon_{\mathrm{TM},0}} = \frac{\phi}{\varepsilon_{\mathrm{A}}} + \frac{1-\phi}{\varepsilon_{\mathrm{B}}}, \text{ for TM wave} \tag{9.83}$$

The results of the effective medium formulation are compared with those of the RCWA in Fig. 9.22, in which the reflectance predicted by the EMA is independent of the polarization. Both of the effective medium formulations cannot predict the radiative properties well at shorter wavelengths. The agreement between effective medium formulations and the RCWA is reasonable in the long-wavelength end, except that the EMA is worse for TE wave. Chen et al. [31] performed a detailed study on the effects of temperature, wavelength, polarization, and angle of incidence on the absorptance of nanoscale patterned wafers for the CMOS technology. They also compared the configuration of combined polycrystalline silicon gates with $SiO_2$ trenches or a $SiO_2$ film. The results demonstrate nanostructures can have a significant impact on the radiative properties in unexpected ways. Hence, further research is much needed to fully understand the effect of complex nanostructures on radiative energy transfer and properties. The effective medium formulations will be further discussed and applied to various inhomogeneous media such as carbon nanotube arrays and metallodielectric photonic crystals in Sect. 9.5.7.

## 9.4 Bidirectional Reflectance Distribution Function (BRDF)

As discussed in Chap. 8, the bidirectional reflectance distribution function (BRDF) is a fundamental radiative property that describes the redistribution of energy reflected from a rough surface and/or an inhomogeneous medium. Knowledge of BRDFs is essential for the analysis of radiative heat transfer between rough surfaces. Because the major heating source in rapid thermal processing is the lamp radiation, knowledge of the radiative properties of materials is important for the thermal budget and temperature control during the process. A challenging problem is for the accurate measurement of wafer temperature based on radiation thermometry, because it is nonintrusive and can achieve fast response. The accuracy of radiation thermometry can be affected by the emittance change and the background radiation, especially when the measured surface is rough, such as the backside of the silicon wafer [41]. The surface roughness affects not only the emittance of the wafer but also the directional distribution of the reflected radiation by scattering. Therefore, a detailed understanding of the directional radiative properties of rough surfaces is essential to model the apparent emittance, considering the background radiation and multiple reflections.

Roughness is a measure of the topographic relief of a surface. It describes features of irregularities on the surface. Some common roughness parameters and functions include rms roughness $\sigma_{\mathrm{rms}}$, *power spectral density* (PSD), autocorrelation length $\tau_{\mathrm{cor}}$, and *slope distribution function* (SDF) [42, 43]. A surface appears to be smooth if the wavelength is much greater than $\sigma_{\mathrm{rms}}$. A highly polished surface can have an rms roughness on the order of nanometers. Some surfaces that look rough to human eyes may appear to be smooth for the far-infrared radiation. The reflection of radiation by rough surfaces is more complicated. For randomly rough surfaces, the scattered energy distribution or the BRDF often exhibits a peak around the direction of specular reflection, an off-specular lobe, and a diffuse component.

The BRDF of a surface can be predicted by solving the Maxwell equations if the surface roughness is fully characterized. The boundary integral method is commonly used to rigorously solve the Maxwell equations by matching the boundary conditions for the electric and magnetic fields. Since the rigorous electromagnetic wave solution generally requires a huge memory with a high-speed CPU, this approach is practically applicable to 1D rough surfaces only, though in some cases, solutions for 2D rough surfaces have been obtained. It is common to use approximation methods, such as the Rayleigh-Rice perturbation theory, the Kirchhoff approximation, and the geometric optics approximation. These approximations are appropriate only within certain ranges of roughness and wavelength.

The geometric illustration for the BRDF definition has been given in Fig. 8.11. The Rayleigh-Rice perturbation theory can be used for relatively smooth surfaces, i.e., for surfaces with $\sigma_{\mathrm{rms}} \cos\theta_{\mathrm{i}}/\lambda < 0.05$, or small particles on surfaces. It is based on a statistical Fourier analysis of the surface and predicts that the BRDF is directly proportional to the PSD and inversely proportional to the fourth power of the wavelength [43]. The Kirchhoff approximation is another physical-optics-based method that is often used to model the surface scattering with wave characteristics, like wave diffraction, by assuming that the radius of the surface curvature is smaller than the wavelength and there is no multiple scattering. The Kirchhoff approximation is applicable when the surface profile is slightly undulating (i.e., without sharp crests and deep valleys). The condition for this approximation to hold is that $\sigma_{\mathrm{rms}}$ must be relatively small compared with $\lambda$ and $\tau_{\mathrm{cor}}$. In the Kirchhoff approximation, the effects of shadowing and multiple scattering, which may be significant at large angles of incidence, are usually neglected. Most studies assumed that the roughness statistics is Gaussian.

The geometric optics approximation (GOA) neglects interference and diffraction effects and treats a rough surface as one with many small facets where an incident ray reflects specularly. Under these assumptions, the ray-tracing technique can be applied to predict the BRDF either with appropriate analytical expressions or with a Monte Carlo method. The shadowing and multiple scattering can be taken into account through a probability density function called shadowing or masking function. Multiple scattering can be incorporated into the geometric optics formulation with the Monte Carlo method. The GOA is applicable to surfaces whose $\sigma_{\rm rms}$ and $\tau_{\rm cor}$ are greater than $\lambda$. There exists a good agreement between the simulation results employing the GOA and the rigorous electromagnetic wave solution. However, the simulation based on geometric optics requires much lesser computational resources and takes much lesser time than that based on the rigorous solution. In the following, the GOA-based analytical formulation and ray-tracing algorithms will be presented, and the results are compared for anisotropic surfaces.

### 9.4.1 The Analytical Model

For the in-plane BRDF (either $\phi_{\rm r} = \phi_{\rm i}$ or $\phi_{\rm r} = \phi_{\rm i} + 180°$), referring to Fig. 8.11, Zhu and Zhang [44] unified several analytical models considering first-order scattering only. The expression of the BRDF is given in the following:

$$f_{\rm r}(\theta_{\rm i}, \phi_{\rm i}, \theta_{\rm r}, \phi_{\rm r}) = \frac{p(\zeta_x, \zeta_y)S(\theta_{\rm i})S(\theta_{\rm r})}{4\cos\theta_{\rm i}\cos\theta_{\rm r}\cos^4\alpha}\rho(n, \theta_0) \tag{9.84}$$

Here, $p(\zeta_x, \zeta_y)$ is the 2D SDF, and $\zeta_x$ and $\zeta_y$ are the slopes in $x$- and $y$-directions, given by

$$\zeta_x = \frac{\partial\zeta}{\partial x} = -\frac{\sin\theta_{\rm i}\cos\phi_{\rm i} + \sin\theta_{\rm r}\cos\phi_{\rm r}}{\cos\theta_{\rm i} + \cos\theta_{\rm r}} \tag{9.85a}$$

and

$$\zeta_y = \frac{\partial\zeta}{\partial y} = -\frac{\sin\theta_{\rm i}\sin\phi_{\rm i} + \sin\theta_{\rm r}\sin\phi_{\rm r}}{\cos\theta_{\rm i} + \cos\theta_{\rm r}} \tag{9.85b}$$

The microfacet reflectance $\rho(n, \theta_0)$, where $n$ is a complex refractive index and $\theta_0$ is the local incidence angle, is calculated from Fresnel's reflection coefficients by averaging over the two polarizations. In the denominator of Eq. (9.84), $\alpha$ is the inclination angle of the microfacet. For $\phi_{\rm r} = \phi_{\rm i}, \alpha = (\theta_{\rm i} + \theta_{\rm r})/2$ and $\theta_0 = |\theta_{\rm i} - \theta_{\rm r}|/2$, while for $\phi_{\rm r} = \phi_{\rm i} + 180°, \alpha = |\theta_{\rm i} - \theta_{\rm r}|/2$ and $\theta_0 = (\theta_{\rm i} + \theta_{\rm r})/2$. A shadowing function $S$ is used in Eq. (9.84) to account for shadowing and re-striking and is a function of the incidence or reflection zenith angles and the rms slope $w$, which equals $\sqrt{2}\sigma_{\rm rms}/\tau$ for Gaussian surfaces. Smith [45] derived a shadowing function based on the Gaussian statistics. The Smith shadowing function is expressed as

$$S(\theta) = \frac{1 - 0.5\mathrm{erfc}(\Gamma)}{1 - 0.5\mathrm{erfc}(\Gamma) + \exp\left(-\Gamma^2\right) / \left(2\sqrt{\pi}\Gamma\right)}, 0 \le \theta \le 90° \quad (9.86)$$

where $\theta$ is the zenith angle of incidence (for shadowing) or reflection (for masking), and $\Gamma = \tan(90° - \theta) \Big/ \left(\sqrt{2}w\right)$. While the expression is simple, the GOA allows calculations for the in-plane BRDF with first-order scattering only.

### 9.4.2 *The Monte Carlo Method*

Lee et al. [46] developed two ray-tracing techniques for modeling the BRDF in the Monte Carlo method, namely, the surface generation method (SGM) and the microfacet slope method (MSM). The major difference lies in how to simulate the rough surfaces. The SGM is the most commonly used ray-tracing method, in which a surface realization (i.e., a numerically generated rough surface) is required prior to tracing the ray bundles. Therefore, the origin and direction of reflection is determined based on the physical location and orientation of the microfacet that the ray strikes. The BRDF is obtained from an ensemble average over a sufficiently large number of surface realizations. On the other hand, the MSM does not need to generate the entire surface a priori. In the MSM, ray tracing is performed by generating a normal vector of a microfacet for each ray bundle, based on the SDF and the direction of the incoming ray [47]. Because a surface profile does not exist in the MSM, the optical path of a propagating ray and whether the ray re-strikes the surface cannot be directly determined. Hence, the MSM relies on a shadowing function, which is the probability that a reflected ray re-strikes another surface facet, to model multiple scattering. Zhu et al. [48] compared the two ray-tracing techniques with rigorous solutions of the electromagnetic wave equation, using the boundary integral method, for dielectric surfaces coated with a thin film. Although the MSM is not applicable for very rough surfaces at oblique incidence, it takes less computational time and has the advantage for multiscale problems, such as light scattering from semitransparent materials, because the MSM algorithm is compatible in both micro- and macroscales.

The *spectral method* is commonly used for surface realization in the SGM by using the power spectrum. The power spectrum can be obtained from the roughness statistics. The autocorrelation function multiplied by $\sigma_{\mathrm{rms}}$ and the PSD are a Fourier-transform pair. A rough surface, defined with the height distribution function and the autocorrelation function, are usually generated with the spectral method, regardless of whether the surface is Gaussian or not. However, it is difficult to generate an anisotropic surface with this method. On the other hand, the surface topographic data from the AFM measurement are stored in a 2D array of the height, which can be conveniently incorporated into the SGM algorithm without using the spectral method. The challenge is how to deal with the trade-off between the measurement area, spatial resolution, noise and artifacts in the AFM measurements, measurement time, and the number of measurements that will produce statistical meaningful results.

The anisotropic SDF can be numerically evaluated as a 2D histogram using topographic data for use in the MSM. A weight function must be included in generating microfacets because, statistically, the incident energy that is intercepted by a microfacet depends not only on the SDF but also on the projected area of the microfacet. The rejection method allows the generation of microfacets, following the weighted SDF, with uniform random numbers. The rejection method is suitable for any type of distribution function as long as a comparison function is appropriately selected. Meanwhile, the Smith shadowing function determines the probability of re-striking, in the MSM.

The polarization state may change upon reflection by a 2D rough surface, because of the random orientation of the microfacets. When the microfacet reflectivity is calculated using Fresnel's reflection coefficients, the change of the polarization state should also be considered. In a 2D rough surface, even though the incident radiation is purely *s* or *p* polarized (TE wave or TM wave, respectively), the radiation incident at the microfacet can have both polarization components in the local coordinates. Furthermore, depolarization may occur upon reflection so that the polarization of the scattered wave is different from that of the incident wave. The geometrical relations between wavevectors and polarization vectors delineate the contribution of each polarization to the reflectivity. As illustrated in Fig. 9.23, unit vectors in the direction of incidence and reflection, i.e., $\mathbf{s}_i$ and $\mathbf{s}_r$, respectively, are defined in the following:

$$\mathbf{s}_i = \begin{pmatrix} -\sin\theta_i \cos\phi_i \\ -\sin\theta_i \sin\phi_i \\ -\cos\theta_i \end{pmatrix} \quad \text{and} \quad \mathbf{s}_r = \begin{pmatrix} \sin\theta_r \cos\phi_r \\ \sin\theta_r \sin\phi_r \\ \cos\theta_r \end{pmatrix} \tag{9.87}$$

The vectors $\mathbf{s}_i$ and $\hat{\mathbf{z}}$ define the plane of incidence in the global coordinates, and the vectors $\mathbf{s}_r$ and $\hat{\mathbf{z}}$ define the plane of reflection. A unit vector $\mathbf{h}_i$ perpendicular and a unit vector $\mathbf{v}_i$ parallel to the plane of incidence characterize the two polarizations of the incident wave. Here, $\mathbf{h}_i$ indicates the electric field for *s* polarization while $\mathbf{v}_i$ the electric field for *p*-polarization. Similarly, $\mathbf{h}_r$ and $\mathbf{v}_r$ represent the two polarizations of the reflected wave. Hence,

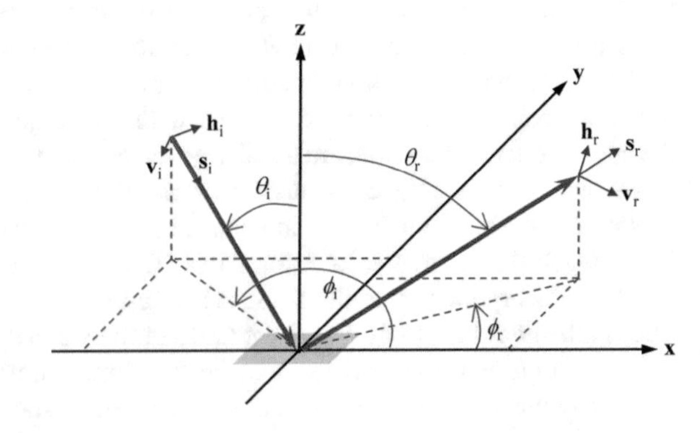


**Fig. 9.23** Schematic of incident and scattered rays. Here, *x*, *y*, and *z* are the global coordinates, where the *x*-*y* plane is the mean plane of a rough surface

$$\mathbf{h}_{\mathrm{i}} = \frac{\hat{\mathbf{z}} \times \mathbf{s}_{\mathrm{i}}}{\left|\hat{\mathbf{z}} \times \mathbf{s}_{\mathrm{i}}\right|} = \begin{pmatrix} \sin\phi_{\mathrm{i}} \\ -\cos\phi_{\mathrm{i}} \\ 0 \end{pmatrix} \quad \text{and} \quad \mathbf{h}_{\mathrm{r}} = \frac{\hat{\mathbf{z}} \times \mathbf{s}_{\mathrm{r}}}{\left|\hat{\mathbf{z}} \times \mathbf{s}_{\mathrm{r}}\right|} = \begin{pmatrix} -\sin\phi_{\mathrm{r}} \\ \cos\phi_{\mathrm{r}} \\ 0 \end{pmatrix} \tag{9.88}$$

$$\mathbf{v}_{\mathrm{i}} = \mathbf{h}_{\mathrm{i}} \times \mathbf{s}_{\mathrm{i}} = \begin{pmatrix} \cos\theta_{\mathrm{i}}\cos\phi_{\mathrm{i}} \\ \cos\theta_{\mathrm{i}}\sin\phi_{\mathrm{i}} \\ -\sin\theta_{\mathrm{i}} \end{pmatrix} \quad \text{and} \quad \mathbf{v}_{\mathrm{r}} = \mathbf{h}_{\mathrm{r}} \times \mathbf{s}_{\mathrm{r}} = \begin{pmatrix} \cos\theta_{\mathrm{r}}\cos\phi_{\mathrm{r}} \\ \cos\theta_{\mathrm{r}}\sin\phi_{\mathrm{r}} \\ -\sin\theta_{\mathrm{r}} \end{pmatrix} \tag{9.89}$$

Calculation of the reflectivity involves two conversions of the polarization components. The $s$- and $p$-polarization components of the incident wave defined in the global coordinates are first converted to their counterparts in the local coordinates. The local polarization components are multiplied by Fresnel's reflection coefficients and then converted to the global components. Accordingly, the microfacet reflectivities for the co- and cross-polarizations can be expressed as follows:

$$\rho_{ss} = \left|(\mathbf{v}_{\mathrm{r}} \cdot \mathbf{s}_{\mathrm{i}})(\mathbf{v}_{\mathrm{i}} \cdot \mathbf{s}_{\mathrm{r}})r_s + (\mathbf{h}_{\mathrm{r}} \cdot \mathbf{s}_{\mathrm{i}})(\mathbf{h}_{\mathrm{i}} \cdot \mathbf{s}_{\mathrm{r}})r_p\right|^2 \Big/ \left|\mathbf{s}_{\mathrm{i}} \times \mathbf{s}_{\mathrm{r}}\right|^4 \tag{9.90a}$$

$$\rho_{sp} = \left|(\mathbf{h}_{\mathrm{r}} \cdot \mathbf{s}_{\mathrm{i}})(\mathbf{v}_{\mathrm{i}} \cdot \mathbf{s}_{\mathrm{r}})r_s - (\mathbf{v}_{\mathrm{r}} \cdot \mathbf{s}_{\mathrm{i}})(\mathbf{h}_{\mathrm{i}} \cdot \mathbf{s}_{\mathrm{r}})r_p\right|^2 \Big/ \left|\mathbf{s}_{\mathrm{i}} \times \mathbf{s}_{\mathrm{r}}\right|^4 \tag{9.90b}$$

$$\rho_{ps} = \left|(\mathbf{v}_{\mathrm{r}} \cdot \mathbf{s}_{\mathrm{i}})(\mathbf{h}_{\mathrm{i}} \cdot \mathbf{s}_{\mathrm{r}})r_s - (\mathbf{h}_{\mathrm{r}} \cdot \mathbf{s}_{\mathrm{i}})(\mathbf{v}_{\mathrm{i}} \cdot \mathbf{s}_{\mathrm{r}})r_p\right|^2 \Big/ \left|\mathbf{s}_{\mathrm{i}} \times \mathbf{s}_{\mathrm{r}}\right|^4 \tag{9.90c}$$

$$\rho_{pp} = \left|(\mathbf{h}_{\mathrm{r}} \cdot \mathbf{s}_{\mathrm{i}})(\mathbf{h}_{\mathrm{i}} \cdot \mathbf{s}_{\mathrm{r}})r_s + (\mathbf{v}_{\mathrm{r}} \cdot \mathbf{s}_{\mathrm{i}})(\mathbf{v}_{\mathrm{i}} \cdot \mathbf{s}_{\mathrm{r}})r_p\right|^2 \Big/ \left|\mathbf{s}_{\mathrm{i}} \times \mathbf{s}_{\mathrm{r}}\right|^4 \tag{9.90d}$$

where $r$ denotes Fresnel's reflection coefficient. The subscripts $s$ and $p$ stand for each polarization. On the left-hand side, the double subscripts indicate the polarization for the incidence and the reflection, respectively.

In terms of the microfacet reflectivities, the reflected energies $G_{\mathrm{r},s}$ and $G_{\mathrm{r},p}$ are related to the incident energies $G_{\mathrm{i},s}$ and $G_{\mathrm{i},p}$ by

$$\begin{bmatrix} G_{r,s} \\ G_{r,p} \end{bmatrix} = \begin{bmatrix} \rho_{ss} & \rho_{ps} \\ \rho_{sp} & \rho_{pp} \end{bmatrix} \begin{bmatrix} G_{i,s} \\ G_{i,p} \end{bmatrix} \tag{9.91}$$

The reflectivity is defined as a ratio of the reflected energy $G_{\mathrm{r}} = G_{\mathrm{r},s} + G_{\mathrm{r},p}$ to the incident energy $G_{\mathrm{i}} = G_{\mathrm{i},s} + G_{\mathrm{i},p}$; thus, it depends on the polarization state of the incident wave. To facilitate the calculation, the incident energy of each ray bundle is set to unity such that $\left(G_{\mathrm{i},s}, G_{\mathrm{i},p}\right) = (1,\ 0)$ for $s$-polarization, $\left(G_{\mathrm{i},s}, G_{\mathrm{i},p}\right) = (0,\ 1)$ for $p$-polarization, and $\left(G_{\mathrm{i},s}, G_{\mathrm{i},p}\right) = (0.5, 0.5)$ for random polarization (i.e., unpolarized incidence). For the first reflection, $G_{\mathrm{r},s}$ and $G_{\mathrm{r},p}$ are calculated from Eq. (9.91). For multiple reflections, the previously reflected energies are substituted for $G_{\mathrm{i},s}$ and $G_{\mathrm{i},p}$, and the next reflected energy is updated according to Eq. (9.91). Each ray bundle is traced until it leaves the surface, and then, the information of its direction and energy for each polarization is stored in a database. Because the energy of the bundle is reduced after each reflection, there is no need to use random numbers to decide whether a ray bundle is reflected at the microfacet or not.

In a special case, when the planes of incidence and reflection are identical, the polarization state is maintained for either *s*- or *p*-polarization if only the first-order scattering has been considered. This means that the vectors $\mathbf{h}_i$ and $\mathbf{h}_r$ are either parallel or antiparallel (refer to Fig. 9.23); consequently, $\mathbf{h}_i \cdot \mathbf{s}_r = 0$ and $\mathbf{h}_r \cdot \mathbf{s}_i = 0$. It can be seen from Eqs. (9.90a)–(9.90d) that $\rho_{sp} = \rho_{ps} = 0$, $\rho_{ss} = |r_s|^2$, and $\rho_{pp} = |r_p|^2$. The corresponding BRDF is called the in-plane BRDF ($\phi_r = \phi_i$ or $\phi_r = \phi_i + 180°$). Nevertheless, the cross-polarization term is nonzero for the in-plane BRDF when multiple scattering is significant. After a large number of ray bundles have been traced, the BRDF can be calculated in terms of the energy of the ray bundles:

$$f_r(\lambda, \theta_i, \phi_i; \theta_r, \phi_r) = \frac{1}{G_i(\theta_i, \phi_i)} \frac{\Delta G_r(\theta_r, \phi_r)}{\cos\theta_r \Delta\Omega_r} \tag{9.92}$$

where $G_i(\theta_i, \phi_i)$ is the total energy of the incident ray bundles, and $\Delta G_r(\theta_r, \phi_r)$ is the energy of the ray bundles leaving the surface within the solid angle $\Delta\Omega_r$, in the direction $(\theta_r, \phi_r)$. The integration of the BRDF yields the directional-hemispherical reflectance. The directional emittance can be obtained according to the conservation of energy and Kirchhoff's law.

### 9.4.3 Surface Characterization

In most studies, surface roughness is assumed to satisfy the Gaussian statistics in the derivation of the BRDF model and for the surface generation in the Monte Carlo simulation. Furthermore, the roughness statistics of 2D rough surfaces is assumed to be isotropic in most publications so that the autocorrelation function is independent of the direction. However, the Gaussian distribution may miss important features of natural and man-made rough surfaces that are strongly anisotropic. Before the invention of the AFM, the surface profile was usually measured with a mechanical profiler that scans the surface line-by-line. Some mechanical stylus profilers can measure rough surfaces with a vertical resolution of a few nanometers. However, the lateral resolution is usually on the order of 1 μm due to the large radius of the stylus probe. Because the radius of curvature of the probe tip is in the range from 5 to 50 nm, AFM can provide detailed information on the topography of a small area on the microrough surfaces, with a vertical resolution of subnanometers and a lateral resolution around 10 nm. The result is stored in an array, containing the height information, $z(m, n)$, where $m = 1, 2, \ldots, M$ and $n = 1, 2, \ldots, N$ are the points along the *x*- and *y*-directions, respectively.

To evaluate the 2D slope distribution $p(\zeta_x, \zeta_y)$, each surface element is determined by the four closest nodes in the data array. The four-node element can be considered as two triangular surfaces with a common side. The surface normals for the two triangles can be averaged to give the mean slope of the surface element such that [44]

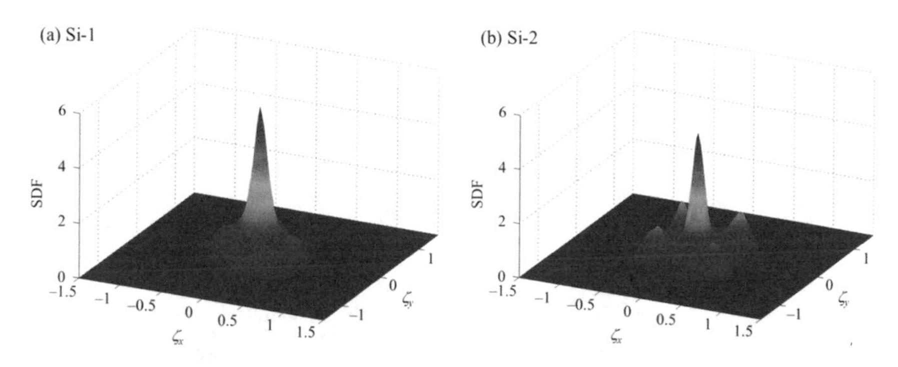


**Fig. 9.24** 2D slope distribution obtained from AFM topographic measurements for two samples: **a** Si-1; **b** Si-2

$$\zeta_x = \frac{z_{m+1,n} - z_{m,n}}{2l} + \frac{z_{m+1,n+1} - z_{m,n+1}}{2l} \tag{9.93a}$$

$$\zeta_y = \frac{z_{m,n+1} - z_{m,n}}{2l} + \frac{z_{m+1,n+1} - z_{m+1,n}}{2l} \tag{9.93b}$$

where $l$ is the lateral distance between adjacent data points. The SDF can be determined by evaluating the slopes of all measured surface elements. For a scan area of $100 \times 100\ \mu\text{m}^2$, the lateral interval $l \approx 0.2\ \mu\text{m}$, when the data are stored in a $512 \times 512$ array.

The 2D SDFs from the AFM measurement in the tapping mode, for two lightly doped < 100 > single-crystal silicon surfaces, are shown in Fig. 9.24 [44]. In the contact mode, lateral or shear forces can distort surface features and reduce the spatial resolution. Thus, deep valleys may not be correctly measured. The AFM scanning performed in the tapping mode with sharper silicon tips allows measuring precipitous slopes. The two SDFs are non-Gaussian and anisotropic, while the anisotropy of Si-1 is not as striking as that of Si-2. The SDF of Si-1 contains only one dominant peak at the center, indicating that a large number of microfacets are only slightly tilted. The SDF of Si-2 also has a dominant peak at the center, though smaller than that of Si-1. Four side peaks can also be seen that are nearly symmetric. These side peaks are associated with the formation of {311} planes, during the chemical etching in the (100) crystalline wafer [44, 46]. The angle between the (100) plane and any of the four (311) planes is $\cos^{-1}(3/\sqrt{11}) = 25.2°$, which is close to the location of the observed side peaks.

### 9.4.4 *Comparison of Modeling with Measurements*

Figure 9.25 compares the predicted BRDFs based on the slope distribution with the BRDFs measured using TAAS at $\lambda = 635$ nm, for Si-2, which is strongly anisotropic

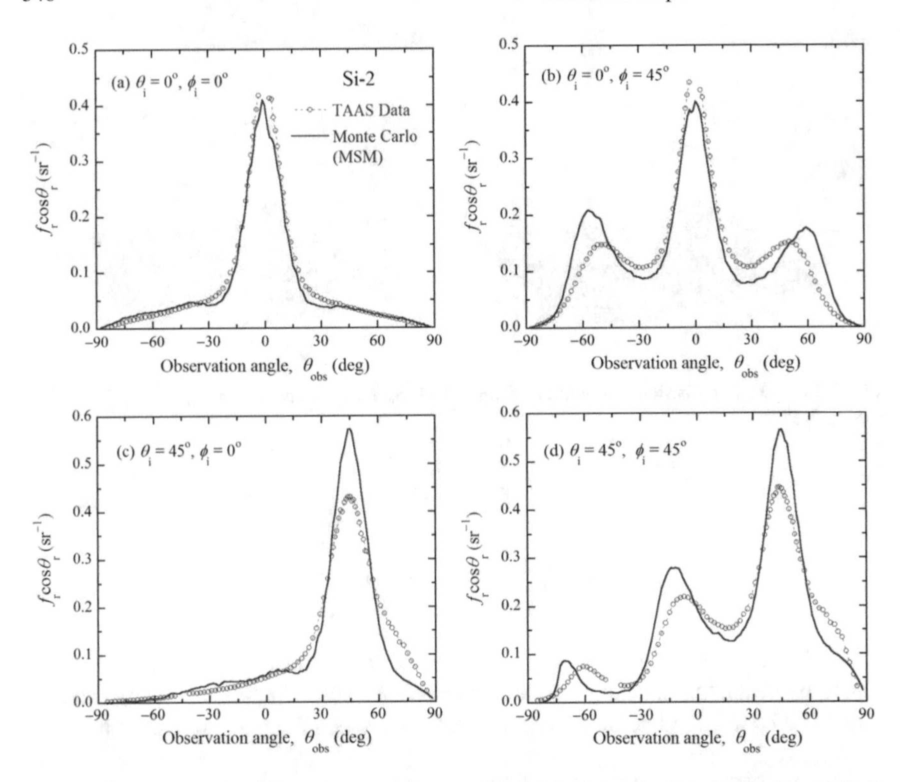


**Fig. 9.25** Comparison of Monte Carlo model based on the MSM and the measured in-plane BRDF for Si-2. The observation angle $\theta_{obs}$ is the same as the reflection polar angle when $\phi_r = \phi_i + \pi$ and negative refraction polar angle when $\phi_r = \phi_i$ [46]

[46]. For clarity, only the prediction using the MSM is presented. The predictions with the SGM and the analytical model yield a similar agreement with experiments. As can be seen from Fig. 9.25, the prediction and the measurement agree well, except near $\theta_{obs} = 0°$, where the measurements cannot be taken within $\pm 2.5°$ and the simulation has a large fluctuation. The simulation captures the general features and trends of the measured BRDF, while some discrepancies exist near the side peaks. For $\theta_i = 0°$ and $\phi_i = 45°$, as shown in Fig. 9.25b, the BRDF contains two large side peaks associated with the side peaks in the SDF for Si-2 at $|\zeta_x| \approx |\zeta_y| \approx 0.38$ in Fig. 9.24b. The Monte Carlo simulations also predict the side peaks located approximately at $\theta_r = 57°$, which deviates somewhat from the measured value of 50°. Based on Snell's law, the inclination angle of microfacets is half of $\theta_r$, at $\theta_i = 0°$. Therefore, the measured side peaks in the BRDF correspond to an inclination angle 25°, which is very close to the angle of 25.2° between any of the four {311} planes and the (100) plane. On the other hand, the predicted side peaks correspond to an inclination angle of 28.5°, which is almost the same as that calculated from the slope at $|\zeta_x| = |\zeta_y| = 0.38$. Consequently, the side peak position obtained from the BRDF measurement is more reliable than that predicted by the Monte Carlo methods using the topographic data from the AFM

measurement. Due to the artifacts in the AFM measurements, the BRDF values are underpredicted when $15° < \theta_r < 50°$ and overpredicted when $50° < \theta_r < 80°$. When $\theta_i = 45°$, the Monte Carlo method overpredicts the specular peak, presumably due to the limitation of geometric optics. The disagreement between the predicted and measured BRDFs, for $60° < \theta_{obs} < 85°$, may be due to the combined result of the artifacts in the AFM measurement, the limitation of the GOA, and multiple scattering. For $\theta_i = 45°$ and $\phi_i = 45°$, a small side peak appears at $\theta_{obs} = -60°$ in the measured curve and at $\theta_{obs} = -71°$ in the predicted curve. This is believed to be due to microfacets with {111} orientation that have an inclination angle of 54.7°. The small side peak should occur around $\theta_{obs} = -64.4°$ based on simple geometric arguments.

Figure 9.26 shows the directional-spectral emittance measured using an integrating sphere coupled with a monochromator [49]. The directional emittance was calculated from the measured the directional-hemispherical reflectance at an incidence angle of approximately 7°. The emittance values calculated from the models based on Gaussian distribution and anisotropic slope distribution are compared with those obtained from experiments. For Si-1, which is nearly isotropic, the difference between the models is small and the agreement with the experiment is excellent. The combined uncertainty in the measurement is estimated to be 0.01, except at $\lambda =$ 1000 nm, where the silicon wafer becomes slightly transparent. For Si-2, however, the Gaussian model underpredicts the emittance and there is a large enhancement of the emittance due to anisotropy, as large as 0.05. The Monte Carlo model, based on the MSM, significantly improves the prediction. Given the fact that the AFM surface topographic measurements may not perfectly match the actual surface slope distribution, an uncertainty of 0.01 has been estimated for the Monte Carlo model. It can be seen that the prediction agrees with the measurement better at short wavelengths, where the geometric optics is more suitable.

The out-of-plane BRDFs of Si-1 and Si-2, coated with a Au film, are calculated with the MSM. The results at $\theta_i = 30°$ are shown in Fig. 9.27 as contour plots in a

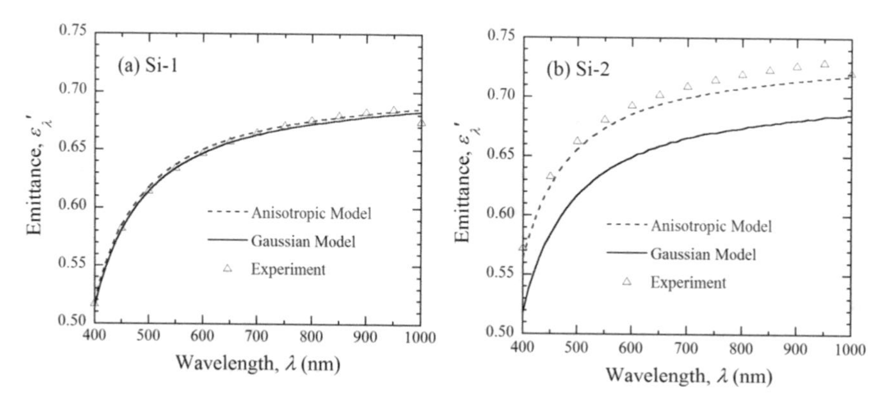


**Fig. 9.26** Comparison of the predicted and measured emittance of Si-1 and Si-2, in a polar angle approximately equal to 7° [49]

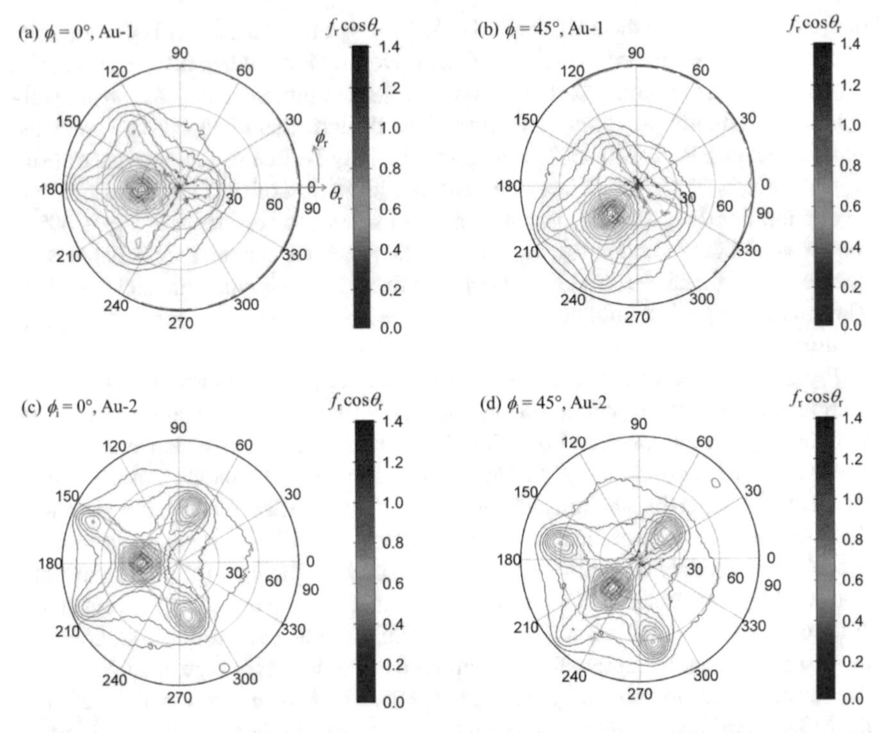


**Fig. 9.27** BRDF predicted by the MSM at $\theta_i = 30°$ for random polarization [46]: **a** Au-1 at $\phi_i = 0°$, **b** Au-1 at $\phi_i = 45°$, **c** Au-2 at $\phi_i = 0°$, and **d** Au-2 at $\phi_i = 45°$. In the polar contour plots, the radial coordinate corresponds to $\theta_r$, and the azimuthal coordinate corresponds to $\phi_r$

polar coordinates system [46]. In these plots, the coated rough surfaces are identified as Au-1 and Au-2, respectively. The radial and azimuthal coordinates correspond to $\theta_r$ and $\phi_r$, and the color contour represents $f_r \cos \theta_r$.

The BRDFs depend little on $\phi_r$ around the specular direction; however, the dependence becomes large as the angular separation from the specular peak increases. The region where the BRDF is independent of $\phi_r$ is broader for Au-1 than for Au-2. The predicted BRDFs for Au-2 display a strong specular reflection peak, together with the four large side peaks associated with {311} planes. In addition, a small side peak associated with a {111} plane appears at large $\theta_r$, as illustrated in Fig. 9.27c at $\phi_r = 294°$ and another in Fig. 9.27d at $\phi_r = 45°$. The results for Si without Au coating are similar except with smaller BRDF values and hence are not shown. The actual magnitudes of the small side peaks may be smaller than those predicted by the MSM, and their positions may shift toward smaller $\theta_r$. Nevertheless, Fig. 9.27 indicates that the Monte Carlo method is an effective technique to study the BRDFs for anisotropic surfaces. A comprehensive review of surface and volume scattering can be found from Zhu et al. [50].

## 9.5 Plasmon, Polariton, and Electromagnetic Surface Wave

In metals, electrons move freely like a gas as discussed previously. This is somewhat similar to the plasma state where a gas is ionized and becomes highly conductive; however, in a solid medium, the ion cores with positive charges are at fixed locations. The collective rapid oscillations of electrons (or electron density) in a plasma or metal are called *plasma oscillation*, which form a longitudinal wave since the field is oscillating in the same direction as the wavevector. The oscillation frequency is known as the plasma frequency, already introduced in Sect. 8.4.2. A *plasmon* is the quantum or quasiparticle of plasma oscillation. Based on the simple Drude model, the dielectric function due to free electrons in metal can be written as

$$\varepsilon(\omega) = 1 - \frac{\omega_{\mathrm{p}}^2}{\omega^2} \tag{9.94}$$

which can be obtained from Eq. (8.118) by neglecting damping and setting $\varepsilon_\infty = 1$. Plasmon can be excited by an oscillating electric field, such as in electron energy loss spectroscopy [51]. According to Eq. (8.3) for a charge neutral medium, $\nabla \cdot \mathbf{D} = \nabla \cdot (\varepsilon_{\mathrm{m}} \mathbf{E}) = 0$. This requires either $\nabla \cdot \mathbf{E} = 0$ or $\varepsilon_{\mathrm{m}} = \varepsilon_0 \varepsilon = 0$. However, when $\varepsilon \neq 0$, the electric field must be perpendicular to the wavevector, and this a transverse wave cannot excite a plasmon or a longitudinal charge oscillation in a bulk. It is therefore only possible to couple a plasmon in a bulk medium with an electromagnetic wave when $\varepsilon = 0$, so that the electric field can have a nonzero component in the wavevector direction. The coupling of a photon and a plasmon is called a *plasmon polariton*. The concept and term of polaritons were introduced in the 1950s to describe the coupling between photons and quasiparticles (such as plasmons, phonons, and excitons) in solids [52].

When $\varepsilon \neq 0$, only transverse electromagnetic waves can exist in the medium. According to Eq. (8.24), the wavevector and the angular frequency are related by

$$k(\omega) = \frac{\omega}{c_0}\sqrt{\varepsilon(\omega)} \quad \text{or} \quad \omega(k) = \frac{c_0 k}{\sqrt{\varepsilon(k)}} \tag{9.95}$$

The above relationship is called the plasmon-polariton dispersion. It suggests that propagating waves can exist in the medium only at frequencies higher than the plasma frequency: $\omega > \omega_{\mathrm{p}}$. In real materials, due to transitions at very high frequencies, the zero permittivity (or near-zero permittivity if damping is considered) occurs at $\omega \approx \omega_{\mathrm{p}}/\sqrt{\varepsilon_\infty}$.

From lattice dynamics, phonons have high-frequency longitudinal and transverse modes, called optical phonons, as discussed in Chap. 6. In ionic crystals, lattice vibrations generate dipole moments with oscillating electric fields. Both the longitudinal modes and transverse modes can interact with infrared photons, forming *phonon polaritons*. The interaction modifies the photon dispersion so the phase speed is frequency-dependent due to the dispersion of the dielectric function. The interaction also modifies the phonon behavior since it allows propagating waves to exist inside the solid in certain spectral regions rather than only at the resonance frequencies for longitudinal and transverse optical phonons: $\omega_{\text{LO}}$ and $\omega_{\text{TO}}$ [53]. Rewrite Eq. (8.123) by neglecting damping, we have

$$\varepsilon(\omega) = \varepsilon_\infty \left( \frac{\omega_{\text{LO}}^2 - \omega^2}{\omega_{\text{TO}}^2 - \omega^2} \right) \tag{9.96}$$

Equation (9.96) gives a low-frequency dielectric constant $\varepsilon(\omega \to 0) = \varepsilon_\infty \omega_{\text{LO}}^2 / \omega_{\text{TO}}^2$. As the frequency increases, the dielectric function increases to a pole at $\omega = \omega_{\text{TO}}$. It becomes negative in the region $\omega_{\text{TO}} < \omega < \omega_{\text{LO}}$ and varies from $-\infty$ to 0 as $\omega$ increases from $\omega_{\text{TO}}$ to $\omega_{\text{LO}}$. Beyond $\omega_{\text{LO}}$, the dielectric function increases from zero to $\varepsilon_\infty$. The interaction of photon and phonon yields two branches where propagating waves can exist inside the solid. Between $\omega_{\text{TO}}$ and $\omega_{\text{LO}}$ the wavevector calculated from Eq. (9.95) is purely imaginary. Subsequently, no propagating waves can exist inside the medium. As discussed in the previous chapter, this region is called the reststrahlen band where the reflectance is very high.

The dispersion curves according to Eq. (9.95) are plotted in Fig. 9.28 for both plasmon polaritons with Eq. (9.94) and phonon polaritons with Eq. (9.96). They are often called bulk (or volume) plasmon and phonon polaritons, respectively. As shown in Fig. 9.28a, the horizontal dashed line is the excitation frequency for a bulk plasmon polariton and the solid line is the dispersion curve when propagating waves can exist ($\omega > \omega_{\text{p}}$). The inclined dash-dotted line shows the linear dispersion when the dielectric function is equal to 1 and is independent of frequency. Similarly, the two horizontal dashed lines in Fig. 9.28b correspond to transverse and longitudinal optical phonon frequencies, respectively. The values of $\varepsilon_\infty = 3$ and $\omega_{\text{LO}}/\omega_{\text{TO}} = 1.5$ are used to produce the plots. The inclined dash-dotted line represents light propagation in a nondispersive dielectric with a (constant) phase speed, $v_{\text{p}} = c_0/\sqrt{\varepsilon_\infty}$. The

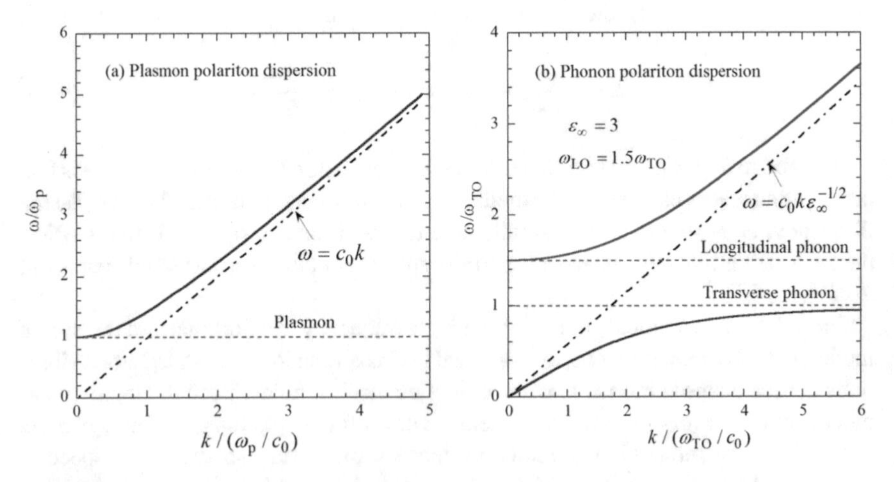


**Fig. 9.28** Dispersion curves for **a** a plasmon polariton and **b** a phonon polariton, according to Eqs. (9.94)–(9.96). Note that normalized coordinates are used

bulk phonon polariton modifies the dispersion and results in a branch below $\omega_{\mathrm{TO}}$ and another branch above $\omega_{\mathrm{LO}}$. The phase speed of electromagnetic waves in the medium is a function of frequency in these two branches where waves can propagate inside the medium.

Many unique phenomena can be explained or better understood by introducing the concept and properties of polariton excitations. Up to this section, the dielectric function has been used in calculating radiative properties without any in-depth discussion on the nature of polaritons. In the following subsections, we will study surface polaritons, coupled polaritons, localized surface plasmon polaritons, magnetic polaritons, graphene plasmons, and hyperbolic polaritons as well. The terminologies used in the literature change over time and may vary to some extent between different authors. This can be rather confusing for researchers who are relatively new to the field. It should be noted that oftentimes it is indeed difficult to clearly distinguish one type versus the other type of polaritons or modes; therefore, it is not surprising that different authors may analyze the problem from different aspects as well as use different notations and terminologies.

### 9.5.1 Surface Plasmon (or Phonon) Polariton

Plasmons can be formed near the surface of the metal, where electrons oscillate along the interface; such plasmons are called *surface plasmons*. The quantum that describes the interaction between a surface plasmon and a photon is called a *surface plasmon polariton* (SPP). The interaction can result in strong absorption or emission of photons in a specific direction, characterized by the wavevector of the electromagnetic wave. SPPs have played an important role in near-field microscopy, nanophotonics, and biomolecular sensor applications [54–56]. Surface plasmons usually occur in the visible or near-infrared region of the electromagnetic wave spectrum for highly conductive metals such as Ag, Al, and Au. It is also possible to excited SPPs at longer wavelengths in doped semiconductors [57].

In some polar dielectric materials, phonons, or bound charges can also interact with the electromagnetic waves in the mid-infrared spectral region to cause resonance effects near the surface; the associate quasiparticle is called *surface phonon polariton* (SPhP). SPhPs have applications in tuning the thermal emission properties [58, 59] and nanoscale nondestructive imaging [60].

The excitation of an SPP or SPhP gives rise to a *surface electromagnetic wave*, which propagates along the interface and decays into both media. It should be noted that in traditional materials, only TM waves can excite SPPs or SPhPs because the electric field must have a component parallel to the charge oscillation. Furthermore, the electric field has a nonzero component both parallel with and perpendicular to the interface. Hence, surface polaritons cannot simply be characterized as a transverse wave or a longitudinal wave. In the following, the basic mechanisms of surface polaritons are presented, with an emphasis on the fields near the interface and the influence on the radiative properties.

As shown in Fig. 9.29a, the charges oscillate along the surface or interface between air and a metal when a surface plasmon polariton is excited. The field associated with a SPP is localized at the surface. Oscillations of the charge in the $x$-$z$ plane result in a magnetic field in the $y$-direction. The amplitude of the field decays away from the interface, as shown in Fig. 9.29b. Note that $\eta = \mathrm{Im}(k_z)$ and both $\eta_1$ and $\eta_2$ are positive. Surface plasmons can be excited by electromagnetic waves and are important for the study of optical properties of metallic materials, especially near the plasma frequency, which usually lies in the ultraviolet. The requirement of evanescent waves on both sides of the interface prohibits the coupling of propagating waves in air directly to surface plasmons. For this reason, surface electromagnetic waves are often regarded as *nonradiative modes.* Another way to understand the excitation condition is that both photons and surface polaritons must have the same frequency (energy) and parallel wavevector component (momentum). However, the momentum of SPPs ($\hbar k_{\mathrm{SPP}}$) is greater than that of photons in free space ($\hbar k_0$). The attenuated total reflectance (ATR) arrangements are commonly used to excite surface plasmons, as illustrated in Fig. 9.30, for (a) prism-air-metal configuration named after A. Otto and (b) the prism-metal-air configuration named after E. Kretschmann and H. Raether. When light is incident from the prism, $k_x$ can be sufficiently large to match the SPP momentum. It is therefore possible for evanescent waves to occur simultaneously in the underneath metallic and air layers. A detailed discussion with historical aspects can be found from Raether [61].

In addition to the requirement of evanescent waves on both sides of the interface, the *polariton dispersion relations* must be satisfied. They are expressed as follows when both media extend to infinity in the $z$-direction:

$$\frac{k_{1z}}{\varepsilon_1} + \frac{k_{2z}}{\varepsilon_2} = 0 \quad \text{for TM wave} \tag{9.97}$$

$$\frac{k_{1z}}{\mu_1} + \frac{k_{2z}}{\mu_2} = 0 \quad \text{for TE wave} \tag{9.98}$$

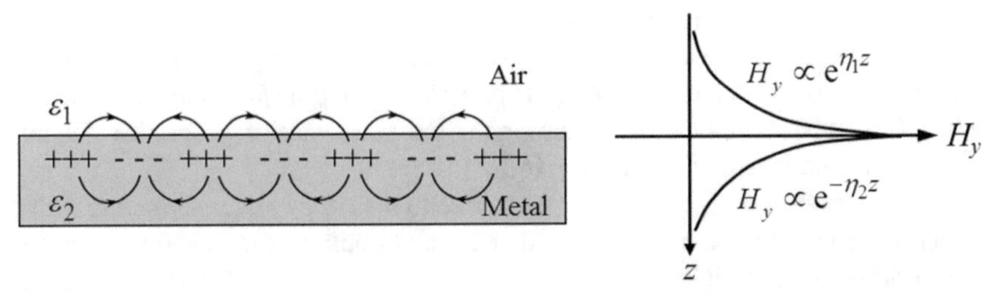


**Fig. 9.29** Illustration of surface plasmon polariton. **a** Charge fluctuations and the magnetic field at the interface between a metal and air. **b** The exponentially decaying field amplitudes away from the interface

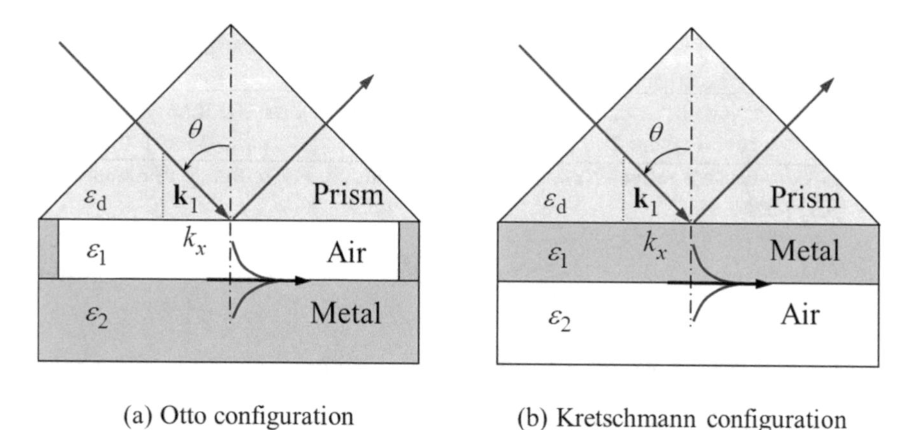


**Fig. 9.30** Typical configurations for coupling electromagnetic waves with surface plasmons using attenuated total reflectance arrangements. **a** The Otto configuration (prism-air-metal). **b** The Kretschmann configuration (prism-metal-air); also called the Kretschmann-Raether configuration. Note that a polar dielectric may be substituted for the metal to excite SPhPs

Let us consider lossless media first. In order for evanescent waves to occur, we must have $k_{1z} = \mathrm{i}\eta_1$ and $k_{2z} = \mathrm{i}\eta_2$ with $\eta_1$ and $\eta_2$ being positive, in order for the field $\mathrm{e}^{\mathrm{i}k_x x - \mathrm{i}k_{1z} z} = \mathrm{e}^{\mathrm{i}k_x x + \eta_1 z}$ to decay toward $z = -\infty$ and $\mathrm{e}^{\mathrm{i}k_x x + \mathrm{i}k_{2z} z} = \mathrm{e}^{\mathrm{i}k_x x - \eta_2 z}$ to decay toward $z = \infty$. This means that the sign of permittivity must be opposite for media 1 and 2 in order to couple a surface plasmon with a TM wave. On the other hand, we will need a magnetic material with negative permeability for a TE wave to excite a surface polariton. In fact, *surface magnon polaritons* have been demonstrated in the microwave region [62]. A magnon is the quantum for the collective motion of magnetic dipoles. As discussed in Chap. 8, negative index materials (NIMs) exhibit simultaneously negative permittivity and permeability in the same frequency region. They are sometimes called double-negative (DNG) materials. Therefore, both TE and TM waves may excite surface polaritons with a NIM, as predicted by Ruppin [63].

Comparing Eqs. (9.97) and (9.98) with Fresnel's reflection coefficients, Eqs. (8.68a) and (8.77a), respectively, it can be seen that the condition for the excitation of surface polaritons is that the denominator of the reflection coefficient be zero. A pole in the reflection coefficient is an indication of a resonance. Therefore, the excitation of a surface plasmon polariton gives rise to a *surface plasmon resonance* (SPR). Taking a TM wave for example, since $k_{1z}^2 = \mu_1\varepsilon_1\omega^2/c^2 - k_x^2$ and $k_{2z}^2 = \mu_2\varepsilon_2\omega^2/c^2 - k_x^2$, we can solve Eq. (9.97) to obtain

$$k_x = \frac{\omega}{c}\sqrt{\frac{\mu_1/\varepsilon_1 - \mu_2/\varepsilon_2}{1/\varepsilon_1^2 - 1/\varepsilon_2^2}} \tag{9.99}$$

Equation (9.99) relates the frequency with the parallel component of the wavevector and it may be considered as another form of the polariton dispersion relation or SPR condition. It should be noted that solutions of this equation are for both $k_{1z}/\varepsilon_1 + k_{2z}/\varepsilon_2 = 0$ and $k_{1z}/\varepsilon_1 - k_{2z}/\varepsilon_2 = 0$, i.e., not only the poles but also the zeros of the Fresnel reflection coefficient are included. For nonmagnetic materials, Eq. (9.99) becomes

$$k_{\rm SPP} = \frac{\omega}{c_0}\sqrt{\frac{\varepsilon_1\varepsilon_2}{\varepsilon_1 + \varepsilon_2}} \tag{9.100}$$

Here, $k_{\rm SPP}$ signifies a surface plasmon polariton resonance or excitation condition. It should be emphasized that it is the parallel component of the wavevector, since SPP propagates along the interface. One should bear in mind that the permittivities are in general functions of the frequency. For a metal with a negative real permittivity ($\omega < \omega_{\rm p}$), the normal component of the wavevector is purely imaginary for any real $k_x$ because $\mu\varepsilon\omega^2/c_0^2 < 0$. Thus, evanescent waves exist in metals regardless of the angle of incidence.

Consider either the Otto configuration or Kretschmann-Raether configuration, shown in Fig. 9.30. We can use the three-layer structure with a middle layer, medium 1, of thickness $d$. According to Eq. (9.7), the reflection coefficient can be expressed as follows:

$$r = \frac{r_{01} + r_{12}{\rm e}^{2{\rm i}\psi}}{1 + r_{01}r_{12}{\rm e}^{2{\rm i}\psi}} = \frac{r_{01} + r_{12}{\rm e}^{-2\eta_1 d}}{1 + r_{01}r_{12}{\rm e}^{-2\eta_1 d}} \tag{9.101}$$

where the subscript 0 signifies the incidence medium, which is the prism, and $\psi = k_{1z}d = {\rm i}\eta_1 d$ is the phase shift. When $d$ is sufficiently large, $\exp(-2\eta_1 d) \ll 1$, and the reflectance $R'_\lambda = rr^* \approx r_{01}r_{01}^*$ is close to unity. When surface polaritons are excited, however, $r_{12}$ increases dramatically and thus it is possible for $r_{12}{\rm e}^{-2\eta_1 d}$ to be of the same magnitude as $r_{01}$, but with opposite phases (i.e., a phase difference of $\pi$). At the condition of surface plasmon resonance, the reflectance $R'_\lambda$ drops suddenly. Let us use an example to illustrate the polariton dispersion curves and the effect on the reflectance in ATR arrangements.

**Example 9.7** Calculate the dispersion relation between Al and air. Calculate the reflectance versus angle of incidence for both the Otto and Kretschmann-Raether configurations at $\lambda = 500$ nm, using Al as the metallic material. Determine the *polariton propagation length* at the wavelength $\lambda = 500$ nm. Assume the prism is made of KBr with $\varepsilon_{\rm d} = 2.46$ and the dielectric function of Al can be described by the Drude model.

**Solution** The Drude model parameters for Al have been given in Example 8.7. Thus, we have $\varepsilon_2(\omega) = 1 - \omega_{\rm p}^2/(\omega^2 + {\rm i}\omega\gamma)$, where the plasma frequency $\omega_p = 2.4 \times 10^{16}$ rad/s and the scattering rate $\gamma = 1.4 \times 10^{14}$ rad/s. One way to calculate the dispersion relation is to assume $\omega$ is real and calculate $k_x(\omega) = k'_x(\omega) + {\rm i}k''_x(\omega)$. The dispersion curves between Al and air ($\varepsilon_{\rm air} = 1$) are usually plotted in an $\omega - k_x$

graph, for the real part of $k_x$ shown in Fig. 9.31a by the solid line. At very low frequencies, the magnitude of $\varepsilon_2$ is so large that $k_x \approx \omega/c_0$. Note that the dash-dotted line with $k_x = \omega/c_0$ represents the *light line*. On the left of this line, there exist propagating waves in air; whereas on the right of the light line, evanescent waves occur in air because $k_x > \omega/c$. The light line can be considered as a wave incident from air at 90° incidence. On the polariton dispersion curve, $k_x$ increases quickly as $\omega$ increases and reaches an asymptote at $\omega = \omega_\mathrm{p}/\sqrt{2}$, when the real part of the dielectric function of Al approaches to–1. Between $\omega_\mathrm{p}/\sqrt{2} < \omega < \omega_p$, the real part of the dielectric function of Al becomes negative with an absolute value less than 1. Therefore, the solution of Eq. (9.100) has a large imaginary part, while the real part of $k_x$ drops to near zero, as reflected by the bending of the dispersion curve toward left and the steep rise upward. Beyond $\omega > \omega_\mathrm{p}$, metal becomes transparent and the real part of the dielectric function becomes positive. Solutions beyond $\omega > \omega_p$ correspond to zeros in the reflection coefficient and thus are not the solutions for Eq. (9.97), which are poles of the reflection coefficient. Notice that the dotted line refers to the light line of the prism. In the shaded region, there exist evanescent waves in air and propagating waves in the prism; as a result, surface plasmons can be coupled to propagating waves in the prism.

The reflectance ($R'_\lambda = rr^*$) is calculated from Eq. (9.101) at the wavelength $\lambda = 500$ nm, corresponding to a wavenumber of 20,000 cm$^{-1}$. As can be seen from Fig. 9.31a, at this frequency, the surface polariton curve is very close to the light line in air. Therefore, the excitation of SPP is expected to be near the critical angle $\theta_\mathrm{c} \approx 39.6°$ between the prism and air. The reflectance would be close to 1 at $\theta > \theta_\mathrm{c}$. However, as shown in Fig. 9.31b, the reflectance drops suddenly around 40° due to SPR. Furthermore, the reflectance dips are very sensitive to the thickness of the middle layer. In the Otto configuration, the air thickness of 900 nm yields a sharp dip. For the Kretschmann-Raether configuration, on the other hand, a metallic film thickness of 24 nm yields a sharp dip in the reflectance. If the Al film exceeds 50 nm,

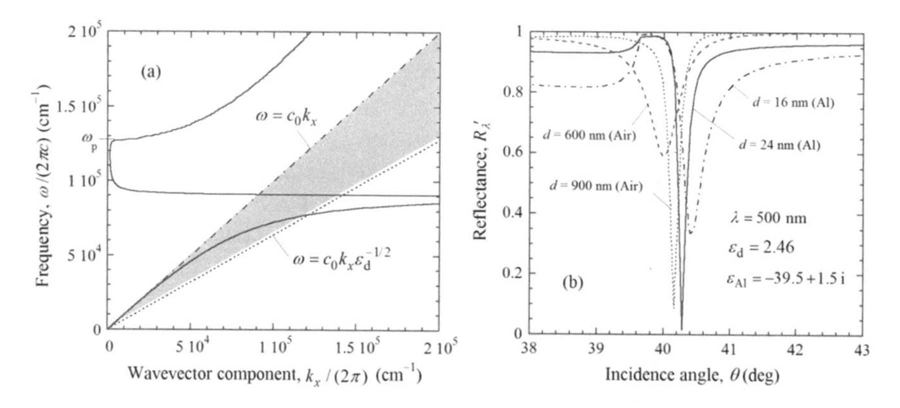


**Fig. 9.31** **a** The dispersion relation of surface plasmon polaritons between Al and air, where $k_x$ is the real part solution of Eq. (9.100). **b** Reflectance in ATR arrangements, either with Al or air as the middle layer

the reflectance is close to 1. Therefore, the location and width of the reflectance minimum depend on the thickness of the middle layer.

When a surface plasmon is excited, the reflectance minimum implies a strong absorption near the metal surface due to coupling of the electromagnetic energy to a surface wave. The propagation length of the surface electromagnetic wave or surface plasmon can be determined using the imaginary part of $k_x$. Note that the field can be expressed as $e^{ik'_x x - k''_x x}$ for surface waves propagating in the positive $x$-direction and as $e^{ik'_x x + k''_x x}$ for surface waves propagating in the negative $x$-direction. The power is proportional to the square of the field amplitude, and the $e^{-1}$ power decaying length or the *polariton propagation length* is given by [61].

$$l_{sp} = 1/(2k''_x) \tag{9.102}$$

Using Eq. (9.100) and the given dielectric functions, we obtain $l_{sp} \approx 80\ \mu m$. Note that the Drude model somewhat underpredicts the imaginary part of the dielectric function. If $\mathrm{Im}(\varepsilon)$ of Al is taken as 10 at $\lambda = 500$ nm, one obtains $l_{sp} \approx 13\ \mu m$, still much longer than the wavelength.

An alternative way to calculate the dispersion is to take the wavevector component $k_x$ as purely real while using a complex frequency $\tilde{\omega} = \omega' + i\omega''$ [61]. In this case, the imaginary part of the frequency represents a temporal decay. In practice, if only the resonance condition is desired, the dispersion solution is often approximated by dropping the imaginary part in the dielectric function, which gives a real function $\omega = \omega(k_x)$ as the dispersion relation. Caution should be taken when the imaginary part of the dielectric function is large. In such a case, a sharp resonance does occur, though the spectrum may contain an extremum close to $\omega = \omega(k_x)$.

It should be mentioned that SPPs or SPhPs can also be excited by using a grating. When light is incident onto a grating at a $k_x$ value that is less than $k_{SPP}$, the parallel wavevector component of reflected and refracted waves depends on the diffraction order according to the Bloch–Floquet condition: $k_{x,j} = k_x + 2\pi j/\Lambda$, where $j$ is the diffraction order and $\Lambda$ is the period of the gratings. For this reason, surface polaritons can be excited along a grating surface in the direction perpendicular to the grooves. To analyze the grating excitation of SPP, the dispersion relation can be folded into the region for $k_x \le \pi/\Lambda$ (first Brillouin zone). As an example, Fig. 9.27a shows the SPP dispersion relation for a binary grating made of Ag with $\Lambda = 1.7\ \mu m$ in the reduced zone scheme. The SPP dispersion curve is very close to the light line. The solid lines are the folded dispersion curves; the dash-dotted lines, which are also folded, correspond to an incidence angle of 30°. The intersections identify the location where surface plasmons can be excited for a TM wave incidence, when the magnetic field is parallel to the grooves.

The reflectance of a shallow grating on Ag is calculated and plotted in Fig. 9.32b at $\theta = 0°$ and 30°. The grating height $d = 100$ nm, and the filling ratio $\phi = 0.65$ (see Fig. 9.21 for the grating geometry). For a TE wave, no drops exist in the reflectance because surface waves cannot be excited. The reflectance is very high for TE waves and has little difference between $\theta = 0°$ and $\theta = 30°$. For a TM wave, the excitation

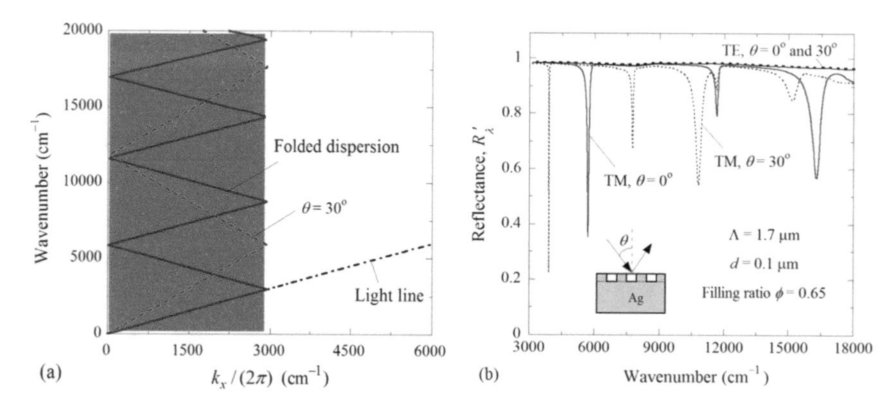


**Fig. 9.32** **a** Dispersion curves for gratings. **b** Reflectance for a Ag grating

of surface polaritons is responsible for the dips in the reflectance. Furthermore, the frequency locations agree well with those predicted by the dispersion curves. Note that at normal incidence, the excitation frequencies are located at the intersections between the dispersion curve and the vertical axis, as shown in Fig. 9.32a. These dips have also been known as Wood's or the Rayleigh-Wood anomalies, when a diffraction order just appears at the grazing angle [64]. The actual resonance frequency may shift slightly from the frequency associated with the appearance or disappearance of a diffraction order, because the dispersion curve is not a straight line [34, 35, 65]. The Rayleigh-Wood anomaly may also occur for gratings whose dielectric functions have a positive real part, i.e., not associated with surface plasmon polaritons. Surface roughness is yet another way to excite surface waves because a rough surface can be considered as a Fourier expansion of multiple periodic components, each acting as a grating. Plasmonic excitations have been used to enhance the performance of photovoltaic cells using corrugated metallic film or a layer of metallic nanoparticles [66].

It should be mentioned that many polar dielectric or semiconductor materials such as MgO, SiC, and GaAs contain a phonon absorption band, called the *reststrahlen band*, where Re($\varepsilon$) is negative and Im($\varepsilon$) is very small. The surface polariton condition described in Eq. (9.97) can be satisfied in the infrared to excite SPhPs [58–60]. In the following discussion of polaritons, the word "metal" is used to signify a material with a negative real permittivity or a negative-$\varepsilon$ material.

## 9.5.2 *Localized Surface Plasmon Resonance*

When surface plasmons are confined to small structures, such as the tip of a scanning microscopic probe, quantum dots or nanoparticles, nanowires, or nanoapertures, they are referred to as *localized plasmons*. The resonance behavior in nanoparticles or quantum dots has enormous applications in chemical sensing and medical diagnoses.

Gustav Mie in 1908 developed a formulation to describe scattering and absorption of small spherical particles. The exact solution is expressed in an infinite series of spherical multipole partial waves [67]. When the radius $r_0$ is much smaller than the wavelength, the *scattering efficiency factor* and the *absorption efficiency factor* of the sphere can be expressed as [1]

$$Q_{\text{sca},\lambda} = \frac{8}{3}\xi^4 \left| \frac{\varepsilon_2 - \varepsilon_1}{\varepsilon_2 + 2\varepsilon_1} \right|^2 \tag{9.103}$$

and

$$Q_{\text{abs},\lambda} = 4\xi \,\text{Im}\left( \frac{\varepsilon_2 - \varepsilon_1}{\varepsilon_2 + 2\varepsilon_1} \right) \tag{9.104}$$

where $\xi = 2\pi r_0/\lambda_1$ is called the size parameter with $\lambda_1$ being the wavelength in the surrounding dielectric medium (matrix) whose dielectric function is $\varepsilon_1$; while $\varepsilon_2$ is the dielectric function of the particle [67]. It should be noted that the efficiency factors may exceed one; this is why the word "factor" is appended after efficiency. Nevertheless, the word factor is often dropped in the literature.

Equations (9.103) and (9.104) are the dipole approximation, also called quasi-static or *electrostatic limit*, of the exact solutions of Maxwell's equations for light scattering by a sphere. While Eq. (9.103) has the same form as the expression of Rayleigh scattering with the $1/\lambda^4$ relationship of the *scattering cross section*, defined as $C_{\text{sca}} = \pi r_0^2 Q_{\text{sca},\lambda}$, the scattering of metallic spheres is distinctly different from that of dielectric spheres because the dielectric function of metals is complex and depends strongly on the wavelength. The scattering cross section is usually a very complex function of the wavelength. This is especially true when the resonance condition $\varepsilon_2 = -2\varepsilon_1$ is satisfied, resulting in a *localized surface plasmon resonance* (LSPR) whose quanta are called *localized SPPs*. Geometric optics completely failed to describe scattering and absorption of small particles. The scattering cross section can be much greater than the actual surface area, especially at the LSPR. Furthermore, the absorbed energy can exceed that of a *blackbody* of the same size [67]. In fact, the blackbody concept is misleading in the subwavelength regime. The actual resonance condition may be complicated for different geometries and coatings, as well as for clusters of particles or nanoparticle aggregates. Numerical methods are frequently used to calculate LSPR in complicated structures such as discrete dipole approximation (DDA), T-matrix method, finite-element method (FEM), boundary element method (BEM), and finite-difference time-domain (FDTD) techniques.

Nanoplasmonics has become one of the most active research fields in the past two decades [68–70]. The use of metal nanoparticles dispersed in a dielectric matrix to create lusterware and stained glass has been known since the 4th century A.D. (e.g., the Lycurgus Cup). Michael Faraday was the first to perform detailed scientific experiments to study the effect of metal particle type and size on the color in the 1850s. For finite size particles, the scattering and absorption efficiencies of spherical particles or shells are expressed as the Mie solution that is a summation of contributions of electric and magnetic dipoles and multipoles [67]. This results in a redshift of LSPR spectrum, and resonance wavelength becomes longer and the absorption peak becomes broader as the particle size increases. LSPR is also very sensitive to the refractive index of the surrounding medium as well as the shape and arrangement of the nanoparticles [71–78]. Resonance phenomena in small particles have been applied to surface-enhanced Raman scattering microscopy and surface-enhanced fluorescence microscopy for single-molecule detection [79, 80]. The study of resonance phenomena in small particles continues to be an active research area because of the applications in biological imaging and molecular sensing [74–80] as well as photothermal therapy [81, 82]. Surface wave scattering and coupling of a cluster or layer of nanoparticles near a substrate surface have also been theoretically and experimentally investigated and may enable tools for characterization of nanoparticle clusters [83, 84].

Plasmonic core-shell nanoparticles have been studied for solar energy harvesting in a fluid because of their broadening absorption band in the visible and near-infrared spectrum with high absorption and extinction efficiencies [85–88]. These nanoparticles have also been applied for local heating, vapor generation, and steam production [89–91]. Manipulating and probing local plasmonic heating have also been demonstrated [92, 93]. Lee et al. [94] theoretically modeled the solar absorptance of a nanofluid containing nanoparticles with different Au nanoshell thicknesses over $SiO_2$ core of different sizes. The nanofluid with blended nanoparticles can significantly enhance solar collector efficiency to 70% with a layer thickness of 1.5 mm at a particle volume fraction as low as 0.05% [94]. Xuan et al. [95] experimentally demonstrated plasmonic nanofluid with $TiO_2$/Ag core-shell nanoparticles (with an outer radius of 30 nm) for solar absorption enhancement. Wang et al. [96] modeled carbon-Au core-shell structures to enhance the solar absorption efficiency for single nanoparticle by taking advantage of the absorptive properties of carbon core. Their results showed an enhancement of solar absorption efficiency that can exceed blackbody limit for the same geometric cross section by a factor of 1.5 with spherical nanoparticles (with an external diameter around 100 nm) and 2.3 with cubic nanoparticles. Further enhancement is also possible with star-shape nanoparticles [97].

For isolated nanoparticles, when LSPR occur, the field decays away from the interface. However, when nanoparticles are placed in close proximity, evanescent waves can be coupled. Furthermore, in a linear chain of nanoparticle, the collective oscillations of the electron clouds inside the nanoparticles can give rise to a propagating wave, forming a 1D waveguide like in a nanowire [98–100]. Nanoparticle plasmonic waveguides can transfer electromagnetic energy beyond the diffraction limit and can enable detection of manipulation at extreme length scale, with potential applications from single-photon light source to photonic integrated circuit [101–103]. Coupled point-dipole approximation is typically used to model the plasmonic interaction and plasmon wave propagation along the nanoparticle chains. When considering the effects of retardation and losses, the dispersion for transverse modes splits into two anticrossing branches. The propagation length, bandwidth, and group velocity of plasmonic waveguide have been extensively studied for both finite and infinite nanoparticle chains, with and without disorder [104–107]. The absorption and scattering in the visible and near-infrared regions for 2D and 3D plasmonic nanoparticle arrays have been modeled [108, 109]. The effects of propagating SPP and SPhP waves and near-field radiative transfer between nanoparticles on the thermal transport in a nanoparticle chain or array have also been investigated [110–113]. Further discussion on near-field thermal radiation will be given in Chap. 10.

### 9.5.3 Polaritons in Thin Films and Layered Structures

Polaritons can exist on both surfaces of a thin film, resulting in a standing wave inside the film, as shown in Fig. 9.33. Economou [114] performed a detailed investigation of different configurations of a thin-film structure. An essential requirement for coupled surface polaritons to occur is the existence of evanescent waves that decay in both media 1 and 3. Such a method was used in Sect. 9.2.4 for obtaining the mode equation for waveguides. A more convenient method to derive the polariton relations is to set the denominator of the reflection coefficient to zero. For the configuration shown in Fig. 9.33, similar to Eqs. (9.7) or (9.101), we can write $r = \frac{r_{12}+r_{23}e^{2ik_{2z}d}}{1+r_{12}r_{23}e^{2ik_{2z}d}}$, whose poles are at $1 + r_{12}r_{23}e^{2ik_{2z}d} = 0$. A further extension using the Fresnel coefficients for TM waves gives,

$$\tanh(ik_{2z}d)\left(\frac{k_{2z}^2}{\varepsilon_2^2} + \frac{k_{1z}k_{3z}}{\varepsilon_1\varepsilon_3}\right) = \frac{k_{2z}}{\varepsilon_2}\left(\frac{k_{1z}}{\varepsilon_1} + \frac{k_{3z}}{\varepsilon_3}\right) \tag{9.105}$$

which is the polariton dispersion relation for a slab sandwiched between two semi-infinite media. Because $\tanh(ik_{2z}d) = i\tan(k_{2z}d)$, when medium 3 is identical to medium 1, Eq. (9.105) is identical to the mode equation of a planar waveguide given in Eq. (9.52). Attention should also be paid to the different meanings of the subscripts in Eqs. (9.52) and (9.105). For the coupled surface polariton, however, $k_{2z}$ is purely imaginary if loss is neglected. In the case of $\varepsilon_3 = \varepsilon_1$ and $\mu_3 = \mu_1$, Eq. (9.105) can be rewritten into two equations [115]:

$$\frac{k_{1z}}{\varepsilon_1} + \frac{k_{2z}}{\varepsilon_2}\tanh\left(\frac{k_{2z}d}{2i}\right) = 0 \tag{9.106a}$$

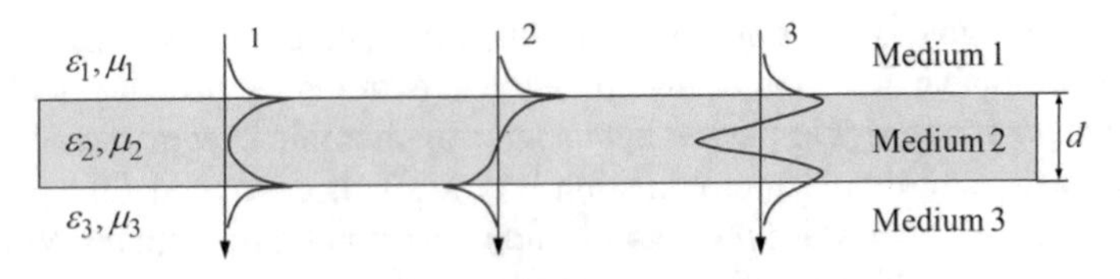


**Fig. 9.33** Illustration of polaritons in a slab. 1—symmetric mode coupled surface polaritons; 2—antisymmetric mode coupled surface polaritons; and 3—bulk polariton

$$\frac{k_{1z}}{\varepsilon_1} + \frac{k_{2z}}{\varepsilon_2}\coth\left(\frac{k_{2z}d}{2\mathrm{i}}\right) = 0 \tag{9.106b}$$

Each of them gives a dispersion curve, and the field distribution can be illustrated in Fig. 9.33 for case 1: a lower frequency symmetric mode, where the surface charges are symmetric and the magnetic fields at the interfaces are in phase, and case 2: a higher frequency antisymmetric mode, where the surface charges are asymmetric with respect to the middle plane and the magnetic fields at the interfaces are out of phase. Due to the coupling of surface waves (which are two evanescent waves decaying toward opposite directions), the field inside medium 2 resembles a standing wave. It should also be noted that when $d \to \infty$, both Eqs. (9.106a) and (9.106b) reduce to the surface polariton equation between two semi-infinite media.

The preceding discussion is also applicable to TE waves. The only change is to replace $\varepsilon$'s by $\mu$'s in the Fresnel reflection coefficients and hence the dispersion relations. If medium 2 is a metal with a negative real permittivity ($\varepsilon_2 < 0$) and media 1 and 3 are dielectric, evanescent waves must exist in the dielectric and coupled surface polaritons can interact only with TM waves for $k_x > \max\left(\sqrt{\varepsilon_1}\omega/c_0, \sqrt{\varepsilon_3}\omega/c_0\right)$. If medium 2 is a NIM, i.e., $\varepsilon_2 < 0$, $\mu_2 < 0$, in the frequency of interest, both TE and TM waves can excite coupled surface polaritons.

If a dielectric is placed as medium 2 between two metallic media, 1 and 3, resonance is also possible, even though $k_{2z}$ is real, since $k_{1z}$ and $k_{3z}$ are imaginary in media 1 and 3. A standing wave can be formed in medium 2, which is a waveguide mode discussed previously. Such an excitation is identified as a bulk polariton or a guided mode polariton [63, 114–117]. The field distribution for a bulk polariton is illustrated in Fig. 9.33 as case 3. Both TE and TM waves can excite bulk polaritons when $k_x > k_0$. When $k_x \le k_0$, propagating waves can exist in vacuum and the structure is essentially a Fabry–Perot resonance cavity [16]. Furthermore, more polariton modes may exist if the thickness $d$ is large enough [115, 117]. Each polariton dispersion line corresponds to an order of the waveguide modes as discussed in Sect. 9.2.4. As in dielectric waveguides, the metal cladding can be replaced by a dielectric material of smaller refractive index.

Park et al. [115] identified a polariton regime map for a NIM slab sandwiched between two different dielectrics, one of which is a vacuum, as shown in Fig. 9.34a. The NIM is represented by the permittivity and permeability functions given in Eqs. (8.135) and (8.136). For developing the dispersion curves, the damping terms is typically assumed to be zero; therefore,

$$\varepsilon_2(\omega) = 1 - \frac{\omega_\mathrm{p}^2}{\omega^2} \text{ and } \mu_2(\omega) = 1 - \frac{F\omega^2}{\omega^2 - \omega_0^2} \tag{9.107}$$

Figure 9.34b represents the regimes with $F = 0.56$ and $\omega_0 = 0.4\omega_\mathrm{p}$ shown in the $\omega - k_x$ graph, where both $k_x$ and $\omega$ are normalized with respect to $\omega_\mathrm{p}$. Note that no polaritons can be excited for $\omega > \omega_\mathrm{p}$ because both $\varepsilon_2$ and $\mu_2$ are positive. In the shaded region for $0.4\omega_\mathrm{p} < \omega < 0.6\omega_\mathrm{p}$, both $\varepsilon_2$ and $\mu_2$ are negative, and this entails

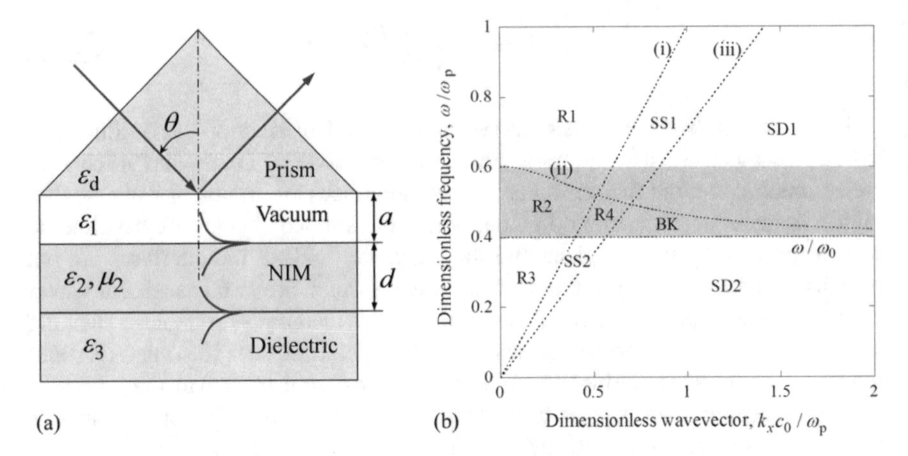


**Fig. 9.34** Illustration of polaritons in a NIM slab. **a** ATR arrangement. **b** Regimes of surface and bulk polaritons [115]

a NIM region. Four dotted lines (i), (ii), (iii), and $\omega = \omega_0$ separate nine different regions. Lines (i), (ii), and (iii) correspond to $\omega = k_x c_0 \sqrt{\varepsilon\mu}$ for media 1, 2, and 3, respectively. If the two dielectrics are identical, lines (i) and (iii) will merge and the regions in between will be eliminated. Notice that the condition for $\omega = k_x c_0 \sqrt{\varepsilon\mu}$ corresponds to $k_z = 0$ in any given medium. In the regions on the left of line (i), $k_x$ is too small to excite any evanescent waves in media 1 and 3; hence, no polaritons can exist in regions R1, R2, and R3. In regions between lines (i) and (iii), an evanescent wave appears in medium 1 whilst a propagating wave exists in medium 3. A surface polariton may exist only at the interface between media 1 and 2, and energy may be transmitted from the prism into medium 3 through *photon tunneling* which will be further discussed in Chap. 10. In regions on the right of line (iii), evanescent waves emerge in both media 1 and 3; hence, surface polaritons may exist at dual boundaries, and several bulk polaritons may also exist. Unlike plasmon polariton or phonon polariton discussed in the beginning of Sect. 9.5, here, the bulk polariton dispersion is expressed in terms of $\omega - k_x$ relations and may be considered as a planar bulk polariton.

In the upper regions of line (ii), evanescent waves exist in the NIM layer. In the shaded area, surface polaritons may be observed in region SS1 at a single boundary and in region SD1 at dual boundaries of the NIM slab, for both polarizations. Surface polaritons may also exist in regions SS1 and SD1 above the shaded area only for TM waves. On the other hand, in regions between the lines $\omega = \omega_0$ and (ii), propagating waves exist in the NIM layer because $k_{2z} > 0$. Therefore, no polaritons may exist in region R4, whereas bulk polaritons can occur in region BK. Below the line $\omega = \omega_0$, medium 2 behaves like a normal metal because $\varepsilon_2 < 0$ and $\mu_2 > 0$ surface polaritons may occur only for TM waves at a single boundary in region SS2 and both boundaries in region SD2.

The reflection coefficient for the four-layer structure shown in Fig. 9.35a can be expressed as follows:

$$r = \frac{Y_{01}X_{12}X_{23}e^{-i\phi_1} + X_{01}X_{12}Y_{23}e^{i\phi_1} + Y_{01}Y_{12}Y_{23}e^{-i\phi_2} + X_{01}Y_{12}X_{23}e^{i\phi_2}}{X_{01}X_{12}X_{23}e^{-i\phi_1} + Y_{01}X_{12}Y_{23}e^{i\phi_1} + X_{01}Y_{12}Y_{23}e^{-i\phi_2} + Y_{01}Y_{12}X_{23}e^{i\phi_2}} \quad (9.108)$$

where $X_{ij} = 1 + \frac{k_{jz}\varepsilon_i}{k_{iz}\varepsilon_j}$ and $Y_{ij} = 1 - \frac{k_{jz}\varepsilon_i}{k_{iz}\varepsilon_j}$ for TM waves, $X_{ij} = 1 + \frac{k_{jz}\mu_i}{k_{iz}\mu_j}$ and $Y_{ij} = 1 - \frac{k_{jz}\mu_i}{k_{iz}\mu_j}$ for TE waves, and $\phi_1 = k_{1z}a + k_{2z}d$ and $\phi_2 = k_{1z}a - k_{2z}d$ are the phase terms. This analytical expression may be used as an alternate to the matrix formulation for the calculation of $r$, and the reflectance $R = rr^*$ as well, of the four-layer structure.

Figure 9.35 shows the calculated reflectance spectra for different NIM layer thicknesses, normalized to $\lambda_p = 2\pi c_0/\omega_p$ , for both TM wave (solid curves) and TE

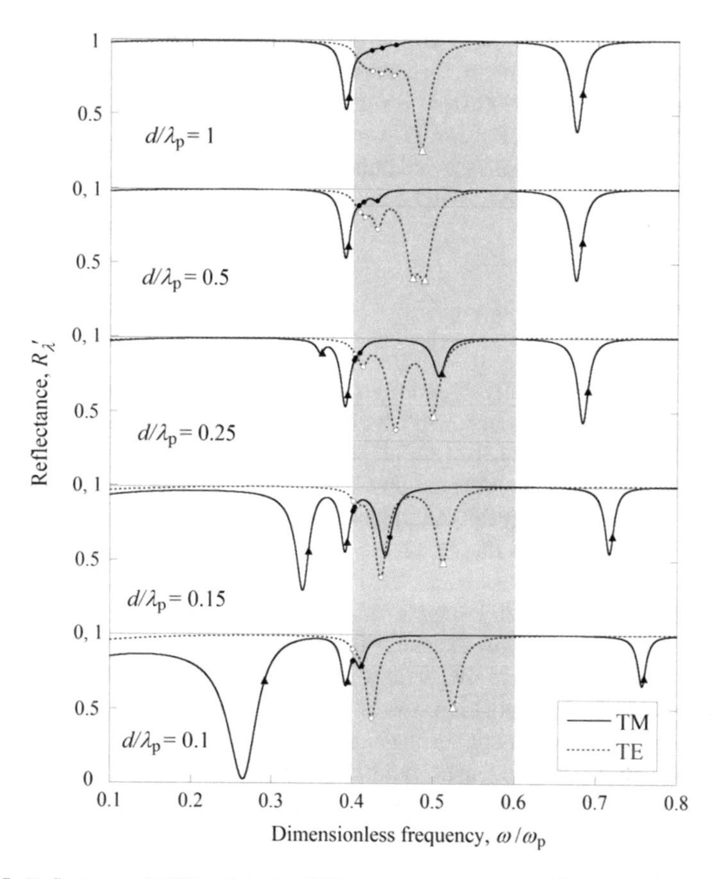


**Fig. 9.35** Reflectance of NIM slab in the ATR arrangement shown in Fig. 9.34a for both TM and TE waves at $\theta = 60°$ [115]

wave (dotted curves). The permittivity and the permeability of the NIM are modeled with $F = 0.56$, $\omega_0 = 0.4\omega_p$, and damping coefficients $\gamma_e = \gamma_m = 0.012\omega_p$ using Eqs. (8.135) and (8.136). The thickness of the vacuum layer is assumed to be $a = 0.25\lambda_p$. For the prism, $\varepsilon_d = 6$, and for the dielectric, $\varepsilon_3 = 2$. The incidence angle is kept at $\theta = 60°$ so that only evanescent waves exist in medium 3. The corresponding regions are SD1, BK, and SD2 in Fig. 9.34b. The shaded region in Fig. 9.35 corresponds to the frequency region where the refractive index is negative.

Several dips, due to surface and bulk polaritons, can be clearly seen in the reflectance spectra. Triangular and circular marks (filled for TM wave and unfilled for TE wave) represent surface and bulk polariton resonance frequencies that are obtained from the polariton dispersion relations in the lossless case. While damping terms affect the width of the dips, it is the vacuum gap distance $a$ that affects the location of the reflectance dips strongly. For $d/\lambda_p = 0.5$ and TE waves, there are three bulk polaritons in $0.4 < \omega/\omega_p < 0.45$ and two surface polariton curves in $0.45 < \omega/\omega_p < 0.5$. When the NIM layer thickness is reduced, the surface polariton of the lower frequency, in the pass band, is converted into a bulk polariton, while the other bulk polaritons are compressed to the vicinity of $\omega_0$ and have little effect on the reflectance. The transition from a surface polariton to a bulk polariton occurs at $d/\lambda_p$ between 0.25 and 0.5 for TE waves, and between 0.15 and 0.25 for TM waves. It is clear that both surface and bulk polaritons affect the radiative properties significantly. Further examples will be provided in Sect. 9.6.

### 9.5.4 Magnetic Polariton

Metamaterials are artificially ordered structures that display physical properties rarely observed in naturally occurring materials. Electromagnetic metamaterials are synthesized or unconventional materials with exotic electric and magnetic properties, such as a negative index material as discussed before. In the optical and even microwave frequencies, very few natural materials have a magnetic response and their permeability is the same as that of vacuum (relative permeability $\mu = 1$). Conventional magnetic materials have either ferromagnetic or antiferromagnetic resonances usually in the radio frequency region [62, 118]. Pendry et al. [119] first proposed to use metallic subwavelength structures to enable magnetic resonance and theoretically calculated the effective magnetic permeability for the split-ring resonators as given in Eq. (8.136). The split-ring structure is shown in Fig. 9.36a, along with other structures that exhibit magnetic resonances.

When a time-varying magnetic field is parallel to the axis of a spiral coil of metal wire, an induced magnetic field will occur due to the resultant current in the coil according to Lenz's law. Diamagnetism can also occur with the split-ring and other resonators shown in Fig. 9.36. In order to scale the diamagnetic response to the near infrared, the single split-ring and U-shape cells have also been employed. With these artificial structures, sometimes called *magnetic atoms*, electromagnetic waves can interact with materials via both the electric and magnetic fields [120].

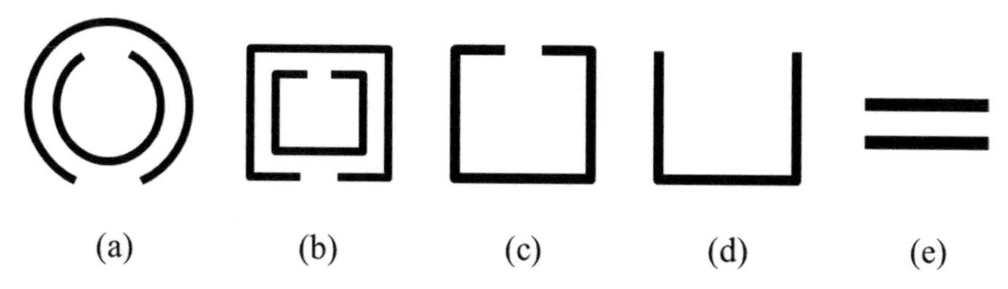


**Fig. 9.36** Structures made of conducting split ring(s) of circular and straight lines, U shape, and paired wires/strips that can result in diamagnetism, when the magnetic field is normal to the plane of these structures. These unit cells are sometimes called magnetic atoms

In fact, diamagnetic response can occur with a pair of conducting wires due to the antiparallel currents induced by a time-varying magnetic field perpendicular to the plane of the wires. Metamaterials have been demonstrated in the near infrared and for red light with paired metal strips. Another successful variation is the so-called fishnet double-layer structure, which has been used to produce NIM in the infrared region.

Since the effective magnetic permeability can be negative in certain frequency region, surface polaritons can be excited with these artificial magnetic materials in the optical frequency region. These resonance excitations are frequently called magnetic plasmon polaritons or *magnetic polaritons* (MPs), though diamagnetism behavior is very different from the quantification of electron spin waves (i.e., magnons). Since subwavelength metallic structure is needed, the magnetic polariton is undoubtedly related to plasmonic resonances. The excitation of MPs can enable tailoring of spectral radiative properties, and the dispersion curves are quite different from SPP/SPhP. The resonance frequency of the fundamental mode (first order) MP can be predicted using an equivalent electric circuit model, which usually contains capacitive and inductive elements for a unit cell of the structure. Some basic background is given next.

Consider two parallel metal strips separated by a dielectric gap as shown in Fig. 9.37a. When the fundamental mode of MP is excited, there exists an oscillating electric current around the structure that produces a transverse oscillating magnetic field. The charge distributions near the metal surface are illustrated in Fig. 9.37a, which results in a current flow along the effective capacitor-inductor circuit (neglecting resistive element), as shown in Fig. 9.37b. Assume the metal strip length is $h$ and depth in the direction perpendicular to the paper is $l$. The skin thickness $\delta$ is usually determined by the field penetration depth. Consider a dielectric gap width $b$ with a dielectric constant $\varepsilon_\text{d}$, which may be set to 1 for vacuum or air. The gap capacitance is split into two in the $LC$ circuit model and expressed as

$$C = c_1 \frac{\varepsilon_0 \varepsilon_\text{d} hl}{2b} \tag{9.109}$$

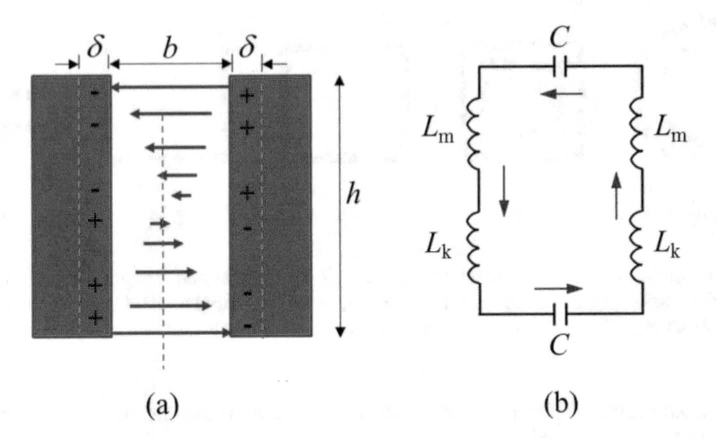


**Fig. 9.37** **a** Schematic of the charge distribution at the surface of two parallel conducting plates and the electric field in the dielectric gap; **b** the equivalent *LC* circuit model for magnetic resonance, where the arrows indicating electric current flow loop which produces a magnetic field perpendicular to the paper

where $c_1$ is a numerical factor near 0.5 that accounts for the nonuniform charge distribution at the metal surfaces [121]. The mutual inductance between the two plates can be written as

$$L_{\mathrm{m}} = \mu_0 \frac{hb}{2l} \tag{9.110}$$

while the kinetic inductance due to drift electrons can be expressed as

$$L_{\mathrm{k}} = -\frac{\varepsilon'}{\omega^2 |\varepsilon|^2} \frac{h}{\varepsilon_0 \delta l} \approx \frac{h}{\varepsilon_0 \omega_{\mathrm{p}}^2 \delta l} \tag{9.111}$$

Here, $\varepsilon'$ is the real part of the dielectric function $\varepsilon$ of the plate material. If the metal plate is thinner than the field penetration depth, $\delta$ should be taken as the thickness of the plate. In writing the last expression, the dielectric function has been approximated as $\varepsilon(\omega) \approx -\omega_{\mathrm{p}}^2/\omega^2$. The total impedance can be expressed as

$$Z_{\mathrm{tot}}(\omega) = 2\mathrm{i}\left[\omega(L_{\mathrm{m}} + L_{\mathrm{k}}) - (\omega C)^{-1}\right] \tag{9.112}$$

The resonance frequency can be obtained by setting $Z_{\mathrm{tot}} = 0$; hence,

$$\omega_{\mathrm{R}} = [(L_{\mathrm{m}} + L_{\mathrm{k}})C]^{-1/2} \tag{9.113}$$

The actual *LC* circuit may be more complicated due to additional coupling between the unit cells and can become rather complicated for complex structures

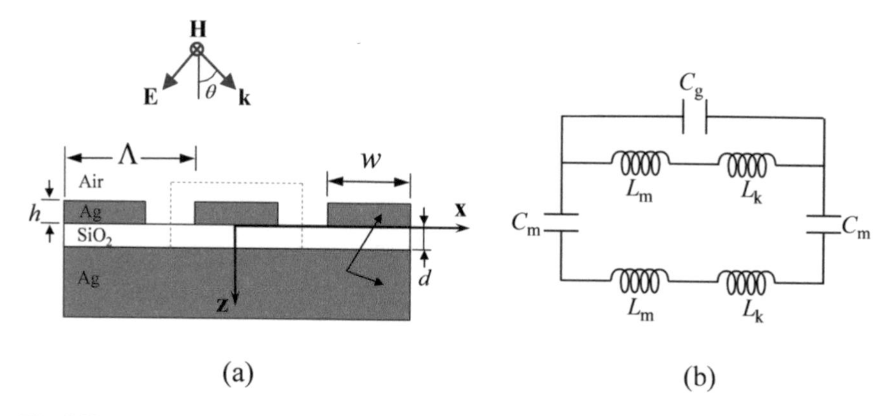


**Fig. 9.38** **a** Schematic of a 1D periodic structure when the metal gratings are separated from the base plane by a dielectric layer; **b** the equivalent circuit showing capacitors in the dielectric as well as gap capacitances of a unit cell as illustrated on the left panel

[121, 122]. In the following, we use an example to illustrate the MP dispersion, field distribution, and the effect on the reflectance of a grating structure.

Consider the structure depicted in Fig. 9.38a made of periodic silver strips with a dielectric spacer deposited on a silver film, which may be assumed opaque. Without the spacer, it is simply a binary Ag grating. The geometry of the one-dimensional grating geometry is represented by a period $\Lambda$, strip width $w$, and thickness $h$. The thickness of dielectric spacer is denoted by $d$. A linearly polarized TM wave is incident from air at an incidence angle $\theta$. The oscillating magnetic field in the $y$-direction between the metal strip and the base plane can cause antiparallel currents in the metal strip and the metal surface.

Using RCWA simulation with 101 Fourier components and the frequency-dependent dielectric functions of Ag and $SiO_2$ from Palik [2], The reflectance is calculated and shown in Fig. 9.39 in the wavenumber range from 3,000 to 20,000 $cm^{-1}$ at $\theta = 25°$. The grating parameters are $\Lambda = 500\,nm$, $w = 250$ nm (filling ratio of 0.5), and $h = d = 20$ nm. For the sake of comparison, the reflectance of a simple grating is shown as the dotted curve, by setting $d = 0$ without changing the other parameters. It is well known that gratings can support SPP according to Eq. (9.100) and can result in sharp reflectance dip at $\bar{\nu} = 13{,}780\,cm^{-1}$. On the other hand, if a 20 nm $SiO_2$ spacer is added, several additional reflectance dips also show up.

These reflectance dips are attributed to the excitation of magnetic resonances. The diamagnetic response is then coupled to the metallic film to cause a surface magnetic polariton with a fundamental mode at the wave number around 5,670 $cm^{-1}$. Magnetic polaritons of the second and higher order harmonics can also be excited. Therefore, the reflectance spectrum with spacer exhibits dips at $\bar{\nu} = 5,670$, 11,490, and 16,095 $cm^{-1}$, corresponding, respectively, to the fundamental, second, and third harmonic modes. Unlike SPP whose resonance frequency is very sensitive to the incidence angle, MPs resonance frequency depends on weakly on the incidence

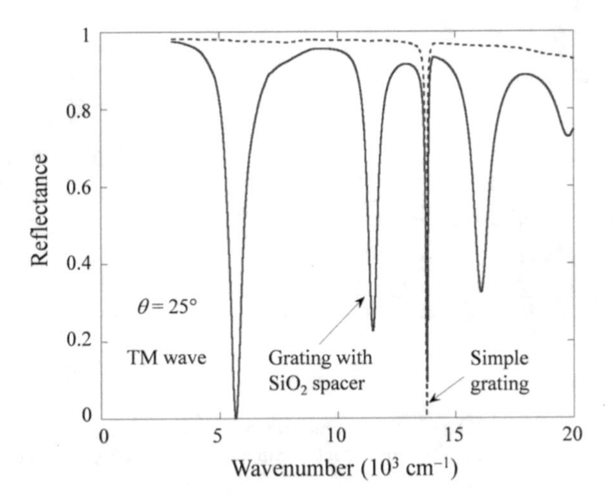


**Fig. 9.39** Calculated reflectance of the grating structure shown in Fig. 9.38 with and without the dielectric spacer using RCWA [123]

angle up to a certain value. Furthermore, SPP is more sensitive to the grating period, while MPs are more sensitive to the strip width [123].

Since the metal film forming the base plane is opaque, one can regard the reflectance dip as the emittance peak according to Kirchhoff's law. The contour plots of the spectral-directional emittance $\varepsilon_{\nu,\theta}$ are shown in Fig. 9.40a for the simple grating and Fig. 9.40a for the grating with spacer by folding the dispersion curves. Darker colors represent lower emittance, whereas brighter colors correspond to higher emittance (reflectance dip). The region outside the light line on the lower right corner is left blank. The emissivity is greatly enhanced when surface plasmons are excited.

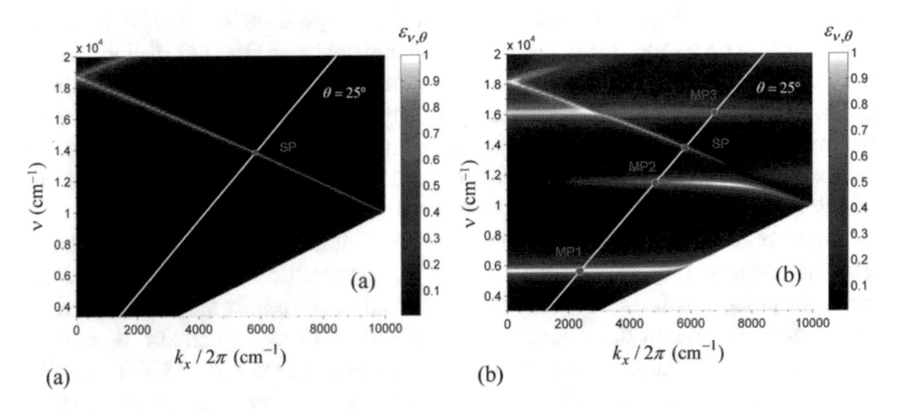


**Fig. 9.40** Contour plot of the spectral-directional emittance of **a** the simple grating and **b** the Ag grating and film separated by a $SiO_2$ spacer [123]. At an incidence angle $\theta = 25°$, the surface plasmon polariton is labeled as SP, while the magnetic polaritons are labeled as MP1, MP2, and MP3 for the fundamental, second, and third harmonic modes, respectively

The branch at $\bar{\nu} < 18{,}200\,\text{cm}^{-1}$ corresponds polaritons coupled with the –1 diffraction order and the high-frequency branch is associated with the +1 diffraction order. The intersection of the surface plasmon dispersion line and the inclined white line, representing $\theta = 25°$, is marked as SP that corresponds to the reflectance dip shown in Fig. 9.39. In general, the resonance condition of SPP depends strongly on both frequency and the incidence angle; thus, the emittance peak exhibits both spectral and directional selectivity.

It can be seen from Fig. 9.40b that the excitation of MPs gives several nearly horizontal bright dispersion bands. The even-order magnetic polaritons (such as the second harmonic mode) can only be excited at oblique incidence, whereas the odd-order magnetic polaritons can be excited at normal incidence. Furthermore, SPPs can strongly interact with MPs to give rise to crossing and anticrossing mode coupling [124]. As a result, the interaction of SPP and MP can result in either enhancement or suppression of the emittance. The magnetic polaritons are localized in the vicinity of metal strips and are not significantly coupled with each other. Hence, the emittance peak resulted from MP exhibits diffuse characteristic (independent of the direction) that is desirable for thermophotovoltaic emitters [125, 126].

The physical mechanism of the magnetic resonance can be better understood by plotting the magnetic field distribution and electric current flow, as shown in Fig. 9.41 for $\theta = 25°$ at the MP1, MP2, and MP3 resonance frequency. Here, the $z$-axis is pointed upwards so that the Ag strips appear to be below the Ag film. The background contour represents the logarithmic values of the square of the magnetic field magnitude, and the arrows indicate the electric field vectors. Antiparallel currents in the metallic strips and film confine strong magnetic field inside the dielectric spacer. The considered structure acts similarly to the metal strip pairs regarding the magnetic field distribution. The number of resonances corresponds well with the number of order of MPs. The corresponding electric field distribution further confirms the magnetic induction around the antinodes of the magnetic field distribution. Hence, the effective permeability of the considered structure exhibits a resonance like dispersion according to the electric and magnetic fields distribution in the dielectric spacer. Although not shown here, SPPs generate enhanced magnetic field very close to the interface between the dielectric and metal film, as well as along the interface between metal strips and air.

MPs have been used to explain resonance transmittance in a periodic strip array, double strip gratings, as well as patterned metallic structures over a metal ground plane separated by a thin dielectric layer [121–130]. The fundamental modes or microcavity resonance in deep gratings can also be accurately modeled using the LC circuit and this has been demonstrated recently [131, 132]. Resistive elements can also be introduced [133]. Furthermore, the metal materials may be replaced by polar dielectric and in the region where $\varepsilon' < 0$, MPs can be excited due to coupled SPhPs [134]. Due to the complexity involved and different point of view, the magnetic resonances or polaritons discussed here are often identified by other names such as LSPR, metamaterial, gap plasmons, etc. for controlling radiative properties [135–138]. Gap plasmon (or gap surface plasmon) refers to the plasmonic resonance occurring in a structure where a dielectric is sandwiched between an array

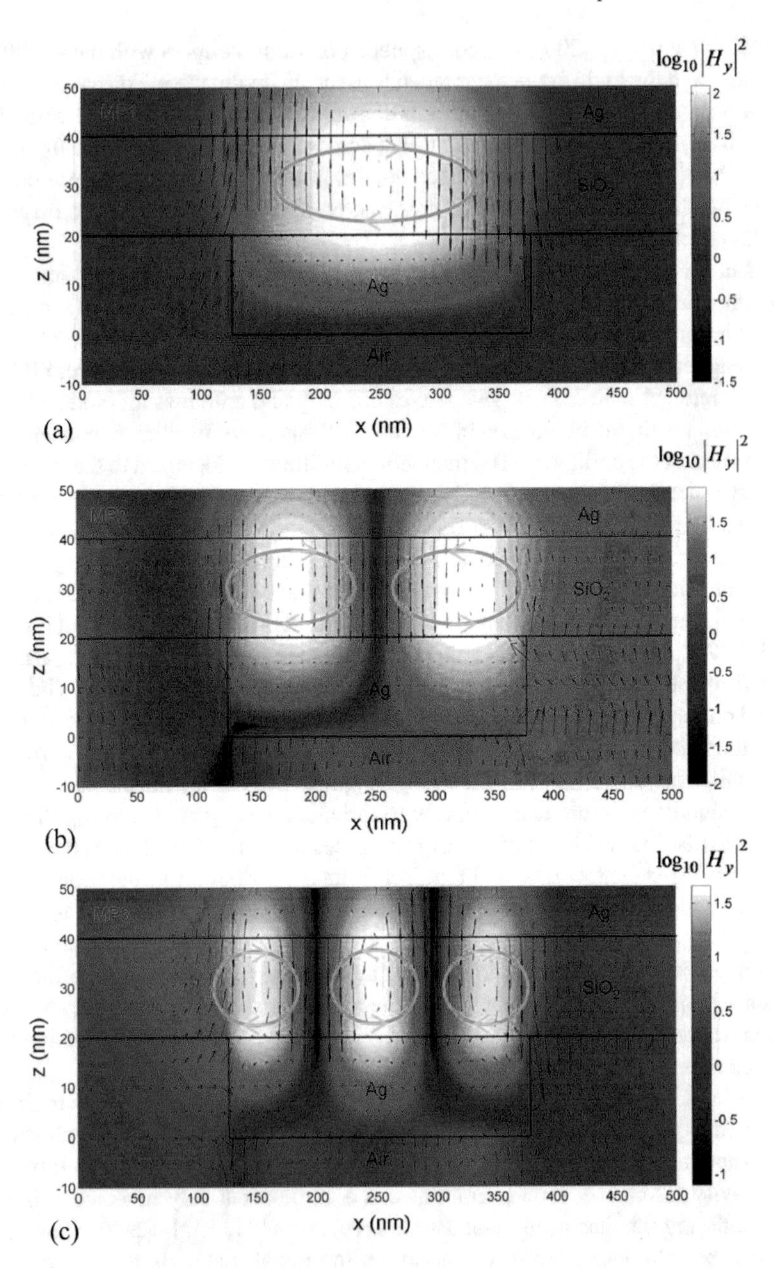


**Fig. 9.41** Contour shows the magnitude strength and the arrows illustrate electric field when MPs are excited at $\theta = 25°$ [123]. The loops indicate electric current flow. **a** MP1: $\bar{\nu} = 5,670\,\text{cm}^{-1}$; **b** MP2: $\bar{\nu} = 11,490\,\text{cm}^{-1}$; **c** MP3: $\bar{\nu} = 16,095\,\text{cm}^{-1}$

of subwavelength metal elements and a metal ground plane [137], or between two uniform metal films [125, 138]. It should be noted that the electric field distribution does not always form a close loop when MPs are excited. However, when both the conductive current and displacement current are considered, the electric current always forms a closed loop [126]. The local energy density and dissipation can also be calculated; hence, the local absorption distribution can be determined [139]. The application of MPs and MP coupled with other resonances will be given in forthcoming sections.

### *9.5.5 Graphene: Optical Properties and Graphene Plasmon*

Since the discovery of a simple method for isolating graphene through the exfoliation of graphite in 2004, which leads to the 2010 Nobel Prize in Physics to Andre Geim and Konstantin Novoselov, graphene has been extensively studied for potential applications in nanoelectronics, optoelectronics, plasmonics, transformation optics, and energy conversion. Nowadays, high-quality graphene and few-layer graphene can be fabricated by chemical vapor deposition (CVD) on metal (e.g., Cu or Ni) surfaces and then transferred to other substrates. Unlike conventional metals, the electrons in graphene are massless quasiparticles that exhibit a linear energy-momentum dispersion governed by the Dirac equation for 2D relativistic fermions. As such, graphene offers certain exotic characteristics such as the extremely high mobility, a universal conductance in the optical frequency region, and unique plasmonic characteristics with 2D graphene patches and ribbons. Furthermore, the infrared conductance of graphene can be tuned by chemical doping or voltage gating, leading to promising high-speed photodetectors as well as optical modulators and antennas [140, 141]. Researchers have also developed scattering-type scanning near-field optics for characterizing graphene plasmons (GPs) with nanometer spatial resolution [141, 142]. This section introduces sheet conductivity of graphene, as well as the associated optical properties and plasmons in a single layer of graphene, a dielectric adjacent to two graphene sheets, and patterned a graphene ribbon array, as schematically shown in Fig. 9.42.

As shown in Fig. 9.42a, graphene atoms form a honeycomb structure along the plane for which each carbon atom is adjacent to three carbon atoms, forming strong covalent bonds with $sp^2$ hybridization ($\sigma$-bonds). The remaining *p*-orbital electron in each carbon atom dangles above and below the sheet and is highly mobile. As discussed in Chap. 6, these electrons form a large conjugated $\pi$-bond system that is responsible for the unique electrical and optical characteristics of graphene. Graphene can be characterized as a 2D material with a sheet conductivity that needs to be modeled based on its Dirac band structure. Alternatively, it may be thought as an ultrathin slice of anisotropic material that has a high in-plane conductivity and behaves like a dielectric when the electric field is perpendicular to the plane.

The sheet conductivity is defined as the electric conductance for a square sheet and its unit is siemens [S] or [$\Omega^{-1}$]. The contribution to graphene optical conductivity

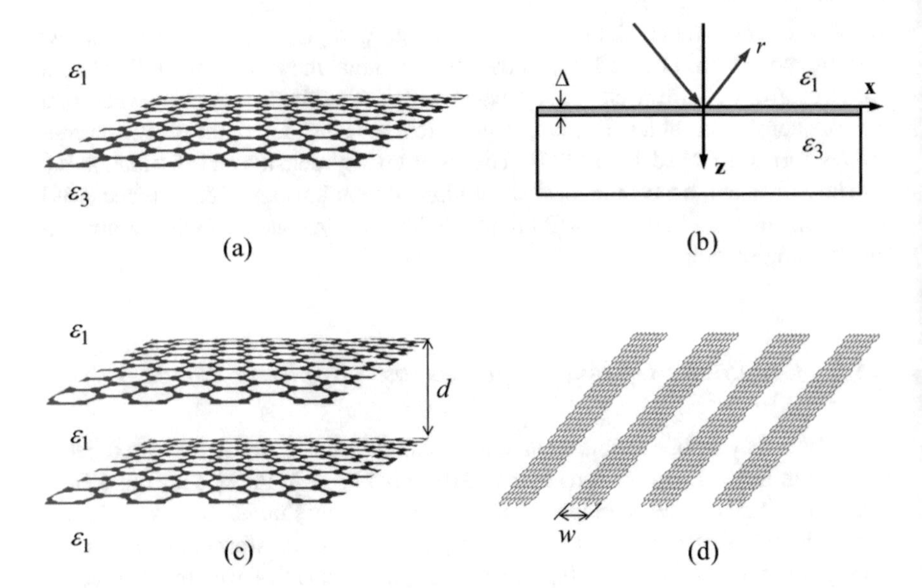


**Fig. 9.42** Illustration of graphene plasmons. **a** A monolayer graphene sheet sandwiched between two (dielectric) media. **b** A plane wave incident from medium 1 onto a graphene layer adjacent to a substrate. **c** Two parallel graphene sheets separated by a distance *d* inside a homogeneous media. **d** A periodic array of graphene ribbon whose width is *w*

includes the intraband transition and interband transition. The sheet conductivity may be written as $\sigma_s = \sigma_D + \sigma_I$, where $\sigma_D$ stands for the Drude-like free electron contribution and $\sigma_I$ is due to interband transition. The conductivity can be modeled as a complex function of frequency, temperature, scattering rate, and chemical potential (i.e., Fermi energy). The conductivity components can be derived from the Kubo formula and expressed as follows [143]:

$$\sigma_D = \frac{i}{\omega + i/\tau}\frac{2e^2 k_B T}{\pi\hbar^2}\ln\left[2\cosh\left(\frac{\mu}{2k_B T}\right)\right] \tag{9.114}$$

and

$$\sigma_I = \frac{e^2}{4\hbar}\left[G\left(\frac{\hbar\omega}{2}\right) + i\frac{4\hbar\omega}{\pi}\int_0^\infty \frac{G(\zeta) - G(\hbar\omega/2)}{(\hbar\omega)^2 - 4\zeta^2}\mathrm{d}\zeta\right] \tag{9.115}$$

where $G(\zeta) = \sinh(\zeta/k_B T)/[\cosh(\mu/k_B T) + \cosh(\zeta/k_B T)]$, $e$ is the electron charge, and $\mu$ is the chemical potential (i.e., Fermi level) of graphene. Near room temperature, the scattering rate $(1/\tau)$ is on the order of $10^{13}$ rad/s. The chemical potential depends on the doping and can be tuned by electrostatic gating. Note that the intraband conductivity has a Drude-like term and dominates the conductivity at low frequencies, while the interband conductivity dominates at high frequencies. It can be shown that when the frequency falls in the near infrared to visible region, $\sigma_s \approx \sigma_I \approx g_0 = e^2/(4\hbar)$, which is independent of the frequency. This is called the universal optical conductance, which has real part only and gives graphene a wavelength-independent optical absorptivity of about 2.3% per layer. The conductivity being a real value also suggests that no plasmonic response can exist in graphene from the ultraviolet to the near infrared. In the mid- and far-infrared region, when the Drude-like term dominates, graphene can support surface plasmons. When $\mu \gg k_B T$, Eq. (9.114) reduces to

$$\sigma_D \approx \frac{i}{\omega + i/\tau} \frac{e^2 \mu}{\pi \hbar^2} \tag{9.116}$$

which is very much the same as the Drude model. Both the chemical potential and scattering rate are related to the carrier concentration. Under the same limit,

$$\sigma_I = \begin{cases} g_0\left(1 - \frac{i}{\pi} \ln \frac{\hbar\omega + 2\mu}{\hbar\omega - 2\mu}\right), & \hbar\omega > 2\mu \\ g_0\left(-\frac{i}{\pi} \ln \frac{2\mu + \hbar\omega}{2\mu - \hbar\omega}\right), & \hbar\omega < 2\mu \end{cases} \tag{9.117}$$

Note that in Eqs. (9.114) to (9.117), the chemical potential $\mu$ takes the absolute value. Equation (9.117) suggests that the real part of $\sigma_I$ is a step function, while the imaginary part has a negative dip around the photon energy of $2\mu$ .

The calculated sheet conductivity related to $g_0$ is plotted in Fig. 9.43 at $T$ = 300 K using Eqs. (9.114) and (9.115) for several $\mu$ values by assuming $\tau = 10^{-13}$ s. It can be seen that the real part is $g_0$ at short wavelength and drops at photon energy close to $2\mu$, where the imaginary part is negative. At longer wavelength, $\sigma_D$ dominates and both the real and imaginary part increases with wavelength. At very long wavelength in the microwave and radio wave region, $\text{Re}(\sigma_D) \gg \text{Im}(\sigma_D)$ and graphene becomes a very good 2D conductor. It should be mentioned that when $\hbar\omega > 3\,\text{eV}$ or $\lambda < 0.4\,\mu\text{m}$, the linear dispersion at the K point of the Brillouin

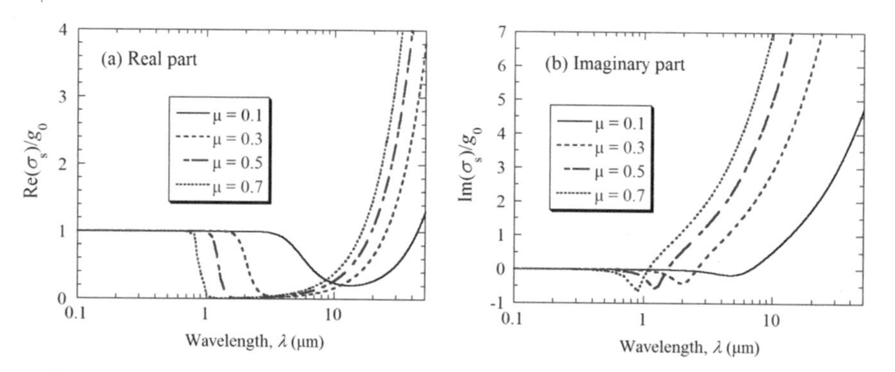


**Fig. 9.43** **a** Real part and **b** imaginary part of the sheet conductivity of graphene calculated using Eqs. (9.114) and (9.115) at 300 K for different chemical potentials

zone breaks down. Hence, the real part of the sheet conductivity first increases with the frequency, reaches a peak in the UV region due to excitons at the saddle-point singularity at the M point of the Brillouin zone, and then drops [144].

In calculating the radiative properties, graphene may be treated as either a sheet conductivity $\sigma_s(\omega)$ or an ultrathin layer with a thickness $\Delta = 0.335\,\text{nm}$, which is the interlayer distance in graphite, and an in-plane dielectric function

$$\varepsilon(\omega) = \varepsilon_h + i\frac{\sigma_s(\omega)}{\omega\varepsilon_0\Delta} \tag{9.118}$$

where $\varepsilon_h$ is the dielectric function of the host medium. The out-of-plane dielectric function of graphene may be taken as $\varepsilon_h$ [145].

When graphene is treated as a sheet conductor, Fresnel's coefficients need to be modified. Consider the structure shown in Fig. 9.42b, we have for $p$-polarization [146]

$$r_p = \frac{\frac{\varepsilon_3}{k_{3z}} - \frac{\varepsilon_1}{k_{1z}} + \frac{\sigma_s}{\omega\varepsilon_0}}{\frac{\varepsilon_3}{k_{3z}} + \frac{\varepsilon_1}{k_{1z}} + \frac{\sigma_s}{\omega\varepsilon_0}} \quad \text{and} \quad t_p = \frac{2\frac{\varepsilon_3}{k_{3z}}}{\frac{\varepsilon_3}{k_{3z}} + \frac{\varepsilon_1}{k_{1z}} + \frac{\sigma_s}{\omega\varepsilon_0}} \tag{9.119a}$$

For $s$-polarization, taking the relative permittivity of the upper and lower media as $\mu_1$ and $\mu_3$, respectively, we have

$$r_s = \frac{\frac{k_{1z}}{\mu_1} - \frac{k_{3z}}{\mu_3} - \sigma_s\omega\mu_0}{\frac{k_{1z}}{\mu_1} + \frac{k_{3z}}{\mu_3} + \sigma_s\omega\mu_0} \quad \text{and} \quad t_s = \frac{2\frac{k_{1z}}{\mu_1}}{\frac{k_{1z}}{\mu_1} + \frac{k_{3z}}{\mu_3} + \sigma_s\omega\mu_0} \tag{9.119b}$$

Note that surface plasmon with a single graphene sheet can be excited for both TM wave and TE wave even if $\mu_1 = \mu_3 = 1$. For TM wave, the dispersion of the graphene surface plasmon is given by

$$\frac{\varepsilon_1}{\sqrt{k_x^2 - \varepsilon_1 k_0^2}} + \frac{\varepsilon_3}{\sqrt{k_x^2 - \varepsilon_3 k_0^2}} = -i\frac{\sigma_s}{\omega\varepsilon_0} \tag{9.120a}$$

If $\varepsilon_3 = \varepsilon_1$, it can be shown that

$$k_x = k_{\text{GSP}} = k_0\sqrt{\varepsilon_1 - \frac{4\varepsilon_0\varepsilon_1^2}{\sigma_s^2\mu_0}} \tag{9.120b}$$

where $k_0 = \omega/c_0$ and the subscript GSP stands for *graphene surface plasmon*. Using Eq. (9.116) for $\sigma_D$ by neglecting $\sigma_I$ ($\hbar\omega \ll 2\mu$), it can be shown that $k_{\text{GSP}} \approx a_1(\omega^2 + i\omega/\tau)$ where $a_1 = 2\pi\hbar^2\varepsilon_0/(e^2\mu)$ depends on the chemical potential. The imaginary part determines the propagation length. Graphene plasmon for TE wave can also be excited when $\text{Im}(\sigma_s) < 0$, which cannot be satisfied by conventional materials [147]. The detail of TE wave GSP is not discussed here as the spectral region is somewhat limited as can be seen from Fig. 9.43b.

For the two layers of graphene separated by a dielectric as shown in Fig. 9.42c, the dispersion includes a symmetric branch and an antisymmetric branch, governed by [148]

$$1 + \frac{\sigma_s k_{1z}}{\omega \varepsilon_0 \varepsilon_1} = \coth\left(\frac{\mathrm{i} k_{1z} d}{2}\right), \text{ symmetric branch} \tag{9.121a}$$

$$1 + \frac{\sigma_s k_{1z}}{\omega \varepsilon_0 \varepsilon_1} = \tanh\left(\frac{\mathrm{i} k_{1z} d}{2}\right), \text{ antisymmetric branch} \tag{9.121b}$$

Note that when coupled surface plasmons are excited, $k_{1z} = \mathrm{i}\sqrt{k_{\mathrm{GSP}}^2 - \varepsilon_1 k_0^2}$ is imaginary. Like coupled SPPs between two metal films, these equations can be solved to find the dispersion relations. When $d \to \infty$, both Eqs. (9.121a) and (9.121b) reduces to Eq. (9.120b). Figure 9.44 illustrates the dispersion curves in the mid-infrared. Large parallel wavevector is needed to excite GSPs. Coupled GSPs can enhance near-field radiative transfer between two graphene sheets [146, 149] and will be discussed in Chap. 10.

For graphene ribbon array as shown in Fig. 9.42d when the magnetic field is parallel to the ribbon (transverse magnetic wave), Fabry–Perot-type resonance across the ribbon occurs that gives the so-called graphene ribbon plasmon. This is due to that the wave is reflected back and forth at the edge of the ribbon with a phase shift $\phi_{\mathrm{edge}}$. For a graphene ribbon width $w$, the resonance of the $m$th order can be expressed as [150]

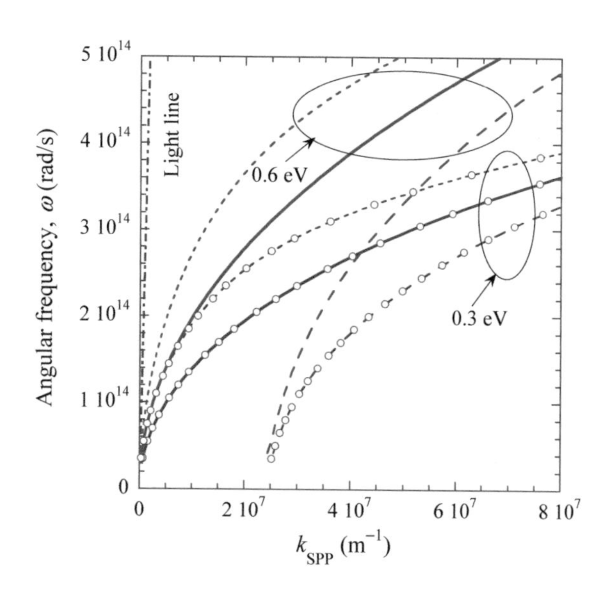


**Fig. 9.44** GSP dispersion relations for a single layer (solid lines) and two layers separated by $d = 20$ nm (dashed lines above and below each solid line). The upper and lower branches will meet at an apex if the frequency and wavevector are sufficiently large (not shown). The graphene chemical potentials are $\mu = 0.3$ eV and 0.6 eV with a scattering rate of $6.28 \times 10^{13}$ rad/s

$$w\mathrm{Re}(k_{\mathrm{GSP}}) + \phi_{\mathrm{edge}} = m\pi \tag{9.122}$$

where $m = 0$ represents the fundamental modes. Substituting the expression of $k_{\mathrm{GSP}}$ for a monolayer of graphene in vacuum, we obtain

$$\omega_m = \frac{e}{\hbar}\sqrt{\frac{\mu(m\pi - \phi_{\mathrm{edge}})}{2\pi\varepsilon_0 w}} \tag{9.123}$$

The phase shift has been found to be $\phi_{\mathrm{edge}} = \pi/4$ for free-standing ribbons in vacuum [151] and $\phi_{\mathrm{edge}} = -\pi$ if the edge is adjacent to a metal [152]. The ribbon polariton is proportional to $\sqrt{\mu/w}$. Since the ribbon array can be considered as a 1D grating, plasmonic modes can be excited within the light line. The ribbon modes are essentially discrete points on the folded dispersion line. The resulting ribbon modes according to Eq. (9.123) are independent of the wavevector $k_x$ or horizontal lines. Graphene plasmons can often be coupled to other types of polaritons such as magnetic polaritons or phonon polaritons to affect the radiative properties [59, 146, 152]. Figure 9.45 illustrates the ribbon plasmon polaritons and its coupling with the magnetic polariton. The contour plots are based on RCWA simulation, while the ribbon polariton dispersion curves are predicted with Eq. (9.123) using different $\phi_{\mathrm{edge}}$ depending on whether the ribbon is adjacent to vacuum or Ag. Only odd orders $m = 1, 3, 5\ldots$ of ribbon plasmon polaritons can be excited at normal incidence. Furthermore, the MP in deep grating can significantly enhance the absorptance of graphene as well as structure to achieve nearly perfect absorption. Besides, the MP resonance condition (fundamental mode) can be well predicted by an equivalent *LC* circuit model [152].

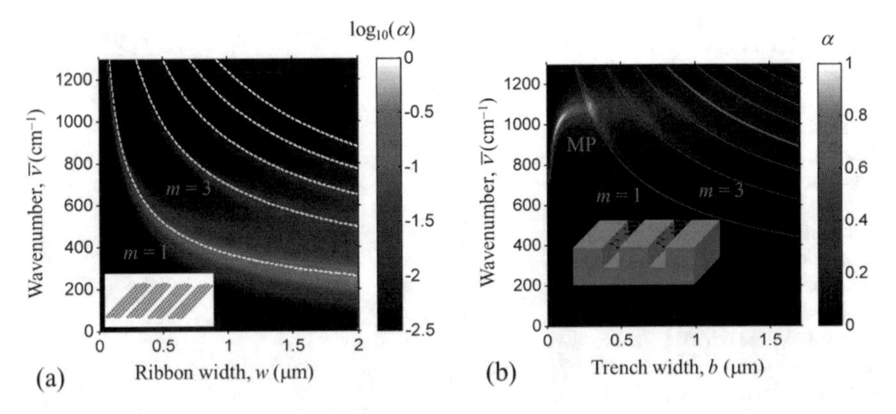


**Fig. 9.45** Calculated absorptance at normal incidence using RCWA of **a** a graphene ribbon array; **b** a graphene ribbon located in the trench of a Ag grating so that $w = b$ [152]. Note that the period for both structures is 4 μm. For the Ag grating, the depth is 2 μm. The chemical potential and relaxation time of graphene are taken as 3 eV and $10^{-13}$ s, respectively

### 9.5.6 Hyperbolic (Plasmon or Phonon) Polariton

A composite with repeating alternative layers of metal and dielectric forms a unique type of photonic crystal called hyperbolic metamaterial. When the thicknesses of individual layers are much smaller than the wavelength, the effective dielectric behavior depends on whether the electric field is parallel with or perpendicular to the interfaces. Such a subwavelength multilayer structure can be modeled as a uniaxial crystal with a dielectric tensor. The distinct feature of a hyperbolic dispersion is that the two dielectric components have different signs. The excitation of plasmon polaritons between the two metal films plays a key role in the exotic optical properties of metallodielectric multilayers and is therefore called a volume (bulk) plasmon polariton or hyperbolic (plasmon) polaritons [138, 153, 154]. Thanks to the anisotropic dielectric or magnetic behavior, hyperbolic metamaterials can enable negative refraction, form waveguides, facilitate far-field diffraction limited imaging (so called optical hyperlens), as well as enhance spontaneous emission and near-field thermal radiation [146, 149, 155, 156].

Figure 9.46 shows a multilayer structure with alternative metal or dielectric layers arranged periodically. For simplicity, assume both materials are nonmagnetic so that the permeability is the same as free space. When the optic axis is aligned with the $z$-axis, the dielectric tensor can be expressed as

$$\bar{\bar{\varepsilon}} = \begin{pmatrix} \varepsilon_{\mathrm{O}} & 0 & 0 \\ 0 & \varepsilon_{\mathrm{O}} & 0 \\ 0 & 0 & \varepsilon_{\mathrm{E}} \end{pmatrix} \tag{9.124}$$

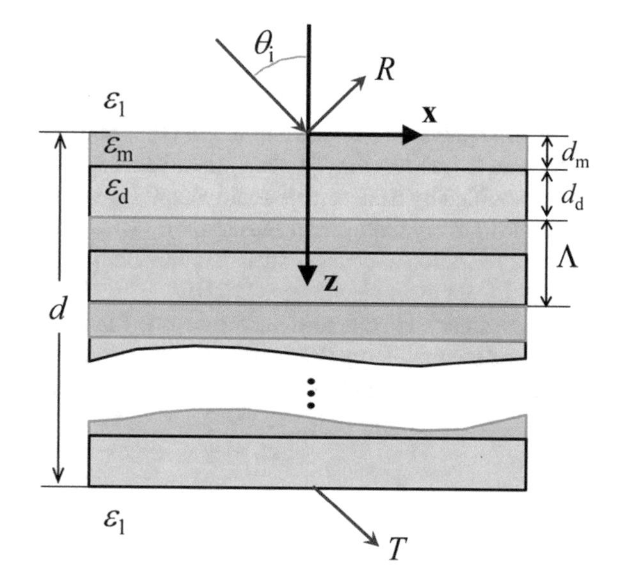


**Fig. 9.46** Illustration of wave incident on a metallodielectric medium with a total thickness $d$, the upper and lower medium are semi-infinite lossless dielectric. Here, $d_{\mathrm{m}}$ and $d_{\mathrm{d}}$ are the thickness of the metal layer and dielectric layer, respectively, and $\Lambda$ is the period of the unit cell

where the (complex) dielectric function for electric field perpendicular to the optic axis is called the ordinary component $\varepsilon_{\mathrm{O}}$ and that for the electric field parallel with the optic axis is called the extraordinary component $\varepsilon_{\mathrm{E}}$. The detail of wave propagation in anisotropic media is beyond the scope of this text. However, it will be shown that when the optic axis is parallel with one of the coordinates, only minor modifications in Fresnel's coefficients need be made to calculate the reflectance, transmittance, as well as the field distribution inside a layer of homogeneous uniaxial medium.

In the coordinates shown in Fig. 9.46, for TE wave, since the electric field is parallel to the $y$-axis and perpendicular to the optic axis ($z$-axis), the uniaxial medium can be treated as an isotropic medium using its ordinary dielectric function $\varepsilon_{\mathrm{O}}$. For a TM wave in the case of oblique incidence, both $\varepsilon_{\mathrm{O}}$ and $\varepsilon_{\mathrm{E}}$ affect the reflection, refraction, and wavevector. Assume medium 1 is a lossless dielectric medium with a dielectric function $\varepsilon_1$, and the angle of incidence is $\theta_{\mathrm{i}}$, we have $k_{1z} = \sqrt{\varepsilon_1 k_0^2 - k_x^2}$, where $k_x = \sqrt{\varepsilon_1} k_0 \sin\theta_1 = \sqrt{\varepsilon_1}\omega/c_0 \sin\theta_1$ is the same in all media. Taking the metamaterial as a homogeneous medium 2, it can be shown that [157–160]

$$k_{2z} = \sqrt{k_0^2 \varepsilon_{\mathrm{O}} - k_x^2 \varepsilon_{\mathrm{O}}/\varepsilon_{\mathrm{E}}} \tag{9.125}$$

Then, the Fresnel coefficients can be modified as [146, 159]

$$r_{12,p} = \frac{\frac{k_{1z}}{\varepsilon_1} - \frac{k_{2z}}{\varepsilon_{\mathrm{O}}}}{\frac{k_{1z}}{\varepsilon_1} + \frac{k_{2z}}{\varepsilon_{\mathrm{O}}}} = \frac{\frac{\varepsilon_{\mathrm{O}}}{k_{2z}} - \frac{\varepsilon_1}{k_{1z}}}{\frac{\varepsilon_{\mathrm{O}}}{k_{2z}} + \frac{\varepsilon_1}{k_{1z}}} \tag{9.126a}$$

$$t_{12,p} = \frac{2\frac{k_{1z}}{\varepsilon_1}}{\frac{k_{1z}}{\varepsilon_1} + \frac{k_{2z}}{\varepsilon_{\mathrm{O}}}} = \frac{2\frac{\varepsilon_{\mathrm{O}}}{k_{2z}}}{\frac{\varepsilon_{\mathrm{O}}}{k_{2z}} + \frac{\varepsilon_1}{k_{1z}}} \tag{9.126b}$$

Furthermore, the phase shift in medium 2 can be expressed by $\psi = k_{2z}d$, where $d$ is the total thickness of medium 2. This allows an extension of the calculation of wave propagation in a uniaxial thin film [159, 160]. It should be noted that when the optic axis is tilted in the $x$-$z$ plane, the dielectric tensor given in Eq. (9.124) needs to be modified by a coordinate rotation and $k_{2z}$ has different values for forward and backward propagation [161]. Furthermore, when the optic axis is in the $y$-$z$ plane, cross-polarization may occur such that the polarization of the reflected wave may be different from the incident wave [162].

The ordinary and extraordinary dielectric functions can be obtained based on the effective medium theory (EMT) based on the dielectric function of the metal $\varepsilon_{\mathrm{m}}$ and dielectric $\varepsilon_{\mathrm{d}}$, similar to Eqs. (9.82) and (9.83) as follows:

$$\varepsilon_{\mathrm{O}} = \phi\varepsilon_{\mathrm{m}} + (1-\phi)\varepsilon_{\mathrm{d}} \tag{9.127}$$

and $1/\varepsilon_{\mathrm{E}} = \phi/\varepsilon_{\mathrm{m}} + (1-\phi)/\varepsilon_{\mathrm{d}}$, which can be expressed as

$$\varepsilon_{\mathrm{E}} = \frac{\varepsilon_{\mathrm{d}}\varepsilon_{\mathrm{m}}}{\phi\varepsilon_{\mathrm{d}} + (1-\phi)\varepsilon_{\mathrm{m}}} \tag{9.128}$$

where $\phi = d_m/\Lambda = d_m/(d_m + d_d)$ is the metal volume fraction or filling ratio.

It should be noted that doped semiconductors may be substituted for the metal layer in Fig. 9.46 to excite hyperbolic plasmon polariton (HPP) in the mid-infrared region. This is because the plasma frequency of free carriers in heavily doped semiconductors typically falls in the near- to mid-infrared region as discussed in Chap. 8. Furthermore, a polar material may also be used since near the real part of the dielectric function is negative near the reststrahlen band, giving rise to hyperbolic phonon polaritons (HPhPs). As an example, Fig. 9.47 shows the effective dielectric functions (real part only) of two multilayer structures [155]. One of them uses *n*-doped Si with a doping concentration of $10^{20}$ cm$^{-1}$, which has a plasma frequency $\omega_p = 1.09 \times 10^{15}$ rad/s (corresponding to a wavelength of 1.73 μm), and Ge for which the dielectric function in the infrared is taken as a constant $\varepsilon_d = 16$ [2]. The other structure uses SiC whose dielectric function is given in Eq. (8.123) with a filling ratio of 0.3 and Ge. As shown in Fig. 9.47a for the D-Si/Ge multilayer, $\varepsilon'_O < 0$ at $\omega < 1.82 \times 10^{14}$ rad/s, while $\varepsilon'_E < 0$ at $2.3 \times 10^{14}$ rad/s $< \omega < 2.9 \times 10^{14}$ rad/s. These are identified as type II and I hyperbolic bands, respectively. Similarly, in the SiC/Ge multilayer, two hyperbolic bands exist in the mid-infrared spectral region that fall in the SiC reststrahlen band from 1.49 to $1.83 \times 10^4$ rad/s.

To better illustrate optical hyperbolicity, Eq. (9.125) may be rewritten in a more general dispersion relation of a uniaxial medium as follows:

$$\frac{k_x^2 + k_y^2}{\varepsilon_E} + \frac{k_z^2}{\varepsilon_O} = \frac{\omega^2}{c_0^2} \tag{9.129}$$

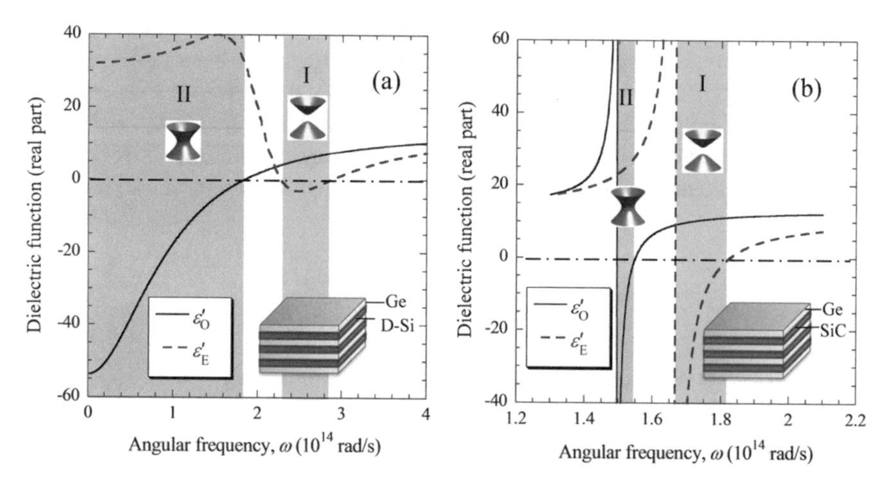


**Fig. 9.47** Real part of the dielectric function calculated from EMT [155]: **a** doped Si and Ge with $\phi = 0.5$; **b** SiC and Ge with $\phi = 0.3$ for SiC. The multilayer structures are illustrated by the insets. The hyperbolic band are shaded and indicated by the hyperbola

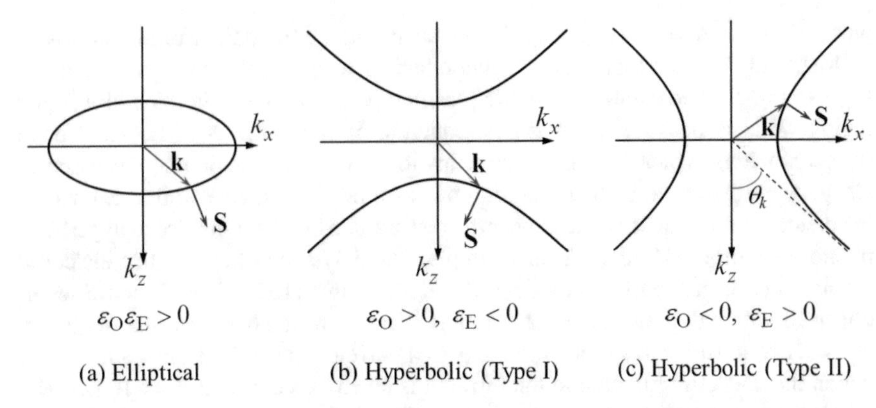


**Fig. 9.48** Isofrequency curves for extraordinary waves in uniaxial media: **a** elliptical, **b** type I hyperbolic, and **c** type II hyperbolic. The directions of wavevector and Poynting vector are indicated. The curves can be rotated about the z-axis to form 3D isofrequency surfaces

For an isotropic medium when $\varepsilon_E = \varepsilon_O$, it is the same as Eq. (9.95). Therefore, Eq. (9.129) may be identified as the polariton dispersion relation for extraordinary waves. The constant-frequency curves depend on the signs of $\varepsilon_O$ and $\varepsilon_E$, as illustrated in Fig. 9.48 for $k_y = 0$ in the lossless case. For lossy material, usually only $\varepsilon_O'$ and $\varepsilon_E'$ are taken into consideration. For hyperbolic dispersion, waves can propagate inside the medium at very large $k_x$. In the asymptotic limit ($k_x \gg k_0$), $\tan(\theta_k) = k_x/\mathrm{Re}(k_z) \approx \sqrt{-\varepsilon_E'/\varepsilon_O'}$. Furthermore, the Poynting vector is always perpendicular to the isofrequency surface.

It can be seen from Fig. 9.48 that for type I hyperbolic metamaterial, the $x$-component of the Poynting vector $S_x$ is negative, suggestion bending the light in the wrong way. This is not surprising since even for a metal whose $\varepsilon'$ is negative, $S_x$ becomes negative. The advantage of using a multilayer is that it can reduce the loss so that light can propagate through many metal layers whose total thickness is significantly greater than the penetration depth. The Poynting vector refraction angle is defined as

$$\theta_S = \tan^{-1}\left(\frac{S_x}{S_z}\right) = \tan^{-1}\frac{\mathrm{Re}(k_x/\varepsilon_E)}{\mathrm{Re}(k_z/\varepsilon_O)} \tag{9.130}$$

at the interface, where $S_z$ is always positive. When $k_x \gg k_0$, $\theta_S = \tan^{-1}\sqrt{-\varepsilon_O'/\varepsilon_E'}$. When the hyperbolic medium is semitransparent, the Poynting vector in medium 2 is affected by both the forward and backward propagating wave and must be calculated based on the total field.

For type II hyperbolic metamaterial, $k_z < 0$. While negative refraction cannot be observed for top incidence shown in Fig. 9.47, negative refraction can occur for side incidence [145]. Furthermore, the metal layers can be substituted by graphene sheets. It should be noted that when $\varepsilon_O < 0$, ordinary waves cannot propagate inside the medium but SPP can be excited near the surface of the medium when it is adjacent to a dielectric.

**Example 9.8** Prove Eq. (9.125) for a uniaxial medium, omitting subscript 2, for a TM wave.

**Solution** Note that the medium is a nonmagnetic medium whose dielectric tensor is given by Eq. (9.124), so that the optic axis is in the $z$-direction. The Maxwell equation in terms of wavevector Eq. (8.26) should be modified as $\mathbf{k} \times \mathbf{H} = -\omega\varepsilon_0\bar{\bar{\varepsilon}}\mathbf{E}$. For TM wave incidence, $k_y = 0$ and $\mathbf{H} = H_y\hat{\mathbf{y}}$. Hence, we have $k_x H_y = -\omega\varepsilon_0\varepsilon_{\rm O} E_x$, $E_y = 0$, and $k_z H_y = \omega\varepsilon_0\varepsilon_{\rm E} E_z$. Plugging the electric field components into $\mathbf{k} \times \mathbf{E} = \omega\mu_0\mathbf{H}$, which is Eq. (8.25), we obtain

$$k_z^2\varepsilon_{\rm E} + k_x^2\varepsilon_{\rm O} = (\omega/c_0)^2\varepsilon_{\rm O}\varepsilon_{\rm E}$$

This proves Eq. (9.125); since the plane of incidence can be rotated about the optic axis, we can also prove that Eq. (9.129) is for a plane wave propagating in the uniaxial medium whose magnetic field is perpendicular to the plane of incidence. Readers can find the derivations for more general cases from [157, 158]. Since $\mathbf{k} \cdot \mathbf{D} = \mathbf{k} \cdot (\bar{\bar{\varepsilon}}\mathbf{E}) = 0$, $\mathbf{k} \perp \mathbf{D}$ always holds. However, the electric field $\mathbf{E}$ and the wavevector $\mathbf{k}$ may not be perpendicular to each other.

In general, it is known that for EMT to be applicable, the wavelength should be much greater than the period of the unit cell. Zhang and Zhang [160] considered the wave propagation and radiative properties for several multilayer structures. Consider a metallodielectric multilayers with Ag and $TiO_2$ (rutile) whose refractive index is around 3 ($\varepsilon_{\rm d} \approx 9$) in the visible and ultraviolet. Let $d = 480$ nm and $\phi = 0.5$. The Poynting vector can be traced in each layer based on the combined field for the forward and backward waves, forming so-called energy streamlines [67, 155]. Figure 9.49 shows the energy streamlines for $\theta_{\rm i} = 40°$ at $\lambda = 400$ nm and 600 nm, respectively. Note that $\varepsilon'_{\rm O}$ is negative at wavelengths longer than 440 nm, due to the large negative value of $\varepsilon'_{\rm Ag}$. As the wavelength is reduced to below 440 nm, while $\varepsilon'_{\rm Ag} < 0$, its magnitude becomes smaller than $\varepsilon_{\rm d}$, so that $\varepsilon'_{\rm O}$ becomes positive while $\varepsilon'_{\rm E}$ becomes negative, according to Eqs. (9.27) and (9.128).

In Fig. 9.49, the dots show calculations based on the effective medium theory. The solid lines are for $d_m = d_d = 12$ nm (20 periods) and the dashed lines are for $d_m = d_d = 30$ nm (8 periods); they are calculated using the transfer matrix method discussed in Sect. 9.2.2. Note that the streamlines are curved in the incident medium ($\varepsilon_1 = 1$) due to interference with the reflected wave. As the thickness of the layer is reduced, the results are closer to the EMT. At $\lambda = 400$ nm the transmittance calculated using EMT is 53%. On the other hand, the transmittance calculated using the exact solution is 42% for $\Lambda = 24$ nm and 15% for $\Lambda = 60$ nm. This suggests that EMT can help qualitatively understand the underlying physics but care should be taken in using EMT for accurate calculation of the radiative properties [160]. Loss is high at $\lambda = 600$ nm, resulting relatively large deviation and negligible transmittance.

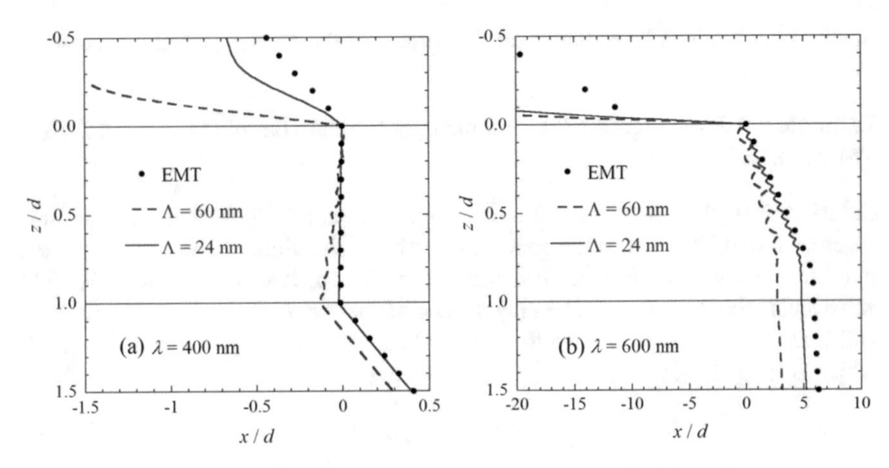


**Fig. 9.49** Energy streamlines for a Ag-$TiO_2$ metallodielectric hyperbolic medium [160]: **a** $\lambda = 400$ nm with type I hyperbolicity. **b** $\lambda = 600$ nm with type II hyperbolicity

The validity of EMT can be further analyzed based on the photonic crystal dispersion relation [158], based on the Bloch theorem.

$$\cos(k_{\mathrm{b}}\Lambda) = \cos(\gamma_{\mathrm{m}}d_{\mathrm{m}})\cos(\gamma_{\mathrm{d}}d_{\mathrm{d}}) - \frac{1}{2}\left(\frac{\varepsilon_{\mathrm{m}}\gamma_{\mathrm{d}}}{\varepsilon_{\mathrm{d}}\gamma_{\mathrm{m}}} + \frac{\varepsilon_{\mathrm{d}}\gamma_{\mathrm{m}}}{\varepsilon_{\mathrm{m}}\gamma_{\mathrm{d}}}\right)\sin(\gamma_{\mathrm{m}}d_{\mathrm{m}})\sin(\gamma_{\mathrm{d}}d_{\mathrm{d}}) \tag{9.131}$$

Here, $k_{\mathrm{b}}$ is the Bloch wavevector of the 1D PC, $\gamma_{\mathrm{m}} = \sqrt{\varepsilon_{\mathrm{m}}k_0^2 - \beta^2}$ or $\gamma_{\mathrm{d}} = \sqrt{\varepsilon_{\mathrm{d}}k_0^2 - \beta^2}$ is the $z$-component of the wavevector in metal and dielectric, respectively, and $\beta = \sqrt{k_x^2 + k_y^2}$ is the parallel wavevector component. Under the limit $k_{\mathrm{b}}\Lambda \ll 1$, $\gamma_{\mathrm{m}}d_{\mathrm{m}} \ll 1$, and $\gamma_{\mathrm{d}}d_{\mathrm{d}} \ll 1$, we can use the approximation $\cos(x) \approx 1 - x^2/2$ and $\sin(x) \approx x$. Equation (9.131) can be reorganized using Eqs. (9.127) and (9.128) to obtain [153, 158]:

$$\frac{\beta^2}{\varepsilon_{\mathrm{E}}} + \frac{k_{\mathrm{b}}^2}{\varepsilon_{\mathrm{O}}} = \frac{\omega^2}{c_0^2} \tag{9.132}$$

which is exactly the dispersion relation for a uniaxial medium when $k_{\mathrm{b}} = k_z$. The practical criteria are $\gamma_{\mathrm{m}}d_{\mathrm{m}} < \pi/4$, $\gamma_{\mathrm{d}}d_{\mathrm{d}} < \pi/4$, and $k_z\Lambda < 1$ [160, 163]. However, more strict criteria may be necessary to accurately predict the radiative properties and near-field radiation.

Certain naturally occurring materials may also support hyperbolic polaritons, such as graphite, YBCO superconductor, hexagonal boron nitride (hBN), $Bi_2Te_3$, etc. [35, 59, 164, 165]. These materials may enable a large range of thermal applications coherent thermal emission to energy harvesting. In addition to planar multilayers, arrays of nanowires, carbon nanotubes, and nanoholes can also support hyperbolic polaritons [120, 149, 159], as discussed in the following.

### 9.5.7 General Effective Medium Theory

Arrays of nanowires and nanoholes are illustrated in Fig. 9.50, assuming the optical properties are isotropic in the *x*-*y* plane so the subwavelength structure forms an effective uniaxial medium whose optic axis is parallel to the *z*-axis. For nanowires, the matrix is assumed to be a dielectric with a dielectric function $\varepsilon_d$ (which may be vacuum or air), and the filler is assumed to be a metallic or some polar materials with a dielectric constant of $\varepsilon_m$. The volume filling fraction or ratio for the wire $\phi$. It is expected that for a low filling ratio, the behavior for electric field perpendicular to the optic axis should be dielectric and for electric field parallel to the optic axis should be dilute metallic. For nanoholes, the filling ratio is based on the holes, the filler is a dielectric ($\varepsilon_d$) inside a metallic matrix ($\varepsilon_m$). The effective medium behavior of hole arrays may be complicated and magnetic resonance may also be excited. In this section, the general effective medium theory is present, without considering magnetic response, and the multilayers and periodic arrays are viewed as special cases.

The effective medium theory has been studied and compared by many researchers. In essence, it uses mean field and seeks an average permittivity based on the permittivities and volume fractions of individual constituents. The Maxwell-Garnett theory assumes one constituent as the host and all other constituents as embedded grains that are spatially separated, and therefore, it is valid for dilute systems (i.e., relatively low volume fractions of the filling constituents). On the other hand, the Bruggeman approximation treats all constituents equally as grains imbedded in an otherwise

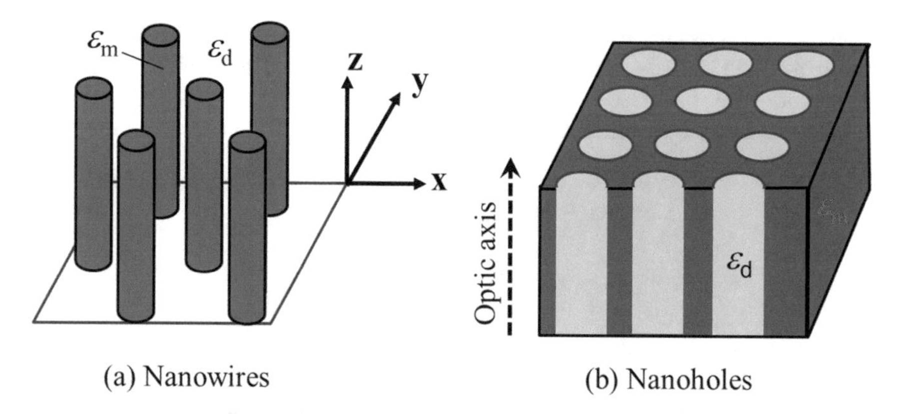


**Fig. 9.50** Schematic drawings of arrays of **a** nanowires and **b** nanoholes. The actual dimensions may be large as long as they are much smaller than the wavelength

homogenous "effective" medium which is assumed to possess the average properties of the composite. Furthermore, the shape of the filler can play a major role for different polarization. Therefore, a geometric factor $g$ can be included whose value depends on the geometry and polarization. The general expression of the effective medium theory for the dielectric function can be expressed as [67, 166]:

$$\frac{\varepsilon_{\mathrm{eff}}-\varepsilon_{\mathrm{h}}}{\varepsilon_{\mathrm{h}}+g(\varepsilon_{\mathrm{eff}}-\varepsilon_{h})}=\phi_1\frac{\varepsilon_1-\varepsilon_{\mathrm{h}}}{\varepsilon_{\mathrm{h}}+g(\varepsilon_1-\varepsilon_{\mathrm{h}})}+\phi_2\frac{\varepsilon_2-\varepsilon_{\mathrm{h}}}{\varepsilon_{\mathrm{h}}+g(\varepsilon_2-\varepsilon_{\mathrm{h}})}+\ldots \tag{9.133}$$

For the Bruggeman approximation, the host medium is the resulting effective medium, thus, $\varepsilon_{\mathrm{h}}=\varepsilon_{\mathrm{eff}}$. In this case, the left side becomes zero and the right side is a function in terms of $\varepsilon_{\mathrm{eff}}$ whose zeros are the solutions of $\varepsilon_{\mathrm{eff}}$. For a two-constituent system with spherical particles or randomly oriented particles, $g = 1/3$, and we obtain Eq. (9.81) which is the traditional effective medium approximation [38]. The Maxwell-Garnett theory treats one of the constituents as the host. Suppose there are $N$ types of constituents and $\phi_N$ is greater than 50%. Taking the $N$th constituent as the host, Eq. (9.133) can be rewritten as follows:

$$\frac{\varepsilon_{\mathrm{eff}}-\varepsilon_N}{\varepsilon_N+g(\varepsilon_{\mathrm{eff}}-\varepsilon_N)}=\sum_{j=1}^{N-1}\phi_j\frac{\varepsilon_j-\varepsilon_N}{\varepsilon_N+g(\varepsilon_j-\varepsilon_N)} \tag{9.134}$$

Since $0 \le g \le 1$, this equation allows an explicit expression for $\varepsilon_{\mathrm{eff}}$. Next, we will discuss some simple systems with two components only.

As discussed by Bohren and Hoffman [67], the summation of all three components of $g$ must be 1. For a long wire when the electric field is polarized along the wire, $g = 0$. It follows that for aligned wire or hole arrays, $g_{\mathrm{E}} = 0$ and $g_{\mathrm{O}} = 0.5$ since there are two in-plane directions. This ensures that the sum of the three components of $g$ is 1. In the case of layered structures, $g_{\mathrm{E}} = 1$ and $g_{\mathrm{O}} = 0$. It can be shown that both the Maxwell-Garnett and Bruggeman formulations give the same results for a two-component system when $g = 0$ or $g = 1$. Therefore, Eq. (9.127) and (9.128) can be obtained from the general EMT formula Eq. (9.133). This is left as an exercise in Problem 9.39.

With the Maxwell-Garnett theory, it can be shown that for nanowire arrays,

$$\varepsilon_{\mathrm{O}}=\varepsilon_{\mathrm{d}}\frac{\varepsilon_{\mathrm{m}}(1+\phi)+\varepsilon_{\mathrm{d}}(1-\phi)}{\varepsilon_{\mathrm{m}}(1-\phi)+\varepsilon_{\mathrm{d}}(1+\phi)} \tag{9.135a}$$

$$\varepsilon_{\mathrm{E}}=\phi\varepsilon_{\mathrm{m}}+(1-\phi)\varepsilon_{\mathrm{d}} \tag{9.135b}$$

where $\phi$ is the filling ratio of metal. For air or vacuum, we can set $\varepsilon_{\mathrm{d}} = 1$. The expression for nanoholes can be obtained by switching $\varepsilon_{\mathrm{m}}$ and $\varepsilon_{\mathrm{d}}$ in Eq. (9.135), and let $\phi$ be the volume fraction of the void (holes). It should be noted that the Bruggeman approximation gives a different expression for $\varepsilon_{\mathrm{O}}$ that requires finding the roots of a quadratic equation (usually only one solution is physical). When the volume fraction exceeds 0.3, it may be more appropriate to use the Bruggeman approximation.

Both metal nanowires and nanoholes and semiconductor nanowires and nanoholes have been shown to exhibit exotic optical properties, including hyperbolic dispersion and negative refraction [57, 149, 156, 159, 161, 167]. It should be noted that Mie resonance dielectric particles, especially with phonon resonances like SiC can be embedded in another dielectric to enable both electrical and magnetic resonances [168].

The similar concept can be applied to aligned carbon nanotubes (CNT). However, since CNT is rolled graphene sheet, the effective dielectric functions for a vertically aligned CNT array in air are usually expressed in terms of the ordinary ($\varepsilon_\perp$) and extraordinary ($\varepsilon_\parallel$) components of graphite using a coordinate transform [169]:

$$\varepsilon_{\rm O} = \frac{(1+\phi)\varepsilon_\parallel + (1-\phi)\sqrt{\varepsilon_\parallel/\varepsilon_\perp}}{(1-\phi)\varepsilon_\parallel + (1+\phi)\sqrt{\varepsilon_\parallel/\varepsilon_\perp}} \tag{9.136a}$$

$$\varepsilon_{\rm E} = \phi(\varepsilon_\parallel - 1) + 1 \tag{9.136b}$$

The volume fraction can be estimated by the ratio of the density of the CNT array to that of graphite and is usually between 2 and 15% [170]. However, fabricated nanowires are often misaligned or entangled. To account for these imperfections, an alignment parameter is introduced such that the dielectric functions are modified as follows [171, 172]:

$$\varepsilon_{\rm O,mod} = x\varepsilon_{\rm O} + (1-x)\varepsilon_{\rm E} \tag{9.137a}$$

$$\varepsilon_{\rm E,mod} = x\varepsilon_{\rm E} + (1-x)\varepsilon_{\rm O} \tag{9.137b}$$

Aligned CNTs exhibit high absorptance from the visible to infrared wavelength region due to interband transitions and free-electron absorption, as well as impedance matching with air [170–172]. Furthermore, CNT arrays form type I hyperbolic metamaterial that can facilitate light collimating, infrared imaging, solar energy harvesting, and near-field radiation [171–173].

## 9.6 Spectral and Directional Control of Thermal Radiation

There have been numerous studies on the spectral and directional control of thermal radiation such as coherent thermal emission, spectrally selective emitters or absorbers, extraordinary optical transmission, resonance perfect absorption, and wideband perfect absorption. Many of the metamaterial structures have a relatively small total thickness and are fabricated onto a substrate. They are sometimes called metasurfaces when the patterned structure has subwavelength thickness. Some of the metasurfaces can have optical chirality and respond differently for electromagnetic waves with left-handed or right-handed circular polarizations. We will limit our discussion to linearly polarized plane waves and for materials with linear responses. Several examples of using micro/nanostructures and metamaterials for tailoring the far-field radiative properties are discussed in this section.

## 9.6.1 Polariton-Enhanced Transmission

As mentioned previously, polaritons can enhance transmission for metal films, such as with metallodielectric multilayers. The prism-air-metal-prism structure can be used to excite surface polaritons at the interface between air and the metal to enable a larger tunneling transmittance for TM waves. Photon tunneling refers to energy transmission through a vacuum gap or dielectric film when the parallel wavevector exceeds the wavevector in the medium. In such case, the perpendicular wavevector is imaginary and only evanescent waves can exist in the medium. When the thickness of the medium is sufficiently small, energy can still be transmitted through the vacuum gap or dielectric film. This will be extensively studied in Chap. 10. In this section, we consider propagating waves that are incident from air or vacuum. Examples are given for layers with NIMs and with a paired negative-$\varepsilon$ and negative-$\mu$ composite, as well as for a periodic slit array and graphene covered slit arrays.

The transmittance of multilayer structures with alternating vacuum and NIM gaps are shown in Fig. 9.51 for TE wave incidence [174]. The number of layers ($2N$) includes the top and bottom semi-infinite dielectric media with $\varepsilon_\text{d} = 2.25$ which can be considered as prisms. Therefore, for the $2N$ layers, there are $N - 1$ layers of alternating vacuum and NIM between the two prisms. The dielectric function and relative magnetic permeability of the NIM are calculated with Eqs. (8.121) and

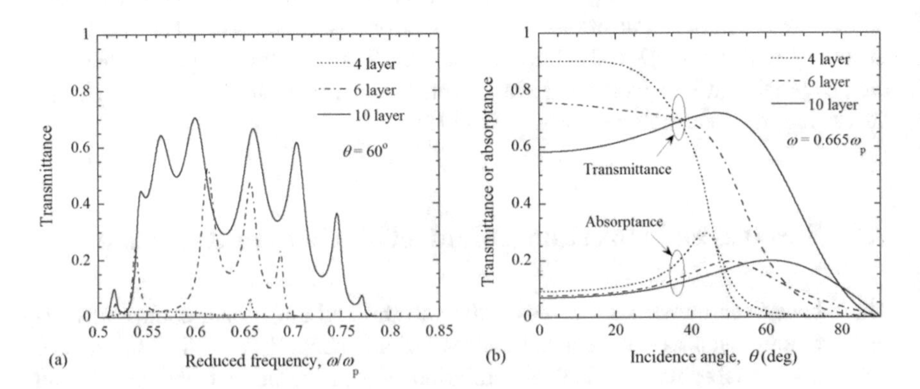


**Fig. 9.51** Radiative properties of multilayer structures with NIMs for TE wave [174]. **a** Transmittance spectra at incidence angle $\theta = 60°$. **b** Transmittance and absorptance as a function of incidence angle at $\omega = 0.665\omega_\text{p}$

(8.122) using $\omega_0 = 0.5\omega_p$, $F = 0.785$, and $\gamma_e = \gamma_m = 0.0025\omega_p$. The total thickness of the NIM is fixed to $0.85\lambda_p = 1.7\pi c/\omega_p$. The thickness of the vacuum layer is the same as that of the NIM layer. The transmittance spectra at an incidence angle $\theta_i = 60°$ are shown in Fig. 9.51a. The tunneling transmittance is greatly enhanced by reducing the individual layer thicknesses while maintaining the same total thickness. The enhanced transmittance is caused by the coupled surface polaritons as well as bulk polaritons. Figure 9.51b illustrates the transmittance and the absorptance as functions of the incidence angle. Note that the critical angle between the prism and vacuum is 41.8°. While the transmittance is slightly reduced as the number of layers increases for propagating waves in vacuum, the tunneling transmittance is greatly enhanced. The absorptance reaches a peak at a certain incidence angle. At large incidence angles, the absorptance also increases as the number of layers increases. Therefore, the enhanced transmittance is associated with a reduction in the reflectance.

A large number of publications have dealt with a paired negative-$\varepsilon$ and negative-$\mu$ bilayer composite and demonstrated unique transmission and emission properties [175, 176]. Since $\varepsilon_1\varepsilon_2 < 0$ and $\mu_1\mu_2 < 0$, when loss is neglected, both $k_{1z}$ and $k_{2z}$ are purely imaginary regardless of $k_x$. For simplicity, let us model the electric and magnetic properties of these two materials with loss using

$$\varepsilon_1(\omega) = 1 - \frac{\omega_p^2}{\omega^2 + \mathrm{i}\omega\gamma_e} \quad \text{and} \quad \mu_2 = 1 \tag{9.138a}$$

and

$$\varepsilon_2(\omega) = \varepsilon_2 \quad \text{and} \quad \mu_2(\omega) = 1 - \frac{F\omega^2}{\omega^2 - \omega_0^2 + \mathrm{i}\omega\gamma_m} \tag{9.138b}$$

where $\varepsilon_2$ is real positive. The dispersion relations for $\varepsilon_2 = 1, 2,$ and 4 are shown in Fig. 9.52a for both polarizations, assuming $\omega_0 = 0.5\omega_p$, $F = 0.785$, and $\gamma_e = \gamma_m = 0$. It can be seen that polaritons can be coupled with propagating waves in air, even at normal incidence. Figure 9.52b shows the transmittance spectra of such a bilayer with loss using $\varepsilon_2 = 4$ and $\gamma_e = \gamma_m = 0.0025\omega_p$, for TE wave incidence from air. The thicknesses are assumed to be $a = 2d = 0.425\lambda_p$. Sharp transmission peaks occur near the surface polariton resonance frequency. It should be noted that, if each individual layer is used, the transmittance is very small since evanescent waves exist in each medium. The calculation of the transmittance for a TM wave is left as an exercise. Jiang et al. [176] examined the resonance transmission of a PC, made of alternating layers of negative-$\varepsilon$ and negative-$\mu$ materials, for potential application of high-$Q$ filters.

The cross coupling of surface plasmon polaritons between corrugated metal films has been studied since the 1970s and employed to enhance light emission from tunnel junctions and light-emitting diodes. The coupled surface polaritons enable the coherent transmission of light through a narrow wavelength region in well-defined directions. A schematic of corrugated or grating-perturbed surfaces is shown in Fig. 9.53a,

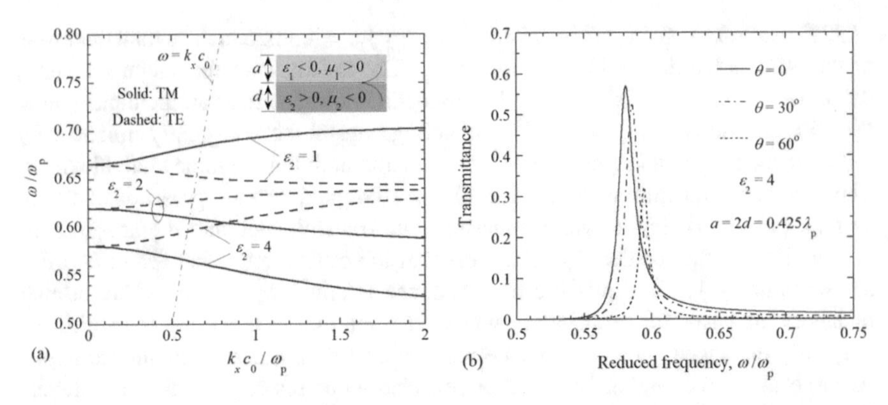


**Fig. 9.52** Dispersion relations and transmittance of a paired negative-$\varepsilon$ and negative-$\mu$ composite in air. **a** Dispersion relations for both polarizations between two semi-infinite lossless media. **b** Spectral transmittance for a TE wave at different angles of incidence

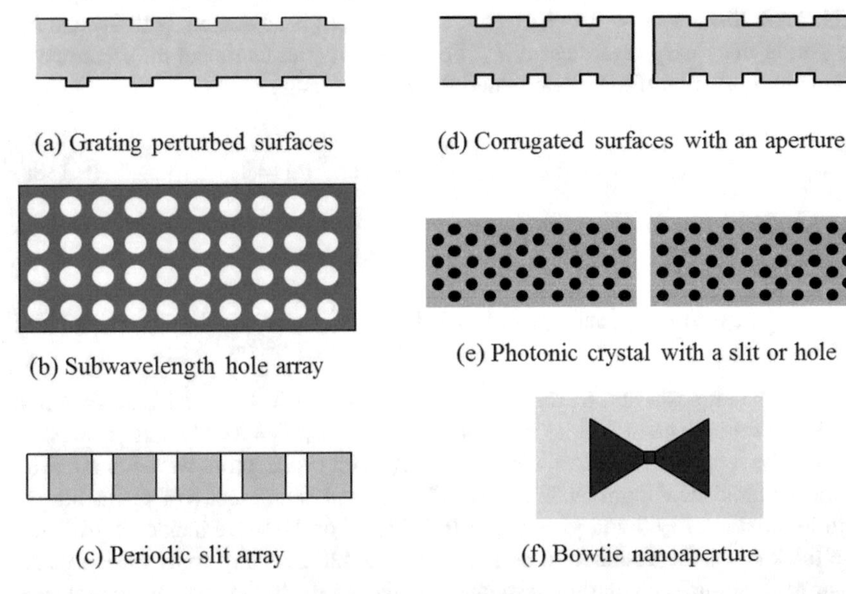


**Fig. 9.53** Various structures for transmission enhancement. **a** Grating or periodically perturbed surfaces for cross coupling of surface plasmons. **b** Subwavelength hole array. **c** 1D periodic slit array in a metal or polar material. **d** Corrugated metallic surfaces with an aperture for directional transmission. **e** Photonic bandgap structure for beaming light. **f** Bowtie nanoaperture for near-field focusing and transmission enhancement

along with some structures that have been extensively investigated for the control of light transmission through nanostructures. The work by Ebbesen and coworkers in 1998 on enhanced transmission of metallic films perforated with subwavelength holes has spurred a keen interest in studying transmission of light through nanostructures [177], including 2D hole arrays and 1D slit arrays, as shown in Figs. 9.53b–d, that include annular aperture and gratings. Coupled and localized SPPs and Fabry–Perot-type resonances have been used to explain the observed enhancement with some success, though quantitative predictions require rather complicated numerical modeling.

Another type of the enhanced transmission configuration is an aperture in corrugated surfaces, as shown in Fig. 9.53d for a metallic film and Fig. 9.53e that uses the bandgap of PCs to beam the light. The corrugated surface serves as a funnel to guide the light into the aperture or slit. In either case, the transmitted light becomes highly directional and the transmittance spectrum exhibits sharp peaks [177]. These structures may be considered as periodic slits with an infinite period or distance of separation. Circularly corrugated surfaces have also been used to funnel light through a subwavelength aperture. Surface plasmon-mediated transmission through single nanoholes and double slits without corrugated surfaces has also been studied. Light transmission through single nanoapertures of different shapes has been of great interest to nanolithography. For example, the bowtie nanoaperture as illustrated in Fig. 9.53f can couple SPPs to the near field and function as a nanoantenna to collect light and focus light to subwavelength spots [178].

Several mechanisms responsible for extraordinary optical transmission (EOT) in periodic 1D slit arrays were described including Wood's anomaly, SPPs or coupled SPPs, EMT, cavity resonance modes, as well as MPs [34, 124, 177, 179]. Chen et al. [179] fabricated 1D Au slit arrays on Si with a period between 800 nm and 1000 nm and measured the mid-infrared transmittance for individual polarization. They demonstrated the effects of Wood's anomaly for both polarizations at the grating-substrate interface. In order for Wood's anomaly to occur, the $z$-component of the wavevector inside the dialectic must be zero for given diffraction order. This gives the dispersion relation regardless of the polarization as

$$k_{\mathrm{W}} = k_x = n_{\mathrm{d}} k_0 - j \frac{2\pi}{\Lambda} \tag{9.139}$$

where $\Lambda$ is the grating period, $n_{\mathrm{d}}$ is the refractive index of the adjacent dielectric, and $j = 0, \pm 1, \pm 2, \ldots$ is the diffraction order. For normal incidence, Eq. (9.139) reduces to $\lambda_j = n_{\mathrm{d}} \Lambda / j$, where $\lambda_j$ is the wavelength in vacuum and $j = 1, 2, \ldots$ For TM waves, Wood's anomaly dispersion is very close to the SPP dispersion and there is essentially no distinction between Wood's anomaly and SPP. However, for TE waves, Wood's anomaly yields slight shift in the transmission spectrum [179]. When the wavelength is much greater than the period, the EMT can help explain the broadband high transmittance for TM wave [34, 179]. Since the electric field is perpendicular to the metal trips, the effective conductivity is very small and the metal gratings exhibit dielectric behavior as predicted by Eq. (9.128). However, for deep metal gratings with narrow slit opening, resonance transmission occurs. This is often explained by the Fabry–Perot-type cavity resonance or organ-pipe mode based on the following equation [180, 181]:

$$2kh_{\text{eff}} = 2\pi m, \quad m = 1, 2, \ldots \tag{9.140}$$

where $h_{\text{eff}} \approx h$ is a modified or effective grating height due to phase shift at the ends. Assuming the phase shift is zero so that $h_{\text{eff}} = h$, Eq. (9.140) gives a fundamental resonance wavelength of $\lambda_{\max} = 2h$. Wang and Zhang [124] used magnetic polariton to explain resonance transmission and absorption for a metal slit array. Because MPs or cavity resonance modes are localized resonances where a strong field exists inside the slit, both the transmittance and absorptance reach a maximum (although with a slightly different locations) about the resonance frequency. The radiative properties for the Ag slit array shown in Fig. 9.54a are plotted in Fig. 9.54b at normal incidence for $\Lambda = 500$ nm, $h = 400$ nm, and $b = 50$ nm. The MP1 resonance wavenumber corresponds to a wavelength of 1.33 μm, which is more than 3 $h$. Wood's anomaly (more accurately, SPP) also enhances absorptance or transmittance at $\lambda = \Lambda = 400$ nm. It should be noted that MPs are insensitive to the incidence angle unlike SPP. Furthermore, the resonance bandwidth for MP is usually broader than that for SPP. The *LC* circuit model, according to Fig. 9.37b and Eq. (9.113), also allows the quantitative prediction of the dependence of resonance frequency on the slit width as demonstrated in Ref. [124].

Liu et al. [129] designed a high-extinction-ratio polarizer in the wavelength region from 1.6 to 2.3 μm that has a transmittance above 89% for TM waves and essentially zero transmittance for TE waves using double-layer slit arrays made of Ag strips separated by a thin dielectric film. The openings of the upper and lower slits are blocked by the metal strips of the other gratings with relative a lateral shift between the upper and lower gratings. The excitation of two MP modes yields a relatively wideband transmittance for TM waves that is insensitive to the incidence angle up to about 60°. The transmission band may be tuned by the geometric parameters [129]. Furthermore, graphene-covered 1D or 2D grating arrays can support MPs in the mid-infrared [182]. In addition to the grating geometry, the transmittance band can be

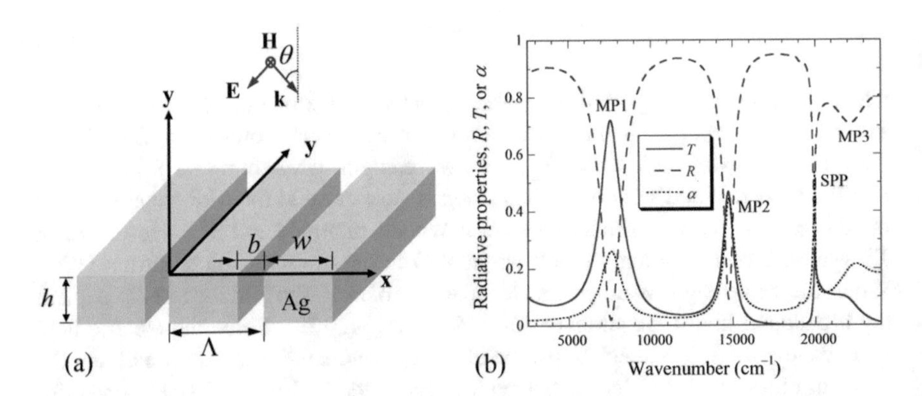


**Fig. 9.54** **a** Schematic of the 1D slit array with Ag gratings. **b** Reflectance ($R$), transmittance ($T$), and absorptance ($\alpha$) spectra at normal incidence for wavenumbers from 2500 cm$^{-1}$ to 25000 cm$^{-1}$ with a grating period $\Lambda = 500$ nm, height $h = 400$ nm, and slit width $b = 50$ nm [124]

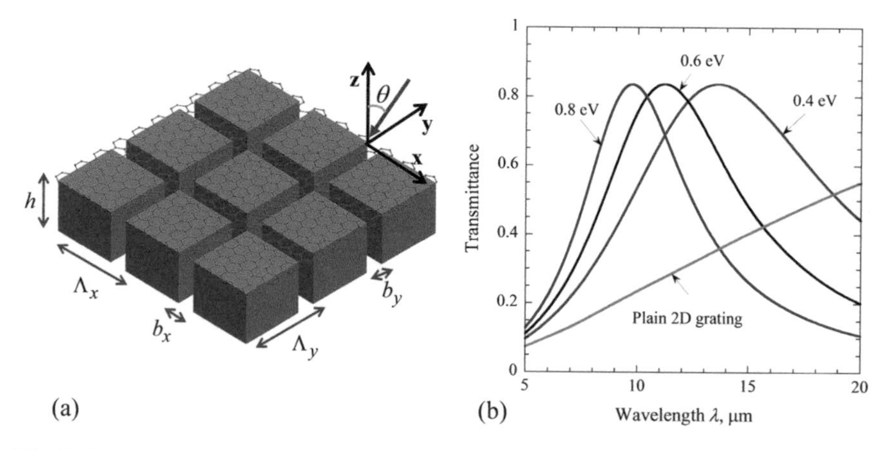


**Fig. 9.55** **a** Schematic of graphene-covered 2D pillar arrays with periods and slit widths in $x$- and $y$-directions as $\Lambda_x$, $\Lambda_y$, $b_x$, $b_y$, respectively, and Ag grating height $h$. **b** Transmittance of plain and graphene-covered 2D pillar arrays at normal incidence for different graphene chemical potentials ($\mu = 0.4$ eV, 0.6 eV, and 0.8 eV). The geometric parameters are $h = 200\,\text{nm}$, $b_x = b_y = 50\,\text{nm}$, and $\Lambda_x = \Lambda_y = 1000\,\text{nm}$ [182]

tuned by the chemical potential of graphene, as shown in Fig. 9.55 for a 2D Ag pillar array. The advantage of the 2D grating is that the transmittance is independent of the polarization at near normal incidence due to the structural symmetry. A commercial finite-difference time-domain (FDTD) software package was used to compute the transmittance. Without graphene, there is a broadband transmission with increasing transmittance toward longer wavelengths. The graphene coverage enables an MP excitation that can be modeled by an equivalent *LC* circuit. The excitation of MP results in a transmittance peak with significantly reduced reflectance [182].

## 9.6.2 Perfect Absorption

Black materials with low reflectivity have numerous applications; for instance, high-efficiency absorbers or emitters for energy conversion, radiometers and bolometers for space-borne infrared systems, calibration standards and backing materials for radiation measurements, and light trappers in optical systems [1, 41, 183, 184]. High absorptance can be achieved using various pigments in painting materials, metallic particles, oxidization (such as anodized aluminum), carbon black, and gas-evaporated gold blacks (ultralow-density pure Au flakes) [183]. This section describes black absorbers made by aligned carbon nanotubes (CNTs), metasurfaces, and with 2D materials such as graphene or hBN.

Vertically aligned CNT films have been shown to exhibit record high visible absorptance and very high absorptance all the way to far-infrared region [172, 184, 185]. The absorption of carbon-related materials in the ultraviolet is mainly due to

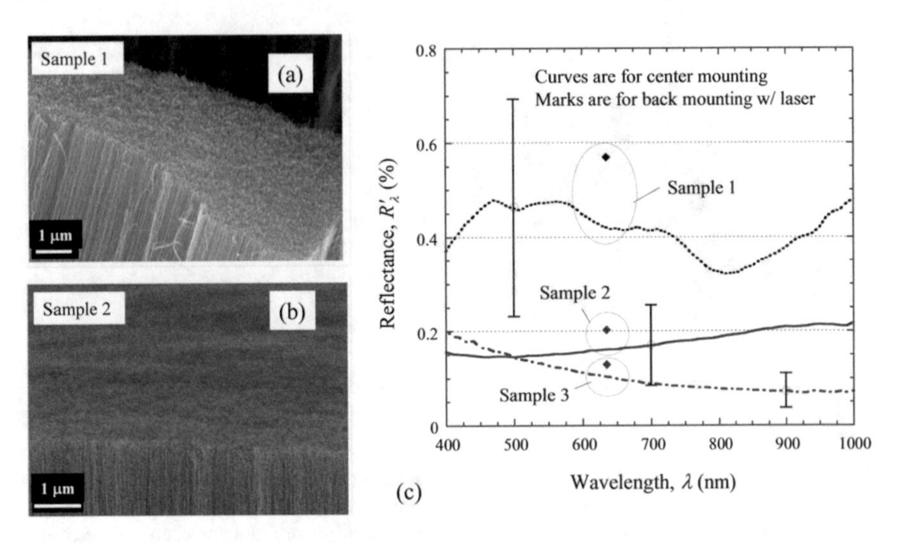


**Fig. 9.56** **a**, **b** SEM images of two vertically aligned CNT arrays identified as Sample 1 and Sample 2. **c** Directional-hemispherical reflectance in percentage at wavelengths from 400 nm to 1000 nm measured with a monochromator and a Si photodetector [185]

the electronic transition between the π and π* bands. As for graphene, the Dirac-like interband transition can have a major impact on the absorption spectrum at high frequencies in infrared and visible region. In the mid- to far-infrared region, absorption is mainly due to the intraband transition in the graphene sheet (like free electrons in metal). Due to the alignment and low density of the CNT arrays, the refractive index is not very far from that of free space, especially for the ordinary direction. On the other hand, the absorption is stronger in the extraordinary direction since the electric field is parallel to the tube direction [170–173]. Subsequently, the absorptance of vertically aligned CNTs can exceed 0.99 or 0.999 in the visible and near-infrared region.

Figure 9.56 shows the directional-hemispherical reflectance at wavelengths from 400 nm to 1000 nm measured for three CNT samples grown on Si. All these samples are opaque and the SEM for Sample 1 and Sample 2 are shown in Fig. 9.56a, b, respectively. In the spectral measurements with a tungsten halogen lamp and monochromator, the sample is mounted at the center of the integrating sphere. For measurement with a laser beam at 635 nm wavelength, the sample is mounted at the back of the integrating sphere. The relative error for the reflectance measurements is estimated to be 30% due to the low reflection signal. The surface of Sample 1 is relatively smooth and it is a specular black, whose reflectance is between 0.3 and 0.5%, suggesting an absorptance from 0.995 to 0.997. The reflectance for Samples 2 and 3 with relatively rough surfaces are even lower, and their absorptance ranges from 0.998 to 0.999 in the measured spectral region [185]. The BRDF of these samples has also been characterized by using high amplification gains to detect the very low reflected signal. The measured results agree well with model predictions based on the EMT described in Eqs. (9.136a), (9.136b) and a surface roughness model according to Eqs. (9.25a) and (9.25b) [185].

The dielectric functions of vertically aligned CNT array considering misalignment can be calculated from Eqs. (9.136a), (9.136b) to (9.137a), (9.137b). Then the modified Fresnel coefficient given in Eq. (9.126a) for reflection upon uniaxial medium can be used to calculate the directional-hemispherical reflectance by ignoring surface roughness. The normal reflectance of semi-infinite CNT films is shown in Fig. 9.57a for varying filling ratio and alignment factors in the wavelength region from 1 μm to 1000 μm. It can be seen that $R'_\lambda$ for $\phi = 0.05$ and $x = 0.99$ and for $\phi = 0.03$ and $x = 0.98$ is less than 0.1% at $\lambda < 25\,\mu\text{m}$ and less than 2% up to $\lambda = 100\,\mu\text{m}$. Beyond $\lambda = 200\,\mu\text{m}$, however, the reflectance increases to around 50%. Experimental measurements of vertically aligned single-wall CNT arrays demonstrated high absorptance of 0.98 for $0.2\,\mu\text{m} < \lambda < 200\,\mu\text{m}$ [186]. This is possibly due to surface texture effect that can create a profile of gradient refractive index to for impedance matching with that of air. Another reason is that CNTs can have different band structures, chemical potentials, as well as scattering rate than what were used in the modeling.

In general, the reflectance increases as the angle of incidence approaches 90°. Therefore, the hemispherical reflectance is expected to be greater than that at normal incidence. The calculated hemispherical absorptance spectra of the CNT arrays with $\phi = 0.05$ and $x = 0.98$ by integrating over the zenith angle are plotted in Fig. 9.57b. In the near- and mid-infrared region for random polarization, the hemispherical absorptance exceeds 0.95 for $\lambda < 9\,\mu\text{m}$ and exceeds 0.90 for $\lambda < 33\,\mu\text{m}$. Interestingly, the absorptance is higher for $s$-polarized incident waves at $\lambda < 40\,\mu\text{m}$. For an isotropic medium, the reflectance for $p$-polarization is smaller than that for $s$-polarization at any incidence angles. This is not always the case for a uniaxial

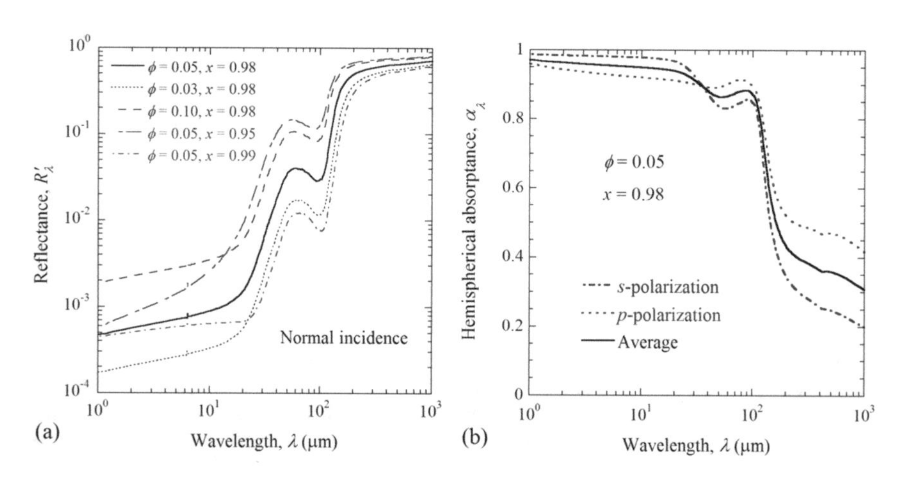


**Fig. 9.57** Predicted radiative properties of vertically aligned CNT arrays. **a** Normal reflectance spectra at wavelengths from 1 μm to 1000 μm, with varying filling ratios and alignment factors. **b** Hemispherical absorptance spectra for $s$- and $p$-polarization, and the average of the two [172]

medium. The reflectance at large incidence angles for CNT arrays can be higher for $p$-polarization than for $s$-polarization. Subsequently, the absorptance is higher for $s$-polarization than for $p$-polarization unless the wavelength exceeds 40 μm, when the reflectance is always higher for $s$-polarization than for $p$-polarization [172]. Aligned doped silicon nanowires can also achieve broadband nearly perfect absorptance in the mid-infrared region for both polarizations [57, 187]. Furthermore, the doped silicon nanowires form a hyperbolic metamaterial and can support negative refraction with low loss in the mid-infrared [187].

Patterning subwavelength metallic elements on a thin dielectric film, which is coated on a metal ground plane, forms a metamaterial (also called metasurface) that can excite magnetic polaritons as discussed previously. The localized magnetic resonance results in an omnidirectional high absorption peak as discussed in Sect. 9.5.4 [120, 123]. The metasurface may be understood to have an effective magnetic permeability due to the diamagnetism [135–137]. By using 2D structures, the absorption peaks can be achieved for incident waves with both $p$- and $s$-polarizations. Different pattern shapes (such as strips, crosses, circular disks, squares, and square rings) and sizes can be formed to control the emission peaks to make it broader or to create multiple peaks [130, 167, 188].

Sakurai et al. [188] studied the effect of patterns on the polarization dependent absorptance. The normal spectral emittance spectra for two structures, shown in the insets, are plotted Fig. 9.58. The FDTD method was used to calculate the infrared reflectance of the two metamaterial absorbers by solving the Maxwell equations. The

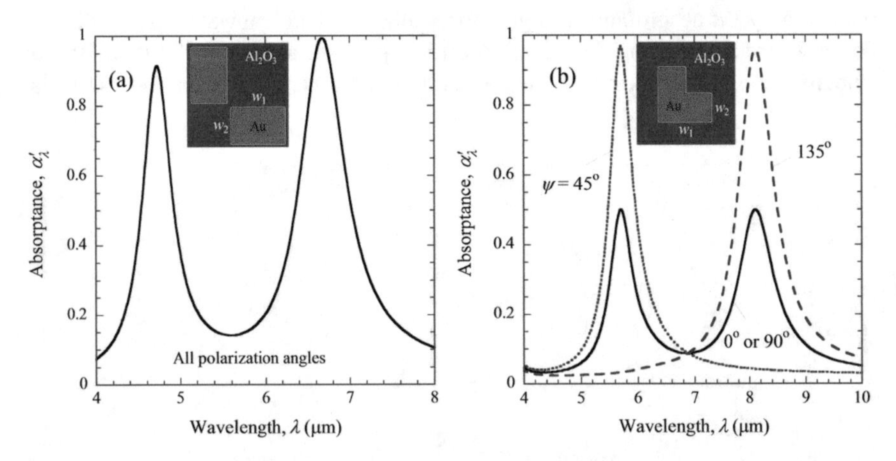


**Fig. 9.58** Absorptance spectra at normal incidence for two different patterns as indicated in the insets. **a** Pattern made of two separate rectangles ($L = 1.7$ μm and $w = 1.1$ μm) for which the absorptance is independent of the polarization angle $\psi$; **b** Pattern made of an L-shape ($L = 1.7$ μm, and $w = 0.85$ μm) for which the absorptance depends on $\psi$. The unit cell period is $\Lambda = 3.2$ μm for either pattern in both the $x$- and $y$-directions. The pattern is made of Au film with a thickness of 100 nm on a 140-nm-thick continuous $Al_2O_3$ film, which is coated on a semi-infinite Au ground plane [188]

optical constants of gold are taken from Palik [2]. For simplicity, losses and dispersion of $Al_2O_3$ are neglected by assuming a constant refractive index $n = 1.57$ based on tabulated value at $\lambda = 6.25$ μm [2]. The polarization angle $\psi$ is defined as the angle between the electric field and the plane of incidence such that $\psi = 0°$ and 90° correspond to TM and TE waves, respectively. The azimuthal angle for normal incidence is set to $\phi = 0°$ or the plane of incidence is the *x*-*z* plane. When $\psi = 0°$, the electric field is parallel with the *x*-direction. In this case, $\psi$ defines the angle between the electric field and the *x*-axis. It can be seen from Fig. 9.58a that absorptance for the double-rectangle patterned metamaterial is enhanced significantly around $\lambda = 4.72$ μm and 6.67 μm. Furthermore, the absorptance is independent of the polarization angle. On the other hand, the polarization effects can be observed in Fig. 9.58b for the L-shape patterned metamaterial. Note that the curves for the cases $\psi = 0°$ and 90° overlap; however, for $\psi = 45°$ and 135°, the absorptance spectra are dramatically different. The peak at $\lambda = 4.72$ μm shows up for $\psi = 45°$ but disappears when $\psi = 135°$. The reverse is true at $\lambda = 6.67$ μm where the peak occurs for $\psi = 135°$ but disappears for $\psi = 45°$. The following focuses on the case of the double-rectangle patterned metamaterial.

The field distributions revealed that each absorption peak in Fig. 9.58a corresponds to a resonance absorption by each rectangle. The *LC* circuit model can be used to predict the absorption peaks due to the excitation of the fundamental mode of MP based on the short and long sides of the rectangle [188]. The geometric symmetry makes the absorption spectra independent of the polarization angle. For the L-shape, two resonance modes are observed at $\psi = 45°$ and 135° with different field distributions. The MP resonance frequency can also be predicted by modifying the effective width of the strip and the capacitance between the patterned structure and the ground plane. At $\psi = 0°$ or 90°, both modes can be excited but with roughly half of the strength, resulting in two peaks with roughly half of the absorptance as shown in Fig. 9.58b. Zhao et al. [188] obtained an analytical expression to describe the polarization dependence (or independence) of periodic micro/nanostructures.

Broadband perfect absorption can also be achieved with metallodielectric multilayers, made into sawtooth gratings [189]. As discussed in Sect. 9.5.6, metallodielectric multilayers form an effective hyperbolic metamaterial. The trapezoidal shape helps impedance matching and can create multiple waveguide modes that trap the incident radiation at different height when the group velocity is zero. Natural hyperbolic materials [59, 164, 165] may also be used to achieve perfect absorption. Figure 9.59 shows the dielectric functions of hBN and the calculated spectral absorptance, demonstrating a mid-infrared perfect absorption band [190].

The sheet of hBN consists of alternating boron and nitrogen atoms in a honeycomb arrangement. Due to its wide bandgap (~ 5.9 eV), hBN is a dielectric material. Due to its 2D layered crystalline structure, hBN has anisotropic lattice vibration modes in the mid-infrared, making it a natural low-loss hyperbolic metamaterial. The in-plane phonon modes ($\omega_{TO,\perp} = 1370\,cm^{-1}$ and $\omega_{LO,\perp} = 1610\,cm^{-1}$) and out-of-plane phonon modes ($\omega_{TO,\parallel} = 780\,cm^{-1}$ and $\omega_{LO,\parallel} = 830\,cm^{-1}$) contribute to the in-plane (**E** lies in the *x*-*y* plane, denoted by $\perp$) and out-of-plane (**E** parallel to the optic

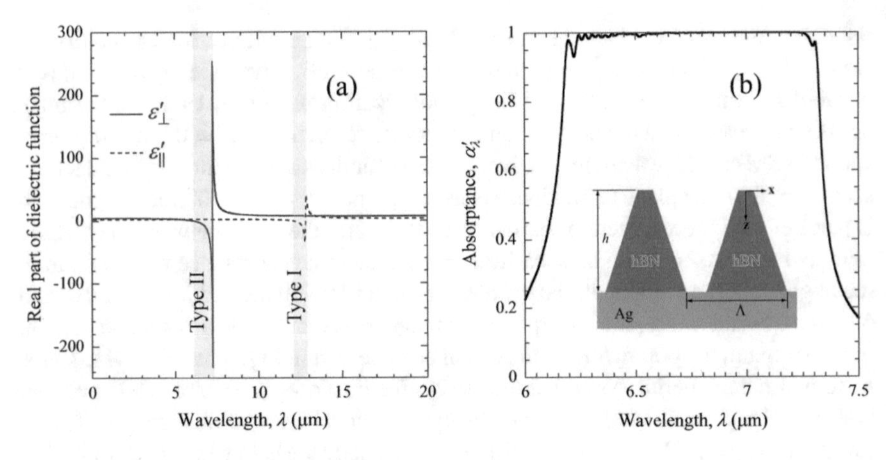


**Fig. 9.59** **a** Real part of the dielectric function of hBN with two hyperbolic bands. **b** Calculated absorptance for TM waves of 1D trapezoidal hBN gratings on an opaque Ag layer [190]. The period $\Lambda = 3\,\mu\text{m}$, height $h = 10\,\mu\text{m}$, and the bases of the trapezoid are 40 nm and 2 μm, respective

axis or the $z$-direction, denoted by$\|$) dielectric functions, which can be written as follows [35, 59]:

$$\varepsilon_\perp = \varepsilon_{\infty,\perp}\left(1 + \frac{\omega_{\text{LO},\perp}^2 - \omega_{\text{TO},\perp}^2}{\omega_{\text{TO},\perp}^2 - i\gamma_\perp\omega - \omega^2}\right) \tag{9.141a}$$

$$\varepsilon_\parallel = \varepsilon_{\infty,\parallel}\left(1 + \frac{\omega_{\text{LO},\parallel}^2 - \omega_{\text{TO},\parallel}^2}{\omega_{\text{TO},\parallel}^2 - i\gamma_\parallel\omega - \omega^2}\right) \tag{9.141b}$$

The remaining parameters used in the model are $\varepsilon_{\infty,\parallel} = 2.95$, $\gamma_\parallel = 4\,\text{cm}^{-1}$, $\varepsilon_{\infty,\perp} = 4.87$, and $\gamma_\perp = 4\,\text{cm}^{-1}$. The actual parameters may vary depending on the defects and isotope fractions [191]. The real part of the dielectric function becomes negative between the TO and LO phonon modes, making the in-plane and out-of-plane dielectric functions of hBN possess opposite signs in either Reststrahlen band as shown in Fig. 9.59b.

The absorptance for TM waves of the designed structure displayed in Fig. 9.59b is close to unity in a relatively broadband for $6.2\,\mu\text{m} < \lambda < 7.3\,\mu\text{m}$ in the type II hyperbolic region. The calculations are based on a modified RCWA considering anisotropic media [35, 190]. Additional calculations not shown here demonstrate that the broadband absorptance is insensitive to the incidence and remains higher than 0.8 even at an incidence angle $\theta = 80°$. Furthermore, the mechanisms can be explained by considering local absorption distribution inside the hBN trapezoidal according to

$$w(x,z) = \frac{1}{2}\varepsilon_0\omega\left(\varepsilon''_\perp|E_x|^2 + \varepsilon''_\parallel|E_z|^2\right) \tag{9.142}$$

where $w$ is the local power dissipation density. The field distribution and dissipation profile demonstrated a slow-light effect and the absorption distribution is strongly wavelength dependent. The absorption in the short wavelength end occurs near the top of the trapezoid and that in the long-wavelength end occurs near the bottom of the trapezoid. Furthermore, due to hyperbolic phonon polaritons (HPhP), resonance perfect absorption may also exist in both trapezoid and square gratings of hBN [190].

For the 1D trapezoidal structure, perfect absorption based on the hyperbolic polaritons can only be achieved for extraordinary waves ($p$-polarization). Wang et al. [165] proposed to use pyramid structures to enable perfect absorption for both polarizations. Using crystalline $Bi_2Te_3$, which exhibits anisotropy in the visible and near-infrared region, perfect absorption was predicted at wavelengths from 300 nm to 2400 nm for harvesting solar energy. Furthermore, with the selected pyramidal structures on Ag substrate, the absorptance was predicted to exceed 99.9% in the whole spectral region [165]. Fabrication of the nanoscale structure period of 200 nm with micrometer (3 μm height) remains a major challenge.

Graphene may be used as ultrafast photodetectors from the visible to the near infrared due to its higher carrier mobility and interband absorption. However, the absorptance of a monolayer graphene sheet is only about 2.3% in this spectral region; this is related to the fine structure constant [192]. Various methods can be used to boost the absorption of graphene, including the use of microcavities, photonic crystals, optical antennas, localized plasmonic resonators, etc. Zhao et al. [193] employed deep metal grating to enhance graphene absorption and to achieve nearly perfect absorption when MPs or SPPs are excited. The structure is graphene-covered deep metal gratings, as shown in Fig. 9.60a and the calculated absorptance spectra with and without graphene are shown in Fig. 9.60b for $p$-polarization at an incidence angle $\theta = 10°$. In the calculation using RCWA, graphene is treated as layer with a finite thickness of 0.3 nm.

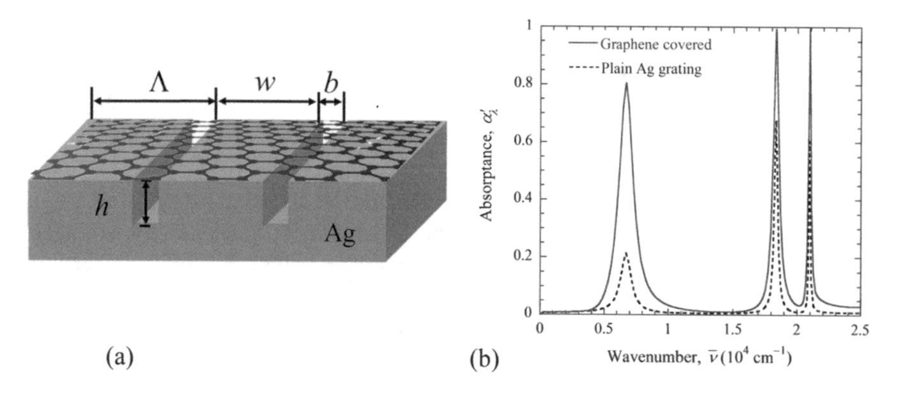


**Fig. 9.60** **a** Schematic of graphene-covered Ag deep grating. **b** Calculated absorptance at an incidence angle $\theta = 10°$ for TM waves, with and without graphene, for $h = 200$ nm, $b = 30$ nm, $\Lambda = 400$ nm, and graphene chemical potential $\mu = 0.3$ eV [193]

Three peaks can be seen in Fig. 9.60b at wavenumbers $\bar{\nu} = 6700\,\text{cm}^{-1}$ (1.49 μm), $18350\,\text{cm}^{-1}$ (545 nm), and $20930\,\text{cm}^{-1}$ (478 nm). These peaks correspond well with MP1, MP2, and SPP in the deep grating. Note that due to interband transitions, graphene behaves as a resistive element with a finite conductivity at wavelength shorter than 2.1 μm. For a plain Ag grating, the absorptance is only 0.21, 0.66, and 0.57 at the resonance of MP1, MP2 and SPP, respectively. With graphene coverage, the absorptance at the corresponding wavenumber is raised to 0.81, 0.99, and 1.0, respectively. Yet, it is interesting to find out the absorption distribution and what fraction is actually absorbed by the graphene.

Zhao et al. [193] calculated the absorption distribution using the power dissipation density as discussed in Eq. (9.142) for isotropic materials as

$$w(x, z) = \frac{1}{2}\varepsilon_0\omega\varepsilon''(x, z)|\mathbf{E}(x, z)|^2 \tag{9.143}$$

Furthermore, the fraction of absorption can be calculated by the ratio of the total absorbed power within a volume $V$ to the incoming power through the exposed surface area $A$ as follows:

$$\alpha = \frac{\int\int\int w(x, z)dV}{\frac{1}{2}c_0\varepsilon_0|\mathbf{E}_{\text{inc}}|^2 A\cos\theta} \tag{9.144}$$

Clearly, the denominator is the product of the Poynting vector and the projected surface area. It was shown that only the suspended graphene contributes to absorption, while the graphene over Ag does not. At $\bar{\nu} = 6700\,\text{cm}^{-1}$ (MP1), the fraction of absorption by graphene is 0.68, which is nearly 30 times greater than a free-standing graphene. The fraction of absorption by Ag grating is 0.13, which is less than the value of 0.21 without the graphene coverage. Applying graphene reduces the field strengths inside the trench region but graphene itself significantly absorbs the incident radiation. At MP2 and SPP resonance frequencies, the fraction of absorption of graphene is 0.77 and 0.80, respectively, while the fraction of absorption of the Ag grating is reduced to 0.22 and 0.20, respectively. In the mid-infrared, graphene can modify the MP resonance frequency since its inductance is not negligible due to intraband transitions [152, 182].

Understanding the distribution of local absorption and electromagnetic energy density in nanostructures are very important for solar cell, local heating, and other optoelectronic devices [139]. In addition to graphene, other 2D materials such as hBN can couple with deep metal gratings to yield perfect absorption peaks, especially when hybrid hyperbolic phonon-plasmon polaritons are excited [35, 59].

### 9.6.3 Tailoring Thermal Emission with Nanostructures

Thermal emission is a spontaneous emission process that occurs from any object. Spontaneous emission from the molecules or atoms can be enhanced or suppressed by micro/nanostructures, such as photonic crystals and nanocavities [21, 194]. It has been clearly demonstrated that the spontaneous emission from light emitters embedded in photonic crystals can be suppressed by the photonic bandgap, whereas the emission efficiency in the direction where resonance modes exist can be enhanced. Thermal radiation emitted from solids is generally manifested as broadband and quasi-isotropic. By introducing thin-film coatings and multilayer structures, the emission spectrum can be significantly modified. Wavelength-selective coatings have been developed since the 1960s for space application and solar collectors. Gratings can also modify the wavelength and angular dependence of thermal emission [58]. These approaches can be generalized to multidimensional complex microstructures, including photonic crystals, for spectral and directional control of spontaneous emission. There are a number of applications that require spectral and directional selection of thermal radiation. Besides space application and solar energy, thermophotovoltaic (TPV) devices utilize a heating source or an emitter around 1500 K to generate electricity based on photovoltaic principle. The efficiency is often limited by the portion of absorption of long-wavelength photons that cannot create electron-hole pairs in the photovoltaic cell [65, 126, 127].

It has been known for a long time that radiative properties, especially the directional and spectral properties, can be modified by surface roughness and structures. Most of the earlier studies dealt with rather simple geometries and did not consider diffraction; see [1, 8]. The emergence of microfabrication and the increased computing capabilities have led to more systematic investigations of the effect of microstructures and material properties on thermal emission and absorption characteristics. Hesketh et al. [181] published a series of studies on the thermal emission from periodically grooved micromachined silicon surfaces. The grooves were 45 μm deep with straight ridges etched on heavily doped *p*-type Si wafers, with a grating period $\Lambda$ ranging from 10 to 22 μm. Thermal emission was measured at temperatures between 300 and 400 °C at wavelengths ranging from 3 to 14 μm. Compared with smooth Si wafers, the grooved surfaces increased the spectral emittance and the observed enhancement was polarization dependent even at normal incidence. The observed resonance emittance enhancement was explained by *organ pipe* resonant modes since geometric optics models largely failed to predict the observed behavior. Heinzel et al. [195] fabricated 2D arrays of tungsten circular pillars as near-infrared emitters and hole arrays on gold films as wavelength-selective filters, for applications in TPV systems. For these structures, the lateral period was between 1 and 2 μm, and the thickness was between 200 and 300 nm. Surface plasmons may play a key role in the wavelength selection. Theoretical modeling based on a 2D RCWA was performed and compared with experiments. The emittance was measured in vacuum at temperatures up to 1700 K with a Fourier-transform spectrometer [195].

Greffet and coworkers [58] showed a strongly coherent thermal emission mediated by surface phonon polaritons using a SiC grating with a period of 6.25 μm and a height of 0.28 μm. They observed *coherent thermal emission* for TM waves within narrowband at wavelengths near 11 μm toward well-confined directions, when heated to 800 K. Spectrally coherent thermal emission means that the emission is confined in a narrow wavelength region for any given direction; this is also referred to as *temporal coherence*, because coherence time and coherence length are interrelated. A nearly monochromatic radiation will have a very long coherence length and time. When the emission at a given wavelength is confined to a narrow angular range, it is referred to as *spatial coherence*, like a collimated beam whose wavefronts do not alter significantly as it travels. At the resonance conditions, surface plasmon or phonon polaritons are coupled with spontaneous emission due to randomly fluctuating charges or dipoles in the thermal field; consequently, thermal emission is enhanced at a particular wavelength and direction. At wavelengths where the surface mode is not excited, the radiation emitted inside the material is either absorbed by the neighboring atoms or reflected back by the material-air interface, yielding a very low emissivity that is typical for metallic materials in the infrared. It should be noted that the bandwidth for spontaneous thermal emission is far greater than that of a laser, which operates under the principle of stimulated emission. Nevertheless, a much longer coherence length than that of blackbody radiation or emission from plain solids could be achieved.

Maruyama et al. [196] fabricated 2D microcavities using Cr-coated Si surfaces and demonstrated discrete thermal emission peaks from these structures as a result of cavity resonances and the enhanced density of states. The structure is illustrated in Fig. 9.61. The resonance wavelengths are identified by the cavity modes (that include both TE and TM waves) that must satisfy

$$\lambda(m, n, p) = 2\left[\left(\frac{m}{a}\right)^2 + \left(\frac{n}{b}\right)^2 + \left(\frac{p}{2h}\right)^2\right]^{-1/2} \tag{9.145}$$

where $m$ or $n = 0, 1, 2, 3$, $p = 0, 1, 3, 5$ (odd number only), and no more than one index can be zero. Equation (9.145) has been verified by experiments [196, 197]. It

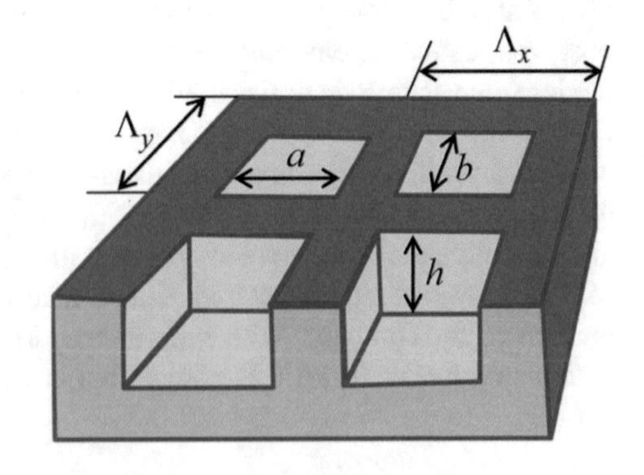


**Fig. 9.61** Illustration of 2D microcavities

should be noted that SPPs can also be observed. The localized cavity modes require that the wavelength to satisfy Eq. (9.145) and is usually not sensitive to the incident directions. On the other hand, as discussed previously, SPPs are very sensitive to the incidence angle.

For 1D deep gratings, resonance cavity mode can occur with $m = 0$ and $p = 1$, so that the fundamental mode is predicted by $\lambda_{max} = 4\,h$. Using the *LC* circuit model, it can be shown that the excitation wavelength for MP1 can be much greater than 4 h [131]. This is mainly due to the large kinetic inductance at smaller gap distances (or coupled SPPs). However, if the structure geometry and wavelengths are scaled to the far-infrared or microwave region, the effect of kinetic inductance is small. The predicted MP1 resonance wavelength is between 4 h and 5 h. Yang et al. [132] demonstrated MP in Al deep gratings. The MP1 resonance wavelength for a 1.44-μm-deep grating was 7.18 μm or 5 h. The values are consistent among the RCWA simulation, *LC* circuit model, and measurements for three samples.

Various types of microstructures have been considered for wavelength selective emitters for TPV applications such as 1D grooves and complex gratings, multilayered films, microcavities, inverse opal and woodpile photonic crystals, perforated hole arrays, metasurfaces, etc. [27, 65, 195–199]. Zhao et al. [127] designed a 2D trilayer structure as a wavelength-selective and polarization-insensitive TPV emitter. The structure is similar to the metasurface structure mentioned earlier, except that the metallic material is tungsten with square patterns as shown in Fig. 9.62a. The simulation was based on a 2D RCWA code to calculate the directional-hemispherical spectral reflectance, and the emittance can then be determined using Kirchhoff's law. The normal emittance spectrum for the 2D structure was compared with those for 1D

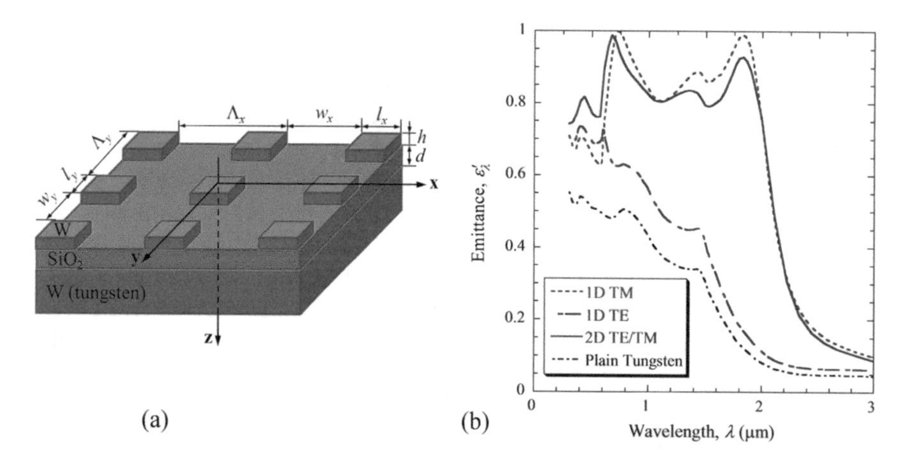


**Fig. 9.62** **a** Schematic of the 2D tungsten grating/thin-film nanostructure. **b** Predicted normal emittance spectra of the 2D structure with $\Lambda_x = \Lambda_y = 600$ nm, $l_x = l_y = 300$ nm, $h = d = 60$ nm [126, 127]. The emittance spectra of the 1D counterpart for both polariton and of a plain tungsten are also shown. Note that for the 1D metamaterial, the tungsten is continuous in the *y*-direction ($l_y = \Lambda_y$)

structure and plain tungsten as shown in Fig. 9.62b. The optical constants were taken from Palik [2]. The objective is to achieve high emittance at wavelength shorter than the bandgap of the TPV cell, which corresponds to $\lambda_g = 2.1\ \mu m$ (for $In_{0.2}Ga_{0.8}Sb$) and to suppress the emittance at wavelengths longer than $\lambda_g$. Note that the normal emittance is independent of polarization for the 2D structure. The emittance spectrum of the 1D grating/thin-film structure for TE waves is similar to that of plain tungsten except for the peak at 0.6 μm, which is due to Wood's anomaly.

The overall emittance at normal direction is the average of those for TE and TM waves. As an example, the normal emittance $\lambda = 1.7\ \mu m$ for the 2D structure is 0.85 and the emittance averaged over the two polarizations for the 1D structure is only 0.58. Therefore, the throughput and efficiency of the TPV system can be significantly improved with the 2D grating/thin-film structure [127]. The emittance spectrum for the 1D structure with TM waves and that for the 2D structures are very similar, both contain two major emission peaks (near 0.7 and 1.8 μm) that do not exist in the spectra for the TE wave and plain tungsten. It has been shown that [127] the emittance peak around $\lambda = 0.7\ \mu m$ is due to the excitation of SPP, while the emittance peak at $\lambda = 1.83\ \mu m$ is due to the coupling of the magnetic resonance inside a micro/nanostructure with the external electromagnetic waves or magnetic plasmon polariton. When the MP is excited, the magnetic field is strongly enhanced in the dielectric layer between the tungsten grating and tungsten substrate. Several small peaks located near 0.4, 0.8, and 1.4 μm are associated with the interband transitions of tungsten [127].

The effect of MPs on the thermal emission has been demonstrated using a trilayer metamaterial as shown in Fig. 9.38a [128]. The structure was fabricated on a Si substrate coated with a 200 nm $SiO_2$ film as a barrier layer using the plasma-enhanced CVD. A 200 nm Au film was thermally evaporated to the sample surface with a 30 nm Ti film as an adhesive layer. After a 185 nm $SiO_2$ spacer was deposited onto the Au film, the Au grating was formed with deep-UV lithography and lift-off process. The emittance was measured using the emissometer setup described in Sect. 8.5.4 (see Fig. 8.30). The measured emittance spectra of one sample with a period $\Lambda = 7\ \mu m$ and Au strip width of 3.5 μm are shown in Fig. 9.63. The thickness of the Au strip array is 170 nm. Two detectors were used to measure different spectral regions: a DTGS pyroelectric detector was used for $1000\ cm^{-1} < \bar{\nu} < 2000\ cm^{-1}$ and a liquid-nitrogen-cooled InSb photodetector was used for $2000\ cm^{-1} < \bar{\nu} < 3000\ cm^{-1}$. Note that the measurements are for *p*-polarization with different emission angles. The peak MP1 can be well predicted by the *LC* circuit model and is insensitive to the incidence angle. MP2 does not show up in the normal direction but can be seen at oblique incidence. This agrees well with the RCWA predictions as shown in Fig. 9.40b, although the materials, geometric parameters, and wavelength region are all different. MP3 and be coupled with SPP to form anticrossing modes that is very sensitive to the emission angle as shown in Fig. 9.63b.

Planar structures can also be used for tailoring the thermal emission spectrum such as the asymmetric Fabry–Perot resonance cavity [16]. Fu et al. [200] proposed to use the paired negative-$\varepsilon$ and negative-$\mu$ bilayer to achieve coherent emission

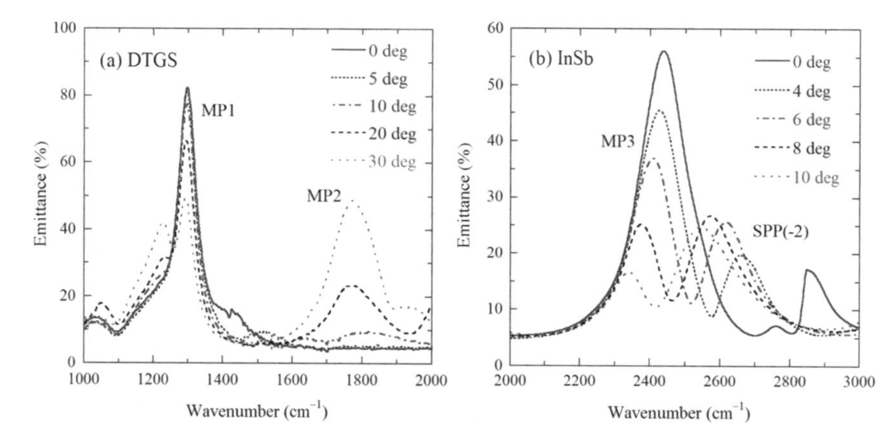


**Fig. 9.63** The emittance for TM wave at 700 K for different directions measured with the high-temperature emissometer based on an FTIR spectrometer using **a** a DTGS detector and **b** an InSb detector [127]

through the excitation of surface polaritons at all angles, for both TE and TM waves, as illustrated in the following example.

**Example 9.9** For a thin metallic-type film, with $\mathrm{Re}(\varepsilon_1) < 0$ and $\mathrm{Re}(\mu_1) > 0$, of thickness *d*, on an opaque magnetic material, with $\mathrm{Re}(\mu_2) < 0$ and $\mathrm{Re}(\varepsilon_2) > 0$, calculate the emittance, using the functions given in Eqs. (9.138). Assume the parameters are $F = 0.785$, $\omega_0 = 0.5\omega_\mathrm{p}$, $d = 0.5\lambda_\mathrm{p} = \pi c/\omega_\mathrm{p}$, $\varepsilon_2 = 4$, and $\gamma_\mathrm{e} = \gamma_\mathrm{m} = 0.002\omega_\mathrm{p}$.

**Solution** Under the lossless conditions, the polariton dispersion relations are the same as shown in Fig. 9.52a for two semi-infinite media. The directional-spectral emittance can be calculated by $\varepsilon'_\lambda = 1 - R'_\lambda$, because the magnetic medium is semi-infinite, where $R'_\lambda$ can be evaluated using Eq. (9.101) for each polarization. The calculation results are shown in Fig. 9.64a at normal direction as well as at $\theta = 45°$ for either TE or TM wave using the reduced frequency. It can be seen that the peak shifts toward lower frequencies for the TM wave and higher frequencies for the TE wave as $\theta$ increases and the center frequency of the peak $\omega_\mathrm{c}$ is in good agreement with the polariton dispersion curves, shown in Fig. 9.52a. The *Q*-factor, defined as $Q = \omega_\mathrm{c}/\delta\omega$ with $\delta\omega$ being the FWHM, is around 100. Figure 9.64b shows the angular distribution of the emission at the center frequencies shown on the left figure. The emission is not diffuse but rather direction selective.

Fu et al. [200] further proposed to use a three-layer structure with a negative-$\varepsilon$ film and a negative-$\mu$ film onto a negative-$\varepsilon$ substrate to achieve a higher *Q* and a spatially coherent source. In such a case, surface polaritons at both sides of the negative-$\mu$ medium can be coupled. A temporally coherent diffuse emitter was also predicted.

Surface electromagnetic waves coupled with a PC can also produce coherent emission characteristics because a PC can support surface modes or surface waves

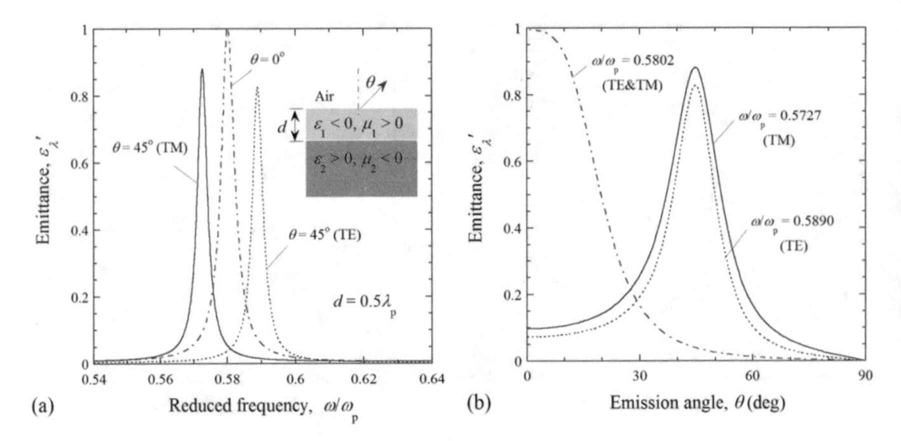


**Fig. 9.64** Emissivity of a negative-$\varepsilon$ layer of thickness $d$ on a semi-infinite negative-$\mu$ layer. **a** Frequency dependence at fixed angles. **b** Angular dependence at fixed frequencies

for both the TM and TE waves in the stop band [17]. If a metallic layer is coated on a 1D PC, surface waves can be excited by a propagating wave in air; this will result in a strong reduction in the reflectance at the resonance frequency. Lee et al. [201] predicted coherent thermal emission based on a modified 1D PC coated with a thin film of SiC. When the thicknesses and dielectric properties are adjusted, surface waves can be excited in the stop band of the PC by radiative waves propagating in air, for either polarization. Subsequently, the emission from the proposed structure contains sharp peaks within a narrow spectral band and toward well-defined directions. The geometry and the electric field distribution are illustrated in Fig. 9.65a, and the field plots are shown in Fig. 9.65b at different frequencies.

A PC is a heterogeneous structure as discussed previously, here, $\varepsilon_a = n_a^2$, $\varepsilon_b = n_b^2$, and $\mu_a = \mu_b = 1$ (nonmagnetic). Therefore, it is inappropriate to define the equivalent $\varepsilon$ and $\mu$ of the PC separately by considering it as a homogeneous medium.

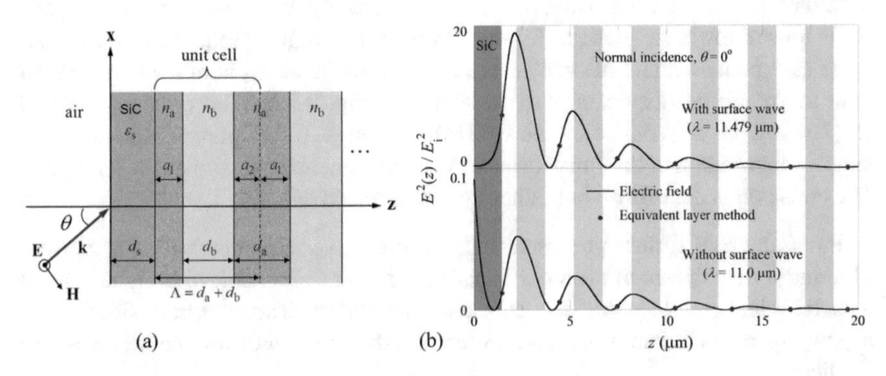


**Fig. 9.65** Schematic of the SiC-coated 1D PC (upper) and the field distributions (lower) for a TE wave incident from air [201]

However, surface waves can be excited at the stop band of the PC because there exists in the PC an *effective evanescent wave*, which is an oscillating field whose amplitude gradually decays to zero as $z$ approaches infinity. The effective evanescent wave does not carry energy into a semi-infinite PC. Note that the wavelength range corresponding to the stop bands of the PC can be scaled by changing the thickness of the unit cell. Here, $\Lambda$ is chosen to be 3 μm in order to approximately match the wavelengths corresponding to the first bandgap of the 1D PC, shown in Fig. 9.19, with the phonon absorption band of SiC. Surface waves can be excited at the SiC-PC interface within the SiC phonon absorption band for both polarizations. By using the equivalent layer method or the supercell method, it is possible to obtain dispersion relations of surface waves between a PC and another medium [202–204].

Figure 9.65b shows the square of the electric field, normalized to the incident field, inside the SiC-PC structure at $\theta = 0$. The real part of the complex electric field is used to show the actual field inside the structure. The solid line represents the field calculated from the matrix formulation described in Sect. 9.2.2. An oscillating field exists inside the PC, and the amplitude of the oscillating field decays gradually with increasing $z$. The dots represent the electric field obtained using the equivalent layer method, which matches the matrix solutions at the boundaries of each unit cell. The upper panel corresponds to the wavelength $\lambda = \lambda_c = 11.479$ μm when a surface wave is excited, and the lower panel corresponds to $\lambda = 11.0$ μm where no surface mode is excited. The field strength at the boundary between SiC and the PC is enhanced by more than an order of magnitude due to excitation of the surface wave. When a surface wave is excited, the incident energy is resonantly transferred to the surface wave, which causes a large absorption in SiC. Because SiC is the only material in the structure that can absorb the incident energy, it is also responsible for the emission of radiation from the SiC-PC structure. It is interesting to note that the maximum electric field is slightly off from the interface between SiC and the PC. If a smooth curve connects all the dots, the magnitude of the electric field will be maximum at the SiC-PC interface and decay gradually deep into the PC. Furthermore, the Poynting vector or the energy flux toward the positive $z$-direction is zero inside the PC at the stop band. Therefore, the effective field inside the PC at the stop band resembles an evanescent wave in a semi-infinite medium. The fact that the field near the SiC-PC interface is greatly enhanced confirms the existence of a surface wave. Further, surface waves at the interface between SiC and the PC can be excited at any angle of incidence and for both polarizations.

Figure 9.66a shows the calculated spectral-directional emittance in the wavelengths between 10.5 and 12.5 μm at $\theta = 0°$, 30°, and 60° for both polarizations [201]. Notice that since the emission peak values depend on the thickness of SiC, $d_s$ can be tuned to maximize the emissivity for any given emission angle and polarization states. Here, the thickness of SiC is set to be $d_s = 1.45$ μm, which results in a near-unity emissivity at $\theta = 60°$ for $s$-polarization. Narrowband emission peaks can be seen and for TE waves, they are centered at $\lambda_c = 11.479$, 11.293, and 10.929 μm for emission angles of 0°, 30°, and 60°, respectively. The spectral emission peaks clearly indicate temporal coherence of the thermal emission. The corresponding quality factor $Q = \lambda_c/\delta\lambda$ are 230, 185, and 133, respectively, which are comparable to

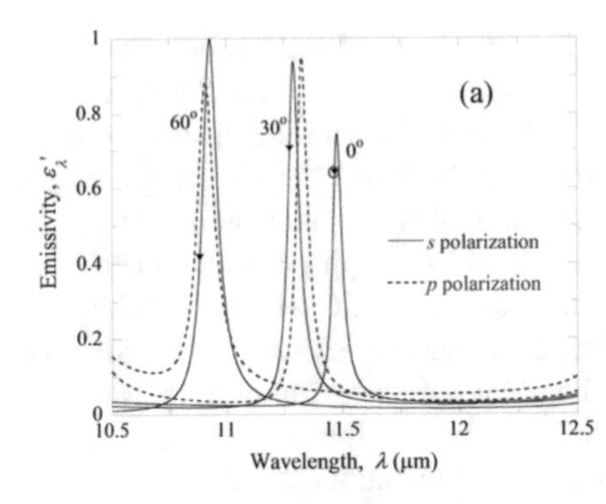


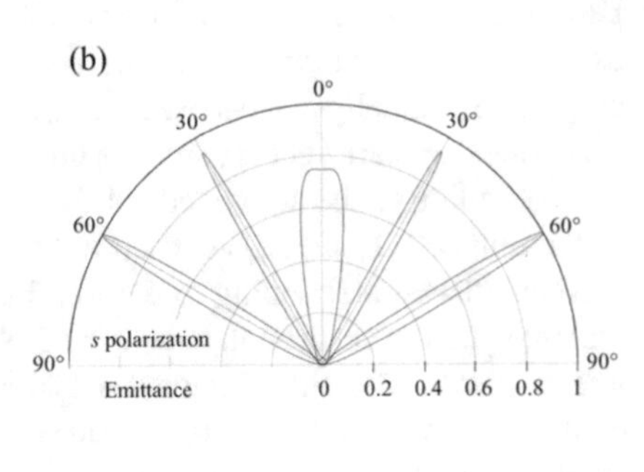


**Fig. 9.66** Calculated directional-spectral emissivity of the SiC-PC structure when surface wave is excited [201]. **a** Spectral dependence at different polar angles for both polarizations. The symbols with circle for TM wave and triangles for TE wave indicate the resonance wavelengths predicted using the dispersion relation. **b** Polar plot of showing the angular distribution of the emissivity at $\lambda_c = 11.479$, 11.293, and 10.929 μm for TE waves

those for SiC gratings [58]. From the solution of the surface wave dispersion relation, assuming no absorption in SiC, the resonance wavelength can be predicted for the given emission angle. These values are also marked as circles for TM waves and triangles for TE waves.

The spatial coherence of the proposed emission source can be seen from the angular distributions of the emissivity, shown in Fig. 9.66b at the three peak wavelengths for TE waves. The emittance is plotted as a polar plot to clearly show the angular lobe into a well-defined direction. However, if one considers the actual source with finite dimensions, due to the axial symmetry of the planar structure, the coherent emission from the SiC-PC structure exhibits (axially symmetric) circular patterns, in contrast to the antenna shape for the grating surfaces. The emittance at each $\lambda_c$ is confined in a very narrow angular region, although the angular spread corresponding to the peak at $\theta = 0°$ is broader than the other two peaks [201].

Lee et al. [204] demonstrated spectral coherence near the wavelength of 1 μm using truncated 1D PC on Ag, which was deposited on a silicon substrate, as displayed in Fig. 9.67. A Ti adhesive layer was first deposited on a Si substrate, followed by a Ag film, which is thick enough to be opaque (semi-infinite). The truncated PC with six unit cells was formed on the Ag film using plasma-enhanced CVD of $SiO_2$ and $Si_3N_4$ layers, as shown in Fig. 9.67a. The refractive index at the wavelength $\lambda = 1$ μm is approximately $n_1 = 1.45$ for $SiO_2$ and $n_2 = 2.0$ for $Si_3N_4$. The thicknesses were obtained from fitting the reflectance dip wavelengths to be $d_1 = d_2 = 153$ nm and the surface termination layer is $d_t = 100$ nm. The dispersion relation can be predicted and agrees with the measured resonances [203].

The bidirectional reflectance of the fabricated sample was measured with a custom-designed three-axis automated scatterometer, as described in Fig. 8.31, at 891 nm wavelength. The measured and calculated reflectances at different incidence angles are shown in Fig. 8.31b for TE waves. The measured reflectance exhibits a very sharp dip with a minimum at $\theta = 54°$ and compares well with the prediction.

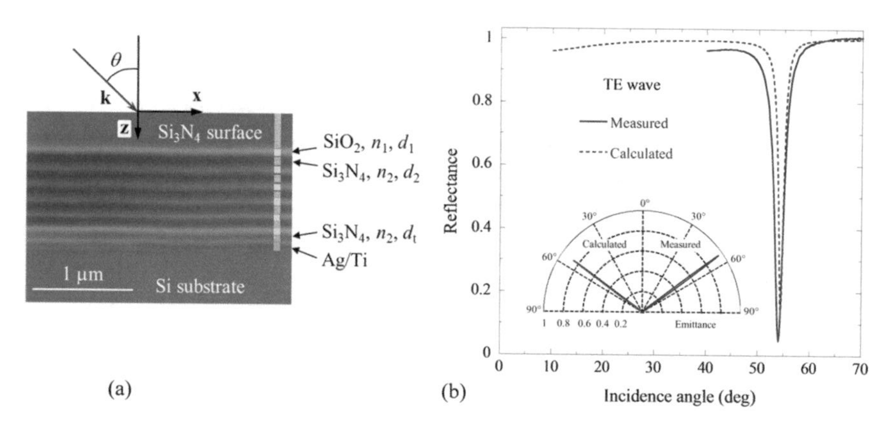


**Fig. 9.67** **a** A cross-sectional image of the fabricated PC-on-Ag structure [204]. **b** Reflectance for TE waves at $\lambda = 891$ nm as a function of $\theta$. The inset shows the spectral-directional emittance calculated from Kirchhoff's law. In the inset, the right half represents the measurement and the left half represents the prediction

The electron scattering rate of Ag film in the fabricated film may be larger than that the sample used for the tabulated optical constants [2]. The additional loss may result in a line broadening of the reflectance dip as well as a reduction of the reflectance at small incidence angles. Surface waves for both TE and TM waves have been demonstrated [203, 204].

The directional emissivity, obtained from the measured reflectance using Kirchhoff's law, is plotted in the vicinity of the peak as an inset in Fig. 9.67b. The right half represents the measurement and the left half represents the prediction. The angular-dependent emissivity exhibits strong directional selectivity. The coherence length given by $L_{\rm coh} = \lambda/(\pi\,\Delta\theta\cos\theta)$ is a measure of the spatial coherence, where $\Delta\theta$ is the full-width-at-half-maximum of the emissivity peak. The estimated $\Delta\theta$ from the measurement is 2.2°, and the corresponding coherence length is 14.1λ [204]. These values are comparable with those obtained from binary gratings [68].

## 9.7 Summary

This chapter provides a detailed treatment of the radiative properties of stratified media based on the electromagnetic wave theory, considering partial coherence, and extended to the discussion of periodic layered structures or 1D photonic crystals. Electromagnetic waveguides and guided mode equations are also introduced. The coupled-wave analysis for periodic gratings is present with some examples on the diffraction in gratings. Numerical models and experimental measurements of surface roughness and BRDF for anisotropic surfaces are summarized.

Built upon previous background, a comprehensive description of plasmonics and polaritonics is given in Sect. 9.5 without invoking details of quantum theory. The topics are evolved gradually from basic definitions of plasmons and phonon polaritons to surface plasmon (or phonon) polaritons, localized SPPs, coupled SPPs, bulk polaritons, magnetic polaritons, graphene plasmons, and hyperbolic polaritons. Some application examples are provided in Sect. 9.6 on how the far-field radiative properties can be tailored using phenomena and theories explained in preceding sections. Further discussions on near-field energy transfer by electromagnetic waves, especially due to evanescent waves, surface plasmons, and various types of polaritons will be given in the next chapter.

## Problems

9.1. A greenhouse looks like a small glasshouse used to grow plants in the winter. Based on the transmittance curve of fused silica ($SiO_2$), shown in Fig. 9.2, explain why glass walls can keep the plants warm in the winter. Discuss the greenhouse effect in the atmosphere. What gases are responsible for the greenhouse effect?

9.2. Calculate the transmittance $T$, the reflectance $R$, and the absorptance $A$ of a thick (without considering interference) silicon wafer (0.5 mm thick) at normal incidence. Plot $T$, $R$, and $A$ versus wavelength, in the range from 2.5 to 25 μm. The refractive index and the extinction coefficient of the doped silicon are given in the following table.

Optical constants of a doped silicon wafer

| Wavelength $\lambda$ (μm) | Refractive index $n$ | Extinction coefficient $\kappa$ |
|---|---|---|
| 2.5 | 3.44 | 0 |
| 5.0 | 3.43 | $1.0 \times 10^{-7}$ |
| 7.5 | 3.42 | $8.4 \times 10^{-5}$ |
| 10.0 | 3.42 | $2.1 \times 10^{-4}$ |
| 12.5 | 3.42 | $4.0 \times 10^{-4}$ |
| 15.0 | 3.42 | $5.0 \times 10^{-4}$ |
| 17.5 | 3.42 | $9.0 \times 10^{-4}$ |
| 20.0 | 3.42 | $1.0 \times 10^{-3}$ |
| 22.5 | 3.42 | $1.1 \times 10^{-3}$ |
| 25.0 | 3.42 | $1.3 \times 10^{-3}$ |

9.3. Calculate and plot the transmittance and reflectance of at $\lambda = 5$ μm as functions of the polar angle $\theta$ for the same silicon wafer described in Problem 9.2. Consider the individual polarizations and their average. Compare your results with those by Zhang et al. (*Infrared Phys. Technol.*, **37**, 539, 1996).

9.4. Using data from the table in Problem 9.2, calculate and plot the normal transmittance of a 100-μm-thick silicon wafer, near 10 μm wavelength, considering interference.

(a) Plot the transmittance in terms of wavelength (μm) with an interval between the data spacing of 0.05 and 0.005 μm, respectively, on one graph.
(b) Plot the transmittance in terms of wavenumber ($\text{cm}^{-1}$) with an interval between the data spacing of 5 and 0.5 $\text{cm}^{-1}$, respectively, on one graph.
(c) What is the fringe-averaged transmittance at 10 μm wavelength?
(d) What is the free spectral range in wavenumber and in wavelength? How will $\Delta\bar{\nu}$ and $\Delta\lambda$ change if the wavelength $\lambda$ is changed to 20 μm?

9.5. For gold, the refractive index at $\lambda = 0.5\,\mu\text{m}$ is $n = 0.916 + \text{i}1.84$, and at $\lambda = 2.0\,\mu\text{m}$ is $n = 0.85 + \text{i}12.6$. Calculate the normal transmittance of a free-standing gold film at these wavelengths for $d = 10$, 20, 50, and 100 nm, using Eqs. (9.2), (9.10) and (9.11). Which equation gives the correct results, and why?

9.6. For the three-layer structure shown in Fig. 9.3, calculate the normal reflectance for $n_1 = 1.45$ (glass), $n_2 = 1$ (air gap), and $n_3 = 2$ (substrate) without any absorption at $\lambda = 1\,\mu\text{m}$. Plot the reflectance as a function of the air gap width $d$. Obtain the analytical formulae of the reflectance maximum and minimum.

9.7. Assume that glass has a refractive index of 1.46 without any absorption in the visible spectrum ($0.4\,\mu\text{m} < \lambda < 0.7\,\mu\text{m}$). Design an antireflection coating (for normal incidence) that will minimize the reflectance from a semi-infinite glass. You need to determine the coating thickness and the refractive index (assuming it is also independent of wavelength). Plot the normal reflectance of the coated glass surface in the spectral range from 0.4 to 0.7 μm. What material will you recommend for use with the desired property?

9.8. To evaluate the effect of antireflection coating for oblique incidence, assume the antireflection coating has a refractive index of 1.21 and a thickness of 114 nm. What will be the reflectance, at 45° and 60°, for each polarization?

9.9. While the extinction coefficient is often related to absorption or loss, it should be noted that when $\kappa \gg n$, it is the real part of the refractive index that is related to the loss. This is because the dielectric function can be expressed as $\varepsilon = \varepsilon' + \text{i}\varepsilon'' = (n^2 - \kappa^2) + \text{i}2n\kappa$ and $\varepsilon''$ is related to the dissipation. For a semi-infinite medium, a purely negative dielectric function means perfect reflection. The effect of $n$ on the absorption by a thin film can be studied by considering a thin film of thickness $d$ with a complex refractive index $n_2 = n + \text{i}\kappa$. For a wavelength of $\lambda = 0.5\,\mu\text{m}$ and at normal incidence, let $d = 30$ nm and $\kappa = 3.0$. Plot the transmittance, the reflectance, and the emittance (which is the same as the absorptance), against the refractive index $n$ ranging from 0.01 to 2. Discuss the effect on $n$ on the absorption.

9.10. Use the dielectric function of SiC given in Example 8.7 to calculate the normal emittance for a SiC film at wavelengths from 9 to 15 μm, for different film thicknesses: $d = 1, 10, 100$, and $1000\,\mu$m. Assume the multiply reflected waves to be perfectly coherent.

9.11. Calculate the emittance as a function of the emission angle for a doped silicon wafer of $200\,\mu$m thickness, at $\lambda = 20\,\mu$m with $n_2 = 3.42 + i0.001$. Consider $p$ and $s$ polarizations separately, and then, take an average. Assume the multiply reflected waves to be perfectly coherent.

9.12. This problem concerns the transmission and reflection of infrared radiation of a YBCO ($YBa_2Cu_3O_7$) film on a thin MgO substrate of $325\,\mu$m thickness, at 300 K and normal incidence. For the YBCO film, use the properties for sample A from Kumar et al. [15]. For MgO, use the Lorentz model in Problem 8.31.

(a) Plot the radiation penetration depth of the YBCO film, $\delta_f(\lambda)$, and that of MgO, $\delta_s(\lambda)$, for $1\,\mu\text{m} < \lambda < 1000\,\mu$m.

(b) Neglecting the interference effect in the MgO substrate, calculate and plot the transmittance $T$, the film-side reflectance $R_f$, and the backside reflectance $R_s$, for $1\,\mu\text{m} < \lambda < 1000\,\mu$m, with different film thicknesses: 0, 30, 48, 70, and 400 nm. Plot $T$, $R_f$, and $R_s$ in terms of both wavelength ($\mu$m) and wavenumber ($\text{cm}^{-1}$).

(c) Repeat the previous calculation, considering the interference effects in the MgO substrate, for $200\,\mu\text{m} < \lambda < 1000\,\mu$m (50 to $10\,\text{cm}^{-1}$). Plot in terms of the wavenumber only. What happens with the interference fringes when the film thickness is 48 nm?

9.13. Calculate the normal transmittance of a 10 $\mu$m film with a refractive index $n = 2.4$ without any absorption in the spectral range from 1000 to 3000 $\text{cm}^{-1}$. One surface of the film is polished, and the other surface has a roughness $\sigma_{rms}$ of 0.10 $\mu$m. How does $\sigma_{rms}$ value affect the transmittance? Compare your result with that shown in Fig. 9.10.

9.14. Reproduce Example 9.2 and Fig. 9.10. Suppose the coherence spectral width $\delta\nu = 1.5\Delta\nu$, where $\Delta\nu$ is the free spectral range. Determine the fringe-averaged transmittance. Explain why the peaks and the valleys flip after fringe averaging.

9.15. Calculate and plot the transmittance of a Fabry–Perot resonance cavity, assuming the medium to be lossless with $n_2 = 2$, $d_2 = 100\,\mu$m, and $R = 0.90$, for normal incidence in the wavenumber region from 950 to 1050 $\text{cm}^{-1}$. What are the free spectral range, the FWHM of the peak, and the $Q$-factor of the resonator? Does the theoretically predicted FWHM match with the plot?

9.16. Group project: A reflectance Fabry–Perot cavity can be constructed by coating a $SiO_2$ film onto a silver substrate first and then a thin silver film onto the $SiO_2$ film. Derive a formula for the reflectance. Based on Kirchhoff's law, one can calculate the emissivity of the structure. Show that the emissivity exhibit sharp peaks close to unity at specific wavelengths for normal incidence. When the wavelength is fixed, calculate the emissivity versus the polar angle for each polarization. Plot and show that there exist angular lobes in the emissivity of such structures. Hint: Choose the thicknesses of the silver film (on the order of 100 nm) and the $SiO_2$ film (on the order of 3000 nm), and the wavelength (around 1 μm). Use the optical constants from Palik [2].

9.17. Develop a MATLAB code for the multilayer radiative properties based on the matrix formulation described in the text for both TE and TM waves. Refer to author's website [36].

9.18. For a planar waveguide with $n_1 = 1.54$ and $n_2 = 1.23$, with a thickness of $d = 200$ nm, how many total modes are there in the waveguide at $\lambda = 635$ nm, $\lambda = 1.55\ \mu$m, and $\lambda = 3.2\ \mu$m?

9.19. Derive Eq. (9.52) by setting the determinant of the characteristic 4×4 matrix to be zero. Sometimes, it is desirable to plot the solutions of Eq. (9.52) in curves that relate $\omega$ to $k_x$. These curves are called waveguide dispersion relations. Given $n_1 = 1.6$, $n_2 = 1.3$, and $d = 300\ \mu$m, plot the dispersion curves for the first four TE modes. Explain why the group velocities are different for different modes, even though the refractive indices are independent of wavelength.

9.20. In an asymmetric dielectric waveguide, the guided region (refractive index $n_1 = 3.5$) is sandwiched between two different materials ($n_2 = 1.5$ and $n_3 = 2.5$). Show that the mode equation can be expressed as $2k_{1z}d + \psi_2 + \psi_3 = 2m\pi$, for $m = 0, 1, 2, \ldots$, where $\psi_2$ and $\psi_3$ are the phase angles upon total internal reflection by media 2 and 3, respectively. If the thickness of the guided region is $d = 3\ \mu$m, find the wavelength region where the fiber is a single-mode fiber ($TE_0$ only). Find the wavelength region where the fiber allows only $TE_0$ and $TM_0$ modes to be guided, i.e., single mode for each polarization.

9.21. Evaluate the plot the photonic band structures of a Bragg reflector made of quarter-wave high- and low-index materials GaAs, $n = 3.49$, and AlAs, $n = 2.95$, around the wavelength of 1064 nm. Optional: Plot the normal reflectance near 1064 nm wavelength with 7, 17, and 27 periods, assuming that the substrate is GaAs.

9.22. Derive Eqs. (9.68) and (9.69).

9.23. Based on Eq. (9.72), show that when the evanescent wave exists, it will decay toward negative $z$. Change the subscript from I to III, and show that when the evanescent wave exists, it will decay toward positive $z$.

9.24. Derive Eqs. (9.79) and (9.80).

9.25. Use different effective medium formulations to compute the effective dielectric function for silicon with a filling ratio $\phi = 1/6$ in air at $\lambda = 300$ nm ($n = 5.0$ and $\kappa = 4.2$), $\lambda = 400$ nm ($n = 5.6$ and $\kappa = 0.39$), $\lambda = 500$ nm ($n = 4.3$ and $\kappa = 0.073$), and $\lambda = 800$ nm ($n = 3.7$ and $\kappa = 0.0066$).

9.26. Consider a grating region consisting of Si, with a filling ratio of 1/6, on a semi-infinite Si substrate. The height of the grating is 50 nm. Calculate the reflectance for normal incidence, using different effective medium formulations at the corresponding wavelengths given in Problem 9.25. Compared your results with those in Fig. 9.22.

9.27. Use RCWA downloaded from author's website [36] to reproduce Fig. 9.22.

9.28. Plot the shadowing function for a Gaussian distribution as a function of the polar angle $\theta$ for the rms slope $w = 0.05, 0.1, 0.2$, and $0.3$.

9.29. Calculate the BRDFs at $\lambda = 0.5$ and $2\,\mu$m based on the analytical model for a gold surface (opaque) with a Gaussian roughness statistics. The SDF is given by $p(\zeta_x, \zeta_y) = \frac{1}{2\pi w} \exp\left(-\frac{\zeta_x^2+\zeta_y^2}{2w^2}\right)$. Use the optical constants from Problem 9.5 and the rms slope $w = 0.1$ and 0.3.

9.30. Comment on the limitations of different analytical models for the BRDF, such as the Rayleigh-Rice perturbation theory, the Kirchhoff approximation, and the geometric optics approximation.

9.31. Calculate the real and imaginary parts of $k_x$ based on the SPP relation given in Eq. (9.100) for Al, like in Example 9.7. What is $k_x'$ for $\lambda = 400$ nm? Assuming that the prism has an index of refraction $n_{\rm d} = 1.53$, find the incidence angle that would yield $k_x = k_x'$. Calculate the reflectance for Al in the ATR arrangements at $\lambda = 400$ nm. Discuss whether the obtained reflectance dip in the angular distribution of the reflectance agrees with that predicted by the surface plasmon polariton. Calculate the polariton propagation length at this wavelength.

9.32. Studies suggest that the surface plasmon dispersion relation described in Eq. (9.100) can be solved by assuming that $k_x$ is real but $\omega = \omega' + {\rm i}\omega''$. The real part of $\omega$ corresponds to the SP resonance frequency, while the imaginary part corresponds to the bandwidth. Develop a computer program to solve $\omega'(k_x)$ and $\omega''(k_x)$ for Al with a thickness 24 nm that is adjacent to the prism with $\varepsilon_{\rm d} = 2.46$. For $\theta = 40.23°$, calculate the reflectance spectrum near the resonance frequency of the surface polariton, and compare the bandwidth with the calculated $\omega''$.

9.33. Examine Fig. 9.32 to confirm whether the surface polariton resonance frequencies predicted by the dispersion relation agree with the reflectance dips for a TM wave incident on a grating. Note that one of the dips in the dotted line ($\theta = 30°$) overlaps with that of the solid line ($\theta = 0°$) near 12,000 cm$^{-1}$. Hence, there are three notable dips in the reflectance at $\theta = 0°$ and five notable dips at $\theta = 30°$.

9.34. Discuss why a nanoparticle can absorb more energy than a blackbody of the same size. Is it possible for a nanoparticle of radius $r_0$ to emit more energy than $4\pi r_0^2 \sigma_{\rm SB} T^4$, where $T$ is the temperature of the spherical particle? Furthermore, is it possible for a nanoaperture to transmit more energy than the product of the incident energy flux (i.e., irradiance) times its area? Why or why not?

9.35. Reproduce some cases in Fig. 9.35 under the same conditions for $a = 0.25\lambda_{\rm p}$ and $d = 0.25\lambda_{\rm p}$. To examine the effect of $a$, recalculate the reflectance spectra with $a = 0.15\lambda_{\rm p}$ and $0.1\lambda_{\rm p}$ for the same $d$. Compare your results with those of Park et al. [115].

9.36. Based on the dielectric function model of SiC, at $\lambda = 11\,\mu$m, $\varepsilon_{\rm s} = -3.256 + 0.208{\rm i}$ and $n_{\rm s} = 0.059 + 1.953{\rm i}$, which correspond to a radiation penetration depth of 0.448 $\mu$m. If a film of SiC with a thickness of $d = 1.8\,\mu$m is sandwiched between two prisms of the same dielectric constant $\varepsilon_{\rm d} = 2.89$, calculate the transmittance as a function of the incidence angle. Considering a prism-air-SiC-prism arrangement, where the width of the air gap is $a =$

5 μm, calculate the transmittance again. You should see a peak near 43° with a transmittance around 0.3 for a TM wave. Verify that the transmittance enhancement is due to surface plasmon excitation, by calculating the angle-dependent reflectance.

9.37. Consider a prism-air-Al-prism configuration with $\varepsilon_d = 2.45$ for both prisms, air gap width $a = 120$ nm, and aluminum thickness $d = 30$ nm. Use the dielectric function of Al from Example 8.7 to calculate the transmittance and the reflectance at $\lambda = 180$ nm for a TM wave at the incidence angle $\theta = 47°$. Discuss the effect of the air gap width.

9.38. Show Eq. (9.132) is indeed an approximation for Eq. (9.131) under the conditions described in the text.

9.39. Show that both the Maxwell-Garnett and Bruggeman formulations as given in Eq. (9.133) yield the same expressions as given in Eqs. (9.127) and (9.128) for metallodielectric multilayers.

9.40. Reproduce Figs. 9.51 and 9.52. Discuss single-negative and double-negative materials.

9.41. Reproduce Fig. 9.54b using the code from the author's webpage [36].

9.42. Reproduce Example 9.9. Discuss the surface polariton effects on both TE and TM waves.

## References


1. J.R. Howell, M.P. Mengüç, R. Siegel, *Thermal Radiation Heat Transfer*, 6th edn. (CRC Press, New York, 2016)
2. E.D. Palik (ed.), *Handbook of the Optical Constants of Solids*, vol. I, II, and III (Academic Press, San Diego, 1998)
3. T.J. Bright, J.I. Watjen, Z.M. Zhang, C. Muratore, A.A. Voevodin, Optical properties of $HfO_2$ thin films deposited by magnetron sputtering: from the visible to the far-infrared. Thin Solid Films **520**, 6793–6802 (2012)
4. I.H. Maltison, Interspecimen comparison of the refractive index of fused silica. J. Opt. Soc. Am. **55**, 1205–1209 (1965)
5. D. Poelman, P.F. Smet, Methods for the determination of the optical constants of thin films from single transmission measurements: a critical review. J. Phys. D Appl. Phys. **36**, 1850–2222 (2003)
6. O.S. Heavens, *Optical Properties of Thin Solid Films* (Dover Publications, New York, 1965)
7. Z. Knittl, *Optics of Thin Films* (Wiley, New York, 1976)
8. M.Q. Brewster, *Thermal Radiative Transfer and Properties* (Wiley, New York, 1992)
9. Z.M. Zhang, Reexamination of the transmittance formulae of a lamina. J. Heat Transfer **119**, 645–647 (1997); Z.M. Zhang, Optical properties of a slightly absorbing film for oblique incidence. Appl. Opt. **38**, 205–207 (1999)
10. L. Mandel, E. Wolf, *Optical Coherence and Quantum Optics* (Cambridge University Press, Cambridge, UK, 1995)
11. G. Chen, C.L. Tien, Partial coherence theory of thin film radiative properties. J. Heat Transfer **114**, 636–643 (1992)
12. K. Fu, P.-f., Hsu, Z.M. Zhang, Unified analytical formulation of thin-film radiative properties including partial coherence. Appl. Opt. **45**, 653–661 (2006)
13. B.J. Lee, V.P. Khuu, Z.M. Zhang, Partially coherent spectral radiative properties of dielectric thin films with rough surfaces. J. Thermophys. Heat Transfer **19**, 360–366 (2005)

14. J.M. Vaughan, *The Fabry–Perot Interferometer: History, Theory, Practice and Applications* (Adam Hilger, Bristol, 1989)
15. A.R. Kumar, Z.M. Zhang, V.A. Boychev, D.B. Tanner, L.R. Vale, D.A. Rudman, Far-infrared transmittance and reflectance of $YBa_2Cu_3O_{7-\delta}$ on Si substrates. J. Heat Transfer **121**, 844–851 (1999); A.R. Kumar, V.A. Boychev, Z.M. Zhang, D.B. Tanner, Fabry–Perot resonators built with $YBa_2Cu_3O_{7-\delta}$ films on Si substrates. J. Heat Transfer **122**, 785–791 (2000)
16. L.P. Wang, B.J. Lee, X.J. Wang, Z. M. Zhang, 2009, Spatial and temporal coherence of thermal radiation in asymmetric Fabry–Perot resonance cavities. Int. J. Heat Mass Transfer **52**, 3024–3031 (2009); L.P. Wang, S. Basu, Z.M. Zhang, Direct measurement of thermal emission from a Fabry–Perot cavity resonator. J. Heat Transfer **134**, 072701 (2012)
17. P. Yeh, *Optical Waves in Layered Media* (Wiley, New York, 1988); also see P. Yeh, A. Yariv, C.S. Hong, Electromagnetic propagation in periodic stratified media. I. General theory. J. Opt. Soc. Am. **67**, 423–438 (1977)
18. C.L. Mitsas, D.I. Siapkas, Generalized matrix method for analysis of coherence and incoherent reflectance and transmittance of multilayer structures with rough surfaces, interfaces, and finite substrates. Appl. Opt. **34**, 1678–1683 (1995)
19. H.A. Haus, *Waves and Fields in Optoelectronics* (Prentice-Hall, Englewood Cliffs, 1984)
20. J.A. Kong, *Electromagnetic Wave Theory*, 2nd edn. (Wiley, New York, 1990)
21. K.J. Vahala, Optical microcavities. Nature **424**, 839–846 (2003)
22. Z. Guo, H. Quan, Energy transfer to optical microcavities with waveguides. J. Heat Transfer **129**, 44–52 (2007)
23. R.F. Cregan, B.J. Mangan, J.C. Knight et al., Single-mode photonic band gap guidance of light in air. Science **285**, 1537–1539 (1999); P.St.J. Russell, Photonic crystal fibers. Science **299**, 358–362 (2003)
24. J.D. Joannopoulos, R.D. Meade, J.N. Winn, *Photonic Crystals* (Princeton University Press, Princeton, 1995)
25. K. Sakoda, *Optical Properties of Photonic Crystals* (Springer, Berlin, 2001)
26. Z.M. Zhang, L.P. Wang, Measurements and modeling of the spectral and directional radiative properties of micro/nanostructured materials. Int. J. Thermophys. **34**, 2209–2242 (2013)
27. J.G. Fleming, S.Y. Lin, I. El-Kady, R. Biswas, K.M. Ho, All-metallic three-dimensional photonic crystals with a large infrared bandgap. Nature **417**, 52–55 (2002); C.H. Seager, M.B. Sinclair, J.G. Fleming, Accurate measurements of thermal radiation from a tungsten photonic lattice. Appl. Phys. Lett. **86**, 244105 (2005)
28. H.A. Macleod, *Thin Film Optical Filters*, 3rd edn. (Institute of Physics, Bristol, 2001)
29. D. Maystre (ed.), *Selected Papers on Diffraction Gratings, SPIE Milestone Series 83* (The International Society for Optical Engineering, Bellingham, 1993)
30. R. Petit (ed.), *Electromagnetic Theory of Gratings* (Springer, Berlin, 1980)
31. Y.B. Chen, Z.M. Zhang, P.J. Timans, Radiative properties of pattered wafers with nanoscale linewidth. J. Heat Transfer **129**, 79–90 (2007)
32. M.G. Moharam, E.B. Grann, D.A. Pommet, T.K. Gaylord, Formulation for stable and efficient implementation of the rigorous coupled-wave analysis of binary gratings. J. Opt. Soc. Am. A **12**, 1068–1076 (1995); M.G. Moharam, D.A. Pommet, E.B. Grann, T.K. Gaylord, Stable implementation of the rigorous coupled-wave analysis for surface-relief gratings: enhanced transmittance matrix approach. J. Opt. Soc. Am. A **12**, 1077–1086 (1995)
33. L.F. Li, Use of Fourier series in the analysis of discontinuous periodic structures. J. Opt. Soc. Am. A **13**, 1870–1876 (1996)
34. B.J. Lee, Y.-B. Chen, Z.M. Zhang, Transmission enhancement through nanoscale metallic slit arrays from the visible to mid-infrared. J. Comput. Theor. Nanosci. **5**, 201–213 (2008)
35. B. Zhao, Z.M. Zhang, Perfect mid-infrared absorption by hybrid phonon-plasmon polaritons in hBN/metal-grating anisotropic structures. Int. J. Heat Mass Transfer **106**, 1025–1034 (2017)
36. Z. M. Zhang, http://zhang-nano.gatech.edu. Last viewed March 25, 2019
37. J.C. Maxwell-Garnett, Colours in metal glasses and in metallic films. Philos. Trans. R. Soc. Lond. A **203**, 385–420 (1904)

38. D.A.G. Bruggeman, Dielectric constant and conductivity of mixtures of isotropic materials. Ann. Phys. (Leipz.) **24**, 636–679 (1935)
39. S.M. Rytov, Electromagnetic properties of a finely stratified medium. Sov. Phys. JETP **2**, 466–475 (1956)
40. P. Lalanne, D. Lemercier-Lalanne, On the effective medium theory of subwavelength periodic structures. J. Mod. Opt. **43**, 2063–2085 (1996)
41. Z.M. Zhang, B.K. Tsai, G. Machin, *Radiometric Temperature Measurements – I. Fundamentals & II. Applications* (Academic Press, Academic Press (an Imprint of Elsevier), Amsterdam, 2009)
42. Z.M. Zhang, C.J. Fu, Q.Z. Zhu, Optical and thermal radiative properties of semiconductors related to micro/nanotechnology. Adv. Heat Transfer **37**, 179–296 (2003)
43. P. Beckmann, A. Spizzichino, *The Scattering of Electromagnetic Waves from Rough Surfaces* (Artech House, Norwood, 1987)
44. Q.Z. Zhu, Z.M. Zhang, Anisotropic slope distribution and bidirectional reflectance of a rough silicon surface. J. Heat Transfer **126**, 985–993 (2004); Q.Z. Zhu, Z.M. Zhang, Correlation of angle-resolved light scattering with the microfacet orientation of rough silicon surfaces. Opt. Eng. **44**, 073601 (2005)
45. B.G. Smith, Geometrical shadowing of a random rough surface. IEEE Trans. Antennas Propag. **15**, 668–671 (1967)
46. H.J. Lee, Y.B. Chen, Z.M. Zhang, Directional radiative properties of anisotropic rough silicon and gold surfaces. Int. J. Heat Mass Transfer **49**, 4482–4495 (2006)
47. Y.H. Zhou, Z.M. Zhang, Radiative properties of semitransparent silicon wafers with rough surfaces. J. Heat Transfer **125**, 462–470 (2003); H.J. Lee, B.J. Lee, Z.M. Zhang, Modeling the radiative properties of semitransparent wafers with rough surfaces and thin-film coatings. J. Quant. Spectrosc. Radiat. Transfer **93**, 185–194 (2005)
48. Q.Z. Zhu, H.J. Lee, Z.M. Zhang, Validity of hybrid models for the bidirectional reflectance of coated rough surfaces. J. Thermophys. Heat Transfer **19**, 548–557 (2005)
49. H.J. Lee, A.C. Bryson, Z.M. Zhang, Measurement and modeling of the emittance of silicon wafers with anisotropic roughness. Int. J. Thermophys. **28**, 918–932 (2007)
50. Q.Z. Zhu, H.J. Lee, Z.M. Zhang, Radiative properties of materials with surface scattering or volume scattering: a review. Front. Energy Power Eng. China **3**, 60–79 (2009)
51. Z.L. Wang, J.M. Cowley, Reflection electron energy loss spectroscopy (REELS): a technique for the study of surfaces. Surf. Sci. **193**, 501–512 (1988)
52. J.J. Hopfield, Theory of the contribution of excitons to the complex dielectric constant of crystals. Phys. Rev. **112**, 1555–1567 (1958)
53. K. Huang, Lattice vibrations and optical waves in ionic crystals. Nature **167**, 779–780 (1951); K. Huang, On the interaction between the radiation field and ionic crystals. Proc. R. Soc. Lond. A **208**, 352–365 (1951)
54. S. Kawata (ed.), *Near-Field Optics and Surface Plasmon Polaritons* (Springer, Berlin, 2001)
55. J. Tominaga, D.P. Tsai (eds.), *Optical Nanotechnologies – The Manipulation of Surface and Local Plasmons* (Springer, Berlin, 2003)
56. J. Homola, S.S. Yee, G. Gauglitz, Surface plasmon resonance sensors: Review. Sensors and Actuators B **54**, 3–15 (1999)
57. X.L. Liu, Z.M. Zhang, Silicon metamaterials for infrared applications, in *Silicon Nanomaterials Sourcebook*, ed. by K.D. Sattler (Taylor&Francis Books, 2017), pp. 345–369
58. J.-J. Greffet, R. Carminati, K. Joulain, J.-P. Mulet, S. Mainguy, Y. Chen, Coherent emission of light by thermal sources. *Nature* **416**, 61–64 (2002); F. Marquier, K. Joulain, J.-P. Mulet, R. Carminati, J.-J. Greffet, Y. Chen, Coherent spontaneous emission of light by thermal sources. Phys. Rev. B **69**, 155412 (2004)
59. B. Zhao, Z.M. Zhang, Design of optical and radiative properties of solids, in *Handbook of Thermal Science and Engineering*, ed. by F.A. Kulachi (Springer Nature, 2017), pp. 1023–1068
60. R. Hillenbrand, T. Taubner, F. Kellmann, Phonon-enhanced light-matter interaction at the nanometer scale. Nature **418**, 159–162 (2002); R. Hillenbrand, Towards phonon photonics:

scattering-type near-field optical microscopy reveals phonon-enhanced near-field interaction. Ultramicroscopy **100**, 421–427 (2004)
61. H. Raether, *Surface Plasmons on Smooth and Rough Surfaces and on Gratings* (Springer, Berlin, 1988)
62. J. Matsuura, M. Fukui, O. Tada, ATR mode of surface magnon polaritons on YIG. Solid State Commun. **45**, 157–160 (1983)
63. R. Rupin, Surface polaritons of a left-handed medium. Phys. Lett. A **277**, 61–64 (2000); R. Rupin, Surface polaritons of a left-handed material slab. J. Phys.: Condens. Matter **13**, 1811–1819 (2001)
64. A. Hessel, A.A. Oliner, A new theory of Wood's anomalies on optical gratings. Appl. Opt. **4**, 1275–1297 (1965)
65. Y.-B. Chen, Z.M. Zhang, Design of tungsten complex gratings for thermophotovoltaic radiators. Opt. Commun. **269**, 411–417 (2007)
66. H.A. Atwater, A. Polman, Plasmonics for improved photovoltaic devices. Nat. Mater. **9**, 205–213 (2010)
67. C.F. Bohren, D.R. Huffman, *Absorption and Scattering of Light by Small Particles* (Wiley, New York, 1983); C.F. Bohren, How can a particle absorb more than the light incident on it? Am. J. Phys. **51**, 323–327 (1983)
68. M.I. Stockman, Nanoplasmonics: past, present, and glimpse into future. Opt. Express **19**, 22029–22106 (2011)
69. N.J. Halas, S. Lal, W.S. Chang, S. Link, P. Nordlander, Plasmons in strongly coupled metallic nanostructures. Chem. Rev. **111**, 3913–3961 (2011)
70. V. Amendola, R. Pilot, M. Frasconi1, O.M. Maragò, M.A. Iatì, Surface plasmon resonance in gold nanoparticles: a review. J. Phys.: Condens. Matter **29**, 203002 (2017)
71. S. Link, M.A. El-Sayed, Spectral properties and relaxation dynamics of surface plasmon electronic oscillations in gold and silver nanodots and nanorods. J. Phys. Chem. B **103**, 8410–8426 (1999)
72. J.P. Kottmann, O.J.F. Martin, D.R. Smith, S. Schultz, Plasmon resonances of silver nanowires with a nonregular cross section. Phys. Rev. B **64**, 235402 (2001)
73. F. Wang, Y.R. Shen, General properties of local plasmons in metal nanostructures. Phys. Rev. Lett. **97**, 206806 (2006)
74. K.L. Kelly, E. Coronado, L.L. Zhao, G.C. Schatz, The optical properties of metal nanoparticles: the influence of size, shape, and dielectric environment. J. Phys. Chem. B **107**, 668–677 (2003)
75. A.J. Haes, S. Zou, G.C. Schatz, R.P. Van Duyne, A nanoscale optical biosensor: the long range distance dependence of the localized surface plasmon resonance of noble metal nanoparticles. J. Phys. Chem. B **108**, 109–116 (2004)
76. J. Chen, B. Wiley, Z.-Y. Li, D. Campbell, F. Saeki, H. Cang, L. Au, J. Lee, X. Li, Y. Xia, Gold nanocages: engineering their structure for biomedical applications. Adv. Mater. **17**, 2255–2261 (2005)
77. P.K. Jain, K.S. Lee, I.H. El-Sayed, M.A. El-Sayed, Calculated absorption and scattering properties of gold nanoparticles of different size, shape, and composition: applications in biological imaging and biomedicine. J. Phys. Chem. B **110**, 7238–7248 (2006)
78. Y. Xia, Y. Xiong, B. Lim, S.E. Skrabalak, Shape-controlled synthesis of metal nanocrystals: simple chemistry meets complex physics? Angew. Chem. Int. Ed. **48**, 60–103 (2009)
79. P. Johansson, H. Xu, M. Käll, Surface-enhanced Raman scattering and fluorescence near metal nanoparticles. Phys. Rev. B **72**, 035427 (2005)
80. B. Sharma, R.R. Frontiera, A.-I. Henry, E. Ringe, R.P. Van Duyne, SERS: materials, applications, and the future. Mater. Today **15**, 16–25 (2012)
81. L.R. Hirsch, R.J. Stafford, J.A. Bankson et al., Nanoshell-mediated near-infrared thermal therapy of tumors under magnetic resonance guidance. Proc. Natl. Acad. Sci. **100**, 13549–13554 (2003)
82. X. Huang, I.H. El-Sayed, W. Qian, M.A. El-Sayed, Cancer cell imaging and photothermal therapy in the near-infrared region by using gold nanorods. J. Am. Chem. Soc. **128**, 2115–2120 (2006)

83. P.P. Venkata, M.M. Aslan, M.P. Mengüç, G. Videen, Surface plasmon scattering by gold nanoparticles and two-dimensional agglomerates. J. Heat Transfer **129**, 60–70 (2007)
84. B.N. Khlebtsov, V.A. Khanadeyev, J. Ye, D.W. Mackowski, G. Borghs, N.G. Khlebtsov, Coupled plasmon resonances in monolayers of metal nanoparticles and nanoshells. Phys. Rev. B **77**, 035440 (2008)
85. J.R. Cole, N.J. Halas, Optimized plasmonic nanoparticle distributions for solar spectrum harvesting. Appl. Phys. Lett. **89**, 153120 (2006)
86. N. Harris, M.J. Ford, M.B. Cortie, Optimization of plasmonic heating by gold nanospheres and nanoshells. J. Phys. Chem. B **110**, 10701–10707 (2006)
87. R. Taylor, P.E. Phelan, T.P. Otanicar, R.J. Adrian, R.S. Prasher, Nanofluid optical property characterization: towards efficient direct-absorption solar collectors. Nanoscale Res. Lett. **6**, 225 (2011)
88. Q.Z. Zhu, Z.M. Zhang, Radiative properties of micro/nanoscale particles in dispersions for photothermal energy conversion, in *Nanoparticle Heat Transfer and Fluid Flow*, ed. by W.J. Minkowycz, E.M. Sparrow, J. Abraham (CRC Press, Boca Raton, 2012), pp. 143–174
89. G. Baffou, R. Quidant, F.J. García de Abajo, Nanoscale control of optical heating in complex plasmonic systems. ACS Nano **4**, 709–716 (2010)
90. R. Taylor, P.E. Phelan, T.P. Otanicar, R.J. Adrian, R.S. Prasher, Vapor generation in a nanoparticle liquid suspension using a focused, continuous laser. Appl. Phys. Lett. **95**, 161907 (2009)
91. O. Neumann, A.S. Urban, J. Day, S. Lal, P. Nordlander, N.J. Halas, Solar vapor generation enabled by nanoparticles. ACS Nano **7**, 42–49 (2013)
92. G. Baffou, C. Girard, R. Quidant, Mapping heat origin in plasmonic structures. Phys. Rev. Lett. **104**, 136805 (2010)
93. Z.J. Coppens, W. Li, D.G. Walker, J.G. Valentine, Mapping heat origin in plasmonic structures probing and controlling photothermal heat generation in plasmonic nanostructures. Nano Lett. **13**, 1023–1028 (2013)
94. B.J. Lee, K. Park, T. Walsh, L. Xu, Radiative heat transfer analysis in plasmonic nanofluids for direct solar thermal absorption. J. Heat Transfer **134**, 021009 (2012)
95. Y.M. Xuan, H.L. Duan, Q. Li, Enhancement of solar energy absorption using a plasmonic nanofluid based on $TiO_2$/Ag composite nanoparticles. RSC Adv. **4**, 16206–16213 (2014)
96. Z.L. Wang, X.J. Quan, Z.M. Zhang, P. Cheng, Optical absorption of carbon-gold core-shell nanoparticles. J. Quant. Spectrosc. Radiat. Transfer **205**, 291–298 (2018)
97. Z.L. Wang, Z.M. Zhang, X.J. Quan, P. Cheng, A numerical study on effects of surrounding medium, materials, and geometry of nanoparticles on solar absorption efficiencies. Int. J. Heat Mass Transfer **116**, 825–832 (2018)
98. M. Quinten, A. Leitner, J.R. Krenn, F.R. Aussenegg, Electromagnetic energy transport via linear chains of silver nanoparticles. Opt. Lett. **23**, 1331–1333 (1998)
99. R.M. Dickson, L.A. Lyon, Unidirectional plasmon propagation in metallic nanowires. J. Phys. Chem. B **104**, 6095–6098 (2000)
100. M.L. Brongersma, J.W. Hartman, H.A. Atwater, Electromagnetic energy transfer and switching in nanoparticle chain arrays below the diffraction limit. Phys. Rev. B **62**, R16356 (2000)
101. S.A. Maier, P.G. Kik, H.A. Atwater, S. Meltzer, E. Harel, B.E. Koel, A.G. Requicha, Local detection of electromagnetic energy transport below the diffraction limit in metal nanoparticle plasmon waveguides. Nat. Mater. **2**, 229–232 (2003)
102. A.F. Koenderink, Plasmon nanoparticle array waveguides for single photon and single plasmon sources. Nano Lett. **9**, 4228–4233 (2009)
103. J.A. Schuller, E.S. Barnard, W. Cai, Y.C. Jun, J.S. White, M.I. Brongersma, Plasmonics for extreme light concentration and manipulation. Nat. Mater. **9**, 193–204 (2010)
104. D.S. Citrin, Plasmon polaritons in finite-length metal nanoparticle chains: The role of chain length unraveled. Nano Lett. **5**, 985–989 (2005)
105. A.F. Koenderink, A. Polman, Complex response and polariton-like dispersion splitting in periodic metal nanoparticle chains. Phys. Rev. B **74**, 033402 (2006)

106. V.A. Markel, A.K. Sarychev, Propagation of surface plasmons in ordered and disordered chains of metal nanospheres. Phys. Rev. B **75**, 085426 (2007)
107. B. Willingham, S. Link, Energy transport in metal nanoparticle chains via sub-radiant plasmon modes. Opt. Express **19**, 6450–6461 (2011)
108. M.B. Ross, C.A. Mirkin, G.C. Schatz, Optical properties of one-, two- and three-dimensional arrays of plasmonic nanostructures. J. Phys. Chem. **120**, 816–830 (2016)
109. G. Baffou, R. Quidant, C. Girard, Thermoplasmonics modeling: a Green's function approach. Phys. Rev. B **82**, 165424 (2010)
110. P. Ben-Abdallah, K. Joulain, J. Drevillon, C. Le Goff, Heat transport through plasmonic interactions in closely spaced metallic nanoparticle chains. Phys. Rev. B **77**, 075417 (2008)
111. J. Ordonez-Miranda, L. Tranchant, S. Gluchko, S. Volz, Energy transport of surface phonon polaritons propagating along a chain of spheroidal nanoparticles. Phys. Rev. B **92**, 115409 (2015)
112. E.J. Tervo, Z.M. Zhang, B.A. Cola, Collective near-field thermal emission from polaritonic nanoparticle arrays. Phys. Rev. Mater. **1**, 015201 (2017)
113. E.J. Tervo, M.E. Gustafson, Z.M. Zhang, B.A. Cola, M.A. Filler, Photonic thermal conduction by infrared plasmonic resonators in semiconductor nanowires. Appl. Phys. Lett. **114**, 163104 (2019)
114. E.N. Economou, Surface plasmons in thin films. Phys. Rev. **182**, 539–554 (1969)
115. K. Park, B.J. Lee, C.J. Fu, Z.M. Zhang, Study of the surface and bulk polaritons with a negative index metamaterial. J. Opt. Soc. Am. B **22**, 1016–1023 (2005)
116. K.L. Kliewer, R. Fuchs, Optical modes if vibration in an ionic crystal slab including retardation. I. Nonradiative region. Phys. Rev. **144**, 495–503 (1966)
117. G.J. Kovacs, G.D. Scott, Optical excitation of surface plasma waves in layered media. Phys. Rev. B **16**, 1297–1311 (1977)
118. M.I. Kaganov, N.B. Pustyl'nik, T.N. Shalaeva, Magnons, magnetic polaritons, magnetostatic waves. Phys. Usp. **40**, 181–224 (1997)
119. J.B. Pendry, A.J. Holden, D.J. Robbins, W.J. Stewart, Magnetism from conductors and enhanced nonlinear phenomena. IEEE Trans. Microw. Theory Tech. **47**, 2075–2084 (1999)
120. W. Cai, V. Shalaev, *Optical Metamaterials — Fundamentals and Applications* (Springer, Berlin, 2010)
121. J. Zhou, E.N. Economon, T. Koschny, C.M. Soukoulis, Unifying approach to left-handed material design. Opt. Lett. **31**, 3620–3622 (2006)
122. N. Engheta, Circuits with light at nanoscales: optical nanocircuits inspired by metamaterials. Science **317**, 1698–1702 (2007)
123. B.J. Lee, L.P. Wang, Z.M. Zhang, Coherent thermal emission by excitation of magnetic polaritons between periodic strips and a metallic film. Opt. Express **16**, 11328–11336 (2008)
124. L.P. Wang, Z.M. Zhang, Resonance transmission or absorption in deep gratings explained by magnetic polaritons. Appl. Phys. Lett. **95**, 111904 (2009)
125. J.X. Chen, P. Wang, Z.M. Zhang, Y.H. Lu, H.H. Ming, The coupling between gap plasmon polariton and magnetic polariton in a metallic-dielectric multilayer structure. Phys. Rev. E **84**(026603), 2011 (2011)
126. L.P. Wang, Z.M. Zhang, Wavelength-selective and diffuse emitter enhanced by magnetic polaritons for thermophotovoltaics. Appl. Phys. Lett. **100**, 063902 (2012)
127. B. Zhao, L.P. Wang, Y. Shuai, Z.M. Zhang, Thermophotovoltaic emitters based on a two-dimensional grating/thin-film nanostructure. Int. J. Heat Mass Transfer **67**, 637–645 (2013)
128. L.P. Wang, Z.M. Zhang, Measurement of coherent thermal emission due to magnetic polaritons in subwavelength microstructures. J. Heat Transfer **135**, 091014 (2013)
129. X.L. Liu, B. Zhao, Z.M. Zhang, Wide-angle near infrared polarizer with extremely high extinction ratio. Opt. Express **21**, 10502–10510 (2013)
130. A. Sakurai, B. Zhao, Z.M. Zhang, Prediction of the resonance condition of metamaterial emitters and absorbers using LC circuit model, in *The 15th International Heat Transfer Conference (IHTC-15), IHTC15-9012*, Kyoto, Japan, 10–15 August 2014

131. B. Zhao, Z.M. Zhang, Study of magnetic polaritons in deep gratings for thermal emission control. J. Quant. Spectrosc. Radiat. Transfer **135**, 81–89 (2014)
132. P. Yang, H. Ye, Z.M. Zhang, Experimental demonstration of the effect of magnetic polaritons on the radiative properties of deep aluminum gratings. J. Heat Transfer **141**, 052702 (2019)
133. A. Sakurai, B. Zhao, Z.M. Zhang, Resonant frequency and bandwidth of metamaterial emitters and absorbers predicted by an RLC circuit model. J. Quant. Spectrosc. Radiat. Transfer **149**, 33–40 (2014)
134. L.P. Wang, Z.M. Zhang, Phonon-mediated magnetic polaritons in the infrared region. Opt. Express **19**, A126–A135 (2011)
135. N. Liu, M. Mesch, T. Weiss, M. Hentschel, H. Giessen, Infrared perfect absorber and its application as plasmonic sensor. Nano Lett. **10**, 2342–2348 (2010)
136. C.M. Watts, X. Liu, W.J. Padilla, Metamaterial electromagnetic wave absorbers. Adv. Mater. **24**, OP98–OP120 (2012)
137. A. Pors, O. Albrektsen, I.P. Radko, S.I. Bozhevolnyi, Gap plasmon-based metasurfaces for total control of reflected light. Sci. Rep. **3**, 2155 (2013); F. Ding, Y. Yang, R.A. Deshpande, S.I. Bozhevolnyi, A review of gap-surface plasmon metasurfaces: fundamentals and applications. Nanophoton **7**, 1129–1156 (2018)
138. I. Avrutsky, I. Salakhutdinov, J. Elser, V. Podolskiy, Highly confined optical modes in nanoscale metal-dielectric multilayers. Phys. Rev. B **75**, 241402R (2007)
139. J.M. Zhao, Z.M. Zhang, Electromagnetic energy storage and power dissipation in nanostructures. J. Quant. Spectrosc. Radiat. Transfer **151**, 49–57 (2015)
140. A.N. Grigorenko, M. Polini, K.S. Novoselov, Graphene plasmons. Nat. Photon. **6**, 749–758 (2012)
141. Z. Fei, A.S. Rodin, G.O. Andreev, W. Bao, A.S. McLeod, M. Wagner, L.M. Zhang, Z. Zhao, M. Thiemens, G. Dominguez, M.M. Fogler, A.H. Castro-Neto, C.N. Lau, F. Keilmann, D.N. Basov, Gate-tuning of graphene plasmons revealed by infrared nano-imaging. Nature **487**, 82–85 (2012)
142. P. Alonso-González, A.Y. Nikitin, F. Golmar, A. Centeno, A. Pesquera, S. Vélez, J. Chen, G. Navickaite, F. Koppens, A. Zurutuza, F. Casanova, L.E. Hueso, R. Hillenbrand, Controlling graphene plasmons with resonant metal antennas and spatial conductivity patterns. Science **344**, 1369–1373 (2014)
143. L.A. Falkovsky, S.S. Pershoguba, Optical far-infrared properties of a graphene monolayer and multilayer. Phys. Rev. B **76**, 153410 (2007)
144. K.F. Mak, J. Shan, T.F. Heinz, Seeing many-body effects in single- and few-layer graphene: observation of two-dimensional saddle-point excitons. Phys. Rev. Lett. **106**, 046401 (2011)
145. R.Z. Zhang, Z.M. Zhang, Tunable positive and negative refraction of infrared radiation in graphene-dielectric multilayers. Appl. Phys. Lett. **107**, 191112 (2015)
146. B. Zhao, Z.M. Zhang, Enhanced photon tunneling by surface plasmon-phonon polaritons in graphene/hBN heterostructures. J. Heat Transfer **139**, 022701 (2017)
147. S.A. Mikhailov, K. Ziegler, New electromagnetic mode in graphene. Phys. Rev. Lett. **99**, 016803 (2007)
148. B. Wang, X. Zhang, X. Yuan, J. Teng, Optical coupling of surface plasmons between graphene sheets. Appl. Phys. Lett. **100**, 131111 (2012)
149. X.L. Liu, R.Z. Zhang, Z.M. Zhang, Near-perfect photon tunneling by hybridizing graphene plasmons and hyperbolic modes. ACS Photon. **1**, 785–789 (2014)
150. A.Y. Nikitin, T. Low, L. Martin-Moreno, Anomalous reflection phase of graphene plasmons and its influence on resonators. Phys. Rev. B **90**, 041407 (2014)
151. L. Du, D. Tang, X. Yuan, Edge-reflection phase directed plasmonic resonances on graphene nano-structures. Opt. Express **22**, 22689–22698 (2014)
152. B. Zhao, Z.M. Zhang, Strong plasmonic coupling between graphene ribbon array and metal gratings. ACS Photon. **2**, 1611–1618 (2015)
153. S.V. Zhukovsky, A.A. Orlov, V.E. Babicheva, A.V. Lavrinenko, J.E. Sipe, Photonic-band-gap engineering for volume plasmon polaritons in multiscale multilayer hyperbolic metamaterials. Phys. Rev. A **90**, 013801 (2014)

154. Z. Jacob, L.V. Alekseyev, E. Narimanov, Optical hyperlens: far-field imaging beyond the diffraction limit. Opt. Express **14**, 8247–8256 (2006)
155. T.J. Bright, X.L. Liu, Z.M. Zhang, Energy streamlines in near-field radiative heat transfer between hyperbolic metamaterials. Opt. Express **22**(S4), A1112–A1127 (2014)
156. A. Poddubny, I. Iorsh, P. Belov, Y. Kivshar, Hyperbolic metamaterials. Nat. Photon. **7**, 958–967 (2013)
157. D.R. Smith, D. Schurig, Electromagnetic wave propagation in media with indefinite permittivity and permeability tensors. Phys. Rev. Lett. **90**, 077405 (2003)
158. J. Schilling, Uniaxial metallo-dielectric metamaterials with scalar positive permeability. Phys. Rev. E **74**, 046618 (2006)
159. H. Wang, X.L. Liu, L.P. Wang, Z.M. Zhang, Anisotropic optical properties of silicon nanowire arrays based on effective medium calculation. Int. J. Therm. Sci. **65**, 62–69 (2013)
160. R.Z. Zhang, Z.M. Zhang, Validity of effective medium theory in multilayered hyperbolic materials. J. Quant. Spectrosc. Radiat. Transfer **197**, 132–140 (2017)
161. X.J. Wang, J.L. Abell, Y.-P. Zhao, Z.M. Zhang, Angle-resolved reflectance of obliquely aligned silver nanorods. Appl. Opt. **51**, 1521–1531 (2012)
162. X.H. Wu, C.J. Fu, Z.M. Zhang, Effect of orientation on the directional and hemispherical emissivity of hyperbolic metamaterials. Int. J. Heat Mass Transfer **135**, 1207–1217 (2019)
163. X.L. Liu, T.J. Bright, Z.M. Zhang, Application conditions of effective medium theory in near-field radiative heat transfer between multilayered metamaterials. J. Heat Transfer **136**, 092703 (2014)
164. E.E. Narimanov, A.V. Kildishev, Metamaterials: naturally hyperbolic. Nat. Photon. **9**, 214–216 (2015)
165. Z.L. Wang, Z.M. Zhang, X.J. Quan, P. Cheng, A perfect absorber design using a natural hyperbolic material for harvesting solar energy. Sol. Energy **159**, 329–336 (2018)
166. D.E. Aspnes, E. Kinsbron, D.D. Bacon, Optical properties of Au: sample effects. Phys. Rev. B **21**, 3290–3299 (1980)
167. Y. Tian, A. Ghanekar, M. Ricci, M. Hyde, O. Gregory, Y. Zheng, A review of tunable wavelength selectivity of metamaterials in near-field and far-field radiative thermal transport. Materials **11**, 000862 (2018)
168. M.S. Wheeler, J.S. Aitchison, M. Mojahedi, Three-dimensional array of dielectric spheres with an isotropic negative permeability at infrared frequencies. Phys. Rev. B **72**, 193103 (2005); M.S. Wheeler, J.S. Aitchison, J.I.L. Chen, G.A. Ozin, M. Mojahedi, Infrared magnetic response in a random silicon carbide micropowder. Phys. Rev. B **79**, 073103 (2009)
169. F.J. García-Vidal, J.M. Pitarke, J.B. Pendry, Effective medium theory of the optical properties of aligned carbon nanotubes. Phys. Rev. Lett. **78**, 4289–4292 (1997)
170. X.J. Wang, J.D. Flicker, B.J. Lee, W.J. Ready, Z.M. Zhang, Visible and near-infrared radiative properties of vertically aligned multi-walled carbon nanotubes. Nanotechnology **20**, 215704 (2009)
171. H. Ye, X.J. Wang, W. Lin, C.P. Wong, Z.M. Zhang, Infrared absorption coefficients of vertically aligned carbon nanotube films. Appl. Phys. Lett. **101**, 141909 (2012); Z.M. Zhang, H. Ye, Measurements of radiative properties of engineered micro/nanostructures. Annu. Rev. Heat Transfer **16**, 351–411 (2013)
172. R.Z. Zhang, X.L. Liu, Z.M. Zhang, Modeling the optical and radiative properties of vertically aligned carbon nanotubes in the infrared region. J. Heat Transfer 137, 091009 (2015); R.Z. Zhang, X.L. Liu, Z.M. Zhang, Near-field radiation between graphene-covered carbon nanotube arrays. AIP Adv. **5**, 053501 (2015)
173. R.Z. Zhang, Z.M. Zhang, Negative refraction and self-collimation in the far infrared with aligned carbon nanotube films. J. Quant. Spectrosc. Radiat. Transfer **158**, 91–100 (2015)
174. C.J. Fu, Z.M. Zhang, D.B. Tanner, Energy transmission by photon tunneling in multilayer structures including negative index materials. J. Heat Transfer **127**, 1046–1052 (2005)
175. A. Alu, N. Engheta, Pairing an epsilon-negative slab with a mu-negative slab: resonance, tunneling and transparency. IEEE Trans. Antennas Propag. **51**, 2558–2571 (2003)

176. H. Jiang, H. Chen, H. Li, Y. Zhang, S. Zhu, Compact high-$Q$ filters based on one dimensional photonic crystals containing single-negative materials. J. Appl. Phys. **98**, 013101 (2005)
177. F.J. Garcia-Vidal, L. Martin-Moreno, T.W. Ebbesen, L. Kuipers, Light passing through subwavelength apertures. Rev. Mod. Phys. **82**, 729–787 (2010)
178. L. Wang, S.M. Uppuluri, E.X. Jin, X. Xu, Nanolithography using high transmission nanoscale bowtie apertures. Nano Lett. **6**, 361–364 (2006)
179. Y.-B. Chen, B.J. Lee, Z.M. Zhang, Infrared radiative properties of submicron metallic slit arrays. J. Heat Transfer **130**, 082404 (2008)
180. F. Marquier, J.-J. Greffet, S. Collin, F. Pardo, J.L. Pelouard, Resonant transmission through a metallic film due to coupled modes. Opt. Express **13**, 70–76 (2005)
181. P.J. Hesketh, J.N. Zemel, B. Gebhart, Organ pipe radiant modes of periodic micromachined silicon surfaces. Nature **324**, 549–551 (1986); P.J. Hesketh, B. Gebhart, J.N. Zemel, Measurements of the spectral and directional emission from microgrooved silicon surfaces. J. Heat Transfer **110**, 680–686 (1998)
182. X.L. Liu, B. Zhao, Z.M. Zhang, Blocking-assisted infrared transmission of subwavelength metallic gratings by graphene. J. Opt. **17**, 035004 (2015)
183. M.J. Persky, Review of black surfaces for space-borne infrared systems. Rev. Sci. Instrum. **70**, 2193–2216 (1999)
184. J. Lehman, C. Yung, N. Tomlin, D. Conklin, M. Stephens, Carbon nanotube-based black coatings. Appl. Phys. Rev. **5**, 011103 (2018)
185. X.J. Wang, L.P. Wang, O.S. Adewuyi, B.A. Cola, Z.M. Zhang, Highly specular carbon nanotube absorbers. Appl. Phys. Lett. **97**, 163116 (2010); X.J. Wang, O.S. Adewuyi, L.P. Wang, B.A. Cola, Z.M. Zhang, Reflectance measurements for black absorbers made of vertically aligned carbon nanotubes, in *SPIE Proceedings*. Reflection, Scattering, and Diffraction from Surfaces II, vol. 7792 (2010), p. 77920R
186. K. Mizuno, J. Ishii, H. Kishida, Y. Hayamizu, S. Yasuda, D.N. Futaba, M. Yumura, K. Hata, A black body absorber from vertically aligned single-walled carbon nanotubes. Proc. Natl. Acad. Sci. USA **106**, 6044–6047 (2009)
187. X.L. Liu, L.P. Wang, Z.M. Zhang, Wideband tunable omnidirectional infrared absorbers based on doped-silicon nanowire arrays. J. Heat Transfer 135, 061602 (2013); X.L. Liu, Z.M. Zhang, Metal-free low-loss negative refraction in the mid-infrared region. Appl. Phys. Lett. **103**, 103101 (2013)
188. A. Sakurai, B. Zhao, Z.M. Zhang, Effect of polarization on dual-band infrared metamaterial emitters or absorbers. J. Quant. Spectrosc. Radiat. Transfer **158**, 111–118 (2015); B. Zhao, A. Sakurai, Z.M. Zhang, Polarization dependence of the reflectance and transmittance of anisotropic metamaterials. J. Thermophys. Heat Transfer **30**, 240–246 (2016)
189. Y. Cui, K.H. Fung, J. Xu, H. Ma, Y. Jin, S. He, N.X. Fang, Ultrabroadband light absorption by a sawtooth anisotropic metamaterial slab. Nano Lett. **12**, 1443–1447 (2012)
190. B. Zhao, Z.M. Zhang, Perfect absorption with trapezoidal gratings made of natural hyperbolic materials. Nanoscale Microscale Thermophys. Eng. **21**, 123–133 (2017); B. Zhao, Z.M. Zhang, Resonance perfect absorption by exciting hyperbolic phonon polaritons in 1D hBN gratings. Opt. Express **25**, 7791–7796 (2017)
191. A.J. Giles, S. Dai, I. Vurgaftman, T. Hoffman, S. Liu, L. Lindsay, C.T. Ellis, N. Assefa, I. Chatzakis, T.L. Reinecke, J.G. Tischler, M.M. Fogler, J.H. Edgar, D.N. Basov, J.D. Caldwell, Ultralow-loss polaritons in isotopically pure boron nitride. Nat. Mater. **17**, 134–139 (2018)
192. R.R. Nair, P. Blake, A.N. Grigorenko, K.S. Novoselov, T.J. Booth, T. Stauber, N.M.R. Peres, A.K. Geim, Fine structure constant defines visual transparency of graphene. Science **320**, 1308 (2008)
193. B. Zhao, J.M. Zhao, Z.M. Zhang, Enhancement of near-infrared absorption in graphene with metal gratings. Appl. Phys. Lett. **105**, 031905 (2014); B. Zhao, J.M. Zhao, Z.M. Zhang, Resonance enhanced absorption in a graphene monolayer by using deep metal gratings. J. Opt. Soc. Am. B **32**, 1176–1185 (2015)
194. S. Noda, M. Fujita, T. Asano, Spontaneous-emission control by photonic crystals and nanocavities. Nat. Photon. **1**, 449–458 (2007)

195. A. Heinzel, V. Boerner, A. Gombert, B. Bläsi, V. Wittwer, J. Luther, Radiation filters and emitters for the NIR based on periodically structured metal surfaces. J. Mod. Opt. **47**, 2399–2419 (2000)
196. S. Maruyama, T. Kashiwa, H. Yugami, M. Esashi, Thermal radiation from two-dimensionally confined modes in microcavities. Appl. Phys. Lett. **79**, 1393–1395 (2001); H. Sai, Y. Kanamori, H. Yugami, Tuning of the thermal radiation spectrum in the near-infrared region by metallic surface microstructures. J. Micromech. Microeng. **15**, S243–S249 (2005)
197. F. Kusunoki, J. Takahara, T. Kobayashi, Quantitative change of resonant peaks in thermal radiation from periodic array of microcavities. Electron. Lett. **39**, 23–24 (2003); F. Kusunoki, T. Kohama, T. Hiroshima, S. Fukumoto, J. Takahara, T. Kobayashi, Narrow-band thermal radiation with low directivity by resonant modes inside tungsten microcavities. Jpn. J. Appl. Phys. **43**, 5253–5258 (2004)
198. S. Basu, Y.-B. Chen, Z.M. Zhang, Microscale radiation in thermophotovoltaic devices – a review. Int. J. Ener. Res. **31**, 689–716 (2007)
199. V. Rinnerbauer, S. Ndao, Y.X. Yeng, W.R. Chan, J.J. Senkevich, J.D. Joannopoulos, M. Soljačić, I. Celanovic, Recent developments in high-temperature photonic crystals for energy conversion. Energy Environ. Sci. **5**, 8815–8823 (2012)
200. C.J. Fu, Z.M. Zhang, D.B. Tanner, Planar heterogeneous structures for coherent emission of radiation. Opt. Lett. **30**, 1873–1875 (2005)
201. B.J. Lee, C.J. Fu, Z.M. Zhang, Coherent thermal emission from one-dimensional photonic crystals. Appl. Phys. Lett. **87**, 071904 (2005); B.J. Lee, Z.M. Zhang, Coherent thermal emission from modified periodic multilayer structures. J. Heat Transfer **129**, 17–26 (2007)
202. J.A. Gaspar-Armenta, F. Villa, Photonic surface-wave excitation: photonic crystal-metal interface. J. Opt. Soc. Am. B **20**, 2349–2354 (2003)
203. B.J. Lee, Z.M. Zhang, Indirect measurements of coherent thermal emission from a truncated photonic crystal structure. J. Thermophys. Heat Transfer **23**, 9–17 (2009)
204. B.J. Lee, Y.-B. Chen, Z.M. Zhang, Surface waves between metallic films and truncated photonic crystals observed with reflectance spectroscopy. Opt. Lett. **33**, 204–206 (2008)

# Chapter 10
# Near-Field Energy Transfer

Thermal radiation has played an important role in incandescent lamps, solar energy utilization, temperature measurements, materials processing, remote sensing for astronomy and space exploration, combustion and furnace design, food processing, cryogenic engineering, as well as numerous agriculture, health, safety, and surveillance applications. Near-field effects can realize emerging technologies, such as superlenses, subwavelength light sources, polariton-assisted biosensors, and energy conversion devices. The control of thermal radiative properties by micro/nanoscale 1D, 2D, and 3D photonic structures has been extensively addressed in previous chapters. Because of the important applications in energy transport and conversion, this chapter focuses on near-field radiative heat transfer between objects in close vicinity. The phenomenon of photon tunneling and the principle of fluctuation–dissipation theorem will be presented, along with recent theoretical and experimental developments.

## 10.1 From Near-Field Optics to Nanoscale Thermal Radiation

Near-field optics and near-field microscopy have played a significant role in nanoscience and nanobiotechnology in the past 30 years and continue to be an active research area, especially when dealing with field localization and resonances in micro/nanostructures, with applications in biochemical sensing and nanolithography. The preceding two chapters have laid the foundation of electromagnetic waves in bulk materials and nanostructures. The present chapter offers a more detailed treatment of the energy transfer by electromagnetic waves in the near field. The applications include nanomanufacturing, energy conversion systems, and nanoelectronics thermal management.



Z. M. Zhang, *Nano/Microscale Heat Transfer*, Mechanical Engineering Series,
https://doi.org/10.1007/978-3-030-45039-7_10

Ernst Abbe in 1873 and Lord Rayleigh in 1879 studied the required angular separation between two objects for their images to be resolved. The resolution of a conventional microscope is diffraction limited such that the smallest resolvable distance is approximately $0.5\lambda/n$, where $\lambda$ is the wavelength in vacuum and $n$ is the refractive index of the medium. Even with immersion oil ($n \approx 1.5$), the imaging sharpness is rather limited to the order of wavelength. The concept of near-field imaging was first described by E. H. Synge in a paper published in Philosophical Magazine in 1928. This work envisioned the use of a subwavelength aperture as small as 10 nm in diameter to introduce light to a specimen (e.g., a stained biological section), placed within 10-nm distance, which could move in its plane with a step size less than 10 nm. By measuring the transmitted light with a photoelectric cell and a microscope, an ultramicroscopic image could be constructed. In a subsequent paper published in 1932 (also in Philosophical Magazine), Synge described the idea of using piezoelectricity in microscopy. Synge's works, however, were largely unnoticed and the idea of near-field imaging was rediscovered many years later. Ash and Nicholls published a paper in Nature in 1972 entitled "Super-resolution aperture scanning microscope." This work experimentally demonstrated near-field imaging with a resolution of $\lambda/60$ using 10-GHz microwave radiation ($\lambda = 3\,\text{cm}$). Readers are referred to Refs. [1, 2] for more details about the early history of near-field microscopy.

In the 1980s, two groups successfully developed near-field microscopes in the visible region. The IBM group in Zurich formed the aperture through a quartz tip coated with a metallic film on its sides [3], whereas the Cornell group used silicon microfabrication to form the aperture [4]. The fabrication process was later improved by using metal-coated tapered optical fibers. In the early 1990s, Nobel Laureate R. Eric Betzig at Bell Labs and his collaborators demonstrated single-molecule detection and data storage capabilities of 45 gigabits per square inch [5]. Nowadays, the near-field scanning optical microscope (NSOM), also known as the scanning near-field optical microscope (SNOM), has become a powerful tool in the study of fundamental space- and time-dependent processes, thermal metrology, and optical manufacturing with a spatial resolution of less than 50 nm. NSOM is usually combined with the atomic force microscope (AFM) for highly controllable movement and position sensing. An alternative approach is to use a metallic AFM tip to couple the far-field radiation with the near-field electromagnetic waves in a subwavelength region underneath the tip. This is the so-called *apertureless NSOM*, which does not require an optical fiber or an aperture. Apertureless tips allow high-intensity laser energy to be focused on nanoscale dimensions for laser-assisted nanothermal manufacturing [1–10].

Figure 10.1 illustrates three typical NSOM designs. The first is an aperture-based setup, where a very small opening is formed on an opaque plate and collimated light is incident from the above. The second is based on a tapered optical fiber whose tip serves as an aperture. The third uses an apertureless metallic sharp tip, which reflects (scatters) the incident laser light. All of the three designs have one thing in common. The light is confined to a narrow region, whose width may be much less than a wavelength. Furthermore, the electromagnetic field within one wavelength distance is very intense and highly collimated. In the near-field region, evanescent

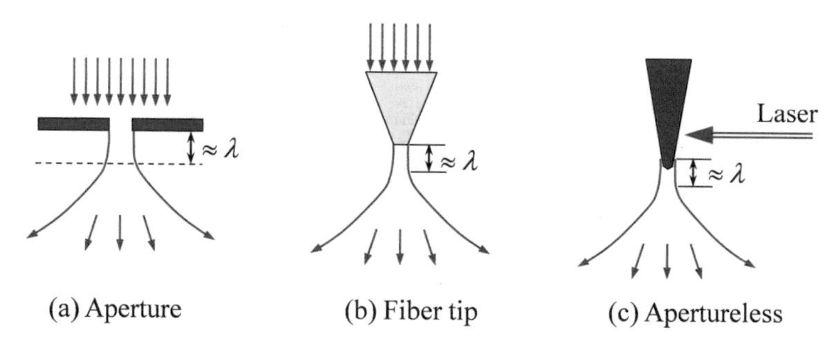


**Fig. 10.1** Schematic illustrations of different NSOM setups. **a** Aperture on an opaque plate. **b** Aperture at the end of a coated optical fiber. **c** Apertureless metallic tip. The opening or the tip is much smaller than the wavelength λ. The electric field is highly collimated in the near field within a distance of λ and diverges as the distance increases

waves dominate. Because the amplitude of an evanescent wave decays exponentially away from the aperture or tip, the far-field, or the radiation field diverges and becomes very weak. Understanding the nature of evanescent waves and the localized fields is essential for the NSOM and other near-field optical devices.

Evanescent waves are also essential in energy transfer between adjacent objects, through photon tunneling, and in surface plasmon polaritons or surface phonon polaritons as discussed in the previous chapter. The concept of Fabry–Perot resonant cavities has also been introduced. Two- and three-dimensional optical cavities and microwave cavities support resonance modes, which are standing waves within the cavity. These devices are important for photonics and optoelectronics. High $Q$-factors can be achieved with microfabricated cavities for quantum electrodynamics (QED), enhancement and suppression of spontaneous emission, and biological and chemical sensing [11–13]. Cavity QED is a field that was initiated in the 1980s to study the spontaneous emission of atoms inside a subwavelength cavity. Both enhancement and inhibition of spontaneous emission have been theoretically and experimentally demonstrated. There have been a large number of publications dealing with spontaneous emission of microstructures since nanostructures may enhance or suppress spontaneous emission [11–13]. As discussed in the previous chapter, surface plasmons may enhance transmission, suppress transmission but enhance absorption, or enhance both transmission and absorption in nanostructures at the same time, depending on the coupling with the resonance and boundary conditions.

If an object is at thermal equilibrium with itself, can it emit more energy (in any spectral range, polarization, and solid angle) to free space (far field) than a blackbody with the identical shape and size at the same temperature? Saying in other words: “Can the emitted intensity from an object exceed the blackbody intensity described by Planck’s law at any particular wavelength and angle of emission?” By using intensity, we are talking about the far field, not the near field. The answer is definitely “yes,” if the overall size of the body is less than the wavelength, and definitely “no” if the overall size of the body is much greater than the wavelength of interest, even though the object is made of subwavelength structures. Rather than considering spontaneous emission toward an empty space, let us consider the thermodynamic equilibrium in an enclosure, where the object is placed inside and is in thermodynamic equilibrium with the enclosure. Generally speaking, stimulated emission is much smaller than stimulated absorption (see Chap. 3, Sect. 3.6), and we can treat the net absorption as stimulated absorption subtracted by stimulated emission. At thermodynamic equilibrium, the net absorbed energy must be the same as the spontaneously emitted energy of any objects inside the cavity. The density of states inside any medium is modified by its electric and magnetic properties. For a medium with a refractive index of $n$, Planck's distribution as given in Eq. (8.44) is proportional to $n^2$. If the refractive index depends on wavelength, the group velocity will be different from the phase velocity and, hence, the equilibrium distribution will further deviate from Planck's law. If absorption is also considered, the equilibrium distribution inside the medium will be completely different. However, *Planck's distribution is always observed in the evacuated region, as long as the location is away from either the object or the walls of the enclosure.* This condition or restriction implies that the enclosure must have enough room for the evacuated region to be much greater than the characteristic wavelength. It is impractical to establish the concept of blackbody for objects and cavities with a size less than the wavelength of interest, as noted by Planck over 100 years ago [14]. It has been known for some time that a large field enhancement exists near the surface when surface polaritons are excited [15]. The enhancement also exists around subwavelength structures [16]. However, the energy density in an evacuated large enclosure at thermal equilibrium is the same as Planck's distribution, except at close vicinity of the objects, including the walls of the enclosure.

Spontaneous emission can be viewed as a coupling of the field inside the material with that outside the material. A small object can couple with the electromagnetic field by bending the energy streamlines or the Poynting vectors (due to the coupling of the incident and emitted fields) toward it, and hence, the object will absorb more energy than a "blackbody" of the same size [16]. Figure 10.2 illustrates qualitatively and somewhat exaggerated interactions of the incident field with a small object and a

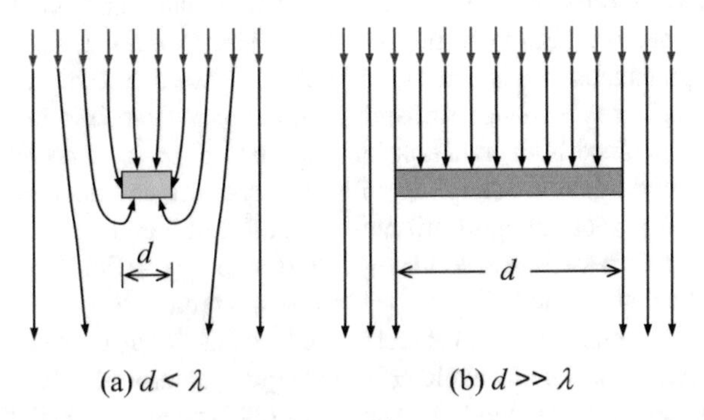


**Fig. 10.2** Schematic drawing of the energy streamlines for an incident plane wave, showing the Poynting vectors of the incident field and the cross-coupling between the incident and scattered fields. **a** A small object with resonance absorption. **b** A large object with subwavelength structures

large object. A small object can perturb the incoming energy streamlines (or the field of the electromagnetic waves) by creating an additional term in the Poynting vector that arises from the coupling between the incident and scattered fields. Therefore, the absorptance and spontaneous emission can be enhanced at the resonance wavelength, which depends on the geometric structure and material's properties. This was briefly discussed in the previous chapter, e.g., Eq. (9.104) for small spheres. On the other hand, for a large body, the incoming energy is limited by the projected area without any geometric enhancement even with surface or volume micro/nanostructures. Because of reflection and transmission, the net absorbed energy is always smaller than the energy incident on the object. Hence, it is not possible for spontaneous emission from a large object or composite to exceed the blackbody intensity in the far field.

Many researchers have demonstrated near-field thermal radiation when objects at different temperatures are separated by distances smaller than the characteristic thermal wavelength [17, 18]. Recently, theoretical calculations and experiments have demonstrated that radiative heat transfer between two nanostructures separated by a distance greater than the characteristic wavelength (i.e., in the far field) can exceed that predicted by Planck's law [19, 20]. The rest of this chapter aims at presenting and explaining the basic theory and formulations for calculation of near-field radiative energy exchange between objects. Following the methodology used in the previous chapters, we will start with simple geometric structures and easy-to-understand materials. Advanced concepts and derivations will be subsequently introduced. Recent advances in theoretical and experimental research and in the applications of nanoscale thermal radiation will be outlined.

## 10.2 Photon Tunneling and Near-Field Radiative Heat Transfer

### *10.2.1 Photon Tunneling by Coupled Evanescent Waves*

In the preceding sections, we have clearly demonstrated that an evanescent wave exists inside the optically rarer medium, which can be air or vacuum, and decays exponentially away from the surface. Furthermore, the evanescent wave or field does not carry energy in the direction normal to the interface. On the other hand, if another optically denser medium is brought to close proximity of the first medium, as shown in Fig. 10.3, energy can be transmitted from the first to the third medium, even though the angle of incidence is greater than the critical angle. This phenomenon, known as *frustrated total internal reflection*, *photon tunneling*, or *radiation tunneling*, is very important for energy transfer between two bodies when the distance of separation is shorter than the dominant wavelength of the emitting source. Frustrated total internal reflection has been known since Newton's time and was theoretically investigated by Hall in 1902 [21]. Cryogenic insulation is a practical example of when

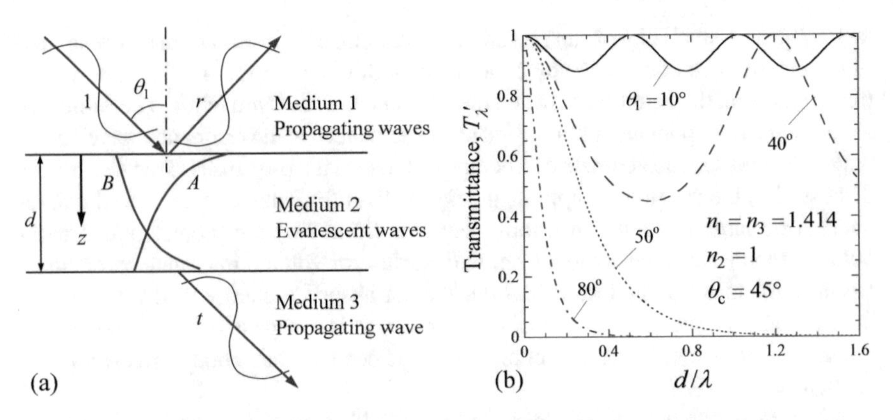


**Fig. 10.3** Illustration of photon tunneling. **a** Schematic drawing of the three layers and fields, where A and B indicate decaying and growing evanescent waves, respectively. **b** Calculated transmittance for a TE wave, assuming $n_1 = n_3 = 1.414$ and $n_2 = 1$. Note the distinct differences between the interference effect and the photon tunneling phenomenon, where the transmittance decreases with increasing $d$ and becomes negligibly small for $d > \lambda$

photon tunneling may be significant [22]. Advances in micro/nanotechnologies have made it possible for the energy transfer by photon tunneling to be appreciable and even dominant at room temperature or above. This may have applications ranging from microscale thermophotovoltaic devices to nanothermal processing and thermal management in nanoelectronics [17, 18, 23–25].

While photon tunneling is analogous to electron tunneling, through a potential barrier, which may be explained by quantum mechanics [26], it can be understood by the coupling of two oppositely decaying evanescent waves [27–29]. Because of the second interface, a backward-decaying evanescent wave is formed inside layer 2, the optical rarer medium. The Poynting vector of the coupled evanescent fields has a nonzero normal component, suggesting that the energy transmission between the media is possible as long as the gap width is smaller than the wavelength. Beyond this wavelength, the field strength of the forward-decaying evanescent wave is too low when it reaches the second interface and the reflected evanescent field is negligible. The matrix formulation discussed in Chap. 9 can be used to calculate the transmittance and the reflectance through the gap (i.e., medium 2) as if there were propagating waves. To illustrate this, consider all three layers are dielectric. Taking the TM wave incidence as an example, let us write the magnetic field inside medium 2 as follows:

$$H_y(x, z) = (A\mathrm{e}^{\mathrm{i}k_{2z}z} + B\mathrm{e}^{-\mathrm{i}k_{2z}z})\mathrm{e}^{\mathrm{i}k_x x}, \quad 0 \le z \le d \tag{10.1}$$

where $A$ and $B$ are determined by the incident field and boundary conditions. When two waves are combined, the Poynting vector of the field $\langle \mathbf{S} \rangle = \frac{1}{2}\mathrm{Re}\left[(\mathbf{E}_1 + \mathbf{E}_2) \times (\mathbf{H}_1^* + \mathbf{H}_2^*)\right]$ has four terms. Two of them can be associated with the power flux of each individual wave, while the other two represent the interaction between the waves. After simplification, the normal component of the Poynting vector can be expressed as

$$\langle S_z \rangle = \frac{k_{2z}}{2\omega\varepsilon_2\varepsilon_0}\left(|A|^2 - |B|^2\right), \text{ when } k_{2z}^2 = k_2^2 - k_x^2 > 0 \tag{10.2a}$$

and

$$\langle S_z \rangle = -\frac{\eta_2}{\omega\varepsilon_2\varepsilon_0}\mathrm{Im}\left(AB^*\right), \text{ when } \eta_2^2 = -k_{2z}^2 = k_x^2 - k_2^2 > 0 \tag{10.2b}$$

Because there is no loss or absorption, $\langle S_z \rangle$ is independent of $z$ in medium 2, and the ratio of $\langle S_z \rangle$ in medium 2 to that of the incidence in medium 1 is the transmittance. When propagating waves exist in medium 2 or the angle of incidence is smaller than the critical angle, interference will occur and the energy flux in the $z$-direction can be represented by the forward- and backward-propagating waves, see Eq. (10.2a). The transmittance oscillates as the thickness of medium 2 is increased. When evanescent waves exist in medium 2 at incidence angles greater than the critical angle, the transmittance is a decaying function of the thickness of medium 2, as shown in Fig. 10.3b. While the individual evanescent wave does not carry energy, the coupling results in energy transfer, as suggested by Eq. (10.2b). Equation (9.10), derived in the previous chapter, can be used to calculate the transmittance. These equations are applicable to arbitrary electric and magnetic properties as long as the medium is isotropic and homogeneous within each layer. The phase shift $\psi$ in these equations is purely imaginary when medium 2 is a dielectric or vacuum.

**Example 10.1** Assuming that the incident field has an amplitude of 1, determine $A$ and $B$ in Eq. (10.1) for $\theta_1 > \theta_c = \sin^{-1}(n_2/n_1)$, when all three media are dielectric with $n_3 = n_1 > n_2$. Find an expression of the tunneling transmittance using real variables only.

**Solution** The tangential field components can be written as follows for the three-layer structure shown in Fig. 10.3a. Let $\eta_2 = \sqrt{k_x^2 - k_2^2} = (2\pi n_1/\lambda)\sqrt{\sin^2\theta_1 - \sin^2\theta_c}$, where λ is the wavelength in vacuum.

$$H_y = \begin{cases} (\mathrm{e}^{\mathrm{i}k_{1z}z} + r\mathrm{e}^{-\mathrm{i}k_{1z}z})\mathrm{e}^{\mathrm{i}k_x x}, & z \le 0 \\ (A\mathrm{e}^{-\eta_2 z} + B\mathrm{e}^{\eta_2 z})\mathrm{e}^{\mathrm{i}k_x x}, & 0 < z \le d \\ t\mathrm{e}^{\mathrm{i}k_{1z}z}\mathrm{e}^{\mathrm{i}k_x x}, & z > d \end{cases} \tag{10.3}$$

$$E_x = \begin{cases} \frac{k_{1z}}{\omega n_1^2\varepsilon_0}(\mathrm{e}^{\mathrm{i}k_{1z}z} - r\mathrm{e}^{-\mathrm{i}k_{1z}z})\mathrm{e}^{\mathrm{i}k_x x}, & z \le 0 \\ \frac{\mathrm{i}\eta_2}{\omega n_2^2\varepsilon_0}(A\mathrm{e}^{-\eta_2 z} - B\mathrm{e}^{\eta_2 z})\mathrm{e}^{\mathrm{i}k_x x}, & 0 < z \le d \\ \frac{k_{1z}}{\omega n_1^2\varepsilon_0}t\mathrm{e}^{\mathrm{i}k_{1z}z}\mathrm{e}^{\mathrm{i}k_x x}, & z > d \end{cases} \tag{10.4}$$

The continuity of the tangential components at the two interfaces allows us to determine $t$, $r$, $A$, and $B$. Note that because the incident field has an amplitude of 1, the preceding equations do not yield a set of homogeneous linear equations as in the case of guided waves. According to Eq. (8.90), $r_\mathrm{p} = \mathrm{e}^{\mathrm{i}\delta}$, where $\delta = -2\alpha$ and $\cot(\alpha) = (k_{1z}/n_1^2)/(\eta_2/n_2^2)$ for a TM wave. Let us rewrite Eqs. (9.7) and (9.8) for the reflection and transmission coefficients as follows:

$$r = \frac{\mathrm{e}^{\mathrm{i}\delta}(1 - \mathrm{e}^{-2\eta_2 d})}{1 - \mathrm{e}^{2\mathrm{i}\delta}\mathrm{e}^{-2\eta_2 d}} \tag{10.5}$$

$$t = \frac{(1 - \mathrm{e}^{2\mathrm{i}\delta})\mathrm{e}^{-\eta_2 d}}{1 - \mathrm{e}^{2\mathrm{i}\delta}\mathrm{e}^{-2\eta_2 d}} \tag{10.6}$$

where we have used the relationship of Fresnel's coefficients and set the phase shift in Eq. (9.6) to $\psi = \mathrm{i}\eta_2 d$. After matching the boundary conditions at $z = d$, we have

$$A = 0.5t[1 - \mathrm{i}\cot(\alpha)] \quad \text{and} \quad B = 0.5t[1 + \mathrm{i}\cot(\alpha)]\mathrm{e}^{-\eta_2 d} \tag{10.7}$$

It can be shown that the normal component of the Poynting vector is the same in media 2 and 3 (see Problem 10.1). The tunneling transmittance becomes

$$T'_\lambda = tt^* = \frac{2[1 - \cos(2\delta)]\mathrm{e}^{-2\eta_2 d}}{1 + \mathrm{e}^{-4\eta_2 d} - 2\cos(2\delta)\mathrm{e}^{-2\eta_2 d}} \tag{10.8a}$$

or

$$T'_\lambda = \frac{\sin^2(\delta)}{\sin^2(\delta) + \sinh^2(\eta_2 d)} \tag{10.8b}$$

Clearly, the tunneling transmittance does not oscillate as $d$ increases; rather, it decreases monotonically from 1 to 0 as $d$ is increased from 0 to infinity. The tunneling transmittance can also be thought as a phonon tunneling probability for the given mode as specified by the polarization and frequency. Equations (10.5)–(10.7) can also be applied to TE waves by setting $\cot(\alpha) = k_{1z}/\eta_2$; this changes the Fresnel reflection coefficient $r_p$ to $r_s$. Equations (10.8a, 10.8b) can be conveniently used for calculating the tunneling transmittance between dielectrics for both polarizations.

### *10.2.2 Thermal Energy Transfer Between Closely Spaced Dielectrics*

Energy exchange between closely spaced dielectric plates can be calculated by integrating Planck's function over all wavelengths as well as over the whole hemisphere using the directional-spectral transmittance. In essence, thermal emission originates from one medium (volumetrically) and is then transmitted through the space (gap) to another medium, where it is absorbed volumetrically. Let us use an example to illustrate the procedure and the effect of photon tunneling and interferences on the near-field thermal radiation.

**Example 10.2** Calculate the hemispherical transmittance between two dielectrics of $n_1 = n_3 = 3$, separated by a vacuum gap $d$ ($n_2 = 1$). Use the results to calculate the radiative energy transfer between the two media, assuming $T_1 = 1000$ K and $T_3 = 300$ K.

**Analysis**. In the far field, we can use the following formula discussed in Chap. 2 (see Example 2.6) to calculate the net radiative heat flux:

$$q''_{13,d\to\infty} = \frac{\sigma_{\mathrm{SB}} T_1^4 - \sigma_{\mathrm{SB}} T_3^4}{1/\varepsilon_1 + 1/\varepsilon_3 - 1} \tag{10.9}$$

The hemispherical emissivity (or emittance) of each surface can be evaluated using Eq. (8.105), which can be rewritten as follows, assuming that the emissivity is independent of the azimuthal angle $\phi$:

$$\varepsilon_\lambda^{\mathrm{h}} = 2\int_0^{\pi/2} \varepsilon'_\lambda(\theta)\cos\theta\sin\theta \mathrm{d}\theta \tag{10.10}$$

One could average the directional-spectral emissivity over the two polarizations. However, the preferable way is to calculate the hemispherical emissivity for each polarization and use it to calculate the net heat flux by taking half of Eq. (10.9). Equation (10.10) can be weighted to the blackbody emissive power and integrated over all wavelengths to obtain the total, hemispherical emissivity according to Eq. (8.106). The heat fluxes calculated for the two polarizations can then be added, resulting in the net heat flux in the far-field limit, as loosely given in Eq. (10.9). Clearly, the calculated far-field heat flux is always smaller than that between two blackbodies given by $q''_{13,\mathrm{BB}} = \sigma_{\mathrm{SB}}(T_1^4 - T_3^4)$ for the parallel-plate configuration. However, the situation will be different in the near field when interference and tunneling effects are important.

**Solution** The hemispherical transmittance can be evaluated in the similar way by an integration over the hemisphere. Note that only a small cone of radiation, originated from medium 1, will result in propagating waves in medium 2. This half cone angle is the critical angle, which is $\theta_{\mathrm{c}} = \sin^{-1}(n_2/n_1) \approx 19.5°$. Thus, we can divide the hemispherical transmittance into two parts to separately evaluate the transmittance. Keeping in mind that the transmittance is defined as the ratio of the transmitted energy to the incident energy, we can sum the two parts to obtain the hemispherical transmittance.

$$T_{\lambda,\mathrm{h}} = T_{\lambda,\mathrm{prop}} + T_{\lambda,\mathrm{evan}} \tag{10.11}$$

where

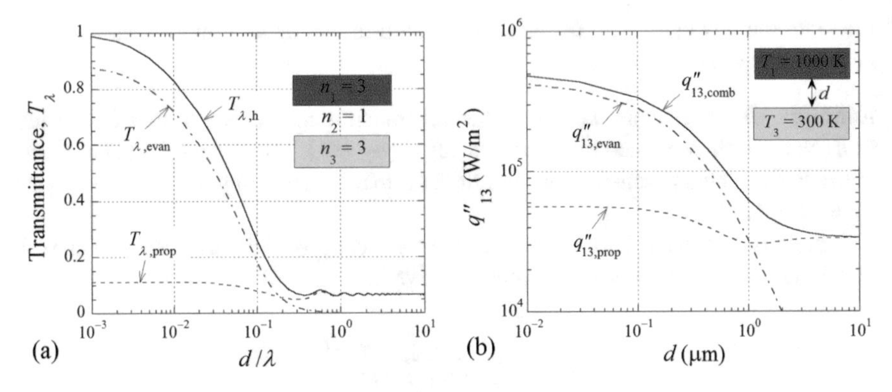


**Fig. 10.4** Radiation heat transfer between dielectric surfaces in close proximity. **a** Contributions to hemispherical transmittance by interference and tunneling, where the transmittance is the average of both polarizations. **b** Net heat flux as a function of the distance of separation

$$T_{\lambda,\mathrm{prop}} = 2\int_0^{\theta_c} T'_\lambda \cos\theta \sin\theta \mathrm{d}\theta \tag{10.12a}$$

and

$$T_{\lambda,\mathrm{evan}} = 2\int_{\theta_c}^{\pi/2} T'_\lambda \cos\theta \sin\theta \mathrm{d}\theta \tag{10.12b}$$

If $n_1 \neq n_3$, $\theta_c$ will depend on whether the incidence is from medium 1 or 3; however, the resulting hemispherical transmittance will remain the same. We can calculate the average transmittance for the two polarizations and the results are shown in Fig. 10.4a. The contribution of propagating waves exhibits some oscillations when $d$ and $\lambda$ are close to each other, but reaches a constant value when $d/\lambda \to 0$ where all waves will be constructively added. At $d/\lambda \gg 1$, the constructive and destructive interferences cancel out so that $T_{\lambda,\mathrm{prop}}$ becomes a constant again. The contribution of evanescent waves becomes important when $d/\lambda < 1$ and starts to dominate over that of the propagating waves when $d/\lambda \ll 1$. When $d/\lambda \to 0$, the evanescent wave or tunneling contributes to nearly 90% of the transmittance when $n_1 = 3$. This explains why photon tunneling is very important for the near-field energy transfer.

Planck's blackbody distribution function, given by Eq. (8.44), can be rewritten for each polarization in media 1 and 3, respectively, as

$$e_{\mathrm{b},\lambda}(\lambda, T_1) = \frac{n_1^2 C_1}{2\lambda^5(\mathrm{e}^{C_2/\lambda T_1} - 1)} \tag{10.13a}$$

and

$$e_{\mathrm{b},\lambda}(\lambda, T_3) = \frac{n_3^2 C_1}{2\lambda^5 (\mathrm{e}^{C_2/\lambda T_3} - 1)} \tag{10.13b}$$

where $\lambda$ in $\mu$m is the wavelength in vacuum, and $C_1 = 3.742 \times 10^8$ W $\mu$m$^4$/m$^2$ and $C_2 = 1.439 \times 10^4$ $\mu$m K are the first and second radiation constants in vacuum. The emissive power in a nondispersive dielectric is increased by a factor of the square of the refractive index, as a result of the increased photon density of states. The factor 2 in the denominator is included because only single polarization has been considered. The net radiation heat flux from medium 1 to 3 is

$$q''_{1\to 3} = \int_0^\infty e_{\mathrm{b},\lambda}(\lambda, T_1) T_{\lambda,\mathrm{h}}(\lambda) \mathrm{d}\lambda \tag{10.14a}$$

and that from medium 3 to 1 is

$$q''_{3\to 1} = \int_0^\infty e_{\mathrm{b},\lambda}(\lambda, T_3) T_{\lambda,\mathrm{h}}(\lambda) \mathrm{d}\lambda \tag{10.14b}$$

where $T_{\lambda,\mathrm{h}}$ is obtained from Eq. (10.11). Hence, the net radiation heat transfer becomes

$$q''_{13} = q''_{1\to 3} - q''_{3\to 1} \tag{10.15}$$

One can also separately substitute the hemispherical transmittance of propagating and evanescent waves to Eqs. (10.14a, 10.14b). Equation (10.15) should be individually applied to TE and TM waves, and then summed together to get the net heat flux. The integration limits can be set such that the lower limit $\lambda_L = 0.1\lambda_{\mathrm{mp}}$ and the upper limit $\lambda_H = 10\lambda_{\mathrm{mp}}$, where $\lambda_{\mathrm{mp}}$ is the wavelength corresponding to the maximum blackbody emissive power at the temperature, as expressed in Eq. (8.45), for the higher temperature medium. The calculated results of the near-field radiative transfer are shown in Fig. 10.4b as a function of the separation distance $d$. Several important observations can be made. (a) When $d \ll \lambda_{\mathrm{mp}}$, the propagating waves result in $q''_{13,\mathrm{prop}} = \sigma_{\mathrm{SB}}(T_1^4 - T_3^4)$ and the evanescent waves result in $q''_{13,\mathrm{evan}} = (n_1^2 - 1)\sigma_{\mathrm{SB}}(T_1^4 - T_3^4)$. The combined net radiation heat transfer is $q''_{13,\mathrm{comb}} = n_1^2\sigma_{\mathrm{SB}}(T_1^4 - T_3^4)$. (b) As the distance increases, the evanescent wave contribution goes down monotonically and becomes negligible when $d = \lambda_{\mathrm{mp}}$, which is about 3 $\mu$m. (c) Due to interference effects, the energy transfer by propagating waves decreases slightly as $d/\lambda$ increases and then reaches the far-field limit, Eq. (10.9), when $d/\lambda \gg 1$.

If the media were conductive, the previous calculations are not appropriate because of the large imaginary part of the refractive index or the dielectric function. In fact, near-field radiation heat transfer can be greatly enhanced with the presence of surface waves or if the media are semiconductors [23–25]. The treatment of these situations requires knowledge of fluctuational electrodynamics, which will be discussed in Sect. 10.4 at length.

### 10.2.3 Resonance Tunneling Through Periodic Dielectric Layers

There exists a photonic analogue of *resonance tunneling* of electrons in double-barrier quantum well structures. The geometry to illustrate resonance photon tunneling is depicted in Fig. 10.5a, with periodic layers of thicknesses $a$ and $b$, like the photonic crystal (PC) structure discussed in Sect. 9.3, and a period $\Lambda = a + b$. For tunneling to occur, the double-prism structure can be used so that light is incident from medium 1 with a refractive index $n_1$. The barrier of thickness $b$ is made of another dielectric with a refractive index $n_2$ that is lower than $n_1$. There are $N$ periods or unit cells in total between the end media. Light is incident at an incidence angle $\theta_1 > \theta_c = \sin^{-1}(n_2/n_1)$. Yeh [27] performed a detailed analysis of this phenomenon and derived the equation of transmittance, which can be expressed as

$$T'_\lambda = \frac{1}{1 + \frac{\sinh^2(\eta b)}{\sin^2(\delta)} \frac{\sin^2(NK\Lambda)}{\sin^2(K\Lambda)}} \tag{10.16}$$

where $K$ is the Bloch wavevector of the PC, $\delta$ is the phase angle upon total internal reflection, and $\eta$ is the imaginary part of the normal component of the wavevector in the lower index dielectric, as defined in Example 10.1. It can be seen that Eq. (10.16) reduces to Eq. (10.8a) for $N = 1$. Note that the layer of thickness $a$ has the same index as the media at the ends. In this case, the transmittance is 1 at $b = 0$, and decreases monotonically with increasing $b$ as discussed previously for the three-layer setup shown in Fig. 10.3.

The following equation can be used to calculate $K\Lambda$:

$$\cos(K\Lambda) = \cos(k_{1z}a)\cosh(\eta b) + \cot(\delta)\sin(k_{1z}a)\sinh(\eta b) \tag{10.17}$$

where $k_{1z}$ is the normal component of the wavevector in medium 1. While $\cos(K\Lambda)$ is real, $K\Lambda$ is in general complex. However, there exist regions or pass bands where $|\cos(K\Lambda)| \le 1$ so that $K\Lambda$ is real. Note that evanescent waves exist in lower index dielectric layers and subsequently, no interference effects should show up in the transmittance in a regular PC in the pass band; see Fig. 9.20. It can be shown that the transmittance expressed in Eq. (10.16) becomes unity when the following equation holds:

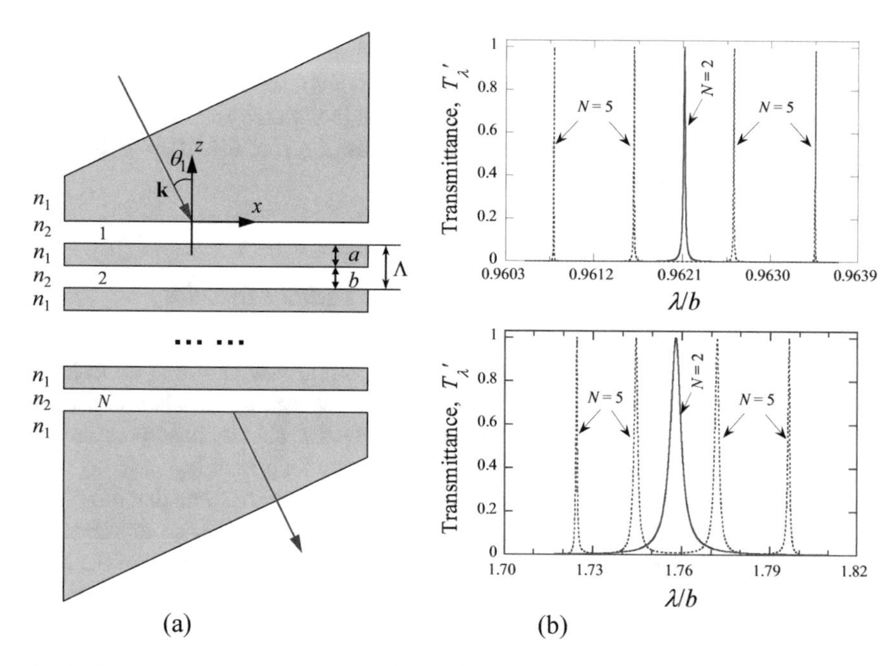


**Fig. 10.5** Illustration of resonance tunneling. **a** Alternate high-index ($n_1$) and low-index ($n_2$) multiple dielectric layers in between two prisms. **b** Calculated transmittance spectra for $N = 2$ and 5, in different wavelength regions, with $n_1 = 3$, $n_2 = 2$, $a = b/2$, and $\theta_1 = 45°$

$$\frac{\sin(NK\Lambda)}{\sin(K\Lambda)} = 0 \tag{10.18}$$

The denominator of this equation cannot be zero; therefore, $K\Lambda \neq m\pi$, $m = 0, \pm 1, \pm 2, \ldots$ It turns out that in each pass band, there exist $(N - 1)$ solutions, for a given combination of $\omega$, $k_x$, and the thicknesses $a$ and $b$. As an example, Fig. 10.5b illustrates the transmittance as a function of $\lambda/b$ when $n_1 = 3$, $n_2 = 2$, $\theta_1 = 45^\circ$, and $a/b = 0.5$. Because of the narrow transmittance peaks, the plot is broken into two panels, each corresponding to a pass band. For $N = 2$, there is only one peak in each pass band, while for $N = 5$, there are four peaks. Yeh [27] showed that the resonance frequencies correspond to the guided modes in the multilayer-waveguide equations. Hence, the fields are highly localized near the higher index layer. Total internal reflection causes very high reflection on the surfaces of the higher index layer and produces resonances similar to those in a Fabry–Perot cavity resonator. It should be noted that extremely sharp transmittance peaks can be obtained when $\lambda$ is close to the gap thickness $b$ (see the upper panel).

Resonance tunneling may have applications as narrow band-pass filters. Due to the guided modes and the localized field, the magnitude of the evanescent wave may be amplified in the forward direction in some region (see Problem 10.4). Similar to the lateral shift by total internal reflection, due to the parallel energy flow in the high-index layer (waveguide), there must be a lateral shift of the transmitted light for finite beams. Little has been reported in the literature about the beam shift and the field distribution in dielectric multilayer structures, when resonance tunneling occurs. When such structures are used in near-field thermal radiation by adding more layers in the vacuum gap between the top layer and the bottom layer of Fig. 10.3a, dispersion and loss often need to be considered.

### *10.2.4 Photon Tunneling with Negative Index Materials*

Negative index materials (NIMs), for which the permittivity and the permeability become negative simultaneously in a given frequency region, can also be used to enhance photon tunneling [28]. The concept of NIMs has already been presented in Sect. 8.4.6. The structure is illustrated in Fig. 10.6a with a pair of layers in between two prisms. One of the layers has a negative refractive index. Assume that one of the layers is vacuum and the other can be described by $\varepsilon = \mu = -1$, so its refractive index is exactly–1. The transmittance becomes unity when the thickness of the NIM layer and that of the vacuum are the same, regardless of the angle of incidence and polarization. Let us use the full notation of $\varepsilon$ and $\mu$ without using the refractive index. The transmission coefficient can be expressed as follows [28]:

$$t = \frac{8}{\xi_1 \mathrm{e}^{-\mathrm{i}\phi_1} + \xi_2 \mathrm{e}^{\mathrm{i}\phi_1} + \xi_3 \mathrm{e}^{-\mathrm{i}\phi_2} + \xi_4 \mathrm{e}^{\mathrm{i}\phi_2}} \tag{10.19}$$

Here, the phase angles $\phi_1$ and $\phi_2$ can be expressed as

$$\phi_1 = k_{2z} d_2 + k_{3z} d_3 \quad \text{and} \quad \phi_2 = k_{2z} d_2 - k_{3z} d_3 \tag{10.20}$$

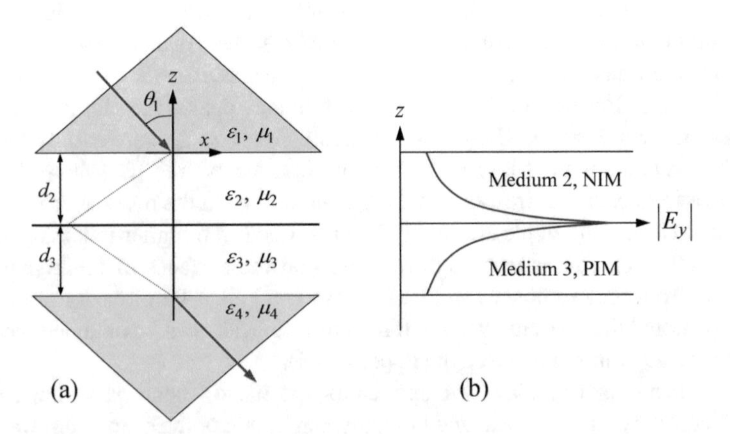


**Fig. 10.6** Photon tunneling with a layer of NIM. **a** The tunneling arrangement. **b** The field distribution in the middle layers for a TE wave

where $d_2$ and $d_3$ are the thicknesses of layers 2 and 3, and $k_{2z}$ and $k_{3z}$ are the normal components of the wavevector in media 2 and 3, respectively. Note that when tunneling occurs, $k_{2z}$ and $k_{3z}$ become purely imaginary for the lossless case, as will be discussed later. For a TE wave, the coefficients in Eq. (10.19) are

$$\xi_1 = \left(1 + \frac{k_{2z}\mu_1}{k_{1z}\mu_2}\right)\left(1 + \frac{k_{3z}\mu_2}{k_{2z}\mu_3}\right)\left(1 + \frac{k_{4z}\mu_3}{k_{3z}\mu_4}\right) \tag{10.21a}$$

$$\xi_2 = \left(1 - \frac{k_{2z}\mu_1}{k_{1z}\mu_2}\right)\left(1 + \frac{k_{3z}\mu_2}{k_{2z}\mu_3}\right)\left(1 - \frac{k_{4z}\mu_3}{k_{3z}\mu_4}\right) \tag{10.21b}$$

$$\xi_3 = \left(1 + \frac{k_{2z}\mu_1}{k_{1z}\mu_2}\right)\left(1 - \frac{k_{3z}\mu_2}{k_{2z}\mu_3}\right)\left(1 - \frac{k_{4z}\mu_3}{k_{3z}\mu_4}\right) \tag{10.21c}$$

and

$$\xi_4 = \left(1 - \frac{k_{2z}\mu_1}{k_{1z}\mu_2}\right)\left(1 - \frac{k_{3z}\mu_2}{k_{2z}\mu_3}\right)\left(1 + \frac{k_{4z}\mu_3}{k_{3z}\mu_4}\right) \tag{10.21d}$$

For a TM wave, the transmission coefficient is defined based on the magnetic fields and the coefficients can be easily obtained by substituting $\varepsilon$'s for $\mu$'s in Eqs. (10.21a)–(10.21d). The sign selection of $k_{lz}$ was mentioned in Sect. 9.2.2 in the discussion of the matrix formulation. Basically, when there exist propagating waves in medium $l$, $k_{lz} = (2\pi n_l/\lambda)\sqrt{1 - (n_1/n_l)^2 \sin^2\theta_1}$ and its sign becomes negative in a NIM. On the other hand, if the waves become evanescent in medium $l$, we use $k_{lz} = \mathrm{i}(2\pi/\lambda)\sqrt{n_1^2 \sin^2\theta_1 - n_l^2} = \mathrm{i}\eta_l$. Here, $\eta_l$ is always positive in a lossless medium, even in a NIM. Assume that the prisms are made of the same materials so that the properties of medium 1 and medium 4 are identical. Furthermore, layer 2 is made of a NIM with index-matching conditions, i.e., $\varepsilon_2 = -\varepsilon_3$ and $\mu_2 = -\mu_3$ so that $n_2 = -n_3$. For propagating waves in the middle layers, $k_{2z} = -k_{3z}$ and $\xi_3 = \xi_4 = 0$, so Eq. (10.19) can be further simplified to

$$t = \frac{1}{\cos(k_{3z}\Delta) - \mathrm{i}Y\sin(k_{3z}\Delta)} \tag{10.22}$$

where $\Delta = d_3 - d_2$, $Y = \frac{1}{2}\left(\frac{k_{3z}\mu_1}{k_{1z}\mu_3} + \frac{k_{1z}\mu_3}{k_{3z}\mu_1}\right)$ for TE waves, and $Y = \frac{1}{2}\left(\frac{k_{3z}\varepsilon_1}{k_{1z}\varepsilon_3} + \frac{k_{1z}\varepsilon_3}{k_{3z}\varepsilon_1}\right)$ for TM waves. Because media 1 and 4 are made of the same material, the transmittance for propagating waves can be written as follows:

$$T'_\lambda = \frac{1}{\cos^2(k_{3z}\Delta) + Y^2\sin^2(k_{3z}\Delta)} \tag{10.23}$$

For evanescent waves, we have $k_{2z} = k_{3z} = \mathrm{i}\eta_3$, where $\eta_3 = (2\pi/\lambda)\sqrt{n_1^2 \sin^2\theta_1 - n_3^2}$. Now that $\xi_1 = \xi_2 = 0$, Eq. (10.19) can be simplified as follows:

$$t = \frac{1}{\cosh(\eta_3\Delta) + \mathrm{i}\cot(\delta)\sinh(\eta_3\Delta)} \tag{10.24}$$

where $\cot(\delta) = \frac{1}{2}\left(\frac{\eta_3\mu_1}{k_{1z}\mu_3} - \frac{k_{1z}\mu_3}{\eta_3\mu_1}\right)$, with $\delta$ being the phase change upon total internal reflection between medium 1 and 2. The transmittance $T'_\lambda = tt^*$ is real and always decreases with increasing $\Delta$, the difference between the layer thicknesses. Although Eqs. (10.22) and (10.24) are identical because $\sin(\mathrm{i}x) = \mathrm{i}\sinh(x)$ and $\cos(\mathrm{i}x) = \cosh(x)$, the use of real variables allows us to observe the variation of transmittance with $\Delta$ easily. When tunneling occurs, the field is highly localized near the interface between the NIM and the PIM layers, as shown in Fig. 10.6b for a TE wave, where the fields are the sum of the forward-decaying and backward-decaying evanescent waves. The amplitude of the evanescent wave in the NIM increases in the direction of energy flow. It can be shown that the amplitude will still increase in medium 2, even though the NIM is placed in layer 3 and layer 2 is a vacuum. This corresponds to another resonance effect, which is associated with the excitation of surface electromagnetic waves or surface polaritons as discussed in Sect. 9.5.1.

The directional and hemispherical transmittances for the structure shown in Fig. 10.6a are illustrated in Fig. 10.7 with the following parameters: $n_1 = n_4 = 1.5$, $n_2 = -1$ ($\varepsilon_2 = \mu_2 = -1$), and $n_3 = 1$ (vacuum). Both the directional and hemispherical transmittances become 1 when $d_3 = d_2$. The hemispherical transmittance has two components, due to propagating and evanescent waves. The effects of loss and dispersion have also been examined [29].

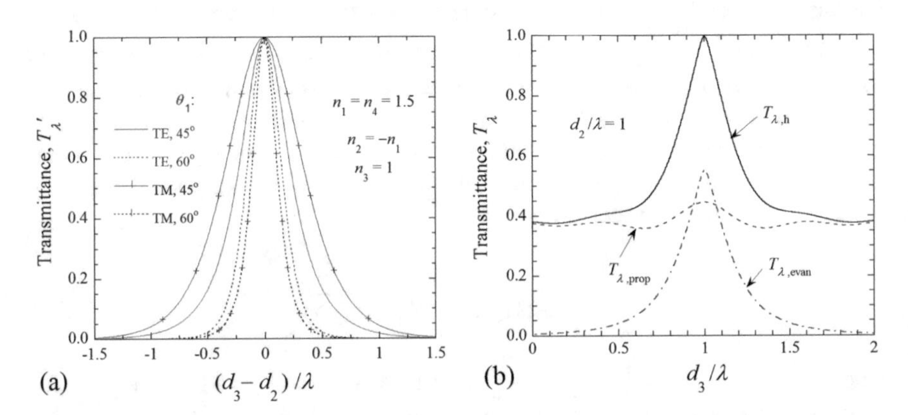


**Fig. 10.7** Transmittance for a four-layer structure with one middle layer being matching-index NIM. **a** Directional transmittance. **b** Hemispherical transmittance

## 10.3 Energy Streamlines and Superlens

As discussed in Chap. 8, Sect. 8.4.6, a NIM or a double-negative material (DNG) forms a *flat lens* that can focus light (see Fig. 8.18). Pendry [30] predicted that a DNG flat lens not only focuses propagating waves but also allows complete transmission of evanescent waves because of an amplifying effect on the evanescent wave amplitude. Furthermore, a single-negative material (SNG) like a Ag film also exhibits focusing properties in the closest proximity. Such a lens is thereafter called a *perfect lens* or *superlens*. Many researchers have been working on the fabrication of micro/nanostructures with tailored electric and magnetic properties. Photonic crystals have also been realized with focusing properties for electromagnetic waves (photons). Researchers have also experimentally demonstrated that a flat Ag lens can focus light at nanoscale distances for nanolithographic applications [31, 32].

While the electromagnetic wave theory describes the tunneling phenomenon and surface polaritons elegantly, the energy ray concept meets a difficulty for coupled evanescent waves because the parallel component of the wavevector for an evanescent wave is so large that no polar angle within the real space can be defined. On the other hand, the Poynting vector can always be defined, and by following the traces, the energy streamline method appears to be a promising technique for analyzing the energy flow directions in the near field. The basic concept developed in a recent study by Zhang and Lee [33] is described next. For convenience, let us consider the layered medium to be oriented along the $x$-direction as shown in Fig. 10.8, where media 1 and 3 are semi-infinite. If the incident wave is a TM wave with an angular frequency $\omega$, the magnetic field in each region is given by

$$H_z(x, y) = \left[A\mathrm{e}^{\mathrm{i}k_x x} + B\mathrm{e}^{-\mathrm{i}k_x x}\right]\mathrm{e}^{\mathrm{i}k_y y} \tag{10.25}$$

where $A$ and $B$ are the coefficients of forward and backward waves at the interface as indicated in Fig. 10.8; $x$ is relative to the origin in media 1 and 2, while in medium

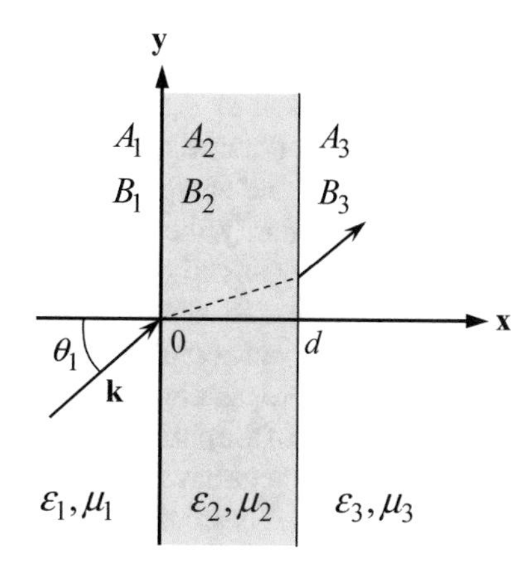


**Fig. 10.8** Schematic of a three-layer structure, where $A_j$ and $B_j$ ($j = 1, 2,$ and 3) are the coefficients of forward and backward waves at the nearest interface

3, $x$ is relative to $d$; and $k_x$ and $k_y$ are the $x$ (normal) and $y$ (parallel) components of the wavevector. Note that $k_x^2 + k_y^2 = \varepsilon\mu\omega^2/c^2$ for this geometry. The components of the time-averaged Poynting vector can be expressed as follows:

$$\langle S_x \rangle = \frac{1}{2\omega\varepsilon_0}\text{Re}\left(\frac{k_x}{\varepsilon}\right)\left[|A|^2 e^{-2k_x'' x} - |B|^2 e^{2k_x'' x}\right] - \frac{1}{\omega\varepsilon_0}\text{Im}\left(\frac{k_x}{\varepsilon}\right)\text{Im}\left(AB^* e^{2ik_x' x}\right) \tag{10.26}$$

$$\langle S_y \rangle = \frac{k_y}{2\omega\varepsilon_0}\text{Re}\left(\frac{1}{\varepsilon}\right)\left[|A|^2 e^{-2k_x'' x} + |B|^2 e^{2k_x'' x}\right] + \frac{k_y}{\omega\varepsilon_0}\text{Re}\left(\frac{1}{\varepsilon}\right)\text{Re}\left(AB^* e^{2ik_x' x}\right) \tag{10.27}$$

Here, $k_x = k_x' + \text{i}k_x''$ is the normal component of the wavevector. Note that the present section uses a slightly different notation from that of preceding sections. The last terms in Eqs. (10.26) and (10.27) arise from the coupling between the forward and backward waves. The direction of $\langle \mathbf{S} \rangle$ of the combined wave can always be defined by a polar angle $\phi = \arctan(\langle S_y \rangle / \langle S_x \rangle)$; in contrast, it is not always possible to define the angle of incidence or refraction $\theta = \arctan(k_y/k_x)$ in the real space. The trajectory of $\langle \mathbf{S} \rangle$ for given values of $\omega$ and $k_y$ may be called an energy streamline, which defines the path of the net energy flow. The matrix formulation can be used to evaluate $A$ and $B$ in each layer by setting $A_1 = 1$ and $B_3 = 0$. Note that the dependence of $\mu$ is implicit in Eqs. (10.26) and (10.27), since $k_x$ is a function of $\mu$, and furthermore, $A_j$ and $B_j$ depend on $k_x$. For TE waves, the magnetic field can be replaced by the electric field in Eq. (10.25), and $\varepsilon$ should be replaced by $\mu$ in Eqs. (10.26) and (10.27).

The energy streamlines in the prism-DNG-prism and prism-SNG-prism configurations are shown in Fig. 10.9, for different incidence angles or $k_y$ values. The energy transport is from the left to the right, and the trajectory of the Poynting vector in the three regions forms a zigzag path, especially when $d \ll \lambda$ . The $x$- and $y$-axes are normalized to the slab thickness $d$. All streamlines are for positive $k_y$ values and pass through the origin. With the dielectric prism ($\varepsilon = 2.25$), the critical angle is $\theta_c = 41.81°$.

Causality requires that $\langle S_x \rangle$ be positive; furthermore, when loss is neglected, $\langle S_x \rangle$ is independent of $x$. Note that $\langle S_y \rangle = \frac{k_y}{2\omega\varepsilon_0}\text{Re}\left(\frac{1}{\varepsilon}\right)|H_z|^2$ is opposite to $k_y$ when $\text{Re}(\varepsilon) < 0$, as in the DNG layer (medium 2). At $\theta_1 = \theta_c$, when the phase refraction angle $\theta_2 = 90°$, the energy refraction angle $\phi_2$ is much less than 90°. In order to remove the singularity, the computation for $\theta_1 = \theta_c$ can be approximated by using an angle that is either slightly greater or slightly smaller than $\theta_c$. Furthermore, the dash-dotted line in the slab separates the propagating-wave streamlines (inside the cone) from the evanescent-wave streamlines (outside the cone). The observation that the energy paths of propagating and evanescent waves are separated by a cone provides a new explanation of the photon tunneling phenomenon based on wave optics. Note that tunneling phenomena have been extensively studied in quantum mechanics regarding the time delay and beam shift, but the results are somewhat controversial [27]. The energy transmittance through the slab, calculated by $T = |A_3/A_1|^2$, is labeled for

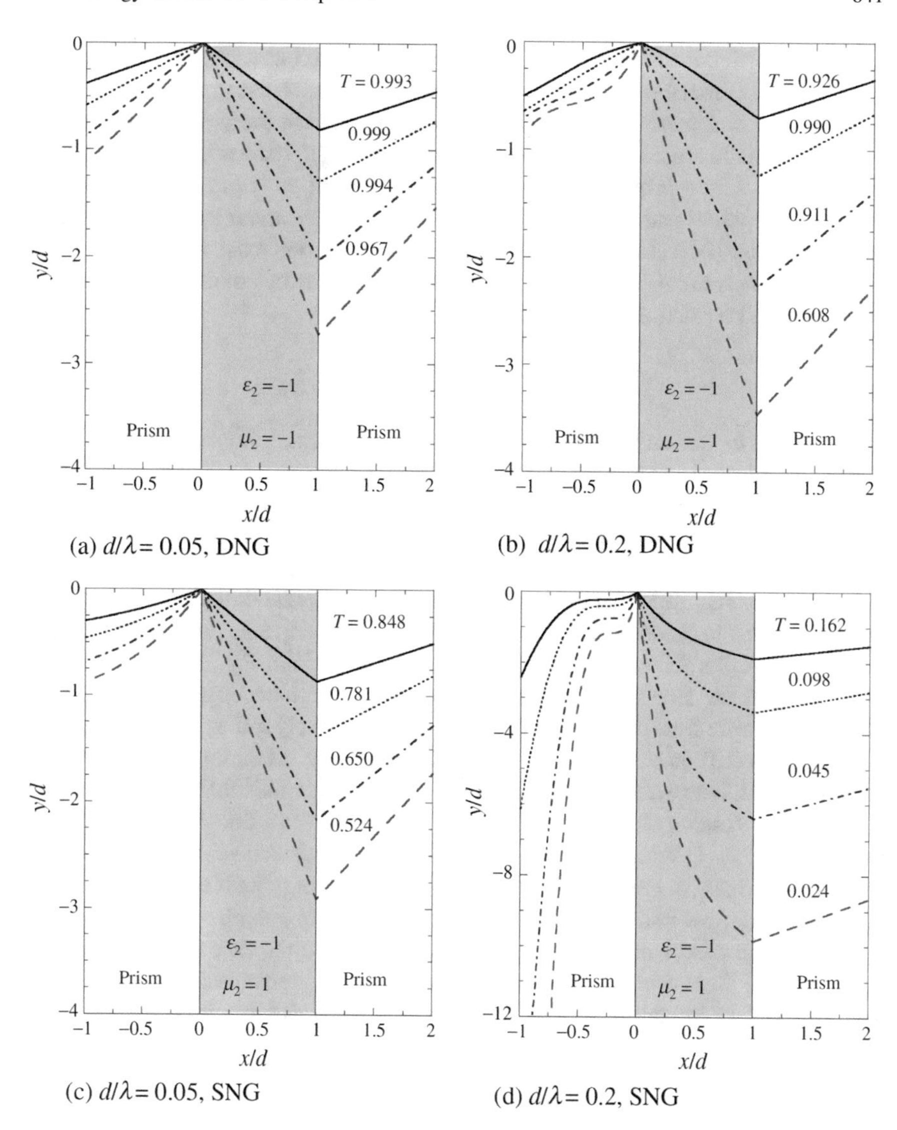


**Fig. 10.9** Energy streamlines for prism-DNG-prism and prism-SNG-prism configurations at various incidence angles [33]: $\theta_1 = 20°$ (solid), 30° (dotted), 41.81° (dash-dotted), and 50° (dashed). The prism has $\varepsilon = 2.25$ and $\mu = 1$, so $\theta_1 = 41.81°$ corresponds to the critical angle for DNG in **a** and **b**. Only evanescent waves exist in medium 2 for SNG in **c** and **d**. The transmittance $T$ from medium 1 to 3 is shown for each incidence angle

each streamline. The tunneling transmittance decreases rapidly as $d$ increases, and the streamlines are curved when $d = \lambda/5$. Figures 10.9c and 10.9d are for a negative-$\varepsilon$ but positive-$\mu$ slab (such as a metal, but lossless). In this case, only evanescent waves exist in the slab because $k_x$ is purely imaginary even at normal incidence.

Energy is carried through medium 2 by coupled evanescent waves, whose path can be completely described by a streamline. The transmittance with a SNG slab is much smaller than that with a DNG slab, and the beam shift in the $y$-direction becomes very large, as illustrated in Fig. 10.9d. Nevertheless, Figs. 10.9a and 10.9c look alike. When $d \ll \lambda$, the propagating waves and evanescent waves are similar because both the sinusoidal and hyperbolic functions are the same under the small-argument approximation [34]. Assume that only propagating waves exist in medium 1, and both $\varepsilon_1$ and $\varepsilon_2$ are real. The following approximations can be obtained for the energy incidence and refraction angles in the limit $d/\lambda \to 0$:

$$\phi_1 = \theta_1 \quad \text{and} \quad \tan \phi_2 = (\varepsilon_1/\varepsilon_2) \tan \phi_1 \tag{10.28}$$

Note that $\mu_2$ does not affect the TM wave results in the *electrostatic limit*, when the distance is much shorter than the wavelength. However, the effect of $\mu_2$ becomes significant when $d/\lambda > 0.1$.

Both positive and negative phase-time shifts were noticed by Li [35] for an optically dense dielectric slab in air without evanescent waves. It is worthwhile to take a look at the streamlines for the vacuum-dielectric-vacuum configuration. For propagating waves, because the second term in Eq. (10.27) depends on $x$, the streamline exhibits wavelike features for $d = \lambda$, as can be seen from Fig. 10.10a, where the solid curve is the streamline and the dashed lines are the traces of the wavevector. The lateral shift of the energy line is determined by point Q rather than P. When $d/\lambda$ is reduced to 0.01 as shown in Fig. 10.10b, the streamline is almost a straight line in each medium. However, point Q becomes closer to the $x$-axis than P, in contrary to Fig. 10.10a. When $d/\lambda \ll 1$, Snell's law determines $\theta_2$ and Eq. (10.28) determines $\phi_2$ . The shift of Q with respect to P depends on the incidence angle, which can be positive or negative. Perhaps the lateral shift of the energy path can be understood by the energy flow parallel to the film as a result of the combined field, similar to the Goos–Hänchen shift. The difference here is due to the fact that a plane wave of infinite width is used to calculate the lateral shift of transmission through a thin film, as well as tunneling. While Poynting vector traces have been presented for transmission through nanoslits as well as for scattering around nanoparticles, the application of the streamline method to planar layers reveals some fundamental and counterintuitive behavior; see Bohren and Huffman [16] and Bashevoy et al. [36]. It appears to be more natural for thermal engineers to deal with energy streamlines rather than evanescent waves. This method allows the visualization of energy flow in the optical near field.

Understanding the energy transport in the subwavelength region has an enormous impact on near-field optics and nanolithography. Figure 10.11 shows the streamlines for the three-layer structure made by Fang et al. [32]. A 35-nm-thick Ag film was evaporated over a polymethyl methacrylate (PMMA) followed by a photoresist (PR) coating. The source is assumed to be at $x = -40$ nm and $y = 0$ inside the PMMA. The properties at $\lambda = 365$ nm are taken from Ref. [32] as follows: $\varepsilon_1 = 2.30 + 0.0014i$ for the PMMA, $\varepsilon_2 = -2.40 + 0.25i$ for Ag, and $\varepsilon_3 = 2.59 + 0.01i$ for the PR. The solid lines are for propagating waves in the PMMA at $\theta_1 = 20°$ and $50°$. The dash-dotted lines correspond to $\theta_1 = 90°$ or $k_y = \text{Re}(k_1) = 2\pi \text{Re}(\sqrt{\varepsilon_1})/\lambda$,

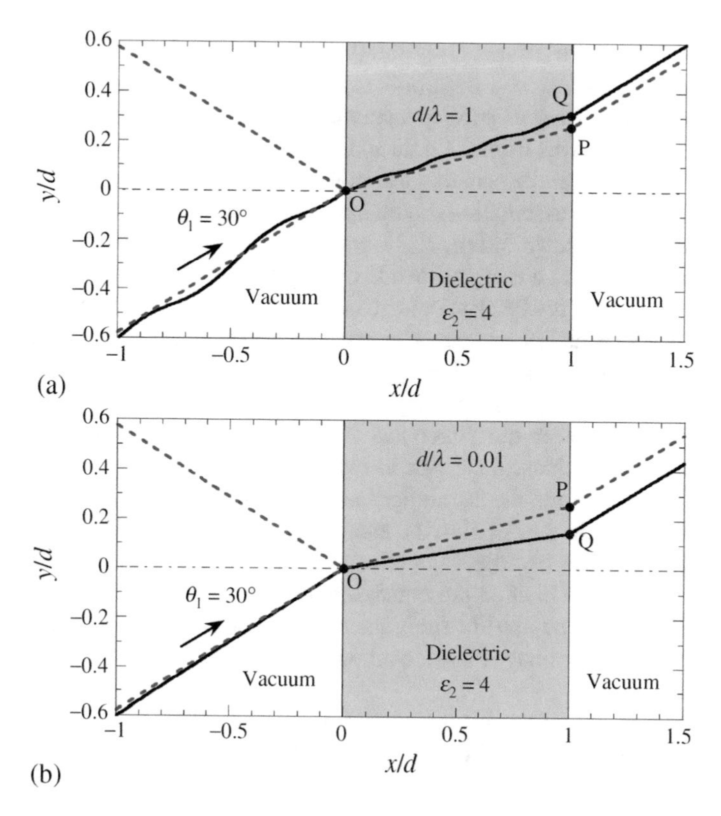


**Fig. 10.10** The streamline for vacuum-dielectric-vacuum configuration at $\theta_1 = 30°$ when **a** $d/\lambda = 1$ and **b** $d/\lambda = 0.01$ [33]. Solid curves are streamlines, and dashed lines are the wavevector direction

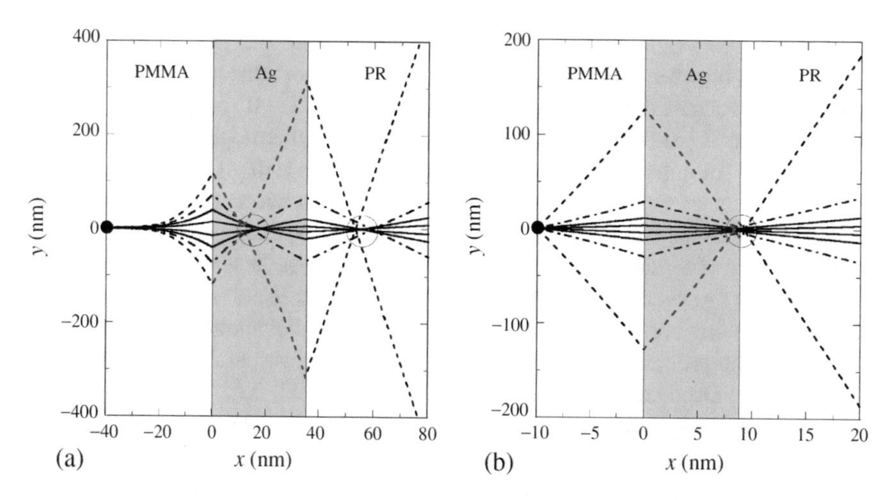


**Fig. 10.11** Energy streamlines for a three-layer structure, showing the imaging features for a silver lens with a thickness of **a** $d = 35$ nm and **b** $d = 8.75$ nm, at the wavelength $\lambda = 365$ nm. The dot represents the source, and circles indicate the foci

where $\varepsilon_1$ is the dielectric function of the PMMA. Outside the cone, defined by the dash-dotted lines with $k_y = 1.06\text{Re}(k_1)$, evanescent waves exist inside the PMMA. Note that evanescent waves exist in vacuum for $\theta_1 > 41.25°$. The use of PMMA allows evanescent waves from the light source with $k_y$ much greater than $\omega/c$ to be transmitted through. In the calculations, both the PMMA and the PR are assumed semi-infinite, and this assumption should have little effect on the imaging properties.

The streamlines shown in Fig. 10.11 are curved (i.e., $\phi_1 \neq \theta_1$). The streamline graph clearly reveals two foci, one inside the Ag film and the other at about 20 nm outside the Ag film in the PR. It should be noticed that the foci are somewhat blurred due to losses. The actual structure fabricated by Fang et al. [32] was more complicated and may require an integration over the wavevector space to fully understand the imaging properties. To examine the proximity limit, the thickness of the Ag film and the distance between the source and the Ag film are fourfold reduced without changing other conditions. As shown in Fig. 10.11b, a single focus is formed near the Ag-PR interface, and the streamlines are nearly straight lines in each medium. Because of the loss in the Ag film, the energy refraction angle in Ag depends on $k_y$ and is slightly greater than that calculated from Eq. (10.28) based on the real parts of $\varepsilon$'s. The streamline method presented here provides information on the paths of light energy and can be used to study lateral beam shifts in photon tunneling and to construct near-field images inside and outside of flat lenses made of a NIM or a silver film.

Further discussion of energy streamlines in near-field thermal radiation for lossy media, multilayers, and hyperbolic metamaterials will be given later.

## 10.4 Radiative Transfer Between Two Semi-Infinite Media

Heat transfer between surfaces placed at extremely short distances has important applications in near-field scanning thermal microscopy [37–40] and thermal rectifier or thermal diode [41–43]. The microscale (near-field) thermophotovoltaic devices offer promise for enhanced performance for energy harvesting [17, 18, 23, 44–47]. The calculation of near-field radiation heat transfer between dielectric materials has already been described in Sect. 10.2.2. However, dispersion and dissipation are unavoidable in real material systems. Nanoscale radiation heat transfer can be further enhanced by several orders of magnitude using lossy materials, especially when surface polaritons are excited [24, 25, 44, 48–50]. Earlier theoretical works were centered on the prediction of the net heat flux between two parallel metallic plates, using a simple Drude model for the dielectric function [51, 52]. While many metals support surface waves through surface plasmon polaritons, the plasma frequencies are usually much higher than the characteristic frequencies of thermal sources. Consequently, the near-field enhancement of thermal radiation between good conductors is not very large. On the other hand, semiconductors and semimetals, with smaller electric conductivities, may greatly enhance radiation heat flux at nanometer scales [25, 49]. The use of polar materials such as SiC allows surface phonon polaritons to be excited, resulting in large near-field radiation heat transfer that is concentrated in a very narrow wavelength band [24, 48, 49]. Analytical solutions of nanoscale energy transfer between a sphere and a surface or between two spheres are also available [48, 51, 52]. Since 2011, numerical methods have also been extensively applied to model near-field radiative heat transfer between nanostructures, including 2D materials [53–61].

This section introduces fluctuational electrodynamics, originally developed by Rytov and coworkers in late 1950s, based on the fluctuation–dissipation theorem [62]. Detailed discussions will be given on the calculation of the near-field thermal radiation between two parallel plates, with an example based on doped silicon and polar dielectric materials. The fluctuation–dissipation theorem has applications in the study of thermal conductivity of nanostructures and has also been used to study van der Waals forces and noncontact friction at nanometer distances [63–65].

### *10.4.1 Fluctuational Electrodynamics*

Consider the geometry shown in Fig. 10.12a, where two homogeneous media, each at equilibrium but with different temperatures, $T_1$ and $T_2$, are separated by a vacuum gap of width $d$, ranging from several tens of micrometers down to 1 nm. For a nonmagnetic and isotropic medium, the complex dielectric function or relative permittivity is the only property needed to fully characterize the optical behavior. Some models of the dielectric function such as the Drude and/or Lorentz models were discussed in Chap. 8. The foundation of fluctuational electrodynamics is the fluctuation–dissipation theorem, under which thermal radiation is assumed to arise from the random movement of charges inside the medium at temperatures exceeding 0 K. The charge movement causes fluctuating electric currents that in turn result in a fluctuating electromagnetic field in space and time. The frequency components of the fluctuating field or current can be analyzed via the correlation function. The

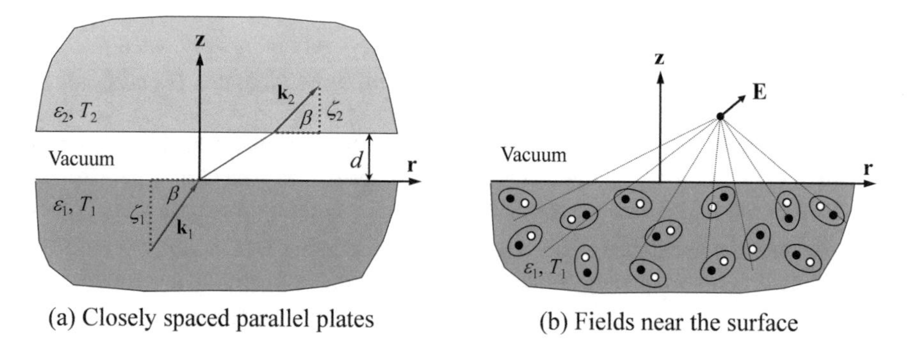


**Fig. 10.12** Schematic drawings for the study of near-field thermal radiation in the cylindrical coordinates. **a** Radiation heat transfer between two parallel plates separated by a vacuum gap. **b** The electric field near the surface due to thermally induced charge fluctuations

movement of charges from an equilibrium position can also be viewed as oscillating dipoles as illustrated in Fig. 10.12b. The electromagnetic field at any location is a superposition of contributions from all of the point sources in the radiating region. The electromagnetic waves deep inside the medium will attenuate due to absorption (i.e., dissipation) inside the medium. The basic assumptions for calculating near-field radiative heat transfer between two parallel plates are as follows: (a) Each medium is semi-infinite and at a thermal equilibrium, presumably due to a sufficiently large thermal conductivity of the solid. (b) Both media are nonmagnetic, isotropic, and homogeneous, so that the frequency-dependent complex dielectric function (relative permittivity) $\varepsilon_1$ or $\varepsilon_2$ is the only material property that characterizes the electrodynamic response and thermally excited dipole emission of medium 1 or 2. (c) Each surface is perfectly smooth, and the two surfaces are parallel to each other. In the structure as illustrated in Fig. 10.12b, thermal emission in the far field as well as the photon local density of states (LDOS) near the surface can be directly calculated via fluctuational electrodynamics [66, 67].

Due to axial symmetry, we can use the cylindrical coordinates so that the spatial variable $\mathbf{x} = \mathbf{r} + \mathbf{z} = r\hat{\mathbf{r}} + z\hat{\mathbf{z}}$. Consider a monochromatic electromagnetic wave propagating from medium 1 to 2. The complex wavevectors in media 1 and 2 are $\mathbf{k}_1$ and $\mathbf{k}_2$, respectively, with $k_1^2 = \varepsilon_1 k_0^2$ and $k_2^2 = \varepsilon_2 k_0^2$, where $k_0 = \omega/c_0 = 2\pi/\lambda$ is the magnitude of the wavevector in vacuum. In general, $\varepsilon_1$ and $\varepsilon_2$ are complex; hence, $k_1$ and $k_2$ should be viewed as complex functions of $\omega$. Only real and positive $\omega$ values are considered so that the wavevector in vacuum $k_0 = \omega/c_0$ is real. The monochromatic plane wave can be expressed in terms of a time- and frequency-dependent field, $\exp(\mathrm{i}\mathbf{k}_j \cdot \mathbf{x} - \mathrm{i}\omega t)$, where $j = 0$, 1, or 2 refers to vacuum, medium 1, or medium 2, respectively. The phase-matching condition requires the parallel components of all three wavevectors to be the same. To simplify the notation, let us use $\beta$ for the parallel component and $\zeta_j$ for the normal component of the wavevector $\mathbf{k}_j$. Thus, $\mathbf{k}_j = \beta\hat{\mathbf{r}} + \zeta_j\hat{\mathbf{z}}$ and $\zeta_j = \sqrt{k_j^2 - \beta^2}$. The spatial dependence of the field in vacuum can be expressed as $\exp(\mathrm{i}\beta r + \mathrm{i}\zeta_0 z)$. Because its amplitude must not change along the $r$-direction, $\beta$ must be real. Keep in mind that both $r$ and $\beta$ are positive in the cylindrical coordinates. The normal component of the wavevector in vacuum $\zeta_0 = \sqrt{k_0^2 - \beta^2}$ will be real when $0 \le \beta \le k_0$ and purely imaginary when $\beta > k_0$. Thus, an evanescent wave exists in vacuum when $\beta > k_0$. Note that $\zeta_1$ and $\zeta_2$ are in general complex.

The random thermal fluctuations produce a spatial-time-dependent electric current density $\mathbf{j}(\mathbf{x}, t)$ inside the medium whose time average is zero. The current density can be decomposed into the frequency domain using the Fourier transform, which gives $\mathbf{j}(\mathbf{x}, \omega)$. With the assistance of the dyadic Green's function $\overline{\overline{\mathbf{G}}}(\mathbf{x}, \mathbf{x}', \omega)$, the induced electric field in the frequency domain can be expressed as a volume integration:

$$\mathbf{E}(\mathbf{x}, \omega) = \mathrm{i}\omega\mu_0 \int\limits_{V'} \overline{\overline{\mathbf{G}}}(\mathbf{x}, \mathbf{x}', \omega) \cdot \mathbf{j}(\mathbf{x}', \omega) d\mathbf{x}' \tag{10.29}$$

where $\mu_0$ is the magnetic permeability of vacuum, and the integral is over the region $V$ that contains fluctuating sources. The physical significance of the Green's function is that it is a transfer function for a current source $\mathbf{j}$ at a location $\mathbf{x}'$ and the resultant electric field $\mathbf{E}$ at $\mathbf{x}$. Mathematically, the dyadic Green's function satisfies the vector Helmholtz equation:

$$\nabla \times \nabla \times \overline{\overline{\mathbf{G}}}(\mathbf{x}, \mathbf{x}', \omega) - k^2 \overline{\overline{\mathbf{G}}}(\mathbf{x}, \mathbf{x}', \omega) = \overline{\overline{\mathbf{I}}}\delta(\mathbf{x} - \mathbf{x}') \tag{10.30}$$

where $k$ is the amplitude of the wavevector at $\mathbf{x}$, and $\overline{\overline{\mathbf{I}}}$ is a unit dyadic. The corresponding magnetic field $\mathbf{H}(\mathbf{x}, \omega)$ can be obtained from the Maxwell equation:

$$\mathbf{H}(\mathbf{x}, \omega) = \frac{1}{\mathrm{i}\omega\mu_0} \nabla \times \mathbf{E}(\mathbf{x}, \omega) \tag{10.31}$$

The spectral energy density of the thermally emitted electromagnetic field in vacuum can be calculated from Eq. (8.19) in terms of the ensemble average. Therefore,

$$u_\omega(\mathbf{x}, \omega) = \frac{\varepsilon_0}{4} \langle \mathbf{E}(\mathbf{x}, \omega) \cdot \mathbf{E}^*(\mathbf{x}, \omega) \rangle + \frac{\mu_0}{4} \langle \mathbf{H}(\mathbf{x}, \omega) \cdot \mathbf{H}^*(\mathbf{x}, \omega) \rangle \tag{10.32}$$

where "< >" denotes the ensemble average of the random currents. The emitted energy flux can be expressed by the ensemble average of the Poynting vector, i.e.,

$$\langle \mathbf{S}(\mathbf{x}, \omega) \rangle = \frac{1}{2} \langle \mathrm{Re}[\mathbf{E}(\mathbf{x}, \omega) \times \mathbf{H}^*(\mathbf{x}, \omega)] \rangle \tag{10.33}$$

To evaluate the ensemble average, the required cross-spatial correlation function between the fluctuating currents at two locations $\mathbf{x}'$ and $\mathbf{x}''$ inside the emitting medium is given as [68]

$$\langle j_m(\mathbf{x}', \omega) j_n^*(\mathbf{x}'', \omega') \rangle = \frac{4\omega\varepsilon_0 \mathrm{Im}(\varepsilon)}{\pi} \Theta(\omega, T) \delta_{mn} \delta(\mathbf{x}' - \mathbf{x}'') \delta(\omega - \omega') \tag{10.34}$$

where $j_m$ ($m = 1, 2,$ or 3) stands for the $x$-, $y$-, or $z$-component of $\mathbf{j}$. The Dirac delta function $\delta(\omega - \omega')$ implies that the spectral quantities with different frequencies are statistically uncorrelated. The term $\delta_{mn}\delta(\mathbf{x}' - \mathbf{x}'')$ comes from the fact that the medium is isotropic and said to be *local* whereby certain properties are only affected by its immediate surroundings. In Eq. (10.34), $\Theta(\omega, T)$ is the mean energy of Planck's oscillator at the frequency $\omega$ in thermal equilibrium and is given by

$$\Theta(\omega, T) = \frac{\hbar\omega}{\exp(\hbar\omega/k_B T) - 1} + \frac{\hbar\omega}{2} \tag{10.35}$$

In Eq. (10.35), the second term $\frac{1}{2}\hbar\omega$ accounts for vacuum fluctuation or zero-point energy [69]. This does not affect the calculation of the net radiative energy exchange since the second term is independent of temperature. However, in the calculation of the Casimir forces, the second term needs to be included [68]. When the second term is dropped, the calculated energy density should be regarded as being relative to the vacuum ground energy density. The local density of states (LDOS) or density of modes $D(z, \omega)$ is defined by the following relation [68]:

$$u_\omega(z, \omega) = D(z, \omega)\Theta(\omega, T) \tag{10.36}$$

The spectral energy density $u_\omega(z, \omega)$ and LDOS are independent of $r$ because of the infinite-plate assumption. The physical significance of $D(z, \omega)$ [$\text{m}^{-3}\,\text{s}\,\text{rad}^{-1}$] is the number of modes per unit angular frequency interval per unit volume. It can become very large when $z \to 0$ (in the proximity of the surface) at certain frequencies. Equation (10.36) assumes that the contribution is only from the medium and did not consider the contribution from free space as well as that reflected by the interface. This omission is justifiable in the near-field regimes because the contribution from free space may be orders of magnitude smaller than that from the medium.

Note that the energy density given in Eq. (10.32) and the Poynting vector defined in Eq. (10.33) can be evaluated by integration over the source region, using the correlation function given in Eq. (10.34). As an example, we can write the component

$$\left\langle E_m(\mathbf{x}, \omega) E_n^*(\mathbf{x}, \omega) \right\rangle = \frac{4\varepsilon_0 \mu_0^2 \omega^3}{\pi} \sum_i \int_{V'} \varepsilon'' \Theta(\omega, T) G_{mi}(\mathbf{x}, \mathbf{x}', \omega) G_{ni}^*(\mathbf{x}, \mathbf{x}', \omega) \mathrm{d}\mathbf{x}' \tag{10.37}$$

where $\varepsilon'' = \mathrm{Im}(\varepsilon)$. The dyadic Green's function may be expressed in terms of a 2D spatial Fourier transfer as [70, 71]

$$\overline{\overline{\mathbf{G}}}(\mathbf{x}, \mathbf{x}', \omega) = \frac{1}{4\pi^2} \int_{-\infty}^{\infty} \int_{-\infty}^{\infty} \bar{\bar{\mathbf{g}}}(\beta, z, z', \omega) \mathrm{e}^{\mathrm{i}\beta(r - r')} \mathrm{d}k_x \mathrm{d}k_y \tag{10.38}$$

After substituting Eq. (10.38) into Eq. (10.37), the integration over $\mathrm{d}x\mathrm{d}y$ is transformed to $\mathrm{d}k_x\mathrm{d}k_y$. The integration in the $k$-space can be converted using cylindrical coordinates for isotropic media. For a homogeneous medium or even for multilayers, the integration over the $z$-direction in the source region (according to Fig. 10.12b) can be evaluated. Usually, a magnetic Green's function is introduced to facilitate the derivations. The results for the simple case with parallel plates as an example are discussed next. More details are given in Sect. 10.5.4.

### 10.4.2 Near-Field Radiative Heat Transfer Between Two Parallel Plates

The Green's function depends on the geometry of the physical system, and for two parallel semi-infinite media sketched in Fig. 10.12a, it takes the following form [24, 68]:

$$\overline{\overline{\mathbf{G}}}(\mathbf{x}, \mathbf{x}', \omega) = \frac{\mathrm{i}}{4\pi} \int_0^\infty \frac{\beta \mathrm{d}\beta}{\zeta_1} \left( \hat{\mathbf{s}} t_s \hat{\mathbf{s}} + \hat{\mathbf{p}}_2 t_p \hat{\mathbf{p}}_1 \right) \mathrm{e}^{\mathrm{i}(\zeta_2 z - \zeta_1 z')} \mathrm{e}^{\mathrm{i}\beta(r-r')} \tag{10.39}$$

where $\mathbf{x} = r\hat{\mathbf{r}} + z\hat{\mathbf{z}}$ and $\mathbf{x}' = r'\hat{\mathbf{r}} + z'\hat{\mathbf{z}}$. Note that $t_s$ and $t_p$ are the transmission coefficients from medium 1 to medium 2 for $s$-and $p$-polarizations, respectively, and can be calculated using Airy's formula given in Eq. (9.8). The unit vectors are $\hat{\mathbf{s}} = \hat{\mathbf{r}} \times \hat{\mathbf{z}}$, $\hat{\mathbf{p}}_1 = \left(\beta\hat{\mathbf{z}} - \zeta_1\hat{\mathbf{r}}\right)/k_1$, and $\hat{\mathbf{p}}_2 = \left(\beta\hat{\mathbf{z}} - \zeta_2\hat{\mathbf{r}}\right)/k_2$. If the interest is to calculate the radiation field from a medium to vacuum, $t_s$ and $t_p$ can be replaced by the Fresnel transmission coefficients between the medium and vacuum. The Poynting vector and energy density can then be calculated using the Green's function by performing integration over the $z$-direction in either region 1 or region 2. Note that the term $\mathrm{e}^{\mathrm{i}\beta(r-r')}$ will drop out when multiplied by its complex conjugate [70]. The local density of states in vacuum near the surface of medium 1 can be expressed in two terms, that is,

$$D(z, \omega) = D_{\mathrm{prop}}(\omega) + D_{\mathrm{evan}}(z, \omega) \tag{10.40}$$

where

$$D_{\mathrm{prop}}(\omega) = \int_0^{k_0} \frac{\omega}{2\pi^2 c_0^2 \zeta_0} \left(2 - \rho_{01}^s - \rho_{01}^p\right) \beta \mathrm{d}\beta \tag{10.41a}$$

and

$$D_{\mathrm{evan}}(z, \omega) = \int_{k_0}^{\infty} \frac{e^{-2z\eta_0}}{2\pi^2 \omega \eta_0} \left[\mathrm{Im}(r_{01}^s) + \mathrm{Im}(r_{01}^p)\right] \beta^3 \mathrm{d}\beta \tag{10.41b}$$

Here, $r_{01}$ is the Fresnel reflection coefficient, $k_0 = \omega/c_0$ is the wavevector in vacuum, and $\rho_{01} = |r_{01}|^2$ is the (far-field) reflectivity at the interface between vacuum and medium 1, $\eta_0 = -\mathrm{i}\zeta_0 = \sqrt{\beta^2 - k_0^2}$, and superscripts $s$ and $p$ signify $s$-polarization and $p$-polarization, respectively. Note that $r_{01}^s = (\zeta_0 - \zeta_1)/(\zeta_0 + \zeta_1)$ and $r_{01}^p = (\zeta_0 - \zeta_1/\varepsilon_1)/(\zeta_0 + \zeta_1/\varepsilon_1)$. It is assumed that medium 2 is either far away or does not exist. Basu et al. [50] calculated the LDOS in the vacuum gap by including medium 2. It should be mentioned that, in deriving Eq. (10.40), the imaginary part of the permittivity of medium 1 in Eq. (10.34) has been combined with other terms. No matter how small $\mathrm{Im}(\varepsilon_1)$ may be, such as for a dielectric, it must not be zero for the semi-infinite assumption to hold. The contribution of propagating waves given by Eq. (10.41a) is independent of $z$ and exists in both near and far fields; whereas the contribution of evanescent waves decreases exponentially with increasing $z$. In the far-field limit, the contribution of propagating waves is responsible for thermal emission. In fact, Eq. (10.41a) contains terms related to the directional-spectral emissivity: $\varepsilon'^{s}_{\omega,1} = 1 - \rho^{s}_{01}$ and $\varepsilon'^{p}_{\omega,1} = 1 - \rho^{p}_{01}$. As it gets closer and closer to the surface, the contribution of evanescent waves near the surface may dominate when $\mathrm{Im}(r^{p}_{01})$ is large, especially in the case when surface phonon polaritons can be excited. Subsequently, extremely large energy densities can exist near the surface at that particular frequency [24, 66].

The $z$-component of the time-averaged Poynting vector can be expressed in the following,

$$\langle S_z(\mathbf{x},\omega)\rangle = \frac{1}{2}\mathrm{Re}\big[\langle E_x(\mathbf{x},\omega)H_y^*(\mathbf{x},\omega)\rangle - \langle E_y(\mathbf{x},\omega)H_x^*(\mathbf{x},\omega)\rangle\big] \tag{10.42}$$

$$q''_{\omega,1-2} = \frac{\Theta(\omega,T_1)}{4\pi^2}\int_0^{\infty} \xi_{12}(\omega,\beta)\beta\,\mathrm{d}\beta \tag{10.43}$$

Therefore, the spectral energy flux from medium 1 to medium 2 can be expressed as where $\xi_{12}(\omega,\beta) = \xi^{p}_{12} + \xi^{s}_{12}$

$$= \frac{16\mathrm{Re}(\varepsilon_1\zeta_1^*)\mathrm{Re}(\varepsilon_2\zeta_2^*)\left|\zeta_0^2 \mathrm{e}^{2\mathrm{i}\zeta_0 d}\right|}{\left|(\varepsilon_1\zeta_0+\zeta_1)(\varepsilon_2\zeta_0+\zeta_2)(1-r^{p}_{01}r^{p}_{02}\mathrm{e}^{2\mathrm{i}\zeta_0 d})\right|^2} + \frac{16\mathrm{Re}(\zeta_1)\mathrm{Re}(\zeta_2)\left|\zeta_0^2 \mathrm{e}^{2\mathrm{i}\zeta_0 d}\right|}{\left|(\zeta_0+\zeta_1)(\zeta_0+\zeta_2)(1-r^{s}_{01}r^{s}_{02}\mathrm{e}^{2\mathrm{i}\zeta_0 d})\right|^2}$$

Here, $\xi^{j}_{12}(\omega,\beta)$, $j = p$ or $s$, is called the *energy transmission coefficient* or *photon tunneling probability* for evanescent waves for the given polarization. Note that $0 \le \xi^{j}_{12}(\omega,\beta) \le 1$. Equation (10.43) includes the contributions from both propagating and evanescent waves. The expression of $q''_{\omega,2-1}$ is readily obtained by replacing $\Theta(\omega,T_1)$ with $\Theta(\omega,T_2)$ since the exchange function is reciprocal, namely, $\xi^{j}_{12}(\omega,\beta) = \xi^{j}_{21}(\omega,\beta)$. The net total energy flux can be calculated by integrating $(q''_{\omega,1-2} - q''_{\omega,2-1})$ over the frequency, viz.

$$q''_{\mathrm{net}} = \frac{1}{4\pi^2}\int_0^{\infty}\int_0^{\infty} [\Theta(\omega,T_1) - \Theta(\omega,T_1)]\xi_{12}(\omega,\beta)\beta\,\mathrm{d}\beta\,\mathrm{d}\omega \tag{10.44}$$

Equation (10.44) provides a way to calculate radiative transfer that is applicable for both the near- and far-field heat transfer. The contribution of evanescent waves when

$\beta > \omega/c_0$ with imaginary $\zeta_0$ reduces as $d$ increases and is negligible when $d$ is on the order of the wavelength. The energy transfer can be separated into contributions of propagating waves and coupled evanescent waves (i.e., photon tunneling). For propagating waves, we have for either $p$-or $s$-polarizations,

$$\xi_{\text{prop}}^{p,s}(\omega,\beta) = \frac{(1-\rho_{01}^{p,s})(1-\rho_{02}^{p,s})}{\left|1-r_{01}^{p,s}r_{02}^{p,s}\mathrm{e}^{2\mathrm{i}\zeta_0 d}\right|^2}, \quad \text{when} \quad \beta < k_0 \tag{10.45}$$

where $r_{0j}$ is the Fresnel coefficients and $\rho_{0j} = r_{0j}r_{0j}^*$ is the reflectivity from vacuum to the $j$th medium. If only the propagating waves are considered as is the case in the far field, we note that $\beta = k_0 \sin\theta = (\omega/c_0)\sin\theta$, where $\theta$ is the polar angle in vacuum. The integration over $\beta$ from 0 to $k_0$ is equivalent to the integration from $\theta = 0$ to $\pi/2$. By averaging the oscillation terms, we can obtain the far-field and incoherent limit when $d \gg \lambda$:

$$\left|1-r_{01}^{p,s}r_{02}^{p,s}\mathrm{e}^{2\mathrm{i}\zeta_0 d}\right|^2 \to \left(1-\rho_{01}^{p,s}\rho_{02}^{p,s}\right) \tag{10.46}$$

Thus, it can also be shown that the inverse of Eq. (10.45) becomes

$$\frac{1-\rho_{01}^{p,s}\rho_{02}^{p,s}}{(1-\rho_{01}^{p,s})(1-\rho_{02}^{p,s})} = \frac{1}{\varepsilon_{\omega,1}^{\prime\, p,s}} + \frac{1}{\varepsilon_{\omega,2}^{\prime\, p,s}} - 1 \tag{10.47}$$

The total energy flux in the far-field limit becomes

$$q''_{\text{net,far}} = \frac{1}{4\pi^2 c_0^2}\int_0^\infty\int_0^{\pi/2}[\Theta(\omega,T_1)-\Theta(\omega,T_2)]\omega^2 \times\left(\frac{1}{1/\varepsilon_{\omega,1}^{\prime p}+1/\varepsilon_{\omega,2}^{\prime p}-1}+\frac{1}{1/\varepsilon_{\omega,1}^{\prime s}+1/\varepsilon_{\omega,2}^{\prime s}-1}\right)\cos\theta\sin\theta\,\mathrm{d}\theta\,\mathrm{d}\omega \tag{10.48}$$

which is similar to the equation found in radiation heat transfer texts cited in previous chapters, except that angular frequency is used here instead of wavelength. While the energy flux includes the contributions of both polarizations, one should integrate the two polarizations separately as done in Eq. (10.48). If the emissivities of the two surfaces have different dependences on the polar angle and polarization status, averaging over the two polarizations to obtain the directional emissivity of each surface may cause some error in the calculation.

For evanescent waves in vacuum when $\beta > k_0$, the photon tunneling probability $\xi$ becomes

$$\xi_{\text{evan}}^{p,s}(\omega,\beta) = \frac{4\,\mathrm{Im}(r_{01}^{p,s})\mathrm{Im}(r_{02}^{p,s})\mathrm{e}^{-2\eta_0 d}}{\left|1-r_{01}^{p,s}r_{02}^{p,s}\mathrm{e}^{-2\eta_0 d}\right|^2}, \quad \text{when } \beta > k_0 \tag{10.49}$$

Clearly, the tunneling probability decays exponentially as the distance of separation $d$ increases.

At the nanometer scale, when near-field radiation dominates, especially for metallic media, doped silicon, or polar materials in the absorption band, the tunneling probability from Eq. (10.49) can be expressed with an approximate formula. Note that at $\beta \gg k_0$, we have $\zeta_1 \approx \zeta_2 \approx \zeta_0 \approx i\beta$. In this case, it can be shown that $r_{01}^s$ and $r_{02}^s$ are negligibly small, and hence, the contribution of TE waves can be ignored. Furthermore, $r_{01}^p \approx (\varepsilon_1 - 1)/(\varepsilon_1 + 1)$ and $r_{02}^p \approx (\varepsilon_2 - 1)/(\varepsilon_2 + 1)$ are independent of $\beta$; therefore,

$$\xi_{\text{evan}}(\omega, \beta) \approx \frac{4\text{Im}(r_{01}^p)\text{Im}(r_{02}^p)\text{e}^{-2\beta d}}{\left|1 - r_{01}^p r_{02}^p \text{e}^{-2\beta d}\right|^2} \tag{10.50}$$

Using the relation: $\text{Im}\left(\frac{\varepsilon-1}{\varepsilon+1}\right) = \frac{2\text{Im}(\varepsilon)}{|\varepsilon+1|^2}$, the spectral heat flux from 1 to 2 in the limit $d \to 0$ can then be expressed as

$$q''_{\omega,1-2} \approx \frac{\Theta(\omega, T_1)}{\pi^2 d^2} \frac{\text{Im}(\varepsilon_1)\text{Im}(\varepsilon_2)}{|(\varepsilon_1 + 1)(\varepsilon_2 + 1)|^2} \int_{x_0}^{\infty} \left|1 - \frac{(\varepsilon_1 - 1)(\varepsilon_2 - 1)}{(\varepsilon_1 + 1)(\varepsilon_2 + 1)}\text{e}^{-x}\right|^{-2} x\text{e}^{-x}\text{d}x$$

where $x_0 = 2k_0 d$. The heat flux will be inversely proportional to $d^2$ in the proximity limit. The integral approaches 1 when $\left|\frac{(\varepsilon_1-1)(\varepsilon_2-1)}{(\varepsilon_1+1)(\varepsilon_2+1)}\right| \ll 1$. Consequently, the net spectral flux becomes [24]

$$q''_{\omega,1-2} - q''_{\omega,2-1} \approx \frac{1}{\pi^2 d^2} \frac{\text{Im}(\varepsilon_1)\text{Im}(\varepsilon_2)}{|(\varepsilon_1 + 1)(\varepsilon_2 + 1)|^2} [\Theta(\omega, T_1) - \Theta(\omega, T_2)] \tag{10.51}$$

When $\beta \gg k_0$, Eq. (10.41b) reduces to

$$D_{\text{evan}}(z, \omega) \approx \frac{1}{\pi^2 \omega} \frac{\text{Im}(\varepsilon_1)}{|\varepsilon_1 + 1|^2} \int_{k_0}^{\infty} \text{e}^{-2\beta z}\beta^2 \text{d}\beta \tag{10.52a}$$

By evaluating the integration and keeping the highest order terms only, one obtains the following asymptotical expression for $z \to 0$ as [66]

$$D_{\text{evan}}(z, \omega) \approx \frac{1}{4\pi^2 \omega z^3} \frac{\text{Im}(\varepsilon_1)}{|\varepsilon_1 + 1|^2} \tag{10.52b}$$

This equation suggests that, as $z$ decreases, the near-field density of states increases with $z^{-3}$ and is localized at the surface. There are questions about when the fluctuation–dissipation theory will fail and when conduction will dominant radiation. In general, the locality and homogeneous assumptions should be valid until the separation spacing approaches interatomic distances or is below about 1 nm [72, 73].

Surface roughness may prevent the two interfaces from reaching interatomic spacing before touching each other at certain locations. Some controversials still remain as evidenced by the recent experiments performed at distances below about 1 nm from different labs [74, 75].

### 10.4.3 *Effect of Surface Plasmon Polaritons (SPPs)*

Radiation heat transfer may be important when the characteristic dimensions are on the nanometer scale. AFM cantilevers with integrated heaters and nanoscale sharp tips made of doped silicon have been developed for thermal writing and reading [76]. These heated cantilever tips may provide local heating for the study of radiative energy transfer between two objects separated by a few nanometers. It is critical to quantitatively predict the near-field radiation heat flux between doped silicon. The dielectric function of doped silicon can be described by the Drude model, considering the effects of temperature and doping level on the concentrations and scattering times of electrons and holes, as described in Sect. 8.4.4 of Chap. 8. Polar dielectric materials may excite phonon polaritons. The excitation of coupled surface plasmon polaritons (SPPs) or coupled surface phonon polaritons (SPhPs) can significantly enhance near-field radiation in a narrowband near the resonance frequency.

To calculate the radiative energy flux, it is essential to perform the integration of the energy transmission coefficient in Eq. (10.42), $\xi_{12}(\omega, \beta)$, over the wavevector $\beta$ ranging from 0 to infinity. Then, the integration of the spectral energy flux can be carried out over all frequencies. Usually, angular frequency is preferred over wavelength for consistency with the electrodynamic formulation. The integration over $\beta$ from 0 to $k_0$ corresponds to radiation heat transfer by propagating waves. In this range, the integrand exhibits highly oscillatory behavior for large $d$. In this regard, Simpson's rule is an effective technique in dealing with oscillatory integrands. The integration for $\beta$ from $k_0$ to infinity corresponds to radiation heat transfer by evanescent waves, and the photon tunneling probability is given as $\xi_{\text{evan}}(\omega, \beta)$ in Eq. (10.49). For small $d$ values, the upper limit $\beta_{\max}$ should be on the order of $1/d$; but for large $d$ values, $1/d$ would be less than $k_0$. A semi-empirical criterion can be used to set $\beta_{\max}$ as $3/d$ or $100k_0$, whichever is larger, to ensure an integration error less than 1%. For materials with strong SPhP resonances like SiC, an even large $\beta_{\max}$ needs to be applied and the limit is set to $\pi/d_c$, where $d_c \approx 0.5$ nm is on the order of the lattice constant [48, 77]. An effective way to perform the integration is to break it into several parts and evaluate each part using Simpson's rule. For example, the integration can be carried out in two parts, $k_0 < \beta < 6k_0$ and $6k_0 < \beta < \beta_{\max}$. A relative difference of 0.1% may be used as the convergence criterion between consecutive iterations. For conventional radiation heat transfer calculations, the lower and upper bounds of the integration over frequency (or wavelength) can be selected such that 99% of the blackbody emissive power falls between the limits. For example, 99% of blackbody radiation emissive power is concentrated between 1.2 and 25 μm at 1000 K, and between 4 and 85 μm at 300 K. Although $\lambda_{\text{mp}}$ predicted by Wien's displacement law does not exactly correspond to the peak in near-field radiation [48, 49, 78], these criteria can give satisfactory results. The enhancement of near-field radiation heat transfer is generally greater at longer wavelengths; as such, the integration should be performed over a much broader spectral region.

Figure 10.13 shows the predicted radiation heat transfer between two silicon plates [25]. Medium 1 is intrinsic silicon at $T_1 = 1000$ K, whereas medium 2 is at $T_2 = 300$ K. Medium 2 is either intrinsic or doped silicon with phosphorus as the donor (*n*-type). For convenience of discussion, sometimes the higher temperature heat source is called the emitter and the lower temperature heat sink is called the receiver. In the calculations, the wavelength region is chosen in the range from approximately 0.94 to 1880 μm ($\omega$ from $10^{12}$ to $2 \times 10^{15}$ rad/s). The dotted line represents the far-field radiation heat flux between two blackbodies, $\sigma_{\text{SB}}(T_1^4 - T_2^4)$, as predicted by the Stefan–Boltzmann law. Wien's displacement law suggests that the dominant wavelength $\lambda_{\text{mp}}$ for the 1000 K emitter is around 3 μm. The energy flux is essentially a constant when the distance $d$ is greater than 10 μm, which is the far-field regime. The net energy flux increases quickly when $d < \lambda_{\text{mp}}$ due to photon tunneling. When medium 2 is intrinsic or lightly doped, i.e., $N_{\text{D2}} < 10^{15}$ cm$^{-3}$, the maximum $q''_{\text{net}}$ is achieved when $d < 50$ nm. The maximum net energy flux is 21.3 times that of the far-field limit and 11.7 times that of blackbodies for intrinsic silicon, as predicted earlier when the silicon plates are treated as dielectrics. On the other hand, $q''_{\text{net}}$ for $N_{\text{D2}} > 10^{16}$ cm$^{-3}$ continues to increase as $d$ is reduced and does not saturate. The heat flux at $d = 1$ nm with $N_{\text{D2}} = 10^{18}$ cm$^{-3}$ is 800 times greater than that between two blackbodies and exceeds that between doped silicon in the far field by more than three orders of magnitude.

If one of the media is a slightly absorbing dielectric, as is the case for silicon with a carrier concentration less than $10^{15}$ cm$^{-3}$, the Fresnel coefficients becomes imaginary beyond the critical angle. There is a propagating wave in the medium and an evanescent wave in vacuum (corresponding to frustrated total internal reflection).

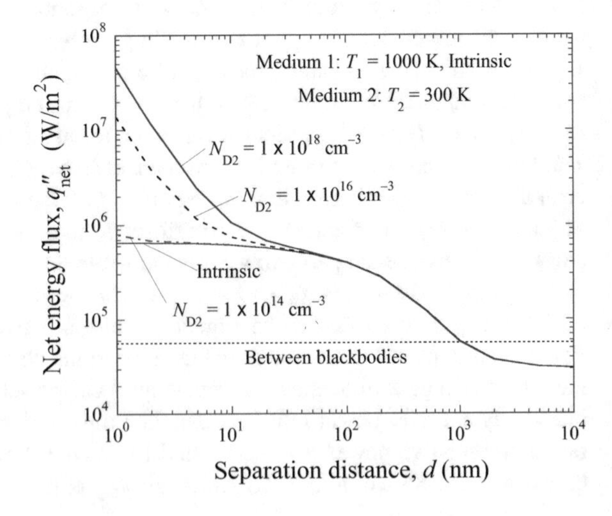


**Fig. 10.13** Net energy flux between an emitter made of intrinsic Si at 1000 K and a receiver made of Si with different doping levels at 300 K [25]

If the refractive index of the dielectric medium is $n$, then $\xi_{\text{evan}}(\omega, \beta)$ is nonzero for $k_0 < \beta < nk_0$. However, because the extinction coefficient $\kappa$ is negligibly small, $\xi_{\text{evan}}(\omega, \beta)$ becomes very small beyond $n\omega/c$ and decays exponentially with increasing $\beta$ . Therefore, for lightly doped silicon, the enhancement is limited to approximately $(n^2-1)\sigma_{\text{SB}}(T_1^4-T_2^4)$ when the near-field flux reaches $q''_{\text{net}} \approx n^2\sigma_{\text{SB}}(T_1^4-T_2^4)$, as discussed previously. Because of the small difference between the refractive indices of the two media, $n$ is used here for both media for simplicity. On the other hand, if $\kappa$ is relatively large, the integration over $\beta > nk_0$ will have a significant contribution and may even dominate the heat flux when $d$ reaches a few nanometers.

The enhancement of near-field heat transfer can be better understood from the energy flux spectra shown in Fig. 10.14. The units of $q''_\omega$ are expressed as W m$^{-2}$ s rad$^{-1}$ rather than J m$^{-2}$ rad$^{-1}$ to keep the integrity of the angular frequency units, i.e., rad/s. Notice that at 1000 K, the carrier concentration is about $10^{18}$ cm$^{-3}$ for intrinsic silicon. The spectral flux between two blackbodies at 1000 and 300 K, calculated from Planck's spectral emissivity power, is also shown for comparison. Interference becomes important at $d = 10$ μm and causes the wavy features in the spectral energy flux. When the receiver is intrinsic, as Fig. 10.14a reveals, the shape of the spectrum is similar for $d < 100$ nm and scaled up with $n^2(\approx 11.7)$ times that between blackbodies. However, the slightly increased $\kappa$ due to phonon or impurity absorption, along with free carrier absorption in the far infrared, can result in an increase in the spectral energy flux at very small distances. These effects are relative small and do not have notable influence on the total heat flux as seen previously in Fig. 10.13.

The near-field spectral flux is greatly enhanced when medium 2 is doped, as can be seen from Fig. 10.14b, especially in the far-infrared region. At $d < 100$ nm, the peak due to near-field enhancement is located around $\lambda = 80$ μm and becomes higher than that due to blackbody emission around $\lambda = 3$ μm. As mentioned previously, the

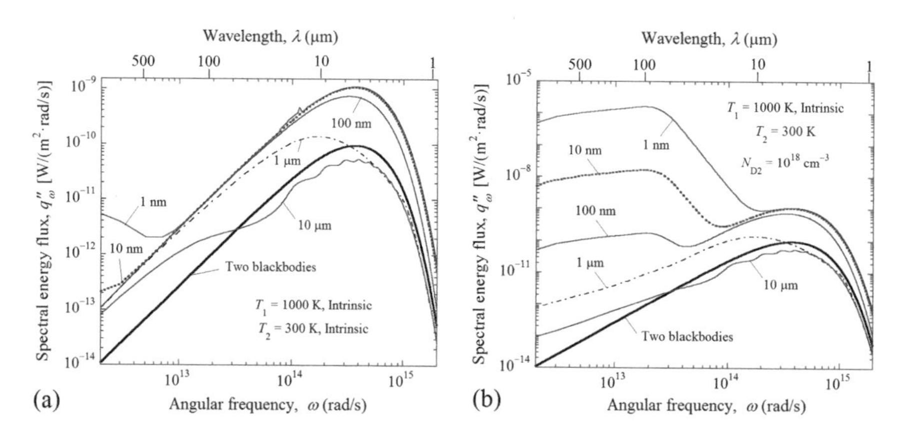


**Fig. 10.14** Spectral energy flux for different separation distances between silicon plates, where medium 1 is always intrinsic, $T_1 = 1000$ K, and $T_2 = 300$ K [25]. **a** Medium 2 is intrinsic. **b** Medium 2 is $n$-type silicon with a donor concentration $N_{\text{D2}} = 10^{18}$ cm$^{-3}$

increased energy flux in the longer wavelengths requires the integration to be carried out over a much broader range than typically done with the blackbody spectrum. It has been shown that for highly conductive materials and polar materials in the resonance region, Eqs. (10.50) through (10.52a, 10.52b) can give excellent approximations in the extreme near field [24, 48, 77]. However, for doped silicon, Eq. (10.50) is not applicable for $\omega > 10^{14}$ rad/s, where the major contribution of evanescent waves comes from $k_0 < \beta < nk_0$, i.e., propagating waves in silicon. Even in the frequency region from $10^{12}$ to $10^{14}$ rad/s, Eqs. (10.50) and (10.51) significantly underpredict the near-field radiation between silicon plates. Therefore, care must be taken in applying the asymptotic expressions.

By comparison with the measured spectral properties of thin films made of heavily doped silicon, Basu et al. [50] obtained more realistic parameters for the carrier concentration and mobility in heavily doped silicon at temperatures from 250 to 400 K and used the modified Drude model to calculate the near-field radiative heat transfer between doped silicon plates. The basic argument was that the model employed by Fu and Zhang [25] gives a much lower majority carrier concentration since only a fraction of the dopant atoms are assumed to be ionized and contribute to the carrier concentration. Figure 10.15a shows the calculated net heat flux for $T_1 = 400$ K and $T_2 = 300$ K with various doping concentrations. In all cases, *n*-type silicon (e.g., doped with phosphorus atoms) is used and it is assumed that both medium 1 and medium 2 have the same doping concentration $N_{D1} = N_{D2}$. Comparing Fig. 10.15a with Fig. 10.13, we see that the enhancement near room temperature is even stronger than at elevated temperatures. As the doping concentration increases from $10^{18}$ to $10^{19}$ cm$^{-3}$, the heat flux for $d < 100$ nm continues to increase but does not change much as the doping concentration is increased to approximately $10^{19}$ cm$^{-3}$. Any further increase in the doping concentration will result in a reduction of the heat flux. The spectral heat flux for the same emitter and receiver temperatures is plotted in Fig. 10.15b for three doping levels at a distance $d = 10$ nm. The peak locations are totally determined by the dielectric functions rather than the blackbody spectrum.

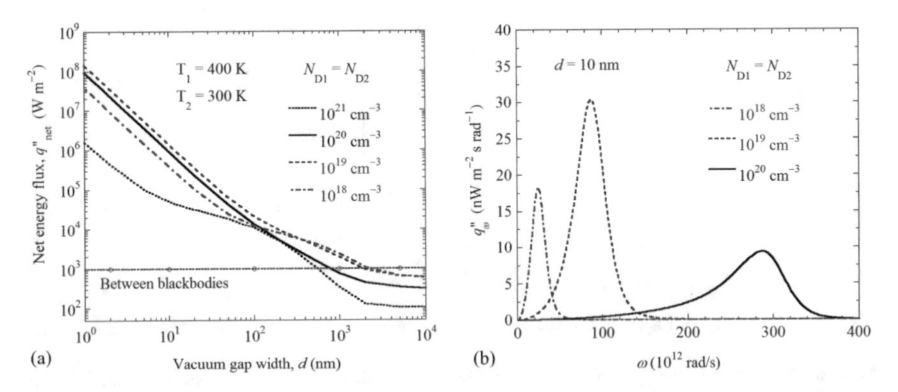


**Fig. 10.15** Radiative heat transfer between heavily doped *n*-type silicon [50]: **a** Net heat flux versus separation distance; **b** spectral heat flux at $d = 100$ nm

The location of the peak shifts toward higher frequencies as the doping concentration increases (more metallic behavior): from $\omega_{\mathrm{m}} = 2.3 \times 10^{13}$ to $7.1 \times 10^{13}$ to $2.5 \times 10^{14}$ rad/s for doping concentrations of $10^{18}$, $10^{19}$, and $10^{20}$ cm$^{-3}$, respectively. These peaks are associated with the coupled SPPs as discussed in Sect. 9.5.3. Wien's displacement does not hold in this case. The resonance frequency for the SPPs is governed by the dielectric function when the real part $\varepsilon'$ approaches −1 [49]. Furthermore, the imaginary part $\varepsilon''$ of the dielectric function determines the width as well as the peak height. A larger $\varepsilon''$ generally gives a broader peak with a smaller height. As the total heat flux equals the area under the curve, the values are similar for $N_{\mathrm{D}} = 10^{19}$ and $10^{20}$ cm$^{-3}$.

The contribution of coupled SPPs to near-field radiation is related to the mode's large $\beta$ values at the peak resonance frequency $\omega_{\mathrm{m}}$ given in Fig. 10.15. This can be clearly seen using the contour plot shown in Fig. 10.16, where the value of $\beta\xi_{12}/2\pi$ is shown as a function of the normalized wavevector $\beta/k_0$ and the angular frequency $\omega$ for $d = 10$ nm. Both the emitter and receiver are made of $10^{20}$ cm$^{-3}$ doped Si. Note that the contribution from TE waves is negligible at this gap spacing for heavily doped Si. The contour plot reveals that the contribution by propagating waves (in the region $\beta/k_0 < 0$) is much smaller than that by evanescent waves. The two dashed curves are calculated from Eq. (9.106) for the symmetric and asymmetric coupled SPP branches. The left branch corresponds to the symmetric mode, while the right branch represents the asymmetric mode. The actual distribution according to the contour plot is much broader; however, the shape of the brighter region generally matches with the dispersion curves. The peak is located near $\omega_{\mathrm{m}} = 2.6 \times 10^{14}$ rad/s and $\beta = 60k_0$. Most of the nanoscale energy transfer is through modes about the resonance frequency and wavevector location. Note that near-field radiation inherently excites coupled SPPs due to the existence of large-$\beta$ modes without using gratings. Furthermore, as $d$ is reduced, the modes with larger $\beta$ values can be excited, resulting in the $1/d^2$ dependence of the total heat flux [79].

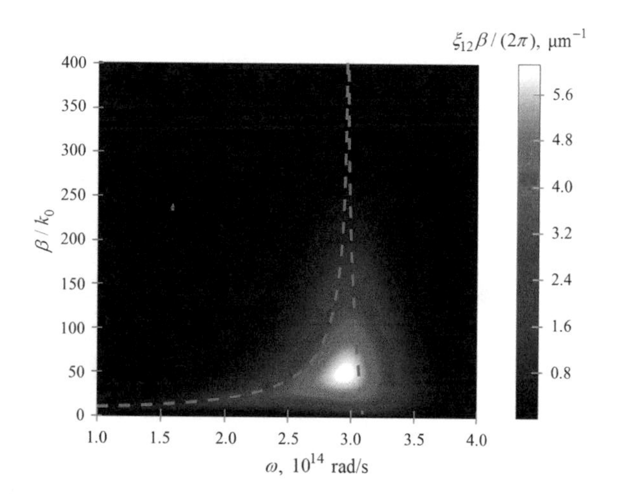


**Fig. 10.16** Contour plot of the function $\xi_{12}(\omega, \beta)\beta$ over $2\pi$ for the case with $N_{\mathrm{D1}} = N_{\mathrm{D2}} = 10^{20}$ cm$^{-3}$ when $d = 10$ nm; the two branches of the SPP dispersion are shown as the dashed curves [50]

While the contribution of TE waves is smaller at very small gap spacings, the contribution needs to be included in moderate gap spacings. The two contributions are separately plotted along with the sum in Fig. 10.17 for two levels of $10^{20}$ and $10^{21}$ cm$^{-3}$, respectively. When the doping concentration is $10^{20}$ cm$^{-3}$ as shown in Fig. 10.17a, the TM wave contribution dominates the net energy transfer, although the TE wave contribution may need to be considered if $d > 100$ nm. For doping concentration of $10^{21}$ cm$^{-3}$, on the contract, the TE wave contribution is greater than the TM wave contribution for $d > 12$ nm. The contribution of TM waves further increases while that of TE waves saturates as $d$ further reduces. Most of the TE wave contributions are limited to smaller $\beta$ values, because both $\text{Im}(r_{01}^s)$ and $\text{Im}(r_{02}^s)$ decrease quickly as $\beta$ increases and become negligible when $\beta > 5k_0$. Similar results have been observed between two metallic surfaces [78].

**Example 10.3** At what distance $d$, would the nanoscale thermal radiation, between two plates at $T_1 = 400$ K and $T_2 = 300$ K, exceed that of heat conduction by air at the pressure $P = 1$ atm? Consider doping concentrations $N_{\text{D1}} = N_{\text{D2}} = 10^{19}$ cm$^{-3}$ for $n$-type silicon.

**Solution** When $d$ is much smaller than the mean free path, which is about 70 nm at standard atmospheric conditions, boundary scattering or ballistic scattering dominates gas conduction. The thermal conductivity decreases linearly as $d$ decreases, whereas the heat flux is independent of $d$ in this regime. Assuming a thermal accommodation coefficient of 1, the heat transfer by gas conduction can be estimated from the theory in Chap. 4, Eq. (4.93), as

$$q''_{\text{cond}} = \frac{c_v(\gamma + 1)P}{(8\pi RT_{\text{m}})^{1/2}}(T_1 - T_2) \tag{10.53}$$

where $R$ is the ideal gas constant, $P$ is the pressure, $T_m = 4T_1T_2/\left(\sqrt{T_1} + \sqrt{T_2}\right)^2$ is a mean temperature, and $c_v$ is the specific heat at constant volume evaluated at $T_{\text{m}}$. The

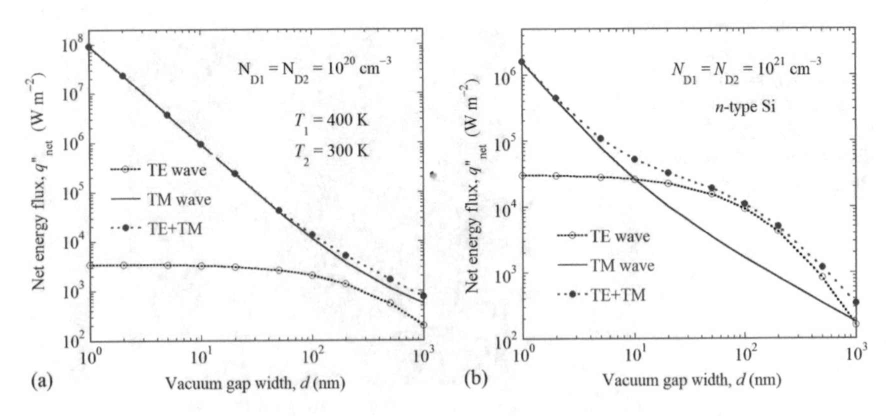


**Fig. 10.17** Polarization dependence of the total heat flux between heavily doped silicon when $T_1 = 400$ K and $T_2 = 300$ K [50]: **a** $N_{\text{D1}} = N_{\text{D2}} = 10^{20}$ cm$^{-3}$ ; **b** $N_{\text{D1}} = N_{\text{D2}} = 10^{21}$ cm$^{-3}$

resulting $q''_{\text{cond}}$ for air at a pressure $P = 1$ atm is approximately $1.1 \times 10^7$ W/m$^2$ for $T_1 = 400$ K and $T_2 = 300$ K. According to Fig. 10.15a, the calculated near-field net energy transfer by radiation is at the same level when $d \approx 3$ nm with heavily doped silicon. At $d = 1$ nm, the near-field radiation heat transfer can be an order of magnitude greater than the heat transfer by air conduction at the atmospheric pressure. Because the conduction heat flux further decreases as the pressure is reduced, nanoscale thermal radiation may dominate the heat transfer process for scanning thermal probes and heated cantilever tips that use heavily doped silicon.

The radiation heat transfer coefficient can be defined as $h_{\text{r}} = q''_{\text{net}}/(T_1 - T_2)$ in analogy to Newton's law of cooling. It can be seen from Fig. 10.15a that for heavily doped silicon, $h_{\text{r}} \sim 10^6$ W/m$^2$ K at $d = 1$ nm and $h_{\text{r}} \sim 10^4$ W/m$^2$ K at $d = 10$ nm. It is important to verify whether the local-equilibrium assumption is valid. Assume that the near-field radiation penetration depth is 100 nm and the thermal conductivity for doped silicon is 100 W/m · K. For a heat flux of $10^9$ W/m$^2$, the temperature drop would be 1 K within the radiation penetration depth. Therefore, the local-equilibrium assumption should still be valid. However, for a wafer of 100-μm thickness, the temperature drop would be 1000 K. The preceding calculations suggest that indeed near-field radiation can be an effective way of heating and cooling. As an alternative to the parallel-plate configuration, it is possible to pattern one of the silicon wafers with a 2D array of truncated cones or pyramids to remove heat locally for thermal control in nanoelectronics, for example. Local cooling based on near-field radiation has also been demonstrated with a $SiO_2$ coated tungsten tip [80].

### *10.4.4 Effect of Surface Phonon Polaritons (SPhPs)*

We will use an example to discuss the effect of SPhPs on near-field radiative heat flux between polar dielectric materials.

**Example 10.4** Calculate the radiative heat transfer coefficient near room temperature between two parallel plates separated by a vacuum gap, as illustrated in Fig. 10.12a for various polar materials. Use polar materials SiC, MgO, and silica (amorphous $SiO_2$), and assume that the emitter and receiver are made of the same material.

**Solution** The dielectric function of SiC has been given in Example 8.8, and that of MgO has been given in Problem 8.31. The dielectric function of amorphous $SiO_2$ can be found from either using the data from Palik's handbook or the Lorentz model. They are also available on the author's webpage. Codes for near-field radiation between parallel plates are also downloadable.

The radiative heat transfer coefficient $h_{\text{r}}$ may be calculated by setting $T_1 = 301$ K and $T_2 = 300$ K, since the value of the net heat flux per degree temperature difference is the same as that of $h_{\text{r}}$ . Another way is to modify Eq. (10.44) to the following

$$h_{\text{r}}(d, T) = \frac{1}{4\pi^2} \int_0^\infty \int_0^\infty \frac{\partial \Theta}{\partial T} \xi_{12}(\omega, \beta) \beta \mathrm{d}\beta \mathrm{d}\omega \tag{10.54}$$

Both methods give essentially the same results. Figure 10.18 shows the calculated $h_r$ for $d$ from 1 nm to 100 μm. It can be seen that the trends are similar and both silica and MgO have slightly higher near-field radiative heat transfer coefficients than SiC. As discussed by Wang et al. [49], there is a trade-off between the height and width of the resonance peak. Furthermore, silica has two phonon resonance modes. The heat flux spectra of the three materials are very different due to the different frequencies of the optical phonons. Details are left as exercises.

For SiC, the enhanced near-field radiation is attributed to the excitation of coupled-SPhP in the mid-infrared near $\lambda = 10.5$ [24, 48, 70]. The peak is very sharp due to the small imaginary part of the dielectric function $\varepsilon''$ at the resonance frequency of phonon polaritons. The enhancement of nanoscale radiation may be understood from the large values of $\xi_{12}$ around the resonance frequency where $\varepsilon''/|1+\varepsilon|^2$ is large in both media. The contour plot of $\xi_{12}\beta/2\pi$ for SiC at $d = 10$ nm is displayed in Fig. 10.19. A very narrow peak can be seen with much larger values of $\beta/k_0$ and $\xi_{12}\beta/2\pi$, as compared with Fig. 10.16. The FWHM bandwidth is only about 0.1 μm, and 80% of the heat flux is within the region $10.4\,\mu\text{m} < \lambda < 10.7\,\mu\text{m}$. The dispersion relations given in Eq. (9.106) are plotted as the white dotted lines, which capture the resonance modes very well. The energy streamlines and field distributions have also been investigated [70, 81].

Figure 10.20 plots the contour of the field distribution and energy streamlines adjacent to the vacuum gap of $d = 100$ nm for a TM wave $\lambda = 10.55\,\mu\text{m}$ [81]. The energy streamlines are calculated for $\beta = 40\,k_0$ (where the peak is located) for $d =$ 100 nm. The region below $z = 0$ is a semi-infinite SiC emitter and that above $z = d$ is a semi-infinite SiC receiver (whose temperature is set to zero). It can be seen that the field varies periodically along the interfaces and the contrast (magnitude) is greatly enhanced near the interfaces due to SPhPs. The energy streamlines are curved

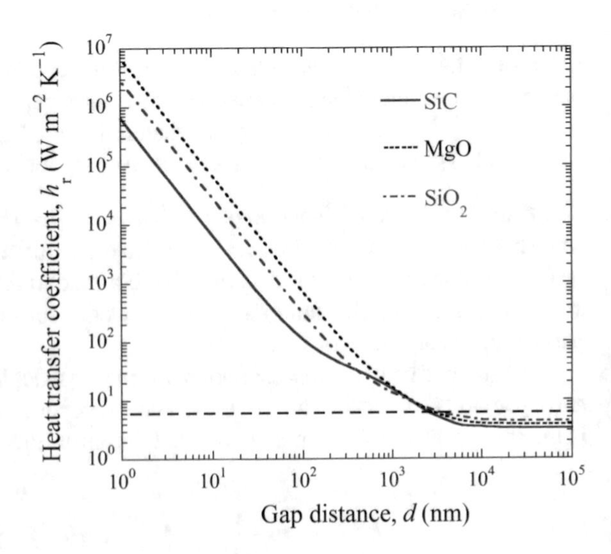


**Fig. 10.18** Radiative heat transfer coefficient as a function of the separation distance for several materials at 300 K

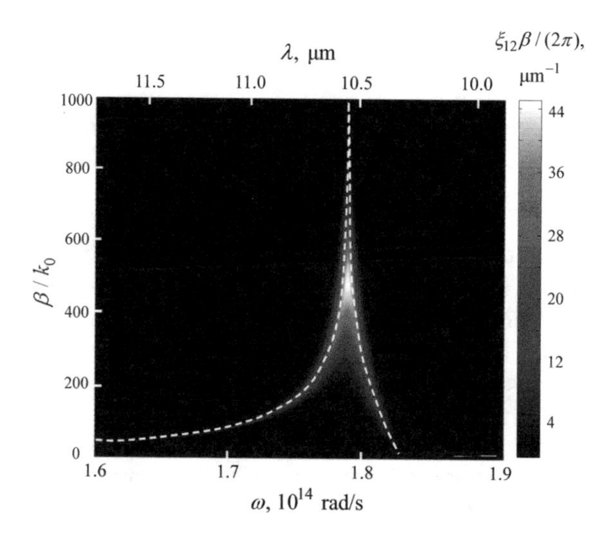


**Fig. 10.19** Contour plot of the function $\xi_{12}(\omega, \beta)\beta$ over $2\pi$ for the case when both the emitter and the receiver are made of SiC with $d = 10$ nm. The dashed lines are the SPhP dispersion curves

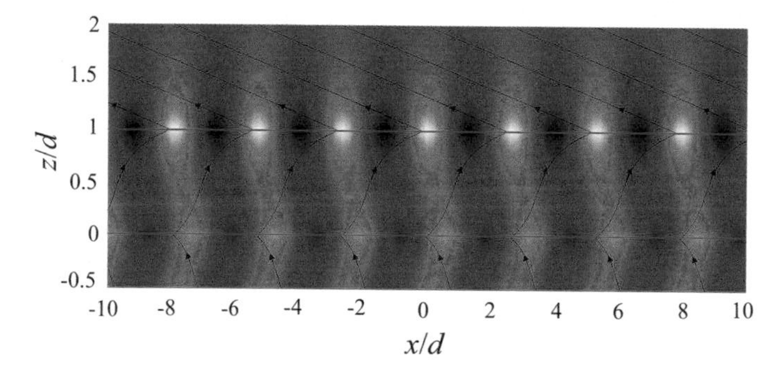


**Fig. 10.20** Contour plot of the field distribution $|H_y|$ and energy streamlines for $d = 100$ nm and $\lambda = 10.55$ μm when both the emitter and receiver are made of SiC [81]. The region $z < 0$ represents the emitter and $z > d$ represents the receiver

due to the coupled forward and backward evanescent waves, except in the receiver where only decaying evanescent waves exist. Negative refraction between SiC and vacuum and between vacuum and SiC can be seen in the energy streamlines since the Poynting vector changes sign due to the negative $\varepsilon'$ value of SiC. This plot also shows that in order for the medium to be approximated as being infinitely extended laterally, the lateral dimension should be at least one order of magnitude greater than the gap spacing.

### 10.4.5 The Landauer-Like Formulism

It is instructive to consider the $k$-space integration for a given frequency in terms of the mode, as illustrated in Fig. 10.21. Each cross or a unit area in the $k$-space represents a mode or energy transfer channel [82, 83]. The integration of $\xi_{12}$ over the $k$-space, or rings determined by the radius $\beta$ in the case of isotropic media, yields the efficiency of energy transfer at the given frequency. Three regions are identified as 1, 2, and 3. In region one, the radius is $k_0$ and when $\beta < k_0$, propagating waves exist in vacuum. For region 2, $k_0 < \beta < nk_0$, where $n > 1$ is the refractive index of the medium, propagating waves exist in the medium such as in a dielectric medium, while photon tunneling occurs via frustrated total internal reflection. These modes are called *frustrating modes* [84]. In the case when both medium 1 and 2 are dielectric with the same refractive index $n$, near-field radiation can contribute up to $(n^2 - 1)$ times the blackbody radiation. For region 3, when surface electromagnetic waves can be excited, the outer radius is determined by $d^{-1}$ so that the total area can be enhanced by $d^{-2}$ if the tunneling probability is close to 1. These modes are called surface modes due to the excitation of coupled SPPs or SPhPs. Following the work of Biehs et al. [83], one may first integrate over all frequencies, along with the energy transmission coefficients, and express the radiative heat transfer coefficient in a Landauer-like formula, as discussed in Sect. 5.6, as follows,

$$h_{\mathrm{r}}(d, T) = \frac{\pi^2 k_{\mathrm{B}}^2 T}{3h} \left( \sum_{j=s,p} \int_0^\infty \frac{1}{2\pi} \tau_{12}^j(\omega, \beta, d) \beta \mathrm{d}\beta \right) \Delta T \tag{10.55}$$

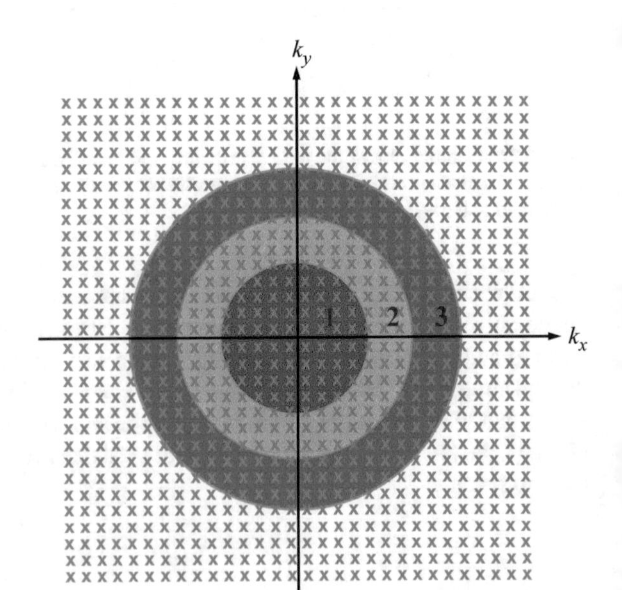


**Fig. 10.21** Illustration of different modes in the $k$-space. The three regions are identified as propagating waves, frustrated modes, and surface modes

Here, the heat source and sink temperatures are assumed to be $T_1 = T + \Delta T$ and $T_2 = T$, and $\tau_{12}^{j}$ is a mean transmission factor that is between 0 and 1, viz.

$$\tau_{12}^{p,s} = \frac{3}{\pi^2} \int_0^\infty f(x) \xi_{12}^{p,s}(x, \beta, d) \mathrm{d}x \tag{10.56}$$

where $x = \hbar\omega/k_{\mathrm{B}}T$ and $f(x) = x^2 \mathrm{e}^x/(\mathrm{e}^x - 1)^2$ [83]. Equations (10.55) and (10.56) offer another view of nanoscale thermal radiation in terms of quantum conductance as for electrons and phonons in the ballistic regime. When anisotropy exists or in a photonic crystal structure, additional modes such as hyperbolic modes can also enhance near-field thermal radiation as discussed in subsequent sections.

## 10.5 Multilayers, Anisotropic Media, and 2D Materials

In this section, near-field radiative transfer between multilayers, hyperbolic metamaterials, and 2D anisotropic materials such as graphene and hexagonal boron nitride are studied. This section will end with a discussion of the dyadic Green's function for multilayer systems.

### *10.5.1 Multilayers and Hyperbolic Modes*

If the emitter and/or receiver are made of a multilayer at a uniform temperature and each layer is isotropic, the net near-field heat flux can still be calculated with Eqs. (10.44), (10.45), and (10.49). However, the reflection coefficients $r_{01}^{p,s}$ and $r_{02}^{p,s}$ must be replaced by the reflection coefficients from vacuum to each multilayered structure, as done in Chap. 9 using the transfer matrix method based on Eq. (9.40), $r = B_1/A_1 = M_{21}/M_{11}$ for either polarization. If the penetration depth in a medium is much smaller than the layer thickness, that layer should be treated as a semi-infinite layer ($N$th layer in Fig. 9.12). The sign of the wavevector $k_{Nz}$ should be such that it decays towards $+\infty$ or $-\infty$ . For a multilayer that is semitransparent, the semi-infinite medium is vacuum. In such a case, the default is to treat the $N$th layer as a blackbody at the same temperature as the adjacent medium. This method is also applicable to magnetic materials for which $\mu_l = \mu_l(\omega)$ in any arbitrary layer $l$ as long as each layer is homogeneous and isotropic. For magnetic materials or materials that can be modeled as an effective homogeneous magnetic medium, surface polaritons can also be excited for TE waves [85–87]. It should be noted that for multilayers with a temperature gradient in the $z$-direction or if one wishes to calculate the heat transfer from a particular layer to another, the multilayer Green's functions are necessary and will be discussed in Sect. 10.5.4.

For uniaxial media whose optic axes are aligned parallel to the $z$-axis, the Fresnel coefficients are the same as with an isotropic medium for $s$-polarization using the ordinary dielectric function. For $p$-polarization, Eqs. (9.125) and (9.126) can be used to calculate the reflection and transmission coefficients. Suppose the two adjacent media are identified as $i$ and $j$, then we have

$$k_{iz} = \sqrt{k_0^2 \varepsilon_{i,\mathrm{O}} - k_x^2 \varepsilon_{i,\mathrm{O}} / \varepsilon_{i,\mathrm{E}}} \tag{10.57a}$$

$$k_{jz} = \sqrt{k_0^2 \varepsilon_{j,\mathrm{O}} - k_x^2 \varepsilon_{j,\mathrm{O}} / \varepsilon_{j,\mathrm{E}}} \tag{10.57b}$$

where subscripts O and E signify ordinary and extraordinary components. For a plane wave incident from medium $i$ to medium $j$,

$$r_{ij}^p = \frac{\dfrac{k_{iz}}{\varepsilon_{i,\mathrm{O}}} - \dfrac{k_{jz}}{\varepsilon_{j,\mathrm{O}}}}{\dfrac{k_{iz}}{\varepsilon_{i,\mathrm{O}}} + \dfrac{k_{jz}}{\varepsilon_{j,\mathrm{O}}}} = \frac{\dfrac{\varepsilon_{j,\mathrm{O}}}{k_{jz}} - \dfrac{\varepsilon_{i,\mathrm{O}}}{k_{iz}}}{\dfrac{\varepsilon_{j,\mathrm{O}}}{k_{jz}} + \dfrac{\varepsilon_{i,\mathrm{O}}}{k_{iz}}} \tag{10.58a}$$

$$t_{ij}^p = \frac{2\dfrac{k_{iz}}{\varepsilon_{i,\mathrm{O}}}}{\dfrac{k_{iz}}{\varepsilon_{i,\mathrm{O}}} + \dfrac{k_{jz}}{\varepsilon_{j,\mathrm{O}}}} = \frac{2\dfrac{\varepsilon_{j,\mathrm{O}}}{k_{jz}}}{\dfrac{\varepsilon_{j,\mathrm{O}}}{k_{jz}} + \dfrac{\varepsilon_{i,\mathrm{O}}}{k_{iz}}} \tag{10.58b}$$

These equations can be combined to the matrix formulation to calculate the refraction coefficients with multiple uniaxial laminae whose optic axes are parallel to the $z$-axis. Formulations for more complicated arrangements such as tilting or rotation will be given in Sect. 10.5.3. While simple, these formulas are especially useful for calculating near-field radiative transfer between two uniaxial media or films. As discussed in Sect. 9.5.6, hyperbolic metamaterials may be realized with artificial multilayers or arrays of aligned nanowires/nanotubes based on the effective medium theory. Furthermore, some naturally existing materials can also exhibit hyperbolicity. Two examples are given next: one is aligned CNT arrays and the other is doped-Si nanostructures, such as nanowires, nanoholes, and multilayers.

**Example 10.5** Calculate the radiative heat transfer between vertically aligned carbon nanotube arrays with an alignment factor $x = 0.98$ and filling ratio $\phi = 0.05$. Assume that the CNTs are sufficiently long so that the CNT arrays may be considered as semi-infinite. Compare the result to those for graphite–graphite and SiC–SiC at different distances. In all cases, the emitter and receiver can be set at $T_1 = 300$ K and $T_2 = 0$ K, respectively. Also plot the spectral heat flux for CNTs and graphite at $d = 10$ nm.

**Solution** The dielectric function of aligned CNT arrays can be modeled using Eqs. (9.136) and (9.137) and references cited there. Graphite is also anisotropic and its dielectric functions were also described in Sect. 9.5.7. The net radiative heat fluxes for the three materials are shown in Fig. 10.22 normalized to $\sigma T_1^4$ for comparison. It can be seen that CNT arrays can significantly enhance the near-field heat transfer, to about ten times that of SiC for 10 nm $< d <$ 100 nm. This is mainly due to the hyperbolic band at low frequencies as can be seen from Fig. 10.22b by

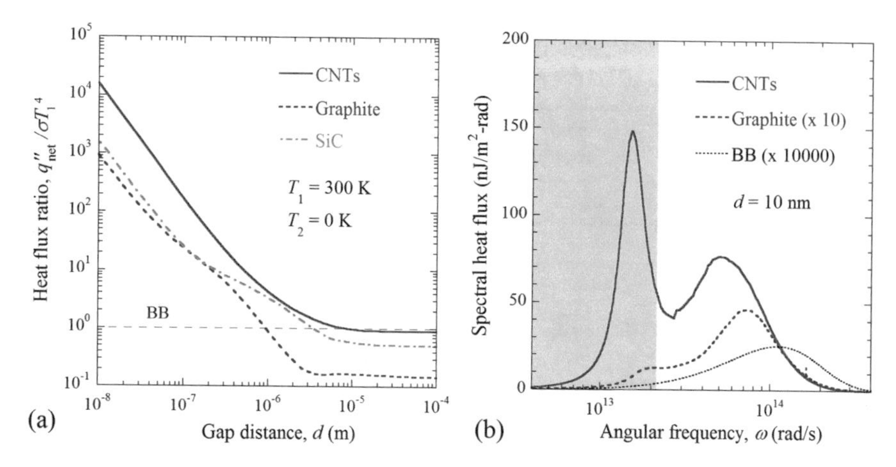


**Fig. 10.22** Near-field radiative heat transfer for aligned CNT, graphite and SiC [88]: **a** distance dependence; **b** spectral heat flux at $d = 10$ nm

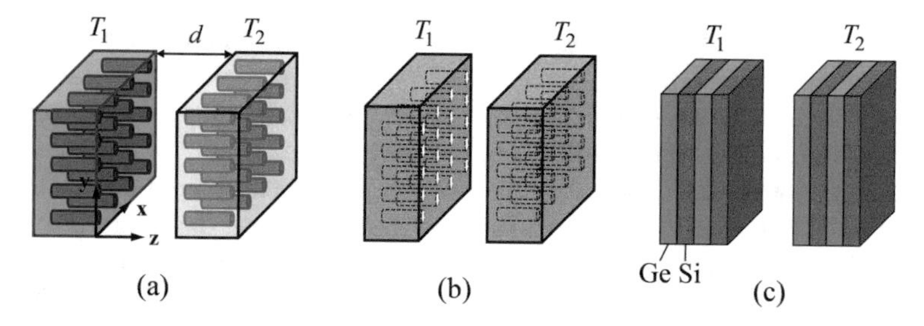


**Fig. 10.23** Configuration of two near-field radiative heat transfer between two semi-infinite nanostructured metamaterials, separated by a vacuum gap at a distance $d$: **a** Doped-S nanowires (D-SiNWs), **b** doped-Si nanoholes (D-SiNHs), and **c** D-Si/Ge multilayers

the peak in the shaded region (hyperbolic band of CNTs). In the heat flux spectra, the spectrum for graphite is magnified by 10 and that for blackbody (BB) is magnified by $10^4$ to make the comparison clear. The near-field radiation with graphite is also significantly enhanced though not as much as that for SiC, even though graphite can support both coupled-SPP and hyperbolic types of resonances [88]. In the far-field, CNT resembles a blackbody with high absorptance as discussed in Chap. 9. On the other hand, graphite resembles metals with a high reflectance, resulting in much lower radiative heat flux in the far field.

**Example 10.6** Consider the three different configurations shown in Fig. 10.23 for D-SiNWs, D-SiNHs, and D-Si/Ge multilayers, for which Ge may be treated as a dielectric with $\varepsilon_d = 16$. (a) Calculate and plot $h_r$ at room temperature as a function of $d$, assuming the Si volume fraction $\phi = 0.05$ for nanowires, $\phi = 0.3$ for nanoholes, and $\phi = 0.4$ for multilayers. (b) Calculate and plot $h_r$ as a function of the volume fraction of Si for $d = 10$ nm.

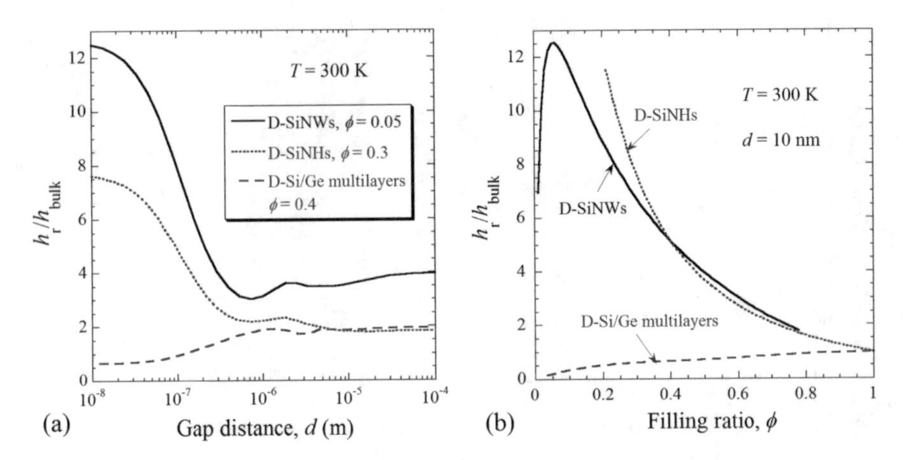


**Fig. 10.24** Radiative heat transfer coefficient for doped Si nanostructures [89]: **a** distance dependence; **b** dependence on the filling ratio. For D-SiNHs, the volume fraction is also for Si

**Solution** The dielectric function of these structures can be modeled as a uniaxial medium with effective medium theory described in the previous chapter, assuming that the structural features are sufficiently small. In general, they should be smaller than the characteristic wavelength and $1/\beta$, where $\beta$ is the parallel wavevector component. Equations (9.135a) and (9.135b) should be used for nanowires and nanoholes, while Eqs. (9.127) and (9.128) should be used for multilayers. The results are plotted in Fig. 10.24 normalized to that of bulk doped Si with $N_D = 10^{20}\,\text{cm}^{-1}$ [50, 89]. The volume fraction of Si in the case of nanoholes has a low limit when the wall thickness is zero. The lower limits of $\phi$ for nanowires and multilayers are arbitrarily set.

It should be noted that when $d$ is from 10 to 100 μm, the heat transfer coefficient for the nanowires is very close to that between two blackbodies. At 10–100 nm gap distances, bulk doped Si can greatly enhance near-field radiation as shown in Fig. 10.15 due to the excitation of coupled SPPs. Nanowires and nanoholes can further enhance nanoscale thermal radiation by an order of magnitude compared with the bulk. The reason is discussed next.

**Discussion**. The hyperbolic modes enable photon tunneling with a large transmission probability in a broad frequency region, as shown in Fig. 10.25a for D-SiNWs. On the other hand, both the hyperbolic modes and low-frequency SPP mode can enhance the photon tunneling probability for D-SiNHs as can be seen from Fig. 10.25b. Compared with Fig. 10.16 for bulk Si (note that the contour plots are rotated by 90°), the region where $\xi_{12}^{p}$ is relatively large becomes much broader. Detailed discussion with further parametric study and contour plots for the multilayer structures can be found from Liu et al. [89].

Bright et al. [71] examined the lateral shift and energy streamlines for hyperbolic materials made of multilayers. Green's functions for anisotropic media must be used to calculate the Poynting vector components. It was shown that the lateral shifts of

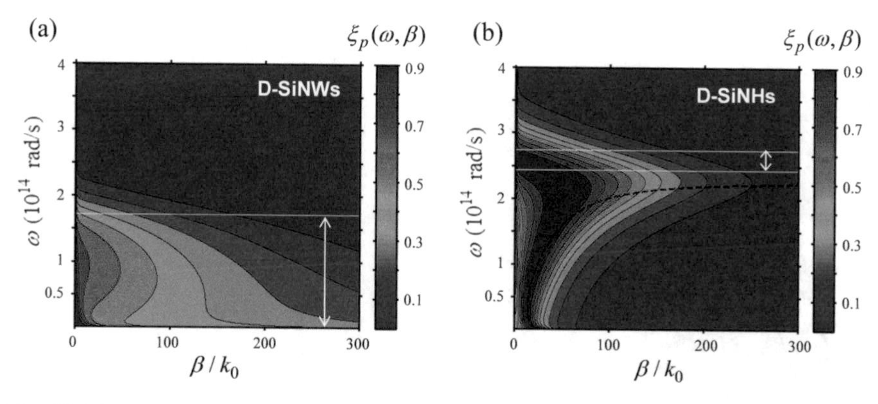


**Fig. 10.25** Energy transmission coefficient contours for $p$-polarization at $d = 10$ nm [89]. The hyperbolic band is indicated by the vertical arrows. **a** D-SiNWs; **b** D-SiNHs, where the black dashed line denotes the lower branch of SPP dispersion. The filling ratios are the same as for Fig. 10.24a

energy streamlines can be 3–4 orders of magnitude that of the gap distances. Care must also be taken in the applicability of EMT when $\beta$ becomes large [90–92]. The general consensus is $d > 1/\beta$. Suppose we are dealing with a wavelength of $\lambda = 19\,\mu\text{m}$ ($\omega \approx 10^{14}$ rad/s), for $\beta = 300\,k_0$, the gap spacing should be greater than 10 nm.

## 10.5.2 *Graphene and Hexagonal Boron Nitride*

Graphene can also be incorporated as either a thin film or a conducting sheet using Eqs. (9.119a) and (9.119b). Furthermore, hBN is a natural hyperbolic material as described in Fig. 9.59. The dielectric functions of hBN can be calculated with Eqs. (9.141a) and (9.141b). The near-field heat transfer for other 2D materials such as $MoS_2$ and black phosphorous (BP) has also been studied [93–96]. This section focuses on graphene sheets, graphene over CNT arrays, hBN films, and graphene over hBN.

Liu et al. [97] studied hybridized graphene plasmon and hyperbolic modes by considering graphene covered D-SiNW arrays, as shown in Fig. 10.26a. The tunneling probability is shown in Fig. 10.26b for three cases: D-SiNWs, graphene, and graphene-covered D-SiNWs. The nanowires are assumed to be sufficiently long for the nanowire array to be treated as semi-infinite. In the case of graphene–graphene, the space extending to infinity is assumed to possess the same temperature as the adjacent graphene sheet. In the calculations, the properties of graphene and doped Si are evaluated at 300 K at a vacuum gap spacing of 200 nm. The chemical potential of graphene is set to be 0.3 eV and relaxation time is $10^{-13}$ s. The filling ratio of the doped-Si nanowire array is $\phi = 0.02$ and the dopant concentration is $N_D = 10^{20}\,\text{cm}^{-1}$. It can

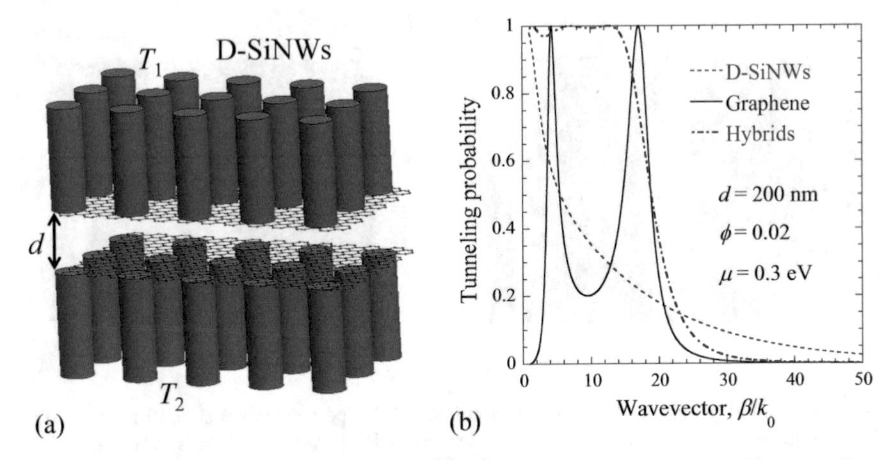


**Fig. 10.26** **a** Schematic of graphene-covered D-SiNWs; **b** Photon tunneling probability $\xi_{12}^{p}$ at $\omega = 5 \times 10^{13}$ rad/s and $d = 200$ nm with two D-SiNWs, graphene sheets, and graphene-covered D-SiNWs (hybrid) [97]

be seen from Fig. 10.26b that $\xi_{12}^{p}$ decreases exponentially with increasing $\beta$ with D-SiNWs only. Due to coupled graphene surface plasmons, there are two peaks for the case with two graphene sheets. When the D-SiNWs are covered with graphene, $\xi_{12}^{p}$ in the hybrid case is close to unity when $\beta < 15k_0$. The near-field radiative transfer can therefore be greatly enhanced with graphene coverage.

Note that the heat transfer coefficient $h_{\mathrm{r}}$ at $d = 200$ nm is 135 W/m$^2$ K for D-SiNWs and is increased to 615 W/m$^2$ K with graphene coverage. The heat transfer coefficient between two blackbodies is $h_{\mathrm{r,BB}} = 4\sigma_{\mathrm{BB}} T^3 \approx 6.1$ W/m$^2$ K. Due to the excitation of graphene plasmons, even with two graphene sheets, $h_{\mathrm{r}} = 454$ W/m$^2$ K. Covering graphene on D-SiNWs can give a heat transfer coefficient greater than the sum of the individual cases.

The enhanced energy transmission coefficient can be clearly seen by the contour plots displayed in Fig. 10.27. The dashed lines on Fig. 10.27b are calculated dispersion curves of coupled GSPs from Eqs. (9.121a) and (9.121b). The effect of

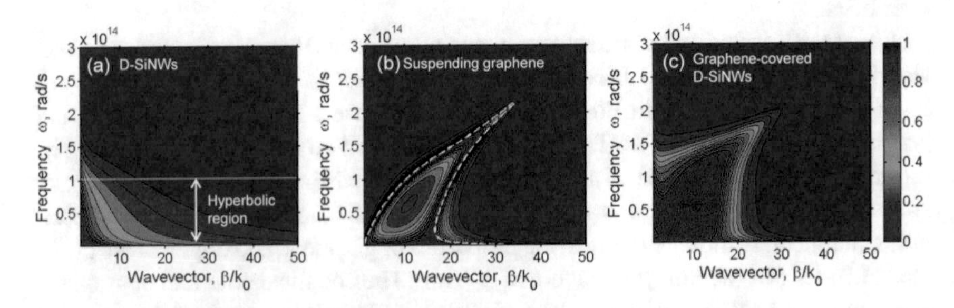


**Fig. 10.27** Energy transmission coefficient $\xi_{12}^{p}(\omega, \beta)$ contours for $p$-polarization at $d = 200$ nm [97]: **a** D-SiNWs; **b** graphene; **c** hybrid, where the scale bar is for all three cases

hyperbolic band at frequencies $\omega < 1.02 \times 10^{14}$ rad/s on D-SiNWs is similar to Fig. 10.25a, although the magnitude of $\beta$ is very different due to the different $d$ values. For suspended graphene, the coupled GSPs can give a large enhancement in $\xi_{12}^{p}$ near the two branches when GSPs are excited. When graphene covers the D-SiNWs for both the emitter and receiver, $\xi_{12}^{p}$ becomes very high at $\omega < 1.5 \times 10^{14}$ rad/s and $\beta < 20k_0$. The existence of D-SiNWs can significantly modify the GSPs, suggesting that a hybridization occurs between the hyperbolic modes and the coupled graphene plasmon modes. All the photons emitted in this regime will be absorbed, which is the blackbody behavior in the near field [83, 97]. The calculated heat transfer coefficient for the hybrid case is close to 80% of the theoretical limit [83]. It should be noted that graphene coverage does not always enhance the near-field radiation and parametric adjustment is also important to maximize the enhancement [97]. Near-field radiation for graphene-covered bulk doped Si [98] and CNT arrays [99] has also been investigated.

**Example 10.7** Calculate the near-field radiative heat flux for graphene-covered hBN films, shown in Fig. 10.28*a*, as a function of $d$. The thickness of hBN is $h = 50\,\mu\text{m}$ for both the emitter and receiver. Assume $T_1 = 300\,\text{K}$ and $T_2$ is very small. For graphene, take $\mu = 0.37\,\text{eV}$ and $1/\tau = 10^{13}$ rad/s.

**Solution** As discussed in Chap. 9, hBN is a natural hyperbolic material whose dielectric tensor can be expressed as $\bar{\bar{\varepsilon}}(\omega) = \text{diag}(\varepsilon_\perp, \varepsilon_\perp, \varepsilon_\parallel)$ using Eq. (9.141). Assume that the dielectric functions of graphene and hBN can be evaluated at 300 K and are independent of temperature. Since hBN is a thin film, some formulation is given below when graphene is treated as a conducting sheet [100, 101]. Since we can assume $T_2 = 0\,\text{K}$, we may write

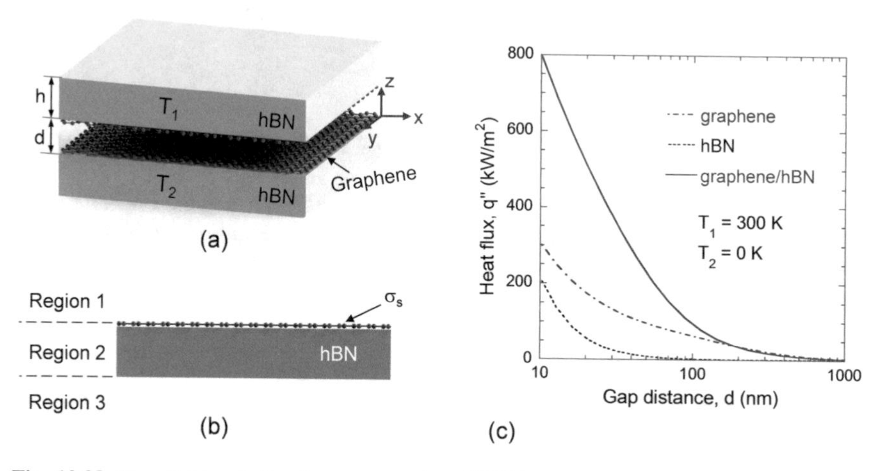


**Fig. 10.28** Near-field radiation between graphene/hBN heterostructures [100]. **a** Schematic of near-field radiation between graphene-covered hBN film structures; **b** Illustration for the reflection coefficient calculation. **c** Radiative heat flux versus gap spacing for the cases with graphene, hBN, and the heterostructure

$$q''_{1\to 2} = \frac{1}{4\pi^2}\int_0^\infty\int_0^\infty \Theta(\omega, T_1)\left[\xi_{12}^s(\omega,\beta) + \xi_{12}^p(\omega,\beta)\right]\beta \mathrm{d}\beta \mathrm{d}\omega \tag{10.59}$$

where

$$\xi_{12}^j(\omega,\beta) = \begin{cases} \dfrac{\left(1-\rho_1^j\right)\left(1-\rho_2^j\right)}{\left|1-r_1^j r_2^j \mathrm{e}^{2\mathrm{i}\zeta_0 d}\right|^2}, & \beta < k_0 \\ \dfrac{4\mathrm{Im}(r_1^j)\mathrm{Im}(r_2^j)e^{-2\eta_0 d}}{\left|1-r_1^j r_2^j e^{-2\eta_0 d}\right|^2}, & \beta > k_0 \end{cases} \tag{10.60}$$

Here, superscript $j$ denotes the polarization status ($s$ or $p$), $r_1$ is the reflection coefficient from vacuum to medium 1 (emitter), $r_2$ is the reflection coefficient from vacuum to medium 2 (receiver), and $\rho_1 = r_1 r_1^*$ and $\rho_2 = r_2 r_2^*$ for each polarization. If the emitter and receiver are made of the same structure whose properties are independent of temperature, $r_2 = r_1$. For graphene-covered hBN film as shown in Fig. 10.28b, the reflection coefficient for either polarization can be expressed as [101]

$$r = r_{12} + \frac{t_{12}t_{21}r_{23}\mathrm{e}^{2\mathrm{i}k_{2z}h}}{1 - r_{21}r_{23}\mathrm{e}^{2\mathrm{i}k_{2z}h}} \tag{10.61}$$

where $k_{2z}$ is for the uniaxial medium whose optic axis is parallel to the $z$-axis. The above expression has the same form as Airy's formula, Eq. (9.7). However, the expression for $k_{2z}$ and the Fresnel coefficients for an anisotropic medium must be used. It can be shown that

$$k_{2z} = \left(\varepsilon_{2,\perp}k_0^2 - \varepsilon_{2,\perp}\beta^2/\varepsilon_{2,\parallel}\right)^{1/2} \quad \text{for } p \text{ polarization} \tag{10.62a}$$

or

$$k_{2z} = \left(\varepsilon_{2,\perp}k_0^2 - \beta^2\right)^{1/2} \quad \text{for } s \text{ polarization} \tag{10.62b}$$

When there is a conducting surface (with negligible thickness), the Fresnel reflection and transmission coefficients between two dielectric materials (which may be anisotropic) can be expressed in a way similar to Eq. (9.119a) and Eq. (9.119b), as follows:

$$r_{\alpha\beta}^p = \frac{\frac{\varepsilon_{\beta,\perp}}{k_{\beta z}} - \frac{\varepsilon_{\alpha,\perp}}{k_{\alpha z}} + \frac{\sigma_s}{\omega\varepsilon_0}}{\frac{\varepsilon_{\beta,\perp}}{k_{\beta z}} + \frac{\varepsilon_{\alpha,\perp}}{k_{\alpha z}} + \frac{\sigma_s}{\omega\varepsilon_0}} \quad \text{and} \quad t_{\alpha\beta}^p = \frac{2\frac{\varepsilon_{\beta\perp}}{k_{\beta z}}}{\frac{\varepsilon_{\beta\perp}}{k_{\beta z}} + \frac{\varepsilon_{\alpha\perp}}{k_{\alpha z}} + \frac{\sigma_s}{\omega\varepsilon_0}} \tag{10.63}$$

where $\alpha, \beta = 1, 2$ or 3 according to Fig. 10.28b. If the medium is a vacuum or an isotropic dielectric, we can set the dielectric functions for different polarizations to be the same. For nonmagnetic materials, the expressions for $s$-polarizations are

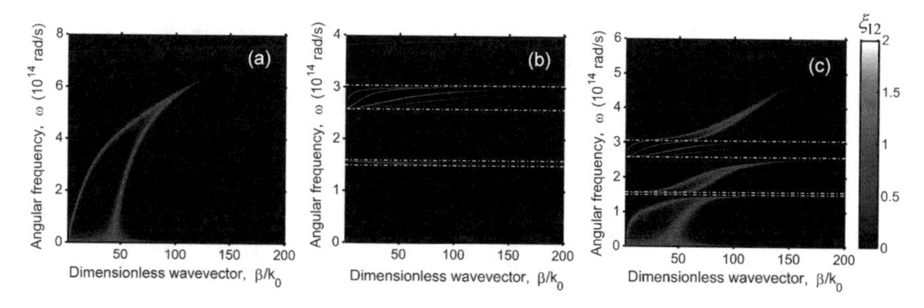


**Fig. 10.29** Contour plot of $\xi_{12}$ for three cases at $d = 20$ nm [100]. **a** Two graphene sheets; **b** Two hBN films; **c** Two graphene/hBN heterostructures

$$r^s_{\alpha\beta} = \frac{k_{\alpha z} - k_{\beta z} - \sigma_s \omega \mu_0}{k_{\alpha z} + k_{\beta z} + \sigma_s \omega \mu_0} \quad \text{and} \quad t^s_{\alpha\beta} = \frac{2k_{\alpha z}}{k_{\alpha z} + k_{\beta z} + \sigma_s \omega \mu_0} \tag{10.64}$$

If there is no graphene, we can set $\sigma_s = 0$ in Eqs. (10.63) and (10.64). By substituting Eqs. (10.62)–(10.64) into Eq. (10.61), we can calculate the reflection coefficients which are needed for evaluating $\xi_{12}$ using Eq. (10.60). Then the heat flux can be calculated with Eq. (10.59). One can also calculate the reflection coefficient when graphene covers both sides of the hBN film. The results are plotted in Fig. 10.28c for the cases with graphene sheets, hBN films, and graphene-covered hBN films (shown in Fig. 10.28b). A dramatic enhancement in near-field radiative flux can be observed with the heterostructure. At $d = 10$ nm, the heat flux is 305 kW/m$^2$ between graphene monolayers and 212 kW/m$^2$ between hBN films ($h = 50$ μm). With the heterostructure, the heat flux is increased to 800 kW/m$^2$, which is 50% more than the sum of the heat fluxes of the individual graphene and hBN cases. When $d$ exceeds about 200 nm, the heat flux for the heterostructures is very close to that between suspended graphene layers and the effect of hBN is negligibly small. The hybridization of the graphene surface plasmon modes with the hBN phonon polariton modes are discussed next.

**Discussion.** The contours of $\xi_{12}$ are shown in Fig. 10.29 for the three cases. Again, the dominant contribution comes from TM waves, though the maximum of the scale is set to 2. The bright bands shown in Fig. 10.29 indicate efficient photon tunneling due to the excitation of different polaritons, corresponding to the dispersion curves where the denominator of $\xi^p_{12}$ in Eq. (10.60) approaches zero. The GSPs result in the two bands in Fig. 10.29a, corresponding to the symmetric (lower frequencies) and asymmetric (higher frequencies) coupled GSPs, similar to Fig. 10.27b. For hBN films shown in Fig. 10.29b, multiple bulk polaritons or waveguide modes exist in each Reststrahlen band, as seen between the horizontal dashed lines. However, when the hBN film is covered by a graphene layer, new branches can be seen in different regions, as shown in Fig. 12.29c. Note that the angular frequency ranges are different in different contours to show the branches clearly. The hBN Reststrahlen bands divide the GSPs into different regions. These hybridized polaritons outside the hBN Reststrahlen bands are called surface plasmon-phonon polaritons (SPPPs), which are due to the coupling between surface plasmons in graphene and phonon polaritons in hBN [100]. Zhao et al. [101] also investigated the near-field radiation with multiple layers of graphene-hBN heterostructures and demonstrated the effect of additional hybridized modes.

### *10.5.3 Anisotropic Media*

Near-field radiative heat transfer between anisotropic media may also be important. Using either effective or naturally existing anisotropic materials may enable various mechanisms to allow the near-field heat flux to be manipulated. Figure 10.30a illustrates a 1D grating structure that may be modeled as an effective uniaxial medium. The grating orientation can be rotated with respect to the $z$-axis so that the angle between the two gratings can be varied to modulate near-field radiation [89, 102–104]. Figure 10.30b represents a general case of uniaxial medium whose optic axis is tilted and also can be rotated. Multilayers are also possible. When the emitter and receiver are individually at thermal equilibrium, the net radiative heat flux can be calculated based on the scattering theory and expressed in a way similar to (10.44) as following.

$$q''_{\text{net}} = \frac{1}{8\pi^3}\int_0^\infty [\Theta(\omega, T_1) - \Theta(\omega, T_2)]\mathrm{d}\omega \int_0^{2\pi}\int_0^\infty \xi(\omega, \beta, \phi)\, \beta \mathrm{d}\beta \mathrm{d}\phi \tag{10.65}$$

where $\phi$ is the azimuthal angle and the energy transmission coefficient $\xi(\omega, \beta, \varphi)$ can be expressed as [89, 102]

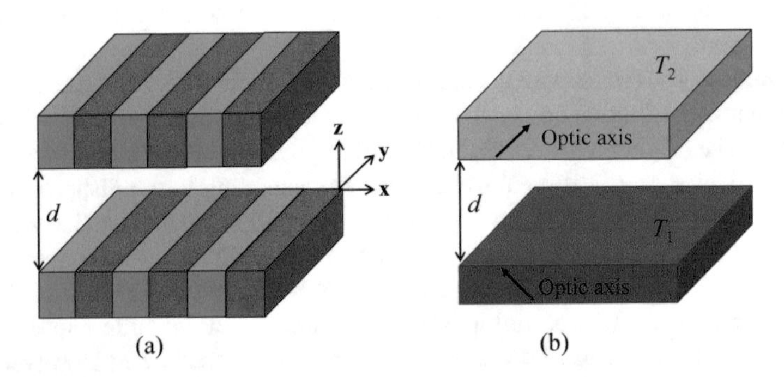


**Fig. 10.30** Schematic of near-field radiation between anisotropic media. **a** 1D gratings made of two materials that can be treated as an effective uniaxial medium; **b** Two arbitrarily oriented uniaxial films separated by a distance $d$

$$\xi(\omega,\beta,\phi) = \begin{cases} \mathrm{Tr}\left[\left(\mathbf{I} - \mathbf{R}_2^\dagger \mathbf{R}_2\right)\mathbf{D}\left(\mathbf{I} - \mathbf{R}_1^\dagger \mathbf{R}_1\right)\mathbf{D}^\dagger\right], & \beta < k_0 \\ \mathrm{Tr}\left[\left(\mathbf{R}_2^\dagger - \mathbf{R}_2\right)\mathbf{D}\left(\mathbf{R}_1 - \mathbf{R}_1^\dagger\right)\mathbf{D}^\dagger\right]\mathrm{e}^{-2\eta_0 d}, & \beta > k_0 \end{cases} \tag{10.66}$$

Note that the superscript "†" denotes a conjugate transpose (Hermitian transpose), Tr(**M**) takes the trace of a matrix **M**, and **I** is a 2 × 2 unit matrix. Here, **R** is the reflection coefficient matrix for reflection from vacuum to medium 1 or 2, respectively, and can be expressed as

$$\mathbf{R}_j = \begin{bmatrix} r_j^{ss} & r_j^{sp} \\ r_j^{ps} & r_j^{pp} \end{bmatrix}, \quad j = 1, 2 \tag{10.67}$$

where the first and second superscripts denote the polarization of the incident and reflected waves, respectively. Note that for isotropic medium, the polarization for the incident and reflected waves is the same so that only co-polarization coefficients $r^{pp}$ and $r^{ss}$ are nonzero, while the cross-polarization terms $r^{ps}$ and $r^{sp}$ are zero. These coefficients can be obtained by using a modified 4 × 4 transfer matrix method for an anisotropic medium or multilayer [104, 105]. The matrix **D** is given by

$$\mathbf{D} = (\mathbf{I} - \mathbf{R}_1 \mathbf{R}_2 \mathrm{e}^{2\mathrm{i}k_0 d})^{-1} \tag{10.68}$$

where –1 signifies matrix inverse.

Biehs et al. [102] calculated heat flux between two Au gratings or SiC gratings and demonstrated a large modulation if the gratings are rotated relative to each other. Large modulation by rotating hBN films and graphene-covered hBN films has also been theoretically predicted [103, 104]. The above formulation can be extended to biaxial anisotropic materials such as black phosphorus [94–96] as long as the reflection coefficient matrix can be obtained using anisotropic optics.

### *10.5.4 Green's Functions for Multilayer Structures*

Green's functions are necessary if the temperatures are nonuniform or if the local fields and LDOS or streamlines are desired. Consider a multilayer structure shown in Fig. 10.31a, consisting of perfectly parallel and smooth interfaces. Each layer is made of an isotropic, homogeneous nonmagnetic material ($\mu = 1$) with a relative permittivity $\varepsilon_l$, $l = 1, 2, \ldots N$. Region 1 extends to negative infinity and Region $N$ extends to positive infinity. The lateral extensions in the $x$- and $y$-direction are both assumed to be $\pm\infty$. For layers from Region 2 to Region $(N - 1)$, the thickness is $d_l = z_l - z_{l-1}$, similar to Fig. 9.12. However, here, each medium can emit radiation according to the fluctuation–dissipation theorem and each medium may absorb radiation. It is assumed that the temperature in each region is uniform and at local equilibrium. This assumption can be relaxed or can be satisfied by dividing the region

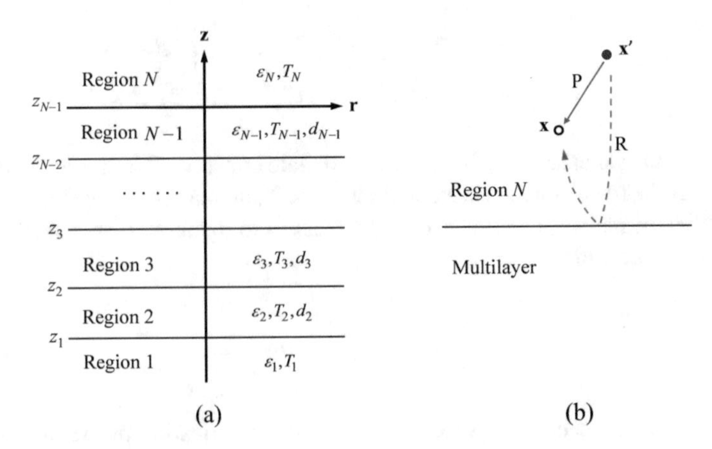


**Fig. 10.31** **a** Schematic of near-field radiation in a multilayered structure expressed in cylindrical coordinates. **b** Schematic of response at **x** from a source at **x**′ with a propagation path (solid) and a reflection path (dashed)

with a temperature gradient to sufficiently small slices. Region 1 or $N$ maybe vacuum or any material that is thick enough to be treated semi-infinite. Some middle layer(s) can be assumed vacuum to study radiative transfer between two-layered structures or even three-body problems [55]. In some calculations, one may set $T_l = 0$ for $l > j$ so that only $z < z_j$ portion can emit radiation. One can evaluate the absorption by each layer when $l > j$ from the media below $z_j$. This can be done if the $z$-component of the Poynting vector due to thermal fluctuation can be evaluated at $z_{l-1}$ and $z_l$ using $S_z(z_{l-1}) - S_z(z_l)$. The dyadic Green's function in different regions can be derived in the following [106–108].

Consider two points $\mathbf{x} = \mathbf{r} + \mathbf{z}$ and $\mathbf{x}' = \mathbf{r}' + \mathbf{z}'$, located in Region $N$, the semi-infinite region of uniform temperature. The response at location $\mathbf{x}$ to a source at location $\mathbf{x}'$ ($\mathbf{x} \neq \mathbf{x}'$) may come from direct propagation, as if in a homogeneous medium everywhere, or after multiple reflections (scattering) by the multilayer. The two paths are illustrated in Fig. 10.31b.

For the direct propagation, the Green's function can be written the same as for a homogeneous infinite medium. The term due to direct propagation or primary wave is also applicable when the two points are located in the same region in any layer. Therefore, we will use a subscript $l = 1, 2, \ldots N$. Using cylindrical coordinates, we have

$$\overline{\overline{\mathbf{G}}}_{\mathrm{P},l}(\mathbf{x}, \mathbf{x}', \omega) = \begin{cases} \dfrac{\mathrm{i}}{8\pi^2} \iint \dfrac{\mathrm{d}k_x \mathrm{d}k_y}{\zeta_l} \left[\hat{\mathbf{e}}(\zeta_l)\hat{\mathbf{e}}(\zeta_l) + \hat{\mathbf{h}}(\zeta_l)\hat{\mathbf{h}}(\zeta_l)\right] \mathrm{e}^{\mathrm{i}\boldsymbol{\beta}\cdot(\mathbf{r}-\mathbf{r}')} \mathrm{e}^{\mathrm{i}\zeta_l(z-z')}, & \text{if } z > z' \\ \dfrac{\mathrm{i}}{8\pi^2} \iint \dfrac{\mathrm{d}k_x \mathrm{d}k_y}{\zeta_l} \left[\hat{\mathbf{e}}(-\zeta_l)\hat{\mathbf{e}}(-\zeta_l) + \hat{\mathbf{h}}(-\zeta_l)\hat{\mathbf{h}}(-\zeta_l)\right] \mathrm{e}^{\mathrm{i}\boldsymbol{\beta}\cdot(\mathbf{r}-\mathbf{r}')} \mathrm{e}^{-\mathrm{i}\zeta_l(z-z')}, & \text{if } z < z' \end{cases} \tag{10.69}$$

where $\boldsymbol{\beta} = k_x\hat{\mathbf{x}} + k_y\hat{\mathbf{y}}$ and $\zeta_l = k_{lz}$ are used to represent $\mathbf{k}_l$ in the cylindrical coordinates. Note that

$$\hat{\mathbf{e}}(\pm\zeta) = \frac{k_y}{\beta}\hat{\mathbf{x}} - \frac{k_x}{\beta}\hat{\mathbf{y}} \tag{10.70a}$$

is the unit vector for TE waves, which is independent of the sign of $\zeta_l$, and

$$\hat{\mathbf{h}}(\pm\zeta) = \mp\frac{\zeta}{k\beta}(k_x\hat{\mathbf{x}} + k_y\hat{\mathbf{y}}) + \frac{\beta}{k}\hat{\mathbf{z}} \tag{10.70b}$$

Note that Eq. (10.69) is similar to Eq. (10.38) but with different notations. The term due to reflection or response by the multilayer can be expressed in Region $N$ as

$$\overline{\overline{\mathbf{G}}}_{\mathrm{R},N}(\mathbf{x},\mathbf{x}',\omega) = \frac{\mathrm{i}}{8\pi^2}\iint \frac{\mathrm{d}k_x\mathrm{d}k_y}{\zeta_l}\begin{bmatrix} R_N^s\hat{\mathbf{e}}(\zeta_l)\hat{\mathbf{e}}(-\zeta_l)\mathrm{e}^{\mathrm{i}\boldsymbol{\beta}\cdot(\mathbf{r}-\mathbf{r}')}\mathrm{e}^{\mathrm{i}\zeta_l(z+z')} \\ +R_N^p\hat{\mathbf{h}}(\zeta_l)\hat{\mathbf{h}}(-\zeta_l)\mathrm{e}^{\mathrm{i}\boldsymbol{\beta}\cdot(\mathbf{r}-\mathbf{r}')}\mathrm{e}^{\mathrm{i}\zeta_l(z+z')}\end{bmatrix},\quad l = N \tag{10.71}$$

where $R_N^s$ and $R_N^p$ are the reflection coefficients of the multilayer for incidence from Region $N$ onto the plane at $z = z_{N-1} = 0$. When $\mathbf{x}'$ is located in Region 1 (semi-infinite), transmission coefficients can be used to replace the reflection coefficients in Eq. (10.71) and other terms should also be replaced since upward waves exist. This is similar to Eq. (10.39) for a three-layer system (two semi-infinite media separated by a vacuum). Suppose $\mathbf{x}$ is still in Region $N$, while the source is from a different layer $l(l = 2, \ldots N-1)$, one needs to consider the upward and downward propagating waves: the upward wave will propagate to $\mathbf{x}$ while the downward wave will be reflected by the lower stratified media and travel upward to reach $\mathbf{x}$ [106]. The field amplitudes at different boundaries can be obtained using the transfer matrix method discussed in Sect. 9.2.2 and used to find the reflection and transmission coefficients. The magnetic Green's function can also be obtained using Maxwell's equations. With Green's functions, we can relate the contribution of the excitation at $\mathbf{x}'$ to the field at location $\mathbf{x}$. This is expressed as a function of the source location at $z'$ and an integration over the parallel wavevector space, as discussed before. According to Eqs. (10.29) and (10.37), to obtain the field or Poynting vector at location $\mathbf{x}$ caused by the source layer, an integration over $z'$ within the layer is needed in addition to integration over $k_x$ and $k_y$. If we wish to find the Poynting vector or LDOS at a particular location $\mathbf{x}$ inside any prescribed layer $l$ due to fluctuating sources in certain region $V'$, which could extend to the whole space or within certain regions only, we need to determine the Green's functions for the source $\mathbf{x}'$ at an arbitrary location and then integrate over $z'$ before integrating over all parallel wavevectors.

In general, when $\mathbf{x}$ is in any layer $l$ and $\mathbf{x}'$ is in any layer $n$, $(l, n = 1, 2, \ldots N)$, we can express the electric Green's function as

$$\overline{\overline{\mathbf{G}}}_{\mathrm{e}}(\mathbf{x},\mathbf{x}',\omega) = \frac{\mathrm{i}}{4\pi}\sum_{j=p,s}\int\frac{\beta\mathrm{d}\beta}{\zeta_n}F^j(\beta,z,z',\omega)\mathrm{e}^{\mathrm{i}\beta(r-r')} \tag{10.72}$$

where

$$F^s(\beta, z, z', \omega) = A^s_{ln} e^{i\zeta_l(z-z_l)-i\zeta_n z'} \hat{\mathbf{e}}^+_l \hat{\mathbf{e}}^+_n + B^s_{ln} e^{-i\zeta_l(z-z_l)-i\zeta_n z'} \hat{\mathbf{e}}^-_l \hat{\mathbf{e}}^+_n$$
$$+ C^s_{ln} e^{i\zeta_l(z-z_l)+i\zeta_n z'} \hat{\mathbf{e}}^+_l \hat{\mathbf{e}}^-_n + D^s_{ln} e^{-i\zeta_l(z-z_l)+i\zeta_n z'} \hat{\mathbf{e}}^-_l \hat{\mathbf{e}}^-_n \quad (10.73)$$

For $p$-polarization, change the superscript from "$s$" to "$p$" and replace "$\hat{\mathbf{e}}$" with "$\hat{\mathbf{h}}$". Here, we have used $\hat{\mathbf{e}}^\pm_l \equiv \hat{\mathbf{e}}(\pm\zeta_l)$ and $\hat{\mathbf{h}}^\pm_l \equiv \hat{\mathbf{h}}(\pm\zeta_l)$ for simplicity. For $n = N$, set $z_n = z_{n-1}$ . The coefficients $A$'s and $B$'s are the amplitudes of the forward and backward waves at layer $l$ due to an upward emitting source located at point $\mathbf{x}'$ ; the coefficients $C$'s and $D$'s are the amplitudes of the forward and backward waves at layer $l$ due to a downward emitting source located at point $\mathbf{x}'$. If the emission is from the lower half space in Region 1 ($n = 1$), then all $C$'s and $D$'s become zero. If the emission is from the upper half space in Region $N$ ($n = N$), then all $A$'s and $B$'s become zero. Suitable matrix formulations or recursive methods can be used to find the coefficients for a given source located anywhere [109–111]. It should be noted that Eq. (10.72) only includes the scattered field. If $l = n$, the electric Green's function should be obtained by adding the contribution of the primary wave Eqs. (10.69)–(10.72).

There are different ways to obtain these coefficients, which are the wave amplitudes at the boundaries due to a point source excitation. Note that the coefficients $A$'s and $B$'s (for an upward emitting source) can be evaluated separately from the coefficients $C$'s and $D$'s (for a downward emitting source). Taking $[A_l, B_l]$, $l = 1, 2, \ldots N$, for example, the transfer matrix method (TMM) explained in Sect. 9.2.2 is a convenient method that can relate the field amplitudes $[A_l, B_l]$ to $[A_{l+1}, B_{l+1}]$. If all the matrices are multiplied, it can relate $[A_1, B_1]$ to $[A_N, B_N]$. When $\beta$ is very large, such as in the case when a surface wave is excited, $k''_{lz} = \mathrm{Im}(\zeta_l)$ tends to be very large. The propagating matrix in the TMM contains both $\exp(ik_{lz}d_l)$ and $\exp(-ik_{lz}d_l)$ terms; the latter will exponentially grow with $k''_{lz}d$ and may approach infinity, resulting in numerical instability in such situations. This problem can be circumvented using a recursive method [109–111]. Francoeur et al. [55] gave a detailed formalism using the scattering matrix ($S$-matrix) method for near-field radiative transfer in stratified media. The $S$-matrix relates the amplitudes of the outgoing (scattered) waves to those of the incident waves. In terms of the coefficients $A_l$ and $B_l$, the $S$-matrix relates $[A_l, B_{l-1}]$ to $[A_{l-1}, B_l]$. Ultimately, it relates $[A_N, B_1]$ to $[A_1, B_N]$ for the whole multilayer. The scattering matrix method is an implicit formulation. Once the source excitation is prescribed, a recursive approach can be used to determine the matrix components [55]. An advantage of using the recursive method or $S$-matrix method is that only $\exp(ik_{lz}d_l)$ terms are involved in the formulation while the $\exp(-ik_{lz}d_l)$ terms that might cause numerical instability are eliminated. The calculation of near-field radiative transfer between different layers is very important for detailed modeling of microscale thermophotovoltaic devices since we need to determine where the photon is absorbed to generate photocurrent [17, 45, 84, 109, 112–115].

The magnetic Green's function can be defined in the following:

$$\mathbf{H}(\mathbf{x}, \omega) = \int\limits_{V'} \overline{\overline{\mathbf{G}}}_{\mathrm{m}}(\mathbf{x}, \mathbf{x}', \omega) \cdot \mathbf{j}(\mathbf{x}', \omega) \mathrm{d}\mathbf{x}' \tag{10.74}$$

Comparing Eq. (10.74) with Eq. (10.29), it can be shown that $\overline{\overline{\mathbf{G}}}_{\mathrm{m}}(\mathbf{x}, \mathbf{x}', \omega) = \nabla_{\mathbf{x}} \times \overline{\overline{\mathbf{G}}}_{\mathrm{e}}(\mathbf{x}, \mathbf{x}', \omega)$, where $\overline{\overline{\mathbf{G}}}_{\mathrm{e}}(\mathbf{x}, \mathbf{x}', \omega)$ is given in Eq. (10.72). Recalling the Poynting vector defined in Eq. (10.33) and the $z$-component Poynting vector in Eq. (10.42), we can write the spectral heat flux

$$q''_w(z) = \frac{k_0^2}{\pi^2} \int\limits_{-\infty}^{\infty} \varepsilon''_l(\omega)\Theta(\omega, T_l)\mathrm{Re}\left[\mathrm{i} \int\limits_0^{\infty} \left(g_{\mathrm{e},x\alpha} g^*_{\mathrm{m},y\alpha} - g_{\mathrm{e},y\alpha} g^*_{\mathrm{m},x\alpha}\right)\beta d\beta\right] \mathrm{d}z' \tag{10.75}$$

Here, $l$ indicates the particular layer where $z'$ is located, and $g_{\alpha\beta}$ ($\alpha, \beta = x, y,$ or $z$) denotes a tensor component of $\bar{\bar{\mathbf{g}}}$ defined according to Eq. (10.38) for the electric and magnetic Green's functions, respectively. It should be noted that in Eq. (10.75),

$$g_{\mathrm{e},x\alpha} g^*_{\mathrm{m},y\alpha} = g_{\mathrm{e},xx} g^*_{\mathrm{m},yx} + g_{\mathrm{e},xy} g^*_{\mathrm{m},yy} + g_{\mathrm{e},xz} g^*_{\mathrm{m},yz} \tag{10.76a}$$

and

$$g_{\mathrm{e},y\alpha} g^*_{\mathrm{m},x\alpha} = g_{\mathrm{e},yx} g^*_{\mathrm{m},xx} + g_{\mathrm{e},yy} g^*_{\mathrm{m},xy} + g_{\mathrm{e},yz} g^*_{\mathrm{m},xz} \tag{10.76b}$$

Zheng and Xuan [86] extended the formulation to stratified media with magnetic materials in which $\mu_l$ are complex functions of frequency. In addition to $\overline{\overline{G}}_{\mathrm{ee}}$ and $\overline{\overline{G}}_{\mathrm{me}}$, which are the electric and magnetic Green's functions due to electric dipoles, one must consider the electric Green's function $\overline{\overline{G}}_{\mathrm{em}}$ and the magnetic Green's function $\overline{\overline{G}}_{\mathrm{mm}}$ due to magnetic dipoles. For two stratified media separated by a vacuum gap, as long as each side is at local equilibrium, the net radiative heat flux can be evaluated based on Eq. (10.59), which is consistent with Eq. (10.65) based on the scattering theory. The additional terms due to magnetic dipoles only affect the Fresnel coefficients. Nevertheless, magnetic metamaterials can excite surface waves for $s$-polarizations that can cause a large heat flux in certain wavelength region [85–87]. It should be noted that the dyadic Green's functions for anisotropic media or multilayer structures have also been formulated though the equations are quite complicated [116–118]. In the case of a uniaxial medium whose optic axis is parallel to the $z$-axis, a slight modification can be made for the $p$-polarization Fresnel coefficients as discussed previously.

## 10.6 Nanostructures and Numerical Methods

Volokitin and Persson [48] obtained an analytical expression for heat transfer between a sphere and a planar half medium (plate) that reduces to the case of dipole-plate solutions when the sphere radius is much smaller than the closest gap distance ($d$) between the sphere and the plate [24, 82]. In the proximity limit, the heat conductance is proportional to $d^{-3}$. Here, the conductance is defined as the heat transfer rate per unit temperature difference between the two objects. For two particles that can be modeled as dipoles (when the radius is much smaller than the center-to-center distance $d$), the thermal conductance between two dipoles varies with $d^{-6}$ as $d$ becomes very small [53]. Narayanaswamy and Chen [54] derived an analytical solution for near-field radiative heat transfer between two spheres using the vector spherical wave (VSW) expansion method, which was extended to cylindrical geometrics and nanorods [119]. In principle, the method is applicable to spheres of any size at arbitrary separation distances. In terms of two spheres, the VSW expansion method begins by expanding the field in terms of the VSWs of each sphere. Then the VSWs of one sphere are re-expanded in terms of the VSWs of the other in order to satisfy the boundary conditions. Recurrence relations for VSWs can be used to reduce the computational demands for calculation of the translation coefficients of each spherical wave function. The convergence criteria have been developed for large sphere and small sphere approximations in terms of the number of VSWs needed in the series. It should be noted that when the sphere is very large and at extremely small distances (i.e., in the proximity limit), the thermal conductance is proportional to $1/d$ due to the curvature effect. It has also been pointed out that for dipole approximations to be valid in the case for two spheres of the same radius $a$, typically, the center-to-center distance $d$ should be at least $3a$ [54, 121, 122]. This criterion is very important for analyzing thermal radiation in dipolar chains and arrays [123] as will be discussed later. When $d > 4a$, that is, the gap in between is greater than the diameter of the sphere, the dipole approximation can be safely applied [124].

Numerical methods are very important for the calculation of near-field thermal radiation between complex structures, including periodic structures [18, 58, 59, 91, 108, 125–134]. This section provides an overview of the commonly used numerical methods, mainly including the scattering theory for periodic gratings, finite-difference time-domain (FDTD), boundary element method (BEM) using fluctuating surface current, and thermal-discrete dipole approximation (T-DDA). An emphasis is placed on the periodic structures with some unique observations.

### *10.6.1 The Scattering Theory for Periodic Structures*

The scattering theory has received significant attention in calculating radiative transfer and the Casimir–Lifshitz force between objects at different temperatures (while each object is at thermal equilibrium) [128, 135]. In principle, if the scattering matrices are known for all objects with arbitrary shapes the radiative heat transfer between the objects can be calculated if their temperatures are known [136, 137]. A convenient way to obtain the scattering matrix for a periodic grating is to use RCWA as discussed in Sect. 9.3.3. The net radiative heat flux between the two grating structures shown in Fig. 10.32 can be expressed as follows.

$$q''_{\text{net}} = \frac{1}{8\pi^3}\int_0^\infty [\Theta(\omega, T_1) - \Theta(\omega, T_2)]\mathrm{d}\omega \int_{-\pi/\Lambda}^{\pi/\Lambda}\int_{-\infty}^{\infty} \xi\left(\omega, k_x, k_y\right)\mathrm{d}k_x\mathrm{d}k_y \tag{10.77}$$

where $\Lambda$ is the grating period in the $x$-direction. It is assumed that the gratings are extended to infinity in the $y$-direction. Thus, the integration over $k_x$ is carried out in the first Brillouin zone. Although rotating gratings or gratings in both $x$- and $y$-directions can be modeled in general, the RCWA calculation for 2D gratings and for conical diffraction may be very time consuming with sufficient orders in order to obtain convergence. According to the scattering theory, the energy transmission coefficient $\xi(\omega, k_x, k_y)$, which includes all polarization states, can be expressed as a trace of the matrix product [138, 139]

$$\xi\left(\omega, k_x, k_y\right) = \mathrm{Tr}\left(\mathbf{D}\mathbf{W}_1\mathbf{D}^\dagger\mathbf{W}_2\right) \tag{10.78}$$

where

$$\mathbf{D} = (\mathbf{I} - \mathbf{S}_1\mathbf{S}_2)^{-1} \tag{10.79}$$

$$\mathbf{W}_1 = \Sigma_{-1}^{\text{pw}} - \mathbf{S}_1\Sigma_{-1}^{\text{pw}}\mathbf{S}_1^\dagger + \mathbf{S}_1\Sigma_{-1}^{\text{ew}} - \Sigma_{-1}^{\text{ew}}\mathbf{S}_1^\dagger \tag{10.80}$$

$$\mathbf{W}_2 = \Sigma_1^{\text{pw}} - \mathbf{S}_2^\dagger\Sigma_1^{\text{pw}}\mathbf{S}_2 + \mathbf{S}_2^\dagger\Sigma_1^{\text{ew}} - \Sigma_1^{\text{ew}}\mathbf{S}_2 \tag{10.81}$$

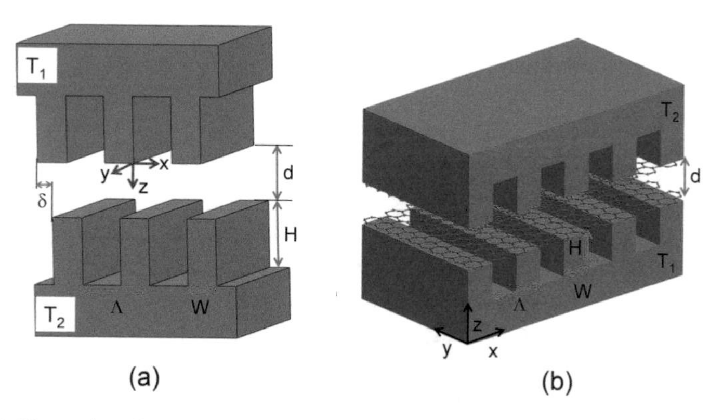


**Fig. 10.32** Illustration of near-field radiative transfer between two grating structures. **a** Two parallel gratings with a lateral displacement; **b** Graphene-covered gratings

Here, † is the Hermitian adjoint (conjugate transpose), $\mathbf{S}_1 = \mathbf{R}_1$, and $\mathbf{S}_2 = \mathrm{e}^{ik_{0z}d}\mathbf{R}_2\mathrm{e}^{ik_{0z}d}$, where $\mathbf{R}_1$ or $\mathbf{R}_2$ is the reflection (scattering) matrix for each grating that can be obtained from RCWA. Due to diffraction by the grating, the $z$-component of wavevector $k_{0z}$ depends on the diffraction order. Therefore, proper mode counting is needed to calculate $\mathbf{S}_2$ [138]. All the matrices are square matrices ($M \times M$). When the highest diffraction order used in the computation is $q$, $M = 2(2q + 1)$. This is because $\mathbf{R}$ can be written similar to Eq. (10.67) consisting of co-polarization and cross-polarization components, except that each corner is a $(2q+1)\times(2q+1)$ matrix that is to be determined using RCWA based on the diffraction order. The symbol Σ in Eqs. (10.80) and (10.81) denotes a matrix operator (rather than summation). Superscripts "pw" and "ew" identify propagating and evanescent modes, while subscript "1" and "–1" indicates forward and backward, respectively. The operator definition can be found in Ref. [139]. Sufficient orders must be used in the RCWA calculations to ensure numerical convergence and accuracy. Guérout et al. [138] demonstrated an enhanced near-field heat transfer between nanocorrugated Au surfaces compared with two flat surfaces when the separation is at the minimum distances between the two gratings. On the other hand, the proximity approximation (PA) considers the heat fluxes between flat surfaces separated by different gap distances and hence always predicts a reduction rather than enhancement. For silica gratings, the PA model was found to be reasonable for aligned gratings but failed to explain the effect of lateral displacement [139].

Liu et al. [140] modeled doped-Si gratings using the scattering theory and compared with the EMT approach and PA model. For the EMT model, the grating is assumed to be an effective uniaxial medium whose optic axis is parallel to the grooves. In the proximity approximation, as shown in Fig. 10.32a, the heat flux between two parallel gratings with a lateral displacement may be expressed as

$$q''_{\mathrm{PA}} = \begin{cases} \frac{\phi\Lambda-\delta}{\Lambda}q''_d + \frac{2\delta}{\Lambda}q''_{d+H} + \frac{\Lambda-\phi\Lambda-\delta}{\Lambda}q''_{d+2H}, & \delta \le \phi\Lambda \\ 2\phi q''_{d+H} + (1-2\phi)q''_{d+2H}, & \phi\Lambda < \delta \le 0.5\Lambda \end{cases} \tag{10.82}$$

for gratings with a volume fraction $\phi \le 0.5$.

$$q''_{\mathrm{PA}} = \begin{cases} \frac{\phi\Lambda-\delta}{\Lambda}q''_d + \frac{2\delta}{\Lambda}q''_{d+H} + \frac{\Lambda-\phi\Lambda-\delta}{\Lambda}q''_{d+2H}, & \delta \le (1-\phi)\Lambda \\ (2\phi-1)q''_d + 2(1-\phi)q''_{d+H}, & (1-\phi)\Lambda < \delta \le 0.5\Lambda \end{cases} \tag{10.83}$$

when $\phi > 0.5$ [140]. In Eqs. (10.82) and (10.83), $q''_d$, $q''_{d+H}$, and $q''_{d+2H}$ are the radiative heat flux for plane–plane configurations at gap distance of $d$, $d + H$, and $d + 2H$, respectively. Due to symmetry, the range of lateral displacement that needs to be considered is from 0 to Λ/2 only.

The near-field radiative heat flux between gratings based on the scattering theory is compared with the predictions from EMT and PA, as shown in Fig. 10.33, and compared with that of bulk silicon. The Si considered here is $n$-type with a doping concentration of $10^{20}$ cm$^{-3}$ and in the modeling, $T_1 = 310$ K and $T_2 = 290$ K. The predicted heat fluxes by EMT and PA are independent of the period for aligned gratings as shown in Fig. 10.33a, where $d = 400$ nm, $\phi = 0.2$, and $H = 1$ μm. When the period is sufficiently large, the PA prediction agrees with the scattering theory (marked as exact) well. As the period decreases, the heat flux predicted by the scattering theory approaches and finally coincides with that by EMT. With decreasing the period and width of the gratings, it becomes difficult for waves to sense the small features and, therefore, homogenizing the grating as an effective medium becomes more reasonable. Corrugating bulk doped silicon helps to enhance the radiative heat flux for small periods.

With respect to the gap distance, as shown in Fig. 10.33b using the same parameters as in Fig. 10.33a, except now the period $\Lambda$ is fixed to 0.2 μm while $d$ is the variable. The agreement between the scattering theory and EMT is very good when $d >$ 0.6 μm. However, EMT has limitations at small gap distances because significant photon tunneling contributions arise from modes having large parallel wavevector ($\beta$) values [92]. Interestingly, with decreasing $d$, the exact solution approaches the PA prediction. Hence, the near-field radiative heat transfer tends to be localized at small gap spacing since the field will be highly confined due to the dominant contribution of high-$k$ modes. For the chosen values of $\phi$, $H$, and $\Lambda$, doped-Si gratings can outperform their bulk counterparts in terms of heat transfer enhancement for $d > 15$ nm. Another interesting observation is that the near-field heat flux for the gratings follows a power law close to $d^{-1}$ for submicron gap spacing rather than the well-known $d^{-2}$ as is the case for both bulk and homogenized media. Liu et al. [140] also showed that the lateral displacement has little effect on the heat transfer for small periods when EMT applies; however, $\delta$ can affect the heat flux in the regime when PA is applicable such as with a large $\Lambda/d$. The applicable regimes also apply to Casimir force calculations [140].

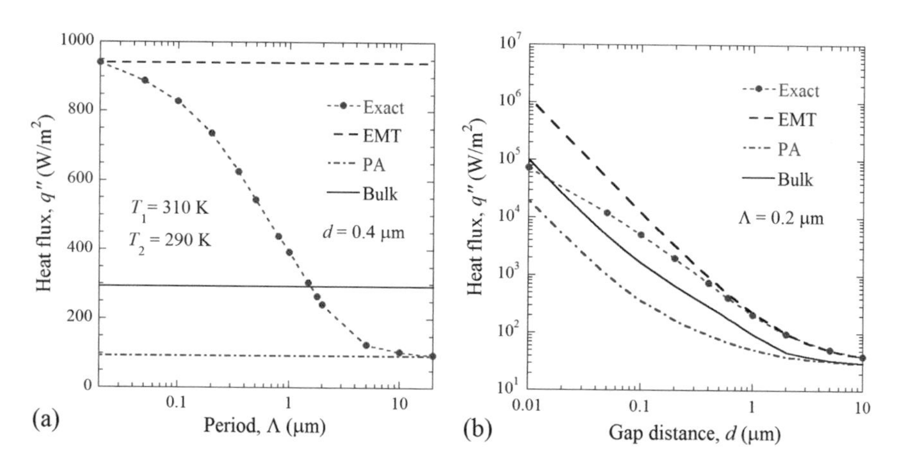


**Fig. 10.33** Comparison of radiative heat flux calculated from the scattering theory (exact) with EMT, PA, and bulk for $T_1 = 310$ K and $T_2 = 290$ K [140]: **a** varying period; **b** varying gap distance. The other parameters are set as $\delta = 0$, $\phi = 0.2$, and $H = 1$ μm

Yang and Wang [141] calculated the near-field radiation for aligned metal gratings using scattering theory with RCWA. Their calculations demonstrated a strong resonance effect due to the excitation of magnetic polaritons (MPs) within the two metal strips at the closest distance ($d \leq 100\,\text{nm}$), rather than due to the cavity resonance (guided modes) in the grooves as discussed in Ref. [138]. The resonance frequency and dispersion can be well accounted for by the MP modes as explained in this work [141].

Liu and Zhang [142] studied near-field radiative heat transfer between graphene-covered silica gratings as shown in Fig. 10.32b. The results are plotted in Fig. 10.34a for varying graphene chemical potentials at a gap distance of 100 nm, in comparison with those for bulk silica and graphene-covered bulk silica. Due to the coupling and hybridization of graphene plasmons with surface phonon polaritons, a maximum heat flux (about four times that for bulk silica) can be achieved with graphene-covered gratings at $\mu = 0.28$ eV. The coupling between graphene and bulk silica results in some enhancement but not as strong as in graphene-covered silica gratings. It should be noted that the heat flux between plain silica gratings is generally lower than that between bulk silica as indicated in Fig. 10.34b. With the parameters chosen and for $d = 100$ nm, PA is a good approximation for plain grating. Therefore, the radiative heat transport or photon tunneling between plain silica gratings tends to be a local phenomenon due to the short lateral propagating length of surface phonon modes [139]. Nevertheless, placing a graphene sheet on top of the grating will make the heat transfer much more effective. The heat flux for graphene-covered corrugated silica is much greater than that without graphene. Besides, the heat flux does not increase linearly with $\phi$ according to PA since PA cannot capture the hybridized plasmonic effects and this underestimates the heat flux for filling ratios less than 1. Detailed

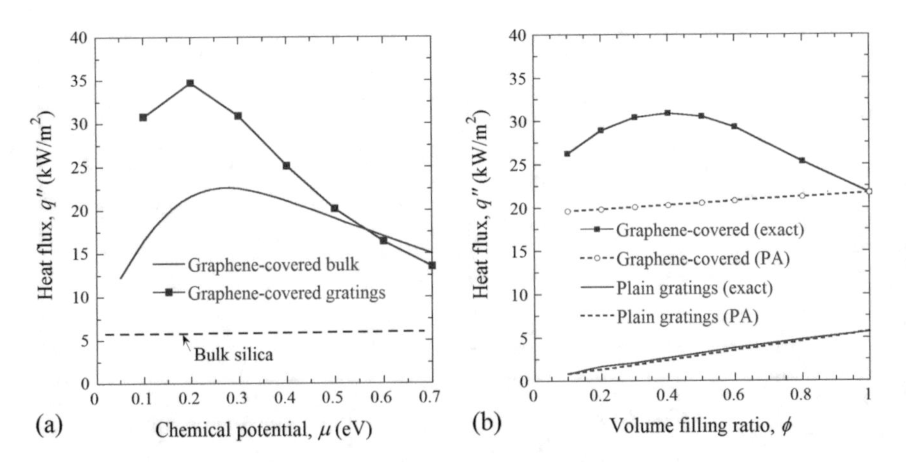


**Fig. 10.34** Comparison of near-field radiative heat flux for graphene-covered silica grating with other structures at $T_1 = 310$ K and $T_2 = 290$ K with a gap distance $d = 100$ nm, period $\Lambda = 500$ nm, and height $H = 500$ nm [142]: **a** Effect of chemical potential with $\phi = 0.4$; **b** Effect of filling ratio with $\mu = 0.3$ eV for covered and uncovered silica gratings

discussions and the contour plots of the transmission coefficient can be found from Ref. [142].

The near-field radiative transfer between two aligned graphene ribbon arrays separated by a vacuum gap of $d$, as shown in Fig. 10.35a has also been investigated [60]. As discussed in Example 10.7 and shown in Fig. 10.28c, near-field radiative heat transfer between two single-layer graphene sheets can be greatly enhanced due to coupled graphene plasmons. By patterning graphene into ribbon arrays (laterally extended to infinity), hyperbolic modes can be supported. Using scattering theory with RCWA or the EMT, the near-field radiative heat transfer is calculated for the emitter $T_1 = 310\,\text{K}$ and receiver $T_2 = 290\,\text{K}$, as shown in Fig. 10.35b. The results are divided between two graphene sheets for the same total area. Note that the graphene area in the ribbon array is only one-fifth that of a graphene sheet with $\phi = W/\Lambda = 0.2$. As expected, EMT failed to predict the heat flux when $d < 50$ nm or so. When $d = 20$ nm, according to the (exact) scattering theory, the heat transfer enhancement is near 14 times with only 20% of the material.

The energy transmission coefficient $\xi$ is plotted in the $k$-space for graphene sheets and for graphene arrays at a fixed frequency $\omega = 5 \times 10^{13}$ rad/s, shown in Fig. 10.36. Coupled graphene surface plasmon (GSP) results in two bright annuluses that can be predicted by the coupled GSPs. The white dashed circle with a radius $\beta = 4.6k_0$ indicates the isofrequency curve of GSPs for a single sheet of graphene in vacuum, which is split into two regions due to the coupling of the two graphene sheets as discussed previously. The regimes with large $\xi$ for graphene ribbon arrays become hyperbolic, where the white dashed lines are the calculated hyperbolic dispersion based on the graphene conductivity and EMT, which is a good approximation even at $d = 50$ nm. By patterning graphene into ribbons, the closed circular isofrequency dispersion is opened to become hyperbolic, leading to broadband singularities in the DOS. The hyperbolic graphene plasmons can couple strongly with extremely high-$k$

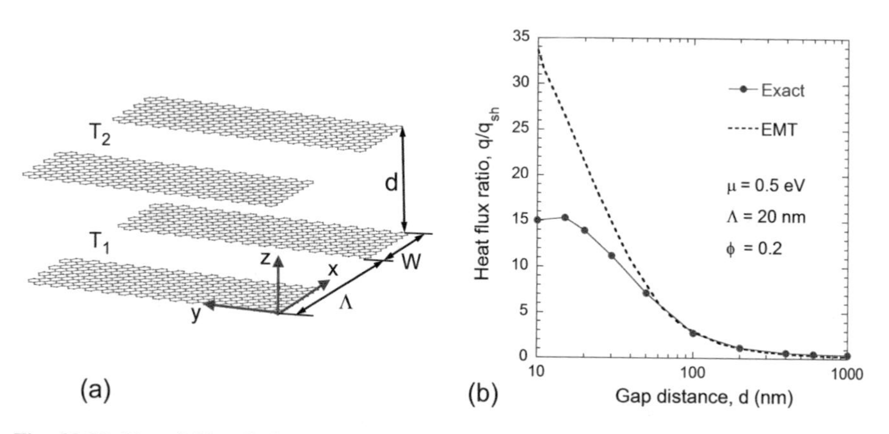


**Fig. 10.35** Near-field radiative transfer between graphene ribbons [60]: **a** schematic of two graphene ribbon arrays separated by a distance $d$. **b** Ratio of the heat transfer rate for two ribbon arrays to that for two graphene sheets near room temperature calculated from the scattering theory (exact) and EMT using parameters given in the figure

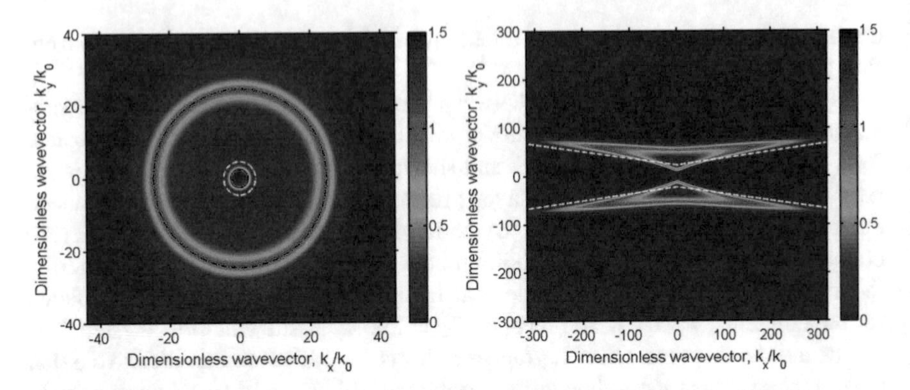


**Fig. 10.36** Contour plots of the energy transmission coefficient contours at $d = 50$ nm and $\omega = 5 \times 10^{13}$ rad/s [60]: **a** graphene sheets in which the dashed circle is for a single-layer graphene plasmon; **b** graphene ribbon arrays showing hyperbolic dispersions with dashed curves

modes, enabling very efficient radiative energy transport, especially at $d < 100$ nm [60].

Liu et al. [143] also investigated nanopatterned black phosphorene arrays in different directions and demonstrated quasi-elliptic coupled SPPs and quasi-hyperbolic coupled SPPs. The scattering theory has also been used to model the performance of a near-field thermophotovoltaic device using tungsten nanogratings [144] and a thermoradiative power generator with a heat sink made of ZrC nanogratings [145].

## 10.6.2 Finite-Difference Time-Domain (FDTD) Method

A finite difference refers to a numerical differentiation that can be used to approximate a derivative. Thus, the finite difference method is a numerical method for solving differential equations by converting them into a set of linear algebraic equations, using a spatial mesh (or grid) scheme. The values at all grid points are solved at an initial time and advanced over small time steps in a leapfrog manner. Since the electromagnetic wave equations involve coupled electric and magnetic vectors that change rapidly with time, it was not so easy to discretize them until 1966 when Kane S. Yee developed a numerical algorithm with the cubic unit cell, known as the Yee space lattice [146]. The Yee space lattices elegantly fill the space in the Cartesian coordinates with an array of contours that satisfy the Ampere's law and Faraday's law. The field components can then be discretized and iteratively solved, while marching in time progressively. Further improvements were made since middle 1970s by many researchers in terms of numerical stability and convergence, treatment of boundary conditions and curved boundaries, convolution method for dispersive materials, near-field to far-field transformation, and various applied sources such as a plane wave, a Gaussian beam, a short pulse, and a dipole emitter [146]. With the advances in computing capabilities and algorithm developments, FDTD has become a powerful numerical method for modeling electromagnetic wave propagation and scattering in broad frequency regions for both very large and very small geometric structures. Many commercial, open source, and homemade FDTD solution packages are available [147].

For a nonmagnetic, isotropic, and homogeneous medium, the 3D Maxwell curl equations can be written in terms of the field components as follows,

$$\frac{\partial H_x}{\partial t} = \frac{1}{\mu}\left(\frac{\partial E_y}{\partial z} - \frac{\partial E_z}{\partial y}\right) \tag{10.84a}$$

$$\frac{\partial H_y}{\partial t} = \frac{1}{\mu}\left(\frac{\partial E_z}{\partial x} - \frac{\partial E_x}{\partial z}\right) \tag{10.84b}$$

$$\frac{\partial H_z}{\partial t} = \frac{1}{\mu}\left(\frac{\partial E_x}{\partial y} - \frac{\partial E_y}{\partial x}\right) \tag{10.84c}$$

and

$$\frac{\partial D_x}{\partial t} = \frac{\partial H_z}{\partial y} - \frac{\partial H_y}{\partial z} - \sigma E_x - J_x^{\mathrm{o}} \tag{10.85a}$$

$$\frac{\partial D_y}{\partial t} = \frac{\partial H_x}{\partial z} - \frac{\partial H_z}{\partial x} - \sigma E_y - J_y^{\mathrm{o}} \tag{10.85b}$$

$$\frac{\partial D_z}{\partial t} = \frac{\partial H_y}{\partial x} - \frac{\partial H_x}{\partial y} - \sigma E_z - J_z^{\mathrm{o}} \tag{10.85c}$$

where $\mathbf{J}^{\mathrm{o}}$ is an internal current source that may be applied to the cell and $\mathbf{J} = \sigma\mathbf{E}$ is the induced current. For a dispersive medium, a Fourier transform needs to be performed to obtain the convolution of the dielectric function. In the time domain,

$$\mathbf{D}(t) = \varepsilon_0\varepsilon_\infty\mathbf{E}(t) + \varepsilon_0\int_0^t \mathbf{E}(t-\tau)\chi(\tau)\mathrm{d}\tau \tag{10.86}$$

where $\varepsilon_\infty$ is the high-frequency dielectric constant such as in the Drude–Lorentz model, and $\chi(t)$ is the time-dependent electric susceptibility, which can be obtained by an inverse Fourier transform of $\chi(\omega) = \varepsilon(\omega) - \varepsilon_\infty$ [146]. Alternatively, the Fourier transform is performed using recursive accumulators such as the piecewise linear recursive convolution [148]. Equations (10.84) to (10.86) can be discretized following the Yee algorithm or some alternative ones and solved iteratively for prescribed geometrics, materials, boundary conditions, and illumination sources. The frequency response and spectral information can be analyzed during the post-processing using a fast Fourier transform (FFT) [147].

Fu and Hsu [149] used FDTD to study the far-field scattering or bidirectional reflectance distribution function (BRDF) from 1D random rough dielectric and metal surfaces. Xuan et al. [150] developed a FDTD code and investigated the BRDF of 2D rough surfaces in the visible and near infrared. In Chap. 9, examples were given on using FDTD to obtain the far-field radiative properties of nanostructures and periodic structures, also see [59, 151]. Most other studies use commercially available software packages.

For near-field radiative heat transfer calculations, the internal source needs to be treated based on the randomly fluctuating current described previously. Lu et al. [125] used the Langevin approach where the fluctuating current term is implemented through the equation of motion to directly calculate thermal emission from photonic crystals. The time-dependent current source can be obtained by a Fourier transform of Eq. (10.34). An alternative fluctuating current source can be developed based on the white noise spectrum where the correlation is instantaneous [58]. This method has been extended to calculate near-field radiative heat transfer between nanostructures [127, 152]. The source current can also be generated using the FFT [153] and this method has been employed to calculate near-field radiative transfer using FDTD [131]. One can treat the fluctuation term as electric dipoles, along each of the three axes, that are distributed in the source region. The corresponding electric or magnetic dyadic Green's function is the ratio of the Fourier transform of the time-dependent electric field or magnetic field to the transform of the current induced by the dipole [132]. A large number of dipoles need to be generated to guarantee sufficient accuracy in the solution. Didari and Mengüç [132, 154] used the Ricker wavelet, which is the second derivative of the Gaussian pulse, as the source term to eliminate the DC term since the time-average of the electric field for a typical pulse is nonzero.

Liu and Shen [91] used FDTD with the Wiener chaos expansion, which is based on Hermite polynomial chaos expansion and can separate the deterministic effects from the randomness. Wen [126] used a finite-difference frequency-domain method to calculate near-field radiation resulting from all of the different eigenmodes of randomly fluctuating thermal currents. Liu and Shen [91] used FDTD to obtain solutions of the Wiener chaos expansion for near-field radiative transfer between split-ring metamaterial and metal-wire arrays. The Wiener chao expansion method combined with FDTD appears to be an efficient technique for spatially incoherent sourses than the brute-force FDTD [155].

As an example, Liu and Zhang [61] used the scattering theory with RCWA and FDTD for calculating near-field radiative heat transfer between metasurfaces made of $n$-type doped-Si ($10^{20}$ cm$^{-3}$) 1D strip arrays and 2D periodic patches, as shown in Fig. 10.37. The numerical results for $d = 100$ nm and $\Lambda = 100$ nm are compared to EMT as well as that for homogeneous films ($\phi = 1$ with the same thickness). The brute-force FDTD was used to calculate the near-field radiation for the 2D metasurfaces by implanting the Ricker wavelet sources using a commercial software package [61]. As shown in Fig. 10.38a, patterning thin films into metasurfaces could increase the radiative heat flux due to the hyperbolic dispersion and coupled SPPs, especially for the 1D structure when the filling ratio is small $\phi < 0.2$. The 2D pattern gives slight heat transfer enhancement for $f < 0.36$, but results in some deterioration for $0.36 < f < 1$. While calculations from EMT can capture the general trend, it significantly overpredicts the heat flux. As shown in Fig. 10.38b, the radiative heat flux of 2D metasurface increases monotonically with the thickness, while the heat

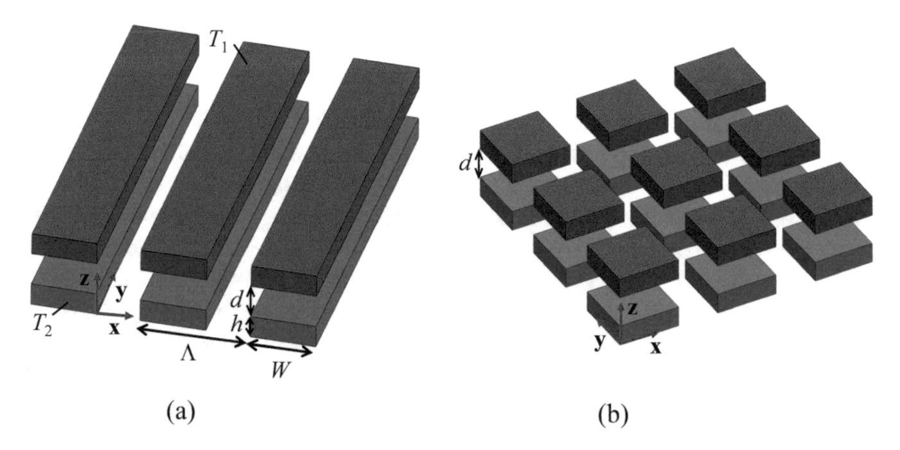


**Fig. 10.37** Illustration of near-field radiative transfer between metasurfaces [61]: **a** 1D thin strips; **b** periodic 2D patches. The filling ratio can be calculated by $\phi = W/\Lambda$ for 1D patterning and $\phi = W^2/\Lambda^2$ for 2D patterning

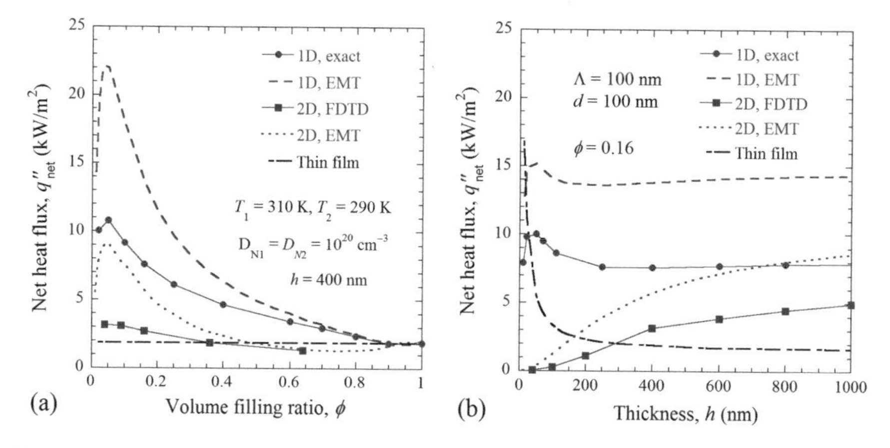


**Fig. 10.38** Calculated heat fluxes for $T_1 = 310$ K and $T_2 = 290$ K with $\Lambda = 100$ nm and $d = 100$ nm [61]: **a** varying the filling ratio while fixing $h = 400$ nm; **b** varying the thickness $h$ with $\phi = 0.16$ for both the 1D and 2D metamaterials. The exact calculation for 1D patterning is based on the scattering theory

flux of 1D metasurface is not so sensitive to the thickness. Interestingly, the radiative heat flux for thin films monotonically increases with decreasing thickness until the thickness is about 3 nm due to the coupling of SPPs in thin film structures. For Si at this doping level, reducing the film thickness to below 50 nm and even to a few nanometers is another way for to enhance radiative energy transfer. Note that due to the large parallel component of the wavevector $\beta$, the penetration depth for nanoscale heat transfer becomes extremely small [156].

### 10.6.3 Boundary Element Method (BEM)

The brute-force FDTD can take a long time for modeling individual nanostructures. For the Wiener chaos expansion, it is challenging to find the proper current modes of the thermal source. Based on the scattering theory, if the field correlator for the isolated object is obtained, then the radiative energy exchange can be calculated [58, 128]. Analytical solutions exist only for limited geometric shapes such as plates, spheres, or even cylinders [128, 129]. McCauley et al. [157] used the boundary element method (BEM) or method of moments to discretize the scattering matrix in the cylindrical wave multipole basis. They have calculated cylinder-plate and tip-plate near-field heat transfer in addition to sphere-plate configurations. Rodriguez et al. [130] developed a fluctuating-surface-current (FSC) formulation for arbitrary geometries using the surface-integral-equation (SIE) formulation within the framework of BEM. In general, the BEM involves a much smaller number of unknowns since the scattering fields are expressed using the fictitious surface current that can be determined by the boundary conditions. The volume integral can be replaced by a surface integral of the fictitious surface currents, which can be discretized using a set of basis functions [158]. These surface currents are arbitrary vector fields that do not need to satisfy any wave equations. The ensemble-averaged spectral heat flux can be written in terms of a trace of the Green's functions.

$$q''_{\omega}(\omega, T_1, T_2) = \frac{1}{2\pi}[\Theta(\omega, T_1) - \Theta(\omega, T_2)]\mathrm{Tr}\left[\frac{\overline{\overline{\mathbf{G}}}_1 + \overline{\overline{\mathbf{G}}}_1^{\dagger}}{2}\overline{\overline{\mathbf{W}}}_{21}^{\dagger}\frac{\overline{\overline{\mathbf{G}}}_2 + \overline{\overline{\mathbf{G}}}_2^{\dagger}}{2}\overline{\overline{\mathbf{W}}}_{21}\right] \tag{10.87}$$

where the dyadic $\overline{\overline{\mathbf{W}}}_{21}$ relates the field incident on object 2 to the equivalent currents at the surface of object 1 and can be calculated using an inverse of the SIE matrix. Rodriguez et al. [130] used this method to calculate radiative transfer between various nanostructured objects, such as finite cylinders, sphere-ring, disks, interlocked rings, cones, and cone-plate configurations. Recently, Nguyen et al. [159] calculated the near-field radiative transfer between a heated AFM tip and a substrate. The shape of the tip was modeled with different geometries such as semi-spheres, sharp cones, and rounded cones. Since the plate was modeled as a large cylinder, the influence of the cylinder size on the numerical results was also examined [159]. Note that the same formulation has been modified and applied for calculation of momentum exchange or Casimir forces [160]. A complete derivation in terms of surface integrals of the tangential components of the Green's functions can be found from Ref. [65]. For inhomogeneous medium, Polimeridis et al. [161] extended the FSC formulation and developed a fluctuating-volume-current (FVC) formulation that incorporates a volume-integral-equation (VIE) matrix. Since both the FSC and FVC use a source term, the scattering matrix does not require the separation of the incoming and outgoing waves like in the scattering theory discussed in Sec. 10.6.1. In the end, the radiative transfer (power) and momentum transfer (Casimir force) are expressed in simple trace formulas involving SIE or VIE and the current correlation matrices. While there are more unknowns using the FVC than the FSC formulation, the FVC method can handle more complex property-structure combinations as those with inhomogeneous permittivities [161].

### *10.6.4 Multiple Dipole Approaches*

Earlier study of heat transfer in multiple dipole systems was motivated by plasmonic nanoparticle arrays and waveguides [162, 163]. Ben-Abdallah and coworkers [122, 164] developed a kinetic theory (KT) approach for a metallic dipole chain (see Fig. 10.39a), based on the dispersion relation of the collective plasmonic modes. Within the framework of the KT approach, the diffusive regime refers to the situation when the propagation length of the polaritonic wave is much shorter than the chain length. Similar to the BTE equation for phonons, the group velocity and mean free path are important parameters entering into the formulation and can be determined from the dispersion relations using $\Lambda_P(\omega) = \left|\tau_P v_{g,P}\right|$ and $v_{g,P} = (\mathrm{d}\omega/\mathrm{d}k)_P$, respectively, where $P$ denotes a particular polarization. The relaxation time or mean free path can be obtained from either the complex wavevector or the complex frequency [165–167]. When the propagation length is much longer than the chain length, plasmon or phonon polaritons can propagate ballistically through the nanoparticle arrays. For 2D and 3D nanoparticle arrays, the dispersion relation may be numerically computed [168]. Recently, Tervo et al. [169] theoretically calculated a photonic thermal conductivity up to 1 W/m K in semiconductor nanowires by infrared plasmonic resonators. For nanoparticle powders with SiC and $SiO_2$, it is still an open question whether phonon polariton effects through nanoscale radiation could play a significant role in the thermal conductivity of the powders.

Baffou et al. [170] extended the discrete-dipole approximation (DDA) with the Green's function and calculated the near-field heating effect in thermoplasmonics without considering the spectral nature of thermal emission. Since 2011, Ben-Abdallah et al. [171, 172] and Messina et al. [173] developed a multiple-dipole

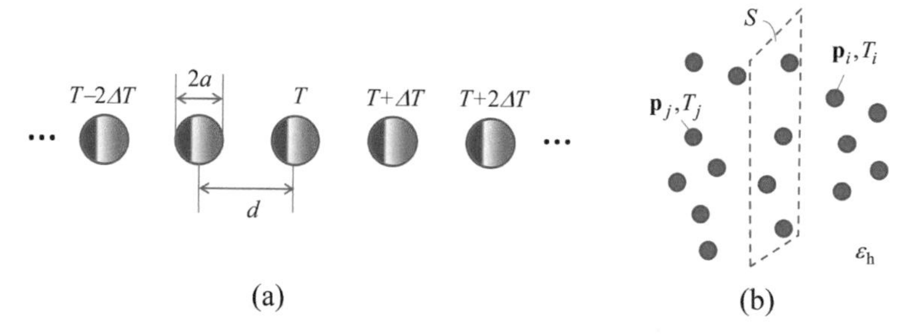


**Fig. 10.39** Schematic of multiple dipole systems: **a** A chain of nanoparticles; **b** $N$ nanoparticles (dipoles) embedded in a dielectric matrix with a relative permittivity $\varepsilon_h$

radiative transfer formalism that incorporates fluctuational electrodynamics to model the emission and absorption by individual particles; each is assumed to be an electric dipole, immersed in a host medium which could also emit thermal radiation. Edalatpour and Francoeur [56] formulated a thermal discrete-dipole approximation (T-DDA) for calculation of radiative transfer between arbitrary 3D objects.

The multiple dipole approaches are summarized below, largely following the derivations given in Tervo et al. [123] with modifications to be consistent with the previous given formulation in the preceding sections of this chapter. Consider a many-dipole system embedded in a host dielectric with a relative permittivity $\varepsilon_h$, shown in Fig. 10.39b. For the dipole approximation to be valid, particles should be sufficiently smaller than the thermal wavelength [16]. Since the thermal wavelength is in the infrared, this is usually valid for particles with diameters below 300 nm. In addition, as mentioned previously, the minimum separation distance between particles needs to be greater than a diameter or at least a radius [54, 121–123].

For an electric dipole located at $\mathbf{x}_j$ with a dipole moment $\mathbf{p}_j$, the induced electric field at a location $\mathbf{x}_i$ can be expressed as follows.

$$\mathbf{E}_{ij}(\mathbf{x}_i, \omega) = \frac{k_0^2}{\varepsilon_0}\overline{\overline{\mathbf{G}_{\mathbf{ij}}^{\mathbf{0}}}}(\mathbf{x}_i, \mathbf{x}_j, \omega) \cdot \mathbf{p}_j(\omega) \tag{10.88}$$

where $k_0 = \omega/c_0$ is the wavevector in vacuum, and the Green's function can be expressed as

$$\overline{\overline{\mathbf{G}_{\mathbf{ij}}^{\mathbf{0}}}}(\mathbf{x}_i, \mathbf{x}_j, \omega) = \left[\overline{\overline{\mathbf{I}}} + \frac{1}{k^2}\nabla \otimes \nabla\right]\frac{\mathrm{e}^{\mathrm{i}kR}}{4\pi R} \tag{10.89}$$

or

$$\overline{\overline{\mathbf{G}_{\mathbf{ij}}^{\mathbf{0}}}}(\mathbf{x}_i, \mathbf{x}_j, \omega) = \frac{\mathrm{e}^{\mathrm{i}kR}}{4\pi R}\left[\left(1 + \frac{\mathrm{i}kR - 1}{k^2R^2}\right)\overline{\overline{\mathbf{I}}} + \frac{3 - 3\mathrm{i}kR - k^2R^2}{k^2R^2}\frac{\mathbf{R} \otimes \mathbf{R}}{R^2}\right] \tag{10.90}$$

Here, $k = k_0\sqrt{\varepsilon_h}$, $\mathbf{R} = \mathbf{x}_i - \mathbf{x}_j$ and $R = \left|\mathbf{x}_i - \mathbf{x}_j\right|$. Note that $R$ and $\mathbf{R}$ depend on both locations which are omitted to make the expression more concise. From the Green's function, we can see that the term $R^{-1}$ dominates the far field. However, as the distance decreases, the near-field terms that vary with $R^{-2}$ and $R^{-3}$ become more and more important. The magnetic field can be written as

$$\mathbf{H}_{ij}(\mathbf{x}_i, \omega) = -\mathrm{i}\omega\nabla \times \overline{\overline{\mathbf{G}_{\mathbf{ij}}^{\mathbf{0}}}}(\mathbf{x}_i, \mathbf{x}_j, \omega) \cdot \mathbf{p}_j(\omega) \tag{10.91}$$

where

$$\nabla \times \overline{\overline{\mathbf{G}_{\mathbf{ij}}^{\mathbf{0}}}}(\mathbf{x}_i, \mathbf{x}_j, \omega) = \frac{\mathrm{e}^{\mathrm{i}kR}}{4\pi R}\left(\mathrm{i} - \frac{1}{kR}\right)\frac{k\mathbf{R} \times \overline{\overline{\mathbf{I}}}}{R} \tag{10.92}$$

In general, $\mathbf{x}_i$ can be any arbitrary location with or without a dipole as long as $R \neq 0$. To study dipole–dipole interactions, we can take $\mathbf{x}_i$ as a dipole location and the total field at $\mathbf{x}_i$ can be expressed in the following [173],

$$\mathbf{E}_i(\mathbf{x}_i, \omega) = \mathbf{E}^{(\mathrm{inc})}(\mathbf{x}_i, \omega) + \sum_{j(\neq i)} \mathbf{E}_{ij}(\mathbf{x}_i, \omega) \tag{10.93}$$

Here, the first term on the right-hand side denotes the incident field without scattering, which is often assumed to be zero to neglect the radiation from the host medium; the summation takes into consideration dipolar scattering of all particles and the field emitted by all other dipoles. In the following, we set $\mathbf{E}^{(\mathrm{inc})} = 0$ for simplicity. According to Eq. (10.88), $\mathbf{E}_{ij}$ is related to the dipole moment $\mathbf{p}_j$ by the dyadic Green's function. The dipole moment includes two parts:

$$\mathbf{p}_j = \mathbf{p}_j^{(\mathrm{fl})} + \mathbf{p}_j^{(\mathrm{ind})} \tag{10.94}$$

where the first term arises from the thermal fluctuation and the second term is due to scattering by all dipoles, viz.

$$\mathbf{p}_j^{(\mathrm{ind})} = \varepsilon_0 \alpha_j \mathbf{E}_j = \varepsilon_0 \alpha_j \sum_{n(\neq j)} \mathbf{E}_{jn} \tag{10.95}$$

Here, $\alpha_j$ is the dressed polarizability that has taken into account the radiative correction [173], which can be expressed as

$$\alpha_j = \frac{\alpha_j^{\mathrm{CM}}}{1 - \mathrm{i}\alpha_j^{\mathrm{CM}} k^3/(6\pi\varepsilon_{\mathrm{h}})} \tag{10.96}$$

where the Clausius–Mossotti polarizability [16] for a spherical particle is

$$\alpha_j^{\mathrm{CM}} = 3\varepsilon_{\mathrm{h}} V_j \frac{\varepsilon_j - \varepsilon_h}{\varepsilon_j + 2\varepsilon_h} \tag{10.97}$$

Plugging Eq. (10.95) into Eq. (10.94) and using Eq. (10.88), we get

$$\mathbf{p}_j = \mathbf{p}_j^{(\mathrm{fl})} + \alpha_j k_0^2 \sum_{n(\neq j)} \overline{\overline{\mathbf{G}_{\mathbf{jn}}^{\mathbf{0}}}}(\mathbf{p}_j, \mathbf{x}_n, \omega)\mathbf{p}_n \tag{10.98}$$

This forms a set of $N$ linear equations that relates the $\{\mathbf{p}_j\}$ to $\left\{\mathbf{p}_j^{(\mathrm{fl})}\right\}$ or vice versa, each is an $N \times 1$ matrix. One can put Eq. (10.98) in the following matrix form:

$$\left\{\mathbf{p}_j^{(\mathrm{fl})}\right\}_{N\times 1} = \{A\}_{N\times N}\{\mathbf{p}_j\}_{N\times 1} \tag{10.99}$$

Note that since the dipole moment is a vector, there are actually $3N \times 3N$ matrix elements to deal with. An equivalent way is to write Eq. (10.98) as follows [123].

$$\mathbf{p}_j = \mathbf{p}_j^{(\mathrm{fl},0)} + \alpha_j^{(0)} k_0^2 \sum_n \overline{\overline{\mathbf{G}_{\mathbf{jn}}^{\mathbf{0}}}}(\mathbf{x}_j, \mathbf{x}_n, \omega)\mathbf{p}_n \tag{10.100}$$

where $\alpha_j^{(0)} = (\varepsilon_j - \varepsilon_\mathrm{h})V_j$ is the bare polarizability and

$$\overline{\overline{\mathbf{G}_{\mathbf{jj}}^{\mathbf{0}}}} = \left(\frac{ik}{6\pi} - \frac{1}{3V_j k^2}\right)\overline{\overline{\mathbf{I}}} \tag{10.101}$$

The introduction of Green's function for $n = j$ takes into account the dressed polarizability. However, the correlation of the fluctuating dipole moments has slightly different expressions. In Eq. (10.100), the correlation function in terms of the dipole moments is expressed in terms of the bare polarizability [123]. The power dissipated (or absorbed) by particle $j$ from particle $n$ can be expressed as

$$\dot{Q}_{nj} = \frac{\omega}{2}\mathrm{Im}\langle \mathbf{p}_j \cdot \mathbf{E}_{jn}^* \rangle \tag{10.102}$$

which can be converted to time domain by

$$\dot{Q}_j(\mathbf{x}_j, t) = \frac{1}{2}\int_0^\infty \int_0^\infty \omega \mathrm{Im}\langle \mathbf{p}_j \cdot \mathbf{E}_j^* \rangle e^{-i(\omega-\omega')t} d\omega d\omega' \tag{10.103}$$

where the dipole moments can be solved in terms of the fluctuating dipole moments, according to Eq. (10.100). Using the fluctuation–dissipation theorem, $\left\langle \mathbf{p}_j \cdot \mathbf{E}_j^* \right\rangle$ can be expressed in terms of $\left\langle p_{i,\alpha}^{(\mathrm{fl},0)}(\omega) p_{j,\beta}^{(\mathrm{fl},0)*}(\omega') \right\rangle$, which can be written as

$$\left\langle p_{j,\alpha}^{(\mathrm{fl},0)}(\omega) p_{n,\beta}^{(\mathrm{fl},0)*}(\omega') \right\rangle = \frac{4}{\omega\pi}\varepsilon_0 \mathrm{Im}(\alpha_j^0)\Theta(\omega, T_j)\delta_{jn}\delta_{\alpha\beta}\delta(\omega - \omega') \tag{10.104}$$

Equation (10.104) can be derived from Eq. (10.34) using $\mathbf{j}_j = -i\omega \mathbf{p}_j V_j$ and $\mathrm{Im}(\alpha_j^0) = V_j \mathrm{Im}(\varepsilon_j)$. The actual calculations are typically performed using matrices. In such case, the complex conjugate should be replaced by the conjugate transpose. After some manipulation, it can be shown that the total energy absorbed by particle $j$ is

$$\dot{Q}_j(\mathbf{x}_j) = \int_0^\infty \mathrm{d}\omega \sum_{n \neq j} \Theta(\omega, T_n)\tau_{nj}(\omega) \tag{10.105}$$

where $\tau_{nj}$ can be expressed as a function of the trace of the coefficient matrix [123, 173]. The heat transfer between two particles can be expressed as

$$\dot{Q}_{nj} = \int_0^\infty \left[\Theta(\omega, T_n) - \Theta(\omega, T_j)\right]\tau_{nj}(\omega)\mathrm{d}\omega \tag{10.106}$$

If the temperature difference is small (near equilibrium), the thermal conductance between two particles can be expressed as

$$G_{nj}(T) = \int_0^\infty \tau_{nj}(\omega)\frac{\partial\Theta}{\partial T}\mathrm{d}\omega \tag{10.107}$$

Considering either a 1D chain or a 2D particle array, the thermal conductivity in the direction $\hat{\mathbf{s}}$ along the temperature gradient at a given location may be calculated from [123]

$$\kappa(T) = \frac{1}{S}\sum_j\sum_n G_{nj}(T)(\mathbf{r}_j - \mathbf{r}_n)\cdot\hat{\mathbf{s}} \tag{10.108}$$

where $S$ is an effective cross-sectional area. Tervo et al. [123] used this to calculate the radiative thermal conductivity of ordered and disordered nanoparticle arrays. Dong et al. [174] incorporated magnetic dipoles into the formulation. Near-field radiation transfer has also been investigated recently for spherical core-shell particles [175] and for graphene wrapped nanoparticles [176].

Note that in the case of only two dipoles in vacuum, it can be shown that in the near-field regime

$$\tau_{12} = \frac{1}{4\pi^3}\left(\frac{3}{d^6} + \frac{k_0^2}{d^4} + \frac{k_0^4}{d^2}\right)\mathrm{Im}(\alpha_1^{\mathrm{CM}})\mathrm{Im}(\alpha_2^{\mathrm{CM}}) \tag{10.109}$$

where $d$ is the center-to-center distance [48, 53]. In the near-field limit, $\tau_{12}$ is proportional to $d^{-6}$ but independent of the frequency; while in far-field limit, $\tau_{12} \propto \omega^2 d^{-2}$. Note that since we are considering subwavelength particles, the far-field result is different from that based on geometric optics. Hence, the energy transfer between two nanoparticles (or any nanostructures) in the far field may exceed that between two ideal blackbodies with the same geometry since the concept of blackbody is not applicable for subwavelength structures [16, 19].

Edalatpour and Francoeur [56] developed a thermal discrete-dipole approximation (T-DDA) for calculation of near-field radiation between closely spaced nanostructures, as shown in Fig. 10.41a, that can be inhomogeneous and each object does not need to be at a thermal equilibrium. The method is based on the well-known DDA method for light scattering calculations [177] to implant fluctuational dipoles

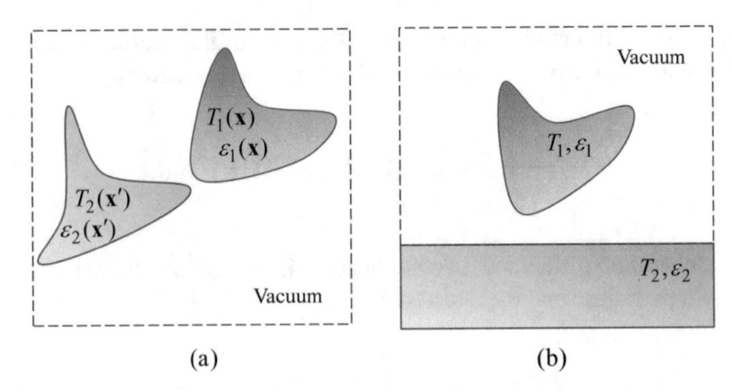


**Fig. 10.40** Schematic of the T-DDA method used for simulating nanostructure near-field radiative transfer: **a** Two arbitrary structures in vacuum; **b** An arbitrary structure above a semi-infinite medium

as described previously in this section. The method is based on discretizing objects of arbitrary shape into cubic subvolumes, each of which can be approximated as a point dipole whose polarizability can be treated using the Clausius–Mossotti model described in Eq. (10.97). The volume integration is then discretized using summations of the dipolar points; and this method can be used to study many-body problems [178]. If the interaction equation with $N$-dipoles is solved and their individual radiative exchanges can be calculated according to Eq. (10.106), then, one can set the receiver temperature to zero (no emission) and add all of the power received by the dipoles within the receiver volume to calculate the absorbed power. Since thermal emission is spatially uncorrelated, one can choose any volume and use the same way to calculate the power absorbed by any subset of dipoles.

The T-DDA method has also been extended to calculating a small object near a semi-infinite planar medium shown in Fig. 10.40b by using Sommerfeld's integrals to evaluate the electric dipole radiation above an infinite plane [179]. As the number of dipoles increases the simulation accuracy improves, while the required computational resources and time grow quickly. Even though only electric dipoles are involved, the DDA-based method takes into account of multipole effects if sufficient subvolumes (dipoles) are used to discretize the structures. The T-DDA method has also been extended to study anisotropic media including magneto-optical materials [180]. All computational methods are time-consuming when dealing with complex problems and each has its own advantages and disadvantages.

## 10.7 Measurements and Applications

Before closing this chapter and the whole book, we would like to review the progress in the development of experiments for measuring near-field radiative heat transfer, as well as the prospective application of nanoscale thermal radiation.

### 10.7.1 Measurements of Near-Field Thermal Radiation

When the first edition of this book was published in 2007, experimental investigations of near-field radiative energy transfer were very limited [108]. Tien and coworkers [181] and Hargreaves [182] were the first to measure the energy flux of two parallel plates at cryogenic temperatures. Domoto et al. [181] used 8.5-cm-diameter copper disks as the emitter and receiver that were coaxially mounted and varied the gap spacing $d$ from 10 μm to 2 mm. Near-field effects were observed when $d < 200$ μm since the temperature is around 10 K. At $d = 10$ μm, the heat flux was enhanced by more than three times compared to the far-field value, though it is still much smaller than that between blackbodies. Hargreaves [182] measured the near-field heat transfer between two chromium-coated plates separated by a vacuum gap from approximately 6 to 1.5 μm. At $d \approx 1.5$ μm and near room temperature, the heat transfer rate was five times greater than that for the far field, approaching 40% of the value between two blackbodies.

More than 20 years later, in an attempt to probe near-field radiation at submicron vacuum gaps, Xu et al. [37] used a scanning tunneling microscope (STM) with a heated indium needle that has a flat tip surface of 100-μm diameter. A thin-film thermocouple with a flat junction area of $160 \times 160$ μm$^2$ was evaporated onto a glass substrate to probe the heat flow. Müller-Hirsch et al. [38] investigated the heat transfer between a tungsten tip and a planar thermocouple on a substrate by cooling the substrate. While the proximity effect was observed at distances down to 10 nm or so, it was difficult to quantitatively determine the absolute heat flux between the tip and the substrate. There were also challenges in accurately measuring the temperatures of the tip and the substrate [37, 38].

Kittel et al. [183] used a scanning thermal microscope tip with a platinum wire inside a glass micropipette which is gold coated to form a thermocouple junction that could be calibrated to indicate the tip temperature. The gold-coated platinum tip with a radius of approximately 60 nm can be resistively heated. Near-field radiative heat transfer has been measured between the tip (which is maintained near 300 K) and a plate whose temperature is lowered to 100 K using liquid nitrogen via a cold finger. The plate is either made of gallium nitride or coated with a gold film for comparison. Near-field thermal radiative transfer was observed for gap spacings of $d = 100$ nm down to about 1 nm. The measured values agree with those predicted from fluctuational electrodynamics for 10 nm $< d <$ 100 nm. However, when $d <$ 10 nm, the measured heat transfer rate saturates and deviates from the predicted trend which continues to increase as $d$ further decreases. Nevertheless, this is a significant development in nanoscale thermal radiation measurements.

Tremendous progress has been made in recent years toward the experimental realization of near-field enhancement of thermal radiation [73–75, 184–209]. Generally speaking, the measurements can be categorized according to their configurations shown in Fig. 10.41 as plate-plate, tip-plate, sphere-plate, and microfabricated suspended structures. The key challenges are how to maintain parallelism and determine the gap distances as well as how to measure the temperatures and heat transfer rate

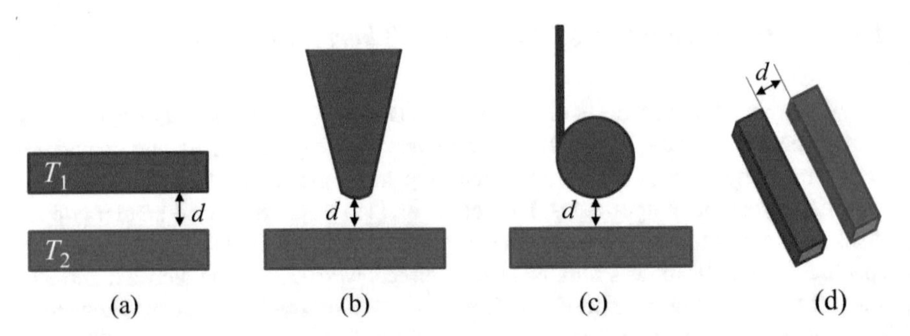


**Fig. 10.41** Various experimental configurations for measuring near-field radiative heat transfer: **a** plate-plate; **b** tip-plate, **c** sphere-plate, and **d** microfabricated suspended structures

accurately. Corrections for heat conduction and far-field thermal radiation are often required. Since all measurements have been performed in an evacuated chamber, convection and gas conduction are usually negligibly small. Basu [108] gave a comprehensive review of the studies done before 2015. A brief survey of the major publications and some examples are given below, along with updated literature since 2015.

Hu et al. [184] measured the near-field heat transfer between two parallel, optical-flat glass (fused quartz) disks of 1.27-cm diameter, separated by sparsely dispensed polystyrene spheres whose nominal diameters are 1 μm. The microspheres serve as spacers with a low thermal conductivity to reduce the conduction heat transfer. The emitter is heated by a heating pad with a temperature controller using a platinum resistance temperature sensor from about 50 to 100 °C. The temperature of the heat sink (receiver) is measured by a thermocouple to be around 24 °C. The heat transfer was measured by a heat flux sensor placed between the receiver and the heat sink. Their results demonstrated a near-field radiative heat transfer exceeding that between two blackbodies by more than 35%, and the measured heat fluxes agree with the predicted values for a vacuum gap of $d = 1.6\,\mu\text{m}$ [184]. More recently, Lang et al. [185] used monodisperse silica nanospheres to reduce the gap spacing down to 150 nm, which was verified with interferometric measurements, and achieved a heat flux that exceeds the blackbody limit by more than an order of magnitude between two 20-mm-diameter disks near room temperature.

Narayanaswamy et al. [186] developed a method for measuring nanoscale radiative heat transfer between a sphere and a plate down to a 100 nm gap spacing, using a biomaterial AFM cantilever, as shown in Fig. 10.42a. The method was further improved using a higher resolution piezoelectric controller that enabled measurements for a vacuum gap down to 30 nm [187]. Silicon dioxide microspheres with different sizes were used with a radius $a = 25$ and 50 μm, respectively. The plate substrate was made of $SiO_2$ or Si or a gold-coated surface. With the configuration of $SiO_2$–$SiO_2$ that can support coupled SPhPs, nearly three orders of magnitude enhancement of the heat transfer coefficient was demonstrated at a gap spacing $d =$

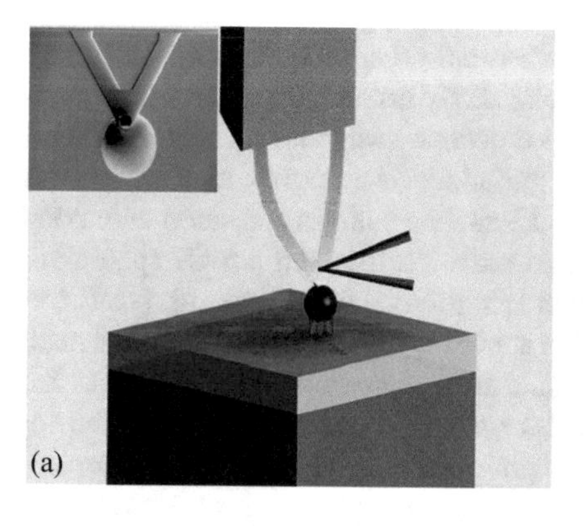


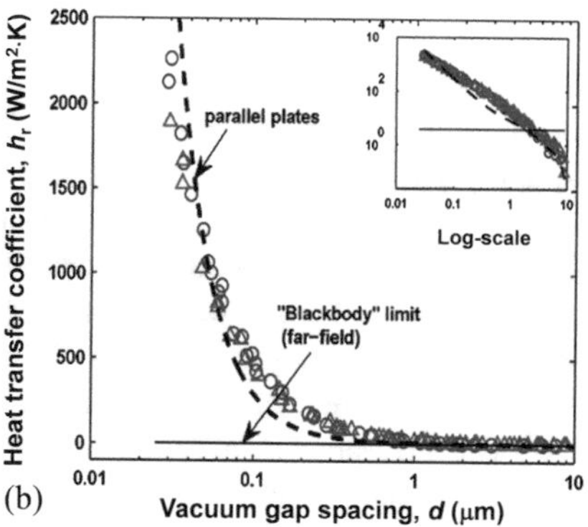


**Fig. 10.42** Experimental setup and measured results for sphere-plate configuration: **a** experimental setup; **b** measured radiative heat transfer coefficient, where the triangular marks are for radius $a =$ 25 μm and circles are for $a = 50$ μm microspheres. Inset is the plot in log–log scale. *Reprinted with permission from Shen et al.* [187]; *copyright (2009) American Chemical Society*

30 nm as shown in Fig. 10.42b. The principles of their measurement technique are briefly described here.

As shown in Fig. 10.42a, the biomaterial cantilever is made by coating a gold film onto a silicon nitride triangular beam which can measure the temperature at the tip, where a silica microsphere is epoxied, due to the different thermal expansion coefficients of Au and $SiN_x$. By measuring the deflection of the cantilever using the focused laser beam and a position sensitive detector (PSD), the temperature of the sphere can be determined. The laser irradiation also heats the tip (or sphere), resulting in bending of the cantilever. The tip temperature can be raised a few tens of degrees relative to the base. The effective thermal conductance of the cantilever was calibrated by measuring the bending with different input powers. As the gap between the sphere and the surface is reduced by moving the substrate using a piezoelectric motion controller, near-field radiation results in a cooling of the sphere that can be measured by the beam deflection. With the vertically mounted cantilever beam, the bending of the cantilever causes negligibly small vertical displacement of the sphere. It should be noted that, due to the small heating power, the base of the cantilever and the flat substrate are passively maintained at ambient temperature. Therefore, both the sphere (emitter) temperature and the near-field radiative transfer can be measured based on the additional bending of the cantilever [186, 187].

Given the relatively large sphere, the near-field radiative heat transfer between the sphere and the plate can be approximated by treating the sphere as slices of flat disks and calculated by integration of the plate-plate heat flux. This is called the proximity approximation or specifically the Derjaguin approximation [54]. As discussed previously, the near-field radiative heat transfer is proportional to $1/d^2$; while upon integration, the radiative heat transfer coefficient is proportional to $1/d$. In order to compare with the flat plate heat transfer coefficient ($h_r$), Shen et al. [187] normalized $h_r$ using an effective surface area of $A_{\rm eff} = \pi R_{\rm eff}^2$, where $R_{\rm eff} = \sqrt{2ad}$.

The resulting heat transfer coefficients between a $SiO_2$ sphere and a $SiO_2$ plate are plotted in Fig. 10.42b, against the far-field blackbody limit ($h_{r,BB} = 3.8\ W/m^2\ K$). It can be seen that $h_r$ increases dramatically as $d$ decreases and the value at $d = 30$ nm is 2230 $W/m^2$ K or about three orders of magnitude greater than the blackbody limit. Surface roughness for these measurements is less than 4 nm as measured with AFM [187]. Shi et al. [188] studied the near-field radiation between a $SiO_2$ sphere and a silicon substrate with various doping concentrations. Later, Shi et al. [189] also investigated the near-field radiation between a microsphere and a metamaterial made of metal (Ni) wire arrays embedded in anodic aluminum oxide (AAO) matrix. The AAO can be removed partially to expose a protruded nanowire array at the top layer of 400 nm height. They found that the protruded case gives the maximum heat conductance while the bare nanowire array gives negligibly small heat conductance, suggesting that metal nanowires can serve as a low-loss waveguide to couple the high-$k$ modes to the hyperbolic metamaterial.

A different method for measuring nanoscale radiative heat transfer between a sphere and a plate was developed by Rousseau et al. [190]. In this work, a biomaterial cantilever was also vertically mounted with a microsphere attached to its tip near the surface, similar to the work of Ref. [186]. The interference pattern of the laser beam reflected from the cantilever was measured with an optical fiber system and used to monitor the deflection of the cantilever. The flat substrate was heated by a few tens of degrees and mounted on a piezoelectric actuator. The temperature of the substrate and that of the end of the cantilever were measured using thermocouples. The microsphere temperature is deduced by analyzing the thermal resistance network and calibration of the thermal conductance based on far-field experiments. Note that the radiative thermal resistance even at $d = 50$ nm still dominates, so the temperature of the sphere is very close to that measured at the end of the cantilever. Silica spheres of radius 20 μm and 11 μm were used with a surface roughness of 40 nm and 150 nm, respectively. For the $a = 20$ nm sphere, measurements showed an enhancement factor that exceeds 400 in radiative heat transfer when $d$ is reduced from 2500 to 30 nm [190]. Later, van Zwol et al. [191, 192] used this setup to measure near-field radiation between a $SiO_2$ sphere and SiC, doped-Si, or graphene-covered substrate, as well as a $VO_2$ (a phase transition material) plate. Quantitative measurements of near-field thermal radiation in the sphere-plate configuration have provided convincing experimental confirmation of the fluctuation–dissipation theorem in predicting near-field energy transfer at gap distances down to 30 nm [186–192].

Even though the sphere-plate experiments have been successfully used in measuring near-field heat transfer at submicron gaps, such a geometrical configuration is not as ideal as the plate–plate geometry. As an example, the effective area $2\pi ad$ for a sphere of radius $a = 50$ μm and a gap spacing $d = 50$ nm is less than 16 $\mu m^2$, which is too small for applications such as near-field thermophotovoltaic devices or rectifiers. In the last 10 years, continuous efforts have been made to experimentally demonstrate the near-field enhancement of energy transfer for parallel-plate geometry with an area greater than a square millimeter [193–202]. Ottens et al. [193] used sapphire ($\alpha$-$Al_2O_3$) plates, each with a 50 × 50 $mm^2$ area and 5-mm thickness, to measure the near-field radiative transfer coefficient from 100 μm down to about 2 μm. Their measured heat transfer coefficients exceed that between blackbodies in the far-field at the smallest gap spacings near room temperature. Si-diode thermometers were used to measure the temperatures of the emitter and receiver. The sapphire surfaces are optically smooth with a specified flatness of 160 nm over the 50 mm lateral displacement. The four corners of the facing surfaces (each with 1 $mm^2$ area) were coated with a 200-nm-thickness copper film to form capacitor plates for gap spacing measurements. Kinematic mirror mounts with stepper motors were used to linearly move the emitter in the $z$-direction as well as to adjust the tip and tilt angular movements [193]. Kralik et al. [194] performed cryogenic near-field radiation measurements between two parallel tungsten-coated plates of 35-mm diameter with separation distances from 500 μm down to about 1 μm. The gap spacing was measured with a capacitance method. They used a plane-parallelism equalizer to maintain parallelism between the emitter and receiver at cryogenic temperatures. They observed heat flux enhancement of nearly four orders of magnitude over the corresponding far-field value (or two orders of magnitude greater than the blackbody limit) when the cold plate temperature is between 10 and 40 K and the hot plate is maintained at 5 K higher at $d \approx 1$ μm [194]. Ijiro and Yamada [195] measured near-field radiative transfer between two 25-mm-diameter $SiO_2$ glass plates at room temperature using an optic-fiber coupled spectrometer to determine the gap spacing at several locations. They used a piezoelectric motor that drives a kinematic mount with a linear stage for both translational motion and tip/tilt angular adjustment to keep the emitter and receiver surface parallel with a vacuum gap distance from 100 μm down to 1 μm. The near-field radiative heat flux is nearly twice that for the far-field case at room temperature. Furthermore, they also fabricated 5 μm × 5 μm × 5 μm microcavities on the glass plate. Interestingly, while the far-field heat flux is enhanced by 20% with the microcavities, the near-field heat flux is actually lower with the microcavities. The far-field enhancement can be explained by the enhanced emittance/absorptance due to guided modes or cavity resonance; while the near-field radiative transfer may be explained by the proximity limit [139, 142]. Nevertheless, when the microcavities are coated with a Au layer, the near-field radiative transfer can be significantly enhanced compared to that between flat Au surfaces [195]; this is consistent with the predictions [138].

In 2015–2016, several groups reported measurements of near-field radiation between parallel plates at distances below 500 nm with surface areas exceeding 1 $mm^2$ using spacers or micropillars between the emitter and receiver [196–199]. Ito et al. [196] fabricated micropillars in truncated square pyramid shape of thicknesses 500, 1000, and 2000 nm to measure the near-field radiative heat transfer between rectangular fused quartz (silica) plates of 19 mm by 8.6 mm. These spacers were etched on the silica surface with a lateral spacingof 1 mm between nearest pillars. A small force (about 1 N) was applied to the plate by compressing a spring with a very low spring constant. The emitter was heated to 5, 10, 15, and 20 K above the receiver temperature of 293 K. A flat heat flux sensor was used to measure the heat flux to the thermoelectrically cooled receiver. The measured heat transfer coefficients are higher than those predicted by theory, presumably due to the contribution of heat conduction through the pad, though quantitative analysis of the heat conduction contribution was not provided [196]. Lim et al. [197] used microfabrication to form a 13.4-mm-long and 0.59-mm-wide strip with a heater as the emitter and a 0.48-mm-wide receiver of the same length, made of doped silicon. The gap was formed by patterned metal spacers, located far away from the heating and receiving strips, and could be varied by adjusting the normal load. The effective surface area is approximately 6.4 $mm^2$. The separation distance, as measured by the capacitance between the emitter and the receiver, can be varied from about 900 nm down to 400 nm. A heat transfer coefficient of 2.9 times the far-field blackbody limit was observed at $d = 400$ nm [197]. Watjen et al. [198] fabricated $SiO_2$ micropillars to separate 10 mm × 10 mm doped-Si plates with various vacuum gaps down to $d = 200$ nm and obtained near-field heat flux more than an order of magnitude higher than the blackbody limit. The experimental setup and measurement results are outlined in the following.

As shown in Fig. 10.43a, the spring presses the stack of layers onto a copper base to form a nearly one-dimensional heat flow path. A heater is epoxied onto a copper plate above the sample using silver grease to ensure good thermal contact. A tiny hole drilled halfway through the side allows a thermocouple to be inserted on each copper plate to measure the hot-side and cold-side temperatures. The thermocouples measure the relative temperatures, while the absolute temperature is measured at the base (heat sink) as $T_0$ using a calibrated silicon diode thermistor. The heat transfer rate from the emitter to the receiver is measured by a heat flux meter (HFM). The sample is placed in the middle and consists of two doped-Si pieces with a sparse array of $SiO_2$ micropillars, as shown in Fig. 10.43b. The pillars with heights ranging from 200 to 800 nm were fabricated using ultraviolet photolithography on a Si wafer and then cut into 10 mm by 10 mm pieces. The patterned plate and a flat (unpatterned) plate of the same size were mated together and pressed by spring loading forces to maintain the necessary gap spacing. Heat conduction was reduced using an array of 1-μm-diameter $SiO_2$ pillars with a relatively large span, i.e., $S = 300$, 400, or 500 μm. It was estimated that more than half of the heat flux is due to radiation when $S$ exceeds about 300 μm [198].

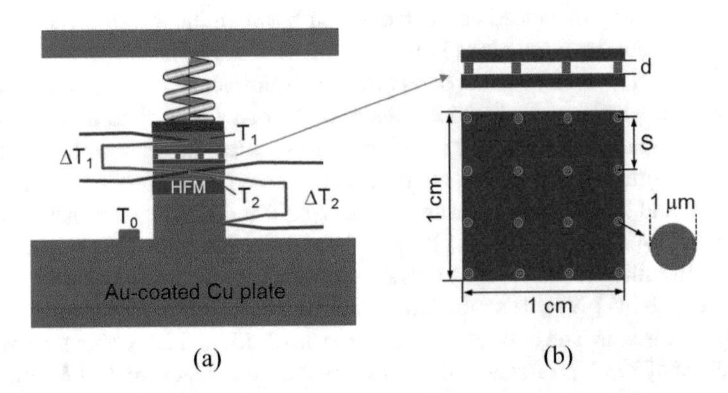


**Fig. 10.43** Illustration of the near-field radiative transfer measurement setup [198]: **a** The measurement stage with a heater, the specimen, temperature sensors, and a heat flux meter (HFM). **b** Schematic of two doped-Si plates separated by $SiO_2$ micropillars

A Fourier-transform infrared spectrometer (FTIR) was used to measure the reflectance of each sample to quantify the gap spacing prior to the heat transfer measurement. Different forces were applied for both the FTIR measurement and the heat transfer measurement, allowing the gap spacing to be adjusted. Contact thermal resistances were also considered in the data analysis [198]. The measured radiative heat transfer rate $q_{rad}$ and heat transfer coefficient $h_r$ are plotted in Figs. 10.44a, 10.45b, respectively. The heat transfer rate for three gap distances is plotted against the temperature difference between the emitter ($T_H$) and receiver ($T_L$). The solid symbols represent the measured results with uncertainty bounds indicated by the error bars. The dashed lines are from the fluctuational electrodynamic calculation with the shaded region indicating the range of uncertainty due to the uncertainty in the gap spacing determination. The large enhancement in nanoscale thermal radiation is attributed to the excitation of coupled surface plasmon polaritons (SPPs).

Figure 10.44b plots the radiative heat transfer coefficient $h_r$ for 14 measurements, as shown by the filled symbols with error bars. The calculated values with the same gap spacings and temperatures are shown as open circles. The solid curve represents the calculated results using the average emitter and receiver temperatures: $\overline{T}_H = 318.5\,\text{K}$ and $\overline{T}_L = 302.3\,\text{K}$. The two dotted lines are the upper and lower bounds of the calculation uncertainty. The blackbody limit is indicated by the dashed horizontal line calculated based on $\overline{T}_H$ and $\overline{T}_L$. It can be seen that $h_r$ increases as $d$ decreases, reaching 81.2 W/m$^2$ K, which is 11 times that of the blackbody limit. The methodology and setup could be further modified and improved for the study of near-field energy conversion and thermal management [198].

Bernardi et al. [199] fabricated a compliant Si membrane structure with 20-μm thickness that supports a 5 mm × 5 mm Si emitter with a thickness of about 0.5 mm in

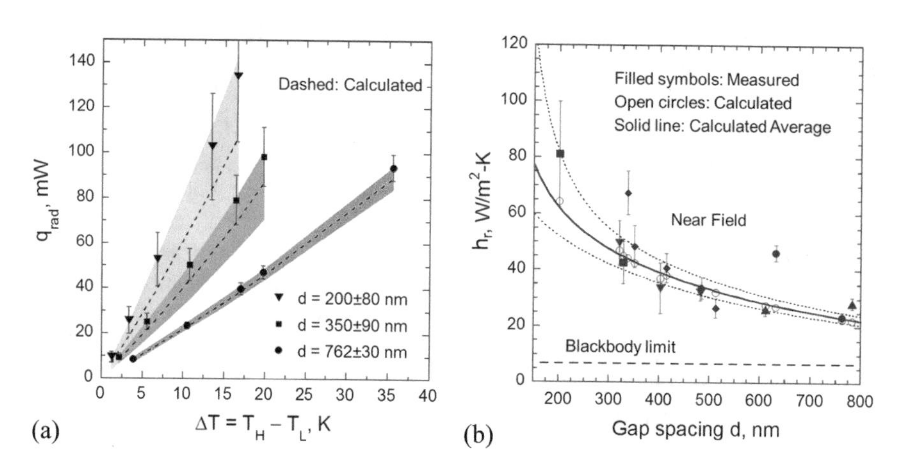


**Fig. 10.44** Near-field radiative transfer measurement results [198]: **a** Radiative heat transfer rate ($q_{rad}$) for three different gap spacings as a function of the temperature difference. **b** Radiative heat transfer coefficient ($h_r$) for 14 measurements at different gap spacings with $\Delta T$ ranging from 10 K to 20 K. Different marks represent different samples and the spacing for some samples was varied by the applied force

the middle. The emitter can be heated and compressed closer to the receiver surface by an applied force. The vacuum gap distance is confined by 3.5-μm-tall SU-8 photoresist posts, with a diameter of 0.25 mm, placed away from the emitter without an applied force. $SiO_2$ stoppers of 150-nm height were fabricated on the receiver plate below the emitter to limit the vacuum gap spacing when a force is applied above the emitter. The radiative heat flux between two intrinsic silicon plates was obtained for temperature differences as large as 120 K at gap distances from 3500 nm down to 150 nm [199]. As discussed previously, nanospheres have also been used to create a gap down to 150 nm between 20-mm-diameter fused silica plates or BK7 glass plates [185]. Yang et al. [200] used four photoresist posts with a height of 3700, 1400, or 430 nm, to create a vacuum gap between the emitter and receiver, with lateral dimensions of 20 mm × 20 mm. They used this setup to measure the near-field radiative transfer between graphene-covered intrinsic silicon substrates with a vacuum gap down to 430 nm and demonstrated enhanced radiative transfer due to graphene surface plasmons. Ying et al. [201] fabricated SU-8 polymer posts to measure the near-field radiative transfer between doped silicon plates with vacuum gaps from about 500 nm down to 190 nm, measured by the capacitance method. Ghashami et al. [202] developed a nanopositioning platform using piezomotors to provide six degrees of freedom with 1-nm translational resolutions in all three axial directions and 1-μrad rotational resolutions in each rotational direction. They measured near-field radiation between crystalline quartz plates from 1200 nm down to 200 nm at temperature differences up to 156 K, and observed more than 40 times enhancement of the blackbody limit in the far field. More recently, DeSutter [203] fabricated micropillars within micrometer-deep pits so that the length of the pillars could be much longer than the vacuum gap spacing to minimize conduction heat transfer. The emitter was heated to about 100 K above the receiver, which was maintained at 300 K. They achieved a vacuum gap spacing down to 110 nm between two doped-Si plates with a surface area of 5.2 × 5.2 $mm^2$ and observed an enhancement of radiative transfer approximately 28.5 times that of the blackbody limit. The recent advances in the plate–plate configuration hold great promise for practical applications of nanoscale thermal radiation.

Using MEMS to fabricate suspended structures is another way for measuring near-field thermal radiation [204–209]. Feng et al. [204] fabricated a MEMS device that contains two suspended membrane islands of 77 μm × 77 μm with Pt heaters and four long beams. The membrane was made of a silicon nitride beam sandwiched between two $SiO_2$ films. They showed that at a vacuum gap spacing of 1 μm, the near-field radiative transfer coefficient is about ten times that of the far field, though still less than heat conduction through the beams. St-Gelais et al. [205] fabricated double nanobeams (referring to Fig. 10.41d) with an electrostatic actuator to tune the gap spacing down to about 200 nm in their first paper and down to about 40 nm in their second paper. They observed a near-field enhancement of nearly two orders of magnitude at $d \approx 42$ nm, with a temperature difference between the two beams exceeding 200 K. The nanobeam is 200 μm long with a height of nearly 500 nm and width of 1–2 μm. Although the effective area is small (on the order of 1 μm), this double-beam MEMS structure can be made on-chip and the gap spacing can be tuned by an applied voltage.

Another group [206–209] developed a nanopositioning platform that uses piezoelectric actuators and high-resolution stepper motors to control the transverse and rotational movements of one of the MEMS structures relative to the other. The fabricated structures for near-field measurements between a sphere and a plate are shown in Fig. 10.45 [206]. Integrated heaters and thermometers were made on both structures, with lateral dimensions on the order of 100 μm. A laser and a PSD were used to detect the mechanical contact.

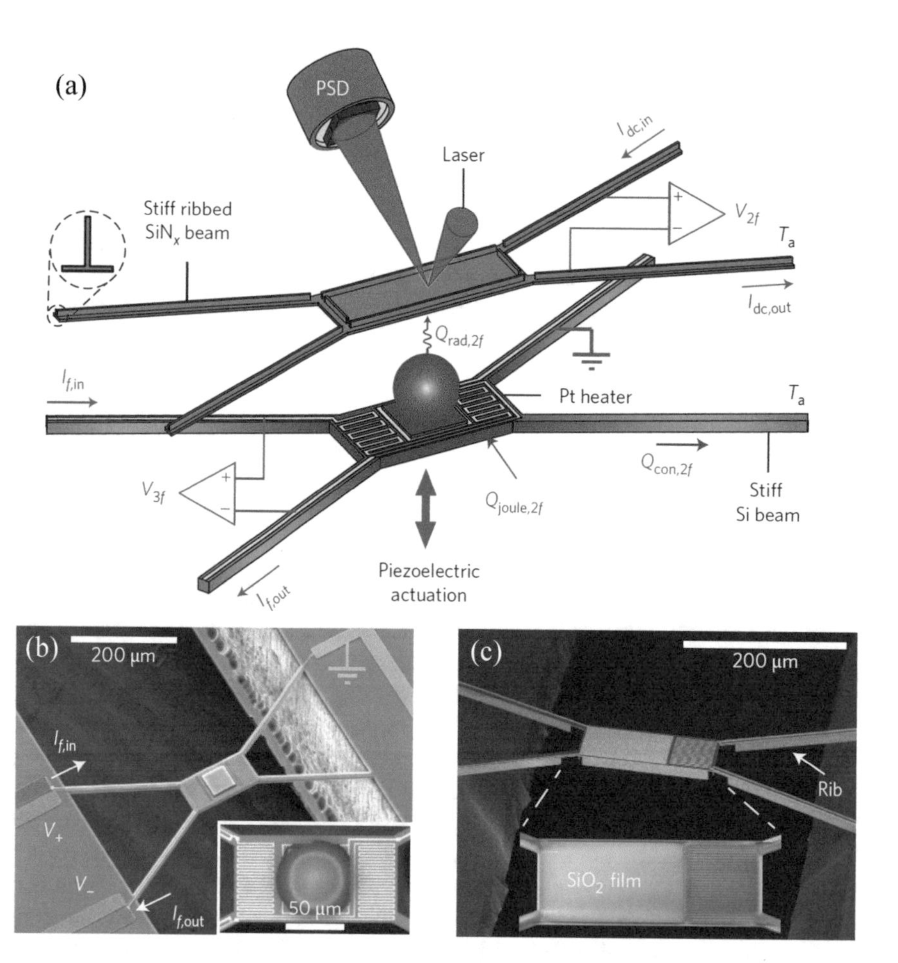


**Fig. 10.45** The nanopositioning platform to place two MEMS structures with nanoscale gaps in between: **a** schematic drawing of the experimental setup; **b** SEM images of the suspended emitter with a sphere and heater/thermometer (inset); **c** SEM images of the receiver structure with a $SiO_2$ film and heater/thermometer (inset). *Reprinted with permission from Song et al.* [206]; *copyright (2015) Springer Nature*

Song et al. [206] measured the near-field radiation in the sphere-plate configuration at distances from 10 μm down to 20 nm using a 53-μm-diameter $SiO_2$ sphere, as shown in Fig. 10.45. Furthermore, they have investigated the effect of flat $SiO_2$ layer thickness by changing it from 50 nm to 3 μm. Their results confirmed the theoretical prediction that the photon penetration depth is largely dependent on the vacuum gap spacing in the near-field regime [156]. A 48 μm by 48 μm square mesa with a height of 20 μm was fabricated for measuring near-field radiation between planar surfaces at separation distances down to near 30 nm [207]. Recently, Fiorino et al. [208] demonstrated nearly 1200-fold enhancement of near-field radiation when the gap spacing is reduced to 25 nm. This setup was also used to build a nanoscale thermophotovoltaic (TPV) device with 40 times enhancement in the output power compared to that of the far field [46]. Lim et al. [209] used MEMS devices (with supporting pads) as done in their previous work [197] for measuring near-field radiative heat transfer between layered hyperbolic metamaterials.

Continuous efforts have also been placed in measuring nanoscale thermal radiation at ultrasmall spacings with tip-based configurations [210, 211]. Probing near-field heat transfer at vacuum gap spacings down to 1 nm and below has been made possible; yet further theoretical and experimental investigations are necessary in order to better understand the interplay between phononic and photonic contributions [72–75, 212, 213].

### *10.7.2 Application Prospects of Nanoscale Thermal Radiation*

In late 1960s and early 1970s, the study of near-field thermal radiation was mainly motivated by space thermal control using multilayer insulation, known as heat shields or radiation shields, where metal-coated polymer films are stacked in close spacing to diminish radiative heat loss from the spacecraft to deep space [22, 181, 182]. The study of near-field radiation in the 1990s was largely driven by the applications of scanning thermal microscopy [37, 38, 82]. Significant developments have been made in the last 10 years in tip-based near-field spectroscopy for imaging microscale thermal sources and for probing the LDOS of thermal emission from nanostructured surfaces [211, 214–216]. In late 1990s and early 2000s, microscale TPV energy conversion devices were proposed and qualitatively demonstrated in a study [23, 217, 218]. Near-field thermophotovoltaics has been a subject of numerous theoretical investigations and some experimental studies using various materials and nanostructures [44–46, 109, 113, 144, 219–222]. Novel radiative energy conversion devices have also been proposed and the performance can be significantly improved at nanoscale separation distances [47, 112, 115, 145, 223–229]. In addition, the enhanced energy density and radiative heat flux associated with the strong evanescent wave coupling in the near field between a sharp tip and a surface have also been used to achieve local heating or cooling and to fabricate nanostructures at the nanometer length scale beyond the diffraction limit [9, 10, 80]. Plasmonic lithography has been developed for maskless nanofabrication and heat-assisted magnetic recording (HAMR) technology [230–233]. In recent years, a large number of scenarios have been envisioned for tuning heat flow using nanoscale thermal radiation with promising applications in thermal management, including noncontact thermal diode or rectifier [41, 42, 192, 234–239], thermal modulator and switcher [102, 103, 240–244], thermal transistor or actuator and interconnector with many bodies [176, 245–252], as well as thermal memory and memristor [253–256]. Some of the devices have already been demonstrated experimentally [238, 239, 242, 244, 252]. Readers can also find comprehensive literature surveys from Refs. [18, 108, 133, 144]. In the following, we highlight the recent research progress and challenges in applying near-field thermal radiation to radiative energy conversion systems.

As discussed in Sect. 6.8.4, thermophotovoltaics is promising for waste heat recovery since there are no moving parts and the device can be compact and portable. A TPV system is based on the same principle as a solar cell, except with a terrestrial heat source that provides mid-infrared photons instead of the sun whose emission is largely in the visible and near infrared. In a TPV receiver, the bandgap of the semiconductor is usually narrower so that photons emitted at longer wavelengths can also be utilized to generate photocurrent. By operating in the near-field regime, it has been predicted that the power output can be significantly enhanced [17, 109]. Near-field TPV systems have recently been demonstrated [46, 222]. Fiorino et al. [46] used a nanopositioning system and achieved a 40-fold enhancement in power output when the separation distance is reduced to 60 nm. The emitter is an 80-μm-diameter Si mesa that can be heated to a temperature up to 655 K. The photovoltaic (PV) cell is made of InAsSb with a bandgap of 0.345 eV (corresponding to a wavelength of 3.6 μm). The maximum power output is 30.2 nW due to the small active area and low efficiency (0.02%). The low efficiency may be due to the low emitter temperature and other imperfections. Inoue et al. [222] fabricated an on-chip TPV device that coupled the frustrated evanescent modes from a 2-μm-thick silicon film to a PV cell made of InGaAs with a bandgap of 0.73 eV (i.e., 1.7 μm). Using spacers, a vacuum gap less than 150 nm was achieved with emitter temperatures above 1000 K. They demonstrated a 10-fold near-field enhancement of the photocurrent with a conversion efficiency close to 1% for an effective area of 500 μm × 500 μm. Continuous efforts and development are needed to further increases the active surface area and conversion efficiency.

From thermodynamic consideration, a low-temperature heat sink (such as a cold sky or outer space) can also be utilized to produce power by using an object at the ambient temperature as the heat "source" [257]. This concept has been further developed into a device called the thermoradiative cell, which is essentially a photodiode (*p-n* junction) on the higher temperature emitter exposed to a lower temperature background [258, 259]. In this case, the $I$–$V$ curve shown in Fig. 6.28b is shifted upward so that a negative open-circuit voltage exists [47]. The system operates with a negative bias voltage and positive photocurrent output, and the output power is equal to $|I \times V|$. The physical mechanism can be interpreted as being due to the negative chemical potential of the *p-n* junction that reduces the number of photons it could emit at frequencies above the bandgap. In the mean time, a current is driven through the circuit to generate photocurrent [227]. By using narrow bandgap semiconductors, the operating temperature of the thermoradiative cell can be relatively low (i.e., 500 K or lower when the surroundings are near or below ambient temperature). The thermoradiative energy converter has also been analyzed and shows great promise to achieve high throughput and conversion efficiency by operating in the near-field regime [223, 224]. Wang et al. [145] theoretically demonstrated enhanced performance of thermoradiative cells using nanoscale gratings as the heat sink due to the excitation of resonance modes.

Light-emitting diodes (LEDs) operate with a high bias voltage that produces photon emission at frequencies greater than the bandgap. This phenomenon can also be explained by the modification of Planck's distribution due to increased photon chemical potential of the *p-n* junction as proposed by Wurfel [260]:

$$I_\omega(\omega, T, \mu) = \left(\frac{\omega}{2\pi c}\right)^2 \frac{\hbar\omega}{\mathrm{e}^{(\hbar\omega-\mu)/k_\mathrm{B}T} - 1}, \quad \text{for } \hbar\omega > E_\mathrm{g} \tag{10.110}$$

The chemical potential for an ideal diode is proportional to the bias voltage and given by $\mu = eV$, where $e$ is the absolute value of the electron charge. When the photon energy is less than the bandgap energy, the distribution function is the same as Planck's distribution at thermal equilibrium, i.e., $\mu = 0$ in Eq. (10.110). Equation (10.110) can also be written in terms of a modified Bose–Einstein distribution function that includes photon chemical potential as

$$f_{\mathrm{BE,mod}}(\omega, T, \mu) = \frac{1}{\mathrm{e}^{(\hbar\omega-\mu)/k_\mathrm{B}T} - 1} \tag{10.111}$$

The modified spectral entropy of a mode can be expressed as [47, 223]

$$s_\omega(\omega, T, \mu) = k_\mathrm{B}[(1+f)\ln(1+f) - f\ln f] \tag{10.112}$$

where $f$ is the modified Bose–Einstein distribution function given in Eq. (10.111).

The principle of LED is called electroluminescence, which can be employed to produce a refrigeration effect to remove heat from a diode at a lower temperature [261]. This can be understood since the intensity of emission for photon energies higher than the bandgap, as calculated from Eq. (10.110) with a positive bias voltage, can be much greater than that at the equilibrium temperature. The challenge is that the net emission at higher frequencies must be sufficiently large to counterbalance the net photon energy transfer at lower frequencies. By operating in the near-field regime with carefully selected materials combinations, the performance can be greatly enhanced [112, 225–227]. The near-field refrigeration effect has recently been demonstrated experimentally [228]. In principle, it is also positive to use a reverse bias (effectively reduce thermal emission) on the hot side to achieve negative luminescent refrigeration with near-field enhancement [262]. The four devices, namely,

TPV, thermoradiative cell, electroluminescent refrigerator, and negative electroluminescent refrigerator can be analyzed coherently using the concept of chemical potential and the modified Planck distribution [47, 227]. In addition, two *p-n* junctions can be used, one on the emitter and one on the receiver to form combined energy conversion systems for achieving either high-efficiency, high throughput power generation [115] or refrigeration [226]. The chemical potential in a real *p-n* junction or *p-i-n* junction is the difference between the electron and hole quasi-Fermi levels. A recent study showed that the reverse saturation current, which is important in modeling TPV devices, can increase significantly in near-field operation [263]. In the previous calculations, it has been assumed that the electron (or hole) population in the conduction (or valence) band has reached a well-established quasi-equilibrium condition. This assumption needs to be further examined considering the coupled charge transport and local photon generation and recombination processes. The spatial variation of the chemical potential needs to be further considered, which can also affect the photoemission and photocurrent generation. While Eq. (10.110) has been used to fit the LED emission spectra [260], quantitative and systematic measurements and comparison of the emission spectra at various temperatures and with varying bias voltages (such as negative bias) are very limited. Challenges also remain in how to reduce the imperfections of the semiconductor materials and the associated nonradiative losses.

## 10.8 Summary

This last chapter of the book describes a field that is clearly associated with radiation heat transfer but has its foundations deeply rooted in physical optics, electrodynamics, quantum mechanics, and perhaps quantum electrodynamics. This field has been advancing rapidly, especially in the past 15 years or so. Significant developments have been made in terms of computation, experiments, as well as novel applications of nanoscale thermal radiation. Recent graduate textbooks on radiation heat transfer have included coverage on near-field thermal radiation. Compared with the first edition published in 2007, this chapter has largely been rewritten to provide comprehensive coverage of the theoretical foundations, numerical methods, and experimental measurement techniques developed in recent years. Extensive examples associated with radiative transfer in micro/nanostructured materials have also been provided. It is expected that more and more prototype demonstrations and practical realizations will emerge in the coming years to harvest near-field radiation in thermal management and energy conversion applications. Fundamental issues that need to be researched further include the delineation of the conduction and radiation regimes at subnanometer separations, as well as the interplay of phonon and photon transports in a vacuum due to quantum fluctuations.

## Problems

10.1 Consider the three-layer structure shown in Fig. 10.3a. Use Eqs. (10.2b) and (10.7) to show that the normal component of the Poynting vector in medium 2 is not a function of $z$ in the case of photon tunneling. Prove that $\langle \mathbf{S} \rangle_z$ in medium 3 is the same as that in medium 2. Can you separate the incident power from the reflected power at the interface between media 2 and 3?

10.2 For thermal radiation originated from medium 1 with a refractive index of $n_1 = 2$ to air, what is the critical angle? If medium 3 is close to medium 1 to form an air gap of width $d$, plot the directional-spectral transmittance at different angles as a function of $d/\lambda$. Calculate the hemispherical transmittance for $s$-polarization, for both propagating and evanescent waves, and plot it against $d/\lambda$.

10.3 Two dielectric materials 1 and 3 are placed in close vicinity at cryogenic temperatures, separated by a vacuum of thickness $d$. Given $n_1 = n_3 = 4$, $T_1 = 4$ K, and $T_3 = 2$ K, calculate the net radiative energy transfer when $d = 1$ mm. Plot the radiative energy transfer from 1 to 3 and that from 3 to 1, as a function of $d$.

10.4 In the resonance tunneling setup shown in Fig. 10.5a, show that for $N = 2$ the transmittance can be expressed as $T'_\lambda = \sin^2(\delta)/[\sin^2(\delta) + 4p^2 \sinh^2(\eta b)]$, where $p = \cos(k_{1z}a)\cosh(\eta b) + \cot(\delta)\sin(k_{1z}a)\sinh(\eta b)$. For $n_1 = 3$, $n_2 = 2$, and $a = b$, find the wavelengths where resonance tunneling occurs. Plot the transmittance spectra for the TM wave, and determine the FWHM of each peak. [Discussion: It is interesting to find out the field distribution and localization in the three middle layers. The amplitude of the evanescent wave may either increase or decrease in the forward direction. One can use the matrix formulation to solve the field distribution to demonstrate the growth of evanescent waves in this arrangement. Discuss the lateral beam shift of the transmitted beam due to the parallel energy flow in the central layer.]

10.5 Derive Eqs. (10.22), (10.23), and (10.24), assuming layer 2 is a PIM and layer 3 is a NIM in Fig. 10.10a, with the same absolute values of refractive index. How will the field distribution in Fig. 10.10b change if the two middle layers switch positions?

10.6 Refer to photon tunneling with negative index layers. Consider two dielectric prisms with refractive index $n_1 = n_5 = 1.5$, sandwiching three middle layers of thicknesses $d_2, d_3,$ and $d_4$. Media 2 and 4 are vacuum with $n_2 = n_4 = 1$, while the middle layer, medium 3, is made of a NIM with $n_3 = -1$, i.e., $\varepsilon_3 = \mu_3 = -1$. Show that when the incidence angle is greater than the critical angle, the transmission coefficient can be expressed as follows: $t = 1/[\coth(\eta_2\Delta) + \mathrm{i}\cot(\delta)\sinh(\eta_2\Delta)]$, where $\eta_2 = \sqrt{k_x^2 - n_2^2\omega^2/c^2}$, $\delta$ is the phase angle upon total internal reflection from medium 1 to 2, and $\Delta = d_2 + d_4 - d_3$. Plot the transmittance as a function of $\Delta/\lambda$, at incidence angles of 45° and 60° for each polarization. Derive the expression for the transmittance of propagating waves. Calculate the hemispherical transmittance for a chosen polarization, and plot it as a function of $\Delta/\lambda$ for both the propagating and evanescent waves in vacuum.

10.7 Reproduce Fig. 10.10, and discuss the features of energy streamlines for the radiative transfer through a dielectric film. Switch the vacuum and the dielectric regions so that the structure becomes dielectric-vacuum-dielectric, with $d/\lambda = 0.1$ and $0.01$. Show the streamlines for both propagation and evanescent waves in air.

10.8 Calculate the nanoscale heat transfer between two SiC plates with a vacuum gap as a function of the gap width. Assume that $T_1 = 600$ K and $T_2 = 300$ K, and use the dielectric function of SiC at room temperature. Plot and discuss the spectral energy transfer in the near field.

10.9 Consider the radiation heat transfer between two plates at $T_1 = 800$ K and $T_2 = 300$ K, separated by a vacuum gap of width $d$. The dielectric function of the plates can be modeled as a Drude model: $\varepsilon_1(\omega) = \varepsilon_2(\omega) = 1 - \frac{\omega_p^2}{\omega^2 + i\omega\gamma}$. Choose different values of $\omega_p$ and $\gamma$ to calculate the near-field and far-field radiation heat transfer. Comment on the effect of each parameter. [Hint: You probably want to set $\omega_p$ in the near infrared, say, at 8000 cm$^{-1}$, and $\gamma \approx 0.01\omega_p$ to start with.]

10.10 Calculate the near-field radiative heat flux at temperatures $T_1 = 350$ K and $T_2 = 250$ K for both SiC and MgO at different gap distances. Plot the heat flux spectra for both materials at $d = 20$ nm. Discuss the contributions of different phonon modes.

10.11 Calculate the near-field radiative heat flux at temperatures $T_1 = 500$ K and $T_2 = 300$ K at $d = 100$ nm for different combinations of emitter and receiver SiC–SiC, SiC–MgO, and MgO–MgO. Plot the heat flux spectra in all three cases. Discuss why dissimilar materials give the smallest heat flux. Do you expect any difference if SiC–MgO is swapped?

10.12 Calculate the radiative heat transfer coefficient at $T = 300$ K between two parallel plates make of SiC as a function of the gap spacing $d$ considering the following cases. As given in Example 8.8, the experimentally obtained scattering rate is $\gamma = 4.76$ cm$^{-1}$. In this homework, you are asked to theoretically explore the effect of $\gamma$ on the radiative heat transfer coefficient by setting $\gamma = 1$ cm$^{-1}$, 4.76 cm$^{-1}$, 10 cm$^{-1}$. Plot the spectral heat transfer coefficient near the resonance frequency.

10.13 Develop a code to plot the contour plots shown in Figs. 10.16 and 10.19.

10.14 Repeat the calculation of Example 10.5 for CNTs with $x = 0.98$ but with several values of the filling ratio ($\phi$): 0.03, 0.05, 0.09, and 0.15.

10.15 Repeat the calculation of Example 10.6 for silicon nanowires (SiNWs) for volume filling ratio $\phi = 0.02, 0.03, 0.05$, and 0.09.

10.16 Repeat the calculation of Example 10.6 for silicon nanoholes (SiNHs) for Si volume fraction $\phi = 0.3, 0.4, 0.6$, and 0.8.

10.17 Calculate the near-field radiative heat flux between two graphene sheets separated by different distances for $\mu = 0.2, 0.3$, and 0.4 eV. Assume $T_1 = 320$ K and $T_2 = 300$ K.

10.18 Reproduce Figs. 10.26b and 10.27. What if the volume fraction and/or the doping level of Si are changed?

10.19 Calculate the near-field radiative heat transfer between graphene-covered CNT arrays using parameters from Ref. [99].

10.20 Reproduce Fig. 10.28b for the three cases. What if the hBN layer thickness is halved or doubled?

10.21 Reproduce Fig. 10.28b for the three cases. Then, evaluate the heat flux when graphene covers both sides of the hBN film as in case 4. Plot them together.

10.22 Obtain the contour plots shown in Fig. 10.29. Discuss hybrid surface plasmon-phonon polaritons. If you integrate the heat flux over the frequency in different regions, which region contributes the most to near-field heat transfer?

10.23 Consider two SiC spheres, each with a diameter of $a = 100$ nm. One is at a temperature $T_1 = 350$ K and the other is at $T_2 = 300$ K. They are placed in vacuum and separated by a center-to-center distance of $d$. Plot the net heat transfer rate as a function of the distance for $3a < d < 7a$.

10.24 Redo Problem 10.23 for $5\ \mu\text{m} < d < 20\ \mu\text{m}$, i.e., in the far-field limit using dipolar approximation. You may also choose $d = 10\ \mu\text{m}$ to plot the spectral heat flux. Calculate the net heat transfer if the spheres were blackbodies. How do the results compare?

10.25 Perform a literature study on the experimental techniques in measuring near-field thermal radiation.

10.26 What are the applications or potential applications of near-field radiative heat transfer?

10.27 Comment on the heat transfer by conduction and by radiation for a powder of $SiO_2$ or SiC nanoparticles.

10.28 In a porous material, when and how will radiation heat transfer exceed conduction heat transfer?

## References

1. L. Novotny, The history of near-field optics. Prog. Opt. **50**, 137–180 (2007)
2. L. Novotny, B. Hecht, *Principles of Nano-Optics*, 2nd edn. (Cambridge Univ. Press, Cambridge, UK, 2012)
3. A. Lewis, M. Isaacson, A. Harootunian, A. Muray, Development of a 500 Å spatial resolution light microscope. Ultramicroscopy **13**, 227–232 (1984)
4. D.W. Pohl, W. Denk, M. Lanz, Optical stethoscopy: image recording with resolution $\lambda/20$. Appl. Phys. Lett. **44**, 651–653 (1984)
5. E. Betzig, R.J. Chichester, Single molecules observed by near-field scanning optical microscopy. Science **262**, 1422–1425 (1993); E. Betzig, J.K. Trautman, R. Wolfe et al., Near-field magneto-optics and high density storage. Appl. Phys. Lett. **61**, 142–144 (1992); E. Betzig, J.K. Trautman, Near-field optics: microscopy, spectroscopy and surface modification beyond the diffraction limit. Science **257**, 189–195 (1992)
6. B. Hecht, B. Sick, U.P. Wild, Scanning near-field optical microscopy with aperture probes: fundamentals and applications. J. Chem. Phys. **112**, 7761–7774 (2000)
7. S. Kawata (ed.), *Near-Field Optics and Surface Plasmon Polaritons* (Springer, Berlin, 2001)

8. J. Tominaga, D.P. Tsai (eds.), *Optical Nanotechnologies – The Manipulation of Surface and Local Plasmons* (Springer, Berlin, 2003)
9. Y.F. Lu, B. Hu, Z.H. Mai, W.J. Wang, W.K. Chim, T.C. Chong, Laser-scanning probe microscope based nanoprocessing of electronics materials. Jpn. J. Appl. Phys. **40**, 4395–4398 (2001)
10. A. Chimmalgi, G.P. Grigoropoulos, K. Komvopoulos, Surface nanostructuring by nano-/femtosecond laser-assisted scanning force microscopy. J. Appl. Phys. **97**, 104319 (2005)
11. H. Mabuchi, A.C. Doherty, Cavity quantum electrodynamics: coherence in context. Science **298**, 1372–1377 (2002)
12. K.J. Vahala, Optical microcavities. Nature **424**, 839–846 (2003)
13. L.A. Blanco, F.J.G. de Abajo, Spontaneous light emission in complex nanostructures. Phys. Rev. B **69**, 205414 (2004)
14. M. Planck, *The Theory of Heat Radiation,* Dover Publications, New York, 1959. [Reproduction of the Masius translation in 1914.]
15. H. Raether, *Surface Plasmons on Smooth and Rough Surfaces and on Gratings* (Springer-Verlag, Berlin, 1988)
16. C.F. Bohren, D.R. Huffman, *Absorption and Scattering of Light by Small Particles* (Wiley, New York, 1983)
17. S. Basu, Z.M. Zhang, C.J. Fu, Review of near-field thermal radiation and its application to energy conversion. Int. J. Energy Res. **33**, 1203–1232 (2009)
18. X.L. Liu, L.P. Wang, Z.M. Zhang, Near-field thermal radiation: recent progress and outlook. Nanos. Micros. Thermophys. Eng. **19**, 98–126 (2015)
19. V. Fernández-Hurtado, A.I. Fernández-Domínguez, J. Feist, F.J. García-Vidal, J.C. Cuevas, Super-Planckian far-feld radiative heat transfer. Phys. Rev. B **97**, 045408 (2018)
20. D. Thompson, L. Zhu, R. Mittapally, S. Sadat, Z. Xing, P. McArdle, M.M. Qazilbash, P. Reddy, E. Meyhofer, Hundred-fold enhancement in far-field radiative heat transfer over the blackbody limit. Nature **561**, 216–221 (2018)
21. E. E. Hall, The penetration of totally reflected light into the rarer medium. Phys. Rev. (Ser. I) **15**, 73–106 (1902)
22. E.G. Cravalho, C.L. Tien, R.P. Caren, Effect of small spacing on radiative transfer between two dielectrics. J. Heat Transfer **89**, 351–358 (1967); C. L. Tien and G. R. Cunnington, Cryogenic insulation heat transfer. Adv. Heat Transfer **9**, 349–417 (1973)
23. M.D. Whale, E.G. Cravalho, Modeling and performance of microscale thermophotovoltaic energy conversion devices. IEEE Trans. Energy Conversion **17**, 130–142 (2002)
24. J.-P. Mulet, K. Joulain, R. Carminati, and J.-J. Greffet, Nanoscale radiative heat transfer between a small particle and a plane surface. Appl. Phys. Lett. **78**, 2931–2933 (2001); *ibid*, Enhanced radiative heat transfer at nanometric distance. Microscale Thermophys. Eng. **6**, 209–222 (2002)
25. C.J. Fu, Z.M. Zhang, Nanoscale radiation heat transfer for silicon at different doping levels. Intl. J. Heat Mass Transfer **49**, 1703–1718 (2006)
26. R.Y. Chiao, A.M. Steinberg, Tunneling times and superluminality. Prog. Opt. **37**, 345–405 (1997)
27. P. Yeh, Resonant tunneling of electromagnetic radiation in superlattice structures. J. Opt. Soc. Am. A **2**, 568–571 (1985); P. Yeh, *Optical Waves in Layered Media* (Wiley, New York, 1988)
28. Z.M. Zhang, C J. Fu, Unusual photon tunneling in the presence of a layer with a negative refractive index. Appl. Phys. Lett. **80**, 1097–1099 (2002); C.J. Fu, Z.M. Zhang, Transmission enhancement using a negative-refraction layer. Microscale Thermophys. Eng. **7**, 221–234 (2003)
29. C.J. Fu, Z.M. Zhang, D.B. Tanner, Energy transmission by photon tunneling in multilayer structures including negative index materials. J. Heat Transfer **127**, 1046–1052 (2005)
30. J.B. Pendry, Negative refraction makes a perfect lens. Phys. Rev. Lett. **85**, 3966–3969 (2000)
31. D.O.S. Melville, R.J. Blaikie, C.R. Wolf, Submicron imaging with a planar silver lens. Appl. Phys. Lett. **84**, 4403–4405 (2004)

32. N. Fang, H. Lee, C. Sun, X. Zhang, Sub-diffraction-limited optical imaging with a silver superlens. Science **308**, 534–537 (2005)
33. Z.M. Zhang, B.J. Lee, Lateral shift in photon tunneling studied by the energy streamline method. Opt. Express **14**, 9963–9970 (2006)
34. A. Alu, N. Engheta, Pairing an epsilon-negative slab with a mu-negative slab: resonance, tunneling and transparency. IEEE Trans. Antennas Propag. **51**, 2558–2571 (2003)
35. C.-F. Li, Negative lateral shift of a light beam transmitted through a dielectric slab and interaction of boundary effects. Phys. Rev. Lett. **91**, 133903 (2003)
36. M.V. Bashevoy, V.A. Fedotov, N.I. Zheludev, Optical whirlpool on an absorbing metallic nanoparticle. Opt. Express **13**, 8372–8379 (2005)
37. J.-B. Xu, K. Läuger, R. Möller, K. Dransfeld, and I. H. Wilson, Heat transfer between two metallic surfaces at small distances. J. Appl. Phys. **76**, 7209–7216 (1994); J.-B. Xu, K. Läuger, K. Dransfeld, and I. H. Wilson, Thermal sensors for investigation of heat transfer in scanning probe microscopy. Rev. Sci. Instrum. **65**, 2262–2266 (1994)
38. W. Müller-Hirsch, A. Kraft, M.T. Hirsch, J. Parisi, A. Kittel, Heat transfer in ultrahigh vacuum scanning thermal microscopy. J. Vac. Sci. Technol. A **17**, 1205–1210 (1999)
39. R. Hillenbrand, T. Taubner, and F. Kellmann, Phonon-enhanced light-matter interaction at the nanometer scale. Nature **418**, 159–162 (2002); R. Hillenbrand, Towards phonon photonics: scattering-type near-field optical microscopy reveals phonon-enhanced near-field interaction. Ultramicroscopy **100**, 421–427 (2004)
40. Y. De Wilde, F. Formanek, R. Carminati, B. Gralak, P.A. Lemoine, K. Joulain, J.-P. Mulet, Y. Chen, J.-J. Greffet, Thermal radiation scanning tunnelling microscopy. Nature **444**, 740–743 (2006)
41. C.R. Otey, W.T. Lau, S. Fan, Thermal rectification through vacuum. Phys. Rev. Lett. **104**, 154301 (2010)
42. L.P. Wang, Z.M. Zhang, Thermal rectification enabled by near-field radiative heat transfer between intrinsic silicon and a dissimilar material. Nanoscale Microscale Thermophys. Eng. **17**, 337–348 (2013)
43. P. Ben-Abdallah and S.-A. Biehs, Phase-change radiative thermal diode. Appl. Phys. Lett. **103**, 191907 (2013); *ibid*, Near-field thermal transistor. Phys. Rev. Lett. **112**, 044301 (2014)
44. A. Narayanaswamy, G. Chen, Surface modes for near field thermophotovoltaics. Appl. Phys. Lett. **82**, 3544–3546 (2003)
45. M. Francoeur, R. Vaillon, M.P. Mengüç, Thermal impacts on the performance of nanoscale-gap thermophotovoltaic power generators. IEEE Trans. Energy Conv. **26**, 686–698 (2011)
46. A. Fiorino, L. Zhu, D. Thompson, R. Mittapally, P. Reddy, E. Meyhofer, Nanogap near-field thermophotovoltaics. Nat. Nanotechnol. **13**, 806–811 (2018)
47. E. Tervo, E. Bagherisereshki, Z.M. Zhang, Near-field radiative thermoelectric energy converters: a review. Front. Energy **12**, 5–21 (2018)
48. A. I. Volokitin and B. N. J. Persson, Radiative heat transfer between nanostructures. Phys. Rev. B **63**, 205404 (2001); *ibid*, Resonance phonon tunneling of the radiative heat transfer. Phys. Rev. B **69**, 045417 (2004)
49. X.J. Wang, S. Basu, Z.M. Zhang, Parametric optimization of dielectric functions for maximizing nanoscale radiative transfer. J. Phys. D Appl. Phys. **42**, 245403 (2009)
50. S. Basu, B.J. Lee, Z.M. Zhang, Infrared radiative properties of heavily doped silicon at room temperature. J. Heat Transfer **132**, 023301 (2010); *ibid*, Near-field radiation calculated with an improved dielectric function model for doped silicon. J. Heat Transfer **132**, 023302 (2010)
51. D. Polder, M. van Hove, Theory of radiative heat transfer between closely spaced bodies. Phys. Rev. B **4**, 3303–3314 (1971)
52. J.J. Loomis, H.J. Maris, Theory of heat transfer by evanescent electromagnetic waves. Phys. Rev. B **50**, 18517–18524 (1994)
53. G. Domingues, S. Volz, K. Joulain, J.-J. Greffet, Heat transfer between two nanoparticles through near field interaction. Phys. Rev. Lett. **94**, 085901 (2005)
54. A. Narayanaswamy, G. Chen, Thermal near-field radiative transfer between two spheres. Phys. Rev. B **77**, 075125 (2008)

55. M. Francoeur, M.P. Mengüç, R. Vaillon, Solution of near-field thermal radiation in one-dimensional layered media using dyadic Green's functions and the scattering matrix method. J. Quant. Spectrosc. Radiat. Transfer **110**, 2002–2018 (2009)
56. S. Edalatpour, M. Francoeur, The thermal Discrete Dipole Approximation (T-DDA) for near-field radiative heat transfer simulations in three-dimensional arbitrary geometries. J. Quant. Spectrosc. Radiat. Transfer **133**, 364–373 (2014)
57. G. Bimonte, T. Emig, M. Kardar, M. Krüger, Nonequilibrium fluctuational quantum electrodynamics: heat radiation, heat transfer, and force. Annu. Rev. Conden. Mat. Phys. **8**, 119–143 (2017)
58. C.R. Otey, L. Zhu, S. Sandhu, S. Fan, Fluctuational electrodynamics calculations of near-field heat transfer in non-planar geometries: a brief overview. J. Quant. Spectrosc. Radiat. Transfer **132**, 3–11 (2014)
59. Y.M. Xuan, An overview of micro/nanoscaled thermal radiation and its applications. Photon. Nanostr. Fundam. Appl. **12**, 93–113 (2014)
60. X.L. Liu, Z.M. Zhang, Giant enhancement of nanoscale thermal radiation based on hyperbolic graphene plasmons. Appl. Phys. Lett. **107**, 143114 (2015)
61. X.L. Liu, Z.M. Zhang, Near-field thermal radiation between metasurfaces. ACS Photon. **2**, 1320–1326 (2015)
62. S.M. Rytov, Correlation theory of thermal fluctuations in an isotropic medium. Sov. Phys. JETP **6**, 130–140 (1958); S.M. Rytov, Yu.A. Kravtsov, V.I. Tatarskii, *Principles of Statistical Radiophysics III: Elements of Random Fields,* vol. 3 (Springer-Verlag, Berlin, 1987) Chap. 3
63. A.L. Volokitin, B.N.J. Persson, Resonant photon tunneling enhancement of the van der Waals friction. Phys. Rev. Lett. **91**, 106101 (2003)
64. J.R. Zurita-Sánchez, J.-J. Greffet, L. Novotny, Friction forces arising from fluctuating thermal fields. Phys. Rev. A **69**, 022902 (2004)
65. Narayanaswamy, Y. Zheng, A Green's function formalism of energy and momentum transfer in fluctuational electrodynamics. J. Quant. Spectrosc. Radiat. Transfer **132**, 12–21 (2014)
66. K. Joulain, R. Carminati, J.-P. Mulet, J.-J. Greffet, Definition and measurement of the local density of electromagnetic states close to an interface. Phys. Rev. B **68**, 245405 (2003)
67. L.P. Wang, S. Basu, Z.M. Zhang, Direct and indirect methods for calculating thermal emission from layered structures with nonuniform temperatures. J. Heat Transfer **133**, 072701 (2011)
68. K. Joulain, J.-P. Mulet, F. Marquier, R. Carminati, J.-J. Greffet, Surface electromagnetic waves thermally excited: radiative heat transfer, coherence properties and Casimir forces revisited in the near field. Surf. Sci. Rep. **57**, 59–112 (2005)
69. P.W. Milonni, M.-L. Shih, Zero-point energy in early quantum theory. Am. J. Phys. **59**, 684–698 (1991)
70. B.J. Lee, Z.M. Zhang, Lateral shifts in near-field thermal radiation with surface phonon polaritons. Nanoscale Microscale Thermophys. Eng. **12**, 238–250 (2008)
71. T.J. Bright, X.L. Liu, Z.M. Zhang, Energy streamlines in near-field radiative heat transfer between hyperbolic metamaterials. Opt. Express **22**, A1112–A1127 (2014)
72. V. Chiloyan, J. Garg, K. Esfarjani, G. Chen, Transition from near-field thermal radiation to phonon heat conduction at sub-nanometre gaps. Nat. Comm. **6**, 6755 (2015)
73. K. Kim, B. Song, V. Fernández-Hurtado et al., Radiative heat transfer in the extreme near field. Nature **528**, 387–391 (2015)
74. K. Kloppstech, N. Könne, S.-A. Biehs, A.W. Rodriguez, L. Worbes, D. Hellmann, A. Kittel, Giant heat transfer in the crossover regime between conduction and radiation. Nat. Commun. **8**, 14475 (2017)
75. L. Cui, W. Jeong, V. Fernández-Hurtado, J. Feist, F.J. García-Vidal, J.C. Cuevas, E. Meyhofer, P. Reddy, Study of radiative heat transfer in Ångström- and nanometre-sized gaps. Nat. Commun. **8**, 14479 (2017)
76. W.P. King, T.W. Kenny, K.E. Goodson, G. Cross, M. Despont, U. Dürig, H. Rothuizen, G.K. Binnig, P. Vettiger, Atomic force microscope cantilevers for combined thermomechanical data writing and reading. Appl. Phys. Lett. **78**, 1300–1302 (2001)

77. S. Basu, Z.M. Zhang, Maximum energy transfer in near-field thermal radiation at nanometer distances. J. Appl. Phys. **105**, 093535 (2009)
78. Z.M. Zhang, X.J. Wang, Unified Wien's displacement law in terms of logarithmic frequency or wavelength scale. J. Thermophys. Heat Transfer **24**, 222–224 (2010)
79. P.O. Chapuis, S. Volz, C. Henkel, K. Joulain, J.-J. Greffet, Effects of spatial dispersion in near-field radiative heat transfer between two parallel metallic surfaces. Phys. Rev. B **77**, 035431 (2008)
80. B. Guha, C. Otey, C.B. Poitras, S. Fan, M. Lipson, Near-field radiative cooling of nanostructures. Nano Lett. **12**, 4546–4550 (2012)
81. B.J. Lee, K. Park, Z.M. Zhang, Energy pathways in nanoscale thermal radiation. App. Phys. Lett. **91**, 153101 (2007)
82. J.B. Pendry, Radiative exchange of heat between nanostructures. J. Phys.: Condens. Matter **11**, 6621–6633 (1999)
83. S.-A. Biehs, E. Rousseau, J.-J. Greffet, Mesoscopic description of radiative heat transfer at the nanoscale. Phys. Rev. Lett. **105**, 234301 (2010)
84. M.P. Bernardi, O. Dupré, E. Blandre, P.-O. Chapuis, R. Vaillon, M. Francoeur, Impacts of propagating, frustrated and surface modes on radiative, electrical and thermal losses in nanoscale-gap thermophotovoltaic power generators. Sci. Rep. **5**, 11626 (2015)
85. K. Joulain, J. Drevillon, P. Ben-Abdallah, Noncontact heat transfer between two metamaterials. Phys. Rev. B **81**, 165119 (2010)
86. Z.H. Zheng, Y.M. Xuan, Theory of near-field radiative heat transfer for stratified magnetic media. Int. J. Heat Mass Transfer **54**, 1101–1110 (2011)
87. M. Francoeur, S. Basu, S.J. Petersen, Electric and magnetic surface polariton mediated near-field radiative heat transfer between metamaterials made of silicon carbide particles. Opt. Express **19**, 18774–18788 (2011)
88. X.L. Liu, R.Z. Zhang, Z.M. Zhang, Near-field thermal radiation between hyperbolic metamaterials: graphite and carbon nanotubes. Appl. Phys. Lett. **103**, 213102 (2013)
89. X.L. Liu, R.Z. Zhang, Z.M. Zhang, Near-field radiative heat transfer with doped-silicon nanostructured metamaterials. Int. J. Heat Mass Transfer **73**, 389–398 (2014)
90. M. Tschikin, S.-A. Biehs, R. Messina, P. Ben-Abdallah, On the limits of the effective description of hyperbolic materials in the presence of surface waves. J. Opt. **15**, 105101 (2013)
91. B. Liu, S. Shen, Broadband near-field radiative thermal emitter/absorber based on hyperbolic metamaterials: direct numerical simulation by the Wiener chaos expansion method. Phys. Rev. B **87**, 115403 (2013)
92. X.L. Liu, T.J. Bright, Z.M. Zhang, Application conditions of effective medium theory in near-field radiative heat transfer between multilayered metamaterials. J. Heat Transfer **136**, 092703 (2014)
93. J. Peng, G. Zhang, B. Li, Thermal management in $MoS_2$ based integrated device using near-field radiation. Appl. Phys. Lett. **107**, 133108 (2015)
94. J. Shen, S. Guo, X.L. Liu, B. Liu, W. Wu, H. He, Super-Planckian thermal radiation enabled by coupled quasi-elliptic 2D black phosphorus plasmons. Appl. Therm. Eng. **144**, 403–410 (2018)
95. Y. Zhang, H.-L. Yi, H.-P. Tan, Near-field radiative heat transfer between black phosphorus sheets via anisotropic surface plasmon polaritons. ACS Photon. **5**, 3739–3747 (2018)
96. L. Ge, Y. Cang, K. Gong, L. Zhou, D. Yu, Y. Luo, Control of near-field radiative heat transfer based on anisotropic 2D materials. AIP Adv. **8**, 085321 (2018)
97. X.L. Liu, R.Z. Zhang, Z.M. Zhang, Near-perfect photon tunneling by hybridizing graphene plasmons and hyperbolic modes. ACS Photon. **1**, 785–789 (2014)
98. M. Lim, S.S. Lee, B.J. Lee, Near-field thermal radiation between graphene-covered doped silicon plates. Opt. Express **21**, 22173–22185 (2013)
99. R.Z. Zhang, X.L. Liu, Z.M. Zhang, Near-field radiation between graphene-covered carbon nanotube arrays. AIP Adv. **5**, 053501 (2015)

100. B. Zhao, Z.M. Zhang, Enhanced photon tunneling by surface plasmon-phonon polaritons in graphene/hBN heterostructures. J. Heat Transfer **139**, 022701 (2017)
101. B. Zhao, B. Guizal, Z.M. Zhang, S. Fan, M. Antezza, Near-field heat transfer between graphene/hBN multilayers. Phys. Rev. B **95**, 245437 (2017)
102. S.-A. Biehs, F.S.S. Rosa, P. Ben-Abdallah, Modulation of near-field heat transfer between two gratings. Appl. Phys. Lett. **98**, 243102 (2011)
103. X.L. Liu, J.D. Shen, Y.M. Xuan, Pattern-free thermal modulator via thermal radiation between Van der Waals materials. J. Quant. Spectrosc. Radiat. Transfer **200**, 100–107 (2017)
104. X.H. Wu, C.J. Fu, Z.M. Zhang, Influence of hBN orientation on the near-field radiative heat transfer between graphene/hBN heterostructures. J. Photon. Energy **9**, 032702 (2018)
105. X.H. Wu, C.J. Fu, Z.M. Zhang, Effect of orientation on the directional and hemispherical emissivity of hyperbolic metamaterials. Int. J. Heat Mass Transfer **135**, 1207–1217 (2019)
106. L. Tsang, J.A. Kong, K.-H. Ding, *Scattering of Electromagnetic Waves: Theories and Applications* (Wiley, New York, 2000)
107. A. Narayanaswamy, G. Chen, Thermal radiation in 1D photonic crystals. J. Quant. Spectrosc. Radiat. Transfer **93**, 175–183 (2005)
108. S. Basu, *Near-Field Radiative Heat Transfer across Nanometer Vacuum Gaps: Fundamentals and Applications* (Elsevier, Amsterdam, 2016)
109. K. Park, S. Basu, W.P. King, Z.M. Zhang, Performance analysis of near-field thermophotovoltaic devices considering absorption distribution. J. Quant. Spectrosc. Radiat. Transfer **109**, 305–316 (2008)
110. G.J. Kovacs, Optical excitation of surface plasmon-polaritons in layered media, in *Electromagnetic Surface Modes*, ed. by A.D. Boardman (Wiley, New York, 1982)
111. W.C. Chew, *Waves and Fields in Inhomogeneous Media* (IEEE Press, New York, 1995)
112. X.L. Liu, Z.M. Zhang, High-performance electroluminescent refrigeration enabled by photon tunneling. Nano Energy **26**, 353–359 (2016)
113. B. Zhao, K. Chen, S. Buddhiraju, G. Bhatt, M. Lipson, S. Fan, High-performance near-field thermophotovoltaics for waste heat recovery. Nano Energy **41**, 344–350 (2017)
114. J. DeSutter, R. Vaillon, M. Francoeur, External luminescence and photon recycling in near-field thermophotovoltaics. Phys. Rev. Appl. **8**, 014030 (2017)
115. B. Zhao, P. Santhanam, K. Chen, S. Buddhiraju, S. Fan, Near-field thermophotonic systems for low-grade waste-heat recovery. Nano Lett. **18**, 5224–5230 (2018)
116. J.K. Lee, J.A. Kong, Dyadic Green's functions for layered anisotropic medium. Electromagnet. **3**, 111–130 (1983)
117. A. Eroglu, Y.H. Lee, J.K. Lee, Dyadic Green's functions for multi-layered uniaxially anisotropic media with arbitrarily oriented optic axes. IET Microwaves Antennas Propag. **5**, 1779–1788 (2011)
118. A. Eroglu, *Wave Propagation and Radiation in Gyrotropic and Anisotropic Media* (Springer, New York, 2010)
119. L.Y. Carrillo, Y. Bayazitoglu, Nanorod near-field radiative heat exchange analysis. J. Quant. Spectrosc. Radiat. Transfer **112**, 412–419 (2011); *ibid*, Sphere approximation for nanorod near-field radiative heat exchange analysis. Nanoscale Microscale Thermophys. Eng. **15**, 195–208 (2011)
120. K. Sasihithlu, A. Narayanaswamy, Convergence of vector spherical wave expansion method applied to near-field radiative transfer. Opt. Express **19**, A772–A785 (2011)
121. S.Y. Park, D. Stroud, Surface-plasmon dispersion relations in chains of metallic nanoparticles: an exact quasistatic calculation. Phys. Rev. B **69**, 125418 (2004)
122. P. Ben-Abdallah, K. Joulain, J. Drevillon, C. Le Goff, Heat transport through plasmonic interactions in closely spaced metallic nanoparticle chains. Phys. Rev. B **77**, 075417 (2008)
123. E.J. Tervo, M. Francoeur, B.A. Cola, Z.M. Zhang, Thermal radiation in systems of many dipoles. Phys. Rev. B **100**, 205422 (2019)
124. K. Park, Z.M. Zhang, Fundamentals and applications of near-field radiative energy transfer. Front. Heat Mass Transfer **4**, 013001 (2013)

125. C. Luo, A. Narayanaswamy, G. Chen, J.D. Joannopoulos, Thermal radiation from photonic crystals: a direct calculation. Phys. Rev. Lett. **93**, 213905 (2004)
126. S. Wen, Direct numerical simulation of near-field thermal radiation based on Wiener Chaos expansion of thermal fluctuating current. J. Heat Transfer **132**, 072704 (2010)
127. A.W. Rodriguez, O. Ilic, P. Bermel, I. Celanovic, J.D. Joannopoulos, M. Soljacic, S.G. Johnson, Frequency-selective near-field radiative heat transfer between photonic crystal slabs: a computational approach for arbitrary geometries and materials. Phys. Rev. Lett. **107**, 114302 (2011)
128. M. Krüger, T. Emig, M. Kardar, Nonequilibrium electromagnetic fluctuations: heat transfer and interactions. Phys. Rev. Lett. **106**, 210404 (2011)
129. C.R. Otey, S. Fan, Numerically exact calculation of electromagnetic heat transfer between a dielectric sphere and plate. Phys. Rev. B **84**, 245431 (2011)
130. A.W. Rodriguez, M.T. Homer Reid, S.G. Johnson, Fluctuating-surface-current formulation of radiative heat transfer for arbitrary geometries. Phys. Rev. B **86**, 220302 (2012); *ibid*, Fluctuating-surface-current formulation of radiative heat transfer: theory and applications. Phys. Rev. B **88**, 054305 (2013)
131. A. Datas, D. Hirashima, K. Hanamura, FDTD simulation of near-field radiative heat transfer between thin films supporting surface phonon polaritons: lessons learned. J. Therm. Sci. Technol. **8**, 91–105 (2013)
132. A. Didari, M.P. Mengüç, Analysis of near-field radiation transfer within nano-gaps using FDTD method. J. Quant. Spectrosc. Radiat. Transfer **146**, 214–226 (2014); *ibid*, Near-field thermal radiation transfer by mesoporous metamaterials. Opt. Express **23**, A1253-A1258 (2015)
133. B. Song, A. Fiorino, E. Meyhofer, P. Reddy, Near-field radiative thermal transport: From theory to experiment. AIP Adv. **5**, 053503 (2015)
134. J.C. Cuevas, F.J. García-Vidal, Radiative heat transfer. ACS Photon. **5**, 3896–3915 (2018)
135. G. Bimonte, Scattering approach to Casimir forces and radiative heat transfer for nanostructured surfaces out of thermal equilibrium. Phys. Rev. A **80**, 042102 (2009)
136. R. Messina, M. Antezza, Scattering-matrix approach to Casimir-Lifshitz force and heat transfer out of thermal equilibrium between arbitrary bodies. Phys. Rev. A **84**, 042102 (2011)
137. M. Krüger, G. Bimonte, T. Emig, M. Kardar, Trace formulas for nonequilibrium Casimir interactions, heat radiation, and heat transfer for arbitrary objects. Phys. Rev. B **86**, 115423 (2012)
138. R. Guérout, J. Lussange, F.S.S. Rosa, J.P. Hugonin, D.A.R. Dalvit, J.-J. Greffet, A. Lambrecht, S. Reynaud, Enhanced radiative heat transfer between nanostructured gold plates. Phys. Rev. B **85**, 180301 (2012)
139. J. Lussange, R. Guérout, F.S.S. Rosa, J.-J. Greffet, A. Lambrecht, S. Reynaud, Radiative heat transfer between two dielectric nanogratings in the scattering approach. Phys. Rev. B **86**, 085432 (2012)
140. X.L. Liu, B. Zhao, Z.M. Zhang, Enhanced near-field thermal radiation and reduced Casimir stiction between doped-Si gratings. Phys. Rev. A **91**, 062510 (2015)
141. Y. Yang, L.P. Wang, Spectrally enhancing near-field radiative transfer between metallic gratings by exciting magnetic polaritons in nanometric vacuum gaps. Phys. Rev. Lett. **117**, 044301 (2016)
142. X.L. Liu, Z.M. Zhang, Graphene-assisted near-field radiative heat transfer between corrugated polar materials. Appl. Phys. Lett. **104**, 251911 (2014)
143. X.L. Liu, J. Shen, Y.M. Xuan, Near-field thermal radiation of nanopatterned black phosphorene mediated by topological transitions of phosphorene plasmons. Nanoscale Microscale Thermophys. Eng. **23**, 188–199 (2019)
144. J.I. Watjen, X.L. Liu, B. Zhao, Z.M. Zhang, A computational simulation of using tungsten gratings in near-field thermophotovoltaic devices. J. Heat Transfer **139**, 052704 (2017)
145. B. Wang, C. Lin, K.H. Teo, Z.M. Zhang, Thermoradiative device enhanced by near-field coupled structures. J. Quant. Spectrosc. Radiat. Transfer **196**, 10–16 (2017)

146. A. Taflove, S.C. Hagness, *Computational Electrodynamics: The Finite-Difference Time-Domain Method*, 3rd edn. (Artech House, Norwood, 2005)
147. S.D. Gedney, *Introduction to the Finite-Difference Time-Domain (FDTD) Method for Electromagnetics* (Morgan & Claypool Publishers, 2011)
148. F.L. Teixeira, Time-domain finite-difference and finite-element methods for Maxwell equations in complex media. IEEE Trans. Antennas Propag. **56**, 2150–2166 (2008)
149. K. Fu, P.F. Hsu, Modeling the radiative properties of microscale random roughness surfaces. J. Heat Transfer **129**, 71–78 (2007); *ibid*, Radiative properties of gold surfaces with one-dimensional microscale Gaussian random roughness. Int. J. Thermophys. **28**, 598–615 (2007)
150. Y.M. Xuan, Y.G. Han, Y. Zhou, Spectral radiative properties of two-dimensional rough surfaces. Int. J. Thermophys. **33**, 2291–2310 (2012)
151. B. Zhao, Z.M. Zhang, Design of optical and radiative properties of surfaces, in *Handbook of Thermal Science and Engineering*, ed. by F.A. Kulacki, et al. (Springer, New York, 2018), pp. 1023–1068
152. D. Lu, A. Das, W. Park, Direct modeling of near field thermal radiation in a metamaterial. Opt. Express **25**, 12999–13009 (2017)
153. J. Andreasen, H. Cao, A. Taflove, P. Kumar, C. Cao, FDTD simulation of thermal noise in open cavities. Phys. Rev. A **77**, 023810 (2008)
154. A. Didari, M.P. Mengüç, A design tool for direct and non-stochastic calculations of near-field radiative transfer in complex structures: the NF-RT-FDTD algorithm. J. Quant. Spectrosc. Radiat. Transfer **197**, 95–105 (2017); *ibid*, A biomimicry design for nanoscale radiative cooling applications inspired by *Morpho didius* butterfly. Sci. Rep. **8**, 16891 (2018)
155. M. Badieirostami, A. Adibi, H.M. Zhou, S.N. Chow, Model for efficient simulation of spatially incoherent right using the Wiener chaos expansion method. Opt. Lett. **32**, 3188–3190 (2007)
156. S. Basu, Z.M. Zhang, Ultrasmall penetration depth in nanoscale thermal radiation. Appl. Phys. Lett. **95**, 133104 (2009)
157. A.P. McCauley, M.T. Homer Reid, M. Krüger, S.G. Johnson, Modeling near-field radiative heat transfer from sharp objects using a general three-dimensional numerical scattering technique. Phys. Rev. B **85**, 165104 (2012)
158. S. Rao, D. Wilton, A. Glisson, Electromagnetic scattering by surfaces of arbitrary shape. IEEE Trans. Antennas Propag. **30**, 409–418 (1982)
159. K. L. Nguyen, O. Merchiers, and P.-O. Chapuis, Near-field radiative heat transfer in scanning thermal microscopy computed with the boundary element method. J. Quant. Spectrosc. Radiat. Transfer **202**, 154–167 (2017)
160. M.T. Homer Reid, J. White, S.G. Johnson, Fluctuating surface currents: an algorithm for efficient prediction of Casimir interactions among arbitrary materials in arbitrary geometries. Phys. Rev. A **88**, 022514 (2013)
161. A.G. Polimeridis, M.T. Homer Reid, W. Jin, S.G. Johnson, J.K. White, A.W. Rodriguez, Fluctuating volume-current formulation of electromagnetic fluctuations in inhomogeneous media: incandescence and luminescence in arbitrary geometries. Phys. Rev. B **92**, 134202 (2015)
162. M. Quinten, A. Leitner, J.R. Krenn, F.R. Aussenegg, Electromagnetic energy transport via linear chains of silver nanoparticles. Opt. Lett. **23**, 1331–1333 (1998)
163. S.A. Maier, M.L. Brongersma, P.G. Kik, S. Meltzer, A.A.G. Requicha, H.A. Atwater, Plasmonics – a route to nanoscale optical devices. Adv. Mater. **13**, 1501–1505 (2001)
164. P. Ben-Abdallah, Heat transfer through near-field interactions in nanofluids. Appl. Phys. Lett. **89**, 224301 (2006)
165. J. Ordonez-Miranda, L. Tranchant, S. Gluchko, S. Volz, Energy transport of surface phonon polaritons propagating along a chain of spheroidal nanoparticles. Phys. Rev. B **92**, 115409 (2015)
166. E.J. Tervo, Z.M. Zhang, B.A. Cola, Collective near-field thermal emission from polaritonic nanoparticle arrays. Phys. Rev. Mater. **1**, 015201 (2017)

167. C. Kathmann, R. Messina, P. Ben-Abdallah, S.-A. Biehs, Limitations of kinetic theory to describe near-field heat exchanges in many-body systems. Phys. Rev. B **98**, 115434 (2018)
168. J. Ordonez-Miranda, L. Tranchant, K. Joulain, Y. Ezzahri, J. Drevillon, S. Volz, Thermal energy transport in a surface phonon-polariton crystal. Phys. Rev. B **93**, 035428 (2016)
169. E.J. Tervo, M.E. Gustafson, Z.M. Zhang, B.A. Cola, M.A. Filler, Photonic thermal conduction by infrared plasmonic resonators in semiconductor nanowires. Appl. Phys. Lett. **114**, 163104 (2019)
170. G. Baffou, R. Quidant, C. Girard, Thermoplasmonics modeling: a Green's function approach. Phys. Rev. B **82**, 165424 (2010)
171. P. Ben-Abdallah, S.-A. Biehs, K. Joulain, Many-body radiative heat transfer theory. Phys. Rev. Lett. **107**, 114301 (2011)
172. P. Ben-Abdallah, R. Messina, S.-A. Biehs, M. Tschikin, K. Joulain, C. Henkel, Heat superdiffusion in plasmonic nanostructure networks. Phys. Rev. Lett. **111**, 174301 (2003)
173. R. Messina, M. Tschikin, S.A. Biehs, P. Ben-Abdallah, Fluctuation-electrodynamic theory and dynamics of heat transfer in systems of multiple dipoles. Phys. Rev. B **88**, 104307 (2013)
174. J. Dong, J.M. Zhao, L.H. Liu, Radiative heat transfer in many-body systems: coupled electric and magnetic dipole approach. Phys. Rev. B **95**, 125411 (2017)
175. J. Chen, C.Y. Zhao, B.X. Wang, Near-field thermal radiative transfer in assembled spherical systems composed of core-shell nanoparticles. J. Quant. Spectrosc. Radiat. Transfer **219**, 304–312 (2018)
176. J.L. Song, Q. Cheng, Z.X. Luo, X.P. Zhou, Z.M. Zhang, Modulation and splitting of three-body radiative heat flux via graphene/SiC core-shell nanoparticles. Int. J. Heat Mass Transfer **140**, 80–87 (2019)
177. B.T. Draine, P.J. Flatau, Discrete-dipole approximation for scattering calculations. J. Opt. Soc. Am. A **11**, 1491–1499 (1994)
178. S. Edalatpour, J. DeSutter, M. Francoeur, Near-field thermal electromagnetic transport: an overview. J. Quant. Spectrosc. Radiat. Transfer **178**, 14–21 (2016)
179. S. Edalatpour, M. Francoeur, Near-field radiative heat transfer between arbitrarily shaped objects and a surface. Phys. Rev. B **94**, 045406 (2016)
180. R.M. Abraham Ekeroth, A. Garcia-Martin, J.C. Cuevas, Thermal discrete dipole approximation for the description of thermal emission and radiative heat transfer of magneto-optical systems. Phys. Rev. B **95**, 235428 (2017)
181. E.G. Cravalho, G.A. Domoto, and C.L. Tien, Measurements of thermal radiation of solids at liquid helium temperatures. in *Progress in Aeronautics and Astronautics,* J.T. Bevans (ed.), vol. 21, pp. 531–542 (1968); G.A. Domoto, R.F. Boehm, and C.L. Tien, Experimental investigation of radiative transfer between metallic surfaces at cryogenic temperatures. J. Heat Transfer **92**, 412–417 (1970)
182. C.M. Hargreaves, Anomalous radiative transfer between closely-spaced bodies. Phys. Lett. **30A**, 491–492 (1969); C.M. Hargreaves, Radiative transfer between closely-spaced bodies. Philips Res. Rep. Suppl. **5**, 1–80 (1973)
183. A. Kittel, W. Muller-Hirsch, J. Parisi, S.A. Biehs, D. Reddig, M. Holthaus, Near-field heat transfer in a scanning thermal microscope. Phys. Rev. Lett. **95**, 224301 (2005)
184. L. Hu, A. Narayanaswamy, X.Y. Chen, G. Chen, Near-field thermal radiation between two closely spaced glass plates exceeding Planck's blackbody radiation law. Appl. Phys. Lett. **92**, 133106 (2008)
185. S. Lang, G. Sharma, S. Molesky, P.U. Kränzien, T. Jalas, Z. Jacob, A.Yu. Petrov, M. Eich, Dynamic measurement of near-field radiative heat transfer. Sci. Rep. **7**, 13916 (2017)
186. A. Narayanaswamy, S. Shen, G. Chen, Near-field radiative heat transfer between a sphere and a substrate. Phys. Rev. B **78**, 115303 (2008)
187. S. Shen, A. Narayanaswamy, G. Chen, Surface phonon polaritons mediated energy transfer between nanoscale gaps. Nano Lett. **9**, 2909–2913 (2009)
188. J. Shi, P. Li, B. Liu, S. Shen, Tuning near-field radiation by doped silicon. Appl. Phys. Lett. **102**, 183114 (2013)

189. J. Shi, B. Liu, P. Li, P., L. Y. Ng, and S. Shen, Near-field energy extraction with hyperbolic metamaterials. Nano Lett. **15**, 1217–1221 (2015)
190. E. Rousseau, A. Siria, G. Jourdan, S. Volz, F. Comin, J. Chevrier, J.-J. Greffet, Radiative heat transfer at the nanoscale. Nat. Photon. **3**, 514–517 (2009)
191. P.J. Van Zwol, S. Thiele, C. Berger, W.A. De Heer, J. Chevrier, Nanoscale radiative heat flow due to surface plasmons in graphene and doped silicon. Phys. Rev. Lett. **109**, 264301 (2012)
192. P.J. Van Zwol, L. Ranno, J. Chevrier, Tuning near-field radiative heat flux through surface excitations with a metal insulator transition. Phys. Rev. Lett. **108**, 234301 (2012)
193. R.S. Ottens, V. Quetschke, S. Wise, A.A. Alemi, R. Lundock, G. Mueller, D.H. Reitze, D.B. Tanner, B.F. Whiting, Near-field radiative heat transfer between macroscopic planar surfaces. Phys. Rev. Lett. **107**, 014301 (2011)
194. T. Kralik, P. Hanzelka, M. Zobac, V. Musilova, T. Fort, M. Horak, Strong near-field enhancement of radiative heat transfer between metallic surfaces. Phys. Rev. Lett. **109**, 224302 (2012)
195. T. Ijiro, N. Yamada, Near-field radiative heat transfer between two parallel $SiO_2$ plates with and without microcavities. Appl. Phys. Lett. **106**, 023103 (2015)
196. K. Ito, A. Miura, H. Iizuka, H. Toshiyoshi, Parallel-plate submicron gap formed by micromachined low-density pillars for near-field radiative heat transfer. Appl. Phys. Lett. **106**, 083504 (2015)
197. M. Lim, S.S. Lee, B.J. Lee, Near-field thermal radiation between doped silicon plates at nanoscale gaps. Phys. Rev. B **91**, 195136 (2015)
198. J.I. Watjen, B. Zhao, Z.M. Zhang, Near-field radiative heat transfer between doped-Si parallel plates separated by a spacing down to 200 nm. Appl. Phys. Lett. **109**, 203112 (2016)
199. M.P. Bernardi, D. Milovich, M. Francoeur, Radiative heat transfer exceeding the blackbody limit between macroscale planar surfaces separated by a nanosize vacuum gap. Nat. Commun. **7**, 12900 (2016)
200. J. Yang, W. Du, Y. Su, Y. Fu, S. Gong, S. He, Y. Ma, Observing of the super-Planckian near-field thermal radiation between graphene sheets. Nat. Commun. **9**, 4033 (2018)
201. X. Ying, P. Sabbaghi, N. Sluder, L.P. Wang, Super-Planckian radiative heat transfer between macroscale surfaces with vacuum gaps down to 190 nm directly created by SU-8 posts and characterized by capacitance method. ACS Photon. **7**, 190–196 (2020)
202. M. Ghashami, H. Geng, T. Kim, N. Iacopino, S.K. Cho, K. Park, Precision measurement of phonon-polaritonic near-field energy transfer between macroscale planar structures under large thermal gradients. Phys. Rev. Lett. **120**, 175901 (2018)
203. J. DeSutter, L. Tang, M. Francoeur, A near-field radiative heat transfer device. Nat. Nanotechnol. **14**, 751–755 (2019)
204. C. Feng, Z. Tang, J. Yu, C. Sun, A MEMS device capable of measuring near-field thermal radiation between membranes. Sensors **13**, 1998–2010 (2013)
205. R. St-Gelais, B. Guha, L. Zhu, S. Fan, and M. Lipson, Demonstration of strong near-field radiative heat transfer between integrated nanostructures. Nano Lett. **14**, 6971–6975 (2014); R. St-Gelais, L. Zhu, S. Fan, and M. Lipson, Near-field radiative heat transfer between parallel structures in the deep subwavelength regime. Nat. Nanotechnol. **11**, 515–519 (2016)
206. B. Song, Y. Ganjeh, S. Sadat et al., Enhancement of near-field radiative heat transfer using polar dielectric thin films. Nat. Nanotechnol. **10**, 253–258 (2015)
207. B. Song, D. Thompson, A. Fiorino, Y. Ganjeh, P. Reddy, E. Meyhofer, Radiative heat conductances between dielectric and metallic parallel plates with nanoscale gaps. Nat. Nanotechnol. **11**, 509–514 (2016)
208. A. Fiorino, D. Thompson, L. Zhu, B. Song, P. Reddy, E. Meyhofer, Giant enhancement in radiative heat transfer in sub-30 nm gaps of plane parallel surfaces. Nano Lett. **18**, 3711–3715 (2018)
209. M. Lim, J. Song, S.S. Lee, B.J. Lee, Tailoring near-field thermal radiation between metallo-dielectric multilayers using coupled surface plasmon polaritons. Nat. Commun. **9**, 4302 (2018)
210. L. Worbes, D. Hellmann, A. Kittel, Enhanced near-field heat flow of a monolayer dielectric island. Phys. Rev. Lett. **110**, 134302 (2013)

211. A. Jarzembski, C. Shaskey, K. Park, Tip-based vibrational spectroscopy for nanoscale analysis of emerging energy materials. Front. Energy **12**, 43–71 (2018)
212. K.Y. Fong, H.-K. Li, R. Zhao, S. Yang, Y. Wang, X. Zhang, Phonon heat transfer across a vacuum through quantum fluctuations. Nature **576**, 243–247 (2019)
213. A. Jarzembski, T. Tokunaga, J. Crossley, J. Yun, C. Shaskey, R. A. Murdick, I. Park, M. Francoeur, and K. Park, "Force-induced acoustic phonon transport across single-digit nanometre vacuum gaps," arXiv preprint arXiv:1904.09383, 2019/4/20
214. F. Huth, M. Schnell, J. Wittborn, N. Ocelic, R. Hillenbrand, Infrared-spectroscopic nanoimaging with a thermal source. Nat. Mater. **10**, 352–356 (2011)
215. A.C. Jones, B.T. O'Callahan, H.U. Yang, M.B. Raschke, The thermal near-field: Coherence, spectroscopy, heat transfer, and optical forces. Prog. Surface Sci. **88**, 349–392 (2013)
216. A. Babuty, K. Joulain, P.-O. Chapuis, J.-J. Greffet, Y. De Wilde, Blackbody spectrum revisited in the near field. Phys. Rev. Lett. **110**, 146103 (2013)
217. R.S. DiMatteo, P. Greiff, S.L. Finberg, K.A. Young-Waithe, H.K.H. Choy, M.M. Masaki, C.G. Fonstad, Enhanced photogeneration of carriers in a semiconductor via coupling across a nonisothermal nanoscale vacuum gap. Appl. Phys. Lett. **79**, 1894–1896 (2001)
218. S. Basu, Y.-B. Chen, Z.M. Zhang, Microscale Radiation in Thermophotovoltaic Devices – A Review. Int. J. Energy Res. **31**, 689–716 (2007)
219. O. Ilic, M. Jablan, J.D. Joannopoulos, I. Celanovic, M. Soljačić, Overcoming the black body limit in plasmonic and graphene near-field thermophotovoltaic systems. Opt. Express **20**, A366–A384 (2012)
220. R. Messina, P. Ben-Abdallah, Graphene-based photovoltaic cells for near-field thermal energy conversion. Sci. Rep. **3**, 1383 (2013)
221. S. Jin, M. Lim, S.S. Lee, B.J. Lee, Hyperbolic metamaterial-based near-field thermophotovoltaic system for hundreds of nanometer vacuum gap. Opt. Express **24**, A635–A649 (2016)
222. T. Inoue, T. Koyama, D.D. Kang, K. Ikeda, T. Asano, S. Noda, One-chip near-field thermophotovoltaic device integrating a thin-film thermal emitter and photovoltaic cell. Nano Lett. **19**, 3948–3952 (2019)
223. W.C. Hsu, J.K. Tong, B. Liao, Y. Huang, S.V. Boriskina, G. Chen, Entropic and near-field improvements of thermoradiative cells. Sci. Rep. **6**, 34837 (2016)
224. C. Lin, B. Wang, K.H. Teo, Z.M. Zhang, Performance comparison between photovoltaic and thermoradiative devices. J. Appl. Phys. **122**, 243103 (2017)
225. K. Chen, P. Santhanam, S. Sandhu, L. Zhu, S. Fan, Heat-flux control and solid-state cooling by regulating chemical potential of photons in near-field electromagnetic heat transfer. Phys. Rev. B **91**, 134301 (2015)
226. K. Chen, T.P. Xiao, P. Santhanam, E. Yablonovitch, S. Fan, High-performance near-field electroluminescent refrigeration device consisting of a GaAs light emitting diode and a Si photovoltaic cell. J. Appl. Phys. **122**, 143104 (2017)
227. C. Lin, B. Wang, K.H. Teo, Z.M. Zhang, A coherent description of thermal radiative devices and its application on the near-field negative electroluminescent cooling. Energy **147**, 177–186 (2018)
228. L. Zhu, A. Fiorino, D. Thompson, R. Mittapally, E. Meyhofer, P. Reddy, Near-field photonic cooling through control of the chemical potential of photons. Nature **566**, 239–244 (2019)
229. A. Datasa, R. Vaillonc, Thermionic-enhanced near-field thermophotovoltaics. Nano Energy **61**, 10–17 (2019)
230. W. Srituravanich, N. Fang, C. Sun, Q. Luo, X. Zhang, Plasmonic nanolithography. Nano Lett. **4**, 1085–1088 (2004)
231. W.A. Challener, C. Peng, A.V. Itagi et al., Heat-assisted magnetic recording by a near-field transducer with efficient optical energy transfer. Nat. Photon. **3**, 220–224 (2009)
232. B.C. Stipe, T.C. Strand, C.C. Poon et al., Magnetic recording at 1.5 $Pb/m^2$ using an integrated plasmonic antenna. Nat. Photon. **4**, 484–488 (2010)
233. N. Zhou, X. Xu, A.T. Hammack, B.C. Stipe, K. Gao, W. Scholz, E.C. Gage, Plasmonic near-field transducer for heat-assisted magnetic recording. Nanophoton. **3**, 141–155 (2014)

234. P. Ben-Abdallah, S.-A. Biehs, Phase-change radiative thermal diode. Appl. Phys. Lett. **103**, 191907 (2013)
235. Y. Yang, S. Basu, L. Wang, Radiation-based near-field thermal rectification with phase transition materials. Appl. Phys. Lett. **103**, 163101 (2013)
236. J.G. Huang, Q. Li, Z.H. Zheng, Y.M. Xuan, Thermal rectification based on thermochromic materials. Int. J. Heat Mass Transfer **67**, 575–580 (2013)
237. A. Ghanekar, J. Ji, Y. Zheng, High-rectification near-field thermal diode using phase change periodic nanostructure. Appl. Phys. Lett. **109**, 123106 (2016)
238. M. Elzouka, S. Ndao, High temperature near-field nanothermomechanical rectification. Sci. Rep. **7**, 44901 (2017)
239. A. Fiorino, A thermal diode based on nanoscale thermal radiation. ACS Nano **12**, 5774–5779 (2018)
240. Y. Yang, S. Basu, L. Wang, Vacuum thermal switch made of phase transition materials considering thin film and substrate effects. J. Quant. Spectrosc. Radiat. Transfer **158**, 69–77 (2015)
241. Y. Yang, L. Wang, Electrically-controlled near-field radiative thermal modulator made of graphene-coated silicon carbide plates. J. Quant. Spectrosc. Radiat. Transfer **197**, 68–75 (2017)
242. K. Ito, K. Nishikawa, A. Miura, H. Toshiyoshi, H. Iizuka, Dynamic modulation of radiative heat transfer beyond the blackbody limit. Nano Lett. **17**, 4347–4353 (2017)
243. G.T. Papadakis, B. Zhao, S. Buddhiraju, S. Fan, Gate-tunable near-field heat transfer. ACS Photon. **6**, 709–719 (2019)
244. N.H. Thomas, M.C. Sherrott, J. Broulliet, H.A. Atwater, A.J. Minnich, Electronic modulation of near-field radiative transfer in graphene field effect heterostructures. Nano Lett. **19**, 3898–3904 (2019)
245. P. Ben-Abdallah, S.-A. Biehs, Near-field thermal transistor. Phys. Rev. Lett. **112**, 044301 (2014)
246. Z.H. Zheng, Y.M. Xuan, Enhancement or suppression of the near-field radiative heat transfer between two materials. Nanoscale Microscale Thermophys. Eng. **15**, 237–251 (2011)
247. R. Messina, M. Antezza, P. Ben-Abdallah, Three-body amplification of photon heat tunneling. Phys. Rev. Lett. **109**, 244302 (2012)
248. B. Liu, Y. Liu, S. Shen, Thermal plasmonic interconnnects in graphene. Phys. Rev. B **90**, 195411 (2014)
249. W. Gu, G.-H. Tang, W.-Q. Tao, Thermal switch and thermal rectification enabled by near-field radiative heat transfer between three slabs. Int. J. Heat Mass Transfer **82**, 429–434 (2015)
250. A. Ghanekar, Y. Tian, M. Ricci, S. Zhang, O. Gregory, Y. Zheng, Near-field thermal rectification devices using phase change periodic nanostructure. Opt. Express **26**, A209–A218 (2018)
251. Y.H. Kan, C.Y. Zhao, Z.M. Zhang, Near-field radiative heat transfer in three-body systems with periodic structures. Phys. Rev. B **99**, 035433 (2019); *ibid*, Enhancement and manipulation of near-field radiative heat transfer using an intermediate modulator. Phys. Rev. Appl. **13**, 014069 (2020)
252. D. Thompson, L. Zhu, E. Meyhofer, and P. Reddy, Nanoscale radiative thermal switching via multi-body effects. Nat. Nanotechnol. **15**, 99–104 (2020)
253. V. Kubytskyi, S.-A. Biehs, P. Ben-Abdallah, Radiative bistability and thermal memory. Phys. Rev. Lett. **113**, 074301 (2014)
254. S.A. Dyakov, J. Dai, M. Yan, M. Qiu, Near field thermal memory based on radiativephase bistability of $VO_2$. J. Phys. D Appl. Phys. **48**, 305104 (2015)
255. K. Ito, K. Nishikawa, H. Iizuka, Multilevel radiative thermal memory realized by the hysteretic metal-insulator transition of vanadium dioxide. Appl. Phys. Lett. **108**, 053507 (2016)
256. J. Ordonez-Miranda, Y. Ezzahri, J.A. Tiburcio-Moreno, K. Joulain, J. Drevillon, Radiative thermal memristor. Phys. Rev. Lett. **123**, 025901 (2019)
257. S.J. Byrnes, R. Blanchard, F. Capasso, Harvesting renewable energy from Earth's mid-infrared emissions. Proc. Nat. Acad. of Sci. (PNAS) **111**, 3927–3932 (2014)

258. R. Strandberg, Theoretical efficiency limits for thermoradiative energy conversion. J. Appl. Phys. **117**, 055105 (2015)
259. P. Santhanam, S. Fan, Thermal-to-electrical energy conversion by diodes under negative illumination. Phys. Rev. B **93**, 161410(R) (2016)
260. P. Wurfel, The chemical potential of radiation. J. Phys. C: Solid State Phys. **15**, 3967–3985 (1982)
261. S.-T. Yen, K.-C. Lee, Analysis of heterostructures for electroluminescent refrigeration and light emitting without heat generation. J. Appl. Phys. **107**, 054513 (2010)
262. K. Chen, P. Santhanam, S. Fan, Near-field enhanced negative luminescent refrigeration. Phys. Rev. Appl. **6**, 024014 (2016)
263. D. Feng, E.J. Tervo, S.K. Yee, Z.M. Zhang, Effect of evanescent waves on the dark current of thermophotovoltaic cells. Nanoscale Microscale Thermophys. Eng. **24**, 1–19 (2020)

# Appendix A
# Physical Constants

Avogadro's constant ($N_{\mathrm{A}}$) $6.022 \times 10^{26}\ \mathrm{kmol}^{-1}$
Universal gas constant ($\bar{R}$) $8.314$ kJ/kmol K
Speed of light in vacuum ($c_0$) $299{,}792{,}458\ \mathrm{m/s}$ (exact)
Boltzmann's constant ($k_{\mathrm{B}}$) $1.381 \times 10^{-23}\ \mathrm{J/K}$
Planck's constant ($h$) $6.626 \times 10^{-34}\ \mathrm{J\,s}$
Stefan-Boltzmann constant ($\sigma_{\mathrm{SB}}$) $5.670 \times 10^{-8}\ \mathrm{W/m^2\,K^4}$
Electron charge (absolute value) ($e$) $1.602 \times 10^{-19}$ C (coulumb)
Electron mass ($m_{\mathrm{e}}$) $9.109 \times 10^{-31}$ kg
Proton mass ($m_{\mathrm{p}}$) $1.673 \times 10^{-27}$ kg
Standard acceleration of gravity ($g_n$) $9.80665\ \mathrm{m/s^2}$ (exact)
Magnetic permeability (vacuum) ($\mu_0$) $4\pi \times 10^{-7}\ \mathrm{N/A^2}$ (exact)
Electrical permittivity (vacuum) ($\varepsilon_0$) $8.854 \times 10^{-12}\ \mathrm{C^2/N\,m^2}$ (or F/m)

1 atm = 760 mm Hg = 101.325 kPa (standard atmosphere, exact)
1 eV = $1.602 \times 10^{-19}$ J (electron volt)

## SI Prefixes

| Power | $10^{-21}$ | $10^{-18}$ | $10^{-15}$ | $10^{-12}$ | $10^{-9}$ | $10^{0}$ | $10^{9}$ | $10^{12}$ | $10^{15}$ | $10^{18}$ | $10^{21}$ |
|---|---|---|---|---|---|---|---|---|---|---|---|
| Prefix | zepto | atto | femto | pico | nano | — | giga | tera | peta | exa | zetta |
| Symbol | z | a | f | p | n | — | G | T | P | E | Z |

Reference: http://physics.nist.gov/cuu/index.html



Z. M. Zhang, *Nano/Microscale Heat Transfer*, Mechanical Engineering Series,
https://doi.org/10.1007/978-3-030-45039-7

# Appendix B
# Mathematical Background

## B.1 Some Useful Formulae

### *B.1.1 Series and Integrals*

Binary equation:

$$(a+b)^N = b^N + Nab^{N-1} + \frac{N(N-1)}{2!}a^2b^{N-2} + \frac{N!}{3!(N-3)!}a^3b^{N-3}$$
$$+ \cdots + Na^{N-1}b + a^N = \sum_{M=0}^{N} \frac{N!}{M!(N-M)!} a^M b^{N-M} \tag{B.1}$$

Geometric series:

$$1 + \mathrm{e}^{-x} + \mathrm{e}^{-2x} + \mathrm{e}^{-3x} + \cdots = \frac{1}{1-\mathrm{e}^{-x}} \quad (x > 0) \tag{B.2}$$

Using the Taylor expansion, we can write

$$\mathrm{e}^x = 1 + x + \frac{x^2}{2!} + \frac{x^3}{3!} + \cdots \tag{B.3}$$

$$\ln(1+x) = x - \frac{x^2}{2} + \frac{x^3}{3} - \frac{x^4}{4} + \cdots \quad (-1 < x < 1) \tag{B.4}$$

Integrate $\int_{-\infty}^{\infty} \mathrm{e}^{-x^2} \mathrm{d}x$. This integral may be evaluated by a transformation from the Cartesian coordinators to polar coordinators:



Z. M. Zhang, *Nano/Microscale Heat Transfer*, Mechanical Engineering Series,
https://doi.org/10.1007/978-3-030-45039-7

$$\left(\int_{-\infty}^{\infty} e^{-x^2} dx\right)\left(\int_{-\infty}^{\infty} e^{-y^2} dy\right) = \iint\limits_{\substack{-\infty<x<\infty \\ -\infty<y<\infty}} e^{-(x^2+y^2)} dx dy$$

$$= \iint\limits_{\substack{0<r<\infty \\ 0<\phi<2\pi}} e^{-r^2} r dr d\phi = \int_0^{\infty} e^{-r^2} 2\pi r dr = \pi \int_0^{\infty} e^{-t} dt = \pi$$

Therefore,

$$\int_{-\infty}^{\infty} e^{-x^2} dx = \sqrt{\pi} \tag{B.5}$$

It can be seen that $\int_{-\infty}^{\infty} e^{-ax^2} dx = \sqrt{\pi/a}$. It should be noticed that $\int_{-\infty}^{\infty} x e^{-ax^2} dx = 0$, but

$$\int_0^{\infty} x e^{-ax^2} dx = \frac{1}{2a}$$

Furthermore

$$\int_0^{\infty} x^{n+2} e^{-ax^2} dx = \frac{n+1}{2a} \int_0^{\infty} x^n e^{-ax^2} dx \quad (n = 0, 1, 2\ldots) \tag{B.6}$$

Another type of important integral equation is the following.

$$\int_0^{\infty} \frac{x^n e^x}{(e^x - 1)^2} dx = n \int_0^{\infty} \frac{x^{n-1}}{e^x - 1} dx \tag{B.7}$$

where

$$\int_0^{\infty} \frac{x^{n-1}}{e^x - 1} dx = (n-1)!\zeta(n) \tag{B.8}$$

Here, $\zeta(n)$ is the Riemann zeta function defined as

$$\zeta(n) = 1 + \frac{1}{2^n} + \frac{1}{3^n} + \frac{1}{4^n} + \cdots\cdots \tag{B.9}$$

The values of $\zeta(n)$ are given below for several $n$ values:

| $n$ | 1 | 2 | 3 | 4 | 5 | 6 | 7 | 8 |
|---|---|---|---|---|---|---|---|---|
| $\zeta(n)$ | $\infty$ | $\frac{\pi^2}{6}$ | 1.202… | $\frac{\pi^4}{90}$ | 1.037… | $\frac{\pi^6}{945}$ | 1.008… | $\frac{\pi^8}{9450}$ |

Examples are $\int_0^\infty \frac{x}{e^x-1}dx = \frac{\pi^2}{6}$, $\int_0^\infty \frac{x^2}{e^x-1}dx = 2.404\ldots$, and $\int_0^\infty \frac{x^3}{e^x-1}dx = \frac{\pi^4}{15}$.

### B.1.2 The Error Function

The error function is defined as

$$\text{erf}(x) = \frac{2}{\sqrt{\pi}} \int_0^x e^{-x^2} dx \tag{B.10}$$

The complementary error function is $\text{erfc}(x) \equiv 1 - \text{erf}(x)$. The error function can only be evaluated numerically. As shown in the table below, erf($x$) changes with $x$ almost linearly for $x < 0.5$ but approaches to unity rapidly as $x$ increases.

| $x$ | 0 | 0.01 | 0.1 | 0.2 | 0.5 | 1 | 2 | 3 | $\infty$ |
|---|---|---|---|---|---|---|---|---|---|
| erf($x$) | 0 | 0.0113 | 0.1125 | 0.2227 | 0.5205 | 0.8427 | 0.9953 | 0.99998 | 1 |

### B.1.3 Stirling's Formula

Stirling's formula is an approximation of the logarithm of a factorial for large numbers. Note that

$$\begin{aligned}\ln x! &= \ln 1 + \ln 2 + \ln 3 + \cdots + \ln x \\ &= \sum_{n=1}^{x} \ln n \approx \int_0^x \ln x\, dx = x \ln x - x + 1 \approx x \ln x - x\end{aligned}$$

More complicated analysis results in the same approximation for large $x$. Stirling's formula is expressed as

$$\ln x! \approx x \ln x - x, \text{ for } x \gg 100 \tag{B.11}$$

The relative error of this approximation is 13.8% for $x = 10$ and less than 1% for $x > 100$. Therefore, it is applicable for very large $x$.

## B.2 The Method of Lagrange Multipliers

The method of Lagrange multipliers is a procedure for determining the maximum/minimum point in a continuous function subject to one or more constraints. Consider a continuous function $f(x_1, x_2, \ldots, x_n)$. At the maximum/minimum,

$$\mathrm{d}f = \sum_{i=1}^{n} \frac{\partial f}{\partial x_i} \mathrm{d}x_i = 0 \tag{B.12}$$

Therefore, if $x_i$ and $x_j (i \neq j)$ are independent, we must have

$$\frac{\partial f}{\partial x_i} = 0, \quad i = 1, 2, \ldots, n \tag{B.13}$$

If they are dependent and related by $m(m < n)$ constraint equations (or constraints), then

$$\psi_j(x_1, x_2, \ldots, x_n) = 0, \; j = 1, 2, \ldots, m \tag{B.14}$$

and

$$\mathrm{d}\psi_j = \sum_{i=1}^{n} \frac{\partial \psi_j}{\partial x_i} \mathrm{d}x_i = 0, \quad j = 1, 2, \ldots, m \tag{B.15}$$

Multiply $\beta_j$ to the $j$th equation in Eq. (B.15) and add to Eq. (B.12), we obtain

$$\sum_{i=1}^{n} \left( \frac{\partial f}{\partial x_i} + \sum_{j=1}^{m} \beta_j \frac{\partial \psi_j}{\partial x_i} \right) dx_i = 0 \tag{B.16}$$

where $\beta_{j's}$ are called Lagrangian multipliers. For Eq. (B.16) to hold, we must have

$$\frac{\partial f}{\partial x_i} + \sum_{j=1}^{m} \beta_j \frac{\partial \psi_j}{\partial x_i} = 0, \quad i = 1, 2, \ldots, n \tag{B.17}$$

The $n$ equations given about allow the determination of $m\beta_{j's}$ and $n - m$ independent variables.

**Example B.1** Determine the positive values of $x$, $y$, and $z$ that will maximize the function $f(x, y, z) = 8xyz$, subject to the constraint $\frac{x^2}{a^2} + \frac{y^2}{b^2} + \frac{z^2}{c^2} = 1$, where $a$, $b$, and $c$ are positive constants.

**Solution** The constraint equation may be rewritten as $\psi(x, y, z) = \frac{x^2}{a^2} + \frac{y^2}{b^2} + \frac{z^2}{c^2} - 1 = 0$. Hence,

$$\mathrm{d}f = 8yz\mathrm{d}x + 8xz\mathrm{d}y + 8xy\mathrm{d}z = 0$$

$$\beta \mathrm{d}\psi = \frac{2\beta x}{a^2}\mathrm{d}x + \frac{2\beta y}{b^2}\mathrm{d}y + \frac{2\beta z}{c^2}\mathrm{d}z = 0$$

Adding these two equations and setting the coefficients to zero, we have $8yz + 2\beta x/a^2 = 0$, $8xz + 2\beta y/b^2 = 0$, and $8xy + 2\beta z/c^2 = 0$; that is, $\beta = -4a^2yz/x$, $\beta = -4b^2xz/y$, and $\beta = -4c^2xy/z$. Dividing the product of the three equations $\beta^3 = -64a^2b^2c^2xyz$ by each equation gives $\beta^2 = 16a^2b^2c^2(x^2/a^2)$, $\beta^2 = 16a^2b^2c^2(y^2/b^2)$, and $\beta^2 = 16a^2b^2c^2(z^2/c^2)$. Solving for and substituting $x^2/a^2$, $y^2/b^2$, and $z^2/c^2$ into the constraint equation, we obtain $\beta = -4abc/\sqrt{3}$. Therefore, $x = a/\sqrt{3}$, $y = b/\sqrt{3}$, and $z = c/\sqrt{3}$. Thus, the maximum of the given function under the specified constraint is $f_{\max} = 8abc/3\sqrt{3} \approx 1.54abc$.

## B.3 Permutation and Combination

This section discusses several permutation and combination problems that are directly related to the derivation of equilibrium distributions of different types of particles, such as molecular gases, electrons in a conductor, electrons and holes in semiconductors, photons in a thermodynamic equilibrium, and phonons in a crystalline solids. One of the important concepts in quantum statistics is related to the indistinguishable nature of particles, also known as identical particles or indiscernible particles. We can understand the indistinguishability by considering identical particles, for which no way exists for us to track their trajectories (location and velocity to be able to discern one from the other). These particles include photons, phonons, electrons, protons, etc.

Case 1. *How many ways can we arrange N distinguishable objects in a row?*

There are $N$ objects to select for the first place, $N - 1$ for the second, $N - 2$ for the third, and so on. The number of permutations of $N$ objects is therefore given by

$$_N\mathrm{P}_N = N! \tag{B.18}$$

Case 2. *How many ways can we arrange N objects out from a group of* g *distinguishable objects* ($N \leq g$)?

An equivalent problem is: *How many ways can we put N distinguishable objects in g distinguishable boxes with a limit that each box can at most have one object ($N \le g$)?* There are $g$ ways of placing the first object, $g - 1$ ways of placing the second, $g - 2$ ways of placing the third, …, and $g - N + 1$ ways of placing the $N$th object. Therefore, the number of permutations of $g$ objects taken $N$ at a time is given by

$$_g\mathrm{P}_N = g(g-1)(g-2)\cdots(g-N+2)(g-N+1) = \frac{g!}{(g-N)!} \qquad \text{(B.19)}$$

Case 3. *How many ways can we put N distinguishable objects into g distinguishable boxes (without regard to order within the boxes)?*

Because each box can contain any number of objects, there are $g$ ways of placing each object. Hence, the number of ways is

$$g^N \qquad \text{(B.20)}$$

Here, $g$ can be smaller than, equal to, or greater than $N$. Note that this is equivalent to the permutation problem with repetition: *How many ways can we arrange N objects taken from g types of objects (each type has more than N identical objects) by allowing repetition?*

**Example B.2** a. How many 4-digit integers can be made from the numbers 1, 2, …, and 9, without allowing repetition? b. Same as (a) but with repetition. c. Same as (b) but with zero.

**Solution** a. There are $9 \times 8 \times 7 \times 6 = 3024$. b. There are $9 \times 9 \times 9 \times 9 = 6561$. c. There are $9 \times 10 \times 10 \times 10 = 9000$, because the first number must not be zero for it to be a 4-digit integer.

**Example B.3** a. How many ways can we put 3 different books on 5 shelves without caring about their order on each shelf? b. Same as (a) but each shelf cannot have more than one book.

**Solution** a. Since each shelf can have any number of books and each book can go to any shelf, the ways to put the books are $5 \times 5 \times 5 = 125$. b. In this case, there are $5 \times 4 \times 3 = 60$ ways only.

Case 4. *How many ways can we choose N from g distinguishable objects without caring about their order ($N \le g$)?*

This is a combination problem. Because the order to arrange the objects is not considered, the number of combinations of $N$ objects taken from a group of $g$ objects is then given by

$$_g\mathrm{C}_N = \frac{g!}{N!(g-N)!} \qquad \text{(B.21)}$$

An equivalent problem is: *How many ways can we put N indistinguishable objects in g distinguishable boxes with a limit of at most one object in each box?* We learned from Case 2 that there are $g!/(g-N)!$ ways of placing $N$ distinguishable objects in $g$ boxes. Now that the $N$ objects are indistinguishable, the number of ways is reduced by a factor of $N!$.

Case 5. *How many ways can we place N distinguishable objects in r distinguishable boxes such that there are N*1 *objects in the first box, N*2 *in the second, ..., and Nr in the* rth *box?*

Because the order within each box is not considered, we must divide the total number of arrangements $N!$ by the number of arrangements in each box, keeping in mind that $N_1 + N_2 + \cdots + N_r = N$. Therefore, the number of ways is

$$\frac{N!}{N_1!N_2! \cdots N_r!} = \frac{N!}{\prod_{i=1}^{r} N_i!} \tag{B.22}$$

Case 6. *How many ways can we place N indistinguishable objects in g distinguishable boxes without limiting the number of objects per box?*

The answer to this problem is less straightforward as compared with previous cases. The order within each box does not matter since the objects are indistinguishable. Let's use a dot for each object and use $g-1$ slashes to separate them into $g$ groups such that:

●●/●●●●●/●/●●●....../ / ●●●

Each arrangement corresponds to one way of placing $N$ indistinguishable objects in $g$ distinguishable boxes. Although the slashes are identical, their order makes the "boxes" distinguishable. Note that the dot and slash are symbols: each occupies one location. The question becomes how many ways to select $g-1$ slash locations out from $N+g-1$ total locations? Said differently, how many ways can we select $N$ dot locations out from $N+g-1$ total locations? The answer is equivalent to the combination problem given in Case 4, except that there are $N+g-1$ total locations, i.e.,

$$\frac{(N+g-1)!}{N!\,(g-1)!} \tag{B.23}$$

The preceding discussions are very important for understanding statistical thermodynamics as discussed in Chap. 3.

## B.4 Events and Probabilities

If an evenly cast coin is tossed, the probability of ending up with a head or tail would each be 0.5. Denote the occurrence of head as event $A$ and that of tail as event $B$, we can write the probability of each event as $p(A) = 0.5$ and $p(B) = 0.5$. In general, the probability of any event is between 0 and 1, i.e.,

$$0 \leq p(A) \leq 1 \tag{B.24}$$

If $p(A) = 0$, it is an impossible event, and if $p(A) = 1$, it is a certain event. If $A^*$ is used for anything but $A$, then $p(A) + p(A^*) = 1$. Two events may be dependent or independent. If we toss the coin twice, the result of the second toss is independent of that of the first. Similarly, if we throw two dice, the result of each die is independent of that of the other. On the other hand, if two balls are drawn sequentially from a box containing 3 red and 4 black balls, the probability of the second ball being red depends upon whether the first ball is red or black. If $A$ and $B$ are independent events, then the probability for both $A$ and $B$ to happen is

$$p(A \text{ and } B) = p(A) \times p(B) \tag{B.25}$$

while the probability of either A or B to happen is

$$p(A \text{ or } B) = p(A) + p(B) - p(A) \times p(B) = 1 - p(A^*) \times p(B^*) \tag{B.26}$$

**Example B.4** What is the probability of getting 7 if two dice are thrown?

**Solution** The numbers on the six faces of each die is 1, 2, 3, 4, 5, and 6. Therefore the total number of combinations is 36. The combinations that yield 7 are (1,6), (2,5), (3,4), (4,3), (5,2), and (6,1). Thus, there are 6 out of 36 combinations that will give a sum of 7. The probability of getting 7 is then $p(7) = 1/6$. It can be shown that the probability of getting 8 is $p(8) = 5/36$.

Consider an experiment for which the probability of event $A$ to occur is $\phi$. For a single trial, the probability is $\phi$ for event $A$ and $1 - \phi$ for anything but $A$. For $N$ trials, the probability for event $A$ to occur $M$ times is given by the following equation:

$$p(M) = {}_N\text{C}_M\, \phi^M (1 - \phi)^{N-M} = \frac{N!}{M!(N-M)!}\phi^M (1 - \phi)^{N-M} \tag{B.27}$$

which is equal to the corresponding coefficient of the binomial equation (B.1) by setting $a = \phi$ and $b = 1 - \phi$.

**Example B.5** Toss three coins, what are the probabilities for getting all tails, one head and two tails, two heads and one tail, and all heads?

**Solution** Here $\phi = 0.5$ and $1 - \phi = 0.5$. Notice that

$$\left(\frac{1}{2}+\frac{1}{2}\right)^3=\left(\frac{1}{2}\right)^3+3\left(\frac{1}{2}\right)^2\left(\frac{1}{2}\right)+3\left(\frac{1}{2}\right)\left(\frac{1}{2}\right)^2+\left(\frac{1}{2}\right)^3=\frac{1}{8}+\frac{3}{8}+\frac{3}{8}+\frac{1}{8}$$

We have $p(0) = 0.125$, $p(1) = 0.375$, $p(2) = 0.375$, and $p(3) = 0.125$.

**Example B.6** Calculate the probability for the number 4 to appear more than twice in six tosses of a fairly weighted die.

**Solution** The probability of 4 to occur in one toss is $\phi = 1/6$. Using Eq. (B.27), we get

$$p(0) = 1 \times \left(\frac{1}{6}\right)^0 \times \left(\frac{5}{6}\right)^6 \approx 0.3349$$

$$p(1) = 6 \times \left(\frac{1}{6}\right)^1 \times \left(\frac{5}{6}\right)^5 \approx 0.4019$$

$$p(2) = 15 \times \left(\frac{1}{6}\right)^2 \times \left(\frac{5}{6}\right)^4 \approx 0.2009$$

Therefore, $p(>2) = 1 - p(0) - p(1) - p(2) \approx 0.0623$.

## B.5 Distribution Functions

Figure B.1 shows a plot of a surface roughness distribution (histogram) measured from an atomic force microscope (AFM) for an unpolished silicon wafer within 50 μm × 1 μm area with a total of 512 × 10 = 5120 data points. The vertical axis records the number of points with height between $x_{i-1}$ and $x_i$. Let $N$ be the total number of data points and $N_i$ the number of points with a height greater or equal to $x_{i-1}$ but less than $x_i$. Then, $N = \sum_i N_i$, and *average* and *variance* (mean-square deviation) are

$$\bar{x}_i = \frac{1}{N}\sum_i x_i N_i \quad \text{and} \quad u_{\text{var}} = \frac{1}{N}\sum_i (x_i - \bar{x}_i)^2 N_i \tag{B.28}$$

The average is the mean surface height, and the square root of the variance is the *root-mean-square* (rms) roughness, respectively. The rms value associated with a set of measurements is called the *standard deviation*. If we randomly pick a point, the probability for it to have a height between $x_{i-1}$ and $x_i$ is

$$p(x_{i-1}, x_i) = N_i/N \tag{B.29}$$

For large $N$, we may expect a continuous distribution function,

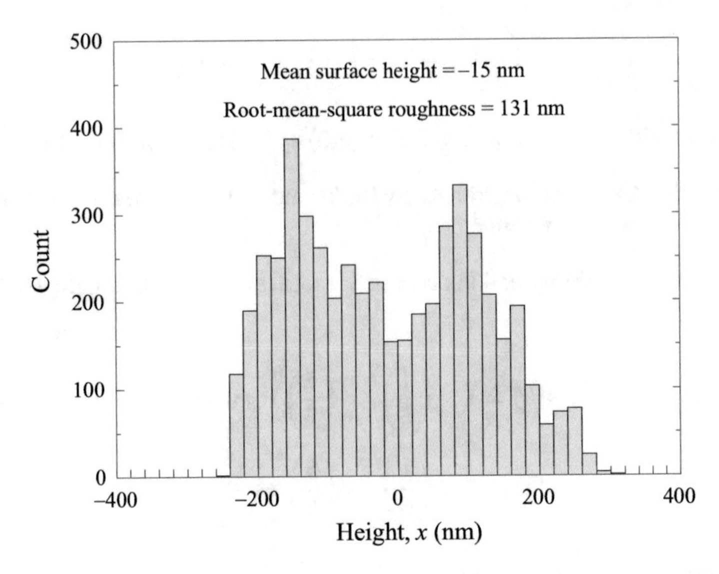


**Figure B.1** Histogram of surface roughness for a silicon surface measured using an AFM

$$f(x) = \lim_{\Delta x_i \to 0} \left( \frac{N_i}{\Delta x_i} \right)$$

where $\Delta x_i = x_i - x_{i-1}$, and $f(x)$ is called a *distribution function*. By definition,

$$\int_{x_{i-1}}^{x_i} f(x)\mathrm{d}x = N_i \quad \text{and} \quad \int_{-\infty}^{\infty} f(x)\mathrm{d}x = N \tag{B.30}$$

The average and variance of the distribution can then be expressed as

$$\bar{x} = \frac{1}{N} \int_{-\infty}^{\infty} xf(x)\mathrm{d}x \quad \text{and} \quad u_{\text{var}} = \frac{1}{N} \int_{-\infty}^{\infty} (x - \bar{x})^2 f(x)\mathrm{d}x \tag{B.31}$$

The average of $x^2$, $\overline{x^2}$is in general different from $u_{\text{var}}$:

$$\overline{x^2} = \frac{1}{N} \int_{-\infty}^{\infty} x^2 f(x)\mathrm{d}x \tag{B.32}$$

The distribution function $f(x)$ may be normalized by dividing $N$ to obtain

$$F(x) \equiv \frac{f(x)}{N} \tag{B.33}$$

where $F(x)$, the normalized distribution function, is called the *probability density function* (PDF). It is related to the probability by,

$$p(x_1, x_2) = \int_{x_1}^{x_2} F(x)\mathrm{d}x,\ p(-\infty, x) = \int_{-\infty}^{x} F(x)\mathrm{d}x, \text{ and } \frac{\mathrm{d}}{\mathrm{d}x} p(-\infty, x) = F(x) \tag{B.34}$$

Note that the function $P(x) = p(-\infty, x)$ defined in Eq. (B.34) is called the *cumulative distribution function* (CDF) whose value varies from 0 to 1. Furthermore, it can be shown that

$$\int_{-\infty}^{\infty} F(x)\mathrm{d}x = 1,\ \int_{-\infty}^{\infty} xF(x)\mathrm{d}x = \bar{x}, \text{ and } \int_{-\infty}^{\infty} (x - \bar{x})^2 F(x)\mathrm{d}x = u_{\mathrm{var}} \tag{B.35}$$

**Example B.7** Under certain conditions, the *x*-component velocity ($U$) of $N$ particles in a fixed volume obeys the following distribution (the *Gaussian distribution* or *normal distribution*):

$$f(U) = A\exp\left(-\frac{U^2}{2\sigma^2}\right)$$

where $U \in (-\infty, \infty)$, and $A$ and $\sigma$ are positive constant. Determine the following: (a) The number of particles $N$ in the volume; (b) The probability density function, $F(U)$; (c) The average velocity ($\bar{U}$); (d) The variance ($u_{\mathrm{var}}$); and (e) the average of $U^2$.

**Solution** Using the definitions and formulations given above, we have

(a) $N = \int_{-\infty}^{\infty} f(U)\mathrm{d}U = \int_{-\infty}^{\infty} A\exp\left(-\frac{U^2}{2\sigma^2}\right)\mathrm{d}U = A\sqrt{2\pi}\sigma$;
(b) $F(U) = \frac{f(U)}{N} = \frac{1}{\sqrt{2\pi}\sigma}\exp\left(-\frac{U^2}{2\sigma^2}\right)$;
(c) $\bar{U} = \int_{-\infty}^{\infty} UF(U)\mathrm{d}U = 0$;
(d) $u_{\mathrm{var}} = \int_{-\infty}^{\infty} U^2F(U)\mathrm{d}U = \sigma^2$; and
(e) $\overline{U^2} = \sigma^2 = u_{\mathrm{var}}$ because $\bar{U} = 0$.

**Discussion** The general form of Gaussian probability density function is

$$F(x) = \frac{1}{\sigma\sqrt{2\pi}}\exp\left(-\frac{(x-\mu)^2}{2\sigma^2}\right)$$

It is a bell-shaped graph centered around $\bar{x} = \mu$ with $u_{\mathrm{var}} = \sigma^2$. It has two inflection points at $x = \mu \pm \sigma$, at which the second order derivative becomes zero. If the Gaussian statistics is used to describe the variations of a set of experimental measurements, the standard deviation $\sigma$ is called the *standard uncertainty*. The probability for a measurement to fall within $|x - \mu| < \sigma$ is 68% and increases to 95% within $|x - \mu| < 2\sigma$. The *expanded uncertainty* is usually defined based on the 95% confidence interval, which is approximately $2\sigma$ for Gaussian statistics.

## B.6 Complex Variables

A complex quantity $z$ may be expressed in terms of a real component $x = \mathrm{Re}(z)$ and an imaginary component $y = \mathrm{Im}(z)$ so that

$$z = x + \mathrm{i}y \tag{B.36}$$

where $\mathrm{i} = \sqrt{-1}$ and $x$ and $y$ are both real. The most convenient way to understand a complex variable is to use the complex plane shown in Fig. B.2. The expression of a complex number is very similar to a two-dimensional vector. Notice that $r = |z| = \sqrt{x^2 + y^2}$ is the magnitude or *complex modulus* and $\phi = \arg|z| = \tan^{-1}(y/x)$ is the phase or *complex argument* of $z$. It is obvious that $x = r\cos\phi$ and $y = r\sin\phi$. By defining

$$\mathrm{e}^{\mathrm{i}\phi} = \cos\phi + \mathrm{i}\sin\phi \tag{B.37}$$

We can also express the complex quantity in terms of its magnitude and phase as follows:

$$z = r\mathrm{e}^{\mathrm{i}\phi} \tag{B.38}$$

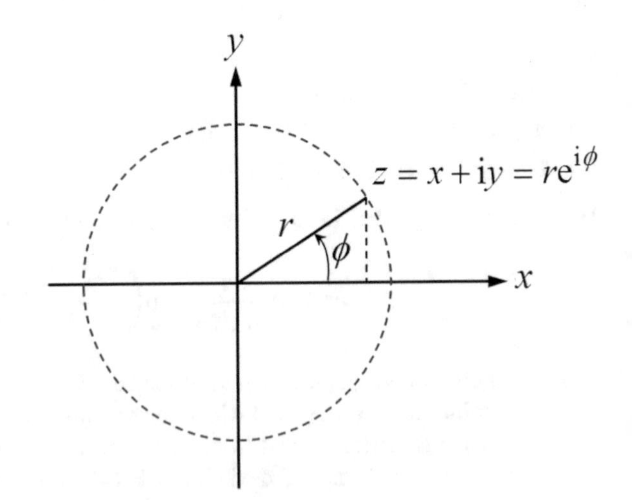


**Figure B.2** Illustration of the complex plane and complex quantity

The complex conjugate is defined as

$$z^* = x - \mathrm{i}y = r\mathrm{e}^{-\mathrm{i}\phi} \tag{B.39}$$

Hence,

$$zz^* = x^2 + y^2 = r^2 = |z|^2 \tag{B.40}$$

Most of the algebra for real variables can be transformed straightforwardly to complex algebra. For example, if $A = A' + \mathrm{i}A'' = r_A \mathrm{e}^{\mathrm{i}\phi_A}$ and $B = B' + \mathrm{i}B'' = r_B \mathrm{e}^{\mathrm{i}\phi_B}$, then

$$A \pm B = (A' \pm B') + \mathrm{i}(A'' + B'') \text{ and } AB = r_A r_B \mathrm{e}^{\mathrm{i}(\phi_A + \phi_B)} \tag{B.41}$$

It can be shown that

$$(A \pm B)^* = A^* + B^* \text{ and } (AB)^* = A^* B^* \tag{B.42}$$

Furthermore,

$$A^n = r^n \mathrm{e}^{\mathrm{i}n\phi} = r^n(\cos n\phi + \mathrm{i} \sin n\phi) \tag{B.43}$$

**Example B.8** Suppose $z = -1 + \mathrm{i}\delta$, where the real number $\delta \ll 1$. First evaluate $y = z^2$ and then evaluate $x = y^{1/2}$.

**Solution** Clearly $z$ is in the second quadrant of the complex plane and $y = 1 - \mathrm{i}2\delta - \delta^2 = (1 - \delta^2) - \mathrm{i}2\delta$ is in the fourth quadrant. Alternatively, we can write $z = \sqrt{1 + \delta^2}\mathrm{e}^{\mathrm{i}\phi}$, where $\phi = \tan^{-1}(-\delta) \approx \pi - \delta$ for small $\delta$. Hence, $y = (1 + \delta^2)\mathrm{e}^{\mathrm{i}2\phi} \approx (1 + \delta^2)\mathrm{e}^{\mathrm{i}(2\pi - 2\delta)} = (1 + \delta^2)\mathrm{e}^{-\mathrm{i}2\delta}$. Finally, $x = y^{1/2} = \sqrt{1 + \delta^2}e^{-\mathrm{i}\delta} \approx 1 - \mathrm{i}\delta = -z$. However, if we use $y \approx (1 + \delta^2)\mathrm{e}^{\mathrm{i}(2\pi - 2\delta)}$, we will end up with $x = \sqrt{z^2} = z$. This example shows that multiple solutions often exist in complex algebra. Which solution should be accepted depends on the particular physical problem. Care must be taken when using a computer to do complex calculations to ensure that the final solution is physical.

Sometimes we may deal with problems involving a complex quantity $z$ with a complex magnitude $\alpha = \alpha' + \mathrm{i}\alpha''$ and a complex phase $\beta = \beta' + \mathrm{i}\beta''$ such that

$$z = \alpha \mathrm{e}^{\mathrm{i}\beta} \tag{B.44}$$

It can be considered as the multiplication of two complex quantities, such that $|z| = \mathrm{e}^{-\beta''}\sqrt{\alpha' 2 + \alpha''^2}$, and $\arg(z) = \arg(\alpha) + \beta'$. Alternatively, we can write $\mathrm{Re}(z) = \alpha' e^{-\beta''} \cos\beta' - \alpha'' e^{-\beta''} \sin\beta'$ and $\mathrm{Im}(z) = \alpha' e^{-\beta''} \sin\beta' + \alpha'' e^{-\beta''} \cos\beta'$. Note that $|\mathrm{e}^{\mathrm{i}\beta}| = \mathrm{e}^{-\beta''}$, which is not equal to 1 unless $\beta''$ is zero.

Complex functions $f = f(z)$ can be defined when $z$ is a complex variable. The differentiation and integration can also be performed. In addition to the difficulty in dealing with multiple solutions, singularities are frequently involved.

## B.7 The Plane Wave Solution

The wave equation is a hyperbolic equation. In the one-dimensional case, it is given as

$$\frac{\partial^2 u}{\partial x^2} = \frac{1}{c^2}\frac{\partial^2 u}{\partial t^2} \tag{B.45}$$

where $x$ is the spatial coordinate and $t$ is the time. It can be verified that the following is a solution of the wave equation:

$$u(x,t) = A\cos(kx - \omega t) \tag{B.46}$$

as long as $k = \omega/c$. Note that any analytic function of $(x \pm ct)$ would satisfy Eq. (B.45). Hence, Eq. (B.46) is only a special solution that we choose to illustrate the nature of the wave equation. Let us further simply the problem by taking only positive values of $A$, $k$, $\omega$, and $c$. Figure B.3 shows the spatial dependence of the wave function at $t = 0$ and $t = \delta t$. Clearly, $A$ is the amplitude of the wave. The period in space, which is the wavelength $\lambda$, is related to $k$ by

$$k = \frac{2\pi}{\lambda} \tag{B.47}$$

Therefore, $k$ is called the wavevector because it is a vector in the 3D coordinates with a magnitude $k$. The wavenumber is defined as the number of waves per unit length, i.e., $\bar{\nu} = 1/\lambda$. From the time dependence, we can see that the period $T = 2\pi/\omega$. The frequency is the number of periods (cycles) per unit time, hence, $\nu = 1/T$, with a unit Hz. Therefore, $\omega = 2\pi\nu$ is called the angular frequency with a unit rad/s. Notice that $\psi = kx - \omega t$ is called the phase. The speed of propagation is determined by the movement of the constant phase position:

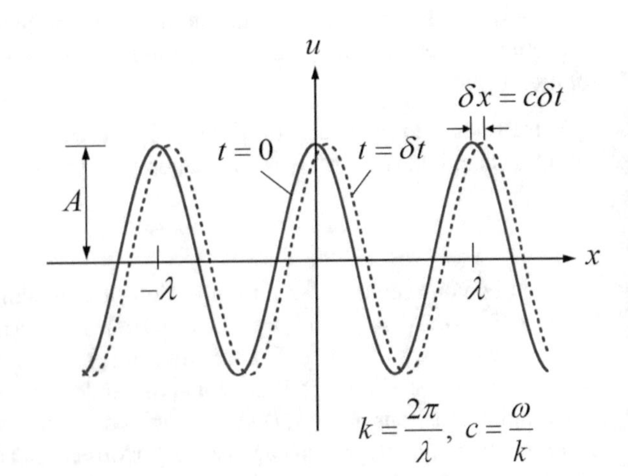


**Figure B.3** Illustration of the wave function and phase speed

$$v_\mathrm{p} = \left(\frac{dx}{dt}\right)_\psi = \frac{\omega}{k} = c \tag{B.48}$$

We have just shown that $c$ is the speed of propagation of the wave or the *phase speed*. In a 3D case, the wave equation is written as

$$\nabla^2 u = \frac{1}{c^2}\frac{\partial^2 u}{\partial t^2} \tag{B.49}$$

where $\nabla^2 = \frac{\partial^2}{\partial x^2} + \frac{\partial^2}{\partial y^2} + \frac{\partial^2}{\partial z^2}$. The solution for a given frequency may be expressed as

$$u(\mathbf{r}, t) = A\mathrm{e}^{\mathrm{i}\mathbf{k}\cdot\mathbf{r} - \mathrm{i}\omega t} \tag{B.50}$$

where $\mathbf{k} = k_x\hat{\mathbf{x}} + k_y\hat{\mathbf{y}} + k_z\hat{\mathbf{z}}$ is called the *wavevector* and its magnitude is $k = 2\pi/\lambda = (k_x^2 + k_y^2 + k_z^2)^{1/2}$. It can be shown that Eq. (B.50) represents a plane wave whose constant phase plane is always perpendicular to **k**, and this wave propagates in the **k** direction. Furthermore, using the plane wave solution, we see that

$$\frac{\partial}{\partial x}u = \mathrm{i}k_x u,\ \frac{\partial}{\partial y}u = \mathrm{i}k_y u, \text{ and } \frac{\partial}{\partial z}u = \mathrm{i}k_x u$$

or the gradient

$$\nabla u = \left(\hat{\mathbf{x}}\frac{\partial}{\partial x} + \hat{\mathbf{y}}\frac{\partial}{\partial y} + \hat{\mathbf{z}}\frac{\partial}{\partial z}\right)u = \mathrm{i}\mathbf{k}\,u \tag{B.51}$$

Similarly,

$$\nabla^2 u = -k^2 u \tag{B.52}$$

In the above discussion, $u$ is treated as a scalar. Frequently, we need to deal with a vector function, such as the electric field **E**, the wave equation can be written as

$$\nabla^2 \mathbf{E} = \frac{1}{c^2}\frac{\partial^2 \mathbf{E}}{\partial t^2} \tag{B.53}$$

Its solution can be expressed as

$$\mathbf{E}(\mathbf{r}, t) = \mathbf{A}\exp(\mathrm{i}\mathbf{k}\cdot\mathbf{r} - \mathrm{i}\omega t) \tag{B.54}$$

where the amplitude **A** is also a vector. It can be shown that

$$\frac{\partial}{\partial x}E_x = \mathrm{i}k_x E_x,\ \frac{\partial}{\partial y}E_y = \mathrm{i}k_y E_y, \text{ and } \frac{\partial}{\partial z}E_z = \mathrm{i}k_x E_z$$

Thus, the divergence is

$$\nabla \cdot \mathbf{E} = \frac{\partial E_x}{\partial x} + \frac{\partial E_y}{\partial y} + \frac{\partial E_z}{\partial y} = \mathrm{i}\mathbf{k} \cdot \mathbf{E} \tag{B.55}$$

the curl is

$$\nabla \times \mathbf{E} = \begin{pmatrix} \hat{\mathbf{x}} & \hat{\mathbf{y}} & \hat{\mathbf{z}} \\ \frac{\partial}{\partial x} & \frac{\partial}{\partial y} & \frac{\partial}{\partial z} \\ E_x & E_y & E_z \end{pmatrix} = \begin{pmatrix} \hat{\mathbf{x}} & \hat{\mathbf{y}} & \hat{\mathbf{z}} \\ ik_x & ik_y & ik_z \\ E_x & E_y & E_z \end{pmatrix} = \mathrm{i}\mathbf{k} \times \mathbf{E} \tag{B.56}$$

and

$$\nabla^2 \mathbf{E} = -k^2 \mathbf{E} \tag{B.57}$$

Equation (B.54) represents a monochromatic wave solution. For polychromatic waves with multiple frequencies, the phase speed is frequency dependent in a dispersive medium. In such as case, waves of different frequency will travel at different speeds. A wave group (or wave packet) contains waves of more than one frequency. The *group velocity* represents the velocity of energy carried by the wave packet and is given by

$$\mathbf{v}_\mathrm{g} = \frac{\mathrm{d}\omega}{\mathrm{d}\mathbf{k}} = \hat{\mathbf{x}}\frac{\mathrm{d}\omega}{\mathrm{d}k_x} + \hat{\mathbf{y}}\frac{\mathrm{d}\omega}{\mathrm{d}k_y} + \hat{\mathbf{z}}\frac{\mathrm{d}\omega}{\mathrm{d}k_z} \tag{B.58}$$

The functional relation $\omega = \omega(\mathbf{k})$ is called a *dispersion relation*. In the 1D case, we can express Eq. (B.58) as

$$v_\mathrm{g} = \mathrm{d}\omega/\mathrm{d}k \tag{B.59}$$

If the phase speed $c = \omega/k$ is constant, we say that the dispersion relation $\omega = ck$ is linear. Subsequently, the group velocity is the same as the phase velocity because $v_\mathrm{g} = \mathrm{d}\omega/\mathrm{d}k = c = \omega/k = v_\mathrm{p}$.

**Example B.9** Consider light propagating in a glass whose refractive index $n = 1.5 + \alpha\omega^2$, where $\alpha$ is a very small coefficient. Find the dispersion relation, the phase speed and group speed as functions of $\omega$.

**Solution** The speed of light in a medium $c = c_0/n$, where $n$ is the refractive index. Therefore, $v_\mathrm{p}(\omega) = \omega/k = c = c_0/(1.5 + \alpha\omega^2)$. The dispersion relation is given by $1.5\omega + \alpha\omega^3 = c_0 k$. The group speed $v_\mathrm{g}(\omega) = (\mathrm{d}k/\mathrm{d}\omega)^{-1} = c_0/(1.5 + 3\alpha\omega^2) = c_0/n_\mathrm{g}$, where $n_\mathrm{g} = n + \omega \mathrm{d}n/\mathrm{d}\omega$ is called the group index.

Some useful vector operators and identities are given below for convenience:

$$\mathbf{A} \cdot \mathbf{B} = \mathbf{B} \cdot \mathbf{A} \tag{B.60}$$

$$\mathbf{A} \times \mathbf{B} = -\mathbf{B} \times \mathbf{A} \tag{B.61}$$

$$\mathbf{A} \times (\mathbf{B} \times \mathbf{C}) = (\mathbf{C} \times \mathbf{B}) \times \mathbf{A} = (\mathbf{A} \cdot \mathbf{C})\mathbf{B} - (\mathbf{A} \cdot \mathbf{B})\mathbf{C} \tag{B.62}$$

$$\mathbf{A} \cdot (\mathbf{B} \times \mathbf{C}) = \mathbf{B} \cdot (\mathbf{C} \times \mathbf{A}) = \mathbf{C} \cdot (\mathbf{A} \times \mathbf{B}) = \det(\mathbf{A}\,\mathbf{B}\,\mathbf{C}) \tag{B.63}$$

$$\nabla \times (\nabla \times \mathbf{A}) = \nabla(\nabla \cdot \mathbf{A}) - \nabla^2 \mathbf{A} \tag{B.64}$$

$$\nabla \cdot (\phi \mathbf{A}) = \phi \nabla \cdot \mathbf{A} + \mathbf{A} \cdot \nabla \phi \tag{B.65}$$

$$\nabla \times (\phi \mathbf{A}) = \phi \nabla \times \mathbf{A} + \nabla \phi \times \mathbf{A} \tag{B.66}$$

$$\nabla \cdot (\mathbf{A} \times \mathbf{B}) = \mathbf{B} \cdot (\nabla \times \mathbf{A}) - \mathbf{A} \cdot (\nabla \times \mathbf{B}) \tag{B.67}$$

$$\nabla \times (\mathbf{A} \times \mathbf{B}) = (\mathbf{B} \cdot \nabla)\mathbf{A} - \mathbf{B}(\nabla \cdot \mathbf{A}) + \mathbf{A}(\nabla \cdot \mathbf{B}) - (\mathbf{A} \cdot \nabla)\mathbf{B} \tag{B.68}$$

and

$$\nabla(\mathbf{A} \cdot \mathbf{B}) = \mathbf{A} \times (\nabla \times \mathbf{B}) + (\mathbf{A} \cdot \nabla)\mathbf{B} + \mathbf{B} \times (\nabla \times \mathbf{A}) + (\mathbf{B} \cdot \nabla)\mathbf{A} \tag{B.69}$$

If $\mathbf{A}$ is a constant matrix, say $\mathbf{A} = \mathbf{K}$ then Eqs. (B.68) and (B.69) reduce respectively to

$$\nabla \times (\mathbf{K} \times \mathbf{B}) = \mathbf{K}(\nabla \cdot \mathbf{B}) - (\mathbf{K} \cdot \nabla)\mathbf{B}$$
$$\nabla(\mathbf{K} \cdot \mathbf{B}) = \mathbf{K} \times (\nabla \times \mathbf{B}) + (\mathbf{K} \cdot \nabla)\mathbf{B}$$

The divergence theorem or Gauss's theorem can be expressed as

$$\iiint_V \nabla \cdot \mathbf{E}\, \mathrm{d}V = \oiint_A \mathbf{E} \cdot \mathbf{n}\, \mathrm{d}A \tag{B.70}$$

It states that the integral of the divergence over the entire volume is equal to the surface integral over the enclosed surface. The curl theorem or Green's theorem states that

$$\iint_A (\nabla \times \mathbf{E}) \cdot \mathbf{n} \mathrm{d}A = \oint_C \mathbf{E} \cdot \mathbf{dr} \tag{B.71}$$

In this equation, it is assumed that $C$ is a closed, piecewise smooth curve that bounds the surface area $A$. The equation converts a surface integration of the curl of a vector to a line integration of the vector. Both the divergence theorem and curl theorem can be considered special cases of Stokes' theorem.

## B.8 The Sommerfeld Expansion

In the free-electron theory discussed in Chap. 5, the integration often includes the Fermi-Dirac function

$$f_{\mathrm{FD}}(\varepsilon, T) = \frac{1}{\mathrm{e}^{(\varepsilon-\mu)/k_{\mathrm{B}}T} + 1} \tag{B.72}$$

where $\varepsilon$ and $\mu$ are the electron energy and chemical potential, $k_{\mathrm{B}}$ is the Boltzmann constant, and $T$ is absolute temperature. Unless the temperature is very high, $k_{\mathrm{B}}T \ll \mu$; note that $\mu$ is a weak function of temperature, $\mu = \mu(T)$. At $T \to 0$ K, the chemical potential is called the Fermi energy $\mu_{\mathrm{F}} = \mu(0)$. However, the chemical potential $\mu$ is often called Fermi level or Fermi energy as well in many texts. At very low temperatures, $f_{\mathrm{FD}}(\varepsilon, 0) \equiv f_{\mathrm{FD}}(\varepsilon, T \to 0) = 1$ for $\epsilon < \mu$, and $f_{\mathrm{FD}}(\varepsilon, 0) = 0$ for $\epsilon > \mu$, as illustrated in Fig. 5.5a. Thus,

$$\int_0^\infty G(\varepsilon) f_{\mathrm{FD}}(\varepsilon, 0) \mathrm{d}\varepsilon = \int_0^{\mu_{\mathrm{F}}} G(\varepsilon) \mathrm{d}\varepsilon \tag{B.73}$$

When $k_{\mathrm{B}}T \ll \mu$, $f_{\mathrm{FD}}(\varepsilon, T)$ is essentially the same as $f_{\mathrm{FD}}(\varepsilon, 0)$, except when $|\varepsilon - \mu| < k_{\mathrm{B}}T$. The integration at intermediate temperatures will be discussed later. The following approximation is often used when $k_{\mathrm{B}}T \ll \mu$:

$$\int_0^{\mu} G(\varepsilon) \mathrm{d}\varepsilon = \int_0^{\mu_{\mathrm{F}}} G(\varepsilon) \mathrm{d}\varepsilon + (\mu - \mu_{\mathrm{F}}) G(\mu_{\mathrm{F}}) + \cdots \tag{B.74}$$

Let us now consider the derivative

$$\frac{\partial}{\partial \varepsilon} f_{\mathrm{FD}}(\varepsilon, T) = -\frac{1}{k_{\mathrm{B}}T} \frac{\mathrm{e}^{(\varepsilon-\mu)/k_{\mathrm{B}}T}}{\left[\mathrm{e}^{(\varepsilon-\mu)/k_{\mathrm{B}}T} + 1\right]} \tag{B.75}$$

The derivative is nonzero only when $|\varepsilon - \mu| < k_{\mathrm{B}}T$. When $T \to 0$, the peak at $\varepsilon = \mu$ goes to infinite. Note that

$$\int_0^\infty \frac{\partial f_{\mathrm{FD}}}{\partial \varepsilon} \mathrm{d}\varepsilon = \int_0^\infty \mathrm{d} f_{\mathrm{FD}} = f_{\mathrm{FD}}|_0^\infty = 0 - 1 = -1$$

Therefore, $\partial f_{\mathrm{FD}}/\partial \varepsilon$ is a Dirac delta function, i.e.,

$$\frac{\partial f_{\mathrm{FD}}}{\partial \varepsilon} \approx -\delta(\varepsilon - \mu), k_{\mathrm{B}}T \ll \mu \tag{B.76}$$

Hence,

$$\int_0^\infty G(\varepsilon)\frac{\partial f_{\mathrm{FD}}}{\partial \varepsilon}\mathrm{d}\varepsilon \approx -G(\mu), k_{\mathrm{B}}T \ll \mu \tag{B.77}$$

The above equation is exact only at absolute zero temperature. A difficulty arises when the integrand contains terms such as $(\varepsilon - \mu)$ or $(\varepsilon - \mu)^2$. In this case, higher-order terms must be retained. Sommerfeld in 1927 developed an expansion to handle the integral. A detailed discussion can be found from the work of J. McDougall and E.C. Stoner, *Phil. Trans. Roy. Soc. Lond.*, Series A, **237**, 67 (1938). The approximations necessary for the free-electron model of metals are discussed below. At intermediate temperatures when $T > 0$ K, Eq. (B.73) can be written in terms of an expansion as follows:

$$\int_0^\infty G(\varepsilon) f_{\mathrm{FD}}\mathrm{d}\varepsilon = \int_0^\mu G(\varepsilon)\mathrm{d}\varepsilon + \frac{\pi^2(k_{\mathrm{B}}T)^2}{6}G'(\mu) + \frac{7\pi^4(k_{\mathrm{B}}T)^4}{360}G'''(\mu) + \cdots \tag{B.78}$$

where $G'(\mu) = \left.\frac{\mathrm{d}G}{\mathrm{d}\varepsilon}\right|_{\varepsilon=\mu}$ and $G'''(\mu) = \left.\frac{\mathrm{d}^3G}{\mathrm{d}\varepsilon^3}\right|_{\varepsilon=\mu}$.

**Example B.10** When $G(0) = 0$, show that

$$\int_0^\infty G(\varepsilon)(\varepsilon - \mu)\left(-\frac{\partial f_{\mathrm{FD}}}{\partial \varepsilon}\right)\mathrm{d}\varepsilon \approx \frac{\pi^2(k_{\mathrm{B}}T)^2}{3}G'(\mu) \tag{B.79}$$

and

$$\int_0^\infty G(\varepsilon)(\varepsilon - \mu)^2\left(-\frac{\partial f_{\mathrm{FD}}}{\partial \varepsilon}\right)\mathrm{d}\varepsilon \approx \frac{\pi^2(k_{\mathrm{B}}T)^2}{3}G(\mu) \tag{B.80}$$

**Solution** We can use Eq. (B.78) by dropping the term with $(k_{\mathrm{B}}T)^4G'''(\mu)$ and higher order terms.

$$\int_0^\infty G(\varepsilon)(\varepsilon - \mu)\left(-\frac{\partial f_{\mathrm{FD}}}{\partial \varepsilon}\right)\mathrm{d}\varepsilon$$

$$= -f_{\mathrm{FD}} G(\varepsilon)(\varepsilon-\mu)|_0^\infty + \int\limits_0^\infty G(\varepsilon) f_{\mathrm{FD}} \mathrm{d}\varepsilon + \int\limits_0^\infty (\varepsilon-\mu) G'(\varepsilon) f_{\mathrm{FD}} \mathrm{d}\varepsilon$$

$$= \int\limits_0^\mu G(\varepsilon)\mathrm{d}\varepsilon + \frac{\pi^2 (k_{\mathrm{B}}T)^2}{6} G'(\mu) + \int\limits_0^\mu (\varepsilon-\mu) G'(\varepsilon)\mathrm{d}\varepsilon + \frac{\pi^2 (k_{\mathrm{B}}T)^2}{6} G'(\mu)$$

$$= \frac{\pi^2 (k_{\mathrm{B}}T)^2}{3} G'(\mu)$$

since $\int_0^\mu (\varepsilon-\mu)G'(\varepsilon)\mathrm{d}\varepsilon = (\varepsilon-\mu)G(\varepsilon)|_0^\mu - \int_0^\mu G(\varepsilon)\mathrm{d}\varepsilon = -\int_0^\mu G(\varepsilon)\mathrm{d}\varepsilon$. This proves Eq. (B.79). The proof of Eq. (B.80) is similar and it is left as an exercise.

Another useful equation is

$$\frac{\partial f_{\mathrm{FD}}}{\partial T} = \frac{\mathrm{e}^{(\varepsilon-\mu)/k_{\mathrm{B}}T}}{\left[\mathrm{e}^{(\varepsilon-\mu)/k_{\mathrm{B}}T}+1\right]^2} \frac{1}{k_{\mathrm{B}}T}\left(-\frac{\varepsilon-\mu}{T} - \frac{\mathrm{d}\mu}{\mathrm{d}T}\right) = -\frac{\partial f_{\mathrm{FD}}}{\partial T}\left(\frac{\varepsilon-\mu}{T} + \frac{\mathrm{d}\mu}{\mathrm{d}T}\right) \tag{B.81}$$

If we neglect $\mathrm{d}\mu/\mathrm{d}T$, then

$$\frac{\partial f_{\mathrm{FD}}}{\partial T} \approx -\frac{\partial f_{\mathrm{FD}}}{\partial \varepsilon}\left(\frac{\varepsilon-\mu}{T}\right) \tag{B.82}$$

# Index




Z. M. Zhang, *Nano/Microscale Heat Transfer*, Mechanical Engineering Series,
https://doi.org/10.1007/978-3-030-45039-7

## C

**D**

## E

**F**

**N**

**O**



**P**

**Q**



**R**

## S

**U**



**V**

## W



## X



## Y



## Z

Printed in the United States
by Baker & Taylor Publisher Services